T3-BPX-008

HELMINTHS, ARTHROPODS AND PROTOZOA OF DOMESTICATED ANIMALS

TO
BILLIE
AND
ANNETTE

HELMINTHS, ARTHROPODS AND PROTOZOA OF DOMESTICATED ANIMALS

Seventh Edition

E. J. L. Soulsby

MA, PhD, MRCVS, DVSM

Professor of Animal Pathology, Department of Clinical Veterinary Medicine, University of Cambridge; formerly Chairman, Department of Pathobiology, and Professor of Parasitology, University of Pennsylvania; Chairman, Graduate Department of Parasitology, Graduate School of Arts and Sciences, University of Pennsylvania

1982

Lea & Febiger · Philadelphia

Published by BAILLIÈRE TINDALL
a division of Cassell Ltd
Greycoat House, 10 Greycoat Place, London SW1P 1SB

First published 1934 as *Veterinary Helminthology and Entomology* by H. O. Mönnig
Fourth edition 1956 and Fifth edition 1962 by Geoffrey Lepage
Sixth edition 1968 as *Helminths, Arthropods and Protozoa of Domesticated Animals* by E. J. L. Soulsby
Reprinted 1969, 1971, 1974, 1976, 1977, 1978
Seventh edition 1982

ISBN 0 7020 0820 6

English Language Book Society Edition 1982

Published in the United States of America
by Lea and Febiger, Philadelphia

ISBN 0 8121 0780 2

Printed in Great Britain by Wm. Clowes, Beccles, Suffolk

British Library Cataloguing in Publication Data

Soulsby, E. J. L.
Helminths, arthropods and protozoa of domesticated animals.—7th ed.
1. Veterinary helminthology
I. Title
636.089'6962 SF810.A3

ISBN 0-7020-0820-6

Contents

Preface

The object of Professor Mönnig's *Veterinary Helminthology and Entomology*, first published in 1934, was 'to give in a scientific way, the most important practical facts of the subject'. To achieve this aim in a contemporary context necessitates an understanding of modern knowledge of the morphology, physiology and biology of the parasites and of the host response to them. This implies that the demarcations between various disciplines become increasingly blurred and in the present edition it has been possible to combine advances in several areas so as to present a more comprehensive understanding of the subject of parasitism than has been possible previously. A further aim has been to provide users of the text with the basis of an understanding for problems they may encounter in the future in addition to supplying existing knowledge of parasitic diseases of domesticated animals. There is every evidence that the problems of the next decade will differ, probably markedly, from those of today. The intensification of livestock farming, and the problems of animal waste disposal in industrialized countries and the new approaches to agriculture in third world nations, much of it relying increasingly on animals, are examples of the new horizons of understanding needed of parasite diseases in the future.

This seventh edition takes note of these needs for the conventional domesticated animals, but in addition it widens the concept of 'domesticated animals'. Thus, accounts of parasites and their associated diseases in fish, both commercially farmed and aquarium bred, are included since the veterinarian is increasingly concerned with fish husbandry and diseases. Parasitic entities of laboratory animals receive increased attention in this edition, to meet the expanded role of the veterinarian in laboratory animal medicine. More information is included on parasites of wild animals and exotic pets since these are now much more within the purview of veterinary care than previously.

The practice has been continued of including complete reference citations. Apart from ensuring some standard of rigorousness in the presentation and avoiding the rigidity of pedantry, this practice provides the reader, especially the student, with a point of departure for further reading on the subject.

This edition also continues to take a global view of parasitic diseases of domesticated animals. The task of compiling accounts of parasitic disease entities in far-away places has been greatly facilitated by communications from numerous individuals offering advice and information about their own countries. It is hoped the compilation of these data and the interpretation of them into the present extensively revised text has resulted in a textbook of veterinary parasitology which can serve both as a practical manual and as a source of reference for veterinarians throughout the world.

E. J. L. Soulsby

June 1982

Acknowledgements

Dr Sheelagh Lloyd has assisted in a major way in the production of this edition. I am particularly grateful to her for her help with the sections on the helminths and arthropods, for the index and for proof reading. There is little doubt that without her assistance, the completion of the revision would have been less thorough than it is, and more delayed than it has been.

The publishers, Baillière Tindall, have been patient and persuasive. They have been sympathetic to last minute revisions and understanding of the various claims and conflicts of time and effort which compete with completion schedules. It has been a particular pleasure to be associated with this publishing company.

Reviewers of the previous (6th) edition have assisted in eliminating errors, have suggested alternate emphasis to various topics and have, in general, been most helpful in their comments. In particular, I would like to thank Dr J. G. Dingle, Dr A. R. Kazam, Professor R. N. Reinecke, Professor Lloyd Whitten and Dr H. McL. Gordon for detailed comments on the sixth edition.

A number of illustrations are new to the seventh edition. Permission to use them was granted as follows:

The front cover picture, a scanning electronmicrograph of the male tail of the nematode *Angiostrongylus costaricensis,* with permission from Dr Ming Ming Wong; Figs 1.1. and 1.2. Mezhdunarodnaya Kniga and the Copyright Agency of the U.S.S.R; Fig. 1.55. Professor E. J. L. Soulsby and Blackwells Scientific Publications; Figs 1.57, 1.58, 1.62, 1.63, Dr Anna Verster and Zeitschift fur Parasitenkunde; Figs 1.88 and 1.89 Dr L. F. Taffs; Fig. 1.110 Dr Edward G. Batte; Fig. 1.117 Dr Herman Zaiman; Fig. 1.161 Dr Lucie Lange and Faculty of Veterinary Science, University of Pretoria, Republic of South Africa; Fig. 1.175 Dr Ming Ming Wong; Figs 1.182 and 1.183 Dr Dickson Despommier; Figs 1.186 and 1.187 Dr E. G. Batte; Figs 2.36, 2.37, 2.40, 2.65, 2.66, 2.68 Dr John Smart (from Castellani and Chalmers); Figs 2.59 and 2.92 Dr John Smart and the Trustees of the British Museum (National History); Figs 2.69, 2.70, 2.71, 2.73, 2.74 Dr William Beesley; Fig. 2.72 Messrs. Oliver and Boyd, Edinburgh; Fig. 2.88 Superintendant of Documents, U.S. Department of Agriculture; Fig. 3.6 Dr Jackie Townsend; Figs 3.24 and 3.25 Dr J. F. Christensen and The Journal of Parasitology; Fig. 3.26 Dr John M. Vettering and The Journal of Parasitology; Fig. 3.27 Dr Jacob Frankel and The Journal of Parasitology; Figs 3.28, 3.29, 3.30, 3.37, 3.39, 3.40 and 3.41 Dr J. P. Dubey and Journal American Veterinary Medical Association; Figs 3.36 A and B Dr Sheelagh Lloyd; Figs 3.36 D and E Dr W. A. Watson; Fig. 3.44 Dr D. F. Mahoney and Academic Press Inc.

To my secretaries, Miss Julia Newham and Miss Clare Blenkinsop, who have attended to the typing, I offer my thanks for their assistance and patience.

And to my wife, with deep appreciation I acknowledge her support and understanding over the years this revision has taken place.

E. J. L. Soulsby

Introduction

Parasitology is a multidisciplinary subject which embraces the fields of biochemistry, physiology, cell biology, immunology and pharmacology, to mention only a few. Rather than attempt an extensive general introduction as many basic biological data as space would allow have been introduced into the text at the appropriate point, with emphasis being placed on those facts which provide a better understanding of the biology of a parasite, its pathogenic effects and control measures against it.

DEFINITIONS

Parasitology is a study of the phenomenon of parasitism. Until recently there has been a desire on the part of authors to separate animal associations into parasitism, symbiosis, commensalism, mutualism, phoresy, etc.

In *commensalism* the relationship is one where one organism benefits nutritionally from another, without at the same time harming the benefactor; in *symbiosis* the original implication was that of ' living together of dissimilar organisms ', the mutually beneficial nature of this association being realized more recently; *mutualism* is essentially symbiosis and the use of the term emphasizes the mutual benefit derived by the members of the association; *phoresy* implies a temporary relationship, usually with no metabolic dependence, in which one organism transports or shelters another. *Parasitism,* on the other hand, has implied a harmful association, the parasite living at the expense of the host.

Such definitions have invariably led to further qualifying definitions (e.g. temporary parasite, obligatory parasite, periodic parasite) and also to debate as to what is to be regarded as a harmful effect. There are numerous species of organisms which cannot, by any stretch of the imagination, be regarded as pathogens and yet, because they are parasites, it has been assumed *à priori* that they must produce some harmful effect.

With the growth of molecular biological studies of ' parasites ' it is clear that there is no longer any justification for retaining strict definitions of animal associations, and certainly none for retaining the idea that a parasite is a form that causes harm to its host. It is more satisfactory to define parasitism as a state in which an organism (the parasite) is metabolically dependent to a greater or lesser extent on another (the host). Within this framework, it would then be useful to designate those forms which are harmful and those which are harmless, or even beneficial, to the host. Two extreme examples might serve to illustrate this: the ciliates of the rumen of the ruminant or the caecum of the horse are metabolically dependent on the host, yet they are far from pathogenic and there is much to indicate that they are beneficial; on the other hand the abomasal worm of sheep, *Haemonchus contortus*, is also metabolically dependent on the host but it is a serious pathogen.

FORMS WHICH ARE PARASITIC

This book deals only with the helminths, arthropods and protozoa which are parasites of domesticated animals and other species. It omits the vast assemblage of other parasitic species which also belong to the three groups mentioned above and which are spread throughout practically every major phylum of the animal kingdom.

The early beginnings of parasitic life must remain obscure since, being invertebrates without exo- or endo-skeletons, they have left no clue to their ancestry. Nevertheless, since they must have evolved with their hosts, the palaeontological history of the host may shed some light on the evolution of parasites. It is probable that early parasitic forms showed little host specificity, but as they became adapted to the host, and evolved with it, the parasites no doubt became increasingly committed to that host.

A recent consideration of the evolution of host–parasite relationships is based on immunological phenomena. This suggests that the older and better adapted parasites have increasingly eliminated antigens (? recognition factors) that were foreign to the host. Ultimately, such parasites became forms that shared a number of common antigens with the host; the degree of antigenic disparity between the host and the parasite determined the success or otherwise of the parasitic organism.

Such a hypothesis could fit into the concept that the pathogenic species of parasites are those which have been parasites for the shortest period of evolutionary time, since they have yet to evolve to the stage where they incite little or no response on the part of the host.

During the evolution of the host–parasite partnership, it is likely that many biological and morphological characters were lost and also that many were gained. The degree to which this took place would depend, amongst other things, on the host parasitized, the site in the host and the environment of the host. Parasites of hosts that are, today, obviously related, and which evolved from a common ancestor, probably underwent similar physiological adaptations as the hosts evolved. It is also likely that the physiological changes were expressed as similar changes in morphology and today, therefore, one would expect a similar series of species in the same habitat in related hosts.

HOST PARASITE SPECIFICITY AND SITE SELECTION

There is increasing evidence that host–parasite compatibility is based on recognition mechanisms which are both host in origin and of exquisite specificity. An example is that of red cell susceptibility to *Plasmodium vivax* infection in man being associated with Duffy blood group determinants which function as erythrocyte receptors for the parasite. Another example is the use of components of the host's complement system for the attachment to and penetration of *Babesia* parasites into erythrocytes. However, the molecular basis of host–parasite specificity is largely unknown and advances in the understanding of this would provide sounder bases for control.

Site selection and the ability to reorganize the local environment are features of parasitism which are under increasing research scrutiny. Intracellular protozoa such as *Toxoplasma* and *Leishmania* species may modify the normal phagocytic mechanisms of macrophages and helminths such as *Trichinella spiralis* markedly alter the local environment to provide for their physiological needs.

An important aspect of host–parasite specificity is the evasion of the host response by parasites. Various mechanisms are used to accomplish this: some modify the affector or effector areas of the immune response, some are concerned with the provision of an antigenic disguise by which parasites acquire host characteristics, while others are able to modulate their surface antigens so rapidly that they readily evade any host immune response which is stimulated by the infection.

BIOCHEMISTRY AND PHYSIOLOGY

To attempt any broad statement on these subjects would be foolhardy, even if such remarks were restricted to the parasites of domestic animals.

Major advances have been made in these fields in recent years and many have been derived from the field of chemotherapy, where an understanding of drug action has provided basic information on the physiology of parasitic organisms. The student is urged to consult the references given at the end of this introduction for information on the biochemical and physiological aspects of parasitism.

EFFECT OF PARASITES ON THEIR HOSTS

There are many species of parasites which are relatively harmless, but there are also many forms which produce pathological changes which may lead to severe ill health or death of the host. The effects are very varied and in many cases represent a combination of several entities. The parasite may compete with the host for food and where this is a specific effect (e.g. competition for vitamin B12 by *Diphyllobothrium latum*) the host may suffer a specific deficiency syndrome (e.g. anaemia in the case of *D. latum* infection). More generally, however, the competition for food is much less well defined. The parasite indirectly may be the cause of decreased food utilization by the host, it may cause a reduced appetite with a concomitant reduction of food intake, an increased passage of food through the digestive tract or a deceased synthesis of protein in skeletal muscle. Changes in the absorptive surface of the intestine may result in marked alterations in the efflux and influx of water and sodium and chloride ions into the bowel and in morphological and biochemical changes in epithelial cells and their microvilli.

The removal of the host's tissues and fluids by parasites is best illustrated by the blood-sucking activities of certain nematodes (e.g. hookworms, *Haemonchus*) and arthropods (e.g. ticks, blood-sucking flies) and in some cases death of the host is directly attributable to excessive loss of blood.

One of the most common effects of parasitism is destruction of the host's tissues. This may be by a mechanical action when, for example, parasites or their larval stages migrate through or multiply in tissues or organs, or when various organs of attachment (e.g. head-spines or teeth, claws, suckers, etc.) are inserted into the tissues as anchors. Destruction may be by pressure as a parasite grows larger (e.g. hydatid, coenurus), or by blockage of ducts such as blood vessels to produce infarction (*Strongylus*), of lymph vessels to produce oedema and elephantiasis (filariasis) or of the intestinal canal to produce necrosis and rupture (ascarids).

Often the destruction of tissues is a secondary effect. It may arise from bacterial infection of lesions caused by a parasite (e.g. bowel ulcers) or by the reaction of the host to the parasite. The latter effect may be due to fibrosis of a lesion (e.g. cirrhosis due to *Fasciola hepatica*), excessive proliferation of epithelium (e.g. *Eimeria stiedai* in the bile ducts of the rabbit), endothelium (e.g. aneurysm caused by *Strongylus*), lymphoid tissue (e.g. leishmaniasis) or the initiation of malignant propensities (e.g. *Spirocerca lupi* in the dog). Tissue damage may also be caused by the immunological response of the host, resulting in necrosis, dermatitis (cercarial dermatitis) or oedema (ascariasis and dictyocauliasis of the lung).

These are a few examples of how pathogenic parasites cause their ill effects; many other examples could be quoted but a comparable list could also be given of those forms for which no adequate explanation of the pathogenic mechanisms can be given. Even with the forms where there is an apparently clear end-result of the parasitism (e.g. anaemia, emaciation or paralysis), often little is known about the chemical pathology and the chronology of the disease process. Anaemia, for example, is a common feature of parasitism and where this is associated with blood-sucking parasites the cause and effect may appear relatively clear. Even in these circumstances, however, it may not be as simple as has been supposed, and in other parasitic diseases a complex series of nutritional, biochemical and pathological conditions inter-react to be manifest as 'anaemia'.

IMMUNOLOGY

The immune response to parasites is an important phenomenon in the pathology of infection and in the control of parasite populations. Practical uses of the immune response have been the development of immunodiagnostic tests and the production of vaccines.

The subject of immunology in parasitic infections has now become so immense that, as with biochemistry and physiology, it would be unwise to attempt to give a proportionate impression of the vast amount of facts in the limited space of this introduction. This would, inevitably, result in abbreviated treatment of major concepts and, in any case, it becomes more and more difficult to present any unified hypothesis, if one exists, of the immunology of parasitism. Consequently, the pertinent facts about immunity to the various species of parasites are presented in the appropriate parts of the text.

GENERAL READING

References are given below to texts which provide an introduction to the various aspects of parasitology dealt with in this book. In addition, the reader's attention is directed to the series *Advances in Parasitology* (Academic Press, London and New York) and *Trends and Perspectives* in the journal *Parasitology* (Cambridge University Press). Other journals including *The Journal of Parasitology* (Journal of the American Society of Parasitologists), *Experimental Parasitology* (Academic Press, New York, London, Toronto and Sydney), *International Journal for Parasitology* (Pergamon Press, Oxford, New York, Paris and Frankfurt), *Parasitology Research* (Zeitschrift für Parasitenkunde) (Springer International), *Veterinary Parasitology* (Elsevier, Amsterdam) and *Parasite Immunology* (Blackwell Scientific Publications, Oxford) are recommended for regular perusal.

GENERAL REFERENCES

Chen, T. C. (1973) *General Parasitology* New York and London: Academic Press.
Cohen, S. & Sadun, E. (1976) *Immunology of Parasitic Infections* Oxford: Blackwell Scientific.
Euzeby, J. (1981) *Diagnostic Experimental des Helminthoses Animals.* Ministere de l'Agriculture, Paris: Informations Techniques des Services Vétérinaires.
Fallis, A. M. (1977) *Parasites, Their World and Ours.* Proc. 18th Symposium of the Royal Society of Canada. Ottawa, Ontario: The Royal Society of Canada.
Levine, N. D. (1968) *Nematode Parasites of Domestic Animals and of Man.* Minneapolis, Minnesota: Burgess.
Levine, N. D. (1973) *Protozoan Parasites of Domestic Animals and of Man*, 2nd ed. Minneapolis, Minnesota: Burgess.
Price, P. W. (1980) *Evolutionary Biology of Parasites.* Princeton, N.J.: Princeton University Press.
Von der Bossche, H. (1980) *The Host–Invader Interplay*, 3rd International Symposium on The Biochemistry of Parasites and Host–Parasite Relationships, Amsterdam: Elsevier North Holland Biomedical Press.
Whitfield, P. J. (1979) *The Biology of Parasitism.* London: Edward Arnold.

I

Helminths

The name helminth is derived from the Greek words *helmins* or *helminthos*, a worm, and is usually applied only to the parasitic and non-parasitic species belonging to the phyla Platyhelminthes (flukes, tapeworms and other flatworms) and Nemathelminthes (roundworms and their relatives). The Annelida (earthworms, leeches) are fundamentally different from both the Platyhelminthes and the Nemathelminthes and are not regarded as helminths, though some (e.g. leeches) may be parasitic and others (e.g. earthworms) may serve as intermediate hosts for helminths.

PHYLUM: PLATYHELMINTHES

Dorsoventrally flattened and usually hermaphrodite worms with solid bodies without a body cavity, all of which, except most of the species of the Class Turbellaria, are parasitic. The organs are embedded in tissue called the parenchyma and the excretory organs are flame-cells. Respiratory and blood-vascular systems are absent. Like the Nemathelminthes, but unlike the Annelida, the Platyhelminthes are not metamerically segmented. The life-history is usually indirect.

CLASS: TURBELLARIA (EDDYWORMS)

A class comprised chiefly of non-parasitic species living in fresh water, the sea or on land, the body having a ciliated covering. The parasitic species are not parasitic in domesticated animals.

CLASS: TREMATODA (FLUKES)

Species have an alimentary canal.

CLASS: EUCESTODA (TRUE CESTODES)

Species have no alimentary canal, food being absorbed through the surface tegument.

CLASS: COTYLODA

This class contains forms previously classified as Cestodaria and includes some new orders. Forms in this class are mainly parasites of fishes.

PHYLUM: NEMATHELMINTHES

CLASS: NEMATODA (ROUNDWORMS)

Cylindrical worms, both ends being usually somewhat pointed. The body is not metamerically segmented. The cuticle, which usually looks smooth to the unaided eye, may show various cuticular structures under magnification, but the metameric rings visible in annelid worms are not present. Beneath the cuticle there is a hypodermis and beneath this a layer of muscle cells of a type not found in any other animals. Down the centre of the cylindrical body runs the alimentary canal, which is a tube usually consisting of a mouth at the anterior end of the worm, a muscular oesophagus and an intestine leading to an anus which is not terminal, so that a short tail is present. Between the muscle cells and the alimentary canal there is a perienteric space filled with fluid. In the body wall there are lateral canals and glands associated with them are regarded as being excretory organs. Flame-cells are not present and cilia are absent from the body. The sexes are in separate individuals. The life-histories are either direct or indirect.

CLASS: NEMATOMORPHA (HAIR WORMS, GORDIACEA)

The adults are non-parasitic, long, threadlike worms living in the water or moist soil. The sexes are separate. The larvae are parasitic in insects, centipedes and millipedes. Neither the adults nor the larvae are parasitic in domesticated animals, but they may be mistaken for parasitic species.

CLASS: ACANTHOCEPHALA (THORNYHEADED WORMS)

Cylindrical worms with a thick cuticle and a retractable proboscis provided with spines or hooks. There is no alimentary canal (cf. Eucestoda). The excretory organs contain bunches of flame-cells. The sexes are separate and the life-histories are indirect.

Trematodes (Phylum: Platyhelminthes)

CLASS: TREMATODA RUDOLPHI, 1808

The bodies of trematodes or flukes are dorsoventrally flattened and are unsegmented and leaflike. All the organs are embedded in a parenchyma, no body cavity being present. Suckers, hooks or clamps attach these species to the exterior or the internal organs of their hosts. A mouth and an alimentary canal are present, but usually there is no anus. The mouth leads into a muscular pharynx, succeeding which is an intestine, and this divides into two branches, which may themselves branch. The branched excretory system has flame cells and it discharges into an excretory bladder which usually has a posterior opening. The reproductive system is hermaphrodite, except in the family Schistosomatidae, the species of which are unisexual. The life-histories are direct (Monogenea) or indirect (Digenea). There are three subclasses (Schell 1970), whereas Yamaguti (1963) considers such groups as orders.

SUBCLASS: MONOGENEA

Species of this subclass are parasitic chiefly on cold-blooded aquatic vertebrates (fishes, amphibia and reptiles) and most of them are ectoparasitic. None of them occur on domesticated animals. The life-histories are, so far as is known, direct.

SUBCLASS: ASPIDOGASTREA

This subclass contains only one family, the Aspidogastridae, the species of which are parasitic in, or on, fishes, turtles, Mollusca or Crustacea, none being parasitic on domesticated animals.

SUBCLASS: DIGENEA

To this subclass belong all the species parasitic in domesticated animals. The life-histories require one, two or more intermediate hosts.

SUBCLASS: MONOGENEA CARUS, 1863

Species of this subclass are parasites of cold-blooded aquatic or amphibious vertebrates (fishes, amphibians, reptiles) and occasionally aquatic invertebrates. They are primarily ectoparasites, particularly of the gills, skin, fins and buccal cavity. Also, they are found in organs communicating directly with the exterior such as the urinary tract. There is a posterior adhesive organ, or haptor, which may bear suckers, clamps and hooks and identification is based primarily on the morphology of this haptor. The parasites are viviparous (Gyrodactylidae) or oviparous and the life cycle is direct. The eggs are usually operculate with prolongations or filaments at one or both poles. The filaments may serve to attach eggs to the host or other objects. The larva or oncomiridium which hatches bears cilia and one or more pairs of eye spots. Once hatched, the oncomiridia have a short free-swimming period within which to find a host on which they reach sexual maturity. Monogenean parasites are widely distributed and can be pathogenic parasites of fishes.

The suborders and families of the species of monogenetic trematodes of major interest are given in the following list, the classification being based on Yamaguti (1963) and Schell (1970):

SUBORDER: MONOPISTHOCOTYLEA Odhner, 1912

Superfamily: GYRODACTYLOIDEA Johnson and Tiegs, 1922

Family: Gyrodactylidae Cobbold, 1864

Genus: *Gyrodactylus* Nordmann, 1832

Superfamily: DACTYLOGYROIDEA Yamaguti, 1963
Family: Dactylogyridae Bychowsky, 1933
Genus: *Dactylogyrus* Diesing, 1850

Superfamily: CAPSALOIDEA Price, 1935
Family: Capsalidae Baird, 1853
Genus: *Benedenia* Diesing, 1858

SUBORDER: POLYOPISTHOCOTYLEA Odhner, 1912
Superfamily: DICLIDOPHOROIDEA Price, 1936
Family: Discocotylidae Price 1936
Genus: *Discocotyle* Diesing, 1850

Superfamily: DIPLOZOOIDEA Yamaguti, 1963
Family: Diploozidae Tripathi, 1959
Genus: *Diplozoon* Nordmann, 1832

SUBORDER: MONOPISTHOCOTYLEA ODHNER, 1912

SUPERFAMILY: GYRODACTYLOIDEA JOHNSTON AND TIEGS, 1922

These parasites are elliptical and flattened on their ventral surface. The posteriorly placed disc-shaped attachment organ bears one or two pairs of large hooks or anchors surrounded by smaller marginal hooklets.

FAMILY: GYRODACTYLIDAE COBBOLD, 1864

Members of this family are small, elongate, viviparous parasites of fishes, amphibians, cephalopods and crustaceans. The posterior haptor is well developed, usually with one pair of anchors and 16 marginal hooklets.

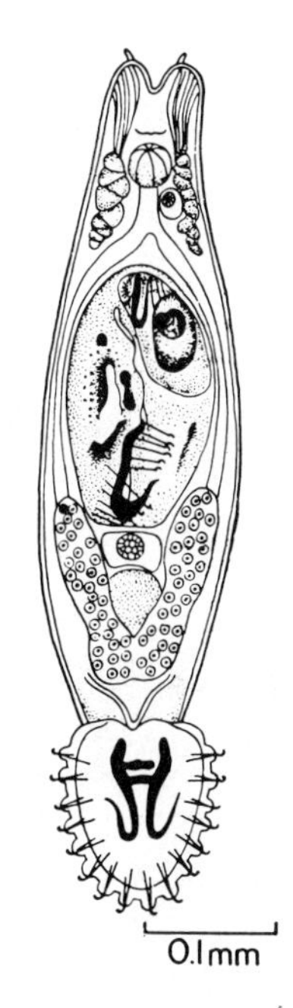

Fig 1.1 Adult *Gyrodactylus elegans*. (*From Dogiel et al. 1961; by courtesy of Oliver and Boyd*)

Genus: Gyrodactylus Nordmann, 1832

Gyrodactylus elegans Nordmann, 1832 (Fig. 1.1) parasitizes the skin, fins and also gills of a wide variety of fish and is a common parasite of trout. It may also affect frogs. It is important in fish culture in Europe, North America and the USSR and also is widespread in Africa and the Middle East. The adults are less than 1 mm in length. They are viviparous and give rise to individuals identical to themselves. The larva forming within the uterus of the adult may, itself, contain a larva and, in all, up to four generations may occur within one individual (serial polyembryony). Further, the parasite just emerged from the mother may immediately produce its own daughter individual. Thus, if conditions are ideal, the parasite numbers can increase very rapidly.

Numerous other *Gyrodactylus* spp. affect the skin, fins and also gills of a wide variety of marine and freshwater teleosts. They can be of importance in fish culture and are common parasites of aquarium fishes.

SUPERFAMILY: DACTYLOGYROIDEA YAMAGUTI, 1963

FAMILY: DACTYLOGYRIDAE BYCHOWSKY, 1933

Members of this family are cosmopolitan in distribution and are important parasites of marine and freshwater fishes. The parasites are oviparous. The posterior haptor bears one or two pairs of anchors and 14 marginal hooklets.

Genus: Dactylogyrus Diesing, 1850

Adults of this genus are up to 2 mm in length and the haptor bears one pair of anchors and 14 marginal hooklets. They are important parasites of the gills of cyprinids, particularly carp, in pond and natural waters. They may also affect bream, goldfish, stickleback, pike and trout.

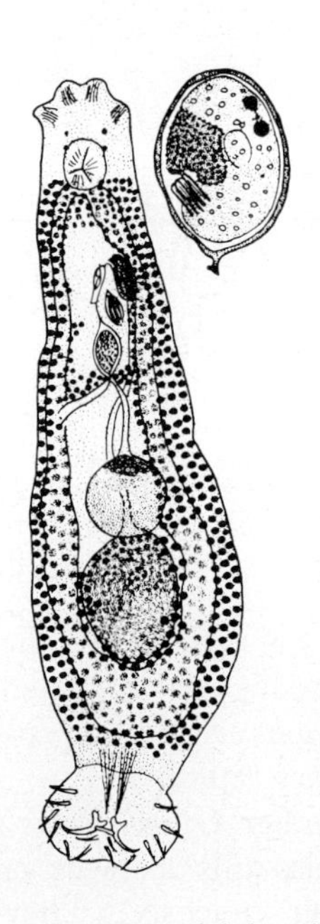

Fig 1.2 *Dactylogyrus vastator*. Adult and egg containing embryo. (*From Dogiel et al. 1961; by courtesy of Oliver and Boyd*)

Dactylogyrus vastator Nybelin, 1924, is particularly important in carp fry in Europe, the USSR and elsewhere, such as in the Middle East. Adults on the gills begin to lay eggs in the spring and reproduction and infection reach a peak in the warmest months when epidemics of infection may occur. In the autumn the infection rate decreases. Reproduction of the few remaining parasites and overwintered eggs continue the cycle in the spring (Bauer 1961). The life-cycle requires one to five days.

Dactylogyrus extensus Mueller and Van Cleave, 1932, affects carp in North America, Europe and the USSR. This parasite is capable of reproduction at lower temperatures than *D. vastator* and the intensity of the infection may rise through the winter, although at a slower rate than in the summer (Bauer 1969). Thus, *D. extensus* has a more northerly distribution than does *D. vastator*.

Other genera of importance in this family are *Neodactylogyrus* Diesing, 1850; *Ancryocephalus*, Creplin, 1839; *Cleidodiscus* Mueller, 1934; *Haliotrema* Johnston & Tiegs, 1922; and *Actinocleidus* Mueller, 1937.

SUPERFAMILY: CAPSALOIDEA PRICE, 1936

FAMILY: CAPSALIDAE BAIRD, 1853

Genus: Benedenia Diesing, 1858

Benedenia spp. affect the oral mucosa and body surface of marine teleosts. Adults are up to 5 mm in length. The posterior disc-shaped haptor is armed with three pairs of dissimilar anchors. *Benedenia* spp. can be important parasites of aquarium fishes and cultured fishes. They are important parasites of mullets in the Middle East.

SUBORDER: POLYOPISTHOCOTYLEA ODHNER, 1912

Adhesive haptor developed as suckers, clamps or anchor complexes. Parasitic on the skin, gills and buccal cavity of fishes, amphibians and reptiles.

SUPERFAMILY: DICLIDOPHOROIDEA PRICE, 1936

FAMILY: DISCOCOTYLIDAE PRICE, 1936

Genus: Discocotyle Diesing, 1850

Discocotyle sagittata Diesing, 1850, has four pairs of equally developed clamps on the posterior adhesive haptor. It is a small but common parasite of the gills of salmonids in North America, Europe and the USSR. *D. sagittata* has been associated with serious gill damage on trout in Great Britain and mass mortality of trout in the USA (Davis 1953).

SUPERFAMILY: DIPLOZOOIDEA YAMAGUTI, 1963

FAMILY: DIPLOZOOIDAE TRIPATHI, 1959

Genus: Diplozoon Nordmann, 1832

Diplozoon paradoxum Nordmann, 1832, is an unusual monogenean parasite on the gills of freshwater fish, particularly cyprinids. Adults lay eggs which are fastened by a long thread to the gills of the fish. Larvae hatch from these eggs as individuals but they must fuse as pairs to reach sexual maturity. The adult parasites are permanently fused as pairs and are about 4–5 mm in length.

Pathogenicity and control of Monogenean trematodes

Monogenean trematodes are widely distributed and are not infrequently pathogenic parasites of fishes. Many species are found on the gills, fins and skin and they can be an important cause of loss in young fish. *Gyrodactylus* spp. are found primarily on the skin while *Dactylogyrus* spp. parasitize the gills. Their life-cycle is direct and it may take less than one to five days. Thus, although monogeneans are not normally pathogenic to fish in their natural habitat, the numbers on fish can reach epidemic proportions in marine aquaria and in cultured fish populations since conditions for reinfection are optimum.

Adult monogeneans feed on mucus, epithelium and sometimes blood. The hooks on the parasite and feeding of the fluke cause frayed gills, rapid breathing and the fish become dull and feeble. Irritation of the fins causes fraying and sluggishness. The skin is irritated and the fish rub against objects causing damage which may lead to bacterial infection. Skin ulcers may result. Colours fade and the fish become dark and slimy, perhaps with black spots. They lose weight and become emaciated. The infected fish becomes more susceptible to predation.

A variety of chemicals are effective in the treatment of monogenean infections in fish. Treatment should be repeated to kill the parasites which hatch from eggs. The following may be used: formalin 1 : 4000 or 1 : 5000 for one hour (the formaldehyde must be free of paraformaldehyde since this is toxic to fish); potassium antimonyl tartrate 150 mg/100 ml; trichlorophon 2–3.5% solution for two or three minutes; ammonia solution 1 : 2000 for 15 minutes followed by methylene blue, 2–4 ml of 1% solution per 3.8 litres; quinacrine hydrochloride 1 g/500 litres; acriflavine 1 g/100 litres; or potassium permanganate 4–5 mg/litre.

SUBCLASS: DIGENEA VAN BENEDEN, 1858

In general the digenetic trematodes are dorsoventrally flattened, some being long and narrow, some leaf-shaped while a few, the amphistomes, have thick fleshy bodies. The schistosomes are long and worm-like.

The cuticle, or more correctly the tegument, may be smooth or spiny but as well as serving as an outer covering. The tegument is a metabolically active surface similar to that of cestodes (p. 87).

The organs of attachment consist of an anterior sucker (oral sucker) placed at the anterior end of the body and ventral sucker or acetabulum usually in the anterior third of the ventral surface, but the position varies and in some forms the ventral sucker may be missing.

The digestive system opens at the mouth which is surrounded by the anterior sucker. Posteriorly there is a muscular pharynx then an oesophagus, this leading into the intestine which usually divides into two blind caeca. A few species of trematodes have an anus. Secondary branching of the caeca may occur. Electronmicroscopy of the gut epithelium (of *F. hepatica*) reveals looped and branching projections, delimited by a triple membrane and having a dense core. The epithelial cells have both absorptive and secretory functions. They are tall and columnar during the secretory phase and after secretion of digestive enzymes etc. they collapse and regeneration gradually occurs. The cells are rich in RNA and active protein synthesis occurs.

The excretory system consists of a bladder which is sac-like in its simplest form, but may have various shapes, and usually opens at the posterior extremity of the body. From this central collecting organ branched tubes run out into the parenchyma, ending in terminal organs, flame-cells, which are characteristic of the flatworms. These cells have a basal cytoplasmic portion which contains the nucleus and bears a number of long cilia that lie in the proximal wide part of the tube which is attached to the cell. The flame-cell is the excretory cell, collecting from its surroundings the waste products to be excreted, while its cilia produce a current in the tubes and so propel these substances to the bladder. The number of flame-cells and the arrangement of their ducts is a diagnostic characteristic.

The nervous system is composed of a circum-oesophageal ring of fibres and paired ganglia, from which three pairs of nerves run forwards and three backwards to all parts of the body. Sense organs as such are not present in the adult, although the free-living larval forms (miracidium and cercaria) may be provided with patches of pigment called 'eye-spots'.

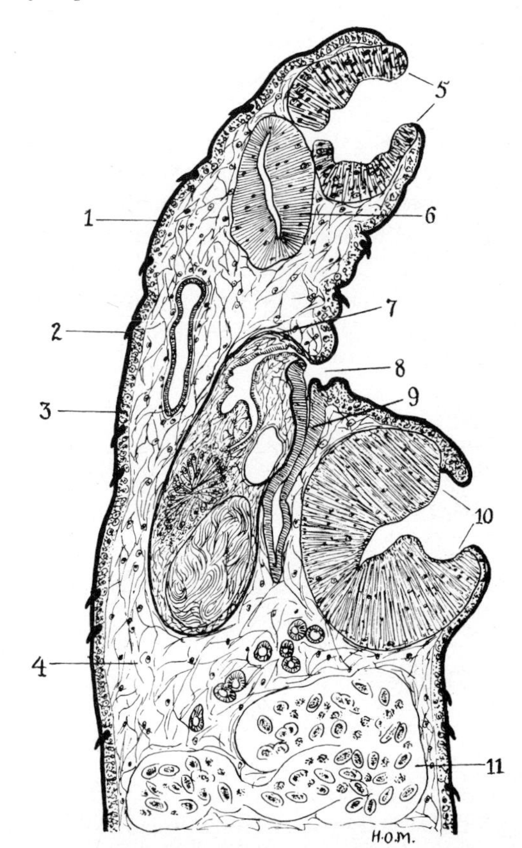

Fig 1.3 Longitudinal section of *Fasciola hepatica*.
1 tegument
2 cuticular spine
3 syncytial cells
4 parenchyma
5 oral sucker
6 pharynx
7 cirrus-sac
8 genital sinus
9 metraterm
10 ventral sucker (acetabulum)
11 uterus with eggs

With the exception of the *Schistosomatidae* and *Dioymozoidae* the digenetic trematodes are hermaphrodite (Fig. 1.4). The male organs consist as a rule of two testes which may be spherical, lobed, branched or divided into a number of smaller bodies. The vasa efferentia unite to form the vas deferens, which usually widens distally and forms a vesiculum seminalis surrounded by the prostate gland, and then ends in the cirrus, a protrusible portion which may be armed with spines. There may be a cirrus-sac enclosing these terminal organs. The genital pore is usually anterior and ventral, but may be posterior or lateral. It is surrounded by a genital sinus or atrium, which is in some species (e.g of the family *Heterophyidae*) developed into a sucker and in which the female pore is also situated. Self-fertilization usually takes place; the cirrus is evaginated and enters the uterus, or the genital sinus is closed and communication is established in this way.

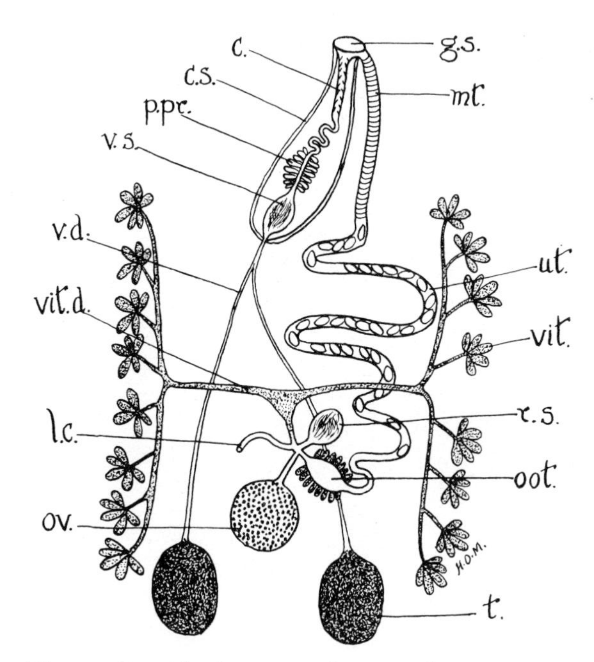

Fig 1.4 Reproductive organs of a trematode.

c	cirrus
cs	cirrus-sac
gs	genital sinus
lc	Laurer's canal
mt	metraterm
oot	ootype surrounded by Mehlis's glands
ov	ovary
ppr	pars prostatica
rs	receptaculum seminis
t	testis
ut	uterus
vd	vas deferens
vs	vesicula seminalis
vit	vitellaria
vitd	vitelline duct

The female organs (Fig. 1.4) consist of an ovary which is usually slightly lobed and discharges the ova into the oviduct. The latter frequently bears a receptaculum seminis and a narrow canal opening on the dorsal surface of the body (Laurer's canal), the function of which is obscure. The insertion of the penis of another fluke into Laurer's canal has been observed, but it is not likely that copulation normally occurs in this way. There is a paired vitelline or yolk gland, consisting usually of a number of follicles situated laterally in the body and discharging into the yolk duct, which joins the oviduct in a special wide portion, the oötype, in which the eggs are formed. The oötype is surrounded by numerous unicellular glands, called Mehlis's glands, which are sometimes collectively called the shell-gland. The shells of the eggs are formed from material contributed by both Mehlis's glands and the vitelline glands. For a discussion of the formation and structure of the egg shell in trematodes and cestodes see Smyth and Clegg (1959). On leaving the oötype the eggs enter the uterus, which may be short or much convoluted and opens at the genital pore. In some species the distal part of the uterus forms a wide and sometimes muscular portion or metraterm. The eggs usually have an operculum, and those of many species develop in the uterus, so that they are ready to hatch when they are laid.

Life-cycle

The eggs of the *Digenea* are usually passed in the faeces of the host and under suitable conditions of moisture and warmth a larva, *miracidium*, hatches. The eggs cannot withstand dessication. Hatching is controlled by a number of factors such as light, temperature and salinity. Five larval stages may occur in the life-cycle; miracidium, sporocyst, redia, cercaria and metacercaria. Daughter sporocysts and rediae may also occur. Mesocercariae, prolonged cercarial stages, may occasionally occur (e.g. in *Alaria* spp). The miracidium is roughly

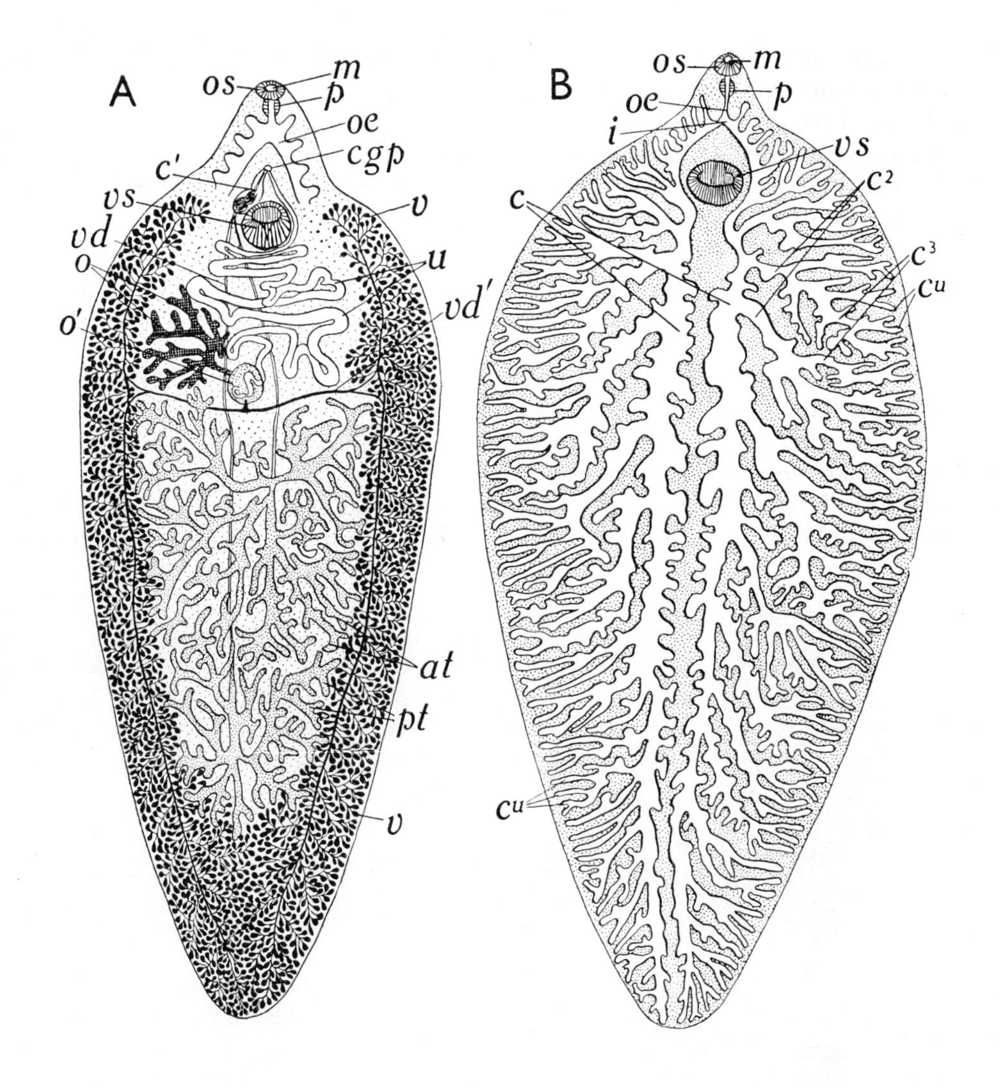

Fig 1.5 *Fasciola hepatica*. A, The reproductive system. B, The alimentary system

os	oral sucker
vs	ventral sucker
m	mouth
p	pharynx
oe	oesophagus
c	intestinal caecum
c^2, c^3, c^u	secondary, tertiary and ultimate branches of the intestinal caeca
cgp	common genital pore
c^I	cirrus
v	vitellaria
vd	vas deferens
vd^1	duct of the vitelline gland
u	uterus
o	ovary
o^1	ootype
at	anterior testis
pt	posterior testis

triangular in shape, the anterior end being broader, and it is usually covered with a ciliated ectoderm and may have an anterior spine for boring into the intermediate host. The intermediate host is a snail, though second intermediate hosts such as other snails or a wide range of other invertebrates, or even vertebrates, may occur in the life-cycle. The miracidium is usually provided with excretory and nervous systems and may have a sac-like gut and an eye-spot. A number of germinal cells are attached to the walls of the body cavity. This larval stage does not feed and further development occurs after it enters a snail. Penetration is assisted by the boring action of the miracidia and probably by enzyme secretions from the apical gland. Following penetration the ciliated coat is lost and the form becomes a *sporocyst*—an undifferentiated mass of cells. Within the sporocyst the germinal cells multiply and produce either daughter sporocysts or rediae, although rediae are not always a feature of all digenetic trematodes and do not occur in the life cycle if daughter sporocysts are formed. One or more generations of rediae may occur. The final stage, the *cercaria*, is produced by the sporocyst or the redia. The redia has an oral sucker, a pharynx, a sac-like intestine, an excretory system and a birth-pore, through which cercariae produced inside it escape. The cercaria has suckers and an intestine like that of the adult, excretory and nervous systems, special glands and sometimes an anterior spine. It is frequently also provided with a

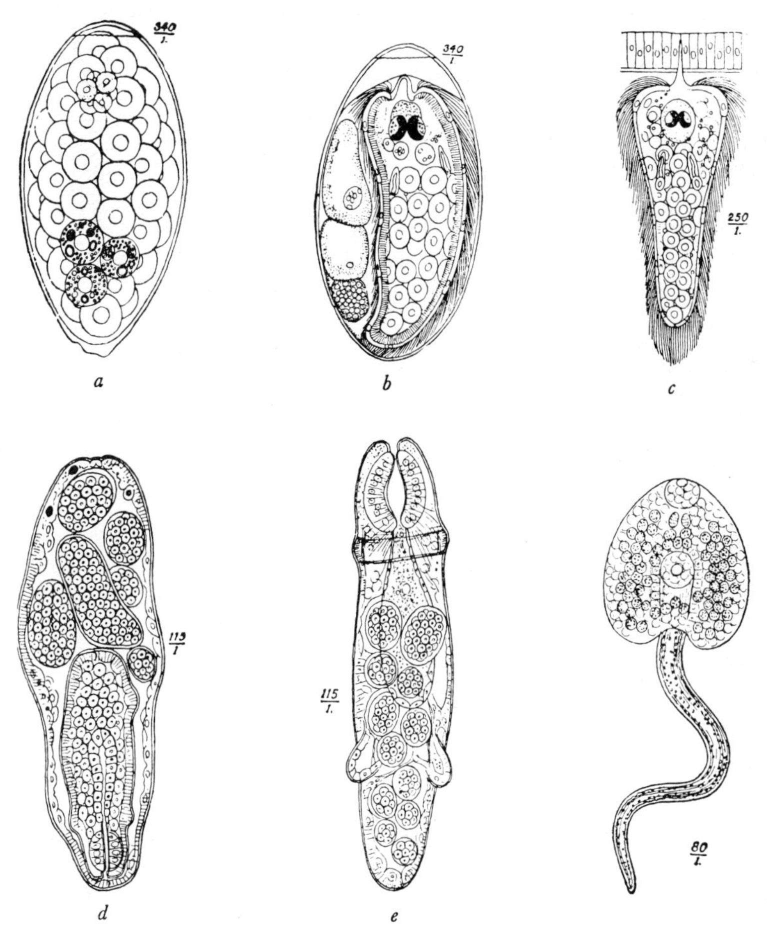

Fig 1.6 Development of *Fasciola hepatica*. *a*, Egg. *b*, Egg containing miracidium. *c*, Miracidium. *d*, Sporocyst. *e*, Redia. *f*, Cercaria. (*From Baylis, after Thomas*)

tail, by which it propels itself through the water after escaping from the snail. The cercaria leaves the snail host actively through an opening or through the tegument or cercariae are expelled passively, in masses. Cercariae usually encyst on or in a second intermediate host or on vegetation. The encysted form undergoes physiological maturation to produce the infective stage, the *metacercaria*.

The metacercaria is the final larval stage and has to reach the definitive host in order to complete its life-cycle. It may enter this host passively on contaminated herbage, in water or on or in a second intermediate host. However, in the family Schistosomatidae cercariae actively penetrate the skin of the definitive host.

The reproductive potential in the trematode life-cycle is enormous and millions of cercariae may be produced from a single miracidium.

When the encysted metacercaria is swallowed by the final host excystation occurs in the intestinal tract and the immature stage migrates to its predilection site. The structure of cercariae varies and they have been given names that refer to details of their anatomical structure. They are discussed by Dawes (1946), Schell (1970) and other authors.

Physiology

Not as much information is available on the physiology of trematodes as on the physiology of cestodes and nematodes, primarily because in vitro culture of these parasites is not as far advanced. The physiology of trematodes is reviewed by Smyth (1966, 1976). The food materials of trematodes comprise blood, mucus and tissues taken into the caeca via the oral sucker. Absorption of amino acids and monosaccharides may also take place through the tegument. Digestion appears to be extracellular in the lumen of the caeca.

The parasitic stages of members of the family Strigeidae have been cultured in vitro but few parasites of economic importance in domestic animals and man have shown in vitro development, although Clegg (1965) had considerable

Table 1.1. Classification of Cercariae according to Lühe (1909)

(A) Body of cercaria with a dorsal, longitudinal undulating fin-fold along the body	*Lophocercaria*
(B) Body without longitudinal membrane	
(I) Tail with two symmetrical, slender caudal appendages which arise separately and are several times longer than the body. Oral aperature at middle of ventral surface. Intestine simple, sac-like	*Gasterostome cercaria*
(II) Caudal appendage variable, absent or distally forked, but never split to the base. Oral aperture anterior. Intestine bifurcate	
(*a*) Ventral sucker absent	*Monostome cercaria*
(*b*) Ventral sucker present	
(1) Ventral sucker posterior, immediately anterior to base of slender tail	*Amphistome cercaria*
(2) Ventral sucker anterior to hind end of body and well separated from base of tail if latter is present	*Distome cercaria*

Distome Cercariae are further classified as follows:

(A) Cercariae single	
(I) Tail well developed	
(*a*) Body retractile into a chamber formed by base of tail	*Cystocercous cercaria*
(*b*) Body not retractile into tail	
(1) Tail not bifurcate	
(*a*) Tail with bristles	
(i) Tail contractile to equal or exceed width of body	*Rhopalocercous cercaria*
(ii) Tail always distinctly narrower than body	*Leptocercous cercaria*
(*b*) Tail with bristles (marine)	*Trichocercous cercaria*
(2) Tail distally bifurcate	*Furcocercous cercaria*
(II) Tail stumpy or absent	
(*a*) Tail stumpy	*Microcercous cercaria*
(*b*) Tail absent	*Cercariaeum*
(B) Cercariae joined into a colony by their tapering tail ends (marine)	*Rat-king cercariae*

success with *Schistosoma mansoni* which showed development to near sexual maturity in vitro.

A more detailed account of the systematics of the Trematoda may be obtained from Yamaguti (1958) and Schell (1970) and the biology from Whitfield (1979).

The following families and genera of the subclass Digenea are dealt with in this volume. The classification is based on Yamaguti (1958) and Schell (1970). The digenetic trematodes of importance in fish are dealt with first and should be consulted with the section dealing with monogenean trematodes of fishes (pp. 8–11).

SUBORDER: PROSOSTOMATA Odhner, 1905

Family: CLINOSTOMIDAE Lühe, 1901
Genus: *Clinostomum*, Leidy, 1856

Family: STRIGEIDAE Railliet, 1919
Genus: *Apatemon* Szidat, 1928
Parastrigea Szidat, 1928
Cotylurus Szidat, 1928

Family: DIPLOSTOMIDAE Poirier, 1886
Genus: *Diplostomum* Nordmann, 1832 (Syn. *Proalaria*)
Posthodiplostomum Dubois, 1938
Neodiplostomum Railliet, 1919
Alaria Schrank, 1788

Family: ALLOCREADIIDAE Stossich, 1903
Genus: *Crepidostomum* Braun, 1900

Family: ACANTHOCOLPIDAE Lühe, 1909
Genus: *Stephanostomum* Looss, 1899

Family: SANGUINICOLIDAE Graff, 1907
Genus: *Sanguinicola* Plehn, 1905

Family: DICROCOELIIDAE Odhner, 1911
Genus: *Dicrocoelium* Dujardin, 1845
Platynosomum Looss, 1907
Athesmia Looss, 1899
Eurytrema Looss, 1907
Concinnum Bhaleras, 1936

Family: HETEROPHYIDAE Odhner, 1914
Genus: *Heterophyes* Cobbold, 1866
Metagonimus Katsurada, 1913
Euryhelmis Poche, 1925
Cryptocotyle Lühe, 1899
Apophallus Lühe, 1909

Family: PROSTHOGONIMIDAE Nicoll, 1924
Genus: *Prosthogonimus* Lühe, 1899

Family: PLAGIORCHIIDAE Ward, 1917
Genus: *Plagiorchis* Lühe, 1899

Family: OPISTHORCHIIDAE Braun, 1901
Genus: *Opisthorchis* Blanchard, 1895
Clonorchis Looss, 1907
Pseudoamphistomum Lühe, 1909
Metorchis Looss, 1899
Parametorchis Skrjabin, 1913

Family: NANOPHYETIDAE Dollfus, 1939
Genus: *Nanophyetus* Chapin, 1927

Family: FASCIOLIDAE Railliet, 1895
Genus: *Fasciola* Linnaeus, 1758
Fascioloides Ward, 1917
Fasciolopsis Looss, 1899
Parafasciolopsis Ejsmont, 1932

Family: ECHINOSTOMATIDAE Poche, 1926
Genus: *Echinostoma* Rudolphi, 1809
Echinoparyphium Dietz, 1909
Hypoderaeum Dietz, 1909
Echinochasmus Dietz, 1909
Isthmiophora Lühe, 1909

Family: NOTOCOTYLIDAE Lühe, 1909
Genus: *Notocotylus* Diesing, 1839
Catatropis Odhner, 1905
Ogmocotyle Skrjabin and Schultz, 1933

Family: BRACHYLAEMIDAE Joyeux and Foley, 1930
Genus: *Postharmostomum* Witenberg, 1923

Family: HASSTILESIIDAE Hall, 1916
Genus: *Skrjabinotrema* Orlov, Erschov and Banadin, 1934

Family: TROGLOTREMATIDAE Odhner, 1914
Genus: *Collyriclum* Kossack, 1911
Troglotrema Odhner, 1924

Family: PARAGONIDAE Dollfus, 1939
Genus: *Paragonimus* Braun, 1899

Family: PARAMPHISTOMATIDAE Fischoeder, 1901
Genus: *Paramphistomum* Fischoeder, 1901
Cotylophoron Stiles and Goldberger, 1910
Calicophoron Näsmark, 1937
Ceylonocotyle Näsmark, 1937
Gigantocotyle Näsmark, 1937
Gastrothylax Poirier, 1883
Fischoederius Stiles and Goldberger, 1910
Carmyerius Stiles and Goldberger, 1910
Gastrodiscus Leuckart, 1877
Gastrodiscoides Leiper, 1913
Pseudodiscus Sonsino, 1895

Family: CYCLOCOELIDAE Kossack, 1911
Genus: *Tracheophilus* Skrjabin, 1913

Family: SCHISTOSOMATIDAE Poche, 1907
Genus: *Schistosoma* Weinland, 1858
Ornithobilharzia Odhner, 1912
Bilharziella Looss, 1899
Heterobilharzia Price, 1929
Bivitellobilharzia Vogel and Manning, 1940
Australobilharzia Johnston, 1917
Gigantobilharzia Odhner, 1910
Trichobilharzia Skrjabin and Zakharow, 1920

FAMILY: CLINOSTOMIDAE LÜHE, 1901

Medium-sized to large trematodes with a flat body. Parasitic in reptiles, birds and mammals.

Genus: Clinostomum Leidy, 1856

Clinostomum complanatum (Rudophi 1819) and **C. marginatum** (Rudolphi 1819) are found in the mouth and throat of piscivorous birds such as herons and bitterns. They are cosmopolitan in distribution.

Life-cycle. Eggs are dropped while the bird is drinking or are swallowed and passed in the faeces. The first intermediate hosts are *Helisoma* and *Lymnaea* spp. (snails). The metacercariae are found subcutaneously and in the muscles of freshwater fish such as bass, perch, trout and other fishes and in frogs. The infection is common in southern Asia and tropical countries.

Pathogenesis. The metacercariae are a cause of unsightly yellow cysts ('yellow grub'). White, coiled parasites, 1.25–4 mm long and 2 mm wide, are found in pinhead to 2.5 mm in diameter cysts. The lesions may occur in gold-fish, other aquarium fish and laboratory fish obtained from their natural habitat, especially tropical countries. A case of laryngopharyngitis due to the ingestion of fish infected with *Clinostomum* spp. has been recorded in man.

Control. See p. 24.

FAMILY: STRIGEIDAE RAILLIET, 1919

These worms are characterized by a constriction which divides the body into an anterior, flattened or cup-shaped portion containing the suckers, and a posterior, cylindrical part containing the reproductive organs. The ventral sucker may be poorly developed or absent and behind it there is usually a special adhesive organ. The genital pore opens posteriorly in a depression or 'bursa copulatrix'. The testes lie tandem, with the ovary anterior to them. The cirrus-sac and pouch are usually absent. The uterus contains relatively few large eggs. The vitellaria are follicular and well developed, either in both parts of the body or only in its posterior part. They are parasites in the alimentary canal, chiefly of birds but some species also occur in mammals. The cercariae are furcocercous, provided with a pharynx, and develop from sporocysts in snails. Metacercariae are usually found in fishes, but also in snails, leeches, etc.

Genus: Apatemon Szidat, 1929

Apatemon gracilis (Rudolphi 1819) (Fig. 1.7) occurs in the intestine of the pigeon, duck and wild duck in Europe, North and South America and the

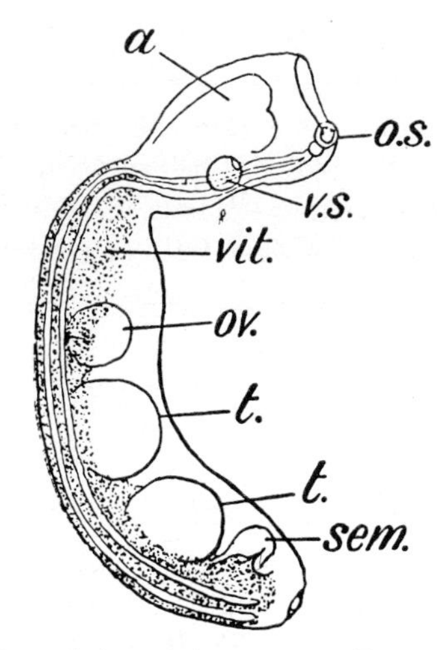

Fig 1.7 Lateral view of *Apatemon gracilis*.

a	adhesive organ
os	oral sucker
ov	ovary
sem	seminal vesicle
t	testes
vit	vitellaria
vs	ventral sucker

Far East. It is 1.5–2.5 mm long by 0.4 mm and is concave dorsally. The anterior cup-shaped part forms about one-third of the total length. It contains an adhesive organ. The cirrus and cirrus-sac are absent. The 'bursa' contains at its base a weak copulatory organ or genital cone. The vitellaria are restricted to the posterior part of the body. The eggs measure 100–110 by 75 μm.

Life-cycle. The eggs are passed in the faeces of the host and the miracidium hatches under favourable conditions in about three weeks. After entering a suitable snail the sporocyst develops. It is very slender and about 20 mm long. The cercariae are formed directly by the sporocysts. The secondary intermediate hosts are the leeches *Haemopis sanguisuga* and *Herpobdella atomaria*. Metacercariae are also found on the viscera and in the eye of fishes (Blair 1976).

Genus: Parastrigea Szidat, 1928

Parastrigea robusta Szidat, 1929, occurs in the intestine of the domestic duck in Europe. It is 2–2.5 mm long and resembles the previous species, but the anterior portion has two large, lateral expansions and a narrow opening, while the vitelline glands are mainly situated in the lateral expansions and the adhesive organ, but also partly in the posterior part of the body. The eggs measure 90–100 by 50 μm.

Life-cycle. Unknown.

A number of other species are also known, especially from ducks in many areas of the world.

Pathogenesis. The parasites attach themselves to the mucosa of the intestine by means of the anterior cup-shaped part of the body, into which they draw a number of villi, constricting them strongly at their bases. The blood vessels in the villi become markedly hyperaemic and burst, discharging the blood into the cavity, from which it is ingested by the worm. The constricted villi then degenerate and are apparently digested by the secretion of a gland in the adhesive organ. Heavy infections are associated with anaemia, a haemorrhagic enteritis and even death.

Diagnosis is made by finding the eggs in the faeces or the worms at autopsy.

Treatment. Carbon tetrachloride has been used in doses of 1–2 ml.

However, anthelmintics effective against trematodes in mammals may be effective, e.g. bithionol, oxyclozanide, albendazole, etc.

Genus: Cotylurus Szidat, 1928

Cotylurus cornutus (Rudolphi, 1808), **C. platycephalus** (Creplin, 1825), **C. flabelliformis** (Faust, 1925) and **C. variegatus** (Creplin, 1825) are found in the small intestine and rectum of domestic and wild ducks, pigeons, gulls and razorbills in Europe, Africa, Asia and North and South America. The adults are small, less than 1.5 mm long. They resemble *Apatemon* spp. but the bursa contains a strong copulatory organ. The eggs of the various species are large, 90–112 by 56–76 μm and unembryonated when laid.

Life-cycle. The miracidia hatch in six to 23 days and penetrate into snails such as *Lymnaea stagnalis, L. palustris* and *Valvata piscinalis*. The cercariae are furcocercous and have a pharynx.

They emerge from the first intermediate host and penetrate into other snails of the same species or other planorbid or physid snails. The metacercariae may develop as hyperparasites when snails are already infected with sporocysts and rediae of other trematodes. Metacercariae of *C. variegatus* have been found in the air bladder, peritoneum and body cavity of fish such as *Perca fluviatalis* and *Acerina cernua* (Odening & Bockhardt 1971). The prepatent period in birds may be four to seven days and the adult parasites are found for up to one month.

Nestling pigeons may become infected through their parents feeding them crop contents.

Pathogenesis. Cotylurus spp. may cause a haemorrhagic enteritis, such having been recorded in razorbills due to *C. platycephalus* (Lowe & Baylis 1934).

FAMILY: DIPLOSTOMATIDAE POIRIER, 1886

These trematodes are similar to the Strigeidae, but the anterior part of the body is more flattened and spatulate and frequently ear-like processes are present on the anterolateral parts of the forebody. The posterior part of the body is cylindrical. Parasites of birds and mammals. The metacercariae are found in fishes and amphibians.

Genus: Diplostomum Nordmann, 1832 (Syn: Proalaria)

Diplostomum spathaceum (Rudolphi, 1819) Olsson, 1876. This form is found in the intestine of a wide variety of gulls in Europe, North America and the USSR. The total length is 2–4 mm. The anterior region is shorter and broader than the posterior, the suckers are small and the ventral sucker is incorporated into an accessory adhesive organ which occupies about one-third of the breadth of the anterior part. The vitellaria occupy most of the posterior region and extend forward on each side of the adhesive organ. Eggs are large (100 by 60 μm) and are embryonated when passed in the faeces.

Life-cycle. The first intermediate hosts are snails of the genus *Lymnaea*. Furcocercariae are produced. These penetrate the skin of fish and migrate to the eye. The metacercariae are flattened, leaf-like, 400 μm long and develop in the lens of the eye. A wide variety of freshwater fish are infected, including cyprinids, salmonids, stickleback and eel.

Pathogenesis. A heavy mortality in black-headed gull chicks due to massive infections of *D. spathaceum* has been recorded by Jennings and Soulsby (1958).

The metacercariae are important parasites of fishes. The cercariae cause haemorrhages when they penetrate the skin and death may result if a large number of cercariae penetrate the skin at one time. In the eye the metacercariae are visible as white spots and the entire eye is white in heavy infections. The fish are blind, there is an increase in intraocular pressure, the cornea may rupture and secondary bacterial or fungal infection may occur. A few parasites may cause damage to young fish, but adult fish may carry up to 50 parasites without blindness.

Control. See p. 24. Van Duijn (1973) reports that the infected eye may be lanced and disinfectant applied. The infection may be controlled with good results on fish breeding farms through the use of copper sulphate (0.5 g/litre water) or malachite green (0.3 g/m^3 water) (Lavrovskii 1977).

Genus: Posthodiplostomum Dubois, 1958

Posthodiplostomum cuticula (Nordmann, 1832) has a distinctly two segmented body with a small acetabulum situated at or near the middle of the anterior part of the body. The adults are found in the intestine of herons and kingfishers in Europe, North America and the USSR.

Life-cycle. The first intermediate hosts are planorbid molluscs. The second intermediate hosts are fish, particularly cyprinids, and the metacercariae may be found in the skin, fins, superficial musculature, cornea, etc.

Pathogenesis. Infection ('black spot disease') can be common in the skin of cyprinids, particularly carp, and can be important in young fish in pond culture and in natural waters. The metacercariae are found in light-coloured cysts of connective tissue around which accumulate pigment cells forming small brown and black spots (0.85–3.8 mm in diameter) in the cutis and underlying muscles. In fishes with little pigment the spots remain a brownish colour. The parasite in the cysts is flat, oval and 1–2 mm long. It bears two suckers and an attachment organ. Heavily infected young fish may die and the presence of infection is aesthetically unpleasing and reduces the commercial value of the fish.

Control. See p. 24.

Posthodiplostomum minimum (MacCallum, 1921) is a common infection of a wide variety of freshwater fish in North America. Non-pigmented white cysts ('white grub'), 1 mm in diameter, are found in the mesentery, kidneys, liver, pericardium and spleen. The definitive host is the heron.

Genus: Neodiplostomum Railliet, 1919

Neodiplostomum perlatum (Ciurea, 1911) occurs in the intestine of piscivorous birds. The metacercariae are found in the skin, fins, muscles and perhaps internal organs of carp. The metacercariae have the appearance of small grey pearls, giving the name 'grey pearl disease'.

Neodiplostomum multicellulata (Miller, 1923) is found in the intestine of herons. The first intermediate hosts are snails in the genus *Physagyrina* and the metacercariae are found in large thin-walled cysts in the liver of bass, bluegill and sunfish. The cysts may destroy the liver.

Genus: Alaria Schrank, 1788

Alaria alata (Goeze, 1782) Hall and Wigdor, 1918 is found in the intestine of dog, cat, fox and also mink in Europe, Africa, Japan, Australia, South and North America. **A. canis** La Rue and Fallis, 1934, **A. americana** Hall & Wigdor, 1918, **A. michiganensis** Hall & Wigdor, 1918, **A. mustelae** Bosma, 1931 and **A. arisaemoides** Augustine & Vribe, 1927 are found in dogs, cats, coyotes, foxes and other animals in North America. Other species such as **A. marcianae** (La Rue, 1917) Walton, 1950 are found in South America. The normal definitive hosts of *Alaria* spp. are wild carnivores and occasionally domestic animals may be infected.

These species are 2–6 mm in length and the flat expanded anterior part is much longer than the posterior cylindrical part. At the anterior lateral corners of the anterior part there are two ear-like tentacles. The suckers are very small and the adhesive organ consists of two long folds with distinct lateral margins. The vitellaria are in the anterior part of the body while the gonads are in the posterior part. The eggs are yellowish brown and measure 98–134 by 62–68 μm.

Life-cycle. Miracidia hatch from the eggs and swim in water, entering freshwater snails such as *Planorbis vortex* and *P. planorbis*. Sporocysts produce cercariae with bifurcated tails. These cercariae then penetrate the second intermediate host, tadpoles and frogs, and the mesocercariae (equivalent to metacercariae) may be found encysted in the muscles of frogs and toads. The definitive host becomes infected on eating these and the fluke undergoes a prolonged migration through the abdominal and thoracic cavities or via the circulation to the lungs and thence to the small intestine via the trachea and pharynx. Paratenic hosts may also be involved in the life-cycle. Mice, rats, snakes and birds may be infected with mesocercariae after eating frogs or toads. Man and other animals may also be infected with mesocercariae in this way. In these paratenic hosts the fluke undergoes the pulmonary part of the migratory cycle. The definitive host may then be infected by eating paratenic hosts (Cuckler 1940). The fluke develops to maturity within ten days.

Pathogenesis. Heavy infections may cause a catharrhal duodenitis (Erlich 1938) but the majority of infections are usually non-pathogenic. Man may act as paratenic host and Fernandes et al. (1976) have recorded a fatal case in man associated with eating inadequately cooked frogs' legs. Several thousand mesocercariae were found in the peritoneal cavity, brain, heart, kidneys, liver, lungs, lymph nodes, pancreas, spinal cord, spleen and stomach. Death resulted from asphyxiation due to extensive pulmonary haemorrhage.

Treatment. Anthelmintics effective against other trematode infections may be of value. Thus, bithionol, niclofolan, praziquantel and niclosamide may be effective. Broad-spectrum benzimidazole anthelmintics, i.e. albendazole, fenbendazole, may also be useful. An organism suspected to be a mesocercaria of *Alaria* in the eye of man was destroyed by laser applications with subsequent improvement in vision (Shea et al. 1973).

FAMILY: ALLOCREADIIDAE STOSSICH, 1973

Members of this family are small to medium-sized trematodes. The vitellaria are follicular and extensive. The testes are usually double and the ovary usually pretesticular. The adults are parasites of the digestive tract, particularly the intestine, of marine and freshwater fishes.

Genus: Crepidostomum Braun, 1900

Crepidostomum spp. are common parasites of the intestine of salmonids and cyprinids, also catfish, eel and other freshwater fish in Europe and North America. The adults are elongate, oval, subcylindrical and unarmed. They measure 1–2 mm in length. The oral sucker is terminal and the ventral sucker is in the anterior half of the body. Two (or four) testes are arranged in tandem in the posterior half of the body and the ovary is pretesticular. The eggs are oval and unembryonated when laid.

Life-cycle. First intermediate hosts are clams or *Lymnaea* snails and the second intermediate hosts are mayfly nymphs, gammarids or crustaceans such as crayfish.

Pathogenesis. Large numbers of adults may cause inflammation of the intestine.

Treatment and control. See p. 24.

FAMILY: ACANTHOCOLPIDAE LÜHE, 1909

Members of this family are elongate distomes with circumoral spines surrounding a small oral sucker. The testes are tandem or diagonal in the hindpart of the body. The ovary is pretesticular. The vitellaria are follicular and in the hindbody. Adults are parasitic in the intestine of fishes.

Genus: Stephanostomum Looss, 1899

Stephanostomum baccatum (Nicoll, 1907) is found in the intestine of marine fishes, particularly flatfish, in the Atlantic. Adults are elongate and covered with spines. The oral sucker is terminal with two rows of circumoral spines. The eggs are large.

Life-cycle. First intermediate hosts are gastropods such as *Buccinum undatum* and *Neptuna antiqua* (Mackenzie & Liversidge 1975). Cercariae emerge and encyst as metacercariae in unpigmented cysts in the muscles of winter flounder, plaice and other pleuronectids.

Pathogenesis. The infections are usually nonpathogenic. Other *Stephanostomum* spp. are found in the intestine of marine fishes throughout the world.

FAMILY: SANGUINICOLIDAE GRAFF, 1907

Digenea, without suckers, inhabiting the vascular system of marine and freshwater fishes. Testes divided into numerous follicles. Ovary median.

Genus: Sanguinicola Plehn, 1905

Sanguinicola inermis Plehn, 1905, and other species are parasitic in the circulatory system of cyprinids in Europe and the USSR and in salmonids in North America. Lanceolate adults, up to 1 mm long, are found in the ventral aorta and large vessels of the gills. The uterus contains only one egg. Eggs are triangular in shape bearing lateral projections and measure 45–75 by 30–45 μm when embryonated. The dimensions of the egg nearly double during embryonation.

Life-cycle. Eggs are discharged, particularly in the summer, into the vascular system and are carried in the branchial capillaries to the gills. Here the miracidia develop over seven or eight weeks, hatch and break out of the gills. Intermediate hosts include *Lymnaea* spp. snails. The cercariae which emerge attach to the gills and skin of fish, penetrate into the circulatory system and reach sexual maturity after about one month.

Pathogenesis. Sanguinicola spp. are important in cultured carp and trout. The acute gill form of the disease is seen in one- and two-year-old carp and trout and may cause severe mortality. The triangular-shaped eggs occlude the branchial capillaries and cause thrombosis and necrosis of the gills. Migrating miracidia produce haemorrhages. Secondary bacterial infection may occur. Infected fish have pale or translucent, flaccid gills and are very sluggish. Large numbers of eggs may be found in the gills and if a large number hatch at one time the extensive trauma and blood loss may be fatal. *Salmo clarki*, experimentally infected with *S. klamathensis* Wales, 1958, have a decreased packed cell volume and oxyhaemoglobin levels. There was an 80% mortality within three months of infection and infected fish weighed less and were shorter than control fish (Evans 1974*a*).

Chronic disease is seen in older fish. Eggs are carried in the blood to other organs, particularly the kidneys, and are trapped in capillaries. The eggs become encapsulated and there is necrosis and hypertrophy of the renal epithelium (Evans 1974*b*). There is resultant nephritis, ascites, exopthalmus and bristling of scales.

Control. Prophylactic measures are similar to those for other trematodes of fishes (see p. 24). Different age groups of fish should be kept separately since older fish may be carriers of the infection. Bithionol (100–120 mg/kg), hexachloroparaxylol (0.5–2.5 g/kg), oxyclozanide (40–50 mg/kg), bromsalan (120–160 mg/kg) or antimosan (15 ml/kg) given two to four times in the feed decrease the intensity of the infection by 30–70% and cure 10–40% of infected fishes.

IMPORTANCE OF DIGENETIC TREMATODES IN FISHES

Adult digenetic trematodes are common in the digestive tract of marine and freshwater fishes. Although they are not usually a cause of serious disease, *Crepidostomum* spp. may cause significant inflamation of the gut. However, it is likely that the presence of digenetic trematodes in the intestines of fishes will cause a retardation in growth similar to that seen in large domesticated animals infected with gastrointestinal nematodes. This would be of economic importance in fish farming. In the future, treatment of infected fish with modern anthelmintics given in the feed with resultant increases in growth rates is an approach to the control of these parasites. Another adult digenetic trematode, *S. inermis*, is resident in the circulatory system of fishes. It can cause significant losses due to acute disease in young cultured carp and salmonids. Also, the parasite may be associated with chronic disease and debility in older fish.

Fish act as second intermediate hosts for a wide variety of digenetic trematodes. The cercariae encyst as metacercariae (black spot, yellow grub, grey pearl, etc.) in the skin, fins and underlying musculature. These infections are of economic importance since their presence is unsightly and decreases the commercial value of the fish. In addition, heavily infected young fish are debilitated, more susceptible to predation and adverse conditions and may die. Further, metacercariae of some species (*Diplostomum*) may be found encysted in the eyes of fish. Here they cause opacities and blindness, affecting the feeding ability of the fish.

Some metacercariae found in fish, whose adult stages are normally parasitic in piscivorous birds and mammals, may infect man. These include *O. tenuicollis* and *C. sinensis* which are widespread in the skin of fish in Asia. *H. heterophyes* is common in the Middle East and Asia in fish in brackish waters while *H. brevicaeca* is found in the Phillipines. Metacercariae of *M. yokogawai* are found in mullet in the Far East and Baltic regions. *C. lingua*, seen occasionally in man in northern Europe, occurs in herring and other inshore marine fish. *Nanophyetus salmincola* may be found occasionally in man.

Control

A variety of control measures may be used. These include the elimination of definitive hosts, piscivorous birds and mammals, from the areas of the fish ponds. Molluscs can be eliminated through the use of molluscicides. Molluscan intermediate hosts can also be killed by draining and drying ponds or through the use of quicklime, chloride of lime, etc. It must be remembered that millions of cercariae may be released from snails in the inlet water to the ponds, and such snails should be eliminated by the use of filters, screens or electric grids across the water inlet.

Adult digenetic trematode infections might be reduced by treatment with drugs such as di-N-butyl tin oxide, bithionol, hexachloroparaxylol, oxyclozanide, bromsalan, antimosan or the broad-spectrum benzimidazole anthelmintics. Little treatment is available for metacercarial infections. The cysts may be excised from aquarium fishes. Infected fish may then be bathed in a picric acid solution for one hour (2–7 ml in 100 litres of water). Infections with metacercariae have been controlled also through the use of copper sulphate or malachite green in the water.

FAMILY: DICROCOELIIDAE ODHNER, 1911

Species of this family are small or medium-sized flukes parasitic in the biliary and pancreatic ducts of amphibia, reptiles, birds and mammals. The body is flattened and elongate, with weak musculature and loose parenchyma through which the internal organs are easily seen. The cuticle often lacks spines. The suckers are not far apart. A pharynx and an oesophagus are present and the intestinal caeca are simple, not quite reaching the posterior end of the body. The excretory bladder is simple and tubular. The testes are situated not far behind the ventral sucker and the ovary is usually behind them. The genital pore opens in the middle line in front of the ventral sucker. The cirrus is small. A Laurer's canal and a small receptaculum seminis are present. The well-developed vitelline glands lie chiefly in the lateral regions of the body. Most of the space behind the genital glands is filled by the many folds of the uterus. The numerous small eggs are deep brown in colour.

Genus: Dicrocoelium Dujardin, 1845

Dicrocoelium dendriticum (Rudolphi, 1819) Looss, 1899 (syn. *D. lanceolatum*), occurs in the bile ducts of the sheep, goat, ox, deer, pig, dog, donkey, hare, rabbit, elk, coypu and rarely man. Hamsters, cotton rats, white rats and guinea-pigs are also susceptible and the Syrian hamster is the most satisfactory laboratory host. It occurs in Europe, Asia, North Africa, North America and less commonly in South America. It has not been reported from Central and South Africa or Australia. The prevalence in some countries may be sporadic or focal. For example, in the British Isles it is restricted mainly to the western islands off Scotland with sporadic occurrences in the north of England and Ireland. The fluke is 6–10 mm long and 1.5–2.5 mm wide. The body is elongate, narrow anteriorly and widest behind the middle. The cuticle is smooth. The oral sucker is smaller than the ventral. The testes are slightly lobed and lie almost tandem, immediately posterior to the ventral sucker, with the ovary directly behind them. The vitelline glands occupy the middle third of the lateral fields. Behind the gonads the central field is occupied by the transverse coils of the uterus, filled with brown eggs (Fig. 1.8). The latter measure 36–45 by 20–30 μm and are operculate and embryonated when laid.

Life-cycle. Two intermediate hosts, a snail and an ant, are required. The two principal snail hosts are *Zebrina detrita* in Europe and *Cionella lubrica* in North America but some 29 other species have been reported to serve as first intermediate hosts (Soulsby 1965). These include the species *Abida frumentum, Ena obscura, Theba carthusiana, Theba fruticicola*, *Helicella ericetorum*, *H. italia*, *Planorbis marginatus*, *P. complanatus*, *Xerophila candidula* and *Arion* spp. but there is no information on the relative roles of these under natural conditions. The miracidia do not hatch out of the eggs until the eggs have been swallowed by the

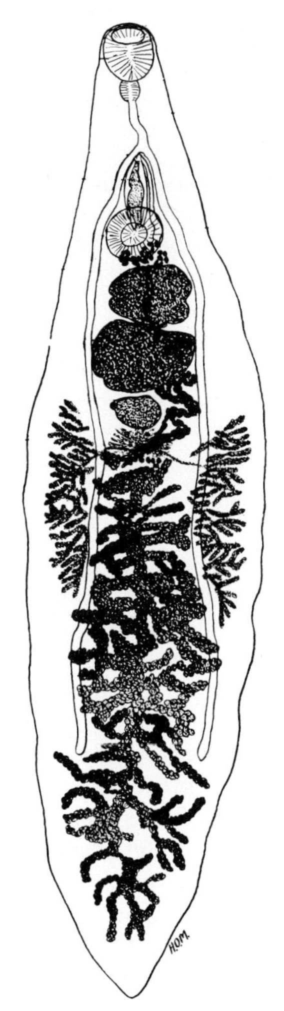

Fig 1.8 Ventral view of *Dicrocoelium dendriticum*.

intermediate host. They hatch in the snail's gut and migrate to the mesenteric gland, where they grow to polymorphous sporocysts lacking a distinct cuticle. These produce a second generation of sporocysts, provided with a cuticle and a birth pore, which in turn produce cercariae (*Cercaria vitrina*). Rediae are not formed. The rate of development is slow; three months or more being required to produce cercariae. The cercariae emerge from the sporocysts only in damp weather following a dry spell and, in the pulmonary chamber of the snail, they clump together in masses called *slime-balls*, in each of which there may be 200–400 cercariae, held together by a sticky, gelatinous substance. These slime-balls are expelled from the snail when a drop in temperature occurs, and adhere to vegetation. The slime-balls are eaten by ants of the genus *Formica*. In the USA, *F. fusca* is concerned, in Europe *F. fusca*, *F. cunicularia*, *F. gagatis* and *F. rufibarbis*, in the Middle East *F. rufibarbis* and in the USSR, *F. fusca*, *F. rufibarbis* and *Proformica nasuta*. Metacercariae are produced in the abdominal cavity, as many as 128 per ant, the time required varying from 26 to 62 days, depending on the environmental temperature. Some metacercariae enter the brain of the ant, causing tetanic spasm of the mouthparts when environmental temperatures are lowered. Affected ants attach to herbage overnight and are available to grazing animals in the early morning. The definitive hosts are infected by swallowing infected ants. It is now generally accepted that cercariae enter the liver via the bile duct. Krull (1958), using experimental infections of hamsters and white mice, observed the metacercariae entering the intestinal opening of the common bile duct and states that they enter the liver by this route rather than by penetration of the host's tissues. He found that the metacercariae may, in these hosts, reach all parts of the biliary system within an hour after their entry through the common bile duct. The young flukes develop in the smaller bile ducts, the older ones go to the larger bile ducts. The prepatent period is 47–54 days and the adults may live for six years or longer (Kirkwood & Peirce 1971).

Pathogenesis and clinical signs. These small flukes penetrate into the fine branches of the bile ducts, in which they lie greatly extended and attached by means of their suckers.

The incidence of infection may be high and for instance may reach 100% in sheep in Yugoslavia. These sheep may contain an average of 2000 *D. dendriticum* but 14 000 adults per sheep is not uncommon (Rukavina 1977). In Spain, 34% of cattle, 23% of sheep and 45% of goats were infected (Manas et al. 1978) while the incidence in cattle in Switzerland was 46% (Marchand 1975). Despite the heavy burdens which may occur in field cases the pathological changes are much less than, for example, those associated with *Fasciola hepatica* infection.

In advanced infection there is extensive cirrhosis and scarring of the liver surface and the bile ducts are markedly distended with large numbers of flukes. Early fibrosis occurs in the portal triads and this later extends in an interlobular and perilobular manner, ultimately producing a condition resembling portal cirrhosis. Marked proliferation of the bile duct glandular epithelium occurs.

The clinical picture in severe cases consists of anaemia, oedema and emaciation, but many cases show no clinical signs.

Treatment. Several anthelmintics have been shown to be effective against *D. dendriticum* though frequently high doses are necessary and these may be uneconomical. Hetolin (*βββ*-tris-4-chlorophenyl propionic acid 4-methylpiperazine hydrochloride) in doses of 19–22 mg/kg is 90% effective (Lämmler 1963). Albendazole, given at 15 mg/kg or 7.5 mg/kg as two doses two to three weeks apart, is also 90% effective. Other benzimidazole anthelmintics are also effective. Cambendazole (25 mg/kg) reduced worm burdens by 95% (Foix 1977) and large doses of fenbendazole (150 mg/kg) and thiabendazole (200–300 mg/kg) were up to 90% effective (Šibalić et al. 1963; Düwel et al. 1975). Similarly, praziquantel (50 mg/kg) was 92% effective, but at lower dose rates of 40 mg/kg and 30 mg/kg it was only 76% and 51% effective (Güralp et al. 1977).

Control. Control measures rely on the treatment of infected animals and the control of snails and ants. However, several factors operate to make control difficult. The eggs of *D. dendriticum* may remain viable for months in soil or faeces and also withstand sub-zero temperatures. The intermediate hosts live in dry areas and, being scattered over the pastures, are less vulnerable than amphibious or aquatic snails to molluscicide treatment. Wild animals are commonly affected: in the USA the woodchuck, cottontail rabbit and deer serve as reservoir hosts for domestic stock and in Europe the rabbit may maintain infection for considerable periods.

Cultivation of pasture to improve soil texture and disrupt ant nests has been suggested and biological control of snails with domestic poultry has been practiced.

Dicrocoelium hospes Looss, 1907, is a closely related species found in the gall bladder of the ox in East and West Africa. In Togo, *Limicolaria* spp. are the first intermediate hosts and the ants *Dorylus* and *Cematogaster* spp. act as second intermediate hosts (Bourgat et al. 1975). In Uganda 80.6% of cows were infected and the incidence appeared to be increasing (Kajubiri & Hohorst 1977).

Genus: Platynosomum Looss, 1907

Platynosomum fastosum (Kossack, 1910) (syn. *P. concinnum*) occurs in the liver and bile ducts of domestic and wild cats in Malaysia, Central and South America, the Caribbean, West Africa, Florida and other southern states of the USA and the Pacific. The incidence of infection can be high and 62 of 100 cats in Honolulu were found to be infected (Chung et al. 1977). Adults measure 4–8 by 1.5–2.5 mm. The testes are horizontal in position and the operculate brownish oval eggs measure 34–50 by 20–35 μm. The eggs are embryonated when laid.

Life-cycle. Eggs are ingested by the snail, *Sublima octona*. Cercariae are produced and these encyst in isopod crustaceans and in lizards, *Anolis cristatellus* being important in Puerto Rico, within which metacercariae are found in the bile ducts. Metacercariae have also been recovered from a toad. Cats become infected by eating lizards. The fluke excysts and migrates up the common bile duct to the bile ducts and gall bladder and develops to maturity in eight to 12 weeks.

Pathogenesis. P. fastosum does not normally cause severe feline disease and frequently only a mild and temporary inappetance associated with hepatic dysfunction may be seen (Taylor & Perri 1977). However, gross liver lesions can be common (Chung et al. 1977) and a marked dilatation of the bile ducts can occur with desquamation of bile duct epithelium. The liver

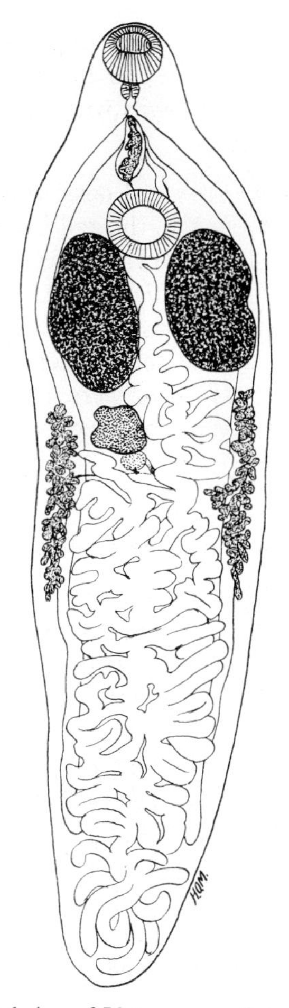

Fig 1.9 Ventral view of *Platynosomum fastosum*.

may be markedly enlarged. Clinical signs are diarrhoea, vomiting, a progressive icterus and possibly death (lizard poisoning) (Leam & Walker 1963). In the terminal stages diarrhoea and vomiting may be continuous.

Treatment. Praziquantel (20 mg/kg) and nitroscanate (100 mg/kg) caused an increase in faecal egg count for two weeks. The egg count then fell to negative, with small numbers of eggs being seen at irregular intervals over the next four months (Evans & Green 1978). Thiabendazole and diamphenethide were ineffective as anthelmintics.

Platynosomum ariestis Travassos, 1918, occurs in the intestine of sheep in Brazil. It is non-pathogenic.

Genus: Athesmia Looss, 1899

Athesmia foxi Goldberger and Crane, 1911, is found in the bile ducts of a variety of monkeys in South America. The adult is long and slender and 8.5 by 0.7 mm. The egg is oval, thick-shelled and operculate and 30 by 19 μm. The first intermediate host is a mollusc. The rest of the life-cycle is unknown. The parasite causes inflammation and enlargement of the bile ducts. It can be common in monkeys imported for experimental purposes.

Genus: Eurytrema Looss, 1907

Eurytrema pancreaticum (Janson, 1889) is found in the pancreatic ducts and more rarely in the bile ducts and the duodenum of sheep, goat, cattle and buffalo in eastern Asia and Brazil and humans in China. It measures 8–16 by 5–8.5 mm. The body is thick and armed with spines, which are often lost in the adult stage. The suckers are large, the oral being the larger of the two. The pharynx is small and the oesophagus short. The testes are horizontal, slightly posterior to the level of the ventral sucker. The genital pore opens just behind the bifurcation of the intestine. The cirrus sac is tubular and reaches back past the anterior margin of the ventral sucker. The ovary is situated near the median line, behind the testes, and the uterus fills the posterior part of the body. The vitelline glands are follicular and are laterally situated. The eggs measure 40–50 by 23–34 μm.

Life-cycle. Tang (1950) found that the land-snails, *Bradybaena* spp. and *Cathaica ravida sieboldtiana*, belonging to the family Fruiticoidolidae, serve as first intermediate hosts of this species. Two generations of sporocysts occur in the snails, the second producing cercariae about five months after infection. Cercariae are extruded onto herbage and are eaten by grasshoppers, *Conocephalus maculatus* and tree crickets, *Oecanthus longicaudus*, which serve as secondary intermediate hosts in Malaysia and the USSR. Metacercariae occur in the haemocele, becoming infective three weeks after infection of the grasshopper. Sheep and goats are infected by inadvertently eating infected grasshoppers, the immature flukes migrating via the pancreatic duct. In sheep and cattle the prepatent period is 80–100 days (Nadykto 1973).

Several other species of this genus have been described from domestic and other ruminants, but it is not clear whether these are all distinct species. *E. ovis*, described by Tubangui in 1925 from sheep in the Philippines, occurs in the perirectal fat of that host. *E. coelomaticum* (Giard and Billet, 1892) is common in the pancreatic ducts of sheep and cattle in Brazil and other parts of South America and occurs in Europe and Asia.

Pathogenesis. Basch (1966) has described the pathological lesions in cattle. A few flukes may elicit little change but usually there is catarrhal inflammation with destruction of duct epithelium. Eggs may penetrate into the walls of ducts causing inflammatory foci and granulomata in which plasma cells and eosinophils predominate. The granulomata are confined to the walls of the ducts and the parenchyma is not affected. Occasionally severe fibrosis may occur producing atrophy of the pancreas but the remaining parenchyma is normal. Severely infected animals may be poor in condition, but no other clinical signs have been definitely ascribed to these parasites.

Treatment. A variety of anthelmintics have been tried, but without success. These include hexachloroparaxylene, hexachlorethane, chlorophos, bithionol, niclofolan and thiabendazole. High doses of albendazole or praziquantel might be effective.

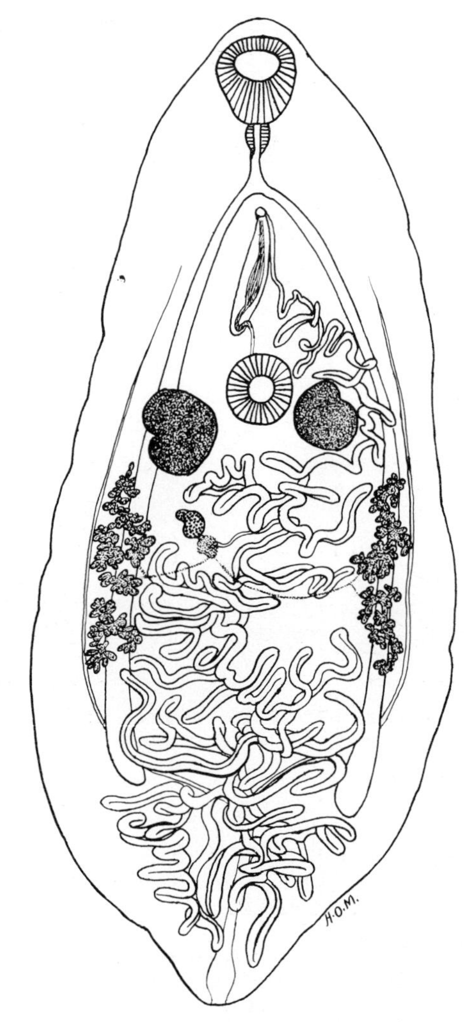

Fig 1.10 Ventral view of *Eurytrema pancreaticum*.

Genus: Concinnum Bhaleras, 1936

Concinnum procyonis Denton, 1942 (syn. *Eurytrema procyonsis*). This species has been reported from the pancreatic ducts, gall bladder and bile ducts of cats, red and grey foxes and racoons in the USA (New York, Connecticut, Maryland, Kentucky). The snail *Mesodon thyroidus* has been infected experimentally and Denton (1944) suggested animals may be infected by ingestion of the snail. However, an arthropod, perhaps a grasshopper, is suggested as a second intermediate host.

Generally this parasite causes no apparent ill health. Parasites are found in the medium sized pancreatic ducts. Periductal fibrosis may produce cord-like ducts and there may be atrophy of glandular acini due to duct fibrosis, otherwise the parenchyma is normal (Sheldon 1966).

Concinnum brumpti Railliet, Henry & Joyeux, 1912, has been recorded from the liver and pancreas of African anthropoid apes.

Concinnum ten (Yamaguti, 1939) has been found in wild carnivores in Japan (Uchida 1976).

FAMILY: HETEROPHYIDAE ODHNER, 1914

Small trematodes, usually not over 2 mm long and wider posteriorly than anteriorly. The body is covered with scale-like spines decreasing in number in the posterior region. The ventral sucker is usually situated near the middle of the body and may be weak or absent. A pharynx and a long oesophagus are present and the intestinal branches reach almost to the posterior end. The genital pore opens close to the ventral sucker and is frequently surrounded by a genital sucker. The testes are oval or slightly lobed, horizontal or diagonal, and situated near to the posterior end of the body. The seminal vesicle is well developed and there is no cirrus-sac. The ovary is oval or slightly lobed, anterior to the testes and median or to the right of the middle. The vitellaria are lateral and usually restricted to the posterior part. The coiled uterus is in the posterior half of the body and contains relatively few eggs. Parasites in the intestines of mammals and birds. Where known, the life-cycle includes two intermediate hosts—snails and fishes or frogs.

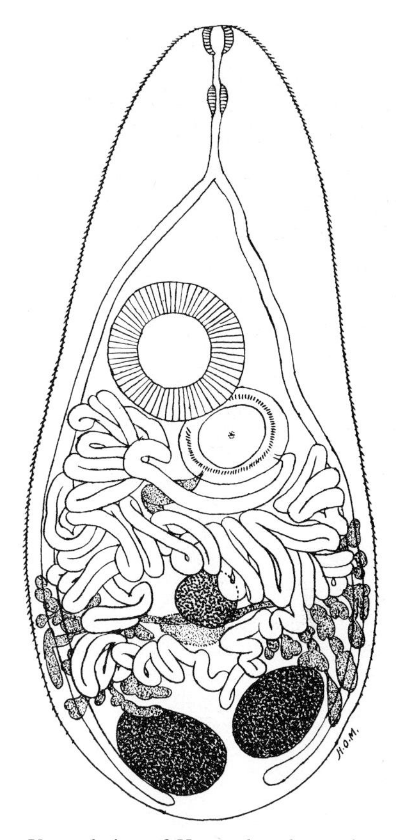

Fig 1.11 Ventral view of *Heterophyes heterophyes*.

Genus: Heterophyes Cobbold, 1866

Heterophyes heterophyes (v. Siebold, 1852) occurs in the small intestine of the dog, cat, fox and man, primarily in the Middle East and east Asia. It measures 1–1.7 by 0.3–0.7 mm and is wider posteriorly than anteriorly. The ventral sucker is situated immediately anterior to the middle and is 0.23 mm wide. The genital sucker lies directly behind it and to one side and bears an incomplete circle of 70–80 small rods. The testes are oval and horizontal in position. The eggs have thick shells; they are light brown in colour, provided with an operculum which fits into a slightly thickened rim of the shell, and measure 26–30 by 15–17 μm.

Life-cycle. The first intermediate hosts are the snails *Pirenella conica* in the Middle East and *Cerithidia cingulata* in Asia which ingest the eggs. The second intermediate host is a fish (*Mugil cephalus*, *M. capito*, *Tilapia nilotica*, *Aphanius fasciatus and Acanthogobius* spp.), in which the metacercaria is encysted. It resembles the adult fluke. Infection takes place through eating infected raw fish. In Egypt the fish is eaten salted as 'fessikh'. The metacercariae live up to seven days in the salted fish. The prepatent period is nine days.

Pathogenesis. The flukes penetrate into the mucosa and may cause slight desquamation, but the pathogenicity of the parasites is so low that they are generally regarded as being practically harmless. In severe human cases diarrhoea may develop; this is usually intermittent and sometimes haemorrhagic. The presence of infection in fish reduces the quality of the flesh and is particularly important since the parasite can infect man. *Mugil capito* caught off the coast of Israel have been shown to contain between 2300 and 6000 metacercariae per gram of fish (Lahav 1974).

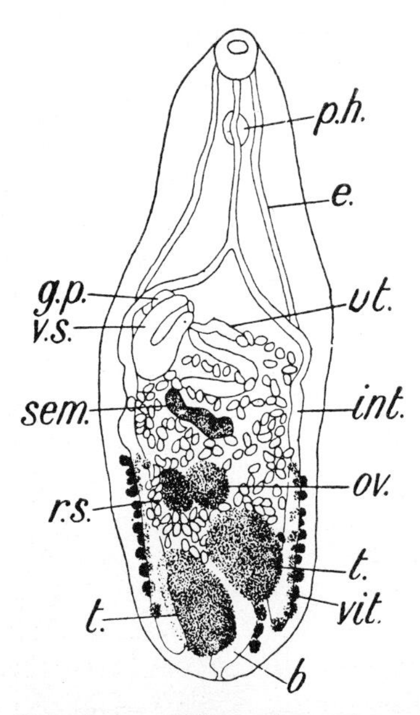

Fig 1.12 Ventral view of *Metagonimus yokogawai*. (*From Baylis, after Ciurea*)

b	excretory bladder
e	excretory canal
gp	genital pore
int	intestinal caecum
ov	ovary
ph	pharynx
rs	receptaculum seminis
sem	seminal vesicle
t	testes
ut	uterus
vit	vitellaria
vs	ventral sucker

Diagnosis is made by finding the eggs in the faeces. They have to be differentiated, especially from the eggs of *Opisthorchis*.

Treatment. See *Metagonimus* sp.

Prevention. The infection can be prevented by using no raw fish as food, while salted fish should not be used within 10 days of salting.

Genus: Metagonimus Katsurada, 1913

Metagonimus yokogawai (Katsurada, 1912) occurs in the small intestine of the dog, cat, pig, man and pelican in eastern Asia and the Balkans. Mice have been infected experimentally. The parasite measures 1–2.5 by 0.4–0.7 mm and is wider posteriorly than anteriorly. The cuticle is armed with spines over the whole body. The ventral sucker is situated to the right of the median line and the genital pore opens in a sinus immediately anterior to it. The sinus and the sucker are surrounded by an elevated muscular ring. The testes are slightly oblique in position. The ovary is median, and the vitelline glands are composed of coarse follicles lying in the posterior parts of the lateral fields. The eggs are very similar to those of *Heterophyes* and measure 27–30 by 15–17 μm.

The first intermediate host is the snail *Semisulcospira libertina* and related species. The second intermediate hosts are several species of fresh-water fish particularly the Cyprinidae (the trout, *Plecoglossus altivelis*, also *Salmo perryi* and *Odontobutis* and *Leuciscus* spp.). The cercariae become encysted under the scales or in the tissue of the gills, fins or tail and the final host infects itself by eating these fish raw.

Pathogenicity etc. is as in the case of *Heterophyes*.

Treatment. Praziquantel at a dose rate of 20 mg/kg given on two consecutive days is the treatment of choice. This anthelmintic is highly effective in man and egg counts become negative following treatment. Niclofolan (two doses of 2 mg/kg) and niclosamide (two doses of 100 mg/kg) decreased egg counts by 89% and 67% respectively (Rim et al. 1978). In another trial 2 g of niclosamide decreased or eliminated egg counts in man (Ahn et al. 1978).

Genus: Euryhelmis Poche, 1925

Euryhelmis squamula (Rudolphi, 1819) occurs in the intestine of the fox, polecat, weasel and mink. The body is broad and flat, measuring about 0.6 by 1.45 mm. The oral sucker is anterior and there are a pharynx and an oesophagus of median length. The intestinal caeca are more or less parallel to the margins of the body. The ventral sucker is situated near the middle and is about half as large as the oral. The testes are lobed in the adult, side by side and posterior in position. The elongate cirrus-sac winds around the right side of the ventral sucker. The genital pore lies

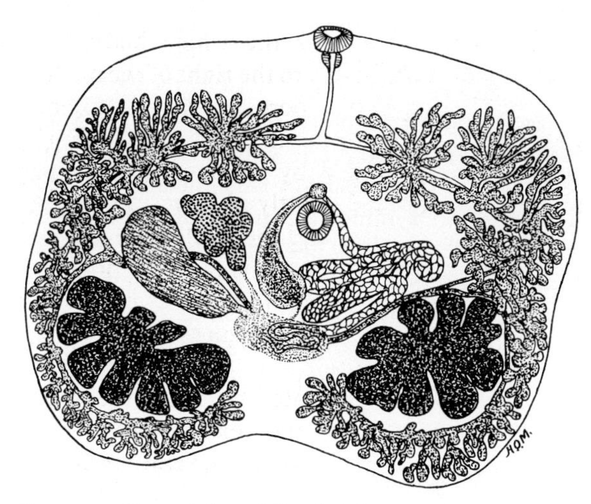

Fig 1.13 Ventral view of *Euryhelmis squamula*.

anterior to the ventral sucker. The ovary is lobed in the adult and lies anterior to the right testis, with a large receptaculum seminis in between. The uterus is coiled and lies mainly between the left testis and the ventral sucker. The vitellaria are well developed and extend along the course of the intestinal caeca. The eggs measure 29–32 by 12–14 μm.

Combes et al. (1974) described the life-cycle of *E. sqamula* in the Pyrenees. The definitive host was *Neomys fodiens*. The first intermediate host was the snail *Bythinella reyniesii* and the parapleurolophocercous cercariae then penetrated and encysted under the skin of young frogs, *Rana temporaria* (not tadpoles). The metacercariae become infective after two months.

Severe infections cause a fatal haemorrhagic enteritis in the mink.

Tetrachlorethylene is effective. It is probable that the benzimidazole anthelmintics, praziquantel, niclosamide and niclofolan, are effective. The animals should be prevented from eating frogs.

Euryhelmis monorchis (Ameel, 1938) occurs in the mink in the USA. It has a single testis on the same side as the ovary and the male genital organs disappear later. Its first intermediate host is the snail *Pomatiopsis lapidaria*; second intermediate hosts are the frogs *Rana clamitans*, *R. pipiens* and *R. palustris*.

Genus: Cryptocotyle Lühe, 1899

Cryptocotyle lingua (Creplin, 1825) Fischoeder, 1903, is common in the intestines of the herring gull, greater and lesser black-backed gull, common tern, kittiwake, razorbill, Slavonic grebe and night-heron. It also occurs in dogs, particularly sled-dogs, man, seal, silver fox, mink and cat. It is found in Europe, Canada, the USA, Siberia and Japan. Christensen and Roth (1949) found it in 17% of dogs in Copenhagen, Cameron (1945) reported it to be the commonest trematode in fox and mink farms in Canada and McTaggert (1958) found it in mink in Scotland.

This species is larger than many heterophyid flukes, having a body shaped like a spatula measuring 0.5–2.0 mm long by 0.2–0.9 mm broad. The cuticle is spiny and the suckers are feeble, the oral one being larger than the ventral one, which is near the middle of the body and is enclosed in the genital sinus. The two branches of the intestine are long and slender. The genital atrium is near the middle of the body. The slightly lobed testes lie side by side or diagonally at the posterior end of the body, the tri-lobed ovary being in front of them and on one side of the middle line. The vitellaria fill all the space outside the intestinal caeca. The uterus has a few wide folds. The eggs are unembryonated when laid and measure 32–50 by 18–25 μm.

Life-cycle. Eggs are eaten by the common shore periwinkle, *Littorina littorea*. Metacercariae encyst in marine fish, particularly those of the families Atherinidae and Pleuronectidae. Tautog, herring, cunner, gudgeon and cod are commonly infected and the metacercariae may remain viable for several years in the fish.

Pathogenesis. Large numbers of flukes cause a marked enteritis associated with degeneration of the epithelium, haemorrhagic erosions and the production of a large quantity of viscid mucus. Cats are less susceptible to the infection, parasites failing to attain their full size in this host. *C. lingua* may be important in farmed trout in that metacercariae form black spots in the skin and superficial musculature.

Cryptocotyle concava (Creplin, 1825) Lühe, 1899, is essentially a parasite of sea birds; however, Christensen and Roth (1949) found it in fox and mink farms and also in 2% of dogs in Denmark. Likewise it has been found in dogs and foxes in USSR, Roumania, North America and Japan.

It is a smaller species than *C. lingua*, measuring 0.5–1.5 mm by 0.35–0.88 mm. The eggs measure 30–40 by 16–20 μm. The molluscan intermediate host *Amnicola longinqua* and metacercariae encyst in fish of the genera *Atherina*, *Gobius*, *Mullus*, *Gasterosteus*, etc.

Cryptocotyle jejuna Nicoll, 1907, resembles *C. concava*, the eggs being 28–36 by 16–19 μm in size, and occurs in Europe. A key for the differentiation of the three species of *Cryptocotyle* is given by Soulsby (1965). *C. jejuna* has been found in dogs experimentally infected by feeding *Gobius melanostomus*.

Both these species produce pathogenic effects similar to those seen in *C. lingua* infection.

Genus: Apophallus Lühe, 1901

Apophallus mühlingi (Jägershiöld, 1899) Lühe, 1909, is normally a parasite of gulls and cormorants in Europe but has also been found in the intestine of cats and dogs. It is a small form, 1.2–1.6 mm long by 0.2–0.23 mm broad. The eggs measure 32 by 18 μm. The life-cycle is incompletely known but metacercariae are found in fish of the Cyprinidae family.

Apophallus donicum Skrjabin and Lindtrop, 1919, occurs in the small intestine of cat, dog, fox, seal and man in eastern Europe and North America. It is a small form, 0.5–1.15 mm in length by 0.2–0.4 mm in breadth. The cuticle is spiny and the testes are large, rounded and lie in the posterior part of the body. The eggs measure 35–40 by 19–24 μm. Rediae and cercariae have been found in the stream snail, *Flumenicola virens*. The metacercariae occur in a wide variety of fish in the family Cyprinidae, such as *Perca*, *Lucioperca*, and *Scardinius* spp. The prepatent period in man is eight days (Neimi & Macy 1974).

FAMILY: PROSTHOGONIMIDAE NICOLL, 1924

These are small distomes, tapering anteriorly and rounded posteriorly. The cuticle is covered with spines. The ventral sucker is in the anterior half. The testes are symmetrical and behind the ventral sucker while the ovary lies between these or dorsal to the ventral sucker. The vitellaria usually lie in grape-like bunches in the lateral fields. The eggs are small. Parasites of birds and rarely mammals.

Genus: Prosthogonimus Lühe, 1899

Prosthogonimus pellucidus (v. Linstow, 1873) (syn. *P. intercalandus*) occurs in the bursa of Fabricius, oviduct and posterior intestine of the fowl, duck and various wild birds. It measures 8–9 by 4–5 mm and is broad posteriorly. It has a pale reddish-yellow colour when fresh. The irregularly oval testes lie horizontal at the middle of the body. The genital pore is situated next to the oral sucker and the cirrus-sac is elongate, extending to near the ventral sucker. The ovary is much lobed and

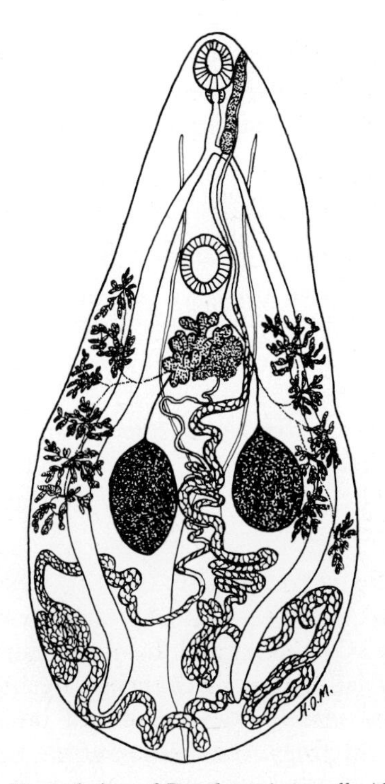

Fig 1.14 Ventral view of *Prosthogonimus pellucidus*.

lies partly dorsal to the ventral sucker. The vitellaria extend from the level of the ventral sucker to the posterior end of the testes. The operculate, dark-brown eggs bear a small spine at the pole opposite to the operculum. They measure 26–32 by 10–15 μm.

Prosthogonimus macrorchis Macy, 1934, is a similar species to the above occurring in the bursa of Fabricius and oviduct of domestic poultry and ducks and also wild birds in North America. It is 5–7 mm in length and the testes are relatively larger than in *P. pellucidus.*

Prosthogonimus ovatus (Rudolphi, 1803) Lühe, 1899, is found in the bursa of Fabricius and oviduct of domestic fowl and geese and a wide variety of wild birds in Europe, Africa and Asia. It is a smaller form than the above two species measuring 3–6 by 1–2 mm. The testes are slightly elongate lying side by side behind the mid part of the body. The ovary is deeply lobed and the eggs are small, 22–24 by 13 μm. Nath (1973) considers *P. ovatus* to be the only valid species of this genus.

Other species include **P. anatinus** Markow, 1902, of domestic ducks in USSR, **P. cuneatus** (Rudolphi, 1809) Braun, 1901, of the swan and **P. oviformis** Strom, 1940, of the duck in Europe.

Life-cycle. Two intermediate hosts are required, the first a water snail and the second the nymphal stage of various species of dragonflies. *Amnicola limosa porata* is the first intermediate host for *P. macrorchis*, *Bithynia teutaculata* serves *P. pellucidus* as such and *B. leachi*, *Gyraulus albus* and *G. gredleri* are the snail hosts for *P. ovatus.* Sporocysts are formed which produce cercariae without undergoing redial development and these, being liberated from the snail, swim about in the water. They are then drawn into the anal openings of dragonfly naiads by the breathing movements of these insects. The tail of the cercaria is lost in the respiratory chamber of the naiad and the metacercaria penetrates into the muscles and encysts in the haemocele of the nymph. Metacercariae may persist in the insect until it is mature and the final host is infected by eating either the adult dragonfly or the nymphal stage.

Several species of dragonflies may serve as hosts. In North America those of the genera *Tetragoneuria*, *Leucorhynia*, *Epicordulia* and *Mesotheronis* are concerned (Macy 1934) and in Europe *Libellula*, *Platycnemis* and *Epicordulia* (Panin 1957). In the final host the liberated immature trematodes migrate to the cloaca and the bursa of Fabricius where they become adult. In the mature fowl, in which the bursa is atrophied, the parasites enter the oviduct.

Pathogenesis. P. pellucidus, as well as other species of this genus, are considered to be the most pathogenic trematode parasites of poultry in Europe and America. Fowls are mainly affected, but occasionally also ducks. In laying birds the movements of the oviduct apparently assist them to enter this organ and here they cause marked irritation, resulting in an acute inflammation of the oviduct, the production of abnormal eggs and discharges of albumen from the cloaca. The irritated oviduct readily performs retroperistaltic movements, causing broken yolks, albumen, bacteria and parasite material to enter the peritoneal cavity, where they set up peritonitis, usually with fatal results.

Clinical signs. The disease is usually seen in spring or early summer. At first the general health is not disturbed, but several hens may begin to lay eggs with soft shells or without any shell at all. They may show a marked tendency to sit on the nest. There may be a discharge from the cloaca in the form of a milky fluid consisting chiefly of lime, which glues the feathers together around the anus. The irritated oviduct passes the eggs through so rapidly that no shell is deposited, although the lime-secreting glands act normally, and this secretion is then discharged separately. The birds become listless, the abdomen is pendulous and the legs are held widely apart in walking. Laying is suspended and the birds are obviously ill. The feathers around the cloaca are soiled with albumen, which is discharged and may contain yellowish-white strands and parasites. If peritonitis develops, the comb and wattles become cyanotic and the birds soon become prostrated and die. An 'aseptic peritonitis' may also be seen, in which case the yolk in the peritoneal cavity becomes inspissated and may obstruct intestinal peristalsis. One to three adult *P. macrorchis* were found in the eggs laid by an infected chicken on four consecutive days (Sardey & Bhilegaonkar 1979).

Post mortem. The oviduct shows varying degrees of inflammation according to the severity of the disease, from a catarrh to a croupous inflammation with a dirty, cheesy mass in the lumen. It may contain broken yolks and, frequently, large concretions of yolk and albumen. The parasites are not easily seen on the mucous membrane. In cases of peritonitis the abdominal cavity contains a dirty fluid and the organs are stuck together by a cheesy mass. Inspissated yolk may be present between the intestines. The serous membranes show a marked congestion and haemorrhages may be present.

Diagnosis. The eggs of the parasites can be found in large numbers in the discharges from the cloaca. In some cases the parasites may already have disappeared, but the disease continues. In such cases the worm eggs can frequently be found in the abdominal cavity at autopsy.

Treatment. No satisfactory treatment is known for removing the parasites from the oviduct. Anthelmintics such as albendazole and praziquantel may be useful.

Prophylaxis. The extermination of snails, as well as preventing the birds from eating dragonflies, as far as possible, is indicated.

FAMILY: PLAGIORCHIIDAE WARD, 1917

These are fairly large trematodes with slender or plump bodies. The cuticle is usually covered with spines. A short prepharynx, a pharynx and an oesophagus are present and the intestinal caeca are of variable length. The genital pore is usually somewhat lateral and lies at varying levels between the two suckers. Testes are entire or lobed, horizontal or tandem. A cirrus-sac is present. The ovary lies in front of the testes and the ventral sucker, to the right of the middle, and may be lobed. The vitellaria lie in the lateral fields. The uterus passes backwards and forwards between the testes. Eggs are numerous with thin shells.

Genus: Plagiorchis Lühe, 1899 (syn. Lepoderma)

Species of this genus have an elongated body that tapers at each end. The genital opening is a little in front of the ventral sucker, usually to the left of the middle line and behind the bifurcation of the intestine. The rounded or oval testes are obliquely behind one another and the rounded ovary is near the hind end of the cirrus-sac, which is on the right side of the ventral sucker and extends behind it.

Plagiorchis megalorchis Rees, 1952, was found by Rees (1952) in Wales in turkey poults and this author found that its first intermediate host is the snail *Lymnaea pereger*, and its second intermediate hosts *Chironomus riparius*, *Culicoides stigma*, *C. nubeculosus* or *Anatopynia varius*. Rees suggests that the normal definitive host is a wild bird, possibly a gull or a heron. She concluded that *P. laricola* Skrjabin, 1924, of gulls and terns reported by Foggie (1937) from turkey poults in Northern Ireland was *P. megalorchis*. Fedorova (1973) considers that *P. megalorchis* is a synonym of *P. laricola* and the nomenclature of the genus *Plagiorchis* is still not settled.

Plagiorchis lutrae Fahmy, 1954, occurs in the otter.

Plagiorchis arcuatus Strom, 1924, occurs in the oviducts of fowl in Germany and USSR. It has been associated with inflammation of the oviduct.

Life-cycle. Plagiorchis cercariae develop after eggs are ingested by *Lymnaea* and *Physa* snails. The second intermediate hosts are invertebrate species belonging to the Mollusca, Insecta or Crustacea. Thus, the first intermediate host of *P. megalorchis* is *Lymnaea pereger* and the second intermediate hosts are the aquatic larval stages of the nematocerans *Chironomous riparius*, *Culicoides stigma* and *C. nubeculosus* (Soulsby 1965). However, plagiochids also tend to abbreviate their life-cycle, either by progenesis or by eliminating the metacescarial stage, and thus the second intermediate host. Encysted metacercariae have been found in sporocysts within the first intermediate host (Grabda-Kazubska 1976).

FAMILY: OPISTHORCHIIDAE BRAUN, 1901

Small to medium-sized flukes, parasitic in the gall-bladder and bile ducts of reptiles, birds and mammals. Usually much flattened with a translucent body which is narrow anteriorly. The suckers are weak and not far apart. A pharynx and an oesophagus are present and the intestinal caeca reach near to the posterior extremity. The excretory bladder has a long stem and short branches. The genital pore opens in the middle line just anterior to the ventral sucker. A cirrus-sac is absent and the tubular seminal vesicle is coiled. The testes lie in the posterior part of the body; they are situated diagonally and are spherical or lobed. The ovary is not far anterior to the testes. The vitelline glands are moderately developed and lie in the lateral fields. The uterine coils usually do not extend behind the ovary. Eggs numerous and light brown.

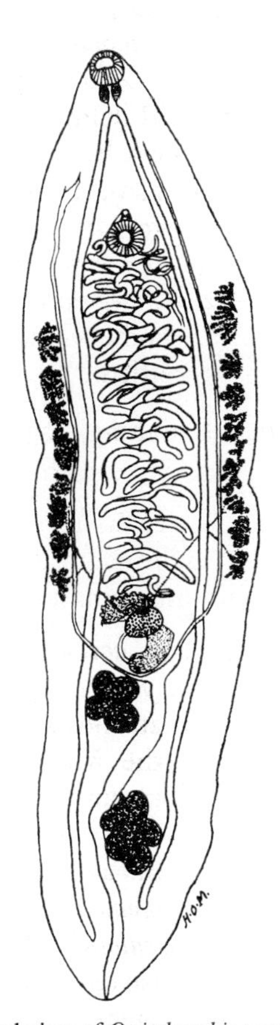

Fig 1.15 Ventral view of *Opisthorchis tenuicollis.*

Genus: Opisthorchis R. Blanchard, 1895

Opisthorchis tenuicollis (Rudolphi, 1819) (syn. *O. felineus* Rivolta, 1884), occurs in the bile ducts, and more rarely in intestine and pancreatic ducts of the dog, cat, fox, pig, Cetacea and man. It has been found in eastern Europe, particularly Poland and the German Democratic Republic, Siberia and other parts of Asia.

Opisthorchis viverrini Stiles and Hassal 1896 occurs in the civet, domestic and wild cats, dogs and man in South East Asia, particularly Thailand and Laos.

They measure 7–12 by 1.5–2.5 mm and have a reddish colour when fresh. The tegument is smooth. The oesophagus is short and the intestinal caeca extend almost to the posterior end. The testes are lobed and the excretory bladder passes between them. The prostate gland and cirrus are absent and there is a weak ejaculatory duct. The ovary is small and lies in the mid-line at the beginning of the posterior third of the body. The vitellaria occupy the middle thirds of the lateral fields; they consist of a series of transversely arranged follicles. The transverse uterine coils do not extend behind the ovary. The eggs measure about 26–30 by 11–15 μm. They have an operculum which fits into a thickened rim of the shell and, when they are laid, they contain a miracidium, the internal structure of which is asymmetrical.

Life-cycle. In the case of *O. tenuicollis* the first intermediate host is the snail *Bithynia leachi* and *B. infata*, and the metacercariae are found in several cyprinid fish (*Leuciscus rutilus*, *Blicca björkna*, *Tinca tinca*, *Idus melanotis*, *Barbus barbus*, *Abramis brama* etc.), into which the cercariae penetrate through the skin and encyst in the subcutaneous tissues, especially at the bases of the fins. Infection of the final host occurs through eating raw infected fish, young flukes migrating via the bile duct to the smaller bile ducts.

With *O. viverrini* the first intermediate hosts include the snails *Bithynia goniomphalus*, *B. funniculata* and *B. laevis* and as second intermediate hosts the cyprinids *Cyclocheilicthus siaja*, *Hampala dispar* and *Puntius orphoides*.

Pathogenesis, clinical signs and diagnosis. These are comparable to those seen in *C. sinensis* infection (see below). Dilatation of bile ducts with adenomatous thickening of the epithelium is common, marked fibrosis occurring in advanced cases. Several cases of carcinoma of the liver or pancreas of cats and man have been ascribed to *O. tenuicollis* eggs.

Treatment and prophylaxis. See *C. sinensis* (below). Experimental work has demonstrated that molluscicides at levels sublethal to *B. inflata* free from *O. tenuicollis* were lethal to infected snails (Beer 1976). This may prove to be of use in the control of the infection in the future.

Control consists of preventing the consumption of raw fish, improvement of sanitation and health education.

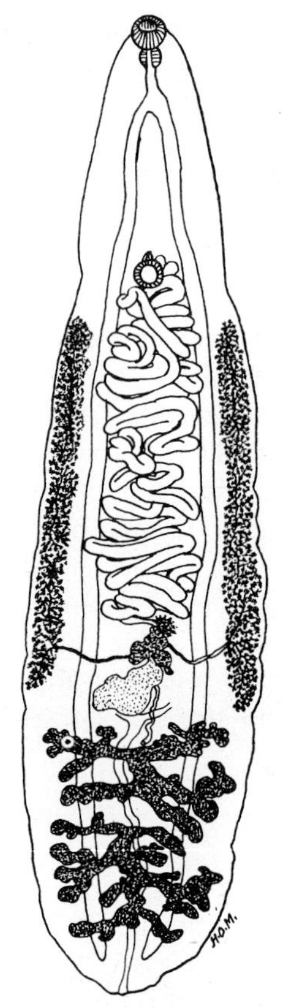

Fig 1.16 Ventral view of *Chlonorchis sinensis.*

Genus: Clonorchis Looss, 1907

Chlonorchis sinensis (Cobbold, 1875) (syn. *O. sinensis*) This species, which is often called the Oriental or Chinese liver fluke, occurs in the bile ducts and sometimes in the pancreatic ducts and the duodenum of man and the dog, cat, pig, weasel, mink, badger. It is common in the south-eastern parts of Asia, China and Japan. The prevalence in man in these areas varies from 30 to 80%. It may reach a size of 25 by 5 mm. It is flat, transparent, wide posteriorly and tapering anteriorly (Fig. 1.16). The testes are much branched and posterior in position; the cuticle is spiny in the young fluke but smooth in the adult. The eggs measure 27–35 by 12–20 μm; they have a thick light brown wall and contain, when they are laid, a miracidium, the internal structure of which is asymmetrical. The operculum of the egg fits into a prominent rim of the shell, while the opposite pole frequently bears a small hook-like structure.

Life-cycle. The eggs normally hatch only after they have been swallowed by the first intermediate host, which may be one of various species of operculated snails, among which are species of the genera *Parafossalurus*, *Bulinus*, *Bithynia*, *Melania* and *Vivipara*. In the snails the miracidium develops into a sporocyst which produces rediae and these in turn produce cercariae which have fairly long tails and elongate bodies with pigmented eye-spots. The second intermediate hosts are fishes belonging to several genera of the family *Cyprinidae*; more than 40 have been reported naturally infected. After breaking out of the snail the cercaria swims about and, on meeting a suitable fish, penetrates partly or completely into the tissues of the fish; losing its tail, it becomes encysted in the fish. Infection of the final host occurs through eating raw, infected fish. The metacercariae are fairly resistant to environmental conditions and may survive two or more months in refrigerated fish. Some metacercariae also survive in salted, dried or pickled fish. In some fish the metacercariae are found only under the scales and animals which are fed with the scales and offal of such fish become infected, while humans who eat the rest of the fish do not. The metacercariae are liberated in the duodenum of the final host and reach the liver by way of the bile duct. Eggs of the fluke are passed out of the final host from the sixteenth day after infection. Adults may survive in the liver for at least 25 years (Attwood & Chou 1978).

Pathogenesis. The severity depends on the intensity of the infection and the vast majority of cases are asymptomatic with few or no changes in the liver.

The worms live in the narrow proximal parts of the bile ducts. They cause a catarrhal cholecystitis, desquamation of bile duct epithelium and increased production of mucus. Widespread periductal fibrosis occurs and occasionally occlusion of the bile ducts with bile stasis and jaundice may be seen. With superimposed bacterial infections there may be cholangitis and cholangiohepatitis. Cholangiocarcinoma may be found concurrently, particularly in severe cases. This is particularly prevalent in Thailand. One-third of patients with *C. sinensis* have parasites in the pancreatic ducts with resultant pancreatitis (Muller 1977).

Clinical signs are not seen except in fairly heavy infections. The symptoms in man include diarrhoea and abdominal pain followed in severe cases by icterus, ascites and other symptoms, resulting from cirrhosis of the liver and derangement of the portal circulation.

Diagnosis is made from the clinical signs, history, geographical area and by finding the eggs in the faeces. The eggs must be differentiated from those of other flukes, especially those of species of the family *Heterophyidae*. The latter, however, contain a symmetrically arranged embryo. Serological tests (complement fixation, passive haemagglutination, immunoelectrophoresis and fluorescent antibody) have been used.

Treatment. No completely satisfactory treatment is yet available. In experimentally infected rabbits hexachlorophene (15 doses of 20 mg/kg) was 100% effective at eliminating adult parasites from the liver; disophenol (30 mg/kg), clioxamide (10 doses of 200 mg/kg) and dithiazanine iodide (six doses of 50 mg/kg) were 99% effective; hexachloroparaxylene (10 doses of 50 mg/kg or five doses of 200 mg/kg) was 97–99.9% effective; and niridazole (10 doses of 100 mg/kg) was 74% effective (Rim et al. 1975).

In man, high doses of some anthelmintics are efficacious but the prolonged use of these drugs may cause side effects. Hexachloroparaxylene (100 mg/kg divided into three doses and given five times on alternate days) is 80–90% effective but temporary side effects may be seen. In addition, this drug is nephrotoxic and neurotoxic and causes anaemia in experimental dogs. Chloroquine (5 mg/kg/day for two months) may be used; side effects are retinopathy and optic neuritis. Dehydroemetine, which has some side effects, is given 30 times on alternate days at 2.5 mg/kg and niclosamide (1–2 mg/kg for two or three days) may be used. These latter drugs reduce egg output but normally only temporarily (Harinasuta 1978). Dithiazinine iodide has been used with success in light infections (Yamaguchi et al. 1962). Niclofolan (1–2 mg/kg for two or three days) is promising and praziquantel (one dose of 50 mg/kg) reduced egg output by 99% (Rim 1978).

Prophylaxis. Thorough cooking or freezing of all fish used as food is essential. This should stop the infection to a great extent, but further measures to eradicate the snail intermediate hosts (which are operculate) and to prevent their infection from human excreta ought also to be considered. General conditions of pisciculture in Asia favour the transmission of infection since animal and human faeces are often added to ponds as fish food. Treatment of night-soil with ammonium sulphate to kill the fluke eggs has been recommended.

Genus: Pseudamphistomum Lühe, 1908

Pseudamphistomum truncatum (Rudolphi, 1819) occurs in the bile ducts of the dog, cat, fox, glutton, seal and sometimes man in Europe, the USSR and India. It measures 2–2.25 by 0.6–0.8 mm. The body is truncate posteriorly and the cuticle is spiny. The posterior ends of the intestinal caeca bend inwards around the testes, which are spherical and almost horizontal. The uterus extends mainly between the testes and the ventral sucker, which is situated near to the middle of the body (Fig. 1.17). The eggs measure 29 by 11 μm.

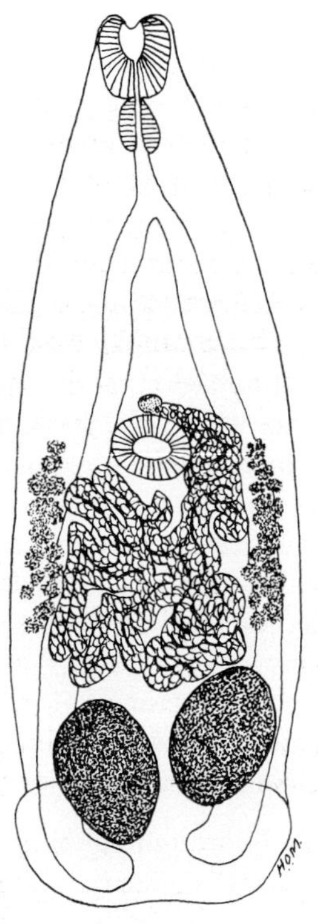

Fig 1.17 Ventral view of *Pseudamphistomum truncatum*.

Life-cycle. The first intermediate host snail is unknown. Schuurmans-Stekhoven (1931) found the metacercariae in the freshwater fish *Leuciscus rutilus*, *Scardinius erythrophthalmus*, *Abramis brama* and *Blicca björkna*.

Pathogenesis. Not well known. Apparently the fluke is not very pathogenic though liver enlargement and bile duct fibrosis have been described in the silver fox. *P. truncatum* has been the cause of losses in seal populations in the Caspian Sea.

Genus: Metorchis Looss, 1899

Metorchis albidus (Braun, 1893) occurs in the gall bladder and bile ducts of dog, cat, fox and grey seal in Europe, North America and the USSR. The parasite may also be found in chickens, ducks, geese and gulls as well as in man. It measures 2.5–6.6 mm by 1.0–1.6 mm; the tegument is spiny and the testes lobed and diagonally placed in the posterior part of the body. The ovary is rounded and just in front of the anterior testis. The genital pore opens in front of the ventral sucker. The vitellaria are restricted in the lateral field between the genital pore and the ovary. Eggs are small, 24–30 by 13–16 μm.

Intermediate hosts are freshwater snails such as *Amnicola limosa* and *Bithynia tentaculata*. Cyprinid fish such as *Leuciscus idus* and *Blicca björkna* are second intermediate hosts. Pathogenic effects are similar to those of *Opisthorchis* spp.

Metorchis conjunctus (Cobbold, 1860) is a similar species in North America occurring in the bile ducts of cat, dog, fox, mink and racoon. It is up to 6 mm in length and the eggs measure 22–32 by 11–18 μm. *Amnicola limosa porosa* serves as first intermediate host and cercariae encyst in the muscles of the common sucker *Catostomus commersoni*.

Genus: Parametorchis Skrjabin, 1913

Parametorchis complexus (Stiles & Hassall, 1894) has been found in the bile ducts of cats and dogs in Maryland and New York, USA. In this species the uterus forms a rosette around the ventral sucker, the vitelline glands are confined to the anterior third of the body and the testes are lobed and lie in tandem in the posterior part of the body. The parasite is 5–10 by 1.5–2 mm and the eggs measure 24 by 12 μm.

The pathogenic effects of the above two liver flukes are similar to the other Opisthorchiidae.

FAMILY: NANOPHYETIDAE DOLLFUS, 1939

These are small, pyriform or elongate trematodes. The oral sucker is terminal and well developed.

The ventral sucker is in the middle third of the body. The testes are symmetrical in the hind body and the ovary pretesticular. It is found in the intestine of mammals, rarely fishes.

Genus: Nanophyetus Chapin, 1927

Nanophyetus salmincola Chapin, 1927 (syn. *Troglotrema salmincola*) occurs in the small intestine of the dog, cat, fox, coyote, racoon, opossum, otter, mink, lynx, etc. in the Pacific north-western area of the USA and in eastern Siberia. Previously the Siberian form was called *N. schikhobalowi* (Skrjabin and Podjapolska, 1931) but it is now regarded as a synonym of *N. salmincola* (Witenberg, 1932). The parasite also infects some piscivorous birds and man. The incidence of infection in man in Siberia can be very high. The biology of *N. salmincola* and the rickettsiae it transmits is reviewed by Millemann and Knapp (1970).

The worms are white or cream in colour. The testes are large and oval and lie ventral to the posterior ends of the intestinal caeca, behind the middle of the body. There is a large cirrus-sac and the genital pore is situated a short distance behind the ventral sucker. The vitellaria are composed of large follicles and lie mainly laterally and dorsally. The ovary is spherical and situated behind and to the right of the ventral sucker.

Eggs measure 52–82 by 32–56 μm; they are yellowish brown, have an indistinct operculum and a small blunt point at the opposite end. They are unembryonated when laid.

Life-cycle. The raccoon and spotted skunk are the principal definitive hosts in the USA. The eggs are passed in the faeces and develop slowly to hatch in three months or longer. The first intermediate hosts are pleuronectid stream snails, *Oxytrema silicula* in the USA and *Semisulcospira laevigata* and *S. cancellata* in Siberia. Sporocysts have not been found but rediae and daughter rediae may be seen.

The liberated cercariae swim free in the water and penetrate through the skin into fish which are the second intermediate hosts. These are primarily fishes in the family Salmonidae (genera *Onchorhynchus*, *Salmo*, *Salvelinus*, *Brachymystax* and *Coregonus*) but also non-Salmonids of the families Cottidae, Cyprinidae and the Pacific Giant Salamander. Metacercariae occur primarily in the kidneys, muscles and fins, but also in almost all the organs. Metacercariae are 0.11–0.25 mm in diameter, they are infective within 10–11 days and may survive five years. They can remain viable in fish held at 3°C for up to five and a half months. The prepatent period in dogs is five to eight days and adults may survive 18–250 days.

Pathogenesis. Fish can tolerate large numbers of parasites but heavy infections can cause death in young rainbow and brook trout as well as in other species. Signs of infection in fish include a decrease in swimming activity, loss of equilibrium, erratic swimming and vertical or horizontal tail curvature in some fish. In dogs the trematodes penetrate deeply into the mucosa of the duodenum or attach to the mucosa of other parts of the small or large intestine. In large numbers a superficial enteritis is produced and this may lead to a haemorrhagic enteritis.

However, the real importance of *N. salmincola* lies in its ability to transmit the rickettsial agents of 'salmon poisoning' and 'Elokomin fluke fever'. 'Salmon poisoning' is caused by *Neorickettsia helminthoeca* or is a disease complex of the two rickettsial agents, while 'Elokomin fluke fever' can occur on its own and is an immunologically distinct rickettsia.

Only Canidae (dogs, foxes, coyotes) are susceptible to 'salmon poisoning'. The incubation period is five to seven days (or longer) and there is a sudden onset of fever and a complete loss of appetite. Within a few days, purulent discharges from the eyes occur, vomiting is marked and there is a profuse diarrhoea, which may be haemorrhagic. Mortality varies from 50 to 90% of infected animals, but recovered animals are immune to reinfection.

'Elokomin fluke fever' has a wider host range, affecting dogs, foxes, coyotes, bears, racoons, ferrets and man. There is a high morbidity and low mortality in dogs. Since this organism may also be part of the 'salmon poisoning' complex the clinical signs are similar. There is fever, lymphadenopathy with an increase in circulating mononuclear cells, emesis, diarrhoea, dehydration and occular discharge.

The rickettsial agents have been found in all stages of the life-cycle of *N. salmincola* and may remain viable for as long as the metacercariae in fish (five years).

Diagnosis. Eggs of *N. salmincola* can be detected in the faeces and rickettsiae can be demonstrated in fluid aspirated from the mandibular lymph node.

Treatment. Since the main disease entity is caused by a rickettsia, anthelmintic compounds will be of little avail. Treatment should involve relief of the emesis, diarrhoea and dehydration. Tetracyclines are effective, as are sulphonamides and chloramphenicol. Gloxazone is more effective against other rickettsial infections than are tetracyclines and could be tried.

Prophylaxis. No infected fish should be fed in a raw or undercooked state. If such fish is accidentally eaten, apomorphine can be given, and it is stated to have prevented the disease when administered as long as three hours after infected fish had been eaten.

FAMILY: FASCIOLIDAE RAILLIET, 1895

These are large flukes parasitic in the bile ducts and intestines of mammals, especially ungulates, with a broad, leaf-shaped body and usually a spiny tegument. The anterior and ventral suckers are close together. A pharynx and a short oesophagus are present and the intestinal caeca are commonly much branched, especially laterally. The excretory bladder is also much branched. The genital pore is median, directly anterior to the ventral sucker. Testes tandem, lobed or branched. Vitellaria strongly developed, filling the lateral fields and extending medially as well. Receptaculum seminis absent. The eggs have thin shells and are operculate.

Genus: Fasciola Linnaeus, 1758

Fasciola hepatica Linnaeus, 1758, occurs in the bile ducts of the sheep, goat, ox and other ruminants, pig, hare, rabbit, beaver, coypu, elephant, horse, dog, cat, kangaroo and man. In the unusual hosts, such as man and the horse, the fluke may be found in the lungs, under the skin or in other situations. The fluke is cosmopolitan in its distribution and is the cause of fascioliasis (liver fluke disease, liver rot), especially in sheep and cattle.

F. hepatica may reach a size of 30 by 13 mm. It is leaf-shaped, broader anteriorly than posteriorly, with an anterior cone-shaped projection which is followed by a pair of broad 'shoulders'. It is greyish-brown in colour, changing to grey when preserved. The ventral sucker is situated at the level of the shoulders and is about as large as the oral. The tegument is armed with sharp spines. The intestinal caeca have numerous branches and extend far back. The testes are much branched, filling the median field in about the second and third quarters of the body. There is a well-developed cirrus and the cirrus-sac also encloses the prostate and seminal vesicle. The ovary is situated to the right of the middle, anterior to the testes, and is branched. The vitelline glands consist of fine follicles filling the lateral fields and the ducts of the follicles unite to form two transverse ducts, which pass inwards to open into a median yolk reservoir, from which a duct passes to the oötype. The uterus lies anterior to the testes. The eggs measure 130–150 by 63–90 μm; and the miracidium develops only after the eggs have been laid (see Fig. 1.5).

Life-cycle. The eggs enter the duodenum with the bile and leave the host in the faeces. The rate of development and the hatching of *F. hepatica* eggs depends on temperature (see below) but at 26°C eggs hatch in about 10–12 days producing the first larval stage, the miracidium (see p. 15). The miracidium is broad anteriorly with a small papilliform protrusion, the tegument is ciliated and the organism has a pair of eye spots. For further development an amphibious snail of the genus *Lymnaea* is required. Throughout the years an increasing number of molluscan intermediate hosts of *F. hepatica* have been reported: these are tabulated by Soulsby (1965). However, many of the snails incriminated are now regarded as being synonymous species and genera (Kendall 1965). The following *Lymnaea* species are considered to be the most important in the transmission of *F. hepatica*: *L. truncatula* is the most important and widespread (Europe, Asia, Africa and North America) intermediate host of *F. hepatica*. In North America the principal species incriminated as an intermediate host of *F. hepatica* is *Lymnaea bulimoides* and in Australia it is *Lymnaea tomentosa*. Other species which have been incriminated in the transmission of *F. hepatica* include *L. viator* and *L. diaphena* (South America), *L. columnella* (Central and North America, Australia and New Zealand) and *L. humilis* (North America). Speciation of these has yet to be conclusively delineated.

Experimental infection of several other species of snails has been attempted and achieved though this does not necessarily imply that the snail will play a major part in the transmission of the infection under natural conditions. Thus *L. palustris*, *L. pereger*, *L. glabra* and *L. stagnalis* may be infected within the first few days of hatching but there is no evidence that these species in their mature form play any substantial part in the natural incidence of the disease. Berghen (1964) successfully infected *L. stagnalis* and *L. palustris* but was unable to obtain free emergence of cercariae from the snails.

The miracidium penetrates actively into the snail, casting off its ciliate covering, and develops into the sporocyst, which reaches a length of over 1 mm. Dawes (1960) has, however, shown that the final stage of penetration into the snail is performed, not by the miracidium, but by the young sporocyst. Studying the penetration of *Fasciola hepatica* into *Lymnaea truncatula* and *F. gigantica* into *L. auricularia*, he found that the miracidium adheres by suction to the epithelial cells of the snail and breaks these down, probably by means of enzymes secreted by the apical organ of the gut, and then casts off, as it enters the snail, its ciliated epithelium, so that the final swift thrust into the snail is effected by the unciliated young sporocyst. Dawes suggests that all the digenetic trematodes enter their intermediate hosts in this manner. Each sporocyst gives rise to five to eight rediae which, when fully developed, are 1–3 mm long. They are characterized by a circular thickening behind the level of the pharynx and a pair of blunt processes at the beginning of the posterior quarter. Daughter rediae may develop under unfavourable conditions, but the next normal generation is one of cercariae (see Fig. 1.6). These leave the snail in four and a half to seven weeks from the time of infection. They have a body of 0.25–0.35 mm long, a tail of twice that length and no eye-spots; the dark, granular, cystogenous glands are conspicuous in the lateral parts of the body. Within a few minutes to two hours the cercariae settle on blades of grass or other plants just below water-level and, after casting off the tail, secrete a covering from the cystogenous glands forming cysts about 0.2 mm in diameter. A small number may encyst at the surface of the water and sink to the bottom. The cercariae are now infective. They are swallowed by the final host with the plants on which they are encysted, or animals, like cattle that walk into the water to drink, may stir up the cercariae lying at the bottom and swallow them.

Development in the vertebrate host. Following ingestion of the metacercariae excystation occurs in the duodenum. Factors concerned with this have been studied by Wikerhauser (1960) and Hughes (1963). The former found that excystation could be induced by treating metacercarial cysts with acid pepsin followed by treatment with trypsin and bile. Hughes confirmed the necessity for pretreatment with acid pepsin and obtained excystation in an artificial intestinal juice composed of trypsin, pancreatin, sodium taurocholate and cholesterol. Hughes found that excystation could occur with cysts no more than two days of age.

Within 24 hours of infection the majority of immature trematodes occur in the abdominal cavity and by four to six days after infection the majority have penetrated the liver capsule and are found migrating in the liver parenchyma. Some young flukes may reach the liver by way of the blood stream but the usual route is via the peritoneal cavity. Migration in the liver occurs for five to six weeks and about seven weeks after infection they begin to enter the main bile ducts; from this time onwards, an increasing number arrive there and reach sexual maturity. From eight weeks onwards, eggs are found in the bile and subsequently in the faeces; however, there is no synchrony in the behaviour of an infection and even though the infection may be patent, a proportion of developmental stages attain maturity later so that over a period of two months a succession of developmental stages reach maturity. Occasionally, especially in cattle, immature flukes may be carried to other organs such as the lungs and in pregnant animals, occasionally parasites may be found in the fetus.

Factors affecting development of the egg. At temperatures below 10°C no development occurs in the egg, but from 10 to 26°C there is an increasing rate of development. At 12°C, 60 days or more are required, at 15°C about 40 days and at 26°C about 12 days. Under field conditions in Great Britain, eggs are unlikely to hatch in less than three weeks (Rowcliffe & Ollerenshaw 1960) and in Australia, the incubation period is 21 days in summer and 90 days in winter (Boray 1963). Eggs may survive for some time at low temperatures and under field conditions accumulation of unhatched eggs may occur over winter. However, these eggs are subjected to adverse conditions and those which survive until the spring usually do not make as significant a contribution as those which are passed by infected animals in the spring (Ollerenshaw 1959).

Factors affecting the development of the parasite in the snail. Little development occurs in the snail below a mean day/night temperature of 10°C, but above this the rate of development increases to about 28°C. The minimum period for completion of the life-cycle is about 21 days at 27°C. Boray (1963) found that at temperatures above 20°C there was an increase in mortality of infected snails and also the infectivity of metacercariae fell markedly above this temperature. From extensive studies, Boray concluded that in Australia the contamination of pasture was evenly distributed throughout the year. In warm weather many cercariae of the parasite may be produced but this is offset by an increased mortality of snails and reduced infectivity of metacercariae.

A relationship between the size of the snail and the number of developing parthenitae was observed by Kendall (1949), larger snails having almost ten times the number of developmental stages than smaller snails, this being a reflection of the amount of food which the snail received. Detailed studies of the effect of population density, food and other factors which affect snails and the production of cercariae are given by Boray (1963). In field studies, Kendall and Ollerenshaw (1963) confirm that the size of snails and not their numbers appears to be the major factor influencing the number of parasites present.

The ability of the snail hosts of *F. hepatica* (*L. truncatula*, *L. tomentosa* and *L. bulimoides*) to undergo aestivation is important in the epidemiology of *F. hepatica* infection. Under field conditions, some snails may survive for several months in dry mud and Taylor (1949) has stated that three out of ten snails may be alive after aestivating under artificial drought conditions for 12 months. Though aestivation of the snail retards the development of *F. hepatica*, the parthenitae of the fluke can survive for at least ten months in aestivating snails. On the return of moist conditions for the snail it grows to maturity very rapidly and similarly the developmental stages of *F. hepatica* undergo rapid development so that within a short time large numbers of cercariae may accumulate on herbage when moist conditions return to the snail habitat.

Ecology of Lymnaea truncatula. This snail is commonly seen in poorly drained land, drainage ditches, areas of seepages of springs or broken drains. Temporary habitats are afforded by muddy gateways, vehicle-wheel ruts, wet and muddy places near drinking troughs and hoof prints of animals on clay soil. Taylor (1949), in confirming the wide distribution of the snail in Great Britain and probably in Europe, noted that concentrated populations generally occurred in places which became dry for periods of weeks or even months. A return to wetness of these habitats provided optimum conditions for the multiplication of snails. The American species, *L. bulimoides*, inhabits similar terrain to *L. truncatula*; however, *L. tomentosa* is an amphibious snail and well adapted for aquatic life. It is found in and around ditches, field dams and similar places. The chances of extension of the area of colonization are greater than with *L. truncatula* and *L. tomentosa* may migrate against the water current at an average speed of migration of 50 cm/hour against a water velocity of 15 cm/second. Snails may float or drift with water currents for long distances. Slightly, but not markedly, acid pH conditions of the soil are preferred by the snail. Under ideal conditions, dense populations of snails may develop and Ollerenshaw (1959) has observed up to 3300 snails/m^2. Detailed consideration has been given to the ecology of *L. truncatula* in Great Britain. In west Scotland, adult snails which survive the winter by hibernation commence laying eggs in the spring and continue to do so throughout the summer, although they die off in the late summer and have disappeared by September/October. The first daughter generation of snails hatching from such eggs grow and produce eggs from late July onwards. This adult population is inhibited in winter but surviving snails form the nucleus of the parent generation the following spring. Development of the second daughter generation commences in late summer and autumn but falling temperatures in autumn inhibit this development and it is not completed until the following spring. Thus, one and a half generations of snails are seen per year. Farther south in Britain the period suitable for development is longer (May to October) and two generations of snail breeding can be completed each year, while in central France three generations of snails per year are possible.

Infection of snails. Ollerenshaw and Rowland (1959) described two annual cycles of infection in snails on the basis of meteorological data and an examination of snail populations and disease incidence in Anglesey. Of these two cycles, summer infection of snails is the most important. Eggs deposited on pasture in spring and early summer hatch and the miracidia infect snails. After a minimum of five weeks development in the snail the infection passes on to the herbage in late summer and autumn. Infection of the definitive host with the metacercariae from this summer infection results in losses beginning in October and continuing throughout winter. Winter infection of snails may also occur. Thus, miracidia which develop from eggs in late summer initiate development in snails at this time. Development of this infection in the snail is inhibited during the winter, but recommences in the spring, so that the infection passes on to the herbage in late spring and early summer. When the metacercariae are ingested by the definitive host, they give rise to disease from July to October. This winter cycle of infection is not as important as the summer cycle of infection probably due to the mortality of infected snails during the winter.

Longevity of metacercariae. Metacercariae have been shown to survive for more than one year under laboratory conditions but, under pasture conditions, it is likely that a dangerous level of infection does not persist for such a long period. Taylor (1949) indicated that a large proportion of metacercariae fell off certain types of herbage within four to six weeks and though such forms may remain viable for some time they are generally unavailable to the grazing animal. In laboratory studies Boray (1963) found that at temperatures of 12–14°C 100% of metacercariae could survive for six months and 5% for ten months. Kendall (1965) reports experiments in which herbage remained infected for periods of between 270 and 340 days. A relative humidity of 70 was necessary for prolonged survival of the metacercariae. Survival of metacercariae below freezing temperatures has been reported. Cercariae may survive on moist hay for eight months. On the other hand, in normal hay-making, it is unlikely that metacercariae will survive the dessication for an extended period. Failure of metacercariae to survive in silage for more than 35 or 57 days has been demonstrated and in Gulf Coast regions of the United States, Olsen (1947) showed that metacercariae were destroyed by heat and drought during the four summer months. On such pastures, sheep did not become infected until the early winter.

More detailed consideration of the relationship between *F. hepatica* and the snail host may be found in Kendall (1965) and Soulsby (1965).

Pathogenesis. The pathological manifestations depend on the number of metacercariae ingested. Since, under natural conditions, i.e. in sheep, there is little evidence of immunity, additional infections are additive and at autopsy a succession of developmental stages may be found in an animal. No appreciable damage is done during passage through the intestinal wall or the peritoneal cavity, the principal lesions occurring in the liver, in either the parenchyma or the bile ducts. Essentially the disease entity can be divided into an acute form and a chronic form.

Acute fascioliasis is less common than the chronic entity and is almost invariably seen in sheep. It is essentially a traumatic hepatitis produced by the simultaneous migration of large numbers of immature trematodes and is seen mainly towards the end of summer when large numbers of cercariae are shed onto the herbage. Taylor (1951) estimated that 10 000 cysts must be given to produce the syndrome in sheep. The most damaging stages are those six to eight weeks of age, these causing extensive destruction of liver parenchyma and marked haemorrhage. If numbers are excessive, rupture of the liver capsule may occur with haemorrhage into the peritoneal cavity. Animals may die within a few days of the onset of clinical signs and in these the liver is enlarged, pale and friable and shows numerous haemorrhagic tracts on the surface and throughout the substance and fibrinous clots on the liver surface and also throughout the peritoneal cavity. At the proximal part of the tract, an immature parasite may be seen, distal to which is a zone of haemorrhage and then a posterior zone of reddish grey material consisting of infiltrated cells. Small flukes 0.7–2 mm can be squeezed from the cut surface or obtained from the excess of peritoneal fluid.

In less acute forms of disease, the liver is covered with migratory tracts but an infiltration of white cells is more in evidence and early fibrosis may be seen. This subacute type may be superimposed on an existing chronic infection and then a more marked cellular response may be seen, possibly indicating a form of immune response to the second infection.

The clinical entity of the acute and subacute forms is seen in animals of all ages and states of nutrition. Death may occur rapidly or after several days. Animals are disinclined to move, are anorexic and show a distended abdomen which is painful to the touch.

A complication of the acute condition is 'black disease', caused by *Clostridium oedematiens novyi*. This is an anaerobe which proliferates in the anaerobic necrotic lesions produced by the immature trematodes. The organism apparently occurs in normal sheep, the clinical entity occurring only after liver damage has been produced by some other agent. Black disease is common in Australia but is also seen in Europe and the USA. Sheep aged two to four years are usually affected. A vaccine is available for the condition.

Chronic fascioliasis is the most common form of the infection in sheep, cattle and other animals (including man) and the major consequence of infection with *F. hepatica* is hepatic fibrosis. The pathology may be divided into hepatic fibrosis and hyperplastic cholangitis (Dargie et al. 1974; Rushton & Murray 1977).

Migration of immature flukes in the liver produces migratory tracts within which there is traumatic destruction of the liver parenchyma, haemorrhage and necrosis. Migration of the flukes also results in thrombus formation in the hepatic veins and liver sinusoids and subsequent obstruction to the blood flow by these thrombi causes an ischaemic, coagulative necrosis in the liver parenchyma. Healing and regeneration of these lesions begins approximately four to six weeks after infection, collagen is laid down and fibrosis occurs. Subsequent contraction of such scar tissue results in a considerable distortion of the hepatic architecture. Probably in an attempt to restore normal hepatic architecture, bands of fibrous tissue then develop to connect the fibrotic migratory tracts to normal tissues in the portal canals, central veins and liver capsule. These tracts subdivide the liver parenchyma into irregular lobules.

Pericellular fibrosis, around single hepatocytes or groups of hepatocytes, and monolobular fibrosis begin to develop 12–20 weeks after infection. Monolobular fibrosis comprises strands of fibrous tissue connecting the portal canals, thus outlining the hepatic lobules. This fibrosis assists in restoring the hepatic architecture and results in a straightening of the hepatic plates counteracting the distortion produced by other hepatic fibrosis. Monolobular fibrosis is greatest, and thick fibrotic tracts are seen, in the ventral lobe of the liver where fluke migration is maximal. However, monolobular fibrosis also occurs in areas distal to fluke migration. In these areas delicate strands of fibrous tissue are laid down. This fibrosis occurs from week 20 onwards, when regeneration of hepatic tissue is greatest, and may be stimulated by the hepatic regeneration (Rushton & Murray 1977).

Fibrous tissue is also laid down in the portal canals. As early as seven days after infection lymphocytes migrate from the hepatic vein into the surrounding tissues and emigration of eosinophils, lymphocytes and macrophages follows. The tissues around the secondary and tertiary portal veins become oedematous and the veins become partially or totally occluded by the pressure of the oedema and cells. Ultimate healing of the inflammatory reaction around the veins with the formation of fibrous tissue may result in a permanent partial occlusion of these veins. This obstruction to the portal blood supply causes a compensatory increase (up to 20-fold) in blood flow in the hepatic artery. The resultant marked increase in intrahepatic blood pressure can cause a compensatory perisinusoidal fibrosis. The hepatic arteries also become thickened and tortuous (Murray 1973).

A hyperplastic cholangitis is caused by the presence of the adult flukes in the bile ducts. At first the epithelium of the bile ducts is hyperplastic, both close to and distal to the sites of fluke residence, and numerous eosinophils and mononuclear cells infiltrate the lamina propria. The spines and suckers of the flukes subsequently denude the bile duct epithelium and organization of the inflammatory reaction results in fibrosis of the lamina propria of the bile duct and surrounding tissues. Movement of flukes through the bile duct tree exacerbates these lesions. Fluke eggs lodged in smaller bile ducts also induce further fibrosis following the formation of granulomatous reactions to such eggs. The hyperplastic biliary mucosa becomes permeable to plasma proteins, particularly albumin, and this, together with the blood-sucking activities of the adult flukes, accounts for the hypoalbuminaemia and hypoproteinaemia evident during infection. In cattle, calcification of the fibrotic lesions may eventually develop and encrustations of calcium are frequently seen, at times forming complete casts of the bile duct and blocking it. The walls of the ducts are commonly calcified in cattle, they protrude markedly from the surface and are difficult to cut with a knife. They resemble the stem of a clay pipe, giving the common name of 'pipe-stem liver' to the infection.

In cattle, too, parasites are often found in other organs, especially the lungs. Here they occur in hazel-nut-sized cysts containing a brownish purulent gelatinous material in which a living, but more frequently a dead and calcified parasite may be found.

Chemical pathology. A profound anaemia and changes in the serum proteins are found in sheep. These are much less marked in cattle.

The anaemia observed in fascioliasis is associated primarily with intrabiliary haemorrhage due to the blood-sucking activity of the adult flukes. The red cell loss per fluke is approximately 0.5 ml/day (Holmes et al. 1968). However, the erythrokinetics of infected sheep have been studied by Berry and Dargie (1978) and haemodilution and intrahepatic haemorrhage also contribute to the anaemia while the ultimate degree of anaemia observed is associated with the animal's erythropoeitic capacity. Thus, Berry and Dargie (1978) demonstrated that a marked increase in plasma volume occurred during the first seven weeks of an experimental infection in sheep. The rapid drop in PVC early in infection coincided with and must be associated with this haemodilution demonstrated by the increase in plasma and blood volumes, although some red cell losses, due to intrahepatic haemorrhage, also occur at this time.

Intrabiliary haemorrhage and consequent loss

of red cells into the intestine occurred eight to nine weeks after infection and thereafter increased in severity producing an anaemia due to blood loss. The resultant progressive loss of iron into the gastrointestinal tract was associated with an increased plasma iron turnover rate and a progressive reduction in plasma iron concentration. Ultimately, an iron-deficiency anaemia may develop (Berry & Dargie 1978).

Marked changes in plasma protein concentrations occur in animals infected with *F. hepatica*. Thus, early in infection, during fluke migration, there is hyperproteinaemia, hyperglobulinaemia and hypoalbuminaemia (Reid 1973). The hypoalbuminaemia is associated with plasma volume expansion and reduced albumin synthesis, the latter probably resulting from the preferential use of amino acids for globulin synthesis and from damage to the liver parenchyma, the major site of albumin synthesis (Dargie & Berry 1979).

As the infection progresses and adult parasites migrate to the bile ducts a marked hypoalbuminaemia, hypoproteinaemia and sometimes hypoglobulinaemia are observed. These changes are associated with the progressive and marked loss of plasma proteins, particularly albumin, into the gastrointestinal tract through the blood-sucking activities of the adult flukes and through leakage of protein through the bile duct epithelium. However, the severity of the hypoalbuminaemia is related not just to the loss of albumin into the intestine, but also to the rate of albumin synthesis and the fractional and total rates of albumin catabolism. These, in turn, are related to the levels of nutrition, appetite and fluke burden (Dargie & Berry 1979). These authors demonstrated that in experimentally infected sheep, those which catabolized the most albumin also synthesized the most albumin and such sheep became the least hypoalbuminaemic and survived the longest. In general the rate of albumin synthesis was closely related to the status of the albumin pools in the body and depletion of these pools, by the loss of albumin into the intestine, acted as a trigger to alter albumin catabolism and to stimulate albumin synthesis. The ability of such infected animals to increase their rate of albumin synthesis was then related to protein intake (dietary protein and degree of inappetance).

The increased synthesis of albumin in these animals probably diverts available amino acids away from other protein metabolism (muscle, milk, wool), thus accounting for the lowered levels of productivity seen in animals infected with *F. hepatica* (Dargie & Berry 1979).

A disease entity in cattle in Great Britain referred to as the fascioliasis/ostertagiasis complex has been reported by Reid et al. (1967). This should be differentiated from Type II ostertagiasis (see p. 223) with which it may readily be confused. The complex occurs from January to March; it may involve both housed and outwintered stock following the first grazing season at average stocking rates during autumn on poorly drained permanent pastures (Type II ostertagiasis is associated with over-stocked pastures grazed during the previous spring and summer). Clinical signs include intermittent soft faeces, occasionally profuse diarrhoea, progressive loss of weight, bottle jaw, pale mucous membranes and morbidity and mortality rates of 30% and 10% respectively. Severe anaemia is always present, plasma pepsinogen may be slightly elevated (compared to marked elevation in Type II ostertagiasis) and total serum proteins are low. At autopsy 200–500 adult *F. hepatica* with 20 000 or more adult and larval stages of *Ostertagia ostertagi* may be found, the abomasal mucosa showing mild to severe hyperplasia. In contrast, in Type II ostertagiasis no or only a few *F. hepatica* may present, and more than 50 000 *O. ostertagi* may occur in an abomasum, which shows severe hyperplasia and oedema.

Clinical signs. In acute cases, in sheep, the animal dies suddenly; blood-stained froth appears at the nostrils and blood is discharged from the anus, as in a case of anthrax. In the chronic cases the first signs are seen at a time when the young worms, burrowing through the liver parenchyma, have reached a fair size. The sheep is off colour and this is followed by an increasing anaemia. There is an increasing lack of vigour which is observed when the animals are caught or driven. The appetite diminishes, the mucous membranes become pale and oedema develops. Oedema is more conspicuous in some breeds than in others and it may appear especially in the intermandibular space, the name 'bottle-jaw' being then given to it. The skin becomes dry and doughy to the touch. The wool is dry and brittle, falling out in patches. The debility, emaciation and general depression increase and there may occasionally be diarrhoea or constipation and slight fever. At this stage, or even earlier, death may occur. The flukes usually live about nine months in the sheep and then die and pass out through the intestine, but some may live up to five years and in one case a survival time of 11 years has been recorded. If the animals recover, the signs gradually abate, but the wool later shows a 'break' in the part grown during the illness and the lesions in the liver are never completely repaired.

In the case of cattle, the most characteristic signs are digestive disturbances. Constipation is marked and the faeces are passed with difficulty, being hard and brittle. Diarrhoea is seen only in the extreme stages. Emaciation increases rapidly, while dullness and weakness soon lead to prostration, especially in calves.

Diagnosis. This is confirmed by finding the eggs in the faeces. They must be distinguished from the eggs of other flukes, especially the large eggs of paramphistomes. The *Fasciola* egg has a yellow shell with an indistinct operculum, and the embryonic cells are also rather indistinct. The paramphistome eggs have transparent shells and distinct opercula; their embryonic cells are clear and there is frequently a small knob at the posterior pole, while the eggs themselves are often larger than those of the liver-fluke.

Treatment. A number of agents may be employed.

Carbon tetrachloride has been used for more than 50 years. It remains of use and, in some areas, is the only drug available for treatment of *F. hepatica* in sheep. It is not recommended for use in cattle. In sheep, a routine dose of 1 ml is satisfactory for strategic control and this may be increased to 5 ml for the control of outbreaks of disease. At these dose rates the anthelmintic is essentially only effective against the adult fluke. In general, sheep tolerate carbon tetrachloride well, though cases of carbon tetrachloride poisoning are by no means uncommon and occasionally several deaths may occur in a flock under treatment. Carbon tetrachloride poisoning in sheep causes liver and kidney disfunction. Intramuscular administration of carbon tetrachloride reduces the risk of toxicity seen with oral administration. For sheep the equivalent of 1–2 ml of carbon tetrachloride in a bland oil such as liquid paraffin has given efficient results, resulting in a high efficacy against mature parasites. In cattle doses of 0.05 ml/kg to a maximum dose of 5–10 ml have been used with the incorporation of a local anaesthetic in the injection. One disadvantage of the intramuscular medication is the production of necrosis or abscesses at the injection site.

Hexachlorethane, 220 mg/kg or up to 400 mg/kg in three to four doses, may have a greater than 90% efficacy against mature flukes and is well tolerated by cattle, although occasionally fatalities may occur. Liver damage appears to be a predisposing factor and the feeding of root crops may increase the sensitivity.

Hexachlorethane may also be used in sheep, dosages varying from 20 to 30 g per animal.

Hexachlorophene may be used for both cattle and sheep. It is given orally or subcutaneously. For sheep a dose of 15–20 mg/kg is generally recommended for routine use; it is greater than 90% effective against mature flukes and is usually well tolerated. For acute fascioliasis 40 mg/kg may be employed. In cattle, 10–20 mg/kg is 90% effective against adult parasites and has some efficacy against late parenchymal stages (Edwards & Parry 1972).

Hetol (1,4-*bis*-trichloromethyl-benzol). Lämmler (1960) suggested a field therapeutic dose of this drug of 150 mg/kg for sheep, this level giving satisfactory results. Behrens (1960) recommended a dose of 5 g to lambs five to six months of age, and 10 g to older sheep. For cattle a dose of 125 mg/kg has been reported to be effective by the above authors. The drug is more effective against adult flukes than immature ones and, in general, the compound is well tolerated by both cattle and sheep at the recommended doses.

Bithionol (30–35 mg/kg) has been used in cattle with a 66–68% efficacy. However, bithionol sulphoxide (40 mg/kg) was 98–100% effective against chronic *F. hepatica* infections in sheep.

Tremendous advances have been made recently in both the efficacy and safety of the anthelmintics available for the treatment of *F. hepatica*. Rafoxanide, nitroxynil, oxyclozanide and diamphenethide are all effective anthelmintics but different efficacies are seen against the mature and immature stages of infection. Thus, *diamphenethide* given at a dose rate of 100 mg/kg was shown to be 100% effective against flukes one to five weeks old, 73–85% effective against seven-week-old parasites and 57-65% effective against nine- to ten-week-old parasites in sheep. Diamphenethide is thus a treatment of choice for acute fascioliasis in sheep. At a dose rate of 150 mg/kg diamphenethide may have a greater than 90% efficacy against adult *F. hepatica* (Čorba & Armour 1973/1974; Rew et al. 1978).

Oxyclozanide is highly effective, up to 100% against mature *F. hepatica* infections in sheep, cattle and goats when given at a dose rate of 15–20 mg/kg to sheep and 10–15 mg/kg to cattle. Three times the dose of oxyclozanide (45 mg/kg) in sheep is effective against immature flukes and may be used for the treatment of acute fascioliasis.

Rafoxanide is also highly effective (over 99%) against mature flukes, up to 98% effective against six-week-old flukes and 50–90% effective against four- to five-week-old immature parasites. In cattle and sheep it is administered at a dose rate of 7.5 mg/kg.

Nitroxynil is given subcutaneously at a dose rate of 10 mg/kg. It has an efficacy of up to 100% against adult flukes in the bile ducts in cattle and sheep and is up to 90% effective against immature flukes over four weeks of age in the parenchyma when given at a dose rate of 15 mg/kg.

Roseby and Boray (1970) demonstrated that *brotianide* had up to a 99% efficacy against 12-week-old flukes and six-week-old flukes in sheep when given at a dose rate of 3.5 mg/kg and 7.1 mg/kg, respectively. *Niclofolan* and *hexachloroparaxylol* have also been shown to be highly effective against mature *F. hepatica* infections.

Some benzimidazole anthelmintics, such as albendazole and oxfendazole, are effective against *F. hepatica* in addition to their efficacy against gastrointestinal nematodes. Thus, *albendazole* at 7.5 mg/kg in sheep and 15 mg/kg in cattle may have a greater than 90% efficacy against adults of *F. hepatica*. Similarly, *oxfendazole* has comparable levels of activity.

Oxyclozanide, 10 mg/kg, has been shown to be effective in the treatment of *F. hepatica* infection in horses (Owen 1977).

Prophylaxis. The control of *F. hepatica* in sheep and cattle is achieved through a combination of the control of the snail intermediate host and the treatment of infected animals.

Snails may be controlled by the elimination of snail habitats, through the use of mulluscicides and by biological control. Snail habitats can be eliminated by improved drainage and problem areas, such as drainage ditches and seepage from springs, can be fenced off. Biological control through the use of ducks and frogs which ingest *L. truncatula* may be tried. These certainly do not eliminate molluscs but ducks may be useful on certain stretches of water. Molluscicides can be highly effective in the control of *F. hepatica*. Copper sulphate solutions of 1 in 100 000 to 1 in 5 million are effective in the destruction of snails and many of their eggs. Also, copper sulphate can be applied as a powder with a sand extender at a rate of 10–35 kg/hectare. Stock should not be grazed on treated pastures until a rain has fallen and the molluscicide can be toxic to fish. *N*-tritylomorpholine is a highly effective molluscicide when applied at 0.45 kg in 680 litres per hectare. Molluscicides are usually applied in spring or mid-summer. The spring application is easy to apply and highly effective, killing off overwintered infected snails and parent snails which would supply the nucleus of the year's breeding population. Midsummer applications kill off infected snails prior to the emergence of the summer infection of *F. hepatica* on to the pasture in late summer. Midsummer applications are not always as effective as spring applications. Molluscicides have also been shown to be effective when applied in the autumn. Use of molluscicides may reduce snail populations by more than 90% and consequently greatly reduce the level of infection in the definitive hosts (Crossland 1976). However, snails have a tremendous reproductive potential and only a few snails may, on occasion, be sufficient to repopulate the pasture and to contaminate the pasture heavily with metacercariae.

Infection with *F. hepatica* can be controlled by anthelmintic treatment of the definitive host. Several programmes of anthelmintic treatment have been outlined. Monthly, or less frequent, treatment of sheep through the winter months has been used. This eliminates the infection ingested in late summer and in autumn and that arising from surviving metacercariae on the pasture in winter. It prevents contamination of the pasture by infected animals in the spring. Armour and Urquhart (1974) demonstrated the effectiveness of a dosage regimen for sheep in the British Isles. Adult sheep were treated in March or April and again in May or June to prevent contamination of the pasture in spring and summer. The animals should be treated again in October to kill the migrating parasites which develop from the summer infection in snails and again in January to eliminate any remaining parasites. Where fascioliasis is not a major problem anthelmintic treatment in May, October and January should suffice.

Cattle may be treated in December and early May to prevent contamination of the pastures in the spring and summer.

Unfortunately, wild animals such as deer and rabbits are frequently infected with *F. hepatica*. Contamination of the pasture by these can maintain the life-cycle in the face of control by the anthelmintic treatment of domestic animals.

Meteorological forecasting of fascioliasis. In Great Britain control of *F. hepatica* can be based on application of molluscicides in the spring or mid-summer in order to kill overwintered snails or snails carrying the summer infection, respectively, plus the use of anthelmintics monthly, or at less frequent intervals, from October to April. However, such extensive dosing is not required every year. Thus, forecasting of fascioliasis eliminates the expense of instituting control measures in years in which disease may not materialize (Ollerenshaw & Rowland 1959; Gibson 1978).

This forecasting is based on an association between meteorological conditions and the incidence of disease. In England and Wales air and soil temperatures are usually above the critical temperature of 10°C required for the development of the snail and the infection in the months of May to October. Thus, during this time surface wetness becomes the critical factor and the 'Mt' system of forecasting the incidence of fascioliasis is based on an estimation of surface wetness using the formula $n(R - P + 5)$ where R is the monthly rainfall in inches, P is the evapotranspiration in inches and n is the number of wet days per month. Wet months produce high values and a value of 100 or more (values over 100 are limited to 100) is optimal for parasite development. Development then becomes dependent on temperature. Temperature is optimal in June, July, August and September and the formula $n(R - P + 5)$ is applied. However, in May and October temperatures are lower and more variable and the rate of parasite development is slower. Thus the index is modified to $n(R - P + 5)/2$ for May and October. The summation of the values for the six months, May to October, gives the seasonal index. In England and Wales a seasonal index of less than 400 forecasts little or no disease; occasional losses may be seen at an index of 400–450, while disease is prevalent when the index is 450 or more. In general, conditions in September and October are constant and the 'Mt' index can be taken at 150 for these months. Thus, disease in sheep in the winter, resulting from summer infection in the snails, can be forecast based on the May to August values. On occasion prevalence can be overestimated if a dry autumn follows a wet summer. In addition, the effect of other factors such as irrigation and the presence of irrigation and drainage ditches must be taken into account. These will allow the life-cycle to be maintained in dry summers.

The critical values of the seasonal index which are indicative of disease may be modified to apply to conditions in various areas of the country. The index has been modified for use in France, the Netherlands and Italy. Thus, in central France, the formula is modified to $n(R - P + 125)/25$ where the rainfall and evapotranspiration are measured in mm and n is the number of rain days with precipitation of over 0.1 mm. The higher temperatures in this area mean that full, not halved, values for the index are calculated for the six months May to October and a critical maximum monthly value of 85 is applied.

In Northern Ireland, where there may be significant variation in the incidence of disease between different areas and where sufficient meteorological data may not be available, the Stormont 'wet-day' fluke forecasting system has been devised (Ross 1978). In general, a 'standard year' incidence of disease (acute or subacute fascioliasis in sheep in particularly wet areas and widespread chronic fascioliasis) is associated with either 12 wet-days (rainfall over 1 mm) per month from June to September, or with 12 wet-days in June, or with 18 wet-days in June to mid-July provided temperatures are optimum from June to mid-July.

Winter infection of snails producing disease in sheep in late summer and autumn may be forecast also. If the summation of the values of 'Mt' for August, September and October is 250 and if either the following May or June is wet then disease is forecast.

These forecasting systems provide the information that is required in June to assist in the decision to use molluscicides, in July to assist in the planning of grazing control and in August to assist in the decision on the use of anthelmintics.

Immunity and immunization. Natural differences in host susceptibility to infection with *F. hepatica* are seen. Thus some animals, such as cattle and pigs, have a moderate to high degree of resistance to primary infection while others, such as sheep, are highly susceptible to infection with *F. hepatica*. Similarly, sheep do not normally develop a protective immune response to reinfection while cattle develop the ability to eliminate primary infections and develop protection against reinfection with *F. hepatica* (van Tiggle 1978).

The ability of cattle to develop protection against reinfection with *F. hepatica* suggested that immunization might be feasible. Nansen (1975) was able to immunize calves with irradiated metacercariae of *F. hepatica*. Six- to seven-month-old calves were immunized with three doses of irradiated metacercariae and, after grazing infected pastures, there was a marked decrease in faecal egg counts and a 71% reduction in the number of flukes recovered from immunized calves as compared to unimmunized animals. In general, the inability to immunize sheep against infection with *F. hepatica* reflects their inability to develop a protective immune response. However, recent experiments by Campbell et al. (1977) and Dineen et al. (1978) demonstrated that the infection of sheep with *Taenia hydatigena* 12 weeks to nine months before infection with *F. hepatica* resulted in protection against infection with *F. hepatica*. Conversely, Hughes et al. (1978) were unable to demonstrate development of cross-protection and this failure may reflect the importance of genetic resistance to disease, since different breeds of sheep were used in these experiments.

Human fascioliasis. Sporadic human cases of fascioliasis occur throughout the world. In Europe and other areas these cases are associated primarily with the eating of watercress contaminated with metacercariae. Human cases are most common after wet summers when watercress or other vegetable beds may become overrun with water draining from wet or swampy animal pastures. In man the presence of adult *F. hepatica* in the bile ducts causes a variety of symptoms: malaise, intermittent fever, weight loss, pain under the right costal margin and often pruritus with eosinophilia. Urticaria with dermatographia may be seen, as may mild jaundice and anaemia. The infection may be diagnosed by the identification of eggs of *F. hepatica* in the faeces or fasting duodenal contents. Serological tests, particularly immunodiffusion and fluorescent antibody tests using *F. hepatica* antigen, may be used.

The infection may be treated surgically, but the treatment of choice is bithionol, at a dose rate of 50 mg/kg on ten alternate days or 1 g three times a day for 15 days. Emetine is also effective and metronidazole (1.5 g/day) has been used.

Adult *F. hepatica* can be found in aberrant sites such as in the lungs and subcutaneously. Here the parasites are found in cysts containing brownish purulent material. They may be removed surgically.

Fasciola gigantica Cobbold, 1885. This is the common liver fluke of domestic stock in Africa; it occurs frequently in Asia, the Pacific islands such as Hawaii, the Philippines, southern USA, southern Europe, European Russia and the Middle East. Mixed infections of *F. hepatica* and *F. gigantica* may occur. For example, in Pakistan Kendall (1954) found mixed infections on the boundaries of highland areas.

F. gigantica resembles *F. hepatica* but is readily recognized by its larger size, being 25–75 mm in length and up to 12 mm in breadth. The anterior cone is smaller than that of *F. hepatica*, the shoulders are not as prominent and the body is more transparent. The eggs measure 156–197 by 90–104 μm.

Life-cycle. The most important intermediate host is *Lymnaea auricularia*. In the Indian subcontinent *L. rufescens* and *L. acuminata* are responsible and in Malaysia, *L. rubiginosa*. The host snail in Africa is *L. natalensis*. The snail hosts of *F. gigantica* are aquatic forms living in fairly large permanent bodies of water which contain abundant vegetation. Still or slightly moving clear water provides the most satisfactory habitat. They occur at sea level and at high altitudes and the snail vectors of *F. gigantica* can survive an amphibious existence but can only aestivate for very short periods.

The development in the snail is comparable to that of *F. hepatica* in *L. truncatula* except that it takes longer. Thus at 26°C eggs of *F. gigantica* hatch in 17 days. In the warm season in East Africa, 75 days are required for development in the snail, this being extended to 175 days in the cold season. Critical studies of the development of the parthenitae of *F. gigantica* were conducted by Dinnik and Dinnik (1964); these authors found that one to six first-generation rediae may develop from a sporocyst of *F. gigantica* at 26°C, each redia producing daughter rediae and then cercariae. Metacercariae encyst on plants under water such as in the aquatic environments of rice fields. Metacercariae may survive for up to four months on stored plants and thus infection may be transmitted by feeding rice straw. In the definitive host the life-cycle is similar to that of *F. hepatica* but longer. *F. gigantica* adults reach the bile ducts, after migration in the liver parenchyma, nine to 12 weeks after infection.

Pathogenesis. This is essentially the same as that of *F. hepatica*, the acute and chronic form of infection occurring in sheep, though in cattle only the chronic form occurs. In endemic areas the infection can be extremely common. For instance in the Tsinchiang province of China 50% of cattle, 45% of goats and 33% of buffalo had eggs of *F. gigantica* in their faeces and in other areas of China up to 80% of slaughtered cattle were infected with the parasite. Similarly, in Iraq, 71% of water buffalo, 27% of cows, 19% of goats and 7% of sheep were found to be infected.

Treatment. Modern anthelmintics which are effective against *F. hepatica* are also efficacious in the treatment of *F. gigantica*. Thus, rafoxanide, 7.5 mg/kg, and oxyclozanide, 15 mg/kg, induced a 97–99% reduction in egg counts in infected sheep (Kadhim & Jabbir 1974). Rafoxanide is 100% effective against adult parasites when administered at dose rates of 2.5–5 mg/kg and 10 mg/kg kills 87% of immature, eight-week-old, *F. gigantica* (Troncy & Vasseau-Martin 1976). Brotianide, 15–20 mg/kg, and niclofolan, 4–6 mg/kg, kill adult *F. gigantica* and over 90% of the migratory stages although they are least effective against the youngest flukes (Karrasch et al. 1975). Niclosamide is 60–98% effective and albendazole could be tried.

Control. Similar treatments are employed as for *F. hepatica*. Control is based on similar principles to *F. hepatica* but since the molluscan vector is an aquatic form, its habitats may be more difficult to treat with molluscicides than for the control of *F. hepatica* snails. In some areas, the use of molluscicides may be contraindicated since the large bodies of water are also important fishing areas. However, *N*-tritylmorpholine, 0.09–0.1 parts per million in a water body in Kenya, eradicated snails for 11 months (Preston & Castelino 1977).

Infection may be avoided by grazing livestock on higher ground and avoiding lakes, swamps and dams though in many parts of the world these are used as watering places for the livestock. A detailed consideration of the control measures applicable in Africa is given by Coyle (1959) who, amongst other things, recommends the piping of water to water-troughs rather than using the dams or lakes themselves for watering places for stock. The use of bore holes and a hydraulic ram to pump water to a higher level than the large body of water is suggested. He also discusses biological control. The control of *F. gigantica* by anthelmintic treatment of infected animals has been studied in the Philippines where 94% of cattle were infected. Strategic control through treatment of all animals reduced infection levels to 8% and 7% in two areas over a 16–24-month period.

Fasciola jacksoni (Stazzi, 1900) occurs in the elephant, in which it produces clinical signs similar to those in ovine fascioliasis. Nitroxynil (10 mg/kg subcutaneously) has been used for treatment (Caple et al. 1978).

Genus: Fascioloides Ward, 1917

Fascioloides magna (Bassi, 1875) occurs in the liver, rarely the lungs, of cattle, horse, sheep and pigs in North America and in cattle, sheep and deer in Europe. The infection is common in wild animals such as the moose (*Alces alces americana*), wapiti (*Cervus canadiensis*), white-tailed deer (*Odocoileus virginianus*), northern white-tailed deer (*O. v. borealis*), mule deer (*O. hemionus columbianus*) and elk (*A. alces*) and is reported in yak and bison in North America. (Davis & Libke 1971). The parasite is found particularly in the Gulf States, the Great Lakes area, western Canada, the West Coast and the Rocky Mountain area. The levels of infection may be high and in Texas *F. magna* was found in 70% of deer, 52% of feral hogs and 38% of cattle (Foreyt & Todd 1972). In the south-eastern states of the USA 12.8% of *O. virginianus* were found to be infected (Pursglove et al. 1977). The parasite was introduced into Europe by the importation of infected animals and the incidence of infection can be high in cattle and sheep in Europe and eastern Europe. The infection is also present in wild ruminants such as fallow deer (*Dama dama*) and sambar (*C. unicolor*) in Italy and red deer (*C. elaphus*) in Italy and Germany.

The worms are oval, with a rounded posterior end, and are thick and flesh-coloured. They measure 23–100 mm long, 11–26 mm broad and 2–4.5 mm thick. There is no distinct anterior cone-like projection. The eggs measure 109–168 by 75–96 μm and have a protoplasmic appendage 4–21 μm in length at the pole opposite the operculum.

Life-cycle. The eggs are passed in the one-celled stage and hatch after four weeks or longer. About seven to eight weeks are required for development in the intermediate hosts, which are the snails *Fossaria parva*, *F. modicella*, *F. modicella rustica*, *Lymnaea bulimoides techella*, *Pseudosuccinea columella*, *Stagnicola palustris nuttalliana* and *S. bulimoides*. *S. palustris nuttalliana* occurs in the stagnant parts of permanent or semi-permanent water which contains large amounts of dead and living vegetation. The parasite does not survive in areas which dry up during the year. *F. parva* occurs in wet, swampy areas and it can aestivate for periods of dryness. *P. columella* is found in pools and streams as are the other species. Metacercariae encyst on vegetation in water but are also very resistant to dessication once the water has receded. On ingestion by the definitive host the parasite probably migrates through the intestinal wall and peritoneal cavity to the liver.

The normal hosts of *F. magna* are considered to be members of the family Cervidae. The white-tailed deer is a common natural host in North America. In deer the young fluke migrates extensively in the liver and the flukes then become encapsulated with connections between the capsule and the bile ducts through which eggs are passed into the bile and thus the faeces. The prepatent period is 30–32 weeks. Each capsule contains two or occasionally three flukes and there appears to be a relationship between fluke pairing and maturation (Foreyt & Todd 1976*a*).

The larger Bovidae, including cattle, bison and yak, are aberrant hosts and following some migration in the liver the *F. magna* become encapsulated in closed cysts. Adults may reach maturity in 32–44 weeks but eggs are not passed out of the liver and into the faeces. Similar closed cysts are found in pigs. Eggs have been reported in the faeces of cattle but these must be rare and probably only associated with severe hepatic destruction (Foreyt & Todd 1976*a*).

In the sheep, which is also an abnormal host, the behaviour of *F. magna* is entirely different. Uninterrupted migration occurs in the liver and encapsulation is rarely seen.

Pathogenesis. In deer the parasites are found in thin-walled, loose vascular fibrous connective tissue cysts up to 4 cm in diameter. Both afferent and efferent bile ducts open into the cavity. Lines of fibrosis in the liver parenchyma resulting from the migratory tracts of the young flukes may be seen.

In the Bovidae and the pig the parasites become enclosed in a thick-walled, fibrous tissue capsule and the bile ducts are occluded. The cysts are filled with a black fluid containing the parasites, bile pigments and eggs. Mature flukes may be found also in calcifying cysts and free in the liver parenchyma. Evidence of migration, haemorrhage, haematoma, infarction, necrosis and fibrosis may be seen but the animals suffer no apparent ill health.

Sheep are usually killed by the uninterrupted migration that occurs. Extensive liver damage, haemorrhage and necrosis result from the unrestricted wandering of the parasite. Adhesive peritonitis may be seen and flukes may be found free in the peritoneal cavity and the lungs.

Characteristically in infections with *F. magna* there is an accumulation of black iron porphyrin pigment in the liver, omentum, kidneys and other internal organs (Foreyt & Todd 1976*a*).

Epidemiology. F. magna is indigenous in parts of North America and Europe. Domestic cattle and sheep become infected when they graze pasture frequented by deer. The presence of infection is dependent on the presence of infected deer since neither cattle nor sheep play an important role in the dissemination of the infection. In the USA 13–60% of white-tailed deer and 58% of elk may be infected and 12% of other deer may be infected in Canada. A high percentage of deer in eastern Europe are infected with *F. magna*.

Treatment. Oxyclozanide (13–28.5 mg/kg) and rafoxanide (12–25 mg/kg) have been used and are effective in the treatment of mature *F. magna* in deer (Foreyt & Todd 1973). However, although rafoxanide (10–15 mg/kg) was effective in cattle, oxyclozanide killed only 27% of the parasites in this host. Similarly, although nitroxynil (11–24 mg/kg) inhibited egg production it did not kill the infection in deer. Clioxanide and diamphenethide were also ineffective in deer (Foreyt & Todd 1974, 1976*b*). Albendazole, 15–35 mg/kg, was more than 90% effective when used to treat calves naturally infected with *Fascioloides magna* (Theodorides 1977).

Control. Sheep and cattle should not be raised on land inhabited by deer. Additional measures include the elimination of snails by molluscicides. However, this presents difficulties because of the different ecological requirements of the snails which may serve as intermediate hosts. Control of Cervidae may be impossible in certain areas and deer should not be moved into a clean area unless they have been demonstrated to be free of infection. Destruction of deer has been practised in a few areas.

Genus: Fasciolopsis Looss, 1899

Fasciolopsis buski (Lankester, 1857) occurs in the small intestine of man and pig, in the south-eastern and far eastern parts of Asia, particularly in China. It is a large, thick-set fluke without shoulders, rather variable in size, but usually measuring about 30–75 by 8–20 mm (Fig. 1.18). The shape is elongate-oval, slightly broader posteriorly than anteriorly. The ventral sucker is situated near the anterior extremity and is much larger than the oral. The cuticle bears spines which are frequently lost. A pharynx and a short oesophagus are present, followed by the unbranched intestinal caeca, which reach almost to the posterior end of the worm. The testes are tandem, branched and posterior in position. The cirrus-sac is long and tubular, opening anterior to the ventral sucker. The ovary is branched, lying to the right of the mid-line. The vitelline glands occupy the lateral fields. The eggs have thin shells with an operculum; they are brown in colour and measure 125–140 by 70–90 μm.

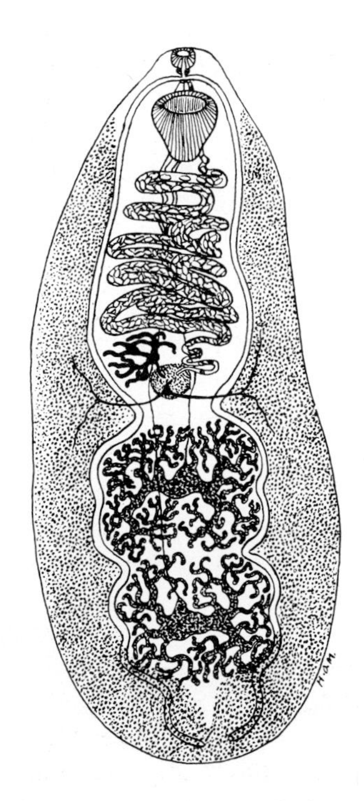

Fig 1.18 Ventral view of *Fasciolopsis buski*.

Life-cycle. Similar to that of *Fasciola hepatica*. Miracidia hatch from eggs in 16–18 days at 30°C and they penetrate flat, spiral-shelled snails of various species of *Planorbis* and *Segmentina*. These snails feed on certain plants—the water calthrop, *Trapa natans* and *T. bicornis*, and the water chestnut, *Eliocharis tuberosa*—which are cultivated for food and are usually fertilized with human night-soil. The cercariae emerge in the summer months and encyst on the tubers or nuts of these plants. These tubers are eaten raw but the outer shell of the nut is commonly peeled off with the teeth particularly by children, in whom the infection is usually heavier than in adults. In some areas the nuts are cooked before they are eaten and then animal infection is usually much higher than human infection. These and possibly also other plants may carry the infection to pigs.

Pathogenesis. The parasite is chiefly of importance as a cause of disease in man though infection is often associated with few or no manifestations of ill health. It attaches itself to the

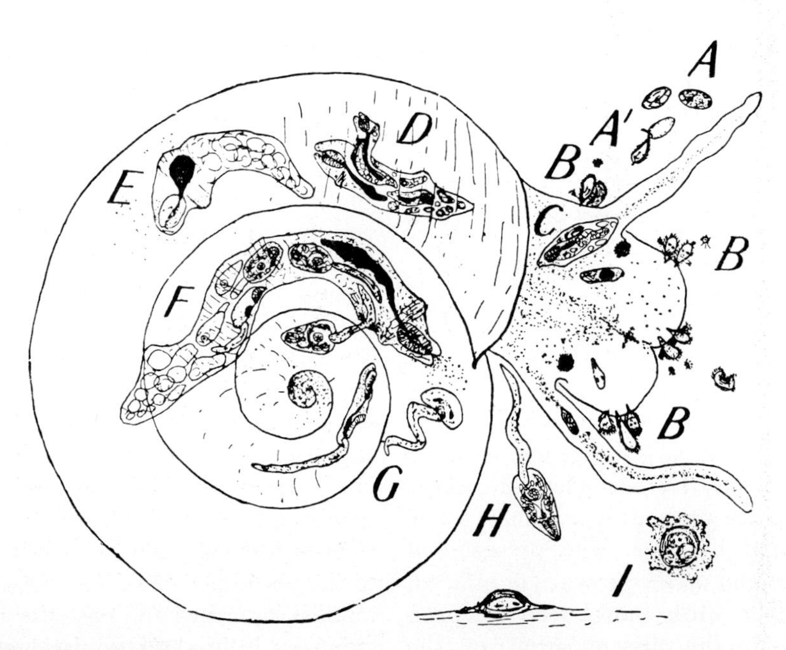

Fig 1.19 Life-cycle of *Fasciolopsis buski*.

- A egg
- A′ miracidium escaping from egg
- B miracidia entering snail
- C sporocyst
- D redia with daughter redia escaping
- E daughter redia
- F older daughter redia with cercariae
- G cercariae
- H cercariae escaping from snail
- I encysted cercariae

intestinal mucosa, causing a local inflammation or severe deep ulcerative lesions in heavy infections, and produces abdominal pain, diarrhoea, oedema and ascites. Infection may be common and up to 39% of children have been found infected in India and Bangladesh (Muttalib & Islam 1975).

Treatment. 1 g of hexylresorcinol is an effective treatment. Tetrachlorethylene is also highly effective and a dose of 0.08–0.14 ml/kg cured 77% of human infections and reduced faecal egg counts by 97%. Niclosamide was less effective and 160 mg/kg reduced faecal egg counts by 46% and eliminated the parasite in only 12% of infected persons. However, niclosamide induced fewer and less severe side effects than did tetrachlorethylene (Suntharasamai et al. 1974). Praziquantel is likely to be effective and niclofolan could be tried.

Prophylaxis is mainly a matter of hygienic disposal of human night-soil and the faeces of pigs. The tubers and nuts of the plants mentioned should not be used as food without at least scalding them in boiling water. Control measures include health education to promote the cooking of water nuts and the avoidance of drinking unfiltered or unboiled water.

Genus: Parafasciolopsis Ejsmont, 1932

Parafasciolopsis fasciolaemorpha Ejsmont, 1932. This form occurs in the gall-bladder and digestive tract of elk and wild goat in the Soviet Union and Poland. It is 3–7.5 mm long by 1–2.5 mm broad, the cuticle is spiny, the anterior sucker 220–285 μm, the ventral sucker 550–850 μm in diameter, the egg is brownish coloured and measures 110–140 by 70–86 μm. The intermediate host is the snail, *Planorbis* (*Coretus*) *corneus*, which is found in stagnant or slow-running deep water in areas where there is much vegetation, such as swamps. Drozdz (1963) states the parasite to be adapted to swampy habitats, deer and elk being infected when they are feeding in such areas. It is unlikely that sheep and cattle will be infected because of the association of the parasite with swampy areas, but heavy infections may be lethal to elk.

FAMILY: ECHINOSTOMATIDAE POCHE, 1926

More or less elongate flukes with a strong ventral sucker situated not far behind the smaller oral sucker. The latter is surrounded dorsally and laterally by a 'head-collar', which bears a single or double row of large spines. The tegument is usually provided with scales or spines. The digestive tract consists of a pharynx, an oesophagus, which nearly reaches the ventral sucker, and simple intestinal caeca which extend to the posterior extremity. The genital pore opens just anterior to the ventral sucker. The testes are entire or lobed, tandem or slightly diagonal, usually situated in the posterior half of the body. A cirrus-sac is present. The ovary is anterior to the testes, median or to the right, and a receptaculum seminis is absent. The vitellaria consist of coarse follicles lying in the lateral fields and frequently extending into the central field behind the testes. Uterus anterior to the ovary, containing relatively large eggs with thin shells. Parasites in the intestine and sometimes the bile ducts of birds and mammals. The life-history is similar to that of *Fasciola hepatica,* but the cercariae frequently enter another snail, an amphibian or a fish, in which they encyst, and the final host becomes infected by ingesting the second intermediate host.

Genus: Echinostoma Rudolphi, 1809

Echinostoma revolutum (Frölich, 1802) occurs in the rectum and caeca of the duck, goose, and other aquatic birds, the partridge, the pigeon and the fowl, as well as in man. It is 10–22 mm long and up to 2.25 mm broad (Fig. 1.20). The head-collar bears 37 spines, of which five on either side form a group of 'corner spines'. The tegument is spiny in the anterior region. The testes are tandem, elongate, oval or slightly lobed, situated behind the middle with the ovary anterior to them. The cirrus-sac lies between the bifurcation of the intestine and the ventral sucker and may extend slightly beyond the anterior margin of the latter. The eggs measure 90–126 by 59–71 μm.

Life-cycle. The eggs hatch after developing under favourable conditions for about three

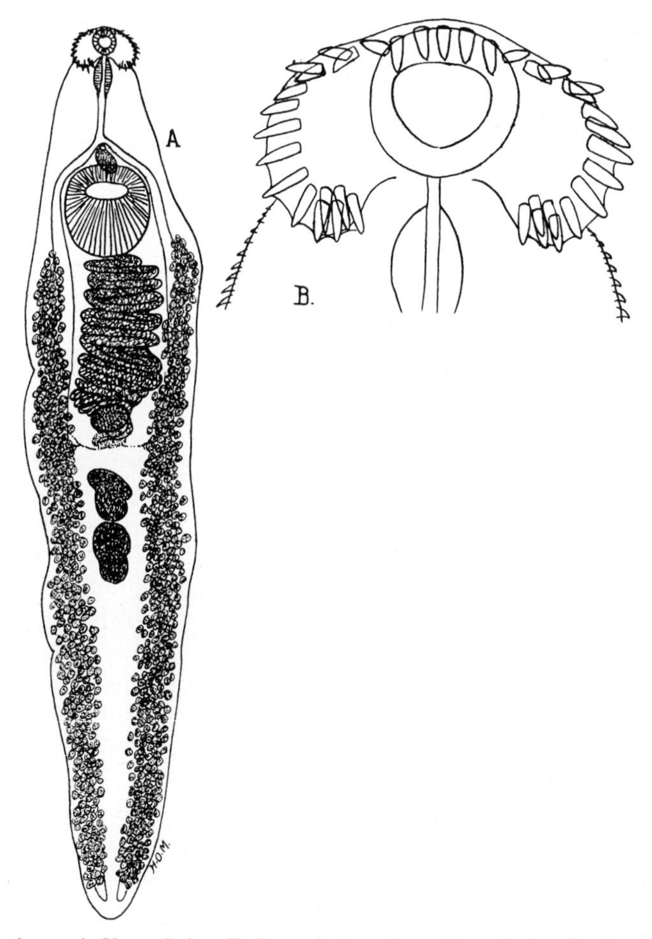

Fig 1.20 *Echinostoma revolutum.* A, Ventral view. B, Ventral view of anterior end, showing head-crown.

weeks and the miracidium penetrates into an intermediate host, *Stagnicola palustris*, *Helisoma trivolvis*, *Physa gyrina*, *P. occidentalis*, *P. oculans*, *Planorbis tenuis*, *Lymnaea stagnalis*, *L. attenuate*, *L.* (*Radix*) *pereger* or *L. swinhoei*. Cercariae are produced in two or three weeks and these either encyst in the snail or escape and enter another of the same or a different species, for instance *Vivipara vivipara*, *Sphaerium corneum*, *Fossaria* spp. or tadpoles of *Rana esculenta*. The final host becomes infected by ingesting these snails and the worms grow to adulthood in 15–19 days.

Pathogenesis. This parasite has generally been regarded as fairly harmless, but heavy infections may cause severe enteritis. Beaver (1937) reported haemorrhagic enteritis 10 days after infection of pigeons. On post mortem 600 flukes were found. Death in pigeons associated with several thousand echinostomes was reported by van Heelsbergen (1927a). In 1975, 3% of a flock of geese were lost following infection with *E. revolutum* and *Notocotylus attenuatus* after they had been moved on to swampy ground. The geese were emaciated, weak and had catarrhal enteritis (Griffiths et al. 1976).

Diagnosis is made by finding the eggs of the worms in the faeces of the host.

Treatment. See below.

Prophylaxis should be directed towards extermination of the snails. Where possible, the birds should have access only to ponds in which the snails can be properly controlled.

Echinostoma paraulum (Dietz, 1909) (syn. *Echinoparyphium paraulum*). Occurs in the small intestine of the duck, the pigeon and man. It measures 6–10.5 by 0.8–1.4 mm. The cuticle bears spines almost to the posterior extremity, but these may be lost and their absence has in some cases caused much confusion. The head-collar, which is continuous across the ventral surface, bears 37 spines: 27 in a double dorsolateral row and at either end five 'corner spines'. The oral sucker measures 0.25–0.3 mm in diameter and the ventral 0.72–0.88 mm; the latter lies at the end of the first quarter of the body. There is a short prepharynx, a pharynx and an oesophagus; the latter is 0.4–0.6 mm long. The testes are tandem and lie in the third quarter of the body; the anterior has frequently three and the posterior four lobes. The cirrus-sac may extend back to the middle of the ventral sucker. The ovary lies just anterior to the testes. The eggs measure about 100 by 70 μm.

Life-cycle. Unknown. The first intermediate host is certainly a snail and some authors suspect fish and snails as the second intermediate host.

Pathogenesis. Krause (1925) and Wetzel (1933) have observed deaths in pigeons caused by this parasite. The birds showed inappetence, thirst, diarrhoea, lassitude and progressive weakness. At post mortem there was atrophy of the breast muscles and catarrhal enteritis with much mucus, becoming haemorrhagic behind the duodenum. The parasites were found chiefly in the middle portion of the intestine and could occur in large numbers.

Treatment. See below.

Prophylaxis. Extermination of the possible intermediate hosts, such as snails, is indicated.

Echinostoma ilocanum (Garrison, 1908) occurs in the intestine of man in the Philippines and South-East Asia. It has also been found in the dog and the Norway rat, the latter serving as a reservoir host. The first intermediate hosts are the snails *Gyraulus convexiusculus*, *G. prashadi* and *Hippeutis umbilicalis*. Cercariae encyst on almost any freshwater mollusc but *Pila luzonica*, *P. conica* and *Viviparus javanicus* are especially important since they are regarded as a delicacy and are eaten raw or at the most with a sprinkling of salt and vinegar. The pathogenic effects consist of inflammatory lesions of the intestinal mucosa at the site of attachment of the worms. Diarrhoea and intestinal colic may occur.

Echinostoma hortense Asada, 1926, occurs in Japan, Korea and Manchuria. The natural definitive hosts are the rat, weasel, marten and dog. Man may be infected and human cases in Japan have been associated with the ingestion of raw green frogs and loach (*Misgurnus*

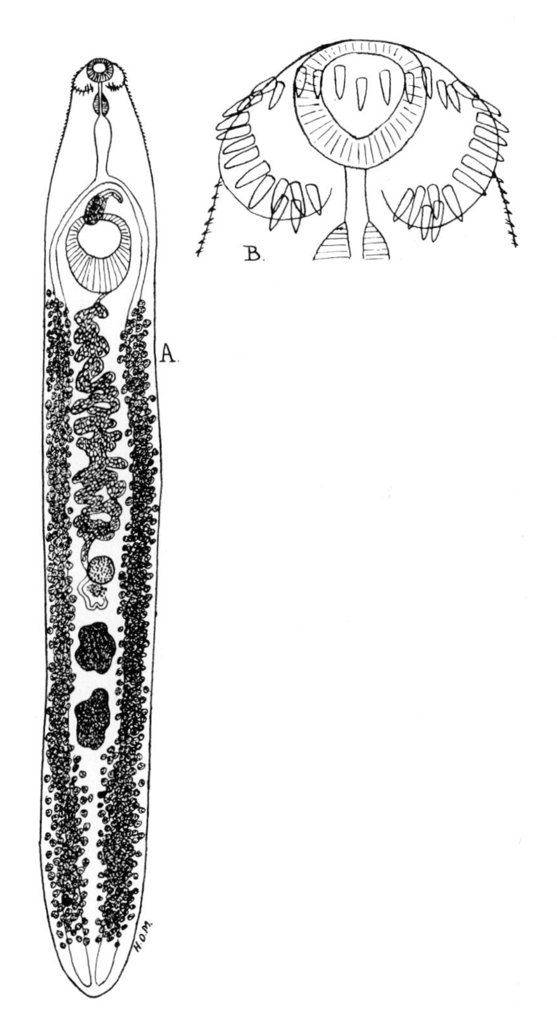

Fig 1.21 *Echinostoma paraulum.* A, Ventral view. B, Ventral view of anterior end showing the head-crown.

anguillicaudatus) which act as second intermediate hosts (Arizono et al. 1976). Metacercariae may be found in 49% of loach and in 30% of *Acheilognathus* (Tani 1976).

Other species of the genus include **E. jassyenese** Léon and Ciurea, 1922, found in the intestine of man in Roumania, **E. lindoensis** Sandground and Bonne, 1940 of rat and man in Indonesia and **E. suinum** Ciurea, 1921, in the intestine of pigs in Roumania and Hungary.

Treatment. Modern anthelmintics may be effective against echinostome infection in domestic animals and man since brotianide (75 mg/kg), oxyclozanide (15–30 mg/kg) and rafoxanide were highly effective against experimental infections with *E. caproni* in mice (Leger & Notteghem 1975). Also, flubendazole (10 or 50 mg/kg/day for five days) was effective in mice (Notteghem et al. 1979). Other benzimidazole anthelmintics and praziquantel should be tried.

Prophylaxis should be directed towards extermination of the snails. Humans should not eat raw snails, frogs or loach.

E. revolutum and other echinostomes may be used for biological control of the intermediate hosts of the schistosomes. Thus, *Biomphalaria glabrata* infected with *Schistosoma mansoni* were more susceptible to infection with echinostomes and then the larval trematodes are antagonistic to each other.

Genus: Echinoparyphium Dietz, 1909

Echinoparyphium recurvatum (v. Linstow, 1873) occurs in the small intestine, especially the duodenum, of the domestic duck, wild duck, fowls and pigeons. The parasite has also been recorded in dogs, cats, rats and man in Malaysia, Indonesia and Egypt. It is up to 4.5 mm long and 0.5–0.8 mm wide (Fig. 1.22). The anterior end is curved ventrad and is armed with spines anterior to the ventral sucker. The head-crown has 45 spines, of which four are corner spines on either side. The ventral sucker is 0.32–0.36 mm wide and situated at the first quarter of the body. Testes oval, tandem, not lobed and in contact with each other.

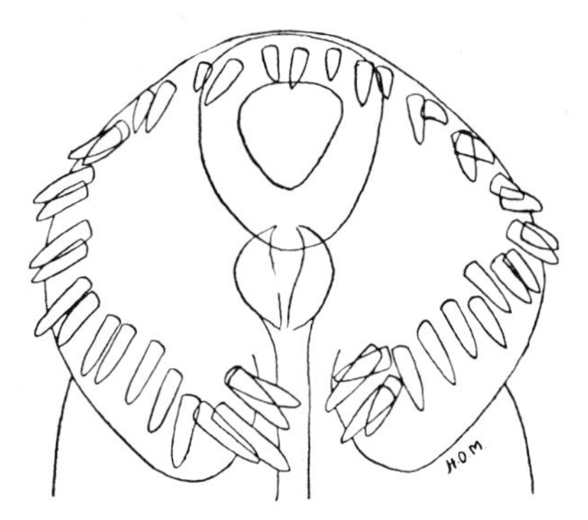

Fig 1.22 Ventral view of the anterior end of *Echinoparyphium recurvatum*, showing the head-crown.

The ovary is transversely oval and the uterus is short, containing three to seven eggs, which measure 108–110 by 81–84 μm.

Life-cycle. The first intermediate hosts are *Lymnaea ovata*, *L. auricularia*, *L. palustris*, *L. stagnalis*, *Planorbis planorbis*, *P. corneus*, *Vivipara vivipara* etc. The frog, *Rana temporaria*, and snails, among which are *Valvata piscinalis* and *Planorbis albus*, act as second intermediate hosts. The cercariae encyst in the digestive glands of the snails and may migrate to other snails. They encyst in the kidneys of the tadpoles and even adult frogs (Harper 1929). The prepatent period is seven to 12 days.

Pathogenesis. Van Heelsbergen (1927*b*) reported emaciation, anaemia and sometimes weakness of the legs in infected fowls. At autopsy a marked enteritis was found, with swelling of the mucosa and excessive mucus in the bowel. In the USA Annereaux (1940) reported a marked enteritis in turkeys due to *E. recurvatum* and in Great Britain Soulsby (1955) found it the cause of death in mute swans on the River Axe in Somerset.

Treatment. See *Echinostoma* spp. above.

Prophylaxis should be directed towards extermination of the snail intermediate hosts and, where possible, the birds should be prevented from ingesting infected frogs. It is obvious that it is impossible to exterminate snails everywhere without upsetting the water fauna and the statement that the snails must be exterminated should be taken and applied with discretion.

Genus: Hypoderaeum Dietz, 1909

Hypoderaeum conoideum (Bloch, 1872) occurs in the posterior part of the small intestine of the duck, goose, swan, wild aquatic birds, fowl and pigeon. It is 5–12 mm long and up to 2 mm broad (Fig. 1.23). The body is elongate and tapers posteriorly. The ventral sucker is relatively large and situated close to the anterior. The head-collar is weakly developed and bears 47–53 (usually 49) spines, of which two on either side form the 'corner spines'. The anterior part of the body is well armed with spines. The oesophagus is very short. Testes are elongate, slightly lobed, tandem, behind the middle. The cirrus-sac is club-shaped, reaching back almost to the posterior margin of the ventral sucker. The eggs measure 95–108 by 61–68 μm.

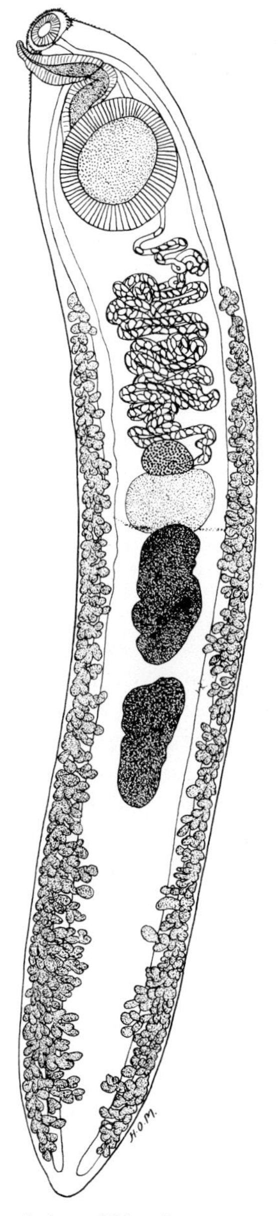

Fig 1.23 Ventral view of *Hypoderaeum conoideum.*

Life-cycle. The first intermediate hosts are *Lymnaea stagnalis*, *L. pereger*, *L. ovata* and *Planorbis corneus*. Vevers (1923) has infected ducks by feeding infected *L. peregra*, showing that the cercariae in the snail are infective. However cercariae usually enter a second intermediate host which carries the infection to the final host. The cercariae have been found in the kidneys of the tadpoles and young forms of the frog *Rana esculenta*.

Pathogenesis. Localized enteritis has been noted in infected ducks (Vevers 1923).

Genus: Echinochasmus Dietz, 1909

Echinochasmus perfoliatus (v. Rátz, 1908) occurs in the intestine of the dog, cat, fox and pig in Europe and Asia. It measures 2–4 by 0.4–1 mm. The head-crown bears 24 spines in a single row and the anterior region of the body is spiny. The ventral sucker is nearly twice as large as the oral one. The testes are large, tandem and situated behind the middle. The ovary lies to the right, anterior to the testes. The cirrus-sac lies anterior to the ventral sucker. The vitellaria extend from in front of the ventral sucker to the posterior end of the body in the lateral fields, also extending inwards behind the testes. The eggs measure 85–105 by 60–75 μm and have a yellow colour.

In Japan the primary intermediate host is the snail *Bulinus striatus japonicus* and several fresh-water fish (*Abramis brama*, *Esox lucius*, *Idus idus*, *Aspius aspius*, *Scardinius erythrophthalmus*, etc.) serve as secondary intermediate hosts.

This parasite may cause severe enteritis.

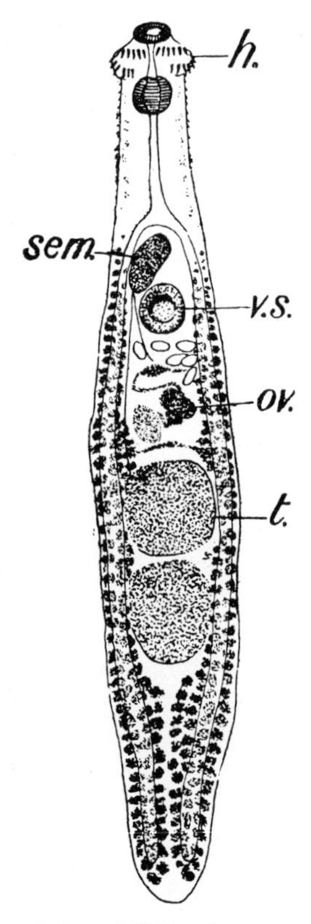

Fig 1.24 Ventral view of *Echinochasmus perfoliatus*. (*From Baylis, after V. Rátz*)

h	head crown
ov	ovary
sem	seminal vesicle
t	testis
vs	ventral sucker

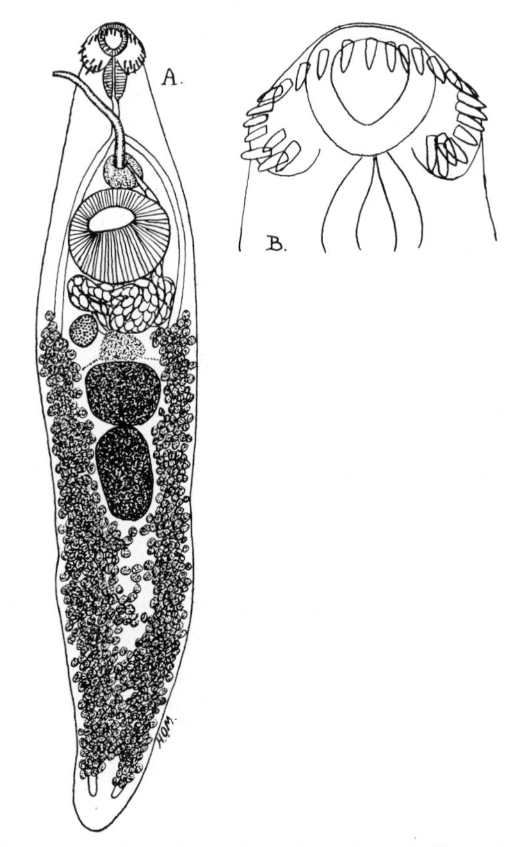

Fig 1.25 *Isthmiophora melis*. A, Ventral view. B, Ventral view of the anterior end showing the head-crown.

The benzimidazole anthelmintics, such as oxyclozanide, rafoxanide and brotianide, may be useful in treatment. Snails should be exterminated where this is possible and raw fish should not be fed to the animal hosts.

Genus: Isthmiophora Lühe 1909

Isthmiophora melis (Schrank, 1788) occurs in the small intestine of the cat, fox, polecat, mink, pine marten, beech marten, badger, otter, weasel and hedgehog (*Erinaceus europaeus*) in Europe. The worm is elongate, measuring 3.5–12 by 1.3–3.3 mm. The head-collar bears a dorsally continuous row of spines, the whole ventral surface is covered with small spines and dorsally they extend to the level of the pharynx. The oral sucker is much smaller than the ventral sucker. The latter lies in the first quarter of the body and the intestine bifurcates near its anterior border. The testes are median, tandem, entire or slightly lobed; the anterior one lies at the middle of the body. The ovary lies anterior to the testes to the right of the mid-line and the receptaculum seminis to the left. The cirrus-sac is well developed and extends dorsally to the ventral sucker. The cirrus is spiny. The vitellaria extend backwards from the level of the receptaculum seminis and almost meet behind the posterior testis. The uterus is short and the eggs measure 120–125 by 91–94 μm.

According to Beaver (1941) the first intermediate host in the United States is the snail *Stagnicola emarginata angulata* and the metacercariae are found in tadpoles.

Treatment is as for *Echinochasmus* spp. Heavy infections may be seen in the polecat without producing clinical signs. The mink, however, is very susceptible to the effects of these worms, which produce a haemorrhagic enteritis in this host.

FAMILY: NOTOCOTYLIDAE LÜHE, 1909

These trematodes have no ventral sucker. The ventral surface of the body is provided with three or five rows of unicellular glands situated in groups. The tegument is armed with fine spines anteriorly and ventrally. A pharynx is absent and the oesophagus is short, while the intestinal caeca extend to the posterior end of the body. The genital pore usually opens directly behind the oral sucker and the cirrus-sac is elongate. The testes are horizontally situated near the posterior end of the body and lateral to the intestinal caeca. The ovary lies between them. The vitelline glands occupy the lateral fields in the posterior part, anterior to the testes. The uterus forms more or less regular transverse coils extending from the ovary to the posterior end of the cirrus-sac. The eggs bear long filaments at both poles. Parasites in the intestine of aquatic birds and mammals.

Genus: Notocotylus Diesing, 1839

Notocotylus attenuatus (Rudolphi, 1809) commonly occurs in the caeca and rectum of the fowl, duck, goose and wild aquatic birds. It measures 2–5 by 0.6–1.5 mm and is narrow anteriorly. There are three rows of ventral glands. The eggs are small, 20 μm long and bear a long filament at either pole.

Other species include **N. impricatus** Szidat, 1935, in domestic poultry in Europe and North America and **N. thienemanni** Szidat and Szidat, 1933, in domestic and wild ducks in Europe.

Life-cycle. The intermediate hosts are the snails *Planorbis rotundatus*, *Lymnaea palustris*, *L. limosa* and *Bulinus japonicus*.

Pathogenesis. *N. attenuatus* is not usually considered to be very pathogenic. However, it has been associated with erosion of the caecal mucosa and disease in geese and ducklings. Infected ducklings were emaciated, had diarrhoea, locomotory ataxia and catarrhal mucoenteritis (Michalski 1977).

Treatment. Oxyclozanide (15–30 mg/kg) is highly effective (Michalski 1977), as is bithional (0.3–0.5 g/kg).

Genus: Catatropis Odhner, 1905

Catatropis verrucosa (Fröhlich, 1789) occurs in the caeca of the fowl, duck, goose and wild aquatic birds. It measures 1–6 by 0.75–2 mm and is reddish in colour. The body is rounded anteriorly and posteriorly. There are three rows of ventral glands each containing eight to 12 glands. The elliptical, reddish eggs measure 18–28 μm in length, not including the filaments, each of which is 160–200 μm long.

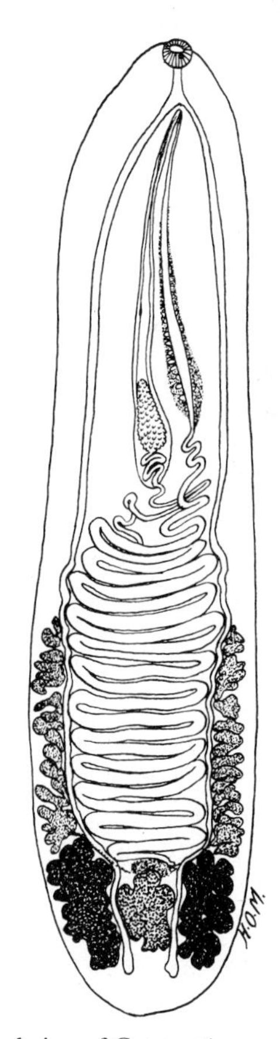

Fig 1.26 Ventral view of *Catatropis verrucosa*.

Szidat (1930) has followed the development through the snail *Planorbis* (*Coretus*) *corneus*. The cercariae have simple tails and three eye-spots. They leave the snail and encyst on water-plants, snails etc., which may be ingested by the final host. The worms become sexually mature within a short time after infection.

Genus: Ogmocotyle Skrjabin and Schulz, 1933

Ogmocotyle indica (Bhalerao, 1942) occurs in sheep, goats and cattle in India. The parasites live in all parts of the digestive tract behind the oesophagus, but particularly the duodenum. Heavy infections are frequent. No pathogenic changes have been ascribed to these parasites.

They are pear-shaped and concave ventrally, measuring 0.8–2.7 by 0.31–0.96 mm. The genital opening is to the left of the midline, a short distance anterior to the middle of the body. The ovary has four distinct lobes. The eggs measure 18–27 by 11–13 μm and carry filaments at the poles.

The Notocotylidae are rarely associated with pathogenic effects.

FAMILY: BRACHYLAEMIDAE JOYEUX AND FOLEY, 1930 (SYN. HARMOSTOMIDAE)

More or less elongate, small or medium-sized trematodes usually with smooth bodies. A prepharynx and oesophagus are present and the intestinal caeca extend to the posterior end of the body. The testes are posterior in position, tandem or slightly diagonal, and the ovary lies between them. The vitellaria are follicular and occupy the lateral fields mostly behind the middle of the body. The genital pore is posterior, median or slightly lateral or even terminal or dorsal. The cirrus-sac contains a cirrus, but the seminal vesicle lies free. Parasitic in the intestine of vertebrates.

Genus: Postharmostomum Witenberg, 1923

Postharmostomum commutatus (Diesing, 1858) occurs in the caeca of the fowl, pheasant, turkey, pigeon and guinea-fowl in southern Europe, North Africa and Indo-China. It measures 3.7–7.5 by 1–2 mm. The body is rounded anteriorly and tapers posteriorly. The ventral sucker lies within the anterior third of the body. The testes are irregularly rounded; the posterior one is median and the anterior one lies to the left of the midline, while the ovary lies to the right. The vitellaria consist of fine follicles extending in the lateral fields forwards from the level of the posterior testis. The uterus has coiled ascending and descending branches, and the genital pore is situated near the anterior border of the anterior testis in the mid-line. The eggs measure 27–32 by 13–18 μm. *Postharmostomum gallinarum* (Witenberg, 1923) found in North Africa, North America and Japan is considered to be synonymous with *P. commutatus*.

The American form utilizes the land snail *Eulota similaris* as an intermediate host (Alicata 1940) though other snails such as *Subulina*, *Euhadra* and *Philomycus* spp. may serve as such. Cercariae after liberation may encyst in the same or other species of snails.

The parasites may cause an inflammation of the caeca.

Postharmostomum suis (Balozet, 1936) occurs in the small intestine of the pig in Tunisia. It is reputed to suck blood, but is apparently not very pathogenic. The intermediate hosts are land snails, especially *Xerophila* species. The eggs are light brown and measure 30–35 by 15–17 μm.

FAMILY: HASSTILESIIDAE HALL, 1916

These are very small oval distomes. The genital pore is submedian or submarginal. The testes are diagonal in the hindbody and the ovary lateral in the testicular zone. Vitellaria are lateral in the anterior half. They are parasites of mammals.

Genus: Skrjabinotrema Orlov, Erschov and Banadin, 1934

Skrjabinotrema ovis Orlov, Ershov and Banadin, 1934. This occurs in the posterior part of the small intestines of sheep in west China and the steppe area of eastern USSR.

The flukes measure 0.79–1.12 by 0.32–0.7 mm. The two oval testes are large and lie diagonally, but touching each other, in the posterior part of the body, the ovary being in front of the right testis. The eggs measure 24–32 by 16–20 μm; they are slightly flattened on one side and have a large operculum at one end and a small appendage at the other. Gvozdev and Soboleva (1973) consider the genera *Skrjabinotrema* and *Hasstilesia* as being synonymous. *Hasstilesia tricolor* (Stiles and Hassal, 1894) occurs in the cottontail and black-tailed jack rabbit in the USA. Heavy infections may cause a catarrhal enteritis.

FAMILY: TROGLOTREMATIDAE ODHNER 1914

These trematodes usually have a fleshy body, flattened or concave ventrally and convex dorsally, with a spiny tegument. The ventral sucker is near the midbody and the oral sucker subterminal. The genital pore is median and posterior to the ventral sucker. The testes are elongate or deeply lobed, usually in the median third of the body. The ovary is usually deeply lobed, submedian and pretesticular. Vitellaria are variable in extent and lie in lateral fields. They are parasites of the skin, frontal sinus, kidney and rarely intestine of birds and carnivores.

Genus: Collyriclum Kossack, 1911

Collyriclum faba (Bremser, 1831) occurs in subcutaneous cysts in the fowl and turkey as well as in several small wild birds like sparrows and starlings. It is found in Asia, Europe and North and South America. It measures 3–5 by 4.5–5.5 mm. It is flattened ventrally, convex dorsally and has a spiny tegument. The oral sucker is small, with a diameter of 0.2–0.45 mm. The ventral sucker is absent. The ovary has three main lobes and each is divided into several smaller lobes. The vitellaria are situated in the anterior half of the body and consist each of about seven large follicles. The very small eggs measure 19–21 by 9–11 μm.

Life-cycle. The first intermediate hosts are probably snails. Metacercariae resembling the adults have been found in dragonfly naiads. Only birds which have access to marshy places become infected.

Pathogenesis and clinical signs. The parasites are found mainly around the cloacal opening and, in heavier infections, also along the abdomen and thorax. They are lodged in subcutaneous cysts, 4–6 mm in diameter. Each cyst has a central opening and contains a pair of the worms, lying with their ventral surfaces apposed. The cysts also contain a black fluid and the eggs, which are discharged through the pore. Heavy infections produce anaemia, emaciation and death. The presence of infection decreases the value of the carcase.

Treatment. The cysts should be opened and the worms extracted, followed by suitable treatment of the wound.

Prophylaxis. The birds are to be kept from marshy places where they become infected.

Genus: Troglotrema Odhner, 1914

Troglotrema acutum (Leuckart, 1842) occurs in the frontal and ethmoidal sinuses of the fox, mink and polecat in Europe. The parasites are whitish in colour and measure about 3.27 by 2.25 mm. The body is thick and rounded anteriorly and has a narrow, tail-like, posterior extremity. The ventral sucker is located just anterior to the middle and is as large as the oral. The testes are entire or slightly lobed and lie just behind the middle. The genital pore opens immediately behind the ventral sucker. The ovary is spherical and lies to the right of the midline close to the ventral sucker. The eggs measure about 80 by 50 μm.

Bythinella dunkeri and *B. alta* snails can be experimentally infected. Cercariae developed after nine months at 10–13°C. The metacercariae encyst in the muscles of *Rana temporaria* (Vogel & Voelker 1978).

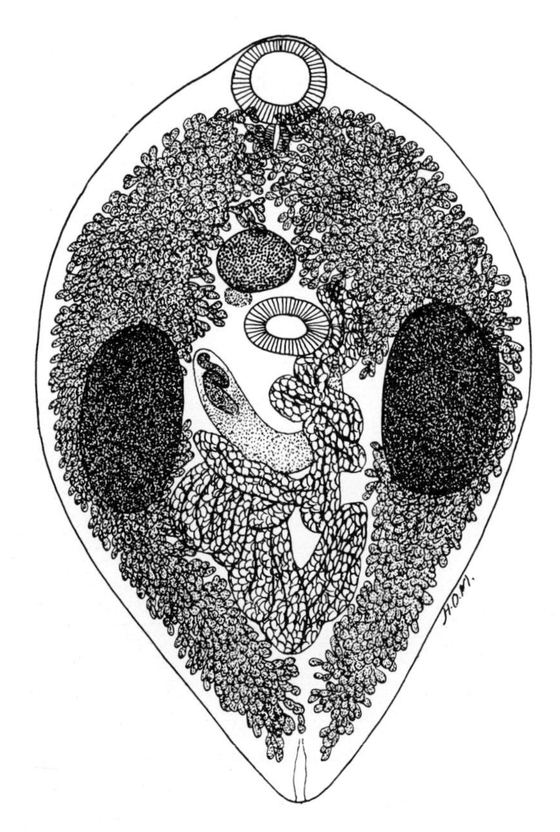

Fig 1.27 Ventral view of *Troglotrema acutum*.

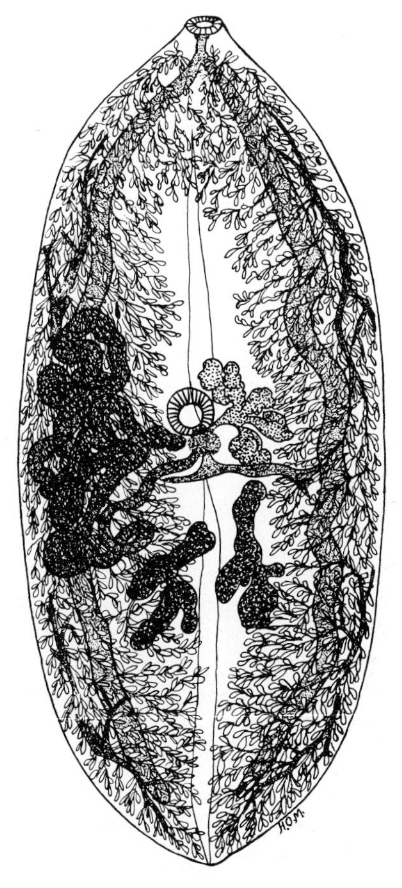

Fig 1.28 Dorsal view of *Paragonimus westermanii*.

The parasites may live in pairs in cysts or, particularly in the fox, they are found freely attached to the mucosa. In the mink and the polecat the parasites cause decalcification and atrophy of the bony walls of the sinuses and eventually perforation. Cysts may break open to the exterior or into the brain cavity.

Treatment is unknown. The animals should be prevented from eating infected frogs.

FAMILY: PARAGONIMIDAE DOLLFUS, 1939

These are ovoid, plump distomes with a spiny tegument and are parasitic in the lungs. The oral sucker is ventroterminal. The ventral sucker is near the midbody and the genital pore is immediately behind this. Testes are in the posterior half of the body and the ovary is pretesticular.

Genus: Paragonimus Braun, 1899

Paragonimus westermanii (Kerbert, 1878), the 'lung-fluke', occurs in the lungs and more rarely in the brain, spinal cord and other organs of the pig, dog, cat, goat, cattle, fox, pine marten, beech marten, mink, musk-rat, wild carnivores and man in China and countries in South-East Asia and the Far East.

P. kellicotti Ward, 1908, is found in wild animals and the cat, dog and pig in North America. The primary host is probably the mink and the muskrat may also be a natural host. Other species of *Paragonimus* are recognized and these include **P. ohirai** Mujazaki, 1939 (China), **P. iloktsuensis** Chen, 1940 (Japan), **P. africanus** Voelker and

Voget, 1965 (Africa), **P. uterobilateralis** Voelker and Voget, 1965 (Africa), **P. caliensis** Little, 1969 (Colombia), **P. peruvianus** Miyazaki, Ibanez and Miraida, 1969 (Peru) and **P. mexicanus** Miyazaki and Ishii, 1968 (Mexico). The comparative morphology of and ecological aspects of infection by these species have been reviewed by Voelker (1973) and Miyazaki (1970). In all cases animals serve as hosts, e.g. dog, mongoose, bush rat, pig, cat, mink etc., according to the geographical area, and man is secondarily infected with this zoonotic parasite.

The adults live in pairs in cysts in the lungs. The parasite is reddish brown in colour and measures 7.5–16 by 4–8 mm. The tegument is covered with spines. The spines of *P. westermanii* are large with bifid tips, while those of *P. kellicotti* are larger and have a number of points each. The ventral sucker is situated slightly anterior to the middle. The eggs are yellowish-brown in colour and measure 75–118 by 42–67 μm; they are provided with an operculum, and the shell is thickened at the pole opposite this.

Life-cycle. The life-cycle has been redescribed by Stromberg and Dubey (1978) and Dubey et al. (1979*b*).

The eggs are laid in the cysts, in which the worms live, and escape through connecting channels into the bronchi or when the cysts rupture. They pass up from the lung with the mucus and may be found in the sputum, which has a characteristic rusty colour. Animals swallow the mucus, so that the eggs are found in the faeces. After development for two to seven weeks (16 days under optimal conditions) the miracidium escapes and penetrates into an aquatic or amphibious snail of the genera *Melania*, *Ampullaria*, *Pomatiopsis*, *Semisulcospira* or *Assiminea* in which sporocysts, rediae and cercariae develop in 78–93 days. The latter have an oval body and a very short tail. After escaping from the snail the cercariae swim about in the water and, on meeting a suitable crab or crayfish, penetrate into it and encyst.

The following genera of crustaceans are known as intermediate hosts: *Astacus*, *Eriocheir*, *Patomon*, *Sesarma*; in Venezuela, *Pseudotelphusa*; in the USA, *Cambarus*. The metacercariae are found in the heart, liver and muscles and mature over six to seven weeks. In the USA there is a seasonal timing of the life-cycle and cercariae enter crayfish in large numbers in late summer or autumn (Stromberg et al. 1978). The final host becomes infected by eating the infected crustacea. Metacercariae which have been released from injured or degenerating crustacea can live in water for three weeks and may be ingested by the definitive host.

After being liberated in the intestine the young fluke penetrates through the wall and wanders through the peritoneal cavity for one to 14 days and then through the diaphragm, entering the lungs from the pleural cavity five to 23 days after infection. They may also enter other organs, such as the brain, from this location. The parasite penetrates the pulmonary parenchyma, a cystic cavity is formed and the parasite matures to the adult stage. Virtually all growth occurs in the lungs. Communication is established with the bronchioles within five weeks and the prepatent period is 30–36 days. The cyst usually contains two parasites surrounded by a purulent fluid mixed with blood and eggs. The interior surface of the cyst is partially epithelialized by cells from the bronchioles (Hoover & Dubey 1978). In cats, 1000–2000 eggs are laid per fluke per day (Stromberg & Dubey 1978).

Pathogenesis and clinical signs. Migrating immature flukes cause an eosinophilic peritonitis, pleuritis and myositis and multifocal pleural haemorrhage. In infected animals there is a chronic bronchiolitis, hyperplasia of bronchiole epithelium and a chronic eosinophilic granulomatous pneumonia associated with degenerating ova in the alveolar tissue (Hoover & Dubey 1978).

Infected animals are lethargic and there is an intermittent cough. Dyspnoea associated with pneumothorax has been seen in one experimentally infected cat. Parasites lodged in cysts in the brain can become problems, as can parasitic cysts in other parts of the body.

Diagnosis of lung cases is readily made by finding eggs in the sputum or faeces. Lesions may be detected by radiography from three to four weeks after infection. The lesions are most frequent in the right caudal lung lobe (Dubey et al. 1978*a*). Diagnosis of parasitic cysts in other parts of the body may be extremely difficult but various serological tests are available (Capron et al. 1965).

Treatment. Albendazole (50–100 mg/kg/day for 14–21 days) decreased the number of eggs shed within eight days, caused morphological degeneration of the adults and reduced the pulmonary lesions in cats (Dubey et al. 1978*b*). Bithionol, 100 mg/kg daily for seven days, or on alternate days for 30 days, is also effective (Macey & Todd 1975; Eliasoff & Harden 1977), as is fenbendazole given as 50 or 100 mg/kg/day in two divided doses for 10–14 days (Dubey et al. 1979*a*). Niclofolan, 1 mg/kg/day for three days or two doses of 2 mg/kg on alternate days, may also be used (Rim et al. 1977). Niclofolan and bithionol are effective in the treatment of infected humans.

Prophylaxis. Freshwater crustacea should not be eaten raw and the extermination of snails should be considered.

FAMILY: PARAMPHISTOMATIDAE FISCHOEDER, 1901

These trematodes are usually thick and circular in transverse section. The ventral (posterior) sucker is situated at or close to the posterior extremity and may be very strongly developed. A large ventral pouch may be present. The anterior sucker sometimes has a pair of posterior pockets. A pharynx is absent, but the oesophagus is present and the intestinal caeca are simple. The tegument is spineless. The genital pore opens ventrally, median, in the anterior third. The testes are frequently lobed and usually anterior to the small ovary. The vitelline glands are lateral and are, as a rule, strongly developed. The uterus runs forwards in the dorsal part of the body and is coiled. Parasites of fishes, amphibia, reptiles, birds and mammals.

In domestic animals, a large number of species have been described from the rumen and reticulum of ruminants and some species occur in the large intestine of ruminants, pigs, equidae and man. The taxonomy of the paramphistomes is complex (Yamaguti 1958; Dawes 1968). Here, the species names used are those listed by Yamaguti (1958).

Check lists of the paramphistomes which occur in cattle and sheep are given by Soulsby (1965). The various genera of the family are discussed below and these are followed by an account of the life-cycles, pathogenicity etc.

Genus: Paramphistomum Fischoeder, 1901

A large number of species have been described from the rumen and reticulum of domestic and wild ruminants. They can be important parasites of cattle, sheep, goats and buffalo, particularly in tropical and subtropical areas.

Paramphistomum cervi (Zeder, 1790) is the commonest species and is found throughout the world. **P. microbothrium** Fischoeder, 1901, is common in Africa and Europe south of Germany and **P. ichikawai** Fukui, 1922, occurs in Australasia. Other species of the genus *Paramphistomum* include **P. gotoi** Fukui, 1922, of cattle in India and Japan; **P. hiberniae** Willmott, 1950, of cattle in Scotland, Ireland and Holland; **P. liorchis** Fischoeder, 1901, of cattle in South and North America (Florida, Louisiana); **P. microbothrioides** Price and McIntosh, 1944, of cattle and sheep in USA; and **P. scotiae** Willmott, 1950, of cattle in Scotland and Ireland.

The colour of live adult specimens is light red. It is one of the 'conical flukes' which are pear-shaped, slightly concave ventrally and converse dorsally, with a large posterior subterminal sucker. The worm measures about 5–13 by 2–5 mm. The genital pore is situated at the end of the anterior third of the body. The testes are slightly lobed and tandem, anterior to the ovary. The vitellaria are in compact groups between the pharynx and the posterior sucker. The eggs measure 114–176 by 73–100 μm.

Genus: Cotylophoron Stiles and Goldberger, 1910

Cotylophoron cotylophorum (Fischoeder, 1901) occurs in the rumen and reticulum of the sheep, goat, cattle and many other ruminants in most parts of the world except the northern temperate region. It closely resembles *P. cervi*, but there is a genital sucker surrounding the genital pore. The eggs measure 125–135 by 61–68 μm. Dawes (1936) names this species *Paramphistomum cotylophorum* (Fischoeder, 1901) and regards the whole genus *Cotylophoron* as being synonymous with the genus *Paramphistomum*.

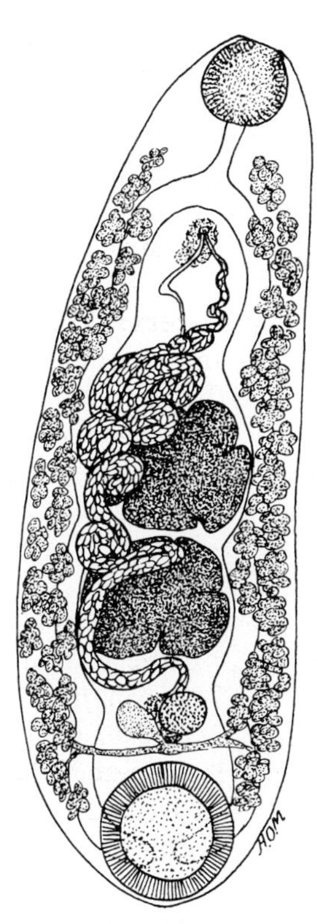

Fig 1.29 Ventral view of *Cotylophoron cotylophorum*.

Genus: Calicophoron Näsmark, 1937

Calicophoron calicophorum (Fischoeder, 1901) occurs in the rumen and reticulum of sheep and cattle in India, Australasia and South Africa. **C. raja** Näsmark, 1937, is found in Africa and **C. cauliorchis** (Stiles and Goldberger, 1910) is found in India and Japan. Eggs measure a mean of 115 by 69 μm.

Genus: Ceylonocotyle Näsmark, 1937

Ceylonocotyle streptocoelium (Fischoeder, 1901) is found in cattle, sheep and antelope in India and Australia. The eggs measure a mean size of 148 by 74 μm. **C. scoliocoelium** (Fischoeder, 1910) is found in India, South-East Asia and Africa.

Genus: Gigantocotyle Näsmark, 1937

Gigantocotyle explanatum Näsmark, 1937, occurs in the bile ducts, gall bladder and duodenum of buffalo, and less commonly, cattle in the Middle East, India and the Far East.

Genus: Gastrothylax Poirier, 1883

Gastrothylax crumenifer (Creplin, 1847) occurs in the rumen and reticulum of sheep, cattle, zebu and buffalo in India, Sri Lanka, China, Asiatic Russia, the Middle East and Europe. It is red when fresh, elongate, circular in transverse section and measures 9–18 by 5 mm. The worms of this genus differ from all other Digenea in having a very large ventral pouch, opening anteriorly and extending over the whole ventral surface up to the posterior sucker, which is large and terminal and has a raised border. The terminal oval sucker is small. The genital pore opens into the pouch, half-way between the pharynx and the intestinal bifurcation. The intestinal caeca end at about the level of the anterior border of the testes, which are lobed and horizontal, with the ovary behind them. The uterus crosses from right to left at about the middle of the body. The eggs measure 115–135 by 66–70 μm.

Genus: Fischoederius Stiles and Goldberger, 1910

Fischoederius elongatus (Poirier, 1883) occurs in the rumen of cattle and other Bovidae in Asia. It is 10–20 mm long and the breadth is about one-quarter of the length. It closely resembles *Gastrothylax*, but one testis lies dorsal to the other and the uterus runs forward in the midline. The intestinal caeca are not widely separated and end a short distance behind the middle of the body. The eggs measure 125–152 by 65–75 μm.

Fischoederius cobboldi (Poirier, 1883) differs from the preceding species in being only 8–10 mm long, while the intestinal caeca end at the posterior border of the posterior testis. This worm occurs in the rumen of cattle, zebu and gayal in Asia. The eggs measure about 110–120 by 60–75 μm.

Genus: Carmyerius Stiles and Goldberger, 1910

Carmyerius spatiosus (Brandes, 1898) occurs in the rumen of cattle, zebu and antelopes in India, Africa and America and is 9–12 mm long. It differs from *Fischoederius* in that the testes are horizontal. The posterior sucker is relatively small and spherical. The intestinal caeca reach the end of the second third of the body. The eggs measure 115–125 by 60–65 μm.

Carmyerius gregarius (Looss, 1896) is found in the buffalo and cattle in India and Africa. It is 7–10 mm long and the intestinal caeca end a short distance behind the middle of the body.

Genus: Gastrodiscus Leuckart, 1877

Gastrodiscus aegyptiacus (Cobbold, 1876) occurs in the large and small intestines of equines, pig and warthog in Africa and India. It is pink in colour when fresh and measures 9–17 by 8–11 mm. There is an anterior, more or less cylindrical, part which is up to 4 mm long and 2.5 mm wide,

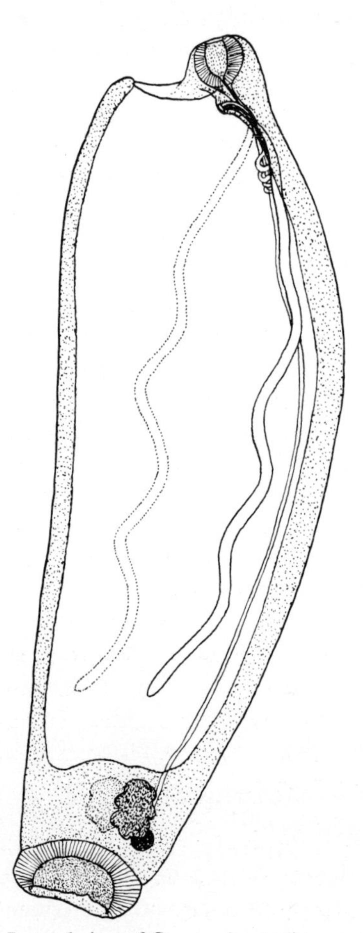

Fig 1.30 Lateral view of *Carmyerius spatiosus*.

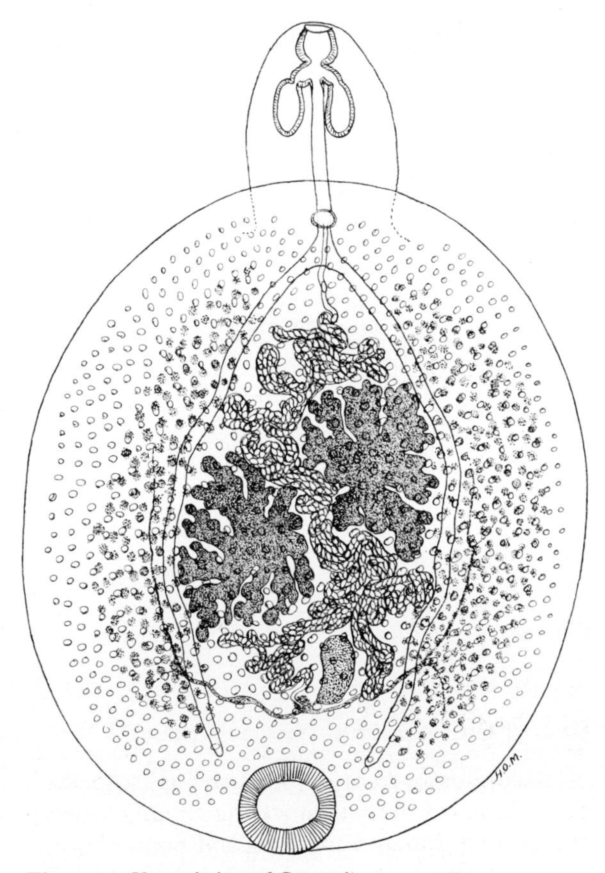

Fig 1.31 Ventral view of *Gastrodiscus aegyptiacus*.

while the rest of the body is saucer-shaped, with the margins curved inwards. The ventral surface is covered by a large number of regularly arranged papillae. The posterior sucker is small and subterminal. The oral sucker has two posterolateral pouches. The intestine branches at the anterior border of the wide portion and the caeca continue to near the hind end of the body. The testes are lobed, slightly diagonal and lie behind the middle with the ovary posterior to them. The vitellaria occupy the lateral fields. The genital pore opens at the level of the intestinal bifurcation. The eggs are oval and measure 131–139 by 78–90 μm.

Gastrodiscus secundus Looss, 1907. This species is found rarely in the colon of the horse and elephant in India.

Genus: Gastrodiscoides Leiper, 1913

Gastrodiscoides hominis (Lewis & McConnell, 1876) Leiper, 1913 (=*Gastrodiscus hominis* (Lewis & McConnell, 1876 (Fischoeder 1902)). The species occurs in the caecum of man and in the colon of the pig in Asia. The pig is the natural host for the parasite. It has also been found in monkeys and field rats.

Genus: Pseudodiscus Sonsino, 1895

Pseudodiscus collinsi (Cobbold, 1875) Stiles and Goldberger, 1910. Varma (1957) found that this species, which occurs in the colon, was the commonest species in horses in India. Other species of the genus occur in the colon of equines.

Life-cycles of paramphistomes

The life-cycle of the different species is generally similar. The composite eggs are clear (unlike those of *Fasciola* spp.), operculate and in the early stages of segmentation when passed in the faeces. The development time to the miracidium varies with the temperature and the species and is approximately 12–21 days.

Liberated miracidia swim in the surrounding water and enter a species of water snail, penetration of the snail occurring by way of the pneumostome and then through the posterior wall of the mantle cavity. However, penetration of the exposed soft parts of the snail may also occur. Young snails are more susceptible than old because the mantle cavity is completely filled with water and the pulmonary aperture permanently open.

Numerous snails have been reported as intermediate hosts, the following genera being of importance in this respect, *Planorbis, Bulinus, Pseudosuccinea, Fossaria, Indoplanorbis, Lymnaea, Pygmanisas, Glyptanisus* and *Cleopatra*.

Further development in the snail is similar in all species and that of *Ceylonocotyle streptocoelium* given by Durie (1953) serves as a typical life-cycle. Following penetration of the mantle cavity the miracidia lose their ciliated covering and by 12 hours an elongate sporocyst (93 by 53 μm) is present. Growth during the next few days is marked and by 11 days the sporocysts are mature and contain a maximum of eight rediae each. The rediae are liberated on the tenth to eleventh day of infection, they undergo marked growth and by the twenty-first day of infection measure 0.5–1 mm in length and contain 15–30 cercariae. Daughter rediae may be formed under certain circumstances.

Cercariae are released from the rediae in an immature state and they require a period of maturation in the snail tissues before being shed. This is 13 days at 27°C (Durie 1953). Mature cercariae are dark brown in colour and possess two distinct eye spots. They are shed during the hours of daylight and discharged within 30 minutes when a snail is stimulated by strong light. Liberated cercariae (*Cercariae pigmentata* Sonsino, 1892) are readily recognized as 'amphistome' because of the presence of anterior and posterior suckers. They are active for several hours, then encyst on herbage or other objects in the water. Encystment is complete in about ten minutes and the new metacercaria gradually darkens to an almost black colour. Such stages remain viable for about three months.

Infection of the final host is by ingestion of the metacercariae with herbage. Excystation occurs in the intestine where the immature paramphistones spend the first part of their vertebrate developmental cycle. They attach to the mucosa primarily in the first 3 m of the small intestine and after six to eight weeks at this site they migrate forward through the reticulum to the rumen, frequently becoming attached along the oesophageal groove. A further few weeks of development are required before maturity is reached. Migration from the small intestine appears to be related to the size of the fluke. For instance, the immature flukes of *P. microbothrium* grow more rapidly in cattle and the majority migrate between days 21 and 35 after infection but migration had only just commenced by day 34 in goats. Further in extremely heavy infections there is a retardation in the size of the flukes in the anterior part of the small intestine and migration to the rumen is delayed in these heavily infected animals (Horak 1967). This delay may be as long as four or five months (Boray 1969). Durie (1953), by experimental infection, reported the prepatent periods of *Ceylonocotyle streptocoelium, Calicophoron calicophorum* and *Paramphistomum ichikawai* in sheep to be 48 days, 49–51 days and 80–95 days respectively.

Epidemiology

Outbreaks of paramphistomiasis generally occur in the drier months. The snail population becomes concentrated around areas of natural water, and these areas, in the dry months, also have the most palatable grazing and thus there is a concentration of cattle, snails and metacercariae over a small area leading to heavy infections. Previous infection and the age of the host afford some protection against reinfection and hence acute disease is usually seen in young animals while older animals, capable of withstanding massive exposure, seed the pastures with eggs (Horak 1971).

Pathogenicity of Paramphistomes

The adult forms in the forestomach are essentially non-pathogenic even though large numbers may be present. At the most there may be a localized loss of rumen papillae. In the case of *Gigantocotyle explanatum* in the bile ducts and gall-bladder there may be a series of superficial haemorrhages indicating the sites of attachment but generally there is no severe pathogenic effect. In very heavy infections the liver may be pale and show a degree of fibrosis.

The immature stages of the paramphistomes in the duodenum and upper ileum are responsible for severe pathological changes. These are embedded in the mucosa and are plug feeders, drawing pieces of the mucosa into the suckers which pinch them off causing necrosis and haemorrhage. In heavy infections a frank, haemorrhagic duodenitis may be produced with immature flukes deeply embeded in the mucosa, sometimes reaching the muscular coat. Histologically there is extensive catarrhal and haemorrhagic inflammation of the duodenum and jejunum with destruction of the intestinal glands, degeneration of the associated lymph nodes and other organs. Associated with these lesions is an anaemia, a hypoproteinaemia, oedema and emaciation.

Clinical signs consist of profuse fluid foetid diarrhoea, marked weakness and frequently death. The animals are thirsty and drink frequently.

In some areas, e.g. India, the Republic of South Africa, Australia, mortality may reach 80–90% and reports have recorded mortalities of 30–40% in cattle and sheep (Boray 1959; Soulsby 1965). *G. aegyptiacus* is normally considered to be non-pathogenic. However, the immature parasite can cause a severe, possibly fatal, hyperacute colitis in horses (Azzie 1975).

Diagnosis

Diagnosis is based on clinical signs, the history of the area and the presence of immature paramphistomes in the fluid faeces. In some circumstances the presence of large number of paramphistome eggs in the faeces is also indicative of the disease since although the pathogenic effects are caused by the immature forms a large number of adult forms may also accompany the immature burden. At post mortem a marked enteritis is evident and large numbers of brownish pink parasites are found on the mucosa and in the intestinal contents. Clinical cases may reveal up to 30 000 immature paramphistomes.

Treatment

Resorantel, rafoxanide, oxyclozanide, niclofolan and a variety of other anthelmintics have been used for the treatment of paramphistomiasis but these anthelmintics may vary in their efficacy against the adult and immature stages. For example, niclosamide (90 mg/kg) was 99.9% effective against the immature stages but only 18% effective against adult paramphistomes in sheep (Boray 1969). Some reports on the efficacy of niclofolan have been variable, but Boray (1969) demonstrated that niclofolan (6 mg/kg) was 96% and 43% effective against immature and mature amphistomes respectively. Bithionol sulphoxide (40 mg/kg) may be up to 100% effective against the immature stages (Horak 1965). Resorantel (65 mg/kg) is reported to be highly effective (100%) against adult amphistomes and 63% effective against the immature stages in calves (Gaenssler 1974). In a field study using naturally infected cattle and buffalo, oxyclozanide, clioxanide and niclosamide were 100%, 90% and 60% effective, respectively (Chhabra & Ball 1976). In this study hexachlorophene and niclofolan were not very effective and were not well tolerated.

Hexachlorophene, dichlorvos, resorantel and oxyclozanide have been used to treat horses infected with *G. aegyptiacus* (Azzie 1975; Roberts et al. 1976).

Control

Since the vectors are water snails, sheep and cattle should be grazed on higher ground, the localized area of water fenced off or treated with molluscicides. Drainage of pools and swamps is a more permanent control measure.

FAMILY: CYCLOCOELIDAE KOSSACK, 1911

These trematodes are medium-sized to large and flattened. The oral sucker is absent and there is usually also no ventral sucker. The mouth is anterior; there is a muscular pharynx and the intestinal caeca are simple or branched and are joined together posteriorly. The genital pore opens a short distance behind the mouth. Testes are diagonal, entire or lobed; the ovary is not lobed and is situated between, or anterior to, the testes. The vitellaria occupy the lateral fields, meeting posteriorly like the intestinal caeca. Parasites of aquatic birds, usually in the body cavity, air-sacs or nasal cavity.

Genus: Tracheophilus Skrjabin, 1913 (Syn: Typhlocoelum)

Tracheophilus cymbius (Diesing, 1850) (syn. *Tracheophilus sisowi*) occurs in the trachea and bronchi of domestic and wild ducks. It measures 6–11.5 by about 3 mm. The testes are not lobed and lie diagonally in the posterior part of the body; the ovary lies at the same level as, or a little in front of, the anterior testis. The eggs measure 122 by 63 μm.

A miracidium hatches from the egg and swims about in the water and contains a single redia (Szidat 1933). When a suitable snail is reached the redia alone enters it. *Helisoma trivolvis* and species of *Planorbis* are the chief intermediate hosts. There is no sporocyst stage. The redia, which settles down near the albuminous gland of the snail, begins to produce small numbers of cercariae after about 11 days. The cercariae have no tail and can be recognized by the complete intestinal ring; they are provided with a ventral sucker and an anterior boring apparatus. They do not leave the snail, but encyst in it in and near the pericardium after having escaped from the redia. The birds become infected by ingesting the infected snails.

The parasites cause obstruction of the trachea and the birds may die of asphyxia.

Albendazole and praziquantel may be of value in treatment.

Birds should be kept away from suspected water and the extermination of snails may be considered.

Two other trematodes which belong to this family and may be harmful are **Tracheophilus cucumerinum** (Rudolphi, 1809) (*T. obovale*), which occurs in the trachea, air-sacs and oesophagus of ducks and related wild birds, causing dyspnoea and asphyxia, and **Hyptiasmus tumidus** (Kossack, 1911) (*H. arcuatus*), which is found in the nasal and orbital sinuses of ducks and geese, causing a catarrh.

FAMILY: SCHISTOSOMATIDAE POCHE, 1907

These are elongate, unisexual and dimorphic trematodes, which inhabit the blood vessels of their hosts. The female is slender and usually longer than the male, and the female of some species is usually carried, especially during copulation, by the latter in a ventral, gutter-like groove, the gynaecophoric canal, formed by the incurved lateral edges of the body. The suckers are weak and close together or absent. There is no pharynx and the intestinal branches usually unite posteriorly to form a single tube which extends to the hind end. The genital pore lies behind the ventral sucker. The testes form four or more lobes, situated anteriorly or posteriorly. The ovary is an elongate, compact organ, lying in front of the posterior union of the intestinal branches. The vitelline gland occupies the part of the body behind the ovary. The eggs are thin-shelled and have no operculum; those of some species have a lateral or terminal spine. They are laid by the females in the small blood vessels of the intestinal wall or the urinary bladder and pass through the tissues, leaving the host with the faeces or urine. The cercariae are furcocercous, without a pharynx, and develop from sporocysts without a redia stage. They enter the host through its skin and do not encyst.

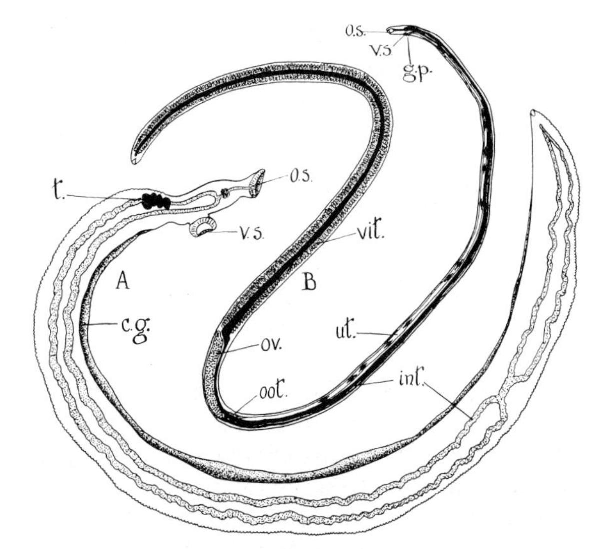

Fig 1.32 Male (A) and female (B) *Schistosoma bovis*.
cg canalis gynaecophorus
gp genital pore
int intestine
os oral sucker
oot ootype
ov ovary
t testes
ut uterus
vs ventral sucker
vit vitellaria

Genus: Schistosoma Weinland, 1858

Schistosoma bovis (Sonsino, 1876) occurs in the portal and mesenteric veins of cattle, sheep and goats in Central, East and West Africa, the Mediterranean area and in the Middle East. Adults are found also in equines, camels, wild ruminants and rodents. *S. bovis* has been found in man but it is less common in man than is *S. mattheei*. It is a serious pathogen of cattle and sometimes sheep.

The males of *S. bovis* are 9–22 mm long and 1–2 mm wide, depending on the degree of folding of the lateral edges. The female is 12–28 mm long. The suckers and the body of the male behind the suckers are armed with minute spines, while the dorsal surface of the male bears small tegumental tubercles. The intestine bifurcates at the level of the ventral sucker and reunites in the female just behind the middle of the body, or sometimes farther back, being continued as a single caecum. In the male the caeca may reunite far back or not at all, or there may be several junctions between them and two posterior caeca. The male has three to six testes in a longitudinal row, situated a short distance behind the ventral sucker. The ovary is about 1 mm long and lies at the middle of the body behind Mehlis's gland. The uterus runs forwards from this point and may contain a large number of eggs. The vitelline gland occupies as a rule the posterior half of the body behind the ovary, but in some females it may fill only the posterior quarter. The eggs are usually spindle-shaped, but small ones are frequently oval, and when passed in the faeces measure on the average 187 by 60 μm; the limits given by various authors are 132–247 by 38–60 μm.

Schistosoma japonicum Katsurada, 1904, occurs in the portal and mesenteric veins of both man and animals in the Far East. A wide range of domestic and wild animals have been found to be infected. These include ruminants, equines, pigs, dogs, cats and rodents. Both man and animals act as maintainance hosts. Four strains of *S. japonicum* occur, the Japanese, Philippine, Chinese and Formosan, and these are recognized both by geographical distribution and by morphological characters. However, a general description of *S. japonicum* is that the male is 9.5–20 mm long and 0.55–0.967 mm wide. The female is 12–26 mm long and about 0.3 mm thick. The suckers lie close together near the anterior end. The tegument is spiny on the suckers and in the gynaecophoric canal. In both sexes the oesophagus is surrounded by a group of glands and the intestine bifurcates before reaching the level of the ventral sucker, again uniting in the last quarter of the body. The testes consist of six to eight lobes in a longitudinal row, lying behind the genital pore, which opens directly posterior to the ventral sucker. The ovary lies behind the middle and the vitelline gland fills the posterior quarter. The oötype is situated directly anterior to the middle, opening into the long, unfolded uterus, which extends forwards to the genital pore. The eggs are passed in the faeces of the host and measure 70–100 by 50–80 μm. They are short, oval, and may have a small lateral spine or knob.

Schistosoma mekongi Voge, Bruckner & Bruce, 1978 occurs in dogs and man in Cambodia. It differs from *S. japonicum* in the size of eggs (40 by 45 μm), the length of the prepatent period in the female host and in its utilization of different snail hosts (*Lithoglyphopsis aperta*). Clinical cases caused by a similar parasite have been reported in Malaysia and they are suspected to be of sylvatic origin.

Schistosoma mattheei (Veglia and Le Roux, 1929) is found in the portal and mesenteric veins, as well as in the veins of the urinogenital tract and stomach, of domestic and wild ruminants, equines, baboons and wild rodents in Central, southern and East Africa. The parasite also infects man with infection rates as high as 40%, but the infection is usually light in man and the parasite usually occurs as a concomitant infection with *S. mansoni* or *S. haematobium*. *S. mattheei* can be differentiated from *S. bovis* on morphological and pathological grounds (Soulsby 1965). Adults measure 17–25 mm but the length of the female is reduced as the infection progresses (Lawrence 1978*a*). The common caecum in the male is 8–12 mm long but its anatomy is very variable. Eggs in the faeces measure 170 to 280 by 72 to 84 μm.

Schistosoma intercalatum Fisher, 1934, is common in the portal and mesenteric veins of domestic and wild ruminants, equines and man in Central Africa. It is considered by some to be synonymous with *S. mattheei* but hybridization studies suggest that it is a different species.

Schistosoma spindale Montgomery, 1906 is found in the mesenteric veins of ruminants and dogs in India and the Far East. Adults measure 5–16 mm in length. The eggs are spindle-shaped and 200 by 70–90 μm with a terminal spine.

Schistosoma indicum Montgomery, 1906 resides in the portal and mesenteric veins of ruminants, equines and camels in Indo-Pakistan. Adults measure 5–22 mm and the eggs in the faeces measure 57–140 by 18–72 μm.

Schistosoma mansoni Sambon, 1907 occurs in the mesenteric veins of man in Africa, South America and the Middle East and humans are the most important definitive host. However, a variety of animals have been found to be naturally infected with *S. mansoni*. These include gerbils and Nile rats in Egypt, rodents in southern Africa and Zaire, various species of rodents and wild mammals and cattle in Brazil and baboons, rodents and dogs in East Africa. Up to 50% of baboons are infected in some areas in East Africa. The significance of these animals in maintaining the infection and in transmission of the infection to man is not known but the levels of infection in animals can be high and it is possible that rodents in South America and the West Indies and baboons in Africa are important in this respect.

Schistosoma haematobium (Bilharz, 1852) occurs in Africa and a locus of infection has been reported in India. Adults reside in the posterior mesenteric arteries and eggs are laid in the walls of the bladder, ureters and urethra. Man is the only significant maintenance host of this species although the infection has been found in animals, e.g. baboons and monkeys in East Africa, rodents in Kenya and southern Africa, pigs in Nigeria and chimpanzees in West Africa. There is no unequivocal evidence that animals play a rôle in the human disease.

Schistosoma nasalis Rao, 1932, closely resembles *S. spindale* and adults measure 5–11 mm in length. Eggs are boomerang shaped and 350–380 by 50–80 μm. The parasites develop in the veins of the nasal mucosa of buffalo, cattle, goats, sheep and horses in the Indian subcontinent. They cause rhinitis and a mucopurulent discharge manifested clinically by signs of coryza, sneezing, dyspnoea and snoring (snoring disease). Infection rates of 40–50% have been recorded in buffalo and cattle. Adult parasites cause dilatation and thrombosis of the veins. The nasal mucosa is studded with granulomas and small abscesses containing eggs. Chronically there is fibrous tissue formation and proliferation of the nasal epithelium. The clinical signs and pathology of infection with *S. nasalis* are less severe in buffalo than in cattle. Buffalo show only pin-head-sized eruptions and congestion of the nasal mucosa suggesting that buffalo are a more suitable definitive host than are cattle (Rajamohanan & Peter 1975).

Schistosoma incognitum Chandler, 1926, (Syn. *S. suis*) occurs in the mesenteric veins of the pig and the dog in the Indian subcontinent. In addition, 60–100% of *Rattus argenifer* caught in rice field dikes were found to be infected in Indonesia (Carney et al. 1977). Infection rates in the pig may be 10% or more and infection is considered to be of economic importance in India. The egg is yellowish-brown, suboval with one side flattened, with a small, stout spine inclining towards the flattened side. In the uterus it measures about 90 by 41 μm.

Schistosoma rodhaini Brumpt, 1931. This species occurs chiefly in rodents, but Deramée et al. (1953) showed that dogs and cats are susceptible to infection with it and natural infections may cause serious disease in dogs in Ruanda.

Other species of the genus include **S. margrebowiei** Le Roux, 1933 of ruminants and zebra in Zambia and the Republic of South Africa and **S. hippopotami** Thurston, 1963 of the hippopotamus in Uganda.

Life-cycle of Schistosoma species

The ovigerous female penetrates deeply into the small vessels of the mucosa or submucosa of the intestine, laying the eggs in the capillaries. From here the eggs pass through the intestinal wall into the intestinal lumen and out in the faeces. When laid the eggs are not fully mature and they continue their development as they pass out in the faeces. In addition, Lawrence (1978*a*) has demonstrated that, in *S. mattheei* infections, a proportion of eggs is laid in the submucosa where they persist for four to five weeks, but are finally broken down and phagocytosed. Also, as the infection progresses, the adults withdraw from the veins in the submucosa into the intestinal veins so eggs are laid in the muscle, adventitia and mesenteric veins. Eggs such as these are carried by the blood to the liver although a proportion continue to be passed in the faeces.

The eggs hatch after contact and dilution of the faeces with water. Agitation ensures optimal hatching. Miracidia infect aquatic snails which are the intermediate hosts. *S. mattheei* infects *Bulinus* (*Physopsis*) *africanus*, *B.* (*P.*) *globosus* and *B.* (*P.*) *nastusus*. *S. bovis* also infects these snails and, in addition, *B. truncatus* has been implicated as an intermediate host (see Malek 1961). The intermediate hosts of *S. japonicum* are species of the genus *Oncomelania* (*O. nosophora, O. quadrasi, O. hupensis* and *O. formosana*) while the intermediate hosts of *S. mansoni* are planorbids of the genus *Biomphalaria*, especially *B. glabrata*. *S. haematobium* develops in *Bulinus truncatus, B. forskali* and *B. obtusispira*, and *S. spindale* develops in snails of the genera *Planorbis, Indoplanorbis* and *Lymnaea*. The miracidia penetrate the tissues of the snail and two generations of sporocysts develop, the second forming the cercariae. The fercocercous cercariae actively emerge from the snail and swim about in the water. Emergence of cercariae from the snail is periodic and those of *S. mansoni* tend to emerge in daylight from 09.00 to 14.00 hours, although emergence is inhibited or partially inhibited at temperatures of below 21°C. Peak shedding of cercariae of *S. mattheei* occurs at

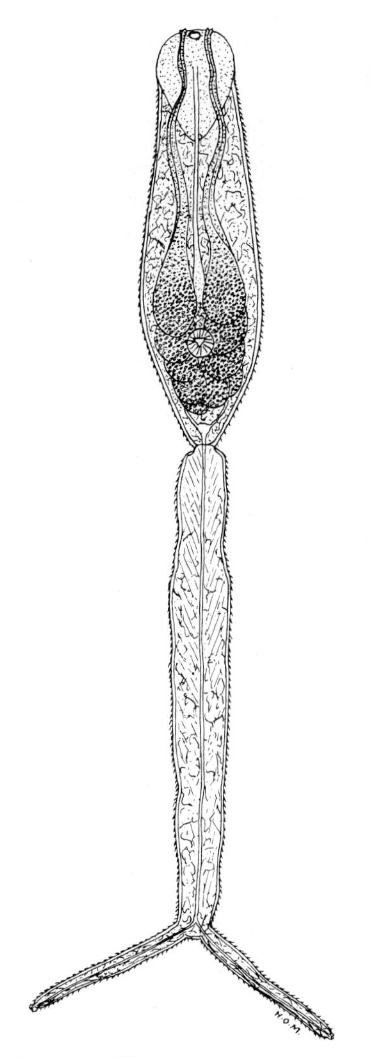

Fig 1.33 Cercaria of *Schistosoma bovis.*

about 17°C and in the eastern Transvaal this occurs between 12.00–14.00 hours in mid-winter and 06.00–08.00 hours in summer.

Development within the snail host is dependant on temperature and varies with the time of year. Thus, the development of *S. mattheei* in *Bulinus* spp. takes a mean of 38 days in mid-summer to a mean of 126 days in winter. The development of *S. mansoni* in *B. glabrata* ranges from 15 to 75 days at 26°C.

The snail hosts are aquatic (although those of *S. japonicum* may be amphibious) and snail survival is best in large, stationary or slow-moving bodies of water. The snails do not survive the drying-up of small bodies of water, but their high biotic potential enables the population to recover rapidly from adverse conditions. The infection is most common in man and animals where the rainfall is highest; there is a marked increase in incidence with increasing water conservation (the building of small and large dams) and with irrigation and with increases in the population density of animals.

Infection of the definitive host is through active skin penetration of the cercariae although cercariae may penetrate the wall of the rumen when swallowed with water. Skin penetration is assisted by the secretions of the cephalic glands which digest the tissues. The cercariae transform into schistosomula which are transported to the lungs via the circulation within four to seven days. They are then carried to the liver, presumably via the blood stream, and from eight days onwards schistosomula are found in the portal vessels of the liver. Pairing of the worms takes place in the portal veins before they leave the liver to reach maturity in the mesenteric veins.

Cattle and horses may become infected when standing in the shallow waters of dams, rivers, etc. during the heat of the day. They and other animals are also infected orally when water tanks and other sources of drinking water become infected with snails and contaminated with faecal material. Prenatal infection with *S. japonicum* has been recorded.

Pathogenesis

Two clinical syndromes are seen in animals infected with schistosomes, an acute intestinal syndrome and a chronic hepatic syndrome. These are described by Lawrence (1978*b,c*).

The *intestinal syndrome* is seen in acute schistosomiasis in heavily infected animals and it develops with patency of the female worms and the passage of large numbers of eggs through the intestinal mucosa. The syndrome begins seven to nine weeks after infection; the duration depends on the intensity of infection and recovery is spontaneous.

Pathological and histopathological examination reveals severe haemorrhagic lesions on the intestinal mucosa, particularly the posterior small intestine and the caecum, although they may extend from the forestomachs to the rectum. The intestinal mucosa is oedematous and covered with haemorrhagic foci 1–10 mm in diameter. Numerous eggs are found in the mucosa and in the blood-stained mucus covering the intestinal surface. The lamina propria is oedematous and infiltrated with eosinophils, lymphocytes, macrophages and plasma cells while the eggs may be found free, in small granulomata or in microabscesses. There is a marked granulomatous response to eggs laid in the submucosa. These eggs gradually disintegrate; there is infiltration with epithelioid cells, fibroblasts and fibrous tissue formation. Long-standing infections are manifested by granulomatous lesions in the mucosa, submucosa and muscularis mucosa; there is prominent fibrous thickening of the intestinal wall and intestinal catarrh.

Adult parasites cause phlebitis in the mesenteric veins and there is proliferation of the tunica intima with partial or complete occlusion of the lumen. As the infection progresses the adult parasites shift away from their predeliction site in the posterior small intestine and anterior large intestine and lesions will be seen in other parts of the intestine as well as in other organs. Adults of *S. bovis* usually are found only in the veins of the intestine, although occasionally lesions will be found in other organs such as the pancreas, lungs and mesenteric lymph nodes. However, adults of *S. mattheei* may be found throughout the intestine from the rumen to the rectum, they may affect the pancreas, mesenteric lymph nodes and lungs. In addition, with *S. mattheei*, lesions in the bladder are striking and common. Pathology in the bladder is seen as linear granulomata or granular patches with associated haemorrhages. Adults of *S. mattheei* and *S. indicum* may, in heavy infections, be found in the pulmonary arteries. The lungs are enlarged, brownish black and contain at first red and then greyish foci or nodules surrounding the adult parasites. Histologically, multiple granulomas surround eggs in the parenchyma (Sharma & Dwivedi 1976).

The *hepatic syndrome* is an immunological disease resulting from the host's cell-mediated immune response to schistosome eggs in the liver. The pathology of this reaction has been studied extensively in mice infected with *S. mansoni* as a model for the pathology seen in human schistosomiasis (Warren 1972, 1977). Eggs swept back in the portal circulation lodge in perisinusoidal interlobular portal venules but these by themselves have little effect on the portal blood supply. However, the immunologically specific host reaction to the eggs leads to extensive damage to the portal vascular system. Soluble antigens escaping through pores in the eggshell sensitize the host and stimulate the accumulation of lymphocytes, macrophages and eosinophils around the egg. The avascular granuloma may reach 100 times the size of the egg. The inflamatory reaction becomes chronic and is characterized by the presence of epithelioid cells, giant cells and fibroblasts. The egg is destroyed and eventual healing is accompanied by the deposition of collagen and fibrous tissue formation. In heavy infections the development and healing of large numbers of these egg granulomas causes massive fibrosis in the portal triads of the liver and the appearance of 'clay-pipe stem' fibrosis.

The development of fibrous scar tissue causes obstruction of the venous portal blood supply. However, there is a compensatory increase in the arterial blood supply and neovascular development of arterial supply in the fibrous tissue. Perfusion of the parenchyma and thus liver function remains, on the whole, unimpaired. However, the obstruction to the portal venous circulation in the liver leads to a congestive splenomegaly. A portal–systemic collateral circulation may develop and oesophageal varices may be seen. Occasionally oesophageal varices may burst, leading to death by haemorrhage. Congestive heart failure may also occur.

Modulation or diminution in the size of the granulomatous response to eggs in the liver has been demonstrated in mice experimentally infected with schistosomes. Thus, peak reactivity and granuloma formation occurred eight weeks after exposure to eggs but a marked diminution of the size of the egg granulomata was seen from weeks 12 to 16 and thereafter. This modulation of the granulomatous response could be transferred with spleen and lymph node cells as well as with serum and thus may be associated with cell-mediated immunity, perhaps suppressor T cells, or humoral immunity.

Modulation of granuloma formation would appear to occur in man since disease is most prominent in 10–14-year-old children while adults in endemic areas, although infected, show fewer clinical manifestations of disease.

Chronic hepatic pathology occurs in schistosome-infected animals, particularly those repeatedly exposed to heavy infections. Granulomatous lesions and periportal inflammation occur, followed by a progressive portal fibrosis. However, the later manifestations of disease in man, splenomegaly and oesophageal varices, are not seen in animals. In contrast to the lesions seen in man the pathology in animals includes the development of lymphoid nodules and follicles around dead parasites in the liver and other organs and in heavy infections there may be massive thrombosis of the portal veins and hepatic infarction.

The prevalence of infection with schistosomes can be high. For instance, up to 92% of cattle in Zimbabwe may be infected with *S. mattheei* and 90% or more of cattle and sheep may be infected with *S. bovis* in the Sudan (Majid et al. 1980). A survey demonstrated that 62% of cattle, 100% of buffalo and 10–12% of sheep and goats were infected with visceral schistosomes in Bangladesh (Islam 1975).

S. bovis and *S. japonicum* are the most pathogenic of the schistosomes of cattle and sheep since as infections with *S. mattheei* progress there is spontaneous elimination of the adult parasites, probably due to an immunological reaction, and thus the pathogenicity of the parasite is limited (Lawrence 1978*a*). The pathogenicity of *S. spindale* is comparable to that of *S. mattheei*.

The mechanism of anaemia in schistosoniasis has been studied by Preston and Dargie (1974*a,b*) and is due to a combination of haemodilution, haemorrhage and dyshaemopoiesis. Initially the drop in packed cell volume seen four to six weeks after infection is due to haemodilution and to a dyshaemopoiesis, the latter being demonstrated by the fact that between four and eight weeks of infection there is a fall in the rate of plasma iron turnover and a slower than normal rate of incorporation of radio-actively labelled iron into the red cells, suggesting an impairment of the marrow reserve capacity. A more severe anaemia is observed beginning six weeks after infection and it is associated with intestinal bleeding, further haemodilution and some impairment of the marrow reserve capacity.

Changes in the nitrogen status and digestive efficiency of sheep infected with *S. mattheei* follow the onset of oviposition (Berry et al. 1973). The decline in nitrogen balance correlated with the severity of other parameters, e.g. diarrhoea, weight loss and hypercatabolism of albumin. The reduced apparent digestibility coefficients of all dietary constituents might be due to the presence of ova in the gastrointestinal tract.

Marked changes occur in the serum protein levels. Dargie et al. (1973) reported that in sheep infected with *S. mattheei* a severe hypoalbuminaemia occurred four to 12 weeks after exposure. An underlying cause was hypercatabolism of albumin due to protein loss into the gastrointestinal tract. Van Zyl (1974) reported an overall increase in total serum proteins in infected sheep, this being due to a net gain in total globulin which compensated for a loss of albumin, and implicated impaired albumin synthesis by the liver as a contributory factor in the reduced albumin levels. The pathogenicity of experimental infections of *S. incognitum* has been reported in dogs by Tewari et al. (1966) and in pigs by Ahluwalia (1972).

Clinical signs

The migration of large numbers of schistosomula through the lungs may cause a temporary cough but this is rare. Acute heavy infections are manifested by profuse diarrhoea or dysentery, dehydration and anorexia. These signs develop at the time of patency seven to nine weeks after infection. Anaemia and hypoalbuminaemia are present, sometimes with oedema. There is a marked decrease in production or loss of weight.

Chronically infected animals are emaciated. There is eosinophilia, anaemia, hypoalbuminaemia and perhaps hypergammaglobulinaemia. Neurological signs may be seen.

Diagnosis

The clinical signs alone will not suffice to arrive at a definite diagnosis, but they should indicate the necessity of faeces examination, which will reveal the eggs of the parasites mixed with blood and mucus. Serological tests have not achieved a level of reliability for use in animals as in man.

Immunity and immunization

Epidemiology studies of *S. bovis* infections of cattle in the Sudan showed a significant fall in age-specific prevalence and intensity, based on faecal egg counts (Majid et al. 1980) and this was demonstrated to be due to naturally acquired resistance to *S. bovis* (Bushara et al. 1980). Cattle from endemic areas withstood challenge infections which proved lethal for control cattle from non-endemic zones. These studies suggested that *S. bovis* infection in cattle could serve as a spontaneously occurring analogue of human schistosomiasis by which homologous and heterologous immunization procedures might be developed.

Success has been achieved in immunization against schistosomiasis in cattle and sheep (Taylor 1980). Irradiated schistosomula of *S. mattheei* and *S. bovis*, when given subcutaneously or intramuscularly to cattle or sheep, induced a greater than 60% level of protection against infection. Of particular importance is the economic advantage that can be attained through immunization against experimental and field infections. Immunized calves had significantly higher growth rates, superior body composition, decreased faecal and tissue egg counts and lower adult worm counts. In addition, immunized calves showed milder histopathological and haematological changes than did unimmunized calves (Taylor 1980).

Similar immunization studies against *S. mansoni* in baboons did not produce significant protection, suggesting that the immunization of man with irradiation-attenuated forms may not be as successful as with ruminant forms. Non-specific immunization of mice with very high doses of BCG, sufficient to induce a disseminated granulomatous response, induced protection against infection with *S. mansoni* but this was ineffective when used to immunize monkeys. However, heterologous immunization by repeated exposure of baboons to *S. rodhaini* and *S. bovis* produced protection against infection with *S. mansoni*. Thus, heterologous immunization might be posssible in human schistosomiasis.

Experimental studies on immunization against *S. incognitum* using ^{60}Co-irradiated cercariae have been reported by Tewari and Ramachandran (1979).

Treatment

The recent development of *praziquantel* seems to provide the drug of choice for the treatment of schistosomiasis in man. Clinical trials to examine its efficacy against *S. mansoni*, *S. haematobium* and *S. japonicum* demonstrate that praziquantel has a greater than 90% efficacy when given as a single oral dose of 60 mg/kg or three doses of 20 mg/kg given four hours apart. Slight side effects (nausea, vomiting, dizziness) may be seen (Davis et al. 1979; Katz et al. 1979; Santos et al. 1979).

In man, other anthelmintics are effective also. *Niridazole* is effective against adult schistosomes but it requires administration for five to ten days and therefore is unsuitable for mass treatment programmes. *Hycanthone*, given as a single oral dose of 1.5–3 mg/kg, has a high efficacy inducing a 96% or greater reduction in egg counts but the drug occasionally has severe side effects and hepatic necrosis and death have been reported. *Lucanthone* is also used but it is given in divided doses. The anthelmintic has at least a 60% efficacy against *S. haematobium* at a dose rate of 60 mg/kg for three days, but its efficacy against *S. mansoni* is lower even when given at a dose rate of 100 mg/kg daily for six to eight days. *Oxamniquine*, 15–20 mg/kg, is 70–100% effective against *S. mansoni*, although it is less effective against *S. mattheei* and *S. haematobium*. Side effects are usually slight. The organophosphate anthelmintic *metrifonate* cures 80–95% of schistosome infections when given at 7.5 mg/kg, the dose being repeated three times at two- to four-week intervals. Possible side effects include nausea and vomiting. *Furapromidium* is used in China and its efficacy against *S. japonicum* may be up to 80%. It is given for 14–20 days at a dose rate of 60 mg/kg/day to adults and 100 mg/kg/day for children.

Generally the therapy of animal schistosomiasis has followed that for the treatment of human infections, but great care must be taken since very variable results have been obtained. Many of the drugs kill the adult schistosomes *en masse* and these then become emboli in the portal veins; portal occlusion and hepatic infarction may result and hepatic failure may occur. The treatment of cattle infected with *S. mattheei* has been discussed by Lawrence (1978*a*). *Tartar emetic, antimosan* and *stibophen* have been effective in the treatment of *S. mattheei* in cattle but their use has been associated with deaths among the treated animals. However, *stibophen*, the sodium salt of antimosan, was very effective in cattle at a dose rate of 7.5 mg/kg given daily for six days. *Lucanthone* is also effective in the treatment of *S. mattheei* in both cattle and sheep; 30 mg/kg given on three alternate days was effective in cattle and moderate efficacy was seen when sheep received 30–50 mg/kg for three days. *Hycanthone* has been used to treat sheep and an intramuscular injection of 3 mg/kg was moderately effective while 6 mg/kg was highly effective. In addition, *niridazole* was effective in sheep at a dose rate of 100 mg/kg for three days. Very variable results have been obtained when *trichlorophon* has been used to treat infected cattle and sheep. Thus, *trichlorophon* was effective against *S. bovis* in cattle when 50–70 mg/kg was given orally on four to six occasions at three-day intervals. However, 75 mg/kg was highly toxic in some treated cattle infected with *S. mattheei*. The lack of preparatory starvation of the animals in the latter experiment may have affected the toxicity. Conversely, the intramuscular injection of trichlorophon in cattle was efficacious when *S. mattheei* infected animals received at least 11 treatments of 10–12 mg/kg given at three- to five-day intervals. *Trichlorophon* given orally to sheep infected with *S. mattheei* at a dose rate of 100–120 mg/kg on four or more occasions three to four days apart was also efficacious but 20 mg/kg given intramuscularly at three- to four-day intervals was toxic.

Nasal schistosomiasis in cattle and buffalo due to *S. nasalis* has been treated successfully with tartar emetic (maximum dose 2 mg/kg) and sodium antimonyl tartrate (1.5 mg/kg twice a day for two days). Trichlorophon was also effective when a dose of 30–40 mg/kg was repeated three times but the anthelmintic was toxic in some cases.

Niridazole, 25 mg/kg, had 100% efficacy against *S. incognitum* in pigs. Piglets required a higher dose, 30–40 mg/kg.

Control

Control of schistosomiasis is based on control of the snail intermediate host and treatment of infected persons and animals.

Biological control has proved effective experimentally but has not yet been shown to be effective in the field. The larval stages of *Echinostoma* spp. are predatory on schistosome larvae within the snail intermediate host. However, the definitive hosts of echinostomes are limited in their distribution which makes their use in nature impractical. Microsporidial protozoa such as *Nosema eurytremae* can cause extensive damage to the intramolluscan stages of schistosomes and other trematodes. Unfortunately only very high levels of microsporidia are effective. Competition between molluscan species such as *Helisoma* spp. has been shown to have some effect in eliminating *B. glabrata* in aquaria, but their competitiveness has not been borne out in the field.

Snail populations can be limited to low levels by the periodic application of molluscicides such as Frescon and Bayluscide to bodies of water. Also, contact between man and animals and snail-infested water should be prevented. The fencing off of lakes and pools and the provision of piped drinking water to troughs aid in preventing infection in animals. In addition, water troughs should be mechanically cleaned periodically. The molluscan intermediate hosts of schistosomes prefer slow-moving or stationary water, so that an increase in the speed of water in irrigation channels will reduce the snail populations. The education of humans at risk, the provision of sanitary facilities and the provision of piped water to houses, laundry units and swimming pools reduces human contact with contaminated water. Before night soil is allowed to contaminate water it should be treated by fermentation for 25–45 days. The heat created is sufficient to kill schistosome eggs. When contact with water cannot be prevented, since farmers and other workers may have to enter water as part of their livelihood, these workers should be provided with protective clothing. Also, repellants such as dibutylphthalate and benzyl benzoate applied to exposed skin may be effective in preventing penetration by schistosome miracidia.

Three types of programme for the control of human infection with *S. mansoni* have been compared on the West Indian island of St Lucia (Jordan 1977). After two years of control the incidence of schistosomiasis in children was reduced from 18.8 to 4.1% when infected persons were treated chemotherapeutically. Snail control with molluscicides reduced infection levels from 22 to 9.8% and the provision of water to houses etc. reduced human contact with water by 90% and the rate of infection in children fell from 22.7 to 11.3%. Chemotherapeutic control was the cheapest of the programmes but it requires the full cooperation of the community.

Genus: Ornithobilharzia Odhner, 1912

In species of this genus, the intestinal caeca tend to anastomose in front of the median posterior intestinal caecum and the ovary forms a spiral in the anterior region of the body, the testes being very numerous and the uterus short, containing only one egg at a time.

Ornithobilharzia bomfordi (Montgomery, 1906) occurs in the mesenteric veins of the zebu (*Bos bubalis*) in India. The male is 6–9 mm long and the female 3–7.3 mm. The male has about 60 testes and the female has a long posterior caecum. The eggs are oval with a posterior spine and measure 100–136 by 44–60 μm. They are found in the faeces of the host.

Ornithobilharzia turkestanicum (Skrjabin, 1913) (syn. *O. turkestanica*) occurs in the mesenteric veins of sheep, goat, camel, horse, donkey, mule, water buffalo, cattle and the cat in Russian Turkestan, Kazakstan, Mongolia, Iraq and France. It is a small species, the male being 4.2–8 mm long and the female 3.4–8 mm. There are 70–80 testes and the ovary is spirally coiled. The uterus contains one egg at a time. The eggs measure 72–77 by 18–26 μm. They have a terminal spine and a short appendage at the other end. It is of little significance in large animals, but leads to debility in sheep and goats, causing marked liver cirrhosis and nodules in the intestines. The intermediate host is the snail *Lymnaea euphratica*.

Genus: Bilharziella Looss, 1899

Bilharziella polonica (Kowalewski, 1896) occurs in the mesenteric and pelvic veins of wild and domestic ducks in Europe and North America. The body is flattened in both sexes and usually lancet-shaped in the posterior half, the gynaecophoric canal being rudimentary. The male is about 4 mm long and 0.52 mm wide and the female is 2.1 mm long and 0.25 mm wide. The short uterus contains one egg at a time. The eggs have a long, narrow, anterior elongation and are swollen posteriorly with a terminal spine; they measure 400 × 100 μm.

Life-cycle. The eggs are laid in the small vessels of the intestinal wall. They gradually penetrate through the wall and are passed out in the faeces. The intermediate host is the snail *Planorbis* (*Coretus*) *corneus*. The cercariae have a pair of pigmented eye-spots. There are three pairs of basophil and three pairs of acidophil cephalic glands. Infection takes place through the mouth and skin.

Pathogenesis. The eggs have been found in the wall of the intestine, where they may occasionally produce inflammatory processes with infiltration of leucocytes and connective tissue proliferation. Eggs may get into the pancreas, spleen and kidneys, but in these organs they die. The parasite is apparently not very pathogenic.

Genus: Heterobilharzia Price, 1929

Species of this genus occur in the mesenteric veins of Canidae and Felidae in North America. Males have 70–83 testes and there is a cirrus sac.

Heterobilharzia americanum Price, 1929, occurs in raccoons, bobcats, occasionally dogs and other mammals (nutria, rabbit) along the coast of the gulf of Mexico. The intermediate hosts are snails of the genera *Lymnaea* (*L. cubensis*), *Pseudosuccinea* (*P. columella*) and *Fossaria*.

Genus: Bivitellobilharzia Vogel and Manning, 1940

Members of this genus occur in the portal veins of elephants. **B. loxodontae** Vogel and Manning, 1940 and **B. nairi** (Mudalier and Ramanujachari, 1948) Dutt and Srivastava, 1955 have been reported.

Genus: Austrobilharzia Johnston, 1917

Species of this genus occur in the blood vessels of water fowl. Males have 18–20 testes and the ovary is an elongated spiral.

Austrobilharzia variglandis Johnston, 1917, occurs in terns and ducks. Marine snails (*Nassa obsoleta* in mainland USA and *Littorina pintado* in Hawaii) serve as intermediate hosts. Various birds can serve as experimental hosts. This species is the cause of schistosome or cercarial dermatitis in man.

Genus: Gigantobilharzia Odhner, 1910

Members of this genus occur in birds. The gynaecophoral canal is narrow and slit-like, the cirrus is absent and suckers are absent.

A number of species occur and a key to them is provided by Brackett (1942) and a list of the species in water fowl is given by McDonald (1969). Freshwater snails of various genera (*Gyraulus*, *Haminoea*, *Physa*, *Anisus*) serve as intermediate hosts in various parts of the world. Cercariae of this genus can cause schistosome dermatitis in man.

Genus: Trichobilharzia Skrjabin and Zakharow, 1920

The gynaecophoral canal is short, shallow and in the form of a widened area of the body; eggs have a terminal spine and the parasites are threadlike. They occur in the mesenteric, renal, cloacal and portal veins of birds including water fowl. A large number of species occurs and these are listed by McDonald (1969). The more common species include *T. ocellata* (La Valette, 1854) Brumpt, 1931 of Europe, Asia and North America; *T. stagnicolae* (Talbot, 1936) McMullen and Beaver, 1945 of *Anas* spp. and *Larus* spp. in North America; and *T. physellae* (Talbot, 1936), McMullen and Beaver, 1945, of *Anas* spp. in Japan, the USA and Canada. Snails of the genera *Lymnaea* and *Physa* serve as intermediate hosts and cercariae cause schistosome dermatitis in man.

Several other genera occur in wild birds. These include **Dendritobilharzia** Skrjabin and Zakharow, 1920 (e.g. *D. pulverulenta* Skrjabin, 1924, of swans, *Anas* spp. and *Fulica*, occurring in the dorsal aorta and found in Europe, Asia and North America) and **Pseudobilharziella** Ejsmont, 1929, of *Anas*, *Nyroca*, *Corvus* and *Cygnus* spp.

The genus **Schistosomatium** Tanabe, 1923, occurs in rodents (*Mus*, *Mustela*, *Sorex*, *Microtas*) in North America and snails of the genera *Lymnaea* and *Physa* spp. serve as intermediate hosts. The species *Schistosomatium douthitti* (Cort, 1914) Price, 1931, has been used extensively in experimental studies of schistosomiasis. It occurs in the mesenteric veins and its cercariae may cause dermatitis.

Cercarial dermatitis or schistome dermatitis, also commonly known as 'swimmers' itch', 'clam-diggers' itch', 'hunters' itch', 'rice-paddy itch', 'lakeside disease' (Japan), 'Badedermatitis' (Germany) or 'gale des nageurs' (France), can be a serious occupational problem where persons have to come into contact with water repeatedly, such as in rice-paddy fields. It is also important in recreational areas throughout the world. It is caused by the penetration of cercariae of non-human schistosomes into the skin of man. Initial exposure to cercariae causes a mild erythema and oedema but repeated exposure causes a severe pruritus and a papular or pustular eruption followed by a severe dermatitis which may persist for several days or several weeks if the lesions become secondarily infected.

Avian schistosomes infecting various water fowl are very important as causes of cercarial dermatitis. *Trichobilharzia* spp. such as *T. ocellata*, *T. physella* and *T. stagnicolae* have been incriminated, as have *Austrobilharzia variglandis* and *Gigantobilharzia* spp. However, the syndrome is also seen with schistosomes infecting domestic and wild mammals and *Heterobilharzia americanum* (racoon, dog) has been incriminated as the cause of 'water dermatitis' in Louisiana (Malek 1961). Buckley (1938) reported that cercariae of *S. spindale* (water buffalo) may be the cause of 'rice-paddy itch' in Malaysia. The condition has been investigated experimentally by Batten (1956).

The problem can be prevented by the use of protective, waterproof clothing when wading in water etc. Also, repellants such as dibutylphthalate and benzyl benzoate may be applied to the skin.

REFERENCES

TREMATODA, GENERAL

Schell, S. B. (1970) *The Trematodes.* Dubuque, Iowa: Brown.
Yamaguti, S. (1963) *Systema Helminthum IV, Monogenea and Aspidogastria.* New York: Interscience.

MONGENEA

Bauer, O. N. (1969) Parasitic diseases of cultured fishes and methods of their prevention and treatment. In *Parasitology of Fishes*, ed. V. A. Dogiel, G. K. Petrushevski & Yu I. Polyanski, pp. 265–298. Edinburgh and London: Oliver and Boyd.
Davis, H. S. (1953) *Culture and Diseases of Game Fishes.* Davis, Calif.: University of California Press.
Schell, S. B. (1970) *The Trematodes*, Dubuque, Iowa: Brown.
Yamaguti, S. (1963) *Systema Helminthum IV, Monogenea and Aspidogastria.* New York: Interscience.

DIGENEA

Clegg, J. A. (1965) *In vitro* culture of *Schistosoma mansoni. Exp. Parasit.*, **16**, 133–147.
Dawes, B. (1946) *The Trematoda—with Special Reference to British and Other European Forms.* London: Cambridge University Press.
Schell, S. B. (1970) *The Trematodes.* Dubuque, Iowa: Brown.
Smyth, J. D. (1966) *The Physiology of Trematodes.* Edinburgh and London: Oliver and Boyd.
Smyth, J. D. (1976) *Introduction to Animal Parasitology.* London: Hodder and Stoughton.
Smyth, J. D. & Clegg, J. A. (1959) Egg shell formation in Trematodes and Cestodes. *Exp. Parasit.*, **8**, 286–323.
Whitfield, P. J. (1979) *The Biology of Parasitism.* London: Edward Arnold.
Yamaguti, S. (1958) *Systema Helminthum Vol. 1. The Digenetic Trematodes of Vertebrates*, Parts 1 and 2. New York: Interscience Publishers.
Yamaguti, S. (1963) *Systema Helminthum, IV. Monogenea and Aspidogastria.* New York: Interscience.

STRIGEIDAE

Blair, D. (1976) Observations on the life-cycle of the strigeoid trematode, *Apatemom* (*Apatemom*) *gracilis* (Rudolphi, 1819), Szidat, 1928. *J. Helminth.*, **50**, 125–132.
Lowe, P. R. & Baylis, H. A. (1934) On a flock of razorbills in Middlesex found to be infested with intestinal flukes with a parasitological report. *Br. Birds*, **28**, 188–190.
Odening, K. & Bockhardt, I. (1971) The life-cycle of the trematode *Cotylurus variegatus* in the lake area of the Spree and Havel rivers. *Biol. Zentralbl.*, **90**, 49–84.

DIPLOSTOMATIDAE

Cuckler, A. C. (1940) Studies on the migration and development of *Alaria* spp. (Trematoda: Strigeata) in the definitive host. *J. Parasit.*, Suppl. **26**, 36.
Ehrlich, I. (1938) Paraziticka fauna pasa s pedrucja grada Zagreba. *Vet. Arh.*, **8**, 531–571.
Fernandes, B. J., Cooper, J. D., Cullen, J. B., Freeman, R. S., Ritchie, A. C., Scott, A. A. & Stuart, P. F. (1976) Systemic infection with *Alaria americana* (Trematoda). *Can. med. Ass. J.*, **115**, 1111–1114.
Jennings, A. R. & Soulsby, E. J. L. (1958) Disease in a colony of blackheaded gulls *Larus ridibundus. Ibis*, **100**, 305–312.
Shea, M., Maberly, A. L., Walters, J., Freeman, R. S. & Fallis, M. (1973) Intraretinal larval trematode. *Trans. Am. Acad. Ophthal. Otolar.*, **77**, 784–791.
Lavrovskii, V. V. (1977) Trout breeding techniques and prophylaxis. *Veterinariya, Moscow*, **5**, 67–69.
Van Duijn, C. (1973) *Diseases of Fishes.* London: Iliffe Books.

ACANTHOCOLPIDAE

Mackenzie, K. & Liversidge, J. M. (1975) Some aspects of the biology of the cercaria and metacercaria of *Stephanostomum baccatum* (Nicoll, 1907) Manter, 1934 (Digenea: Acanthocolpidae). *J. Fish Biol.*, **7**, 247–256.

SANGUINICOLIDAE

Evans, W. A. (1974*a*) Growth, mortality and haematology of cut-throat trout experimentally infected with the blood fluke *Sanguinicola klamathensis*. *J. Wildl. Dis.*, **10**, 341–346.
Evans, W. A. (1974*b*) The histopathology of cut-throat trout experimentally infected with the blood fluke *Sanguinicola klamathensis*. *J. Wildl. Dis.*, **10**, 243–248.

DICROCOELIIDAE

Bourgat, R., Seguin, D. & Bayssade-Dufour, C. (1975) New data on *Dicrocoelium hospes* Looss, 1907: Anatomy of the adult and life cycle. Preliminary note. *Ann. Parasit. hum. comp.*, **50**, 701–713.
Basch, P. F. (1966) Patterns of transmission of the trematode *Eurytrema pancreatum* in Malaysia. *Am. J. vet. Res.*, **27**, 234–240.
Chung, N. Y., Miyahara, A. Y. & Chung, G. (1977) The prevalence of feline liver flukes in the city and county of Honolulu. *J. Am. Animal Hosp. Ass.*, **13**, 258–262.
Denton, J. F. (1944) Studies on the life history of *Eurytrema procynois* Denton, 1942, *J. Parasit.*, **30**, 277–286.
Düwel, D., Kirsch, R. & Reisenleiter, R. (1975) The efficacy of fenbendazole in the control of trematodes and cestodes. *Vet. Rec.*, **97**, 371.
Evans, J. W. & Green, P. E. (1978) Preliminary evaluation of four anthelmintics against the cat liver fluke *Platynosomum concinnum*. *Aust. vet. J.*, **54**, 454–455.
Foix, J. (1977) Treatment of dicrocoeliasis with cambendazole. *Rev. Med. vet.*, **128**, 1111–1119.
Güralp, N., Oguz, T. & Zeybek, H. (1977) Chemotherapeutic trials with Embay 8440 (Praziquantel, Droncit) against *Dicrocoelium dendriticum* in naturally infected sheep. *Ankara Vet. Fakult. Dergisi*, **24**, 85–89.
Kajubiri, V. & Hohorst, W. (1977) Increasing incidence of *Dicrocoelium hospes* (Looss, 1907) (Trematoda: Digenea) in Uganda. *J. Helminth.*, **51**, 212–214.
Kirkwood, A. C. & Peirce, M. A. (1971) The longevity of *Dicrocoelium dendriticum* in sheep. *Res. vet. Sci.*, **12**, 588–589.
Krull, W. H. (1958) The migratory route of the metacercaria of *Dicrocoelium dendriticum* (Rudolphi, 1819) Looss, 1899 (Dicrocoeliidae) in the definite host. *Cornell Vet.*, **48**, 17–24.
Lämmler, G. (1963) Die Experimentalchemotherapie der Dicrocoeliose mit Hetolin. *Dt. tierärztl. Wschr.*, **70**, 373–377.
Leam, G. & Walker, I. E. (1963) The occurrence of *Platynosomum fastosum* in domestic cats in the Bahamas. *Vet. Rec.*, **75**, 46–47.
Manas, A. I., Gomez, G. V., Lozano, M. J., Rodriguez, O. M. & Campos, B. M. (1978) A study of the frequency of dicrocoeliasis in domestic animals in the province of Granada. *Rev. Iberica Parasit.*, **38**, 751–773.
Marchand, A. (1975) Dicrocoeliasis in European ruminants. *Point vet.*, **2**, 6–8.
Nadykto, M. V. (1973) Development of *Eurytrema pancreaticum* (Janson, 1889) (Trematoda: Dicrocoeliidae) in the Primorsk territory. *Parazitologiya*, **7**, 408–417.
Rukavina, J. (1977) Prevalence of dicrocoeliasis in sheep in Bosnia and Hercegovina. *Acta parasit. Jugoslav.*, **8**, 61–63.
Sheldon, W. G. (1966) Pancreatic flukes (*Eurytrema procyonis*) in domestic cats. *J. Am. vet. med. Ass.*, **148**, 251–253.
Sibalic, S., Mladenovic, Z. & Slavica, M. (1963) Effect of thiabendazole on *Dicrocoelium dendriticum* in sheep. *Vet. Glasn*, *17*, 1041–1046.
Soulsby, E. J. L. (1965) *Textbook of Veterinary Clinical Parasitology*. Vol. 1. *Helminths*. Oxford: Blackwell Scientific.
Tabangui, M. A. (1925) Metazoan parasites of Philippine domesticated animals. *Philipp. J. Sci.*, **28**, 11–37.
Tang, C. C. (1950) Studies on the life history of *Eurytrema pancreaticum* Janson, 1889. *J. Parasit.*, **36**, 559–573.
Taylor, D. & Perri, S. F. (1977) Experimental infection of cats with the liver fluke *Platynosomum concinnum*. *Am. J. vet. Res.*, **38**, 51–54.
Uchida, A., Itagaki, H. & Kugi, G. (1976) *Concinnum ten* (Yamaguti, 1939) from carnivorous animals in Japan. *Jap. J. Parasit.*, **25**, 319–323.

HETEROPHYIDAE

Ahn, Y. K., Chung, B. S. & Soh, C. T. (1978) Niclosamide in the treatment of metagonimiasis. *Korean J. Parasit.*, **16**, 65–68.
Cameron, T. W. M. (1945) Fish-carried parasites in Canada. I. Parasites carried by fresh-water fish. *Can. J. comp. Med.*, **9**, 245–254, 283–286, 302–311.
Christensen, N. O. & Roth, H. (1949) Investigation on internal parasites of dogs. *Yearbook: Royal Veterinary and Agricultural College*. Copenhagen, Denmark.
Combes, C., Jourdane, J. & Richard, J. (1974) Studies on the life cycle of *Euryhelmis squamula* (Rudolphi, 1819) a parasite of *Neomys fodiens* in the Pyrenees. *Z. Parasitenk.*, **44**, 81–92.
Lahav, M. (1974) The occurence and control of parasites infesting Mugilidae in fish ponds in Israel. *Bamidgeh*, **26**, 99–103.
McTaggert, H. S. (1958) *Cryptocotyle lingua* in British mink. *Nature, Lond.*, **181**, 651.
Neimi, D. R. & Macy, R. W. (1974) The life cycle and infectivity to man of *Apophallus donicus* (Skrjabin and Lindtrop, 1919) (Trematoda: Heterophyidae) in Oregon. *Proc. Helminth. Soc. Wash.*, **41**, 223–229.
Rim, H. J., Chu, D. S., Lee, J. S., Joo, K. H. & Won, C. Y. (1978) Anthelmintic effects of various drugs against metagonimiasis. *Korean J. Parasit.*, **16**, 117–122.
Soulsby, E. J. L. (1965) *Textbook of Veterinary Clinical Parasitology*, Vol. I, *Helminths*. Oxford: Blackwell Scientific.

PROSTHOGONIMIDAE

Macy, R. W. (1934) Studies on the taxonomy, morphology and biology of *Prosthogonimus macrorchis* Macy, a common oviduct fluke of domestic fowls in North America. *Univ. Minn. agric. Exp. Stn tech. Bull.*, 98.
Nath, D. (1973) Experimental development of *Prosthogonimus ovatus* (Rud, 1803) Lühe, 1899 in common quails, grey partridges and guinea fowls. *Indian vet. J.*, **50**, 465–473.
Panin, W. J. (1957) Variability of the morphological characters and its importance in the systematisation of suckers of the genus *Prosthogonimus* (Lühe, 1909). *Trudy Inst. Zool., Alma-Ata.*, **7**, 170–215.
Sardey, M. R. & Bhilegaonkar, N. G. (1979) Occurence of *Prosthogonimus macrorchis* (Macy, 1934, Syn. *P. rudolphi* Skrjabin, 1919; Witenberg and Eckman, 1939) in the egg of deshi hen. *Indian vet. J.*, **56**, 65–66.

PLAGIORCHIIDAE

Fedorova, O. E. (1973) An analysis of the morphological variations of *Plagiorchis laricola* and its taxonomic position. *Voprosy Zool.*, **3**, 55–58.
Foggie, A. (1937) An outbreak of parasitic necrosis in turkeys caused by *Plagiorchis laricola* (Skrjabin). *J. Helminth.*, **15**, 35–36.
Grabda-Kazubska, B. (1976) Abbreviation of the life cycles in plagiorchid trematodes. *Acta parasit. pol.*, **24**, 125–141.
Rees, F. G. (1952) The structure of the adult and larval stages of *Plagiorchis* (*Multiglandularis*) *megalorchis* n. nom. from the turkey and an experimental demonstration of the life history. *Parasitology*, **42**, 92–113.
Soulsby, E. J. L. (1965) *Textbook of Veterinary Clinical Parasitology*, Vol. 1. *Helminths*. Oxford: Blackwell Scientific.

OPISTHORCHIIDAE

Attwood, H. D. & Chou, S. T. (1978) The longevity of *Clonorchis sinensis*. *Pathology*, **10**, 153–156.

Beer, S. A. (1976) Prospects for the control of *Opisthorchis* infections by treating molluscan populations. *Parazitologia*, **10**, 473–481.

Harinasuta, C. (1978) Diseases of the liver, 3. Liver fluke infections. In *Diseases of Children in the Subtropics and Tropics*, ed. D. B. Jelliffe & J. P. Stanfield, pp. 549–555. London: Edward Arnold.

Muller, R. (1977) Liver flukes. In *Infectious Diseases. A Modern Treatise of Infectious Processes*, ed. P. D. Hoeprich, pp. 654–657. New York: Harper and Row.

Rim, H. J. (1978) The treatment of human clonorchiasis and other trematode infections with Niclofolan (Bilevon, Bayer 9015) and Praziquantel (Embay 8440). *Proc. 4th int. Congr. Parasit. 1978, Warsaw*, Section D, pp. 24–25.

Rim, H. J., Chang, D. S., Hyun, I. & Song, S. D. (1975) Effectiveness of anthelmintic drugs against *Clonorchis sinensis* infection of rabbits. *Korean J. Parasit.*, **13**, 123–132.

Schuurmans-Stekhoven, J. H. (1931) Der zweite Zwischen wirt von *Pseudamphistomum truncatum* (Rud.) nebst Beobachtung über andere Trematoden larven. *Z. Parasitenk.*, **3**, 747–764.

Yamaguchi, T., Uohara, K. & Shinoto, M. (1962) Treatment of *Clonorchis sinensis* with Pankiller (Dithiazanine iodine). *Jap. J. Parasit.*, **11**, 30–38.

NANOPHYETIDAE

Millemann, R. E. & Knapp, S. E. (1970) Biology of *Nanophyetus salmincola* and 'Salmon poisoning' disease. *Adv. Parasit.*, **8**, 1–41.

FASCIOLIDAE: FASCIOLA HEPATICA

Armour, J. & Urquhart, G. M. (1974) Clinical problems of preventative medicine. The control of helminthiasis in ruminants. *Br. vet. J.*, **130**, 99–109.

Behrens, H. (1960) Behandlung des Lebergelbefalls der Schafe mit Hetol. *Dt. tierärztl. Wschr.*, **67**, 467–470.

Berghen, P. (1964) Some Lymnaerdae as intermediate hosts of *Fasciola hepatica* in Belgium. *Exp. Parasit.*, **15**, 118–124.

Berry, C. I. & Dargie, J. D. (1978) Pathophysiology of ovine fascioliasis: the influence of dietary protein and iron on the erythrokinetics of sheep experimentally infected with *Fasciola hepatica*. *Vet. Parasit.*, **4**, 327–339.

Boray, J. (1963) The ecology of *Fasciola hepatica* with particular reference to its intermediate host in Australia. *Proc. 17th Wld vet. Congr. Hannover*, 709–715.

Campbell, N. J., Kelly, J. D., Townsend, R. B. & Dineen, J. K. (1977) The stimulation of resistance in sheep to *Fasciola hepatica* by infection with *Cysticercus tenuicollis*. *Int. J. Parasit.*, **7**, 347–351.

Čorba, J. & Armour, J. (1973/1974) The effect of diamphenetide [β,β-bis (4-acetamidophenyloxy) ethyl ether] on *Fasciola hepatica*. *Helminthologia*, *14/15*, 961–967.

Crossland, N. O. (1976) The effect of the molluscicide *N*-tritylmorpholine on transmission of *Fasciola hepatica*. *Vet. Rec.*, **98**, 45–48.

Dargie, J. D., Armour, J., Rushton, B. & Murray, M. (1974) Immune mechanisms and hepatic fibrosis in fascioliasis. In *Parasitic Zoonoses*, ed. E. J. L. Soulsby, pp. 249–271. New York: Academic Press.

Dargie, J. D. & Berry, C. I. (1979) The hypoalbuminaemia of ovine fascioliasis: the influence of protein intake on the albumin metabolism of infected and of pair-fed control sheep. *Int. J. Parasit.*, **9**, 17–25.

Dawes, B. (1960) A study of the miracidium of *Fasciola hepatica* and an account of the mode of penetration of the sporocyst into *Linnaea truncatula*. In *Lilro Hmenaje Dr. Eduardo Caballero y Caballero*, pp. 95–111. Mexico: Escuela Nacional de Ciencias Biologicas.

Dineen, J. K., Kelly, J. D. & Campbell, N. J. (1978) Further observations on the nature and characteristics of cross protection against *Fasciola hepatica* produced in sheep by prior infection with *Cysticercus tenuicollis*. *Int. J. Parasit.*, **8**, 173–176.

Edwards, C. M. & Parry, T. O. (1972) Treatment of experimentally-produced acute fascioliasis in sheep. *Vet. Rec.*, **90**, 523–526.

Gibson, T. E. (1978) The 'Mt' system for forecasting the incidence of fascioliasis. In *Weather and Parasitic Animal Disease*, ed. T. E. Gibson, Tech. Note 159, pp. 3–5, Geneva: World Meteorological Organization.

Holmes, P. H., Dargie, J. D., MacLean, J. M. & Mulligan, W. (1968) The anaemia of fascioliasis: studies with ^{51}Cr-labelled red cells. *J. comp. Path.*, **78**, 415–420.

Hughes, D. L. (1963) Some studies on the host-parasite relations of *Fasciola hepatica*. PhD Thesis, University of London.

Hughes, D. J., Harness, E. & Doy, T. G. (1978) Failure to demonstrate resistance in goats, sheep and cattle to *Fasciola hepatica* after infection with *Cysticercus tenuicollis*. *Res. vet. Sci.*, **25**, 356–359.

Kendall, S. B. (1949) Nutritional factors affecting the rate of development of *Fasciola hepatica* in *Limnaea truncatula*. *J. Helminth.*, **23**, 179–190.

Kendall, S. B. (1954) Fascioliasis in Pakistan. *Ann. trop. Med. Parasit.*, **48**, 307–313.

Kendall, S. B. (1965) Relationships between the species of *Fasciola* and their molluscan hosts. *Adv. Parasit.*, **3**, 59–98.

Kendall, S. B. & Ollerenshaw, C. B. (1963) The effect of nutrition on the growth of *Fasciola hepatica* in its snail host. *Proc. nutr. Soc.*, **22**, 41–46.

Lämmler, G. (1960) Chemotherapeutische Untersuchungen mit Hetol, einem neuen, hochwirksamen Lebergegelmittel. *Dt. tierärztl. Wschr.*, **67**, 408–413.

Murray, M. (1973) Fascioliasis: pathology. In *Helminth Diseases of Cattle, Sheep and Horses in Europe*, ed. G. M. Urquhart & J. Armour, pp. 81–114. Glasgow: MacLehose.

Nansen, P. (1975) Resistance in cattle to *Fasciola hepatica* induced by X-ray attenuated larvae: results of a controlled field trial. *Res. vet. Sci.*, **19**, 278–283.

Ollerenshaw, C. B. (1959) The ecology of the liver fluke (*Fasciola hepatica*). *Vet. Rec.*, **71**, 957–963.

Ollerenshaw, C. B. & Rowland, L. P. (1959) A method of forecasting the incidence of fascioliasis in Anglesey. *Vet. Rec.*, **71**, 591–598.

Olsen, O. W. (1947) Longevity of metacercariae of *Fasciola hepatica* on pastures in the upper coastal region of Texas and its relationship to liver fluke control. *J. Parasit.*, **31**, 36–42.

Owen, J. M. (1977) Liver fluke infection in horses and ponies. *Equine vet. J.*, **9**, 29–31.

Reid, J. F. S. (1973) Fascioliasis: Clinical aspects and diagnosis. In *Helminth Diseases of Cattle, Sheep and Horses in Europe*, ed. G. M. Urquhart & J. Armour, pp. 81–114. Glasgow: Maclehose.

Reid, J. F. S., Armour, J., Jennings, F. W., Kirkpatrick, K. S. & Urquhart, G. M. (1967) The fascioliasis ostertagiasis complex in young cattle: A guide to diagnosis and therapy. *Vet. Rec.*, **80**, 371–374.

Rew, R. S., Colglazier, M. L. & Enzie, F. D. (1978) Effect of diamfenetide on experimental infections of *Fasciola hepatica* in lambs: anthelmintic and clinical investigations. *J. Parasit.*, **64**, 290–294.

Roseby, F. B. & Boray, J. C. (1970) The anthelmintic efficiency against *Fasciola hepatica* and the toxicity of Bay 4059 in sheep. *Aust. vet. J.*, **46**, 308–310.

Ross, J. G. (1978) Stormont 'wet-day' fluke forecasting. In *Weather and Parasitic Animal Disease*, ed. T. E. Gibson, Tech. Note 159, pp. 14–20, Geneva: World Meteorological Organization.

Rowcliffe, S. A. & Ollerenshaw, C. B. (1960) Observations on the bionomics of the egg of *Fasciola hepatica*. *Ann. trop. Med. Parasit.*, **54**, 172–181.

Rushton, B. & Murray, M. (1977) Hepatic pathology of a primary experimental infection of *Fasciola hepatica* in sheep. *J. comp. Path.*, **87**, 459–470.

Soulsby, E. J. L. (1965) *Textbook of Veterinary Clinical Parasitology*, Vol. 1, *Helminths*. Oxford: Blackwell Scientific Publications.

Taylor, E. L. (1949) The epidemiology of fascioliasis in Britain. *Proc. 14th Int. vet. Congr. London*, **2**, 81–87.

Taylor, E. L. (1951) Parasitic bronchitis in cattle. *Vet. Rec.*, **63**, 859–873.

Wikerhauser, T. (1960) A rapid method for determining the viability of *Fasciola hepatica* metacercariae. *Am. J. vet. Res.*, **21**, 895–897.

Van Tiggle, L. J. (1978) Host-parasite relationship in *Fasciola hepatica* infections. In *Immunopathology and Diagnosis of Liver-Fluke Disease in Ruminants*, pp. 1–64. Rotterdam: Bronder-Offset.

FASCIOLIDAE: FASCIOLA GIGANTICA

Caple, I. W., Jainudeen, M. R., Buick, T. D. & Song, C. Y. (1978) Some clinico-pathological findings in elephants (*Elephas maximus*) infected with *Fasciola jacksoni*. *J. Wildlife Dis.*, **14**, 110–115.
Coyle, T. J. (1959) Control of fascioliasis in Uganda. In *Symposium on Helminthiasis in Domestic Animals*, C.C.T.A. Nairobi, No. 49, pp. 67–80, I.A.C.E.D.
Dinnik, J. A. & Dinnik, N. N. (1964) The influence of temperature on the succession of redial and cercarial generations of *Fasciola gigantica* in a snail host. *Parasitology*, **54**, 59–65.
Kadhim, J. K. & Jabbir, M. H. (1974) Comparative field trial of rafoxanide in sheep with oxyclozanide and hexachlorophene. *J. Egypt. vet. med. Ass.*, **34**, 190—198.
Karrasch, A. W., Horchner, F. & Bohnel, H. (1975) Efficacy of Dirian against *Fasciola gigantica* and paramphistomes in naturally infected cattle from Madagascar. *Berl. Munch. Tierarztl. Wschr.*, **88**, 348–351.
Kendall, S. B. (1954) Fascioliasis in Pakistan. *Ann. trop. Med. Parasit.*, *48*, 307–313.
Preston, J. M. & Castelino, J. B. (1977) A study of the epidemiology of bovine fascioliasis in Kenya and its control using N-trytylmorpholine. *Br. vet. J.*, **133**, 600–608.
Troncy, P. M. & Vasseau-Martin, N. (1976) Rafoxanide for treatment of *Fasciola gigantica* infection in zebu cattle in Chad. *Rev. Élev. med. vet. Pays trop.*, **29**, 31–37.

FASCIOLIDAE: FASCIOLOIDES MAGNA

Davis, J. W. & Libke, K. G. (1971) Trematodes. In *Parasitic Diseases of Wild Mammals*, ed. J. W. Davis & R. C. Anderson, pp. 235–257. Ames: Iowa State University.
Foreyt, W. J. & Todd, A. C. (1972) The occurrence of *Fascioloides magna* and *Fasciola hepatica* together in the livers of naturally infected cattle in south Texas, and the incidence of flukes in cattle, white-tailed deer and feral hogs. *J. Parasit.*, **58**, 1010–1011.
Foreyt, W. J. & Todd, A. C. (1973) Action of oxyclozanide against adult *Fascioloides magna* (Bassi, 1875) infections in white-tailed deer. *J. Parasit.*, **59**, 208–209.
Foreyt, W. J. & Todd, A. C. (1974) Efficacy of rafoxanide and oxyclozanide against *Fascioloides magna* in naturally infected cattle. *Am. J. vet. Res.*, **35**, 375–377.
Foreyt, W. J. & Todd, A. C. (1976*a*) Development of the large American liver fluke, *Fascioloides magna*, in white-tailed deer, cattle and sheep. *J. Parasit.*, **62**, 26–32.
Foreyt, W. J. & Todd, A. C. (1976*b*) Effects of six fasciolicides against *Fascioloides magna* in white-tailed deer. *J. Wildl. Dis.*, **12**, 361–366.
Pursglove, S. R., Prestwood, A. K., Ridgeway, T. R. & Hayes, F. A. (1977) *Fascioloides magna* infection in white-tailed deer of southern United States. *J. Am. vet. Med. Ass.*, **171**, 936–938.
Theodorides, V. J. (1977) The newer benzimadazole anthelmintics. In *Perspectives in the Control of Parasitic Disease in Animals in Europe*, ed. D. W. Jolly & J. M. Somerville, pp. 37–44. London: Unwin.

FASCIOLIDAE: FASCIOLOPSIS AND PARAFASCIOLOPSIS

Drozdz, J. (1963) Naturalne ognisko parafasciolopsozy in Wojewodzt wie Bialostockun. *Wiad. Parazyt.*, **9**, 129–132.
Muttalib, M. A. & Islam, N. (1975) *Fasciolopsis buski* in Bangladesh—a pilot study. *J. trop. Med. Hyg.*, **78**, 135–137.
Suntharasamai, P., Bunnag, D., Tegavanij, S., Harinasuta, T., Migasena, S., Vutikes, S. & Chindanond, D. (1974) Comparative clinical trials of niclosamide and tetrachlorethylene in the treatment of *Fasciolopsis buski* infections. *S.E. Asian J. trop. Med.*, **5**, 556–559.

ECHINOSTOMATIDAE

Annereaux, R. F. (1940) A note on *Echinoparyphium recurvatum* (von Linstow) parasitic in California turkeys. *J. Am. vet. med. Ass.*, **96**, 62–64.
Arizono, N., Uemoto, K., Kondo, K., Matsuno, K., Yoshida, Y., Maeda, T., Yoshida, H., Mutp, K., Inoue, Z. & Takahashi, K. (1976) Studies on *Echinostoma hortense* Asada, 1926, with special reference to human infections. *Jap. J. Parasit.*, **25**, 36–45.
Beaver, P. C. (1937) Experimental studies on *E. revolutum* (Froel.), a fluke from birds and mammals. *Illinois biol. Monogr.*, **15**, 1–96.
Beaver, P. C. (1941) Studies on the life history of *Euparyphium melis* (Trematoda: Echinostomidae). *J. Parasit.*, **27**, 35–44.
Griffiths, H. J., Gonder, E. & Pomeroy, B. S. (1976) An outbreak of trematodiasis in domestic geese. *Avian Dis.*, **20**, 604–606.
Harper, W. F. (1929) On the structure and life history of British fresh-water larval trematodes. *Parasitology*, **21**, 189–219.
Krause, C. (1925) Gehäuftes Sterben bei Tauben durch Echinostomiden. *Berl. tierärztl. Wschr.*, **41**, 262–263.
Leger, N. & Notteghem, M. J. (1975) Study of the flukicide activity of a new compound, brotianide, on *Echinostoma caproni* Richard, 1964. *Ann. Pharm. fr.*, **33**, 273–277.
Notteghem, M. J., Leger, N. & Cavier, R. (1979) A study of the anthelmintic activity of flubendazole on *Echinostoma caproni* Richard, 1964. *Ann. Pharm. fr.*, **37**, 153–156.
Soulsby, E. J. L. (1955) Deaths in swans associated with trematode infection. *Br. vet. J.*, **111**, 498–500.
Tani, S. (1976) Studies on *Echinostoma hortense* (Asada, 1926). (2) The intermediate and final hosts in Akita Prefecture. *Jap. J. Parasit.*, **25**, 461–467.
van Heelsbergen, T. (1927a) Echinostomiasis bij de duif door Echinostoma. *Tijdschr. Diergeneesk.*, **54**, 414–416.
van Heelsbergen, T. (1927b) Echinostomiasis bij kippen door Echinoparyphium. *Tijdschr. Diergeneesk.*, **54**, 413–414.
Vevers, G. M. (1923) Observations on the life-histories of *Hypodaerium conoideum* (Bloch) and *Echinostomum revolution* (Froel): Trematode parasites of the domestic duck. *Ann. appl. Biol.*, **10**, 134–136.
Wetzel, R. (1933) Zum Wirt-Parasitverhältnis des Saugwurmes *Echinoparyphium paraulum* in der Taube. *Dt. tierärztl. Wschr.*, **41**, 772–775.

NOTOCOTYLIDAE; BRACHYLAEMIDAE; HASSTILESIIDAE; TROGLOTREMATIDAE

Alicata, J. E. (1940) The life cycle of *Postharmostomum gallinum*, the cecal fluke of poultry. *J. Parasit.*, **26**, 135–143.
Gvozdev, E. V. & Soboleva, T. N. (1973) Revision of the subfamily Hasstilesiinae (Trematoda: Brachylaemidae). In *Problemy Obshcheĭ i Prikladnoĭ gel' Mintologii*, ed. V. G. Gagarin, pp. 41–48, Moscow: Academy of Sciences.
Michalski, L. (1977) Studies in the efficacy of Zanil in control of *Notocotylus* infection in ducklings. *Wiad. Parazyt.*, **23**, 435–439.
Szidat, L. (1930) Die Parasiten des Hausgeflügels. 4. *Notocotylus* und *Catatropis* Odhner, Zwei, die Blinddärme des Geflügels bewohnende Monostome Trematodengattung, ihre Entwicklung und Übertragung. *Arch. Geflügelk.*, **4**, 105–111.
Vogel, H. & Voelker, J. (1978) On the life-cycle of *Troglotrema acutum*. *Tropenmed. Parasit.*, **29**, 385–405.

PARAGONIMIDAE

Capron, A., Yokogawa, M., Biguet, J., Tsuje, M. & Luffaw, G. (1965) Diagnostic immunologique de la paragonimose humaine. Mise en evidence d'anticorps seriques specifique par immunoelectrophorese. *Bull. Soc. Path. éxot.*, **58**, 474–487.
Dubey, J. P., Hoover, E. A., Stromberg, P. C. & Toussant, M. J. (1978*b*) Albendazole therapy for experimentally induced *Paragonimus kellicotti* infection in cats. *Am. J. vet. Res.*, **39**, 1027–1031.
Dubey, J. P., Miller T. B. & Sharma, S. P. (1979*a*) Fenbendazole for treatment of *Paragonimus kellicotti* infection in dogs. *J. Am. vet. med. Ass.*, **174**, 835–837.
Dubey, J. P., Stromberg, P. C., Toussant, M. J., Hoover, E. A. & Pechman, R. D. (1978*a*) Induced paragonomiasis in cats: Clinical signs and diagnosis. *J. Am. vet. med. Ass.*, **173**, 734–742.
Dubey, J. P., Toussant, M. J., Hoover, E. A., Miller, T. B., Sharma, S. P. & Pechman, R. D. (1979*b*) Experimental *Paragonimus kellicotti* infection in dogs. *Vet. Parasit.*, **5**, 325–337.
Eliasoff, L. B. & Harden, C. R. (1977) Treatment of *Paragonimus kellicotti* infestation. *Feline Pract.*, **7**, 45–47.
Hoover, E. A. & Dubey, J. P. (1978) Pathogenesis of experimental pulmonary paragonimiasis in cats. *Am. J. vet. Res.*, **39**, 1827–1832.
Macey, D. W. & Todd, K. S. (1975) Treatment of canine paragonimiasis with bithionol acetate. *Vet. Med. small Anim. Clin.*, **70**, 57–58.
Miyazaki, I. (1970) Taxonomical and ecological studies on the lung fluke *Paragonimus* in Asia. *H. D. Srivastava. Comm. Vol.*, pp. 155–165.
Rim, H. J., Kim, M. S., Ha, J. H. & Chang, D. S. (1976) Experimental chemotherapeutic effects of niclofolan (Bayer, 9015, Bilevon) in the animals infected with *Paragonimus westermanii* or *P. iloktsuenesis*. *Korean J. Parasit.*, **14**, 140–146.
Stromberg, P. C. & Dubey, J. P. (1978) The life cycle of *Paragonimus kellicotti* in cats. *J. Parasit.*, **64**, 998–1002.
Stromberg, P. C., Toussant, M. J. & Dubey, J. P. (1978) Population biology of *Paragonimus kellicotti* metacercariae in Central Ohio. *Parasitology*, **77**, 13–18.
Voelker, J. (1973) Morphologisch-taxonomische Untersuchungen über *Paragonimus uterobilateralis* (Trematoda, Troglotrematidae) Sowie Beobachtungen über den lebenzyklus und die Verbrütung des Parsiten in Liberia. *Z. tropenmed. Parasit.*, **24**, 1–20.

PARAMPHISTOMATIDAE; CYCLOCOELIDAE

Azzie, M. A. J. (1975) Pathological infection of thoroughbred horses with *Gastrodiscus aegyptiacus*. *J. S. Afr. vet. med. Ass.*, **46**, 77–78.
Boray, J. C. (1959) Studies on intestinal amphistomosis in cattle. *Aust. vet. J.*, **35**, 282–287.
Boray, J. C. (1969) Studies on intestinal paramphistomosis in sheep due to *Paramphistomum ichikawai* Fukui, 1922. *Vet. Med. Rev.*, **4**, 290–308.
Chhabra, R. C. & Ball, H. S. (1976) Efficacy of some drugs against amphistomes in cattle and buffaloes under field conditions in the Punjab. *J. Res. Punjab Agric. Univ.*, **13**, 226–231.
Dawes, B. (1936) On a collection of Paramphistomidae from Malaya, with revision of the genera *Paramphistomum* Fischoeder, 1901 and *Gastrothylax* Poirier, 1883. *Parasitology*, **28**, 330–354.
Dawes, B. (1968) *The Trematoda. With Special Reference to British and Other European Forms.* London: Cambridge University Press.
Durie, P. H. (1953) The paramphistomes (Trematoda) of Australian ruminants. II. The life history of *Ceylonocotyle streoptocoelium* (Fischoeder) Näsmark and of *Paramphistomum ichikawai* Fukui. *Aust. J. Zool.*, **1**, 193–222.
Gaenssler, J. G. (1974) Further trials of the efficacy of Terenol (rosorantel) in cattle and goats in South Africa. *Blue Book for the Veterinary Profession*, **24**, 94–98.
Horak, I. G. (1965) The anthelmintic efficacy of bithionol against *Paramphistomum microbothrium*, *Fasciola* spp. and *Schistosoma mattheei*. *J. S. Afr. vet. med. Ass.*, **36**, 561–566.
Horak, I. G. (1967) Host-parasite relationships of *Paramphistomum microbothrium* Fischoeder, 1901, in experimentally infected ruminants, with particular reference to sheep. *Onderstepoort J. vet. Res.*, **34**, 451–540.
Horak, I. G. (1971) Paramphistomiasis of domestic ruminants. *Adv. Parasit.*, **9**, 33–72.
Roberts, H. M., Adams, J. W. E. & Danks, B. C. (1976) *Gastrodiscus aegyptiacus*: a therapeutic trial. *Rhod. vet. J.*, **6**, 73–76.
Soulsby, E. J. L. (1965) *Textbook of Veterinary Clinical Parasitology. I. Helminths.* Oxford: Blackwell Scientific.
Szidat, L. (1933) Über die Entwicklung und den Infektionmodus von *Tracheophilus sisowi* Skrj. eines Luftröhrenschmarotzers der Enten aus der Trematoden-familie der Zyklozöliden. *Tierärztl. Rdsch.*, **39**, 95–99.
Varma, A. K. (1957) On a collection of paramphistomes from domesticated animals in Bihar. *Indian J. vet. Sci.*, **27**, 67–76.
Yamaguti, S. (1958) *Systema Helminthum Vol. I. The Digenetic Trematodes of Vertebrates*, Parts 1 and 2. New York: Interscience.

SCHISTOSOMATIDAE

Ahluwalia, S. S. (1972) Experimental schistosomiasis incognitum in pigs. *Ind. J. Anim. Sci.*, *42*, 723–729.
Batten, P. J. (1956) The histopathology of swimmer itch. I. The skin lesions of *Schistosomatium douthitti* and *Gigantobilharzia huronensis* in the unsensitized mouse. *Am. J. Path.*, **32**, 363–377.
Berry, C. I., Dargie, J. D. & Preston, J. M. (1973) Pathophysiology of ovine schistosomiasis. IV. Effects of experimental *Schistosoma mattheei* infections on the nitrogen status and digestive efficiency of the host. *J. comp. Path.*, **83**, 559–568.
Brackett, S. (1942) Five new species of avian schistosomes from Wisconsin and Michigan with the life cycle of *Gigantobilharzia gyrauli* (Brackett). *J. Parasit.*, **28**, 25–42.
Buckley, J. J. C. (1938) On a dermatitis in Malays caused by the cercariae of *Schistosoma spindale* Montgomery, 1906. *J. Helminth.*, **14**, 117–120.
Bushara, H. O., Majid, A. A., Saad, A. M., Hussein, M. F., Taylor, M. G., Dargie, J. D., Marshall, T. F. de C., & Nelson, G. S. (1980) Observations on cattle schistomiasis in the Sudan, a study in comparative medicine. II. Experimental demonstration of naturally acquired resistance to *Schistosoma bovis*. *Am. J. trop. Med. Hyg.*, **29**, 442–451.
Carney, W. P., Brown, R. J., Van Peenen, P. P., Purnomo, D., Ibrahim, B. & Koesharjono, R. C. (1967) *Schistosoma incognitum* from Cikuri, West Java, Indonesia. *Int. J. Parasit.*, **7**, 361–366.
Dargie, J. D., MacLean, J. M. & Preston, J. M. (1973) Pathophysiology of Ovine Schistosomiasis. III. Study of plasma protein metabolism in experimental *Schistosoma mattheei* infections. *J. comp. Path.*, **83**, 543–557.
Davis, A., Biles, J. E. & Ulrich, A.-M. (1979) Initial experiences with praziquantel in the treatment of human infections due to *Schistosoma haematobium*. *Bull. Wld Hlth Org.*, **57**, 773–779.
Deramée, O., Thienpont, D., Fain, A. & Jadin, J. (1953) Sur un foyer de bilharziose canine *Schistosoma rodhaini* Brumpt au Raunda-Urundi; note preliminaire. *Ann. Soc. belge Med. trop.*, **33**, 207–209.
Islam, K. S. (1975) Schistosomiasis in domestic ruminants in Bangladesh. *Trop. Anim. Hlth Prod.*, **7**, 244.
Jordan, P. (1977) Schistosomiasis—research to control. *Am. J. trop. Med. Hyg.*, **26**, 877–886.
Katz, N., Rocha, R. S. & Chaves, A. (1979) Preliminary trials with praziquantel in human infections due to *S. mansoni*. *Bull. Wld Hlth Org.*, **57**, 781–785.
Lawrence, J. A. (1978*a*) Bovine schistosomiasis in Southern Africa. *Helminth. Abstr.*, **A47**, 261–270.
Lawrence, J. A. (1978*b*) The pathology of *Schistosoma mattheei* infection in the ox. 1. Lesions attributable to the eggs. *J. comp. Path.*, **88**, 1–14.
Lawrence, J. A. (1978*c*) The pathology of *Schistosoma mattheei* infection in the ox. 2. Lesions attributable to the adult parasite. *J. comp. Path.*, **88**, 15–29.
McDonald, M. E. (1969) *Catalogue of Helminths of Water Fowl (Anatidae)*. Special Scientific Report, Wildlife No. 126. Washington: Bureau of Sport Fisheries and Wildlife.
Malek, E. A. (1961) The biology of mammalian and bird schistosomes. *Bull. Tulane med. Fac.*, **20**, 181–207.

Majid, A. A., de Marshall, T. F., Hussein, M. F., Bushara, H. O., Taylor, M. G., Nelson, G. S. & Dargie, J. D. (1980) Observations on cattle schistosomiasis in the Sudan, a study in comparative medicine. I. Epizootiological observations on *Schistosoma bovis* in the White Nile Province. *Am. J. trop. Med. Hyg.*, **29**, 435–441.
Preston, J. M. & Dargie, J. D. (1974*a*) Pathophysiology of ovine schistosomiasis, V. Onset and development of anaemia in sheep experimentally infected with *Schistosoma mattheei*—studies with ^{51}Cr-labelled erythrocytes. *J. comp. Path.*, **84**, 73–81.
Preston, J. M. & Dargie, J. D. (1974*b*) Pathophysiology of ovine schistosomiasis VI. Onset and development of anaemia in sheep experimentally infected with *Schistosoma mattheei*—ferrokinetic studies. *J. comp. Path.*, **84**, 83–91.
Rajamohanan, K. & Peter, C. T. (1975) Pathology of nasal schistosomiasis in buffaloes. *Karala J. vet. Sci.*, **6**, 94–100.
Santos, A. T., Blas, B. L., Nosenas, J. S., Portillo, G. P., Ortega, O. M., Hayashi, M. & Boehme, K. (1979) Preliminary clinical trials with praziquantel in *Schistosoma japonicum* infections in the Philippines. *Bull. Wld Hlth Org.*, **57**, 793–799.
Sharma, D. N. & Dwivedi, J. N. (1976) Pulmonary schistosomiasis in sheep and goats due to *Schistosoma indicum* in India. *J. comp. Path.*, **86**, 449–454.
Soulsby, E. J. L. (1965) *Textbook of Veterinary Clinical Parasitology, I. Helminths.* Oxford: Blackwell Scientific.
Taylor, M. G. (1980) Vaccines against trematodes. In *Vaccines against Parasites*, ed. A. E. R. Taylor & R. Muller, pp. 115–140. Oxford: Blackwell Scientific.
Tewari, H. C., Dutt, S. C. & Iyer, P. K. R. (1966) Observations on the pathogenicity of experimental infections of *Schistosoma incognitum* Chandler, 1926 in dogs.
Tewari, H. C. & Ramachandran, P. K. (1979) Histopathology of experimental *Schistosoma incognitum* infection in mice following exposure to normal and irradiated cercariae. *J. Helminth.*, *53*, 117–20.
Warren, K. S. (1972) The immunopathogenesis of schistosomiasis: a multi-disciplinary approach. *Trans. R. Soc. trop. Med. Hyg.*, **66**, 417–432.
Warren, K. S. (1977) Modulation of immunopathology and disease in schistosomiasis. *Am. J. trop. Med. Hyg.*, Suppl. **26**, 113–119.
van Zyl, L. C. (1974) Serum protein fractions as determined by cellulose acetate electrophoresis in *Schistosoma mattheei* infested sheep. *Onderstepoort J. vet. Res.*, **41**, 7–14.

Cestodes

Wardle et al. (1974) have divided the cestodes into two classes: COTYLODA and EUCESTODA. The class Cotyloda contains six orders, of which four are discussed in this text. Those described are primarily important as parasites of fish in the plerocercoid or adult stages of the life-cycle, while *Diphyllobothrium* and *Spirometra* species are parasites of man, dogs and cats. The Eucestoda (true cestodes) contains 15 orders, of which seven contain important parasites of domestic animals, man and fish.

Morphology

Tapeworms are hermaphrodite, endoparasitic worms with an elongate flat body and without a body cavity or alimentary canal. They may be a few millimetres to several metres in length. The body consists of a head or *scolex*. This is usually followed by a short unsegmented portion called the *neck* and, in general, the remainder of the body or *strobila* consists of a number of segments or *proglottids* which are separated by transverse constrictions and vary considerably in shape and size. Each proglottid usually contains one or two sets of reproductive organs.

The scolex is usually globular. In the Eucestoda, the scolex is normally provided with four suckers (*acetabula*) or modifications thereof. The suckers may be armed with hooks. A protrusible part, the *rostellum*, often armed, may be present. In the Cotyloda, the scolex may lack true holdfast organs or may have long, narrow, weakly muscular grooves (*bothria*) (or similar devices) on the scolex.

The majority of tapeworms have metameric repetition of their reproductive organs, called *proglottidization*. In the Eucestoda each proglottid is distinctly separated from its neighbours. Proglottids are formed from the neck or growth region and mature as they are pushed away from the scolex. The posterior proglottids, when fully mature, are packed with eggs (*gravid*). In the Cotyloda, segmentation may be absent or indistinct and there may be instantaneous proglottidization of several proglottids at a time (such as from the unsegmented plerocercoid).

The body is covered by a *tegument* composed of a syncytial outer layer formed by the tegumental cells (an inner nucleated layer). The outer cytoplasm is extended into *microtriches* (sometimes referred to as microvilli) which are spine-like processes covered by plasma membrane and containing microtubular structures. The whole acts as the absorptive structure. Further details on tegumental structure and function may be found in Read (1966) and Smyth (1972, 1976).

Muscles lie beneath the syncytial layer and more deeply in the parenchyma-filled body. They divide the body into outer cortical and inner medullary sections. The medulla contains the excretory, nervous and reproductive organs.

The excretory system is a nephridial system with flame cells and efferent canals. There are usually, on either side, two longitudinal canals joined by a loop in the scolex. Each pair may be connected in the posterior portion of the proglottid.

The central part of the nervous system is situated in the scolex and generally consists of a rostellar nerve ring and two lateral nervous ganglia from which six cords run posteriorly. In addition there is a pair of dorsal and ventral nerve cords.

There are one or two sets of reproductive organs per proglottid. The reproductive organs generally mature from the anterior to the posterior proglottids. *Mature* proglottids are those in which the reproductive organs are fully mature and functional. In the Eucestoda, following fertilization of the eggs, the reproductive organs degenerate leaving a gravid proglottid filled with eggs. Gravid proglottids are detached and passed out of the host singly or, occasionally, in chains and the eggs are set free by disintegration of the gravid proglottids (*apolysis*). Or, as has been described for some of the Taeniidea, the eggs are released by the pressure of eggs in the uterus and the muscular activity of the proglottid through an opening called the *thysanus* created in the uterus when the proglottid breaks away from the strobila. In the Cotyloda, the proglottids are not detached and the eggs are normally continuously discharged from the proglottid through a uterine pore and not by apolysis. However, when egg production has ceased, chains of exhausted proglottids may be detached from the strobila (*pseudoapolysis*).

The male reproductive organs usually develop first (protandry or androgyny) and there may be one or, more usually, a large number of testes. These discharge into the vasa efferentia which join and form the vas deferens. This ends in the cirrus surrounded by the cirrus sac. These are usually well developed. The male and female genital pores usually lie close together in the genital sinus on the lateral or ventral surface of the proglottid. Self-fertilization or cross-fertilization between proglottids may occur.

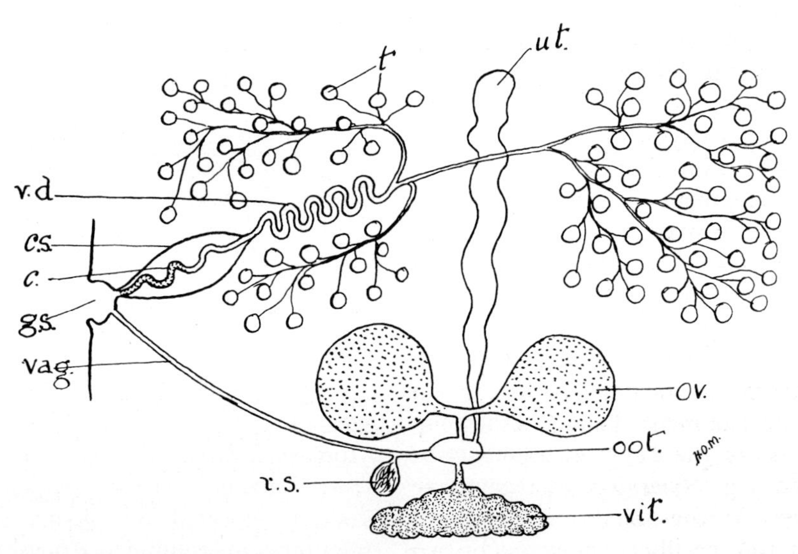

Fig 1.34 Reproductive organs of a cestode.

c	cirrus	t	testes
cs	cirrus-sac	ut	uterus
gs	genital sinus	vag	vagina
oot	ootype	vd	vas deferens
ov	ovary	vit	vitellarium
rs	receptaculum seminis		

The female genital pore leads into the tubular vagina which may bear a seminal vesicle. The vagina ends where it is joined by the oviduct and the vitelline duct, in the oötype, which is surrounded by Mehlis's glands. The ovary is usually bilobed. The vitellaria are usually compact (Eucestoda) or may be follicular (Cotyloda). From the oötype the uterus arises. It may open to the exterior and discharge the eggs (Cotyloda) or it may end blindly and assume various shapes as it fills with eggs (Eucestoda); the uterus may persist as a tube or network of tubes; it may develop diverticula and become sac-like; or it may develop extensive side branching (e.g. *Taenia*). Alternatively, the uterus may degenerate and the eggs, either singly or in groups, pass into protective structures within the proglottids. These include hyaline *egg capsules*, which enclose and protect groups of eggs and are formed from the uterus itself before it degenerates (i.e. *D. caninum*), or *par-uterine organs* which are denser areas of fibromuscular tissue of the parenchyma and are dilatations of the uterus which persist after the uterus has disappeared. Eggs pass from the uterus to the par-uterine organ, which assumes the function of a uterus.

Eggs may be embryonated or unembryonated when passed from the host and the fully embryonated egg contains an *oncosphere*. The oncosphere is bilaterally symmetrical, spherical or oval and is armed with three pairs of hooks. Oncospheral development *in utero* is closely associated with the formation of the embryonic envelopes. Embryonic development of oncospheres and formation of the egg has been reviewed by Rybicka (1966).

Four embryonic envelopes (capsule, outer envelope, inner envelope, oncospheral membrane) are seen during development. The *capsule* (sometimes referred to as the 'egg shell') may be poorly developed (e.g. *Hymenolepis*) or generally absent from eggs in the faeces (e.g. *Taenia*). Alternatively, it may be thick, sclerotinized and operculate (e.g. Diphyllidea) or thin and non-operculate (e.g. Proteocephalidea). The *outer envelope* is formed from two to eight macromeres which surround the oncosphere with cytoplasm. The *embryophore* is the principal and most resistant covering of many oncospheres and it arises from the *inner envelope*. The embryophore occurs in different forms: in parasites such as the Pseudophyllidea and Diphyllidea the embryophore is cellular and bears cilia. The developmental stage which comprises the ciliated embryophore containing the oncosphere is called the *coracidium*. In the Taeniidea, the embryophore is thick and striated, being composed of secreted blocks of a keratin-type protein (Morseth 1966). The embryophore is seen as a thin cellular membrane in the Mesocestoididea. The *oncospheral membrane* is a thin membrane lying beneath the embryophore and surrounding the oncosphere.

Hatching

The process of hatching in cestode eggs describes different events depending on the egg type, since different modes of embryonation have resulted in different egg structures. For instance, in the Diphyllidea and Pseudophyllidea (e.g. *D. latum*) the oncosphere covered by its ciliated embryophore hatches from the outer capsule (shell) under the influence of environmental stimuli such as light. The coracidium is then ingested by the intermediate host and the oncosphere is liberated from the cellular embryophore within the intermediate host, probably under the influence of digestive juices.

In other cestodes (i.e. Taeniidea) the outer capsule is usually absent and hatching and activation of the oncosphere occur in the intestine of the intermediate host and refer primarily to the liberation of the oncosphere from the embryophore. Hatching may be by a combination of mechanical action by the host and the host's digestive enzymes (e.g. *Anoplocephala*) or hatching may be brought about largely by the action of digestive enzymes (e.g. *Taenia*). Activation is a process of stimulation, under physiological conditions, of the hatched oncosphere which then activates and emerges from its oncospheral membrane. The activated oncosphere penetrates the intestinal mucosa and migrates to its site of predilection via the circulatory system (Heath 1971).

Metacestodes. Cestode life-cycles are indirect (except *H. nana*) and require the development of metacestodes (larval stages) in one or more intermediate hosts. Development of the metacestode is reviewed by Šlais (1973).

The common forms of metacestodes which occur in the life-cycles of cestodes of domestic animals and man can be classified as follows:

Procercoid: This is the first metacestode stage in the life-cycle of parasites such as the Pseudophyllidea and Diphyllidea. The procercoid is solid bodied and bears hooks on the cercomer in the posterior region.

Plerocercoid: This follows the procercoid and occurs in the second intermediate host. Plerocercoids are elongate, solid bodied metacestodes which bear an adult scolex (e.g. *D. latum*). Some plerocercoids (e.g. *Schistocephalus*) may show advanced development of the genitalia.

Tetrathyridium: An elongate, solid-bodied metacestode with a deeply invaginated acetabular scolex (e.g. *Mesocestoides*).

Cystercoid: A metacestode with a single non-invaginated scolex withdrawn into a small vesicle with practically no cavity (e.g. *D. caninum*).

Cysticercus: A single scolex invaginated into itself in a large fluid-containing vesicle or bladder (e.g. *T. saginata*).

Strobilocercus: A single scolex, which is not invaginated when fully developed, and is attached to the bladder by a long, segmented strobila (e.g. *T. taeniaeformis*).

Coenurus: A large fluid-containing bladder with a number of invaginated scolices attached to the wall (e.g. *T. multiceps*).

Hydatid: A large fluid-containing bladder which develops other cysts called brood capsules in which the scolices develop (e.g. *Echinococcus*).

The metacestode is passively transferred to the definitive host when the latter ingests the infected intermediate host. The scolex excysts or evaginates, as required, and attaches to the mucosa of the intestine.

Some successes in the *in vitro* culture of cestodes have been reported. *H. nana* has been cultured through its entire life-cycle (Berntzen 1970) while *S. mansonoides* has been cultured through most of its life-cycle (Berntzen & Mueller 1972). Much success has been reported with parasites such as *Schistocephalus* and *Ligula* (see Smyth 1976). In the Taeniidea, a variety of metacestodes have been successfully cultured *in vitro* (Heath 1973) and successful culture of monozoic, but sexually mature, *E. multilocularis* and sexually mature, strobilated *E. granulosus* (sheep strain) has been reported (Smyth & Davies 1974).

The classification of Wardle et al. (1974) is used in the present edition. It differs somewhat from that used in the sixth edition, where two orders, Cyclophyllidea and Pseudophyllidea, were recognized. The present classification is based on two classes, Eucestoda (the true tapeworms, possessing a scolex with four armed or unarmed suckers, segmentation of the body and with a life-cycle including a non-operculate egg) and Cotyloda (the pseudo-tapeworms possessing a scolex with two longitudinal slits and with a life-cycle including an operculated egg). A summary of the orders of interest in these classes is as follows:

CLASS: EUCESTODA Southwell, 1930

ORDER: ANOPLOCEPHALIDEA Wardle, McLeod and Radinovsky, 1974

Family: ANOPLOCEPHALIDAE Blanchard, 1981

Genus: *Anoplocephala* Blanchard, 1848
Paranoplocephala Lühe, 1910
Moniezia Blanchard, 1891
Cittotaenia Riehm, 1881
Bertiella Stiles and Hassal, 1902

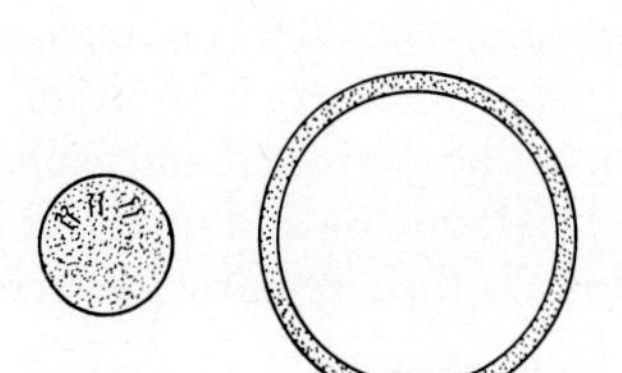

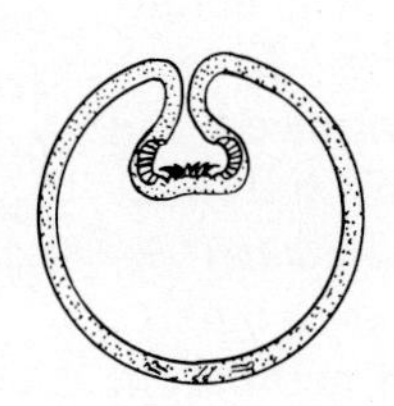

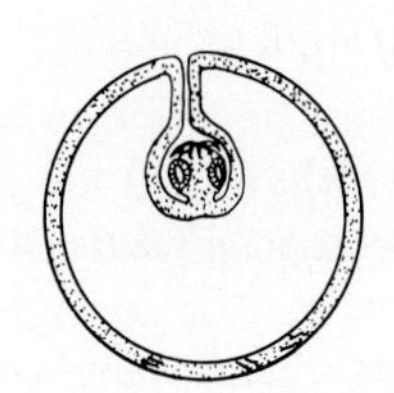

Fig 1.35 Stages in the development of a cysticercus.

Family: THYSANOSOMIDAE Fuhrmann, 1907
Genus: *Avitellina* Gough, 1911
Stilesia Railliet, 1893
Thysanosoma Diesing, 1835
Thysaniezia Skrjabin, 1926

Family: LINSTOWIIDAE Fuhrmann, 1932
Genus: *Inermicapsifer* Janicki, 1910
Atriotaenia Sandground, 1926

ORDER: DAVAINEIDEA Wardle, McLeod and Radinovsky, 1974
Family: DAVAINEIDAE Fuhrmann, 1907
Genus: *Davainea* Blanchard, 1891
Raillietina Fuhrmann, 1920
Cotugnia Diamare, 1893
Houttuynia Fuhrmann, 1920

ORDER: DILEPIDIDEA Wardle, McLeod and Radinovsky, 1974
Family: DILEPIDIDAE Railliet and Henry, 1909
Genus: *Amoebotaenia* Cohn, 1900

Family: DIPYLIDIIDAE Wardle, McLeod and Radinovsky, 1974
Genus: *Choanotaenia* Railliet, 1896
Dipylidium Leuckart, 1863
Diplopylidium Beddard, 1913
Joyeuxiella Fuhrmann, 1935
Metroliasthenes Ransom, 1900

ORDER: HYMENOLEPIDIDEA Wardle, McLeod and Radinovsky, 1974
Family: HYMENOLEPIDIDAE Railliet and Henry, 1909
Genus: *Hymenolepis* Weinland, 1858
Diorchis Clerc, 1903

Family: FIMBRIARIIDAE Wolffhügel, 1898
Genus: *Fimbriaria* Fröhlich, 1802

ORDER: TAENIIDEA Wardle, McLeod and Radinovsky, 1974
Family: TAENIIDAE Ludwig, 1886
Genus: *Taenia* Linnaeus, 1758
Echinococcus Rudolphi, 1801

ORDER: MESOCESTOIDIDEA Wardle, McLeod and Radinovsky, 1974
Family: MESOCESTOIDIDAE Perrier, 1897
Genus: *Mesocestoides* Valliant, 1863
Mesogyna Voge, 1952

ORDER: PROTEOCEPHALIDEA Mola, 1928
Family: PROTEOCEPHALIDAE La Rue, 1911
Genus: *Proteocephalus* Weinland, 1958

CLASS: COTYLODA Wardle, McLeod and Radinovsky, 1974

ORDER: DIPHYLLIDEA Wardle, McLeod and Radinovsky, 1974
Family: DIPHYLLOBOTHRIIDAE Lühe, 1910
Genus: *Diphyllobothrium* Lühe, 1910
Diplogonoporus Loennberg, 1892
Spirometra Mueller, 1937
Ligula Bloch, 1782
Schistocephalus Creplin, 1829

ORDER: PSEUDOPHYLLIDEA Carus, 1863
Family: TRIAENOPHORIDAE Loennberg, 1889
Genus: *Triaenophorus* Rudolphi, 1793

Family: AMPHICOTYLIDAE Ariola, 1899
Genus: *Eubothrium* Nybelin, 1922

ORDER: SPATHEBOTHRIDEA Wardle and McLeod, 1952
Family: CYATHOCEPHALIDAE Nybelin, 1868
Genus: *Cyathocephalus* Kessler, 1868

ORDER: CARYOPHYLLIDEA Beneden (in Olsson, 1893)
Family: CARYOPHYLLAEIDAE Leuckart, 1878
Genus: *Caryophyllaeus* Mueller, 1787

Family: LYTOCESTIDAE Hunter, 1927
Genus: *Khawia* Hsii, 1935

CLASS: EUCESTODA SOUTHWELL, 1930

ORDER: ANOPLOCEPHALIDEA WARDLE, McLEOD AND RADINOVSKY, 1974

The worms of this order have neither rostellum nor hooks. The proglottids are usually wider than long and each has one or two sets of genital organs. The genital pores are marginal. The transverse uterus may persist or be replaced by egg capsules or one or more par-uterine organs. Each egg has three coverings, an outermost vitelline membrane, a middle albuminous coat, and an innermost chitinous membrane, which is frequently pear-shaped, bearing on one side a pair of hooked projections (the *pyriform apparatus*).

FAMILY: ANOPLOCEPHALIDAE, BLANCHARD, 1891

In this family the uterus persists as a transverse tube or network of tubes. The intermediate hosts are mites of the family *Oribatidae.*

Genus: Anoplocephala E. Blanchard, 1848

Anoplocephala perfoliata (Goeze, 1782) occurs in the large and small intestine of horses and donkeys. It is cosmopolitan in distribution and usually it is the most frequently seen species of tapeworm in horses. Adults are up to 5 cm (or 8 cm) in length and 1.2 cm in breadth. The scolex is 2–3 mm in diameter and is provided with a 'lappet' behind each sucker. The proglottids are wider than long, each containing a single set of reproductive organs. The gravid uterus is transverse, large, sac-like and lobed. The eggs measure 65–80 μm in diameter.

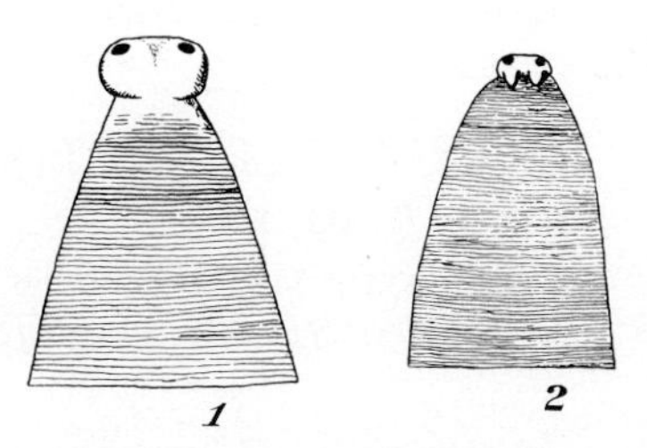

Fig 1.36 1, Anterior end of *Anoplocephala magna.* 2, Anterior end of *A. perfoliata.*

Anoplocephala magna (Abildgaard, 1789) occurs in the small intestine, particularly jejunum and rarely stomach, of horses and donkeys. It may reach 80 cm in length and 2.5 cm in breadth. The scolex is larger, 4–6 mm wide, and without 'lappets'. Eggs measure 50–60 μm in diameter.

Genus: Paranoplocephala Lühe, 1910

Paranoplocephala mamillana (Mehlis, 1831) occurs in the small intestine and occasionally the stomach of the horse. It measures only 6–50 by 4–6 mm. The scolex is narrow, the openings of the suckers are slit-like and situated dorsally and ventrally. Lappets are absent. The eggs measure about 51 by 37 μm.

Life-cycles of Anoplocephala spp.

Oribatid mites serve as intermediate hosts. For *A. perfoliata, Scheloribates laevigatus, S. latipes, Galumna nervosus, Achiperia* spp. and *Ceratozetes* spp. have been experimentally infected; for *A. magna, S. laevigatus* and *S. latipes* are concerned and for *P. mamillana, G. obvious* and *Allogalumna longipluma* are concerned (Sengbusch 1977). Cysticercoids are produced in these mites two to four months after infection. Adult tapeworms are found four to six weeks after ingestion of infected mites with herbage.

Pathogenesis

Light infections in horses produce no clinical signs, but very large numbers may cause ill health, unthriftiness and even death. *A. perfoliata* usually localizes near the ileocaecal orifice and small, dark, depressed, ulcerative lesions may be seen where the scolices are attached to the caecal wall. There may be oedema and occasionally excessive granulation tissue. Rarely, partial occlusion of the ileocaecal orifice may occur. *A. magna*, in very large numbers, can cause a catarrhal or haemorrhagic enteritis. Perforation of the intestine has been recorded in infections with *A. perfoliata* and *A. magna*. *P. mamillana* is seldom responsible for ill health.

Treatment

Micronized mebendazole (15–20 mg/kg) is 96–99% effective against *A. perfoliata* (Kelly & Bain 1975). Bithionol is effective at 7 mg/kg and niclosamide at 88 mg/kg. Safaev (1972) used phenasal (niclosamide) at 200–300 mg/kg. Kamala and dichlorophen have been used.

Genus: Moniezia R. Blanchard, 1891

Moniezia expansa (Rudolphi, 1810) (Fig. 1.37) occurs in the small intestine of sheep, goat, cattle and several other ruminants in most parts of the world. It may reach a length of 600 cm and a width of 1.6 cm. The scolex is 0.36–0.8 mm wide, with prominent suckers. There are neither rostellum nor hooks. The segments are broader than long and each contains two sets of genital organs with marginal genital pores. The ovaries and the vitelline glands form a ring on either side, median to the longitudinal excretory canals, while the testes are distributed throughout the central field. At the posterior border of each proglottid there is a row of rosette-like interproglottidal glands which extend almost across the width of the proglottid. The uterus becomes sac-like when filled with eggs. The eggs are somewhat triangular in shape, containing a well-developed pyriform apparatus, and measure 56–67 μm in diameter.

Moniezia benedeni (Moniez, 1879) (Fig. 1.37) occurs in ruminants, chiefly cattle, and differs from *M. expansa* in being broader (up to 2.6 cm) and in having the interproglottidal glands arranged in a short, continuous row close to the mid-line of the segment. The eggs measure up to 75 μm in diameter.

Life-cycle. Proglottids and eggs are passed in the faeces of infected animals. These proglottids may be eaten by birds which therefore may disseminate the infection. Cysticercoids develop in oribatid mites of the genera *Galumna*, *Oribatula*, *Peloribates*, *Protoscheloribates*, *Scheloribates*,

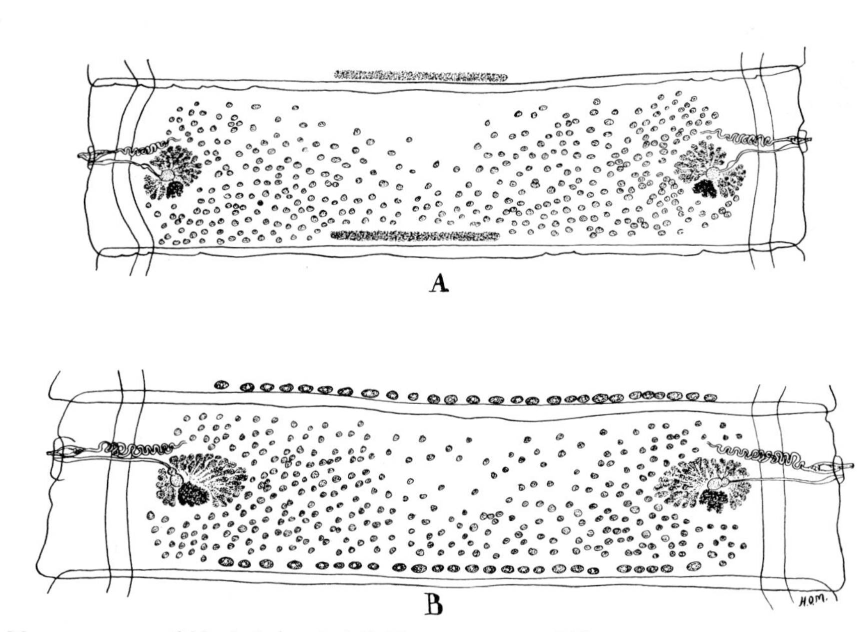

Fig 1.37 A, Mature segment of *Moniezia benedeni*. B, Mature segment of *M. expansa*.

Scutovertex and *Zygoribatula* and others (Sengbusch 1977). Infective stages are produced in approximately four months. Ruminants are infected by the ingestion of infected mites with herbage and the prepatent period is 37–40 days. There is a marked seasonal occurrence of *Moniezia* infection due to mites overwintered on pasture. The parasites are prevalent in young lambs and calves during their first summer on pasture. Lambs become infected very early in life and may pass ripe proglottids when they are six weeks old. Infection is not very common in older animals and, in these, the infections are usually light.

Pathogenesis. As a rule only lambs, kids and calves under six months of age are substantially infected. A wide divergence of opinion exists regarding the pathogenic effects of the *Moniezia* spp. in sheep and cattle. There is little doubt that light infections are of little importance. Some authors, especially from the USSR, ascribe severe pathogenic effects to the tapeworms but workers in the USA (Kates & Goldberg 1951) have failed to detect any serious effect, even from apparently substantial burdens. In heavy infections the intestine may be virtually a solid mass of tapeworms and they can cause diarrhoea and unthriftiness. Obstruction of the intestine has been recorded. In the USSR *Moniezia* spp. infections are considered to be highly pathogenic in lambs. There they cause depressed wool and meat production and are associated with many deaths. A high incidence of enterotoxaemia has been associated with *Moniezia* infections in lambs in the USSR (Vibe 1976). Because of the large size of the cestodes their presence is obvious and frequently the true underlying cause of the parasitism, small trichostrongyles, is overlooked.

Diagnosis. The presence in the faeces of ripe segments, which resemble cooked rice grains and from which *Moniezia* eggs can be identified, indicates the presence of tapeworms.

Treatment. Copper sulphate in combination with phenothiazine or copper sulphate/nicotine sulphate/phenothiazine mixtures have been used for many years and are still used extensively for the treatment of tapeworms and nematodes of sheep. Also, copper sulphate, phenothiazine and salt at a ratio of 1:10:100 available daily during the grazing season may be used prophylactically (Mishareva 1977). Recently, a wide variety of anthelmintics has been shown to be efficacious in the treatment of *Moniezia*. Albendazole (10 mg/kg), fenbendazole (5 mg/kg), cambendazole (20 mg/kg), oxfendazole (5 mg/kg) and praziquantel (15 mg/kg) are effective. Others include bunamidine hydroxynapthoate (25–50 mg/kg); dichlorophen (100 mg/kg); resorantel (65 mg/kg); bithionol (200 mg/kg); and niclosamide (phenasal) (75–150 mg/kg).

Prophylaxis. Animals may be treated in late spring/early summer and again, if required, in the autumn. The level of infected mites on pasture can be controlled by ploughing and reseeding the pastures or by the use of pastures which have not been grazed the previous year.

Genus: Cittotaenia Riehm, 1881

Cittotaenia ctenoides (Railliet, 1890) occurs in the small intestine of the rabbit in Europe. It may grow up to 80 cm long and 1 cm wide. The scolex is about 0.5 mm broad. A short neck is present. The proglottids are all much broader than long and each contains two sets of posteriorly placed genital organs. The legs have a pyriform apparatus and measure about 64 μm in diameter.

Cittotaenia denticulata (Rudolphi, 1804) occurs in the rabbit in Europe. It has no neck and the scolex measures 0.8 mm in diameter.

Cittotaenia pectinata (Goeze, 1782) occurs in hares and rabbits in Europe, Asia and America. Its scolex is about 0.25 mm in diameter and a neck is present.

Life-cycle. The intermediate hosts are the oribatid mites of the genera *Achipteria, Allopelops, Cepheus, Galumna, Scheloribates, Scutovertex, Trichoribates, Xenillus* etc. (Sengbusch 1977).

Pathogenesis. Heavy infections of these tapeworms, especially *C. ctenoides*, frequently cause digestive disturbances, emaciation and even death amongst rabbits.

Treatment. Praziquantel and the benz-imidazole anthelmintics used for the treatment of *Moniezia* should be effective, but there is no detailed information available on their efficacy.

Genus: Bertiella Stiles and Hassal, 1902

Bertiella studeri (Blanchard, 1891) is a common parasite of the small intestine of various primates (rhesus and cynomolgus monkeys, Japanese macaque, mandrills, baboons, gibbons, chimpanzees) and occasionally man in Africa, Asia, Indonesia, the Philippines and Mauritius.

Bertiella mucronata (Meyner, 1895) is a parasite of monkeys in South America. Adults reach 15–45 cm in length. Each proglottid contains a single set of genital organs. Genital pores alternate irregularly. The uterus is a transverse tube in the gravid proglottid. Eggs measure 36–60 μm and have a well developed pyriform apparatus. The intermediate hosts are oribatid mites of the genera *Achipteria, Galumna, Scheloribates* and *Scutovertex* (Sengbusch 1977) and final hosts become infected by accidental ingestion of mites on vegetation.

These parasites cause no apparent signs of disease or lesions. Diagnosis is made by finding the proglottids or characteristic eggs in the faeces.

Treatment. Drugs active against tapeworms of dogs and cats are indicated but additionally those effective against the anoplocephalides of ruminants are likely to be of value. Since transmission requires free-living mites, it is unlikely that this parasite will constitute a problem in primate colonies.

FAMILY: THYSANOSOMIDAE FUHRMANN, 1907

In this family the gravid uterus disappears and is replaced by par-uterine organs or capsules.

Genus: Avitellina Gough, 1911

Avitellina centripunctata (Rivolta, 1874) occurs in the small intestine of sheep and other ruminants in Europe, Africa and Asia. **A. chalmersi** Woodland, 1927, and **A. goughi**, Woodland, 1927, are found primarily in sheep in Africa and Asia; **A. tatia** Bahlerao, 1936, is found in goats in India. Adults reach 3 m in

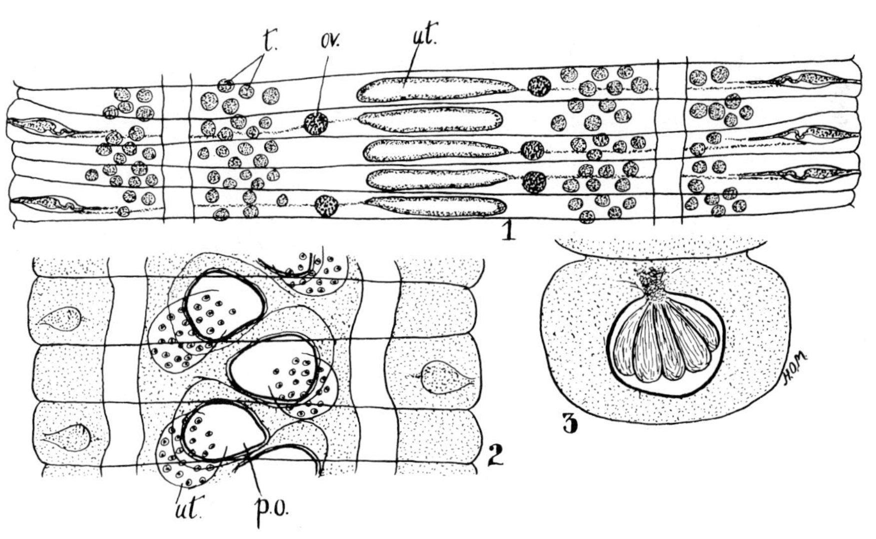

Fig 1.38 *Avitellina centripunctata.* 1, Mature proglottides: 2, Gravid proglottides, eggs passing into paruterine organs; 3, Ripe proglottis, par-uterine organ with eggs in bunches.
ov ovary
po par-uterine organ
t testes
ut uterus

length and are about 3 mm wide. They are almost cylindrical at the posterior end. Proglottids are short and not distinctly segmented. The genitalia are single and the pores alternate irregularly. The uterus lies transversely in the middle portion of the proglottid and eggs then pass into a thick-walled par-uterine organ in which the eggs lie in capsules. The uterus and par-uterine organs show as an opaque line in the medial portion of the proglottids. The eggs, which have no pyriform apparatus, measure 21–45 μm.

Life-cycle. Psocids (bark lice, dustlice, book lice) may serve as intermediate hosts.

Pathogenesis. Probably similar to that of the *Moniezia* spp., but this genus is not considered to be as pathogenic as *Moniezia* in the USSR.

Treatment. Bithionol, praziquantel, cambendazole and phenasal (niclosamide) are available for use. Few studies have been done on the treatment of *Avitellina* spp. infection but anthelmintics effective against *Moniezia* spp. would probably be useful.

Genus: Stilesia Railliet, 1893

Stilesia hepatica Wolffhügel, 1903 occurs in the bile ducts of sheep, goats, cattle and wild ruminants in Africa and Asia. It is an extremely common infection and can be found in 90–100% of sheep in many areas of Africa. Adults are 20–50 cm long and up to 3 mm wide. The scolex has prominent suckers and there is a broad neck. The proglottids are short. The genital organs are single and the genital pores alternate irregularly. The uterus is long, transverse and dumbell-shaped and the eggs pass into two par-uterine organs, each containing about 30 eggs. The ovoid eggs have no pyriform apparatus. They measure 26 by 16–19 μm. The life-cycle is not known, but probably involves oribatid mites.

Pathogenesis. S. hepatica occurs in animals of all ages. It is practically non-pathogenic; extremely heavy infections are often seen in apparently perfectly healthy sheep. Although the bile ducts may be almost occluded, or even form sac-like dilatations filled with worms, no icterus or other clinical signs are seen. In affected livers there may be slight cirrhosis and the walls of the bile duct are usually thickened. Such livers are condemned at meat inspection and, in some parts of the world where *S. hepatica* is present, all sheep livers are automatically condemned since such a high proportion are infected with the parasite.

Treatment. Praziquantel, given at 15 mg/kg, has been used to treat infections with *S. hepatica* (Dey-Hazra 1976).

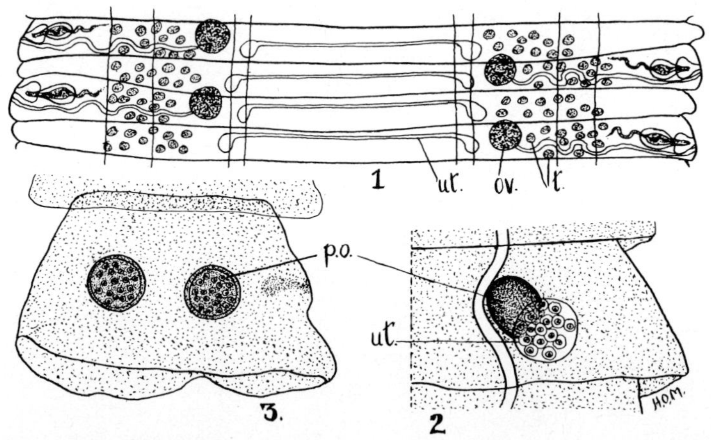

Fig 1.39 *Stilesia hepatica.* 1, Mature proglottides. 2, Half of gravid proglottis, eggs passing into paruterine organ. 3, 'Ripe' proglottis, eggs in par-uterine organs.
ov ovary
po par-uterine organ
t testes
ut uterus

Stilesia globipunctata (Rivolta, 1874) occurs in the small intestine of the sheep and goat as well as cattle and other ruminants in Europe, Africa and Asia. Adults are 45–60 cm long and up to 2.5 mm wide.

Life-cycle. Oribatid mites have been incriminated in the life-cycle of *S. globipunctata* in Chad and India. Psocids may well be involved.

Pathogenesis. S. globipunctata attaches primarily at the junction of the duodenum and jejunum. The scolex of the immature forms penetrates the mucous membrane and there is nodule formation, proliferative inflammation, cell infiltration and epithelial desquamation (Arnjadi 1971). The scolex and anterior proglottids are embedded in the nodule while the posterior proglottids are free in the lumen. Infections can result in death.

Treatment. Praziquantel (15 mg/kg) appears to be highly effective. Other anthelmintics tested by Bankov (1976) were not as effective in removing the scolices.

Genus: Thysanosoma Diesing, 1835

Thysanosoma actinioides Diesing, 1835, the 'fringed tapeworm', occurs in the bile ducts, pancreatic ducts and small intestine of sheep, cattle and deer in America, especially the western parts of the USA and also in South America, but it does not occur outside the western hemisphere. It measures 15–30 cm by 8 mm. The scolex is up to 1.5 mm wide. The segments are short and conspicuously fringed posteriorly. Each segment contains two sets of genital organs and the testes lie in the median field. The uterus is a single undulating tube and the eggs then pass into several par-uterine organs, each usually containing six to 12 eggs. The eggs have no pyriform apparatus.

Life-cycle. Work by Allen (1959) demonstrated that cysticercoids could be recovered from laboratory reared psocids which had been fed egg capsules containing oncospheres. As yet it has not been possible to induce infections in cattle or sheep with infected psocids.

Pathogenesis. The pathogenicity of this parasite has probably been overestimated. The symptoms of selenium (loco) poisoning and other diseases have been ascribed to it in the past (Christenson 1931). It may partly obstruct the flow of bile and pancreatic juice and cause digestive disorders and unthriftiness. The presence of infection causes condemnation of livers at meat inspection; 46% of sheep in Texas have been recorded as being infected.

Treatment. Bithionol (200 mg/kg), niclosamide (400–600 mg/kg) or cambendazole (100 mg/kg) may be used. Praziquantel is likely to be effective.

Prophylaxis. Similar to that for the *Moniezia* spp.

The pathology, pathogenicity and control of *T. actinioides* has been reviewed by Allen (1973).

Genus: Thysaniezia Skrjabin, 1926 (syn. Helictometra)

Thysaniezia giardi (Moniez, 1879) occurs in the small intestine of sheep, goat and cattle in Europe, USSR, Africa and America. It grows to

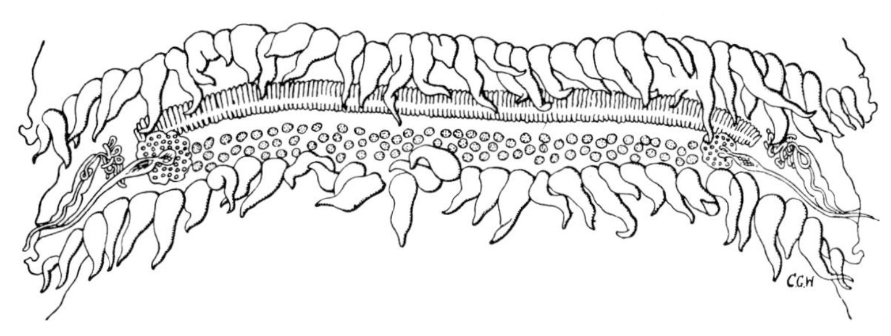

Fig 1.40 Mature proglottis of *Thysanosoma actinioides*. (*After Fuhrmann in Kükenthal*)

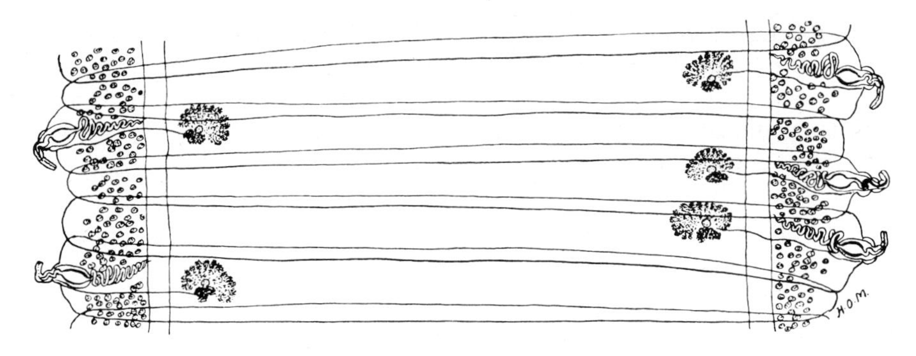

Fig 1.41 Mature proglottides of *Thysaniezia giardi*.

about 200 cm long and 12 mm wide, the width varying greatly in different specimens. The segments are short and each contains a single set of genital organs, very rarely two, the pores alternating irregularly. The testes are lateral to the excretory canals. The side of the segment which contains the cirrus-sac bulges out, thus giving the margin of the worm an irregular appearance. The eggs, which are devoid of a pyriform apparatus, pass from the uterus into a large number of small par-uterine organs.

Life-cycle. Psocids have been infected experimentally (see Svadzhian 1963) but these do not induce infections in sheep. Oribatid mites have been incriminated as intermediate hosts but this now seems unlikely.

Pathogenesis. This worm occurs in young and adult animals, but is rarely seen in numbers sufficiently large to produce clinical signs. It is the most frequent tapeworm of adult cattle in the Republic of South Africa.

Treatment. Bithionol at 125–150 mg/kg is effective. Anthelmintics which are used to treat *Moniezia* are likely to be effective against *Thysaniezia*.

FAMILY: LINSTOWIIDAE FUHRMANN, 1932

In this family the gravid uterus is replaced by egg capsules.

Genus: Inermicapsifer Janicki, 1910

Species in this genus are found in rodents and hyracoids primarily in Africa. **I. madagascariensis** (Davaine, 1870) and **I. cubensis** (Kouré, 1938) have been recorded in humans but no clinical signs have been associated with the infection. Niclosamide and mebendazole are effective anthelmintics (Horstmann et al. 1978).

Genus: Atriotaenia Sandground, 1926

Proglottids wider than long, craspedote, genital pores irregularly alternating. Genital atrium may form large sucker-like organ.

Atriotaenia procyonis Spassky, 1951, is a parasite of mustelids and procyonids in North America, South America, Asia and Europe. It is a common parasite of *P. lotor* in the USA. Beetles serve as intermediate hosts.

Other species in this family in procyonids include **Oschmarenia pedunculata** Chandler, 1952 in *Mephitis* sp. in the USA, **O. wallacei** (Chandler, 1952) from skunks in Minnesota and **O. oklahomensis** Spassky, 1951 from skunks and raccoons in North America.

ORDER: DAVAINEIDEA WARDLE, McLEOD AND RADINOVSKY, 1974

These are small to medium-sized cestodes. The rostellum is retractable and armed with numerous hammer-shaped hooks. The suckers are usually armed. Genital organs are usually single. The uterus may persist or the eggs may pass into egg capsules or into a par-uterine organ.

FAMILY: DAVAINEIDAE FUHRMANN, 1907

In this family the uterus is replaced by egg capsules. Adults are parasitic in mammals or birds.

Genus: Davainea Blanchard, 1891

Davainea proglottina (Davaine, 1860) occurs in the duodenum of the fowl, pigeon and other gallinaceous birds in most parts of the world. Adults are microscopic (0.5–3 mm) with only four to nine proglottids. The rostellum and suckers are armed. Genital pores alternate regularly. The eggs (28–40 μm in diameter) lie singly in parenchymatous capsules in the gravid proglottid.

Life-cycle. Gravid proglottids are found in the faeces primarily in the afternoon or night. The eggs hatch after being swallowed by various species of gastropod molluscs (e.g. *Limax*, *Arion*, *Cepoea*, *Agriolimax* spp.) and a cysticercoid develops in approximately three weeks. After ingestion of the infected mollusc the prepatent period in the fowl is about 14 days.

Pathogenesis and treatment. See p. 101.

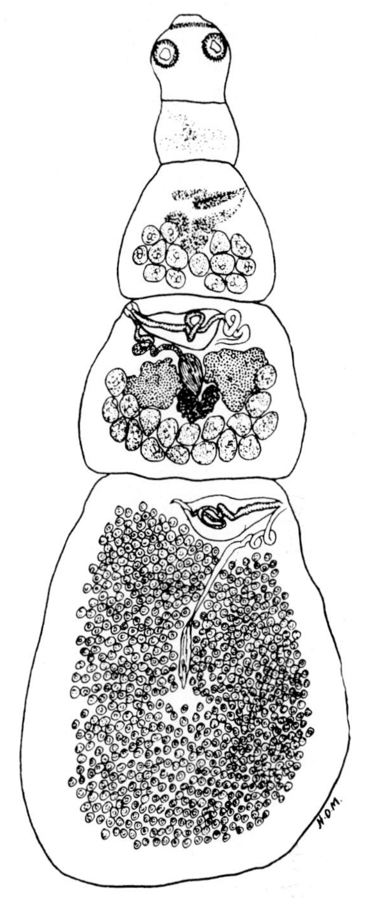

Fig 1.42 Complete specimen of *Davainea proglottina*.

Genus: Raillietina Fuhrmann, 1920

About 200 species are described in this genus and it has been subdivided into several subgenera. A large number of species are found in birds but many are not common.

Raillietina tetragona (Molin, 1858) occurs in the posterior half of the small intestine of the chicken, guinea fowl, pigeon and other birds. It is cosmopolitan in distribution. It is one of the

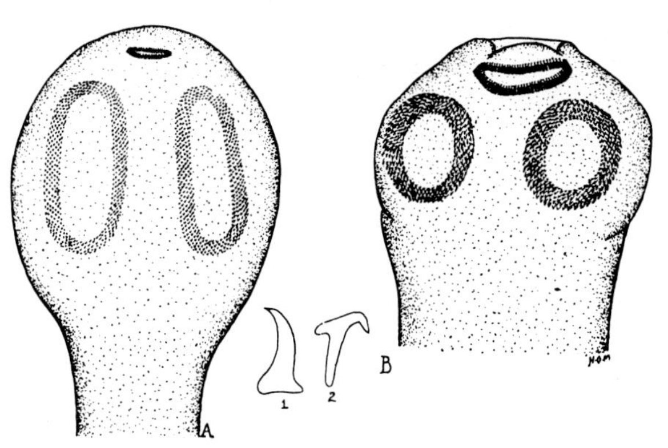

Fig 1.43 A, Scolex of *Raillietina tetragona*. B, Scolex of *R. echinobothrida*. 1, Hook from sucker. 2, Hook from rostellum of B.

largest of the fowl tapeworms and adults reach up to 25 cm in length.

The scolex is smaller than that of *R. echinobothridia*. The rostellum is armed with one or two rows of hooks and the suckers are oval and armed. The genital pores are usually unilateral and the eggs are found in egg capsules each containing six to 12 eggs. The eggs are 25–50 μm in diameter.

Life-cycle. Cysticercoids occur in ants of the genera *Pheidole* and *Tetramorium*. The prepatent period in the chicken is 13 days to three weeks.

Pathogenicity and treatment. See p. 101.

Raillietina echinobothridia (Megnin, 1880) also occurs in the small intestine of the chicken and turkey in most parts of the world. In shape and size it resembles *R. tetragona*. It is distinguished by a more heavily armed rostellum with two rows of hooks and the suckers are circular in outline. Gravid proglottids frequently separate in the middle forming small 'windows' in the strobila.

Life-cycle. The ants *Tetramorium caespitum* and *Pheidole vinelandica* serve as hosts for the cysticercoids in North America. In Europe the ants *T. caespitum*, *T. semilaeve* and *P. pallidula* have been incriminated as vectors. The prepatent period is 20 days.

Pathogenecity and treatment. See p. 101.

Raillietina cesticillus (Molin, 1858) is very common throughout the world in domestic poultry. It is 4 cm, rarely 15 cm, long. It has a large scolex with a wide rostellum armed with 400–500 small hooks. The suckers are inconspicuous and unarmed. The eggs, 75–88 μm in diameter, occur singly in egg capsules.

Life-cycle. The intermediate hosts are beetles of the genera *Calathus*, *Amara*, *Pterostichus*, *Bradycellus*, *Harpalus*, *Poecilus*, *Zabrus* (in Europe) and *Anisotarsus*, *Choeridium*, *Cratacanthus*, *Calathus*, *Stenolaphus*, *Stenocellus*, *Amara* and *Selenophorus* (in North America). Development in the fowl takes 20 days.

Pathogenicity and treatment. See p. 101.

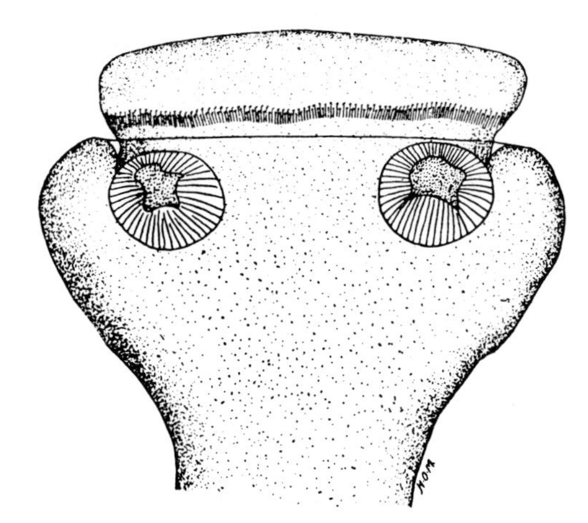

Fig 1.44 Scolex of *Raillietina cesticillus*.

Raillietina georgiensis (Reid and Nugara, 1961) occurs in the wild turkey and domestic turkey in the southern states of the USA. Chickens appear to be refractory to infection. It is a large, robust tapeworm, 15–36 cm in length. Rostellar hooks number 220–260 and are 17–23 μm in length. The intermediate host probably is the ant *Pheidole vinelandica* which has been found naturally infected with cysticercoids on poultry ranges.

Other species of the genus include: **R. williamsi** (Fuhrmann, 1932) of wild turkeys in the USA; **R. magninumida** (Jones, 1939) of guinea-fowl in the USA, cystercoids of the species developing in ground and dung beetles; and **R. ransomi** (Williams, 1931) of the domestic turkey in Florida and wild turkey elsewhere in the USA. A more detailed description of these species is given by Reid (1962).

Genus: Cotugnia Diamare, 1893

Parasites of this genus are found chiefly in tropical areas. They possess two sets of reproductive organs per proglottid. Eggs occur singly in egg capsules.

Cotugnia digonopora (Pasquale, 1890) occurs in the small intestine of the fowl in Europe, Africa and Asia. It is up to 107 mm long. Like the other species of this genus, it has two sets of genital organs in each segment. The rostellum bears two rows of small hooks and the suckers are unarmed.

The life-cycle is unknown. For pathogenicity and treatment, see below.

Cotugnia fastigata (Meggit, 1920) occurs in the duck in Burma and **C. cuneata** (Meggit, 1924) in the pigeon in Burma and India.

Genus: Houttuynia Fuhrmann, 1920

Houttuynia struthionis (Houttuyn, 1773) occurs in the small intestine of the ostrich and the South American rhea. The worms grow up to 60 cm long and 9 mm wide. The scolex is 1–2 mm wide and bears a double row of about 160 large and small hooks. The genital pores are unilateral. In the gravid segments the eggs are contained in parenchymatous capsules, about 15–25 in each.

The life-cycle is unknown.

The parasite is seen especially in ostrich chicks, causing unthriftiness, emaciation and sometimes diarrhoea. The adult birds are frequently carriers of the infection, but rarely show any symptoms.

ORDER: DILEPIDIDEA WARDLE, McLEOD AND RADINOVSKY, 1974

In this order the retractable rostellum is usually armed with one, two or more crowns of rose-thorn-shaped hooks. The suckers may be armed. Genital organs are single or double. The uterus may be sac-like or the eggs may pass into egg capsules or a par-uterine organ. Adults are parasitic in birds and mammals.

FAMILY: DILEPIDIDAE RAILLIET AND HENRY, 1909

In this family the gravid uterus persists as a transverse sac.

Genus: Amoebotaenia Cohn, 1900

Amoebotaenia cuneata (Linstow, 1872) (= *A. sphenoides*) occurs throughout the world in the small intestine of domestic fowl. It is small, up to 4 mm long, and roughly triangular in shape. The rostellum is armed. The genital pores usually alternate irregularly at the extreme anterior end of the proglottid margin. The uterus is sac-like and slightly lobed.

Life-cycle. The intermediate hosts are earthworms of the genera *Eisenia*, *Pheretina*, *Ocnerodrilus* and *Allolobophora*, in which the cysticercoid develops in about 14 days. Fowls acquire the infection frequently after rains when the earthworms come to the surface. The worms grow adult in the fowl in four weeks.

Pathogenesis and treatment. See below.

FAMILY: DIPYLIDIIDAE WARDLE, McLEOD AND RADINOVSKY, 1974

In this family the gravid uterus is replaced by egg capsules containing one or more eggs.

Genus: Choanotaenia Railliet, 1896

Choantaenia infundibulum (Bloch, 1779) occurs in the upper half of the small intestine of the fowl and turkey. It may reach 23 cm in length and the segments are markedly wider posteriorly than anteriorly, giving the worm a characteristic shape. The rostellum is armed with 16–20 slender hooks. Genital pores alternate regularly. The uterus is sac-like. However, the proglottids leave the body before they are completely gravid and there is doubt as to whether the uterus is replaced by egg capsules. The eggs have distinct elongate filaments.

Life-cycle. The intermediate hosts are the house fly, *Musca domestica*, and beetles of the genera *Geotrupes*, *Aphodius*, *Calathus* and *Tribolium*.

Pathogenesis and treatment. See below.

TAPEWORM INFECTIONS OF POULTRY

A large number of cestode species have been reported from domestic poultry and wild birds throughout the world and cestode infections are still common in poultry raised on range or in

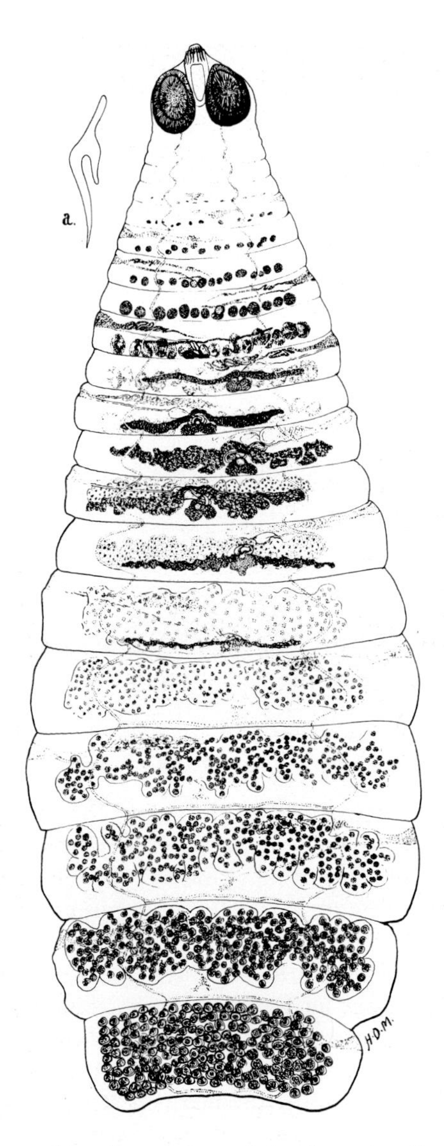

Fig 1.45 Complete specimen of *Amoebotaenia*. a, Rostellar hook, much enlarged.

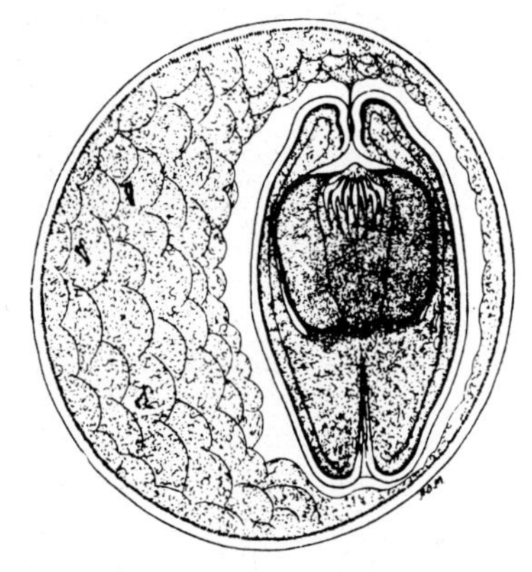

Fig 1.46 Mature cysticercoid of *Amoebotaenia cuneata*.

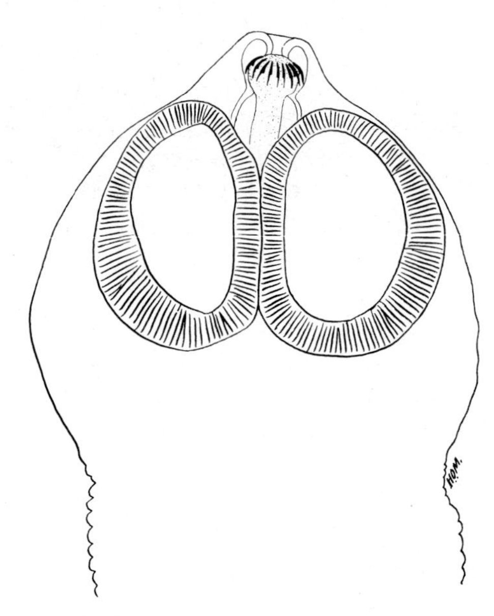

Fig 1.47 Scolex of *Choanotaenia infundibulum*.

backyard flocks. Where intensive management is practised the prevalence of tapeworms has decreased. However, under intensive management, cestodes, particularly *C. infundibulum*, for which flies are the intermediate hosts, may occur in poultry raised in open houses. Also, *R. cesticillus* may survive owing to beetles which breed in litter (Reid 1962).

Pathogenesis

D. proglottina is highly pathogenic and may occur in large numbers. The parasite penetrates deeply within the villi and in heavy infections cause necrosis and a haemorrhagic enteritis is seen. The infection may be fatal. Chronic infections are characterized by reduced growth rate, emaciation and weakness. Of the *Raillietina* species *R. echinobothridia* is the most pathogenic. Nodules may be formed at the site of attachment and a hyperplastic enteritis can occur. *R. tetragona* is less

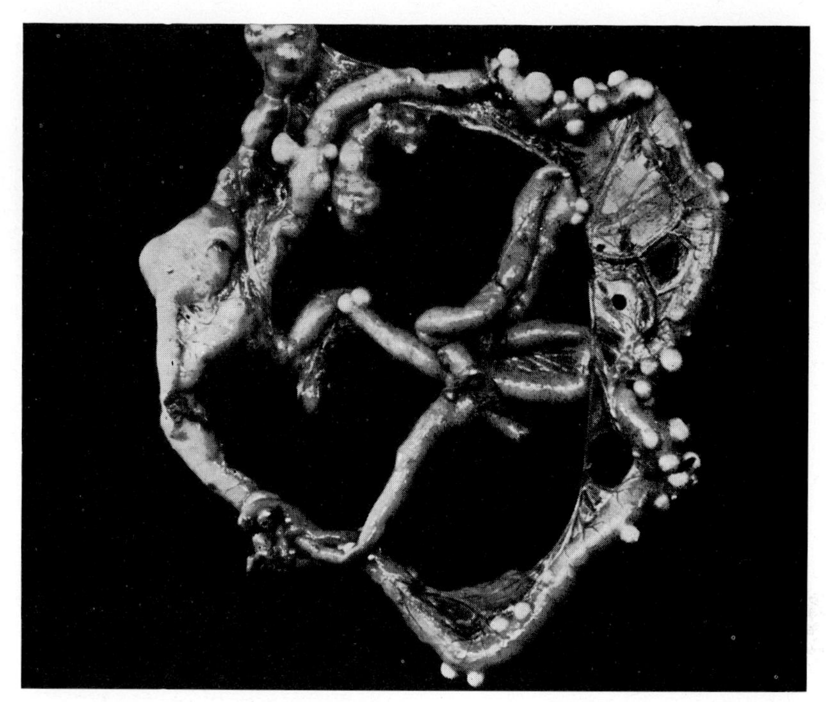

Fig 1.48 Nodular tapeworm disease of small intestine of fowl caused by *Raillietina echinobothrida*. (*K. D. Downham*)

pathogenic but may cause weight loss and decreased production. Other species are not normally harmful unless the infection is extremely heavy. However, infections could cause decreased production in intensively managed birds.

Diagnosis

Although gravid proglottids are passed in the faeces, diagnosis is usually made at necropsy. Representative members of the flock are examined. Since some cestodes are very small mucosal scrapings should be examined microscopically.

Treatment and prophylaxis

Treatment must be associated with control measures directed against the intermediate hosts. Butynorate (75–150 mg/kg) is widely used. Niclosamide and hexachlorophene are also effective. The latter can depress egg production. Praziquantel and the broad-spectrum benzimidazole anthelmintics such as albendazole and oxfendazole may be effective but have yet to be assessed.

Control measures should be directed against the intermediate hosts. Insecticides may be used for flies and ants and metaldehyde bait for slugs. Beetles, earthworms and crustacea are more difficult to control but alternation of ranges may be helpful.

Genus: Dipylidium Leuckart, 1863

Dipylidium caninum (Linnaeus, 1758) occurs in the small intestine of the dog, cat, fox and occasionally man, particularly children. It is the commonest tapeworm of the dog in most parts of the world and has a world-wide distribution. The parasite may be up to 50 cm long. The retractable rostellum has three or four rows of rose-thorn-shaped hooks. Each proglottid contains two sets of genital organs and the ovary and vitelline gland form a mass on either side resembling a bunch of grapes. Eggs lie in egg capsules, each containing up to 30 eggs. Mature and particularly gravid proglottids have a characteristic elongate, oval shape, resembling cucumber seeds.

Several other species of the genus exist. **D. sexcoronatum** von Ratz, 1900 occurs principally in the cat and appears to be a distinct

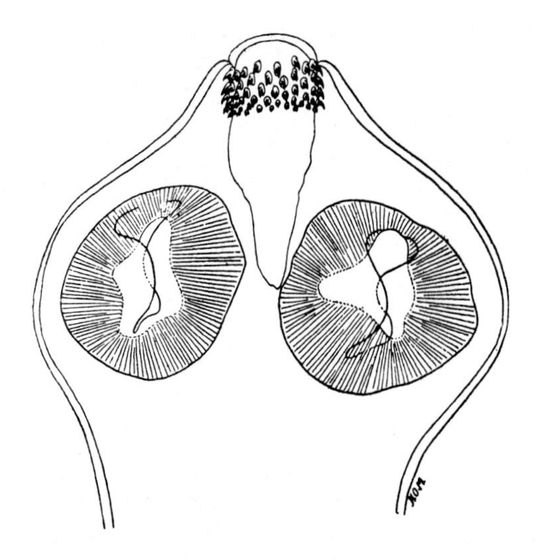

Fig 1.49 Scolex of *Dipylidium caninum*.

species while **D. gracile** Millzner, 1926, **D. compactum** Millzner, 1926, **D. diffusum** Millzner, 1926 and **D. buencaminoi** Tubangui, 1925 may be synonyms of *D. caninum*.

Life-cycle. The gravid proglottids are voided in the faeces or they may leave the host spontaneously and crawl about actively disseminating eggs. The intermediate hosts are fleas (*Ctenocephalides canis*, *C. felix*, *Pulex irritans*) and the dog louse, *Trichodectes canis*, has been incriminated. Larval fleas ingest eggs and the cysticercoids develop in the adult flea. The definitive hosts are infected by swallowing infected fleas. Human infections are probably due to the accidental ingestion of infected fleas when children play with dogs or cats.

Pathogenesis and treatment. See p. 124.

Parasites in the genera **Joyeuxiella** Fuhrmann, 1935, and **Diplopylidium** Beddard, 1913, occur in cats and dogs. Infection can be common in the Middle East, Africa and Australasia. In these genera the egg capsules contain only one egg. Two intermediate hosts are apparently required for the life-cycle. The first intermediate hosts are beetles and the second intermediate hosts are reptiles or small mammals.

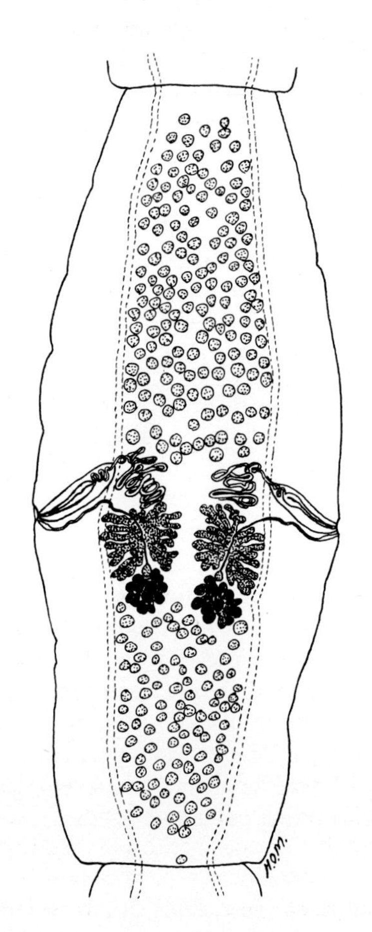

Fig 1.50 Mature proglottis of *Dipylidium caninum*.

Fig 1.51 Rostellum of *Joyeuxiella fuhrmanni*.

Genus: Metroliasthes Ransom, 1900

Metroliasthes lucida Ransom, 1900, is a rather rare parasite occurring in the small intestine of the fowl and turkey in North America, India and Africa. It is about 20 mm long and 1.5 mm wide. The scolex is devoid of a rostellum and hooks. The genital pores are single, irregularly alternating and often prominent. There are 30–40 testes in each proglottid. In the gravid segments the eggs pass into a large par-uterine organ. The intermediate hosts are grasshoppers of the genera *Chorthippus*, *Paroxya* and *Melanoplus*.

ORDER: HYMENOLEPIDIDEA WARDLE, McLEOD AND RADINOVSKY, 1974

Cestodes in this order are small to medium in size, with four unarmed suckers and a retractable rostellum bearing a single circle of hooks. There is a single set of reproductive organs per proglottid and the genital pores are unilateral. Each egg has three delicate, enclosing membranes. The intermediate hosts are arthropods in which cysticercoids develop and adults are found in birds and mammals.

FAMILY: HYMENOLEPIDIDAE RAILLIET AND HENRY, 1909

Genus: Hymenolepis Weinland, 1858

This genus contains a large number of species which occur chiefly in domestic animals and wild birds. The worms are usually narrow and thread-like in appearance and they are rather difficult to distinguish. Details may be found in Wardle et al. (1974).

Hymenolepis nana (von Siebold, 1852) (= *H. fraterna*). The dwarf tapeworm infects rodents, simian primates and man. It is the most common tapeworm of man in the tropics and subtropics and is common in wild and laboratory rodents. Adults are slender and 25–40 mm long. The egg is oval and 44–62 by 30–55 μm. The oncosphere possesses three pairs of hooks. In man the life-cycle is direct and cysticercoids develop in the villi of the small intestine before emerging to develop to maturity in the lumen. The prepatent period is 16 days. Autoinfection has also been recorded. In rodents the life-cycle can be direct or indirect with flour beetles or fleas as intermediate hosts.

Pathogenesis. Heavy infections in man may cause a variety of abdominal symptoms (anorexia, vomiting, diarrhoea, pain). Heavily infected mice may show retarded growth or weight loss.

Treatment and prophylaxis. Infections are easily spread and thus control is difficult. In laboratory rodent colonies repeated treatment and strict sanitary measures are effective. Caesarean-derivation and barrier-maintenance will eradicate the parasite. Niclosamide (100–200 mg/kg) has been effective in the rat and hamster, but regeneration after treatment has been recorded. Bunamidine hydrochloride (200 mg/kg), thiabendazole (300 mg/kg or 0.3% in the feed for 14 days) and mebendazole (30 mg/kg daily for three days) have been effective and other benzimidazoles are worthy of use. Uredofos and praziquantel (25 mg/kg) are effective.

In man, niclosamide and paromomycin given for five to seven days are 90% effective. Praziquantel has been shown to be effective.

Hymenolepis diminuta Rudolphi, 1819, is common in wild rodents and has been recorded in man. The adults are 2–6 cm long. Intermediate hosts are primarily fleas and flour beetles although numerous other species of insects have been recorded.

Hymenolepis microstoma Dujardin, 1845, is an uncommon parasite of rodents. Adult worms are found in the gall-bladder, bile ducts and duodenum. Praziquantel or bithionol may be effective anthelmintics for the treatment of *H. microstoma*.

Hymenolepis carioca de Magalhaes, 1898. This apparently non-pathogenic parasite is common in fowl in the USA. The adult is slender and up to 8 cm long. The intermediate hosts are dung beetles and sometimes flour beetles or *Stomoxys calcitrans.*

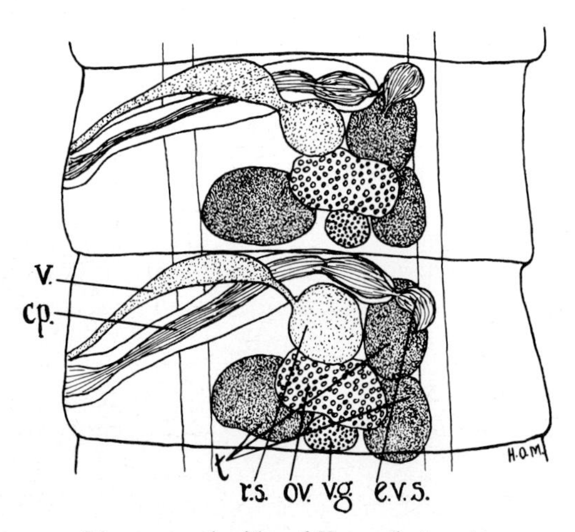

Fig 1.52 Mature proglottides of *Hymenolepis carioca*.
cp cirrhus pouch
evs external vesiculum seminalis
ov ovary
rs receptaculum seminis
t testes
v vagina
vg vitelline gland

Hymenolepis lanceolata Weinland, 1858 is a large species occurring in ducks and geese and is one of the most harmful parasites of this group. Intermediate hosts are aquatic crustaceans.

Hymenolepis cantaniana Polonio, 1860 is fairly common in the small intestine of the chicken and other birds in the USA, Europe and Africa.

Diorchis nyrocae Yamaguti, 1935 is a common cestode of ducks. The intermediate hosts are ostracod or copepod crustacea.

FAMILY: FIMBRIARIIDAE WOLFFHÜGEL, 1898

Genus: Fimbriaria Fröhlich, 1802

Fimbriaria fasciolaris (Pallas, 1781) is an uncommon parasite. It is found throughout the world in the small intestine of the chicken and

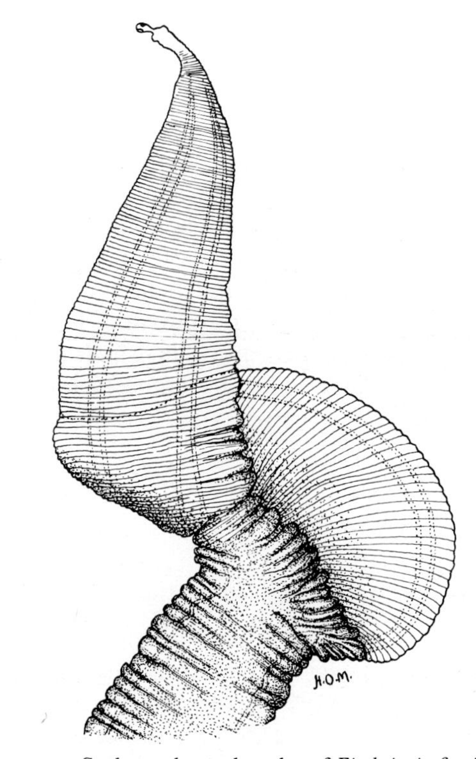

Fig 1.53 Scolex and pseudoscolex of *Fimbriaria fasciolaris.*

anseriform birds. The anterior portion of the body of this cestode forms a folded expansion or 'pseudoscolex' by which the parasite attaches itself.

The cysticercoid has been found in the copepods, *Diaptomus vulgaris* and *Cyclops* spp.

ORDER: TAENIIDEA WARDLE, McLEOD AND RADINOVSKY, 1974

The *Taeniidea* are usually large tapeworms. The gravid proglottids are longer than they are wide. The rostellum may be absent, but is usually present and armed with a double row of small and large hooks (Fig. 1.54). The genital pores are single and irregularly alternating. There are, as a

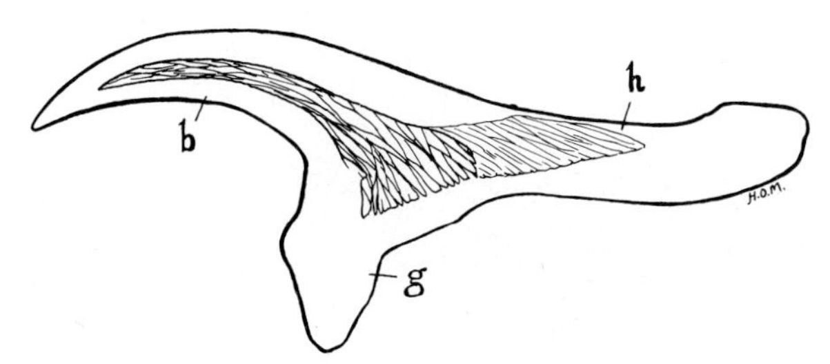

Fig 1.54 Rostellar hook of *Taenia* spp.
b blade
g guard
h handle

rule, a large number of testes and the ovary is situated in the posterior part of the proglottid. The uterus has a median, longitudinal stem and lateral branches. The development of the egg is described by Rybicka (1966). There is an outer envelope and capsule which is usually lost in faecal eggs. Inside this, the inner envelope develops into the embryophore which is made of blocks and gives the egg its characteristically brown, radiated appearance. Within these, the oncospheral membrane surrounds the hexacanth embryo or oncosphere. The structure of the adult stages is described in detail by Verster (1969). The metacestode is a cysticercus, a strobilocercus, a coenurus or an echinococcus (hyatid) cyst.

FAMILY: TAENIIDAE LUDWIG, 1886

Genus: Taenia Linnaeus, 1758

Taenia saginata Goeze, 1782 (= *Taeniarhynchus saginata*) occurs in the small intestine of man (the only definitive host) and the metacestode (cysticercus) is found in cattle although other ruminants will serve as intermediate hosts (i.e. llama, reindeer). Detailed studies of cysterceri found in African wild ruminants such as the giraffe, the wildebeest and the antelope have indicated that these animals are not normally hosts for *T. saginata* metacestodes. However, where farming of game animals is undertaken, this situation may change and, in the absence of wild game taeniids, animals may become susceptible to *T. saginata*. Deliberate infection of game animals with eggs of *T. saginata* has resulted in metacestode development (Stevenson, unpublished). The role of the water buffalo requires further investigation (Sachs 1969). It is generally accepted that man is not normally a host for the larval stage (unlike *T. solium*); however, there are a few reports of the metacestode of *T. saginata* occurring in man (Pawlowski & Schultz 1972).

T. saginata has a cosmopolitan distribution and the infection is particularly important in Africa

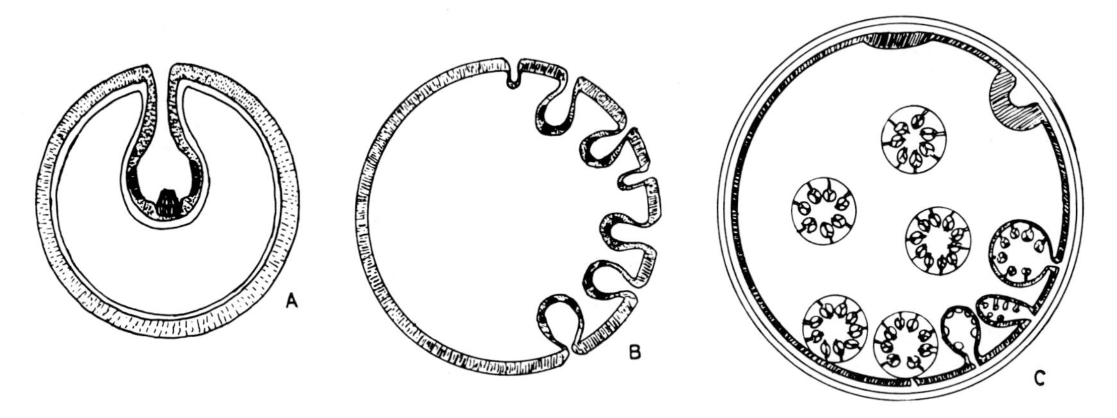

Fig 1.55 Larval stages of taeniid cestodes. A, Cysticercus, single invagination and scolex. B, Coenurus, multiple invaginations each with a scolex. C, Hydatid, development of brood capsules each with a number of scolices. (*After Soulsby 1965*)

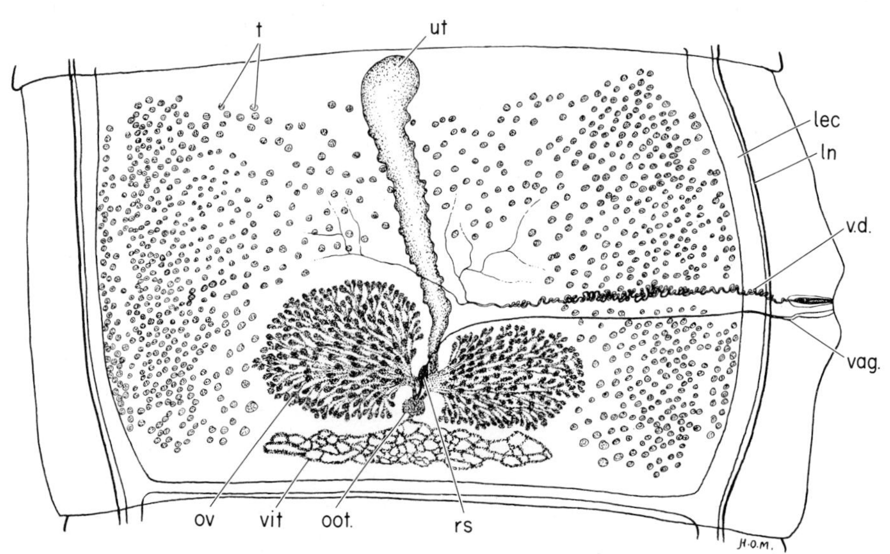

Fig 1.56 Mature proglottid of *Taenia* spp.

lec	longitudinal excretory canal	t	testes
ln	lateral nerve	ut	uterus
oot	oötype and Mehlis' gland	vag	vagina
ov	ovary	vit	vitellarium
rs	receptaculum seminis	vd	vas deferens

and South America and in some Mediterranean countries. It is present, although less significant, in most other countries of the world. Stoll (1947) estimated that 389 million people were infected with the parasite. The prevalence is increasing. The adult worm is 4–8 m, rarely up to 25 m, long and will survive many years. The scolex has four suckers and is without a rostellum or hooks. The eggs are 46–50 by 39–41 μm in diameter. Gravid proglottids contain about 80 000 eggs and the uterus has 14–32 lateral branches. The number of lateral branches of the uterus has been used as a diagnostic feature to differentiate the proglottids of *T. saginata* and *T. solium*. This is unreliable and the presence (*T. saginata*) or absence (*T. solium*) of a vaginal sphincter muscle is a more reliable criterion. This and other distinguishing features are described by Verster (1969); thus the cirrus pouch of *T. saginata* does not extend to the excretory vessels while that of *T. solium* does; the ovary of *T. saginata* is bilobed while three lobes are present in *T. solium*; the eggs are oval in *T. saginata* and spherical in *T. solium*; and the gravid segments leave the host spontaneously in the case of *T. saginata* but not of *T. solium*. The presence or absence of an armed rostellum is diagnostic.

Life-cycle. About ten proglottids are shed daily in the faeces or they migrate spontaneously out of the anus. The proglottids are motile and will migrate a few centimetres over the body, clothes, bed or ground, shedding eggs in the process. The eggs may remain viable for several weeks or months in sewage, in rivers and on pasture (Pawlowski & Schultz 1972). Infection of cattle occurs in a variety of ways. Neonatal calves may be infected when handled by infected persons, who also disperse eggs on pastures. In the USA epizootics of bovine cysticercosis have been recorded in feedlots (Dewhurst et al. 1967), owing to infected workers defaecating in silos, irrigation ditches and hayfields (Slonka et al. 1975). In developed countries there has been an increase in the prevalence of infection in cattle grazing on camping grounds and along major travel arteries. Sewage is an important route of dissemination of eggs. Eggs can pass through sewage treatment

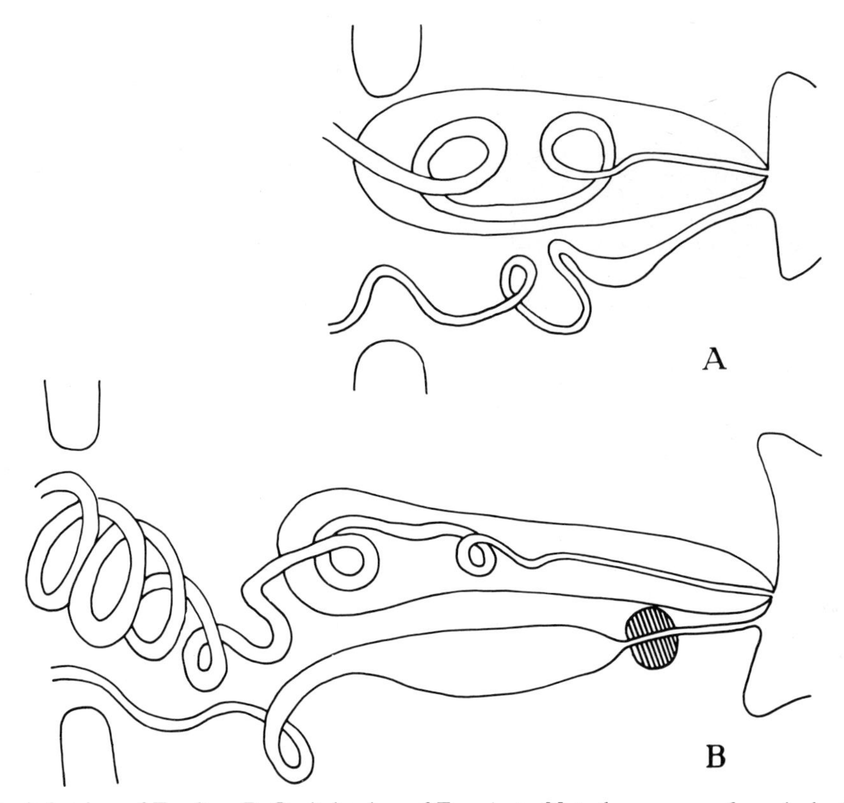

Fig 1.57 A, Genital atrium of *T. solium*. B, Genital atrium of *T. saginata*. Note the presence of a vaginal sphincter muscle in *T. saginata*. (*After Verster 1967*)

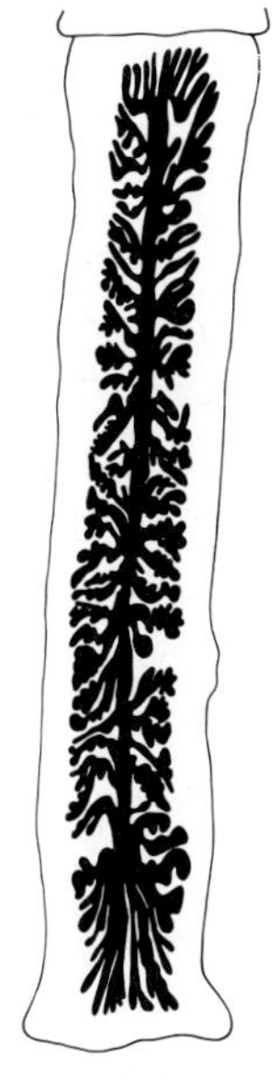

Fig 1.58 Gravid segment of *T. saginata*. (*After Verster 1967*)

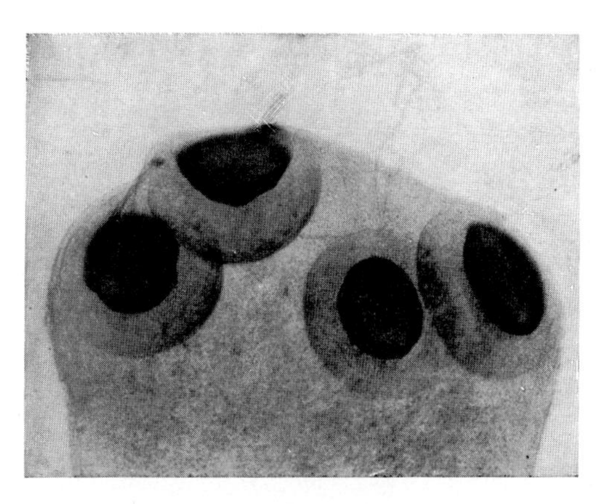

Fig 1.59 Scolex of *Taenia saginata* showing elliptical suckers and absence of rostellum and hooklets. × 80. (*Dr J. F. Brailsford*)

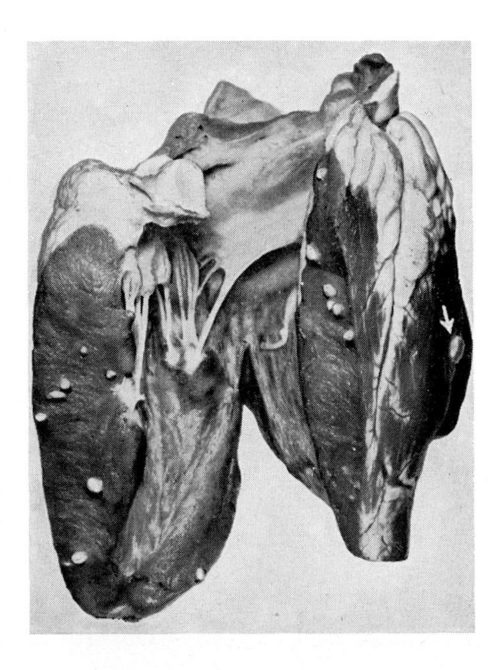

Fig 1.60 Degenerated cysticerci of *T. saginata* in wall of left ventricle of ox heart. One cyst, indicated by arrow, is visible beneath the epicardium. (*E. F. McCleery*)

plants and be released into rivers with the effluent, contaminate pasture treated with sludge or be disseminated by birds feeding on raw sewage (Silverman & Griffiths 1955). Eggs may remain viable for 71 days in liquid manure, 16 days in city sewage, 33 days in river water and 159 days on pastures (Jepsen & Roth 1949). Australian workers, quoted by Seddon (1950), found that the eggs may remain alive on pastures for at least eight weeks and on dry sunny pastures for $14\frac{1}{2}$ weeks.

When the eggs are ingested by cattle the oncosphere hatches and activates under the influence of gastric and intestinal juices and penetrates the intestinal mucosa to reach the general circulation. The embryos are disseminated throughout the body and develop in skeletal and cardiac muscle, but also in fat and organs. The heart and masseters appear to have the highest density and it has been stated in the past that the muscles of predilection are the masseters, heart, diaphragm and tongue, but usually cysticerci are spread throughout the musculature. A

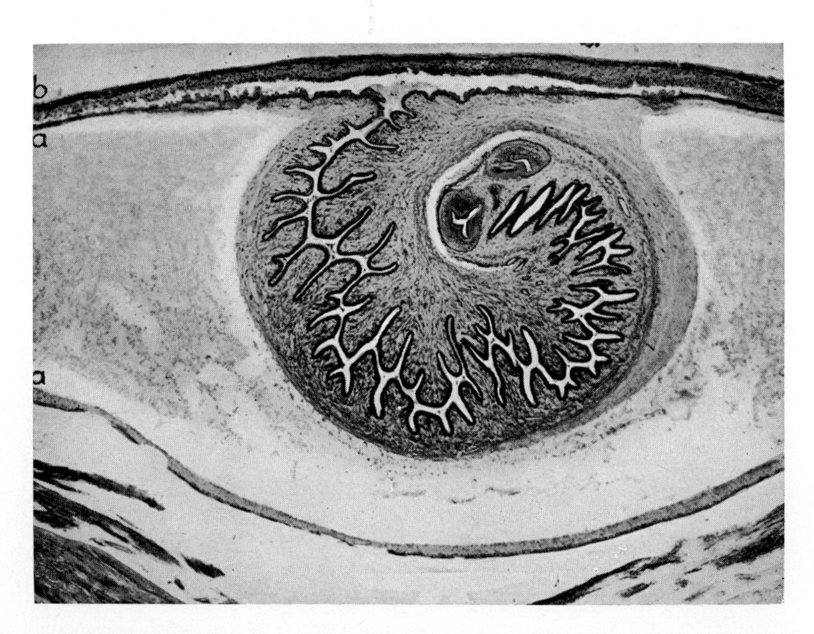

Fig 1.61 Microphotograph of cysticercus of *T. saginata* showing invaginated scolex with two suckers apparent; (a, a) bladder wall; (b) connective tissue capsule. × 28. (*H. Thornton*)

cysticercus (*Cysticercus bovis*) develops and becomes infective in about ten weeks and remains viable for up to nine months or longer. It occurs in a tissue capsule. Man is infected by the ingestion of raw or undercooked, infected (measley) beef. Gravid proglottids begin to be passed by man approximately 100 days after infection.

Cysticerci begin to degenerate four to six months after infection and by nine months a large number may be dead. Penfold (1937) demonstrated that in heavy artificial infections almost all cysticerci were dead in nine months; however, the longevity of the cysticerci is dependent on the degree of infection and the age of the animal at infection (Soulsby 1965). A proportion of cysticerci can remain viable for a prolonged period (perhaps for the life of the host) following infection of neonatal calves.

Prenatal infections have been recorded, but are not common (McManus 1960). It is possible that such infections are confused with early neonatal infections.

Taenia solium Linnaeus, 1758 occurs in the small intestine of man and experimentally it has been established in the gibbon and the chacma baboon. The pig (and wild boar) are the main hosts of the metacestode. Canine infections are rare and play little or no part in transmission, except where dogs are a source of human food. Of particular importance is the fact that man can act as an intermediate host as well as a final host.

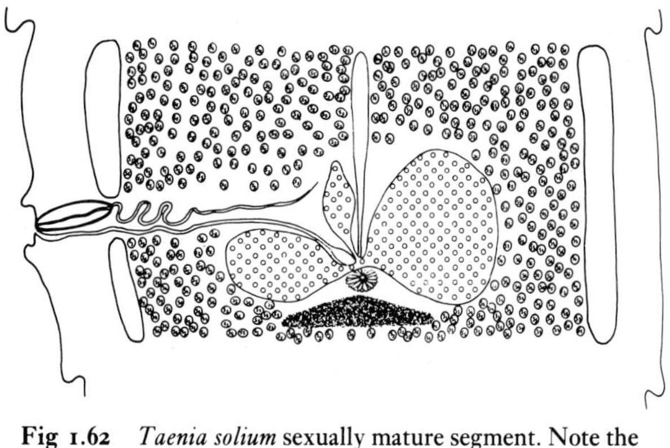

The adult is 3–5 m or up to 8 m long and can survive for up to 25 years. The scolex bears a rostellum which has two rows of hooks. The gravid proglottids are 10–12 mm long by 5–6 mm wide and the uterus has seven to 16 lateral branches. The gravid proglottids, each containing about 40 000 eggs, do not leave the host spontaneously and are voided in the faeces, frequently in chains. The eggs are 26–34 μm in diameter. (See p. 108) for differentiation between *T. solium* and *T. saginata*.)

Life-cycle. The life-cycle is similar to that of *T. saginata* except that pigs act as the intermediate host. The cysticerci (*C. cellulosae*) develop primarily in skeletal and cardiac muscle. The fully developed cysticercus measures up to 20 by 10 mm and is infective after about nine to ten weeks. The life-cycle is completed and man is infected when he eats raw or undercooked, infected (measley) pork. Man may act as an intermediate host and becomes infected with cysticerci of *T. solium* by ingestion of eggs in contaminated food or from dirty hands. Auto-infection by reverse peristalsis of the intestine, once considered to be important, is now largely discounted as an important source of human infection. Cysticerci in man develop primarily in the subcutaneous tissue but second

Fig 1.62 *Taenia solium* sexually mature segment. Note the three-lobed ovary. (*After Verster 1967*)

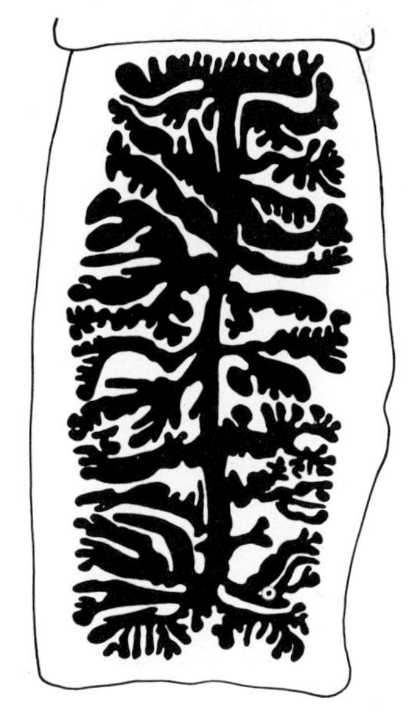

Fig 1.63 Gravid segment of *T. solium*. (*After Verster 1967*)

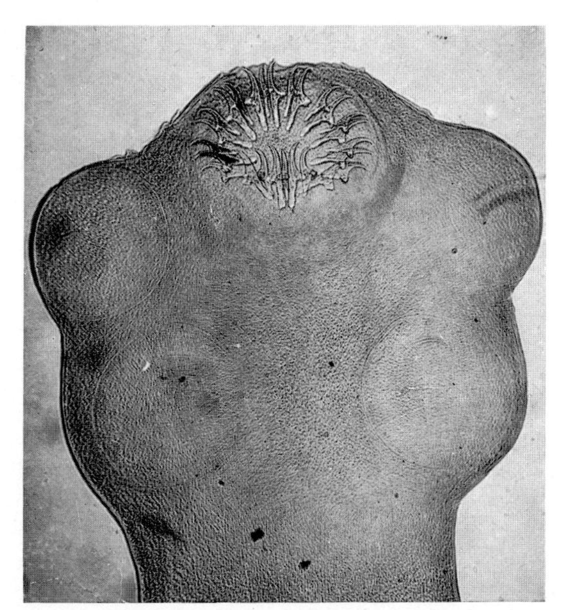

Fig 1.64 Microphotograph of scolex of cysticercus of *T. solium* showing four suckers and rostellum of hooklets. × 100. (*Dr J. F. Brailsford*)

in order of importance are the brain and ocular tissue. In sites such as these tremendous damage can result. In the brain the parasite usually develops in the ventricles and may become racemose (proliferative) in character. Infection causes pain, paralysis, epileptiform attacks and may be fatal. In Latin America as a whole, Schenone et al. (1973) have estimated that at least 300 000 individuals suffer neurocysticercosis per year. In Mexico, 1.9% of human autopsies show cysticercosis (Flisser et al. 1979).

Infection with *T. solium* is important in pork-eating countries, is mainly restricted to regions of low socio-economic development and is endemic in Latin America, southern Africa, south-east Asia and the Indian subcontinent. Infection is common in areas where villages are not supplied with sanitary facilities and where the pigs run loose scavenging for food, with ready access to human faecal material. The cysticerci are found in the pig primarily in the masseters, heart, tongue and shoulders, but will be disseminated throughout the body. The longevity of cysticerci is not known, but the young age at which pigs are slaughtered means that the majority of cysts in pork would be viable.

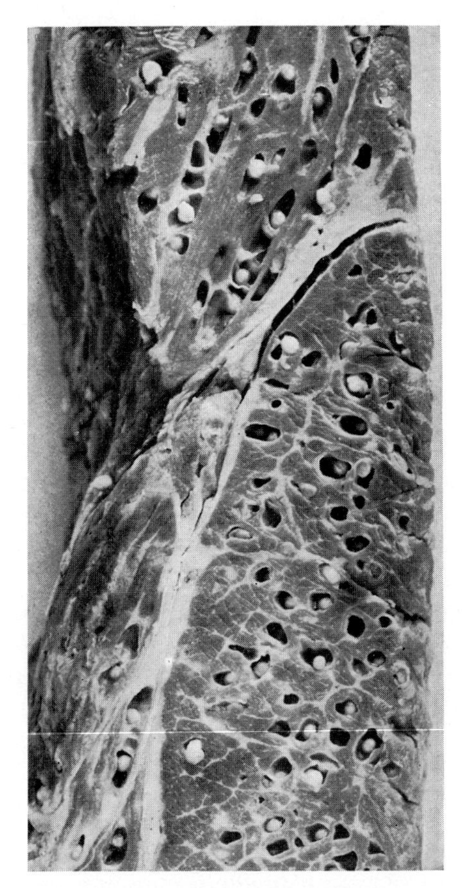

Fig 1.65 Pork showing extensive invasion of muscle by cysticerci of *T. solium*.

Pigs may acquire massive infections, because the gravid segments of *T. solium*, unlike those of *T. saginata*, are not active and may remain in and about the faeces, so that the eggs may be concentrated in these.

Clinical signs. Adult *T. saginata* and *T. solium* in man may cause a variety of non-specific abdominal symptoms such as diarrhoea, constipation and epigastric pain (Pawlowski & Schultz 1972). Neurological symptoms may be evident in persons infected with cysticerci of *T. solium*. Epileptic seizures associated with cerebral cysticercosis have led to severe burns in the mountain people of Irian Jaya (Subianto et al. 1978). Infections in pigs and cattle are usually asymptomatic. Extremely heavy experimental infections in cattle have resulted in myositis, myocarditis and signs of muscular stiffness or weakness. This occurs soon after infection and the animal recovers spontaneously. Owing to their public health importance, bovine and swine cysticercoses have major economic significance resulting from the condemnation or downgrading of infected carcases.

Diagnosis. Perianal swabs for eggs are a useful technique for identifying persons infected with *T. saginata*. Identification of the adult parasite in man is based on the morphological features of the scolex and proglottids. Diagnosis of cysticercosis in pigs and cattle is usually made at meat inspection. Ante-mortem diagnosis by serological methods is not yet feasible since these are not sufficiently specific or sensitive.

Treatment. For adult infections of *T. saginata* and *T. solium* niclosamide, given at a dose rate of 2 g, is effective, but the scolex is partially digested and often not recovered. Paromomycin (5 mg/kg), quinacrine (7–10 mg/kg), inorganic tin compounds and arecoline are effective but not as well tolerated. Treatment with praziquantel (10 mg/kg) is effective. Drugs that cause vomiting (i.e. quinacrine) should be avoided for the treatment of *T. solium*. Treatment of human cysticercosis is by surgical removal of the offending lesion. The prognosis is not good.

Prophylaxis involves treatment of infected persons, public education and hygiene and proper meat inspection. Meat inspection procedures vary in different countries but may involve examination of the muscles exposed when the carcase is split and through multiple incisions into the heart, masseters and perhaps tongue and muscles of the shoulder. However, meat inspection is estimated to detect less than 50% of infected carcasses and inspected carcasses are often drawn from the section of the animal population which is the least likely to be infected (Hammerberg et al. 1978). Infected carcasses may be condemned for human consumption or treated by freezing at −10°C for ten days to two weeks or cooking at 50–60°C. The length of time for which beef carcasses must be frozen in order to kill cysticerci has been reported by Hilwig et al. (1978). All cysticerci were killed after 15 days at −5°C, nine days at −10°C and six days at −15 to −30°C; 24-week-old cysticerci were less susceptible to the lethal effects of freezing than were 16- and 12-week-old cysticerci. The controversy as to the treatment of a carcass in which a single cyst has been detected at meat inspection has been largely resolved by Juranek et al. (1976). Carcasses showing a single dead or viable cyst by routine inspection are likely to contain viable or dead cysts elsewhere. Thus these carcases should be cooked or frozen instead of having the cysticercus removed.

Treatment with compounds such as albendazole, mebendazole and praziquantel and immunoprophylaxis against infections with metacestodes in the intermediate hosts is discussed on p. 126.

Taenia hydatigena Pallas, 1766 is cosmopolitan in distribution and occurs in the small intestine of dogs, wolves and other wild carnivores. The intermediate hosts are domestic and wild ruminants, particularly sheep. The pig may also be infected. Adults are 75–500 cm long and have two rows of 26 and 46 rostellar hooks. Gravid proglottids measure approximately 12 by 6 mm and the uterus has six to ten lateral branches. The eggs are oval and 36–39 by 31–35 μm.

Life-cycle. Ingested eggs hatch in the small intestine and oncospheres reach the liver via the blood. Here the embryos break out of the portal vessels and migrate in the liver parenchyma for up to 30 days. They cause haemorrhagic tracts which later undergo fibrosis. The developing cysticerci migrate into the peritoneal cavity from about 18 days to four weeks after infection. They reach maturity here between days 34 and 53 and are found attached to the greater omentum, intestinal mesentery and the serosal surface of organs. Mature cysticerci (*C. tenuicollis*) are up to 6 cm long and contain a single scolex invaginated into a long neck. Carnivores become infected by ingestion of the cysticercus. The prepatent period in the dog is 51 days and dogs may remain infected for a year or more.

Pathogenesis. The prevalence of infection in sheep is high but the levels of infection are often

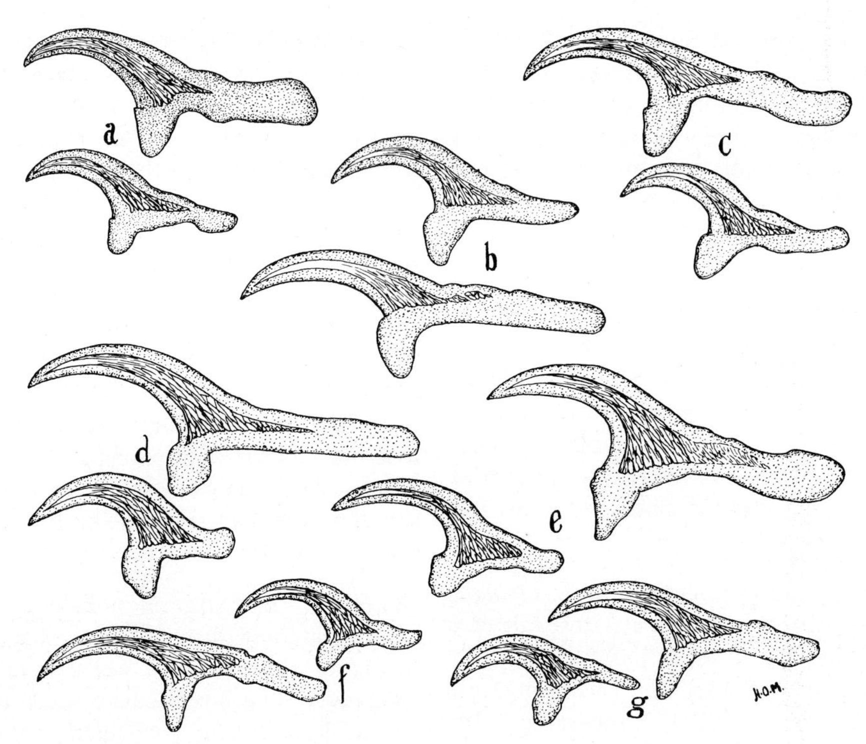

Fig 1.66 Rostellar hooks of *Taenia* species. a, *T. solium*. b, *T. hydatigena*. c, *T. ovis*. d, *T. pisiformis*. e, *T. taeniaeformis*. f, *T. multiceps*. g, *T. serialis*. All × 100 except e × 50.

Fig 1.67 Cysticercus of *T. hydatigena* with the head everted; natural size.

low. Migration of cysticerci in the liver may cause haemorrhagic and fibrotic tracts and viable, caseated or calcified cysticerci may be present. This may result in condemnation of the liver at slaughter. It is possible that infected lambs show some unthriftiness. Heavy infections occasionally occur in young lambs and traumatic hepatitis leading to death may result. This condition must be differentiated from acute fascioliasis and an immature cysticercus is readily differentiated from an immature fluke. Mature cysticerci in the peritoneal cavity usually cause no harm.

Pathogenesis in the dog is discussed on p. 124.

Treatment and prophylaxis. See p. 124.

Taenia ovis (Cobbold, 1869) is a common tapeworm of dogs and wild carnivores in most parts of the world. The intermediate hosts are primarily sheep and also goats. Adults are up to 1 or 2 cm long and the uterus has 11–20 lateral branches. The eggs are oval and 19–31 by 24–26 μm in diameter.

Life-cycle. The life-cycle is similar to that of *T. saginata*. Oncospheres are disseminated in the general circulation and develop in the skeletal and cardiac musculature. The heart, diaphragm and masseters appear to be predilection sites. Mature cysticerci (*C. ovis*) are up to 6 mm long and are infective at 46 days. The prepatent period in dogs is about 60 days.

Pathogenesis. Clinical signs are not normally seen in infected animals. The parasite is of importance as a cause of financial loss to the meat industry because of the unaesthetic appearance of the cysts in meat. For instance, shipments of boneless mutton from Australia to the United States may be rejected if metacestodes of *T. ovis* are detected.

Pathogenesis in dogs is discussed on p. 124.

Treatment and prophylaxis. See p. 124.

Taenia krabbei Moniez, 1789 is a tapeworm of wild carnivores (e.g. *Canis latrans*, *C. lupus*, *Lynx rufus*) and the dog occurring in northern countries. The intermediate stage, *Cysticercus tarandi*, is found in the muscles of reindeer, gazelle and other wild ruminants. The worm is about 26 cm long or longer. There are 26–34 hooks, the large ones 148–170 μm long and the small ones 85–120 μm. The mature segments are much broader than long and the organs are compressed and transversely elongated. The uterus has nine to ten lateral branches on either side.

Taenia pisiformis (Bloch, 1780) (syn. *T. serrata*) occurs in the small intestine of the dog, fox, several wild Carnivora and rarely the cat. The intermediate hosts are lagomorphs, primarily rabbits and hares, and rodents. It may grow up to 200 cm long. The rostellum bears 34–48 hooks in two rows, the large hooks 225–294 μm long and the small ones 132–177 μm. The gravid segments measure 8–10 by 4–5 mm, and the uterus has eight to 14 lateral branches on either side. The eggs are slightly oval and 43–53 by 43–49 μm.

Life-cycle. The life-cycle is similar to that of *T. hydatigena*, but the intermediate hosts are chiefly rabbits and hares. The young stages, after having developed in the liver for about 15–30 days, penetrate through the parenchyma of this organ and the adult bladderworm (*Cysticercus pisiformis*) is found in the peritoneal cavity attached to the viscera. It is a small cyst, about the size of a pea.

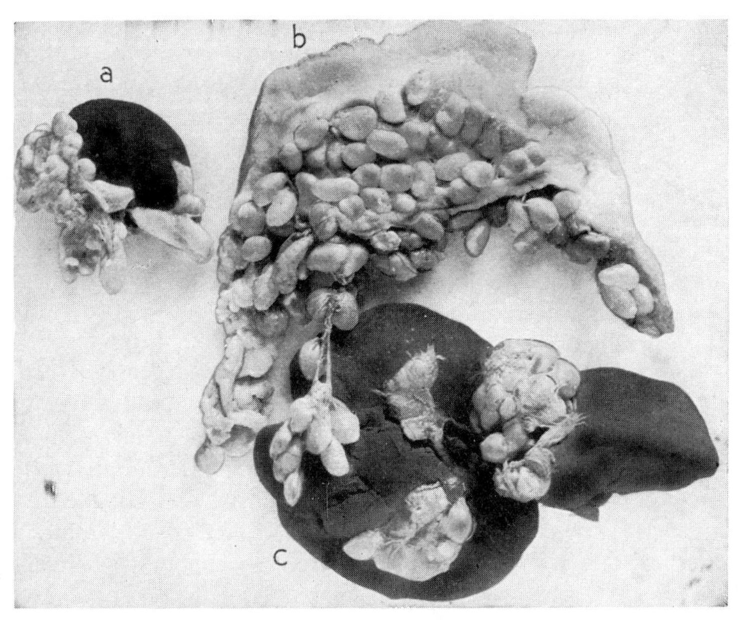

Fig 1.68 Abdominal viscera of rabbit showing subserous distribution of cysts of metacestodes of *T. pisiformis* in (a) kidney; (b) omentum; (c) liver. (*Dr J. F. Brailsford*)

Pathogenesis. In experimental infections severe damage to the liver and death can result. Light infections may result in digestive disturbances and loss of condition. For pathogenesis in dogs, see p. 124.

Treatment and prophylaxis. See p. 124.

Taenia taeniaeformis (Batsch, 1786) (= *Hydatigera taeniaeformis*) is a parasite with a cosmopolitan distribution which is found in the small intestine of cats and related carnivores (i.e. stoats, lynx). The genus *Hydatigera* has been established for this and other species which have a strobilocercus as a larval stage. However, it is now common to assign the species to the genus *Taenia* and not the genus *Hydatigera*. The intermediate hosts are rodents and occasionally lagomorphs. Adults are up to 60 cm long, characteristically lack a neck and have bell-shaped posterior proglottids. The scolex is large and prominent with two rows of rostellar hooks. The uterus has five to nine lateral branches. Eggs are 31–36 μm in diameter.

Life-cycle. Development of the metacestode (*Cysticercus fasciolaris*) occurs in the livers of rodents. An urban and a sylvatic cycle occur. The urban cycle involves the domestic cat and house and field mice and rats, whereas the sylvatic cycle in North America involves bobcats (*Lynx rufus*) and wild rodents (e.g. *Ondatra* spp.). By 30 days after the ingestion of eggs, an invaginated cysticercus develops. On day 42 the scolex evaginates and becomes connected to the bladder by a segmented strobila, thus resembling a small tapeworm. This larval stage is known as a strobilocercus. Strobilocerci are infective for the cat at 60 days. On ingestion by the cat the posterior portion of the strobila is digested and the scolex attaches to the intestinal wall. The prepatent period of infection in the cat is 36–42 days. Cats may remain infected for up to two years.

Pathogenesis. See p. 124.

Treatment and prophylaxis. See p. 124.

A large number of other taeniid cestodes occur in the small intestine of wild carnivores, but in many cases their life-cycle is not known. These include: **Taenia brauni** Setti, 1897 of dogs and jackals in tropical and southern Africa, the larval stage is a coenurus (see below) being found in Muridae and Hystricidae and has also been reported from man where it occurs in the subcutaneous tissue and the lungs, brain and eyes; **T. bubesi** Ortlepp, 1938 of the lion (*Panthera leo*); **T. crocutae** Mettrick and Beverley-Burton 1961 of the spotted hyena (*Crocuta crocuta*); **T. crassiceps** (Zeder, 1800) adults in foxes, coyotes etc. and cysticercus in various rodents in North America and may also occur in Europe; **T. erythraea** Setti, 1897 of the black-backed jackal (*Canis mesomelas*); **T. gongamai** Ortlepp, 1938 and **T. hlosei** Ortlepp, 1938 of the lion and cheetah (*Acinonyx jubatus*) in the Republic of South Africa; **T. hyaenae** Baer, 1924 of various species of hyena in Central and southern Africa, with cysticerci in various antelopes; **T. laticollis** Rudolphi, 1819, in various carnivores in North America and Central and southern Africa with larval stages occurring in lagomorphs and rodents; **T. lycaontis** Baer and Fain, 1955 of the hunting dog (*Lycaontis pictus*) in Central and East Africa; **T. macrocystis** (Diesing, 1850) Lühe, 1910 of the lynx and coyote in North America with larval stages in *Lepus californicus*; **T. martis** (Zeder, 1803) Freeman 1956 in *Martes* spp. in Europe and North America and cysticerci occur in *Clethrionomys* (vole); **T. mustela** Gmelin, 1790, in *Martes* and *Mustela* species in Europe and North America with larval stages in moles (*Talpa* spp.) and various rodents; **T. omissa** Lühe, 1910 of the cougar (*Felis concolor*) and larvae occur in various deer; **T. parva** Baer, 1926 in *Genetta* spp. and larval stages in various rodents in Europe and Africa; **T. polyacantha** Leuckart 1856 occurs in foxes in Alaska and larval stages occur in microtine rodents; **T. regis** Baer, 1923 in *Panthera leo* (lion); **T. rileyi** Loewen, 1929, of the lynx in North America, larval stages occurring in the mesentery of a variety of rodents; and **T. twitchelli** Schwartz, 1924 in wolverines and larval stages in the lungs and pleural cavity of porcupines in North America (various rodents may be infected experimentally).

Details of the morphology and distribution of these species may be found in Leiby and Dyer (1971) and Verster (1969).

Taenia multiceps (Leske, 1780) (= *Multiceps multiceps*). Previously this species was placed in the genus *Multiceps* Goeze, 1782, features of which were that the rostellar hooks were large with a sinuous handle and the larval stage was a coenurus (a bladderworm with numerous single tapeworm heads attached to the inner wall). Esch and Self (1965) concluded that the morphological criteria used to separate *Multiceps* and *Taenia* are not valid and the only constant difference is the larval stage.

Adults occur in the small intestine of the dog, fox, coyote and jackal throughout the world. The intermediate stage, a coenurus, develops in the brain of the sheep and other ungulates and has been recorded in man. Adults are 40–100 cm in length. The uterus has 14–20 lateral branches. Eggs have a diameter of 29–37 μm.

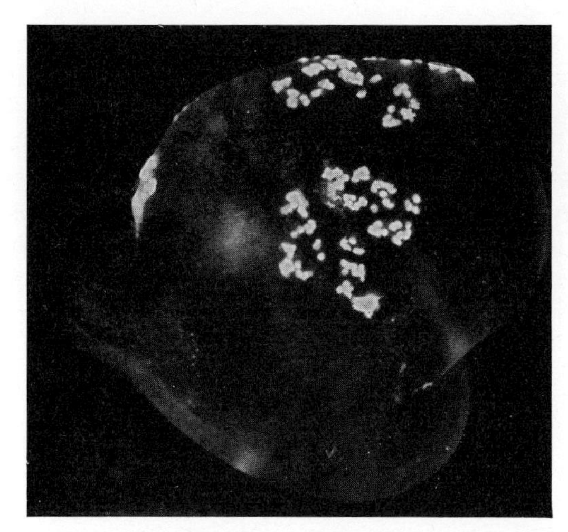

Fig 1.69 *Taenia multiceps* cyst removed from sheep brain. Numerous scolices are seen as white clusters on the inner wall of the cyst. × 2.

Life-cycle. The metacestode (*Coenurus cerebralis*) develops in the brain and spinal cord of sheep, but in the goat the cysts may also reach maturity in other organs, intramuscularly and subcutaneously. The agent of infection in the goat has been referred to as *T. gaigeri* Hall, 1916 but the difference in habitat of the larval stage appears to be related to the species of host and not the parasite. The developing larval stages, on reaching the brain, first migrate in the brain and spinal tissue leaving tortuous yellowish-grey to reddish streaks. As the parasite matures it develops into a large fluid-containing cyst, 5 cm or more in diameter. The cyst contains several hundred protoscolices invaginated in clusters on the cyst wall. The coenurus is infective after six to eight months (Fig. 1.69).

Pathogenesis and clinical signs. In lambs, if a large number of immature stages migrate in the brain, an acute meningoencephalitis may develop. More commonly infection is chronic and is associated with the presence of one or two coenuri in the brain four to six months after infection. There is an increasing degree of destruction of brain tissue as the coenurus develops. The neurological clinical signs are often referred to as 'gid' or 'staggers' and are dependant on the location of the cyst in the central nervous system. Most frequently the cyst is situated in the parietal region on the surface of one of the cerebral hemispheres. The animal holds its head to one side and turns in a circle towards the affected side. It may be blind in the eye on the opposite side. If the parasite is situated in the anterior part of the brain, the head is held against the chest and the animal steps high, or it may walk in a straight line until it meets an obstacle and remains motionless for a

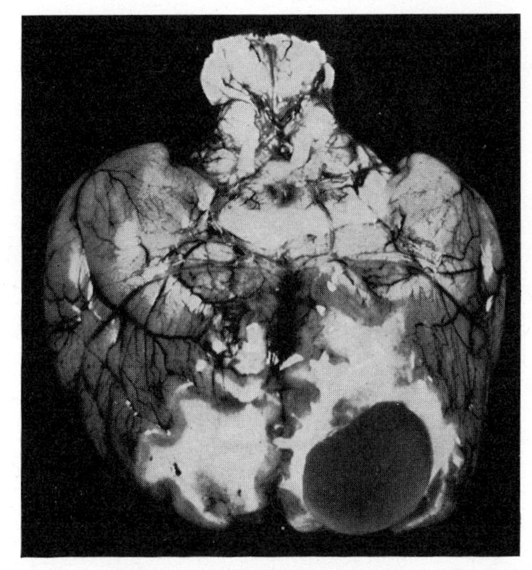

Fig 1.70 Brain of sheep affected with sturdy. The base of the brain is uppermost, and one large *Taenia multiceps* cyst can be seen in the region of the olfactory bulb of the left cerebral hemisphere.

time. Movements which tend to be the reverse of the above are seen if the cyst is in a ventricle. If in the cerebellum, the animal is hyperaesthetic, has a jerky or staggering gait and may become prostrate. In spinal cord the parasite can cause progressive paresis of one or both hind limbs. The clinical signs may be intermittent, but progressively the animal ceases to feed and emaciation develops.

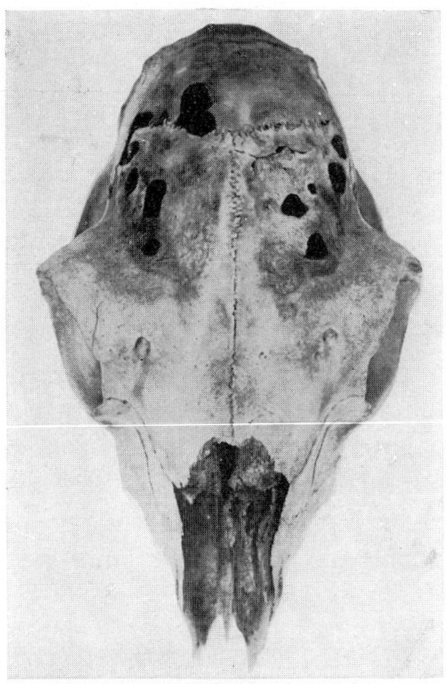

Fig 1.71 Skull of sheep showing perforation caused by *Taenia multiceps*.

If the parasite is situated on the surface of the brain, the skull undergoes pressure atrophy, even to the extent of perforation, so that on examination the affected part can be localized, as it yields to pressure. Such manipulation causes pain and may incite a spell of clinical signs.

Diagnosis. Differential diagnosis is difficult since a specific serological test is not yet available and frequently diagnosis is not made antemortem. Ronshina (1953) states that an early diagnosis (prior to the development of clinical signs) can be made on ophthalmoscopic examination of the eye.

Treatment. In many cases treatment is useless. If the coenurus is located on the surface of the brain, surgical removal is possible. Albendazole has been used on one occasion with apparent helpful effect.

Taenia serialis (Gervais, 1847) is a tapeworm of the dog and fox which has a cosmopolitan distribution. Intermediate hosts are lagomorphs in which a coenurus (*C. serialis*) develops in the subcutaneous and intramuscular connective tissues. It grows to a length of 72 cm and the scolex bears two rows of 26–32 hooks. The large hooks are 135–175 μm long and the small ones 78–120 μm. The gravid uterus has 20–25 lateral branches on either side. The elliptical eggs measure 31–34 by 29–30 μm. The coenurus may be 4 cm or larger

Fig 1.72 Rabbit extensively affected with *Taenia serialis* cysts. (*Crown copyright*)

and the numerous protoscolices form radiating rows. *T. serialis* is regarded by some to be synonymous with *T. multiceps*. For example Clapham (1942) suggests that various species of the genus *Multiceps* be assigned to *T. multiceps*. She envisages that the coenurus stage can occur in the CNS, connective tissue, abdominal cavity and elsewhere, the form adopted in any host being an individual characteristic.

Genus: Echinococcus Rudolphi, 1801

Speciation in the genus *Echinococcus* has been complicated by attempts to assign species or subspecies status to strains of parasites. This has been discussed by Smyth (1969, 1977). At present four species of the genus *Echinococcus* are accepted and regarded as valid taxonomically. These are *Echinococcus granulosus* (Batsch, 1786), *Echinococcus multilocularis* (Leuckart, 1863), *Echinococcus oligarthrus* (Diesing, 1863) and *Echinococcus vogeli* (Rausch and Bernstein, 1972). These four species are morphologically distinct in both the adult and the larval stages. The morphological criteria that are used to differentiate the adult stages of *Echinococcus* spp. include the form of the strobila, the position of the genital pore in the mature and gravid segments, the rostellar hooks, the number of testes and the shape of the uterus. Smyth and Smyth (1964) suggest that 'a continuum of "species" exists between *E. granulosus* and *E. multilocularis*'.

Echinococcus granulosus (Batsch, 1786) is found in the small intestine of carnivores (particularly the dog) and the metacestode (hydatid cyst) is found in a wide variety of ungulates and man. The parasite has a cosmopolitan distribution. Adults are 2–7 mm long and usually possess three or four proglottids (rarely up to six). The penultimate proglottid is mature and the terminal proglottid is gravid and is usually about half the length of the worm. The rostellum has two rows of hooks. The ovary is kidney-shaped. Genital pores alternate irregularly and normally open in the posterior half of the mature and gravid proglottids. The uterus of the gravid proglottid has well-developed diverticula. The gravid proglottid usually disintegrates in the intestine so that only eggs and not proglottids are found in the faeces. The eggs are typical taeniid eggs and measure 32–36 by 25–30 μm.

Life-cycle. Eggs are passed in the faeces of the carnivore, they are immediately infective and on ingestion by ungulates the oncosphere penetrates an intestinal venule or lymphatic lacteal (Heath 1971) to reach the liver or lungs (although other organs can be infected). The hydatid cyst develops slowly over several months. Hydatid cysts are commonly 5–10 cm in diameter although larger ones have been recorded, principally in man; for example a cyst 50 cm in diameter and containing about 16 litres of fluid has been recorded. The hydatid cyst is usually unilocular and is composed of a fairly thick outer concentrically laminated membrane and within this is a granular germinal membrane. From this, brood capsules, each containing protoscolices, develop about five months after infection. At this time the cyst is infective for the definitive host. The brood capsules may become detached and float free in the cyst fluid, being called hydatid sand. Occasionally daughter cysts develop within the hydatid cyst and, if a cyst is ruptured, protoscolices and brood capsules can develop into other external daughter cysts. External daughter cysts may also be formed if a piece of germinal membrane becomes enclosed in a laminated layer. The life-cycle is completed when a dog ingests protoscolices. These evaginate, penetrate deeply between the villi into the crypts of Lieberkuhn and develop to maturity in about 47 days. Dogs may remain infected for about two years. In heavy infections the intestine becomes carpeted with worms.

Not all hydatid cysts produce brood capsules or protoscolices. Thus, they may be sterile. For instance, Thompson (1977) found 27% of horse hydatid cysts and 51% of sheep hydatid cysts to be sterile. Cysts in cattle are frequently sterile and pigs, although not commonly infected, usually have sterile cysts. Sterility of cysts is also associated with the age of the host upon infection.

Hydatid cysts are found primarily in the lungs of sheep where they are frequently multilocular.

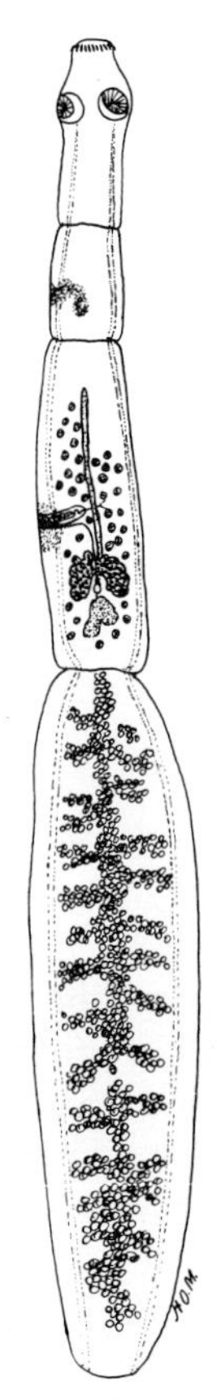

Fig 1.73 Entire specimen of *Echinococcus granulosus*.

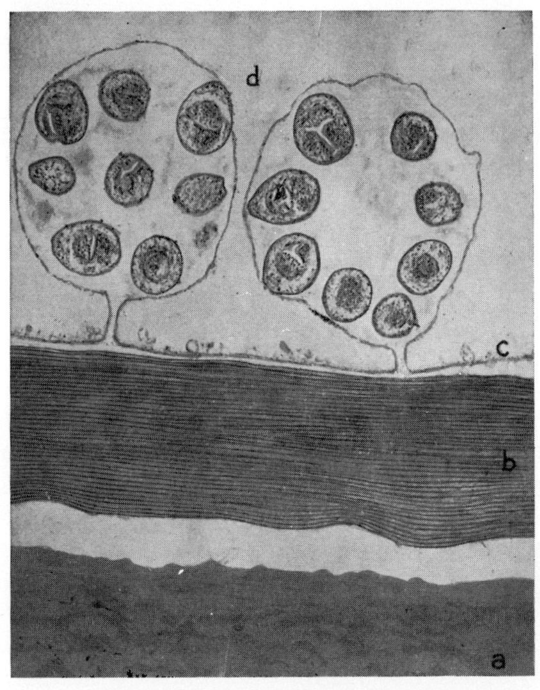

Fig 1.74 Microphotograph of wall of a fertile hydatid from sheep liver. a, Connective tissue capsule formed by reaction of host. b, External cuticular membrane. c, Internal germinal layer attached to which are. d, Two brood capsules each containing eight scolices. × 50.

They are found in both the liver and the lungs of pigs, but primarily in the livers of horse and cattle. In horses the hydatid cysts are usually unilocular. In man hydatid cysts are found in a wide variety of organs.

Hydatid cyst fluid is pale yellow with 17–200 mg protein/100 ml. Cyst fluid has a striking similarity to the serum of the host and it contains immunoglobulin. In particular, anti-complementary substances are found in cyst fluid

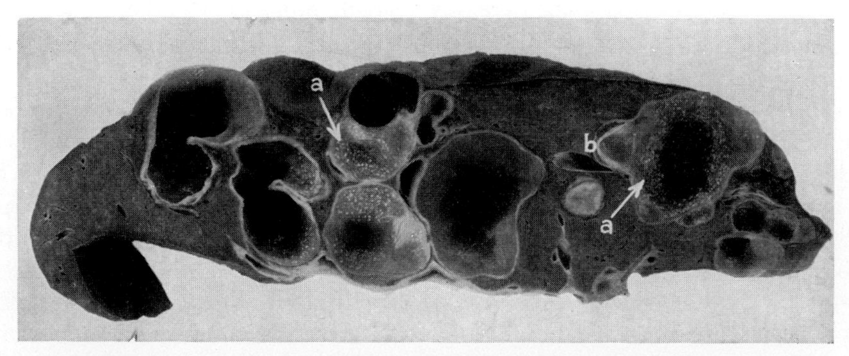

Fig 1.75 Section of sheep liver showing fertile hydatid cysts. The white spots on the germinal layer at (a, a) are brood capsules and at (b) a growing cyst is assuming an irregular shape as it orientates round a bile duct.

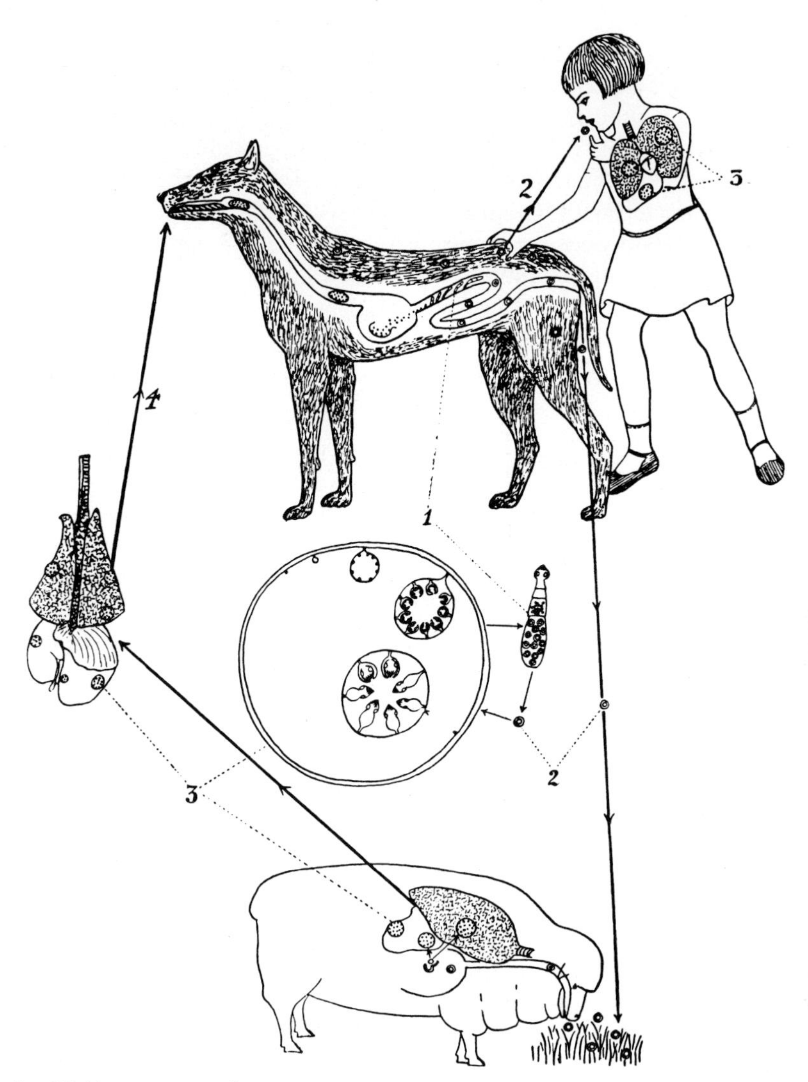

Fig 1.76 Life-cycle of *Echinococcus granulosus*.

and calcareous corpuscles (Kassis & Tanner 1976). These may be associated with the prolonged survival of hydatid cysts (see p. 127).

Host range and distribution. E. granulosus has a broad geographical distribution; it is cosmopolitan in distribution with highly endemic areas in the USSR, the Mediterranean countries and areas of Africa, Latin America and Australia. The larval stage has been found in a wide range of mammals including primates, marsupials, ruminants, lagomorphs and artiodactyls. Occasional natural infection with the larval stage has been reported in carnivores and avian hosts.

A number of strains of *E. granulosus* appear to exist. Smyth and Smyth (1964) describe the occurrence of self-fertilization by *Echinococcus* and hence the tremendous biotic potential of the metacestode leads to the expression of mutants or strains. Clearly a horse strain, involving hounds and horses, and a sheep strain, involving dogs and sheep, exist in the UK (Smyth 1977; Thomson 1978). These differ morphologically and biochemically. Further, the sheep strain, but not the horse strain, develops to sexual maturity when cultured *in vitro* (Smyth & Davies 1974; Smyth et al. 1980). The horse strain appears to be host specific and is widespread in Ireland in the absence of ovine hydatidosis (Hatch 1975). Cranley (1982) reported 48% of horses infected with hydatids in Great Britain. The horse strain does not appear to infect man and the rhesus monkey cannot be infected (Thompson & Smyth 1976). However, the sheep strain does infect man.

Apart from the sheep/dog and horse/dog strains mentioned above, there is evidence that strains exist which utilize camel/dog, pig/dog, buffalo/dog, goat/dog, cattle/dog and man/dog cycles. Further, other strains exist with sylvatic cycles and these include moose/wolf in North America, wallaby/dingo in Australia, deer/jackal in Sri Lanka, deer/coyote in California and hare/fox in Argentina. Studies of isoenzyme patterns in hydatids may be helpful in differentiating the origin of the strain of parasite (McManus & Smyth 1979).

Echinococcus multilocularis Leukart, 1863, (Syn. *Alveococcus multilocularis*) is found in the northern hemisphere, central and eastern Europe and particularly in Canada and the USSR. Adults are smaller than *E. granulosus*, being 1.2–4.5 mm in length. Other morphological differences are reviewed by Smyth (1964) and include the presence of two to six proglottids; a difference in the shape and number of rostellar hooks (14–34, small 20–21 μm, large 25–35 μm); a lack of lateral pouching of the uterus in the gravid proglottids; the genital pore is nearer the anterior of the proglottid; the ante-penultimate segment is characteristically mature; and testes are 14 to 35 in number. Adults occur primarily in foxes. The prepatent period is less than 30 days. Alveolar hydatid cysts, which are multivesicular infiltrative forms, are found principally in microtine rodents (field mouse, vole, shrew, ground squirrel) and man may also be infected. The species is perpetuated mainly by a sylvatic cycle involving foxes and cricetid rodents. A major source of infection for man is through fruits and vegetables that have been contaminated by the faeces of foxes and occasionally cats and dogs. Vegetables, strawberries, grapes and fallen fruits may be contaminated with eggs of *Echinococcus* and *Taenia* species. In Alaska, the close association of man with his sledge dogs and hunting dogs permits contamination of food and drink and trappers may become infected when the pelt of a fox is being removed. Dogs may be important final hosts in northern USSR and team dogs in Alaska (Rausch & Schiller 1956) and Eckert et al. (1974) found the parasite in dog and cat in southern Germany, an area where the parasite was endemic in red foxes.

Echinococcus oligarthus (Diesing, 1863). The adult worm possesses three segments, the penultimate of which is mature. It measures 1.9–2.9 mm. The genital pore is anterior to the middle in mature segments and approximately at the middle in gravid segments. The number of testes varies from 15 to 46. The gravid uterus is sac-like. The rostellar hooks number from 26 to 40 (large and small hooks respectively being 43–60 μm and 28–45 μm).

The hydatid stage is polycystic. The cysts have a tendency to become septate and to form multichambered growths.

E. oligarthus occurs in Central and South America, and occurs in wild felids including the puma (*Felis concolor*), jaguar (*Felis onca*), jaguarundi (*Felis yagouaroundi*) and Geoffroy's cat (*Felis geoffroyi*) as definitive hosts. The agouti (*Dasyprocta*) and possibly other rodents serve as intermediate hosts. Human infection with this species has not been reported.

Echinococcus vogeli (Rausch and Bernstein, 1972). The adult worm is between 3.9 and 5.6 mm in length; it has three segments and the penultimate segment is mature, the genital pore being situated posterior to the middle of both the mature and gravid segment. There are between 50 and 67 testes, the greater proportion being in the anterior half of the segment. The gravid uterus has no lateral branches or sacculations and is characterized by being relatively long and tubular in form. Rostellar hooks 28–36; large and small hooks respectively measure 49–57 μm and 30–47 μm.

The larval stage is similar to that of *E. oligarthus* and the two species can be distinguished by comparing the dimensions of the rostellar hooks of the protoscolex. The large and small protoscolex rostellar hooks of *E. oligarthus* are 25.9–37.9 μm (average 33.4 μm) and 22.6–29.5 μm (average 25.45 μm) respectively while those of *E. vogeli* are 39.1–43.9 μm (average 41.64 μm) and 30.4–36.5 μm (average 33.6 μm) respectively.

E. vogeli occurs in central and northern South America, with the bush dogs (*Speothos venaticus*) and the domestic dogs as definitive host, and pacas (*Cuniculus paca*) and possibly other rodents as intermediate hosts. The larval stage develops in man and causes a polycystic form of hydatid disease.

Subspecies of Echinococcus. A number of other species of the genus have been described but at present only the species mentioned above are accepted as valid. However, there are two currently accepted subspecies of *E. granulosus*, *E.g. granulosus* and *E.g. canadensis*. The former is infective to dogs and domestic ungulates, and was believed to have originated in Europe and to have been spread throughout the world by early settlers. *E.g. canadensis* is the indigenous species in the arctic region of North America and utilizes wolves and wild ruminants in its life-cycle.

E.m. multilocularis is a subspecies of *E. multilocularis* and is infective to dogs and foxes and also probably cats, with small rodents acting as intermediate hosts. Its distribution is restricted to that of its rodent hosts in central Europe and Russia. *E.m. sibiricencis* utilizes the arctic fox and rodents and is found in North America.

No intraspecific variants of *E. oligarthus* and *E. vogeli* have so far been described. It should be noted that not all workers agree on the question of subspeciation.

Pathogenesis. The adult tapeworm is comparatively harmless to the dog although in large numbers enteritis may be seen.

The pathogenicity of the hydatid cyst depends on the severity of the infection and the organ in which it is situated. In domestic animals clinical signs are not commonly seen despite heavy infections. However, human hydatidosis is often associated with clinical signs and the function of the affected organ is frequently impaired. This is of particular importance if the brain or heart is involved. Rupture of a cyst can lead to fatal anaphylactic shock. Also, the released brood capsules and protoscolices may develop into numerous daughter cysts.

Alveolar hydatidosis due to *E. multilocularis* is serious; growth is peripheral and invasive and, like a malignant tumour, metastases often occur.

Diagnosis. Under normal conditions of faecal examination, the eggs of *Echinococcus* cannot be differentiated from those of the *Taenia* spp. To confirm a diagnosis the adult parasites must be demonstrated and a dog must be purged orally with arecoline hydrobromide (1–2 mg/kg) and the mucous portion of the purged material (usually the last passed) is examined microscopically for tapeworms. Dogs which fail to purge can be given a half dose of arecoline one hour later.

Diagnosis of hydatidosis in domestic animals is rarely made ante-mortem. Immunodiagnostic tests are common in human medicine and have been reviewed by Matossian (1977). Tests based upon Arc 5, demonstrable by immunoelectrophoresis (Capron et al. 1970) and double diffusion (Varela-Diaz et al. 1976), are particularly valuable. Radiographic diagnosis is also used.

Treatment. The treatment of *E. granulosus* infection in the dog is discussed on p. 124. Hydatid cysts in man are usually treated surgically. After exposure of the cyst, the cyst fluid is aspirated and a 2.5–10% formalin solution is injected to kill the germinal membrane and any remaining protoscolices. However, this procedure is hazardous since spillage of the cyst fluid, with dissemination of cysts, and leakage of formalin can occur. Kassis and Tanner (1976) propose an alternative technique of the substitution of fresh serum with high complement activity, preferably the patient's own serum, for the formalin. These authors suggest that lysis of the protoscolices remaining in the cyst will occur.

Mebendazole has been used experimentally for the treatment of infections with *E. granulosus* and *E. multilocularis* in laboratory and domestic animals. Oral mebendazole therapy has also been successful in the treatment of humans infected with *Echinococcus* species (Ammann et al. 1979; Wilson et al. 1978). Other compounds are under study (Schantz et al. 1982).

Prevention and control. Programmes for prevention and control are in progress in several countries. The Falkland Islands, New Zealand and Tasmania represent island models and Argentina, Uruguay, Chile, Peru, Bulgaria and parts of the USSR represent continental models. The parasite has been eliminated from Iceland (Beard 1963). Such programmes have been reviewed by Gemmell (1979) and Schwabe (1979); the World Health Organization (1981) has produced guidelines which should be consulted in detail when prevention and control programmes are contemplated.

TAPEWORM INFECTIONS IN DOGS AND CATS

Pathogenesis. Normally, infections with adult cestodes are not very harmful to dogs and cats. However, heavy infections in young animals can cause non-specific abdominal symptoms. There may be diarrhoea or constipation and the animal may have an unthrifty, pot-bellied appearance. Rarely, obstruction of the intestine may occur. Gravid proglottids, particularly those of *D. caninum*, migrate out of the anal sphincter and in the perianal area. The irritation may cause the dog to drag its anus over the ground. However, similar behaviour is associated with pruritus ani and impacted anal glands. Also, proglottids may drop off the animal and migrate over chairs, floors or clothes for some minutes. This is aesthetically unpleasing for the owner.

Diagnosis. Because of the vague clinical signs of most tapeworm infections, many infections are diagnosed on the presence of proglottids in the faeces, on the perianal area or on furniture etc. Differentiation is based primarily on an examination of these proglottids. The characteristic egg capsules of *D. caninum* contain up to 30 eggs. These egg capsules may rupture releasing single eggs which should not be confused with the single eggs within an egg capsule of *Diplopylidium* or *Joyeuxiella* species. Proglottids of *Mesocestoides* species have a single par-uterine organ filled with oncospheres. Proglottids of *Taenia* species are differentiated on the presence of single eggs with a thick, radially striated embryophore. Eggs may also be found in the faeces by sedimentation or flotation techniques. Proglottids of *Echinococcus* species undergo apolysis within the intestine and only eggs are found in the faeces. *Echinococcus* eggs cannot readily be differentiated from those of the *Taenia* species. Thus, if infection with *Echinococcus* is suspected, care should be taken in handling the faecal material. A positive diagnosis is made following purgation of the animal with arecoline hydrobromide (1–2 mg/kg) and the evacuated faeces, especially the mucous portion, are examined microscopically for adult worms. Arecoline acetarsol may be used in cats. Infections with *D. latum* and *Spirometra* species are diagnosed on the presence of the operculate eggs in the faeces using sedimentation techniques.

Treatment. A wide variety of anthelmintics is available for the treatment of cestode infections in dogs and cats. Many should be preceded or succeeded by an overnight fast.

Arecoline hydrobromide (1–2 mg/kg) is effective against most cestodes in dogs, including *E. granulosus*, although it is ineffective against *Mesocestoides* species. The anthelmintic is contraindicated for cats. The antidote is atropine.

Arecoline acetarsol (5 mg/kg) affects *Taenia* species and *D. caninum* and has some effect against *Echinococcus* in dogs and cats. It is unsafe for cats under six months and dogs under three months of age.

Bithionol (200 mg/kg) is effective against the *Taenia* species infecting dogs and cats.

Bunamidine hydrochloride (25–50 mg/kg) is widely used for the treatment of most cestodes, including *E. granulosus*, infecting dogs and cats. The efficacy against *E. granulosus* is increased if two doses of 50 mg/kg are given 48 hours apart.

Dichlorophen (0.3 mg/kg for a dog or 0.1–0.2 mg/kg for a cat) has efficacy against *Taenia* species and *D. caninum* but not *E. granulosus*. It has relatively low toxicity.

Hexachlorophene (15 mg/kg) may be used in the dog.

Mebendazole (animals under 2 kg, 100 mg twice a day for five days; over 2 kg, 200 mg twice a day for five days). Micronized mebendazole is efficacious against both nematodes and cestodes. It is not as effective against *E. granulosus* as against some other cestode genera.

Niclosamide (100–150 mg/kg) is widely used against *D. caninum* and *Taenia* species in dogs and cats. It is not used for *E. granulosus*. The anthelmintic has a wide margin of safety.

Nitroscanate (50 mg/kg) has activity against both nematodes and cestodes. Its effectiveness against cestodes, particularly against *E. granulosus*, would appear to be increased if the anthelmintic is given in a micronized form (Gemmell et al. 1977).

Praziquantel (5 mg/kg) has a wide therapeutic index and is extremely effective against both immature stages and adults of *Taenia* species, *D. caninum*, *Mesocestoides* species and *E. granulosus* and *E. multilocularis* (Thakur et al. 1978). It is probably the drug of choice for the treatment of cestode infections, particularly *Echinococcus*. It does not have any ovicidal activity (Thakur et al. 1979) and hence eggs will continue to be shed and contaminate the environment for several days after treatment. More eggs than usual may thus be introduced into the environment and extra care should be taken because of this potential hazard.

Uredofos, *Diuredosan* (50 mg/kg) has broad-spectrum activity against both cestodes and nematodes in dogs and cats. It is effective against *Taenia* species and *D. caninum*, but less effective against *E. granulosus*.

Streptothricin antibiotics such as SQ21 704 from *Streptomyces griseocarneus* at 50 mg/kg are effective against *Taenia* species and *D. caninum*. Vomiting and diarrhoea can be frequent sequelae of its experimental use (Gemmell et al. 1978).

Control. Since the life-cycles of the cestodes are indirect, prevention of tapeworm infections may be accomplished by the control of the intermediate hosts in association with a regular treatment programme. Dogs should not be fed raw offal or meat from slaughtered animals. Animals must be prevented from catching and eating small wild animals (rats, mice and rabbits) containing the metacestodes of species such as *T. pisiformis* and *T. taeniaeformis*. Fleas should be eliminated from kennels and catteries to control *D. caninum*. Raw fish flesh and offal should not be fed to dogs and cats in order to prevent infection with *D. latum*. However, control of *Mesocestoides* and *Spirometra* species is difficult because of the wide range of intermediate hosts.

At present, control of infection with metacestodes in the intermediate host by means of vaccination seems feasible (see p. 126). This, in turn, would control infection with certain species of cestodes in the dog. Attempts to control infection with adult cestodes in the dog by immunization have met with limited success. Movsevijan et al. (1968) have demonstrated that a degree of resistance can be induced in dogs immunized with irradiated protoscolices of *E. granulosus*. Similarly, Herd et al. (1975) have shown a significant suppression of egg production in dogs immunized with secretory antigens derived from adult *E. granulosus* grown *in vitro*.

Metacestode infections in the intermediate host

Epidemiology. The epidemiology of infection of the intermediate host with metacestodes in the order Taeniidea has been examined extensively by Gemmell and is reviewed by Gemmell and Johnstone (1977). The incidence and prevalence of infection in the intermediate host are affected by a large number of factors. These include:

1. *Level of environmental contamination*. The infected definitive host passes thousands of eggs daily and these may be transmitted to the intermediate host directly, in feed or water or via pasture. When infected definitive hosts have close contact with pasture high levels of contamination develop rapidly. If susceptible animals are exposed to this infection a 'cysticercosis storm' can result. This has also been recorded as a result of close contact between infected humans and feed-lot cattle, and where attendants with contaminated hands teach calves to drink milk from a bucket.

2. *Dispersion of eggs*. On pasture there is rapid dispersion of eggs from the site of contamination for distances of up to 80 m. This dispersion is radial and irrespective of the prevailing winds. The eggs may be dispersed by invertebrates, such as flies (especially blowflies), beetles and earthworms, and by flooding water.

3. *Egg survival.* Eggs will remain viable for periods of over six months, but there is a gradual reduction in the survival rate and evidence that older, senescent eggs are less capable of initiating a viable infection. Senescent oncospheres probably penetrate the tissues but do not develop to maturity. Therefore, they may in fact act as immunizing agents.

4. *Age of host.* When one- to three-week-old lambs, with their mothers, are placed on contaminated pastures they become only lightly infected with *T. ovis.* The lambs are picking up few eggs and/or they are protected by passively transferred immunity. Lambs five weeks or more in age become more heavily infected. This is consistent with a waning of passively transferred immunity and the fact that they are now grazing to a significant extent. However, lambs raised from birth on contaminated pasture gradually develop a strong resistance to reinfection. They may remain lightly infected when compared to susceptible animals introduced to the pasture later in life.

Calves in endemic areas are frequently infected with *T. saginata* in the neonatal period. They do not respond immunologically and this neonatal infection may remain viable for a prolonged period, perhaps for the life of the host. This could contribute to the transmission of the parasite in endemic areas.

5. *Immune response of the host.* Animals develop a marked protective immune response to reinfection with metacestodes. The longevity of this immunity in the absence of reinfection is about one year but it does serve to limit infection and can be exploited by immunization.

6. *Heterologous infection.* Some cross-protection between *Taenia* speceis has been demonstrated experimentally in immunization studies. However, in nature and by means of two egg feeding experiments, it has been demonstrated that heterologous species may exist in the same host and in the same environment.

Treatment. Albendazole, mebendazole and praziquantel have been used experimentally for the treatment of infections with a number of metacestodes. Albendazole (50 mg/kg) given orally was effective against mature metacestodes of *T. saginata* in cattle (Lloyd et al. 1978). However, albendazole given at this rate to naturally infected animals in Kenya failed to reduce the number of viable cysts in treated animals, compared with controls (Stevenson et al. 1981). Praziquantel (four daily doses or a single dose of 50 mg/kg) was highly effective against mature, but less effective against immature, cysticerci (Gallie & Sewell 1978). Mebendazole (50 mg/kg for 14 days administered orally) and praziquantel (50 mg/kg administered subcutaneously) were effective in the treatment of *T. hydatigena* and *T. ovis* in sheep. The lethal effect of both drugs would appear to be related to the biomass of the infection since the anthelmintics were more effective in animals which were lightly infected with *T. hydatigena* than in those more heavily infected with the parasite (Heath & Lawrence 1978).

Praziquantel had no effect on the metacestode of *E. granulosus*, but mebendazole retarded the growth of the metacestodes of this parasite (Heath & Lawrence 1978). Similarly, mebendazole (25 mg/kg for 10 days on two occasions) was effective against the larval stage of *E. granulosus* in pigs. These anthelmintics are likely to be effective against other metacestode infections.

Immunoprophylaxis. Considerable progress has been made in recent years towards the immunization of sheep and cattle against infection with the metacestodes of *T. hydatigena*, *T. ovis* and *T. saginata.* Rickard and his coworkers have demonstrated that secretory/excretory (S/E) antigens, collected from oncospheres of *T. hydatigena* and *T. ovis* cultured *in vitro*, were effective immunizing agents in sheep (Rickard & Adolph 1976). Similarly, these antigens, when administered to pregnant ewes, induced a passive transfer of immunity to the lambs (Rickard & Arundel 1974). In addition, calves can be actively immunized against infection with *T. saginata* using the S/E antigens produced by the oncospheres of *T. saginata* cultured *in vitro* (Rickard & Adolph 1976). Since, in endemic areas, neonatal calves become infected with metacestodes of *T. saginata* and are not capable of responding immunologically to this infection, it is important to protect these animals by passively transferred immunity. Immunization of periparturient cows either intramuscularly or via the intramammary route, and successful passive transfer of immunity to the calves, has now shown that this approach is feasible (Lloyd & Soulsby 1976; Lloyd 1979).

For further progress to be made in immunization against the metacestodes of *T. saginata* and *E. granulosus* alternative sources of antigen are required. This is necessary to avoid the need for human donors of immunizing material and the need to handle eggs and oncospheres of *E. granulosus*. Heterologous immunization is possible since it has been shown that immunization of cattle by intramuscular injection of eggs of *T. hydatigena* will induce partial protection against infection with *T. saginata* (Wikerhauser et al. 1971) The secretory and excretory antigens from oncospheres of *T. taeniaeformis* are capable of inducing a high level of protection against infection with the heterologous parasite, *T. saginata*. This protection is seen following active immunization of calves and following immunization of periparturient cows with passive transfer of immunity to the calves (Lloyd 1979).

Evasion of the immune response. Metacestodes derived from a primary infection survive in the tissues despite the fact that the animals are resistant to reinfection. This is analogous to concomitant immunity in schistosomiasis. Hammerberg and Williams (1978*a*,*b*) have demonstrated that this evasion of the immune response depends, at least in part, on evasion of complement-mediated effector mechanisms. Fluids, extracts and the secretory products of metacestodes are capable of converting C_3 and can deplete haemolytic complement both *in vivo* and *in vitro*. This reaction is not immunological. The active substance would appear to be a polysulphated proteoglycan which may be on the surface of the parasite. Although there is no evidence for systemic consumption of complement, Hammerberg and Williams postulate that local consumption of complement around the metacestode *in vivo* contributes to its evasion of the immune response. Thus, mature metacestodes from a primary infection may evade the action of complement-mediated antibody attack while young metacestodes from succeeding infections are susceptible to this attack.

ORDER: MESOCESTOIDIDEA WARDLE, McLEOD AND RADINOVSKY, 1974

FAMILY: MESOCESTOIDIDAE PERRIER, 1897

Genus: Mesocestoides Valliant, 1863

These are small to medium-sized cestodes. The scolex is unarmed and has four elongate oval suckers and no rostellum. Each proglottid contains a single set of reproductive organs and the genital pore opens on the dorsal surface. In gravid proglottids the oncospheres pass from the uterus into a par-uterine organ. *Mesocestoides* species are parasitic in mammals and charadriiform birds. The parasites are common in a wide variety of carnivores including the dog, birds and occasionally man in Europe, Asia, North America and Africa. Speciation of *Mesocestoides* is controversial but several species are listed by Wardle et al.

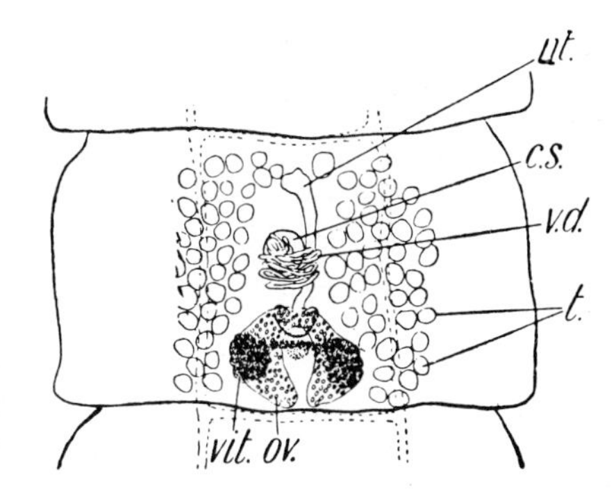

Fig 1.77 Dorsal view of mature segment of *Mesocestoides lineatus*. (*After Baylis*)
cs cirrus-sac
ov ovary
t testes
ut uterus
vd vas deferens
vit vitellarium

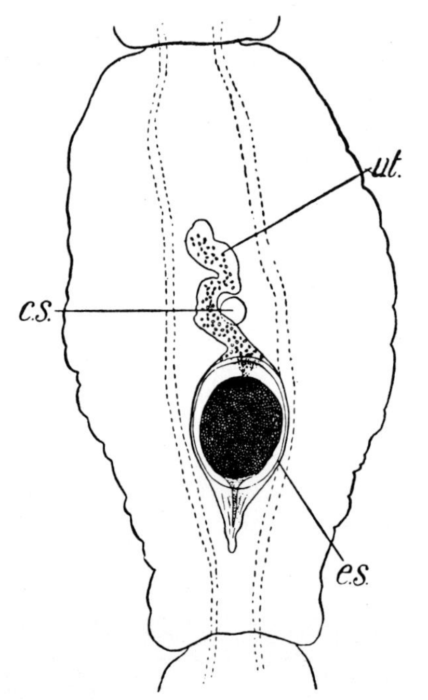

Fig 1.78 Dorsal view of gravid segment of *Mesocestoides lineatus*, (*After Baylis*)
cs cirrus-sac
es egg-sac
ut uterus

(1974). These include **M. variabilis** Mueller, 1927, **M. lineatus** Goeze, 1782, and **M. corti** Hoeppli, 1925.

Life-cycle. The life-cycle of *Mesocestoides* is not fully known but would appear to require two intermediate hosts. Successful experimental infections of coprophagous oribatid mites, in which cysticercoids develop, would suggest that these are the first intermediate hosts. When the infected mite is eaten by a second intermediate host (amphibia, reptiles, birds and mammals such as rodents, dogs and cats) a tetrathyridium is formed. The tetrathyridia are 1 cm or more long and are found in the serous cavities, particularly the peritoneal cavity, and in the liver and lungs. The flat tetrathyridia have an invaginated scolex bearing four suckers. Tetrathyridia will multiply asexually by longitudinal splitting of the parent scolex and massive infections in snakes, lizards and other animals are recorded. When tetrathyridia are ingested by the definitive host they become adult in 16–20 days. However, it has been demonstrated that when tetrathyridia are fed to definitive hosts such as dogs and skunks they can reproduce asexually in the intestine. They may also reach the peritoneal cavity and multiply within it. This asexual reproduction in the definitive host can result in massive infections with *Mesocestiodes*. Tetrathyridia, experimentally injected into the peritoneal cavity, are capable of migrating to the intestine and developing to adults. Presumably, this could occur within the life-cycle.

Pathogenesis. Intestinal infections in the dog are usually harmless, but severe diarrhoea can occur in human intestinal infections.

Infections with tetrathyridia in dogs and cats can cause a severe parasitic peritonitis and ascites.

Treatment. See p. 124.

Genus: Mesogyna Voge, 1952

Par-uterine organ absent; ovary and vitelline glands midline, posteriorly.

Mesogyna hepatica Voge, 1952 occurs in the smaller bile ducts or blood vessels of the liver of *Vulpis macrotis* (kitfox) in California. It is approximately 15 mm long by 0.5 mm. Transmission is probably similar to *Mesocestoides* spp.

ORDER: PROTEOCEPHALIDEA MOLA, 1928

FAMILY: PROTEOCEPHALIDAE LA RUE, 1911

Genus: Proteocephalus Weinland, 1958

Parasites in this genus have an unarmed scolex and four suckers. A fifth apical sucker may be present. The eggs are globular and embryonated when passed in the faeces. The adults are found primarily in fresh-water fishes.

Proteocephalus ambloplitis (Leidy, 1887), the 'bass tapeworm', is of much concern in fish culture in many parts of the world. The life-cycle has been detailed by Fisher and Freeman (1969). It may involve three hosts. Copepods in which procercoids develop are the first intermediate hosts. Plerocercoids are found in the body cavity of the second intermediate hosts, centrarchid and percid fishes. The definitive host is the bass, in which adults develop. However, bass may also act as second intermediate hosts. Procercoids, ingested by bass, migrate to the viscera, particularly gonads, and encyst as plerocercoids. Following stimulation by an increase in temperature, the plerocercoids are then capable of migrating to the intestine of the same bass where they develop to maturity. The occurrence of adults in the intestine is seasonal and peaks in late spring and early summer. The fibrosis caused by the plerocercoids in the gonads affects reproductive potential and may cause sterility.

CLASS: COTYLODA WARDLE, McLEOD AND RADINOVSKY, 1974

ORDER: DIPHYLLIDEA WARDLE, McLEOD AND RADINOVSKY, 1974

This new order contains one family—Diphyllobothriidae—and has as type genus and genotype *Diphyllobothrium latum*. Species in this order are medium to large tapeworms. The scolex has, instead of suckers, narrow, deep, weakly muscular grooves called bothria. Typically the bothria are dorsal and ventral, but they may be inconspicuous or absent. The scolex is unarmed. There is only one set of hermaphrodite reproductive organs in each proglottid. The numerous testes and vitellaria are scattered, the ovary is bilobed and the gravid uterus is a spiral tube. There is a permanent uterine pore usually opening on the ventral side of the proglottid. The eggs are operculate and unembryonated when laid. They may be mistaken for those of trematodes. The life-cycle includes a free-living *coracidium* (a ciliated embryo); a *procercoid* occurring in the first intermediate host, copedid crustaceans; a *plerocercoid* found in the second intermediate hosts, fish; and the definitive hosts (amphibia, reptiles, birds or mammals) contain the adult stage.

FAMILY: DIPHYLLOBOTHRIIDAE LÜHE, 1910

Genus: Diphyllobothrium Lühe, 1910

Diphyllobothrium latum Lühe, 1910 occurs in the small intestine of man, dog, cat, pig, polar bear and other fish-eating mammals in many parts of the world especially the Baltic regions, the Great Lakes area of North America and the USSR. Adults are 2–12 m long or longer, yellowish-grey in colour with dark, central markings caused by the uterus and eggs. Numerous testes and vitellaria lie in the lateral regions of the proglottid and the bilobed ovary is central. The rosette-shaped uterus is central and opens ventrally in the uterine pore, lying immediately behind the genital pore. The eggs, which are passed continuously into the faeces of the host, are light brown, operculate, 67–71 by 40–51 μm and have rounded ends.

Life-cycle. The eggs develop for several weeks to produce a coracidium. This is a six-hooked oncosphere covered with a ciliated embryophore. The motile coracidium hatches into the water and dies fairly soon unless ingested by a suitable immature cyclopid or diaptomid copepod such as *Diaptomus gracilis*. The first larval stage, a procercoid, develops in the haemocele in two to three weeks. Infection of the second intermediate hosts, fresh-water fish, is by ingestion of infected copepods. The parasite, in the viscera or musculature, develops into an elongate solid-bodied plerocercoid with a head resembling the adult. Several species of fresh-water fish can act as second intermediate hosts but the heaviest infections are found in large predatory fish (i.e.

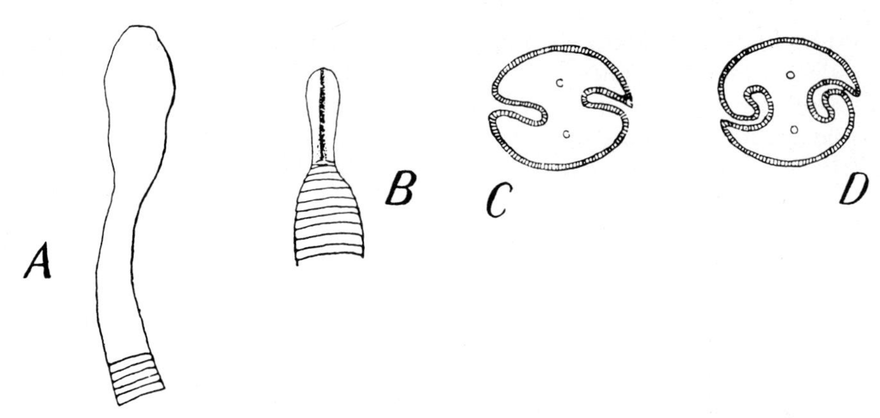

Fig 1.79 Scolex of *Diphyllobothrium latum*. *A*, Lateral view. *B*, Dorsal view. *C*, *D*, Transverse sections at different levels. (*From Baylis, after Stein*)

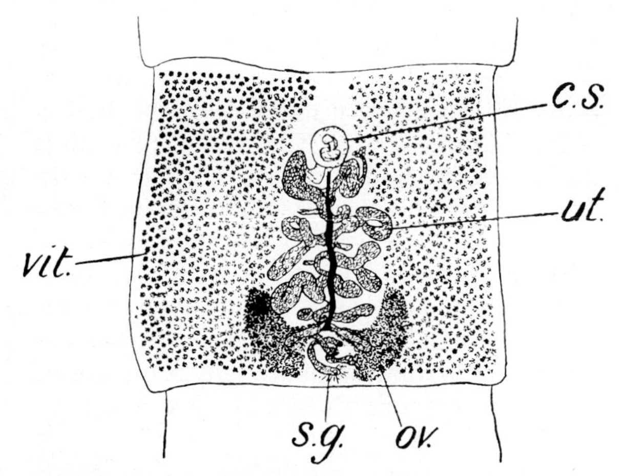

Fig 1.80 Mature segment of *Diphyllobothrium latum*. (*From Baylis, after Stephens*)
cs cirrus-sac
ov ovary
sg shell-gland
ut uterus
vit vitellaria

pike, walleye, barbel, perch, trout). These are infected by eating copepods and smaller infected fish whereupon the plerocercoid transfers into their tissues. The plerocercoids are long-lived and accumulate in the tissues. The definitive host is infected by eating raw fish. The prepatent period in the dog is three to four weeks.

Infection in man occurs in areas where cultural habits include the eating of raw or lightly pickled fish or caviare. Wild fish-eating mammals are readily infected and the incidence of infection in pigs and dogs can be high in areas where they are offered raw fish offal.

Pathogenesis in man. Infection in man can cause non-specific abdominal symptoms. In a small percentage of cases, occurring particularly in the Baltic regions, a macrocytic hypochromic anaemia develops resulting from a competition between parasite and host for vitamin B12. This occurs when the parasite is situated higher in the intestine than usual.

Diagnosis in man is based on symptoms and the presence of the characteristic operculate eggs in the faeces.

Treatment and prophylaxis. Treatment is similar to that described for *T. saginata* and *T. solium*. Praziquantel (25 mg/kg), niclosamide and quinacrine would be the drugs of choice.

Prophylaxis involves the freezing or cooking of fish. Animals should not be fed raw fish. Precautions should be taken against raw sewage reaching fresh water lakes in endemic areas.

Other species of *Diphyllobothrium* have been recorded in man. These include **D. dentriticus** Nitsch, 1824 (normal definitive host gulls); **D. dalliae** Rausch, 1956; **D. pacificum** Baer, 1969 (sea lion); **D. strictum** Markowski, 1952; and **D. minus** Cholodkovsky, 1916, the last two being considered by some to be synonymous with *D. dendriticus*.

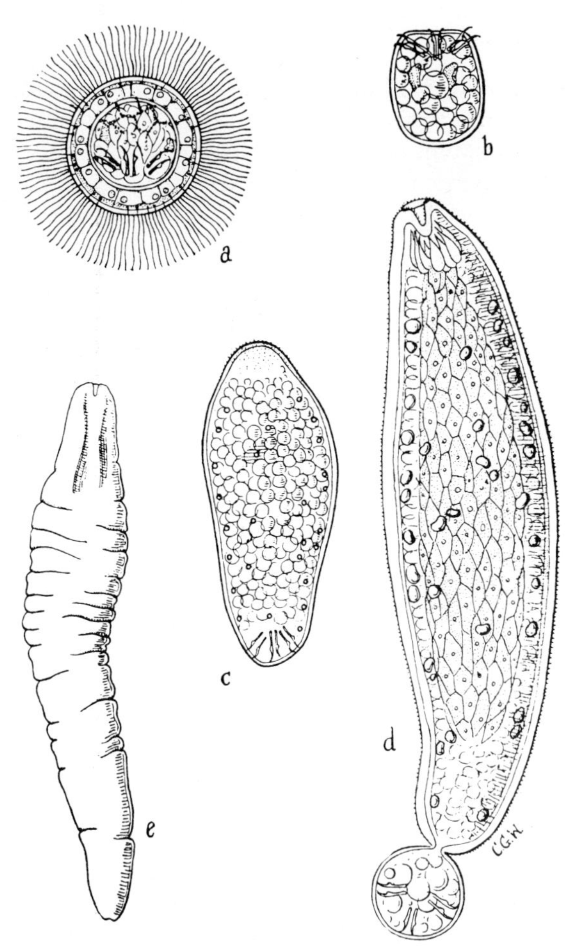

Fig 1.81 Developmental stages of *Diphyllobothrium latum*. *a*, Coracidium. *b*, Onchosphere from body cavity of *Cyclops*, five days after passage through intestinal wall. *c*, Young procercoid from body cavity of *Cyclops*. *d*, Mature procercoid. *e*, Young plerocercoid from body-cavity of pike.

Diplogonoporus grandis Blanchard, 1894 has been found in humans eating marine fish from Japanese waters. The normal definitive host is probably a seal.

Pathogenesis in fish. Plerocercoids of *D. latum* and other species of *Diphyllobothrium* are found lying in or encysted in the viscera and musculature of marine and fresh-water fish. Plerocercoids of *D. latum* are found in a wide variety of fresh-water fishes while those of other *Diphyllobothrium* species are found typically in salmonids and coregonids. Migrating larvae in small fishes can cause much damage with adhesions, sterility and even mortality. Their presence is unaesthetic and affects the market value of the fish (Roberts 1978).

Genus: Spirometra Mueller, 1937

Parasites in this genus are small to medium-sized tapeworms which are morphologically very similar to *Diphyllobothrium* species. They differ in that the uterus is spiraled and not rosette-form in arrangement; the uterus and vagina open separately on the ventral surface of the proglottid; the eggs are pointed, rather than rounded, at each end. The adults are found in wild carnivores and in domestic cats and dogs.

The systematics of *Spirometra* are still controversial but several species are recognized by Wardle et al. (1974). These include, **S. mansonoides** Mueller, 1935, which is found in cats, dogs and racoons in North America and is recorded in South America; **S. mansoni** Joyeux and Houdemer, 1927, which occurs in cats and dogs in the Far East and South America; **S. erinacei** Faust, Campbell and Kellog, 1929, which occurs in cats and dogs in Australia and the Far East; and **S. felis** Southwell, 1928 which has been found in large Felidae in zoos.

Life-cycle. The life-cycle of *S. mansonoides* has been extensively studied and is described by Mueller (1974). Pinkish-coloured adults, which may survive for several years, are found in the jejunum of cats, dogs and racoons. The eggs are passed continuously in the faeces and develop to produce a coracidium which hatches and is swallowed by the early life-cycle stages of fresh-water crustaceans in the genus *Cyclops*. The mature procercoid is reached in 10–14 days. Plerocercoids or *spargana* occur in the second intermediate hosts. The latter are principally water snakes, but tadpoles, other amphibia, alligators, birds and mammals, including man, may also be infected. The plerocercoids are white, ribbon-like structures which may reach several centimetres in length. The life-cycle is completed when cats ingest infected second intermediate hosts. The prepatent period is 10–30 days. Cats, dogs and racoons may be infected with both plerocercoids and adults and the plerocercoids in the tissues of cats and dogs probably can migrate to the gut lumen to continue development in the same host.

Pathogenesis. Adult *Spirometra* species are rarely pathogenic but the plerocercoids are of public health importance as a cause of sparganosis in man. Humans may be infected in three ways:

1. By the accidental ingestion of crustacea infected with procercoids. The procercoids migrate to the subcutaneous tissues and musculature and develop as plerocercoids.

2. By the ingestion of plerocercoids in second intermediate hosts. Pigs, particularly feral pigs raised for human consumption, in Australia and the Far East may be infected with plerocercoids of *S. erinacei* or other species (Bearup 1953). The plerocercoids or spargana, which are easily mistaken for nerves, are found in the connective tissue of the muscles, particularly of the abdomen and hind-legs, and also under the peritoneum, pericardium and pleura. On ingestion by humans (unsuitable as definitive hosts) the plerocercoids could migrate to the tissues and reestablish.

3. When infected frog or snake flesh is used as a dressing for wounds or eyes the plerocercoids may migrate into the human flesh or eye.

The spargana in man migrate, primarily subcutaneously, causing inflammation, urticaria, oedema and eosinophilia. Spargana in man are found principally in the subcutaneous tissues and muscles and are often in the periorbital area.

Treatment. Treatment of humans is by surgical removal. For the treatment of dogs and cats, see p. 124.

Genus: Ligula Bloch, 1782

Ligulosis caused by plerocercoids of parasites such as **Ligula intestinalis** Goeze, 1782, is a common and economically important disease of fresh-water fishes, particularly those in lakes and reservoirs, in most parts of the world. The life-cycle is similar to that of *D. latum*. Adult ligulids up to 28 cm in length are found in piscivorous diving and wading birds. The life-span of the adult tapeworms is only a few days. Procercoids occur in copepods and plerocercoids are found principally in cyprinid fishes. The plerocercoids are long, up to 40 cm, slender, unsegmented and found in the body cavity of fish, particularly carp. The weight of the parasite tissue may exceed that of the host. The large plerocercoids displace internal organs causing functional disturbances and infected fish lose condition and have swollen abdomens. There is atrophy of the gonads due to a derangement in hormonal balance. Control involves the removal of infected fish and suppression of piscivorous birds on fish farms.

Genus: Schistocephalus Creplin, 1829

Schistocephalus solidus Mueller, 1776. Adults are found in the intestine of a wide variety of birds and *S. solidus* has been associated with fatal infections in ducks. The life-cycle is similar to that of *Ligula* species with copepod first intermediate hosts and fish second intermediate hosts. The plerocercoids are large and have a triangular, notched scolex. Pathogenicity is fairly similar to that of *Ligula* species and has been described by Arme and Owen (1967). The parasite has been introduced into fish aquaria through the feeding of live crustaceans.

ORDER: PSEUDOPHYLLIDEA CARUS, 1863

FAMILY: TRIAENOPHORIDAE LOENNBERG, 1793

Genus: Triaenophorus Rudolphi, 1793

Adult *Triaenophorus* species occur in the intestine of predatory fish, particularly pike, in many areas of the world such as the USSR, Europe and North America. The scolex is armed with trident-shaped hooks on the apical disc and the bothria are shallow and pyriform. Copepods are the first intermediate host. The second intermediate hosts are a variety of fish species, particularly trout, and the plerocercoids, the most pathogenic stage, are found encysted in the viscera and musculature. Plerocercoids in the musculature reduce the commercial value of the fish. Also, *T. nodulosus* causes destruction of liver tissue in trout and deaths may occur in heavily infected young trout within one or two weeks of infection. The presence of pike is necessary for the condition to occur on trout farms and control involves the drainage of ponds and the removal of pike.

FAMILY: AMPHICOTYLIDAE ARIOLA, 1899

Genus: Eubothrium Nybelin, 1922

Adults in this genus are small and have simple circular bothria and are parasitic in teleosts. They are particularly common in wild and cultivated salmonids in lakes and reservoirs in Europe and North America. Adults may be 50 cm or more in length and the life-cycle involves either only a copepod intermediate host or copepods and fish (sticklebacks) may act as intermediate hosts.

ORDER: SPATHEBOTHRIIDEA WARDLE AND McLEOD, 1952

FAMILY: CYATHOCEPHALIDAE NYBELIN, 1868

Genus: Cyathocephalus Kessler, 1868

These parasites have a funnel-shaped apical sucking organ on the scolex. The strobila is without marked external segmentation but contains 20–45 sets of reproductive organs. The adults are sexually precocious neotenic procercoids.

Cyathocephalus truncatus (Pallas, 1781) is a common parasite of white fish and trout in northern Europe, the USSR and North America. It is a parasite of the pyloric caeca and in large numbers can be of major pathogenic importance. Heavy infections cause retardation in growth and slow maturation of the ovaries. Sometimes, particularly in the spawning season, the parasite can cause mortality.

ORDER: CARYOPHYLLIDEA BENEDEN (IN OLSSON, 1893)

Parasites in this order are small and have a single set of reproductive organs in an unsegmented body. The life-cycle involves aquatic oligochaetes as intermediate hosts and the adults are found in the alimentary tract of certain families of fresh water fishes in most parts of the world.

FAMILY: CARYOPHYLLAEIDAE LEUCKART, 1878

Genus: Caryophyllaeus Mueller, 1787

Caryophyllaeus fimbriceps Chlopina, 1924, and **C. laticeps** (Pallas, 1781) are important parasites of natural and cultured carp in many areas of the world, particularly Europe and the USSR. The infection in carp is seasonal and the numbers of adults in the intestine peak in spring. There is a decrease in the prevalence and intensity of infection in the autumn. These species are important parasites of carp up to two years of age and cause a sharp deterioration in the condition of the fish. Mortality may be high. The intermediate hosts are oligochaete annelids. Control therefore involves reducing the number of aquatic oligochaetes and this may be accomplished by draining and drying the ponds.

FAMILY: LYTOCESTIDAE HUNTER, 1927

Genus: Khawia Hsü, 1935

Khawia sinensis Hsü, 1935 is found in the Far East and Europe. The life-cycle is similar to that of *Caryophyllaeus* species. Adults in the intestine of carp cause a haemorrhagic enteritis.

CESTODE INFECTIONS OF FISHES

Cestode infections are very common in fish in nature and can be of importance in cultured fish. They are not of particular importance in aquaria fish. Fish may be infected with adult cestodes residing in the intestine and pyloric caeca. These adult infections are often of little importance, but heavy infections with parasites such as *C. truncatus* and *C. fimbriceps* may result in enteritis and a retardation of growth and even death of the fish. Fish are also infected with the plerocercoid stage of a variety of cestodes. Plerocercoids in the musculature are aesthetically unpleasing and although the majority are not of public health importance they are of economic importance since their presence can decrease the commercial and sporting value of the fish. The infection of fish with plerocercoids of *D. latum* is of public health importance. Plerocercoids migrating and encysting in the viscera can be of major importance causing adhesions, disturbance in metabolism, destruction of the gonads (resulting in decreased sexual potential or sterility), loss of condition and perhaps mortality. The cestodes of fishes are described in more detail in Hoffman (1967), Dogiel et al. (1961) and Needham and Wootten (1978).

Treatment. Anthelmintic treatment of cestode infections in fish is not far advanced. Some anthelmintics that have been administered orally are phenothiazine (4 g/fish for three days); di-n butyl tin oxide (25 g/100 kg for three days); phenasal; and niclosamide. Praziquantel and many of the benzimidazole anthelmintics are worthy of trial as additives in the feed. Further details may be found in Hoffman and Meyer (1974).

Prophylaxis. Control of parasitic diseases of cultured fish is described by Needham and Wootten (1978). Fry should be raised separately from older, possibly carrier fish, and efforts should be made to control the intermediate hosts of the cestodes. Water inflow should be supplied by a source from which wild fish and piscivorous birds are excluded. Carrier wild fish and infected crustacea can be screened from the water inflow by mesh or sand filters, respectively. Intermediate hosts such as copepods and oligochaetes may be removed from ponds by periodic drainage and treatment with quicklime or rotenone. Piscivorous birds should be kept away from ponds by screens or wire stretched over the ponds.

REFERENCES

CESTODES—GENERAL

Berntzen, A. K. (1970) Continuous axenic culture of the successive stages of the life cycle of *Hymenolepis nana*. *J. Parasit.*, **56**, (Sect II, Part 2) 397.
Berntzen, A. K. & Mueller, J. F. (1972) *In vitro* cultivation of *Spirometra* spp. (Cestoda) from the plerocercoid to the gravid adult. *J. Parasit.*, **58**, 750–752.
Heath, D. D. (1971) The migration of oncospheres of *Taenia pisiformis*, *T. serialis* and *Echinococcus granulosus* within the intermediate host. *Int. J. Parasit.*, **1**, 145–152.
Heath, D. D. (1973) An improved technique for the *in vitro* culture of taeniid larvae. *Int. J. Parasit.*, **3**, 481–484.
Morseth, D. J. (1966) Chemical composition of embryonic blocks of *Taenia hydatigena*, *Taenia ovis* and *Taenia pisiformis* eggs. *Exp. Parasit.*, **18**, 347–354.
Read, C. P. (1966) Nutrition of intestinal helminths. In: *Biology of Parasites*, ed. E. J. L. Soulsby, pp. 101–126. New York: Academic Press.
Rybicka, K. (1966) Embryogenesis in cestodes. *Adv. Parasit.*, **4**, 107–186.
Šlais, J. (1973) Functional morphology of cestode larvae. *Adv. Parasit.*, **11**, 396–480.
Smyth, J. D. (1972) Changes in the digestive-absorptive surface of cestodes during larval adult differentiation. *Symp. Br. Soc. Parasit.*, **10**, 41–70.
Smyth, J. D. (1976) *Introduction to Animal Parasitology*, p. 466. London: Hodder and Stoughton.
Smyth, J. D. & Davies, Z. (1974) *In vitro* culture of the strobilar stage of *Echinococcus granulosus* (sheep strain); a review of basic problems and results. *Int. J. Parasit.*, **4**, 631–644.
Wardle, R. A., McLeod, J. A. & Radinovsky, S. (1974) *Advances in the Zoology of Tapeworms*, 1950–1970, p. 270. Minneapolis: University of Minnesota Press.

ANOPLOCEPHALIDEA

Allen, R. W. (1959) Preliminary notes on the larval development of the fringed tapeworm of sheep *Thysanosoma actinioides* Diesing, 1834, in Psocids (Psocoptera: Corodentia). *J. Parasit.*, **45**, 537–538.
Allen, R. W. (1973) The biology of *Thysanosoma actinioides* (Cestoda: Anoplocephalidae) a parasite of domestic and wild ruminants. *Bull. agric. Expt. Stn, New Mexico State Univ.*, **604**, 69.
Arnjadi, A. R. (1971) Studies on histopathology of *Stilesia globipunctata* infections in Iran. *Vet. Rec.*, **88**, 486–488.
Bankov, D. (1976) Diagnosis and treatment of *Stilesia* infection in sheep. *VetMed. Nauk.*, **13**, 28–36.
Christenson, R. O. (1931) An analysis of reputed pathogenicity of *Thysanosoma actinioides* in adult sheep. *J. agric. Res.*, **42**, 245–249.
Dey-Hazra, A. (1976) The efficacy of Droncit (R) praziquantel against tapeworm infections in dog and cat. *Vet. Med. Rev.*, **2**, 134–141.
Horstmann, R., Bienzle, U., Kern, P. & Voelker, J. (1978) Tapeworm infestation with *Inermicapsifer inadagascariensis. Tropenmed. Parasit.*, **29**, 406–408.
Kates, K. C. & Goldberg, A. (1951) The pathogenicity of the common sheep tapeworm, *Moniezia expansa. Proc. Helminth Soc. Wash.*, **18**, 87–101.
Kelly, J. D. & Bain, S. A. (1975) Critical test evaluation of micronized mebendazole against *Anoplocephala perfoliata* in the horse. *N.Z. vet. J.*, **23**, 229–232.
Mishareva, T. E. (1977) Special features of the control of helminth infections on industrial sheep farms. *Veterinariya, Kiev*, **45**, 67.
Reid, W. M. (1962) Chicken and turkey tapeworms. *Handbook to Aid in Identification and Control of Tapeworms Found in the United States of America*, p. 71. Athens, Georgia. Georgia Agricultural Experiment Stations..
Safaev, Y. S. (1972) Efficacy of phenasal against anoplocephalids in horses. *Veterinariya, Moscow*, **49**, 68–69.
Sengbusch, H. G. (1977) Review of oribatid mite–anoplocephalan tapeworm relationships (Acari; Oribatei; Cestoda; Anoplocephalidae). *Proc. Symp. east. Branch Ent. Soc. Am.*, 87–102.
Svadzhian, P. K. (1963) Development of *Thysaneizia giardi* (Moniez, 1879) in the bodies of insects of the order of Psocids (Psocoptera). *Dokl. Akad. Nauk armyan, SSSR*, **36**, 303–306.
Vibe, P. P. (1976) Parasitic coenurosis in sheep. *Veterinariya, Moscow*, 6, 58–60.

TAENIIDEA

Amman, R., Akovbiantz, A. & Eckert, J. (1979) Chemotherapie der Echinokokkose des menschen mit Mebendazol (Vermox). *Schweiz. med. Wschr.*, **109**, 148–151.
Beard, T. C. (1973) The elimination of echinococcosis from Iceland. *Bull. Wld Hlth Org.*, **48**, 653–660.
Capron, A., Yarzabal, L., Vernes, A. & Fruit, J. (1970) Le diagnostic immunologique de l'echinoccose humaine. *Path. Biol.*, **18**, 357–365.
Clapham, P. A. (1942) On identifying *Multiceps* spp. by measurement of the large hook. *J. Helminth.*, **20**, 31–40.
Cranley, J. J. (1982) Survey of equine hydatidosis in Great Britain. *Equine vet. J.*, **14**, 153–157.
Dewhurst, L. W., Cramer, J. D. & Sheldon, J. J. (1967) An analysis of current inspection procedures for detecting bovine cystericercosis. *J. Am. vet. med. Ass.*, **150**, 412–417.
Eckert, J., Muller, B. & Partridge, A. J. (1974) The domestic cat and dog as natural definitive hosts of *Echinococcus (Alveococcus) multilocularis* in Southern Federal Republic of Germany. *Tropenmed. Parasit.*, **25**, 334–337.
Esch, G. W. & Self, J. T. (1965) A critical study of the taxonomy of *Taenia pisiformis* Bloch, 1780, *Multiceps multiceps* Leske, 1780 and *Hydatigera taeniaeformis* Batsch, 1786. *J. Parasit.*, **51**, 932–937.
Flisser, A., Perez-Montford, R. & Larralde, C. (1979) The immunology of human and animal cysticercosis: a review. *Bull. Wld Hlth Org.*, **57**, 839–856.
Gallie, G. J. & Sewell, M. M. H. (1978) The efficacy of praziquantel against the cysticerci of *T. saginata* in calves. *Trop. Anim. Hlth Prod.*, **10**, 36–38.
Gemmell, M. A. (1979) Hydatidosis control: a global view. *Aust. vet. J.*, **55**, 118–125.
Gemmell, M. A. & Johnstone, P. D. (1977) Experimental epidemiology of hydatidosis and cysticercosis. *Adv. Parasit.*, **15**, 311–369.
Gemmell, M. A., Johnstone, P. D. & Oudemans, G. (1977) The effect of micronized nitroscanate on *Echinococcus granulosus* and *Taenia hydatigena* infections in dogs. *Res. vet. Sci.*, **22**, 391–392.
Gemmell, M. A., Johnstone, P. D. & Oudemans, G. (1978) The effect of an antibiotic of the streptothricin family against *Echinococcus granulosus* and *Taenia hydatigena* infections in dogs. *Res. vet. Sci.*, **25**, 109–110.
Hammerberg, B. & Williams, J. F. (1978*a*) Interaction between *Taenia taeniaeformis* and the complement system. *J. Immunol.*, **120**, 1033–1038.
Hammerberg, B. & Williams, J. F. (1978*b*) Physicochemical characterization of complement-interacting factors from *Taenia taeniaeformis. J. Immunol.*, **120**, 1039–1045.
Hammerberg, B., MacInnis, G. A. & Hyler, T. (1978) *Taenia saginata* in grazing steers in Virginia. *J. Am. vet. med. Ass.*, **173**, 1462–1464.
Hatch, C. (1975) Observations on the epidemiology of equine hydatidosis in Ireland. *Irish vet. J.*, **29**, 155–157.
Heath, D. D. (1971) The migration of oncospheres of *Taenia pisiformis*, *T. serialis* and *Echinococcus granulosus* within the intermediate host. *Int. J. Parasit.*, **1**, 145–152.
Heath, D. D. & Lawrence, S. B. (1978) The effect of mebendazole and praziquantel on the cysts of *Echinococcus granulosus*, *Taenia hydatigena* and *T. ovis* in sheep. *N.Z. vet. J.*, **26**, 11–15.
Herd, R. P., Chappel, R. J. & Biddell, D. (1975) Immunization of dogs against *Echinococcus granulosus* using worm secretory antigens. *Int. J. Parasit.*, **5**, 395–399.
Hilwig, R. W., Cramer, J. D. & Forsyth, K. S. (1978) Freezing times and temperatures required to kill cysticerci of *Taenia saginata* in beef. *Vet. Parasit.*, **4**, 215–219.
Jepsen, A. & Roth, H. (1949) Epizootiology of *Cysticercus bovis*—resistance of the eggs of *Taenia saginata. Proc. 14th int. vet. Congr. Lond.*, **2**, 43–50.
Juranek, D. D., Forbes, L. S. & Keller, U. (1976) *Taenia saginata* cysticerci in the muscles of beef cattle. *Am. J. vet. Res.*, **37**, 785–789.
Kassis, A. I. & Tanner, C. E. (1976) The role of complement in hydatid disease. *In vitro* studies. *Int. J. Parasit.*, **6**, 25–35.
Leiby, P. D. & Dyer, W. G. (1971) Cyclophyclidean tapeworms of wild carnivora. In: *Parasitic Diseases of Wild Mammals*, ed. J. W. Davis & R. C. Anderson, pp. 174–234, Ames: Iowa State University Press.
Lloyd, S. (1979) Homologous and heterologous immunization against the metacestodes of *Taenia saginata* and *Taenia taeniaeformis* in cattle and mice. *Z. Parasitenk.*, **60**, 87–96.
Lloyd, S. & Soulsby, E. J. L. (1976) Passive transfer of immunity to neonatal calves against the metacestodes of *Taenia saginata. Vet. Parasit.*, **2**, 355–362.
Lloyd, S., Soulsby, E. J. L. & Theodorides, V. J. (1978) Effect of albendazole on the metacestodes of *Taenia saginata* in calves. *Experientia*, **34**, 723–724.
McManus, D. (1960) Prenatal infection of calves with *Cysticercus bovis. Vet. Rec.*, **72**, 847–848.
McManus, D. P. & Smyth, J. D. (1979) Isoelectric focusing of some enzymes from *Echinococcus granulosus* (horse and sheep strains) and *E. multilocularis. Trans. R. Soc. trop. Med. Hyg.*, **73**, 259–265.
Matossian, R. M. (1977) The immunological diagnosis of human hydatid disease. *Trans. R. Soc. trop. Med. Hyg.*, **71**, 101–104.
Movsesijan, M., Sokolic, A. & Mladenovic, Z. (1968) Studies on the immunological potentiality of irradiated *Echinococcus granulosus* forms: immunization experiments in dogs. *Br. vet. J.*, **124**, 425–432.
Pawlowski, Z. & Schultz, M. G. (1972) Taeniasis and cysticercosis (*Taenia saginata*). *Adv. Parasit.*, **10**, 269–343.
Pawlowski, Z., Kozakiewicz, B. & Zatonski, J. (1976) Effect of mebendazole on hydatid cysts in pigs. *Vet. Parasit.*, **2**, 299–302.
Penfold, H. B. (1937) The life history of *Cysticercus bovis* in the tissues of the ox. *Med. J. Aust.*, **1**, 579–583.
Rausch, R. & Schiller, E. (1956) Studies on the helminth fauna of Alaska. XXV. The ecology and public health significance of *Echinococcus sibiricensis*, Rausch and Schiller 1954 on St. Lawrence Island. *J. Parasit.* **46**, 395–419.
Rickard, M. D. & Adolph, A. J. (1976) Vaccination of calves against *Taenia saginata* infection using a 'parasite-free' vaccine. *Vet. Parasit.*, **1**, 389–392.
Rickard, M. D. & Arundel, J. H. (1974) Passive protection of lambs against infection with *Taenia ovis* via colostrum. *Aust. vet. J.*, **50**, 22–24.
Rickard, M. D., White, J. B. & Boddington, E. B. (1976) Vaccination of lambs against infection with *Taenia ovis. Aust. vet. J.*, **52**, 209–214.
Ronshina, G. I. (1953) The early diagnosis of coenuriasis in sheep. *Izadatelstvo Akad. Nauk.*, 587–597.
Rybicka, K. (1966) Embryogenesis in cestodes. *Adv. Parasit.*, **4**, 107–186.
Sachs, R. (1969) Untersuchungen zur Artbeststimmung und Differenzierung der Muskelfinnen ostafrikanischer Wildtiere. *Z. Tropenmed. Parasit.*, **20**, 39–50.

Schantz, P. M., van den Bossche, H. & Eckert, J. (1982) Chemotherapy for larval echinococcosis in animals and humans. Report of a workshop. *Z. Parasitenk*, **67**, 5–26.
Schenone, H., Ramirez, R. & Rojas, A. (1973) Aspectos epidemiologicos de la neurocisticercosis en America Latina. *Boln Chile Parasit.*, **28**, 61–72.
Schwabe, C. W. (1979) Epidemiological aspects of the planning and evaluation of hydatid disease control. *Aust. vet. J.*, **55**, 109–117.
Seddon, H. R. (1950) *Diseases of Animals in Australia, Part I, Helminth Infestations*, Publication No. 5. Commonwealth of Australia, Department of Health, Service Publications (Division of Veterinary Hygiene) No. 6.
Silverman, P. H. & Griffiths, R. B. (1955) A review of methods of sewage disposal in Great Britain, with special reference to the epizootiology of *Cysticercus bovis*. *Ann. trop. Med. Parasit.*, **49**, 436–450.
Slonka, G. F., Moulthrop, J. I., Dewhurst, L. W., Hotchkiss, P. M., Vallaza, B. & Schultz, M. G. (1975) An epizootic of bovine cysticercosis. *J. Am. vet. Med. Ass.*, **166**, 678–681.
Smyth, J. D. (1964) The biology of the hydatid organisms. *Adv. Parasit.*, **2**, 169–219.
Smyth, J. D. (1969) The biology of the hydatid organisms. *Adv. Parasit.*, **7**, 327–347.
Smyth, J. D. (1974) Occurrence of physiological strains of *Echinococcus granulosus* demonstrated by *in vitro* culture of protoscolices from sheep and horse hydatid cysts. *Int. J. Parasit.*, **4**, 443–445.
Smyth, J. D. (1977) Strain difference in *Echinococcus granulosus*, with special reference to the status of equine hydatidossi in the United Kingdom. *Trans. R. Soc. trop. Med., Hyg.*, **71**, 93–100.
Smyth, J. D. & Davies, Z. (1974) *In vitro* culture of the strobilar stage of *Echinococcus granulosus* (sheep strain): a review of the basic problems and results. *Int. J. Parasit.*, **4**, 631–644.
Smyth, J. D., McManus, D., Barrett, N. J., Bryceson, A. & Cowie, A. G. A. (1980) *In vitro* culture of human hydatid material. *Lancet*, **1**, 202–203.
Smyth, J. D. & Smyth, M. M. (1964) Natural and experimental hosts of *Echinococcus granulosus* and *E. multilocularis*, with comments on the genetics of speciation in the genus *Echinococcus*. *Parasitology*, **54**, 493–514.
Soulsby, E. J. L. (1965) *Textbook of Veterinary Clinical Parasitology*, I. *Helminths*, p. 1120. Oxford: Blackwell Scientific.
Stevenson, P., Holmes, P. W. & Muturi, J. M. (1981) Effect of albendazole on *Taenia saginata* cysticerci in naturally infected cattle. *Vet. Rec.*, **109**, 82.
Stoll, N. R. (1947) This wormy world. *J. Parasit.*, **33**, 1–18.
Subianto, D. B., Tumada, L. R., & Margono, S. S. (1978) Burns and epileptic fits associated with cysticercosis in mountain people of Irian Jaya. *Trop. geogr. Med.*, **30**, 275–278.
Thakur, A. S., Prezioso, U. & Marchevsky, N. (1979) *Echinococcus granulosus*: ovicidal activity of praziquantel and bunamidine hydrochloride. *Exp. Parasit.*, **17**, 131–133.
Thakur, A. S., Prezioso, U. & Marchevsky, N. (1978) Efficacy of Droncit against *Echinococcus granulosus* infection in dogs. *Am. J. vet. Res.*, **39**, 859–860.
Thompson, R. C. A. (1977) Hydatidosis in Great Britain. *Helminth. Abstracts*, Series A, **46**, 837–861.
Thompson, R. C. A. (1978) Aspects of speciation in *Echinococcus granulosus*. *Vet. Parasit.*, **4**, 121–125.
Thompson, R. C. A. & Smyth, J. D. (1976) Attempted infection of the rhesus monkey (*Macaca mulatta*) with the British horse strain of *Echinococcus granulosus*. *J. Helminth.*, **50**, 175–177.
Varela-Diaz, V. M., Coltori, E. A., Ricardes, M. I. & Prezioso, U. (1976) Evaluation of immunodiagnostic techniques for the detection of human hydatid cyst carriers in field studies. *Am. J. trop. Med. Hyg.*, **25**, 617–622.
Verster, A. (1969) A taxonomic revision of the genus *Taenia* Linnaeus, 1758 s. str. *Onderstepoort J. vet. Res.*, **37**, 3–58.
Wikerhauser, T., Zukovic, X. & Dzakula, N. (1971) *Taenia saginata* and *T. hydatigena*: intramuscular vaccination of calves with oncospheres. *Exp. Parasit.*, **30**, 36–40.
Wilson, J. F., Davidson, M. & Rausch, R. L. (1978) A clinical trial of mebendazole in the treatment of alveolar hydatid disease. *Am. Rev. resp. Dis.*, **118**, 747–757.
World Health Organization (1981) *Guidelines on Echinococcosis/Hydatidosis Surveillance, Prevention and Control*, ed. J. Eckert, M. A. Gemmell & E. J. L. Soulsby. Geneva.

MESOCESTOIDIDEA; PROTEOCEPHALIDEA

Fisher, J. & Freeman, R. S. (1969) Penetration of parenteral plerocercoids of *Proteocephalus ambloplitis* (Leidy) into the gut of small mouth bass. *J. Parasit.*, **55**, 766–774.
Wardle, R. A., McLeod, J. A. & Radinovsky, S. (1974) *Advances in the Zoology of Tapeworms*, 1950–1970, p. 274. Minneapolis: University of Minnesota Press.

COTYLODA

Arme, C. & Owen, R. w. (1967) Infections of the three-spined stickleback *Gasterosteus aculeatus* (L.) with the plerocercoid larvae of *Schistocephalus solidus* (Mull.) with special reference to pathological effects. *Parasitology*, **57**, 301–314.
Bearup, A. J. (1953) Life history of a spirometrid tapeworm, causing sparganosis in feral pigs. *Aust. vet. J.*, **29**, 217–224.
Dogiel, V. A., Petrushevski, G. K. & Polyanski, Ya. I. (1958) *Parasitology of Fishes*, trans. Z. Kabata, p. 384. Edinburgh: Oliver and Boyd.
Hoffman, G. L. (1967) *Parasites of North American Fresh Water Fishes*, p. 486. University of California Press.
Hoffman, G. L. & Meyer, F. P. (1974) *Parasites of Freshwater Fishes: A Review of their Control and Treatment*, p. 224. Neptune, N. J.: T. F. H. Publishers.
Mueller, J. F. (1974) The biology of *Spirometra*. *J. Parasit.*, **60**, 3–14.
Needham, T. & Wootten, R. (1978) The parasitology of teleosts. In: *Fish Pathology*, ed. R. J. Roberts, pp. 144–182. London: Baillière Tindall.
Wardle, R. A., McLeod, J. A. & Radinovsky, S. (1974) *Advances in the Zoology of Tapeworms*, 1950–1970, p. 274. Minneapolis: University of Minnesota Press.
Roberts, R. J. (ed.) (1978) *Fish Pathology*, London: Baillière Tindall.

Nematodes

CLASS: NEMATODA RUDOLPHI, 1808

The Nematodes are free-living or parasitic, unsegmented worms, usually cylindrical and elongate in shape. An alimentary canal is present. With a few exceptions the sexes are separate and the life-cycle may be direct or include an intermediate host.

Morphology

The shape of the body is elongate, cylindrical and tapering at the extremities. A few exceptions occur; for instance, the females of *Tetrameres*, which swell up after copulation, becoming almost spherical, and those of *Simondsia*, in which the posterior part of the body also assumes a globular shape.

The body is unsegmented, but the cuticle which forms the covering is usually provided with circular annulations not readily visible to the naked eye, or it may be smooth or have longitudinal striations. The cuticula is relatively thick in nematodes and is continuous with the cuticular lining of the buccal cavity, the oesophagus, the rectum and the distal portions of the genital ducts. It may form special adhesive structures; for instance, hooks (*Rictularia*, *Tetrameres* males), simple or more complicated thickenings (*Gongylonema*, *Acuaria*) or a cephalic collar (*Physaloptera*). Many species have lateral flattened cuticular expansions, called *alae*, especially in the cervical region (*Toxascaris, Physocephalus, Oesophagostomum*), and in most species the males bear cuticular expansions at the posterior extremity. Electron microscopy studies of the cuticle of various nematodes show, in general, a number of layers consisting of an outer membrane, a cortical layer, a matrix and fibre layers. These layers are divisible into about nine separate layers. There are a number of differences between species (see Lee 1965).

The cuticle is formed by an underlying subcuticular layer, called the *hypodermis*. This usually consists of cells in the free-living forms and of a syncytium containing a number of nuclei in the parasitic forms. This layer forms four longitudinal thickenings on the inner aspect, situated dorsally, ventrally and laterally and known as the 'longitudinal lines'. The lateral lines contain the longitudinal canals of the excretory system. The cuticle is not significantly permeable to small molecules of nutritional significance (e.g. glucose or amino acids) though the basis of this determination is based mainly on *Ascaris*. In *Ascaridia galli* it has been demonstrated that glucose and alanine do enter through the cuticle and much needs to be done regarding the nutritional significance of the cuticle in other groups. The cuticle is, however, penetrated by certain anthelmintics.

The muscular layer, which follows next and lines the body cavity, consists of a number of cells having a basal contractile portion which is transversely striated, and a cytoplasmic portion which contains the nucleus and is connected to the nerve trunks running in the dorsal or ventral line. The muscular layer is divided into four quadrants by the longitudinal lines.

The mouth is anterior, sometimes subdorsal or subventral, and is usually surrounded by lips. The original forms apparently had three lips, one dorsal and two ventral, each bearing two sensory

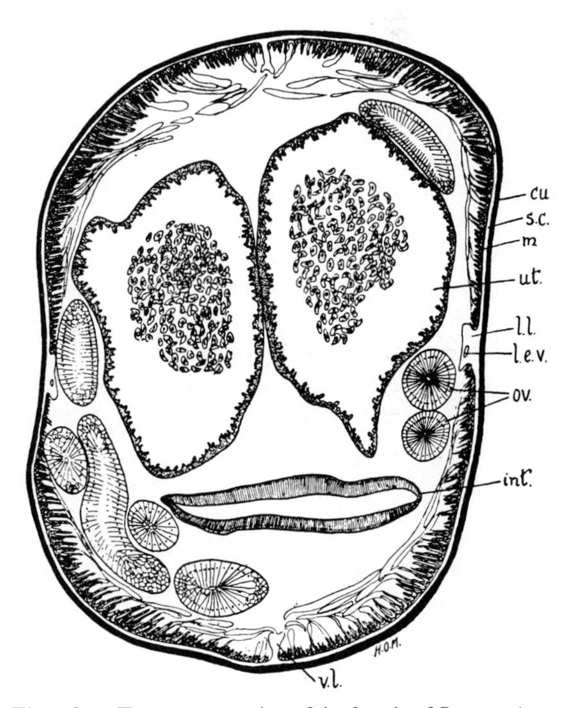

Fig 1.82 Transverse section of the female of *Parascaris equorum*.

cu cuticula
int intestine
lev longitudinal excretory vessel
ll lateral line
m muscular layer
ov ovary
sc subcuticula
ut uterus
vl ventral line

papillae. This arrangement still persists in most free-living and some parasitic forms (Ascarids). In other forms there are two lips, each bearing three papillae, and these lips may be subdivided into three parts each (Spirurids) or may disappear completely (Strongyles, Filarids) and papillae only remain around the mouth opening. Though difficult to see in some parasitic forms, a pair of depressions occur on each side of the head end, lying lateral or posterior lateral to other specialized head structures. These are the *amphids* and are chemoreceptors. They are well supplied with nerve fibres and associated with a glandular structure. Comparable organs, the *phasmids*, are placed in the posterior extremity in many forms. These occur posterior to the anus and glands are usually associated with the organelles. The possession, or otherwise, of *phasmids* has been the basis of the classification of nematodes into the *Phasmidia* (with) or *Aphasmidia* (without). This system of classification is no longer generally in use.

In the forms without lips, secondary structures may develop in their place. The *Strongylidae*, for instance, develop *leaf-crowns* which consist of a large number of fine, pointed processes which arise from the rim of the mouth opening (external leaf-crown) or the rim of the buccal capsule (internal leaf-crown). The mouth may lead into a *buccal capsule*, which has thick cuticular walls and may contain special tooth-like structures, or into a *pharynx*, which is usually cylindrical and surrounded by muscular tissue, or directly into the *oesophagus*. The oesophagus of nematodes shows variations of structure that are used for the classification of species. It is a strongly muscular organ with a triradiate lumen which is thickly lined with cuticle and divides the wall into one dorsal and two subventral sectors, corresponding to the primitive arrangement of the lips. The wall of the oesophagus contains three *oesophageal glands*, one in each sector, which secrete digestive enzymes. The dorsal gland opens into the mouth and the others into the lumen of the oesophagus. Posteriorly the oesophagus may have a bulbar swelling called the *oesophageal bulb*, which contains a valvular apparatus (e.g. *Heterakis*). The muscles of the oesophagus dilate its lumen, so that it sucks in liquid food which is passed into the intestine. In the blood-sucking Ancylostomidae its pumping pulsations may achieve a rate of 120 or more a minute. In order to prevent regurgitation the oesophagus may be separated from the intestine by three valves. In other species the posterior part of the oesophagus may be non-muscular and has a structure that is possibly glandular; it is called the *ventriculus* (e.g. Anisakinae). In the non-parasitic first larvae of many species of nematodes and the individuals of the non-parasitic generations of the Rhabditida described below the oesophagus has a club-shaped anterior portion connected by a narrow neck with a pear-shaped posterior bulb. An oesophagus of this type is called a *rhabditiform* oesophagus to distinguish it from the type of oesophagus which is club-shaped without a posterior bulb, the latter being called a *filariform* oesophagus. A filariform oesophagus is found in the second and later larvae of nematodes and in the individuals of parasitic generations of the Rhabditida. The intestine is a simple tube with a non-muscular wall composed of a single layer of columnar cells standing on a basal membrane. It leads into the rectum, which is lined with cuticle and into which the genital duct opens in the male; the latter, therefore, has a cloaca.

The gut plays an important role in absorption of nutrients. Ultrastructural studies show that the gut cells possess microvilli which are concerned in the absorption of nutrients. The role of the gut in nutrition has been reviewed by Wright (1977).

The part of the body behind the anal or cloacal opening is called the tail.

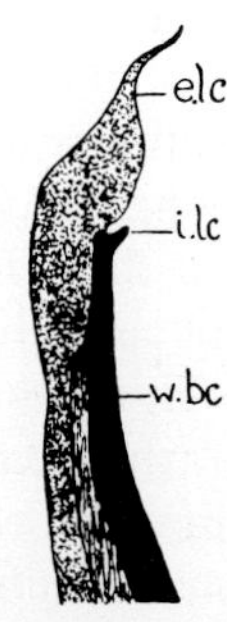

Fig 1.83 Section through the rim of the buccal capsule of *Strongylus equinus*.
elc external leaf crown
ilc internal leaf crown
wbc wall of buccal capsule

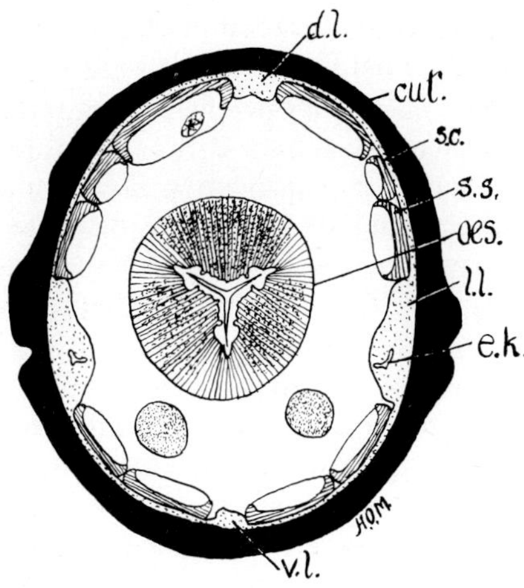

Fig 1.84 Section through the oesophageal region of *Strongylus equinus.*

cut	cuticula	mc	muscle cell
dl	dorsal line	oes	oesophagus
ex	excretory canal	sc	subcuticula
ll	lateral line	vl	ventral line

The region between the layer of muscle-cells and the alimentary canal is called the *perienteric* space. It is filled by tissue, composed of very few cells, in which there are large intracellular spaces containing the *perienteric fluid*, which contains proteins, fats, glucose, enzymes and, in some species, haemoglobin. In *Ascaris* four large cells with numerous processes ending in terminal knobs, called the stellate cells or *phagocytic organs*, phagocytose bacteria and other material experimentally introduced into the perienteric space. There are similar cells in other species (e.g. *Strongylus equinus*) and also free phagocytic cells in the perienteric fluid.

The excretory system, which is also osmoregulatory, opens by a ventral pore situated a short distance behind the anterior extremity and consists of a pair of unbranched lateral longitudinal vessels embedded in the hypodermis which do not end in flame-cells as the excretory tubes of flatworms do. The function of the lateral canals is not fully known. In *Ascaris* these canals are joined by a transverse canal behind the perioesophageal nerve ring to form an H-shaped system. Variations of this type occur. The right-hand canal may, for instance, be absent in Anisakinae and there are no lateral canals in the Trichuroidea and Dioctophymatoidea.

The nervous system consists of a number of ganglia connected by fibres, forming the *nerve ring* which surrounds the oesophagus. From this central organ six nerve trunks arise anteriorly and posteriorly, the principal ones running in the dorsal and ventral lines. Solitary ganglia occur in other parts of the body; for instance, an anal ganglion. The sense organs are:

1. Around the mouth, the amphids, the chemoreceptors, as already described.

2. A pair called the *cervical papillae* or *deirids*, which project through the cuticle at the sides of the oesophagus and are probably tactile (e.g. in *Oesophagostomum*.

3. Frequently a lateral or subdorsal pair near the middle of the body.

4. On the posterior extremity of the male, the genital papillae, which are mostly paired and may have long stalks supporting the caudal alae or the copulatory bursae of the strongyles. In the latter case the stalks contain muscle fibres and the papillae are arranged according to the definite system described below. The female may also bear a pair of papillae on the tail or in the region of the vulva.

5. The phasmids, also chemoreceptors, as already described.

The sexes are usually separate. In the parasitic nematodes the function of reproduction is markedly developed and there is much variation in the genital organs. Sexual dimorphism is sometimes very marked. Apart from other differences, the males are frequently smaller than the females.

The male organs are composed of a single testis in the parasitic and most free-living forms, a vas deferens, sometimes a seminal vesicle and a muscular ejaculatory duct which opens into the cloaca. In most species there are one or two spicules lying in sheaths which also open into the cloaca. These organs are cuticular, often pigmented; they vary in shape and size and are of great value in determination of the species. They serve during copulation for attachment and probably also to expand the vagina and direct the flow of sperms. The spicules are moved by special muscles and in many cases the wall of the cloaca is provided with cuticular thickenings which guide the spicules. Such a thickening on the dorsal wall is called a gubernaculum. Less frequently there is one on the ventral wall, known as a telamon.

In the female the vulva was originally posterior and there were two uteri and two ovaries running forward (prodelph). Some free-living and also some parasitic forms have only one uterus and ovary. On the other hand, the uteri may be subdivided, so that forms with up to 16 uteri are known. The vulva may be found in various positions, even quite close to the anterior extremity (*Oxyuridae*, *Filariidae*) and the uteri may run in opposite directions (amphidelph) or both run backwards (opisthodelph). The ovary is proximally a solid, cylindrical organ containing a number of cells, which divide to form the ova and further on arrange themselves around a central rhachis, from which they later become detached. They pass through an oviduct to the seminal receptacle, a small dilated part of the organ in which spermatozoa are stored and fertilization takes place. This is followed by the uterus, in which the egg-shells are formed and the embryo may develop. The uteri are usually connected to the vagina by muscular ovijectors, or one for both uteri, lined with cuticle like the vagina. The vulva is usually situated on the ventral surface. Some nematodes are oviparous, others are ovoviviparous or viviparous. The eggs vary greatly in shape and size, a fact which is of great importance in making a specific diagnosis by faeces examination. The parasitic nematodes are very prolific and a female may lay several thousand eggs per day.

Development

The original egg cell divides into two, then four and so on, and the embryo passes through a morula stage and later through a 'tadpole' stage, in which the anterior end is broad and the embryo is bent double. Eventually the larva is fully formed and ready to hatch. Normally four moults or ecdyses, in which the whole cuticle is shed and replaced by a new one, take place before the adult stage is reached, but in some cases a moult may be omitted. Each period between two moults consists of two phases: one in which the worm feeds and grows and the second during which it becomes inactive, or lethargic, while structural changes take place in the body in preparation for the next moult (the lethargus). The larval worm becomes infective for the final host as a rule after the second moult, and in those species which are free-living up to this stage the cuticle of the second moult is usually retained as a protective sheath until the worm has entered its host. The infective stage may, in certain species, be reached in the egg-shell. In species which use an intermediate host, the infective larva develops inside this host.

Infection of the host in which the adult nematodes occur may therefore be effected by (*a*) an active, non-parasitic third larva which enters the host through its mouth (e.g. *Strongylus*); (*b*) a passive infective egg containing a second or third infective larva (e.g. *Ascaris*); or (*c*) an intermediate host in which the infective larva develops. In these instances the intermediate host is either eaten by the definitive host (e.g. *Metastrongylus*) or it conveys the infective larva to the definitive host and the infective larva then penetrates through the skin of the definitive host (e.g. *Filarioidea*).

The third and fourth ecdyses take place in the final host, after which the worms are in the adult stage and grow to maturity.

Various types of life-cycles are found among the nematodes, depending to some extent on the degree of adaptation to a parasitic existence that has been reached. The most specialized species (e.g. *Trichinella*) have no period of free existence at all. The life-cycles may therefore be classified as follows:

1. *Without an intermediate host*:

(*a*) Eggs hatch outside the host and larvae are free-living for a time; infective larvae are active, e.g. most *Strongylidae* and *Trichostrongylidae*. Entry into the host is through the mouth with food and water, but the infective larvae of some species can penetrate the host's skin as well as entering through its mouth (*Ancylostoma*, *Bunostomum*).

(*b*) Eggs develop outside the host but do not hatch there; infective larvae are passive inside the egg. Entry into the host occurs only through the mouth, e.g. *Ascaridae*.

2. *With an intermediate host*:

(*a*) Eggs hatch or the worms are viviparous and the larva enters the intermediate host after a short free existence, e.g. *Metastrongylidae*, *Habronema* spp. Intermediate host is eaten by the definitive host.

(*b*) Eggs do not hatch and are ingested by the intermediate host, e.g. *Spiruroidea*. Intermediate host is eaten by the definitive host.

(*c*) The worms are viviparous and the larvae enter the blood of the host, from which they are taken up by a blood-sucking intermediate host, inside which the infective larva develops. When the intermediate host sucks the blood of the definitive host, the infective larvae break out of the proboscis of the intermediate host and penetrate into the definitive host through its skin, e.g. *Filarioidea*.

After having entered the final host many nematodes migrate through the body before settling down to their normal habitat, and some of them do much harm in a mechanical way during this process.

Classification

In the previous edition of this book the classification of the nematodes was based on that originally proposed by Chitwood (1933) and Chitwood and Chitwood (1937) which divided the nematodes into two major groups, *Phasmidia* and *Aphasmidia*. The Russian system of classification was also adopted. Now the classification follows that of Yamaguti (1961), Chitwood (1969) and the *Keys to Nematode Parasites of Vertebrates*, edited by Anderson et al. (1974). These changes affect the higher taxonomic ranks; they have little effect at the family or generic level.

PHYLUM: Nemathelminthes Schneider, 1873
CLASS: NEMATODA Rudolphi, 1808
Subclass: Secernentea Dougherty, 1958

ORDER: ASCARIDIDA Skrjabin & Schulz, 1940
SUPERFAMILY: ASCARIDOIDEA Railliet & Henry, 1915
Family: *Ascarididae* Baird, 1853
Anisakidae Skrjabin & Karokhin, 1945

SUPERFAMILY: OXYUROIDEA Railliet, 1916
Family: *Oxyuridae* Cobbold, 1864
Kathlaniidae Travassos, 1918

SUPERFAMILY: SUBULUROIDEA Travassos, 1930
Family: *Heterakidae* Railliet & Henry, 1914
Subuluridae York & Maplestone, 1926

ORDER: RHABDITIDA Chitwood, 1933
SUPERFAMILY: RHABDITOIDEA Travassos, 1920
Family: *Rhabditidae* Micoletzy, 1922
Strongyloididae Chitwood & McIntosh, 1934

ORDER: STRONGYLIDA Molin, 1861
SUPERFAMILY: STRONGYLOIDEA Weinland, 1958
Family: *Strongylidae* Baird, 1853
Trichonematidae Witenberg, 1925
Amidostomidae Baylis & Daubney, 1926
Stephanuridae Travassos & Vogelsang, 1933
Syngamidae Leiper, 1912

SUPERFAMILY: ANCYLOSTOMATOIDEA Chabaud, 1965
Family: *Ancylostomatidae* Looss, 1905

SUPERFAMILY: TRICHOSTRONGYLOIDEA Cram, 1927
Family: *Trichostrongylidae* Leiper, 1912
Ollulanidae Skrjabin & Schikhobalova, 1952
Dictyocaulidae Skrjabin, 1941

SUPERFAMILY: METASTRONGYLOIDEA Lane, 1917
Family: *Metastrongylidae* Leiper, 1908
Protostrongylidae Leiper, 1926
Filaroididae Schulz, 1951
Skrjabingyidae Kontrimavichus, Delamure and Boev, 1976
Crenosomatidae Schulz, 1951

ORDER: SPIRURIDA Chitwood, 1933
SUPERFAMILY: SPIRUROIDEA Railliet & Henry, 1915
Family: *Spiruridae* Oerley, 1885
Thelaziidae Railliet, 1916
Acuaridae Seurat, 1913
Tetrameridae Travassos, 1924

SUPERFAMILY: PHYSALOPTEROIDEA Sobolev, 1949
Family: *Physalopteridae* Leiper, 1909
Gnathostomatidae Railliet, 1895

SUPERFAMILY: FILAROIDEA Weinland, 1958
Family: *Filariidae* Claus, 1885
Setariidae Skrjabin & Schikhobalova, 1945
Onchocercidae Chabaud and Anderson, 1959

SUPERFAMILY: DRACUNCULOIDEA Cameron, 1954
Family: *Dracunculidae* Leiper, 1912

Subclass: Adenophorea Chitwood, 1958

ORDER: ENOPLIDA Schuurmans, Stethoven and Delenetz, 1933
SUPERFAMILY: TRICHUROIDEA Railliet, 1916
Family: *Trichinellidae* Ward, 1907
Trichuridae Railliet, 1915
Capillariidae Neveu-Lemaire, 1936

SUPERFAMILY: DIOCTOPHYMATOIDEA Railliet, 1916
Family: *Dioctophymatidae* Railliet, 1915
Soboliphymatidae Petrov, 1930

SUPERFAMILY: MERMITHOIDEA Wülker, 1934
Family: *Mermithidae* Braun, 1883
Tetradonematidae Cobb, 1919

REFERENCES

NEMATODES—GENERAL

Anderson, R. C., Chabaud, A. G. & Willmott, S. (1974) *CIH Keys to the Nematode Parasites of Vertebrates. No. 1 General Introduction*, p. 17. Farnham Royal: Commonwealth Agricultural Bureaux.
Chitwood, B. G. (1933) A revised classification of the Nematoda. *J. Parasit.*, **20**, 131.
Chitwood, M. B. (1969) The systematics and biology of some parasitic nematodes. In: *Chemical Zoology*, ed. M. Florkin & B. T. Scheer, Vol. III, pp. 223–244. New York and London: Academic Press.
Chitwood, B. G. & Chitwood, M. G. (1937) *An Introduction to Nematology*, sect. 1, part 1, Baltimore, Md: Monumental Printing.
Lee, D. L. (1965) The cuticle of adult *Nippostrongylus brasiliensis*. *Parasitology*, **55**, 173–181.
Read, C. P. (1966) Nutrition of intestinal helminths. In: *The Biology of Parasites*, ed. E. J. L. Soulsby, pp. 101–125. New York and London: Academic Press.
Wright, K. A. (1977) Structural studies of digestion in some helminth parasites. In: *Parasites, Their World and Ours*, ed. A. M. Fallis, pp. 195–231. Ottawa: Royal Society of Canada.
Yamaguti, S. (1961) *Systema Helminthum*, Vol. III, *The Nematodes of Vertebrates*, p. 1261. New York and London: Interscience.

SUBCLASS: SECERNENTEA DOUGHERTY, 1958 (FORMERLY PHASMIDIA)

Nematodes with phasmids present. Amphids pore-like and labial in position. Males commonly possess caudal alae or copulatory bursae.

ORDER: ASCARIDIDA SKRJABIN & SCHULZ, 1940

This order has probably arisen from Rhabditoidea. They possess three large lips and caudal alae, when present, are laterally placed.

SUPERFAMILY: ASCARIDOIDEA RAILLIET & HENRY, 1915

Mostly large nematodes. Mouth surrounded by three large lips; no buccal capsule, oesophagus usually lacks posterior bulb; intestine may have caeca; tail of female blunt, of male frequently coiled; two spicules in the male; life-cycle may be direct or indirect.

FAMILY: ASCARIDIDAE BAIRD, 1853

The genera of importance are *Ascaris*, *Parascaris*, *Toxascaris* and *Toxocara*.

These are usually relatively large worms with three well-developed lips, one dorsal and two subventral, each of which usually bears two papillae. Between the bases of these lips there may be smaller lips, the interlabia. The inner surface of each lip may bear a dentigerous ridge of small teeth. There is no buccal capsule or pharynx. The oesophagus is usually club-shaped, muscular and without a posterior bulb. The tail of the male is usually without well-developed caudal alae, but it usually bears numerous caudal papillae. The male has paired spicules and the vulva of the female is in front of the middle of the body. The females are oviparous and produce a large number of eggs, which are usually unsegmented when they are laid. The eggs are oval or subglobular and the shell is in most cases thick.

Hatching mechanisms in Ascarids. The processes involved in the hatching of eggs consist of a stimulus from the host which acts on a 'receptor' in the infective egg, this causes a resumption of development from the previous stage of resting and the secretion of 'hatching fluid'. Hatching fluid contains various enzymes which attack the layers of the egg shell and subsequently the infective larval stage emerges.

Studies with *Ascaris lumbricoides*, *Toxocara cati* and *Ascaridia galli* have indicated that dissolved carbon dioxide and undissociated carbonic acid comprise the host stimulus, though strongly reducing conditions and a satisfactory pH are also necessary. Under *in vitro* conditions the reducing conditions can be produced by cysteine, glutathionine, sodium dithionite or sulphur dioxide (Fairbairn 1961). The optimal concentration of undissociated carbonic acid plus dissolved carbon dioxide is of the order of $0.25–0.5 \times 10^{-3}$ M at pH 7.3. Hatching fluid of *Ascaris suum* contains an esterase and a chitinase which attack the lipid and chitin in the egg shell. One of the first effects after stimulation is an increased permeability of the vitelline membrane which allows the enzymes to reach the shell (Rogers 1966).

Genus: Ascaris Linnaeus, 1758

Ascaris suum Goeze, 1782. This species is cosmopolitan in distribution, occurring in the pig. Immature specimens, sometimes described as *A. ovis*, are occasionally found in sheep and cattle and the parasite has also been reported from certain squirrels and the dog. Adults may develop in rabbits under experimental conditions (Berger et al. 1961).

For many years this species was considered synonymous with the human parasite *A. lumbricoides* Linnaeus, 1758; however there is now evidence that they are distinct species. Sprent (1952) has described morphological differences in the denticulation of the lips of the two forms, Ansel and Thibaut (1973) reported differences in the axial truncations of the lips and epidemiological studies in areas where human and swine ascariasis is common failed to indicate any evidence of cross-infection. However, the chromosomes are identical, gametogenesis is similar and patent infections with the human form can be induced in pigs under appropriate circumstances. In a summary of much of the evidence, Taffs (1961) concluded that the two forms should be considered as distinct, but further study is necessary to clarify the situation.

The males measure 15–25 cm by about 3 mm and the females up to 41 cm by 5 mm. The dorsal lip bears two double papillae and each ventrolateral lip one double subventral and a small lateral papilla. Each lip bears on its inner surface a row of minute denticles. The spicules of the male are about 2 mm long and stout. There are a large number of precloacal papillae. The vulva opens near the end of the first third of the body. The eggs are oval, measuring 50–75 by 40–50 μm. They have thick shells, the albuminous layer bears prominent projections and they are brownish-yellow in colour.

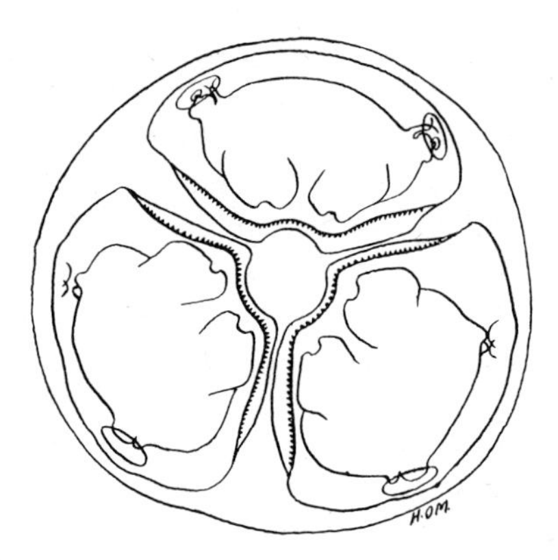

Fig 1.85 Anterior view of the head of *Ascaris suum*.

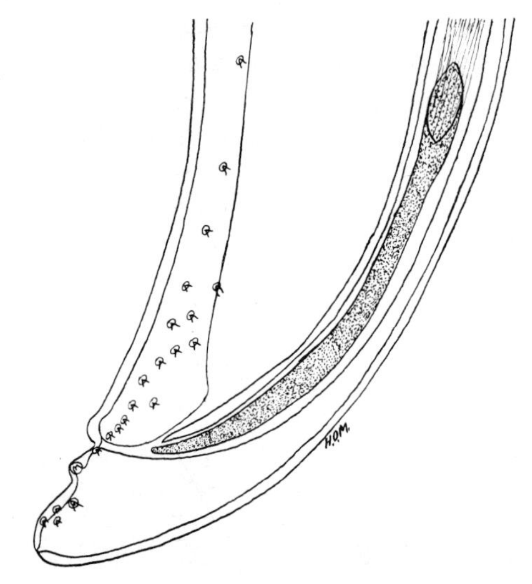

Fig 1.86 Lateral view of the hind end of *Ascaris suum*, showing one spicule.

Life-cycle. A female may lay as many as 200 000 eggs per day though some authors suggest that up to 2 million eggs may be produced daily. The eggs are passed in the faeces of the host and develop to the infective stage in 10 days or longer, depending on the temperature. The eggs are very resistant to adverse conditions, like drying or freezing, and to chemicals, and they may remain viable for as long as five years, or perhaps longer, but hot, dry conditions, such as those prevailing in sandy soil with direct sunlight, kill them in a few weeks. It has been generally accepted that the ascarid life-cycle involves a second-stage larva as the infective form, unlike the majority of other nematodes. Hence in the case of *A. suum* it was presumed that a second-stage larva, with the retained sheath of the first-stage larva, was the infective stage. However, Aranjo (1972) has demonstrated that two ecdyses occur before eggs hatch and has suggested that the infective stage is a third-stage larva. The larva rarely hatch and infection usually takes place through ingestion of the eggs with food or water or from the soiled skin of the mother in the case of sucking pigs.

The ingested eggs hatch in the intestine and the larvae burrow into the wall of the gut. They may pass through into the peritoneal cavity and thence to the liver; but the majority reach this organ by way of the hepatoportal blood stream. They may arrive in the liver 24 hours after the eggs have been ingested, or even earlier. From the liver they are carried by the blood through the heart to the lungs, where they are arrested in the capillaries, although some may pass through into the arterial circulation and reach other organs like the spleen and kidneys. The majority of larvae are recognizable as third-stage larvae between the fourth and fifth day after infection. This may represent a moult from second-stage to third-stage, or a loss of two retained sheaths by an infective third-stage larva (Aranjo 1972). At this time many larvae are still in the liver though a good proportion are migrating to the lungs and may be in the lungs. A period of marked growth and development occurs and this is also associated with the migration of larvae to the lungs.

Larvae break out of the alveolar capillary into the alveolus and pass through the alveolar duct to the small bronchioles and then gradually ascend the bronchial tree. This is a 'tracheal' route of migration (Sprent 1959). As the infection continues, larger numbers of larvae are found more and more towards the anterior end of the large bronchi and trachea. Larvae then migrate from the trachea to the pharynx when they are swallowed and third-stage larvae arrive in the intestine seven to eight days after infection. Douvres et al (1969) state that the moult to the fourth stage occurs about the tenth day, in the intestine, this being contrary to the views of Roberts (1934) who stated that this moult occurred in the respiratory system and that only fourth-stage larvae were able to survive the acid environment of the stomach. At this time larvae measure 1.2–1.4 mm. Large numbers of fourth-stage larvae are present in the small intestine between the fourteenth and twenty-first day after infection and by 21 days they measure 4.5–6.5 mm. The moult to the fifth stage, or young adult, occurs 21-29 days after infection. Maturity occurs after 50–55 days and eggs appear in the faeces at 60–62 days.

Though ascariasis is extremely common in swine, it has proved difficult to establish experimental patent infections in these animals. Various studies have suggested that small doses of eggs are more likely to result in patent infections than large (several thousand eggs). This has been confirmed by Andersen et al. (1973) who demonstrated an inverse relationship between the dose of eggs and the number of adult worms in the intestine. While a mean establishment of adults of 64% resulted from a dosage of 50 eggs per piglet, a mean of 2.9% occurred with a dose of 500 eggs and a mean of 0.013% resulted with doses of 1000 eggs and higher. There was no relationship between the age of the animal and establishment of adult worms.

A detailed account of the morphology and sizes of different developmental stages is given by Soulsby (1965).

The eggs of *A. suum* will hatch and the larvae migrate in many animal species, including man. Much work has been done with *A. suum* infections in guinea-pigs, rabbits, rats and mice and in these the migratory cycle is much the same as in the pig. There are slight differences in size of larvae from various animals, thus larvae from mice are smaller than those from swine. Normally *A. suum* does not mature in such animals though Berger et al. (1961) have reported this in the rabbit.

Migration of *A. suum* in the human occurs also but patent infections do not appear to take place commonly, although accidental human infection in laboratory workers using *A. suum* has been noticed. The development of *A. lumbricoides* in swine has received more attention and patent infections of the human form have been produced. deBoer (1935) produced it in young piglets, this work being confirmed by Soulsby (1961) using baby pigs deprived of colostrum.

There is no evidence for prenatal infection in *A. suum*.

Pathogenesis. During the migratory period the larvae may do much damage if the infection is heavy (Roneus 1966). Destruction of tissue and haemorrhage may occur in the liver, especially around the intralobular veins, but the most important lesions are produced in the lungs, where the larvae cause numerous small haemorrhages into the alveoli and bronchioles, followed by desquamation of the alveolar epithelium, oedema and infiltration of the surrounding pulmonary parenchyma with eosinophils and other cells. In heavy infections death from severe lung damage may occur six to 15 days after infection. Under field conditions, however, it is unlikely that a single large lethal dose would be acquired. Nevertheless marked pathological changes are seen in the lungs and these are due to repeated infections, the lesions of oedema, emphysema and haemorrhage being due to a hypersensitive state resembling asthma.

There is a lack of critical information on the pathogenic effects of the adult parasites. The worms may be so numerous that they may become twisted into bundles in such a way that intestinal obstruction occurs. Ascarids may wander and enter the stomach and be vomited, or pass up the bile duct into the liver, causing biliary stasis, or blockage of the bile ducts; they may even perforate the intestine and produce peritonitis.

In children heavily infected with *A. lumbricoides* Tripathy et al. (1972) reported impaired nitrogen retention and fat and *d*-xylose absorption. Jejunal biopsies revealed broadening and shortening of villi, a decrease in the cyst–villus ratio and cellular infiltration of the lamina propria. These reverted to normal after removal of the worms.

Studies of adult infection in pigs experimentally infected with 15-day-old larvae (Stephenson et al. 1980) revealed an increase in gut weight correlated with the number of worms present and due to hypertrophy of the tunica muscularis. Non-significant changes in nitrogen and fat absorption occurred, but nevertheless these indicated that *A. suum* interferes with the absorption of protein, fat and carbohydrate, leading to decreased energy

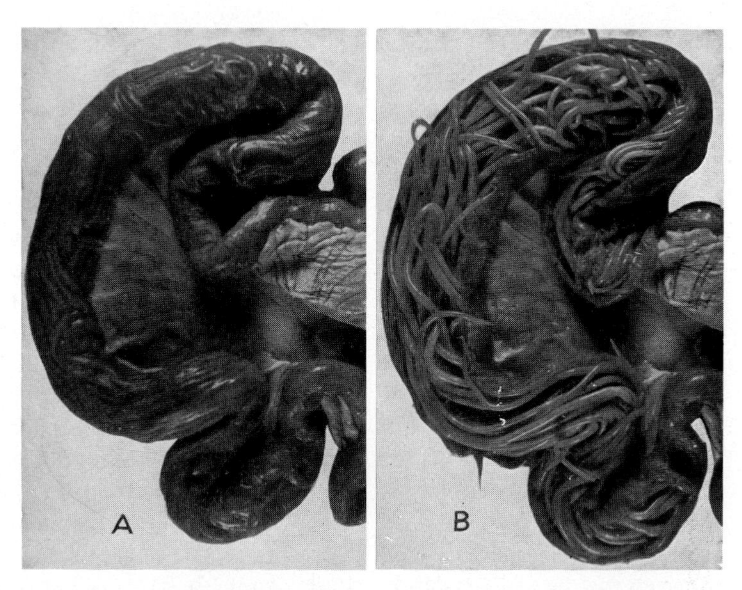

Fig 1.87 Small intestine of a pig infected with *Ascaris suum*. A, Unopened intestine. B, Same intestine opened.

intake, this being more serious when animals were fed low protein diets. The contribution of *A. lumbricoides* to malnutrition in children is reviewed by Stephenson (1980).

Clinical signs. The clinical signs of ascariasis in pigs depend on the severity of the infection. Young pigs are chiefly affected. Newborn pigs which become heavily infected may show signs of pneumonia, especially a cough and exudate into the lungs. In less severe cases the animals cough and their growth is stunted. Heavy infections with adult worms produce diarrhoea and this has a marked effect on their growth rate.

The migration of *A. suum* larvae in the pig may enhance latent infections of enzootic (viral) pneumonia.

Post mortem. The liver shows varying degrees of fibrosis, which may be localized in the form of 'milk spots'. These are usually whitish in colour but may become haemorrhagic, indicating a more recent nature. In chronic infections the liver may be markedly fibrotic. Anderson et al. (1973) have demonstrated that infection of neonatal piglets does not provoke eosinophilia, which is seen in older infected animals. However, 'milk spots' on the livers of infected animals occurred at all ages. Varying degrees of pneumonia or bronchitis may be found or only a number of petechial haemorrhages in the lungs. The larvae can be found by pressing small bits of lung tissue between two slides and examining these under a low magnification or by teasing up a portion of the organ in warm saline. Small haemorrhagic or necrotic foci may also be seen in other organs. In the intestine worms of various ages may be found and some may have become lodged in the bile ducts.

Hepatic lesions in lambs have been ascribed to *A. suum*. These lesions may be mistaken for those caused by *Taenia hydatigena* or *Echinococcus* spp. and are associated with the grazing of sheep on pastures previously grazed by pigs. Pasture may remain infected following cultivation and cropping (Mitchell & Linklater 1980).

Diagnosis. During the early stages of the disease the pulmonary signs will indicate the possible aetiological factor and larvae may be found in the sputum. *Ascaris* eggs will be found in the faeces of the older pigs. Unfertilized eggs are frequently seen and are significant when only female worms are present in large numbers. Apparently fertilization has to be repeated at intervals and unfertilized eggs may be laid even when males are present. Such eggs are variable in shape, elongate or triangular, and contain numerous vacuoles and large granules.

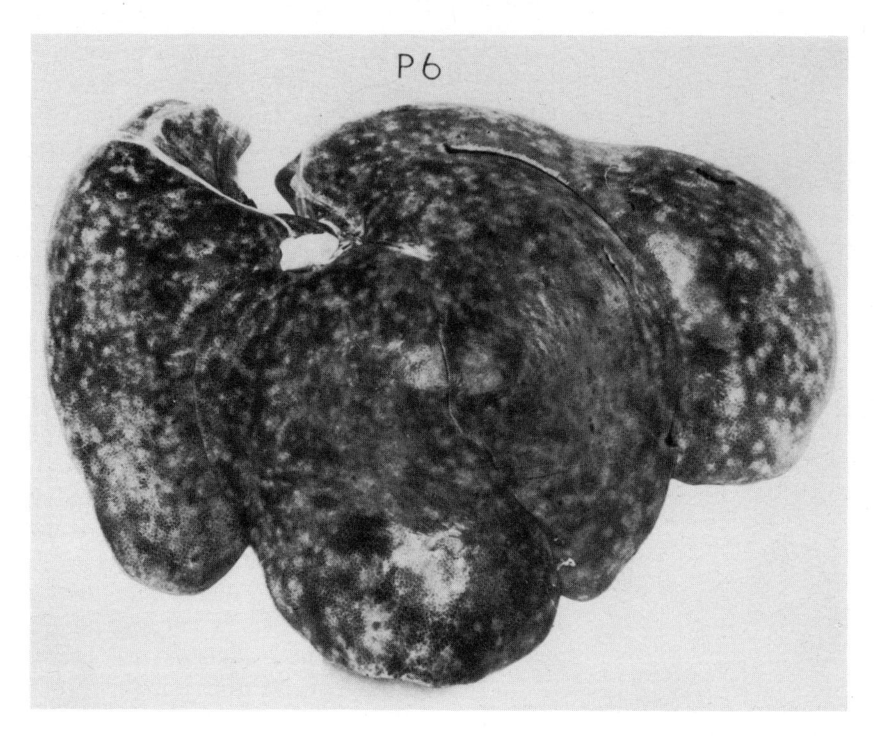

Fig 1.88 Liver of pig infected on several occasions with eggs of *Ascaris suum*. Numerous focal areas of fibrosis, milk spots, are evident. (*Photograph by courtesy Dr L. F. Taffs*)

Treatment. The imidazole and benzimidazole anthelmintics are the compounds of choice for ascarid infections in pigs. They are available for oral use as food additives; some can be injected and some are effective against migrating larvae.

Levamisole (levamisole hydrochloride) is a broad-spectrum anthelmintic effective against *A. suum* and the majority of other important nematodes of pigs. It may be given by subcutaneous injection at a dose rate of 7.5 mg/kg, administered as a drench at the rate of 8 mg/kg or offered in the feed at 0.72 g/kg for herd treatment.

Tetramisole (tetramisole hydrochloride) is active against mature and immature ascarids as well as other common pig parasites. It is given orally at 15 mg/kg in the feed. As a feed additive, 5 kg are added to 500 kg of feed which is calculated to give a dose of 250 g/25 kg.

Parbendazole, a broad-spectrum anthelmintic effective against the majority of important pig nematodes, is given orally at a dose of 30 mg/kg.

Fenbendazole is a broad-spectrum anthelmintic which also has an ovicidal effect. It may be administered as a single oral dose of 5 mg/kg or as a mass medicant in the feed.

Cambendazole, a broad-spectrum anthelmintic, is usually applied as a 'top dressing' to feed to give a minimal dose of 20 mg/kg.

Dichlorvos (2,2 dichlorvinyl dimethylphosphate) may be given at the rate of 10 mg/kg for sows or 40 mg/kg for weaning pigs, usually mixed with the feed.

Morantel tartrate is effective against a spectrum of parasites of pigs. For weaners 5 mg/kg is recommended mixed in the feed given at eight to ten weeks of age. For older animals, e.g. sows and boars, a dose of 12.5 mg/kg in the feed is recommended.

Piperazine compounds have previously been widely used for the removal of ascarids in swine. A number of salts are available (e.g. piperazine citrate, adipate, dihydrochloride, etc.) and all have

Fig 1.89 Lungs of pig ten days after administration of large dose of *Ascaris suum* eggs. Numerous petechial haemorrhages are evident over the surface of the lungs. (*Photograph by courtesy Dr L. F. Taffs*)

a wide margin of safety. They are usually administered in the food at a dose rate of 100–400 mg/kg.

Prophylaxis. Connan (1977) has demonstrated that eggs of *A. suum* may survive for extended periods under conditions prevailing in a pig house. Further, eggs shed into the environment over a period from September to May in Great Britain became infective more or less synchronously in July. Hence infective eggs may be far removed in time from the donor of the eggs. The eggs survive best in damp soil or in dirty buildings. When pastures become heavily contaminated it is necessary to put them through rotational cultivation for a period of some years. Contaminated pig pens can be rendered safe by the use of a solution of hot caustic soda or live steam.

The most important preventive measures are those concerned with the protection of the young pigs immediately after birth or later. A very satisfactory system, which was devised by Ransom in the USA is known as the MacLean County System. The sow is treated for ascarids a little time before farrowing and then within a few days of farrowing is thoroughly washed and scrubbed in order to remove any eggs adhering to the body, and is then placed in a farrowing pen. The latter has a concrete floor and has been prepared by thoroughly scrubbing the floor and walls with boiling water, soda and a hard broom. Within ten days of farrowing the sow and her litter are moved to an *Ascaris*-free field planted with a suitable crop. After weaning the sow is removed and the young pigs grow up able to resist the major effects of infection. In modern intensive pig production units, ascariasis is not a major parasitic disease problem: nevertheless it may become so if pigs are allowed access to pasture at regular or occasional intervals.

Ascaris columnaris Leidy, 1856 occurs in the small intestine of the skunk, stoat, weasel and Siberian polecat (*Putorius eversmanni*). The males are up to 9 cm and the females up to 22.5 cm long. The spicules are about 0.4 mm long. The vulva is situated at the end of the first quarter of the body. The eggs have finely pitted shells and are subglobular, measuring 88–90 by 66–68 μm.

Mice act as intermediate hosts. Sprent (1955) has shown that the larvae of *A. columnaris* and also of *A. devosi* cause an encephalitis in mice.

Genus: Parascaris Yorke & Maplestone, 1926

Parascaris equorum (Goeze, 1782) Yorke & Maplestone, 1926 (syn. *Ascaris megalocephala*, *Ascaris equorum*) occurs in the small intestine of equines, including the zebra, and perhaps also cattle. The males are 15–28 cm long and the females up to 50 cm by 8 mm. This is a rigid, stout worm with a large head. The three main lips are conspicuous, separated by three small intermediate lips and divided into anterior and posterior portions by horizontal grooves on their

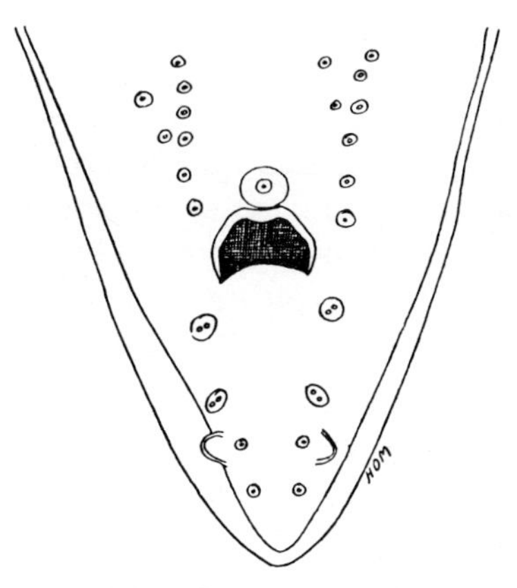

Fig 1.90 Ventral view of the hind end of the male *Parascaris equorum*.

medial surfaces. The male tail has small lateral alae. There are two double and three single pairs of postcloacal papillae. A single median papilla occurs on the anterior border of the cloaca. The spicules are about 2–2.5 mm long. The vulva is situated at the end of the first quarter of the body. The eggs are subglobular with a thick, pitted albuminous layer and measure 90–100 μm in diameter.

Life-cycle. Similar to that of *A. suum*. The worms reach maturity 80–83 days after infection (Clayton & Duncan 1977).

Pathogenicity and clinical signs. Foals three to nine months of age especially suffer from this parasite, sometimes massive burdens being present (more than 1000 worms). Occasionally, patent infections are found in mature horses, but these are of little clinical consequence. A significant age-immunity develops by six months (Clayton 1978).

In heavy infections coughing and circulating eosinophilia are features of the prepatent period (Clayton & Duncan 1977). Adult worms cause a catarrhal enteritis which produces diarrhoea which may be foetid in odour and pale in colour. Flatulence is common. There is general malaise and debility and the coat is harsh. Animals may become pot-bellied. Complications may occur owing to the migration of the adult worms to aberrant sites such as the bile duct; they may also penetrate the bowel wall and cause localized or generalized peritonitis or may 'ball up' and cause an obstruction. Weight gains are depressed by infections and infected foals have a lower total body albumin and serum albumin than worm-free foals. This is ascribed to a depression of albumin synthesis (Clayton 1978). Unthriftiness is a function of the number of worms in the intestine and not a consequence of hepatic or pulmonary damage caused by larval migration (Clayton 1978).

Treatment. The majority of modern anthelmintics used in horses are effective against adult ascarids in foals. These include thiabendazole (44 mg/kg), mebendazole (10 mg/kg), fenbendazole (7.5 mg/kg), cambendazole (20 mg/kg), dichlorvos (26–52 mg/kg) and haloxon (50–70 mg/kg). There are few data on the effect of these and other compounds on the migrating larvae. For good control foals should be treated at one month of age and thereafter at intervals of four to six weeks (Clayton 1978).

Prophylaxis. Attention should be paid to the foals at the time of birth and the foal should be run with its mother in a clean paddock. Boxes in which mares foal should be thoroughly cleaned before the event. Brood mares should be treated for ascarids before foaling. Stables should be cleaned frequently and clean water and food supplied in such a way that contamination is not likely to occur. Manure disposal plays an important part since practically all worm eggs and larvae will be killed by the heat generated during the fermentation of the manure.

Genus: Toxascaris Leiper, 1907

The oesophagus lacks a posterior muscular bulb. The anterior part of the body is provided with large cervical alae and is bent dorsad. The cervical alae of species of the genus, and also those of the genus *Toxocara* described below, give their anterior ends an arrow-like appearance. For this reason they are sometimes called arrow-worms or arrow-headed worms of cats and dogs.

Toxascaris leonina (v. Linstow, 1902) Leiper, 1907 (syn. *T. limbata*) occurs in the small intestine of dog, cat, fox and wild Felidae and Canidae in most parts of the world. The males are up to 7 cm long and the females up to 10 cm. The female genital organs lie behind the level of the vulva. The tail of the male is simple and the spicules are 0.7–0.15 mm long. The eggs are slightly oval, with smooth sides, and measure 75–85 by 60–75 μm.

Life-cycle. The infective stage is the egg containing a second-stage larva. This stage, under optimal conditions outside the host, is reached in three to six days. After ingestion and hatching, second-stage larvae enter the wall of the intestine and remain in this site and stage for about two weeks. Moulting to third-stage larvae commences about 11 days after infection and is followed fairly quickly by a moult to the fourth larval stage. Fourth-stage larvae are present in numbers three to five weeks after infection and may measure up to 8 mm in length. At this stage they are in the mucosa and the lumen of the intestine. Fifth stage larvae are produced about six weeks after infection and eggs are produced from 74 days onwards (Sprent 1959).

No migration of larvae occurs, as compared with *Toxocara canis* (see below).

Larvae of *T. leonina* may occur in mice. In this animal, third-stage larvae are distributed in many tissues and if an infected mouse is eaten by a dog or cat, the larvae are digested from the mouse tissues and develop to maturity in the wall and lumen of the intestine of the final host.

Whereas larvae in the dog and cat are restricted to the intestine, in the mouse they migrate out of this site being distributed all over the body. Sprent (1959) considers this indicative that intermediate hosts are fully utilized in the life-history of the parasite.

Pathogenesis. See Ascariasis in Dogs and Cats.

Toxascaris transfuga (Rudolphi, 1819) occurs in *Ursus* species in many parts of the world.

Other genera of interest in this group of ascarids are **Lagochilascaris** Leiper, 1909 and **Baylisascaris** Sprent, 1968 parasites of wild felines and didelphoids and carnivores and rodents respectively. *Lagochilascaris minor* Leiper, 1909 has been found in subcutaneous abscesses in man in Surinam and Trinidad (Winkel & Treurniet 1956).

Genus: Toxocara Stiles, 1905

Members of this genus have a posterior granular ventriculus at the base of the oesophagus. Intestinal caeca are absent, as are interlabia. Several species occur, the most important of which are *T. canis, T. cati* and *T. vitulorum.*

Toxocara canis (Werner, 1782) occurs in the small intestine of the dog and fox. It is larger than *T. leonina*, the males being up to 10 cm long and the females up to 18 cm. Large cervical alae are present and the body is anteriorly bent ventrad. The female genital organs extend anteriorly and posteriorly to the vulvar region. The male tail has a terminal narrow appendage and caudal alae. The spicules are 0.75–0.95 mm long. The eggs are subglobular, with thick, finely pitted shells, and measure about 90 by 75 μm.

Life-cycle. The life-cycle of *T. canis* is complex and according to the age of the host may involve prenatal (transuterine) and colostral (lactogenic)

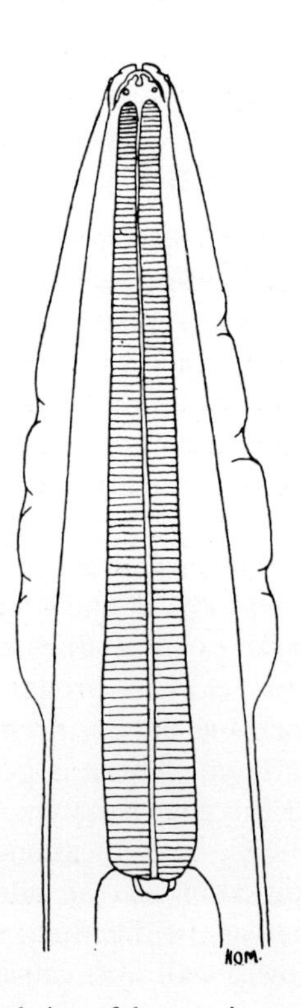

Fig 1.91 Dorsal view of the anterior end of *Toxocara canis.*

transmission, direct transmission or paratenic host transmission. The life-cycle exemplifies both the somatic and the tracheal routes of ascarid migration.

In puppies of a few weeks to less than three months of age the following tracheal migration development cycle occurs (Webster 1958; Sprent 1958). Eggs reach the infective stage in 10–15 days under optimal conditions. Following ingestion they hatch in the duodenum, second-stage larvae (possibly third-stage, according to Aranjo 1972), penetrate the intestinal wall and pass via the lymph stream to mesenteric nodes and then by the portal blood to the liver, the majority reaching this organ by two days after infection. They then pass to the lungs, via the hepatic vein, heart and pulmonary artery, reaching a peak here at about five days after infection, when they measure 800–950 μm in length. They then pass to the tracheal side of the lungs, migrating into the alveoli, the bronchioles and the trachea, eventually being swallowed to reach the stomach on the tenth day. Third-stage larvae are produced in the lungs, trachea or oesophagus and fourth-stage larvae occur in the small intestine about two weeks after ingestion of eggs. The final moult to the adult occurs between three and four weeks after infection and patency occurs at four to five weeks.

As puppies grow older there is a decreasing tendency for the tracheal (above) type of development to be followed and somatic migration occurs, with larvae migrating to various tissues and organs and remaining there as second-stage larvae (Koutz et al. 1966). The age at which tracheal migration ceases and somatic commences appears to vary considerably. This may be a function of the breed and sex of the dog, previous exposure to the parasite and/or the dose of eggs. Greve (1971) reported that the inability to induce patent infections in adult dogs was not related to prior exposure since ascarid-naive dogs failed to produce patent infections when infected with *T. canis* eggs. However Dubey (1979) was able to produce patent infections in dogs aged seven to 52 months when they were fed 100 eggs each of *T. canis*. It is clear that it is unwise to place a strict age limit on when somatic migration supervenes in canine toxocariasis and it is probable that while in the majority of dogs this occurs within the first three months of age, some are capable of supporting tracheal type migrations even as adults.

The somatic type of migration is exemplified when infective eggs of *T. canis* are ingested by an adult bitch. Eight days after infection, second-stage larvae are to be found in various tissues of the body (e.g. liver, lungs, kidneys) and at this stage they have undergone no development (Sprent 1958). Such larvae become resident in the somatic tissues of the adult dog and remain there for some time. The subsequent history of these larvae is not completely clear but during pregnancy they are mobilized and migrate to the fetus giving rise to prenatal infection. Douglas and Baker (1959) noted that this mobilization did not occur before the forty-second day of pregnancy and also that larvae had to be acquired by the pregnant bitch at least 14 days prior to this if prenatal infection were to occur. Not all larvae are mobilized at each pregnancy and some may remain to undergo the process at subsequent pregnancies. The duration of larvae in bitches may be long and animals infected for up to 385 days may still be capable of transmitting infection to puppies. The factor(s) which induce mobilization and migration are unclear but probably there is a hormonal basis involved.

When larvae reach the liver of the fetus they moult to become third-stage larvae and at birth of the puppy, third-stage larvae are present in the lungs and they continue to appear here during the first week of life. The moult to the fourth stage takes place in the first week of life when larvae are in the lungs or, subsequently, the stomach. By the end of the second week after birth, larvae moult to the fifth stage, being 5–7 mm in length at this time. Subsequently, growth is rapid and the adult form may be present by the end of the third week, though an increasing number mature in the next week or two. Patency of prenatal infections varies from 23 to 40 days after birth.

Neonatal infection of puppies from the nursing bitch also occurs through the transmammary route of infection (Stone & Girardeau 1967). Larvae are passed to suckling pups via the colostrum and develop directly to adult worms in the intestine of the puppy. The relative importance of this route of infection has yet to be assessed, but it may be as important as the prenatal mode of infection.

Several workers have reported the occurrence of eggs in the faeces of bitches shortly after parturition. Douglas and Baker (1959) ascribe this to a 'weakening of the immunity' at parturition which permits larvae to pass through the lungs and complete their development in the intestine. However, Sprent (1961) states it may be due to the habit of the bitch of eating the faeces of the puppies and, by so doing, ingesting immature worms which are shed in the faeces of the puppy. These undergo no migration in the bitch and mature in her intestine. Such post-parturient infections are eliminated within a few weeks of the termination of lactation.

An additional mode of infection utilizes the predatory habits of the canine host. Infective eggs ingested by rodents produce second-stage larvae resident in various tissues and organs of these paratenic hosts. Such larvae resume development when the rodent is eaten by a carnivore and the parasites develop, without migration, to the adult stage in the intestine.

Toxocara cati (Schrank, 1788) Brumpt, 1927 (syn. *T. mystax*) (Zeder, 1800) occurs in the small intestine of the cat and wild Felidae. The cervical alae are very broad and are striated. The males are 3–6 cm and the females 4–10 cm long. The spicules are 1.63–2.08 mm long. The egg shell is pitted similarly to the egg of *T. canis* and the diameter of the eggs is 65–75 μm.

Life-cycle. Sprent (1956) has provided a detailed account of the life-cycle. Prenatal infection does not occur, paratenic hosts play an important role in the life-cycle and transmammary transmission occurs. Infection occurs by the ingestion of eggs containing an infective second-stage larva. For the first two days, larvae are found in the stomach wall where they measure 360–460 μm. By the third day some are found in the liver and lungs and by the fifth day they are to be found in the lungs and tracheal washings. Larvae which have passed through the tracheal route are again found in the stomach wall by the tenth day, though many are still to be found in the lungs. By the twenty-first day the number in the stomach wall has greatly increased and larvae are also found in the stomach contents and intestinal contents. Subsequently, the number of larvae in the intestinal contents increases while the number in the lungs and stomach decreases.

The migration is accomplished by larvae in the second stage and third-stage larvae do not occur until larvae have returned to the digestive tract. The majority of third-stage larvae occur in the stomach wall while fourth-stage larvae occur in the stomach contents, the bowel wall and bowel contents.

Rodents play an important part in the life-cycle as paratenic hosts. Second-stage larvae from infective eggs become encapsulated in various organs and tissues, chiefly the liver, where they remain for several months. Upon ingestion by a cat, the larvae are released and develop in the stomach wall to third-stage larvae, remaining here for six days then they re-enter the stomach contents as fourth stage larvae on the thirteenth day and reach maturity in the small intestine from 21 days onwards. No migration via the liver and lungs occurs.

As well as rodents acting as paratenic hosts second-stage larvae may be found in the tissues of earthworms, cockroaches, chickens, sheep and other animals fed infective eggs. Sprent (1956) considers the parasite to be well adapted to the Felidae in that the larger members of the family may acquire the infection by predation on ruminants etc., while the smaller members may acquire the parasite by ingestion of small rodents or invertebrates.

Though prenatal infection with *T. cati* does not occur, transmammary infection is common (Swerczeh et al. 1971). It is possible that the majority of infections in kittens are derived from the milk of infected queens. Larvae occur in the milk throughout lactation in experimentally infected queens and larvae appear in the mammary gland shortly after infection with eggs or are transported to the mammary gland after an extended stay in other tissues before lactation commences. Larvae acquired in this manner by kittens do not undergo migration and behave similarly to those acquired from a paratenic host.

ASCARIASIS IN DOGS, CATS AND FUR-BEARING ANIMALS

Pathogenesis. Heavy infections are most commonly seen in kennels and catteries and, under conditions of poor hygiene, heavy infection of young animals may occur.

Heavy prenatal infection with *T. canis* may lead to the death of whole litters of puppies. Though the migration of larvae through the lungs of the newborn puppy may cause pneumonia, this is uncommon and the more usual effect is a progressive malaise associated with vomiting and diarrhoea as worms mature in the stomach and intestine. Eventually each meal is rejected and the puppy often becomes covered with stale vomit; it may suffer from inhalation pneumonia and, in all, presents an abject picture of despair. Death frequently occurs two to three weeks after birth.

In less severe infection of *T. canis* and in infections of *T. leonina* and *T. cati* there is general unthriftiness, a pot-bellied appearance, intermittent diarrhoea and possibly anaemia.

Adult worms, on occasion, migrate to aberrant sites, such as the bile duct or through the bowel wall, and in these cases the pathogenesis depends on the site of the worm.

Nervous disorders are frequently associated with roundworm infection of dogs. The mechanisms responsible for these effects have yet to be clarified; however, they may result from irritation of the bowel by the worms or may be due to focal lesions in the central nervous system on the death of aberrant larvae.

Clinical signs. The animals are unthrifty and either pot-bellied or with a tucked-up abdomen; the coat is dull and harsh; there is usually emaciation, often anaemia, restlessness and diarrhoea or constipation. Death due to acute intestinal obstruction may occur. In the case of foxes the most dangerous period is during the first few weeks of life, especially from the second to the fourth. Extreme pot-belliedness of fox pups may be associated with the presence of ascarids in the peritoneal cavity.

Diagnosis is made on the basis of the clinical signs and is confirmed by finding the eggs in the faeces.

Treatment. Adult ascarids in dogs and cats are susceptible to a range of anthelmintics. Larval stages in the tissues are much less susceptible and if a drug is active against larval stages, it must frequently be given in a markedly increased dose.

Salts of *piperazine*, which are well tolerated by dogs and cats, are highly effective against *Toxocara* and *Toxascaris* spp. Piperazine adipate at a dose of 100 mg/kg is highly effective against adult forms and at the dose of 200 mg/kg will remove immature worms from puppies one to two weeks of age. This allows for the control of prenatally acquired infections. Other salts, e.g. piperazine dihydrochloride and citrate, are also effective at this dose. The drug is well tolerated and can be given to very young animals. There is no effect by the piperazine salts against larvae in the tissues of the bitch.

Diethylcarbamazine, though one of the older compounds, is highly effective in a single dose of 50 mg/kg against the dog and cat ascarids. It should not be given to dogs with patent infections of *Dirofilaria immitis.*

Thenium [(N:N-dimethyl-N-2-phenoxyethyl-N-2-′thenylammonium) *p*-chlorobenzene sulphonate], variously combined with piperazine or arecoline *p*-stibonobenzoic acid, is given as a single dose of 100 mg/kg. Puppies weighing less than 2.5 kg should not be treated with the drug.

Dichlorvos is highly effective against ascarids and other intestinal nematodes. Doses of 12–15 mg/kg are used, as either a single or a split dose. For cats and puppies a dose of 11 mg/kg is adequate. Dichlorvos is an organophosphate and is a cholinesterase inhibitor. It should not be used in dogs weighing less than 1 kg, in dogs or cats under 10 days of age or in greyhounds or whippets which seem to have an increased sensitivity to the compound. In addition, its use is contraindicated in dogs with impaired liver function, circulatory distress, or patent infections of *Dirofilaria immitis.*

Toluene (methylbenzene) may be formulated with dichlorophene or arecoline, in various proportions, to produce a broad-spectrum compound active against nematodes and cestodes. Doses of 100–200 mg/kg are effective against adult ascarids in dogs; solid food and milk should be withheld for at least 18 hours prior to and four hours after treatment. It should not be given to weak or debilitated animals.

Dichlorophen is variously formulated with toluene and/or arecoline and is given in standard capsule form, according to the size of the dog.

Trichlorphon formulated with atropine is effective at 75 mg/kg for ascarids and other roundworms of dogs, as well as various ecto-parasites. It should not be given to cats, grey-hounds or sick or debilitated animals.

Pyrantel pamoate may be used at a dose of 5 mg/kg and is effective also against hookworms of dogs.

Nitroscanate (4-nitro-4′-isothiocyano-diphenge-ether) is a broad-spectrum anthelmintic highly effective against ascarids, other intestinal nematodes and adult cestodes of dogs. It is safe to use in very young puppies and in pregnant or lactating bitches. Particle size is important and a size of 2–3 μm is highly active. A single dose of 50 mg/kg will remove 86% of immature and 96–100% of adult *Toxocara canis*. Mild vomiting may occur at about 60 mg/kg and/or reversible tranquillizing effects may occur in a few dogs.

Mebendazole, a benzimidazole compound, is highly effective against ascarids and is given twice a day over two days at a dose of 10 mg/kg. It is very safe.

Fenbendazole, a benzimidazole, is 95–100% effective against adult ascarids of dogs and cats with a single dose of 100 mg/kg or this dose divided over five days. Dubey (1979) reported that dogs given 50 mg/kg per day for 14 days harboured significantly fewer larvae in muscles and viscera than non-treated dogs, all having been infected with 10 000 *Toxocara canis* eggs.

Prophylaxis. Good hygiene is essential in kennels and catteries. With *T. leonina* and *T. cati* where oral infection leads directly to a patent infection the parasites can be eliminated by regular treatment of dogs or cats and by frequent thorough cleansing of the premises. Earth exercise areas should be either fenced off or made impervious.

With *T. canis*, in which prenatal infection plays an important part in the life-cycle, the situation is different and more difficult. Transmammary infection with *T. canis* and *T. cati* presents similar problems and the bitch and the queen may harbour ‘dormant’ larvae in her tissues for several months or years, transmitting the infection to several litters. Immediate tactical control consists of recognizing and anticipating prenatal and transmammary infections and treating puppies within two weeks of birth. The evidence that fenbendazole may reduce the number of larvae in the tissues of bitches (Dubey 1979) indicates that such treatment may prevent prenatal infection. More long-term control consists of regular treatments for adult worms to lower or eliminate contamination of the environment and also measures to eliminate an established environmental contamination. This is best achieved by providing impervious surfaces to the kennels so that they may be thoroughly and regularly cleaned. The eggs of the dog and cat ascarids may remain viable for several months, consequently a cursory or token disinfection is of little value.

Since rodents may play an important part in the life-cycles of the parasites, these should be exterminated from the kennels and catteries.

Visceral larva migrans

This condition of children is mainly caused by the larvae of *T. canis*, though the larval stages of *T. leonina*, *T. cati*, *Capillaria hepatica* (of rodents) and *Lagochilascaris minor* (of wild felines) etc. have also been incriminated. Petter (1960) has compiled a list of hosts of ascarids which may be responsible for causing the visceral larva migrans syndrome. The entity is characterized by chronic granulomatous (usually eosinophilic) lesions, associated with larvae of the above parasites, in the inner organs of children, especially the liver, lungs, brain, sometimes the eye and also elsewhere. In the child, larvae migrate in the ‘somatic’ manner, as they would for example in a rodent, and on repeated infection large numbers may occur in the body of the child. Beaver (1966) has reported on one case which had 300 larvae per gram of liver. A related entity, *cutaneous larva migrans*, caused by canine hookworm larvae in man is described on p. 206.

The pathological entity consists of an enlarged liver with eosinophilic granulomatous lesions, hepatomegaly, pulmonary infiltration, intermittent fever, loss of weight, loss of appetite and a persistent cough. However, the clinical picture varies greatly though a relatively constant feature is a high (50%) and persistent circulating eosinophilia. The eye lesions caused by these larvae have received considerable attention in recent years, especially since they often resemble a retinoblastoma. A mistaken diagnosis may result, and on several occasions has resulted, in unnecessary enucleation of the eyeball. The criteria used for a diagnosis of visceral larva migrans include (*a*) leucocytosis, a white blood count of greater than 10 000/mm^3; (*b*) eosinophilia greater than 10%; (*c*) an anti-A isohaemagglutinin titre of 1 : 400 or more and an anti-B titre of 1 : 200 or more; (*d*) IgG and IgM levels greater than two standard deviations above the normal for age and sex; and (*e*) hepatomegaly (Glickman et al. 1978).

The condition is most usually seen in children one to five years of age. Children of this age frequently adopt the habit of dirt eating and where soil is heavily contaminated with *Toxocara* eggs (e.g. soil around doorsteps etc.) the ingestion of even moderate amounts of soil may result in the intake of large numbers of infective eggs. Since it is common to give young puppies to children as playmates, a special hazard may arise since it is the young puppy which is preferentially infected with *T. canis*. However, doorstep and garden soil contaminated by domestic pets is not the sole or, in fact, the main danger. There is a much wider public health problem which has yet to be recognized by the general public at large. It is the extensive fouling of public parks, playgrounds and sidewalks with the faeces of domestic pets, especially in large cities (Jacobs et al. 1977). Though modern-day man has largely solved the problems of mass hygiene, he has yet to provide a solution to the disposal of the excreta of his pets. The public health importance of human *Toxocara* infection in Great Britain has been reviewed by Woodruff (1976). Some 2% of the adult population of the UK possess antibodies to *Toxocara*, though the majority of the sensitizations are asymptomatic. Additionally, a majority have had no direct contact with dogs or cats through pet ownership. Higher prevalencies of seropositiveness are found in certain groups of individuals, suggesting a greater exposure to infection or the involvement of the parasite in the disease process. Thus, as compared with normal healthy individuals of which 2% are seropositive, those with choroidoretinitis, uveitis, hepatomegaly and asthma showed 8.8%, 10%, 29% and 17% positive reactions, respectively. The clinical importance is difficult to estimate since infection is not notifiable. However, in the Hospital for Tropical Diseases, London, 20–30 patients with toxocaral eye lesions are treated annually.

The specific diagnosis of visceral larva migrans is based on the demonstration of the lesions and the larvae in biopsy material. Immunodiagnostic tests are a valuable adjunct to diagnosis (Glickman et al. 1978). A detailed account of visceral larva migrans is given by Beaver (1956, 1966).

Visceral larva migrans also occurs in other animals. Sprent (1955) has extensively studied the migration of dog and cat ascarids in experimental animals. Done et al. (1960) reported brain and spinal cord lesions in pigs infected with *T. canis*: clinical signs of illness occurring about 22 days after infection were associated with encapsulation and death of larvae. Roneus (1963) has reported on the migration of *T. cati* larvae in pigs.

Toxocara vitulorum (Goeze, 1782) Travassos, 1927 (syn. *Neoascaris vitulorum*) occurs in the small intestine of cattle, zebu and the Indian buffalo, and is found in many parts of the world. It has also been recorded from sheep and goats (Warren 1971). The males measure up to 25 cm by 5 mm and the females 30 cm by 6 mm. The cuticle is not as thick as that of other large ascarids and these worms therefore have a soft, translucent appearance. The body does not taper much towards the extremities. There are three lips, broad at the base and narrow anteriorly. The oesophagus is 3–4.5 mm long and has a posterior, granular ventriculus. The tail of the male usually forms a small spike-like appendage. There are about five pairs of post-cloacal papillae; the anterior pair is large and double. The pre-cloacal papillae are variable in number. The spicules are 0.99–1.25 mm long. The vulva is situated about one-eighth of the body length from the anterior end. The eggs are subglobular, provided with a finely pitted albuminous layer and measure 75–95 by 60–75 μm.

Life-cycle. Adults of *T. vitulorum* are found almost exclusively in calves and prenatal and transmammary infections constitute the major sources of parasites for calves; hence the life-cycle resembles that of *T. canis.*

Eggs become infective under optimal conditions in 15 days and may survive for extended periods, like other ascarid eggs. Ingestion of embryonated eggs by neonates, juveniles or adults does not lead directly to a patent infection. Instead larvae are distributed to various tissues and organs, remaining dormant in the site until the latter part of pregnancy in the cow. At this time prenatal infection occurs. Larvae also migrate to the mammary gland and pass out in the milk and, on ingestion by calves, produce adult worms in the intestine.

The sequence of events in prenatal infection in natural and experimental hosts has been described by Warren (1971), Mozgovoi and Slinkhov (1971) and Mozgovoi and Shakhmatova (1969). Larvae in the tissues after oral infection with eggs are considered to be third-stage larvae. From the eighth month of pregnancy onwards they recommence migration, reach the placenta and pass to the amniotic fluid. The fetus is then infected by the ingestion of larvae which reach the stomach and then the intestine and where they mature to adult worms after the birth of the calf. It is presumed that during the migration after the eighth month of pregnancy the larvae also pass to the mammary gland. At birth they occur in the colostrum and on ingestion by a calf develop to adult worms in approximately four weeks (Warren 1971). Egg production is very high and may reach 8×10^6 eggs per female worm per day and egg counts of 100 000 epg or more are not uncommon. However patency is short and natural expulsion of adult worms commences as early as 38 days after birth and by four to six months no adult parasites remain.

Pathogenicity. Light infections may pass unnoticed in calves but clinical signs are seen with worm burdens of 70–500 per calf. The predominant clinical signs are diarrhoea and steatorrhoea. They may be accompanied by colic, signs resembling intestinal obstruction and the presence of mud-coloured, evil-smelling faeces. Emaciation and, later, death, may occur. The parasite has been reported to be a serious pathogen in Africa, the Philippines, Sri Lanka and India; in the latter two countries it contributes significantly to calfhood mortality among buffalos. However, it also occurs in Europe (e.g. Belgium, Germany, Italy, Hungary, USSR), Australia and North America, though in non-tropical areas it is not considered to be a serious pathogen. Infection may be transferred to other localities by cows and imported Limousin cattle have been suggested as a source of infection in Belgian livestock.

Diagnosis can be made by finding the eggs in faeces.

Treatment. Several of the anthelmintics generally used for trichostrongyle infections in cattle are also effective against *T. vitulorum* in calves. These include piperazine, morantel, neguvon, fenbendazole and levamisole, used at doses similar to those for the trichostrongyles (see p. 247). No information is available on the action of these compounds on the larvae in the cow.

FAMILY: ANISAKIDAE SKRJABIN & KAROKHIN, 1945

Genera of importance in this family include *Anisakis*, *Porrocaecum* and *Contracaecum.* Members of the family possess one or more of the following features at the base of the oesophagus: oesophagus with cylindrical ventriculus but intestinal caecum absent (*Anisakis*); posterior ventriculus and anterior caecum projecting forward along the oesophagus (*Porrocaecum*); or posterior ventriculus, anterior caecum and posterior appendix projecting backwards from the ventriculus (*Contracaecum*) (Hartwich 1974). They are parasites of marine mammals, birds, fishes, reptiles and, accidentally, man.

Genus: Porrocaecum Raillet & Henry, 1912

Porrocaecum crassum (Deslongchamps, 1824) occurs in the intestine of domestic and wild ducks. The male is 12–30 mm in length and the female 40–55 mm. The worms are reddish-white in colour. A short anterior caecum arises from the gut; the tail of the male is conical and there are no caudal alae. The eggs are ellipsoidal, reticulated and measure 85 by 110 μm.

Life-cycle. The life-cycle is similar to that of *Porrocaecum ensicaudatum* (Zeder, 1800), a parasite of passerine birds. Levine (1957) infected earthworms of the genera *Lumbricus* and *Octolasium* with eggs of *P. ensicaudatum* and larvae were found in the ventral blood vessels and hearts. Invertebrates fed to passerines released larvae which penetrated the horny layer of the gizzard and then migrated to the duodenum. They penetrated the wall of the intestine, moulted to third-stage larvae in the wall and subsequently developed to maturity after emergence from the wall.

Pathogenesis. Reduced weight gain, diarrhoea and foamy discharges from the nose have been reported in infected ducks. Anaemia associated with inhibition of haemopoiesis has also been reported (Ryzhkova 1953).

Treatment. Anthelmintics active against *Ascaridia* and *Heterakis* are indicated, i.e. piperazine given in the feed at a rate of one part to 350 of feed, giving an intake of 133–300 mg/kg. Various salts of piperazine (e.g. citrate, adipate, hexahydrate) are probably effective. Thiabendazole and other benzimidazoles may be of use.

Other species include **P. depressum** (Zeder, 1800) Bayliss, 1920 in birds of prey, **P. angusticolle** Bayliss and Daubney, 1922 of raptors and **P. aridae** Bayliss, 1936 of the heron.

Genus: Contracaecum Railliet & Henry, 1912

Contracaecum spiculigerum (Rudolphi, 1809) occurs in ducks, geese, swans and a wide variety of waterfowl. The male is 32–45 mm in length and the female 24–64 mm. An oesophageal appendix and an anterior caecum are present. The eggs are spherical and 50–52 μm in diameter. Huizinga (1965) has reported that the life-cycle involves a copepod (*Cyclops*) in which second-stage larvae occur. Various species of fish ingest infected copepods and third-stage larvae penetrate the intestinal wall of the fish and encyst in the body cavity. Birds acquire infection by eating infected fish.

Parasites are found in the lumen or the wall of the proventriculus. Denudation of the mucosa occurs and the extent of this is dependent on the number of worms present. In fish, worms may burrow into the mucosa of the intestine and may also invade the musculature, necessitating rejection of the fish at inspection.

No critical information is available concerning treatment, though drugs active against the other ascarids of poultry will probably be effective.

Other species in the genus include **C. microcephalum** Bayliss, 1920 of the heron and *Anas* spp. and **C. osculatum** Rudolphi, 1802 of seals and other fish-eating mammals.

Genus: Anisakis Dujardin, 1845

Parasites of this genus occur in marine animals. An intestinal caecum is absent; the anterior region of the lips form medial, bilobed processes. Larval stages occur in a variety of marine fishes but may also be found in anadromous fish in fresh water. Ingestion of raw or semi-raw fish containing such larval stages by humans may result in anisakiasis (herring worm disease), in which anisakine nematodes either persist in the alimentary canal or penetrate into the intestinal tissue. In addition to the genus *Anisakis* (the herring worm) the genus **Phocanema** Myers, 1959 (codfish worm) is concerned in such human infection.

Other genera, e.g. **Terranova** Leiper and Atkins, 1914, adults of which occur in elasmobranch fishes, **Thysanascaris** Dollfus, 1933 (e.g. *T. adunca* of sardines) and **Contracaecum** (e.g. *C. osculatum* of seals and larval stages in the anchovy) may possibly be concerned in the human condition.

Clinical signs are variable and non-specific, being associated with an eosinophilic granuloma in the intestinal or stomach wall. A single larva has been received in the majority of cases, being coughed up or found in the stool. Mild cases are associated with slight irritation of the bowel. Severe cases may be mis-diagnosed as ulceration or a malignant tumour.

The condition has been reported mainly from the Netherlands, Germany, Scandinavia and Japan, often associated with the consumption of lightly salted ('green') herrings, the musculature of which contains the larval anisakids. Preventive measures include the evisceration of fish as soon as possible after catch (to prevent additional larvae migrating into the muscles) and freezing fish at $-20°$ for 60 hours if thorough cooking is not intended prior to consumption (WHO 1979).

Various immunodiagnostic tests have been developed to detect the condition (Soulsby 1976).

SUPERFAMILY: OXYUROIDEA RAILLIET, 1916

Nematodes with the ventrolateral papillae rudimentary or absent. Males with two, one or no spicules. Oesophagus possessing a posterior bulb. Petter and Quentin (1976) have provided a key to the genera of the Oxyuroidea. Four families are recognized: Pharyngodonidae Travassos, 1919, Heteroxynematidae (Skrjabin and Shikhobalova, 1948), Kathlaniidae Travassos, 1918 and Oxyuridae Cobbold, 1864. The last is of major interest in this superfamily.

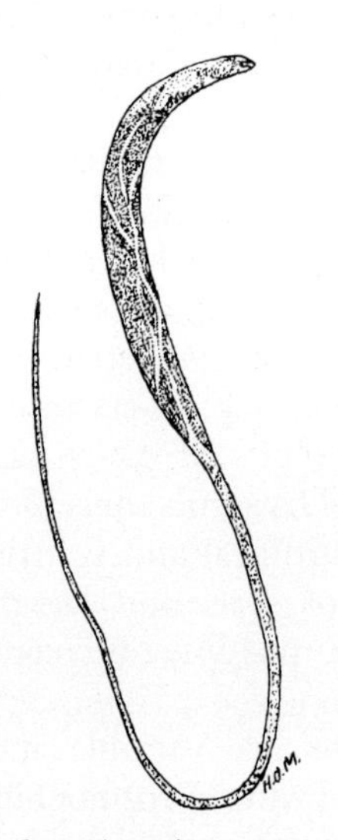

Fig 1.92 Female *Oxyuris equi.*

FAMILY: OXYURIDAE COBBOLD, 1864

These are medium-sized or small worms with three inconspicuous lips. The oesophagus has a well-developed posterior bulb. The male bears a number of large papillae around the cloacal opening. The females are usually much larger than the males and have long, tapering tails. The vulva is situated near the anterior end of the body. The eggs are usually flattened on one side and development takes place without the need for an intermediate host.

Genus: Oxyuris Rudolphi, 1803

Oxyuris equi (Schrank, 1788) (syn. *O. curvula*, *O. mastigodes*) occurs in the large intestine of equines in all parts of the world. The male is 9–12 mm long and the female up to 150 mm. The oesophagus is narrow at the middle and the bulb is not distinctly marked off. The male has one pin-shaped spicule which is 120–150 μm long and the tail bears two pairs of large and a few small papillae. The young females are almost white in colour, slightly curved and have relatively short, pointed tails. The mature females have a slatey-grey or brownish colour and narrow tails which may be more than three times as long as the rest of

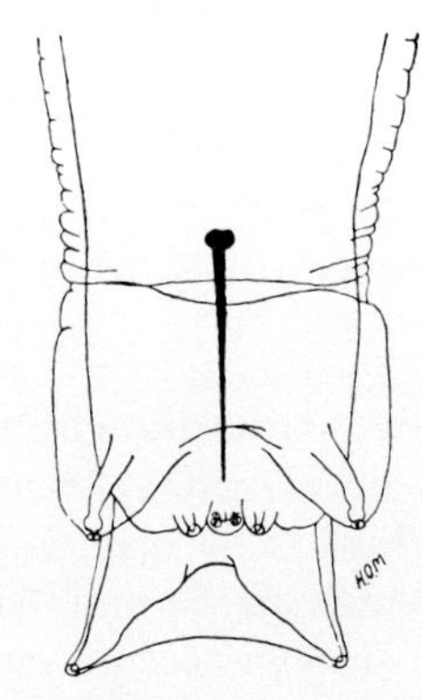

Fig 1.93 Ventral view of the hind end of *Oxyuris equi.*

the body. The eggs are elongate, slightly flattened on one side, provided with a plug at one pole, and measure about 90 by 42 μm.

Life-cycle. The males and young females inhabit the caecum and large colon. After fertilization the mature females wander down to the rectum and crawl out through the anal opening. Eggs are laid in clusters on the skin in the perineal region. Development of the egg is rapid, reaching the infective stage in three to five days. The infective stage may be reached on the perineal region or, more usually, the egg falls off to the ground. Eggs probably survive for several weeks in moist surroundings but dessication is rapidly lethal.

Infection is by ingestion of the infective eggs on fodder and bedding. Infective larvae are liberated in the small intestine and third-stage larvae are found in the mucosal crypts of the ventral colon and caecum. Fourth-stage larvae are produced about eight to ten days after infection: these possess a large buccal capsule and browse on the mucosa. The sexually mature adult stage is reached about four to five months after infection.

Pathogenesis. The fourth-stage larva feeds on the intestinal mucosa of the host. The adult worms are, however, not found attached and probably feed on the intestinal contents. The chief feature of oxyuriasis in equines is the anal pruritus produced by the egg-laying females.

Clinical signs. The irritation caused by the anal pruritus produces restlessness and improper feeding, which results in loss of condition and a dull coat. The animal rubs the base of its tail against any suitable object, causing the hairs to break off and the tail to acquire an ungroomed 'rat-tailed' appearance.

Diagnosis. The clinical signs should lead to an examination of the perineal region, where cream-coloured masses of eggs will be found. These should be removed and identified under the microscope. The condition should be differentiated from mange and anal pruritus due to other causes.

Treatment. Regular anthelmintic treatment for strongyles in horses usually keeps *O. equi* infection under control. However, since the pre-patent period is extended and the majority of drugs do not have a high efficacy against the immature stages, occasionally the parasites may pose a problem. Mebendazole (5–10 mg/kg), cambendazole (20 mg/kg) and dichlorvos (26–52 mg/kg) are the drugs of choice for adult parasites, though other compounds in general use for horses (e.g. thiabendazole, fenbendazole, oxibendazole and pyrantel) have an acceptable level of activity. New arrivals at a stable should be treated routinely to eliminate the infection.

Control of *O. equi* depends on good hygiene in stables. Bedding should be removed frequently and feeding appliances constructed so that they are not contaminated by bedding. A clean supply of water should be available.

Other species of the genus include **O. poculum** von Linstow, 1904 of the horse in Sri Lanka, **O. karamoja** Baylis, 1939 of the rhinoceros and **O. tenuicauda** von Linstow, 1901 in *Equus burchelli* in Africa.

Genus: Enterobius Leach, 1853

Enterobius vermicularis (Linnaeus, 1785). This is the human pinworm or seat worm. As well as the human, it may occur in higher primates such as the chimpanzee, gibbon and marmosets. It does not occur in the dog or cat.

The worms are cream-coloured and slender, the male measuring 2–5 mm and the female 8–13 mm. The female has a long pointed tail: otherwise the parasites resemble *Oxyuris equi*.

Adult worms occur in the caecum, appendix and ascending colon. Gravid females migrate posteriorly and deposit eggs on the perianal and perineal regions. The eggs become infective within a day or so and further infection is by ingestion of infective embryonated eggs, mature worms being produced about two months after infection.

The gravid females produce an intense pruritus. This causes restlessness, insomnia and various effects on behaviour including inattention, lack of cooperation and possibly a feeling of shame and inferiority.

Infection in the simians is not usually serious but the pruritus may lead to aggressive behaviour. Animals may be treated with dichlorvos (8–9 mg/kg daily for two days), pyrvinium pamoate (5 mg/kg twice at an interval of 14 days), piperazine (50 mg/kg for seven days) or thiabendazole (50 mg/kg for two days).

Genus: Passalurus Dujardin, 1845

Passalurus ambiguus (Rudolphi, 1819) occurs in caecum and colon of rabbits, hares and other lagomorphs. Male 4.3–5 mm, female 9–11 mm long. The oesophagus has a prebulbar swelling and a strong bulb. The male tail has a whip-like appendix and small caudal alae supported by papillae. The spicule is simple and 0.09–0.12 mm long. The female has a tapering tail 3.4–4.5 mm long and the cuticle of its distal extremity is marked with about 40 circular striations. The vulva opens 1.54–1.89 mm from the anterior extremity. The eggs are flattened on one side and measure 95–103 by 43 μm.

Development is direct and infection occurs through the ingestion of infective eggs. The young stages are found in the mucosa of the small intestine and the caecum.

These worms sometimes occur in enormous numbers in young rabbits, but appear to be relatively harmless. Should treatment be required, probably piperazine or dichlorvos will be effective.

Genus: Skrjabinema Werestchajin, 1926

This genus contains several species of small worms measuring about 3–8 mm in length, which occur in the caeca of ruminants. They have three large, complicated lips and three small intermediate lips. The oesophagus is cylindrical and terminates in a large spherical bulb. The tail of the male is bluntly rounded and has a cuticular caudal expansion supported by two pairs of processes. There is a single spicule in the male. The life-cycle is direct. Eggs are fully embryonated when deposited by the female on the perianal skin. Eggs drop off and are ingested with feed and water. They hatch in the small intestine and migrate to the large intestine where they mature in about 25 days after infection (Schad 1957).

Skrjabinema ovis (Skrjabin, 1915) has been found in the sheep, goat and antelopes in several countries, **S. alata** (Mönnig, 1932) in the sheep in southern Africa, **S. africana** (Mönnig, 1932) in the steinbock in southern Africa and **S. caprae** Schad, 1959 in the goat in New Mexico. **Skrjabinema tarandi** Skrjabin and Mitskevich, 1930 occurs in reindeer and caribou in Siberia and north-western North America. These worms are not pathogenic, but they may be mistaken for young forms of other nematodes, such as *Oesophagostomum columbianum*.

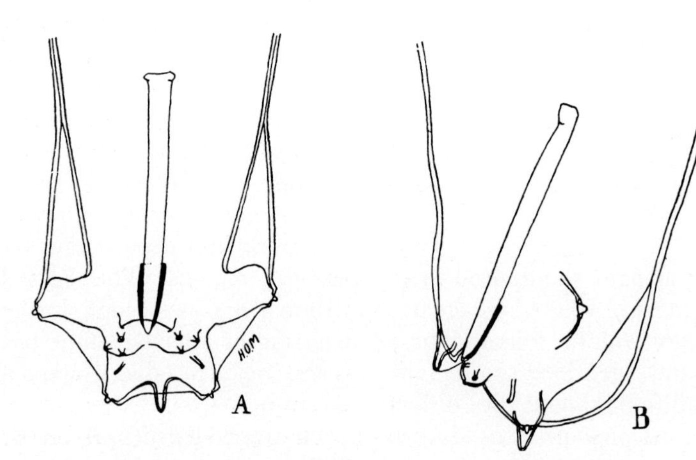

Fig 1.94 Ventral view (A) and lateral view (B) of the hind end of the male *Skrjabinema ovis*.

Genus: Syphacia Seurat, 1916

Members of the genus occur in the large intestine of various rodents. **Syphacia obvelata** (Rudolphi 1802) Seurat 1916 is the mouse pinworm occurring in the house mouse throughout the world. The parasites are small (males 1.1–1.5 mm; females 3.4–5 mm) and the eggs measure 135 by 45 μm. Other species of the genus include **S. muris** Yamaguti, 1941, the pinworm of rats (males 1.2–1.3 mm; females 2.8–3.4 mm) and several other species in wild rodents in various parts of the world.

Worms are pointed and white in colour; there are three distinct lips without a buccal capsule. The oesophagus is oxyurid, having a prebulbar swelling and a posterior globular bulb. These are small cervical alae. There is a single, slender long spicule in the male.

The life-cycle is direct. Females deposit embryonated eggs in the colon and on the perianal skin. These are infective within a few hours and infection occurs from ingestion of eggs from the perianal contaminations, or in contaminated food (Chan 1952). Gravid females occur on the ninth day of infection.

Even in heavy infections there are few or no pathogenic effects. Reduced weight gain and growth have been suggested as a sequel to infection (Hoag & Meier 1966). Diagnosis is based on the detection of eggs in faeces and worms in the caecum and colon. The following anthelmintics are effective: piperazine adipate in the drinking water (4–7 mg/ml for three to ten days), dithiazanine iodide (0.1 mg/g food for seven days), pyrvinium pamoate (1.6 mg/kg food or 0.8 mg/litre of water for 30 days) and dichlorvos (0.5 mg/g food for one day). Eradication is difficult and usually only colonies derived by caesarean section and barrier sustained can be kept free of the infection.

Genus: Aspiculuris Schulz, 1924

Members of this genus are pinworms of rodents. There are three lips round the mouth but no vestibule. Broad cervical alae occur; the tail of the male is conical and a spicule and gubernaculum are absent.

Aspiculuris tetraptera (Nitzsch, 1821) Schulz, 1924 occurs in the large intestine of mice and other rodents, including rats, on a world wide basis. Males are 2–4 mm in length and females 3–4 mm. Eggs are symmetrical, 85 by 37 μm.

The life-cycle is direct. Eggs are passed in the faeces and the infective stage is reached in about six days. Infection is by ingestion of eggs and the prepatent period is about 23 days. Negligible pathogenicity is associated with the infection.

Treatment and control are as for *Syphacia* spp.

Genus: Dermatoxys Schnider, 1866

This genus occurs in the caecum and colon of cotton-tail rabbits and hares and jack rabbits in North America and Brazil.

Dermatoxys veligera (Rudolphi, 1819) Schneider, 1966 is 8–11 mm in the male and 16–17 mm in the female. Eggs are thick-walled, asymmetrical, operculate and measure 110 by 50 μm. There are three lips and a short mouth-vestibule containing three teeth. Cervical alae are present as are caudal alae in the male. There is a single spicule. The female tail is long and tapering, ovaries are marked and deeply coloured. The life-cycle is probably direct. Pathogenesis is unknown.

Genus: Tachygonetria Weal, 1862

This is one of the commonest genera of the large intestine of tortoises. Large numbers may occur and indeed several genera of oxyurids and numerous species may be found in heavy infections. No ill-effect is associated with these infections (Flynn 1973).

FAMILY: KATHLANIIDAE TRAVASSOS, 1918

Oxyuroidea with the isthmus of the oesophagus sub-spherical: oesophagus terminated by a bulb. Parasites of the large intestine of equines, apes and tortoises.

Genus: Probstmayria Ransom, 1907

Probstmayria vivipara (Probstmayr, 1865). This is a minute nematode living in the colon of horses. It measures 2–2.9 mm in length. The females are viviparous and give birth to larvae almost as large as the adults. As a result of this almost unique method of reproduction, infections may be enormous, but the worms are not known to be pathogenic.

Leiperenia leiperi Khalil, 1922 is a similar species in the intestine of the hippopotamus and the African elephant. **Leiperenia galebi** Khalil, 1922 occurs in the Indian elephant.

SUPERFAMILY: SUBULUROIDEA TRAVASSOS, 1930

FAMILY: HETERAKIDAE RAILLIET & HENRY, 1914

Medium sized to small worms with three lips around the mouth, a small buccal cavity and pharynx. Lateral alae extending down the body.

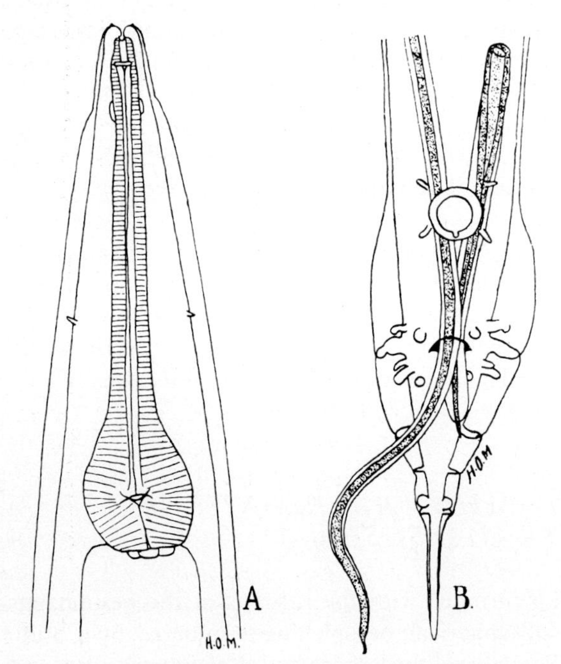

Fig 1.95 *Heterakis gallinarum*. A, Dorsal view of anterior end. B, Ventral view of hind end of male.

Oesophagus in three parts—a short pharynx, a cylindrical, middle part and a bulbous posterior part with a valvular apparatus. A pre-anal sucker at the tail end of the male which has a chitinous rim. Many anal papillae.

Genus: Heterakis Dujardin, 1845

Heterakis gallinarum (Schrank, 1788) Madsen, 1949 (syn. *H. papillosa*, *H. vesicularis*, *H. gallinae*) occurs in the caeca of the fowl, guinea-fowl, pea-fowl, turkey, duck, goose and numerous other birds. The male is 7–13 mm long and the female 10–15 mm. There are large lateral alae extending some distance down the sides of the body. The oesophagus has a strong posterior bulb. The tail of the male is provided with large alae, a prominent, circular, pre-cloacal sucker and twelve pairs of papillae. The spicules are unequal, the right being slender and 2 mm long, while the left has broad alae and measures 0.65–0.7 mm. The vulva opens directly behind the middle of the body. The eggs have thick, smooth shells; they measure 65–80 × 35–46 μm and are unsegmented when laid.

Life-cycle. The eggs develop in the open and reach the infective second larval stage in 14 days at 27° but usually development is longer and may take several weeks at lower temperatures. The eggs are very resistant and may remain viable in the soil for months. When the host swallows an infective egg the larva hatches in the intestine after one or two hours. Up to about the fourth day the young worms are rather closely associated with the caecal mucosa and some injury to the glandular epithelium may occur (Baker 1933). Osipov (1957) considers that second-stage larvae spend two to five days in the glandular epithelium before continuing their development in the lumen. They moult to the third stage on the sixth day after infection, the fourth stage on the tenth day and the fifth stage on the fifteenth day. The first eggs are passed in the faeces of the bird after 24–30 days. Earthworms may serve as transport hosts. The parasites occurred as second-stage larvae in earthworms and infections occurred in poultry which ate infected earthworms (Lund et al. 1966).

Pathogenesis and clinical signs. The direct effects of *H. gallinarum* are slight and only in heavy infections may there be thickening of the caecal mucosa with a number of petechial haemorrhages on the surface. Even so, no marked ill effects are ascribable to infection. With *H. isolonche*, on the other hand, marked lesions may be produced in the caecum of the pheasant. These consist of a nodular typhlitis which leads to diarrhoea, wasting, emaciation and death. All stages of *H. isolonche* may be found in the lesions. *H. beramporia* may also be found in nodules in the caeca.

The principal economic importance of *H. gallinarum* lies in its role as a carrier of *Histomonas meleagridis*, the causal agent of blackhead (enterohepatitis) of turkeys (see p. 567). The protozoon may remain viable in the egg of *H. gallinarum* for a long time, perhaps as long as the egg remains viable. It is also thought that the shelter of the helminth egg allows passage of the protozoan through the anterior part of the digestive tract, which is normally lethal to the blackhead organism.

Diagnosis is made by finding the eggs in the faeces. Caecal faeces have to be examinaed and the eggs must be differentiated from those of *Ascaridia galli* and other related worms.

Treatment and prevention. Phenothiazine is effective at a dose of 1 g per bird. One part of phenothiazine to 60 of feed, given for six hours after an overnight fast, should produce 80–100% clearance. Piperazine is less effective but a mixture of phenothiazine and piperazine may be used to eliminate *Heterakis* and *Ascaridia* infections. 1 g of 7:1 phenothiazine/piperazine mixture will remove 90% or more of each genus. Hygromycin B as a 0.25% mix in the feed is highly effective. Mebendazole is effective against a range of poultry nematodes; 2 g are given in 28 kg of feed. Tetramisole has a similar range of activity; a 10% solution is given in drinking water. Haloxon is given at a rate of 30 g per 50 g of feed; this drug is toxic for geese.

For prevention, strict sanitation of poultry houses and yards is essential.

Other species of this genus which occur in avian hosts include:

Heterakis brevispiculum Gendre, 1911. The spicules are equal, 0.4 mm long and each has a barb near the tip. This species affects chicken and guinea-fowl in South America, Puerto Rico and Africa.

Heterakis isolonche von Linstow, 1906, occurs in the caecum of pheasant, quail and other gallinaceous birds, being distributed globally. Males are 6–12 mm long and females 9–12 mm. The spicules are asymmetrical and there is a perianal sucker 70–150 μm in diameter. The eggs are 65–75 by 37–46 μm.

Heterakis dispar (Shrank, 1790) occurs in the goose and duck. The males are 11–18 mm long and the females 16–23 mm. The subequal spicules are 40–50 μm long.

Other species of *Heterakis* include: **Heterakis beramporia** Lane, 1914 of chicken in South and South-east Asia and the Pacific area; **H. indica** Maplestone, 1931 of the chicken in India; **H. linganensis** Li, 1933 of the chicken in China; **H. meleagris** Hsü 1957 of the turkey in China; and **H. pavonis** Maplestone 1932 of peafowl and pheasant in India and China.

A heterakid of rodents, **Heterakis spumosa** Schneider, 1866, occurs in the caecum of rats throughout the world. Males measure 3.5–8 mm and females 6.8–8 mm. Eggs are 59 by 40 μm and the life-cycle is direct and similar to that of *H. gallinarum*.

Genus: Ascaridia Dujardin, 1845

Ascaridia galli (Schrank, 1788) (syn. *A. lineata*, *A. perspicillum*) occurs in the small intestine of the fowl, guinea-fowl, turkey, goose, and various wild birds in most parts of the world. Male 50–76 mm, female 72–116 mm long. There are three large lips and the oesophagus has no posterior bulb. The tail of the male has small alae and bears ten pairs of papillae, most of which are short and thick. There is a circular precloacal sucker with a thick cuticular rim. The spicules are sub-equal, 1–2.4 mm long.

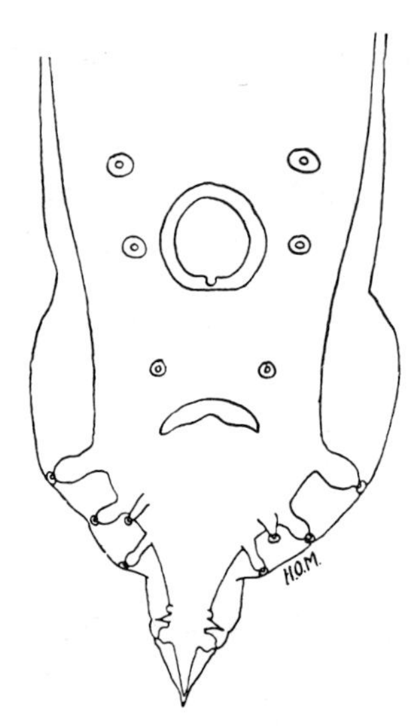

Fig 1.96 Ventral view of the hind end of male *Ascaridia galli*.

The eggs are oval, with smooth shells, and are unsegmented when laid. They measure 73–92 by 45–57 μm.

Life-cycle. The eggs are passed in the faeces of the host and develop in the open, reaching the infective stage in about ten days, or longer. The egg then contains a fully developed second stage larva and is fairly resistant to adverse conditions. The eggs can remain viable for over three months in shaded places, but are rapidly killed by dry, hot weather, even when they are 15 cm deep under the soil exposed to sunlight. Infection takes place by ingestion of the eggs with food or water. Earthworms may ingest the eggs and may, when they are swallowed by the birds, transmit the infection mechanically.

The eggs hatch in the intestine of the host and the larvae live for the first eight days or so in the lumen of the intestine. The majority are then found in the intestinal mucosa from the eighth to the seventeenth day. Subsequently the larvae re-enter the lumen and reach maturity in six to eight weeks, this depending on the age of the chicken. The moult to the third larval stage occurs about eight days after infection, that to the fourth stage at 14–15 days. These moults may be delayed if larvae spend too long in the tissues (Tugwell & Ackert 1952).

Pathogenicity. Young birds are more susceptible to infection than adult birds or others that have had a previous infection. Dietary deficiencies, such as those of vitamins A, B and B12, various minerals and proteins, predispose to heavier infections. Chickens over three months of age are more resistant to infection and this may be associated with a marked increase in goblet cells in the gut mucosa about this time.

Pathogenesis and clinical signs. The most serious infections occur in chickens one to three months of age. Marked lesions may be produced when large numbers of the young parasites penetrate into the duodenal mucosa. They cause haemorrhage and enteritis and the birds become anaemic and suffer from diarrhoea. The birds become unthrifty, markedly emaciated, generally weak and egg production is decreased. In heavy infections intestinal obstruction may occur.

Post mortem. A haemorrhagic enteritis may be seen and larval worms, which are about 7 mm long, are found in the mucosa. In other cases the carcass is emaciated and anaemic and the worms are found in the intestine. Occasionally viable or calcified parasites may be found in the albumin portion of eggs.

Diagnosis can be made by finding the eggs in the faeces or the worms in the intestine at autopsy.

Treatment. The piperazine compounds are highly effective against *A. galli* infections. Several salts may be used: they are given in the feed or drinking water. Thus piperazine adipate at a dose rate of 300–440 mg/kg in the feed is 94–100% efficient; 440 mg of piperazine citrate per litre of water for 24 hours has a similar efficacy and piperazine carbodithioic acid is effective at similar dose rates. The drug has no effect on growth or egg production at therapeutic levels.

Phenothiazine is variable in its effect and up to 2200 mg/kg must be given for even moderate efficacy. It is often used in conjunction with piperazine for the joint control of *Heterakis* and *Ascaridia*. Mebendazole, tetramisole and haloxon at dose rates for *Heterakis* spp (above) are effective against *Ascaridia* spp.

Hygromycin B at the rate of 8 g/tonne of feed administered for 8 weeks is highly effective in controlling *A. galli* infection.

Prophylaxis. Special attention should be paid to the young birds. When birds are kept out of doors, young birds should be separated from the old and the poultry runs should be well drained. Rotation of poultry runs is highly desirable.

Heavy burdens of *A. galli* may occur in birds kept in deep-litter houses, especially when excess moisture occurs. Attention should be paid to ventilation, feeding troughs and drinking water appliances. Periodically, the litter around the feeding and water area should be mixed with dry litter in other parts of the house.

Prior to each new batch of chickens being placed in the litter house, the litter should be stacked for several days to allow heating and sterilization.

Ascaridia columbae (Gmelin, 1790) (syn. *A. maculosa*) occurs in domestic and wild pigeons. It is, like the previous species, a large worm, the male being 16–70 mm and the female 20–95 mm long. The eggs measure 80–90 by 40–50 μm. All development appears to take place in the small intestine and the prepatent period is 37–42 days. Pathogenesis appears limited since heavy burdens can occur without apparent ill effect. However, focal liver lesions have been reported by Wehr and Shalkop (1963); these are eosinophic granuloma.

Other members of the genus *Ascaridia* include:

Ascaridia dissimilis Pefes Vigueras, 1931 of the small intestine of domestic and wild turkeys in North America and Europe. It is smaller than *A. galli* (males 35–52 mm, females 50–75 mm). The life-cycle is similar to that of *A. galli*, as are treatment and control measures.

Ascaridia compar (Schrank, 1790) occurs world-wide in various gallinaceous game birds (e.g. grouse, ptarmigan, partridge, pheasant). Males: 36–43 mm; Females: 84–95 mm. Eggs are 90 by 60 μm. The developmental cycle and pathogenesis are similar to *A. galli.*

Ascaridia numidae Leiper, 1908 occurs in the small intestine of the guinea fowl in Africa and North America.

Ascaridia razia Akhtar, 1937 is found in the domestic pigeon in India.

Genus: Paraspidodera Travassos, 1914

Paraspidodera uncinata (Rudolphi, 1819) Travassos, 1914, occurs in the large intestine of the guinea-pig throughout the world and of the agouti in South America. Males 11–22 mm, females 16–27 mm; eggs are 43 by 31 μm and heterakid in appearance. Spicules are equal in length and there is a preanal sucker. The parasite is not associated with pathogenic effects.

Genus: Pseudoaspidodera Baylis and Daubney, 1922

Pseudaspidodera pavonis Baylis and Daubney, 1922 occurs in the caecum of peafowl in India. Males 6 mm and females 7 mm in length and eggs 70 by 40 μm. Spicules are dissimilar and a preanal sucker is present. No pathology is associated with infection.

A related species, **Pseudaspidoderoides jnanendre** Freitas 1956, is similar to *P. pavonis* and also occurs in the peafowl in India.

FAMILY: SUBULURIDAE YORK AND MAPLESTONE, 1926

Subuluroidea with mouth with no lips or lips poorly visible. Buccal capsule present and frequently teeth are present in it. Oesophagus with a posterior bulb. Pre-anal sucker present, slit-like, without a chitinous rim.

Genus: Subulura Molin, 1860

Subulura brumpti (Lopez Neyra, 1922) Cram, 1926 occurs in the caeca of the fowl, turkey, guinea-fowl and wild related birds in Africa, North and South America and Asia. The males are 6.9–10 mm long and the females 9–17.5 mm. Lateral alae are present. The small buccal capsule has three teeth at its base. The oesophagus has a small swelling posteriorly, followed by a deep constriction and then a spherical bulb. The tail of the male is provided with large lateral alae and is curved ventrad. The pre-cloacal sucker is an elongate slit, surrounded by radiating muscle

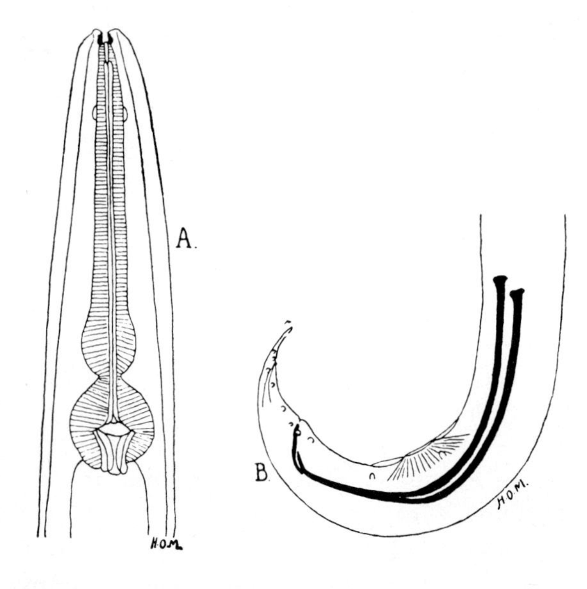

Fig 1.97 *Subularia brumpti*. A, Dorsal view of anterior end. B, Lateral view of hind end of male.

fibres. There are ten pairs of small caudal papillae. The spicules are equal, alate and 1.3–1.5 mm long. The vulva is situated just anterior to the middle of the body. The eggs are subglobular with smooth shells, and contain a fully developed embryo when laid. They measure 52–64 by 41–49 μm.

The intermediate hosts are various beetles of the genera *Blaps*, *Gonocephalum* and *Dermestes* and the cockroach *Blatella germanica*.

Pathogenicity is apparently not marked.

S. differens (Sonsino, 1890) is a similar form occurring in the fowl and guinea-fowl in Southern Europe, Africa and Brazil. **S. strongylina** (Rudolphi, 1819) occurs in the caecum of chicken, guinea fowl and various wild birds in North America. **S. suctoria** (Molin, 1860) is a parasite of the caecum of the chicken and various wild birds in Africa and South America; **S. minetti** Bhalerao, 1941 occurs in the chicken in India.

REFERENCES

ASCARIDOIDEA

Anderson, S., Jørgensen, R. J., Nansen, P. & Nielsen, K. (1973) Experimental *Ascaris suum* infections in piglets. *Acta path. microbiol. scand.* (B), **81**, 650–656.

Ansel, M. & Thibaut, M. (1973) Value of the specific distinction between *Ascaris lumbricoides*, Linné, 1783 and *Ascaris suum* Goeze 1782. *Int. J. Parasit.*, **3**, 317–319.

Aranjo, P. (1972) Observacoes pertinentes as primeiras ecdises de larvas de *Ascaris lumbricoides*, *A. suum* and *Toxocara canis*. *Rev. Inst. Med. trop. S. Paulo*, **14**, 83–90.

Beaver, P. C. (1956) Larva migrans. *Exp. Parasit.*, **5**, 587–621.

Beaver, P. C. (1966) Zoonoses with relation to parasites of veterinary importance. In: *Biology of Parasites*, ed. E. J. L. Soulsby, pp. 215–226. New York: Academic Press.

Berger, H., Wood, I. B. & Willey, C. H. (1961) Observations on the development and egg production of *Ascaris suum* in rabbits. *J. Parasit.*, Suppl. **47**, 15.

deBoer, E. (1935) Experimental onderzoek betreffend *Ascaris lumbrocoides* van mensch en varken. *Tijdschr. Diergeneesk.*, **62**, 965–973.

Clayton, H. M. (1978) The pathogenesis of equine ascariasis. *Proc. Am. Ass. vet. Parasit.*, **2**.

Clayton, H. M. & Duncan, J. L. (1977) Experimental *Parascaris equorum* infection in foals. *Res. vet. Sci.*, **23**, 109–114.

Connan, R. M. (1977) Ascariasis: The development of eggs of *Ascaris suum* under the conditions prevailing in a pig house. *Vet. Rec.*, **100**, 421–422.

Done, J. T., Richardson, M. D. & Gibson, T. E. (1960) Experimental visceral larva migrans in the pig. *Res. vet. Sci.*, **1**, 133–151.

Douglas, J. R. & Baker, N. R. (1959) The chronology of experimental intrauterine infections with *Toxocara canis* (Werner, 1782) in the dog. *J. Parasit.*, **45**, Suppl., 43–44.

Douvres, F. W., Tromba, F. G. & Malakatus, G. M. (1969) Morphogenesis and migration of *Ascaris suum* larvae developing to fourth stage in swine. *J. Parasit.*, **55**, 689–712.

Dubey, J. P. (1979) Effect of fenbendazole on *Toxocara canis* larvae in tissues of infected dogs. *Am. J. vet. Res.*, **40**, 698–699.

Fairbairn, D. (1961) The in vitro hatching of *Ascaris lumbricoides* eggs. *Can. J. Zool.*, **39**, 153–162.

Glickman, L., Schantz, P., Dombroske, R. & Cypess, R. (1978) Evaluation of serodiagnostic tests for visceral larva migrants. *Am. J. trop. Med. Hyg.*, **27**, 492–498.

Greve, J. H. (1971) Age resistance to *Toxocara canis* in ascarid-free dogs. *Am. J. vet. Res.*, **32**, 1185–1192.

Hartwich, G. (1974) Keys to the genera of Ascaroidae. In: *CIH Keys of the Nematode Parasites of Vertebrates*, ed. R. C. Anderson, A. G. Chaband & S. Willmott, pp. 1–15. Farnham Royal: Commonwealth Agricultural Bureaux.

Huizinga, H. W. (1965) Studies on the life cycle and histopathology of the nematode parasite *Contracaecum spiculigerum* in fish eating birds. Northeast Fish Wildlife Conf. Harrisburg, Pennsylvania, Jan. 1965, p. 17.

Jacobs, D. E., Pegg, E. J. & Stevenson, P. (1977) Helminths of British dogs: *Toxocara canis*—a veterinary prospective. *J. small Anim. Pract.*, **18**, 79–92.

Koutz, F. R., Groves, H. F. & Scothorn, M. W. (1966) The prenatal migration of *Toxocara canis* larvae and their relationship to infection in pregnant bitches and in pups. *Am. J. vet. Res.*, **27**, 789–795.

Levine, N. L. (1957) Life history studies on *Porrocaecum ensidicaudatum*, an avian nematode. *J. Parasit.*, 43, 47–48.

Mitchell, G. B. B. & Linklater, K. A. (1980) Condemnation of sheep livers due to ascariasis. *Vet. Rec.*, **107**, 70.

Mozgovoi, A. A. & Shakhmatova, V. I. (1969) On the study of the life cycle of *Neoascaris vitulorum* (Ascaridata: Anisakidae), nematode of ruminants. *Trud. Gel'mint. Lab. Akad Nauk. USSR*, 17, 95–103.

Mozgovoi, A. A. & Slinkhov, R. M. (1971) *Neoascaris vitulorum* of ruminants. *Veterinarya*, 47, 59–61.

Petter, C. (1960) Etude zoologique de la larva migrans. *Ann. Parasit. hum. comp.*, **35**, 118–137.
Roberts, F. H. (1934) The large roundworm of pigs, *Ascaris lumbricoides* L. 1758. Its life history in Queensland, economic influence and control. *Bull. Queensland Dept. Agric. Stock Animal Hlth Stat. Yeerongpilly*, **1**, 1–81.
Rogers, W. P. (1966) Exsheathment and hatching mechanisms in helminths. In: *Biology of Parasites*, ed. E. J. L. Soulsby, pp. 30–40. New York: Academic Press.
Roneus, O. (1963) Parasitic liver lesions in swine experimentally produced by visceral larva migrans of *Toxocara cati*. *Acta vet. scand.*, **4**, 170–196.
Roneus, O. (1966) Studies on the aetiology and pathogenesis of white spots in the liver of pigs. *Acta vet. scand.*, Suppl. **1**, 16.
Ryzhkova, K. M. (1953) Reservoid parasitism among helminths [In Russian]. *Tr. Gel'mintol. Lab. Akad. Nauk, S.S.R.*, **7**, 200.
Soulsby, E. J. L. (1961) Unpublished observations.
Soulsby, E. J. L. (1965) *Textbook of Veterinary Clinical Parasitology, Vol. 1, Helminths.* Oxford: Blackwell Scientific.
Soulsby, E. J. L. (1976) Serodiagnosis of other helminth infections. In: *Immunology of Parasitic Infections*, ed. S. Cohen & E. H. Sadun, pp. 152–161. Oxford: Blackwell Scientific.
Sprent, J. F. A. (1952) Anatomical distinction between human and pig strains of *Ascaris*. *Nature, Lond.*, **170**, 627–628.
Sprent, J. F. A. (1955) On the invasion of the central nervous system in ascariasis. *Parasitology*, **45**, 41–55.
Sprent, J. F. A. (1956) The life history and development of *Toxocara cati*. *Parasitology*, **46**, 54–78.
Sprent, J. F. A. (1958) Observations on the development of *Toxocara canis* (Werner, 1782) in the dog. *Parasitology*, **48**, 184–209.
Sprent, J. F. A. (1959) The life history and development of *Toxocara leonina* (von Linstow, 1902) in dog and cat. *Parasitology*, **49**, 330–371.
Sprent, J. F. A. (1961) Post-parturient of the bitch with *Toxocara canis*. *J. Parasit.*, **47**, 284.
Stephenson, L. S. (1980) The contribution of *Ascaris lumbricoides* to malnutrition in children. *Parasitology*, **81**, 221–233.
Stephenson, L. S., Pond, W. G., Nesheim, M. C., Krook, L. P. & Crompton, D. W. T. (1980) *Ascaris suum*: nutrition absorption, growth and intestinal pathology in young pigs experimentally infected with 15 day old larvae. *Exp. Parasit.*, **49**, 15–25.
Stone, W. M. & Girardeau, M. H. (1967) Transmammary passage of infectious stage nematode larvae. *Vet. Med. small Anim. Clin.*, **62**, 252–253.
Swerczeh, T. W., Nielsen, S. W. & Helmboldt, C. G. (1971) Transmammary passage of *Toxocara cati* in the cat. *Am. J. vet. Res.*, **32**, 89–92.
Taffs, L. F. (1961) Immunological studies on experimental infection of pigs with *Ascaris suum* (Goetz 1782). An introduction with a review of the literature and the demonstration of complement-fixing antibodies in the serum. *J. Helminth.*, **35**, 319–344.
Tripathy, K., Duque, E., Bolaños, O., Lotero, H. & Mayoral, L. G. (1972) Malabsorption syndrome in ascariasis. *Am. J. clin. Nutr.*, **25**, 1276–1280.
Warren, E. G. (1971) Observations on the migration and development of *Toxocara vitulorum* in natural and experimental hosts. *Int. J. Parasit.*, **1**, 85–99.
Webster, G. A. (1958) On prenatal infection and the migration of *Toxocara canis* Werner 1782 in dogs. *Can. J. Zool.*, **36**, 435–440.
Winckel, W. E. F. & Treurniet, A. E. (1956) Infestation with *Lagochilascaris minor* (Leiper) in man. *Doc. Med. Geogr. trop.*, **8**, 23–28.
Woodruff, A. W. (1976) Toxocariasis a public health problem. *Environ. Hlth*, **1**, 29.
World Health Organization (1979) Parasitic Zoonoses. *Wld Hlth Org. tech. Rep. Ser.*, **637**, 107.

OXYUROIDEA AND SUBULUROIDEA

Baker, A. D. (1933) Some observations on the development of the caecal worm *Heterakis gallinae* (Gmelin, 1790; Freeborn, 1923), in the domestic fowl. *Sci. Agric.*, **13**, 356–363.
Chan, K.-F. (1952) Life cycle studies on the nematode *Syphacia obvelata*. *Am. J. Hyg.*, **56**, 14–21.
Flynn, R. J. (1973) *Parasites of Laboratory Animals*, p. 884. Ames: Iowa State University Press.
Hoag, W. G. & Meier, H. (1966) Infectious diseases. In: *Biology of the Laboratory Mouse*, ed. E. L. Green, pp. 589–600. New York: McGraw-Hill.
Lund, E. E., Wehr, E. E. and Ellis, D. J. (1966) Earthworm transmission of *Heterakis* and *Histomonas* to turkeys and chickens. *J. Parasit.*, **52**, 899–902.
Osipov, A. N. (1957) Survival of *Heterakis gallinarum* ova in winter. *Trudy Moskovsk. Vet. Akad.*, **19**, 350–355.
Petter, A. J. & Quentin, J.-C. (1976) Keys to the genera of the Oxyuroidea. In: *CIH Keys to the Nematode Parasites of Vertebrates*, ed. R. C. Anderson, A. G. Chabaud & S. Willmott. Farnham Royal: Commonwealth Agricultural Bureaux.
Schad, G. A. (1959) A revision of the North American species of the genus *Skjabinema* (Nematoda Oxyuroidea). *Proc. Helminth. Soc. Wash.*, **26**, 138–147.
Tugwell, R. L. & Ackert, J. E. (1952) On the tissue phase of the life cycle of the fowl nematode *Ascaridia galli* (Schrank). *J. Parasit.*, **38**, 277–288.
Wehr, E. E. & Shalkop, W. T. (1963) *Ascaridia columbae* infection in pigeons: a histopathologic study of liver lesions. *Avian Dis.*, **7**, 206–211.

ORDER: RHABDITIDA CHITWOOD, 1933

SUPERFAMILY: RHABDITOIDEA TRAVASSOS, 1920

Forms with an oesophagus with a long anterior cylindrical portion (sometimes a median bulbar swelling) and a posterior bulb with a valvular apparatus. Cephalic papillae consisting of an inner circle of six and an outer circle of ten. Oral stylet absent. Excretory system symmetrical and H-shaped: mainly free-living forms, or forms parasitic in invertebrates, amphibians, reptiles etc.

FAMILY: RHABDITIDAE MICOLETZSKY, 1922

Mainly free-living forms, some parasitic in arthropods, earthworms, molluscs etc. Small buccal cavity, three or six lips. Oesophagus rhabditiform, i.e. an anterior wide portion, then a narrow shorter portion and terminated by a spherical posterior bulb with valves. Females oviparous or viviparous. Larval stages may be found on the skin of animals, in wounds and in some cases may penetrate further. Adults and larvae may contaminate cultures and lead to erroneous diagnoses.

Genus: Pelodera Schneider, 1866

Pelodera strongyloides (Scheider, 1860) Scheider, 1866 (syn. *Rhabditis strongyloides*) is a free-living nematode (male 0.9–1.2 mm, female 1.3–1.5 mm) which may in rare cases invade the skin, but probably only when the latter is already damaged. Chitwood (1932) described the infection in a dog in the USA. The skin of the affected areas was red, denuded, partly covered with crusts, and there were pustules surrounded by red zones, as well as nodules. The pustules contained *Rhabditis* larvae, 596–600 μm long. The infection has been reported in dogs in Europe and in other animals (cattle) in the USA. Animals probably become infected by lying on wet bedding. Lesions clear up spontaneously when animals are removed from the source of contamination.

Genus: Rhabditis Dujardin, 1844

Free-living forms occurring in decaying vegetable matter. They may contaminate faecal cultures and occur on the coat of animals, particularly those with loose stools or diarrhoea. Pseudoparasitism due to these forms in the faeces or urine has been observed on rare occasions.

Rhabditis axei Dougherty, 1955 is a common form, males 0.9 mm and females 1.4 mm long. Adults and larvae have a long slender tail and a rhabiditiform oesophagus.

Rhabditis bovis Kreis, 1964 is associated with bovine parasitic otitis in East Africa. The male is 0.8 mm and the female 1.05 mm in length, larvae 0.35–0.61 mm long. The female is viviparous (Kreis 1964).

Jibbo (1966) has described the condition of parasitic otitis, which may be widespread in East Africa. It is recognized by a putrid odour leading to secondary invasion by myiasis-causing diptera. Conditions which facilitate nematode infection of the ear are unclear, but heat and humidity, excessive glandular secretion and the long external auditory meatus of the zebu may be responsible.

Diagnosis is based on clinical signs and the demonstration of large numbers of rhabditid nematodes in the exudate from the ear. Treatments reported to be effective include benzene hexachloride, 3.9% solution introduced into the auditory meatus, DDVP used as a spray in early cases and a mixture of terramycin and lorexane (1 to 64). Dipping yards and night paddocks should be kept dry and manure sprayed with insecticides such as BHC or dieldrin (Jibbo 1966).

Rhabditis macrocerca and **R. clavopapillata** were reported by Kreis and Faust (1933) as saprophytic forms contaminating the hair of dogs and monkeys, particularly those animals which had loose bowel movements.

Rhabditis gingivalis Stefanski, 1954 (syn. *Tricephalobus gingivalis*) was isolated from a granuloma of the gum of a horse in Poland.

Micronema deletrix Anderson and Bemrick 1965 is a saprophytic species usually associated with decaying vegetation. The parasite is very small; the females measure 250 μm in length and are ovigerous. The eggs measure 32–46 by 10 μm. It has been isolated from a bilateral nasal tumour and in maxillary granuloma in horses in Minnesota and elsewhere (Anderson & Bemick 1965; Johnson & Johnson 1966). The parasite has also been reported from the central nervous system and kidney of a horse. The clinical signs resembled a viral encephalitis and the lesions consisted of a vasculitis, haemorrhagic necrosis and malacia (Rubin & Woodward 1974).

FAMILY: STRONGYLOIDIDAE CHITWOOD & McINTOSH, 1934

Free-living generation saprophytic, parasitic generation in gut of vertebrates. Free-living generation with oesophagus with a valvulated bulb. Parasitic generation with markedly elongated cylindrical oesophagus. Heterogenetic.

Genus: Strongyloides Grassi, 1879

This genus contains several species which are parasites in domestic animals. The parasitic forms are parthenogenetic and their eggs may give rise, outside the host, directly to infective larvae of another parasitic generation or to a free-living generation of males and females. The oesophagus in the free-living generation is rhabditiform. The vulva is near the middle of the body, the eggs are few, but large, and have thin shells. This non-parasitic generation produces a parasitic generation. The oesophagus of the parasitic generation is not rhabditiform but is cylindrical, without a posterior bulb (filariform). The infective larvae of the parasitic generation are able to penetrate through the skin of the host and pass with the blood to the lungs, thence up the trachea to the pharynx and on to the intestine. The adult parasitic worms are characterized by their female genital organs and by the relatively long oesophagus.

Strongyloides papillosus (Wedl, 1856) Ransom, 1911 occurs in the small intestine of the sheep, goat, cattle, rabbit and wild ruminants. Similar worms have been found in various fur-bearing animals, including the mink. It is 3.5–6 mm long and 0.05–0.06 mm thick. The oesophagus is 0.6–0.8 mm long. The eggs have rather blunt ends and thin shells. They measure 40–60 by 20–25 μm and contain fully developed embryos when passed in the faeces of the host.

Strongyloides westeri Ihle, 1917 occurs in the small intestine of the horse, pig and zebra. It is up to 9 mm long and 0.08–0.95 mm thick. The oesophagus is 1.2–1.5 mm long and the eggs measure 40–52 by 32–40 μm.

Strongyloides stercoralis (Bavay, 1876) occurs in the small intestine of man, various other primates, dog, fox and cat. The parasitic female is about 2.2 mm long and 0.034 mm thick. The parasitic male is 0.7 mm long. The oesophagus is 0.6 mm long. The eggs measure 50–58 by 30–34 μm, but most usually rhabditiform larvae are found in fresh faeces.

Strongyloides cati Rogers, 1939 (syn. *S. planiceps*) occurs in the small intestine of the cat. It is 2.37–3.33 mm long. The eggs measure 57.6–64 by 23–40 μm and are poorly developed when passed in the faeces.

Strongyloides ransomi Schwartz & Alicata, 1930 occurs in the small intestine of the pig. It is 3.33–4.49 mm long. The eggs measure 45–55 by 26–35 μm.

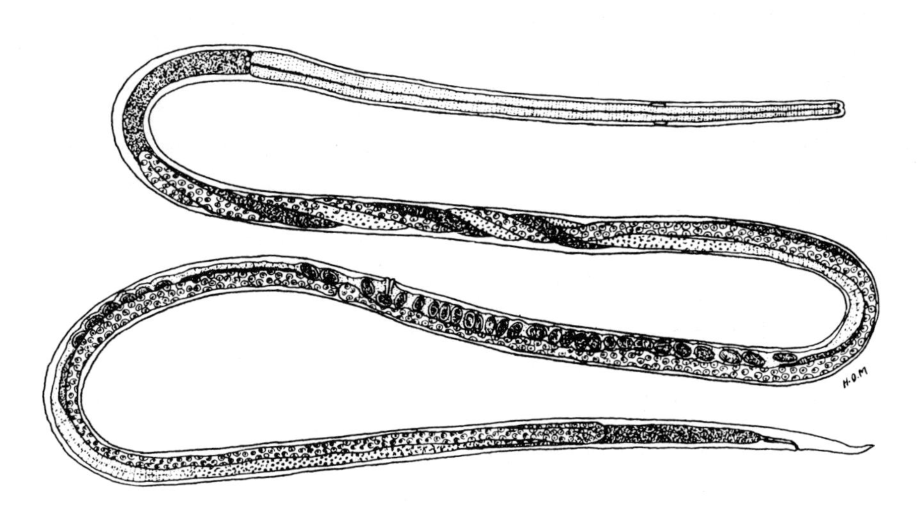

Fig 1.98 Female *Strongyloides westeri.*

Strongyloides avium Cram, 1929 occurs in the small intestine and the caeca of the fowl, turkey and some wild birds. It is 2.2 mm long. The oesophagus is 0.7 mm long and the eggs measure 52–56 by 36–40 μm.

Other species of the genus include:

Strongyloides fuelleborni von Linstow, 1905 of the small intestine of primates (e.g. chimpanzee, baboon, macaques) (*S. simiae* Hung and Hoeppli, 1923 is considered a synonym of *S. fuelleborni*); **S. procyonis** Little, 1966 of raccoons in Louisiana; **S. ratti** Sandground, 1925 in rats; **S. tumefaciens** Price and Dikmans, 1941, associated with tumours of the large intestine of cats in the southern USA and **S. venezuelensis** Brumpt 1934 of rats in North and South America.

Life-cycles

The life-cycle of members of the genus is different from all other nematodes in that completely parasitic and completely free-living cycles occur and a combination of both can occur. The parthenogenetic female is found buried in the mucosa of the small intestine. This form is triploid in character and produces thin-shelled transparent eggs which are passed in the faeces, except in the case of *S. stercoralis* where the eggs hatch in the intestine and first stage larvae are found in the faeces. The first-stage larvae may either develop directly to become third-stage infective larvae (homogonic cycle) or develop to free-living males and females which may subsequently produce infective larvae (heterogonic cycle). When environmental conditions are satisfactory (e.g. warmth, humidity etc.) the heterogonic cycle predominates but when environmental conditions are unfavourable the homogonic cycle predominates.

In the heterogonic cycle the first-stage larvae are rapidly transformed so that within 48 hours sexually mature free-living males and females occur. Following copulation, the free-living female produces eggs which hatch in a few hours and these larvae metamorphose to become infective larvae. There is evidence that only one generation of larvae is produced by the free-living female (Stewart 1963); copulation may occur several times and up to 35 eggs are produced after each mating, a total of about 180 eggs being produced per worm (Premvati 1958).

In the homogonic cycle, first stage larvae metamorphose rapidly to become infective larvae, as little as 24 hours being required for this at 27°C.

Studies on the genetics of *Strongyloides* by Chang and Graham (1957) and Little (1962) have indicated that the parasitic female has a triploid number of chromosomes, the free-living female a diploid number, the free-living male being haploid and the infective larva triploid. Chang and Graham (1957) described three types of eggs from the parthenogenetic female, some triploid, some diploid and some haploid. Little (1962) showed that haploid eggs developed to free-living males and he suggested that the diploid eggs produced either infective larvae (directly) or free-living females. It would seem, however, that a triploid egg would be necessary in the life-cycle if adult parasitic females were to arise from the direct life-cycle.

It is considered that all stages are produced initially, this being determined genetically in the egg, and the success or otherwise of each developmental stage is determined by the environmental conditions. Thus, when conditions are adverse, only the first-stage larvae, which are triploid, survive to produce infective larvae, but when conditions are favourable, larvae of all three genetic types survive.

Infection of the vertebrate host is mainly by skin penetration, though oral infection may occur. Penetration of the mucosa of the mouth or oesophagus at the time of oral infection may lead to systemic migration. Larvae reach a skin capillary or venule and are carried by the blood to the lungs. Here they break out into the alveoli, migrate up the smaller bronchioles to the bronchi and trachea and then descend the oesophagus to the intestine where they mature. The prepatent period is five to seven days.

Prenatal infection has been described for *S. ransomi* in pigs (Stone 1964; Zajicek 1969) and *S. papillosus* in cattle (Pfeiffer 1962). Infection by milk-borne larvae has been reported for *S. ransomi* in pigs (Moncol & Batte 1966), *S. papillosus* in sheep and cattle (Lyons et al. 1970) and *S. westeri* in horses (Lyons et al. 1973). After prenatal and colostral infection no systemic migration occurs and in the case of *S. ransomi* patent infections occur four to six days after feeding on infective colostrum (see Stone & Smith 1973).

In human infection hyperinfection and autoinfection may occur. In the former, first stage larvae in the bowel metamorphose to infective larvae, these penetrate the bowel wall and undergo a lung migration as before. In autoinfection the transformed larvae are voided in the faeces, penetrate the skin of the perianal and perineal regions and continue the lung migration as before. There is little evidence for these modes of infection with the *Strongyloides* spp. of animals.

Pathogenicity and clinical signs

Experimental studies of the pathogenicity of *S. papillosus* by Turner (1959) showed that exposure of lambs to 100 000 or more larvae caused death in 13–41 days. Pathological changes included erosion of the intestinal mucosa and fluid gut contents; the clinical signs consisted of anorexia, loss of weight, diarrhoea and a moderate anaemia. Field outbreaks of disease were associated with a catarrhal enteritis of the upper small intestine but fatalities were few. The larvae of *S. papillosus* are associated with the introduction of the organisms of 'foot rot' into the skin around the feet of sheep (Beveridge 1934).

Severe infections of *S. stercoralis* may occur in dogs, especially in puppies. The condition is most commonly seen in summer when the weather is hot and humid and is frequently a kennel problem. Lesions consist of a catarrhal inflammation of the small intestine while in severe infections there may be necrosis and sloughing of the mucosa. Dogs show moderate to severe diarrhoea which may be blood stained. Dehydration, followed by death, may occur.

Pathological manifestations of *S. ransomi* are usually seen in young suckling pigs, infection being acquired orally from infective larvae adhering to the udder or teats, from the skin penetration by larvae in the soil or litter or, more commonly, by colostral infection. Heavy infections can occur under environmental circumstances which would preclude a free-living phase of the life-cycle.

After colostral infection patent infections were apparent in four days. A protein-losing enteropathy is produced (Enigk & Dey-Hazra 1975) and mortality in young piglets may reach 50%. The chief clinical signs are initially anorexia, then diarrhoea, which soon becomes continuous and frequently haemorrhagic (Dey-Hazra et al. 1977).

Skin lesions may be seen (Ippen 1953) but pulmonary disorders are not frequent in natural outbreaks of the disease. Nevertheless, they can be produced experimentally (Supperer & Pfeifer 1960).

S. westeri in foals produces diarrhoea which may be acute (Enigk et al. 1974*a*). The parasite may be responsible for a high incidence of scouring in nursing foals. Foals usually develop a satisfactory immunity to the infection at 15–23 weeks after birth but in the donkey heavy infections have been recorded at the age of nine to 12 months (Pande & Rao 1960).

Immunology of Strongyloides infections

Essentially a few infections lead to a marked immunity and in domestic animals this is exemplified by the fact that only young animals are severely affected by the parasite.

Diagnosis

Diagnosis is made by demonstrating the eggs or larvae (dogs) in the faeces.

Treatment

For sheep, thiabendazole at a dose rate of 75 mg/kg orally is highly effective. Other compounds which may be used are Dowco 105 (*O*-methyl-*O*-(4-tert.-butyl-2-chlorophenyl)ethylphosphoramidothioate) at 200 mg/kg; Bayer 21/199 (coumaphos, Co-Ral (*O*,*O*-diethyl-*O*-(3-chloro-4-methyl-7-coumarinyl)phosphorothioate) at a rate of 25 mg/kg; and haloxon at a rate of 30–55 mg/kg.

In pigs, thiabendazole is highly effective at a dose rate of 50 mg/kg mixed in the food. Levamisole, 5–10 mg/kg, is also effective.

For dogs, diethylcarbamazine (100 mg/kg), dithiazanine (ten daily doses of 5 mg/kg) and pyvinium pamoate (20 mg/kg per day for five days, five days rest, then repeat the treatment for a further five days) may all be used. Thibendazole at a dose of 50–75 mg/kg is highly effective.

For horses, cambendazole is the drug of choice, given at a dose rate of 20 mg/kg. Fenbendazole at a dose rate of 50 mg/kg is also highly effective.

Prophylaxis

The infective larvae are not resistant to dessication and infection can be prevented by providing clean dry quarters for the animals. Since prenatal and transcolostral infections may occur, they should be anticipated and treated before clinical manifestations of the disease can occur. Enigk et al. (1974*b*) reported that mebendazole at a dose rate of 72–104 mg/kg given over a period of 12–14 days prior to parturition in the pig reduced the number of larvae in milk. A total dose of levamisole of 140 mg/kg, given similarly, had the same effect.

REFERENCES

RHABDITIDA

Anderson, R. V. & Bemick, W. J. (1965) *Micronema deletrix* n. sp., a saprophagous nematode inhabiting a nasal tumour of a horse. *Proc. helminth. Soc. Wash.*, **32**, 74–75.
Beveridge, W. I. B. (1934) Foot-rot in sheep. Skin penetration of *Strongyloides* larvae as a predisposing factor. *Aust. vet. J.*, **10**, 43–51.
Chang, P. C. H. & Graham, G. L. (1957) Parasitism, parthenogenesis and polyploidy; the life cycle of *Strongyloides papillosus. J. Parasit.*, **43**, Suppl. 13.
Chitwood, B. G. (1932) The association of *Rhabditis strongyloides* with dermatitis in dogs. *N. Am. Vet.*, **13**, 35–40.
Dey-Hazra, A., Giese, W. & Enigk, K. (1977) Der gastrointestinale Plasma- und Plasmaproteinverlust beim *Strongyloides ransomi*-Befall des Schweines vor und nach Behandlung mit Thiabendazol. *Dt. tierarztl. Wschr.*, **79**, 421–424.
Enigk, K. & Dey-Hazra, A. (1975) Intestinal plasma and blood loss in piglets infected with *Strongyloides ransomi. Vet. Parasit.*, **1**, 69–75.
Enigk, K., Dey-Hazra, A. & Batke, J. (1974*a*) Zur Klinischen Bedeuteing und Behandlung des galaktogen erworbenen *Strongyloides*—Befalls der Fohlen. *Dt. tierärztl. Wschr.*, **81**, 605–607.
Enigk, K., Weingärtner, E. & Schmelzle, H. M. (1974*b*) Zur Chemoprophylaxe der galaktogenen *Strongyloides*-infection beim Schwein. *Zbl. vet. Med. B.*, **21**, 413–425.
Ippen, R. (1953) Zur Pathogenitat des *Strongyloides ransomi* unter besonderer Berücksuchtigung seines Sitzes in der Schleimhaut der Darmwand. *Arch. exp. vet. Med.*, **7**, 36–57.
Jibbo, J. M. C. (1966) Bovine parasitic otitis. *Bull. epizoot. Dis. Afr.*, **14**, 59–63.
Johnson, K. H. & Johnson, D. W. (1966) Granulomas associated with *Micronema delatrix* in the maxillae of a horse. *J. Am. vet. med. Ass.*, **149**, 155–159.
Kreis, H. A. (1964) Ein neuer Nematode aus dem äusseren Gehörgang von Zeburindern in Ostafrika, *Rhabditis bovis* n. sp. (Rhabditidoidea; Rhabditidae). *Schweizer Arch. Tierheilk.*, **106**, 372–377.
Kreis, H. A. & Faust, E. C. (1933) Two new species of Rhabditis (*Rhabditis macrocerca* and *R. clavopapillata*) associated with dogs and monkeys in experimental Strongyloides studies. *Trans. Am. Microsc. Soc.*, **52**, 162–172.
Little, M. D. (1962) Experimental studies on the life cycle of *Strongyloides. J. Parasit.*, **48**, 41.
Lyons, E. T., Drudge, J. H. & Tolliver, S. C. (1970) *Strongyloides* larvae in milk of sheep and cattle. *Mod. Vet. Pract.*, **51**, 65–68.
Lyons, E. T., Drudge, J. H. & Tolliver, S. C. (1973) On the cycle of *Strongyloides westeri* in the equine. *J. Parasit.*, **59**, 780–787.
Moncol, D. J. & Battle, E. G. (1966) Transcolostral infection of newborn pigs with *Strongyloides ransomi. Vet. Med. small Anim. Clin.*, **61**, 583–586.
Pande, B. P. & Rao, P. (1960) The nematode genus *Strongyloides* Grassi 1879 in Indian livestock. I. Observations on natural infections in the donkey (*Equus asinus*). *Br. vet. J.*, **116**, 281–283.
Pfeiffer, H. (1962) Die pranatale Invasion von *Strongyloides papillosus* beim Rind. *Z. Parasitenk.*, **22**, 104–105.
Premvati (1958) Studies on *Strongyloides* of primates. III. Observations on the free-living generations of *S. fülleborni. Can. J. Zool.*, **36**, 447–457.
Rubin, H. L. & Woodward, J. C. (1974) Equine infection with *Micronema delatrix. J. Am. vet. med. Ass.*, **165**, 256–258.
Stewart, T. B. (1963) Environmental factors affecting the survival and development of *Strongyloides ransomi* with special reference to its free-living stages. Ph.D. Thesis, University of Illinois, Urbana, Illinois.
Stone, W. M. (1964) *Strongyloides ransomi* prenatal infection in swine. *J. Parasit.*, **50**, 568.
Stone, W. M. & Smith, F. W. (1973) Infection of mammalian hosts by milk-borne nematode larvae: A review. *Exp. Parasit.*, **34**, 306–312.
Supperer, R., Pfeiffer, H. (1960) Ueber die Strongyloidose der Kalber. *Wien tierärztl. Mschr.*, **47**, 361–368.
Turner, J. H. (1959) Experimental strongyloidiasis in sheep and goats. I. Single infections. *Am. J. vet. Res.*, **20**, 102–110.
Zajicek, D. (1969) Two cases of prenatal strongyloidosis. *Vet. Med. (Praha)*, **14**, 329–332.

ORDER: STRONGYLIDA MOLIN, 1861

Nematodes with six, three or no lips, usually small if present. Corona radiata (leaf-crowns) may be present. Female reproductive system well developed, uterus with well-developed muscular ovejectors. Males with bursa and rays usually well developed. Oesophagus club-shaped in adult parasites.

Superfamilies of importance in this suborder include Strongyloidea, Ancylostomatoidea, Trichostrongyloidea and Metastrongyloidea.

SUPERFAMILY: STRONGYLOIDEA WEINLAND, 1858

Worms with mouth well developed; the oral opening is often surrounded by corona radiata. Teeth or cutting plates may occur in the buccal cavity. Copulatory bursa on the posterior end of the male worms are well developed. This structure consists of cuticular alae which usually form two lateral lobes and a dorsal lobe, enclosing the posterior extremity, and are supported by modified caudal papillae, known as the 'bursal rays'. These rays contain muscle-fibres and are arranged in a definite order. There are two ventral rays (a ventroventral and a lateroventral), three lateral rays (an anterolateral or externolateral, a mediolateral and a posterolateral) and a set of dorsal rays (usually comprising an externodorsal on either side of the single or divided dorsal ray) (Fig. 1.99). The hind end of the male enclosed in the bursa is called the 'genital cone'. There are usually two equal spicules; a gubernaculum, as well as a telamon, may be present. The families of importance in this superfamily include *Strongylidae, Trichonematidae, Syngamidae* and *Stephanuridae*.

FAMILY: STRONGYLIDAE BAIRD, 1853

There is a well-developed globoid buccal capsule on the dorsal wall of which there may be a median thickening, called the *dorsal gutter*, which carries the duct of the dorsal oesophageal gland. The anterior margin of the buccal capsule usually bears leaf-like cuticular structures called the *leaf-crowns* or *corona radiata*. There may be an external leaf-crown round the mouth opening and an internal leaf-crown on the inner wall of the buccal capsule a little further back. The supposed resemblance of these fringes to a palisade gave origin to the term 'palisade worms' formerly given to these species. The anterior margin of the buccal capsule does not bear teeth or cutting plates, but teeth may be present in the depth of the buccal capsule. The male bursa is strongly developed and has typical rays. The life-cycle is direct in all known cases. Genera of importance include *Strongylus, Triodontophorus, Craterostomum, Oesophagodontus, Codiostomum* and *Chabertia*.

Genus: Strongylus Müller, 1780

Members of this genus are sometimes placed in the three genera *Strongylus, Alfortia* or *Delafondia* or in subgenera with these names. However, it is more convenient to consider them in a single genus.

Strongylus equinus Müller, 1780 occurs in the caecum and colon of equines, including the zebra. The worms are fairly rigid and dark grey in colour; sometimes the red colour of the blood in the intestine can be seen. The male is 26–35 mm long and the female 38–47 mm by about 2 mm thick. The head end is not marked off from the rest of the body. The buccal capsule is oval in outline and there are external and internal leaf-crowns. At the base of the buccal capsule there is a large dorsal tooth with a bifid tip and two smaller subventral teeth. The dorsal oesophageal gland opens into the buccal capsule through a number of pores situated in a thickened ridge, the dorsal gutter, formed by the wall of the buccal capsule. The male has two

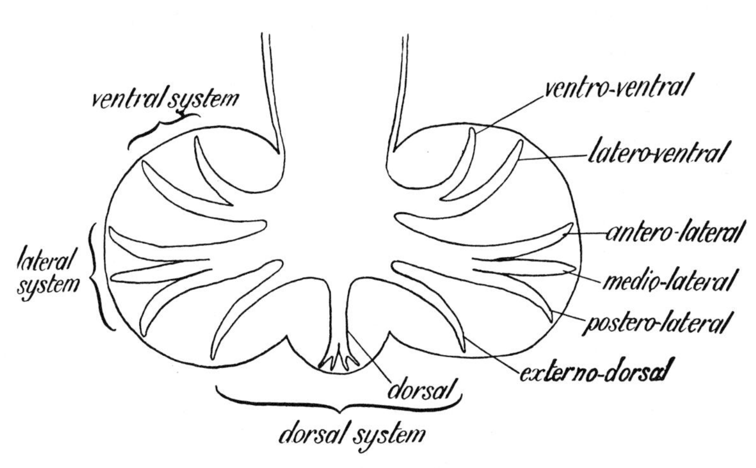

Fig 1.99 Copulatory bursa of the Strongylidae.

simple, slender spicules. The vulva lies 12–14 mm from the posterior extremity. The eggs are oval, thin-shelled, segmenting when laid, and measure 70–85 by 40–47 μm.

Strongylus edentatus (Looss, 1900) (syn. *Alfortia edentatus*) also occurs in the large intestine of equines. The male is 23–28 mm long and the female 33–44 mm by about 2 mm broad. This worm resembles *S. equinus* macroscopically, but the head is somewhat wider than the following portion of the body. The buccal capsule is wider anteriorly than at the middle and contains no teeth.

Strongylus vulgaris (Looss, 1900) (syn. *Delafondia vulgaris*) occurs in the large intestine of equines. The male is 14–16 mm long and the female 20–24 mm by about 1.4 mm thick. This worm is distinctly smaller than the two preceding species. The buccal capsule is roughly oval and contains two ear-shaped dorsal teeth at its base. The elements of the external leaf-crown are fringed at their distal extremities.

Strongylus asini Boulenger, 1920 occurs in the large intestine of the ass and wild equids in East Africa. Males are 18–32 mm long and females 30–42 mm long.

Strongylus tremletti Round, 1962 occurs in the rhinoceros in East Africa.

Life-cycles

Bionomics of strongyle larvae. The eggs of the parasites are passed in the faeces in the early stages of segmentation. The egg is thin shelled, composed of an outer chitinous shell and an inner delicate vitelline membrane. Usually there is a wide fluid cavity between the inner membrane and the cell mass. The shape of the egg is that of a regular ellipse. Embryonation commences immediately but is dependent on suitable environmental conditions such as moisture, oxygen and a favourable temperature. At about 26°C a first stage larva is produced in 20–24 hours; this hatches from the egg to become a free-living stage. Development to the first larval stage may be inhibited by several factors, temperature and lack of moisture being the two major ones. At temperatures below 7.2°C development is extremely slow and the majority of eggs fail to develop to the pre-hatch stage. However, these which do so may hatch if the temperature is raised to above 9°C. Eggs which have not undergone embryonation do not readily survive temperatures below 0°C but if they have reached the pre-hatch stage they may survive for several weeks at low temperatures. Dessication is generally lethal to eggs which have not undergone development to the pre-hatch stage; however, those which have so developed may remain viable for several weeks in a state of resistant dormancy. Upon the return of

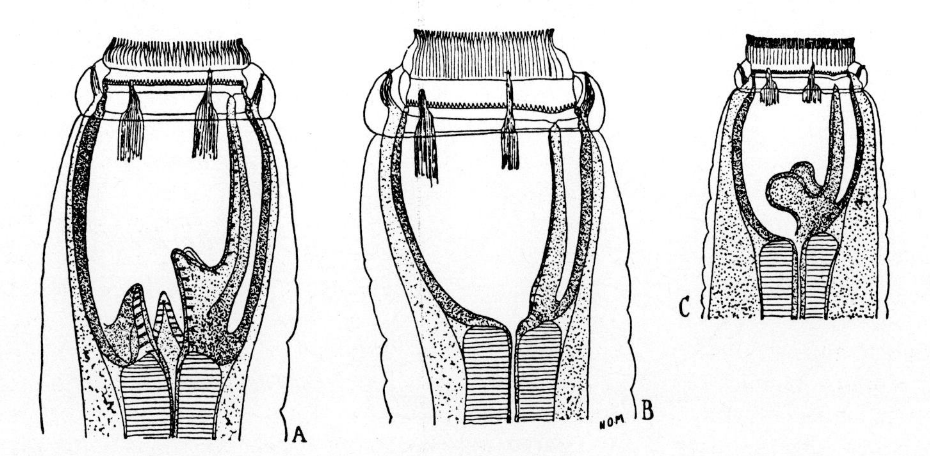

Fig 1.100 Dorsolateral view of the anterior end of various *Strongylus* spp. A, *S. equinus*. B, *S. edentatus*. C, *S. vulgaris*.

moist conditions, hatching can occur within a few minutes.

After hatching from the egg, the larva is in the first stage and is characterized by having a rhabditiform oesophagus. It feeds mainly on bacteria and grows, but soon enters a lethargic state in preparation for the first moult, from which the second-stage larva emerges. This has a less rhabditiform oesophagus than the first-stage larva. The process of feeding and growth is repeated, followed by lethargy. The old cuticle is separated off, but is not shed; it remains as a sheath round the third larva. This sheathed, third larva is the only larval stage that can infect a new host and is therefore called the infective larva. It has a club-shaped oesophagus.

The behaviour of the infective larva is different from that of the earlier stages. It does not feed and exists on the reserve food granules stored in its intestinal cells. As soon as these are exhausted the larva dies. The larva does not actively enter the host, but is swallowed with herbage, or sometimes water, and its habits are such that they increase the possibility of finding a host. These habits are the normal responses of a number of external stimuli, and the following are the most important:

1. The larva is negatively geotropic and it crawls up blades of grass or other herbage.

2. It is positively phototropic to a mild light, but is repelled by strong sunlight. The larva will therefore crawl up blades of grass only in the early morning, towards the evening and at other times of the day in dull weather. At night some of the larvae may descend to the soil. Moisture is necessary for these migrations, as the larvae are unable to crawl on a dry surface, but a very thin film of water suffices.

3. There is a certain amount of response to heat; migration is more active in warm than in cold weather.

A detailed consideration of nematode movement and factors affecting it is given by Croll (1975).

The sheath affords some protection against adverse conditions. One of the most lethal factors is dessication, larvae failing to survive more than a few days when this is marked. However, on pasture the local microclimate at the soil surface may not be as dry as the general environment and larvae may survive for much longer under these conditions. Some larvae may penetrate the soil, where they survive more readily than on the surface. In loose, sandy soil they are able to move more easily and to penetrate deeper than in fine clay soil, so that from this point of view sandy soil is favourable. In water the larvae sink to the bottom and may live for a month or more, depending on the temperature and the presence of other organisms which appear to affect them.

Taking all these factors together, the length of life of a larva in a pasture is favourably affected by moisture, shade and a relatively low temperature. Since the larvae do not feed and have only a limited amount of food reserves, conditions favourable for migration, like warmth, marked daily fluctuations in the intensity of light and loose soil, lead to rapid exhaustion and death. In general, under such conditions, as well as during dry seasons, the larvae will not live longer than about three months, but some may live for a year or longer in a cool climate where sufficient moisture is available in the soil. More detailed considerations of the ecology of the pre-infective and infective stages of strongyles may be found in Soulsby (1965), Levine (1963, 1968) and Ogbourne and Duncan (1977).

Infection is by the ingestion of infective larvae. Liberation of the infective larva from the retained sheath of the second-stage larva (exsheathment) occurs in the small intestine. It is probable that the mechanism of this is comparable to the exsheathment of the trichostrongyle larvae (see p. 215). Poynter (1956) has shown that the infective larvae of several species of horse strongyles could be induced to exsheath by exposing them to fresh equine duodenal contents at 38°C.

Life-cycle of S. equinus. Exsheathed infective larvae penetrate the mucosa of the caecum and colon and enter the subserosa where they cause the formation of nodules. Eleven days after infection, fourth-stage larvae occur in the nodules and these migrate to the peritoneal cavity and then to the liver in which they wander for a further six to eight weeks. Between two and four months after infection larvae leave the liver via the hepatic ligaments and pass via the pancreas to the peritoneal cavity. The moult to the fifth larval stage occurs about 118 days after infection and a permanent buccal capsule is produced. The route taken to the caecum and colon is unknown but probably direct penetration of the wall of these organs occurs. The prepatent period is about 260 days (Wetzel 1941, 1942; Wetzel & Vogelsang 1954).

Life-cycle of S. edentatus. Infective larvae enter the wall of the intestine and pass to the liver via the portal system. In the liver, fourth-stage larvae are produced about 11–18 days after infection. Such fourth-stage forms may migrate in the liver for up to nine weeks and then they pass between the peritoneal layers of the hepatic ligaments to reach the parietal peritoneal region in the right abdominal flank. Late fourth- and early fifth-stage larvae are found in this site in association with haemorrhagic nodules which vary in size from one to several centimetres in diameter. Larvae are found here up to about three months after infection, but they then migrate between the layers of the mesocolon to the walls of the caecum and colon, here again causing haemorrhagic nodules. Such nodules are seen three to five months after infection. Eventually the young adult forms pass to the lumen and become mature. Eggs are produced about 300–320 days after infection (Wetzel 1952; Wetzel & Keesten 1956).

Life-cycle of S. vulgaris. Over the years there has been considerable controversy about the migratory route of the larvae of *S. vulgaris*. This has arisen because of the frequent widespread arterial lesions caused by the larvae, which in some cases, occurred as far anterior in the arterial system as the origin of the aorta.

A detailed consideration of the various theories put forward by various authors is given by Soulsby (1965). However, studies of experimental infections (Enigk 1950; Drudge et al. 1966; Duncan & Pirie 1972) have shown the following developmental cycle. Infective larvae penetrate the intestinal wall where, about eight days after infection, fourth stage larvae are produced. Such fourth-stage forms penetrate the intima of the submucosal arterioles and migrate in these vessels towards the cranial mesenteric artery. They are to be found here from 14 days after infection onwards associated with thrombi and later aneurysms. Starting about 45 days after infection fourth-stage larvae pass back via the arterial system to the submucosa of the caecum and colon, and here become fifth-stage larvae about three months after infection. They then enter the lumen and reach maturity, egg production occurring about six to seven months after infection. Some larvae may linger as fourth- or fifth-stage forms in the aneurysms in the cranial mesenteric artery for several weeks after the main population has returned to the large bowel. Lesions elsewhere in the arterial system may be due to aberrant migration of a few larvae.

Pathogenesis

The pathogenesis of the three *Strongylus* species will be dealt with here, but since *Strongylus* spp. infections are almost always combined with *Trichonema* spp. infection, a general account of strongyle infection of horses, together with treatment and control measures, is given on p. 180.

In their adult forms all three *Strongylus* spp. attach themselves to the mucosa of the large intestine and suck blood. In heavy infections this results in an anaemia of the normochromic, normocytic type which is associated with reduced red cell survival and an increased rate of albumen catabolism (Duncan & Dargie 1975). Lesions produced by the adult worms consist of small haemorrhagic ulcers indicating the site of attachment. Some may become confluent to produce an ulcerous patch. They are however superficial, unlike the deep ulcers produced by the *Triodontophorus* spp. These ulcers are more numerous than worms, suggesting that parasites periodically move to new sites of attachment.

The larval stages of the *Strongylus* spp. may be responsible for severe pathogenic effects. The pathogenesis of the larval stages of *S. vulgaris* has been reviewed by Ogbourne and Duncan (1977). Primary lesions occurred in the wall of the small and large intestine, caused by the penetration of third stage larvae and then migration to the submucosa. Subsequently the fourth and fifth larval stages of *S. vulgaris* are responsible for severe lesions in the arterial system from the aortic valves to the iliac arteries, though the majority of the lesions occur in the region of the cranial mesenteric artery and the arteries which derive from it. Extensive irregular inflammatory lesions occur in the media of the affected arteries producing an endarteritis and the formation of thrombi. Larval stages may be found embedded in the thrombus. At times thrombus formation may be marked, being large, soft friable structures extending for several centimetres in the arterial system. Detachment of such thrombi may lead to a rapidly fatal event, especially if they are situated at the anterior end of the arterial system. With the formation of the thrombus, especially in the cranial mesenteric artery, a thickening of the arterial wall occurs and progressive dilatation begins due to degeneration of elastic fibres. Ultimately a large dilated mass occurs. After migration of the fifth-stage larvae back to the gut lumen, arterial lesions resolve and by nine months the lesions have healed (Duncan & Pirie 1972).

The consequences of such lesions are varied. Detachment of large anterior thrombi may lead to catastrophic events such as occlusion of a coronary artery or the brachiocephalic trunk (Farrelly 1954). Infarction of the iliac artery may lead to temporary lameness while thrombosis of a testicular artery may lead to passive congestion of one or both testicles. Infarction of the kidney has been described. A diarrhoeic syndrome associated with ulceration of the colon and caecum resulting from thrombo-embolism caused by larvae of *S. vulgaris* has been reported by Greatorex (1975), and Merritt et al. (1975) ascribe the effects of verminous arteritis to altered circulation and local irritation.

An association between cranial mesenteric artery aneurysms and colic has long been suggested. Indeed Enigk (1951) produced death in five animals due to haemorrhagic or anaemic infarction of the small and large intestine by giving 800–8000 infective larvae. Similar results were obtained by Drudge et al. (1966). In natural cases, it is unlikely that the massive embolism necessary to produce acute fatal infarction would commonly occur and since the aneurysm of the cranial mesenteric artery is produced slowly, it is likely that collateral circulations will be well established. Nevertheless, Wheat (1975) has suggested that arterial lesions due to *S. vulgaris* are the most common cause of colic. Olt (1932) has suggested that colic may be due to the pressure of the cranial mesenteric aneurysm on associated nerve plexuses. Haematological and biochemical changes consequent on infection have been described by Amborski et al. (1974) and Round (1970). A moderate normocytic, normochromic anaemia occurs during the early phases of experimental infection; there is a marked increase in β-globulin levels 15–20 weeks after infection. Immunoglobulin T is the principal component of the β-globulin increase (Patton et al. 1978).

The pathogenicity of *S. equinus* larvae has been described by Wetzel (1941). Fatal infections caused by 4000 larvae were associated with haemorrhagic tracts in the liver and pancreas. Clinical signs consisted of colic, anorexia and general malaise. Five hundred infective larvae produced no marked clinical signs.

The larval stages of *S. edentatus* may be responsible for serious ill health. In acute experimental infections (3000–75 000 larvae) there is a marked peritonitis, acute toxaemia, jaundice and fever. The peritoneal cavity contains a large amount of haemorrhagic fluid and numerous haemorrhages and fibrinous deposits occur on the peritoneum. Larvae which have migrated to the subperitoneal region of the right flank produce haemorrhagic nodules. These reach their fullest development about three to five months after infection and in severe infections the abdominal cavity is dotted with such nodules. Some may break down, leading to a serious and, at times, fatal intra-peritoneal haemorrhage. At other times, an acute peritonitis occurs which may later turn septic.

Genus: Triodontophorus Looss, 1902

The species of this genus vary from 9 to 25 mm in length. They occur in the large bowel of equids. The buccal capsule is subglobular and rather thick-walled. It has three pairs of teeth at its base and a well-developed dorsal gutter. The spicules of the male end in small hooks and the vulva of the female is near the posterior end of the body. The developmental cycle of the *Triodontophorus* spp. is unknown but it is likely that development is restricted to the bowel mucosa. Species of the genus include:

Triodontophorus serratus Looss, 1902; horse, ass, mule and zebra, world wide in distribution.

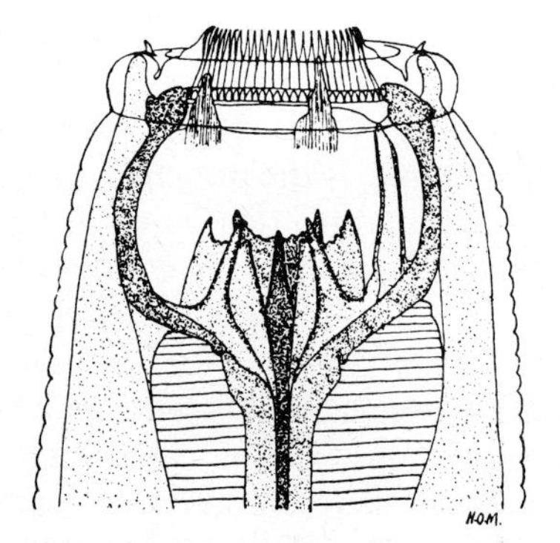

Fig 1.101 Dorsolateral view of the anterior end of *Triodontophorus serratus.*

Triodontophorus brevicauda Boulenger, 1916; horse and ass; worldwide.

Triodontophorus minor Looss, 1902; donkey, worldwide.

Triodontophorus tenuicollis Boulenger, 1916; horse, worldwide.

T. tenuicollis causes the formation of ulcers, which at times may be deep and haemorrhagic, in the right dorsal colon to which bunches of the worms may be found attached.

Genus: Craterostomum Boulenger, 1920

Species of this genus resemble *Triodontophorus*, but they have no teeth in the buccal capsule and the vulva of the female is further forward. Two species may be found in the large intestines of equines. These are: **C. acuticaudatum** Boulenger, 1920 of horse, ass, mule and other equids, being worldwide in distribution; and **C. tenuicauda** Boulenger, 1920 of the pony in India. The developmental cycle is unknown.

Genus: Oesophagodontus Railliet & Henry, 1902

The single species of this genus, **O. robustus** (Giles, 1892), is rather rare in the large intestines of equines. The male is 15–16 mm and the female 19–22 mm long. There is a slight constriction between the anterior end and the rest of the body. The goblet-shaped buccal capsule has a posterior circular ridge and at its base are three tooth-like folds which do not project into the buccal capsule. There is no dorsal gutter. The developmental cycle probably resembles that of the *Trichonema* spp.

Other members of the family *Strongylidae* that occur in non-equid hosts are dealt with after the genera of the family *Trichonematidae* that occur in equids (see p. 183).

FAMILY: TRICHONEMATIDAE WITENBERG, 1925

These strongyles have a short cylindrical or annular buccal capsule. The dorsal gutters are short and do not reach the anterior border of the buccal capsule. Leaf crowns are present.

Various authors have assigned genera in the family Trichonematidae to the subfamily Cyathostominae and have discarded the genus *Trichonema*, this being replaced by four separate genera (*Cyathostomum*, *Cylicocyclus*, *Cylicodontophorus* and *Cylicostephanus*) (e.g. Lichtenfels 1975). This arrangement will be followed in the present edition of this book.

Genus: Cyathostomum Molin, 1861

The buccal capsule of this genus is relatively short, thin-walled and without teeth; a dorsal gutter is not present. The spicules of the males have barbed tips. The vulva of the female is near to the anus. The species is 5–12 mm in length.

The type species is **C. tetracanthum** (Mehlis, 1831) Molin 1861, which occurs in the caecum, sometimes the colon, of horses and other equids throughout the world. **C. coronatum** Looss, 1910 is found in the large intestine of various equids throughout the world. **C. labiatum**

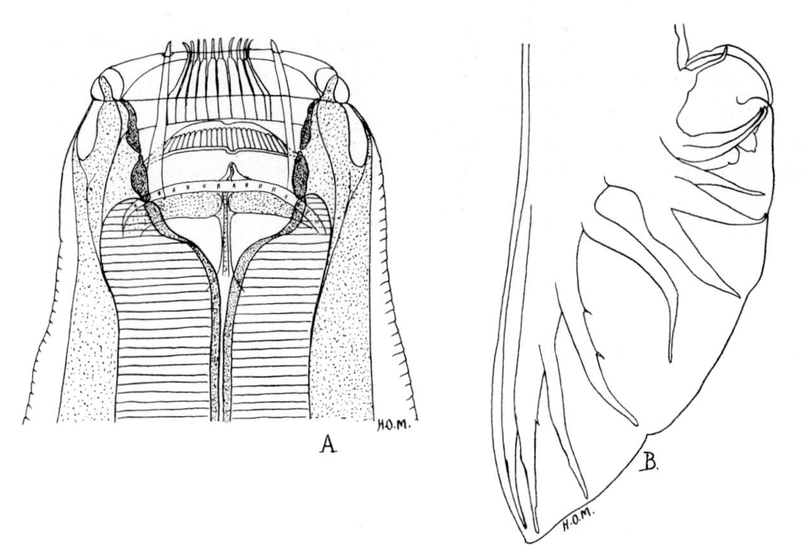

Fig 1.102 *Cyathostomum tetracanthum*. A, Dorsal view of anterior end. B, Lateral view of hind end of male.

(Looss, 1902) occurs in horses and asses globally. **C. labratum** Looss, 1910 occurs in the caecum of asses throughout the world. Other species are **C. ornatum** (Kotlán, 1919) and **C. sagittatum** (Kotlán, 1929) of horses and asses in Europe.

Genus: Cylicocyclus Ihle, 1922

This genus is similar to *Cyathostomum*. The internal leaf-crowns are small and rod-like. The buccal capsule is short, with thin walls. A dorsal gutter is not usually present in the buccal capsule. The species is 10–25 mm long.

There are several species of this genus, of which the following are the most common: **C. nassatus** (Looss, 1900) Chaves, 1930, of the caecum and colon of equids throughout the world; **C. leptostomum** (Kotlán, 1920) Chaves, 1930, of horses and asses; **C. insigne** (Boulenger, 1917) Chaves, 1930, of equids; and **C. ultrajectinus** (Ihle, 1920) Ershov, 1939, of the colon of horses in North America and Europe.

Genus: Cylicodontophorus Ihle, 1922

These are small to medium-sized nematodes, 7–14 mm long. The mouth collar is high, the internal leaf-crown elements long and broad and the buccal capsule short and thick-walled. The type species is **C. bicoronatus** (Looss, 1900) Cram, 1924, of the caecum and colon of horses throughout the world. Two other species, **C. euproctus** (Boulenger, 1917) Cram, 1924 and **C. mettami** (Leiper, 1913) Foster, 1936, occur in the caecum and colon of horses throughout the world.

Genus: Cylicostephanus Ihle, 1922

This is a small genus, 4–10 mm long. The mouth collar is depressed and the elements of the internal leaf-crowns are thin and rod-like. **C. calicatus** (Looss, 1900) Cram, 1924 is the type species and is found in the caecum and colon of horses and asses throughout the world. Other species which occur in the caecum and colon of horses include **C. barbatus** (Smit and Notosoediro, 1923) Cram, 1925, of horses in Java; **C. hybridus** (Kotlán, 1920) Cram, 1924; **C. goldi** (Boulenger, 1917) Lichtenfels, 1975; and **C. longibarsatus** (Yorke and Macfie, 1918) Cram, 1924.

Genus: Poteriostomum Quiel, 1919

Species of this genus resemble *Cyathostomum*, but the externodorsal ray and the dorsal ray of the male bursa arise from a common trunk. The buccal cavity is broader than deep. The elements of the internal leaf-crown are long, broad and acutely tipped; those of the external leaf-crown are numerous, short and thin. They are 12–18 mm in length, **P. imparidentatum** Quiel, 1919 and **P. ratzii** (Kotlán, 1919) Ihle, 1920 occur in the caecum and colon of domestic and wild equids throughout the world.

Genus: Gyalocephalus Looss, 1900

Species of this genus have a short, thick-walled buccal capsule at the base of which are thin, triangular, chitinoid plates which line the very large oesophageal funnel, from which three teeth project into the buccal capsule. There is no dorsal gutter. One species, **G. capitatus** Looss, 1900, is rather rare in the large intestine of equines. The male is 7–8.5 mm and the female 8.5–11 mm long.

Other genera of limited distribution include **Caballonema** Abuladze, 1937, of horses in the USSR, **Cylindropharynx** Leiper, 1911 of zebra and **Sinostrongylus** Hsiung and Chao, 1949 of horses in China.

Illustrated keys to genera and species of the strongyles of horses, with emphasis on North American forms, have been prepared by Lichtenfels (1975).

Life-cycles

Little experimental work has been done on the life-cycles of the cyathostomes in horses. Larvae appear to undergo all their development in the caecum and colon. Larval stages invade the mucosa, become encapsulated and later return to the lumen of the large intestine to reach maturity. The duration of this stay in the mucosa varies greatly from species to species. Some have a prepatent period of about three months while others may remain encapsulated for several months (Wetzel & Vogelsang 1954). Soulsby (1965) has given details of the different larval stages which occur in the mucosa and Ogbourne (1978) has summarized the information of the life-cycles. In general, cyathostomes can mature as early as two months after infection, though Round (1969) reported patency as early as five to six weeks in experimental infections.

Pathogenesis

These parasites are often found in very large numbers in the caecum and colon of horses. Ogbourne (1976) has recorded more than 1 million worms in a horse. Though several authors suggest that adult worms do little harm (reviewed by Ogbourne 1978) heavy infections produce a disquamative catarrhal enteritis. The nodules caused by the larval stages may be so numerous that it is difficult to find an area of normal mucosa. The pathological changes caused by larvae are summarized by Ogbourne (1978).

Clinical signs

Since the cyathostomes invariably coexist with other strongyles, especially *Strongylus* spp., it has been difficult to discern the effects attributable to the *Cyathostominae*. Acute verminous enteritis has been reported by various authors (e.g. Cuillé et al. 1913; Velichkin 1952; Chiejina & Mason 1977; Jeggo & Sewell 1977) involving massive mucosal infections with larval stages and especially their emergence in late winter and spring. Profuse diarrhoea, colic and occasionally death are reported with this condition. Post-mortem examination showed severe inflammation of the caecum and colon with numerous larvae beneath the mucosa (Mirck 1977).

STRONGYLE INFECTION OF EQUINES

Though some of the equine strongyles are blood suckers and some are not, the infections are usually mixed and consequently the general clinical signs can be considered to be caused by all the worms collectively. Specific clinical signs may arise due to the larval stages of the *Strongylus* spp. (see p. 177). Large numbers of strongyles may be present in foals. Eggs of the worms found in the faeces of foals may come from at least two sources: from worms parasitic in the foals, though these usually do not appear until the foals are seven to eight weeks old or eggs may be found in the faeces of foals younger than this, even in the meconium, but these are usually derived from the faeces of the dam licked or eaten by the foals.

It is possible that foals may be infected before birth by larvae, especially those of *S. vulgaris*, present in the blood, but the occurrence of prenatal infection has not been demonstrated experimentally. Older equines develop resistance to reinfection and some may carry heavy infections without serious effects; nevertheless they may be dangerous sources of infection to younger animals. It has been estimated that a horse passing out 1000 eggs per gram of its faeces may pass out 30 million eggs a day.

In general the clinical signs develop slowly. The faeces become soft and have a bad odour. Later diarrhoea develops, the appetite diminishes, the animals become emaciated, easily exhausted and the coat is rough. Anaemia develops and may become marked and associated with oedematous swellings on the abdomen and the legs. Various degrees of the disease are seen, depending on the number of the parasites present and the condition and food of the animal. Death may occur in severe cases.

Strongyle burdens of horses tend to show a seasonal incidence which is a reflection of the pasture burdens of infective larvae. Duncan (1974) demonstrated in Great Britain that mares showed a marked increase in strongyle egg output in late spring and early summer which was reflected by high pasture larval counts in September and October. After this, with egg output in mares low and environmental conditions being unfavourable for development of free-living larval stages, there is a sharp decline in pasture larvae. Infective larvae acquired in late summer lead to adult worm populations which predominate in the spring. Indeed there appears to be an annual turnover in the adult worm population, new adults being established from larvae which have over-wintered as arrested larvae in the gut wall (Ogbourne 1976). Though infective larvae may survive over winter on pasture, they do not constitute a major source of infection for horses the following spring and such larvae do not survive beyond June the following year.

Diagnosis

The eggs of the various strongyle species of the large intestine of horses cannot be readily distinguished from one another and if accurate identification is required faecal cultures should be made to obtain third-stage larvae. These may then be identified using appropriate keys (Soulsby 1965). As a rule the presence of oval, thin-shelled, strongyle-type eggs is sufficient for diagnosis of infection and the presence of 1000 epg of faeces, or more, is evidence that removal of the worms is necessary.

The presence of an aneurysm in the cranial mesenteric artery may be determined in small horses by rectal palpation. The aneurysm may be a large pulsating body varying in size up to 6–7 cm in diameter. Where fibrosis has occurred pulsations may be difficult to recognize. Blood vessels are thickened and firm and irregularities in outline may be detected. When aneurysms are associated with colic, abdominal auscultation reveals hypermobility of the intestines and vertebral percussion over the lumbar and sacral vertebrae may evince pain. Paracentesis may reveal a turbid peritoneal fluid containing neutrophils and eosinophils (Greatorex 1977).

Treatment

A number of anthelmintics are available, the majority having a high level of activity against adult strongyles of horses (*Strongylus* spp. and cyathostomes) and some also are active against the larval stages of these helminths (Duncan 1976).

Phenothiazine. A safe dose is 30–35 g per animal. It is highly effective against cyathostomes but not against *Strongylus* spp. at this level. Higher doses may be toxic to horses, causing anaemia and haemoglobinuria. Though phenothiazine has generally been replaced by other compounds, it is nevertheless a useful anthelmintic.

Piperazine salts, both the adipate and the citrate, are highly effective against *Parascaris equorum* at a dose of 220 mg/kg but show variable action against the strongyles. Piperazine may be formulated with other compounds to produce a broader spectrum of activity.

Thiabendazole, at 44 mg/kg, is effective against all horse large intestinal strongyles. At a dose of 440 mg/kg given twice it shows activity against the migrating stages of *Strongylus* spp.

Mebendazole, at 10 mg/kg, is effective against adult strongyles of horse.

Fenbendazole, at 7.5 mg/kg, is highly effective against adult strongyles. In addition, at 50 mg/kg it is effective in cases of diarrhoea in horses due to prepatent strongyle infection (Jeggo & Sewell 1977) and at 60 mg/kg shows efficacy against migrating stages of *S. vulgaris*.

Cambendazole, at 20 mg/kg, is effective against adult strongyles.

Oxibendazole is given at a dose of 5–10 mg/kg and is effective against strongyles.

Pyrantel emboate, a broad-spectrum drug, is effective at 19 mg/kg.

Dichlorvos, at 26–32 mg/kg, has a broad spectrum and may be given in the feed.

Haloxon is a broad-spectrum organophosphate and is used at 50–70 mg/kg. It may be given in the feed.

Avermectin B1a is reported to be effective against larvae of *S. vulgaris* in the anterior mesenteric artery.

Lloyd (1980) has recommended a general programme of treatment for the control of parasites of horses involving the administration of anthelmintics every two months. For example, 15 January, drugs in Group A; 15 March, drugs in Group B or C; 15 May, drugs in Group A; 15 July, drugs in group A (or B or C); 15 September, drugs in Group A; 15 November, drugs in Group B which have a high efficacy against *Gastrophilus intestinalis*.

Group A anthelmintics have very good efficacy against the adults and some efficacy against the immature stages of the large strongyles (*Strongylus vulgaris*, *S. edentatus* and *S. equinus*) as well as the small strongyles. On the whole they are effective against *Parascaris equorum* and *Oxyuris equi*, affecting both adults and immature stages (mebendazole, thiabendazole, cambendazole, fenbendazole).

Group B anthelmintics are an effective treatment for *Gastrophilus* spp. and the strongyles as well as *P. equorum* and *O. equi*. They are organophosphate anthelmintics (dichlorvos and trichlorofon/phenothiazine/piperazine mixture) and are unrelated to the benzimadazole drugs.

Group C anthelmintics are unrelated to the benzimidazoles and can be used to help prevent the development of parasite resistance against the benzimadazoles (or to treat already resistant strains). They are effective against the small strongyles, large strongyles (with less effect against *S. edentatus*), *P. equorum* and *O. equi* (levamisole/piperazine mixture, pyrantel tartate, pyrantel pamoate).

Regular treatment greatly reduces the egg output and hence the number of infective larvae on pastures. New animals should be treated on arrival with a broad-spectrum anthelmintic and isolated for 48–72 hours. The most suitable drugs for this are mebendazole, cambendazole and dichlorvos. Where the stocking rate is high (e.g. in studs, riding stables etc.) treatment should be given every six weeks through spring, summer and autumn and every two months during winter. Where there are light stocking rates the treatment may be reduced to once every three months.

For the treatment of colic due to verminous aneurysm, Greatorex (1977) has recommended intravenous injection of 6% dextran in 5% dextrose as an antithrombotic agent. Initially 500–1500 ml of dextran 70 (2.5 mg/kg) is given on three successive days followed by 500 ml every four days until nine injections have been given over 27 days.

Control

Apart from the routine use of anthelmintics (see above) the general principles of pasture management for the control of parasitic nematodes enumerated for the control of ruminant parasites are applicable to horse parasites. Pastures should not be overstocked or overgrazed. Special attention should be given to young horses which are more susceptible than the older ones. If possible clean pasture should be available for them. Proper disposal of manure, by allowing it to ferment in heaps, will kill the eggs and larvae by the heat of fermentation. Under crowded conditions manual removal of faeces from pastures is very effective. Alternate grazing of horse pastures with sheep or cattle will assist in reducing pasture burdens of larvae since horse strongyles do not develop in ruminants.

REFERENCES

STRONGYLIDAE AND TRICHONEMATIDAE OF EQUINES

Amborksi, G. F., Bello, T. R. & Torbert, B. J. (1974) Host response to experimentally induced infections of *Strongylus vulgaris* in parasite-free and naturally infected ponies. *Am. J. vet. Res.*, **35**, 1181–1188.

Chiejina, S. N. & Mason, J. A. (1977) Immature stages of *Trichonema* spp. as a cause of diarrhoea in adult horses in spring. *Vet. Rec.*, **100**, 360–361.

Croll, N. A. (1975) Behavioural analysis of nematode movement. *Adv. Parasit.*, **13**, 71–122.

Cuillé, J., Marotel, G. & Roquet, M. (1913) Nouvelle et grave enterite vermineuse du cheval: la cylicostomose larvaire. *Bull. Mém. Soc. Sci. vet. Lyon*, **16**, 172–185.

Drudge, J. H., Lyons, E. T. & Szanto, J. (1966) Pathogenesis of migrating stages of helminths with special reference to *Strongylus* species. In: *Biology of Parasites*, ed. E. J. L. Soulsby, pp. 199–214. New York: Academic Press.

Duncan, J. L. (1974) Field studies on the epidemiology of mixed strongyle infection in the horse. *Vet. Rec.*, **94**, 337–345.

Duncan, J. L. (1976) The anthelmintic treatment of horses. *Vet. Rec.*, **98**, 233–235.

Duncan, J. L. & Dargie, J. D. (1975) The pathogenesis and control of strongyle infection in the horse. *J. S. Afr. vet. Ass.*, **46**, 81–85.

Duncan, J. L. & Pirie, H. M. (1972) The life cycle of *Strongylus vulgaris* in foals. *Res. vet. Sci.*, **18**, 82–93.

Enigk, K. (1950) Zur Entwicklung von *Strongylus vulgaris* (Nematoden) In Wirtstier. *Z. Tropenmed. Parasit.*, **2**, 287–306.

Farrelly, B. T. (1954) The pathogenesis and significance of parasitic endoarteritis and thrombosis in the ascending aorta of the horse. *Vet. Rec.*, **66**, 53–61.

Greatorex, J. C. (1975) Diarrhoea in horses associated with ulceration of the colon and caecum resulting from *Strongylus vulgaris* larval migration. *Vet. Rec.*, **97**, 221–225.

Greatorex, J. C. (1977) Diagnosis and treatment of 'Verminous aneurysm' formation in the horse. *Vet. Rec.*, **101**, 184–189.

Jeggo, M. H. & Sewell, M. M. H. (1977) Treatment of pre-patent equine strongyliasis. *Vet. Rec.*, **101**, 187.

Levine, N. D. (1963) Weather, climate and bionomics of ruminant nematode larvae. *Adv. vet. Sci.*, 8, 215–261.

Levine, N. D. (1968) *Nematode Parasites of Domestic Animals and of Man.* Minneapolis: Burgess.

Lichtenfels, J. R. (1975) Helminths of domestic equids. *Proc. helminth. Soc. Wash.*, **42** (Special Issue), 92.

Lloyd, S. (1980) Unpublished data.

Merritt, A. M., Bolton, J. R. & Cimprich, R. (1975) Differential diagnosis of diarrhoea in horses over six months of age. *J. S. Afr. vet. Ass.*, **46**, 73–76.

Mirck, M. H. (1977) Cyathostominose: een vorm van ernstige strongylidose. *Tijdschr. Diergensk.*, **102**, 932–934.

Ogbourne, C. P. (1976) The prevalence, relative abundance and site distribution of nematodes of the subfamily Cyathostominae in horses killed in Britain. *J. Helminth.*, **50**, 203–214.

Ogbourne, C. P. (1978) Pathogenesis of cyathostome (Trichonema) infections of the horse. A review. *Commonw. Inst. Helminth. misc. Publ.*, 5.

Ogbourne, C. P. & Duncan, J. L. (1977) *Strongylus vulgaris* in the horse: its biology and veterinary importance. *Commonw. Inst. Helminth. misc. Publ.*, 4.

Olt, A. (1932) Das Aneurysma Verminosum des Pferdes und seine unbekannten Beziehungen zur Kolik. *Dt. tierärztl. Wschr.*, **40**, 326–332.

Patton, S., Mock, R. E., Drudge, J. A., Morgan, D. (1978) Increase of immunoglobulin T circulation in ponies as a response to experimental infection with the nematode *Strongylus vulgaris. Am. J. vet. Res.*, **39**, 19–23.

Poynter, D. (1956) Effect of a coliform organism (*Escherichia*) on the second ecdysis of nematode larvae parasitic in the horse. *Nature, Lond.*, **177**, 481–482.

Round, M. C. (1969) The prepatent period of some horse nematodes determined by experimental infection. *J. Helminth.*, **43**, 185–192.

Round, M. C. (1970) The development of strongyles in horses and the associated serum protein changes. In *Equine Infectious Diseases*, ed. J. T. Bryans & M. Gerber, vol. 2, pp. 290–303. Basel: Karger.

Soulsby, E. J. L. (1965) *Textbook of Veterinary Clinical Parasitology, Vol. I, Helminths.* Oxford: Blackwell Scientific.

Velichkin, P. A. (1952) Nodular trichonematosis of the large intestine in stud horses [in Russian]. *Izvest. Mosk. Zootekh. Inst. Konev.*, **7**, 18–25.

Wetzel, R. (1941) Zur Entwicklung des grossen Palisadenwürmer (*Strongylus equinus*) im Pferd. *Arch. Wiss Prakt. Tierheilk*, **76**, 81–118.

Wetzel, R. (1942) Über die Entwicklungsdauer der Palisadenwürmer in Körper des Pferdes und ihre praktische Auswertung. *Dt. tierärztl. Wschr.*, **50**, 443–444.

Wetzel, R. (1952) Die Entwicklungsdauer (Prepatent periode) von *Strongylus endentatus* in Pferd. *Dt. tierärztl. Wschr.*, **59**, 129–130.

Wetzel, R. & Keesten, W. (1956) Die Leberphase der Entwicklung von *Strongylus edentatus. Wien. tierärztl. Mschr.*, **43**, 664–672.

Wetzel, R. & Vogelsang, E. G. (1954) Helmintiasis intestinal del equino. *Rev. med. vet. parasit. Caracas*, **13**, 17–25.

Wheat, J. D. (1975) Causes of colic and types requiring surgical intervention. *J. S. Afr. vet. Ass.*, **46**, 95–98.

NON-EQUINE STRONGYLIDAE AND TRICHONEMATIDAE

GENUS IN BIRDS

Genus: Codiostomum Railliet & Henry, 1911

Codiostomum struthionis Horst, 1885 occurs in the large intestine of the ostrich. The male is about 13 mm long and the female 17 mm. The buccal capsule is subglobular, strongly chitinized and provided with external and internal leaf-crowns, but there are no teeth. The dorsal gutter is well developed and reaches the anterior margin of the buccal capsule. The male bursa has a large projecting dorsal lobe and the vulva of the female is situated close to the anus.

The life-cycle is unknown, but probably direct.

Little is known about its pathogenicity but the parasite must be considered a dangerous one, especially when infections are heavy.

GENUS IN PIGS

Genus: Bourgelatia Railliet, Henry & Bauche, 1919

Bourgelatia diducta Railliet, Henry & Bauche, 1919 occurs in the caecum and colon of the pig in India, South-East Asia and Java. The male is 9–12 mm long and the female about 11–13.5 mm. The mouth is directed straight forwards. The buccal capsule is cylindrical and shallow and its thick wall is divided into an anterior and a posterior portion, the latter being continuous with the lining of the wide oesophageal funnel. The external leaf-crown

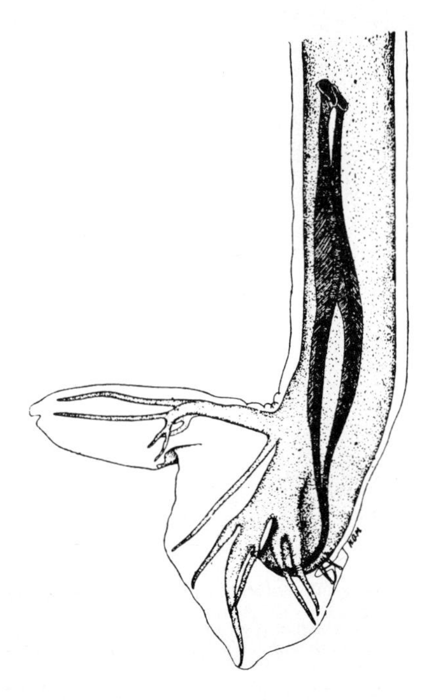

Fig 1.103 Lateral view of the hind end of the male *Codiostomum struthionis.*

has twenty-one long elements projecting from the oral aperture and the internal leaf-crown has about twice as many elements. The spicules are equal, alate and about 1.3 mm long. The vulva opens near the anus. The posterior end of the female is straight and ends in a sharp point. The eggs measure 58–77 by 36–42 μm.

The life-cycle is probably direct.

Very little is known about the effects of this parasite on its host.

GENERA IN ELEPHANTS

Genus: Choniangium Railliet, Henry & Bauche, 1914

The mouth is directed anterodorsally, the buccal capsule is spacious but long and narrow. Cuticular prominences project into it. There are no internal leaf-crowns. Two species occur in the caecum of the Indian elephant (van der Westhuysen 1938): **Choniangium epistomum** Railliet, Henry and Bauche, 1914 and **C. magnostomum** van der Westhuysen, 1938.

Genus: Decrusia Lane, 1914

The anterior end of this genus is bent dorsally. The buccal cavity is cup-like and contains two subventral teeth. A dorsal groove is present, as are external and internal leaf-crowns. One species, **Decrusia additicta** (Railliet, Henry and Bauche, 1914) Lane, 1914, occurs in the large intestine of the Indian elephant.

Genus: Equinurbia Lane, 1914

The mouth is directed dorsally; the buccal capsule is almost spherical. Internal leaf-crowns are absent; teeth absent. **Equinurbia sipunculiformis** (Baird, 1859) Lane, 1914 occurs in the caecum of the Indian elephant.

Genus: Khalilia Neveu-Lemaire, 1924

This is a small strongyle. The buccal capsule is short; the mouth directed forward, the oesophagus stout and hour-glass-shaped. **Khalilia pileata** (Railliet, Henry and Bauche, 1914) Ogden, 1966 occurs in the caecum of the Indian elephant. Males are 9–11 mm and females 11.5–14 mm long. **K. buta** (Vuylsteke, 1953) Popova, 1958 occurs in the large intestine of the African elephant as does **K. sameera** (Khalil, 1922) Neveu-Lemaire, 1924.

Genus: Murshidia Lane, 1914

These are slender worms 20–30 mm long; the mouth is directed forward with a prominent mouth-collar. The buccal capsule is cylindrical, with or without teeth. The type species is **M. murshida** Lane, 1914 in the caecum of the Indian elephant. A number of other species occur in the Indian elephant or the African elephant, but not in both.

Genus: Quilonia Lane, 1914

This species is similar to *Murshidia* spp., slender with a mouth-collar. The buccal capsule is short and annular; one to seven teeth project into the capsule. Several species occur, some in the Indian elephant **Q. renniei** (Railliet, Henry and Joyeux, 1913) Lane, 1914; **Q. travancra**, Lane, 1914 and many in the African elephant (e.g. **Q. africana** Lane, 1921; **Q. uganda,** Khalil, 1922; **Q. ethiopica,** Khalil, 1922). A key to the species is given by van der Westhuysen (1938) and a check list is provided by Round (1968).

The life-cycles of these various genera in elephants are unknown, but they are likely to follow those of comparable genera in the horse. No information is available regarding pathogenicity or treatment, however the benzimidazoles are probably effective.

GENERA IN PRIMATES

Genus: Ternidens Railliet & Henry, 1909

These are strongyles of the large intestine of primates. The buccal capsule is subglobular. Three teeth occur in the oesophageal funnel. There are two sets of leaf-crowns, a transverse cervical groove and an indistinct cephalic vesicle.

Ternidens deminatus Railliet and Henry, 1909 occurs in the large intestine of various primates including chimpanzee, gorilla, various macaques and, occasionally, man. Males are 9.5 mm long and females 12–16 mm. Spicules are slender, equal and 1.15 mm long. The eggs are strongyle type, 62–72 μm by 36–40 μm.

The developmental cycle is unknown but it is probably similar to that of *Oesophagostomum* spp.

Pathogenic effects include anaemia in heavy infections due to the blood-sucking activities of the worm. Nodules may be formed in the large intestine (Flynn 1973).

It is probable that the benzimidazoles, normally used for ruminants, would be effective for the treatment of heavy infections.

GENERA IN RUMINANTS

Genus: Chabertia Railliet & Henry, 1909

Chabertia ovina (Gmelin, 1790) occurs in the colon of sheep, goats, cattle and a number of other ruminants throughout the world. Males are 13–14 mm long and females 17–20 mm long. The anterior end is curved slightly ventral and the large buccal capsule opens anteroventrally. The oral

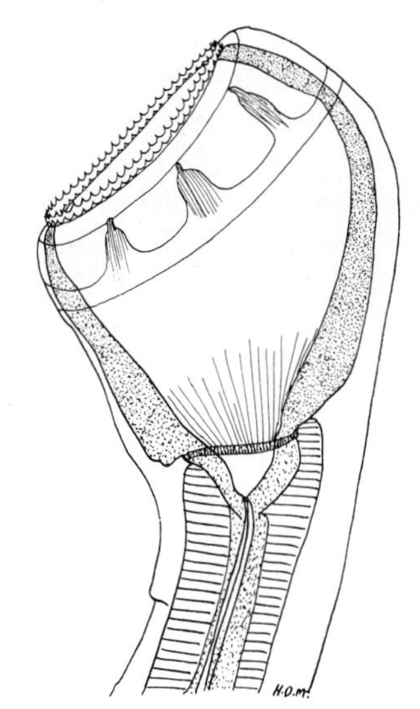

Fig 1.104 Lateral view of the anterior end of *Chabertia ovina*.

aperture is surrounded by a double row of small cuticular elements representing the leaf-crowns. There is a shallow ventral cervical groove, and anterior to it a slightly inflated cephalic vesicle. The male bursa is well developed and the spicules are 1.3–1.7 mm long, with a gubernaculum. The vulva of the female opens about 0.4 mm from the posterior extremity. The eggs measure 90–105 by 50–55 μm.

Life-cycle. This is direct. The sheath of the infective larva has a relatively long tail. Infection occurs per os. Herd (1972) has described the life-cycle. Third-stage larvae undergo an extensive histotropic phase in the wall of the small intestine prior to the third ecdysis seven to eight days after infection. Up to 26 days may elapse before developmental stages reach the colon. Fourth-stage larvae develop mainly in the lumen of the caecum. The fourth ecdysis occurs, on average, about 24 days after infection. Immature adults then pass to the colon, where patency occurs 49 days after infection.

Pathogenesis. The adult worms attach themselves firmly to the mucosa of the colon by means of their buccal capsules and then draw in a plug of mucosa, chiefly the granular layer, which is digested by the secretions of the oesophageal glands of the worm. The worms probably suck blood by accident only, when a blood vessel is ruptured. The adjoining parts of the mucosa show an increased activity of goblet cells and infiltration with lymphocytes and eosinophils. Clinical signs in severely affected animals include a marked diarrhoea with much blood and mucus. At autopsy the worms are found attached to the mucosa of the colon, which is congested, swollen and covered with mucus in severe cases; punctiform haemorrhages may be present. In severe infections sheep lose condition and become anaemic and die. *Chabertia* infections may be responsible for a specific reduction of wool growth in sheep.

Diagnosis is made by finding the eggs in the faeces and by identification of the larvae in faecal cultures.

Treatment. All the benzimidazoles in common use for ruminant gastrointestinal helminths are effective against *C. ovina*. Hence treatment for gastrointestinal nematodes also clears out *C. ovina*.

Genus: Oesophagostomum Molin, 1861

Members of this genus have a cylindrical buccal capsule, usually narrow. Leaf-crowns are present. There is a ventral cervical groove near the anterior end, anterior to which the cuticle is dilated to form a cephalic vesicle. Species are parasites in the small intestine and the large intestine of cattle, sheep, pigs and primates. These nematodes are often referred to as nodular worms, owing to the fact that several species cause nodule formation on the wall of the intestine.

NODULAR WORMS OF SHEEP, CATTLE AND RELATED SPECIES

Oesophagostomum columbianum (Curtice, 1890) Stossich, 1899
Oesophagostomum venulosum (Rudolphi 1809) Railliet, 1896
Oesophagostomum asperum Railliet and Henry, 1913
Oesophagostomum multifoliatum Daubney and Hudson, 1932
Oesophagostomum okapi Leiper 1935
Oesophagostomum walkeri Mönnig 1932
Oesophagostomum radiatum (Rudolphi, 1803)

Oesophagostomum columbianum occurs in the colon of the sheep, goat, camel and a number of wild antelopes. Generally, it is world-wide in distribution, though it is more common in tropical and subtropical areas. It is absent from Great Britain and the west coast of North America. It has been recorded once from cattle but in this case the identification of the worms may have been at fault. The male is 12–16.5 mm long and the female 15–21.5 mm by about 0.45 mm wide. There are large cervical alae which produce a marked dorsal curvature of the anterior part of the body. The cuticle forms a mouth-collar which is fairly high and separated from the rest of the body by a constriction. A cervical groove extends around the ventral surface of the lateral aspects of the body, about 0.25 mm from the anterior end. The cuticle anterior to this groove is inflated to form a cephalic vesicle. Immediately behind the cervical groove the cervical alae arise and their anterior extremities are pierced by cervical papillae. The buccal capsule is shallow, the external leaf-crown consists of 20–24 elements and the internal has two small elements to each of the external. The male bursa is well developed and there are two equal, alate

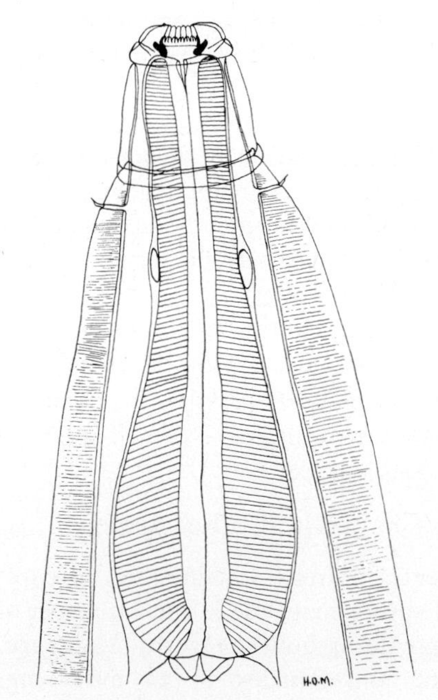

Fig 1.105 Dorsal view of the anterior end of *Oesophagostomum columbianum*.

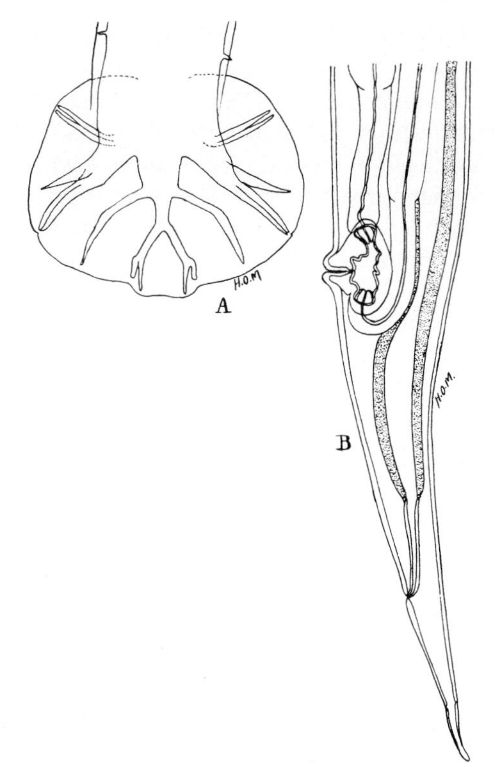

Fig 1.106 *Oesophagostomum columbianum.* A, Bursa of male. B, Lateral view of hind end of female.

spicules, 0.77–0.86 mm long. The tail of the female tapers to a fine point. The vulva is situated about 0.8 mm anterior to the anus. The vagina is very short, transverse, leading into the kidney-shaped 'pars ejectrix' of the ovijectors. The eggs have thin shells and are laid in the eight to sixteen cell stage. They measure 73–89 by 34–45 μm.

Life-cycle. The eggs are passed in the faeces of the host and the development and bionomics of the free stages are similar to those of the *Strongylus* spp. The infective stage is reached under optimum conditions in six to seven days. None of the pre-infective stages are resistant to dessication. After ingestion, infective larvae exsheath in the small intestine and larvae, within one day after infection, penetrate into the wall of the intestine, anywhere from the pylorus to the rectum, being coiled up against the muscularis mucosa and causing cysts to form. Here the third ecdysis takes place at four days after infection and the larvae grow to a length of about 1.5–2.5 mm. They now have a subglobular buccal capsule with a dorsal tooth at the base, and the cervical groove is conspicuous. Normally they return to the lumen of the gut after five to seven days and pass to the colon, where they grow to adult after the fourth ecdysis. The first eggs are passed in the faeces of the host 41 days after infection. A few larvae may remain in the mucosa for extended times in lambs and especially in previously exposed adult animals.

Pathogenesis. In lambs, or in older sheep that have no previous exposure to the parasite, the larvae incite practically no reaction by their migration into the mucosa, so that eventually a large number of adult worms can be found in the colon while there are no nodules in the wall of the intestine. In other cases, owing to previous sensitization, the larvae pass into the submucosa and a marked reaction takes place in the form of a localized inflammation around each larva. Leucocytes, especially eosinophils, and foreign body giant cells collect around the parasite and the focus becomes encapsulated by fibroblasts. The larvae may stay in these nodules for about three months, and when the contents caseate and calcify the parasite either dies or leaves the nodule, and then frequently wanders about between the muscle fibres, leaving behind it a narrow canal filled with material similar to that found in the nodules. Although the nodules usually have a small opening through which pus is discharged into the intestine, the large majority of these larvae do not find their way back into the lumen. In such cases the intestinal wall may therefore show numerous nodules and tracks, while the colon contains few adult worms.

O. columbianum is a serious pathogen of sheep, 200–300 adult worms constituting a severe infection for young sheep. Extensive nodular formation in both the small and large intestine seriously interferes with absorption, bowel movement and digestion. The nodules are frequently suppurative and may rupture to the peritoneal surface causing peritonitis and multiple adhesions. Though the adult worms do not suck blood they cause a marked thickening of the bowel wall, congestion and a large production of mucus. Infections have a profound effect on appetite and growth and also wool growth (Gordon 1950). The anorexia, accompanied by depressed utilization of feed, and the intake of water are dose dependent. After infection with 500 larvae depressed feed intake and utilization are apparent, but with 2000–5000 larvae the effects are severe and there is a marked diarrhoea associated with loss of weight which begins within a week of infection and is most severe by the fifth week (Dobson 1967*a*). Though total serum proteins may be slightly decreased during infection, there is a marked hypoalbuminaemia with an increase in $\beta 2$-globulin and γ-globulin, both attributable to the immune response (Dobson 1967*b*).

On *post mortem* there is a marked emaciation and an almost complete absence of fat. In heavy initial infections a large number of adult worms is seen, the mucous membrane is thickened, reddish and covered with mucus in which the worms are embedded. After repeated infections the ileum and colon may be thickly studded with nodules of various sizes, some having been converted to abscesses and containing a green to yellowish pus or caseous material.

Clinical signs. In lambs the first sign is a marked and persistent diarrhoea, which results in exhaustion and death unless the animals are removed from the infected pasture. The faeces usually have a dark green colour and contain much mucus and sometimes blood. This diarrhoea begins on the sixth day after a severe infection and coincides with the time when the larvae leave the nodules. In more chronic cases there may be an initial diarrhoea, later followed by constipation and occasional spells of diarrhoea. The animal shows progressive emaciation and general weakness. The skin becomes dry and the wool is unthrifty. The characteristic picture of chronic oesophagostomiasis in sheep is that of extreme emaciation and cachexia with atrophy of the muscles, ending in complete prostration for 1–3 days and death.

Diagnosis. Examination of the diarrhoeic faeces may show the fourth-stage larvae in acute cases or the presence of eggs in other cases. Chronic cases without adult parasites can be diagnosed only tentatively and an autopsy of a selected animal may be required. The eggs of *O. columbianum* cannot easily be differentiated from those of many other gastrointestinal nematodes of sheep and faecal cultures have to be made and examined when the larvae have reached the infective stage. The infective larvae of *O. columbianum* and other species of this genus have a sheath provided with a long, whip-like tail, while the tail of the larva itself is much shorter and ends in a simple point (Fig. 136).

Treatment. All the benzimidazole anthelmintics and other broad-spectrum anthelmintics, such as levamisole, morantel and phenothiazine, in doses used for other gastrointestinal nematodes of sheep, are highly effective against *O. columbianum*.

Oesophagostomum venulosum occurs in the colon of sheep, goat, deer and camel. The male is 11–16 mm and the female 13–24 mm long. There are no lateral cervical alae. The cervical papillae are situated behind the level of the oesophagus. The external leaf-crown consists of 18 and the internal leaf-crown of 36 elements. The spicules of the male are 1.1–1.5 mm long.

Life-cycle. This is similar to that of *O. columbianum* (Goldberg 1951). Following ingestion of infective larvae, encysted forms occur in the wall of the small intestine and these moult to the fourth-stage four days after infection. They re-enter the intestine, pass to the large bowel, moult to the adult stage 13–16 days after infection and patency commences 28–31 days after infection.

Pathogenesis. Though *O. venulosum* is very similar to *O. columbianum* and occurs in the same site in the host, the pathogenic effects are quite different. *Oesophagostomum venulosum* is relatively harmless: infection seldom produces nodule formation. Even in heavy experimental infections the clinical effects are of a low order (Goldberg 1952).

Treatment. As for *O. columbianum*.

Oesophagostomum aspersum occurs in the large intestine of the goat and sheep in Central America (Panama) and Asia. Males are 12–13 mm long and females 15–17 mm. The cephalic vesicle is inflated. *O. indicum* reported from sheep in India is a synonym of *O. aspersum*.

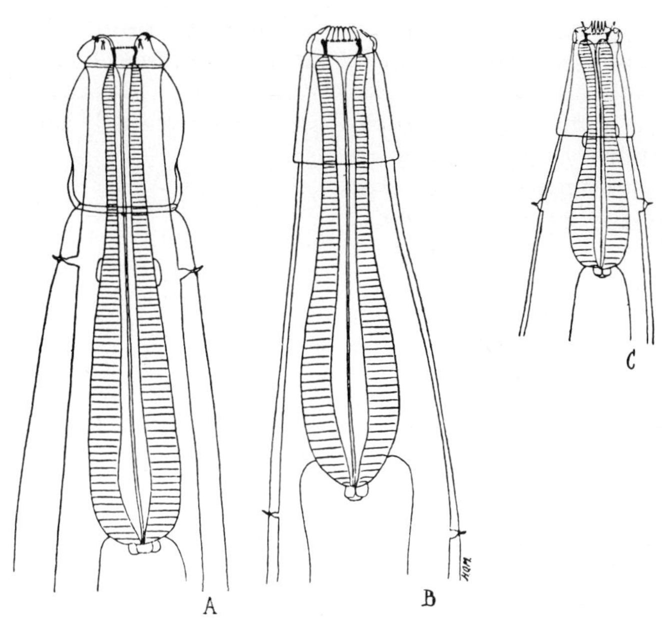

Fig 1.107 Dorsal view of the anterior end of various species of *Oesophagostomum*. A, *O. radiatum*. B, *O. venulosum*. C, *O. dentatum*.

Oesophagostomum multifoliatum occurs in the large intestine of sheep and goats in East Africa. Males are 12–14 mm long and females 14–17 mm. No information is available on the life-cycle or the pathogenic effects of these two species.

Oesophagostomum okapi and **O. walkeri** occur in the okapi and eland respectively.

Oesophagostomum radiatum (Rudolphi, 1803) occurs in the colon of cattle, zebu and water buffalo throughout the world. The male is 14–17 mm long and the female 16–22 mm. This species is characterized by a rounded mouth-collar, a large cephalic vesicle which is constricted behind its middle and the absence of an external leaf-crown. The internal leaf-crown consists of 36–40 minute elements. The vagina is short, as in *O. columbianum*. The spicules are 0.7–0.8 mm long. Eggs are 70–76 by 36–40 μm and strongyle-like. Following ingestion, infective larvae exsheath in the small intestine and enter the wall of both small and large intestines, where they moult to the fourth larval stage five to seven days after infection. They return to the lumen seven to 14 days after infection, pass to the large intestine and moult 17–22 days after infection to the adult stage. Patency is reached 32–42 days after infection (Roberts et al. 1962).

Pathogenesis. *O. radiatum* is one of the more pathogenic species of helminths of cattle when it is present in large numbers. In the acute form of disease there is inflammation of the small and large intestine and black foetid diarrhoeic faeces are passed. The chronic disease may occur in young stock (in which it may be fatal) and in old ones (which usually recover). Extensive nodule formation (pimply gut) occurs, affecting the whole of the intestinal tract. This is associated first with intermittent diarrhoea and later with continuous purging resulting in emaciation, prostration and often death in young animals. Bremner (1961) has shown that the most severe effects were associated with the early fifth-stage larvae. Anorexia was particularly important and this was especially evident from the fourth week of infection onwards. Anaemia of the normochromic normocytic type and hypoproteinaemia, due to a protein-losing enteropathy, are features of the disease.

Treatment. The benzimidazole compounds are highly effective (e.g. thiabendazole, parbendazole, cambendazole, fenbendazole, albendazole and oxfendazole). Where their use is licensed for use in cattle, doses of 60–100 mg/kg are used. Other compounds available include phenothiazine (30 g for calves, 60 g for cattle), piperazine salts (7–15 mg/50 kg) and levamisole (7.5 mg/kg) given by subcutaneous injection. Supportive medication may be necessary in severe cases of diarrhoea.

NODULAR WORMS OF PIGS

Oesophagostomum dentatum (Rudolphi, 1903) Molin, 1861.

Oesophagostomum brevicaudum Schwartz and Alicata, 1930

Oesophagostomum georgianum Schwartz and Alicata, 1930

Oesophagostomum quadrispinulatum (Marcone, 1901) Alicata, 1935

Oesophagostomum granatensis Herrera, 1958

Oesophagostomum dentatum occurs in the large intestine of pigs and peccaries throughout the world. The males are 8–10 mm long and the females 11–14 mm. The cephalic vesicle is prominent, but cervical alae are practically absent. The cervical papillae are towards the posterior end of the oesophagus. The submedian head papillae project forward conspicuously, as do the nine elements of the external leaf-crown. The internal leaf-crown has 18 elements. The spicules are 1.15–1.3 mm long. The eggs measure 35–45 by 60–80 μm.

The life-cycle is similar to that of other members of the genus. Infective larvae exsheath in the small intestine and enter the mucosa of the large intestine, causing small nodules. Larvae re-enter the lumen of the large intestine six to seven days later, having moulted to the fourth-stage at four days after infection. Patency is reached at about 49 days after infection (Kotlán 1948).

Oesophagostomum brevicaudum occurs in the large intestine of pigs in the USA and India. Males are 6–7 mm long and the females 6.5–8.5 mm. External leaf crowns have 14–16 elements and internal leaf crowns 28–32 elements. Spicules are 1.0–1.2 mm long. Females have a short, dorsally bent tail. Eggs are 30–45 by 52–67 μm. The life-cycle is similar to that of *O. dentatum.*

Oesophagostomum georgianum occurs in the large intestine of pigs in the southern USA. It is similar to *O. brevicaudum*. Females are 10 mm long with a ventrally curved tail.

Oesophagostomum quadrispinulatum (syn. *O. longicaudum*) occurs in the large intestine of pigs in North America, Europe, South America and the Philippines. It is similar to *O. dentatum*. The oesophagus is more slender than *O. dentatum* and the tail of the female is 486 μm in length compared with 256 μm in the case of *O. dentatum* (Kendall et al. 1977). The life-cycle is direct, fourth-stage larvae occur four days after infection, adults are present in the large intestine on day 14 and eggs are produced on day 33. The occurrence of cysts containing third-stage larvae is uncommon in *O. quadrispinulatum* infection and Kendall et al. (1977) ascribe this response to infection with *O. dentatum*.

Oesophagostomum granatensis has been described from the pig in Spain, but there is doubt whether it is specifically distinct from *O. dentatum* (Graber et al. 1970).

Other species of the genus which occur in swine include **O. maplestonei** Schwartz, 1931 of the pig in India, **O. rousseloti** Diaouré, 1964 of the pig in Zaire and **O. hsiungi** Ling, 1959 of the pig in China.

Pathogenesis. Nodule formation caused by the larval stages of *O. dentatum* is responsible for enteritis, anorexia and blood-stained faeces (Davidson & Taffs 1965). Severe infections may cause death. The infection is considered to be an important constraint on pig production in the USA. Diagnosis is based on the detection of characteristic eggs in the faeces. A postparturient increase in egg output may occur (Connan 1966), which may also be associated with loss of weight and reduced milk yield.

Treatment. Several compounds are available. Thiabendazole (100 mg/kg) is highly effective; levamisole (8 mg/kg) is given in the water or feed; cambendazole (20–40 mg/kg) is completely effective; parbendazole (20–30 mg/kg in the feed) is completely effective; dichlorvos (40 mg/kg) is completely effective, as are haloxon (35–50 mg/kg) and oxfendazole (4.5 mg/kg) (Kingsbury et al. 1981). Piperazine salts given in drinking water (110 mg/kg) are highly effective.

NODULAR WORMS OF PRIMATES

Oesophagostomum aculeatum (Linstow, 1879)
Oesophagostomum bifurcum (Creplin, 1849)
Oesophagostomum stephanostomum Stossich, 1904

Oesophagostomum aculeatum (syn. *O. apiostomum*) occurs in the large intestine of macaques (*M. irus*, *M. sinica* etc.) and the capuchin monkey in South-East Asia and Indonesia. It was common in macaques and cynomolgus monkeys (e.g. 70% of rhesus) imported to the USA for experimental purposes. It is also reported from imported chimpanzees. Males are 8–10 mm long and females 8.5 mm. Cervical papillae behind middle of oesophagus; three teeth in oesophageal funnel; external leaf crown composed of 10 pointed elements alternating with minute spines. Spicules are 1.15 mm long and eggs 27–40 by 60–63 μm. The developmental cycle is similar to that of other members of the genus. Larvae enter the wall of the caecum, cause nodule formation and then mature in the lumen of the large intestine 30–40 days after infection.

Oesophagostomum bifurcatum occurs in the small and large intestine of various primates (*Papio*, *Cercopithecus*, *Macaca* species) in Africa and Asia. It is common in imported monkeys. A few human infections have been reported. It is similar to *O. aculeatum*; males and females are 8–10 mm long, spicules less than 1.15 mm and eggs similar in size to *O. aculeatum*. The life-cycle similar to that of *O. aculeatum*.

Oesophagostomum stephanostomum occurs in the large intestine of gorilla, chimpanzee and various other monkeys in Africa and South America. Human infection has been reported. Similar in size to the other species of primates; leaf crown composed of 38 elements; eggs 40–55 by 60–80 μm. Life cycle similar to that of *O. aculeatum*. Heavy infections are responsible for diarrhoea, loss of weight and death. In animals under stress in captivity, moderate burdens may assume clinical importance. Thus Rousselot and Pellissier (1952) reported that *O. stephanostomum* was a common cause of death of captive gorillas. At necropsy nodules are seen in the wall of the large intestine and may be white or black due to haemorrhage (Vickers 1969).

Diagnosis is based on the clinical signs and the presence of eggs in the faeces; these should be differentiated from the eggs of *Ancylostoma* and *Ternidens* spp.

Thiabendazole, 100 mg/kg, given orally, i.e. in the food, is effective. It should be repeated at two weeks. It is likely that other benzimidazole compounds will be effective. For control, improved hygiene and management will prevent transmission of infection in imported animals.

REFERENCES

NON-EQUID STRONGYLIDAE AND TRICHONEMATIDAE

Bremner, K. C. (1961) A study of the pathogenic factors in experimental bovine oesophagostomosis. I. An assessment of the importance of anorexia. *Aust. J. agric. Res.*, **2**, 498–512.

Connan, R. M. (1966) A post-parturient rise of faecal nematode egg counts in sows. *Vet. Rec.*, **79**, 156–157.

Davidson, J. B. & Taffs, L. F. (1965) Gastro-intestinal parasites in pigs. *Vet. Rec.*, **77**, 403.

Dobson, C. (1967*a*) The effects of different doses of *Oesophagostomum columbianum* larvae on the body weight, intake and digesibility of feed and water intake of sheep. *Aust. vet. J.*, **43**, 291–296.

Dobson, C. (1967*b*) Pathological changes associated with *Oesophagostomum columbianum* infestations in sheep: serum protein changes after first infestation. *Aust. J. agric. Res.*, **18**, 821–831.

Flynn, R. J. (1973) *Parasites of Laboratory Animals*. Ames: Iowa State University Press.

Goldberg, A. (1951) Life history of *Oesophagostomum venulosum*, a nematode parasite of sheep and goats. *Proc. helminth. Soc. Wash.*, **18**, 36–47.

Goldberg, A. (1952) Effects of the nematode *Oesophagostomum venulosum* on sheep and goats. *J. Parasit.*, **38**, 35–47.

Gordon, H. McL. (1950) Some aspects of parasitic gastroenteritis of sheep. *Aust. vet. J.*, **26**, 14–28, 46–52, 65–72, 93–98.

Graber, M., Raynaud, J. P. & Euzeby, J. (1970) Les oesophagostomes due porc en France. *Bull. Soc. Sci. vet. med. Lyon*, **72**, 423–442.

Herd, R. P. (1971) The parasitic life cycle of *Chabertia ovina* (Fabricius, 1788) in sheep. *Int. J. Parasit.*, **1**, 189–199.

Kendall, S. B., Small, A. J. & Phipps, L. P. (1977) *Oesophagostomum* species in pigs in England. I. *Oesophagostomum quadrispinalatum* description and life history. *J. comp. Path.*, **87**, 223–229.
Kingsbury, P. A., Rowlands, D. ap T. & Reid, J. F. S. (1981) Anthelmintic activity of oxfendazole in pigs. *Vet. Rec.*, **108**, 10–11.
Kotlán, A. (1948) Studies on the life-history and pathological significance of *Oesophagostomum* spp. of the domestic pig. *Acta vet. Hung.*, **1**, 14–30.
Roberts, F. H. S., Elek, P. & Keith, R. K. (1962) Studies on resistance in calves to experimental infections with the nodular worm *Oesophagostomum radiatum* (Rudolphi, 1803) Raieliet, 1808. *Aust. J. Agric. res.*, **13**, 551–573.
Round, M. C. (1968) Check list of the helminth parasites of African mammals. Tech. Comm. No. 38. Farnham Royal: Commonwealth Agricultural Bureaux.
Rousselot, R. & Pellissier, A. (1952) Pathologie du gorille. *Bull. Soc. Path. exot.*, **45**, 565–574.
Vickers, J. H. (1969) Disease of primates affecting the choice of species for toxological studies. *Ann. N.Y. Acad. Sci.*, **162**, 659–672.
van der Westhuysen, O. P. (1938) A monograph of the Helminth Parasites of the Elephant. *Onderstepoort J. vet. Sci. Anim. Indust.*, **10**, 49–190.

FAMILY: AMIDOSTOMIDAE BAYLISS & DAUBNEY, 1926

Species of the family are parasitic in the mucosa of the gizzard, proventriculus and crop of geese and ducks.

Genus: Amidostomum Railliet & Henry, 1909

The shallow, broad buccal cavity has no leaf-crowns. The oesophagus is lined by three longitudinal ridges or plates. The male spicules are short, with bifurcate or trifurcate ends. The vulva of the female is in the posterior part of the body.

Amidostomum anseris (Zeder, 1800) (syn. *A. nodulosum*) occurs world-wide in domestic and wild geese and ducks in the mucosa of the gizzard and sometimes also in the proventriculus and the oesophagus. The worms are slender and reddish in colour. The male is 10–17 mm long and the female 12–24 mm long. The buccal capsule is short, wide and thick-walled, with three pointed teeth at its base. The equal spicules are 0.2–0.3 mm long and each ends in two branches. The vulva is situated at the posterior fifth of the body and may be covered by a flap. The eggs measure 100–110 by 50–60 μm and contain a segmenting embryo when laid.

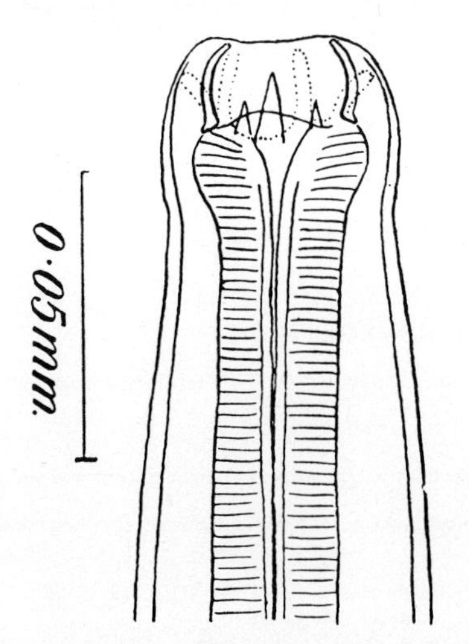

Fig 1.108 Dorsal view of the anterior end of *Amidostomum anseris*.

The life-cycle is direct, like that of other strongyles. Kobulej (1956) has stated that development to the infective third larval stage occurs inside the egg. However, Cowan and Herman (1956) showed that infective larvae could develop outside the egg. Development to the infective stage is rapid, under ideal conditions infective larvae being produced in as little as 48 hours. Infection is by ingestion and larvae remain in the lumen of the gizzard for a few days, then enter the submucosa, reaching patency in as little as 14 days in young goslings, but usually 25 days in older birds.

The worms are very pathogenic to young geese, while the adult birds may act as carriers of the infection without showing clinical signs. The parasites burrow into the mucous and submucous tissues of the gizzard and proventriculus, sucking blood and causing marked irritation, inflammatory changes and haemorrhage. In severe cases there is extensive necrosis of the gizzard, and the horny lining is a reddish brown friable mass, in which numerous worms are embedded.

Heavily infected birds lose their appetite and become emaciated, anaemic, weak and easily tired. Diarrhoea is not unusual. Death is often preceded by prostration.

Diagnosis is best made by autopsy of a selected case.

Previously carbon tetrachloride (2–3 ml in a mixture of flour and water) was used. However, more recent anthelmintics, including the benzimidazole compounds, are quite effective. Thus tetramisole (30–40 mg/kg), levamisole (25 ml/kg in drinking-water) and cambendazole (60–80 mg/kg) are available. Pyrantel tartrate (50 mg/kg) and trichlorphon (75 mg/kg) are also effective.

General hygienic measures are indicated and special attention should be given to young ducks and geese. Treatment of the adult birds before the breeding season is recommended.

Other members of the genus include **Amidostomum skrjabini** Boulenger, 1926, of the gizzard of domestic and wild duck throughout the world; **A. cygni** Wehr, 1933, of the crop of swan in North America; **A. acutum** Seurat, 1918, of the gizzard of teal, eider and scoter; and **A. simile** Freitas and Mendonca, 1954 of the small intestine of the swan in Belgium. The developmental cycle and pathogenesis of these are similar to those of *A. anseris*. Third-stage larvae may develop within the egg in 40 hours in the case of *A. skrjabini* (Leiby & Olsen 1965) and these hatch and reach the infective stage within five days. However, Pande et al. (1964) reported the development to the third stage outside the egg shell.

Genus: Epomidiostomum Skrjabin, 1916

Some authors (e.g. Skrjabin 1952) assign this genus to the family Trichostrongylidae. However, species parasitize the gizzard and proventriculus of ducks and geese similarly to *Amidostomum* species and it is convenient to consider the genus here.

The buccal capsule contains no teeth; the distinctive feature is the presence of 'epaulettes', being cuticular thickenings at the anterior end. Males are 10 mm and females 16 mm long.

Epomidiostomum uncinatum (Lundahl, 1848) Seurat, 1918 (syn. *E. anatinum*) occurs in the gizzard of domestic and wild ducks throughout the world. Other species include: **E. orispinum** Seurat, 1918 of the oesophagus and proventriculus of geese and ducks in North America; **E. skrjabini** Petrov, 1926, oesophagus and proventriculus of domestic and wild geese in the USSR; and **E. vogelsangi** Travassos, 1937 of the swan, in Argentina.

The life-cycle is similar to that of *Amidostomum* spp., third-stage larvae developing within the egg and reaching the infective stage shortly after hatching. On infection, larvae penetrate the lining of the gizzard and reach patency 16–24 days later. The pathogenic effects are similar to those produced by *Amidostomum* spp. (Leiby & Olsen 1965). Treatment is as for *Amidostomum* infection.

FAMILY: STEPHANURIDAE TRAVASSOS & VOGELSANG, 1933

Nematodes with a cup-shaped buccal capsule, containing teeth. Vulva near anus. Parasites of kidney and perirenal tissues. Genus of importance *Stephanurus*.

Genus: Stephanurus Diesing, 1839

Stephanurus dentatus Diesing, 1839, the 'kidney-worm' of swine, occurs in the perirenal fat, the pelvis of the kidney and the walls of the ureters, and as an erratic parasite in the liver or other abdominal organs and sometimes the thoracic organs, as well as the spinal canal, of the pig; it is seen rarely in the liver of cattle and has also been reported from a donkey. The parasite is

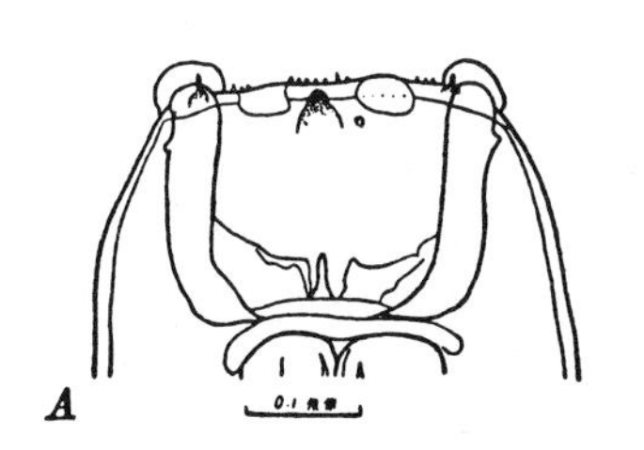

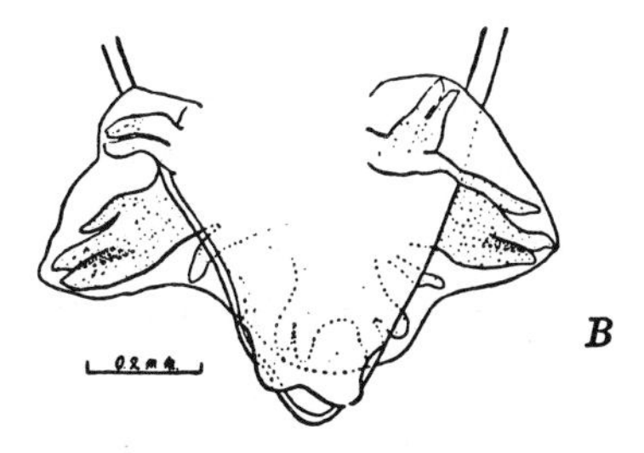

Fig 1.109 *Stephanurus dentatus*. A, Anterior end. B, Ventral view of the posterior end of the male.

widely distributed in tropical and subtropical countries. The male is 20–30 mm long and the female 30–45 mm. The worms are stout, the female being about 2 mm broad, and the internal organs are partly visible through the cuticle. The buccal capsule is cup-shaped and thick-walled with six variable cusped teeth at its base. Its rim bears a leaf-crown of small elements and six external cuticular thickenings or 'epaulettes', of which the dorsal and ventral are the most prominent. The male bursa is small and its rays are short. The two spicules are equal and measure 0.66–1 mm in length. The vulva is situated close to the anus. The eggs are ellipsoidal and thin-shelled and measure 43–70 by 90–120 μm.

Life-cycle. Normally the adult worms are lodged in or near the kidney in cysts which communicate with the ureters, and eggs are passed out in the urine of the host. At this stage the embryo consists of about 32–64 cells. The development of the pre-infective stages is similar to that of *Strongylus* spp. At an optimum temperature of 26°C, eggs hatch after 24–36 hours and the larvae reach the infective stage about four days later after two ecdyses, retaining the second skin as a sheath. The eggs and all larval stages are rapidly killed by freezing and dryness. The infective larvae can live in moist surroundings up to five months, but the majority die off after two or three months.

Infection of the host occurs per os or through the skin. Earthworms (e.g. *Eisenia foetida*) may serve as transport hosts. Infective larvae accumulate in masses in amoebocytes in the earthworm and probably can survive here for several weeks or months (Batte et al. 1960). The sheath of the infective larva is cast soon after infection has occurred and the third ecdysis takes place about 70 hours later, either in the wall of the stomach after oral infection or in the skin and the abdominal muscles after percutaneous infection. The fourth-stage larva possesses a buccal capsule. From both ports of entry the larvae reach the liver: after oral infection via the portal vessels in three days or longer, and after skin penetration via the lungs and the systemic circulation in eight to 40 days. They wander about underneath the liver capsule and eventually, three months or more after infection, penetrate through the latter into the peritoneal cavity. After having reached the perirenal tissues they perforate the walls of the ureters and produce a cyst which continues to communicate with the ureter through a fine canal.

The migrating larvae have a marked tendency to penetrate soft tissues and many of them go astray. After percutaneous infection some remain in the pulmonary capillaries and become encapsulated in the lungs, or they may wander further and reach the pleural cavity and other thoracic organs. In the peritoneal cavity they do not all reach the perirenal tissues, but penetrate into other organs such as the spleen, psoas muscles etc. Penetration of the placenta and prenatal infection has been reported (Batte et al. 1966).

Pathogenesis. Percutaneous infection causes the formation of nodules in the skin, with oedema and enlargement of the superficial lymph glands. These lesions disappear after three or four weeks or longer. The migrating larvae produce lesions of an acute inflammatory nature, especially in the liver. Abscess formation, extensive liver cirrhosis and multiple adhesions may occur (Batte et al. 1966). Aberrant migration of larvae may lead to lesions in the spinal cord. The adult parasite itself is not markedly pathogenic and is found in cysts varying from 0.5 to 4 cm in diameter, each cyst usually containing a pair of worms embedded in green pus. Cysts may occur in the kidney tissue. The ureter is thickened and in chronic cases may be almost occluded.

Clinical signs. The temporary subcutaneous nodules seen in the early stages of the infection may affect the host; precrural nodules, for instance, may cause stiffness of the leg. Posterior paralysis has been ascribed to the parasite. The general effects are a depressed growth rate, loss of appetite and later emaciation. Where cirrhosis of the liver is marked ascites may be present. The infection is a herd problem and the general picture is one of lack of growth and wasting in the pig herd.

Post-mortem. Decomposition takes place rapidly in animals that have died from the results of

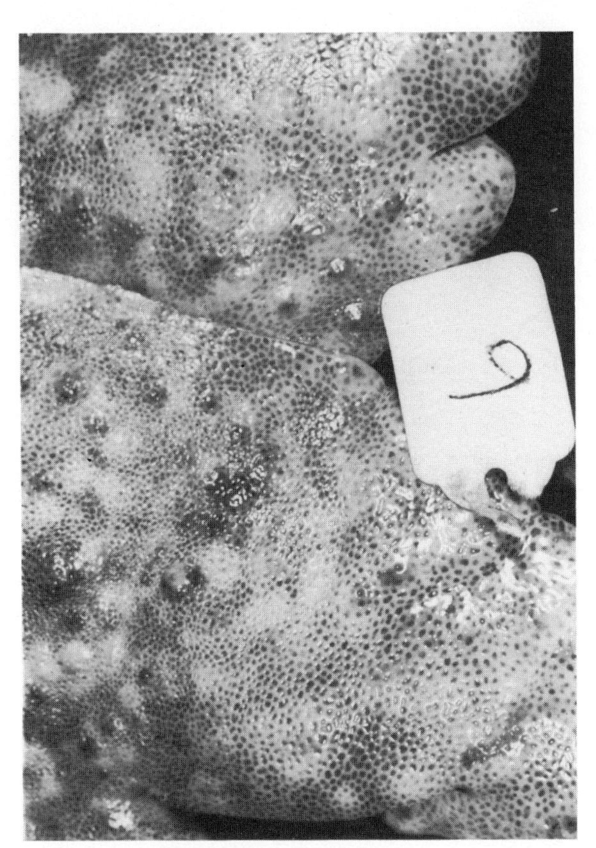

Fig 1.110 Liver cirrhosis associated with the migration of the larval stages of *Stephanurus dentatus* in the pig. (*E. C. Batte*)

this infection. Young and even adult worms may be found in cysts or abscesses in the lungs and other thoracic organs or free in the pleural cavity. The liver is enlarged and its surface is uneven owing to irregular tracks and scars, which are also found deeper in the parenchyma. Cirrhosis may be marked and ascites is then usually present. The perirenal tissues usually show a certain degree of hypertrophy and the surface of the kidneys may bear small scars of healed abscesses. The worms are found in and around the kidneys and ureters, as described above. The portal and mesenteric lymph vessels are enlarged, while in older cases they are indurated and have uneven surfaces.

Diagnosis can be made by finding the eggs of the parasite in the urine of the pig if mature worms are present and in communication with the ureters. In other cases a definite diagnosis can be made at autopsy only. Tromba and Baisden (1960) have described a gel precipitin test for the diagnosis of non-patent *S. dentatus* infection.

Treatment. No satisfactory treatment is available for *S. dentatus* infection. Egerton (1961) has reported that thiabendazole, when incorporated in the food at the rate of 0.1–0.4%, is effective in inhibiting the migration of *S. dentatus* larvae.

Prophylaxis is largely a matter of hygiene. It has been found in the USA that the swine sanitation system used for the control of *Ascaris suum* brought about a considerable decrease in the incidence of *S. dentatus*. A hard, bare strip, 1.5–2 m wide, is provided along the edge of the pasture and the shelters and troughs are placed on a bare patch 10 m wide at the end of the field, since the pigs mostly urinate in these places. Adequate drainage of pig pens and yards is essential. Thus muddy areas and pig wallows should be drained and filled in. Pigs should be fed on a concrete apron which should be cleaned regularly.

Alicata (1953) has reported that the treatment of soil with Polyborate (a mixture of sodium pentaborate tetrahydrate and sodium tetraborate pentahydrate) is successful under laboratory and field conditions in destroying the eggs and larvae of *S. dentatus*. It is applied at the rate of 2.5 kg in 12 litres of water per 10 m^2. Each treatment lasts for about 30 days. The toxicity of Polyborate for pigs is low and they may be allowed access to treated soil without ill-effect. Segregation of clean, especially young, animals from infected ones and periodic unstocking of pens, which are then allowed to dry thoroughly in sunlight, are important measures.

Since patency of *S. dentatus* does not usually occur until animals are two years old a policy of 'gilts-only' breeding in which gilts are bred only once and then sold before kidney worms develop to maturity, is advocated by Stewart et al. (1964).

FAMILY: SYNGAMIDAE LEIPER, 1912

Nematodes with a cup-shaped buccal capsule, without leaf-crowns. Teeth may be present.

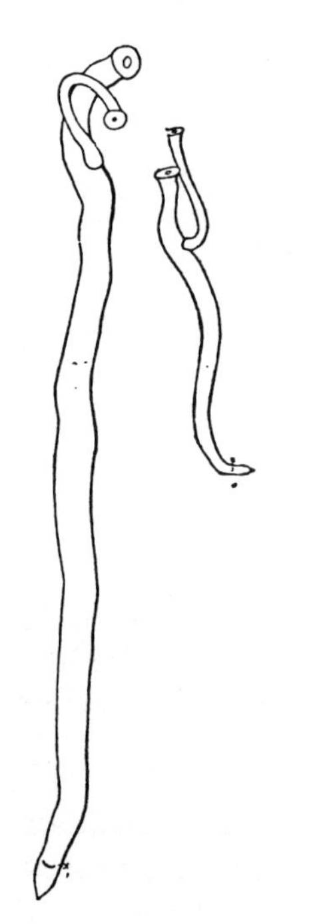

Fig 1.111 Outline of two *Syngamus trachea*. *Left*, Gravid female. *Right*, Immature female.

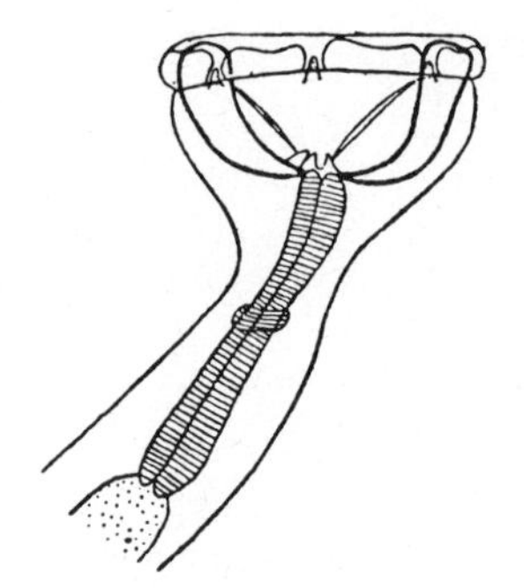

Fig 1.112 Anterior end of *Syngamus trachea*.

Vulva in anterior part of the body. Parasites of respiratory tract.

Genera of importance include *Syngamus*, *Cyathostoma* and *Mammomonogamus*.

Genus: Syngamus v. Siebold, 1836

Syngamus trachea (Montagu, 1811), the 'gapeworm', occurs throughout the world in the trachea of the turkey, fowl, pheasant, guinea-fowl, goose and various wild birds. The parasites are bright red in colour when fresh and the sexes are found permanently in copulation. The male is 2–6 mm long and the female 5–20 mm. The mouth opening is wide, without leaf-crowns, and the buccal capsule is cup-shaped, bearing six to ten small teeth at its base. The male bursa has short, stout rays. The spicules are 53 × 82 μm long, equal and simple in shape. The eggs, which measure 70–100 by 43–46 μm, have a thickened operculum at either pole and are ejected from underneath the male bursa in the 16-celled stage.

Syngamus skrjabinomorpha Ryzhikov, 1949 occurs in the trachea of the domestic goose and chicken in the USSR. Various other species occur in wild birds throughout the world.

Life-cycle. The eggs of the worms are usually coughed up and swallowed by the host and they pass out in the faeces. The infective larva develops inside the egg, requiring, under optimal conditions of moisture and temperature, about three days, but under field conditions usually one to two weeks. The larva moults twice in the egg; it retains the cuticle of the previous stage as a sheath and has a short, pointed tail and a relatively long oesophagus. The infective larva may hatch from the egg and if it does so it soon becomes inactive and shows no negative geotropism; further it is not able to resist desiccation. Alternatively, hatching may not occur; the infective larva is then more protected. Infection occurs on ingestion of the egg or free infective larva. The infective larva may be swallowed by earthworms, snails, slugs, flies and other arthropods, in which they become encysted, and here they may live for several months or even years. Such invertebrates may become heavily infected and act as important transport hosts. According to Morgan and Clapham (1934), passage through earthworms renders the larvae more highly infective, enabling strains from wild birds to pass to chicks more readily than otherwise.

The larvae are carried by the blood to the lungs, which they may reach in six hours. In the lungs they are found in the alveoli. A further ecdysis occurs on the third day after infection, after which the sexes can be differentiated. The final ecdysis occurs on the fourth or fifth day and the young worms migrate to the larger bronchi, where copulation takes place. The trachea is reached as early as the seventh day and patency occurs 17–20 days after infection.

Pathogenesis. In heavy infections migration through the lungs may cause ecchymoses, oedema and even lobar pneumonia. In the trachea the worms attach themselves to the mucosa and suck blood, with consequent catarrhal tracheitis and the secretion of much mucus. The mucus may occlude the air passages causing difficulty in breathing. The males become deeply embedded with their anterior ends in the wall of the trachea of turkeys, causing the development of nodules.

Clinical signs. Gapeworm disease chiefly affects young birds. Adult hens are not usually infected and guinea-fowl are little affected at any age. Turkeys are susceptible to infection at any age and some consider them to be the natural hosts of *S. trachea*. Heavy burdens may occur in turkeys kept in straw yards. They may, however, be dangerous carriers of the disease, from which other birds may acquire infection. Hen-chicks, goslings and artificially reared game birds, especially pheasants, suffer most from the parasites. The characteristic signs of 'gapes' are those of dyspnoea and asphyxia, occurring in spasms on account of the accumulation of mucus in the trachea. The bird shakes and tosses its head about and it may cough, or it extends the neck, opens the beak and performs gaping movements. Death results from asphyxia during such an attack or from progressive emaciation, anaemia and weakness caused by the parasite.

Post-mortem. The carcass is emaciated and anaemic and the worms are found in the posterior part of the trachea, attached to the mucosa and surrounded by mucus which may be streaked with blood.

Diagnosis can be made from the clinical signs and by finding in the faeces the characteristic eggs, which must be differentiated from those of *Capillaria*. The diagnosis will be further confirmed by finding the worms at the autopsy of a selected case.

Treatment. Thiabendazole at a dose of 0.3–1.5 g/kg orally has been used successfully to treat chickens and turkeys. Continuous feeding of 0.05% thiabendazole in the food for seven days is effective in pheasant and partridge and 6 mg/kg of food for 48 days will prevent infection in pheasants. Mebendazole and fenbendazole are effective at levels of 0.01% in the feed for seven to 14 days.

Prevention and control. Severe outbreaks of gapeworm disease are less likely to occur if the birds are not kept for long periods on the same ground. Moist localities where earthworms, slugs and snails occur should be avoided if possible. Rearing pens, including strawyards, should be kept dry and access by wild birds controlled. Rooks and starlings are important in this respect. Turkeys should not be kept on the same ground as chickens or even near to them. Infected pens or yards may be treated with D-D (a mixture of 1,3-dichloropropylene and 1,2-dichloropropane) or Dowfume W-40 (42% ethylene debromide in petroleum extendor) at the rate of 1.5 litres/16 m^2 (Herman & Kramer 1950).

Genus: Cyathostoma E. Blanchard, 1849

Members of this genus occur in the trachea and bronchi and in some cases in the nasal and orbital cavities of geese, ducks, swans and gulls. Worms are not permanently in copulo. The buccal capsule is cup-shaped and has six or seven teeth at its base. The bursae are wide and deep and the spicules filamentous.

Cyathosoma bronchialis (Muehlig, 1884) occurs in the trachea and bronchi of the goose, duck and swan in Europe and North America. The male is 4–5.8 mm long and the female 16–31 mm. The eggs measure 74–83 by 49–62 μm.

The life-cycle is similar to that of *S. trachea*; the infective third-stage larva is reached within the egg. Following ingestion larvae migrate to the bronchi and trachea through the peritoneal cavity and air sacs. Worms occur in the trachea six days after infection (Fernando et al. 1973). Earthworms may be important in the life-cycle.

The parasite is very harmful, especially to young geese and ducks. Even two or three worms may be responsible for death by asphyxiation.

Treatment and prophylaxis are as for *S. trachea*.

Other species of this genus include **C. lari** Blanchard, 1849 of the gull; **C. brantae** Cram, 1928 of the trachea of gulls in North America; and **C. variegatum** Chapin, 1825, which occurs in storks and cranes.

Genus: Mammomonogamus Ryjikov, 1948

Parasites in this genus are similar to *Syngamus*, with the male and female joined in copulo. A cervical papilla is present and the buccal capsule has no cuticular crown. The eggs are not operculate. They are parasites of the trachea and nasal sinuses of mammals.

Mammomonogamus laryngeus (Railliet, 1899) Ryjikov, 1948 occurs in the larynx of cattle and water buffalo, occasionally goat and deer, in India, Malaysia, Vietnam and South America. Human infections have been reported from the Caribbean and South America. The worms are red in colour and joined in copulo. The males are 3–3.5 mm long and the females 8.5–10 mm. The eggs are ellipsoidal, 42–45 by 75–85 μm. The life-cycle is assumed to be similar to that of *S. trachea*. The parasite is not a serious pathogen. The attachment of several worms to the larynx may cause coughing and some loss of condition. In man, coughing, with haemoptysis, may occur.

Mammomonogamus nasicola (von Linstow, 1899) occurs in the nasal cavities of cattle, sheep, goats and deer in Brazil, West Indies, Turkestan and Cameroon. Some authors consider it a synonym of *O. laryngeus*. Males are 4–6 mm long and females 11–23 mm; eggs are 54 by 98 μm. Life-cycle and pathogenesis are unknown.

Other species of the genus include **M. auris** (Faust and Tang, 1934) Ryjikov, 1948 in the trachea of the puma in Brazil; **M. ierei** (Buckley, 1934) Ryjikov, 1948 in the nasal cavity of cats in the West Indies; **M. indicus** (Mönnig, 1932) Ryjikov, 1948 of the pharynx of the Indian elephant; **M. loxodontus** Vuylsteke, 1935) Ryjikov, 1948 of the trachea of the African elephant and **M. mcgaughei** (Seneviratne, 1954) of the pharynx, nasal and frontal sinuses of the cat in Sri Lanka.

SUPERFAMILY: ANCYLOSTOMATOIDEA CHABAUD, 1965

The buccal capsule is subglobular, lips and leaf crowns are absent; oral opening unarmed or with teeth and cutting plates.

FAMILY: ANCYLOSTOMATIDAE LOOSS, 1905

Strongylida with a well-developed buccal capsule, which is devoid of leaf-crowns, but is armed on its *ventral* margin either with teeth or with chitinous cutting plates. The anterior extremity is usually bent in a dorsal direction. The male bursa is normally developed. Most species are voracious blood-suckers and are parasitic in the small intestine. Two subfamilies occur: *Ancylostominae* and *Necatorinae*.

SUBFAMILY: ANCYLOSTOMINAE STEPHENS, 1916

The buccal capsule bears on its ventral margin one to four pairs of teeth. Inside the buccal capsule there are two dorsal teeth. The dorsal gutter does not project into the cavity of the buccal cavity to form a dorsal cone. Species are *Ancylostoma*, *Agriostomum*.

SUBFAMILY: NECATORINAE LANE, 1917

The ventral margin of the buccal capsule bears, instead of teeth, cutting plates (semi-lunes), but these are usually absent from *Globocephalus*. There are also subventral teeth inside the buccal capsule. Dorsal teeth are absent, but small subdorsal (lateral) teeth may be present. The end of the dorsal gutter carrying the duct of the dorsal oesophageal gland projects into the buccal capsule as a prominence called the *dorsal cone*.

Species are *Necator*, *Bunostomum*, *Gaigeria*, *Globocephalus*, *Uncinaria*, *Grammocephalus*, *Bathmostomum*.

SUBFAMILY: ANCYLOSTOMINAE STEPHENS, 1916

Genus: Ancylostoma (Dubini, 1843) Creplin, 1845

Ancylostoma caninum (Ercolani, 1859) Hall, 1913 occurs in the small intestine of the dog, fox, wolf, coyote and other wild carnivores, and very rarely also in man. It is cosmopolitan in distribution, being common in tropical and subtropical zones in North America, Australia and Asia. It is rare in Great Britain, usually being found in imported animals only. The male is 10–12 mm long and the female 14–16 mm long. The worms are fairly rigid and grey or reddish in colour, depending on the presence of blood in the alimentary canal. The anterior end is bent dorsad and the oral aperture is directed anterodorsally. The buccal capsule is deep. There is no dorsal cone. The dorsal gutter ends in a deep notch on the dorsal (posterior) margin of the buccal capsule, the ventral margin of which bears three teeth on either side. In the depth of the capsule there is a pair of triangular dorsal teeth and a pair of centrolateral teeth. The male bursa is well developed and the spicules are 0.8–0.95 mm long. The vulva is situated near the junction of the second and last thirds of the body. The uteri and ovaries form numerous transverse coils in the body. The eggs measure 56–75 by 34–47 μm and contain about eight cells when passed in the faeces.

Ancylostoma tubaeforme (Zeder, 1800). The normal hookworm of the cat. For many years it has been accepted that *A. caninum* occurs in the cat but studies by Biocca (1954) in Italy, by Burrows (1962) in the USA, and others, demonstrated *A. tubaeforme* to be a valid species. The two species are not interchangeable between the two hosts. Apparently cosmopolitan in distribution, but absent from Great Britain. The male is 9.5–11.0 mm, female 12.0–15 mm. Mouth capsule similar to *A. caninum*, but teeth on the ventral margin slightly larger. Spicules of the male are larger than 1 mm (1.23–1.4 mm) (compare *A. caninum*). Eggs are 55–75 by 34.4–44.7 μm.

Ancylostoma braziliense Gomez de Faria, 1910 occurs in the small intestine of the dog, cat, fox and several wild Canidae and sometimes man. It occurs in most tropical and subtropical countries. It is slightly smaller than *A. caninum*,

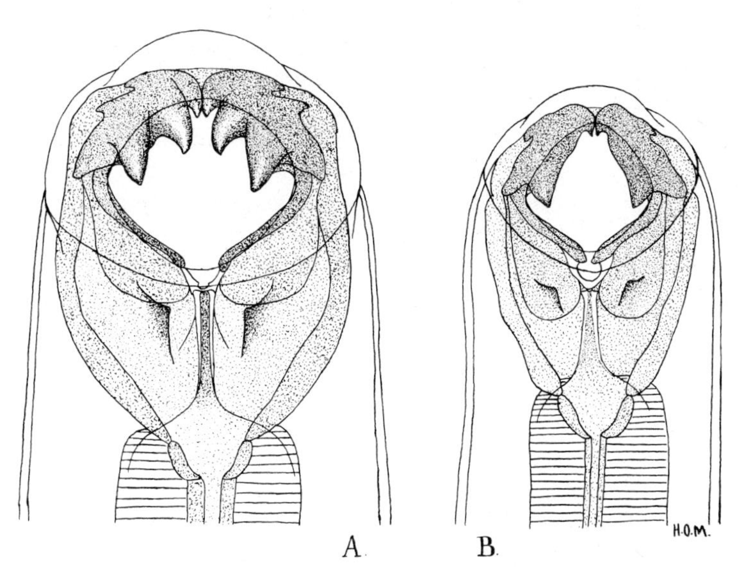

Fig 1.113 Dorsal view of the anterior end of *Ancylostoma caninum* (A) and *Ancylostoma braziliense* (B).

the males measuring 6–7.75 mm and the females 7–10 mm. It can be differentiated from the previous species by the fact that its ventral teeth consist of a large and a small one on either side. The eggs measure 75–45 μm.

Ancylostoma ceylanicum (Looss, 1911) Leiper, 1915 occurs in the dog, cat, civet cat, tiger and occasionally man in Sri Lanka, Malaysia and other parts of Asia. For a long time considered to be a synonym of *A. braziliense*, it differs from it by features of the buccal capsule and bursa. The inner pair of ventral teeth in the mouth are larger than those *A. braziliense* and the origin and direction of the rays of the bursa differ (Biocca 1951).

Ancylostoma duodenale Dubini, 1843, one of the hookworms of man in Europe, Africa, western Asia, China and Japan, has also been recorded from certain wild Carnivora and the pig. Experimental infection of young dogs and cats has been successful, but the parasite does not naturally occur in these hosts. The ventral teeth of this worm consist of two large and one small one on either side.

Other species in the genus include **A. kusimaensis** Nayayoshi, 1955 of the dog and badger in Japan and **A. paraduodenale** Biocca, 1951 of the small intestine of lion and other carnivores in Africa.

Life-cycle of Ancylostoma spp. in the dog

Adult female hookworms produce an average of 16 000 eggs per day, though the number is inversely proportional to the number of adults present (Krupp 1961). The pre-infective stages develop and behave biologically similarly to those of *Strongylus* spp., but none are resistant to desiccation, so that they are therefore found only in moist surroundings. The most suitable is a slightly sandy, moist soil; clay and gravel are not suitable. The optimal temperature for development is between 23 and 30°C for *A. caninum* while that for *A. ceylanicum* and *A. braziliense* is slightly higher. At these temperatures the infective stage larva is reached within a week, but where temperatures are low development is more prolonged (e.g. 22 days at 15°C; nine days at 17°C). Infection of a new host occurs either by the ingestion of infective larvae or by skin penetration by them. A variety of developmental pathways may occur following infection:

1. Oral infection (e.g. experimentally by stomach tube or gelatine capsule) may lead to development directly to adult worms; when larvae are placed in the mouth a proportion will penetrate the buccal and pharyngeal epithelia and undergo migration as if skin penetration had occurred.
2. Skin penetration leads to migration to the lungs and, by 'tracheal migration', to the intestine. Subsequent maturation may occur or, in other animals, there may be somatic migration of larvae to, followed by dormancy in, the musculature.
3. Prenatal infection of the fetus by intra-uterine infection.
4. Colostral or lactogenic infection of pups by the passage of larvae through milk to the suckling puppy.

After oral infection larvae which do not migrate systemically enter the gastric glands or the crypts of Lieberkühn for a few days and then return to the lumen where they moult to the fourth stage (about three days after infection). Patency is reached 15–18 days after infection in young dogs (Herrick 1928) and worms may persist for an average of six months. However, Schad (1977) has reported that developmental inhibition of parasitic larval stages may occur in the small intestine when infective (pre-parasitic) larvae are exposed to sudden chilling and resulting worm burdens in dogs may consist of 60–70% arrested larval stages. Kelly et al. (1976) have also reported arrested development of larvae in the intestine. The ability to arrest is determined largely by a physiological change in the larvae induced by chilling, while the immunological status of the dog determines the number of arrested larvae that establish rather than the capacity to arrest.

In puppies up to approximately three months of age, larvae which penetrate the skin or oral mucous membrane (aided by collagenase-like and other enzymes) reach blood vessels or lymphatic vessels and are carried by these via the venous system or thoracic duct to the heart and lungs.

Larvae pass into the alveoli and migrate up the bronchioles, bronchi and trachea and are then swallowed and mature in the small intestine. The moult to the fourth larval stage occurs after the larvae leave the alveoli (48 hours) and fourth-stage larvae are found in the intestine in large numbers up to the fourth day after infection. The fourth moult to the immature adult occurs on the sixth day, reproductive organs are evident in the adult worms on the twelfth day and mature worms occur by 17 days after infection.

However, in older animals, even those which have not been infected previously, fewer larvae mature and incoming infective larvae follow a somatic migratory route and become dormant in the muscles. For example, in 11-month-old hookworm-naive dogs 53% and 81% of an infection in males and females respectively failed to mature. The specific site of storage was identified as the muscle tissue by Lee et al. (1975). Stoye (1973) has shown in helminth-naive bitches that third-stage larvae of *A. caninum* survive in the musculature for at least 240 days. It would appear that such larvae constitute a reservoir for the population of the mammary gland at the onset of lactation or the repopulation of the intestine when existing adult worm burdens are eliminated. No information is available on the fate of larvae which would arrest in the intestine after chilling in the infective stage should they gain access by skin penetration.

Prenatal infection as a result of intrauterine infection was previously considered to be a common route of infection (Foster 1932). However, the demonstration of the transfer of larvae by colostrum and milk has necessitated a re-evaluation of this and Miller (1970*a*) has reported that under experimental circumstances less than 2% of neonatal infections can be ascribed to intrauterine infection. Miller (1971) was unable to induce prenatal infection in puppies with *A. braziliense*.

When prenatal infection does occur, presumably larvae enter the blood stream of pregnant bitches and pass to the placenta and enter the fetus. The relationship of dormant muscle larvae to intrauterine infection of the fetus has yet to be determined.

Colostral or lactogenic infection with *A. caninum* was reported by Enigk and Stoye (1967) and Stone and Girardeau (1968) and reviewed by Stone and Smith (1973). The designation 'colostral' may be inappropriate in that larvae may be recovered from the milk of bitches up to 20 days after whelping. Miller (1971) has been unable to demonstrate this route of infection in *A. braziliense* or *Uncinaria stenocephala* (see below). The origin of larvae responsible for lactogenic infection, presumably, is the pool of dormant muscle larvae; however, more studies are necessary to establish this.

An additional source of infection, the significance of which is unknown, is that of paratenic hosts bearing infective larvae. Rodents, for example, may accumulate third-stage larvae in their tissues, which when eaten will lead to patent infections (Miller 1970*b*).

The life-cycles of *A. braziliense*, *A. ceylanicum* and *A. tubaeforme* are presumed, in general, to be similar to that of *A. caninum*, except in the instances mentioned above (e.g. failure to induce prenatal infection in *A. braziliense*). More work is necessary to delineate their detailed life-cycles. With *A. ceylanicum* the prepatent period in dogs and cats is 14–17 days (Yoshida 1968), but in man it is reported to be 18–26 days (Yoshida et al. 1972).

The life-cycle of *A. duodenale* is similar to that of *A. caninum*, in that oral and percutaneous infection may occur, the latter being the more common route of infection. Prepatent periods for the infection in man have varied (Miller 1979) (e.g. 38–41 days; 45–74 days; 43–162 days) and may be very extended, since Schad et al. (1973) have observed a strain of *A. duodenale* from India that undergoes arrested development, which is an adaptation to a seasonally unfavourable external environment. This strain has been maintained in dogs for at least six generations by Schad (1979). *Ancylostoma duodenale* also will develop in various primates (monkey, chimpanzee) and Miller (1979) has reviewed work on this aspect.

Genus: Agriostomum Railliet, 1902

Agriostomum vryburgi Railliet, 1902 occurs in the small intestine of the zebu (*Bos indicus*) and

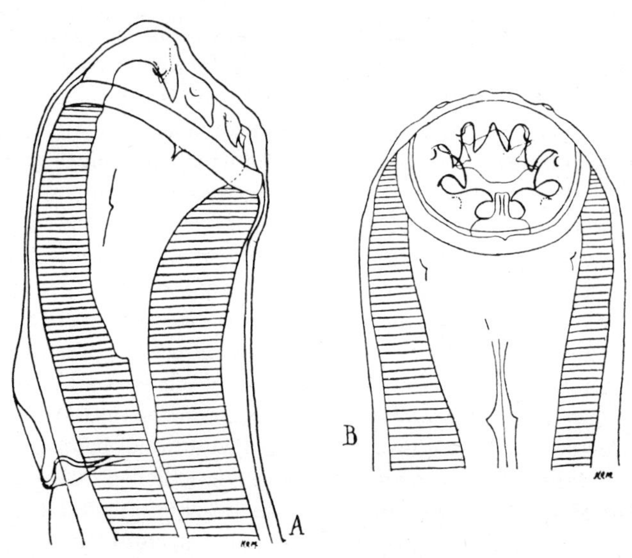

Fig 1.114 Anterior end of *Agriostomum vryburgi*. A, Lateral view. B, Dorsal view.

the ox in India and Sumatra. Males are 9.2–11 mm long and females 13.5–15.5 mm. The buccal capsule opens anterodorsally and is relatively shallow. It is followed by a very large oesophageal funnel which contains two small subventral lancets. The oral margin is provided with four pairs of large teeth and a rudimentary leaf-crown. Spicules are equal, 0.83–0.87 mm long and accompanied by a gubernaculum. The vulva is posterior and eggs measure 125–195 by 60–92 μm.

The life-cycle is probably direct.

Little is known about pathogenicity, but the parasites probably cause anaemia and diarrhoea in heavy infections.

Other members of the genus occur in various antelope species in Africa (e.g. **A. cursoni** Mönnig, 1932; **A. equidentatum** Mönnig, 1929; **A. gungunis** Le Roux, 1929).

SUBFAMILY: NECATORINAE LANE, 1917

Genus: Necator Stiles, 1903

Species of this genus have one pair of ventral cutting plates and, at the base of the buccal capsule, one pair of ventral teeth and one pair of subdorsal teeth. The termination of the dorsal gutter projects into the buccal capsule as a prominence called the *dorsal cone*.

Necator americanus (Stiles, 1902) is a common hookworm of man, occurring in most warm climates, especially in America and Africa. It has also been recorded from the dog and the pig. Good experimental infections have been achieved in the dog by Miller (1966) using corticosteroid treatment. The parasite is relatively unimportant from the veterinary aspect, but it is

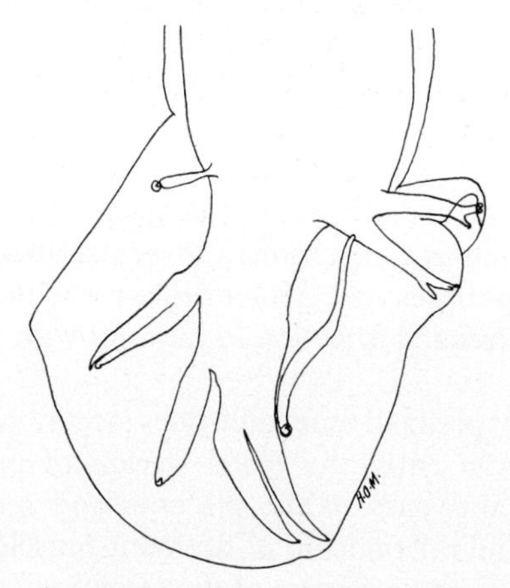

Fig 1.115 Lateral view of hind end of male *Necator americanus*.

a cause of hookworm disease in man in India, the Far East, Australia and the southern United States, to which countries it probably spread from Africa. In the United States it is the predominant human hookworm.

N. suillus Ackert & Payne, 1922, reported from the pig in Trinidad and possibly from South Brazil, is now considered to be *N. americanus*.

Genus: Uncinaria Fröhlich, 1789

Uncinaria stenocephala (Railliet, 1884) is a hookworm found in the dog, cat and fox in temperate climates, for example Europe and North America. The male is 5–8.5 mm long and the female 7–12 mm. There is a pair of chitinous plates at the ventral border of the large, funnel-shaped buccal capsule. Near the base of the buccal capsule there is a pair of subventral teeth. There are no dorsal teeth in the buccal capsule. The dorsal cone does not project into the buccal capsule. The male bursa is well developed and has a short dorsal lobe and two large and separate lateral lobes. The externodorsal rays arise at the base of the dorsal ray, which is cleft for about half its length, the two branches being bidigitate or tridigitate. The anterolateral ray diverges from the other lateral rays. The spicules are slender and 0.64–0.76 mm long. The eggs resemble those of *A. caninum* but are slightly longer and decidedly stouter, being 65–80 by 40–50 μm.

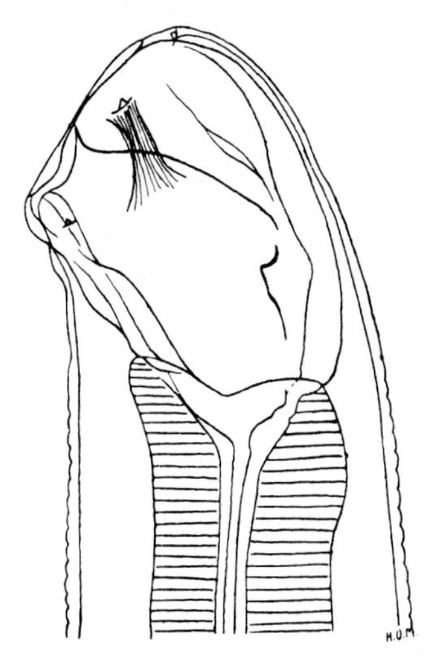

Fig 1.116 Lateral view of anterior end of *Uncinaria stenocephala*.

In general the life-cycle resembles that of *A. caninum*, except that oral infection appears to be the normal and most successful route and few larvae reach the intestine after percutaneous infection (Gibbs 1958). Miller (1971) was unable to produce prenatal or colostral infection in newborn puppies with *U. stenocephala*. Gravid females are present in the intestine of dogs 15 days after infection and eggs are passed in the faeces one to two days afterwards.

Uncinaria criniformis (Goeze, 1782) Railliet, 1899 occurs in the badger and fox in Europe.

Uncinaria lucasi Stiles and Hassall, 1901 occurs in the fur seal in Alaska. Studies of the life-cycle of this parasite established the concept of lactogenic transmission of hookworms (Olsen & Lyons 1965).

ANCYLOSTOMIASIS AND UNCINARIASIS IN DOGS, CATS AND FOXES

Pathogenesis

Miller (1971) has given details of the pathogenesis of the four species of canine hookworm. He emphasizes that the disease syndrome differs for each species of hookworm, that caused by *A. caninum* being the most severe.

Anaemia is the principal consequence of *A. caninum* infection and this is related to the blood loss in the intestine which is associated with the feeding habits of the adult parasites. Further, the severity of the clinical disease is related to the intensity of infection, age, nutritional status, iron reserves and presence of acquired immunity. Initially, the anaemia is normocytic and normochromic but as the animal becomes iron-deficient a microcytic hypochromic anaemia supervenes.

Animals most severely affected by *A. caninum* are puppies which have acquired substantial burdens of worms by the lactogenic route. Puppies of the smaller breeds suffer relatively more severely than those of larger breeds, but in all newborn puppies iron reserves are marginal and milk is a poor source of iron. Older animals are more able to cope with the blood loss, which could be at a rate of as much as one-quarter of the total circulating erythrocyte volume (Miller 1971). Estimates vary for the mean daily blood loss per worm, however; the early estimates by Wells (1931) of 0.8 ml per worm per day now appear to be too high and Miller (1971) suggests a loss of 0.01–0.09 ml per worm per day, depending on the intensity of infection.

Blood loss commences about the eighth day of infection, coincident with the fourth moult and the development of the adult buccal capsule. A distinct diphasic rhythm of blood loss occurs. Thus peak loss is associated with the rapid growth of the maturing worm (10–15 days after infection) and the maximum output of eggs after the twentieth day of infection (Miller 1971).

Death of puppies from anaemia usually occurs between 10 and 24 days after a single primary infection. With older puppies with adequate iron reserves there is a rapid erythropoietic response which compensates for the blood loss. Miller (1971) stresses that hookworm disease caused by *A. caninum* is an acute condition and there is little evidence that other factors, such as malabsorption, contribute to it.

Other sequelae of *A. caninum* infection include diarrhoea, which is seen as early as the fourth day of infection when fourth-stage larvae reach the intestine and by the eighth day fresh blood is mixed with watery mucus (Miller 1971). Skin lesions associated with percutaneous infection range from moist eczma to ulcerations. Damage may be particularly severe on the feet and is often aggravated by licking and biting affected areas. Such lesions often appear after rain and when dogs are allowed on wet grass or sand in endemic areas. Occasionally very heavy infections may lead to pulmonary damage one to five days after infection. A haemorrhagic pneumonitis is produced by larvae as they leave the pulmonary circulation to enter the alveoli. In fatal cases, lungs are consolidated and covered with multiple haemorrhages.

Chronic infection with *A. caninum* is characterized by reduced appetite, poor growth and poor coat condition. In silver foxes the poor coat is a major economic effect of the infection.

Little is known of the pathogenic effects of *A. tubaeforme* infection in cats. Heavy infections are likely to produce effects similar to those of *A. caninum* in dogs, but usually a chronic infection is evident, associated with recurrent diarrhoea, poor growth and poor coat.

Infection with *A. braziliense* is not characterized by anaemia and blood loss is insignificant. Miller (1971) estimates the mean daily blood loss per worm as 0.001 ml. Hypoproteinaemia may be evident in heavy infections (e.g. 500 or more worms) and this arises probably through a protein-losing enteropathy. Digestive upsets and diarrhoea may occur.

Similarly, infection with *U. stenocephala* is not associated with anaemia and the daily blood loss per worm has been estimated at 0.0003 ml (Miller 1971). However, infection induces diarrhoea, which may be severe, and hypoproteinemia, protein plasma levels being reduced by 10% (Miller 1971).

Infection with *A. ceylanicum* results in a blood loss of approximately 0.014 ml per worm per day. However, this parasite is not usually associated with clinical infections in dogs.

Immunity

Miller (1971) has reviewed the immunology of canine infection and also (1978) gives an account of the development, marketing and eventual discontinuance of a gamma-irradiated third-stage larval vaccine.

It has been known for some time that dogs become immune to hookworm infection. Graded doses of larvae given subcutaneously or orally induce this, egg counts rising following the initial infections to reach a maximum about two months after infection. Thereafter there is a marked fall in egg counts and large numbers of adult worms are passed in the faeces (McCoy 1931). McCoy regarded this curative phenomenon as a 'crisis' and it is comparable to the phenomenon of 'self-cure' seen in *Haemonchus contortus* infection. Where infections are built up too rapidly death occurs before the immune response can control the infection. Natural age resistance occurs and is first evident in bitches at eight months of age and in dogs at 11 months (Miller 1965).

Extensive studies by Miller have shown that double vaccination by the subcutaneous inoculation of 1000 infective larvae previously exposed to 40 kr of irradiation protected pups against severe challenge and prevented completely the expected morbidity and mortality associated with heavy challenge by normal larvae (Miller 1971). Vaccination may be started as early as 72 hours after birth and maternal antibody and prenatal-colostral infection do not interfere with the efficacy of the vaccine. However, in practice the vaccine was used following the removal by anthelmintic of colostral acquired infection. The duration of immunity was at least seven months in the absence of further exposure and in practice, with the onset of age resistance at this time, repeat vaccination was unnecessary. An unexpected bonus of the vaccine was that it induced protection against *A. braziliense* and *U. stenocephala*. However, the vaccine failed to live up to economic expectations and marketing ceased in 1975.

Clinical signs

Ancylostomiasis and uncinariasis occur frequently in summer, especially in animals that are confined on a relatively small area of moist ground, like dogs in kennels and foxes in runs. Prenatal or colostral infection may cause severe anaemia with coma and death within three weeks of birth in the case of *A. caninum* infection. In foxes the disease occurs usually when they are two to six months old. It may be acute and rapidly fatal in susceptible animals, while others may develop a marked degree of resistance to the effects of infection. The chief clinical sign is anaemia, accompanied by hydraemia, sometimes oedema, general weakness and emaciation. In the later stages of the disease the blood changes may include eosinophilia. Growth is stunted and the coat becomes dry and harsh; in foxes the fur is poor. Itching of the skin and areas of dermatitis, caused by penetration of the skin by the larvae of the worms, may be observed. The faeces are often diarrhoeic and contain bloody mucus, or they may be of a tarry nature. Death is as a rule preceded by marked weakness and extreme paleness of the mucous membranes.

Post-mortem

Anaemia and cachexia are conspicuous, while oedema and ascites are frequently seen. The liver has a light brown colour and shows fatty changes. The intestinal contents are haemorrhagic. The mucosa is usually swollen, covered with mucus, and shows numerous small red bite-marks of the worms. The latter are found attached to the mucosa, or sometimes free, and their colour grey or reddish, depending on the amount of blood in their intestines.

Diagnosis

Diagnosis is made from the clinical signs and confirmed by finding large numbers (several thousand per gram of faeces) of the characteristic strongyle-type eggs in the faeces. In acute prenatal and colostral infections a severe anaemia may develop before eggs are demonstrable in the faeces of the puppy.

Treatment

In addition to the use of the specific anthelmintics mentioned below, severely affected animals may require blood transfusions or at least some form of readily assimilable iron therapy. Further supportive treatment should include protein rich foods.

When dogs are treated for hookworm infection, note should be taken of the ability of dormant or hypobiotic larvae to repopulate the intestine (see p. 201). This may lead to a false conclusion of drug resistance on the part of the parasite. In such cases treatment should be repeated, if necessary, using the same compound.

Tetrachloroethylene was introduced in the mid 1920s and has since been used extensively for the treatment of hookworms. At a dosage rate of 0.2 ml/kg it is 99% efficient. Preparative treatment consists of an overnight fast before dosing and is followed by a saline laxative. It is seldom used now.

Bephenium compounds such as bephenium chloride, bromide, iodide or hydroxynaphthoate in a single dose of more than 20 mg/kg show 99.4% efficiency. A related compound, *thenium* (thenium *p*-chlorobenzene sulphonate) is highly effective against *A. caninum* and *U. stenocephala* at a rate of 200–250 mg/kg twice daily.

Disophenol (2 : 6-diiodo-4-nitrophenol) is highly effective. A single subcutaneous dose of 7.5 mg/kg is highly efficient against adult *A. caninum* and *A. braziliense* but 10 mg/kg is necessary for good action against *U. stenocephala*.

Dichlorvos at a dose of 12–15 mg/kg is used. The drug is a cholinesterase inhibitor and should not be used simultaneously with or within a few days before or after treatment with other cholinesterase-inhibiting drugs, with other anthelmintics, muscle relaxants, tranquilizers, or modified live-virus vaccines. Dogs showing signs of severe constipation, impaired liver function, circulatory failure or infectious disease should not be treated.

Methylbenzene (toluene) may be formulated with dichlorophen in capsules for dogs of varying weight. Fasting for 18 hours prior to treatment and for four hours afterwards is required.

Styrylpyridinium may be formulated with diethylcarbamazine (6.6 mg with 2.6 mg/kg respectively) and given daily for one week eliminates adult hookworms from dogs. A similar dose prevented infection in puppies given oral doses of infective larvae daily for 11 days.

Tetramisole at single doses of 7.5 and 10 mg/kg by subcutaneous injection or 20 mg/kg orally is very effective against *U. stenocephala* and *A. caninum*.

Mebendazole at an oral dose of 40 mg/kg or two to five doses of 10 mg/kg is effective against the major species of dog hookworms and *Thiabendazole* at a dose of 20 mg/kg orally is also effective.

Fenbendazole at 20 mg/kg as a single dose is effective against immature and mature *A. caninum* and five doses of 20 mg/kg or one dose of 100 mg/kg are 100% effective. For infection in the cat a dose of 30 mg/kg is highly effective.

Nitroscanate at an oral dose of 50 mg/kg is highly effective against *A. caninum* and *U. stenocephala*.

Prophylaxis

The preinfective stages are not resistant to desiccation, so that ground and pens on which susceptible animals are kept should be as dry as possible and faeces should be removed at short intervals. The floors of kennels may be treated with common salt or sodium borate (2 kg/10 m^2) which help to kill the larvae. Where possible the floors of kennels and exercise yards should be made impervious by concrete or similar material.

CUTANEOUS LARVA MIGRANS

This condition may be compared with *visceral larva migrans* (p. 154). It occurs in man and other hosts and is caused by the larvae of nematodes which enter the skin and migrate in it, causing papules and inflamed tracks, sometimes with thickening of the skin and pruritus. The larvae of *Ancylostoma braziliense* most frequently cause it, but other nematodes whose larvae may cause it are *Uncinaria stenocephala*, *Ancylostoma caninum* (experimentally), *Ancylostoma duodenale*, *Necator americanus*, *Bunostomum phlebotomum*, and species of *Strongyloides* and *Gnathostoma* (Beaver 1956). The severity of skin reactions is related to the degree of exposure to infective larvae. When extended exposure to *A. caninum* has occurred, for example, re-exposure leads to papule formation, oedema and a markedly pruritic lesion. Occasionally hookworm larvae may reach the lungs and they have also been reported in opacities of the cornea (Nadbath & Lawlor 1965; Beaver 1966).

Genus: Bunostomum Railliet, 1902 (syn. Monodontus, Molin, 1861)

Bunostomum trigonocephalum (Rudolphi, 1808) is a hookworm which occurs in the small intestine (ileum and jejunum) of sheep and goats in many parts of the world and in Scottish red deer. It has also been recorded from cattle, but the accuracy of this record seems to be doubtful. The male is 12–17 mm long and the female 19–26 mm. The anterior end is bent in a dorsal direction, so that the buccal capsule opens anterodorsally; it is relatively large and bears at its ventral margin a pair of chitinous plates. Near its base is a pair of small subventral lancets. The dorsal gutter, carrying the duct of the dorsal oesophageal gland, ends in a large dorsal cone, which projects into the buccal cavity. There are no

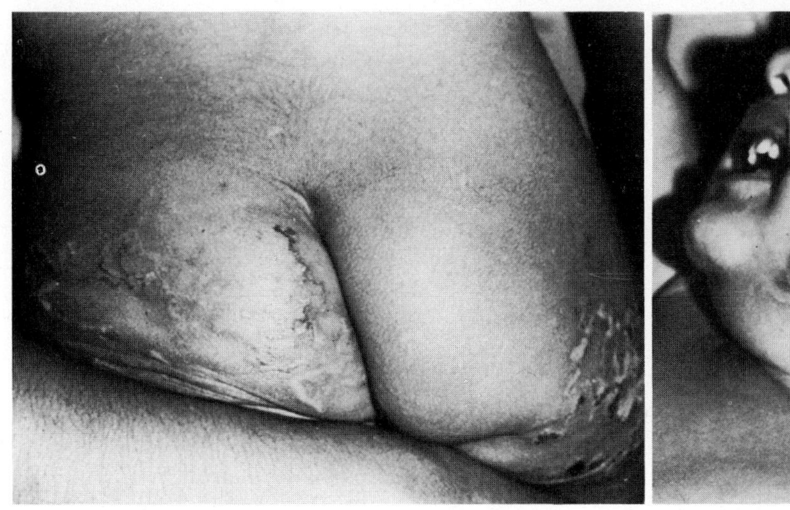

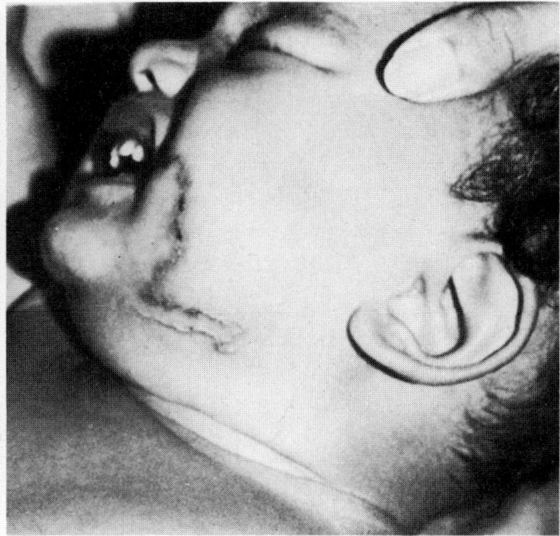

Fig 1.117 Cutaneous larva migrans in a child.

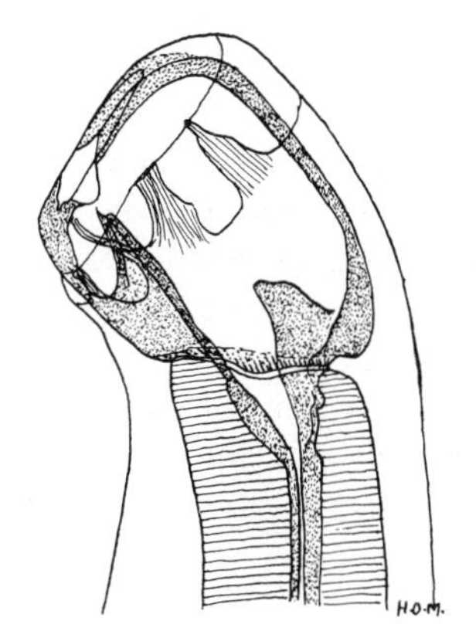

Fig 1.118 Lateral view of the anterior end of *Bunostomum trigonocephalum*.

dorsal teeth in the buccal capsule. The bursa is well developed and has an asymmetrical dorsal lobe. The right externodorsal ray arises higher up on the dorsal stem and is longer than the left, which arises near the bifurcation of the dorsal ray, which divides into two tridigitate branches. The spicules are slender, alate and 0.6–0.64 mm long. The eggs measure 79–97 by 47–50 μm (usually 92 by 50 μm); the ends are bluntly rounded and the embryonic cells are darkly granulated.

The development is direct. Infection of the host occurs through the mouth or skin. Following skin penetration the larvae pass to the lungs, where the third ecdysis occurs. The fourth-stage larvae, which have a buccal capsule, reach the intestine again after 11 days and the first eggs are passed 30–56 days after infection.

The parasite is more important in warmer climates than cold ones. Usually infection occurs along with other gastrointestinal strongyles and the hookworms contribute to the general effects of parasitism. As in the case of ancylostomiasis (see p. 203) the adult worms attach themselves to the intestinal mucosa and suck blood.

When substantial burdens of hookworms occur, the main clinical signs are progressive anaemia, with associated changes in the blood picture, hydraemia and oedema, which shows especially in the intermandibular region as a 'bottle-jaw'. Diarrhoea is not infrequent and the faeces may be dark in colour due to the presence of altered blood pigments. Death is frequently preceded by complete prostration.

The post mortem picture is very similar to that of ancylostomiasis in the dog. In addition, hydrothorax and effusion of fluid into the pericardium are commonly seen.

Diagnosis is made partly from the clinical signs, but the infection must be differentiated from other worm infections that cause anaemia by the identification of the eggs in the faeces or the larvae following culture of the faeces.

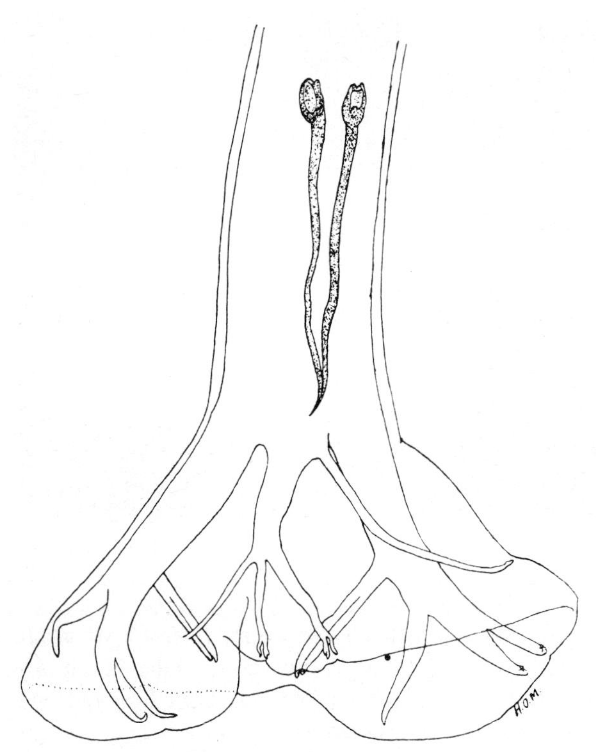

Fig 1.119 Dorsal view of bursa and spicules of male *Bunostomum trigonocephalum.*

The majority of anthelmintics in current use for gastrointestinal nematodes of sheep are highly effective against *B. trigonocephalum.* These include thiabendazole (75 mg/kg), parbendazole (15 mg/kg), fenbendazole (5 mg/kg), cambendazole (20 mg/kg), albendazole (7.5 mg/kg) and thiophanate (50 mg/kg).

The infective larvae cannot resist dryness, so that the infection is found invariably on permanently or occasionally moist pastures. Infection can be controlled by avoiding such places, which should be drained if possible. Around watering troughs the ground should be kept hard and dry or treated frequently with liberal applications of salt.

Bunostomum phlebotomum (Railliet, 1900) is widely distributed and occurs in the small intestine, mainly in the duodenum, of cattle and the zebu. It has also been recorded from sheep, but the accuracy of this is doubtful. The male is 10–18 mm long and the female 24–28 mm. It closely resembles the preceding species, but can be differentiated by its shorter dorsal cone, by the presence of two pairs of subventral lancets in the buccal capsule and the longer male spicules, which measure 3.5–4 mm. The eggs measure about 106 by 46 μm. They have blunt ends and darkly pigmented embryonic cells, so that they can be differentiated from other worm eggs in the faeces.

The life-cycle is similar to that of *B. trigonocephalum.*

This species is a serious pathogen in many parts of the world, e.g. Africa, Australia and southern and mid-western states of USA. However, it may also cause serious ill health in cattle in Europe (Soulsby et al. 1955). In stabled cattle itching of the legs, probably caused by the entry of the skin-penetrating larvae through the skin, which makes the animals stamp their feet and lick their legs, may occur. Diarrhoea, anaemia and marked weakness, especially in calves, are the clinical signs. Submandibular oedema with emaciation may be seen in heavy infections.

Diagnosis is confirmed by finding the eggs in the faeces or by the identification of infective larvae cultured in faeces.

As with the sheep hookworm, the anthelmintics in common use for gastrointestinal nematodes of cattle are effective against *B. phlebotomum.*

Prophylaxis is as in the case of *B. trigonocephalum.* Stabled cattle should be protected by hygienic measures, especially by frequent removal of faeces, by keeping the floors and bedding of yards dry, and preventing contamination of the food and water.

Genus: Gaigeria Railliet and Henry, 1910

Gaigeria pachyscelis Railliet & Henry, 1910, the only known species of this genus, is a hookworm which occurs in the duodenum of sheep and goats in the Indian subcontinent, Africa and Indonesia. The male is up to 20 mm long and the female up to 30 mm. In general it resembles *Bunostomum trigonocephalum.* The buccal capsule contains a large dorsal cone, but no dorsal tooth,

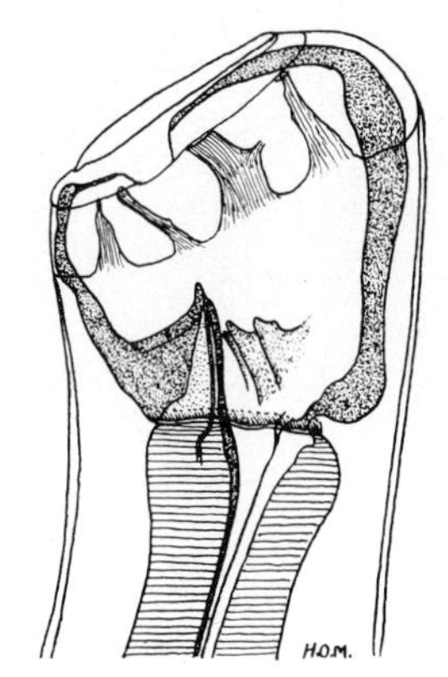

Fig 1.120 Lateral view of the anterior end of *Gaigeria pachyscelis.*

and a pair of subventral lancets which have several cusps each. The male bursa has small lateral lobes joined together ventrally and a voluminous dorsal lobe. The anterolateral ray is short and blunt and it is separated widely from the other lateral rays. The externodorsal rays arise from the main stem of the dorsal ray, which is cleft for about a quarter of its length, the two short branches ending in three very small digitations. The spicules are slender with recurved unbarbed ends; they are 1.25–1.33 mm long. The eggs measure 105–129 by 50–55 μm and have blunt ends.

The life-cycle is direct and similar to that of other hookworms. The infective larvae are sheathed and resemble those of ***Bunostomum trigonocephalum.*** They are not resistant to desiccation. According to Ortlepp (1937) infection occurs only through the skin. The larvae reach the lungs via the blood, where the third ecdysis occurs, and they remain for about 13 days. The fourth-stage larva, which has a globular buccal capsule with a dorsal cone and a pair of subcentral lancets, passes up the bronchi and trachea to the pharynx and is swallowed, reaching the intestine, where the fourth ecdysis occurs and the worms grow adult in about 10 weeks from the time of infection.

The pathogenicity has been studied by Hart and Wagner (1971). The worms are marked blood-suckers and about 24 are stated to be sufficient to cause death. The parasite is more pathogenic to merinos than to the hairy breeds of sheep, such as Persians. In severe infections death occurs suddenly, the only sign being anaemia. In more chronic cases the usual signs of hookworm infection are seen. The worms are found attached to the mucosa of the first part of the small intestine, frequently in groups of two or three, the groups being surrounded by a quantity of fresh blood. Observations made in south-west Africa indicate that the larvae, penetrating

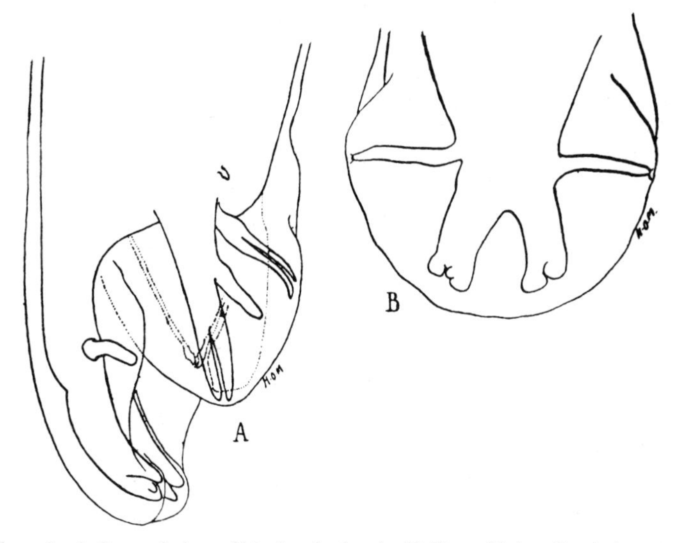

Fig 1.121 *Gaigeria pachyscelis.* A, Lateral view of hind end of male. B, Dorsal lobe of male bursa with dorsal rays.

through the skin of the feet of sheep, may be instrumental in introducing the organisms of foot-rot.

Diagnosis can be made by finding the characteristic eggs in the faeces.

Treatment is as for *Bunostomum trigonocephalum*.

Moist pastures and all moist ground should be avoided. Since infection occurs via the skin, it is essential that the surroundings of watering troughs should be kept dry or be treated with salt. In The Republic of South Africa the distribution of the parasite is closely associated with sandy soil.

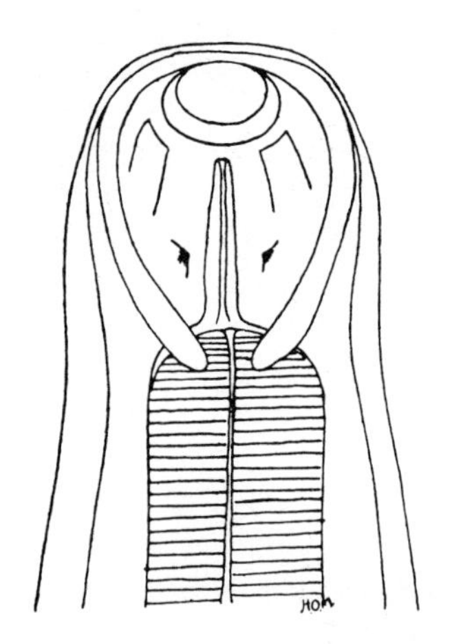

Fig 1.122 Dorsal view of the anterior end of *Globocephalus urosubulatus*.

Genus: Globocephalus Molin, 1861

Several species of this genus occur in the small intestine of the pig. The genus occurs in Europe, Africa, North and South America and the East. The worms are about 4–8 mm long and fairly stout. The mouth opens subdorsally and the buccal capsule is globular or funnel-shaped with an external chitinoid ring. There are neither leaf-crowns nor teeth at the oral margin. Near the base of the buccal capsule a pair of subventral teeth is usually present, but may be small or absent in some specimens. The dorsal gutter is prominent, extending almost to the oral margin. The male bursa is well developed, the spicules are slender and a gubernaculum is present.

The life-cycle is probably direct, but unknown.

Very little is known about pathogenicity but heavy infections are likely to cause anaemia. For treatment, it is likely that compounds active for canine ancylostomiasis would be useful.

Globocephalus longemucronatus Molin, 1861 occurs in the intestine of pig in Europe, Africa and the East (Japan, Philippines). Males are 5.7 mm and females 6–8 mm long. **G. samoensis** (Lane, 1922) occurs in the small intestine of the pig and other suidae in Samoa, New Guinea and Japan. Males are 4.5–6 mm long and females 5.2–6.4 mm. **G. urosubulatus** (Alessandrini, 1909) occurs in the domestic pig in North and South America, the Pacific area, the Indian Subcontinent and Europe. Males are 4.5–5.5 mm long and females 5–6 mm; eggs are 52–56 by 36–35 μm. **G. versteri** Ortlepp, 1964 occurs in the bush-pig in Africa.

Genus: Bathmostomum Railliet and Henry, 1909

The genus resembles other ucinarids, but the wall of the buccal capsule is markedly fissured giving the capsule a ridged appearance. A pair of subventral teeth occur.

Bathmostomum sangeri (Cobbold, 1879) Railliet and Henry, 1909 occurs in the caecum of the Indian elephant. Males are 15–16 mm long and females 20 mm; eggs are 45 by 30 μm.

Genus: Grammocephalus Railliet and Henry, 1909

The buccal capsule is infundibuliform with a pair of triangular lateral lancets and a pair of subventral lancets. A dorsal cone is present. An intestinal diverticulum is present running forward alongside the oesophagus. Parasites of elephants and rhinoceros.

Grammocephalus clathratus (Bairs, 1868) Railliet and Henry, 1910 occurs in the bile ducts of the African elephant. Males are 45–52 mm and females 36 mm; eggs are 50 by 35 μm. **G. inter-**

medius Neveu-Lemaire, 1924 occurs in the bile ducts of the rhinoceros in Kenya and Ethiopia. **G. hybridatus** van der Westhuysen, 1938 is found in the bile ducts of the Indian elephant. Males are 37 mm long and females 37 mm; eggs are 58–63 by 29–38 μm. **G. varedatus** Lane, 1921 is found in the bile ducts of the Indian elephant. Males are 55 mm and females 47 mm; eggs are 68 by 37 μm.

The life-cycles of the genera *Bathmostomum* and *Grammocephalus* are unknown, but probably are similar to that of *Bunostomum*. The pathogenesis is probably similar to that of other hookworms and Sutherland and O'Sullivan (1950) have recorded a fatal infection in an Indian elephant due to *B. sangeri*. It is probable that treatment with the benzimidazole compounds used in cattle will be effective.

REFERENCES

AMIDOSTOMIDAE, STEPHANURIDAE AND SYNGAMIDAE

Alicata, J. E. (1953) Observations on the lethal action of polyborate on swine kidney worm (*Stephanurus dentatus*) larvae in soil. *Am. J. vet. Res.*, **14**, 563–570.

Batte, E. G., Harkema, R. & Osborne, J. C. (1960) Observations on the life cycle and pathogenicity of the swine kidney worm (*Stephanurus dentatus*). *J. Am. vet. med. Ass.*, **136**, 622–625.

Batte, E. G., Moncol, D. J. & Barber, C. W. (1966) Prenatal infection with the swine kidney worm (*Stephanurus dentatus*). *J. Am. vet. med. Ass.*, **149**, 758–765.

Cowan, A. B. & Herman, C. M. (1956) Winter losses of Canada geese at Pea Island, North Carolina. *Proc. SE Ass. Game Fish Com. Mtg*, 1955, 172–174.

Egerton, J. R. (1961) The effect of thiabendazole upon *Ascaris* and *Stephanurus* infections. *J. Parasit.*, **47** (Suppl.), 37.

Fernando, M. A., Hoover, J. J. & Ogungbade, S. G. (1973) The migration and development of *Cyathostoma bronchialis* in geese. *J. Parasit.*, **59**, 759–764.

Herman, C. M. & Kramer, R. (1950) Control of gapeworm infection in game farm birds. *Calif. Fish Game*, **36**, 13–17.

Kobulej, T. (1956) Beiträge zur Biologie des *Amidostomum anseris* (Zeder, 1800). *Acta vet. Hung.*, **6**, 429–449.

Leiby, P. D. & Olsen, O. W. (1965) Life history of nematodes of the genera *Amidostomum* (Strongyloidae) and *Epomidiostomum* (Trichostrongyloidae) occurring in the gizzards of water fowl. *Proc. Helminth Soc. Wash.*, **32**, 32–49.

Morgan, D. O. & Clapham, P. A. (1934) Some observations on gapeworm in poultry and game birds. *J. Helminth.*, **12**, 63–70.

Pande, B. P., Bhatia, B. B. & Dubey, J. P. (1964) On the development of free-living stages of *Amidostomum skrjabini*. A pathogenic nematode in domestic duck. *Curr. Sci. Bangalore*, **33**, 278–279.

Skrjabin, K. I. (1952) *Key to Parasitic Nematodes, Vol. III, Strongylata*. Academy of Sciences of the USSR, Moscow 1952. Israel Program for Scientific Translations, U.S. Dept. Commerce, Office of Technical Services, Washington, D.C.

Stewart, T. B., Hale, O. M. & Andrews, J. S. (1964) Eradication of the swine kidney worm, *Stephanurus dentatus* from experimental pastures by herd management. *Am. J. vet. Res.*, **25**, 1141–1150.

Tromba, E. G. & Baisdon, L. A. (1960) Diagnosis of experimental stephanuriasis in swine by a double diffusion agar precipitin technique. *J. Parasit.*, **46** (Suppl.), 29.

ANCYLOSTOMATIDAE

Beaver, P. C. (1956) Parasitological reviews. Larva migrans. *Exp. Parasit.*, **5**, 587–621.

Beaver, P. C. (1966) Zoonoses, with particular reference to parasites of veterinary importance. In: *Biology of Parasites*, ed. E. J. L. Soulsby, pp. 215–225. New York: Academic Press.

Biocca, E. (1951) On *Ancylostoma brasiliensis* (de Faria, 1910) and its morphological differentiation from *A. ceylanicum* (Looss, 1911). *J. Helminth.*, **25**, 1–10.

Biocca, E. (1954) Ridescrizione di *Ancylostoma tubaeforme* (Zeder, 1800) parassita del gatto, considerato erroneamente sinonimo di *Ancylostoma caninum* (Ercolani, 1859) Parassita del cane. *Riv. Parassit.*, **15**, 267–278.

Burrows, R. B. (1962) Comparative morphology of *Ancylostoma tubaeforme* (Zeder, 1800) and *Ancylostoma caninum* (Erocolani, 1859). *J. Parasit.*, **48**, 715–718.

Enigk, K. & Stoye, M. (1967) Untersuchungen uber den Infektionsweg von *Ancylostoma caninum* Ereolani 1959 (Ancylostomidae) beim Hund. *Kongrbch. Dt. TropenMed. ges.*, **v**, 101–113.

Foster, A. O. (1932) Prenatal infection with the dog hookworm *Ancylostoma caninum*. *J. Parasit.*, **19**, 112–118.

Gibbs, H. C. (1958) On the gross and microscopic lesions produced by the adults and larvae of *Dochmoides stenocephala* (Railliet, 1884) (Ancylostomidae: Nematoda). *Can. J. comp. Med.*, **22**, 382–382

Hart, R. J. & Wagner, A. M. (1971) The pathological physiology of *Gaigeria pachyscelis* infestation. *Onderstpoort J. vet. Res.*, **38**, 111–116.

Herrick, C. A. (1928) A quantitative study of infections with *Ancylostoma caninum* in dogs. *Am. J. Hyg.*, **8**, 125–157.

Kelly, J. D., Thompson, H. G., Chow, D. C. M. & Whitlock, H. V. (1976) Arrested development of larval *Ancylostoma caninum* in the gastro-intestinal tract. *N. Z. vet. J.*, **24**, 93–94.

Krupp, I. M. (1961) Effect of crowding and of super infection on habitat selection and egg production in *Ancylostoma caninum*. *J. Parasit.*, **47**, 957–961.

Lee, K. T., Little, M. D. & Beaver, P. C. (1975) Habitat of *Ancylostoma caninum* in some mammalian hosts. *J. Parasit.*, **61**, 589–598.

McCoy, O. R. (1931) Immunity reactions of the dog against hookworm (*Ancylostoma caninum*) under conditions of repeated infection. *Am. J. Hyg.*, **14**, 268–303.

Miller, T. A. (1965) Influence of age and sex on susceptibility of dogs to primary infection with *Ancylostoma caninum*. *J. Parasit.*, **51**, 701–704.

Miller, T. A. (1966) Comparison of the immunogenic efficiencies of normal and X-irradiated *Ancylostoma caninum* larvae in dogs. *J. Parasit.*, **52**, 612–519.

Miller, T. A. (1970*a*) Prenatal-colostral infection of pups with *Ancylostoma caninum*. *J. Parasit.*, 56[4], Sect. II, 239.

Miller, T. A. (1970*b*) Potential transport hosts and the life cycles of canine and feline hookworms. *J. Parasit.*, **56** (Suppl.), 238.

Miller, T. A. (1971) Vaccination against the canine hookworm diseases. *Adv. Parasit.*, **9**, 153–183.

Miller, T. A. (1978) Industrial development and field use of the canine hookworm vaccine. *Adv. Parasit.*, **16**, 333–342.

Miller, T. A. (1979) Hookworm infection in man. *Adv. Parasit.*, **17**, 315–384.

Nadbath, B. P. & Lawlor, P. P. (1965) Nematode (*Ancylostoma*) in the cornea. *Am. J. Ophthalmol.*, **59**, 486–490.

Olsen, O. W. & Lyons, E. T. (1965) Life cycle of *Uncinaria lucasi* Stiles, 1901 (Nematode Ancylostomatidae) of fur seals, *Callorhinus ursinus* Linn on the Pribilof Islands, Alaska. *J. Parasit.*, **51**, 689–700.

Ortlepp, R. J. (1937) Observations on the morphology and life-history of *Gaigeria pachyselis* (Raill. and Henry, 1919): A hookworm parasite of sheep and goats. *Onderstepoort J. vet. Sci.*, **8**, 183–212.
Schad, G. A. (1977) The role of arrested development in the regulation of nematode populations. In: *Regulation of Parasite Populations*, ed. G. W. Esch, pp. 111–167. New York: Academic Press.
Schad, G. A. (1979) *Ancylostoma duodenale*: Maintenance through six generations of helminth-naive pups. *Exp. Parasit.*, **47**, 246–253.
Schad, G. A., Chowdhury, A. B., Dean, C. G., Kochar, V. K., Nawalinski, T. A., Thomas, J. & Tonascia, J. A. (1973) Arrested development in human hookworm infections: Adaptation to a seasonally unfavourable external environment. *Science, N.Y.*, **180**, 502–504.
Soulsby, E. J. L., Venn, J. A. J. & Green, K. N. (1955) Hookworm disease in British cattle. *Vet. Rec.*, **67**, 1124–1126.
Stone, W. M. & Girardeau, M. H. (1968) Transmammary passage of *Ancylostoma caninum* larvae in dogs. *J. Parasit.*, **54**, 426–429.
Stone, W. M. & Smith, F. W. (1973) Infection of mammalian hosts by milk borne nematode larvae: A review. *Exp. Parasit.*, *34*, 306–312.
Stoye, M. (1973) Untersuchungen über die moglichkeit pränataler und galaktogener infektionen mit *Ancylostoma caninum* Ercolani 1859 (Ancylostomidae beim Hunde). *Zbl. VetMed.*, **B20**, 1–39.
Sutherland, A. K., O'Sullivan, P. J. & Ohman, A. F. S. (1950) Helminthiasis in an elephant. *Aust. vet. J.*, **26**, 88–90.
Yoshida, Y. (1968) Pathobiological studies on *Ancylostoma ceylanicum* infection. *8th int. Congr. trop. Med. Malariol.*, Teheran, 170–171.
Yoshida, Y., Okamoto, K. & Chin, J. K. (1972) Experimental infection of man with *Ancylostoma ceylanicum*. *Chin. J. Microbiol.*, **4**, 157–167.
Wells, H. S. (1931) Observations on the blood sucking activities of the hookworm *Ancylostoma caninum*. *J. Parasit.*, **17**, 167–182.

SUPERFAMILY: TRICHOSTRONGYLOIDEA CRAM, 1927

Stoma reduced or rudimentary. Corona radiata absent. Lips six or three or absent. Thin-bodied, bursa well developed, rarely reduced. Families of importance include *Trichostrongylidae*, *Ollulanidae*, *Dictyocaulidae*.

FAMILY: TRICHOSTRONGYLIDAE LEIPER, 1912

Mostly small forms in which the buccal capsule is absent or very small and is devoid of leaf-crowns, and usually bears no teeth. The male bursa is well developed, with large lateral lobes and a small dorsal lobe. Adults parasitic in the alimentary canals of sheep, cattle, equines and other vertebrates. Genera of importance include *Trichostrongylus*, *Graphidium*, *Ostertagia*, *Marshallagia*, *Cooperia*, *Nematodirus*, *Haemonchus* and *Mecistocirrus*.

Genus: Trichostrongylus Looss, 1905

The species of this genus are small, slender, pale reddish-brown worms without a specially developed head-end. There is no buccal capsule. The excretory pore is usually situated in a conspicuous ventral notch near the anterior extremity. The male bursa has long lateral lobes, while the dorsal lobe is not well defined. The ventral rays of the male bursa are separated widely and the ventroventral ray is conspicuously thinner than the lateroventral, which runs parallel with the lateral rays. The posterolateral ray diverges from the other lateral rays and lies near to the externodorsal ray. The dorsal ray is slender and cleft near its tip into two branches which have short digitations. The spicules are stout, ridged and pigmented brown, and a gubernaculum is present. The eggs are oval, thin-shelled and segmenting when they are laid.

Trichostrongylus colubriformis (Giles, 1892) (syn. *T. instabilis*) occurs in the anterior portion of the small intestine and sometimes also in the abomasum of sheep, goat, cattle, camel and various antelopes. It has also been recorded from the rabbit (*Lepus californicus melanotis*), the pig, the dog and man. Males are 4–5.5 mm long and females 5–7 mm. Spicules equal, 0.135–0.156 mm long. The eggs measure 79–101 by 39–47 μm.

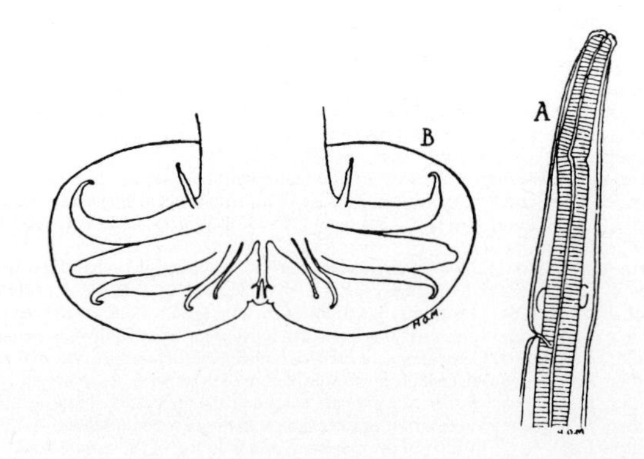

Fig 1.123 *Trichostrongylus colubriformis* (syn. *T. instabilis*). A, Lateral view of anterior end. B, Bursa of male opened out.

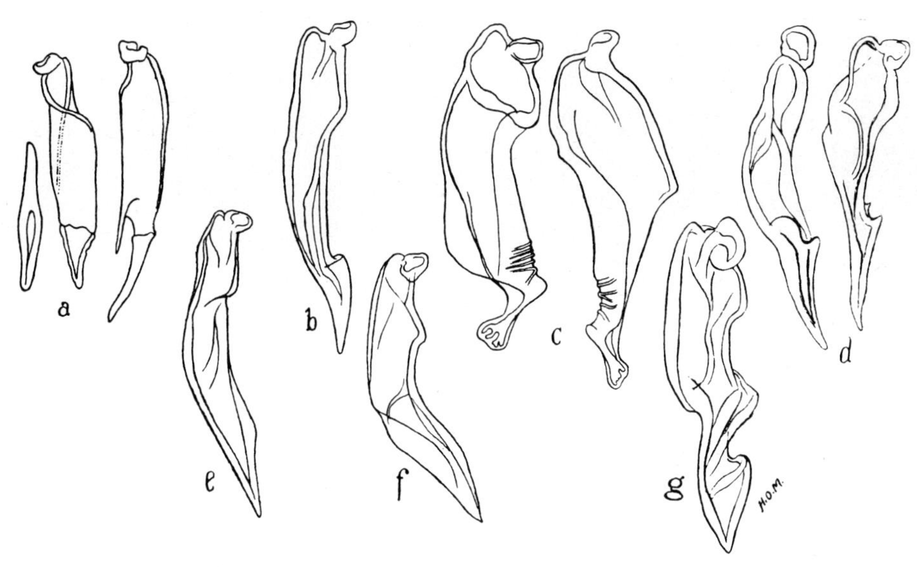

Fig 1.124 Spicules of some species of *Trichostrongylus*. a, *T. axei*. b, *T. colubriformis*. c, *T. rugatus*. d, *T. falculatus*. e, *T. vitrinus*. f, *T. capricola*. g, *T. probolarus*. All are drawn to the same scale. Both spicules are shown only where they are unequal and dissimilar.

Trichostrongylus falculatus Ransom, 1911 occurs in the small intestine of sheep and goat and some antelopes in southern Africa and Australia. Males are 4.5–5.6 mm long. Spicules subequal, about 0.1 mm long. Females have not been described.

Trichostrongylus vitrinus Looss, 1905 occurs in the small intestine of sheep, goat, deer and occasionally pig, rabbit, camel and man throughout the world. Spicules are equal, 0.16–0.17 mm long. Eggs measure 93–118 by 41–52 μm. Males are 4–7 mm long and females 5–8 mm.

Trichostrongylus capricola Ransom, 1907 occurs in the small intestine of sheep and goat. Spicules equal, 0.13–0.145 mm long. Males are 3.5–5.5 mm long and females 5–6.4 mm.

Trichostrongylus probolurus (Railliet, 1896) occurs in the small intestine of sheep, goat, camel and, accidentally, man. Spicules equal, 0.126–0.134 mm long.

Trichostrongylus axei (Cobbold, 1879) (syn. *T. extenuatus*) occurs in the abomasum of sheep, goats, cattle, deer and wild antelope and in the stomach of pig, horse, donkey and man. Males are 2.5–6 mm long and females 3.5–8 mm. The spicules are unequal and dissimilar; the right is 0.085–0.095 mm long and the left 0.11–0.15 mm. Eggs measure 79–92 by 31–41 μm.

Trichostrongylus rugatus Mönnig, 1925 occurs in the small intestine of sheep and goat in southern Africa and Australia. Males are 4–6.6 mm long and females 5.8–7.3 mm. Spicules are unequal and dissimilar; the right is 0.137–0.145 mm long and the left 0.141–0.152 mm.

Trichostrongylus longispicularis Gordon, 1933 occurs in sheep in Australia and in cattle in

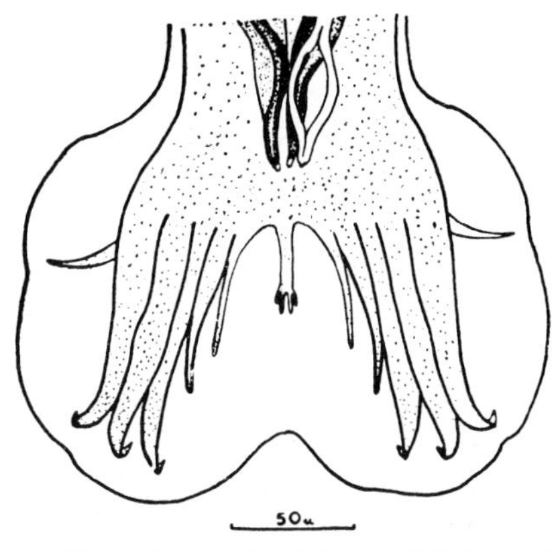

Fig 1.125 Ventral view of male bursa of *T. longispicularis*.

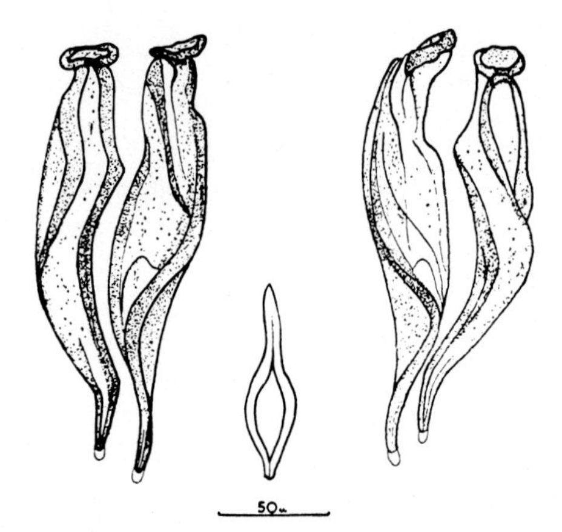

Fig 1.126 Spicules and gubernaculum of *T. longispicularis*.

Australia, Europe and America. The male is 5.5 mm in length. Spicules are subequal, the right being 0.175–0.180 mm and the left 0.19 mm. Both spicules end bluntly, are rounded at the tip and have a tongue-like semitransparent membrane projecting from the tip.

Trichostrongylus drepanoformis Sommerville, 1959 has been described from sheep in Australia. Spicules are bent near the middle to form an angle of 90°.

Trichostrongylus hamatis Daubney, 1933 occurs in the intestine of sheep and steinbok (*Raphicerus campestris*) in East Africa.

Trichostrongylus skrjabini Kalantarian, 1930 is found in the sheep, moufflon and roe deer in the USSR.

Trichostrongylus orientalis Jimbo, 1914 is principally a parasite of the small intestine of man and only occasionally ruminants (sheep). It occurs in Asia and the middle East, especially where nightsoil is used as a fertilizer.

Trichostrongylus retortaeformis (Zeder, 1800) occurs in the small intestine, rarely the stomach, of the rabbit, hare and goat. Males are 5–7 mm long and females 6–9 mm. The spicules are 0.12–0.14 mm long. The eggs measure 85–91 by 46–56 μm.

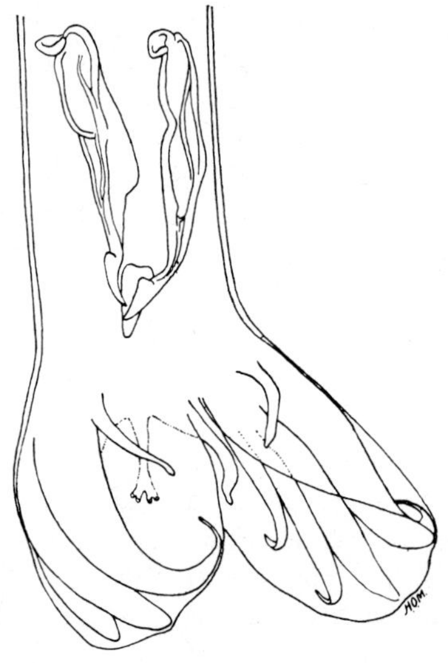

Fig 1.127 Ventral view of hind end of male *Trichostrongylus retortaeformis*.

Trichostrongylus affinus Graybill, 1924 occurs in the small intestine of the rabbit and occasionally sheep and other ruminants.

Trichostrongylus tenuis (Mehlis, 1846) occurs in Europe, Asia and North America in the caeca and small intestine of domestic and wild ducks and geese, fowl, guinea-fowl, turkey, pheasant and partridge. Males are 5–6.5 mm long and females 7.3–9 mm. The spicules are 0.13–0.15 mm long and curved.

Life-cycle of Trichostrongylus spp.

The eggs which are passed in the faeces of the host are strongyle-like, being thin-shelled and in the eight to 32 cell (blastomere) stage. Their development to first stage larvae and subsequently to infective larvae is comparable to that of the horse strongyles (see p. 174). Thus the infective third-stage larva is produced in four to six days under optimal conditions (27°C, O_2, H_2O). Lower temperatures greatly delay development and below 9°C no development occurs. Migration of larvae on blades of grass occurs when light intensity is about 62 foot-candles; moisture favours migration but more than 0.12 ml of water per square centimetre of soil hinders movement. Consequently the greatest numbers of larvae are on blades of grass in the early morning and early evening when temperature, humidity and light intensity are favourable. Survival of *Trichostrongylus* spp. infective larvae is dealt with under Epidemiology of *Trichostrongylus* spp.

Infection is by ingestion of infective larvae on herbage.

Exsheathment. Following ingestion of infective larvae, completing of the second moult, or exsheathment, must occur before the parasitic cycle can begin. This consists of the shedding of the retained sheath of the second-stage larva and the physiology of this has been extensively studied by Rogers (1966; 1970) and Rogers and Brooks (1978). Essentially the process consists of two stages: the first is a host stimulus which causes larvae to secrete 'exsheathing fluid' and the second is the attack by exsheathing fluid on the sheath so that a break in it occurs and larvae, aided by their own movements, are able to escape from it.

The host stimulus consists of the unionized components of bicarbonate–carbon dioxide buffer, undissociated CO_2 and dissolved gaseous CO_2. 70% exsheathment of *Trichostrongylus axei* infective larvae occurred in three hours at 37°C at pH 7.3 in 0.02 M sodium dithionite when the total concentration of carbonic acid plus dissolved gaseous CO_2 was about 0.5×10^{-3} M. With *H. contortus* larvae 1.5×10^{-3} M was necessary to give the same results.

The main effect on larvae occurs within about 30 minutes and the exsheathing fluid attacks an encircling area about 20 μm from the anterior end of the larva. This first appears as a refractile line across the sheath, consisting of a swelling and separation of the sheath in that area. Ultimately the anterior end is detached as a cap and the larva then wriggles out of the sheath.

The mechanism by which exsheathing fluid disrupts the sheath is enzymatic in nature and is due to a specific leucine aminopeptidase (LAP) which attacks its substrate localized to a narrow area on the inside of the sheath. Rogers and Brooks (1978) suggest that a lipase may also be concerned in the process. Rogers (1966) has demonstrated that a low concentration of iodine inhibits exsheathment of larvae of *H. contortus* and *T. colubriformis* and such inhibition can be reversed by subsequent treatment with hydrogen sulphide water.

Parasitic development. The parasitic development of *T. colubriformis* and *T. axei* in cattle was described by Douvres (1957). Parasitic third-stage larvae are found in the abomasum or small intestine two to five days after infection; fourth-stage larvae occur about seven days after infection and fifth-stage larvae are found 15 days after infection. The prepatent period is about 20 days. The morphological features of the various developmental stages are described by Douvres (1957).

Epidemiology of Trichostrongylus spp. larvae

The epidemiology of the gastrointestinal trichostrongyles (*sensu lata*) of sheep and cattle is discussed on p. 240. In common with the other trichostrongyle parasites the development and survival of the free-living stages of the *Trichostrongylus* spp. is dependent on the weather and on pasture conditions. In general, the infective stage is produced in four to six days under optimal conditions at 27°C. The minimum temperature for development is between 10 and 15°C. In Great Britain infective larvae may develop in two weeks, but usually longer, depending on the weather at the time of contamination of the pasture. The most rapid development occurs in summer with peak larval burdens on pasture being reached in six to eight weeks, resulting in heavy infections in lambs from September onwards (Boag & Thomas 1977). The infective larvae, although more susceptible to cold than those of the *Ostertagia* spp., do show a limited ability to overwinter and larvae on pasture may not die out until the following April to June (Gibson & Everett 1967; Boag & Thomas 1970). The ability of larvae to overwinter can result in the appearance of *Trichostrongylus* spp. parasites in lambs in June and July. Conversely, larvae are unable to survive high temperatures and low humidity and a soil temperature of 21–27°C has been shown to be inhibitory for *T. axei* (Callinan 1978). However, Wharton (1982) has reported that infective larvae of *T. colubriformis* readily survive desiccation (e.g. 50% survival for 58–164 days at 33–98% relative humidity at 20°C), whereas second-stage larvae were very susceptible to desiccation. In Western Australia, in winter rainfall areas, infective larval development may take four to 28 days and the maximum yield of larvae on pasture results from eggs deposited in autumn and winter. These larvae are able to persist on herbage and soil for up to 208 days but none survive the Australian dry summer weather. In fact, the length of survival of the larvae is dependent on the time of deposition of the eggs and the occurrence of the first summer weather (Callinan 1978, 1979). Similarly, free-living larvae do not survive the dry season (October to May) in northern Nigeria and the parasite has adapted itself to survive this period within the host (Ogunsusi & Eysker 1979). In contrast, in summer rainfall zones in Australia pasture larval populations are highest in summer, with minimum levels being seen in late winter and spring. This is reflected by infection levels in lambs which increase through the summer, autumn and winter to reach a maximum in spring (Anderson et al. 1978).

Trichostrongylus spp. larvae may undergo inhibition of development within the host and Eysker (1978) demonstrated that the majority of *Trichostrongylus* spp. parasites in adult ewes in the Netherlands were present overwinter as inhibited larvae. However, the levels of inhibition of larvae in all ages of animals in Nigeria were much lower and in both Nigeria and Australia *Trichostrongylus* spp. parasites appear to survive adverse environmental conditions primarily as adult worms within the host (Ogunsusi & Eysker 1979; Anderson et al. 1978).

Pathogenesis

The pathological changes caused by *T. axei* in the stomach of the horse and abomasum of cattle have been described by Leland et al. (1961) and Ross et al. (1967). Histotrophic migration of the larvae occurs and all stages of development are found between the stomach or abomasal epithelium and basement membrane in the calf and within dilated gastric glands in the horse. The infection causes hyperaemia of the mucosa which progresses to a catarrhal inflammation with necrosis and erosion or ulceration of the epithelium. In calves, numerous parasites may be found associated with, or partially embedded in the mucosa in, raised plaque lesions comprising greyish flat areas with sharply demarcated borders ('ringworm lesions'). In the horse hypertrophic gastritis with pedunculated and polypoid lesions may develop, and these are associated with a protein-losing gastropathy but these proliferative changes were associated with heavy infections and were seen only after the infections had progressed for a considerable period of time (one year).

Parasites such as *T. colubriformis*, *T. vitrinis* and *T. retortaeformis* all cause similar pathology in the anterior small intestine. Parasites in all stages of development are found in tunnels beneath the epithelial cells but on the epithelial side of the basement membrane. Displacement and disruption of epithelial cells leave parts of the parasites exposed (Taylor & Pearson 1979). The lamina propria becomes thickened, oedematous and infiltrated with inflamatory cells and the increased permeability of capillaries and venules plus open junctions between the epithelial cells accounts for the loss of plasma proteins into the intestine and the hypoalbuminaemia which develops (Barker 1973*a*). Subtotal to severe villous atrophy may develop in infected areas. The mucosa becomes flattened or composed of irregular masses or ridges while increased mitosis in the intestinal crypts causes protuberant collars of cells surrounding the crypt mouths (Barker 1973*b*). The apices of the epithelial cells are rounded, with sparse, irregular or knob-like microvilli, and there is a deficiency in brush border enzyme activity (Barker 1973*b*; Coop & Angus 1975).

Infection with *Trichostrongylus* spp. results in a protein-losing gastroenteropathy and hypoalbuminaemia develops. In experimental infections, protein loss commenced on day 10 (Barker 1973*b*) and the rate of loss was directly related to larval dose (Symons & Steel 1978). Examination of protein metabolism in infected guinea-pigs revealed decreased skeletal muscle protein synthesis (in that concentrations of RNA/mg DNA fell progressively from day four of infection which was associated with reduced levels of RNA polymerase) but a faster rate of liver protein synthesis (Symons & Jones 1978; Jones & Symons 1978), the latter presumably associated with synthesis of plasma proteins in an attempt to maintain homeostasis. The loss of serum proteins is a factor in anorexia resulting in reduced food intake and reduced feed conversion efficiency. Intestinal parasitism may also influence gastric secretions through the production of secretin by the duodenal mucosa and the influence of secretin on gastrin. For example, in the intestinal phase of *Trichinella spiralis* gastric secretion is impaired due to reduced production of secretin (Dembinski et al. 1979). A correlation between anorexia and plasma concentration of cholecystokinin (CCK) in sheep infected with *T. colubriformis* has been reported by Symons and Hennessy (1981). Mucosal cells of the small intestine secrete CCK and nematode parasitism appears to stimulate secretion of CCK, and increased plasma levels act on the appropriate centre in the brain to depress appetite. An additional factor is that of reduced utilization of food consumed. Thus Coop et al. (1976), while noting a 20% reduction in food intake in trichostrongylosis in lambs, also found that the body weight gain per kilogram of feed consumed was significantly less in parasitized animals compared with controls. Coop et al. (1976) have shown, too, that bone growth is arrested and osteoporosis is present in infected animals, these effects being attributed to low plasma phosphorus concentrations. Absorption of phosphorus and calcium is depressed (Symons & Steel 1978) and hypophosphataemia (Coop 1981) is a feature in infected sheep. The uptake of selenium is decreased in intestinal parasitism in sheep (Horak et al. 1968). Therefore, the overall effects of infection are depressed growth rate, depressed production and stunting of skeletal growth.

It has been estimated that about 2000 worms are necessary to produce marked clinical signs in a year-old sheep, and many more may be needed to produce fatal results, but the effects of the worm vary according to the age and nutrition of the host.

Clinical signs

In horses heavy infections with *T. axei* may cause signs of gastrointestinal disturbance, but in such cases other worms present in the alimentary canal may be additional causes of the signs observed. The same may be true of cattle. In sheep and goats young animals are especially susceptible. In southern Africa trichostrongylosis is a disease of Persian sheep especially, merinos being much less susceptible. When a severe infection is acquired within a short time the disease may be acute and may rapidly lead to death. Such animals usually show neither emaciation nor anaemia, but become weak in the legs and are unable to stand shortly before they die. In more chronic cases the appetite is variable, emaciation occurs, the skin becomes dry and there may be alternating constipation and diarrhoea; if anaemia is noticeable, it is mild. In Australia the parasites cause serious losses in young merino sheep which pass dark, diarrhoeic faeces, the worm being popularly known as the 'black scours worm'.

Post mortem

In acute cases the carcass shows no lesions other than those in the intestine. The mucosa of this organ is swollen, especially in the duodenum, sometimes slightly haemorrhagic, and it may be covered with mucus. The worms can be found by scraping the mucosa into a glass dish of water or by smearing scrapings over a glass plate, which then is held up to the light or examined with a hand lens.

In chronic cases the carcass is emaciated and the liver may show fatty changes. The intestinal mucosa may be thickened, inflamed and ulcerated. There may be flattened, bright red, confluent or focal areas sharply demarcated from the surrounding normal mucosa. Histological examination reveals villous atrophy and the other changes described under pathogenesis.

Immunology of Trichostrongylus spp. infection

Infection induces immunity to re-infection. Both the intake of infective larvae and the presence of adult worms are concerned with the immune response and the immunity is specific, at least at the generic level, in that sheep immune to *H. contortus* are susceptible to *Trichostrongylus* spp. and *vice versa* although some degree of antigen-sharing occurs between some *Trichostrongylus* spp. (Dineen et al. 1977). Re-infection of sheep with infective larvae induces self cure (see p. 236). The effector mechanism of expulsion is not species-specific since challenge infection of sheep immunized against *T. colubriformis* results in a 98–100% level of protection against mixed infections of *T. colubriformis*, *T. vitrinis* and *Nematodirus spathiger* (Dineen et al. 1977). Jarrett et al. (1960), using X-irradiated larvae of *T. colubriformis*, have been able to induce a high degree of resistance to challenge infection; thus double vaccination with irradiated infective larvae reduced the number of worms developing from a challenge infection of 10 000 larvae by approximately 97% and immunized lambs are resistant to both impulse (40 000 larvae) and sequential (2000 larvae each week-day) infection (Gregg & Dineen 1978). However, although high levels of protection develop in 10-month-old lambs, the immunization of young lambs (three months old) is less successful and produces only a partial protection or an absence of response (Gregg et al. 1978). Recently, Windon and Dineen (1980) have demonstrated that when young lambs are immunized and then challenged against infection with *T. colubriformis*, these lambs may be divided on the basis of faecal egg counts into 'responder' and 'non-responder' lambs. The ability of such lambs to respond to immunization against infection with *T. colubriformis* is inherited genetically (Dineen & Windon 1980).

Protective immunity against infection with *T. colubriformis* in guinea-pigs may be adoptively transferred by means of lymphocytes (Adams & Rothwell 1980) and 'transfer factor' transfers a significant degree of resistance against *T. axei* and *T. colubriformis* to recipient sheep (Ross & Halliday 1979*a*, *b*) (see also Soulsby 1979).

Diagnosis

This must be confirmed by making faeces cultures and identifying the infective larvae (see Fig. 1.136).

Prophylaxis and treatment

The general preventive measures described below (p. 250) should be applied. Treatment is discussed on pp. 247–50.

TRICHOSTRONGYLOSIS IN AVIAN HOSTS

The eggs are passed in the faeces of the bird, and, under favourable conditions, the larvae develop to the infective stage as do those of other strongyles. The worms mature in the birds in seven days after infection.

Severe infections may cause a haemorrhagic typhlitis with diarrhoea. Later the birds suffer from loss of appetite, emaciation and anaemia. This parasite is associated with 'grouse disease', an entity which decimated the red grouse population in Scotland in the 19th century (Cobbold 1873). However, the parasite can also cause severe disease in goslings and quail.

Genus: Graphidium Railliet & Henry, 1909

Graphidium strigosum (Dujardin, 1845) occurs in the stomach and small intestine of the rabbit and hare in Europe. Males are 8–16 mm long and females 11–20 mm. The body cuticle bears 40–60 longitudinal ridges. The male bursa has large lateral lobes and a small dorsal lobe.

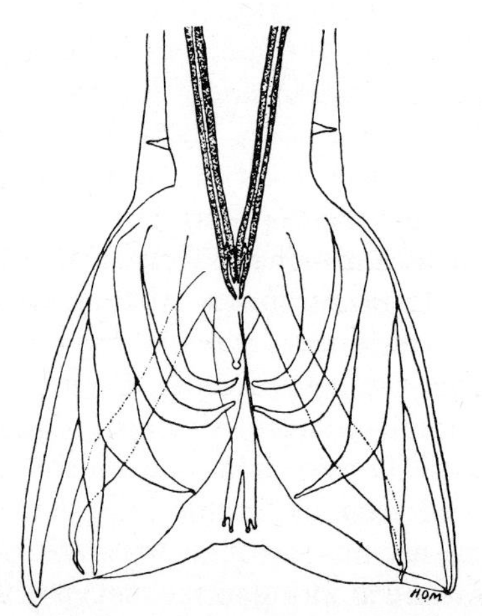

Fig 1.128 Ventral view of hind end of male *Graphidium strigosum*.

The spicules are 1.1–2.4 mm long, slender, and each ends distally in several points. The vulva opens 1.14–3.28 mm from the posterior extremity. The eggs measure 98–106 by 50–58 μm.

The life-cycle is direct.

The worms penetrate deeply into the wall of the stomach or intestine. In some cases no signs are seen even with severe infections, while in other cases anaemia, cachexia and even death may result.

No satisfactory anthelmintic has been determined but probably the benzimidazoles would be of value. For prophylaxis hygienic measures are indicated, such as frequent removal of faeces and dryness of quarters.

Genus: Obeliscoides Graybill, 1924

Obeliscoides cuniculi (Graybill, 1923) occurs in the stomach of rabbits in the USA. Males are 10–14 mm long and females 15–18 mm. Spicules are brown in colour, 0.44–0.47 mm long and are bifurcated at the distal end, each bifurcation ending in a hook. Eggs 76–86 by 44–45 μm.

The life-cycle is direct. Prepatent period about 19 days.

Except in heavy infections this parasite appears to cause no harm.

Genus: Ostertagia Ransom, 1907

The species of this genus, which occur in the abomasum and rarely the small intestine of sheep, goats, cattle and other ruminants, are usually known as brown stomach-worms, because they have this colour when they are fresh. The worms are slender. The cuticle of the anterior extremity may be slightly inflated, transversely striated, and the 'head' is not more than 25 μm wide. The rest of the body cuticle bears 25–35 longitudinal ridges and has no transverse striations. The male bursa has lateral and dorsal lobes and an accessory bursal membrane situated anteriorly on the dorsal side. The spicules are pigmented brown, relatively short, and end posteriorly in two or three processes. The vulva of the female may be covered by a small anterior flap. The systematics and biology of the genus *Ostertagia* sens. lat. have been reviewed recently at a workshop held at the 3rd European multicolloquium of parasitology in Cambridge (Jansen & Gibbons 1981).

Ostertagia ostertagi (Stiles, 1892) occurs in the abomasum of cattle, to which it is strongly adapted, and goats, rarely sheep and horse. Males are 6.5–7.5 mm long and females 8.3–9.2 mm. The spicules are 0.22–0.23 mm long and each ends in three bluntly-hooked processes. The vulva opens in the posterior fifth of the body and is covered by a flap. The eggs measure 80–85 by 40–45 μm.

Ostertagia circumcincta (Stadelmann, 1894) occurs in the abomasum of sheep and goats, being strongly adapted to these species (Borgsteede 1981). Males are 7.5–8.5 mm long and females 9.8–12.2 mm. The spicules are slender, 0.28–0.32 mm long. Each ends in a large knobbed and a small, acute process. The vulva is usually covered by a flap and opens in the last fifth of the body. The eggs measure 80–100 by 40–50 μm. Near the tip of the female tail there is a thickened band

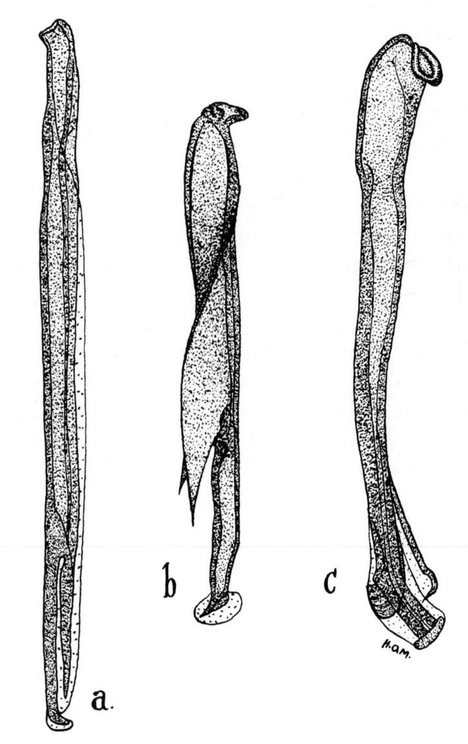

Fig 1.129 Spicules of some common species of *Ostertagia*. a, *O. circumcincta*. b, *O. trifurcata*. c, *O. ostertagi*.

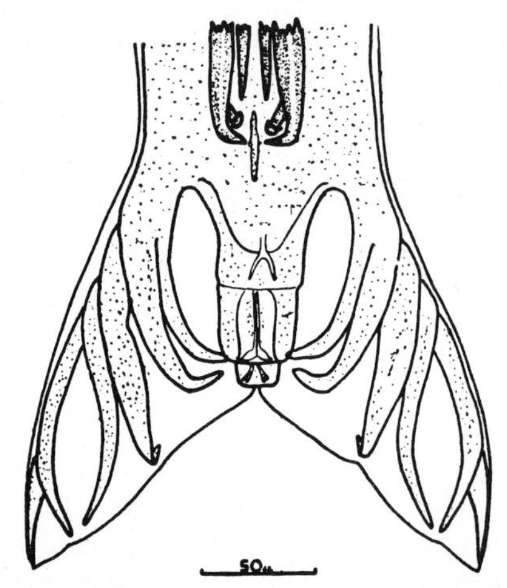

Fig 1.130 Ventral view of male bursa of *Ostertagia lyrata*.

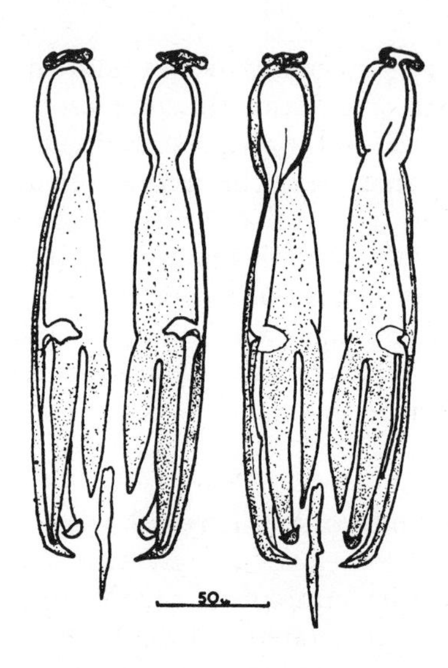

Fig. 1.131 Spicules and gubernaculum of *Ostertagia lyrata*: ventral view on the left; dorsal view on the right.

which bears four to five transverse striations. A similar band is also seen in some other species of this genus.

Ostertagia trifurcata Ransom, 1907 is more adapted to the abomasum of sheep and goats but also occurs in cattle (Borgsteede 1981). The male is 6.5–7 mm long. The spicules are about 0.18 mm long and each ends in a stout, knobbed tip, while just behind the middle two sharp spurs are given off medially.

Ostertagia lyrata Sjöberg, 1926 (syn. *Skrjabinagia lyrata; Grosspiculagia lyrata*) occurs almost exclusively in cattle in Europe, Africa and North America. It is very similar to *O. ostertagi*.

Ostertagia leptospicularis Asadov 1953 is generally considered to be a parasite of cervids in Europe, though it has also been reported from cattle. However, recently it has been reported as a dominant species in bovine ostertagiasis (Al Saqur et al. 1980) and it may increase the severity of disease when associated with *O. ostertagi*. Some authors consider *O. crimensis* Kadenazii and Andreva, 1956, which has been found in cattle and deer in Europe and New Zealand, to be a synonym of *O. leptospicularis*.

Several other species of *Ostertagia* may occur in sheep or cattle and these include: **O. pinnata** Daubney, 1933 (sheep in India, East Africa, Europe); **O. bisonis** Chapin, 1925 (wild ruminants and cattle in North America, the USSR and China); **O. orloffi** Sankin, 1930 (Barbary sheep, cattle and deer in Europe, the USSR and North America); **O. hamata** Mönnig, 1932 (springbok, South Africa); and **O. (Grosspiculagia) podjapolskyi** Schultz, Andreeva and Kadezii, 1954 (cattle, sheep and moufflon in Europe and New Zealand).

The genus **Teladorsagia** Andreeva and Satubaldini, 1954 resembles *Ostertagia* but has a well developed genital cone and lacks an accessory bursal membrane. One species, **Teladorsagia davtiani** Andreevea and Satubaldin, 1954, occurs in sheep and goats in the USSR, the USA and Europe.

Life-cycles

Life-cycles of *Ostertagia* are direct and essentially similar to those of the other Trichostrongylidae. Ingested third-stage larvae exsheath in the rumen and then penetrate the gastric glands in the abomasal mucosa. The third and fourth moults occur in the gastric glands and adult parasites emerge 18–21 days after infection unless hypobiosis (see below) has occurred. The prepatent period is approximately three weeks.

Epidemiology of Ostertagia spp. infection

The epidemiology of *Ostertagia ostertagi* has been extensively studied in Great Britain (Armour et al. 1974). The development and survival of the free-living stages are essentially similar to the other gastrointestinal trichostrongyles. Thus the speed of development of eggs to third-stage larvae is dependent on an average air temperature of over 10°C. Development begins slowly in the spring, peaks in mid-summer and then becomes slower until the autumn; eggs laid in late autumn do not reach the third larval stage until the following spring.

Ostertagia species larvae are fairly resistant to cold and infective larvae, arising from eggs deposited in the previous grazing season, may survive overwinter on pasture until May. Recent data have demonstrated that larvae may survive, presumably in the soil, for an extended period and emerge on to the pasture from July onwards (Bairden et al. 1979). Infective larvae are not markedly resistant to desiccation.

In Europe, ingestion of overwintered larvae in the spring by susceptible stock results in egg-laying adult stages within three weeks. When eggs derived from these, plus eggs derived from the spring or post-parturient rise in faecal egg counts, are deposited on pasture they result in the accumulation of large numbers of infective larvae on pasture from mid-July onwards. Ingestion of sufficient numbers of these larvae results in type I disease in calves during their first summer at pasture. However, should a dry summer supervene, the migration of these larvae from the faecal pad, which acts as a reservoir for the larvae, on to pasture is delayed, and when a dry summer is followed by a wet autumn large numbers of larvae will migrate on to the pasture in the autumn. Under these conditions Type I disease is less likely to occur but an increase in the incidence of Type II disease results.

As autumn progresses an increasing proportion of the larvae ingested do not undergo immediate development to the adult stage but become arrested, as fourth-stage larvae, in the gastric glands. Under appropriate environmental conditions, termed 'conditioning', the free-living stages are influenced so that their development is arrested in the host. In a temperate climate the primary stimulus to the induction of this arrested development is decreasing temperature (Armour & Bruce 1974). This hypobiosis is not permanent and Michel et al. (1976) presented results to suggest that larvae resumed development at a rate of approximately 500–700 a day until nearly all the remaining larvae resumed development together in March/April. The resumption of development of large numbers of larvae could occur earlier in animals with a history and development suggestive of impaired resistance. Armour and Bruce (1974) showed little development of the inhibited larvae until 16–18 weeks after infection or 23 weeks after the larvae were conditioned. The resumption of development of large numbers of inhibited larvae in a wave or succeeding waves in late winter or early spring results in Type II disease. Type II disease is primarily significant in immature cattle but under certain circumstances it may occur in adult cattle (Selman et al. 1976).

Conditioned larvae which remain on pasture also undergo deconditioning. Thus, on pasture, the proportion of larvae capable of undergoing arrest when ingested by the host increases until about December and then decreases to zero by March/April (Michel et al. 1978*a*). The speed of deconditioning is enhanced by an abrupt increase in temperature.

In addition to conditioning of the larvae by environmental conditions, age, previous experience of infection and reproductive status of the host play a role in the proportion of the challenge infection which is arrested within the host (Michel et al. 1979).

The epidemiology of the *Ostertagia* spp. of sheep is similar. For instance, in north-east England, considerable numbers of infective larvae overwinter on the pasture and initiate infection in the spring. Peak larval numbers on pasture occur from mid to late summer and a proportion of parasites survive the winter as inhibited larvae within the host, although a significant number may also survive as adults (Waller & Thomas 1978).

The epidemiology of *Ostertagia* spp. in both cattle and sheep in areas of the world with a temperate climate is similar to that described above but different climates produce differences in the epidemiology. Thus, in areas of the world where hot summers prevail, the infection may survive the hot, unfavourable environmental conditions of summer as inhibited larvae within the host. In Texas, adult parasites of *O. ostertagi* in cattle are seen from November to May, with peak levels of infection occurring in January and February reflecting the levels of pasture larvae. Larvae conditioned for arrested development are acquired from January through March (Craig 1979). Similarly, in winter rainfall zones of Australia, the availability of larvae of the *Ostertagia* spp. of sheep follows the trend of rainfall. Maximal numbers of larvae are available on pasture in late winter. This number declines rapidly in spring and few larvae survive the hot, dry months of summer (Anderson et al. 1978). Lambs become infected in the autumn and winter and autumn-born lambs, infected in late winter, may carry substantial burdens of parasites on through the summer months. Spring-born lambs, infected earlier in the season, have developed substantial acquired resistance by this time.

Pathogenesis

The morphological changes and functional consequences of infection with *O. ostertagi* in calves have been described by Murray et al. (1970). This study demonstrated that the main pathogenic effect of infection with *O. ostertagi* is to reduce the functional gastric gland mass and large areas of the gastric mucosa may be affected.

The pathological changes can be divided into three phases. In the first phase, up to 17 days after infection, lesions are produced by the developing larvae in the gastric glands and morphological changes are confined to the parasitized glands. In infected glands, mature differentiated cells (mucous, zymogenic, parietal) are replaced by undifferentiated cells which develop into tall columnar mucus-secreting cells. Major morphological and functional changes are seen in the second phase, 17–35 days after infection, and are associated with the emergence of adult parasites from the gastric glands and the occurrence of striking changes in the surrounding glands. The enlarged parasitized gland stretches the surrounding glands and stimulates rapid division of epithelial cells so that affected gastric glands become lined with flattened, undifferentiated cells. This lack of differentiation affects particularly the parietal cells and any parietal cells present are not fully differentiated and not functionally active. The rapid cell division results in marked hyperplasia and thickening of the mucosa and junctional complexes between many of the cells are absent. The zona occludens between the cells is shortened or the plasmalemmata are pulled apart.

These morphological changes account for the marked biochemical changes seen at this time and for the clinical signs of the infection. There is a marked increase in abomasal pH since hydrochloric acid is not produced by such undifferentiated parietal cells. Any pepsinogen produced is not converted to pepsin, thus interfering with digestion. The breakdown of cellular junctional complexes results in leakage of macromolecules across the abomasal mucosa (pepsinogen into the plasma exceeding 3 IU tyrosine in severe cases and plasma proteins into the abomasum). This loss of plasma proteins, resultant hypoalbuminaemia and high catabolic rate of albumin develop in a similar manner to that seen in fascioliasis and haemonchosis (Dargie 1975). An increase in viable bacteria in the abomasum has been shown to occur at this time. Affected animals have marked anorexia and severe diarrhoea.

Macroscopically the lesion produced is a raised circular nodule, 2–3 mm in diameter, with a central orifice opening into the parasitized gland. The nodule is due to the extensive cell division and hyperplasia of the mucosa. In heavy infections the whole mucosa is hyperplastic with the characteristic 'morocco leather' appearance.

The third phase is associated with a gradual loss of adult parasites after day 35 of infection and there is a gradual return to a structurally and functionally normal gastric mucosa by days 63–70 of infection.

Similar but more extensive lesions occur in Type II ostertagiasis.

The need to differentiate the fascioliasis–ostertagiasis complex from Type II ostertagiasis has been noted on p. 46. Severe anaemia is always present and plasma pepsinogen levels may be slightly elevated, compared to the marked elevation seen in Type II ostertagiasis.

Clinical signs

Two clinical manifestations of the disease are seen. The first, Type I ostertagiasis, occurs in calves during their first summer at pasture in temperate climates. It occurs from July to October and is associated with the presence of large numbers of adult parasites. There is abomasitis with oedema and necrosis, decreased albumin levels, reduction of appetite and profuse watery diarrhoea, which is often a bright green colour due to the failure of the abomasum to denature chlorophyll. Morbidity is high but mortality is usually low, provided that treatment is instituted.

Type II ostertagiasis consists of a clear clinical entity seen in late winter or early spring in housed or out-wintered cattle after the first grazing season (Selman et al. 1976). It is associated with the emergence of large numbers of inhibited larvae from the gastric glands. However, Type II ostertagiasis may be seen occasionally in adult cattle. It is characterized by a severe chronic diarrhoea and emaciation and frequently ends fatally (Martin et al. 1957), although morbidity is usually low. In some animals more than 200 000 parasites may be found, 60% of which may be immature. The abomasal mucosa is greatly thickened and oedematous; at times there may be superficial necrosis along with an inflammatory exudate. Marked oedema of the abomasum may be seen in older animals. There is marked reduction of serum proteins and subcutaneous oedema may be evident, owing to the loss of serum albumin (Mulligan et al. 1963). Levels of plasma pepsinogen may exceed 3 IU of tyrosine.

Diagnosis

Diagnosis is based on the grazing history and the clinical signs seen in infected animals. Faecal egg counts may be high in Type I disease, often exceeding 1000 epg in severe cases, but may be extremely low or even negative in animals suffering from Type II disease. The elevated levels of plasma pepsinogen may be an aid to diagnosis. In severe cases values may exceed 3 IU of tyrosine (compared with normal levels of less than 1 IU of tyrosine seen in young animals that have not experienced infection; in older animals normal levels may reach 1.5–2 IU tyrosine/litre). Michel et al. (1978*b*) have cautioned against categoric use of plasma pepsinogen levels in diagnosis and gave examples where values may be well above the accepted levels for clinical ostertagiasis in normally growing animals following treatment for ostertagiasis.

Larval counts on herbage may be a guide to diagnosis, particularly when the clinical picture is indefinite, as for example in adult animals (i.e. dairy cows). Levels of infective larvae (Fig. 1.136) of 100/kg of dried herbage have been associated with reduced growth rates, while levels in excess of 1000/kg are associated with clinical disease.

Diagnosis is often confirmed by the response to treatment with benzimidazole anthelmintics: Type I disease usually responds rapidly and within 48 hours appetite has returned and abomasal changes regress rapidly. The Type II disease responds less dramatically and more than one treatment is often required.

Treatment

Treatment of this infection is discussed on p. 247.

Control

The control of ostertagiasis in cattle in Great Britain by means of anthelmintic treatment and pasture management has been described by Armour (1970) and the Technical Development Committee, BVA (1977). Type I disease in susceptible calves may be prevented by the provision of uncontaminated pasture or new ley in the spring. Alternatively, calves grazed on pasture on which overwintered larvae are available in the spring should be treated with anthelmintic and moved to clean pasture in mid-July prior to the peak of larval availability on the pasture. Previously, it was considered that the pasture would be free of overwintered larvae by mid-summer and would thus supply suitable new grazing for susceptible animals. However, recent data have demonstrated that there may be a steady increase in the numbers of infective larvae on aftermath from July onwards. Sufficient larvae may be present to result in Type I disease in calves in late August or September (Bairden et al. 1979). These larvae appear to survive within the soil (Armour et al. 1981), from where they presumably migrate vertically to the surface (Fincher & Stewart 1979) or reach the surface through the aid of hosts such as earthworms (Gronvold 1979). Immature cattle should not be returned to contaminated grazing in autumn or winter, so as to prevent the development of ostertagiasis Type II. In addition, Type II disease may occasionally occur in adult cattle if they are grazed on contaminated pasture in the autumn (Selman et al. 1976).

In the absence of clean pasture, which can be a problem on small farms, animals at risk may be treated at monthly intervals until mid-July. Animals could then be treated with an anthelmintic effective against the larval stages when housed in the winter (see p. 250).

Useful control of bovine ostertagiasis has been achieved by mixed grazing of sheep and cattle. The *Ostertagia* species are fairly host-specific (Borgsteede 1981).

Alternate grazing systems may also be employed. One, developed by the East of Scotland School of Agriculture, utilizes a three-year rotation of cattle, sheep and hay or silage. Rotational grazing in which susceptible calves precede previously exposed older animals results in better herbage utilization and lower rates of contamination of pasture by infective larvae.

Genus: Marshallagia (Orloff, 1933 Travassos, 1937

Marshallagia marshalli (Ransom, 1907) Orloff, 1933. Found in the abomasum and rarely duodenum of sheep, goats and wild ruminants such as antelopes and bighorn sheep primarily in tropical and subtropical climates as far north as southern Europe, the western USA, India and the USSR. Wild ruminants serve as important hosts of the parasite. It is similar to the *Ostertagia* spp. Males are 10–13 mm long and females 12–20 mm. Characterized by a long, slender dorsal ray, 280–400 μm long, which is bifurcated near the tip. Spicules 0.25–0.28 mm long and yellowish-brown in colour: they are split into three processes at the tip. The eggs are large, measuring 160–200 by 75–100 μm, and resemble those of *N. battus*, but the morula is in a more advanced state of division. Further, the geographical distribution of the two parasites differs.

Other species of the genus include: **M. orientalis** (Bhalerao, 1932) Travassos, 1937, hill goat (*Capra sibirica*), India; **M. mongolica** Schumakovitch, 1938, abomasum of sheep, goat and camel, Outer Mongolia; **M. schikhobalovi** Altaev, 1954, sheep, USSR; and **M. dentispicularis** Assadov, 1954, sheep, USSR.

Life-cycle. The life-cycle is similar to that of the *Ostertagia* species, although second-stage larvae appear to hatch from the eggs.The larvae penetrate the gastric mucosa forming whitish-grey nodules, 2–4 mm in diameter. Each nodule may contain two, three or more parasites.The larvae emerge from the nodules 15–18 days after infection and the prepatent period is up to three weeks. Arrested development of larvae may occur and it has been observed that larvae developing in the pyloric region of the abomasum may persist much longer (up to one year) than those developing in the fundic or cardiac regions.

Species of this genus are not known to be serious pathogens.

Genus: Camelostrongylus Orloff, 1933

This genus resembles *Ostertagia* except that the bursa has two large lateral lobes; the spicules are equal, long, narrow and denticulated.

Camelostrongylus mentulatus (Railliet and Henry, 1909) Orloff, 1933 can be an extremely common, but usually non-pathogenic, agent in the abomasum and small intestine of camels. The parasite also infects sheep, goats, antelope and llama. It is found in areas such as the Middle East and Australia. It resembles *O. ostertagi* in size and the eggs are 75–85 by 40–50 μm. Infection produces morphological and functional changes in the abomasum similar to those caused by the *Ostertagia spp.* (Beveridge et al. 1974).

Other genera related to *Ostertagia* include:

Pseudostertagia Orloff, 1933 (*P. bullosa*, Orloff, 1933, abomasum of sheep, bighorn sheep and Barbary sheep and pronghorn antelope in the USA).

Skrjabinagia (Kassimov, 1942) Altaev, 1952 (*S. popovi*, (Kassimov, 1942) Altaev, 1952 and *S. dagestanica* Altaev, 1952 in the small intestine of sheep in USSR; *S. boevi* Bryan, Bainbridge and Kerr 1976, large and small intestine of water buffalo and cattle in the Northern Territory of Australia; *S. kolchida* (Popova, 1937) Andreeva, 1956, abomasum of deer in central Europe).

Spiculopteragia Orloff, 1933 (*S. spiculoptera* Orloff, 1933; abomasum of sheep and various deer in Europe and USSR; *S. peruviana* Guerrero and Chavez, 1964, abomasum of llama, alpaca and vicuña in Peru; *S. boehmi* Gebauer, 1931, moufflon and deer in central Europe).

Longistrongylus Le Roux, 1931 (*L. albifrontis* (Mönnig, 1931) Travassos, 1937, abomasum of blesbok and springbok in southern Africa; *L. meyeri* Le Roux, 1931, abomasum of Thomson's gazelle and African black buffalo in southern Africa).

Genus: Cooperia Ransom, 1907

Species of this genus, which are usually found in the small intestine, rarely in the abomasum, of ruminants, are relatively small worms, of a reddish colour when fresh. The cuticle of the anterior extremity frequently forms a cephalic swelling and the rest of the body cuticle bears 14–16 longitudinal ridges which are transversely striated. The male bursa has a small dorsal lobe. The lateroventral ray is thicker than the ventroventral and divergent from it, but its tip again approaches the latter. The posterolateral is slender and the externodorsal usually arises from the base of the dorsal stem. The spicules are stout, relatively short, pigmented brown, and usually have a ridged, wing-like expansion at the middle. An accessory piece is absent. The vulva may be covered by a flap and is situated behind the middle of the body.

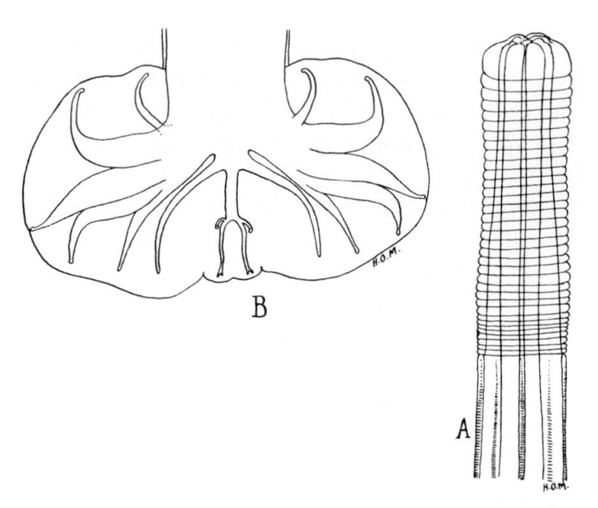

Fig 1.132 *Cooperia curticei*. A, Anterior end. B, Bursa of male opened out.

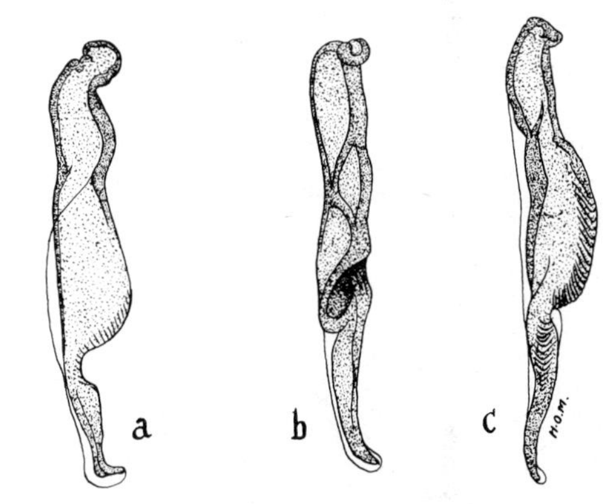

Fig 1.133 Species of some common species of *Cooperia*. a, *C. curticei*. b, *C. punctata*. c, *C. pectinata*. Not drawn to the same scale.

Cooperia curticei (Railliet, 1893) occurs in sheep and goats and rarely cattle. Males are 4.5–5.4 mm and females 5.8–6.2 mm long. Spicules 0.135–0.145 mm long.

Cooperia punctata (v. Linstow, 1907) occurs in cattle and rarely in sheep. Males are 4.7–5.9 mm long and females 5.7–7.5 mm. Spicules 0.12–0.15 mm long.

Cooperia pectinata (Ransom, 1907) occurs in cattle and rarely in sheep. The male is 7 mm long and the female 7.5–9 mm. Spicules are 0.24–0.28 mm long.

Cooperia oncophora (Railliet, 1898) occurs primarily in cattle and also in sheep and rarely the horse. Males are 5.5–9 mm long and females 6–8 mm. Spicules are 0.24–0.3 mm long.

Cooperia surnabada Antipin, 1931 (syn. *C. mcmasteri*, Gordon, 1932) is a parasite of cattle, sheep and camel in Britain, Australia and North America. The male measures 6.8 mm and the female 7.9 mm. In general it resembles *C. oncophora* but the bursa is larger and the rays of the bursa longer and thinner. The spicules are also thinner, 0.27 mm in length and end in a bifurcation, the external part of which ends in a small conical expansion.

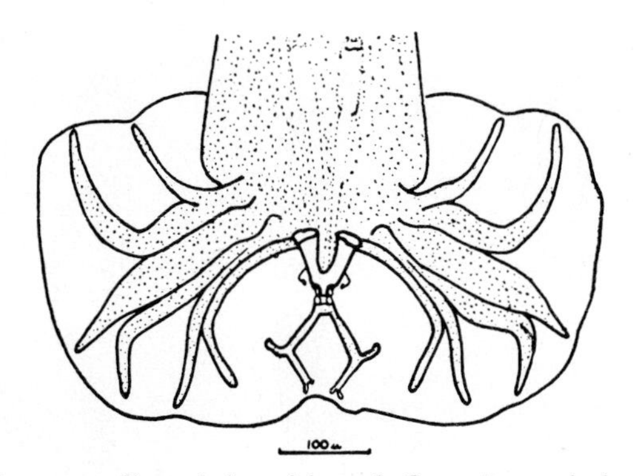

Fig 1.134 Ventral view of the male *Cooperia surnabada*.

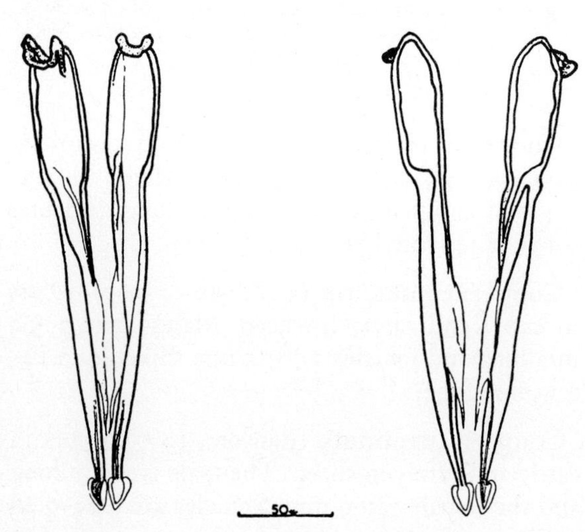

Fig 1.135 Spicules of *Cooperia surnabada*.

Cooperia spatulata Baylis, 1938 occurs in the small intestine of cattle and sheep in Malaysia, East Africa, Australia and North America (Florida, Georgia, Mississippi, Hawaii). Males are 6–8 mm long and females 6–7.8 mm. Spicules are 0.225–0.29 mm and eggs 68–82 by 34–42 μm.

Other species of *Cooperia* include **C. bisonis** Cram, 1925 in the abomasum and small intestine of bison and pronghorn antelope in the USA and of several wild ruminants in Africa. It occurs rarely in cattle and sheep.

The life-cycle of *Cooperia* spp. is direct and in general similar to that of *Trichostrongylus* spp. Infection of the host occurs by ingestion. The infective larva has a pointed tail and is surrounded by a sheath with a medium-sized tail. The prepatent period of *C. curticei* is 14 days, that of *C. oncophora* about three weeks.

The epidemiology of infection is similar to that of the other gastrointestinal trichostrongyles (see p. 000). Infective larvae may survive on pasture for nine to 26 weeks and there may be considerable winter survival of the infective stages. In addition, the parasites are able to undergo arrested development in a similar manner to *O. ostertagi* in order to survive adverse environmental conditions.

The worms penetrate into the mucosa of the small intestine. A light infection is of no consequence, but young cattle and sheep may be severely affected by heavy infections, which are usually acquired on moist pastures. The clinical signs and lesions are similar to those of trichostrongylosis.

Diagnosis has to be made by faeces culture and the identification of the infective larvae (Fig. 1.136). The eggs of species of the genus *Cooperia* cannot be specifically identified in the faeces.

For treatment and prophylaxis see pp. 247–52.

Genus: Paracooperia Travassos, 1935

Similar to *Cooperia*; spicules divide into three processes, one of which is large and alate. Occur in ruminants.

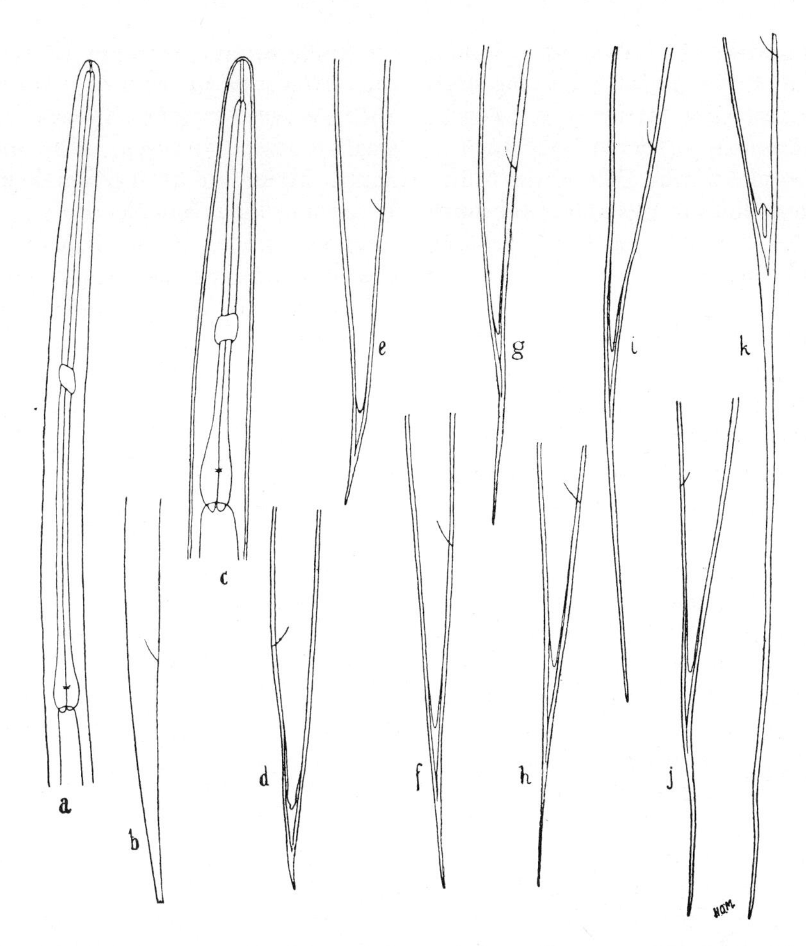

Fig 1.136 Infective larvae of some nematodes of sheep. *a*, *Strongyloides papillosus*, oesophageal region; *b*, hind end of same; *c*, *Trichostrongylus*, oesophageal region; *d*, hind end of same; *e*, *Ostertagia*, hind end; *f*, *Cooperia*, hind end; *g*, *Haemonchus*, hind end; *h*, *Bunostomum*, hind end; *i*, *Oesophagostomum*, hind end; *j*, *Chabertia*, hind end; *k*, *Nematodirus*, hind end. (All drawn to same scale)

	Total length	Oeso-phagus	Tail of larva	Tail of sheath
Strongyloides papillosus (filariform larva)	0.6	0.23	0.09	Sheath absent
Trichostrongylus colubriformis	0.69	0.165	0.06	0.094
Ostertagia spp.	0.84	0.16	0.075	0.112
Cooperia spp.	0.78	0.16	0.067	0.124
Haemonchus contortus	0.69	0.136	0.06	0.142
Bunostomum trigonocephalum	0.57	0.16	0.06	0.140
Chabertia ovina	0.73	0.165	0.064	0.165
Oesophagostomum columbianum	0.79	0.16	0.07	0.214
Nematodirus spathiger	1.1	0.225	0.056	0.326

Paracooperia nodulosa (Schwartz, 1928) Travassos, 1937, (syn. *P. matoffi*) occurs in the small intestine and occasionally caecum and colon of buffalo and occasionally other animals, such as the zebra, in Asia and Africa. The larvae of this parasite occur in nodules in the wall of the small intestine. *P. nodulosa* may be very pathogenic in young buffalo in India.

Genus: Nematodirus Ransom, 1907

The species of this genus are relatively long worms with a filiform anterior portion. They have an inflated cuticle around the anterior end and about 14–18 longitudinal ridges on the body cuticle. The anterior part of the body is thinner than the posterior part. The male bursa has elongate lateral lobes which are covered internally by rounded or oval cuticular bosses, while the dorsal lobe with its supporting rays is split in two and each half is attached to a lateral lobe. The spicules are long and slender and their tips are fused together. The ventral rays are parallel and close together. Except in *N. battus*, the mediolateral and posterolateral rays lie close together, except at their tips. The tail of the female is short and truncate, with a slender terminal appendage. The vulva opens at the posterior third of the body. The eggs are so large that their size readily distinguishes them from those of other trichostrongylid species usually found in farm mammals. Eggs passed in the faeces of the host contain about eight cells.

Nematodirus spathiger (Railliet, 1896), frequently confused with *N. filicollis* (Rudolph, 1802), is the commonest species and occurs in the small intestine of sheep, cattle and other ruminants. Males are 10–15 mm long and females 15–23 mm long. Spicules are 0.7–1.21 mm long, terminating in a spoon-shaped expansion. The eggs measure 175–260 by 106–110 μm and contain an embryo of about eight cells when passed by the host.

Nematodirus battus Crofton and Thomas, 1954. This species of sheep was discovered independently by Crofton and Thomas (1954) in England and Morgan (unpublished) in Scotland. The male is 10–16 mm long and the female 15–24 mm. The spicules of the male are 0.85–0.95 mm long and meet only at the tips in a flattened, bluntly pointed projection. The mediolateral and posterolateral rays of the male bursa are divergent, not parallel as they are in *N. spathiger* and *N. filicollis*. The eggs measure 152–182 by 67–77 μm and have a brown shell.

Nematodirus filicollis (Rudolphi, 1802) occurs in the small intestine of sheep, cattle, goats and deer and is cosmopolitan in distribution. It is essentially similar to *N. spathiger* from which it can be differentiated by the spicules which end in a narrow, pointed enlargement, this being distinct.

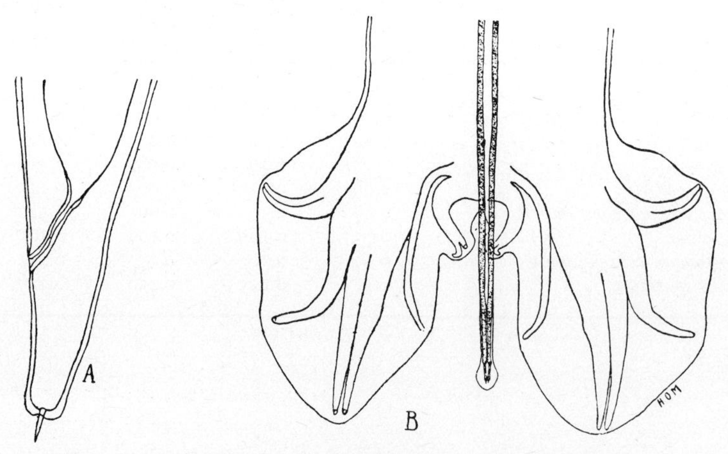

Fig 1.137 *Nematodirus spathiger*. A, Hind end of female. B, Hind end of male with bursa opened out.

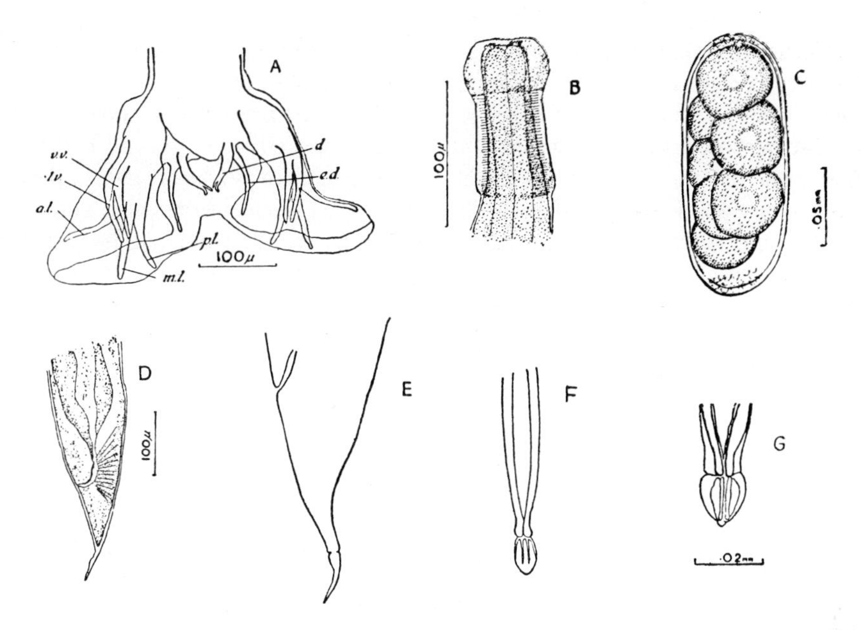

Fig 1.138 *Nematodirus battus.* A, bursa. B, Anterior end. C, Egg. D, E, Posterior end of the female. F, G, Tips of the long spicule of the male

Nematodirus helvetianus May, 1920, is chiefly a parasite of cattle in Europe (Switzerland, Great Britain) and the USA. It also occurs in sheep in these areas. The male is 11–17 mm and the female 18–25 mm. The dorsal lobe of the bursa is not separated from the lateral lobes; the spicules measure 0.9–1.25 mm and at the distal end they form a point, the enclosing membrane being lanceolate. The eggs measure 160–230 by 85–121 μm.

Other species of *Nematodirus* which occur include **N. abnormalis** May, 1920, in sheep, goats, camels in North America, Europe, the USSR, Asia and Australia; **N. oiratianus** Rajevskaja, 1929, in the camel and wild ruminants in Europe and Russia; **N. leporis** Chandler, 1924, and **N. aspinosus** Schultz, 1931, occur in rabbits and hares in North America and the USSR. A number of species occur in wild ruminants. These include: **N. tarandi** Hadwin, 1922, **N. odocoilei**, Beckland and Walker, 1967, **N. andreevi** Sutabaldin, 1954, **N. lamae** Becklund, 1963 and **N. hsuei** Liang, Ma and Lin, 1958 (sheep in China).

Life-cycle and epidemiology of N. battus

Eggs passed in the faeces develop slowly under field conditions in Great Britain and in two to three months the infective third larval stage within the egg is reached in late summer, but development may continue until between November and March. Although these eggs do have a low infectivity when fed to lambs (Gibson 1958) they are largely on the soil surface and unavailable for ingestion. The availability of the infection on pasture is dependent on hatching and the translation of infective larvae to the herbage and this does not occur until the following spring. The larvae become sensitized to hatch by prolonged exposure to cold conditions and the stimulus to hatching is the rise in soil temperature which occurs in spring. For example, Thomas and Stevens (1960) found that when larvae of *N. battus* in eggs were exposed to temperatures of 2–3°C the number of larvae which hatched when the larvae were returned to 21°C increased as the exposure to the low temperature increased, a maximum hatching occurring after seven months exposure to low temperature. Once triggered by appropriate temperature and moisture conditions there is a mass hatch of infective larvae in spring and these accumulate on herbage. The hatched larvae survive only a few weeks and there is then a rapid decline in pasture levels. Thus, the level of infection in lambs is dependent on the hatching date and lambing date. If weather conditions are such that hatching is early and prior to the time at which the lambs are grazing then the prevalence of infection in lambs that year will be low. Conversely, should spring be late, the eggs hatch at a time when the lambs are grazing and heavy infections result.

Infective third-stage larvae, when ingested, penetrate the intestinal mucosa between the villi and moult to the fourth stage by day four. Many leave the mucosa between days four and six but others are still in the mucosa on day 10 and by day 10 the majority have moulted to the fifth stage before emerging from the mucosa. The prepatent period is 15 days (Mapes & Coop 1972). Infection in the lambs is short-lived and a large proportion of adult parasites survive only a few weeks and faecal egg counts then drop rapidly. Elimination of the parasites appears to be dose-dependent since experimental infections of 60 000 larvae in lambs were eliminated by days 24–28 of infection while an infection with 20 000 larvae persisted at least 72 days (Lee & Martin 1976). The lambs then become highly resistant to reinfection. Thus, there is only one generation of parasites per year and eggs passed by a current year's lambs overwinter to hatch the following year and infect the next year's lambs. It is essentially a lamb-to-lamb transmitted infection. However, older lambs and adult animals may carry light infections of *N. battus* and thus may contribute to pasture contamination.

Life-cycle and epidemiology of other Nematodirus spp.

The parasitic life-cycle of *N. filicollis*, *N. spathiger* and *N. helvetianus* is similar to that of *N. battus*. The prepatent period of infection with *N. filicollis* and *N. spathiger* is two to three weeks and that of *N. helvetianus* is three weeks. Rapid expulsion occurs in heavy infections but is delayed in lighter infections (Samizadeh-Yazd & Todd 1979).

However, preparasitic development on pasture differs markedly. *N. filicollis* is less important as a pathogen than *N. battus* since, although hatching is delayed, it occurs over an extended period beginning in autumn and increasing steadily through winter to reach a small spring peak (Boag & Thomas 1975). *N. spathiger* and *N. helvetianus* eggs do not show delayed hatching and their epidemiology is more nearly similar to that described for *Trichostrongylus* spp. under parasitic gastroenteritis (see p. 240).

Clinical signs

Clinical signs in lambs infected with *N. battus* are first seen during the prepatent period and they are associated with the emergence of larvae from the mucosa. Inappetence occurs and within a few days (10–11 days after infection) there is a sudden onset of acute enteritis. Anorexia, severe blackish-green and then yellowish diarrhoea, dehydration and prostration occur. There may be a considerable mortality (up to 30%) in lambs six to 10 weeks of age. After about three weeks there is an improvement in the clinical condition of the animals, a rapid drop in faecal egg counts and surviving lambs are highly resistant to reinfection. The clinical effects of the other *Nematodirus* spp. infection are comparable to trichostrongylosis.

Pathogenesis and pathology

The parasites penetrate the intestinal mucosa causing extensive destruction and tunnelling (Samizadeh-Yazd & Todd 1979) but the posterior portion of the parasite may be protruded into the lumen of the intestine. In lambs, by day 16 of an infection with 60 000 infective larvae of *N. battus* the surface area of the villi becomes reduced. Villous atrophy, similar to that seen during infection with *Trichostrongylus* spp., follows by day 20. At this time the parasites become enclosed in mucus-like material, which may be associated with the rejection that occurs at this time in such heavy infections. The structure of the mucosa returns to normal by day 32 (Martin & Lee 1976).

Carcasses of affected lambs are dehydrated and emaciated and there may be acute inflammation of the mucosa of the whole small intestine. Large numbers of *N. battus*, particularly immature stages, will be present in the intestine. However, a large proportion of immature stages may be lost from the small intestine between days two and four and adults from the mucosal surface between days 12 and 16. Thus very variable worm counts may be seen (Mapes & Coop 1972).

Diagnosis

Acute enteritis occurring in late April or May in lambs grazing pastures grazed by lambs the preceeding year is suggestive of infection with *N. battus*. Clinical signs often occur during prepatency of the infection and thus infection can be confirmed only on post-mortem examination in these cases. The other species of *Nematodirus* contribute to the general picture of parasitic gastroenteritis. Patent infections are readily recognized by the number of large strongyle-type eggs in the faeces.

Control

Lambs should not be raised on pasture occupied by lambs the previous spring. Ploughing and reseeding of pastures reduce the level of infection but not absolutely. Similarly, if a pasture is not grazed by lambs one year the levels of infection decrease markedly but a small proportion of the free-living stages can persist for two years and these low levels of infection can be escalated in following years if lambs are grazed on such a pasture. Alternatively, control may be based on the ability to forecast the time of hatching of eggs of *N. battus*. Ollerenshaw and Smith (1966) and Smith and Thomas (1972) demonstrated that, on the basis of the mean soil temperature taken at a depth of 30 cm for the period 1 to 20 March, it is possible by late March each year to forecast the peak time of hatch and thus predict the period of six to eight weeks each year when the pastures would be heavily contaminated with larvae of *N. battus*. At this time lambs may be moved off the pasture or they may be treated prophylactically with an anthelmintic given two or three times at three-week intervals. Normally rainfall is not computed in the forecasting since it is generally adequate to permit hatching. However, if the weather is unusually dry then hatching will be delayed and a discrepancy between forecasting and prevalence can exist.

Treatment

See p. 247

Genus: Nematodirella Yorke and Maplestone, 1926

This resembles *Nematodirus*. The anterior part of the body is narrow; the spicules are equal, very long, up to half the length of the body. Three species occur in domestic and wild ruminants: **N. longispiculata** Yorke and Maplestone, 1926 (small intestine of sheep, goat and other ruminants in USSR, China, northern Europe); **N. cameli** Rajewskaya and Badinin, 1933 (in camel and reindeer in USSR); and **N. dromedarii** Douvres and Lucker, 1958 in dromedary in Europe. The eggs of all these species are large, 230–270 by 110–140 μm. The life-cycle is probably similar to that of *Nematodirus* spp.

Genus: Haemonchus Cobb, 1898

This is an important genus of the abomasum of various ruminants, frequently associated with disease. Parasites are 10–30 mm long. They possess a small buccal cavity with a slender tooth or lancet; cervical papillae are prominent; the bursa is large, especially the lateral lobes, and the dorsal lobe is small and asymmetrical. The vulva is posterior in the female and knobs, flaps or linguiform processes are present or absent in the vulvar region. Substantial variation may occur in morphological features and the genus appears in a state of evolutionary flux. Gibbons (1979) has reviewed the genus and concludes that nine species are valid.

Haemonchus contortus (Rudolphi, 1803) occurs in the abomasum of sheep, goats, cattle and numerous other ruminants in most parts of the world. It is commonly known as the 'stomach-worm' or 'wireworm' of ruminants, and is one of their most pathogenic parasites.

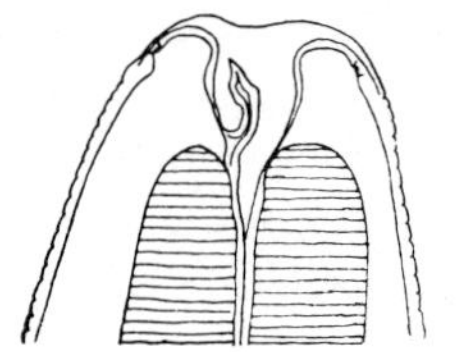

Fig 1.139 Lateral view of anterior end of *Haemonchus contortus.*

Males are 10–20 mm long and females 18–30 mm. The male has an even reddish colour, while in the female the white ovaries are spirally wound around the red intestine, producing the appearance of a barber's pole. The cervical papillae are prominent and spine-like. A small buccal cavity is present, containing a dorsal lancet. The male bursa has elongate lateral lobes supported by long, slender rays; the small dorsal lobe is asymmetrically situated against the left lateral lobe and supported by a Y-shaped dorsal ray. The spicules are 0.46–0.506 mm long, each provided with a small barb near its extremity. The vulva of the female is usually covered by a linguiform process (vulva flap), which is usually large and very prominent, but may be reduced to a small knob-like structure in some specimens. The eggs measure 70–85 by 41–48 μm, and those passed in the faeces of the host contain an embryo divided into 16–32 cells.

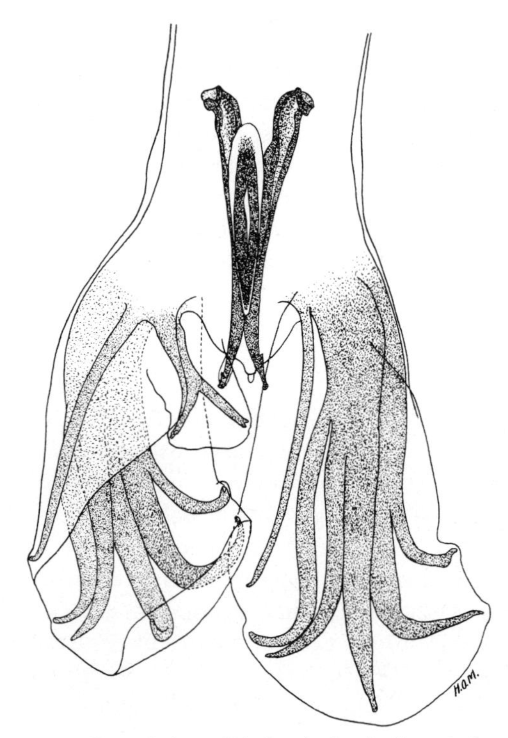

Fig 1.140 Dorsal view of hind end of male *Haemonchus contortus.*

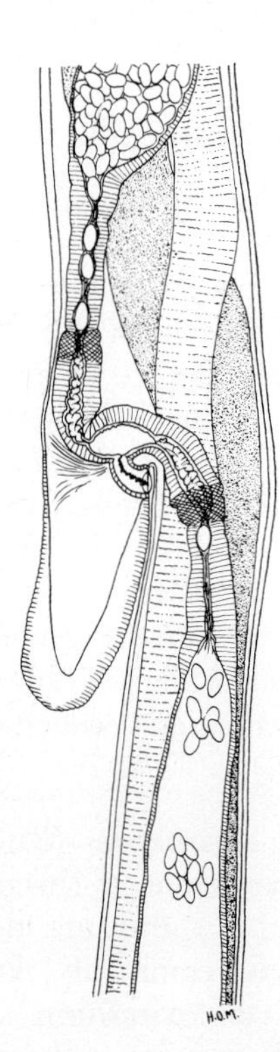

Fig 1.141 Vulvar region of female *Haemonchus contortus.*

Das and Whitlock (1960) studied the comparative morphology of *Haemonchus* spp. from various geographic regions. Consistent morphological differences suggested that two new subspecies and a new variety of *Haemonchus contortus* be created, namely *Haemonchus contortus contortus* (Australia), *H. contortus cayugensis* (New York State, USA) and *H. contortus* var. *utkalensis* (Orissa, India). Subsequently, other subspecies were recognized: *H. contortus bangalorensis* (India), *H. contortus hispanicus* (Spain) and *H. contortus kentuckiensis* (Kentucky). Daskalov (1972) indicated that the host was largely responsible for the development of the various forms of *H. contortus*, the different proportions in a natural mixed population being due to geographical and ecological factors. On the basis of this and other considerations, Gibbons (1979) concluded that the various subspecies and varieties of *H. contortus* are synonyms of *H. contortus*. However, Le Jambre (1981), on the basis of hybridization studies, has concluded that there are distinct ecotypic differences in the genus *Haemonchus* and he argues in favour of retaining the subspeciation of the genus.

Haemonchus placei (Place, 1893) Ransom, 1911. Studies in Australia by Roberts et al. (1954) led to the conclusion that the cattle and sheep forms of *Haemonchus* were distinct and represented two species, *H. contortus* in sheep and *H. placei* in cattle. Studies by Bremner (1955) on the chromosomes (2N = 11 male; 12 female) indicated differences in the X chromosome in the two species, in the cattle species it measured 8 μm whereas in the sheep form it was similar to the autosomes and measured 3 μm. Subsequently, Herlich et al. (1958) demonstrated the occurrence of the two species in the USA. The differential features were based on mean spicule lengths, the distance between the barbs and the tip of the spicules and differences in size and activity of infective larvae. Thus mean spicule lengths are *H. contortus* 398–431 μm, *H. placei* 454–470 μm; barb to tip of spicule left *H. contortus* 41–46 μm, *H. placei* 52–54 μm; right *H. contortus* 21–40 μm, *H. placei* 27–37 μm. F_1 males resulting from mating between *H. contortus* males and *H. placei* females are sterile and sterility also occurs in F_2 males of matings between *H. contortus* females and *H. placei* males (Le Jambre 1979). However, Gibbons (1979) considers such morphological characters unreliable; for example vulvar flaps may vary with host, age and host resistance. She considers that differences are those applicable between strains and does not accept *H. placei* as a valid species, but regards it as a synonym of *H. contortus*.

Haemonchus similis Travassos, 1914. This species has been reported from cattle and deer in Florida, Louisiana and Texas, and also occurs in cattle in Europe and Brazil. It differs from *H. contortus* in that the terminal processes of dorsal ray are longer and the spicules are shorter, being 139–334 μm long.

Haemonchus longistipes Railliet and Henry, 1909 occurs in the camel and dromedary in North Africa and India. The spicules are much longer than those of the other species, being a mean of 625 μm.

A number of other species are recorded from wild ruminants. These include: **H. bedfordi** Le Roux, 1929 (abomasum, African buffalo, various gazelle); **H. dinniki** Sachs, Gibbons and Leveno, 1973 (abomasum, various gazelle); **H. krugeri** Ortlepp, 1964 (abomasum, impala); **H. lawrenci** Sandground, 1933 (small intestine, duiker); **H. mitchelli** Le Roux, 1929 (abomasum, various gazelle, eland, oryx); **H. vegliai** Le Roux, 1929 (abomasum, oryx, antelope).

Life-cycles of Haemonchus spp.

The preparasitic development of *H. contortus* is very similar to that of other strongyles. Detailed studies of the embryonation of eggs and the development of larvae of *H. contortus* have been carried out by Dinaburg (1944), Silverman and Campbell (1959) and Dinnik and Dinnik (1958) and the ecology of larvae on pasture has been reviewed by Crofton (1963) and Levine (1963). Variations in temperature requirements for the various subspecies of the genus are discussed by Le Jambre (1981).

Under satisfactory environmental conditions infective larvae are reached in four to six days. Low temperatures retard development and below 9°C little or no development takes place. Eggs which have reached the 'prehatch' stage are more resistant to adverse conditions and can survive freezing and desiccation more readily than other stages. However, the eggs and infective larvae of *H. contortus* are intolerant of desiccation and low temperatures. Thus, in summer rainfall regions and in areas with mild winters larval availability on pasture tends to increase in the late spring, reach maximal levels after mid-summer and decline through the winter. In general, there is little winter survival. Conversely, in winter rainfall areas the larvae do not survive the hot, dry summer period, though infective larvae do have a considerable ability to survive successive cycles of desiccation and rehydration (Wharton 1982).

H. contortus survives such adverse environmental conditions (cold weather and hot, dry weather) through arrested development within the host. In the temperate climate of Great Britain, the pattern of inhibition of development in *H. contortus* is similar to that in *Ostertagia* spp., although it appears earlier. In East Anglia, for example, significant numbers of arrested larvae are present in sheep in August. This number may reach 80% by September and may be 100% from October to March (Connan 1971). In the north-east of England inhibition may occur as early as July (57%) and may be virtually 100% by September (Walker & Thomas 1975). In contrast, in Nigeria, larvae of *H. contortus* undergo arrested development to survive the dry season. Inhibited larvae represent 14% of the *H. contortus* burden in June to July, 55% in August and 95% by October (Ogunsusi & Eysker 1979). The stimulus for arrested development in *H. contortus* infective larvae, unlike those of *Ostertagia* spp., may not be declining temperatures, since, for example, *H. contortus* larvae may undergo arrested development as early as July in Great Britain. Variations in *in vitro* culture conditions suggest that the development of larvae under 'wet' as opposed to 'dry' conditions stimulates the development of arrest. Thus, arrested development was much more prevalent in cultures with a water content of 77 g/litre as opposed to cultures with 58 g/litre water content (Connan 1978). The number of larvae undergoing arrested development also increased with the length of storage of the cultures.

Resumption of development of arrested larvae occurs at a time when environmental conditions are suitable for the survival of the free-living stages and is possibly associated with a seasonal stimulus. Larvae resume development primarily in April in Great Britain (Connan 1978), resulting in the spring rise in faecal egg counts, and at the beginning of the rainy season in Nigeria (Ogunsusi 1979). The superimposition of parturition on this process permits the maturation of these larvae with a resultant marked periparturient rise in faecal egg counts.

The larvae of *H. placei* of cattle behave similarly to those of *H. contortus* of sheep, except that the faecal pad of cattle may serve as a reservoir for larvae. Durie (1961) has reported that larvae may survive here for several months and ultimately be released when the faecal pad is sufficiently moistened by rain.

Following ingestion of infective larvae exsheathment occurs in the rumen and parasitic larval stages migrate to the abomasum and penetrate between the gastric epithelial cells from which they emerge as fourth-stage larvae. The prepatent period for *H. contortus* in sheep is an average of 15 days and for *H. placei* in cattle between 26 and 28 days. Details of the parasitic developmental stages of *H. contortus* of sheep are given in the classic study by Veglia (1915) and those of the parasite in cattle by Bremner (1956).

Pathogenesis

The principal feature of *Haemonchus* spp. infection is anaemia. Both the adult and the fourth larval stages of *H. contortus* in sheep and *H. placei* in cattle suck blood and, in addition, move and leave wounds which haemorrhage into the abomasum. The average blood loss has been calculated at 0.05 ml/parasite/day and blood first appears in the faeces six to 12 days after infection (Clark et al. 1962).

Dargie and Allonby (1975) have investigated the development of anaemia in sheep heavily infected with *H. contortus* using radio-isotopic techniques (Cr^{51} labelled erythrocytes and Fe^{59} labelled iron) and demonstrated that the anaemia developed in three stages. In the first stage, seven to 25 days after infection, the packed cell volume (PCV) of infected sheep fell rapidly from 33 to 22% while serum iron remained normal. This rapid drop in PCV was due to a time lag between the loss of blood and the activation of the erythropoietic system of the host to compensate for this blood loss. In the second phase of the anaemia, lasting for six to 14 weeks, the PCV was maintained at a steady, but lower than normal, level despite the continued blood loss, since infected sheep compensated for the blood loss through and increase in erythrocyte production (approximately three-fold). During this time there was an increase in plasma iron turnover and a marked loss of iron in the faeces. The sheep had a limited capacity to reabsorb iron from the intestine, with only an average of 11% being reabsorbed, although sheep that suffered the greatest haemorrhage had the greatest capacity to reabsorb iron. Eventually, severe depletion of the iron reserves occurred, affected sheep exhibited low serum iron and low bone marrow reserves of iron and then the third phase, that of anaemia, developed. This was manifested by a rapid drop in PCV resulting from a dyshaemopoiesis due to iron deficiency.

Infected animals lose large quantities of serum proteins into the gut with the mean daily faecal 'clearance' of plasma being recorded as 210–340 ml/day when estimated from the loss of I^{131}-labelled polyvinylpyrrolidone (Dargie 1975). In consequence, the fractional catabolic rate of albumin is markedly increased. However, serum albumin concentration can remain at a steady level for at least several weeks and this must be associated with a marked increase in albumin synthesis until the animals' metabolic reserves are depleted.

Clinical signs

The clinical signs of haemonchosis may be divided into three syndromes: hyperacute, acute and chronic.

Hyperacute haemonchosis is uncommon but it may be seen when susceptible animals are exposed to a sudden massive infection. The extremely large number of parasites causes a rapidly developing severe anaemia, dark-coloured faeces and sudden death from acute blood loss. There is a severe haemorrhagic gastritis. Death may occur in the prepatent period of such heavy infections.

Acute haemonchosis is seen primarily when young susceptible animals become heavily infected. The anaemia may develop fairly rapidly, but there is expansion of the erythropoietic response of the bone marrow. The anaemia is accompanied by hypoproteinaemia and oedema (i.e. bottle jaw) and deaths occur. Faecal egg counts are usually high—up to 100 000 epg. The carcase shows generalized oedema and anaemia and 1000–10 000 parasites may be present in the abomasum.

Chronic haemonchosis is extremely common and of considerable economic importance. The disease is due to chronic infection with a fairly low number of parasites (100–1000). Morbidity is 100% but mortality is low. Affected animals are weak, unthrifty and emaciated. Anaemia and hypoproteinaemia may or may not be severe, depending on the erythropoietic capacity of the animal, the iron reserves remaining and the nutritional metabolic reserves of the host. Faecal egg counts may at times be less than 2000 epg. On post mortem examination there is hyperplastic gastritis and chronic expansion of the bone marrow.

The post mortem signs will vary depending on the clinical syndrome of haemonchosis. In general the mucous membranes and the skin are pale, while the blood has a watery appearance. The internal organs are also markedly pale. Hydrothorax, fluid in the pericardium and ascites are usually conspicuous and extreme cachexia is present, the fat being replaced by a gelatinous tissue. The liver has a light-brown colour; it is fragile and shows fatty changes. The abomasum contains reddish-brown fluid ingesta and a large number of worms that are readily seen and that swim about actively if the carcase is still warm. The mucosa is swollen and covered with small red 'bite-marks' of the parasites. Occasionally shallow ulcers with ragged edges are found, and a number of the worms may be firmly attached with their anterior extremities in these ulcers. The intestine may contain a few worms which are being passed out by the host.

Genetic aspects of Haemonchus spp. infection

The influence of breed on the susceptibility of sheep (and goats) to infection with *H. contortus* has been studied by various workers. Thus Knight et al. (1973) reported that Navaho lambs supported fewer worms than Suffolk, Rambouillet, Targhee and Corriedale lambs, while Florida native lambs were reported to be more resistant than Rambouillet lambs to *H. contortus* infection (Bradley et al. 1973). Todd et al. (1978) compared Targhee and Targhee–Barbados Black-Belly cross lambs using experimental infections and were unable to demonstrate any significant differences ($P > 0.05$) in weight gain, packed cell volumes, eggs per gram of faeces or numbers of worms between the two groups. In Yugoslavia, adult Ciyaja and Merino Prekos sheep were more resistant to infection with *H. contortus* than were Merino Karkas sheep (Cvetković et al. 1973).

Preston and Allonby (1978, 1979) compared the susceptibility to *H. contortus* infection of six breeds of sheep and three of goats in East Africa. The indigenous Red Masai was the most resistant breed of sheep, the Hampshire Down being least resistant and the Blackhead Persian, Merino, Dorper and Corriedale breeds being intermediate. The breed differences were related to differences in the establishment of infection (Preston & Allonby 1979) and to the relative ability of breeds to elicit an immune response. This increased resistance of the indigenous East African sheep may be the result of natural genetic selection in an endemic environment; for example, Red Masai sheep were shown to achieve higher mean levels of anti-larval IgA antibodies than Merinos. On the other hand, with goats, which are browsers in their natural environment, the genetic pressure to develop resistance to *H. contortus* in the indigenous breeds might be precluded.

Differences in susceptibility to *H. contortus* are greater between breeds than between haemoglobin phenotypes within a breed. The influence of the latter has been studied in detail by Altaif and Dargie (1978*a*, *b*) using Scottish Blackface and Finn Dorset sheep. Homozygous haemoglobin type A sheep of both breeds showed lower worm burdens and reduced effects of parasitic infections than animals homozygous for haemoglobin type B. Genetic resistance operated mainly against worm establishment and that this was controlled by the immune response elicited. However, there was a lack of association between 'self-cure' (see below) and haemoglobin phenotype within either breed, though Allonby and Urquhart (1976) had suggested that HbA Merinos underwent self-cure more frequently and effectively than those of HbB.

Further evidence that the genetic resistance of the host to infection with *H. contortus* has an immunological basis is evident from studies of the superiority of HbAA sheep as compared to HbBB sheep to produce better antibody responses when immunized with unrelated antigens i.e. human serum albumin (Cuperlović et al. 1978). Further, Riffkin and Dobson (1979) were able to predict resistance to *H. contortus* in Merino × Border Leicester sheep by means of the *in vitro* responsiveness to antigen of peripheral blood lymphocytes. Preinfection responses to lymphocytes were correlated with resistance to infection in that high responder sheep passed fewer eggs, had lower worm burdens and an earlier onset of the self-cure reaction than did low responder sheep. Lymphocyte responsiveness and resistance was inherited in a predictable manner although the inheritance of lymphocyte responsiveness was related to the response with third-stage larval antigen and not with adult antigen.

Immunology of Haemonchus infections

Probably one of the best known phenomena of immunity to helminths is the 'self-cure' reaction which occurs in *H. contortus* infection. This reaction which results in the loss of a burden of parasites can be induced in suitably infected and sensitized sheep by a challenge dose of infective larvae, the reaction being initiated when the challenge larvae moult from the third to the fourth larval stage (Soulsby & Stewart 1960). The self-cure reaction is more likely to occur in sheep which have experienced several suitably spaced doses of larvae rather than those which carry an initial infection. A latent period of six to seven weeks is required between initial infection and administration of the infection which induces the self-cure reaction. Self-cure is accompanied by a transient rise in blood histamine, an increase in the complement-fixing antibody titre and intense mucosal oedema in the abomasum (Stewart 1953). Sheep failing to show self-cure of the infection showed no rise in circulating histamine. The self-cure reaction could be induced by the injection of exsheathed *H. contortus* larvae into that abomasum as well as by the administration of ensheathed larvae by mouth. The reaction is dependent on antigens associated with living larvae and which act locally, since the intraperitoneal injection of living larvae, while producing a good antibody response, fails to induce self-cure; similarly dead infective larvae given by mouth fail to induce it.

The association of self-cure with the moulting period of the challenge dose of larvae was demonstrated by Soulsby and Stewart (1960) directly by examining the abomasal population of parasites at the time of self-cure and indirectly by obtaining serological evidence of a marked reaction to exsheathing fluid at the time of self-cure. A more detailed consideration of the self-cure reaction is given by Soulsby (1966).

The reaction is not entirely specific since challenge with *H. contortus* larvae will induce self-cure of a *Trichostrongylus* spp. infection: however, the reverse, namely challenge with *Trichostrongylus* larvae, does not induce self-cure of a *H. contortus* infection. The self-cure reaction is not solely an experimental entity. It is an important mechanism for terminating natural gastrointestinal parasitism in sheep in Great Britain (Soulsby 1957) and in Australia natural burdens of *H. contortus* are terminated by the process, especially after rain when the intake of infective larvae provides the stimulus for the reaction. However, self-cure may also occur on lush pasture in the absence of re-infection. This might be attributable to an 'anthelmintic substance' or an 'allergic substance' in freshly growing grass or to physiological alterations in the abomasum (Allonby & Urquhart 1973). Both host and parasite genetic factors may influence the occurrence of the self-cure reaction. *H. contortus cayugensis* does not induce a self-cure reaction (Whitlock 1966) and the self-cure reaction was shown to be more effective and more frequent in Merino HbAA sheep than in HbBB Merinos in East Africa (Allonby & Urquhart 1976).

Self-cure and protection against infection are not necessarily interrelated and Gordon (1968) summarized the main responses that occurred following reinfection with *H. contortus*: (*a*) self-cure and protection against the new infection; (*b*) self-cure and establishment of the new infection; (*c*) no self-cure and establishment of the new infection; and (*d*) no self-cure and no establishment of the new infection.

A substantial degree of protective immunity may be produced against infection with *H. contortus*. Immunization of sheep over six months of age with irradiated larvae induces a significant resistance to challenge infection (Mulligan et al. 1961). However, animals younger than six months of age respond poorly, or not at all, to immunization against *H. contortus* (Manton et al. 1962; Benitez-Usher et al. 1977). This unresponsiveness may be immunologically mediated and *in vitro* lymphocyte responsiveness of lambs to antigen of *H. contortus* does not reach adult levels until the lambs are several months of age.

Resistance to *H. contortus* in sheep may be controlled genetically and marked differences in resistance to infection have been demonstrated in different breeds of sheep as discussed on p. 235. Thus Preston and Allonby (1978) concluded that resistance may be the result of natural selection in an endemic environment. Haemoglobin type has also been associated with resistance to infection with *H. contortus* (Altaif & Dargie 1978) and the superiority of HbAA sheep as compared to HbBB sheep may be immunologically mediated (Cuperlović et al. 1978) (see p. 236).

Lactation has a profound influence on the hosts' response to infection with *H. contortus*. Periparturient relaxation in immunity is manifest by increased susceptibility to newly acquired infection, activation of inhibited larvae, maturation of the parasites to adulthood and increased fecundity of parasites (O'Sullivan & Donald 1973; Connan 1976; Soulsby 1977; Shubber et al. 1981). This periparturient rise in faecal egg output acts as a source of infection for the new generation of animals. The periparturient relaxation in immunity has been associated with hormonal changes and the lactogenic hormone, prolactin, appears important in this phenomenon. In addition, antigen- and phytomitogen-induced blastogenesis of peripheral blood lymphocytes is markedly reduced in late pregnancy and early lactation and is associated with a significant increase in faecal egg output at that time (Chen & Soulsby 1976).

The immune response to *H. placei* in cattle differs from that induced by *H. contortus* in sheep. Roberts (1957) was unable to induce self-cure. He demonstrated that a marked resistance to re-infection was acquired quite rapidly, even after a single dose of larvae. The immunity is manifested by a marked inhibition of development of fourth-stage larvae in the abomasal mucosa. Such inhibited larvae can persist for a considerable time and are apparently inhibited by the presence of a population of adult worms. The removal of such adult worms by anthelmintic treatment may allow the inhibited larvae to develop to maturity and this new population may be seriously pathogenic.

Diagnosis

The clinical signs alone may lead to a suspicion of haemonchosis, and these, supported by high faecal egg counts (including identification of larvae in faecal culture), should be sufficient to establish a diagnosis. However, a definitive diagnosis can only be made by autopsy of a representative clinical case from the flock.

Treatment

See p. 247.

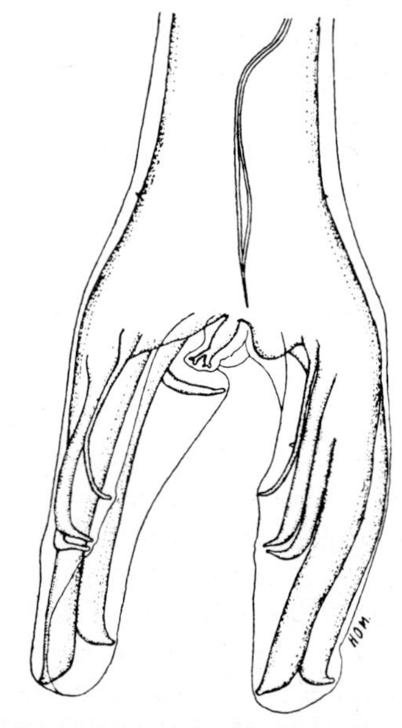

Fig 1.142 Dorsal view of the hind end of male *Mecistocirrus digitatus.*

Genus: Mecistocirrus Railliet & Henry, 1912

Mecistocirrus digitatus (von Linstow, 1906) occurs in the abomasum of sheep, goat, cattle, zebu, buffalo and the stomach of the pig and rarely man in eastern countries and in Central America. Males are up to 31 mm long and females up to 43 mm. The cuticle bears about 30 longitudinal ridges. The cervical papillae are prominent and the small buccal capsule contains a lancet as in *Haemonchus*. The female resembles that of the latter genus in having the ovaries wound spirally around the intestine, but is readily distinguished from it by the position of the vulva which is situated about 0.6–0.9 mm from the tip of the tail and the absence of a vulvar flap. The male bursa has a small, symmetrical dorsal lobe; the ventroventral ray is small, while the lateroventral and anterolateral rays are conspicuously longer than the other rays. The spicules are long and slender, 3.8–7 mm long, and united together for almost their whole length. The eggs measure 95–120 by 56–60 μm.

The life-cycle is direct, the prepatent period being about 60 days (Fernando 1965). In this infection the fourth larval stage is of long duration, lasting from the ninth to the twenty-eighth day of infection.

M. digitatus is an important pathogen of buffalo, cattle and goats in endemic areas, the effects of infection being similar to those of *H. contortus*.

GASTROINTESTINAL NEMATODE INFECTION IN RUMINANTS

Pathogenesis

This has been reviewed by Soulsby (1976), Symons and Steel (1979) and Dargie (1981).

The pathology caused by gastrointestinal parasites is varied. Thus infection with *Ostertagia* spp. is associated with morphological and functional destruction of the gastric glands of the abomasum. The primary pathology caused by *Haemonchus contortus* and *Mecistocirrus digitatus* is marked haemorrhage through wounds in the abomasal mucosa. Infections such as *Trichostrongylus* spp. and *Nematodirus* spp. cause villous atrophy and adults of *Oesophagostomum* spp. and *Chabertia ovina* in the large intestine cause ulceration and haemorrhage. These lesions have been described previously.

In general, anorexia and thus reduced feed intake are observed in parasitized animals. This contributes to the poor weight gains and lowered production seen in parasitized animals. Further, in addition to the loss of whole blood brought about by the blood-sucking activities of haematophagous nematodes, gastrointestinal parasites cause a protein-losing gastroenteropathy (see p. 222). The extensive proliferation of epithelial cells in the parasitized gastrointestinal tract results in replacement of differentiated functional cells with immature non-functional cells with imperfectly formed intercellular junctional complexes (Murray et al. 1970, 1971). This results in leakage of macromolecules through the mucosa into the intestine. The protein loss is also contributed to by the intestinal lymphangiectasia which may be seen (Nielsen & Anderson 1967). Leakage of proteins into the intestine and resultant hypoproteinaemia, particularly hypoalbuminaemia, has a profound effect on protein metabolism. The kinetics of the hypoalbuminaemia are similar to those described for anaemia. In the initial phase there is a fairly rapid fall (20–30%) in serum albumin levels. In the second phase the host compensates for the protein loss by adjusting endogenous protein catabolism. These alterations in protein metabolism are reflected by an increase in the incorporation of amino acids into liver protein synthesis, associated with increased plasma protein synthesis, in an attempt by the animal to maintain homeostasis. Consequently, during this second phase, although albumin levels may be below normal, they do remain stable. However, at the same time reduced skeletal muscle protein synthesis and reduced incorporation of amino acids in wool follicles is seen (Symons & Jones 1975, 1978). The altered protein metabolism and reduced feed intake accompanied by changes in cholecystokinin levels (see p. 217) are associated also with impaired utilization of nutrients for growth and production. In the third phase, exhaustion of body reserves results in a severe hypoalbuminaemia. Anaemia may be seen after prolonged infection with non-blood-sucking gastrointestinal nematodes. The anaemia probably results from a deficiency in the amino acids required for haemoglobin synthesis.

Increased urea synthesis, possibly metabolized from ammonia arising from the deamination of amino acids, elevated levels of plasma urea and increased excretion of urinary nitrogen contribute to the inferior nitrogen balance of parasitized animals (Roseby & Leng 1974). Local dysfunction occurs in the parasitized portion of the gastrointestinal tract. Thus, in ostertagiasis there is decreased production of HCl by the parietal cells and a resultant lack of conversion of pepsinogen to pepsin (Murray et al. 1970). This may interfere with protein digestion. In infections of the small intestine associated with villous atrophy there is a deficiency in brush border enzymes (i.e. alkaline phosphatase, maltase, dipeptidase) (Coop & Angus 1975). However, malabsorption *per se* does not play a major role in production loss by parasitized animals (Symons 1976). Although reduced digestion and absorption of proteins, fats and sugars may occur at the site of parasitism, a compensatory increase in digestion and absorption is seen at a distal site in the intestine. However, animals in the field will be infected with a large number of parasites affecting different parts of the gastrointestinal tract. The presence of each of these species may prevent the compensatory absorption and exacerbate the effects of the other species.

Gastrointestinal parasitism also induces disorders in mineral metabolism. Reduced calcium and/or phosphorus and magnesium absorption, reflected by reduced deposition of these elements in skeletal bone, results in reduced bone growth in young animals (Reveron et al. 1974; Sykes et al. 1975). Skeletal development may be limited further by reduced availability of protein and reduced skeletal growth has long-term implications for production (Sykes et al. 1977).

Economic effects

The effects of parasitism on production are well documented. The anorexia and reduced feed intake, loss of blood and plasma proteins into the gastrointestinal tract, alterations in protein metabolism, depressed levels of minerals, depressed activity of some intestinal enzymes and diarrhoea all contribute to depressed weight gain, wool growth and milk production.

Reduced skeletal growth brought about by mineral deficiencies affects growth rates, since skeletal size ultimately determines the capacity of the growing animal to accumulate muscle (Sykes et al. 1977). Further, reduced levels of amino acid incorporation in muscle protein results in reduced weight gains; even weight loss may occur. Wool production is suppressed as a result of the reduced incorporation of amino acids into protein by wool follicles. The quality of the wool is reduced and a break in wool growth may be seen in heavily parasitized animals.

Clinical parasitism has a marked effect on milk production. Recently, however, the economic importance of subclinical parasitism and its effects on milk production in dairy cows have been demonstrated. Thus, Bliss and Todd (1976) demonstrated the economic advantages of anthelmintic treatment of adult dairy cattle at the time of parturition and again in mid-lactation. Field trials demonstrated increased milk production (more than 200 kg/lactation) after treatment of dairy cows passing fewer than 10 epg of faeces. Milk production was suppressed in cows given 200 000 trichostrongylid larvae when the larvae were administered in the first 90 days of lactation. A significant increase in strongyle egg output was noted in beef cows at calving and anthelmintic treatment at calving markedly reduced post-calving egg counts (Hammerberg & Lamm 1980). Borgsteede (1978) has shown that, in Europe, the species contributing to this lactational rise in egg output in the first two weeks after parturition of dairy animals are *Ostertagia* spp., *Trichostrongylus* spp. and *Cooperia oncophora*, while a second peak of egg output which occurred four weeks later consisted of *C. oncophora*, *Oesophagostomum*, *Haemonchus* and *Bunostomum* spp.

Worm burdens of dairy heifers in England and Wales were investigated by Hong et al. (1981) who showed that over 60% of 143 animals had less than 10 000 worms and only four exceeded 100 000 (mean 16 285). *Ostertagia ostertagi* was always present and formed the major population of the total. *Trichostrongylus axei* was also a common species while other species of the genus *Ostertagia* were usually present but in small numbers. Arrested early fourth-stage larvae were highest in winter and declined during spring.

Such studies indicate there is an increasing tendency for adult animals to carry substantial burdens of nematodes. Possibly this is related to the efficacy of control measures for parasitic gastroenteritis in young animals, with the result that older animals have little experience of, and immunity to, infection. The positive effects of anthelmintic treatment have not been universal and this might be associated with different methods of husbandry (Barger 1979). For instance anthelmintic treatment of housed dairy cattle might be expected to be more efficacious than the treatment of cows on pasture since the latter will be continually subjected to reinfection. The effects of anthelmintic treatment were greatest in the most productive dairy herds (Bliss & Todd 1973). These results may reflect the greater appetite, and therefore greater larval intake, of high-yielding animals (Fox & Jacobs 1981) or may reflect the effects of low levels of parasites on the high demands for energy in the high-yielding dairy cow. It is likely that anthelmintic treatment will be the most effective in high-yielding dairy cattle fed for peak production. Conversely, anthelmintic treatment may be efficacious in poorly managed, heavily parasitized cows.

Little information is available regarding the effects of anthelmintic treatment on milk production in the goat. A single study demonstrated that the treatment of goats with thiabendazole resulted in a 17.6% increase in milk production over a three-week period (Farizy 1971). In sheep, clinical or subclinical parasitism associated with the periparturient relaxation in protective immunity might be expected to suppress milk production. This, in turn, might be reflected by reduced growth rates of the suckled lambs.

Epidemiology

The bionomics of the free-living stages of trichostrongylid nematodes have been reviewed by Levine (1978). The primary factors affecting the development and survival of the eggs and larvae are temperature and moisture and different parasites vary in their ability to survive extremes of temperature and humidity. Thus, *Haemonchus* spp. and *Oesophagostomum columbianum* predominate in hot climates while *Trichostronylus* spp., *Ostertagia* spp. and *Oesophagostomum venulosum* predominate in warm climates. In general the third-stage larva is the least susceptible to adverse environmental conditions. This is followed by the embryonated egg, the unembryonated egg, first-stage larva and then the second-stage larva, although Todd et al. (1976) demonstrated that the unembryonated egg of *H. contortus* was more susceptible to high or low temperatures than were the second- and first-stage larvae.

The free-living stages are killed by excessive heat or cold but the effects of fluctuating temperatures on the parasites are not fully known. Hsu and Levine (1977) demonstrated that although maximum development occurred under constant conditions of temperature and relative humidity (RH) (100%), larvae of *H. contortus* and to a lesser extent *T. colubriformis* developed when temperatures were cycled between 20 and 35°C at 100% RH.

Surface soil moisture is important and a minimum amount is required for development to take place, but the effect of fluctuating moisture content is not known. However, the larvae of both *H. contortus* and *T. colubriformis* were able to develop in small numbers if the RH was cycled between 70 and 100% (mean 84–89%) in contrast to the fact that constant 85% RH did not permit development of larvae of *H. contortus* and only a few infective larvae of *T. colubriformis* were able to develop under the latter conditions (Hsu & Levine 1977). Infective larvae of *H. contortus* do survive repeated desiccation and rehydration, although other larvae may not be able to survive as well (see Wharton 1982). Surface soil moisture is affected by such factors as the amount, frequency and type of precipitation, the rate of evapotranspiration from the soil and vegetation and the soil type.

Translation of infective larvae on to the herbage is also dependent on moisture and temperature. The optimum temperature for migration is related to survival but at times when the temperature is optimal little migration occurs in the absence of adequate moisture. Rose (1962) found that sufficient rainfall or moisture to penetrate the dung pat was required to create suitable conditions for migration. However, through the summer, even in fairly dry weather, if temperatures are not too high, infective larvae can develop and survive within the faecal pat. This acts as a reservoir for the larvae until adequate moisture stimulates their release. There is a logarithmic decrease in the number of larvae found as the distance from the faecal pat increases and the majority of larvae of *H. contortus* were found within 20 cm of the faecal pat (Skinner & Todd 1980). Other factors that aid in the translation of larvae onto the herbage include wind and rain, these serving to disintegrate faecal pats. For example, Boswell and Smith (1976), using a fluorescent marker in cattle faeces, demonstrated distribution of faecal material over an area of pasture of more than one hundred times the original area of the faecal pats. The distribution of larvae may well follow the distribution of the marker. Further, the role of invertebrates such as earthworms and dung beetles as transport hosts of the free-living stages of the helminths is not yet fully known, but they may aid the distribution of the parasites by moving faecal material mechanically (Gronwold 1979). Larvae of the trichostrongylid nematodes may be distributed by spores of the fungus *Pilobolus* (Bizzell & Ciordia 1965) and psychodid flies have been incriminated in the dissemination of infection (Jacobs et al. 1968).

Once on herbage, the survival of the infective larvae is once more dependent on temperature and moisture. Larvae such as *H. contortus* can survive repeated desiccation and rehydration and desiccated larvae of *H. contortus* become infective when moisture is replaced, for instance after the falling of dew. This may be true for other species also. Infective larvae also appear to be able to survive adverse conditions such as cold (overwinter) in the soil. Thus, Nelson (1977) found significant numbers of *Dictyocaulus viviparus* larvae from 7.5 cm downwards in the soil in early spring in Great Britain. As temperatures rise these larvae presumably migrate vertically through the soil (Fincher & Stewart 1979) and in warmer weather no larvae are found in the deeper layers of the soil but larvae are detected near the soil surface and on the herbage. Indeed, survival in the soil may be prolonged and larvae from eggs deposited the previous autumn may emerge on to aftermath following the removal of hay or a silage crop in mid-July (Bairden et al. 1979).

In areas with temperate climates two sources of pasture contamination are available in the spring (Thomas 1973; Michel 1976; Armour 1980; Brunsdon 1980). The first is the spring rise and the post-parturient rise in faecal egg counts arising from the maturation of inhibited or hypobiotic larvae (see below) in the ewe. This contamination by ewes produces a peak in larval availability on pasture from mid-summer and when ingested by lambs results in heavy infections capable of producing disease in late July, August and September. In England the eggs passed by these lambs are deposited on pasture when temperatures are falling and normally give rise only to a lower wave of infective larvae. This second peak is not a frequent cause of disease in England but it is an important source of larvae which overwinter on the pasture. In New Zealand, on the other hand, this second peak is larger than the first and can cause considerable disease (Vlassoff 1973). The larvae on pasture in the autumn also become conditioned and undergo arrested development when ingested by the host and supply a source of infection in the following year.

The second source of pasture contamination is the presence of overwintered larvae on pastures grazed by infected animals the previous summer and autumn. These overwintered larvae remain viable on pasture until May or June and are immediately available for ingestion when lambs or calves are released on the pastures in the spring. Infection accumulates in such animals and they begin passing eggs within three weeks of their introduction to the pasture. The animals may suffer from subclinical disease as a result of ingestion of these overwintered larvae. The eggs passed by these infected animals in April, May and June, together with the eggs passed by ewes in the periparturient period, will result in a peak of larvae on pasture in mid-July and consequently a second wave of infection in animals in mid to late summer and autumn. In addition, the presence of overwintered larvae on pasture could increase the magnitude of the periparturient rise in faecal egg counts since periparturient animals are more susceptible to newly acquired infection (Shubber et al. 1981).

A similar epidemiological pattern with a limited number of parasite generations per year is described for other areas in the world. These include New Zealand and parts of Australia with a temperate climate (Donald & Walker 1973; Vlassoff 1973). Conversely, in winter rainfall areas in Australia larvae are abundant on the pasture in the winter and early spring. These larvae are probably derived mainly from eggs deposited in the late summer and autumn. The faecal egg output of adult sheep is highest in late summer and autumn and conditions are more favourable for larval development in autumn than in winter. A rapid decrease in larval availability occurs in early summer (September/October) associated with a rise in the mean maximum temperature and decrease in moisture so that larvae virtually disappear from pasture in the hot, dry summer months (Anderson 1972, 1973). Again, an increasing number of ingested larvae undergo arrested development in late winter, spring and early summer.

In tropical areas larval availability on pasture coincides with the rainy season. Larval numbers on pasture decrease at the beginning of the dry season and arrested larvae accumulate in the host at this time (Hart 1964; Ogunsussi & Eysker 1979). In Nigeria termination of arrested development occurs in April, the increase in faecal egg counts being coincidental with the arrival of the heavy rains and conditions suitable for the development of the free-living stages (Ogunsussi 1979). In areas of the world where summer temperatures and rainfall are high, several generations of parasites a year are possible (Armour 1980).

The epidemiology and pattern of larval contamination will be altered by local variations in weather. For instance, a dry summer followed by a wet autumn in Great Britain can delay the translation of larvae on to the herbage. The moisture content in the centre of a bovine faecal pat can remain sufficiently high for several weeks to support development of the infective stages but larvae do not migrate on to the pasture until wet weather ensues. In conditions such as this few larvae are available for ingestion in the summer and parasitic gastroenteritis due to the presence of large numbers of adult parasites in the gastrointestinal tract of susceptible animals is less likely to occur. However, with the onset of autumn rainfall large numbers of larvae will be released on to the herbage. These will be conditioned by environmental factors and large numbers may become arrested when ingested by grazing animals. As a result Type II ostertagiasis will be more common and, in particular, the magnitude of the spring and post-parturient rise will be increased, resulting in heavier levels of pasture contamination in the following year.

As the grazing season progresses there is a definite succession of parasite species on the pasture and within the host. This may be associated with the relative ability of the larvae of certain species to overwinter on pasture or to be transmitted in the periparturient rise in faecal egg counts; the rate at which the free-living stages develop; and differences in the resistance of the host against a particular species of parasite.

Hypobiosis in trichostrongylosis

Inhibited or arrested development, or hypobiosis, has been reviewed in a symposium edited by Borgsteede et al. (1978). It is a phenomenon describing the temporary cessation of development of nematodes at a precise point in early parasite development and serves to synchronize the development of the parasite with events in the host and in the environment. In domestic animals, gastrointestinal nematodes such as *Ostertagia* spp., *H. contortus*, *Trichostrongylus* spp., *C. oncophora*, *C. ovina*, *N. helvetianus* and *H. rubidus* are capable of undergoing arrested development within the host. The phenomenon is also seen in *D. viviparus* and *D. filaria* of ruminants, in *A. duodenale* of man and in a variety of other nematodes.

The accumulation of significant populations of inhibited larvae within the host coincides with the onset of environmental conditions which are adverse to the survival of the free-living stages. There is evidence that the degree of inhibition is related to the adversity of the environmental conditions (Armour 1978). For instance, there is a high magnitude of inhibition in areas with extremely cold winters or very arid, hot summers whereas in the comparatively mild winters of Great Britain the order of magnitude is only 50–90% and in hot, humid climates a low level of inhibition is seen.

Varied stimuli serve to initiate inhibition or to condition the infective larvae in such a way that their development in the host is arrested. The stimulus may be associated with host factors, or be parasite-related, i.e. genetic, or be environmental. In temperate areas arrested development of *O. ostertagi* and *O. circumcincta* has been linked with low or declining temperatures in the autumn (Armour & Bruce 1974). However, in western Australia, Nigeria and Israel these parasites undergo arrest to survive the hot, arid summer (Hart 1964; Anderson 1973; Shimshony 1974). There is some evidence for cold conditioning in the arrest of development of *H. contortus* in Canada and New Zealand (Blitz & Gibbs 1972; McKenna 1973). In contrast, in Great Britain *H. contortus* undergoes arrest from mid-summer, so temperature and photoperiod are presumably minimally involved (Walker & Thomas 1975; Connan 1975). Connan (1978) demonstrated *in vitro* the influence of culture period and water content of the cultures on increasing the percentage of larvae capable of undergoing arrest and Walker and Thomas (1975) have suggested that arrested development is a genetically programmed and normal part of the life-cycle of *H. contortus* in Great Britain.

The percentage arrest in development seen in different parasites varies. Thus a high proportion of *H. contortus* survive adverse conditions as inhibited larvae within the host. In contrast, *T. axei* survive primarily as adult parasites within the host, although arrested development does occur also. The strain of parasite may also influence percentage arrest within the host. For instance, the 'Weybridge strain' of *O. ostertagi* may have lost the capability to respond to cold conditioning although it has not lost the capability to undergo arrest. There may also be a relationship between the parasite generation and the ability of its progeny to undergo arrest. The progeny of adults which have developed from overwintered pasture larvae show a lesser propensity for inhibited development than the progeny of parasites that have suffered arrest (Michel 1978). In experimental infection, the number of larvae administered may be a factor in the development of arrest and may be important in infections such as *Graphidium strigosum* in rabbits (Martin et al. 1957).

Arrested development may also be seen as a consequence of host resistance. For instance, in addition to environmental factors, host-mediated factors may play a role in the induction of arrest in *O. ostertagi*. Thus, previous experience of infection may initiate arrested development of *O. ostertagi* and innate age resistance can also be expressed in older animals (Michel et al. 1979). Acquired resistance may be of importance in the development of arrest in other gastrointestinal parasites of domesticated animals and is of primary importance in *T. retortaeformis* in rabbits.

Inhibition of development or hypobiosis is not permanent and resumption of development appears to be timed to occur when environmental conditions are suitable for the development of the free-living stages. The mechanism of stimulation for resumption of development is not clear but the stimulus is probably associated with a diversity of factors. Resumption of development may be spontaneous and occur synchronously or asynchronously. It may be associated with host reproductive factors and loss of immunity associated with removal or loss of the adult parasites. Stress, poor nutrition and concurrent disease may be associated in the release of hypobiotic larvae from the dormant state.

Resumption of development of *O. ostertagi* larvae appears to occur spontaneously and Armour and Bruce (1974) demonstrated that this occurred about 23 weeks after conditioning of the larvae on pasture, although Michel et al. (1976) showed a linear decrease in the number of arrested larvae at the rate of 600–700/day until March/April when the majority of the remaining inhibited larvae developed synchronously.

Deconditioning also occurs spontaneously in infective larvae maintained in vitro if the storage conditions are kept constant and it can be enhanced by an abrupt increase in temperature. Armour and Bruce (1974) attributed this decline to mortality of conditioned larvae but Michel et al. (1975) did not observe preferential death of such larvae. Under field conditions in Great Britain, the proportion of larvae which become arrested when ingested by the host increases until about December, but by March few or none do so.

The spontaneous and constant development of 600–700 inhibited larvae a day might, perhaps, be attributable to a regulatory mechanism dependent on the presence of adult worms while the resumption of development of the bulk of the inhibited larvae in March/April could, perhaps, be the end of a programmed sequence of events beginning with conditioning and ending with maturation (Michel 1978). However, Type II ostertagiasis resulting from the maturation of large numbers of inhibited larvae can occur at any time from December to May and therefore is presumably influenced by factors such as loss of immunity, pregnancy and lactation and poor nutrition. Outbreaks of Type II ostertagiasis tend to be seen in unthrifty animals, and the stress of calving or periparturient immunosuppression may be associated with Type II ostertagiasis, which has been recorded in recently calved heifers (Wedderburn 1970).

Maturation of inhibited larvae of *H. contortus* in sheep and the resultant spring rise in faecal egg counts has been related to a loss of immunity due to the limited exposure to infection in winter, to stress, to immunosuppression associated with the late stages of pregnancy and lactation and to the frequent occurrence of poor nutrition at the end of winter. However, recent data suggest that maturation of the inhibited larvae may be seasonally related. Connan (1978) demonstrated that, although maturation of *H. contortus* occurred over a two- to three-month period, peak maturation occurred in April irrespective of the lambing date in February and the infection of the sheep on pasture. This suggests that maturation is a response to a seasonal stimulus, though spontaneous development at a specific time after conditioning on pasture cannot be excluded. In non-lactating sheep, many of the larvae are rejected before they develop to the adult stage but often these parasites mature at a time of nutritional stress and they may result in disease in affected animals.

Periparturient rise in faecal egg counts in trichostrongylosis

Animals often show an increased faecal egg count, the periparturient rise in faecal egg counts, beginning in late pregnancy and rising to a peak in early lactation. Increased faecal egg counts of *Oesophagostomum* spp. and *H. rubidus* are seen in lactating sows (Connan 1967) and several species of gastrointestinal nematodes are involved in the phenomenon in sheep. That lactation is associated with this periparturient increase in faecal egg counts is demonstrated by the abrupt fall in faecal egg counts that is seen when piglets are weaned or lactation in ewes is prevented by removal of their lambs at birth (Connan 1967; O'Sullivan & Donald 1973). There is evidence that this periparturient increase in faecal egg counts results from a temporary relaxation in immunity and this may be influenced by endocrinological changes. Prolactin has received much attention in this respect and prolactin secretion in the ewe follows the pattern of the host's increased susceptibility to infection with gastrointestinal helminths (Connan 1973). In addition, the injection into ewes of diethylstilboestrol results in a rise in plasma prolactin and is accompanied by an increase in faecal egg counts. Immunologically, there is a marked suppression in phytomitogen and specific *H. contortus* antigen cell-mediated immunity responses in late pregnancy and during lactation (Chen & Soulsby 1976) and cell transfer studies in guinea-pigs have demonstrated that the differentiation of sensitized lymphocytes to effector cells is inhibited during lactation (Dineen & Kelly 1972).

This periparturient relaxation in immunity is manifest by the maturation of inhibited or hypobiotic larvae and, in particular, these larvae complete their development to adulthood in the periparturient animal. That parturition and lactation amplify the effects of an on-going spring rise in faecal egg counts is suggested by the results of Cvetković et al. (1971) and others. In these studies maximum post-parturient faecal egg counts occurred only if lambing was timed to occur at the period during which a spring rise in faecal egg counts was occurring in non-reproductive animals. Lambing before or after this optimum period was not associated with a periparturient rise in faecal egg counts. In addition, there is an increased fecundity of the parasites present in lactating animals and there is also a marked increase in susceptibility to newly acquired infection in the periparturient period (Shubber et al. 1981).

While these increased burdens of parasites, particularly *H. contortus* in lactating ewes, may result in subclinical or clinical disease during lactation, the phenomenon ensures the contamination of pasture with infective stages and transmission of infection to the new generation of animals.

Host genetic factors influencing trichostrongylosis in ruminants

Genetic resistance to disease is of increasing importance in livestock development programmes and the influence of genetic constitution on the susceptibility of animals to gastrointestinal helminths is increasingly under investigation. The topic has been reviewed recently by Frisch (1981) with respect to parasites of cattle in tropical areas. The host genetic factors associated with *H. contortus* infection have been discussed on p. 235. Early work by Whitlock and Madsen (1958) demonstrated that, in sheep, the inheritance of resistance to trichostrongylosis was complex but highly heritable. Since then a marked difference in resistance to infection with *H. contortus* has been demonstrated in different breeds of sheep (see p. 235) and similar differences have been noted in resistance to infection with *Ostertagia* in sheep. Thus Targhee and Panama breeds have been shown to be more resistant to infection than were Hampshire, Suffolk and Rambouillet breeds (Scrivner 1964).

Haemoglobin type has also been shown to influence resistance against infection with helminths although results have been variable. For example, haemoglobin AA sheep were more resistant to infection with *H. contortus* and *O. circumcincta* than sheep with HbBB genotype (Altaif &

Dargie 1978). This may be associated with a greater ability of HbAA sheep to withstand the stress of anaemia due to a higher affinity for oxygen of HbAA or its conversion to HbC. In addition, Romney Marsh and Southdown sheep in Australia are predominantly HbA whereas the original stock in Great Britain are predominantly HbB. Evans and Blunt (1961) have suggested there has been genetic selection for resistance to haemonchosis in Australian sheep.

Genetic aspects of resistance can be demonstrated in lambs vaccinated with irradiated *T. colubriformis* larvae and challenged with normal larvae. Lambs showed a marked bimodal distribution in response to vaccination and could be divided into 'responders' and 'non-responders' (Windon et al. 1980). This response was genetically determined and progeny of responder sires were more responsive than were progeny of non-responder sires as judged by faecal egg counts and by worm counts (Dineen & Windon 1980). Peripheral blood lymphocytes from responder progeny were more responsive *in vitro* to antigen and also mitogens than were the lymphocytes from non-responder animals. This increased lymphocyte responsiveness was evident only after immunization (Dineen & Windon 1980) which is in contrast to the increased preinfection lymphocyte response detected by Riffkin and Dobson (1979) in responder animals.

Immunological aspects of trichostrongylosis

Specific aspects of the immune response as it pertains to *H. contortus* and *Trichostrongylus* spp. are dealt with under the generic descriptions of these parasites (see pp. 217 and 236). More general aspects of the immune response and gastrointestinal nematode infection are reviewed by Soulsby (1979).

Apart from protective immune responses mediated by lymphoid cells and antibodies, other important considerations are survival mechanisms used by parasites in immunologically competent hosts and the failure of immune effector mechanisms in the young ruminant and in the lactating animal. Neonatal immune unresponsiveness to gastrointestinal helminths is especially important and contributes to the high morbidity and mortality of young animals. Reasons postulated for this unresponsiveness included a hierarchical development of responsiveness to parasite antigens, colostral transfer of tolerogenic factors and induced suppression by high doses of infective larvae.

Periparturient relaxation of immunity is an important factor in the epidemiology of a number of parasitic infections. The basic immunological defect has yet to be identified, but is likely to be common to the numerous parasitic infections in which the phenomenon has been identified. In fact the relaxation of immunity may be somewhat local, affecting the intestinal tract, since with *Nematospiroides dubius* infection in mice while marked relaxation of immunity occurred to the gut forms, lactating mice were capable of inducing a marked passive transfer of immunity to their offspring (Shubber et al. 1981). With the gastrointestinal nematodes of ruminants, there is no evidence that passive transfer of immunity plays a role in the control of infections.

Studies in laboratory animals suggest that the mechanism of expulsion of intestinal nematodes involves a specific immunological event (cellular or antibody-mediated immunity response) which triggers non-specific factors such as release of pharmacologically active mediators of inflamation. These act directly or indirectly on the parasite, resulting in expulsion (Dineen & Kelly 1976; Rothwell et al. 1971). The lack of specificity of the effector mechanisms is demonstrated by the fact that challenge of a *H. contortus* infection will induce self-cure not only to *H. contortus* but also to other abomasal parasites and also to *Trichostrongylus* spp. infection in the small intestine (Stewart 1955). The division of the response into specific and non-specific components has been emphasized in experiments studying protection against infection with gastrointestinal nematodes (Dineen et al. 1977). Sheep immunized against infection with *T. colubriformis* were markedly protected against infection with *T. colubriformis*, showed some degree of protection against infection with the closely related species, *T. vitrinus*, but were not protected against infection when challenged with *N. spathiger*. In contrast, when similarly immunized sheep were challenged with a mixed infection of all three species a highly significant (98–100%) protection was evident against all three species. These results demonstrate the requirement for a specific antigenic trigger of the rejection mechanism but the lack of specificity of the terminal effector mechanism.

Various immunoglobulin classes of antibody occur during infection. Immunoglobulin A antibodies are detectable early in infection and IgE antibodies accumulate at the intestinal mucosal surface, being transported there by mast cells. An exponential increase in mast cells may occur in the intestinal mucosa (in rats infected with *Nippostrongylus brasiliensis*) and these, through sensitization by specific IgE antibody and subsequent secretion of vasoactive amines upon contact with parasite antigen, play an important role in the elimination of a parasite burden. Mucus production may also be important in this.

The role of the eosinophil and eosinophilia (long regarded as pathognomonic of helminth infection) has received detailed consideration recently. Adherence of eosinophils to various developmental stages and their degranulation upon parasite surfaces with the release of various enzymes has provided a firmer understanding of effector mechanisms in helminth immunity.

ANTHELMINTIC TREATMENT OF PARASITIC GASTROENTERITIS

Benzimidazoles

Thiabendazole, given orally in a wide variety of formulations, has a wide margin of safety. It is given at a dose rate of 44 mg/kg to sheep and 66 mg/kg to cattle. 110 mg/kg is recommended for the treatment of heavy infections in cattle. Thiabendazole has high efficacy against the adult stages and is effective against the immature stages of gastrointestinal nematodes. It is only moderately effective against adult and immature *Nematodirus* spp., *Strongyloides* spp. and *Bunostomum* spp. It is recommended at 88 mg/kg for lambs with nematodiriasis.

Parbendazole has moderate efficacy against *Nematodirus* and *Bunostomum* spp., although the efficacy increases with increased dose rates. Parbendazole is highly effective in the treatment of the other gastrointestinal nematodes of ruminants.

Cambendazole is highly effective against gastrointestinal nematodes of ruminants, although less efficacious against *Bunostomum* spp. and *O. columbianum*. It also shows activity against *Moniezia* spp. and some activity against *Dictyocaulus* spp.

Mebendazole (15 mg/kg) is intended for removal of immature and adult gastrointestinal nematodes in ruminants. It is also effective in the treatment of *Moniezia* and *Dictyocaulus* spp.

Fenbendazole, given at a dose rate of 7.5 mg/kg to cattle and 5 mg/kg to sheep and goats, is highly efficacious in the treatment of adult and immature gastrointestinal nematodes. It is also ovicidal, killing eggs already in the gut at the time of treatment. It is a broad-spectrum anthelmintic with efficacy against lungworms, *Moniezia* spp. and at higher dose rates against *F. hepatica* and *D. dendriticum*.

Oxfendazole at 4.5–5 mg/kg is a broad-spectrum anthelmintic with high efficacy and ovicidal activity against mature and immature gastrointestinal nematodes, *Dictyocaulus* spp. and *Moniezia* spp. Higher dose rates are required for efficacy against *F. hepatica*.

Albendazole at 5 mg/kg or 7.5 mg/kg is highly efficacious in the treatment of gastrointestinal nematodes, *Moniezia* and *Dictyocaulus* spp. in sheep and cattle, respectively. 7.5 mg/kg and 15 mg/kg are recommended for the treatment of *F. hepatica* in sheep and cattle, respectively. The anthelmintic is ovicidal.

Febantel is metabolized *in vivo* to a benzimidazole methyl-carbamate. At 5–7.5 mg/kg it shows efficacy against mature and immature gastrointestinal nematodes and also affects *Dictyocaulus* spp. lungworms.

Thiophanate undergoes cyclization *in vivo* to a benzimidazole ethyl-carbamate. It has broad-spectrum activity against gastrointestinal nematodes with some efficacy against *Dictyocaulus* at 50–100 mg/kg. Lower doses (1–10 mg/kg) given daily decrease worm burdens, reduce egg output and are ovicidal.

Organophosphates

Haloxon, 37–64 mg/kg, is given orally for the treatment of gastrointestinal nematodes of cattle, sheep and goats. It has low toxicity but delayed neurotoxicity has occasionally been reported after its use.

Crufomate, 16–35 mg/kg in cattle and 40–90 mg/kg in sheep, has activity against the gastrointestinal nematodes *Haemonchus*, *Ostertagia*, *Trichostrongylus* and *Cooperia* spp.

Coumaphos is used as a top dressing on the feed of cattle at a dose rate of 2 mg/kg/day for six days. It is effective against adult *Haemonchus*, *Ostertagia*, *Trichostrongylus* and *Cooperia* spp.

Levamisole

Levamisole, 7.5 mg/kg, is highly effective against both mature and immature gastrointestinal nematodes and lungworms. It may be administered as a drench, in the feed or by subcutaneous injection. The margin of safety is not great after subcutaneous administration to sheep and particularly goats. Overdosage results in signs of cholinergic toxicity, particularly in goats.

Experimentally levamisole combined with a clostridial vaccine heightened the antibody response to the vaccine in sheep and also retained its anthelmintic activity (Hogarth-Scott et al. 1980; Forsyth & Wynne-Jones 1980).

Morantel tartrate

Morantel tartrate at 10 mg/kg has high efficacy against adult and immature stages of the gastrointestinal nematodes, including *N. battus*, in cattle and sheep. Also, low level administration in the feed (1.5 mg/kg/day) results in reduced faecal egg counts and post mortem worm burdens in cattle and sheep. Experimentally, prophylactic administration of morantel tartrate to cattle for the first two and a half months of the grazing season reduced pasture larval numbers by 85% (Pott et al. 1979). Recently, a sustained-release bolus designed to release a sustained level of morantel tartrate over a three-month period has been developed. Administration of the bolus to calves in early spring reduced trichostrongylid egg output by 78% with a consequent 77% reduction in the peak of larval contamination on pasture (Jacobs et al. 1981). A 71% reduction in worm burdens acquired over a grazing season with a mean improvement in weight gain of 24 kg was reported by Armour et al. (1981) using a morantel sustained release bolus. In lactating grazing adult dairy cows given a morantel sustained-release bolus, Bliss et al. (1982) reported a highly significant improvement in milk production, milk fat and protein content compared with untreated animals.

Phenothiazine

Micronized phenothiazine given orally to cattle at 220–440 mg/kg (maximum dose 40–80 g) and sheep at up to 600 mg/kg has good activity against adult gastrointestinal nematodes. Prophylactic daily administration (2 g/day for cattle) reduces the egg-laying capacity of female nematodes and reduces egg hatchability. Various formulations of phenothizine with lead arsenate, copper sulphate and nicotine sulphate are still widely used in various parts of the world for the control of gastrointestinal nematodes.

Avermectins

The avermectins (i.e. Ivermectin) are a new class of drugs with broad-spectrum anthelmintic, insecticidal and acaricidal properties. They are fermentation products produced by *Streptomyces avermitilis*. At 50–200 μg/kg the avermectins have excellent efficacy against immature and adult gastrointestinal nematodes of ruminants. Armour et al. (1980) reported that oral dosage with 100 μg or subcutaneous injection with 100–200 μg removed all adults and inhibited larvae of the common gastrointestinal nematodes of cattle in the UK.

Narrow-spectrum anthelmintics

Napthalophos, 17–47 mg/kg, is an organophosphate anthelmintic active against *H. contortus* and *T. colubriformis*. Salicylanilides and substitued phenols show variable activity against *H. contortus*, i.e. *disophenol* given at 7.5 mg/kg. *Closantel*, 5 mg/kg, shows high efficacy against immature and mature *H. contortus*. Closantel also has anti-trematode activity and has activity against a variety of arthropods.

Resistance to anthelmintics

Anthelmintic resistance in trichostrongylid nematodes has been reviewed by Le Jambre (1978) and Prichard et al. (1980). Drudge et al., in 1957, first reported anthelmintic resistance in *H. contortus* to phenothiazine. In 1964, three years after the commercial release of thiabendazole, a resistant strain of *H. contortus* was reported (Drudge et al. 1964). Since then anthelmintic resistance to the majority of anthelmintics has been shown to develop, in the field and/or in the laboratory, in a variety of nematodes. Anthelmintic resistance has been reported in *H. contortus*, *Trichostrongylus* spp., *Ostertagia* spp., small strongyles in horses and a variety of other parasites.

Prichard et al. (1980) classified the major anthelmintics into four groups:

Group 1: benzimidazoles and probenzimidazoles (febantel and thiophanate)
Group 2: levamisole and morantel
Group 3: salicylanilides and substituted nitrophenols
Group 4: organophosphates

Cross-resistance within these groups occurs. For instance, thiabendazole- and parbendazole-resistant parasites have been found to be cross-resistant to other benzimidazole anthelmintics (Berger 1975; Hall et al. 1978). No cross-resistance has been demonstrated between the groups but multiple resistance, induced by multiple selection with anthelmintics in the same or different groups, has been demonstrated. Indeed, when parasites, *H. contortus*, *O. circumcincta* and *T. colubriformis*, were selected by frequent treatment (at less than one-month intervals) with three anthelmintics (thiabendazole, morantel tartrate and levamisole), the parasites developed resistance to all three anthelmintics as rapidly as they developed resistance when selected by a single drug (Le Jambre 1978). In fact, rapid alternation of anthelmintics may increase the rate at which resistance develops.

Information on the development of resistance has been drawn primarily from studies on insecticide resistance. It is thought that the genes expressing resistance are normally at a low frequency in a population but the frequency of these genes increases since the survivors of treatment (resistant) make a greater contribution to succeeding generations. The intensity of the selection pressure depends on such factors as frequency of treatment, dose rate and infection pressure (Prichard et al. 1980). When selection pressure is removed theoretically the parasite population should return towards susceptibility. However, reversion appears to be slow and resistance in such a parasite developed very rapidly upon re-selection (Kelly & Hall 1979).

Little is known about the way resistant nematodes survive the effect of anthelmintics. Resistance is not absolute but the LD_{50} or LD_{90} is higher in a resistant strain than in a non-resistant strain of a parasite. Experiments with radiolabelled fenbendazole and thiabendazole showed that benzimidazole-resistant *H. contortus* and *T. colubriformis* did not exclude the anthelmintics although fenbendazole was incorporated to a greater extent by susceptible parasites (Pritchard et al. 1978*b*). However, benzimidazoles inhibit the fumarate reductase system in nematodes and inhibition of this system is less pronounced in benzimidazole resistant *H. contortus* (Prichard 1973). Energy production in resistant *H. contortus* in the presence of thiabendazole was aided by the production of ethanol, bypassing the fumarate reductase system (Rew 1978).

Anthelmintic resistance in nematodes has great epidemiological significance and there is also evidence that resistant parasites are more fecund, are more pathogenic, have increased establishment rates in the host and have increased survival of the free-living stages (Kelly et al. 1977). Resistance is diagnosed by faecal egg counts and post mortem worm counts after treatment. Benzimidazole resistance may be detected by examining embryonation and hatching of eggs since eggs of resistant strains will embryonate and hatch in higher concentrations of anthelmintics than do eggs of non-resistant strains (Le Jambre 1976). Little information is available on methods of delaying the development of resistance. Narrow-spectrum anthelmintics should be used if a broad-spectrum anthelmintic is not required, in order to reduce selection pressure on the latter. Frequent rotation of anthelmintics, particularly within a single generation of parasites, should be avoided, However, if only a single anthelmintic is used an extremely high level of resistance will develop eventually and resistance by less pathogenic parasites may not be detected in the face of lack of resistance by a pathogenic parasite resulting in undetected production losses. Slow rotation may be effective in reducing the rate of selection (Prichard et al. 1980).

Anthelmintic treatment of hypobiotic larvae

Until recently no anthelmintic was available for the destruction of hypobiotic larvae of *O. ostertagi* and other parasites. However, Duncan et al. (1976) demonstrated that fenbendazole, given at a dose rate of 7.5 mg/kg, was highly effective (97.5%) against the hypobiotic fourth-stage larvae of *O. ostertagi* in cattle. Other anthelmintics such as albendazole (7.5 mg/kg), oxfendazole (5 mg/kg), Ivermectin (100–200 μg/kg) and thiophanate (100–200 mg/kg) have been shown to be effective. Such anthelmintics have also been demonstrated to be highly effective against hypobiotic larvae of *Ostertagia*, *Haemonchus*, *Trichostrongylus* and other species in cattle and sheep.

Unfortunately variable results have been seen in a number of studies and thus treatment cannot yet be considered a fully reliable method of control. The variability has not yet been explained although a number of hypotheses have been formulated. Michel (1967) suggested that inhibited larvae were not susceptible to the effects of anthelmintics, owing to their depressed metabolism and resultant reduced uptake of anthelmintics. However, Prichard et al. (1978*b*) demonstrated similar incorporation of ^{14}C-labelled thiabendazole into both adults and inhibited larvae of *O. ostertagi*. Variations in the degree of hypobiosis and therefore levels of metabolic rate could be important and this could be dependent on a variety of factors. These include the length of time the larvae have been inhibited, interactions between the season and host factors on the larvae and the aetiology of larval conditioning (environmental or host-induced). The larvae could also undergo cycles of metabolic activity. However, anthelmintics have been shown to be equally effective at different times of the year, i.e. November and January (Duncan et al. 1978). Variability between individual animals within a group may suggest that host factors are important. A possible explanation was closure of the oesophageal groove and direct passage of the anthelmintic into the abomasum with resultant reduced efficacy. Kelly et al. (1977) were unable to demonstrate significant differences in efficacy of anthelmintics when administered intraruminally or directly into the abomasum via artificially induced closure of the oesophageal groove. The half-life of anthelmintics could be important if, because of their reduced energy demands, inhibited larvae are able to survive short-term but not long-term anthelmintic treatment. Prolonged treatment, such as fenbendazole 1 mg/kg/day for 10 days and thiophanate five daily doses of 20 mg/kg, has been shown to be effective (McBeath et al. 1977; Duncan et al. 1979).

CONTROL OF PARASITIC GASTOENTERITIS IN RUMINANTS

The epidemiology and control of gastrointestinal nematodes of ruminants has been the subject of many reviews. These include Gordon (1973), Michel (1976), Urquhart and Armour (1973), Donald et al. (1978), Morley and Donald (1980) and Brunsdon (1980). Control may be based on grazing management, anthelmintic treatment and integrated systems incorporating these two facets. However, differences in epidemiology under different climatic conditions require different approaches for control.

Control based on management incorporates the knowledge of the life-cycles, larval ecology and epidemiology of the gastrointestinal parasites. It is influenced by grazing management: provision of clean pastures; alternate grazing by other species of host; alternate grazing by immunologically resistant hosts of the same species; stocking rate; and timing of reproductive events. Control based on management factors will be aided by the strategic use of anthelmintics. Alternatively, particularly where permanent pastures are utilized, farmers may rely solely on anthelmintics for control, treatments being given as often as every three to four weeks, but this is likely to be uneconomic.

In temperate climates, i.e. Great Britain, two sources of contamination are available for spring-born lambs: overwintered larvae and the periparturient rise in faecal egg counts. Of these, the latter is the most important. If ewes and lambs are grazing clean pasture in the spring, then anthelmintic treatment of the ewe in the periparturient period is effective in reducing the mid-summer peak of larvae on pasture. However, the periparturient ewe is susceptible to newly acquired infection. Thus, when ewes are grazing contaminated pasture in the spring a single anthelmintic treatment will not eliminate the periparturient rise in faecal egg counts. Lambs ingesting overwintered larvae in the spring also contribute to the subsequent mid-summer rise in larvae on pasture. Under these circumstances, the lambs are treated with anthelmintics and moved to clean pasture in early to mid-summer to avoid the peak of pasture larval contamination in mid- to late summer. Clean pastures are provided by silage /or hay aftermath or pasture not grazed during the spring. Anthelmintic treatment without movement will have little effect.

Under New Zealand conditions, the second autumn peak of larvae on pasture, arising from eggs deposited by lambs in the summer and autumn, is the important cause of clinical trichostrongylosis in autumn and winter. Two periods are important in control. Lambs are treated and moved in early to mid-summer at weaning Subsequently they are treated at the beginning of autumn and, in order to avoid the autumn peak of larvae on pasture, are moved to pasture not grazed that summer by lambs. In winter rainfall regions of Australia, and probably other areas with hot, arid summers unsuitable for larval development, movement and treatment of sheep with anthelmintics at the beginning and near the end of the hot, arid period achieves satisfactory control. This rationale is associated with the fact that the late spring and summer are the critical periods in the life-cycle of the parasites involved.

In the British Isles, larvae that have overwintered on pasture are the major source of infection for calves and the use of clean pasture in the spring for calves will markedly reduce the levels of parasitic infection. When calves are grazed on contaminated pasture in the spring they should be treated and moved to clean pasture in mid-July to avoid the peak of larvae on pasture. Pastures not grazed since the end of winter are normally used. To prevent ingestion of conditioned larvae the animals should not be returned to contaminated pasture in the autumn. In other areas where there is a longer period suitable for larval development, treatment and movement to safe pasture on two occasions may suffice and the control measures are reversed where larvae undergo inhibition in the spring before the hot, dry summer months.

Production of safe pastures may be achieved by a variety of means.

Silage and hay aftermath. Overwintered larvae normally disappear from pasture by May or June and the pasture may be considered safe for grazing after the removal of a crop of hay. Recently, however, larvae of *O. ostertagi* and *D. viviparous* have been shown to survive in the soil and sufficient larvae may emerge on to the pasture to cause disease in August/September.

Pasture resting is seldom a suitable strategy. Stock would need to be withheld from pasture for at least six months under cool, moist conditions and two months under hot, arid conditions for larval concentration to be reduced to a reasonably low level.

Variation in timing of reproductive events. The immunosuppressive effect of late pregnancy and lactation amplifies the spring rise in faecal egg counts. The periparturient rise in faecal egg counts may be reduced, but not eliminated, if lambing is timed so that it does not overlap with the spring maturation of inhibited larvae. Since the ewe is the major source of infection for the lamb, early weaning should reduce worm burdens in the early-weaned lambs as compared to the unweaned lambs.

Stocking rate. Increased stocking rate has numerous, often conflicting, effects on levels of infection on pasture. Increased stocking rate will increase contamination of the pasture. If the mass of herbage is reduced by the increased stocking rate then infective stages will be more accessible. However, the reduced plant cover may alter the micro-environment and expose the free-living stages to conditions less favourable for their development and survival. Reduced plant cover may decrease consumption of pasture by the animals but may lower the nutritional status of the host.

Alternate grazing with hosts of different species. The cattle/sheep alternation is the most common. Both species are susceptible to parasites such as *T. axei* and *H. contortus*. *Cooperia* and the intestinal *Trichostrongylus* spp. have a reduced infectivity and patency in the alternate host and there is little cross transmission of *Ostertagia*, *Nematodirus*, *Bunostomum* and *Oesophagostomum* spp. However, genetic adaptation is always a possibility. Alternate grazing can offer substantial benefits with the exceptions of *T. axei* and perhaps *H. contortus/placei*. Cattle or sheep/horse alternate grazing is highly effective with the exception of infection with *T. axei*.

Alternate grazing with hosts of the same species. Older animals are usually fairly resistant to infection with helminths. These resistant animals will reduce the numbers of infective larvae on pasture while their faecal egg output will be lower than that of young susceptible animals. This is particularly true of cattle but less so of sheep. Total contamination by the adult animals could still be sufficient to result in subclinical or clinical disease in young susceptible animals. Further, the ingestion of large numbers of infective larvae could result in disease, particularly ostertagiasis Type II, in adult animals. Grazing adult resistant stock alongside young susceptible stock may also dilute contamination to an acceptable level.

Ploughing, reseeding and burning of pastures can reduce or markedly decrease levels of contamination. Heavy applications of some fertilizers may have some larvicidal effects.

Anthelmintic treatment. Repeated use of anthelmintics in spring and early summer, 'in-feed' incorporation of anthelmintics or administration of a slow-release anthelmintic bolus can reduce peak summer levels of larvae on pasture. Strategic use of anthelmintics in other climatic regions should also be effective.

Use of slurry. Although their viability is low, trichostrongylid eggs can survive in slurry and application of slurry to pasture may alter the epidemiological pattern of the helminths. Slurry application in winter can result in raised larval counts on pasture in spring and early summer.

Forecasting the incidence of parasitic gastroenteritis

Weather plays a dominant role in determining the timing and size of peak larval contamination on pasture. Nematodiriasis, due to *N. battus*, in lambs is successfully forecast on the basis of temperature in the spring. Forecasting the incidence of parasitic gastroenteritis in lambs and calves is more complex owing, in part, to the variety of species involved in the disease. Various studies have shown a relationship between the onset of gastroenteritis and meteorological data, particularly surface moisture (Vlassoff 1975; Ollerenshaw et al. 1978; Thomas & Starr 1978). Correlation may be improved by taking into consideration meteorological data at the end of the previous grazing season (Ollerenshaw et al. 1978).

Thomas and Starr (1978), in showing a correlation between climate and time of summer peak in pasture infective larvae, developed the concept of a 'wet score' allocated to 12-hourly rainfall figures. A 'critical index' of 440 units of wetness was shown to be necessary before the pasture larval peak was reached. They suggested a 'warning index' of 350–380 wetness units, permitting a prediction to be made of the onset of major infection in lambs.

Although forecasting of parasitic gastroenteritis is not yet precise, it will aid in the development and timing of various strategic control measures.

REFERENCES

TRICHOSTRONGYLIDAE: TRICHOSTRONGYLUS

Adams, D. B. & Rothwell, T. L. W. (1980) The role of lymphocytes in immunological memory for resistance to infection by *Trichostrongylus colubriformis* in guinea pigs. *Cell. Immunol.*, **55**, 1–11.

Anderson, N., Dash, K. M., Donald, A. D., Southcott, W. H. & Waller, P. J. (1978) *Epidemiology and Control of Gastrointestinal Parasites of Sheep in Australia*, ed. A. D. Donald, W. H. Southcott, & J. K. Dineen, pp. 23–51. Commonwealth Scientific and Industrial Research Organization, Australia.

Barker, I. K. (1973*a*) A study of the pathogenesis of *Trichostrongylus colubriformis* in lambs with observations on the contribution of gastro-intestinal plasma loss. *Int. J. Parasit.*, **3**, 743–757.

Barker, I. K. (1973*b*) Scanning electron microscopy of the duodenal mucosa of lambs infected with *Trichostrongylus colubriformis. Parasitology*, **67**, 307–314.

Boag, B. & Thomas, R. J. (1970) The development and survival of free-living stages of *Trichostrongylus colubriformis* and *Ostertagia circumcincta* on pasture. *Res. vet. Sci.*, **11**, 380–381.

Boag, B. & Thomas, R. J. (1977) Epidemiological studies on gastro-intestinal nematode parasites of sheep: The seasonal number of generations and succession of species. *Res. vet. Sci.*, **22**, 62–67.

Callinan, A. P. L. (1978) The ecology of the free-living stages of *Trichostrongylus aexi. Int. J. Parasit.*, **8**, 453–456.

Callinan, A. P. L. (1979) The ecology of the free-living stages of *Trichostrongylus vitrinus. Int. J. Parasit.*, **9**, 133–136.

Cobbold, T. S. (1873) *The Grouse Disease: A Statement of Facts Tending to Prove the Parasite Origin of the Epidemic*. London.

Coop, R. L. (1981) Feed intake and utilization by the parasitized ruminant. In: *Isotypes and Radiation in Parasitology IV*. Vienna: IAEA.

Coop, R. L. & Angus, K. W. (1975) The effect of continuous doses of *Trichostrongylus colubriformis* larvae on the intestinal mucosa of sheep and on liver vitamin A concentration. *Parasitology*, **70**, 1–9.

Coop, R. L., Sykes, A. R. & Angus, K. W. (1976) Subclinical trichostrongylosis in growing lambs produced by continuous larval dosing. The effect on performance and certain plasma constituents. *Res. vet. Sci.*, **21**, 253–258.

Dembinski, A. B., Johnson, L. R. & Castro, G. A. (1979) Influence of parasitism on secretion—inhibited gastric secretion. *Am. J. trop. Med. Hyg.*, **28**, 854–859.

Dineen, J. K., Gregg, P., Windon, R. G., Donald, A. D. & Kelly, J. D. (1977) The role of immunologically specific and non-specific components of resistance in cross-protection to intestinal nematodes. *Int. J. Parasit.*, **7**, 211–215.

Dineen, J. K. & Windon, R. G. (1980) The effect of sire selection on the response of lambs to vaccination with irradiated *Trichostrongylus colubriformis* larvae. *Int. J. Parasit.*, **10**, 189–196.

Douvres, F. W. (1957) The morphogenesis of the parasitic stages of *Trichostrongylus aexi* and *Trichostrongylus colubriformis*, nematode parasites of cattle. *Proc. helminth. Soc. Wash.*, **24**, 4–11.

Eysker, M. (1978) Inhibition of development of *Trichostrongylus* spp. as third stage larvae in sheep. *Vet. Parasit.*, **4**, 29–33.

Gibson, T. E. & Everett, G. (1967) The ecology of the free-living stages of *Trichostrongylus colubriformis. Parasitology*, **57**, 533–547.

Gregg, P. & Dineen, K. J. (1978) The response of sheep vaccinated with irradiated *Trichostrongylus colubriformis* larvae to impulse and sequential challenge with normal larvae. *Vet. Parasit.*, **4**, 49–53.

Gregg, P., Dineen, J. K., Rothwell, T. L. W. & Kelly, J. D. (1978) The effect of age in the response of sheep to vaccination with irradiated *Trichostrongylus colubriformis* larvae. *Vet. Parasit.*, **4**, 35–48.

Horak, I. G., Clark, R. & Gray, R. S. (1968) The pathophysiology of helminth infestations. III. *Trichostrongylus colubriformis. Onderstpoort. J. vet. Res.*, **35**, 195–223.

Horton, G. M. J. (1977) Selenium and vitamin E for lambs with trichostrongylosis. *J. Anim. Sci.*, **45**, 891–895.

Jarrett, W. F. H., Jennings, F. W., McIntyre, W. I. M. & Sharp, N. C. C. (1960) Resistance to *Trichostrongylus colubriformis* produced by X-irradiated larvae. *Vet. Rec.*, **72**, 884.

Jones, W. O. & Symons, L. E. A. (1978) Protein metabolism, 6. *Trichostrongylus colubriformis*: skeletal muscle protein catabolism in primary and secondary infections of the guinea pig. *Exp. Parasit.*, **44**, 92–99.

Leland, S. E., Drudge, J. H., Wyant, Z. N. & Elam, G. W. (1961) Studies on *Trichostrongylus axei* (Cobbold, 1879) VII. Some quantitative and pathologic aspects of natural and experimental infections in the horse. *Am. J. vet. Res.*, **22**, 128–138.

Ogunsusi, R. A. & Eysker, M. (1979) Inhibited development of trichostrongylids of sheep in Northern Nigeria. *Res. vet. Sci.*, **26**, 108–110.

Rogers, W. P. (1966) Exsheathment and hatching mechanisms in helminths. In *Biology of Parasites*, ed. E. J. L. Soulsby, pp. 33–39. New York: Academic Press.

Rogers, W. P. (1970) The function of leucine aminopeptidase in exsheathing fluid. *J. Parasit.*, **56**, 138–143.

Rogers, W. P. & Brooks, F. (1978) Leucine aminopeptidase in exsheathing fluid of North America and Australian *Haemonchus contortus. Int. J. Parasit.*, **8**, 55–58.

Ross, J. G. & Halliday, W. G. (1979*a*) Investigations of 'transfer factor' activity in the transfer of immunity to *Trichostrongylus azei* infections in sheep. *Res. vet. Sci.*, **26**, 41–46.

Ross, J. G. & Halliday, W. G. (1979*b*) Investigations of 'transfer factor' activity in immunity to *Ostertagia circumcincta* and *Trichostrongylus colubriformis* infections in sheep. *Int. J. Parasit.*, **9**, 281–284.

Ross, J. G., Purcell, D. A., Dow, C. & Todd, J. R. (1967) Experimental infections of calves with *Trichostrongylus axei*; the course and development of infection and lesions in low level infections. *Res. vet. Sci.*, **8**, 201–206.

Soulsby, E. J. L. (1979) The immune system and helminth infection in domestic species. *Adv. vet. Sci. comp. Med.*, **23**, 71–102.

Symons, L. E. A. & Hennessy, D. R. (1981) Cholecystokinin and anorexia in sheep infected by the intestinal nematode *Trichostrongylus colubriformis. Int. J. Parasit.*, **11**, 55–58.

Symons, L. E. A. & Jones, W. O. (1978) Protein metabolism, 5. *Trichostrongylus colubriformis*: changes of host body mass and protein synthesis in guinea pigs with light to heavy infections. *Exp. Parasit.*, **44**, 7–13.

Symons, L. E. A. & Steel, J. W. (1978) Pathogenesis of the loss of production in gastrointestinal parasitism. In: *The Epidemiology and Control of Gastrointestinal Parasites of Sheep in Australia*, A. D. Donald, W. H. Southcott & J. K. Dineen, pp. 9–22. Commonwealth Scientific and Industrial Research Organization, Australia.

Taylor, S. M. & Pearson, G. R. (1979) *Trichostrongylus vitrinus* in sheep. I. The location of nematodes during parasitic development and associated pathological changes in the small intestine. *J. comp. Path.*, **89**, 397–403.

Wharton, D. A. (1982) The survival of desiccation by the free-living stages of *Trichostrongylus colubriformis* (Nematoda: Trichostrongylidae). *Parasitology*, **84**, 455–462.

Windon, R. G. & Dineen, J. K. (1980) The segregation of lambs into 'responders' and 'non-responders': response to vaccination with irradiated *Trichostrongylus colubriformis* larvae before weaning. *Int. J. Parasit.*, **10**, 65–73.

TRICHOSTRONGYLIDAE: OSTERTAGIA AND RELATED GENERA

Al Saqur, I., Armour, J., Bairden, K., Dunn, A. M. & Jennings, F. W. (1980) *Ostertagia leptospicularis* Asadov 1953 as a pathogen in British cattle. *Vet. Rec.*, **107**, 511.

Anderson, N., Dash, K. M., Donald, A. D., Southcott, W. H. & Waller, P. J. (1979) Epidemiology and control of nematode infections. In: *The Epidemiology and Control of Gastrointestinal Parasites of Sheep in Australia*, ed. A. D. Donald, W. H. Southcott, & J. K. Dineen, pp. 23–51. Commonwealth Scientific and Industrial Research Organization, Australia.

Armour, J. (1970) Bovine ostertagiasis: a review. *Vet. Rec.*, **86**, 184–189.

Armour, J., Bairden, K., Al Saqur, I. M. & Duncan, J. L. (1981) The role of soil as a potential reservoir for infective larvae of *Ostertagia ostertagi*. In: *The Epidemiology and Control of Nematodiasis in Cattle*, ed. P. Nansen, R. J. Jørgensen & E. J. L. Soulsby, pp. 277–286. The Hague: Martinus Nijhoff.
Armour, J. & Bruce, R. G. (1974) Inhibited development in *Ostertagia ostertagi* infections—a diapause phenomenon in a nematode. *Parasitology*, **69**, 161–174.
Armour, J., Jennings, F. W., Murray, M. & Selman, I. (1974) Bovine ostertagiasis: Clinical aspects, pathogenesis, epidemiology and control. In: *Helminth Diseases of Cattle, Sheep and Horses in Europe*, ed. G. M. Urquhart & J. Armour, pp. 11–16. Glasgow: Maclehose.
Bairden, K., Parkins, J. J. & Armour, J. (1979) Bovine ostertagiasis: a changing epidemiological pattern? *Vet. Rec.*, **105**, 33–35.
Beveridge, I., Barker, I. K., Rickard, M. D. & Burton, J. D. (1974) Experimental infection of sheep with *Camelostrongylus mentulatus* and associated gastritis. *Aust. vet. J.*, **50**, 36–37.
Borgsteede, F. H. M. (1981) Experimental cross-infections with gastrointestinal nematodes of sheep and cattle. *Z. Parasitenk.*, **65**, 1–10.
Craig, T. M. (1979) Seasonal transmission of bovine gastrointestinal nematodes in the Texas Gulf coast. *J. Am. vet. med. Ass.*, **174**, 844–847.
Dargie, J. D. (1975) Application of radioisotopic techniques to the study of red cells and plasma protein metabolism in helminth diseases of sheep. *Symp. Br. Soc. Parasit.*, **13**, 1–26.
Fincher, G. T. & Stewart, T. B. (1979) Vertical migration by nematode larvae of cattle parasites through soil. *Proc. Helminth. Soc. Wash.*, **46**, 43–46.
Gronvold, J. (1979) On the possible role of earthworms in the transmission of *Ostertagia ostertagi* third-stage larvae from faeces to soil. *J. Parasit.*, **65**, 831–832.
Jansen, J. & Gibbons, L. M. (1981) Workshop No. 14. Systematics and biology of *Ostertagia* Sens. lat. (Nematoda: Trichostrongylidae). *Parasitology*, **82**, 175–189.
Martin, W. B., Thomas, B. A. C. & Urquhart, G. M. (1957) Chronic diarrhoea in housed cattle due to atypical parasitic gastritis. *Vet. Rec.*, **69**, 736–739.
Michel, J. P., Lancaster, M. B. & Hong, C. (1976) The resumed development of arrested *Ostertagia ostertagi* in experimentally infected calves. *J. comp. Path.*, **86**, 615–619.
Michel, J. P., Lancaster, M. B. & Hong, C. (1978*a*) Arrested development of *Ostertagia ostertagi* and *Cooperia oncophora*: Effect of the time of year on the conditioning and deconditioning of infective larvae. *J. comp. Path.*, **88**, 131–136.
Michel, J. F., Lancaster, M. B. & Hong, C. (1979) The effect of age acquired resistance, pregnancy and lactation on some reactions of cattle to infection with *Ostertagia ostertagi*. *Parasitology*, **79**, 157–168.
Michel, J. F., Lancaster, M. B., Hong, C. & Berrett, S. (1978*b*) Plasma pepsinogen levels in some experimental infections of *Ostertagia ostertagi* in cattle. *Vet. Rec.*, **103**, 370–373.
Mulligan, W., Dalton, R. G. & Anderson, N. (1963) Ostertagiasis in cattle. *Vet. Rec.*, **75**, 1014.
Murray, M., Jennings, F. W. & Armour, J. (1970) Bovine ostertagiasis: structure function and mode of differentiation of the bovine gastric mucosa and kinetics of the worm loss. *Res. vet. Sci.*, **11**, 417–427.
Selman, I. E., Reid, J. F. S. & Armour, J. (1976) Type II ostertagiasis in adult cattle. *Vet. Rec.*, **99**, 141–143.
Technical Development Committee, BVA (1977) The control of bovine ostertagiasis. *Vet. Rec.*, **101**, 11–13.
Waller, P. J. & Thomas, R. J. (1978) Nematode parasitism in sheep in north-east England: The epidemiology of *Ostertagia* species. *Int. J. Parasit.*, 8, 275–283.

TRICHOSTRONGYLIDAE: NEMATODIRUS

Boag, B. & Thomas, R. J. (1975) Epidemiological studies on *Nematodirus* species in sheep. *Res. vet. Sci.*, **19**, 263–268.
Crofton, H. D. & Thomas, R. J. (1954) A further description of *Nematodirus battus* Crofton and Thomas 1951. *J. Helminth.*, **28**, 119–112.
Gibson, T. B. (1958) The role of the egg as the infective stage of the nematodes *Nematodirus battus* and *Nematodirus filicollis*. *Vet. Rec.*, **70**, 496–497.
Lee, D. L. & Martin, J. (1976) Changes in *Nematodirus battus* associated with the development of immunity to this nematode in lambs. In: *Biochemistry of Parasites and Host–Parasite Relationships*, ed. van den Bossche, H. Amsterdam: Elsevier North Holland.
Mapes, C. J. & Coop, R. L. (1972) The development of single infections of *Nematodirus battus* in lambs. *Parasitology*, **64**, 197–216.
Martin, J. & Lee, D. L. (1980) *Nematodirus battus*: scanning electron microscope studies of the duodenal mucosa of infected lambs. *Parasitology*, **81**, 573–578.
Ollerenshaw, C. B. & Smith, L. P. 1966) An empirical approach to forecasting the incidence of nematodiriasis over England and Wales. *Vet. Rec.*, **79**, 536–540.
Samizadeh-Yazd, A. & Todd, A. C. (1979) Observation on the pathogenic effects of *Nematodirus helvetianus* in dairy calves. *Am. J. vet. Res.*, **40**, 48–51.
Smith, L. P. & Thomas, R. J. (1972) Forecasting the spring hatch of *Nematodirus battus* by the use of soil temperature data. *Vet. Rec.*, **90**, 388–392.
Thomas, R. J. & Stevens, A. J. (1960) Ecological studies on the development of the pasture stages of *N. filicollis*, nematode parasites of sheep. *Parasitology*, **50**, 31–49.

TRICHOSTRONGYLIDAE: HAEMONCHUS AND RELATED GENERA

Allonby, E. W. & Urquhart, G. M. (1973) Self cure of *Haemonchus contortus* under field conditions. *Parasitology*, **66**, 43–53.
Allonby, E. W. & Urquhart, G. M. (1976) A possible relationship between haemonchosis and haemoglobin polymorphism in Merino sheep in Kenya. *Res. vet. Sci.*, **20**, 212–214.
Altaif, K. I. & Dargie, J. D. (1978*a*) Genetic resistance to helminths. The influence of breed and haemoglobin type on the response of sheep to primary infections with *Haemonchus contortus*. *Parasitology*, **77**, 161–175.
Altaif, K. I. & Dargie, J. D. (1978*b*) Genetic resistance to helminths. The influence of breed and haemoglobin type on the response of sheep to reinfection with *Haemonchus contortus*. *Parasitology*, **77**, 177–187.
Benitez-Usher, C., Armour, J., Duncan, J. L., Urquhart, G. M. & Gettinby, G. (1977) A study of some factors influencing the immunization of sheep against *Haemonchus contortus* using attendated larvae. *Vet. Parasit.*, **3**, 327–342.
Bradley, R. E., Radhakrishnan, C. V. & Patilkulkarni, V. G. (1973) Response in Florida native and Rambouillet lambs exposed to one and two oral doses of *Haemonchus contortus*. *Am. J. vet. Res.*, **34**, 729–735.
Bremner, K. C. (1955) Cytological polymorphism in the nematode *Haemonchus contortus* (Rudolphi, 1803) Cobb, 1898. *Nature, Lond.*, **174**, 704–705.
Bremner, K. C. (1956) The parasitic life cycle of *Haemonchus placei* (Place) Ransom (Nematoda: Trichostrongylidae). *Aust. J. Zool.*, **4**, 146–151.
Chen, P. & Soulsby, E. J. L. (1976) *Haemonchus contortus* infection in ewes: Blatogenic responses of peripheral blood leucocytes to third stage larval antigen. *Int. J. Parasit.*, **6**, 135–141.
Clark, C. H., Kiesel, G. K. & Goby, C. H. (1962) Measurements of blood loss caused by *Haemonchus contortus* infection in sheep. *Am. J. vet. Res.*, **23**, 977–980.
Connan, R. M. (1971) The seasonal incidence of inhibition of development in *Haemonchus contortus*. *Res. vet. Sci.*, **12**, 272–274.
Connan, R. M. (1976) Effect of lactation on the immune response to gastrointestinal nematodes. *Vet. Rec.*, **99**, 476–477.
Connan, R. M. (1978) Arrested development in *Haemonchus contortus*. In: *Arrested Development of Nematodes in Sheep and Cattle*, ed. F. H. M. Borgsteede, J. Armour & J. Jansen, pp. 53–62. Lelystad: Central Veterinary Institute.
Crofton, H. D. (1963) *Nematode Parasite Population in Sheep and on Pasture*. St.Albans, England: Commonwealth Bureau of Helminthology.
Cuperlovič, K., Altaif, K. I. & Dargie, J. D. (1978) Genetic resistance to helminths: A possible relationship between haemoglobin type and the immune responses of sheep to non-parasitic antigens. *Res. vet. Sci.*, **25**, 125–126.
Cvetković, L. J., Lepojev, O. & Vulic, I. (1973) Resistance to natural infection with gastrointestinal strongyles in Yogoslav breeds of sheep. *Vet. Glasnik.*, **27**, 867–872.
Dargie, J. D. (1975) Application of radioisotope techniques to the study of red cell and plasma protein metabolism in helminth diseases of sheep. *Symp. Br. Soc. Parasit.*, **13**, 1–26.
Dargie, J. M. & Allonby, E. W. (1975) Pathophysiology of single and challenge infection of *Haemonchus contortus* in Merino sheep: Studies on red cell kinetics and the 'self cure' phenomenon. *Int. J. Parasit.*, **5**, 147–157.
Das, K. M. & Whitlock, J. H. (1960) Subspeciation in *Haemonchus contortus* (Rudolphi 1983) Nematoda, Trichostrongyloidea. *Cornell Vet.*, **50**, 182–197.
Daskalov, P. (1972) *Haemonchus contortus* factors determining the polymorphism of linguiform females. *Exp. Parasit.*, **32**, 364–368.

Dinaburg, A. G. (1944) Developmental and survival under outdoor conditions of eggs and larvae of the common ruminant stomach worm *Haemonchus contortus*. *J. agric. Res.*, **69**, 421–433.

Dinnik, J. A. & Dinnik, N. N. (1958) Observations on the development of *Haemonchus contortus* larvae under field conditions in the Kenya highlands. *Bull. epizoot. Dis. Afr.*, **6**, 11–21.

Durie, P. H. (1961) Parasitic gastro-enteritis of cattle: the distribution and survival of infective strongyle larvae on pasture. *Aust. J. agric. Res.*, **12**, 1200–1211.

Fernando, S. T. (1965) The life cycle of *Mecistocirrus digitatus*, a trichostrongylid parasite of ruminants. *J. Parasit.*, **51**, 156–163.

Gibbons, L. M. (1979) Revision of the genus *Haemonchus* Cobb, 1989 (Nematoda: Trichostrongylidae). *Systematic Parasit.*, **1**, 3–24.

Gordon, H. M. (1968) Self cure reaction. In: *The Reaction of the Host to Parasitism*, ed. E. J. L. Soulsby, pp. 174–190. Marburg: Elwert.

Herlich, H., Porter, D. A. & Knight, R. A. (1958) A study of *Haemonchus* in cattle and sheep. *Am. J. vet. Res.*, **19**, 866–872.

Knight, R. A., Vegirs, H. H. & Glimp, H. A. (1973) Effects of breed and date of birth of lambs on gastrointestinal nematode infections. *Am. J. vet. Res.*, **34**, 323–327.

Le Jambre, L. F. (1979) Hybridization studies of *Haemonchus contortus* (Rudolphi, 1893) and *H. placei* (Place, 1893) (Nematoda: Trichostrongylidae). *Int. J. Parasit.*, **9**, 455–463.

Le Jambre, L. F. (1981) Hybridization studies with Australian *Haemonchus placei* (Place, 1893), *Haemonchus contortus cayaguensis* (Das & Whitlock 1960) and *Haemonchus contortus* (Rudolphi, 1803) from Louisiana. *Int. J. Parasit.*, **11**, 323–330.

Levine, N. D. (1963) Weather climate and bionomics of ruminant nematode larvae. *Adv. vet. Sci.*, **8**, 215.

Manton, V. J. A., Peacock, R., Poynter, D., Silverman, P. H. & Terry, R. J. (1962) The influence of age on naturally acquired resistance to *Haemonchus contortus*. *Res. vet. Sci.*, **3**, 308–313.

Mulligan, W., Gordon, H. McL., Stewart, D. F. & Wagland, D. F. (1961) The use of irradiated larvae as immunizing agents in *Haemonchus contortus* and *Trichostrongylus colubriformis* infections in sheep. *Aust. J. agric. Res.*, **12**, 1175–1187.

Ogunsusi, R. A. (1979) Termination of arrested development of trichostrongyles of sheep in northern Nigeria. *Res. vet. Sci.*, **26**, 189–192.

Ogunsusi, R. A. & Eysker, M. (1979) Inhibited development of trichostrongylids of sheep in northern Nigeria. *Res. vet. Sci.*, **26**, 108–110.

O'Sullivan, B. M. & Donald, A. (1973) Responses to infection with *Haemonchus contortus* and *Trichostrongylus colubriformis* in ewes of different reproductive status. *Int. J. Parasit.*, **3**, 521–530.

Preston, J. M. & Allonby, E. W. (1978) The influence of breed on the susceptibility of sheep and goats to a single experimental infection with *Haemonchus contortus*. *Vet. Rec.*, **103**, 509–512.

Preston, J. M. & Allonby, E. W. (1979) The influence of breed on the susceptibility of sheep to *Haemonchus contortus* infection. *Res. vet. Sci.*, **26**, 134–139.

Riffkin, G. G. & Dobson, C. (1979) Predicting resistance of sheep to *Haemonchus contortus* infections. *Vet. Parasit.*, **5**, 365–378.

Roberts, F. H. S. (1957) Reactions of calves to infestation with the stomach worm *Haemonchus placei* (Place, 1893) Ransom 1911. *Aust. J. agric. Res.*, **8**, 740–767.

Roberts, F. H. S., Turner, H. N. & McKevett, M. (1954) On the specific distinctness of the ovine and bovine 'strains' of *Haemonchus contortus* (Rudolphi) Cobb (Nematoda: Trichostrongylidae). *Aust. J. Zool.*, **2**, 275–295.

Shubber, A. H., Lloyd, S. & Soulsby, E. J. L. (1981) Infection with gastrointestinal helminths. Effect of lactation and maternal transfer of immunity. *Z. Parasitenk*, **65**, 181–189.

Silverman, P. H. & Campbell, J. A. (1959) Studies on parasitic worms of sheep in Scotland. I. Embryonic and larval development of *Haemonchus contortus* at constant conditions. *Parasitology*, **49**, 23–38.

Soulsby, E. J. L. (1957) Studies on the serological response in sheep to naturally acquired gastro-intestinal nematodes. II. Responses in a low ground flock. *J. Helminth.*, **31**, 145–160.

Soulsby, E. J. L. (1966) The mechanisms of immunity to gastro-intestinal nematodes. *Biology of Parasites*, pp. 255–276 New York: Academic Press.

Soulsby, E. J. L. (1977) Parasites and domestic animals: Host factors in pathogenesis. In: *Parasites: Their World and Ours*, Proc. 18th Symp. Royal Soc. Canada ed. A. M. Fallis, pp. 25–47. Ottowa: Royal Society of Canada.

Soulsby, E. J. L. & Stewart, D. F. (1960) Serological studies of the self cure reaction in sheep infected with *Haemonchus contortus*. *Aust. J. agric. Res.*, **11**, 595–603.

Stewart, D. F. (1953) Studies on resistance of sheep to infestation with *Haemonchus contortus* and *Trichostrongylus* spp. and on the immunological reactions of sheep exposed to infestation. V. The nature of the self cure phenomenon. *Aust. J. agric. Res.*, **4**, 100–117.

Todd, K. S., Maisfield, M. E. & Levine, N. D. (1978) *Haemonchus contortus* infections in Targhee and Targhee-Barbados Black-Belly Cross Lambs. *Am. J. vet. Res.*, **39**, 865–866.

Veglia, F. (1915) *The Anatomy and Life History of Haemonchus contortus (Rud.)*, 3rd and 4th Reports of the Director of Veterinary Research, pp. 347–500. South Africa: Department of Agriculture.

Walker, P. J. & Thomas, R. J. (1975) Field studies of inhibition of *Haemonchus contortus* in sheep. *Parasitology*, **71**, 285–291.

Whitlock, J. H. (1966) The environmental biology of a nematode. In: *Biology of Parasites*, ed. E. J. L. Soulsby, pp. 185–197. New York: Academic Press.

TRICHOSTRONGYLIDAE: TRICHOSTRONGYLOSIS; PATHOGENESIS

Barger, I. A. (1979) Milk production of grazing dairy cattle after a single anthelmintic treatment. *Aust. vet. J.*, **55**, 68–70.

Bliss, D. H. & Todd, A. C. (1973) Milk production by Wisconsin dairy cattle after worming with Baymix. *Vet. Med. small Anim. Clin.*, **68**, 1034–1038.

Bliss, D. H. & Todd, A. C. (1976) Milk production by Vermont dairy cattle after deworming. *Vet. Med. small Anim. Clin.*, **71**, 1251–1254.

Borgsteede, F. H. M. (1978) Observations on the post-parturient rise of nematode egg output in cattle. *Vet. Parasit.*, **4**, 385–391.

Coop, R. L. & Angus, K. W. (1975) The effect of continuous doses of *Trichostrongylus colubriformis* larvae on the intestinal mucosa of sheep and on liver vitamin A concentration. *Parasitology*, **70**, 1–9.

Dargie, J. D. (1981) Pathophysiology and helminth parasites. In: *Immunology and Pathogenesis of Parasitic Infections in Ruminants and the Influence of Genetic and Nutritional Factors*. Vienna: International Atomic Energy Agency.

Farizy, P. (1971) Enterent d'un traitment anthelmintique au thiabendazole chez la chevre en lactation. *Recl Méd. vét. Alfort*, **146**, 251–260.

Fox, M. T. & Jacobs, D. E. (1981) Observations on the epidemiology and pathogenecity of nematode infections in adult dairy cattle in Great Britain. In: *The Epidemiology and Control of Nematodiasis in Cattle*, ed. P. Nansen, R. J. Jørgensen & E. J. L. Soulsby, pp. 87–100. The Hague: Martinus Nijhoff.

Hammerberg, B. & Lamm, W. D. (1980) Changes in periparturient fecal egg counts in beef cows calving in the spring. *Am. J. vet. Res.*, **41**, 1686–1689.

Hong, C., Lancaster, M. B. & Michel, J. F. (1981) Worm burdens of dairy heifers in England and Wales. *Vet. Rec.*, **109**, 12–14.

Murray, M., Jarrett, W. F. H., Jennings, F. W. & Miller, H. R. P. (1971) Structural changes associated with increased permeability of parasitized mucous membranes to macromolecules. In: *Pathology of Parasitic Diseases*, ed. S. M. Gaafar, pp. 197–200. Indiana: Purdue University Studies.

Murray, M., Jennings, F. W. & Armour, J. (1970) Bovine ostertagiasis: structure, function and mode of differentiation of the bovine gastric mucosa and kinetics of the worm loss. *Res. vet. Sci.*, **11**, 417–427.

Nielsen, K. & Anderson, S. (1967) Intestinal lymphangiectasia in cattle. *Nord. Vet Med.*, **19**, 31–35.

Reveron, A. E., Topps, J. H. & Gelman, A. L. (1974) Mineral metabolism and skeletal development of lambs affected by *Trichostrongylus colubriformis*. *Res. vet. Sci.*, **14**, 310–319.

Roseby, F. B. & Leng, R. A. (1974) Effects of *Trichostrongylus colubriformis* (Nematoda) on the nutrition and metabolism of sheep, II. Metabolism of urea. *Aust. J. agric. Res.*, **25**, 363–367.

Soulsby, E. J. L. (1976) *Pathophysiology of Parasitic Infection*. New York: Academic Press.

Sykes, A. R., Coop, R. L. & Angus, K. W. (1975) Experimental production of osteoporosis in growing lambs by continuous dosing with *Trichostrongylus colubriformis* larvae. *J. comp. Path.*, **85**, 549–559.

Sykes, A. R., Coop, R. L. & Angus, K. W. (1977) The influence of chronic *Ostertagia* circumcincta infection in the skeleton of growing sheep. *J. comp. Path.*, **87**, 521–529.

Symons, L. E. A. (1976) Malabsorption. In: *Pathophysiology of Parasitic Infection*, ed. E. J. L. Soulsby, pp. 11–21. New York: Academic Press.

Symons, L. E. A. & Jones, W. O. (1975) Skeletal muscle, liver and wool protein synthesis by sheep infected by the nematode *Trichostrongylus colubriformis*. *Aust. J. agric. Res.*, **26**, 1063–1072.
Symons, L. E. A. & Jones, W. O. (1978) Protein metabolism, 5. *Trichostrongylus colubriformis*: changes of host body mass and protein synthesis in guinea-pigs with light to heavy infections. *Exp. Parasit.*, **44**, 7–13.
Symons, L. E. A. & Steel, J. W. (1979) Pathogenesis of the loss of production in gastrointestinal parasitism. In: *The Epidemiology and Control of Gastrointestinal Parasites of Sheep in Australia*, ed. A. D. Donald, W. H. Southcott & J. K. Dineen, pp. 9–22. Commonwealth Scientific and Industrial Research Organization, Australia.

TRICHOSTRONGYLIDAE: EPIDEMIOLOGY OF TRICHOSTRONGYLOSIS

Anderson, N. (1972) Trichostrongylid infections of sheep in a winter rainfall region. I. Epizootiological studies in the Western district of Victoria 1966–1967. *Aust. J. agric. Res.*, **23**, 1113–1129.
Anderson, N. (1973) Trichostrongylid infections of sheep in a winter rainfall region. II. Epizootiological studies in the western district of Victoria 1967–1968. *Aust. J. agric. Res.*, **24**, 599–611.
Armour, J. (1980) The epidemiology of helminth disease in farm animals. *Vet. Parasit.*, **6**, 7–46.
Bairden, K., Parkins, J. J. & Armour, J. (1979) Bovine ostertagiasis: a changing epidemiological pattern? *Vet. Rec.*, **105**, 33–35.
Bizzell, W. E. & Ciordia, H. (1965) Dissemination of infective larvae of trichostrongylid parasites of ruminants from faeces to pasture by the fungus *Pilobolus* spp. *J. Parasit.*, **51**, 184.
Boswell, C. C. & Smith, A. (1976) The use of a fluorescent pigment to record the distribution by cattle of traces of faeces from dung pats. *J. Br. Grassld Soc.*, **31**, 135–136.
Brunsdon, R. V. (1980) Principles of helminth control. *Vet. Parasit.*, **6**, 185–215.
Donald, A. D. & Walker, P. (1973) Gastro-intestinal nematode parasite populations in ewes and lambs and the origin and time course of infective larval availability on pastures. *Int. J. Parasit.*, **3**, 219–233.
Fincher, G. T. & Stewart, T. B. (1979) Vertical migration by nematode larvae of cattle parasites through soil. *Proc. Helminth. Soc. Wash.*, **46**, 43–46.
Gronwold, J. (1979) On the possible role of earthworms in the transmission of *Ostertagia ostertagi* third-stage larvae from faeces to soil. *J. Parasit.*, **65**, 831–832.
Hart, J. A. (1964) Observations on the dry season strongyle infestations of zebu cattle in northern Nigeria. *Br. vet. J.*, **120**, 87–95.
Hsu, C.-K. & Levine, N. D. (1977) Degree-day concept in development of infective larvae of *Haemonchus contortus* and *Trichostrongylus colubriformis* under constant and cyclic conditions. *Am. J. vet. Res.*, 38, 1115–1119.
Jacobs, D. E., Tod, M. E., Dunn, A. M. & Walker, J. (1968) Farm-to-farm transmission of porcine oesophagostomiasis. *Vet. Rec.*, **82**, 57.
Levine, N. D. (1978) The influence of weather on the bionomics of the free-living stage of nematodes. In: *Weather and Parasitic Animal Disease*, ed. T. E. Gibson, technical note no. 159, pp. 51–57. Geneva: World Meteorological Organization.
Michel, J. F. (1976) The epidemiology and control of some nematode infections in grazing animals. *Adv. Parasit.*, **14**, 355–397.
Michel, J. F., Lancaster, M. B. & Hong, C. (1979) The effect of age, acquired resistance, pregnancy and lactation on some reactions of cattle to infection with *Ostertagia ostertagi*. *Parasitology*, **79**, 157–168.
Nelson, A. M. R. (1977) Where do lungworms go in wintertime? *Vet. Rec.*, **101**, 248.
Ogunsussi, R. A. (1979) Termination of arrested development of the trichostrongyles of sheep in northern Nigeria. *Res. vet. Sci.*, 26, 189–192.
Ogunsussi, R. A. & Eysker, M. (1979) Inhibited development of trichostrongylids of sheep in northern Nigeria. *Res. vet. Sci.*, **26**, 108–110.
Robinson, J. (1962) *Pilobolus* spp. and the translation of the infective larvae of *Dictyocaulus viviparus* from faeces to pasture. *Nature, Lond.*, **193**, 353–354.
Rose, J. H. (1962) Further observations on the free-living stages of *Ostertagia ostertagi* in cattle. *J. comp. Path.*, **72**, 11–18.
Skinner, W. D. & Todd, K. S. (1980) Lateral migration of *Haemonchus contortus* larvae on pasture. *Am. J. vet. Res.*, **41**, 395–398.
Shubber, A. H., Lloyd, S. & Soulsby, E. J. L. (1981) Infection with gastrointestinal helminths. Effect of lactation and passive transfer of immunity. *Z. Parasitenk*, **65**, 181–189.
Thomas, R. J. (1973) Ovine parasitic gastroenteritis. Epidemiology and control in hill sheep. Epidemiology and control in lowground sheep. In: *Helminth Diseases of Cattle, Sheep and Horses in Europe*, ed. G. M. Urquhart & J. Armour, pp. 41–49. Glasgow: Maclehose.
Todd, K. S., Levine, N. D. & Boatman, P. A. (1976) Effect of temperature on survival of free-living stages of *Haemonchus contortus*. *Am. J. vet. Res.*, **37**, 991–992.
Vlassoff, A. (1973) Seasonal incidence of infective trichostrongyle larvae on pasture grazed by lambs. *N.Z.J. agric. Res.*, **1**, 293–301.
Wharton, D. A. (1982) The survival of desiccation by the free-living stages of *Trichostrongylus colubriformis* (Nematoda: Trichostrongylidae). *Parasitology*, **84**, 455–462.

TRICHOSTRONGYLIDAE: HYPOBIOSIS; PERIPARTURIENT RISE

Armour, J. (1978) The effect of weather factors on the inhibition of larval development in parasitic nematodes. In: *Weather and Parasitic Animal Disease*, ed. T. E. Gibson, technical note no. 159. Geneva: World Meteorological Organization.
Armour, J. & Bruce, R. G. (1974) The inhibition of development of *Ostertagia ostertagi*—a diapause phenomenon in a nematode. *Parasitology*, **69**, 161–174.
Anderson, N. (1973) Trichostrongylid infections of sheep in a winter rainfall region. II. Epizootiological studies in the western district of Victoria, 1967–1968. *Aust. J. agric. Res.*, **24**, 598–611.
Blitz, N. M. & Gibbs, H. C. (1972) Studies on the arrested development of *Haemonchus contortus* in sheep. I. The induction of arrested development. *Int. J. Parasit.*, **2**, 5–12.
Borgsteede, F. H. M., Armour, J. & Jansen, J. (1978) *Arrested Development of Nematodes in Sheep and Cattle*. Lelystad: Central Veterinary Institute.
Chen, P. & Soulsby, E. J. L. (1976) *Haemonchus contortus* infection in ewes: blastogenic responses of peripheral blood leukocytes to third stage larval antigen *Int. J. Parasit.*, **6**, 135–141.
Connan, R. M. (1967) Observations on the epidemiology of parasitic gastroenteritis due to *Oesophagostomum* spp. and *Hyostrongylus rubidus* in the pig. *Vet. Rec.*, **80**, 424–429.
Connan, R. M. (1973) Ovine parasitic gastroenteritis. The spring rise in faecal egg count. In: *Helminth Diseases of Cattle, Sheep and Horses in Europe*, ed. G. M. Urquhart & J. Armour, pp. 36–41. Glasgow: Maclehose.
Connan, R. M. (1975) Inhibited development in *Haemonchus contortus*. *Parasitology*, **71**, 239–246.
Connan, R. M. (1978) Arrested development in *Haemonchus contortus*. In: *Arrested Development of Nematodes in Sheep and Cattle*, ed. F. H. M. Borgsteede, J. Armour & J. Jansen, pp. 52–63. Lelystad: Central Veterinary Institute.
Cvetković, L., Golosin, R. & Kosanović, M. (1971) Seasonal fluctuations in the trichostrongylid worm egg counts in the faeces of unmated ewes and ewes which lambed at different months of the year. *Acta vet. Beograd*, **21**, 77–88.
Dineen, J. K. & Kelly, J. D. (1972) The suppression of rejection of *Nippostrongylus brasiliensis* in lactating rats: the nature of the immunological defect. *Immunology*, **22**, 1–12.
Hart, J. A. (1964) Observations on the dry season strongyle infestations of zebu cattle in northern Nigeria. *Br. vet. J.*, **120**, 87–95.
McKenna, P. B. (1973) The effect of storage on the infectivity of parasitic development of third-stage *Haemonchus contortus* larvae in sheep. *Res. vet. Sci.*, **14**, 312–316.
Martin, W. B., Thomas, B. A. C. & Urquhart, G. M. (1957) Chronic diarrhoea in housed cattle due to atypical parasitic gastritis. *Vet. Rec.*, **69**, 736–739.
Michel, J. F. (1978) Topical themes in the study of arrested development. In: *Arrested Development of Nematodes in Sheep and Cattle*, ed. F. H. M. Borgsteede, J. Armour & J. Jansen, pp. 7–17. Lelystad: Central Veterinary Institute.
Michel, J. F., Lancaster, M. B. & Hong, C. (1975) Arrested development of *Ostertagia ostertagi* and *Cooperia oncophora*: The effect of temperature at the free-living third stage. *J. comp. Path.*, **85**, 133–138.

Michel, J. F., Lancaster, M. B. & Hong, C. (1976) The resumed development of arrested *Ostertagia ostertagi* in experimentally infected calves. *J. comp. Path.*, **86**, 615–618.
Michel, J. F., Lancaster, M. B. & Hong, C. (1979) The effect of age, acquired resistance, pregnancy and lactation on some reactions of cattle to infection with *Ostertagia ostertagi. Parasitology*, **79**, 157–168.
O'Sullivan, B. M. & Donald A. O. (1973) Responses to infection with *Haemonchus contortus* and *Trichostrongylus colubriformis* in ewes of different reproductive status. *Int. J. Parasit.*, **3**, 521–530.
Riffkin, G. G. & Dobson, C. (1979) Predicting resistance of sheep to *Haemonchus contortus infection. Vet. Parasit.*, **5**, 365–378.
Shimshony, A. (1974) Observations on parasitic gastroenteritis in goats in northern Israel. *Refuah Vet.*, **31**, 63–75.
Shubber, A. H., Lloyd, S. & Soulsby, E. J. L. (1981) Infection with gastrointestinal helminths. Effect of lactation and maternal transfer of immunity. *Z. Parasitenk*, **65**, 181–189.
Walker, P. J. & Thomas, R. J. (1975) Field studies on inhibition of *Haemonchus contortus* in sheep. *Parasitology*, **71**, 285–291.
Wedderburn, J. F. (1970) Ostertagiasis in adult cattle: a clinical report of an outbreak in the field. *N. Z. vet. J.*, **18**, 168–170.

TRICHOSTRONGYLIDAE: HOST GENETIC FACTORS

Altaif, K. I. & Dargie, J. D. (1978) Genetic resistance to helminths. The influence of breed and haemoglobin type on the response of sheep to re-infection with *Haemonchus contortus. Parasitology*, **77**, 177–187.
Dineen, J. K. & Windon, R. G. (1980) The effect of sire selection on the response of lambs to vaccination with irradiated *Trichostrongylus colubriformis* larvae. *Int. J. Parasit.*, **10**, 189–196.
Evans, J. V. & Blunt, M. H. (1961) Variation in the gene frequencies of potassium and haemoglobin types in Romney Marsh and Southdown sheep established away from their native environment. *Aust. J. biol. Sci.*, **14**, 100–108.
Frisch, J. E. (1981) Factors affecting resistance to ecto- and endo-parasites of cattle in tropical areas and the implications for selection. In: *Genetic and Nutritional Influences on the Immunology and Pathogenesis of Parsitic Infections in Ruminants.* Vienna: International Atomic Energy Agency.
Scrivner, L. H. (1964) Breed resistance to ostertagiasis in sheep. *J. Am. vet. med. Ass.*, **144**, 883–887.
Whitlock, J. H. & Madsen, H. (1958) The inheritance of resistance to trichostrongylidosis in sheep II. Observations on the genetic mechanism in trichostrongylidosis. *Cornell Vet.*, **48**, 134–145.
Windon, R. G., Dineen, J. K. & Kelly, J. D. (1980) The segregation of lambs into 'responders' and 'non-responders': response to vaccination with irradiated *Trichostrongylus colubriformis* larvae before weaning. *Int. J. Parasit.*, **10**, 65–73.

TRICHOSTRONGYLIDAE: IMMUNOLOGICAL ASPECTS OF TRICHOSTRONGYLOSIS

Dineen, J. K., Gregg, P., Windon, R. G., Donald, A. D. & Kelly, J. D. (1977) The role of immunologically specific and non-specific components of resistance in cross-protection to intestinal nematodes. *Int. J. Parasit.*, **7**, 211–215.
Dineen, J. K. & Kelly, J. D. (1976) The levels of prostaglandins in the small intestine of rats during primary and secondary infection with *Nippostrongylus brasiliensis. Int. Arch. Allergy appl. Immunol.*, **51**, 429–440.
Rothwell, T. L. W., Dineen, J. K. & Love, R. J. (1971) The role of pharmacologically-active amines in resistance to *Trichostrongylus colubriformis* in the guinea-pig. *Immunology*, **21**, 925–938.
Shubber, A. H., Lloyd, S. & Soulsby, E. J. L. (1981) Infection with gastrointestinal helminths: Effect of lactation and maternal transfer of immunity. *Z. Parasitenk*, **65**, 181–189.
Soulsby, E. J. L. (1979) The immune system and helminth infection in domestic species. *Adv. vet. Sci. comp. Med.*, **23**, 71–102.
Stewart, D. F. (1955) 'Self cure' reaction in nematode infestations in sheep. *Nature, Lond.*, **176**, 1273–1274.

TRICHOSTRONGYLIDAE: ANTHELMINTIC TREATMENT

Armour, J., Bairden, K., Duncan, J. L., Jones, R. M. & Bliss, D. H. (1981) Studies on the control of bovine ostertagiasis using a morantel sustained release bolus. *Vet. Rec.*, **108**, 532–535.
Armour, J., Bairden, K. & Preston, J. M. (1980) Anthelmintic efficacy of ivermectin against naturally acquired bovine gastrointestinal nematodes. *Vet. Rec.*, **107**, 226–227.
Berger, J. (1975) The resistance of a field strain of *Haemonchus contortus* to five benzimidazole anthelmintics in current use. *J. S. Afr. vet. Ass.*, **46**, 369–372.
Bliss, D. H., Jones, R. M. & Conder, D. R. (1982) Epidemiology and control of gastrointestinal parasitism in lactating, grazing adult dairy cows using a morantel sustained release bolus. *Vet. Rec.*, **110**, 141–144.
Drudge, J. H., Leland, S. E. & Wyant, Z. N. (1957) Strain variation in the response of sheep nematodes to the action of phenothiazine I. Studies on pure infections of *Haemonchus contortus. Am. J. vet. Res.*, **18**, 317–325.
Drudge, J. H., Szanto, J., Wyant, Z. N. & Elam, R. W. (1964) Field studies on parasite control in sheep: comparison of thiabendazole, ruelene and phenothiazine. *Am. J. vet. Res.*, **25**, 1512–1518.
Duncan, J. L., Armour, J. & Bairden, K. (1978) Autumn and winter fenbendazole treatment against inhibited 4th stage *Ostertagia ostertagi* larvae in cattle. *Vet. Rec.*, **103**, 211–212.
Duncan, J. L., Armour, J., Bairden, K. & Baines, D. M. (1979) The efficacy of thiophanate against gastrointestinal nematodes of cattle including inhibited larvae of *Ostertagia ostertagi. Vet. Rec.*, **105**, 444–445.
Duncan, J. L., Armour, J., Bairden, K., Jennings, F. W. & Urquhart, G. M. (1976) The successful removal of inhibited fourth stage *Ostertagia ostertagi* larvae by fenbendazole. *Vet. Rec.*, **98**, 342.
Forsyth, B. A. & Wynne-Jones, M. (1980) Levamisole vaccine combinations. II. Retained anthelmintic efficacy. *Aust. vet. J.*, **56**, 292–295.
Hall, C. A., Kelly, J. D., Campbell, N. J., Whitlock, H. V. & Martin, I. C. A. (1978) The dose response of several benzimidazole anthelmintics against resistant strains of *Haemonchus contortus* and *Trichostrongylus colubriformis* selected with thiabendazole. *Res. vet. Sci.*, **25**, 364–367.
Hogarth-Scott, R. S., Liardet, D. M. & Morris, P. J. (1980) Levamizole vaccine combinations. I. Heightened antibody response. *Aust. vet. J.*, **56**, 285–291.
Jacobs, D. E., Fox, M. T., Walker, M. J., Jones, R. M. & Bliss, D. H. (1981) Field evaluation of a new method for the prophylaxis of parasitic gastroenteritis in calves. *Vet. Rec.*, **108**, 274–276.
Kelly, J. D. & Hall, C. A. (1979) Resistance of animal helminths to anthelmintics. *Adv. Pharmac. Chemother.*, **16**, 89–128.
Kelly, J. D., Hall, C. A., Whitlock, H. V., Thompson, H. G., Campbell, N. J. & Martin, I. C. A. (1977) The effect of route of administration on the anthelmintic efficacy of benzimidazole anthelmintics in sheep infected with strains of *Haemonchus contortus* and *Trichostrongylus colubriformis* resistant or susceptible to thiabendazole. *Res. vet. Sci.*, **22**, 161–168.
Kelly, J. D., Whitlock, H. V., Thompson, H. G., Hall, C. A., Martin, I. C. A. & Le Jambre, L. F. (1979) Physiological characteristics of free-living and parasite stages of strains of *Haemonchus contortus*, susceptible or resistant to benzimidazole anthelmintics. *Res. vet. Sci.*, **25**, 376–385.
Le Jambre, L. F. (1976) Egg hatch as an *in vitro* assay of thiabendazole resistance in nematodes. *Vet. Parasit.*, **2**, 385–391.
Le Jambre, L. F. (1978) Anthelmintic resistance in gastrointestinal nematodes of sheep. In: *The Epidemiology and Control of Gastrointestinal Parasites of Sheep in Australia*, ed. A. D. Donald, W. H. Southcott & J. K. Dineen, pp. 109–120. Melbourne: CSIRO.
McBeath, D. G., Best, J. M. J., Preston, N. K. & Thompson, F. (1977) The treatment of ostertagiasis Type II in cattle, using fenbendazole in feed blocks. *Vet. Rec.*, **101**, 285–286.
Michel, J. F. (1967) Methods of testing anthelmintics. *Vet. Rec.*, **80**, 336.

Pott, J. M., Jones, R. M. & Cornwell, R. L. (1979) Observations on parasitic gastroenteritis and bronchitis in grazing calves: effect of low level feed incorporation of morantel in early season. *Int. J. Parasit.*, **9**, 153–157.
Prichard, R. K. (1973) The fumerate reductase reaction of *Haemonchus contortus* and the mode of action of some anthelmintics. *Int. J. Parasit.*, **3**, 409–417.
Prichard, R. K., Donald, A. D., Dash, K. M., Hennessy, D. R. (1978*a*) Factors involved in the relative anthelmintic tolerance of arrested 4th stage larvae of *Ostertagia ostertagi*. *Vet. Rec.*, **102**, 382.
Prichard, R. K., Hall, C. A., Kelly, J. D., Martin, I. C. A. & Donald, A. D. (1980) The problem of anthelmintic resistance in nematodes. *Aust. vet. J.*, **56**, 239–250.
Prichard, R. K., Kelly, J. D. & Thompson, H. G. (1978*b*) The effect of benzimidazole resistance and route of administration on the uptake of fenbendazole and thiabendazole by *Haemonchus contortus* and *Trichostrongylus colubriformis* in sheep. *Vet. Parasit.*, **4**, 243–255.
Rew, R. S. (1978) Mode of action of common anthelmintics. *J. vet. Pharmac. Ther.*, **1**, 183–198.

TRICHOSTRONGYLIDAE: CONTROL OF TRICHOSTRONGYLOSIS

Brunsdon, R. V. (1980) Principles of helminth control. *Vet. Parasit.*, **6**, 185–215.
Donald, A. D., Southcott, W. H. & Dineen, J. K. (1978) *The Epidemiology and Control of Gastrointestinal Parasites of Sheep in Australia*. Melbourne: Commonwealth Scientific and Industrial Research Organization Publications.
Gordon, H. M. (1973) Epidemiology and control of gastrointestinal nematodes of ruminants. *Adv. vet. Sci. comp. Med.*, **17**, 395–437.
Michel, J. F. (1976) The epidemiology and control of some nematode infections in grazing animals. *Adv. Parasit.*, **14**, 355–397.
Morley, F. H. W. & Donald, A. D. (1980) Farm management and systems of helminth control. *Vet. Parasit.*, **6**, 105–134.
Ollerenshaw, C. B., Graham, E. G., & Smith, L. B. (1978) Forecasting the incidence of parasitic gastroenteritis in lambs in England and Wales. *Vet. Rec.*, **103**, 461–465.
Thomas, R. J. & Starr, J. R. (1978) Forecasting the peak of gastrointestinal nematode infection in lambs. *Vet. Rec.*, **103**, 465–468.
Urquhart, G. M. & Armour, J. (1973) *Helminth Diseases of Cattle, Sheep and Horses in Europe*. Glasgow: Maclehose.
Vlassoff, A. (1975) The role of climate in the epidemiology of gastrointestinal nematode parasites of sheep and cattle. In: *Symposium on Meteorology and Food Production*, pp. 171–176. Wellington: New Zealand Meteorological Service.

Genus: Hyostrongylus Hall, 1921

Hyostrongylus rubidus (Hassall & Stiles, 1892) occurs in the stomach of the pig in many countries. The male is 4–7 mm long and the female 5–10 mm. The worms are slender and reddish when they are fresh. The body cuticle is transversely striated and also bears 40–45 longitudinal striations. The bursa is well developed, but the dorsal lobe is small. The spicules are 0.13 mm long. The vulva is situated 1.3–1.7 mm anterior to the anus. The eggs measure 71–78 by 35–42 μm.

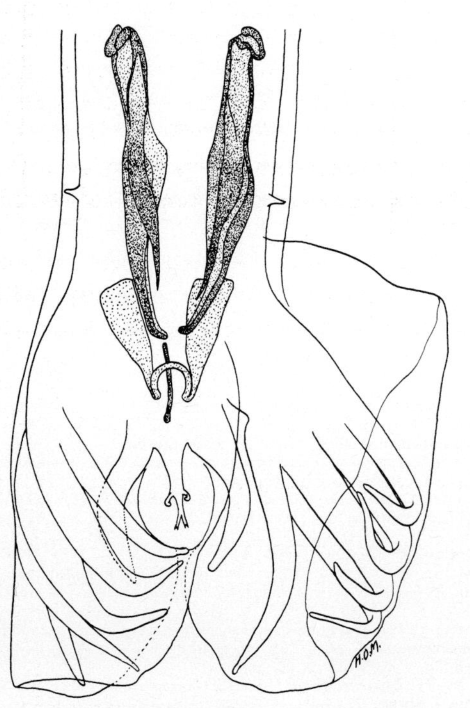

Fig 1.143 Ventral view of the hind end of the male *Hyostrongylus rubidus*.

Life-cycle. The eggs hatch, at ordinary temperatures, in 39 hours and the larvae develop to the infective stage in seven days. They are not very resistant to drying or low temperatures. Infection occurs per os and not through the skin. The worms reach maturity in 17–19 days. Kotlán (1949, 1960) studied the histotropic larval phases in the stomachs of pigs and rabbits and found that the larvae penetrate into the pits of the gastric glands, moving about in these until they become adults, this histotropic phase requiring 13–14 days. Some adults then return to the lumen of the stomach, but others may remain in the gastric glands for several months, causing dilatation of the glands and the formation of nodules 2–6 mm in diameter.

Pathogenicity. Various surveys show a prevalence of 10–60%. The parasites burrow into the gastric mucosa and suck blood. They may be present in small numbers without causing any ill effects, but often their presence is associated with a marked gastritis and in some cases marked ulceration. The animals lose condition rapidly and become weak. The appetite may vary, but the animals are usually thirsty. Diarrhoea occurs and the faeces may be dark in colour.

Experimental infections with doses up to 500 000 infective larvae produce mild fever, inappetence, diarrhoea and reduced weight gain (Castelino et al. 1970). An elevation of gastric pH (up to 6.5) in field cases of hyostrongylosis was reported by Davidson et al. (1967). Indeed, pathophysiological changes in the stomach closely resemble those of the abomasum in ostertagiasis of ruminants (p. 222) and provide an explanation for the disease process. Elevated gastric pH coincides with the time when most marked clinical signs are evident and egg-laying is at its height (Titchner et al. 1974). There is a loss of differentiation of cells of the gastric glands and a reduction in the number of parietal cells. Whereas in ostertagiasis raised gastric pH leads to diarrhoea, in hyostrongylosis vomiting is the more likely sequel and Titchner et al. (1974) suggest that this may be a major factor in gastrointestinal protein loss.

Post-mortem. The degree of emaciation of the carcass depends on the severity of the case. The main lesions are found in the stomach, varying from hyperaemia and catarrh in mild cases to croupous gastritis. In the latter the mucosa is thickened and shows areas covered with yellow, corrugated pseudo-membranes. The worms are found in these membranes as well as in the mucosa.

Diagnosis. A tentative diagnosis can be made by finding the eggs of the worm in the faeces, but it can be definitely confirmed only by autopsy of a selected case.

Treatment. A 97% efficacy was seen following treatment with dichlorvos (17 mg/kg) (Herbert et al. 1975); fenbendazole at a dose rate of 5 mg/kg had a 78%, 96% and 100% efficacy against experimentally administered 5-, 16- and 42-day-old infections, respectively (Kirsch & Düwel 1975); levamisole (7 mg/kg) was more than 90% effective against the adult parasites and had a 40–60% efficacy against the immature stages (Probert et al. 1973); thiabendazole (66–100 mg/kg) and cambendazole (15–30 mg/kg) have also been shown to be effective. Low level dosage with thiabendazole of pigs at pasture (0.05% in feed from three to eight weeks and 0.01% from eight weeks until 10 days before slaughter) prevented *H. rubidus* infection (Taffs & Davidson 1967).

Prophylaxis. Hygienic measures, particularly frequent removal of faeces from sties and effective drainage in the runs or paddocks, are indicated.

Genus: Ornithostrongylus Travassos, 1914

Ornithostrongylus quadriradiatus (Stevenson, 1904) occurs in the crop, proventriculus and small intestine of the pigeon throughout the world. The male is 9–12 mm long and the female 18–24 mm. The fresh worms are red. The cuticle of the head is slightly inflated and the body cuticle has a number of longitudinal striations. In the male bursa the ventral rays are close together and the dorsal is fairly short. The spicules are 0.15–0.16 mm long, each ending in three pointed processes. A telamon is present; it is roughly cross-shaped, the two arms forming an incomplete ring through which the spicules pass. The vulva opens about 5 mm from the posterior extremity. The eggs measure 70–75 by 38–40 μm.

The life-cycle is direct, like that of other Trichostrongylidae.

The worms burrow into the intestinal mucosa and severe infections cause a catarrhal enteritis. They are also blood-suckers and in severe infections a haemorrhagic enteritis with ulceration and necrosis may be seen. The parasite may be responsible for heavy losses in breeding establishments.

Phenothiazine at the rate of 0.4 g/kg has been used with success.

Tetramisole, 50 mg/kg, is 50–71% effective and 0.05% thiabendazole in the feed for 10 days causes a marked reduction in worm burden (Leibovitz 1962). Other benzimidazoles are likely to be effective.

A general improvement in hygiene is indicated in prevention. Pigeon lofts should be cleaned out

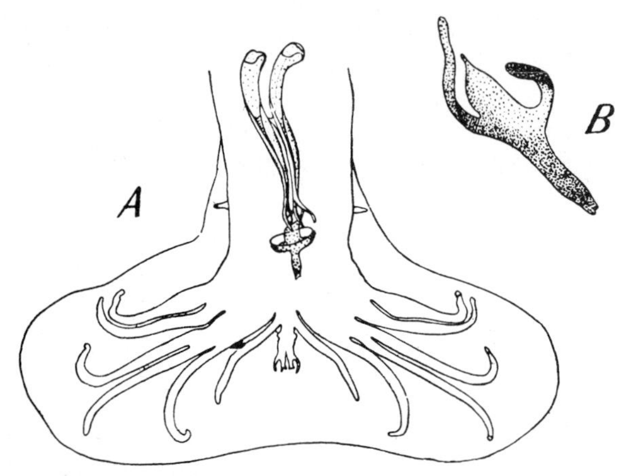

Fig 1.144 Ventral views of *Ornithostrongylus quadriradiatus*. A, Bursa. B, 'Telamon'.

regularly. Where ornamental pigeons and doves are hand fed, the place of feeding should be changed regularly to avoid an accumulation of eggs and infective larvae.

Genus: Libyostrongylus Lane, 1923

Libyostrongylus douglassii (Cobbold, 1882) occurs in the proventriculus of the ostrich in South Africa. The male is about 4–6 mm long and the female 5–6 mm. The colour of fresh specimens is yellowish-red. The male bursa is well developed and the dorsal ray is long and cleft in its distal half, forming three small branches on either side. The spicules are 0.14–0.158 mm long, each ending in a large and a small spine. The vulva opens 0.8 mm from the posterior extremity. The eggs measure 59–74by 36–44 μm.

The eggs are passed in the faeces of the host and the development and bionomics of the free-living stages resemble those of other strongyle species. The infective larval stage is reached under optimal conditions in 60 hours. The infective larva is about 0.745 mm long, including the sheath, and morphologically closely resembles the larvae of *Trichostrongylus* spp. Eggs containing fully formed embryos can resist desiccation as long as three years and the infective larvae remain viable under dry conditions for nine months or longer. Infection of the host occurs per os. The worms develop to maturity in the ostrich in about 33 days, the first eggs being passed in the faeces on the thirty-sixth day.

The young parasites penetrate deeply into the lumina of the glands in the proventriculus. The adults live in the surface epithelium of the organ, sucking blood and causing severe irritation.

Chicks are most susceptible to infection; adult birds, although susceptible, suffer much less from the parasite, especially when they are well fed. The birds become anaemic, weak, emaciated and stunted in growth. Severe losses of chicks may be experienced.

Apart from emaciation and anaemia the characteristic lesions at post mortem are found in the proventriculus. The mucosa is swollen and covered with an excessive amount of tough mucus. The mucosa is desquamated in patches and pseudomembranes may be present, covering haemorrhagic areas. The worms are found in and underneath the mucus and the pseudomembranes.

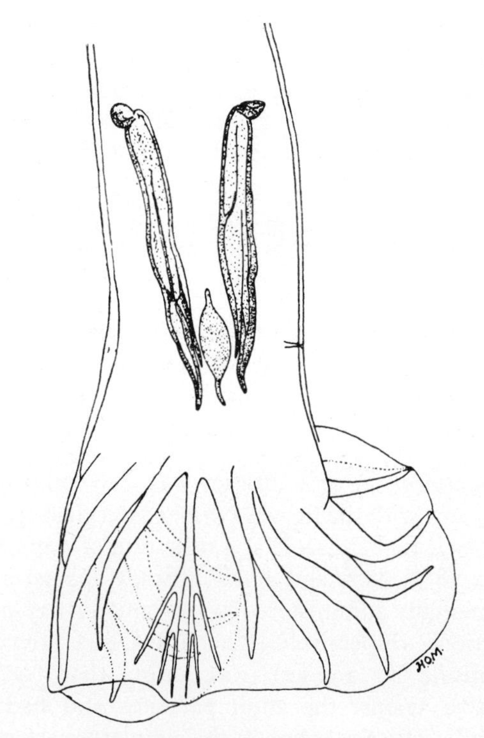

Fig 1.145 Dorsal view of the hind end of the male *Libyostrongylus douglassii*.

The clinical signs will lead to a tentative diagnosis, which must be confirmed by finding the infective larvae in faeces cultures or by autopsy of a selected bird. The eggs cannot be differentiated readily from those of *Codiostomum struthionis*.

No satisfactory treatment is known, but it is possible that the benzimidazole compounds are of use.

Since birds and runs may remain infected for long periods, the chicks should be kept away from the adult birds in clean runs and fed with crops grown on clean lands. If hens and chicks are kept together, the droppings of the adult birds should be removed every 24–48 hours. Infected adult birds suffer little effect from the parasites when they are well fed.

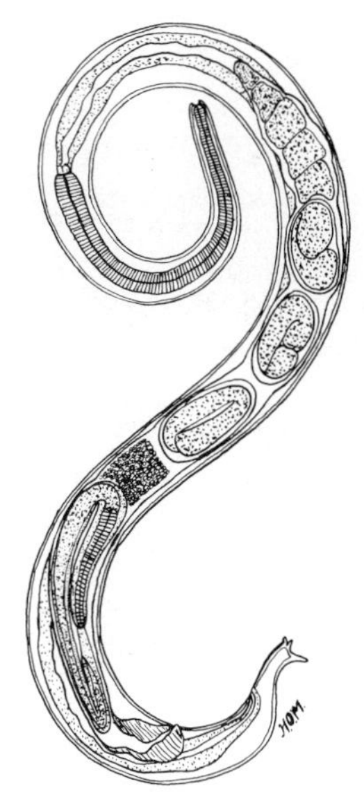

Fig 1.146 Female *Ollulanus tricuspis*.

FAMILY: OLLULANIDAE SKRJABIN & SCHIKHOBALOVA, 1952

Trichostrongyloidea in which the female has two, three or more caudal processes (mucrones). Parasites oviparous or viviparous. Genus of importance *Ollulanus*.

Genus: Ollulanus Leuckart, 1965

Ollulanus tricuspis Leuckart, 1865 occurs in the stomach of the cat, fox, wild *Felidae* and the pig in Europe, North America and Australia. The male is 0.7–0.8 mm long and the female 0.8–1 mm. A small buccal cavity is present. The male bursa is well developed and the spicules, which are 0.046–0.057 mm long and stout, are each split into two for a considerable distance. The tail of the female ends in three or more short cusps. The vulva is situated in the posterior part of the body and there is only one uterus and ovary.

The worms are viviparous and the larvae develop in the uterus of the female to the third larval stage. The infection is spread through the vomit of an infected animal which is eaten by another susceptible one. In addition the life-cycle may be completed endogenously and heavy infections may result.

In a survey of the infection in cats in a rural area of Germany, Hasslinger and Trah (1981) found 38% of 542 cats infected. An average infection of 1500 worms per cat was noted. The most satisfactory diagnostic technique is to examine vomit after the administration of an emetic. Faecal examination is of little value.

The parasite is considered to be relatively harmless to cats, although the worms burrow into the gastric mucosa, causing slight erosions and increased secretion of mucus. A chronic gastritis has been reported by Hänichen and Hasslinger (1977) and chronic catarrhal gastritis and emaciation in the pig have been ascribed to the parasite (Kotlán & Mocsy (1933).

It is possible the benzimidazole anthelmintics will be an effective treatment. Control is dependent on hygiene.

REFERENCES

TRICHOSTRONGYLOIDEA: HYOSTRONGYLUS, ORNITHOSTRONGULUS

Castelino, J. B., Herbert, I. V. & Lean, I. J. (1970) The live weight gain of growing pigs experimentally infected with massive doses of *Hyostrongylus rubidus* (Nematoda) larvae. *Br. vet. J.*, **126**, 579–582.

Davidson, J. B., Murray, M. & Sutherland, I. H. (1967) Observations on the clinical pathology of natural strongyle infestations in the pig, and their control with special reference to *Hyostrongylus rubidus*. In *The Host Response to Parasitism*, ed. E. J. L. Soulsby, pp. 9–23. Marberg: Elwert.

Hänichen, T. & Hasslinger, M.-A. (1977) Chronische Gastritis durch *Ollulanus tricuspus* (Leukart, 1865) beim einer Katze. *Berl. Münch. tierärztl. Mschr.*, **90**, 58–62.

Hasslinger, M.-A. & Trah, M. (1981) Unterschungen zur Verbreitung und zum Nachweis des Magenwurmes der Katze, *Ollularus tricuspus* (Leukart, 1865). *Berl. Munch. tierärztl. Mschr.*, **94**, 235–238.

Herbert, I. V., Jacobs, D. E. & Probert, A. J. (1975) An investigation of the activity of a dichlorvos formulation, Atgard C, against mature *Hyostrongylus rubidus*, the red stomach worm, in growing pigs. *Shell Anim. Hlth Reprint Series*, No. AHR/END 13.

Kirsch, R. & Düwel, D. (1975) Laboratory investigations on pigs with the new anthelmintic fenbendazole. *Res. vet. Sci.*, **19**, 327–329.

Kotlán, A. (1949) On the histotropic phase of the parasitic larvae of *Hyostrongylus rubidus*. *Acta vet. Hung.*, **1**, 76–82.

Kotlán, A. (1960) *Helminthologie*, Budapest: Akademiai Kiado.

Kotlán, A. & Mocsy, J. V. (1933) Ollulanus tricuspus Leuk. als Ursache einer chronische Magenwurmseuche bein Schweine. *Dt. tierärztl. Wschr.*, **41**, 689–692.

Leibovitz, L. (1962) Thiabendazole therapy of pigeons affected with *Ornithostrongylus quadrinadiatus*. *Avian Dis.*, **6**, 380–384.

Probert, A. J., Smith, B. D. S. & Herbert, I. V. (1973) The efficiency of orally and subcutaneously administered levamisole against mature and immature stages of *Hyostrongylus rubidus* (Hassal and Stiles, 1892), the stomach worm of pigs. *Vet. Rec.*, **93**, 302–306.

Taffs, L. F. & Davidson, J. B. (1967) Low level thiabendazole in the control of worm parasites of pigs. *Vet. Rec.*, **81**, 426–435.

Titchner, R. N., Herbert, I. V. Probert, A. J. & Axford, R. F. E. (1974) Observations on the stomach pH in pigs following massive infections with *Hyostrongylus rubidus* larvae. *J. comp. Path.*, **84**, 127–131.

FAMILY: DICTYOCAULIDAE SKRJABIN, 1941

It has been common to classify nematodes in this group in the family Metastrongylidae and thus include them with the other nematodes which are found in the air passages or blood vessels of the lungs. The placing of the genus *Dictyocaulus* in the superfamily Trichostrongyloidea rather than the Metastrongyloidea (see later) is based on the consideration that members of the genus do not require intermediate hosts and thus are 'geohelminths' whereas members of the Metastrongyloidea, as far as is known, require intermediate hosts and are 'biohelminths', according to Russian workers. Members of this family occur in the respiratory passages of the lungs; the bursa is well formed and bursal rays well developed, some may be fused at the distal part. Spicules short and reticulated. The life-cycle is direct. The only genus of importance is *Dictyocaulus*.

Genus: Dictyocaulus Railliet and Henry, 1907

Dictyocaulus filaria (Rudolphi, 1809) occurs in the bronchi of sheep, goats and some wild ruminants. It has a world-wide distribution and causes serious losses in countries in eastern Europe and India. The male is 3–8 cm and the female 5–10 cm long. The worms have a milk-white colour and the intestine shows as a dark line. There are four very small lips and a very small, shallow buccal capsule. In the male bursa the medio- and posterolateral rays are fused together except at their tips; the externodorsals arise separately and the dorsal ray is cleft right from its base. The spicules are stout, dark-brown, boot-shaped and 0.4–0.64 mm long. The vulva is situated not far behind the middle of the body. The eggs measure 112–138 by 69–90 μm and contain fully formed larvae when laid.

Life-cycle. The eggs may hatch in the lungs, but are usually coughed up and swallowed, and first-stage larvae hatch while they pass through the alimentary tract of the host. Some eggs may be expelled in the nasal discharge or sputum. The first-stage larva passed in the faeces is 0.55–0.58 mm long and can be easily recognized by the presence of a small cuticular knob at the anterior extremity and numerous brownish food granules in the intestinal cells. The free stages do not feed, but exist on these food granules. After one or two days the larva reaches the second stage, but does not cast the old cuticle until the third or infective stage is reached, so that the latter is for some time enclosed in two sheaths. The first is then cast while the second is retained for protection. Under ideal conditions the infective stage is reached in six or seven days.

Infection of the host occurs per os. Larvae penetrate into the intestinal wall within three days and pass via the lymph vessels to the mesenteric lymph glands, where they develop and perform the third ecdysis about four days after infection. In the fourth stage males and females can be distinguished. The worms now pass via the lymph and blood vessels to the lungs, where they are arrested in the capillaries and break through into the air passages. Development to maturity in the bronchi of the host takes about four weeks.

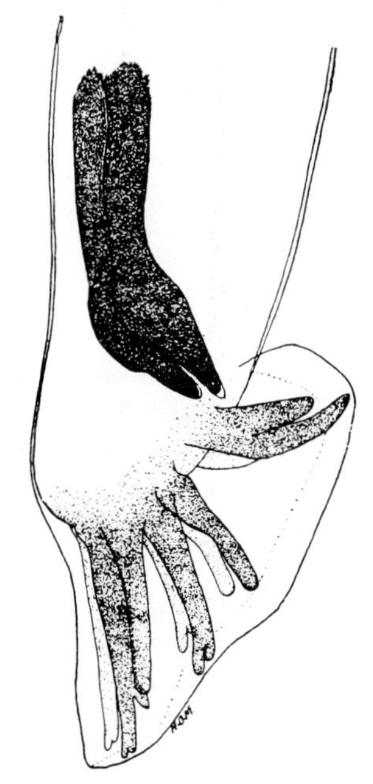

Fig 1.147 Lateral view of hind end of male *Dictyocaulus filaria*.

Epidemiology. The larvae require moisture for development and at a temperature of 27°C they reach the infective stage in six or seven days. In England, in summer, development may take four to seven weeks. Infective larvae may survive over winter to the following March, April and May providing a source of infection for spring born lambs. Survival time of the larvae is shorter in spring and summer than in autumn and winter, thus pasture burdens of larvae increase in the autumn and the heaviest infections in lambs occur in the autumn (Gallie & Nunns 1976).

Pathogenesis. The worms live in the small bronchi, producing a catarrhal parasitic bronchitis. The inflammatory process spreads to the surrounding peribronchial tissues and the exudate frequently passes back into the bronchioles and alveoli, causing atelectasis and catarrh, or pneumonia. Secondary bacterial infection may lead to more extensive areas of pneumonia. Severe infections resemble those produced by *Dictyocaulus viviparus* in cattle (see p. 266).

Clinical signs. Young animals are chiefly affected, but the disease may occur at all ages and is usually chronic. The animals may cough and a tenacious mucus exudes from the nostrils, but cough is not always present and it is not a reliable guide to the severity of the infection. Dyspnoea is usually obvious, the respiration is more rapid than normal and abnormal lung sounds can be heard on auscultation. The temperature is not elevated unless pneumonia develops.

Post-mortem. The lungs show atelectatic areas of variable size. The bronchi in the affected parts contain the worms and a large amount of mucus, which is mixed with blood and is slighly opaque due to desquamated epithelial cells, leucocytes and the eggs of the worms. The bronchial mucosa and the peribronchial tissues are inflamed and infiltrated with leucocytes. Localized and cone-shaped areas of pneumonia, accompanied by atelectasis and compensatory emphysema, may be present. In some cases proliferation of the bronchial epithelium has occurred.

Diagnosis is made by finding the first-stage larvae in the fresh faeces. Eggs may be found in the sputum or nasal discharge, but their absence is not significant.

Treatment. See p. 267.

Prophylaxis. The animals must be removed from infected ground, placed on dry pastures and supplied with clean drinking water. Moist pastures must be avoided, while dry pastures are fairly safe, because the infective larvae are not very resistant to dryness. The general prophylactic measures described on p. 267 for the bovine lungworm are also important. The larvae may

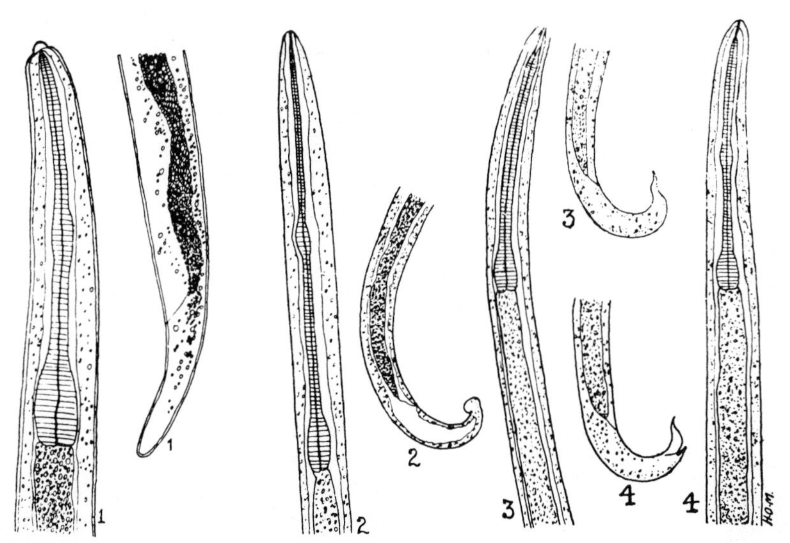

Fig 1.148 Larval stages of lungworms as found in the fresh faeces of hosts. (anterior and posterior ends). 1, *Dictyocaulus filaria*. 2, *Metastrongylus apri*. 3, *Protostrongylus rufescens*. 4, *Muellerius capillaris*.

live through the winter in cold climates, and the infection is acquired either from overwintered larvae or from those shed by older animals, which show no clinical signs. Older animals should therefore not be grazed together with the young stock. A vaccine consisting of irradiation-attenuated infective larvae is available in some countries, such as Yugoslavia and India.

Dictyocaulus viviparus (Bloch, 1782) occurs in the bronchi of cattle, deer, reindeer, buffalo and camel and has a cosmopolitan distribution. The male is 4–5.5 cm long and the female 6–8 cm. The worm closely resembles the preceding species, but the medio- and posterolateral rays are completely fused and the spicules are only 0.195–0.215 mm long. The eggs measure 82–88 by 33–38 μm.

Life-cycle. Similar to that of *D. filaria*. The infective stage is reached in about four days. The first-stage larva found in the fresh faeces of the host is 0.3–0.36 mm long and has no anterior knob, but the intestinal cells also contain numerous brownish granules.

The developmental cycle in the bovine is essentially the same as that of *D. filaria* of sheep. Thus infective larvae exsheath in the small intestine, penetrate the bowel wall and are carried to the local mesenteric lymph nodes. Here they moult to fourth-stage larvae and then continue their migration to the lungs via the thoracic duct and right heart. Jarrett and Sharp (1963) were unable to recover larvae from the lungs earlier than the seventh day of infection and under 'normal circumstances' fourth-stage larvae did not appear in the lungs until the thirteenth to fifteenth day of infection. Fifth-stage larvae are produced in the lungs by the fifteenth day and sexual maturity is reached on the twenty-second day of infection.

In massive infections (240 000 larvae) third-stage larvae may be found in the lungs as early as 24 hours after infection (Poynter et al. 1960); however, it would appear that this is a feature of massive infections only, in either the bovine or the guinea-pig, which may also serve as a host for the parasite.

The longevity of the parasite in the lungs may be prolonged in the case of a few worms; however, the majority of the burden is expelled within 50–70 days of infection. A factor of epidemiological importance, particularly from the point of view of the persistence of infection in a herd, is the survival of the parasite in an inhibited form in the lungs for several months.

Epidemiology. Parasitic bronchitis is particularly important in temperate areas with a sufficiently high rainfall to prevent dessication of the larvae. Disease due to *D. viviparus* is seen primarily in young calves during their first summer at pasture and outbreaks of disease are seen most commonly in late July, August and September, although they may occur from June until November in the northern hemisphere. Older animals generally have a strong acquired immunity but this may wane in the absence of reinfection and adult animals may be susceptible to massive larval challenge.

Three sources of infection exist in spring. Larvae may overwinter on pasture or the infection may survive the winter in the host either as adults or as inhibited immature stages.

Larvae deposited on pasture the previous autumn and winter can overwinter in sufficient numbers to initiate infection and cause disease in susceptible stock in the spring. In addition it is possible that larvae may survive on pasture until at least mid-August since calves grazing aftermath following the removal of a hay crop may show clinical signs of infection even though the pastures were last grazed by infected animals the previous autumn (Duncan et al. 1979; Oakley 1979). It has been suggested that the larvae survive in dung pads or in the soil. The survival of infective larvae in the soil is supported by the experiments of Duncan et al. (1979). Pasture grazed the previous autumn when grazed for 17 days in early May by two tracer calves produced in those calves only 0 and 12 parasites. However, when another group of calves was immediately grazed on the pasture they developed clinical signs of parasitic bronchitis within 42 days. These results suggest that the infective larvae overwintered in the soil and emerged on to the pasture in mid to late May. It has been demonstrated that infective larvae of *Ostertagia ostertagi* can be recovered from soil, root mat and herbage (Armour et al. 1981).

The presence of infection on fields of aftermath may also be associated with the emergence of infective larvae onto the pasture from the soil (Duncan et al. 1979). However, the importance of contamination from adjacent fields through windborne field-to-field transmission of larvae by *Pilobolus* sporangiae has been suggested by Jørgensen (1981). In these experiments infection was present on aftermath of an experimental field which had not been grazed for the previous five years although the field was lightly contaminated with trichostrongyle larvae originating from liquid manure fertilization each spring. Infective larvae were found on this field prior to it being grazed at a time which coincided with heavy pasture contamination on an adjacent infected field, suggesting the occurence of *Pilobolus* transmission.

The importance of *Pilobolus* in the translation of larvae of *D. viviparus* to the herbage was demonstrated by Robinson (1962). The infective larvae of *D. viviparus* are relatively inactive and are frequently found coiled up, showing very little movement. Consequently there is little migration of larvae from the faecal pads onto herbage, except during heavy rainfall, and larvae which do reach the herbage are capable of limited vertical migration only. However, Robinson (1962) reported that the fungus *Pilobolus* may accumulate larvae of *D. viviparus* on the upper surface of the sporangium, which when it explodes may propel the larvae as far as 3 m (10 ft)! The fungus is very common in cattle faeces; surveys in Britain having shown it to be present in 95% of samples examined. Jørgensen (1981) also reported that few *D. viviparus* larvae were translated from faecal pads in the absence of *Pilobus* fungi, their absence markedly reducing (8 : 1) the pasture infectivity for susceptible calves. Translation of larvae on to herbage is also aided by conditions (i.e. parasitic gastroenteritis) which lead to loose faeces or diarrhoea.

A second source of pasture infection in the spring is the small numbers of adult *D. viviparus* which may survive in the lungs for six months or more. For instance, Cunningham et al. (1956) demonstrated that 0–4.5% of adult animals and 9–41% of yearlings were infected with *D. viviparus* when examined in January to May at slaughter-houses in Scotland.

D. viviparus also appears capable of surviving winter as inhibited late fourth-stage and early fifth-stage larvae in the host. These larvae resume their development in spring with resultant contamination of pastures (Gupta & Gibbs 1975).

Pathogenesis and clinical signs. Extensive accounts of the pathology and pathogenesis of bovine parasitic bronchitis have been published by Jarrett et al. (1957, 1960*a*) which should be consulted in their entirety for full appreciation. A summary of the major points of the pathogenesis is given by Soulsby (1965) and Urquhart et al. (1973) which are further summarized below. The major pathogenic phases are the prepatent phase in the lungs, the patent phase and the post-patent phase. The prepatent phase is associated with blockage of many respiratory bronchioles with an eosinophilic exudate and collapse of alveoli. This, clinically, is associated with the onset of tachypnoea and coughing. Emphysema may develop. The patent phase lasting from days 25 to 55 is associated with adult parasites in the bronchi and trachea. There is severe damage to the epithelium of these organs, marked exudation into the bronchi and blockage of air passages. In addition, aspiration of eggs and larvae into the bronchioles and alveoli occurs, leading to consolidation of lobules. Lesions which become obvious in the late prepatent stage of the infection, namely epithelialization of the alveolar epithelium and 'hyaline membrane' formation, remain and may become more marked. The animals show dyspnoea and coughing, with rapid loss of condition. Harsh respiratory sounds with ronchi and emphysematous crackling can be heard. Complications include pulmonary oedema and emphysema though secondary bacterial infection is not common.

By 50 days, if the animals survive, the post-patent phase commences, this being a process of recovery. Clinically the respiratory rate decreases, coughing is less frequent and weight gain is resumed. Severe epithelialization may persist in some animals for some time but by 90 days worms are usually no longer present and the lesions remaining consist of peribronchial fibrosis and epithelialization of a few alveoli surrounding some of the bronchi. In some animals a sudden exacerbation of dyspnoea is seen. This condition is often fatal and is characterized by proliferation of the alveolar epithelialization to involve entire lung lobes.

Mild and transient clinical signs may be seen when an immune animal receives a large challenge infection. Larvae which reach the lungs are destroyed by the immune response with the formation of lymphoreticular granulomas, nodules about 5 mm in diameter and bronchiolar obstruction.

It has been suggested that a possible complication of *D. viviparus* infection is acute pulmonary emphysema and that this is related to the condition of 'fog fever'. However, Breeze et al. (1973) failed to obtain experimental proof that lungworms are concerned with the condition and conclude that there is no evidence to support the idea that the condition is a hypersensitive response to lungworms.

Immunity. The immunology of *D. viviparus* infection is reviewed by Soulsby (1965). In summary, essentially, an initial infection may lead to the fairly rapid acquisition of immunity, so much so that Michel (1962) found a considerable measure of immunity as early as 10 days after an initial infection.

Artificial immunization has been markedly successful using X-irradiated infective larvae and a commercial vaccine is available in Europe, having been developed from the work of Jarrett et al. (1960*b*). This vaccine which consists of two doses of 1000 irradiated larvae given at an interval of a month has been used for two decades with outstanding success. Animals are normally immunized when two months of age or older and exposure to infection is avoided until two weeks after the second dose. However, immunization of calves when three and seven weeks of age can be effective (Benitez-Usher et al. 1976). The protective immune response wanes in the absence of re-infection but exposure to infection after vaccination maintains the levels of immunity and this occurs in endemic areas.

Diagnosis. This is based on the clinical signs of bronchitis, rapid breathing, coughing, etc., and the demonstration of larvae in the faeces. Usually parasitic bronchitis, husk or hoose, is a herd problem, seen especially in young calves and the first indication of it may be an increased incidence of coughing in animals which have recently been placed on pasture. Other pneumonic conditions may be confused with it, e.g. epizootic bronchitis or 'virus pneumonia', *Pasteurella* infection, and 'cuffing pneumonia'.

Treatment. Cyanacethydrazide, 15 mg/kg subcutaneously or 17.5 mg/kg orally, has shown activity against adult parasites but has failed to give consistent results.

Diethycarbamazine, given intramuscularly at 22 mg/kg/day for three days or as a single dose of 50 mg/kg, is primarily effective against the immature parasites and will suppress early infection but not patent infection.

Methyridine given subcutaneously or intraperitoneally at a dose rate of 200 mg/kg shows good activity.

Tetramisole, 15 mg/kg given orally or parenterally, and *levamisole*, 7.5 mg/kg given parenterally, show excellent efficacy. It has been suggested that these are more effective against immature rather than adult parasites but this is disputed.

Morantel, given as a sustained-release bolus to cattle, was effective in the control of *D. viviparus* as well as gastrointestinal parasites (Jacobs et al. 1982).

A number of benzimidazole anthelmintics show excellent efficacy in the treatment of parasitic bronchitis. These include *fenbendazole* (5 mg/kg), *albendazole* (7.5 mg/kg) and *oxfendazole* (4.5 mg/kg) given orally.

In addition the efficacy of fenbendazole against the inhibited stages of *D. viviparus* has been examined by Inderbitzen and Eckert (1978) and Pfeiffer (1978). A single oral dose of 7.5 or 10 mg/kg had an efficacy of 70–87% while fenbendazole, given daily for five days at a dose rate of 1.5 or 2 mg/kg, reduced the number of inhibited larvae in infected cattle by 99.6%.

Exacerbation of the clinical signs and some mortality has been associated with the treatment of parasitic bronchitis. Jarrett et al. (1980) have reported the appearance of new lesions a few days following treatment of infected animals with levamisole or benzimidazole compounds. These were destructive lesions of the bronchial and bronchiolar walls leading to chronic occlusive bronchitis, acute alveolar epithelialization and severe oedema of the intralobular septa and peribronchial tissue. It has been suggested that this is related to the fact that the benzimidazole anthelmintics rapidly kill the worms in situ and if large numbers of parasites are present the products of disintegrating worms and larvae can cause severe histopathological lesions. Conversely, levamisole paralyses the parasites and perhaps allows them to be removed more slowly by physiological mechanisms and with less disintegration (McEwan et al. 1979; Oakley 1981). However, Urquhart (1981) warns that it would be premature to claim that one drug is safer than another on the present evidence.

Prophylaxis. Grazing management should be improved, especially to provide clean pasture for young calves. Animals continuously exposed to infection are at little risk provided the rate of acquisition of the infection is sufficient to stimulate a satisfactory immunity and not enough to cause clinical illness.

With the advent of more effective drugs control by anthelmintic treatment in mid-summer and movement to clean pasture has been practiced. This is no longer necessarily effective since sufficient larvae to cause disease in cattle may be present on 'clean' pastures, such as hay aftermath.

Reliance on treatment for prophylaxis and control is complicated by the unpredictability of natural infection and reinfection. The superimposition of reinfection on existing pulmonary lesions may lead to severe morbidity and mortality. Because of the uncertainties of control by therapy, vaccination, using the irradiated larval vaccine, referred to above, offers the only reliable and effective method of control (Urquhart et al. 1981).

Dictyocaulus arnfieldi (Cobbold, 1884) occurs in the bronchi of the horse, donkey, zebra and tapir and is cosmopolitan in distribution. The infection may be very common in donkeys. The male is up to 36 mm long and the female up to 60 mm. The medio- and posterolateral rays of the male bursa are fused for about half their length. The spicules are 0.2–0.24 mm long. The eggs measure 80–100 by 50–60 μm.

The life-cycle and pathogenicity of *D. arnfieldi* have been reviewed by Round (1976). The life-cycle is similar to that of *D. filaria* but most eggs do not hatch before being passed in the faeces although they hatch within a few hours of being passed. The prepatent period in experimentally infected donkeys was shown to be 12–14 weeks although it has been reported previously to be 39–40 days. Generally it has been believed that donkeys are the natural hosts for *D. arnfieldi* and the parasites may remain alive in donkeys for at least five years. Usually horses are thought to become infected by contact with donkeys and in horses the parasite frequently does not reach patency. However, in the Soviet Union the prevalence of infection in horses may reach 70% in the absence of donkeys and up to 25% of thoroughbred mares on studs in Newmarket, England, were infected as adjudged by the presence of larvae in their faeces (Round 1976). The length of patency in horses is between six and eight weeks.

As a rule this parasite is thought to be not very pathogenic and the donkey, assumed to be the natural host of the parasite, generally suffers little, even from heavy infections, as judged by large numbers of larvae in the faeces (Rose et al. 1970). Non-patent infections in horses may be associated with clinical signs such as coughing, increased respiratory rate and nasal discharge (Round 1972, 1976). In such cases diagnosis cannot be made by faecal examination, but coincident grazing with donkeys may suggest dictyocauliasis.

Thiabendazole, two doses of 440 mg/kg with 24 hours between treatments (Round 1972); mebendazole, 20 mg/kg/day for five days (Clayton & Neave 1979), and fenbendazole, a single dose of 50 mg/kg (Tiefenbach 1976), are effective in the treatment of infection in horses and donkeys. However Urch and Allen (1980) reported that up to 30 mg/kg of fenbendazole failed to remove adult lungworm infection in donkeys.

Dictyocaulus cameli Boev, 1952 has been described from the bronchi of the camel in Europe and Asia. However, many authors consider it a synonym of *D. viviparus*. Similarly, **D. eckerti** Skrjabin, 1931, found in the bronchi of various deer in Europe, Asia and North America, is considered by many to be a synonym of *D. viviparus*.

REFERENCES

TRICHOSTRONGYLOIDEA, DICTYOCAULIDAE

Armour, J., Bairden, K., Al Saqur, I. M. & Duncan, J. L. (1981) The role of the soil as a potential reservoir for infective larvae of *Ostertagia ostertagi*. In *The Epidemiology and Control of Nematodiasis in Cattle: Proceedings of CEC Workshop on Epidemiology and Control of Nematodiasis in Cattle*, ed. P. Nansen, R. J. Jørgensen & E. J. L. Soulsby, pp. 277–286. The Hague: Martinus Nijhoff.

Benitez-Usher, C., Armour, J. & Urquhart, G. M. (1976) Studies on immunization of suckling calves with Dictol. *Vet. Parasit.*, **2**, 209–222.

Breeze, R. G., Pirie, H. M., Selman, I. E. & Wiseman, A. (1973) Fog fever and *Dictyocaulus viviparus*. In *Helminth Diseases of Cattle, Sheep and Horses in Europe*, ed. G. M. Urquhart & J. Armour, pp. 54–8. Glasgow: Maclehose.

Clayton, M. M. & Neave, R. M. S. (1979) Efficacy of mebendazole against *Dictyocaulus arnfieldi* in the donkey. *Vet. Rec.*, **104**, 571–572.

Cunningham, M. P., Jarrett, W. F. H., McIntyre, W. I. M. & Urquhart, G. M. (1956) The carrier animal in bovine parasitic bronchitis. *Vet. Rec.*, **68**, 141–143.

Duncan, J. L., Armour, J., Bairden, K., Urquhart, G. M. & Jørgensen, R. J. (1979) Studies on the epidemiology of bovine parasitic bronchitis. *Vet. Rec.*, **104**, 274–278.

Gallie, G. J. & Nunns, V. J. (1976) The bionomics of the free living larvae and the transmission of *Dictyocaulus* filaria between lambs in North-East England. *J. Helminth.*, **50**, 79–89.

Gupta, R. P. & Gibbs, H. C. (1975) Infection patterns of *Dictyocaulus viviparus* in calves. *Can. vet. J.*, **16**, 102–108.

Inderbitzen, F. & Eckert, J. (1978) The action of fenbendazole (Panacur®) against inhibited stages of *Dictyocaulus viviparus* and *Ostertagia ostertagi* in calves. *Berl. Munch. tierärztl. Wschr.*, **91**, 395–399.

Jacobs, D. E., Fox, M. T., Jones, R. M. & Bliss, D. H. (1982) Control of bovine parasitic gastroenteritis and parasitic bronchitis in a rotational grazing system using the morantel sustained-release bolus. *Vet. Rec.*, **110**, 399–402.

Jarrett, W. F. H., Jennings, F. W., McIntyre, W. I. M., Mulligan, W., Sharp, N. C. C. & Urquhart, G. M. (1960*a*) Symposium on husk. I. The disease process. *Vet. Rec.*, **72**, 1066–1067, 1068, 1086–1087.

Jarrett, W. F. H., Jennings, F. W., McIntyre, W. I. M., Mulligan, W. & Urquhart, G. M. (1960*b*) Immunological studies on *Dictyocaulus viviparus* infection. Immunity produced by the administration of irradiated larvae. *Immunology*, **3**, 145–151.

Jarrett, W. F. H., McIntyre, W. I. M. & Urquhart, G. M. (1957) Husk in cattle. A review of a year's work. *Vet. Rec.*, **66**, 665–676.

Jarrett, W. F. H. & Sharp, N. C. C. (1963) Vaccination against parasitic disease. Reactions in vaccinated and immune hosts in *Dictyocalus viviparus* infection. *J. Parasit.*, **49**, 177–189

Jarrett, W. F. H., Urquhart, G. M. & Bairden, K. (1980) Treatment of bovine parasitic bronchitis. *Vet. Rec.*, **106**, 135.

Jørgensen, R. J. (1981) Recent Danish studies on the epidemiology of bovine parasitic bronchitis. In *The Epidemiology and Control of Nematodiasis in Cattle: Proceedings of CEC Workshop on Epidemiology and Control of Nematodiasis in Cattle*, ed. P. Nansen, R. J. Jørgensen & E. J. L. Soulsby, pp. 215–239. The Hague: Martinus Nijhoff.
McEwan, A. D., Oakley, G. A. & Robinson, M. (1979) Effects of anthelmintics on the pathology of *Dictyocaulus viviparus* infection in cattle. *Vet. Rec.*, **105**, 15–16.
Michel, J. F. (1962) Studies on resistance to *Dictyocaulus* infection. IV. The rate of acquisition of protective immunity in infection of *D. viviparus*. *J. comp. Path.*, **72**, 281–285.
Oakley, G. A. (1979) Delayed development of *Dictyocaulus viviparus* infection. *Vet. Rec.*, **104**, 460.
Oakley, G. A. (1981) Speed of action of some anthelmintics against *Dictyocaulus viviparus*. *Vet. Rec.*, **108**, 172.
Pfeiffer, H. (1978) Efficacy of fenbendazole against arrested lungworms of cattle after repeated administration of small doses. *Wein. tierärztl. Mschr.*, **65**, 343.
Poynter, D., Jones, B. V., Nelson, A. M. R., Peacock, R., Robinson, J., Silverman, P. H. & Terry, R. J. (1960) Recent experiences with vaccination. *Vet. Rec.*, **72**, 1087–1090.
Robinson, J. (1962) *Pilobolus* spp. and the translation of the infective larvae of *Dictyocaulus viviparus* from faeces to pastures. *Nature, Lond.*, **193**, 353–354.
Rose, M. A., Round, M. C. & Beveridge, W. I. B. (1970) Influenza in horses and donkeys in Britain. *Vet. Rec.*, **86**, 768–769.
Round, M. C. (1972) Natural history of lungworm infection of equidae. PhD Dissertation, University of Cambridge.
Round, M. C. (1976) Lungworm infection (*Dictyocaulus arnfieldi*) of horses and donkeys. *Vet. Rec.*, **99**, 393–395.
Soulsby, E. J. L. (1965) *Textbook of Veterinary Clinical Parasitology*. Vol. I *Helminths*. Oxford: Blackwell Scientific.
Tiefenbach, B. (1976) Panacur—weitweite klinische Prüfung eines neuen Breitband Anthelminthikums. *Die Blauen Heft Nr.*, **55**, 204–218.
Urch, D. L. & Allen, W. R. (1980) Studies on fenbendazole for treating lung and intestinal parasites in horses and donkeys. *Equine vet. J.*, **12**, 74–77.
Urquhart, G. M. (1981) Speed of action of some anthelmintics against *Dictyocaulus viviparus*. *Vet. Rec.*, **108**, 132.
Urquhart, G. M., Jarrett, W. F. H., Bairden, K. & Bonazzi, E. F. (1981) Control of parasitic bronchitis in calves: Vaccination or treatment. *Vet. Rec.*, **108**, 180–182.
Urquhart, G. M., Jarrett, W. F. H. & McIntyre, W. I. M. (1973) Bovine dictyocauliasis. Pathology, clinical signs, epidemiology, treatment and control. In *Helminth Diseases of Cattle, Sheep and Horses in Europe*, ed. G. M. Urquhart & J. Armour, pp. 23–30. Glasgow: Maclehose.

SUPERFAMILY: METASTRONGYLOIDEA LANE, 1917

Stoma reduced or rudimentary, frequently six lips around mouth. Usually thin-bodied. Bursa more or less reduced or even absent, bursal rays fused to varying degrees. Parasites of the respiratory passages and blood vessels of the lungs. Where life-cycle is known, intermediate hosts are required. Families of importance include *Metastrongylidae*, *Protostrongylidae*, *Crenosomatidae*, *Filaroididae*.

FAMILY: METASTRONGYLIDAE LEIPER, 1908

Dorsal ray much reduced. Spicules long and filiform. Genus of importance: *Metastrongylus*.

Genus: Metastrongylus Molin, 1861

Metastrongulus elongatus (Dujardin, 1846) (*syn. M. apri*) occurs in the bronchi and bronchioles of the pig, and wild pigs, and has also been recorded from the sheep, deer, ox and other ruminants and accidentally in man. Its distribution is cosmopolitan. The male is up to 25 mm long and the female up to 58 mm. The worms are white and have six small lips or papillae around the oral aperture. The male bursa is relatively small; the anterolateral ray is large and has a swollen tip; the mediolateral and posterolateral rays are fused and the dorsal rays are much reduced. The spicules are filiform, 4–4.2 mm long and end in a single hook each. The posterior end of the female is flexed ventrad. The vulva opens near the anus and the vagina is 2 mm long. The eggs measure 45–57 by 38–41 μm, have thick, rough shells and contain a fully developed embryo when laid.

The eggs are passed in the faeces of the host and may hatch soon thereafter or only after they have been swallowed by the intermediate host. The first-stage larva is 0.25–0.3 mm long; its intestinal cells are filled with opaque granules, the hind end is strongly curved and the tip of the

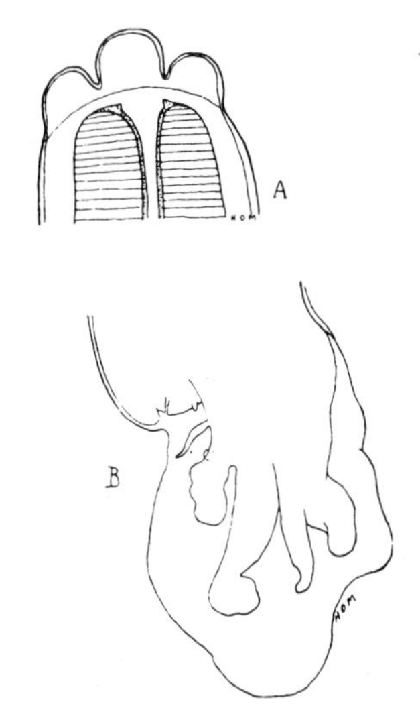

Fig 1.149 Lateral view of *Metastrongylus elongatus*. A, Anterior end. B, Bursa of male.

tail is bluntly rounded or swollen. The larvae may live up to three months in moist surroundings, but are not infective and can proceed with their development only after they have been ingested by a suitable species of earthworm. The following species are concerned in various countries: *Lumbricus terrestris*, *L. rubellus*, *Diplocardia* sp., *Eisenia austriaca*, *E. lönnbergi*, *Dendrobaena rubida*, *Pheretima hupiensis*, *Helodrilus foetidus*, and *H. caliginosus*. Rose (1959) studied the development and resistance of the eggs in the soil and their development in the intermediate hosts. The larvae develop in the blood vessels in the walls of the oesophagus and proventriculus of the intermediate host and in blood spaces outside these organs and reach the infective stage in about 10 days, after performing two ecdyses and retaining the second skin as a sheath. They grow to about 0.52 mm and when they are infective they concentrate in the blood vessels of the earthworm. The latter does not suffer from even very severe infections and the larvae can pass the winter in the earthworm. Under experimental conditions larvae remained viable in *E. foetida* for up to seven years (Kolevatova 1974). They do not escape spontaneously from the intermediate host, but if the earthworm is hurt or dies, the liberated larvae are able to live in moist soil for about two weeks. Pigs become infected by ingesting infected earthworms or accidentally liberated infective larvae. In the pig the development is similar to that of *D. filaria* in the sheep, the larvae passing through the mesenteric lymphatic glands, where they moult once and then reach the lungs, where they grow adult after a further moult. The first eggs are laid after about 24 days.

Metastrongulus pudendotectus (Wostokow, 1905) (syn. *M. brevivaginatus*, *Choerostrongylus pudendotectus*) also occurs in the pig and wild boar in the majority of countries of the world. The male is 16–18 mm long and the female 19–37 mm. It differs from the preceding species mainly in having a larger bursa, spicules only 1.2 mm long and provided with double hooks, and the vagina 0.5 mm long. The tail of the female is straight and a swelling covers the vulva and anus. The eggs measure 57–63 by 39–42 μm.

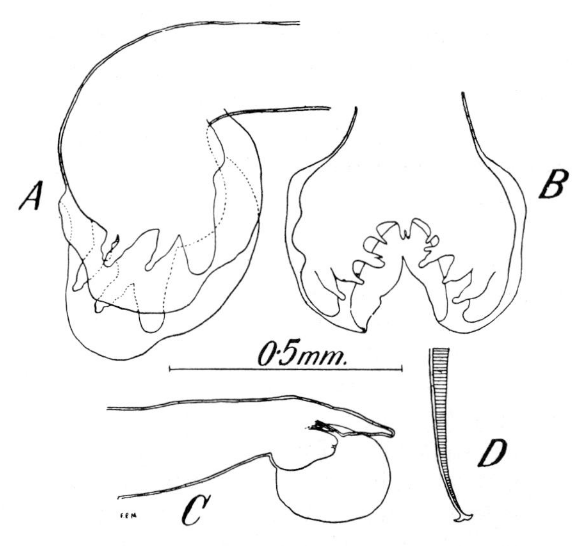

Fig 1.150 Metastrongylus pudendotectus. A, Lateral view of bursa. B, Dorsal view of bursa. C, Posterior end of female. D, Tip of spicule. The scale refers to A and B only.

The life-cycle is similar to that of *M. elongatus*.

Metastrongylus salmi Gedoelst, 1923 occurs in the pig and wild boar in Zaire, south-east Asia, the Pacific area, South America and the United States. The spicules are 2–2.1 mm long and the vagina of the female 1.5 mm. The life-cycle is probably similar to that of *M. elongatus*.

Metastrongylus madagascariensis Chabaud and Grétillat, 1956 was reported from the pig in Madagascar. Males are 9–12 mm long and females 22–26 mm. Embryonated eggs are 42–60 μm.

The highest incidence of infection is in four to six-month-old pigs and up to 50–60% of lungs may be affected on infected premises. In general the disease entity is similar to that caused by *D. filaria*, but the pig lungworms are not as pathogenic as those which occur in ruminants. In young pigs a marked verminous bronchitis and pneumonia may sometimes be seen; these are possibly due to secondary bacterial infection. As a rule, however, the parasites cause mainly loss of condition and retarded growth, which is rather important in pigs. The parasites may sometimes die in the small bronchioles and give rise to the formation of nodules, which must be differentiated from tuberculous nodules at autopsy or at meat inspection.

Shope (1941, 1943) found that the virus of swine influenza is carried by the larvae of pig lungworms and persists in them, when they are carried by earthworms, for 32 months. Pigs infected with such larvae acquire influenza from them, and he showed (1958) that the larvae can carry the virus of swine fever to pigs in a form which requires some form of stress to provoke this virus to pathogenicity. Mackenzie (1958) found that *M. elongatus* had a negligible effect on pigs experimentally infected with it, but that lesions resembling those of epizootic pneumonia could be produced. In a later report Mackenzie (1963) reported that the clinical signs and lesions of epizootic pneumonia were enhanced by *Metastrongylus* infection. It has been suggested that *Metastrongylus* spp. transmit the viruses of Teschen disease and as mentioned above, swine fever. However, Wallace (1977) reviewed the literature on the transmission of swine influenza virus by lungworms and concluded that the studies did not support an essential role for *Metastrongylus* larvae in viral transmission. Similarly, *Metastrongylus* larvae did not appear to transfer the agents of mycoplasmal pneumonia.

Diagnosis is made by demonstrating the embryonating egg in the fresh faeces.

Treatment is comparable to that for *Dictyocaulus viviparus* (see p. 267). Levamisole (15 mg/kg) is highly effective; tetramisole (15 mg/kg) is similarly effective and fenbendazole (20–30 mg/kg) shows high activity.

Infected pigs should be kept on dry ground or in sties with concrete floors and their faeces should be disposed of in such a way that they do not spread the infection. Clean and young pigs should be run on clean fields. Infected paddocks and fields may remain infected for a considerable time, since the intermediate stage can live in the earthworm for an unknown period. A 3% solution of carbathion applied to soil kills earth worms.

FAMILY: PROTOSTRONGYLIDAE LEIPER, 1926

Hair-like forms occurring in the alveoli, bronchioles and parenchyma of the lungs of various species of mammals. Bursa much reduced or absent. Rays of bursa reduced. Spicules marked often with membranous expansions. Gubernaculum and telamon well developed. Genera of importance include *Protostrongylus, Muellerius, Cystocaulus, Spiculocaulus, Bicaulus* etc.

Genus: Protostrongylus Kamensky, 1905

The taxonomy of this genus is discussed by Schultz et al. (1933) and Dougherty (1944, 1951).

Protostrongylus rufescens (Leuckart, 1865) occurs in the small bronchioles of sheep, goats and deer in Europe, Africa, Australia and North America. The male is 16–28 mm long and the female 25–35 mm. The worms are slender and reddish in colour. The bursa is short and strengthened dorsolaterally by a chitinous plate on either side. The ventral, lateral and externodorsal rays are present, but the dorsal is a very thick trunk which bears six papillae on its ventral surface. The spicules are about 0.26 mm long, tubular, with broad, membranous expansions. A gubernaculum is present, as well as a strongly developed telamon. The latter is pigmented brown in its posterior portion, where it forms two arms, each provided distally with a number of teeth. The vulva opens near the anus. The eggs are unsegmented when they are laid and measure 75–120 by 45–82 μm.

Other species in sheep and goats include **P. skrjabini** (Boev, 1936) and **P. kochi** (Schultz, Orlov and Kutass, 1933) (although the latter may be a synonym of *P. rufescens*), in sheep and goats in eastern Europe and the USSR; **P. stilesi** Dikmans, 1931 and **P. rushi** Dickmans, 1937, which can be very common in the lungs of bighorn sheep in the USA; **P. hobmaieri** (Schultz, Orlov and Kutass, 1933); **P. davtiani** (Sawina, 1940) and **P. brevispiculum** Mikacic, 1939. Forrester (1971) has provided a detailed account of this genus in bighorn sheep in North America.

Life-cycle. The eggs develop in the lungs of the host and the first-stage larva which is passed in the faeces is 0.25–0.32 mm long. The tip of its tail has a wavy outline, but is devoid of a dorsal spine (cf. *Muellerius*). For its further development the larva requires an intermediate host, which may be one of several species of snail of the genera *Helicella*, *Theba*, *Abida*, *Zebrina*, *Arianta* etc. First-stage larvae penetrate the foot of the snail. The development to the infective stage requires 12–14 days and two ecdyses are performed. The final host becomes infected by swallowing the snail with its food and the larvae pass to the lungs of the host via the mesenteric lymphatic glands, in which the third ecdysis takes place. The prepatent period has been recorded as 30–37 days. Transplacental transmission occurs and larvae have been found in the livers and lungs of fetuses and newborn lambs (Forrester & Senger 1964).

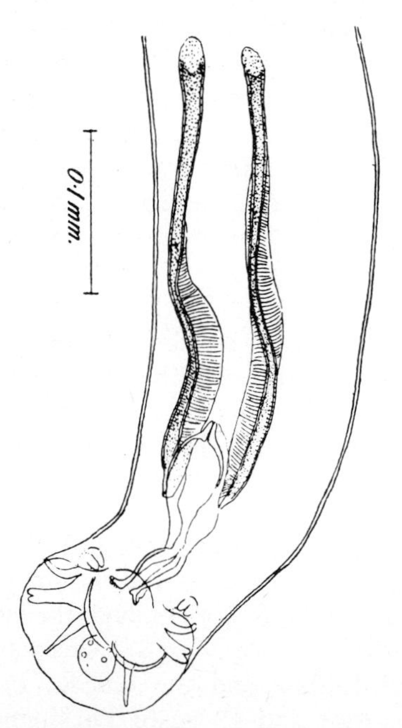

Fig 1.151 Ventral view of posterior end of male *Cystocaulus nigrescens*.

Pathogenesis. The worms live in the small bronchioles, where they produce a local area of inflammation. The resulting exudate fills those alveoli which are situated distally to the site of the parasites, and the inflammatory process spreads to the peribronchial tissues. The affected alveolar and bronchial epithelium is desquamated, blood vessels are occluded and an infiltration with round cells and proliferation of connective tissue takes place in the area. The result is a small focus of lobular pneumonia, roughly conical in shape and yellowish-grey in colour. The pleura at the base of the focus may be involved in a fibrinous pleuritis. The number of such foci in the lungs depends on the number of parasites present. As a rule the animals show no definite clinical signs, although severe infections would undoubtedly affect the general health, and may cause death and the weakened lungs are susceptible to bacterial invasion which may produce acute pneumonia. The association between lungworm and *Pasteurella* organisms is discussed by Forrester (1971).

Diagnosis can be made by finding the larvae in the faeces.

Treatment. Emetine hydrochloride has been used extensively for the Protostrongylidae in the USSR. Cyanacethydrazide has been reported to have good effect against *P. rufescens* (Walley 1957) and Kassai (1958) found diethylcarbamazine phosphate active against *Protostrongylus* spp. Levamisole, 20 mg/kg, and ditrazine, 100 mg/kg, both given subcutaneously, and fenbendazole, given orally at 20–80 mg/kg are recommended. Albendazole at a dose of 5 mg/kg has 85% efficacy. These anthelmintics produce a marked decrease in faecal larval counts but more information is needed as to whether they kill the adult parasites or only temporarily suppress egg production.

Prophylaxis. Where the extermination of the snail intermediate host is possible this should be carried out. Special attention must be paid to the lambs; they should not be run on pastures previously used for infected animals. Pastures may remain infected for a considerable period, as the infective larvae are protected in the snails. The snails creep up plants in the early morning and evening and in rainy weather and the animals should therefore not be allowed to graze at such times, particularly in the autumn, when the infection most frequently occurs.

Other species of the genus include **P. pulmonalis** (syn. *P. terminalis*) Goble and Dougherty, 1943 in the bronchioles of rabbit and hare in Europe; **P. tauricus** Schultz and Kadekacii, 1949 of hares in the USSR; **P. boughtoni** Goble and Dougherty, 1943 in the bronchi and bronchioles of cottontail rabbits and hares in North America; **P. sylvilagi** Scott 1943 in cottontail and jackrabbits in Wyoming; and **P. oryctolagi** Babos, 1955 of rabbits in Hungary.

Genus: Cystocaulus Schulz, Orloff and Kutass, 1933

Cystocaulus nigrescens (Jerke, 1911) occurs in the lung parenchyma and in subpleural nodules of sheep and goats. It is common in sheep in eastern Europe and the USSR; it has been reported from Great Britain, the Middle East and probably occurs elsewhere in sheep raising areas of the world. Both sexes are threadlike, the male 8–9 cm and the female 13–16 cm. The spicules measure 275–379 μm, the gubernaculum is 120–174 μm in length and the crura are dark brown in colour.

Cystocaulus ocreatus (Railliet and Henry, 1907) possibly is a synonym of *C. nigrescens* and is cosmopolitan in distribution.

The developmental cycle is similar to that of *P. rufescens* and includes land snails of the genera *Helix*, *Helicella*, *Theba*, *Cepaea* and *Monacha*. The prepatent period is 25–28 days.

The pathogenic effects of *C. nigrescens* infection are similar to those of the *Prostostrongylus* spp.

Fenbendazole at a dose of 20–80 mg/kg marked reduces the faecal larval count (Eslami & Anwar 1976).

Genus: Muellerius Cameron, 1927

Muellerius capillaris (Müller, 1889) occurs in the lungs of sheep, goats and chamois and is worldwide in distribution. It is probably the commonest lungworm of sheep in Europe. Male

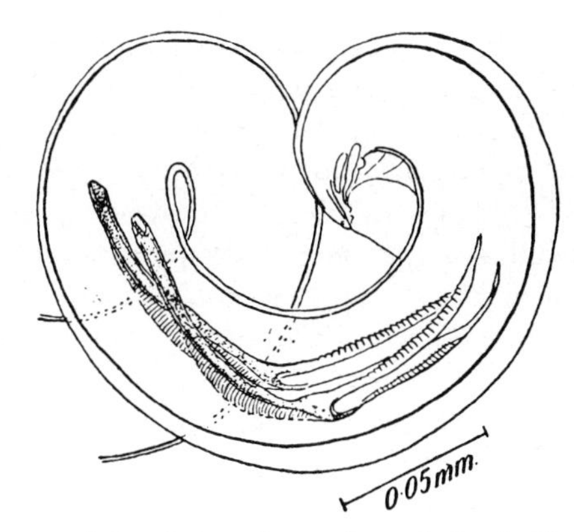

Fig 1.152 Lateral view of posterior end of male *Muellerius capillaris*.

12–14 mm long and female 19–23 mm. The posterior end of the male is spirally coiled and there is no bursa, but a number of papillae surround the cloacal opening. The spicules are 0.15 mm long; they are curved and each consists of a proximal half which is alate and two distal serrated arms ending in sharp points. The vulva opens close to the anus and has a small cuticular swelling on its posterior border. The eggs measure about 100 by 20 μm and are unsegmented when laid.

Life-cycle. The eggs develop in the lungs of the host and the first-stage larvae are passed in the faeces. They are 0.23–0.3 mm long. The oesophagus has two swellings, one near the middle and the other at the distal extremity. The tail of the larva has an undulating tip and a dorsal spine (cf. *Protostrongylus*). These larvae can resist a fair amount of drying, are most active at relatively low temperatures (17°–27°C) and are not killed by freezing. For further development they require a snail intermediate host, into which they penetrate through the foot, or they may be swallowed. A large number of species of molluscs are known to be suitable intermediate hosts; for instance, species of the nude slugs *Limax*, *Agriolimax* and *Arion*, and the snails *Helix* and *Succinea*. The development in the snail and in the sheep is very similar to that of *Protostrongylus*. The infective larvae can live in the snail probably for as long as the snail lives, and for up to a week after the death of the snail.

Pathogenesis. This parasite is usually not found in lambs or kids under six months of age. Thereafter the prevalence increases with age and in goats may reach 100% by three years of age (Lloyd & Soulsby 1978). The adult worms live in the alveoli and the pulmonary parenchyma, especially in the subpleural tissue. The worms produce greyish nodules up to 2 cm in diameter. These consist of necrotic masses, resulting from the degeneration of accumulated leucocytes and pulmonary tissues, and they are surrounded by a wall of connective tissue and a few giant cells. These nodules may calcify. Smaller separate foci are formed by the eggs which are surrounded by leucocytes and epithelioid cells, while the adjoining pulmonary tissue is hyperaemic and the alveoli become filled with round cells. After the egg has hatched the reaction subsides and the lesion may heal. An adenoma-like proliferation of the bronchial epithelium is seen in some cases.

The infected animal as a rule shows no clinical signs, but heavy infections weaken the lungs and assist in reducing the general health and resistance of the host. Secondary bacterial infection of the nodules is not uncommon and at times they coalesce to form septic lesions. In goats, *M. capillaris* often causes a widespread interstitial pneumonia. It is probable that *M. capillaris* may be pathogenic in heavily infected goats, producing moderate (dyspnoea and persistent cough) to severe (pneumonia) clinical signs.

Diagnosis. This can be made by finding the larvae in fresh faeces.

Prophylaxis. This is the same as for *Protostrongylus*. The nude slugs can be more easily controlled than shelled snails and spreading of lime has been recommended for this purpose.

Treatment. Emetine hydrochloride, 8–9 mg/kg, is said to be effective as are diethylcarbamazine and cyanacethydrazide although these drugs probably only cause a temporary decrease in the faecal larval counts. Repeated, high level doses of oxfendazole, albendazole or fenbendazole are likely to be the most effective anthelmintics. However, Halhead (1968) reported deaths in a proportion of ewes after treatment with tetramisole; all dead animals had heavy burdens of *M. capillaris*. Albendazole, 5 mg/kg, gave an average of 89% efficacy against adult and larval stages of *Muellerius capillaris* and *Protostrongylus* spp. (Cordero-del-Campillo et al. 1980).

Genus: Spiculocaulus Schulz, Orlov and Kutass, 1933

Protostrongylids in which the bursa is well developed; the gubernaculum is elaborate and the spicules are long, up to 1 mm in length. **S. leuckarti** Schulz, Orlov and Kutass, 1933 is found in sheep and ibex in the USSR; **S. austriacus** Dougherty and Goble, 1946 occurs in goats and ibex in Europe and the USSR; **S. kwongi** Dougherty and Goble, 1946 occurs in sheep and goats in China; **S. orloffi** Boev and Murzina, 1948, occurs in sheep, goats and ibex in the USSR.

Genus: Bicaulus Schulz and Boev, 1940

Dougherty considers the genus synonymous with *Varestrongylus* Bhalerao, 1932. The parasites occur in eastern Europe.

Bicaulus schulzi (Boev and Wolf, 1938) (syn. *Varestrongylus sinicus*, Dikmans, 1945, *V. schulzi* Dougherty and Goble, 1946). Thread-like forms in the lung parenchyma of sheep and goats. Male 12–15 mm, female 22–25 mm. Spicules 333–378 μm; gubernaculum 190–210 μm. Developmental cycle similar to *Protostrongylus* spp.

Bicaulus sagittatus (Müller, 1891). In *Cervus elaphus* in Europe.

Genus: Neostrongylus Gebauer, 1932

Neostrongylus linearis (Marotel, 1913). A small worm, male 5–8 mm, female 13–15 mm. Spicules unequal in size, one 320–360 μm, the other 160–180 μm. In sheep and goats in Europe and Middle East. The intermediate hosts are molluscs and the prepatent period in sheep is 60–70 days.

Genus: Elaphostrongylus Cameron, 1931

Protostrongylidae occurring in deer. Several species have previously been included in this genus and other genera, e.g. *Pneumostrongylus* and *Parelaphostrongylus* have been regarded by some authors as congeneric with *Elaphostrongylus*. Members of these various genera occur in the lungs and blood vessels of the central nervous system of deer and at times are associated with neurological disorders.

Elaphostrongylus cervi Cameron, 1931 occurs in the intermuscular connective tissue primarily of the breast, thorax and back, as well as the central nervous system of red deer (*Cervus elaphus*), roe deer and fallow deer in Europe and eastern Europe. The parasite has also been found in the wapiti (*C. canadensis*) in New Zealand (Sutherland 1976) and in woodland caribou (*Rangifer tarandus caribou*) in Canada (Lankester & Northcott 1979).

Elaphostrongylus panticola Labimow, 1945 occurs in the brain of *Cervus elaphus* and *Cervus nippon* in the USSR and other parts of Asia.

Elaphostrongylus rangiferi Mitskievitsch, 1964, occurs in reindeer, *R. tarandus*, in Sweden. The taxonomic position of these latter two species is not clear and some authors think that they may be synonymous with *E. cervi*.

Adults in the muscles lay unembryonated eggs in the small blood vessels. These are carried to the lungs where larvae develop, hatch and are then passed up the bronchi and trachea, eventually to be passed in the faeces. Larvae measure 0.3–0.4 mm and have a dorsal spine on the tail. Development of larvae occurs in *Trichia hispida* and *Succinea putris* and takes 27–35 days (Anderson 1971).

The pathology of infection has been reviewed by Anderson (1963, 1964). Parasites in the central nervous system may cause meningoencephalitis and spinal demyelination with consequent ataxia, paralysis and perhaps death (Sutherland 1976). Adults in the muscles are non-pathogenic but trimming of the muscles may be necessary at slaughter. The presence of eggs and larvae in the lungs can cause a verminous pneumonia. The incidence of infection may be high.

Cyanacethydrazide (34 mg/kg) given on two or three occasions; thiabendazole (50 mg/kg for seven days); diethylcarbamazine (ditrazine) phosphate; and phenothiazine all cause some decrease in faecal larvae.

Genus: Parelaphostrongylus Boev & Schulz, 1950

Parelaphostrongylus tenuis (Dougherty, 1945) Pryadko and Boev, 1971 (syn. *Pneumostrongylus tenuis, Odocoileostrongylus tenuis, Elaphostrongylus tenuis, Neurofilaria cornelliensis*). Adults inhabit the cranial venous sinuses and subdural spaces of white-tailed deer (*Odocoileus virginianus*) in North America. The prevalence of infection may be up to 80% and in the white-tailed deer the parasite is non-pathogenic. The parasite also infects, and causes neurological disease in, elk (*Cervus canadensis*), moose (*Alces alces*), caribou (*Rangifer tarandus*), black-tailed deer (*O. hemionus columbicanus*), red deer (*Cervus elaphus*), llama, sheep and goats. Adults are 39–91 mm in length; larvae, which occur in the faeces, measure 348 μm and have a dorsal spine.

Life-cycle. Eggs deposited on the meninges develop, hatch, enter the circulation and are carried to the lungs, while eggs laid in the venous circulation are carried to the lungs where they embryonate and hatch. They then pass up the bronchi and trachea and are swallowed, eventually appearing in the faeces. The intermediate hosts are a variety of terrestrial slugs and snails, i.e. *Arion circumscriptus, Deroceras graciles, D. reticulatum, D. laeve, Cionella lubrica, Zonitoides nitidus* and *Z. arboreus*. Development to the infective stage takes three or four weeks. Infection is by ingestion of infected snails. In the definitive host larvae reach the spinal cord in 10 days and develop in the dorsal horns of the grey matter for 20–30 days. Adults leave the neural parenchyma for the spinal subdural space and migrate in this to the cranium. They enter the venous sinuses by penetrating the dura mater. The prepatent period is about three months (Anderson 1965, 1971).

Pathogenesis. Clinical signs are not normally seen in the natural definitive host, white-tailed deer, but often-fatal neurological disease is commonly seen in abnormal hosts. The parasite is the cause of 'moose disease' in North America and it may reach serious proportions when overpopulation of deer and moose occur. The lesions in experimentally infected wapiti include microcavitation and traumatic damage, particularly in the dorsal horns of the grey matter of the spinal cord (Anderson 1971; Anderson et al. 1966).

Parelaphostrongylus odocoilei (Hobmaier and Hobmaier, 1934) Boev and Schultz, 1950, occurs in connective tissue lining lymphatics and blood vessels of the skeletal musculature beneath the spine, in the abdomen and in the upper parts of the legs. It is found in mule deer (*Odocoileus hemionus hemionus*), black-tailed deer (*O. h. columbianus*) and moose (*Alces alces*). White-tailed deer are refractory to infection. Adults are 24–56 μm in length; larvae 367–378 mm in length have a dorsal caudal spine and are found in the faeces.

Eggs are carried in the circulation to the lungs where they embryonate and are then passed into the faeces. The intermediate hosts are terrestrial gastropods, e.g. *Agriolimax* spp. (Anderson 1971).

Parelaphostrongylus andersoni Prestwood, 1972. Adults, 20–31 mm in length, occur in the musculature of white-tailed deer (*O. virginianus*) in the south-eastern USA. They are found particularly in the longissimus dorsi (Prestwood 1972).

FAMILY: FILAROIDIDAE SCHULZ, 1951

Metastrongyloidea with a small or rudimentary bursa, bursa sometimes absent. Lateral and ventral rays short or digitiform. Parasites of the respiratory system of mammals. Genera of importance: *Filaroides*, *Aelurostrongylus*, *Angiostrongylus*, etc.

Genus: Filaroides v. Beneden, 1858 (syn. Oslerus)

Filaroides osleri (Cobbold, 1879) occurs in nodules in the trachea and bronchi and rarely in the lungs of the dog. The worm is not frequently seen, although it has been found in the USA, Great Britain, India, the Republic of South Africa and New Zealand. However, infection rates may be high in dogs kept under kennel conditions. Wild dogs and dingoes may be infected and 24 of 94 *Canis latrans* in Texas were found to be infected (Pence 1978). Male 5 mm long and slender, female 9–15 mm long and stouter. The hind end of the male is rounded and bears a few papillae. The spicules are short and slightly unequal. The vulva is close to the anus and the worms lay eggs with thin shells, containing larvae and measuring about 80 by 50 μm. The larva has a short S-shaped tail, is 232–266 μm in length and the oesophagus, which is unclear, is about one-quarter of the length.

Life-cycle. The life-cycle is direct and has been described by Dorrington (1968). *F. osleri* (and *F. hirthi*, see below) is an unusual nematode in that no period of development is required outside the definitive host. First-stage larvae found in the saliva or faeces are immediately infective for the definitive host. Infection is acquired early by puppies usually of less than six weeks of age. Infection is probably transmitted from a bitch to her pups when she is licking and cleaning them. Regurgitative feeding at about the time of weaning also may be of importance in the transmission of infection amongst wild canine populations. Infection by faecal contamination is also possible. Upon ingestion larvae migrate via the lymphatic and venous portal systems to the heart and lungs whereupon they migrate up the bronchi to their predilection site, the tracheal bifurcation. The prepatent period is about 10 weeks.

Pathogenesis. The worms live in or under the mucosa of the trachea or bronchi and cause the development of granulomas. These are greyish-white or pink in colour, usually less than 1 cm in diameter, polypoid or sessile, with cavities in which the worms are lodged. In severe infections many haemorrhagic wart-like lesions may cover the area of the bifurcation of the trachea. The nodules develop slowly and following an experimental infection a 2 mm nodule was visible 92 days after infection. At this stage, first stage larvae were found in nearby mucus. By day 252 the nodule had increased to 6 mm in size and a second nodule, 4 mm in diameter, had appeared (Polley & Creighton 1977).

Clinical signs. The clinical signs depend on the severity of the infection and the size and number of the granulomas. Young dogs are chiefly affected and the disease is chronic but occasionally may be fatal. The most marked clinical sign is a rasping persistent cough. Respiratory distress may be present and loss of appetite and emaciation may be present. Mortality may reach 75% in some infected litters.

Diagnosis can be made by bronchoscopy or by finding the larvae in the sputum. Larvae may be found in the faeces; however they are neither plentiful nor highly active and must be looked for very carefully.

In addition the females do not lay eggs continuously. In long-standing cases, thoracic radiography may be useful.

Treatment. Malherbe (1954) obtained good results with 5 ml of either lithium or antimony thiomalate or Stibophen given to dogs for nine to 22 weeks, which was not toxic, and also with a combination of either of these with diethylcarbamazine given daily for seven days, the latter being not effective by itself. Thiacetarsamide (1 ml/kg) given daily by intravenous injection for 21 days has been reported to be effective. Surgical removal by curettage or with biopsy forceps relieves the respiratory distress but does not eliminate the infection. Dorrington (1968) recommends the use of thiocetarsamide (1 mg/5 kg) followed six weeks later by surgical removal of the nodules.

Thiabendazole (64–140 mg/kg/day divided into two doses a day and given for 23 days) has been effective. Faecal larvae were eliminated within 15 days and the nodules regressed within 10 weeks (Bennett & Beresford-Jones 1973). However, in New Zealand a dog was treated by three courses of anthelmintics. First, thiabendazole (64 mg/kg/day for 21 days), followed later by levamisole (10 mg/kg for eight days) and by fenbendazole (100 mg/kg given twice at a three-day interval). When the dog was killed 196 days later live adults were still present in the tracheal nodules (Jones et al. 1977).

Control. The absence of a highly effective treatment and the direct life-cycle makes the control of *F. osleri* in kennels difficult. Good hygiene should be helpful. However, only hand-rearing or foster-rearing of pups on uninfected bitches will prevent the infection. All animals being admitted into a breeding kennels should be examined for infection.

Filaroides hirthi Georgi and Anderson, 1975, occurs in the lung parenchyma of dogs. Pence (1978) questions the validity of this species. The infection was first reported in experimental beagles in New York (Hirthi & Hottendorf 1973). Infection can be common in beagle rearing establishments and, in addition, the parasite has since been recorded in a Yorkshire terrier in Texas (Craig et al. 1978). Adult males are 2.3–3.2 mm long and females 6.6–13.0 mm. *F. hirthi* may be differentiated from *F. milksi* by its smaller size; the spicules are broader, shorter and have broader knobs for the attachment of muscles (Georgi & Anderson 1975). Females are ovoviviparous and the first-stage larvae which hatch in the lungs are 240–290 μm in length.

Life-cycle. The life-cycle is similar to that of *F. osleri* and is described by Georgi et al. (1979*a*). First-stage larvae are immediately infective when passed in the saliva, vomitus or faeces, no development outside the host being required. Upon ingestion, larvae reach the lungs via the portal and mesenteric lymphatic circulation as early as six hours after infection. Development of all four larval stages and the adults occurs in the lungs and the prepatent period of infection is five weeks. Transmission must be similar to that of *F. osleri*; pups are likely to be infected when four or five weeks of age and, in a beagle colony, 50% of the animals were infected by 17 weeks of age (Georgi et al. 1976). It is possible that autoinfection occurs (larvae infecting the same host when passing from the lungs through the intestine) since larvae were found in the mesenteric lymph node for up to 169 days after a single infection (Georgi et al. 1977).

Pathogenesis. Infection with *F. hirthi* does not usually cause clinical disease. The adults occur in nests in the lung parenchyma and there is a focal granulomatous reaction to the parasites. Of major importance is the fact that a broad spectrum of pulmonary changes occurs, some of which mimic drug-induced and neoplastic changes (Hirthi & Hottendorf 1973).

Diagnosis. Zinc sulphate flotation (which is 100 times more effective than Baermann examination) will detect larvae in heavy experimental infections, but diagnosis of natural low-grade infections is rarely achieved (Georgi et al. 1977). Radiographical examination reveals numerous linear and miliary interstitial parenchyma infiltrates in the lung lobes.

Treatment and control. Albendazole, 100 mg/kg in two divided doses a day for five days, killed the majority of adult parasites and larvae were not produced by the remaining females (Georgi et al. 1978). Treatment caused an extensive tissue reaction to the dead parasites and it is therefore not feasible to treat beagles destined for experimental use. However, it is possible that control of the infection in breeding colonies may be achieved by anthelmintic treatment, since Georgi et al. (1979*b*) have demonstrated that it is possible to raise pups free of infection when the bitches are given two courses of treatment with albendazole prior to parturition. In addition, for control, puppies could be hand-reared or foster-reared on uninfected bitches.

Filaroides martis (Werner, 1782) occurs in mustelidae (*Mustela* and *Martes* spp.) in Europe and North America. It is a common parasite of mink in North America. Adult parasites live in the connective tissue surrounding the bronchi and pulmonary arteries. The females are viviparous and the larvae are coughed up, swallowed and passed in the faeces. The life-cycle is indirect, development occurring in gastropods (e.g. *Helix*, *Succinea*, *Zonitoides*) in 16–18 days. In mink larvae remain in the gastric mucosa and submucosa for five to eight days before migrating along the adventitia of arteries to the dorsal aorta from whence they transfer to the adventitia of the pulmonary arteries and follow these to the lung parenchyma (Stockdale & Anderson 1970).

Other species of the genus include **F. milksi** Whitlock, 1956, which occurs in the lung parenchyma and bronchioles of dogs in Europe and North America; **F. (Anafilaroides) rostratus** Gerichter, 1949, which may cause a tracheobronchitis in cats in Asia, Middle East and USA; **F. bronchialis** (Gmelin, 1790) which occurs in mink and polecats; **F. cebus** (Gebauer, 1933) and **F. gordius** (Travassos, 1921), found in the lungs of capuchin and squirrel monkeys; and **F. pilbarensis** (Spratt, 1979) in marsupials in Australia.

Genus: Aelurostrongylus Cameron, 1927

Aelurostrongylus abstrusus (Railliet, 1898) occurs in the lungs of the cat in most parts of the world. The male measures up to 7.5 mm and the female is 9.86 mm long. The male bursa is short and the lobes are not distinct. All the bursal rays can be distinguished; the dorsal forms two stout branches. The spicules are simple and 0.13–0.15 mm long. The vulva opens near the posterior extremity and the eggs measure about 80 by 70 μm.

Life-cycle. The adult worms live in the terminal respiratory bronchioles and alveolar ducts. Eggs are forced into the alveolar ducts and into the adjacent alveoli, forming small nodules. Eggs on the periphery of the lesion hatch first and larvae escape into the air passages, which they ascend, and they are passed out in the faeces of the host. They are about 0.36 mm long and the tail has an undulating appendage of variable shape and usually a dorsal projection. The larvae live only about two weeks in the free state. For further development they require snails and slugs—mainly *Epiphragmophora* spp. and also *Agriolimax agrestis*, *A. columbianus*, *Helix asperus*,

Helminthoglypta californiensis, *H. nickliniana*, *H. arrosa*—as intermediate hosts in which two ecdyses occur.

Various auxiliary or transport hosts, such as rodents, frogs, lizards and birds, which eat infected snails and in which the worm larvae encyst, may aid in infecting cats. In the cat the larvae migrate from the stomach to the lungs via the peritoneal and thoracic cavities and may reach the lungs within 24 hours. The prepatent period is approximately one month and the infection persists for some four to nine months although occasional evidence of adult worms was still seen 24 months after infection (Hamilton 1970; Scott 1973).

Pathogenesis. The prevalence of infection varies from 1% to 26% and although it is a common parasite it does not often cause clinical disease. The pathology is reviewed by Scott (1973). The typical lesions are subpleural nodules which are firm, raised and greyish in colour. These vary in diameter from 1 to 10 mm and they may become confluent, forming larger lesions. In heavy infections, which tend to be fatal, there may be creamy yellow areas on the lungs and the pleural cavity may be filled with a thick milky fluid rich in eggs and larvae. Incision of the lungs produces a milky exudate in acute cases but in the chronic infection calcification is common (Hamilton 1963).

Radiological studies reveal bronchial, alveolar and miliary interstitial disease. The infection causes smooth muscle hypertrophy and hyperplasia of the tunica media of small arteries and hypertrophy of the alveolar and bronchiolar smooth muscle. Thus, Hamilton (1970) considers that the so-called spontaneous pulmonary arteriopathy with pulmonary arterial hypertrophy and hyperplasia seen in 3–70% of cats is in fact due to infection with *A. abstrusus*.

Clinical signs consist of a chronic cough with gradual wasting. Hamilton (1963) reports prolonged respiratory trouble with coughing, sneezing and a nasal discharge. There is dyspnoea, polypnoea with increased lung sounds and rales.

In heavy infections the animal may cough and suffer from diarrhoea and emaciation, which may be followed by death, or recovery may take place. In very severe infections the simultaneous deposition of a large number of eggs in the lungs may cause sudden death.

Diagnosis can be made by finding the larvae in the faeces.

Treatment. Levamisole in doses of 15–100 mg/kg given on alternate days for five to six treatments suppresses the clinical signs and larvae disappear from the faeces. However, West et al. (1977), when treating cheetahs in Dublin Zoo with 12 mg/kg of levamisole, found that even four courses of such treatment induced only a temporary fall in faecal larvae and larvae reappeared one month after treatment. Treatment with albendazole might be of value.

Prophylaxis is in most cases impracticable, since it would imply preventing cats from catching mice, lizards etc.

Genus: Perostrongylus Schlegel, 1933

Perostrongylus pridhami (Anderson, 1963) occurs in the lungs of wild mink (*Mustela vison*) in Canada. Dual infections of *P. pridhami* and *F. martis* are frequently found and the larvae of these two parasites are indistinguishable. The life-cycle is indirect, with a wide range of aquatic and terrestrial gastropods acting as intermediate hosts and mice, frogs, passerine birds and fish as paratenic hosts (Anderson 1971).

Perostrongylus falciformis (Schlegel, 1934) is found in badgers (*Meles m. meles* and *M. m. tauricus*) in Europe. Its life-cycle is similar to that of *P. pridhami* (Anderson 1971).

Genus: Angiostrongylus Kamensky, 1905

Angiostrongylus vasorum (Baillet, 1866) (syn. *Haemostrongylus vasorum*) occurs in the pulmonary artery and rarely in the right ventricle of the dog and fox in Europe, the USSR and South America. The male is 14–18 mm long and the female 18–25 mm and relatively stout. The bursa is small, but all the rays can be distinguished; the ventrals are fused for most of their length and the dorsal ray is stout with short terminal branches.

The vulva is situated in the posterior half of the body. The eggs measure 70–80 by 40–50 μm and are unsegmented when laid.

The eggs are arrested in the lung capillaries, where they develop and hatch. The larvae escape into the air passages and are passed in the faeces of the host. They are 330–360 μm long. The snail *Arion rufus* acts as the intermediate host and infective third-stage larvae develop in 17 days. Infection is by ingestion and in dogs the larvae migrate from the stomach and intestine to the mesenteric lymph node where they are found for four or five days and during which time they moult twice. Parasites are found in the liver on day 8 and by days 9 and 10 are found in the right ventricle and pulmonary artery. Sexual maturity is reached 33–36 days after infection (Guilhon & Cens 1973). Paratenic hosts may occur in the life-cycle but are not yet known.

The parasites and their eggs cause obliteration of the small pulmonary arterial branches through inflammation of the vascular walls. This later leads to perivascular sclerosis and pulmonary emphysema. Nodules of various sizes, consisting of partly organized thrombi, may be present in the lungs. Hypertrophy of the heart and congestion of the liver with consequent ascites follow on the lung disturbances. Affected animals suffer from dyspnoea and may die of cardiac insufficiency.

The worms are usually found at autopsy, but in cases of dyspnoea a diagnosis might be found by finding the larvae in the sputum or fresh faeces. The larva is about 330 μm in length, it has a small cephalic button, the tail is pointed and possesses an undulation with a dorsal appendage.

Levamisole, 10 mg/kg given subcutaneously on three consecutive days, has been used with effect.

Snail control would be advised but the probable involvement of paratenic hosts in the life-cycle makes control of the infection difficult.

Angiostrongylus cantonensis (Chen, 1935) Dougherty, 1946. This the rat lung-worm. It is a relatively common parasite of the lungs of rats in Australia, various Pacific Islands, Malaysia, Taiwan and other parts of the Far East, India and Egypt. Its importance lies in the fact that in the human it may be the cause of eosinophilic meningitis, which may be serious and also, at times, fatal. The parasite is filiform, 17–25 mm in length.

Life-cycle. In rats larvae migrate to the brain and are found in neural parenchyma for up to two weeks. They spend a further two weeks in the subarachnoid space and then migrate via the venous system to the pulmonary artery. The prepatent period is 42–45 days. The eggs lodge in the smaller pulmonary vessels, embryonate in about six days and hatch; the larvae, 270–300 μm long, are coughed up, swallowed and passed in the faeces. The intermediate hosts are terrestrial, aquatic and amphibious gastropods, such as *Vaginulus plebeius* and *Laevicaulus alta* in Australia and *Achatina fulica* and *Bradybaena circulus* in Japan. Development to the infective third-stage larva occurs in 17 days. A wide variety of paratenic hosts occur; these include freshwater prawns (*Macrobrachium lar*), land crabs (*Ocypode ceratophthalma*, *Cardisoma hirtepes*), coconut crab (*Birgus latro*) and planarians (*Geoplana septemlineata*) (Alicata & Jindrak 1970). In Japan frogs and toads have been found to be infected and larvae will survive in frogs for at least 10 weeks (Asato et al. 1978).

Pathogenesis. The infection is largely inapparent in rats although neurological or respiratory signs may occasionally be seen. In other mammals the migration of the parasite is not completed and it remains and causes damage in the central nervous system. Eosinophilic meningoencephalitis and associated neurological symptoms occur in man in southern Asia and the Pacific area (Rosen et al. 1962). Infection of man occurs from the consumption of raw or undercooked intermediate or paratenic hosts. An ascending paralysis due to infection with *A. cantonensis* occurs in puppies in Australia (Mason et al. 1976).

Diagnosis. Infection in man and animals other than the rat must be diagnosed by serological tests such as the ELISA.

Treatment. Thiabendazole, levamisole and prednisolone have been shown to be beneficial in the treatment of man. In rats, mebendazole (6.25

mg/kg for five days) and levamisole (12.5 mg/kg for five days) had a greater than 95% efficacy against *A. cantonensis* larvae (Lämmler & Weidner 1975). High doses of cambendazole, fenbendazole and parbendazole were also effective.

Angiostrongylus costaricensis Morera and Céspedes, 1971, occurs in the subserosal arteries of the caecum and in the cranial mesenteric artery of wild rodents in areas of Central and South America. The intermediate hosts are the slugs, *Vaginulus plebeius*. This parasite causes eosinophilic granulomata in the intestine of man, which may at times be fatal.

A large number of other *Angiostrongylus* species occur in wild animals. These include **A. dujardini** Drozdz and Dobey, 1970, which occurs in the pulmonary arteries and right heart of *Apodemus sylvanticus* and *Clethrionomys glareolus* in Mediterranean areas; **A. mackerrasae** Bhaibulaya, 1968, in rats in Australia; **A. (Rodentocaulus) ondatrae** (Schultz, Orlov & Kutass, 1933) in the muskrat, (*Ondantra zibethica*) in USSR; **A. gubernaculatus** Dougherty, 1946, which occurs in badgers and striped skunks in North America; **A. schmidti** Kinella, 1971, from the rice rat (*Oryzomys palustris*) in the USA; and **A. blarini** Ogren, 1954, and **A. michiganensis** Ash, 1967, in shrews (*Blarina brevicauda* and *Sorex cinereus cinereus*, respectively) in the USA. Drozdz (1970) has proposed that the latter species and others which occur in insectivores should be placed in the genus *Stefanskostrongylus* Drozdz, 1970. Since *A. cantonensis* does not occur in Europe and other areas, but eosinophilic meningitis may be seen there, it is possible that some of the above species could cause the zoonosis in man. However, many (i.e. *A. dujardini*) do not have a predilection for the CNS and are therefore probably unimportant in man (Drozdz & Doby 1971).

Genus: Gurltia Wolffhügel, 1933

Gurltia paralysans Wolffhügel, 1933 occurs in thigh veins of *Felidae* in South America. Male 12 mm, female 20–23 mm. It may cause lameness of the posterior extremity.

Genus: Parafilaroides Dougherty, 1946

Filaroid parasites of pinnipeds. Small, buccal capsule rudimentary; bursa absent.

Parafilaroides gymnurus Dougherty, 1946, occurs in the lungs of the harbour seal (*Phoca vitulina*) in Europe.

Parafilaroides decorus Dougherty and Herman, 1947 occurs in the lungs of the sea lion (*Zalophus californianus*) where it causes lesions similar to those seen in aelurostrongylosis in cats. The intermediate host is a fish, the opal eye fish (*Girella nigricans*) serving as such in the San Nicholas Island rookery (Dailey 1970). This lungworm is responsible for much of the mortality seen in captive sea lions (Dailey & Brownell 1972). Diagnosis is based on finding first-stage larvae in the faeces.

Other species include **P. nanus** Dougherty and Herman, 1947 and **P. prolificus** Dougherty and Herman, 1947, both of the lungs of Stellar's sea lion (*Eumetopias jubatus*) in California. No detailed studies have been done regarding treatment of these infections, but possibly the benzimidazoles would be effective, given at a similar dose rate for the lungworms of pigs.

FAMILY: SKRJABINGYLIDAE KONTRIMAVICHUS, DELYAMURE AND BOEV, 1976

Genus: Skrjabingylus Petrov, 1927

Skrjabingylus nasicola (Leukart, 1842) occurs in the nasal sinuses of mink (*Mustela vison*), polecat and fox in North America, Europe and the USSR. The parasite is 7–10 mm in the male and 18–22 mm in the female. The spicules are similar, being 220–230 μm in length and the gubernaculum is marked.

Other species include **Skrjabingylus chitwoodorum** (Hill, 1939) in skunks (*Mephitis* spp.) in North America; **Skrjabingylus petrowi** Bazanor, 1936, in *Martes* spp. in the USSR; and

Skrjabingylus magnus Webster, 1965, in *Mephitis mephitis* in Canada.

Larvae passed in the faeces range in size, depending on the species, from 375 to 520 μm and have a small terminal spine. The intermediate hosts are terrestrial gastropods and paratenic hosts include frogs, toads, snakes, fish and mice. In the definitive host development occurs in the wall of the stomach and small intestine. The young adults then migrate via the vertebral column, subarachnoid space and olfactory nerves to the nasal sinuses (Lankester 1972). Here they cause decalcification, discoloration and deformity of the sinuses. The incidence of infection may be high. *Skrjabingylus* spp. has been found in the subarachnoid space of a cat in Germany.

Genus: Metathelazia Skinker, 1931

These are parasite of the lungs of carnivores, insectivores and primates. Several species occur, including **M. californica** Skinker, 1931, from various *Felidae*; **M. felis** (Vogel, 1928), from *Felis pardalis*; **M. multipapillata** Gerichter, 1948, from *Erinaceus europaeus*; and **M. ascaroides** Dougherty, 1943 from the langur monkey in Africa.

Genus: Vogeloides Orlov, Davtian and Lubimov, 1933

Some authors consider this a synonym of the genus *Metathelazia*.

Vogeloides massinoi (Davtian, 1933) Dougherty, 1952 occurs in cats in the USSR and USA.

Genus: Pneumospirura Wu and Hu, 1938

Pneumospirura capsulata Gerichter, 1948 occurs in *Meles meles*. **P. bassarisci** Pence and Stone, 1977, is found in the bronchioles of the ringtail, *Bassaricas astutus*, in the USA; **P. rodentium** Wertheim and Giladi, 1977, reported from the lungs of *Gerbillus dasyurus* and *Meriones crassus* in Israel. Werthein and Chabaud (1977) consider the genus *Pneumospirura* synonymous with the genus *Metathelazia*.

FAMILY: CRENOSOMATIDAE SCHULZ, 1951

Fairly short worms, cuticle at the anterior end thrown into ring formation. Bursa well developed. Parasites of lungs of carnivores and insectivores. Genera of importance: *Crenosoma*, *Troglostrongylus*.

Genus: Crenosoma Molin, 1861

Crenosoma vulpis (Dujardin, 1845) occurs in the bronchi and sometimes also the trachea of the fox. It is an important parasite of foxes and it also affects the dog, wolf, raccoon, badger and wolverine and it and other species have been reported in the black bear (*Urus americanus*).

The male is 3.5–8 mm long and the female 12–15 mm. The cuticle of the anterior end of both sexes has 18–26 overlapping circular folds, which bear small spines. The ventral rays are partly fused and also the medio- and posterolaterals; the externodorsals arise separately; the

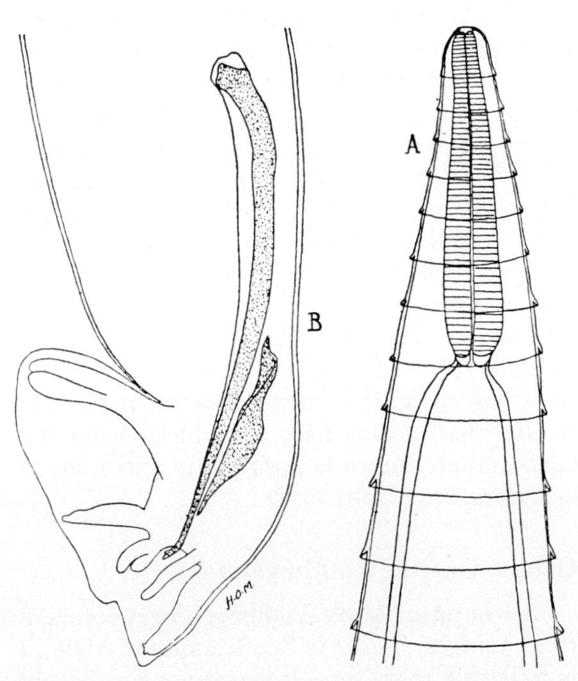

Fig 1.153 *Crenosoma vulpis.* A, Dorsal view of anterior end. B, Lateral view of hind end of male.

dorsal is stout and bears two small processes distally. The spicules are about 0.37 mm long and each has a slender dorsal spur in the posterior third. A gubernaculum is present. The vulva is situated near the middle of the body. The worms are ovoviviparous and produce larvae which are 0.265–0.33 mm long and have a straight tail. The oesophagus of the larva is 0.105–0.113 mm long and the tail 0.035–0.036 mm. There is a fairly transparent vestibulum, and the intestine consists of 14–16 cells.

Wetzel (1940) found that the larvae penetrate into the foot of land snails or slugs (*Helix pomatia*, *Cepea hortensis*, *C. nemoralis*, *Arianta arbustorum*, *Agriolimax agrestis*, *Arion hortensis*, *A. circumscriptus*) in which they reach the infective stage in 16–17 days. When such snails are eaten by the final host the larvae pass via the lymphatic glands and hepatic circulation to the lungs, where they grow adult in 21 days.

A similar disease entity is produced to that seen with *Capillaria aerophila* (see p. 340), consisting of a rhinotracheitis, bronchitis and a nasal discharge. In heavy infections a bronchopneumonia may be a complication.

The disease should be suspected on silver fox farms when the prevalence of chronic tracheobronchitis is substantial. Larvae may be found in the faeces or nasal discharge and the condition should be differentiated from that caused by *Capillaria aerophila*, though the two frequently occur together.

Levamisole as an oral dose of 8 mg/kg was 100% effective while diethycarbamazine, 8 mg/kg twice a day for three days, showed a 92% efficacy (Stockdale & Smart 1975).

The runs of foxes should be kept dry and free of grass in order to produce conditions which are unsuitable for the intermediate host. Molluscicides may be applied to soil, or animals may be kept on wire floors to avoid contact with the earth.

Other *Crenosoma* species include **Crenosoma petrowi** Morosov, 1939, in *Martes* and *Mustela* spp. and the black bear in North America and the USSR; **Crenosoma mephiditis** Hobmaier, 1940, in the skunk, *Mephitis mephitis* in North America; and **Crenosoma striatum** (Zeder, 1800) in hedgehogs in Europe and the USSR. The latter causes respiratory signs and together with *Bordetella bronchiseptica* has been associated with deaths of hedgehogs. The infection may be treated with two subcutaneous injections of tetramisole (0.2 ml of a 1% solution/100 g) given with an interval of one day.

Genus: Troglostrongylus Vevers, 1923 (syn. Bronchostrongylus Cameron, 1931)

Troglostrongylus subcrenatus (Railliet & Henry, 1913). This species occurs in the lungs of wild felidae and the domestic cat in Africa, Italy and the USA. The adult worm superficially resembles *Aelurostrongylus abstrusus* and Fitzsimmons (1961) gives the differential features by which these two species may be distinguished. Its molluscan intermediate hosts are species of the genera *Helicella*, *Chondrula*, *Monacla*, *Retinella* and *Limax* and mice may act as transport hosts of the infective larvae. The mice eat the molluscs and may carry viable infective larvae in cysts on the surface of their lungs for at least 120 days. Cats infect themselves by eating the mice.

REFERENCES

METASTRONGYLOIDEA: METASTRONGYLIDAE, PROTOSTRONGYLIDAE, CRENOSOMATIDAE AND FILAROIDIDAE

Alicata, J. E. & Jindrak, K. (1970) *Angiostongylosis in the Pacific and South-east Asia*. Springfield, Ill.: Charles C. Thomas.

Anderson, R. C. (1963) The incidence, development and experimental transmission of *Pneumostrongylus tenuis* (Dougherty) (Metastrongyloidea: Protostrongylidae) of the meninges of the white-tail deer (*Odocoileus virginianus borealis*) in Ontario. *Can. J. Zool.*, 41, 775–792.

Anderson, R. C. (1964) Neurologic disease in moose infected experimentally with *Pneumostrongylus tenuis* from white-tailed deer. *Path. vet.*, 1, 289–322.

Anderson, R. C. (1965) The development of *Pneumostrongylus tenuis* in the central nervous system of white-tailed deer. *Path. vet.*, **2**, 360–379.
Anderson, R. C. (1971) Metastrongylid lungworms. In *Parasitic Diseases of Wild Mammals*, ed. J. W. Davis & R. C. Anderson, pp. 81–126. Ames: Iowa State University Press.
Anderson, R. C., Lankester, M. W. & Strelive, U. R. (1966) Further experimental studies of *Pneumostrongylus tenuis* in cervids. *Can. J. Zool.*, **44**, 851–861.
Asato, R., Sato, Y. & Otsuru, M. (1978) The occurrence of *Angiostrongylus cantonesis* in toads and frogs in Okinawa Prefecture, Japan. *Jap. J. Parasit.*, **27**, 1–8.
Bennett, D. & Beresford-Jones, W. P. (1973) Treatment of *Filaroides osleri* in a 16-month-old male Yorkshire Terrier with thiabendazole. *Vet. Rec.*, **92**, 226–227.
Cordero-del-Campillo, M., Rojo-Vasquez, F. A. & Diez-Banos, P. (1980) Albendazole against protostrongylid infestations in sheep. *Vet. Rec.*, **106**, 458.
Craig, T. M., Brown, T. W., Shefstad, D. K. & Williams, G. D. (1978) Fatal *Filaroides hirthi* infection in a dog. *J. Am. vet. med. Ass.*, **172**, 1096–1098.
Dailey, M. D. (1970) The transmission of *Parafilaroidis decorus* (Nematoda: metastrongyloidea) in the California sea lion (*Zalophus californianus*). *Proc. Helminth. Soc. Wash.*, **37**, 215–222.
Dailey, M. D. & Brownell, R. I. (1972) A checklist of marine mammal parasites. In *Mammals of the Sea: Biology and Medicine*, ed. S. H. Ridgway, pp 528–589. Springfield, Ill.: Charles C. Thomas.
Dorrington, J. E. (1968) Studies on *Filaroides osler* infestation in dogs. *Onderstepoort J. vet. Res.*, **35**, 225–286.
Dougherty, E. C. (1944) The genus *Metastrongylus* (Molin, 1861) (Nematoda: Metastrongylidae). *Proc. Helminth. Soc. Wash.*, **11**, 66–73.
Dougherty, E. C. (1951) A further revision in the classification of the family Metastrongylidae (Leiper, 1909) (Phylum Nematoda). *Parasitology*, **41**, 91–96.
Drozdz, J. (1970) Revision of the systematics of the genus *Angiostrongylus* Kamensky 1905 (Nematoda: Metastrongyloidea). *Ann. Parasit. hum. comp.*, **45**, 829–837.
Drozdz, J. & Doby, J. M. (1971) Morphology, migration and life-cycle of *Angiostrongylus* (*Parastrongylus*) *dujardini* Drozdz & Doby 1970 (Nematoda: Metastrongyloidea) in its definitive hosts. *Bull. Soc. Sci. Bretagne*, **45**, 229–239.
Eslami, A. H. & Anwar, M. (1976) Activity of fenbendazole against lungworms in naturally infected sheep. *Vet. Rec.*, **99**, 129.
Fitzsimmons, W. M. (1961) *Bronchostrongylus subcrenatus* (Railliet & Henry, 1913) a new parasite recorded from the domestic cat. *Vet. Rec.*, **73**, 101–102.
Forrester, D. J. (1971) Bighorn sheep lungworm-pneumonia complex. In *Parasitic Diseases of Wild Mammals*, ed. J. W. Davis & R. C. Anderson, pp. 158–173. Ames, Iowa: Iowa State University Press.
Forrester, D. J. & Senger, C. M. (1964) Prenatal infection of bighorn sheep with protostrongylid lungworms. *Nature, Lond.*, **201**, 1051.
Georgi, J. R. & Anderson, R. C. (1975) *Filaroides hirthi* sp. n. (Nematoda: Metastrongylidea) from the lung of the dog. *J. Parasit.*, **61**, 337–339.
Georgi, J. R., Fahnestock, G. R., Bohm, M. F. K. & Adsit, J. C. (1979*a*) The migration and development of *Filaroides hirthi* larvae in dogs. *Parasitology*, **79**, 39–47.
Georgi, J. R., Fleming, W. J., Hirth, R. S. & Cleveland, D. J. (1976) Preliminary investigations of the life history of *Filaroides hirthi* Georgi & Anderson, 1975. *Cornell Vet.*, **66**, 309–323.
Georgi, J. R., Georgi, M. E. & Cleveland, D. J. (1977) Patency and transmission of *Filaroides hirthi* infection. *Parasitology*, **75**, 251–257.
Georgi, J. R., Georgi, M. E., Fahnestock, G. R. & Theorides, V. J. (1979*b*) Transmission and control of *Filaroides hirthi* lungworm infection in dogs. *Am. J. vet. Res.*, **40**, 829–831.
Georgi, J. R., Slauson, D. O. & Theodorides, V. J. (1978) Anthelmintic activity of Albendazole against *Filaroides hirthi* lung worms in dogs. *Am. J. vet. Res.*, **39**, 803–806.
Guilhon, J. & Cens, B. (1973) *Angiostrongylus vasorum* (Baillet, 1866). Morphological and biological study. *Ann. Parasit. hum. comp.*, 48, 567–596.
Halhead, W. A. (1968) Tetramisole toxicity in sheep infected with *Muellerius capillaris. Vet. Rec.*, **83**, 58.
Hamilton, J. M. (1963) *Aelurostrongylus abstrusus* infestation of the cat. *Vet. Rec.*, **75**, 417–422.
Hamilton, J. M. (1970) The influence of infestation by *Aelurostrongylus abstrusus* on the pulmonary vasculature of the cat. *Br. vet. J.*, **126**, 202–209.
Hirthi, R. S. & Hottendorf, G. H. (1973) Lesions produced by a new lungworm in beagle dogs. *Vet. Path.*, **10**, 385–407.
Jones, B. R., Clark, W. T., Collins, G. H. & Johnstone, A. C. (1977) *Filaroides osleri* in a dog. *N. Z. vet. J.*, **25**, 103–104.
Kassai, T. (1958) A juhod tudofergessegeinek orvoslasa ditrazinofosafattal. *Magy. Allatorv. Lapja*, **13**, 9–13.
Kolevatova, A. I. (1974) Preservation of infectivity of *Metastrongylus elongatus* larvae in the earthworm intermediate host. *Parazitologiya*, **8**, 49–52.
Lämmler, G. & Weidner, E. (1975) The larvicidal activity of anthelmintics against *Angiostrongylus cantonensis. Berl. Münch. tierärzt. Wsch.*, **88**, 152–156.
Lankester, M. W. (1972) The biology of *Skrjabingylus* spp. (Metastrongyloidea: Pseudaliidae) in Mustelids (Mustelidae). *Diss. Abstr. Int.*, **32B**, 4420.
Lankester, M. W. & Northcott, T. H. (1979) *Elaphostrongylus cervi* Cameron 1931 (Nematoda: Metastrongyloidea) in caribou (*Rangifer tarandus caribou*) of Newfoundland. *Can. J. Zool.*, **57**, 1384–1392.
Lloyd, S. & Soulsby, E. J. L. (1978) Survey of parasites in dairy goats. *Am. J. vet. Res.*, **39**, 1057–1059.
MacKenzie, A. (1958) Studies on lungworm infection of pigs. II. Lesions in experimental infections. *Vet. Rec.*, **70**, 903–906.
MacKenzie, A. (1963) Experimental observations on lungworm infection together with virus pneumonia in pigs. *Vet. Rec.*, **75**, 114–116.
Malherbe, N. D. (1954) The chemotherapy of *Filaroides osleri* (Cobbold 1879) infestation in dogs. A progress report. *J. S. Afr. vet. med. Ass.*, **25**, 9–12.
Mason, K. V., Prescott, C. W., Kelly, W. R. & Waddell, A. H. (1976) Granulomatous encephalomyelitis due to *Angiostrongylus cantonensis. Aust. vet. J.*, **52**, 295.
Pence, D. B. (1978) Notes on two species of *Filaroides* (Nematoda: Filaroididae) from carnivores in Texas. *Proc. Helminth. Soc. Wash.*, **45**, 103–110.
Polley, L. & Creighton, S. R. (1977) Experimental direct transmission of the lungworm *Filaroides osleri* in dogs. *Vet. Rec.*, **100**, 136–137.
Prestwood, A. K. (1972) *Parelaphostrongylus andersoni* sp. n. (Metastrongyloidea: Protostrongylidae) from the musculature of white-tailed deer (*Odocoileus virginianus*). *J. Parasit.*, **58**, 897–902.
Rose, J. H. (1959) *Metastrongylus apri*, the pig lungworm. Observations on the free-living embryonated egg and the larvae in the intermediate host. *Parasitology*, **49**, 439–447.
Rosen, L., Chappell, R., Laqueur, G. L., Wallace, G. D. & Weinstein, P. P. (1962) Eosinophilic meningitis caused by a metastrongylid lungworm of the rat. *J. Am. med. Ass.*, **179**, 620–624.
Schultz, R. S., Orloff, I. W. & Kutass, A. J. (1933) Zur Systematik der Subfamilie *Synthetocaulinae* Skrj. 1932 nebst Beschreibung einiger neuer Gattungen and Arten. *Zool. Anz.*, **102**, 303–310.
Scott, D. W. (1973) Current knowledge of aelurostrongylosis in the cat. *Cornell Vet.*, **63**, 483–500.
Shope, R. E. (1941) The swine lungworm as a reservoir and intermediate host for swine influenza virus. I. The presence of swine influenza virus in healthy and susceptible pigs. II. The transmission of swine influenza virus by the swine lungworm. *J. exp. Med.*, **74**, 41–68.
Shope, R. E. (1943) The swine lungworm as a reservoir and intermediate host for swine influenza virus. III. Factors influencing transmission of the virus and the provocation of influenza. *J. exp. Med.*, **77**, 111–138.
Shope, R. E. (1958) The swine lungworm as a reservoir and intermediate host for hog cholera virus. I. The provocation of masked hog cholera virus in lung worm infested swine by *Ascaris* larvae. *J. exp. Med.*, **107**, 609–622.
Stockdale, P. H.G. & Anderson, R. C. (1970) The development, route of migration and pathogenesis of *Filaroides martis* in mink. *J. Parasit.*, 550–558.
Stockdale, P. H.G. & Smart, M. E. (1975) Treatment of crenasomiasis in dogs. *Res. vet. Sci.*, **18**, 178–181.
Sutherland, R. J. (1976) *Elaphostrongylus cervi* in cervids in New Zealand. I. The gross and histological lesions in red deer (*Cervus elaphus*). *N. Z. vet. J.*, **24**, 263–266.
Wallace, G. W. (1977) Swine influenza and lungworms. *J. infect. Dis.*, **135**, 490–492.
Walley, J. K. (1957) A new drug for the treatment of lungworms in domestic animals. *Vet. Rec.*, **69**, 815–824, 850–853.
Wertheim, G. & Chabaud, A. G. (1977) Scanning electron microscopy of the cephalic structures of Pneumospiruridae (Thalazioidea—Nematoda). Revision of the family. *Ann. Parasit. hum. comp.*, **52**, 647–657.
West, B., Wilson, P. & Hatch, C. (1977) *Aelurostrongylus abstrusus* infection in the cheetah. *J. Helminth.*, **51**, 210–211.
Wetzel, R. (1940) Zur Biologie des Fuchslungenwurmes *Crenosoma vulpis* I. Mitteilung. *Arch. wiss. prakt. Tierheilk.*, **75**, 445–460.

ORDER: SPIRURIDA CHITWOOD, 1933

Oesophagus essentially divided into two regions, an anterior muscular and a posterior glandular region; ventrolateral cephalic papillae absent. Parasites of vertebrates, development of larval stages in arthropods.

SUPERFAMILY: SPIRUROIDEA RAILLIET & HENRY, 1915

Nematodes with two lateral lips which may be further subdivided. A pharynx or cylindrical buccal capsule is usually present. The oesophagus consists of a short, anterior, muscular portion followed by a longer, wide, glandular portion. The hind end of the male is usually spirally coiled and bears lateral alae and papillae. The spicules are, as a rule, unequal and dissimilar. The vulva opens near the middle of the body in most cases, but its position is variable. The eggs are usually thick-shelled and contain larvae when laid. The adults are parasites of vertebrates, frequently living in the lumen or the wall of the stomach. The life-cycle as a rule includes an intermediate host which is an arthropod. Families of importance: *Spiruridae*, *Thelaziidae*, *Tetrameridae*, *Acuariidae*.

FAMILY: SPIRURIDAE OERLEY, 1885

With the typical characteristics of the superfamily. Genera of importance: *Habronema*, *Draschia* and *Hartertia*.

Genus: Habronema Diesing, 1861

Habronema muscae Carter, 1861 occurs in the stomach of equines. Male 8–14 mm long and female 13–22 mm. There are two lateral lips, each being trilobed. The pharynx is cylindrical and provided with a thick cuticular lining. The male has wide caudal alae, four pairs of precloacal papillae and one or two papillae behind the cloaca. The cloacal region of the body is covered with small cuticular bosses or ridges. The left spicule is slender and 2.5 mm long, the right stouter and only 0.5 mm long with a ratio of 5 : 1. The vulva is situated near the middle of the body and opens dorsolaterally, the vagina running for some distance in the body wall. The eggs have thin shells and measure 40–50 by 10–12 μm. Eggs or larvae may be found in the faeces.

Habronema majus Schneider, 1866 (syn. *H. microstoma*) is also a parasite in the stomach of equines and closely resembles the preceding species. It occurs throughout the world. It is larger, males being 16–22 mm long and females 15–22 mm. The pharynx contains a dorsal and a ventral tooth in its anterior part. The male has four pairs of precloacal papillae. The left spicule measures 0.76–0.8 mm and the right 0.35–0.38 mm, with a ratio of 2 : 1.

Genus: Draschia Chitwood and Wehr, 1935

Draschia megastoma Chitwood and Wehr, 1934 (syn. *Habronema megastoma*) occurs in nodules in the stomach wall, rarely free, in the stomach of equines throughout the world. It can be easily recognized by its 'head', which is constricted off from the body. The pseudolabia are unlobed. The pharynx is funnel-shaped. The male has four pairs of precloacal papillae. The left spicule is 0.46 mm long and the right 0.24. The worms are ovoviviparous.

Life-cycles. The embryonated eggs may hatch in the stomach or intestine and larvae or eggs, according to species, are passed in the faeces of the host and are ingested by the maggots of flies which develop in the manure. *H. muscae* and *D. megastoma* develop in the housefly *Musca domestica*, while *H. majus* develops in the stable fly *Stomoxys calcitrans*. The worms reach the infective stage in the maggots at about the time when the latter pupate. In the adult fly the larvae occur free in the haemocoel and pass forwards into the proboscis. The larvae are deposited on the lips, nostrils and wounds of the horse when the fly feeds. In the case of *H. majus* in *S. calcitrans* the

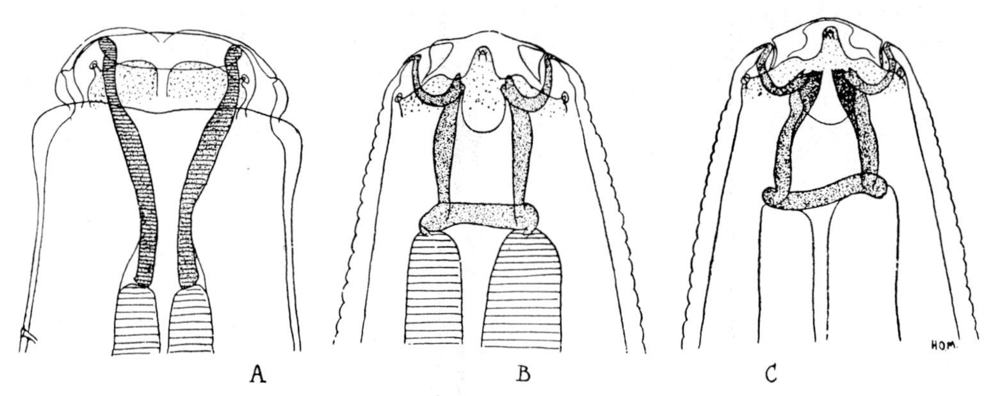

Fig 1.154 Lateral views of the heads of equine species of *Habronema* and *Draschia* spp. A, *D. megastoma*. B, *H. muscae*. C, *H. majus*.

infective larvae interfere with the ability of the fly to penetrate the skin with the proboscis and it reverts to an imbibing method of feeding, obtaining nourishment from moist surfaces such as the lips, nostrils and wounds of horses. It is probable that equines become infected by swallowing flies that fall into water or food. The larvae are liberated into the stomach, where they grow to maturity in about two months.

Pathogenesis. D. megastoma produces large fibrous 'tumours' in the fundus region of the stomach wall. These have one or more openings and a number of cavities in which the worms are located, together with caseous or necrotic material. Upon infection the parasites penetrate into the submucosa of the stomach, forming nodules surrounded by a thin wall of granulation tissue with cellular infiltration. The individual nodules gradually fuse to form the large tumours. These tumours may protrude into the lumen of the stomach sufficiently to interfere mechanically with the function of the stomach, but are otherwise not particularly harmful. The other two species of the genus *Habronema* occur free in the stomach and may penetrate into the mucosa. They produce an irritation which leads to a chronic catarrhal gastritis with the formation of much mucus. *H. majus* may produce ulcers of the stomach.

Cutaneous habronemiasis, also known as 'summer sores', 'bursati', 'granular dermatitis' etc., is caused by *Habronema* and *Draschia* larvae which are deposited in existing wounds by infected flies. The condition is seen in Europe, Australia, Africa, the East, North America and the USSR.

All three species of *Habronema* and *Draschia* may be concerned in the condition but *D. megastoma* appears to be the most important. Generally larvae are deposited by flies on wounds; however, larvae may also penetrate the apparently intact skin (Nishiyama 1958).

Habronema spp. larvae may also be associated with a granular conjunctivitis, this being seen in areas where cutaneous habronemiasis is prevalent. The lesion is seen on the inner canthus of the eye on the nictitating membrane or the skin surrounding the eye and is in the form of a wart-like lesion. Rarely the whole conjunctiva is a mass of granulation tissue.

Habronema spp. may also be found in the lungs, being associated with fibrotic nodules 0.5–2 cm in diameter which develop around the finer bronchioles, being essentially a nodular peribronchitis. The mode of entry of third-stage larvae into the lungs is unknown.

Clinical signs. Gastric habronemiasis is not known to cause marked clinical signs, but it is probably the cause of some cases of chronic gastritis. If the tumours of *D. megastoma* occur near the pylorus they may interfere with the closing of the sphincter and cause digestive disorders.

The lesions of cutaneous habronemiasis occur mainly in warm countries and during the summer. They are seen on those parts of the body that are liable to be injured, such as the legs, withers, sheath and canthus of the eye. They vary in size and have an uneven surface which consists of a soft, brownish-red material and which covers a mass of firmer granulations. The wounds show a tendency to increase in size and do not respond to ordinary treatment until the following winter, when they often heal spontaneously. In some cases chronic lesions develop in the form of granulomata which may reach a considerable size.

Diagnosis. The gastric infection is difficult to diagnose because few eggs or larvae are passed and they are not readily found in the faeces. Some worms or larvae may be found by gastric lavage through a stomach tube. Experimentally, *M. domestica* and *S. calcitrans* larvae may be fed on the faeces and later examined for the presence of infection.

The tumourous masses of *D. megastoma* and the adults of *H. muscae* and *H. majus* are easily seen on post mortem examination. When the gastric contents are lifted away from the stomach wall the latter nematodes will be observed on the surface of the contents and lining the mucosa. They must be differentiated from *Trichostrongylus axei*, which are smaller.

Treatment. A traditional treatment for the infection is that the animal should be starved overnight and in the morning 8–10 litres of a 2% solution of sodium bicarbonate at body temperature are administered into the stomach by means of a stomach tube, in order to loosen the mucus in which the worms lie. The fluid should be withdrawn if possible, but usually it passes into the intestine. Carbon bisulphide is then administered through the tube at the rate of 5 ml/100 kg body weight, followed by a small quantity of water to wash down the drug. This treatment is effective for *H. muscae* and *H. majus*. Duncan et al. (1977) have reported that fenbendazole in doses of 15–60 mg/kg may be effective for *Habronema* spp. However, *Draschia* spp. in tumours may be difficult to treat. Organophosphate anthelmintics in equine use, such as trichlorphon and dichlorvos, may be effective against *Habronema* spp. Neither phenothiazine nor piperazine is effective.

For cutaneous and ocular habronemiasis súrgical removal may be useful, especially when the lesions has become pedunculated. The operation site should be adequately covered to prevent further deposition of larvae by flies.

Topical application of chromic acid in a 10% solution two or three times, apart from killing the larvae, forms a thick crust which protects the lesions from further attack. Radiotherapy has been used in France; a course of 200–400 roentgens is applied weekly for four or more weeks.

Cryosurgery is also effective. The lesion is frozen and thawed twice and the treatment may be repeated if necessary. Organophosphates given systemically and by topical application, the latter with or without DMSO, can also be effective.

Genus: Cyrnea Chabaud, 1958

Several species occur in the gizzard and proventriculus of birds. Lateral pseudolabia are present. The oesophagus is long. The tail of the male is not spiralled. Caudal alae are well developed. The eggs are thick-shelled and embryonated.

Cockroaches serve as intermediate hosts, larvae developing in them to the infective stage in 18–20 days. The prepatent period in birds is about 40 days.

Little or no pathology is associated with these species (Ruff 1978).

Cyrnea piliata (Walton, 1927) are small yellowish nematodes up to 13 mm in length which are found in the proventriculus of a variety of birds.

Cyrnea colini Cram, 1927 occurs in quails, grouse, prairie chicken, wild and domestic turkey, etc. in the USA. It occurs at the junction of the proventriculus and gizzard.

Genus: Spirura Blanchard 1849

The posterior part of the body is spirally twisted and thicker than the anterior part. Eggs are thick-shelled and embryonated when laid. The worms are parasites of the alimentary canal of various rodents, insectivores and carnivores.

Spirura talpae Blanchard, 1849 occurs in *Talpa europaea* and *Epimys rattus* in Europe and Africa.

Spirura rytipleurites (Deslongchamps, 1824) has two varieties, *S. r. rytiplanites* (Deslongchamps, 1824) and *S. r. seurati* Chabaud, 1954. The former occurs in the stomach of cat and fox in Europe and Africa and the latter in the hedgehog in North Africa. Intermediate hosts are cockroaches and other beetles. Nothing is known about the pathogenic effects of the parasite.

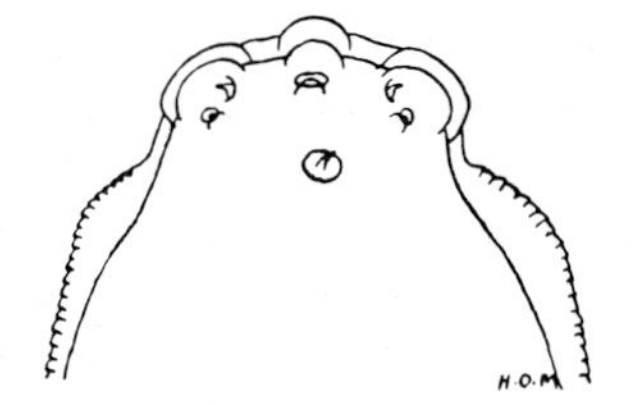

Fig 1.155 Lateral view of anterior end of *Hartertia gallinarum*.

Genus: Protospirura Seurat, 1914

Worms of this genus occur in the stomach of various felids and rodents. They are very similar to those of the genus *Mastophorus* (see below).

Protospirura numidia Seurat, 1914 occurs in *Felis ocreata*, *Canis latrans* and various *Peromyscus* spp. **P. bestianum** Kreis, 1953 occurs in *Martes martes* and **P. muricola** Gedoelst, 1916 is found in a range of rodents as well as several primates (e.g. gorilla, *Aotus* and *Cebus* monkeys). Members of the genus develop in cockroaches and infection is by ingestion of an infected invertebrate.

Genus: Mastophorus Diesing, 1853

Members of this genus are spirurids very similar to *Protospirura*. They have a spiral tail and caudal alae; the eggs are oval, thick-shelled and 56 by 30 μm.

Mastophorus muris Chitwood, 1938 occurs in the stomach of a wide range of rodents all over the world. It has been found in the cat and coyote, probably as accidental infections. Intermediate hosts include various beetles (e.g. *Tenebrio*).

Genus: Hartertia Seurat, 1915

Hartertia gallinarum (Theiler, 1919) occurs in the small intestine of fowl and wild bustards in Africa, particularly South and West Africa, Europe and Asia. The male is 28–40 mm long and the female 60–100 mm. Macroscopically the worms closely resemble *Ascaridia galli*. They have two lateral lips, each divided into three lobes. The male hind end has lateral alae, ventral cuticular bosses, four pairs of pre-cloacal and two pairs of post-cloacal papillae. The anterior lip of the cloaca bears a cuticular thickening, which may resemble a papilla. The left spicule is 2.3 mm long, acutely pointed and provided with four large barbs, while the right is 0.63 mm long and ends bluntly. The vulva opens in the anterior third of the body. The eggs have thick shells; they measure 45–53 by 27–33 μm and contain fully developed embryos when laid.

The eggs are passed in the faeces of the host and for further development have to be ingested by certain termites (*Macrohodothermes mossambicus transvaalensis*). Only the workers become infected. The larvae develop in the body cavity of the termite and reach a length of 1 cm or more. The termites do not appear to suffer from the

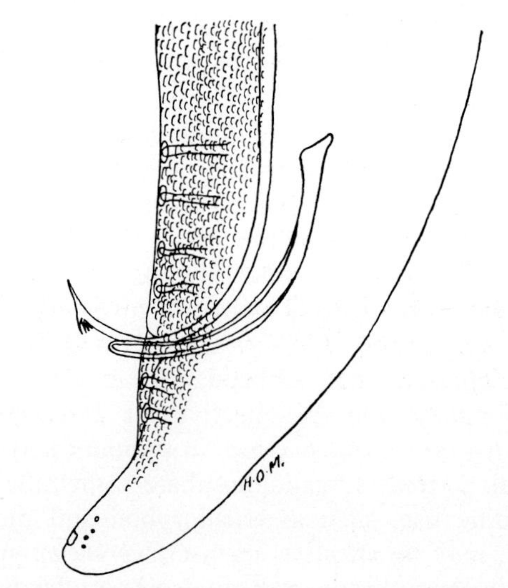

Fig 1.156 Lateral view of hind end of male *Hartertia gallinarum*.

parasites. Fowls become infected by ingesting infected termites and the worms grow to maturity in the bird in three weeks.

Depending on the severity of the infection, the parasites produce various degrees of emaciation, weakness, diarrhoea, decrease of egg production or death.

Diagnosis can be made by finding the eggs in the faeces, but it may not always be possible to distinguish them from eggs of other spirurids occuring in fowls. Diagnosis should therefore be confirmed by autopsy of a selected case.

Treatment as in the case of *Ascaridia galli*. All termites in the vicinity of the fowl runs should be destroyed.

Genus: Streptopharagus Blanc, 1912

Streptopharagus armatus Blanc, 1912 and **S. pigmentatus** (Linstow, 1897) are spirurids which occur in the stomach of monkeys and apes. The eggs are thick-shelled, embryonated when passed in the faeces and measure 28–38 by 17–22 μm. Flynn (1973) reports that *S. armatus* is common in the USA, Japan and Africa. Adults were found in 10.5% of cynomolgous monkeys and 14.3% of Japanese macaques. *S. pigmentatus* occurs in Africa, Asia, Europe and the USA and it is common in rhesus monkeys.

FAMILY: THELAZIIDAE RAILLIET, 1916

Spiruroidea with no pseudolabia; mouth capsule present. Hind end of male with many pre- and post-anal papillae. Spicules unequal. Parasites of the conjunctival sac, lacrimal duct and digestive tract of birds and mammals. Genera of importance include *Thelazia*, *Oxyspirura*, *Spirocerca*, *Ascarops*, *Physocephalus*, *Simondsia* and *Gongylonema*.

Genus: Thelazia Bosc, 1819

Parasites of the conjunctival sac or lacrimal duct of mammals and birds. Body may be transversely striated.

Thelazia rhodesii (Desmarest, 1828) occurs primarily in cattle, but also in sheep, goats and buffaloes and is cosmopolitan in distribution. Milky-white worms, males 8–12 mm long and females 12–18 mm. The cuticle bears prominent transverse striations. The male has about 14 pairs of precloacal and three pairs of post-cloacal papillae. The spicules are 0.75–0.85 and 0.115–0.13 mm long.

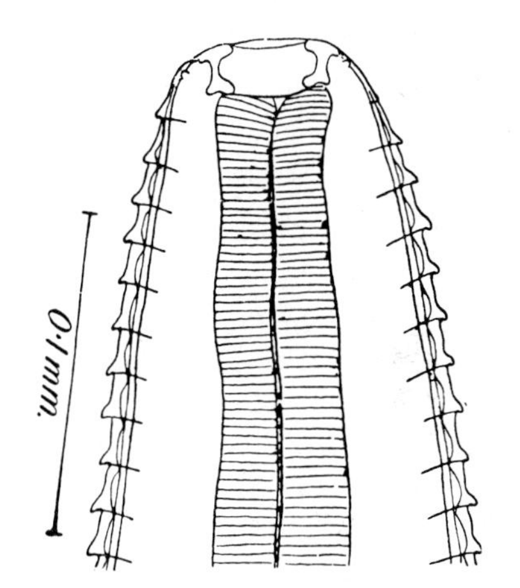

Fig 1.157 Lateral view of anterior end of female *Thelazia rhodesii*.

Thelazia gulosa Railliet & Henry, 1910 occurs in cattle in most parts of the world.

Thelazia alfortensis Railliet & Henry, 1910 occurs in cattle in Europe, but is considered to be a synonym of *T. gulosa*.

Thelazia lacrymalis (Gurlt, 1831) occurs in the horse in most parts of the world.

Thelazia skrjabini Ershov, 1928 occurs in cattle in Europe, Asia and North America.

Thelazia callipaeda Railliet & Henry, 1910 occurs under the nictitating membrane of the dog in the Far East and has also been reported from the rabbit and man. Males are 7–11.5 mm long and females 7–17 mm. The cuticle has fine transverse striations. The male has five pairs and one single pre-cloacal papilla and two pairs of post-cloacals. The left spicule is about 12 times as long as the right. The vulva is situated in the oesophageal region. The eggs when laid contain fully developed larvae which, soon after laying, extend themselves as well as the shells and so become sheathed larvae.

Thelazia californiensis Price, 1930 occurs in sheep, deer, cat, dog and also man in the USA.

Thelazia leesei Railliet and Henry, 1910 has been reported from the dromedary in the USSR and Asia.

Other species of the genus include, **T. bubalis** Ramanujachari and Alwar, 1952 in the conjunctival sac of the water buffalo in India and **T. erschowi** Oserskaja, 1931 in the conjunctival sac of the pig in the Ural region of the USSR.

Life-cycle. Klesov (1950) and Krastin (1950, 1952) have shown that the intermediate hosts of *T. rhodesii* are *Musca larvipara* and *M. convexifrons*, that those of *T. gulosa* are *Musca larvipara* and *M. amica* and that that of *T. skrjabini* is possibly *M. amica*. *Musca oseris* transmits *T. lachrymalis* in the USSR while *M. autumnalis* appears to be an important vector in the USA. Thus, 0.5–13.2% were found to be infected with *Thelazia* spp. in Massachusetts (Geden & Stoffolano 1977). *Fannia* spp. are thought to be more important than *Musca* spp. in the transmission of *T. californiensis* (Weinmann et al. 1974). Klesov (1950) found that the first larvae of *T. rhodesii* enter the gut of the fly from the eye secretions of the definitive host and penetrate to the ovarian follicles of the fly, where they develop, becoming second larvae 3.6–4 mm long, which grow and moult to become third, infective larvae. The third larvae leave the ovarian follicles and migrate to the mouth parts of the fly, from which they are transferred to cattle. The infective larvae are 5.06–7.9 mm long and their development in the fly requires 15–30 days. When infected flies were allowed access to calves, and when infective larvae were experimentally introduced into calves, adult *T. rhodesii* appeared in 20–25 days. The infective larvae of *T. gulosa* are smaller; when they were put on the conjunctiva of a calf, adult *T. gulosa* appeared seven days later. Krastin (1950) obtained infective larvae of *T. gulosa* from *Musca amica* and placed them on the eye region of a calf; six weeks later he obtained one adult *T. gulosa*. Transmission is seasonal and does not occur in the winter months due to the inactivity of the flies.

Pathogenesis. The infection may be common and 5–30% of cattle in Poland were found to be infected as were 37–42% of cattle in Surrey, England (Arbuckle & Khalil 1976) and 20–33% of cattle in the south-eastern USA. Up to 90 parasites have been found in one eye. In England and the USA surveys have demonstrated the presence of infection in 28–38% of horses. Adult parasites are found behind and in the nictitating membrane, lacrimal and nasolacrimal ducts and on the surface of the conjunctiva. However, in many cases eyeworms have practically no pathogenic effect on the host, especially in the larger animals. In South Africa, *T. rhodesii* is found frequently in calves without any lesions and in Great Britain lesions were found in only 4.3% of infected eyes. In some cases lesions of keratitis, ophthalmia etc. ascribed to these parasites are probably due to other causes; however, there can be no doubt that the parasites do cause disease of the eye. Lesions may occur in one or both eyes; initially there is a mild conjunctivitis which may progress to congestion of the conjunctiva and the cornea. As the condition becomes more serious the cornea becomes cloudy, there is marked lacrimation and the affected eye becomes markedly swollen and covered with exudate and pus.

Without treatment a progressive keratitis occurs, there is ulceration of the cornea leading to protusion of the contents of the anterior chamber.

T. callipaeda and *T. californiensis* in man cause conjunctivitis, pain and excess lacrimation.

Diagnosis. A definitive diagnosis is made by the detection of the parasites in the conjunctival sac. It may be necessary to instil local anaesthetic to allow manipulation. Examination of the lacrimal secretions may reveal eggs or first-stage larvae.

Treatment. Removal of the adult parasites with fine forceps, using local anaesthesia, is helpful. Halpin and Kirkly (1962) have reported that a subcutaneous dose of 20 ml of methyridine produces rapid recovery from *Thelazia* infection, as do tetramisole (15 mg/kg) and levamisole (5 mg/kg) given orally or parenterally. These drugs are secreted through the lacrimal glands. However, Michalski (1976) found 2 ml of levamisole injected in the subconjunctival sac to be more effective than levamisole given orally. Eye salves containing 4% morantel tartrate or 1% levamisole have also been used with success.

Genus: Oxyspirura v. Drasche in Stossich, 1897

Parasites of the eyes of birds.

Oxspirura mansoni (Cobbold, 1879) occurs under the nictitating membrane of the fowl, turkey and pea-fowl in many countries. Males are 10–16 mm long and females 12–19 mm. The cuticle is smooth and the pharynx roughly resembles an hourglass in shape. The male tail is curved ventrad and bears no alae. There are four pairs of pre-cloacal and two pairs of post-cloacal papillae. The left spicule is slender, 3–3.5 mm long, and the right 0.2–0.22 mm long and stout. The vulva is situated in the posterior part of the body and measures 50–65 by 45 μm.

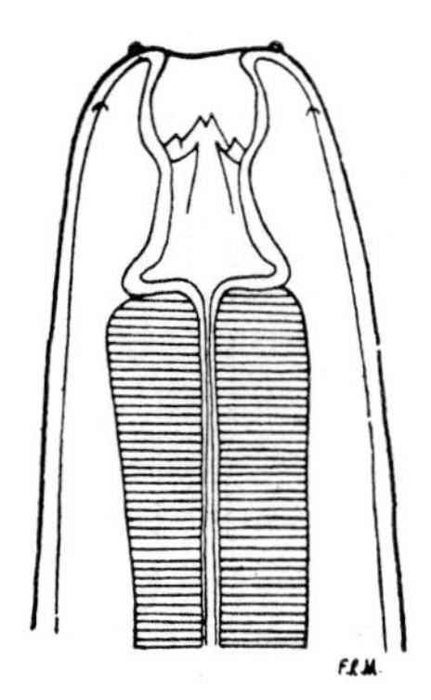

Fig 1.158 Ventral view of anterior end of *Oxyspirura mansoni*.

Oxyspirura parvorum Sweet, 1910, described from the fowl in Australia, is possibly identical with the preceding species.

Oxyspirura petrowi Skrjabin, 1929, occurs in many species of birds and a large number of other *Oxyspirura* spp. have been reported in wild birds in most countries of the world.

Life-cycles. The eggs of the parasite pass down the lacrimal ducts and out in the faeces of the bird. The intermediate stages develop in the cockroach *Pycnoscelus surinamensis*, and the fowls acquire the parasite by ingesting infected cockroaches. The larvae escape from the intermediate host after it has been ingested and apparently wander up the oesophagus, pharynx and lacrimal duct to the eye, as they have been found there 20 minutes after cockroaches were fed to the chicks.

Pathogenesis. Affected birds have ophthalmitis with inflamed, watery eyes. The nictitating membrane is swollen. The eyes are irritated and the birds scratch them. White cheesy material collects under the eyelids and eventually the eyeball may be destroyed.

Treatment. One to three drops in the eye of a 10% solution of tetramisole kills the adult parasites, as does tetramisole given orally at a dose rate of 40 mg/kg.

Prophylaxis. In accordance with the life-cycles given above, prophylaxis should include general hygiene measures and control of the cockroaches and beetles mentioned above. In the case of poultry, cockroaches should be exterminated.

Genus: Spirocerca Railliet and Henry, 1911

Spirocerca lupi (Rudolphi, 1809) (syn. *S. sanguinolenta*) occurs in the walls of the oesophagus, stomach and aorta, and more rarely free in the stomach and in other organs of the dog, fox, wolf, jackal, coyote and wild Felidae such as the lynx and snow leopard. Parasites have also been found in ruminants, such as the goat, and in donkeys.

The worms are usually coiled in a spiral and have a pink colour. Males are 30–54 mm long and females 54–80 mm and rather stout. The lips are trilobed and the pharynx is short. The male tail bears lateral alae, four pairs and one unpaired

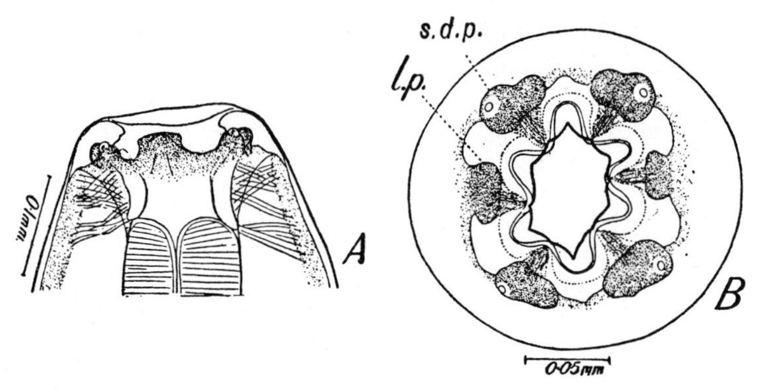

Fig 1.159 Anterior end of female *Spirocerca lupi*. A, Lateral view. B, *En face* view.
l.p. lateral papilla
s.d.p. subdorsal papilla

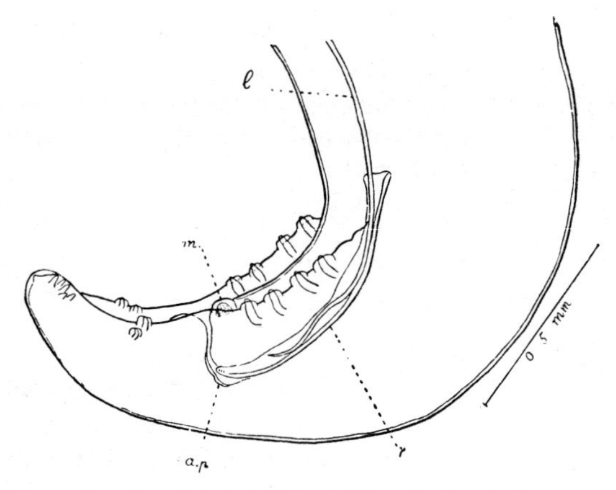

Fig 1.160 Lateral view of posterior end of male *Spirocerca lupi*.
a.p. accessory pore
l. left spicule
m. median preanal papilla
r. right spicule

median pre-cloacal papilla and two pairs of post-cloacal papillae, while a group of minute papillae is situated near to the tip of the tail. The left spicule is 2.45–2.8 mm long, the right 0.475–0.75 mm. The eggs have thick shells; they measure 30–37 by 11–15 μm and contain larvae when laid.

Spirocerca arctica Petrow, 1927 occurs in the dog, fox and wolf (*Vulpes lagopus*) in Northern Russia.

Life-cycle. The eggs of *S. lupi* are passed in the faeces of the host and hatch only after they have been ingested by a suitable coprophagous beetle (*Scarabeus sacer*, *Akis*, *Atenchus*, *Geotrupes*, *Gymnopleurus*, *Cauthon* spp. and others). The larvae develop to the infective stage and become encysted in the beetle, chiefly on the tracheal tubes. If such beetles are swallowed by an unsuitable host, the larval worms become encysted again in the oesophagus, mesentery or other organ of this host. Such cysts have been found in numerous amphibia, reptiles, domestic and wild birds and small mammals such as hedgehogs, mice and rabbits. The larvae can transfer from one paratenic host to another. The final host may become infected by ingesting either beetles that contain infective larvae or other animals in which the encysted forms occur. Larvae encysted in the viscera of chickens may be particularly important in the transmission of infection to dogs when these are fed with discarded viscera. On being liberated in the stomach the larvae penetrate into the stomach wall and, reaching the arteries, migrate in the walls of the gastric and gastroepiploic arteries to the coeliac artery and thence to the aorta, reaching the upper thoracic aorta, about half way between the diaphragm and aortic arch, in about three weeks. After two and a half to three months in the aorta the majority migrate to the oesophagus between 102 and 124 days after infection, traversing the connective tissue of the thoracic cavity in this area. Some may enter veins and reach other organs. The prepatent period of *S. lupi* is five to six months.

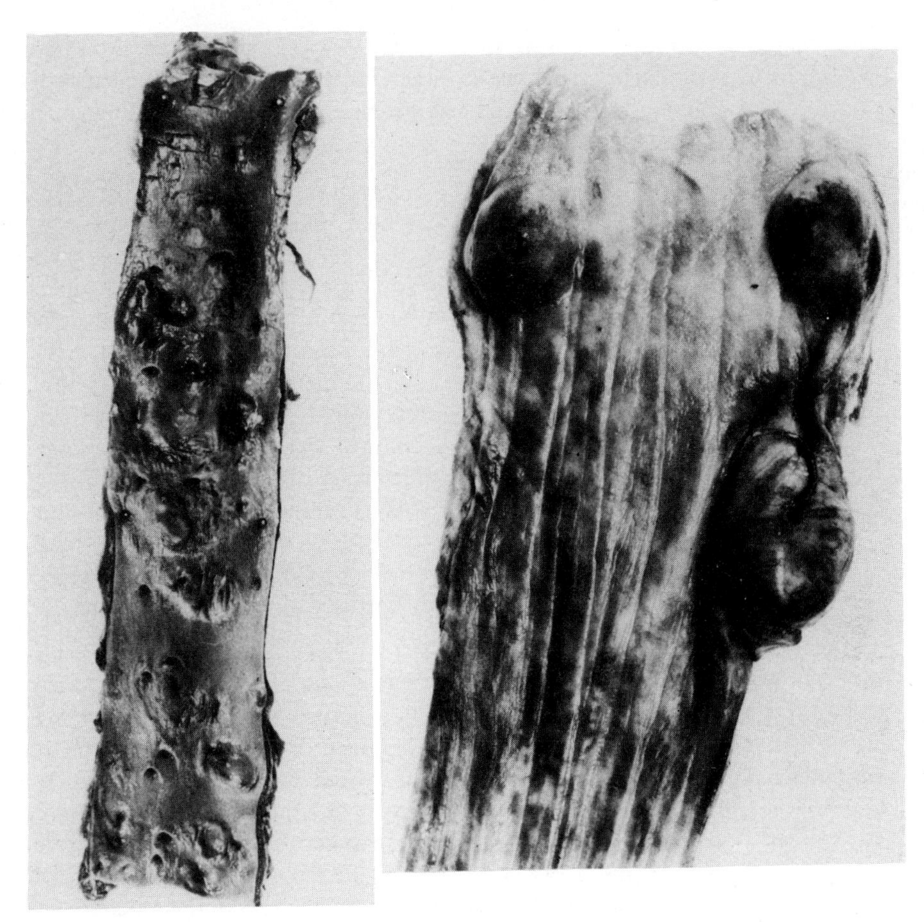

Fig 1.161 *Spirocerca lupi* lesions in dog. *Left*, Aneurysm formation in the wall of the aorta due to larvae stages of *S. lupi*. *Right*, Nodule formation in wall of oesophagus containing adult worms. (*Dr Lucie Long and Faculty of Veterinary Science, University of Pretoria, Republic of South Africa*)

Pathogenesis. Infection with *S. lupi* may be very common in tropical areas often being found in 10 to 80 or even 100% of dogs in these areas. Also, Pence and Stone (1978) in Texas found aortic lesions in 82% of coyotes (*Canis latrans*), 35% of bobcats (*Felis rufus*), one of five grey foxes (*Urocyon cinereoargenteus*) and one of two red foxes (*Vulpes vulpes*). The importance of coyotes as reservoir hosts is not yet known since adults were found in the oesophagus of only 7% of these animals. The presence of adults in the oesophagus of coyotes may, however, be seasonal. The migrating larvae produce haemorrhages, inflammatory reactions and necrosis, as well as purulent streaks or abscesses in the tissues in which they penetrate. These lesions heal rapidly after the larvae have passed on, but stenosis of the vessels may remain. The adult parasites produce nodules in the oesophagus, stomach and aorta, which contain cavities that harbour one or more of the worms.

The lesions and scarring of the thoracic aorta are pathognomonic for infection with *S. lupi*. The intima is rough and granular and eosinophilic granulomatous reactions form nodules around parasitic cavitations in the media. There is degeneration of the elastic tissue and replacement by collagen with eventual calcification and ossification of muscle. All this results in stenosis of the aorta or formation of aneurysms with possible rupture of the vessel. In severe infections the nodular mass in the oesophageal wall may become large and pedunculated and protrude into the lumen of the oesophagus. If marked this mass leads to an interference with digestion and occasionally persistent vomiting and emaciation.

A serious complication of *S. lupi* infection is that it may be an important factor in the development of a malignant tumor in the oesophagus (Bailey 1963). The fibroblasts in the oesophageal nodule are of an embryonal nature but the mechanism of transformation of the oesophageal granuloma to a sarcoma (fibrosarcoma or osteosarcoma) is not known. The majority of cases of oesophageal sarcoma have so far been associated with *S. lupi* infection. The neoplasm may be extensive and may metastasize to the lungs and elsewhere. The neoplasm has been found mainly in hounds, setters and pointers, but its general incidence in other breeds has yet to be determined.

A long-standing complication is hypertrophic pulmonary osteoarthropathy of the long bones. Spondylitis of the adjacent posterior thoracic vertebrae may be seen, but the factors responsible for its development are not known. It may be due to direct irritation of the periosteal tissue by parasites in this region. Also, blockage of the intervertebral arteries by lesions in the aorta may contribute to the development of spondylitis.

Other complications of *S. lupi* infection include pyaemic nephritis, derived from septic foci in the oesophageal lesions and aplastic anaemia.

Clinical signs. The oesophageal lesions may cause interference with deglutition, respiration and circulation. These produce vomiting which is at times so persistent that the animal is unable to retain its food and loses condition rapidly. Sometimes worms may be passed in the vomit.

Haemorrhage from the oesophageal lesions may cause anaemia and haemoptysis may be seen. Rupture of the surrounding weakened oesophageal wall and escape of ingesta into the pleural cavity causes pleuritis.

In less serious cases the lesions cause difficulty in swallowing or interfere mechanically with the action of the stomach. Aortic infection is usually not observed until sudden death is caused by rupture. Where a neoplasm has developed there is general wasting and emaciation, perhaps thickening of the long bones in long-standing cases, and more specific signs depending on the site of the metastases of the tumour.

Diagnosis. Infections in the alimentary canal can be diagnosed by finding the eggs in the faeces or the vomit, but eggs are not passed unless the granuloma has acquired an opening into the lumen of the organ. It may be difficult to differentiate this infection from one with *Physaloptera* in the stomach. The eggs are similar to those of other spirurids and are heavier than, for example, strongyle eggs, and require solutions of high specific gravity (e.g. sodium nitrate) (SG 1.360).

Aortic infection can be diagnosed only tentatively on account of the presence of stenosis or aneurysm. Contrast radiography is useful to demonstrate abnormalities in the oesophagus and aorta. Endoscopy may be of use.

Treatment. McGaughey (1950) and Rao (1953) found that 20 mg/kg of diethylcarbamazine, given daily, caused the signs to disappear in four to ten days. However, this treatment may cause only suppression of egg production and not elimination of the worms (Vaidyanathan 1952). Darne and Webb (1964) found disophenol (2,6-di-iodo-4-nitrophenol) effective in doses of 1 ml/5 kg body weight. Levamisole and albendazole may be of value. When the condition has not progressed too far, thoracotomy and resection of the lesions in the oesophagus may be effective.

Prophylaxis. Infected animals should be isolated and their vomit and faeces disposed of, so that the infection is not allowed to spread. Healthy animals should be prevented as far as possible from eating dung beetles, frogs, mice, hedgehogs, lizards etc., which may carry the encysted larvae.

Genus: Ascarops v. Beneden, 1873 (syn. Arduenna)

Ascarops strongylina (Rudolphi, 1819) occurs in the stomach of the pig, feral pig and wild boar in most parts of the world. The male is 10–15 mm long and the female 16–22 mm and red in colour.

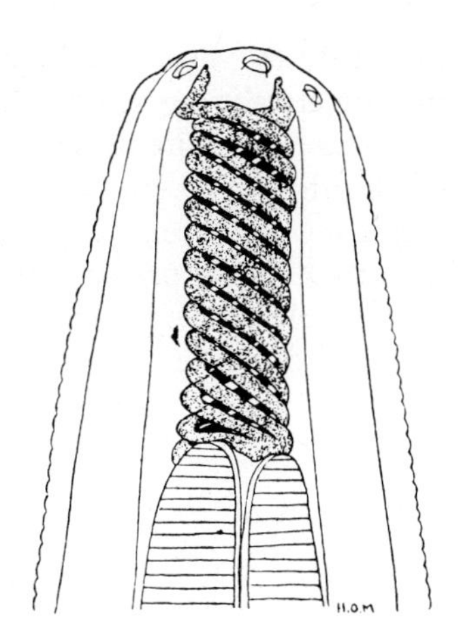

Fig 1.162 Lateral view of anterior end of *Ascarops strongylina*.

A cervical ala is present only on the left side of the body. The pharynx is 0.083–0.098 mm long and its wall is strengthened by thickenings in the form of a triple or quadruple spiral. The right caudal ala of the male is about twice as large as the left and there are four pairs of pre-cloacal papillae and one pair of post-cloacals, all situated asymmetrically. The left spicule is 2.24–2.95 mm long and the right 0.46–0.62 mm. The eggs measure 34–39 by 20 μm and have thick shells surrounded by a thin membrane which produces an irregular outline. They contain larvae when laid.

Ascarops dentata (von Linstow, 1904) is a larger species than the preceding one, the male being 25 mm and the female 55 mm long. It occurs in pigs in Indo-China and Malaysia. The buccal capsule is provided with a pair of teeth anteriorly.

Genus: Physocephalus Diesing, 1861

Physocephalus sexalatus (Molin, 1860) occurs in the stomach of the pig and has also been recorded from some other animals, for instance, hares and rabbits. The male is 6–13 mm long and the female 13–22.5 mm. The cuticle of the anterior extremity is slightly inflated in the region of the pharynx. This inflation is followed by three cervical alae on either side. The cervical papillae are very asymmetrically situated. The mouth is small and toothless. The pharynx is 0.263–0.315 mm long and its walls are strengthened by a single spiral thickening which breaks up into complete rings in the middle portion. The caudal alae of the male are narrow and symmetrical, and there are four pairs of pre-cloacal papillae and the same number of post-cloacal papillae. The left spicule measures 2.1–2.25 mm and the right 0.3–0.4 mm. The eggs have thick shells, contain larvae when laid and measure 34–39 by 15–17 μm.

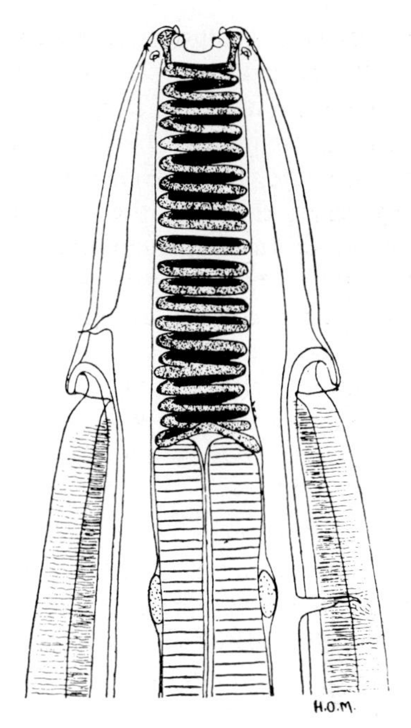

Fig 1.163 Dorsal view of the anterior end of *Physocephalus sexalatus*.

Physocephalus cristatus (Seurat, 1912) occurs in the dromedary.

Life-cycles of Ascarops and Physocephalus spp. Eggs are passed in the faeces of the host where they are swallowed by coprophagous beetles. A large number of beetles are known to act as intermediate hosts, including species of *Scarabeus*, *Phanaeus*, *Gymnopleurus*, *Geotrupes* and *Onthophagus*. In Czechoslovakia larvae of *A. strongylina* and *P. sexalatus* were found in 85% and 26% of *Geotrupes stercorarius* and 44% and 9.1% of *G. s. monticola*, respectively, when these were collected from enclosures in which wild pigs were kept (Páv & Zajiček 1972). The larvae develop in these to the infective stage in 28 days or more. Pigs become infected by eating such beetles. Also, as in the case of *Spirocerca* the larvae become encysted, chiefly in the walls of the alimentary canal, in unsuitable hosts which swallow infected beetles. Cram (1936) found that birds which had eaten beetles on a field used for pigs harboured numerous cysts containing these larvae, but their health was not affected. In the pig the larvae penetrate deeply into the mucosa of the stomach and reach the adult stage in about six weeks.

Pathogenesis. Ascarops and *Physocephalus* are common stomach worms of pigs, although the prevalence of infection varies greatly. However, surveys have demonstrated the infection in 56 and 44% of pigs in Georgia, USA, 84 and 67% of pigs in Hissar, India, and 51 and 67% of pigs examined in Czechoslovakia. The parasites cause no marked disturbance unless they occur in fairly large numbers or the resistance of the animal is weakened by another factor. The worms then irritate the mucosa and produce an inflammation.

Clinical signs. Affected animals, especially young ones, show signs of a chronic or acute gastritis. They lose their appetite, but usually have a marked thirst. Growth is retarded or emaciation and even death may occur.

Post mortem, the gastric contents are small in quantity and there is much mucus. The mucosa, particularly in the fundus region, is reddened and swollen, or it may be covered with pseudomembranes, underneath which the tissues are markedly reddened and ulcerated. The worms are found free or partly embedded in the mucosa.

Diagnosis can be made by finding the eggs of the worms in the faeces.

Treatment. Of the older anthelmintics, *carbon disulphide* at a dose of 0.1 ml/kg given after 36–48 hours starvation is 83–100% effective; *sodium fluoride* is effective against *A. strongylina* and *P. sexalatus* as a 1% mixture in food. Of the more recent anthelmintics, dichlorvos is effective and the benzimadazole compounds such as oxibendazole should be tried.

Prophylaxis. Infected pigs should not be allowed to disseminate the eggs with their faeces and the latter should therefore be disposed of in a suitable way. Healthy pigs should be prevented as far as possible from ingesting dung beetles.The measures advocated for the prevention of *Ascaris* will be effective against the spirurids of pigs to a certain extent, but dung beetles may fly a substantial distance.

Genus: Simondsia Cobbold, 1864

Simondsia paradoxa (Cobbold, 1864) occurs in the stomach of pigs in Europe and Asia. It has also been reported in the stomach of a warthog in the Republic of Mali. The worm is peculiar in that the posterior parts of the bodies of the females, which are globular in shape, are lodged in small cysts in the stomach wall, while the anterior, slender portions protrude. The gravid female is 15 mm long and has lateral cervical alae and a large dorsal and a large ventral tooth. The eggs are oval or ellipsoidal and are 20–29 μm long. The male has a spirally-coiled tail and a cylindrical body, 12–15 mm long, and occurs free or partly embedded in the mucosa. The life-cycle is unknown. The females in mucosal crypts cause nodules and chronic gastritis.

Genus: Gongylonema Molin, 1857

Gongylonema pulchrum Molin, 1857 (syn. *G. scutatum*) occurs in the sheep, goat, cattle, pig, zebu, buffalo and less frequently the horse, camel, donkey and wild boar. It also occurs in man, particularly in the oral epithelium but also subcutaneously. It is found in most parts of the world. The worm inhabits the oesophagus, where it lies in zigzag fashion embedded in the mucosa or submucosa. In ruminants it may also occur in the rumen. Males are up to 62 mm long and females up to 145 mm. The cuticle of the anterior end bears a number of round or oval thickenings of various sizes. The cervical alae are well developed. The lips are small and there is a short pharynx with simple walls. The tail of the male is alate, somewhat asymmetrical, and bears a number of papillae which are also asymmetrically

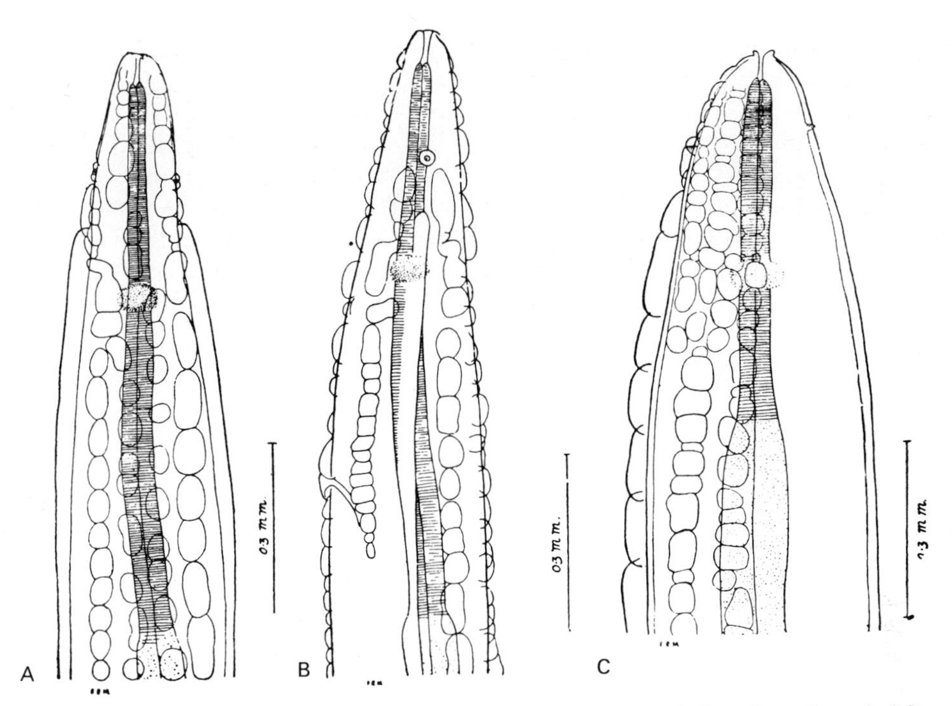

Fig 1.164 A, Dorsal view of the anterior end of *Gongylonema pulchrum*. B, Lateral view of anterior end of *Gongylonema pulchrum*. C, Dorsal view of the anterior end of *Gongylonema verrucosum*.

arranged. The left spicule is slender and 4–23 mm long. The right is 0.084–0.18 mm long and stout. A gubernaculum is present. The vulva opens posteriorly and the eggs measure 50–70 by 25–37 μm.

Life-cycle. The eggs of the parasite are passed in the faeces of the host and hatch after they have been swallowed by coprophagous beetles of the genera *Aphodius*, *Onthophagus*, *Blaps*, *Caccobius*, and others (over 70 species have been incriminated), in which the larvae develop to the infective stage in about 30 days. The small cockroach *Blatella germanica* can be infected experimentally. Infection of the final host takes place by ingestion of the infected beetles. Larvae will emerge spontaneously from cockroaches which fall into water, but it is improbable that such larvae would be important as a source of infection. The route of migration is not known in all animals but in the guinea-pig, Alicata (1935) found larvae embedded in the wall of the gastro-oesophageal region. He suggested that larvae excysted in the stomach and then migrated anteriorly to the oral cavity, finally migrating to the wall of the oesophagus.

Gongylonema verrucosum (Giles, 1892) occurs in the rumen of sheep, goat, cattle, deer and zebu in India, the USA and South Africa. The worms have a reddish colour when fresh. Males are 32–41 mm long and females 70–95 mm. This speceis has a festooned cervical ala as well as cuticular bosses on the left side only. The left spicule is 9.5–10.5 mm long and the right 0.26–0.32 mm.

Gongylonema mönnigi Baylis, 1926 occurs in the rumen of the sheep and goat in South Africa. Males are 42 mm long and females 102–113 mm. It resembles *G. verrucosum*, but the cervical ala is not festooned. The left spicule is about 15 mm long and the right 0.26 mm. The gubernaculum also differs from that of the preceding species.

Gongylonema ingluvicola Ransom, 1904 and **G. crami** Smit, 1927 occur in the crop of the

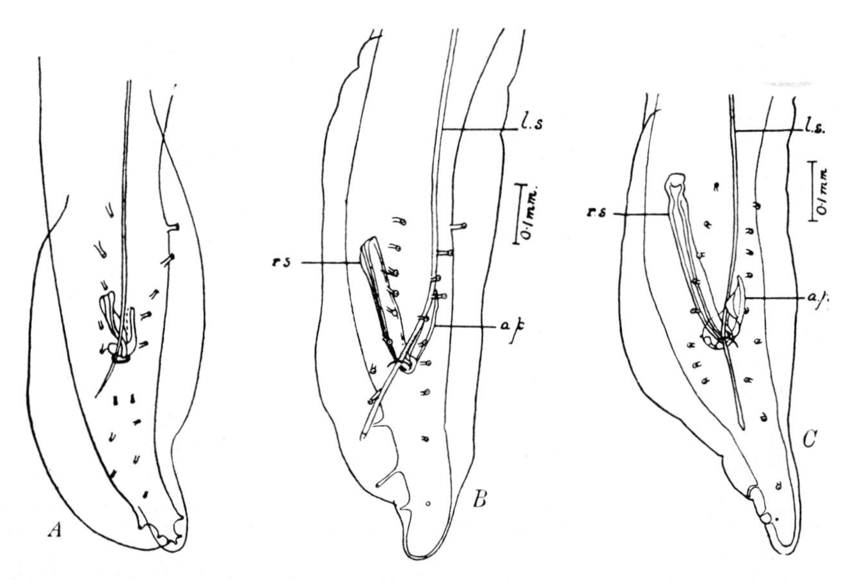

Fig 1.165 Ventral views of the posterior end of male *Gongylonema pulchrum* (A), *G. mönnigi* (B) and *G. verrucosum* (C).
a.p. accessory piece
l.s. left spicule
r.s. right spicule

fowl in North America, Asia (India, Philippines, Taiwan), Europe and Australia. **G. sumani** Bhalerao, 1933 occurs in the crop of the domestic fowl in India (Uttar Pradesh). The eggs measure about 58 by 35 μm. *Blatella germanica* may be infected with this worm.

The adult forms are not markedly pathogenic. The parasites lie embedded in the epithelium. There is a slight chronic inflamatory reaction with hypertrophy and cornification of the epithelium. *G. ingluvicola* may burrow into the crop and is said to cause severe lesions in heavy infections. No treatment is known.

FAMILY: ACUARIIDAE SEURAT, 1913

The cuticle of the anterior part of the body is ornamented with 'cordons'—cuticular ridges or grooves—or epaulette-like thickenings. The cordons may be either recurrent, i.e. they run down the body and turn back forwards again, or non-recurrent, i.e. they do not do this. They may or may not anastomose. For further details see Cram (1927). The lips are usually small and triangular in shape and the pharynx is cylindrical. Parasites in the walls of the gizzard, proventriculus, oesophagus or crop of birds.

Genus: Cheilospirura Diesing, 1861

Cheilospirura hamulosa (Diesing, 1851) (Syn. *Acuaria hamulosa*) occurs in the gizzard of the fowl and turkey in most countries, especially those of Asia, the Americas and Europe. Males are 10–14 mm long and females 16–29 mm. The cordons are double cuticular ridges with an irregular outline and extend far back along the body. The male has four pairs of pre-cloacal and six pairs of post-cloacal papillae. The left spicule is slender and 1.63–1.8 mm long, the right flattened and 0.23–0.25 mm long. The vulva is situated just behind the middle and the eggs measure 40–45 by 24–27 μm and are embryonated when passed in the faeces.

The eggs pass out in the faeces of the host and hatch after they have been swallowed by the intermediate hosts, which are grasshoppers (*Melanoplus*, *Oxya nitidula*, *Spathosternum prasiniferum*), various beetles and weevils. In these hosts the infective larva develops in three weeks and the final host acquires the infection by ingesting such insects. The prepatent period is three weeks.

The parasites live underneath the horny lining of the gizzard, producing soft nodules in the musculature and thus weakening it.

Mild infections are usually not noticed, and the location of the worms does not allow them to be seen at autopsy. Severe infections produce emaciation, droopiness, weakness and anaemia. Cases have been described in which the gizzard was weakened to such a degree that a rupture in the form of a large sac developed.

In mild infections the worms are noticed post mortem only if the horny lining of the gizzard is removed. They are found in soft, yellowish-red nodules which are most frequently seen in the thinner parts of the wall. In severe cases the horny lining may be partly destroyed, and the worms are found below the necrotic material in the altered musculature.

Several species of *Cheilospirura* may occur in fowls and turkeys and their eggs are difficult to distinguish from one another. Spirurid eggs in the faeces would lead to a tentative diagnosis, which has to be confirmed at autopsy of a selected case.

No satisfactory remedy is known, but carbon tetrachloride, tetrachloroethylene and oil of chenopodium have been recommended in the past. However, dichlorvos, levamisole and the benzimidazole anthelmintics should be tried.

Prophylaxis is difficult in the case of birds that run free, since they cannot be prevented from eating the intermediate hosts. When the infection is troublesome the birds have to be kept confined on bare ground and the entrance of insects into the runs, as well as meal beetles and weevils into the food, must be prevented.

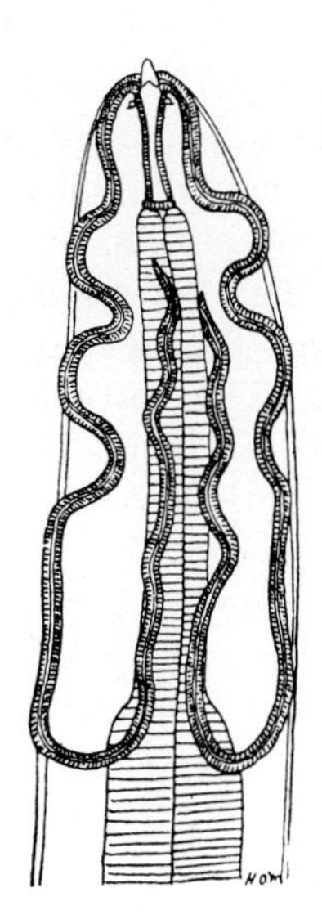

Fig 1.166 Lateral view of the anterior end of *Dispharynx (Acuaria) spiralis.*

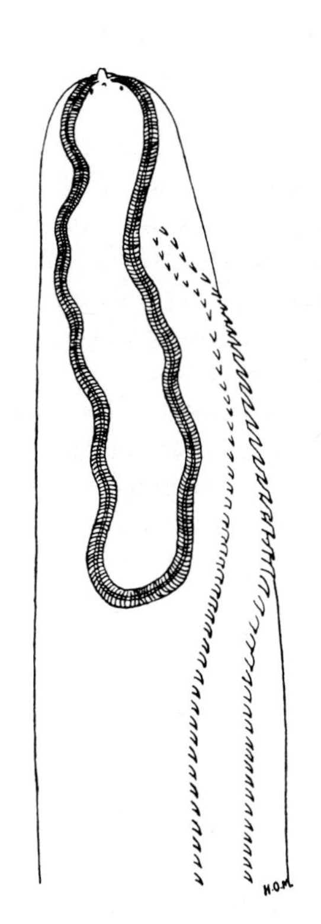

Fig 1.167 Lateral view of anterior end of *Echinuria (Acuaria) uncinata.*

Genus: Dispharynx Railliet, Henry and Sisoff, 1912

Dispharynx spiralis (Molin, 1858) (Syn: *Acuaria spiralis*) occurs in the walls of the proventriculus and oesophagus, more rarely in the intestine, of the fowl, turkey, pigeon, guinea-fowl, pheasant and other birds in many countries especially those of Africa, Asia and the Americas.

Males are 7–8.3 mm long and females 9–10.2 mm. The cordons have a sinuous course and are recurrent, but do not anastomose. The male has four pairs of pre-cloacal and five pairs of post-cloacal papillae. The left spicule is slender and 0.4–0.52 mm long, the right 0.15–0.2 mm long and boat-shaped. The vulva is situated in the posterior part of the body. The eggs measure 33–40 by 18–25 μm; they have thick shells and contain larvae when laid.

The eggs hatch after having been swallowed by the intermediate host, which is an isopod (*Porcellio laevis*, *P. scaber* and *Armadillidum vulgare*) also *Oxya nitidula* and *Spathosternum prasiniferum*. The larvae develop in the body cavity of the isopod, and when the latter is eaten by a suitable bird the parasite grows to maturity in it.

The effects of the parasites vary with the severity of the infection. If worms are present they apparently do not penetrate into the mucosa, which may then show inflammation and thickening only; but in severe infections deep ulcers, in which the anterior extremities of the worms are embedded, may be seen in the proventriculus. There is an extensive destruction of the glands of this organ and a marked cellular infiltration of the underlying tissues. The affected birds, particularly young ones, rapidly lose weight in spite of a voracious appetite and become very weak and anaemic. This parasite may cause severe losses.

Diagnosis is as in the case of *C. hamulosa*.

No satisfactory remedy is known. Tetrachloroethylene or carbon tetrachloride have been recommended in the past. The newer compounds suggested for *C. hamulosa* (above) should be tried.

Where this parasite is troublesome the birds should be confined on bare ground and measures taken to combat the intermediate hosts.

Genus: Echinuria Soloviev, 1912

Echinuria uncinata (Rudolphi, 1819) (syn. *Acuaria uncinata*) occurs in the oesophagus, proventriculus, gizzard and small intestine of the duck, goose, swan and wild aquatic birds. Cosmopolitan. Males are 8–10 mm long and females 12–18.5 mm. The cordons are non-recurrent and they anastomose in pairs. The cuticle bears also four longitudinal rows of spines. The left spicule is 0.706 mm long and the right 0.208 mm. There are four pairs of pre-cloacal and four pairs of post-cloacal papillae; the pre-cloacal papillae stand in two groups of two on either side. The vulva is situated near the posterior extremity and the eggs measure 37 by 20 μm.

The eggs are passed in the faeces of the bird and are swallowed by *Daphnia pulex* and *Gammarus*, in which they hatch and the infective larvae develop. When suitable aquatic birds ingest the infected intermediate hosts the parasites develop to maturity.

The worms penetrate into the wall of the alimentary canal, causing a marked inflammation and the formation of nodules which have a caseous content. In the proventriculus and the gizzard the nodules may become so large that they cause mechanical interference and even obstruction to the passage of food. The birds become dull, the feathers are ruffled and feeding may stop completely. Sudden death is known to occur.

Diagnosis is as in the case of *C. hamulosa*.

Treatment is unknown but probably the compounds suggested for *C. hamulosa* (above) should be tried.

Prophylaxis will obviously be difficult; however, exclusion of wild water fowl from breeding areas and drainage of stagnant pools or treatment of the water with insecticides to kill the *Daphnia* and *Gammarus* are appropriate control measures.

Several other species of related parasites occur in poultry and occasionally produce disease. Those described above are the commonest and the most important.

FAMILY: TETRAMERIDAE TRAVASSOS, 1924

Species of this family are related to the *Acuariidae*, but have no cordons. They show marked sexual dimorphism, the male being white and filiform, with or without spines on the cuticle and tail end, and the female globular (*Tetrameres*) or coiled (*Microtetrameres*).

Genus: Tetrameres Creplin, 1846

This genus is remarkable for the fact that the mature female is almost spherical in shape, blood-red in colour and lies embedded in the proventricular glands of birds. The male is slender and its body cuticle is usually armed with four rows of spines; it is mostly found free in the lumen of the proventriculus, but it may follow the female into the gland temporarily for copulation.

Tetrameres americana Cram, 1927 occurs in the proventriculus of the fowl and turkey, and has been recorded from the USA and South Africa. Males are 5–5.5 mm long and females 3.5–4.5 by 3 mm. The female is subspherical and has four deep grooves in the regions of the longitudinal lines, while the anterior and posterior extremities project as conical appendages. The eggs are 50–60 µm by 30 µm, thick-shelled and embryonated when laid.

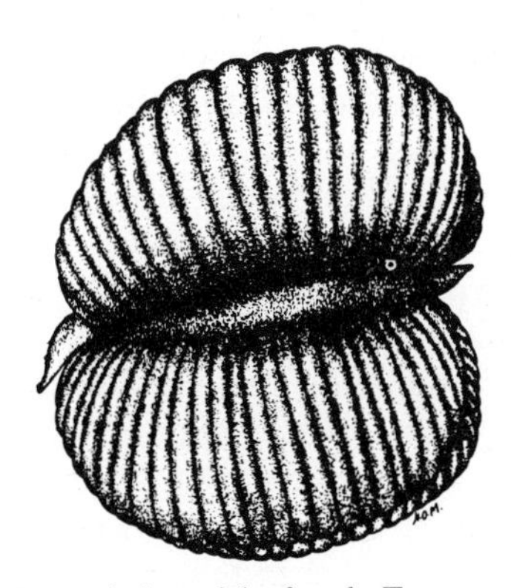

Fig 1.168 Lateral view of the female *Tetrameres americana*.

The eggs are passed in the faeces of the bird and hatch after they have been swallowed by a suitable orthopteran insect (*Melanoplus femurrubrum*, *M. differentialis* and *Blatella germanica*). Infection of the final host occurs through ingestion of the infected intermediate host and, according to Cram (1927), the males and females migrate into the proventricular glands, where they copulate. The males then leave the glands and die. The females contain eggs 35 days after infection, but reach their full size only after three months.

Tetrameres fissispina (Diesing, 1861) occurs in the duck, pigeon, fowl, turkey and wild aquatic birds and has a wide distribution. Males are 3–6 mm long and females 2.5–6 by 1–3.5 mm. The intermediate hosts are the water crustacea *Daphnia pulex* and *Gammarus pulex*, and in the final host the sexes are stated to copulate before the females migrate into the glands.

Tetrameres crami Swales, 1933 occurs in domestic and wild ducks in North America. Males are up to 4.1 mm long and females 1.5–3.5 by 1.2–2.2 mm. The intermediate hosts are amphipods: *Gammarus fasciatus* and *Hyalella knickerbockeri*.

Tetrameres confusa Travassos, 1919 occurs in the proventriculus of fowl pigeon and other birds in Brazil. Intermediate hosts are probably similar to those for *T. fissispina*.

Tetrameres mohtedai Bahlerao and Rao, 1944 occurs in fowl in India and south-east Asia. The intermediate hosts are cockroaches and grasshoppers such as *Spathosternum prasiniferum* and *Oxya nitidula* but Lim (1975) demonstrated that *Setamorpha nutella* is an important intermediate host. The larvae of this small moth develop in stored grain but also in faecal material in litter where they ingest eggs of the nematode. Fowl become infected through ingesting the larvae or the adult moths which hide beneath the surface of the litter during the day.

Tetrameres pattersoni (Cram, 1933) occurs in quail and the intermediate hosts are grasshoppers and cockroaches.

The females suck blood, but the greatest damage is done when the young worms migrate into the wall of the proventriculus, causing marked irritation and inflammation, which may kill chicks. The birds are anaemic and emaciated. At autopsy the adult females can be seen from the outside of the proventriculus as dark objects in the dept of the tissues. On post mortem, the wall of the proventriculus is thickened; there is extensive glandular necrosis and exfoliation of the proventriculus wall.

Piperazine adipate and carbon tetrachloride given for at least three days are effective against the immature and adult worms respectively. The anthelmintics may be given in combination. Iodophene (0.02 mg/kg for three to four days) is highly effective against both adults and immatures. Levamisole, haloxon and thiabendazole are all fairly effective.

Birds confined on bare ground will be less liable to acquire the parasites than those running free. Young chicks particularly should be prevented from eating the intermediate hosts and can be reared on wire floors for this purpose. Screens should be used in poultry houses to keep out moths which are attracted to lights.

SUPERFAMILY: PHYSALOPTEROIDEA SOBOLEV, 1949

Spirurida with pseudolabia well developed; pseudolabia armed with one or more teeth. Families of importance: *Physalopteridae*, *Gnathostomatidae*.

FAMILY: PHYSALOPTERIDAE LEIPER, 1909

The cuticle usually forms a collar-like projection around the anterior extremity. The lips are simple and bear small teeth on their medial surfaces. A pharynx is absent. Genus of importance: *Physaloptera*.

Genus: Physaloptera Rudolphi, 1819

Physaloptera praeputialis von Linstow, 1889 occurs in the stomach of the cat and wild *Felidae* in most parts of the world. Males are 13–40 mm long and females 15–48 mm. The worms are stout and the cuticle in both sexes is posteriorly extended to form a sheath which projects beyond the caudal end of the body. In the fertilized female the vulva is covered by a conspicuous ring of brown cement material. Each lip bears at the middle of its free edge a set of three flattened, internal teeth, and externally to these a single conical external tooth of about the same height as the internals. The male tail bears large lateral alae, joined together anteriorly across the ventral surface. There are four pairs of pedunculated papillae, three sessile papillae on the anterior lip of the cloaca and five pairs of sessile post-cloacal papillae. The left spicule is 1–1.2 mm long and the right 0.84–0.9 mm. The eggs measure 49–58 by 30–34 μm.

Several other species of *Physaloptera* have been described from domestic animals and wild animals. The following may be particularly noted: **Physaloptera rara** Hall and Wigdor, 1918 is known from the stomach and duodenum of dogs and wild Canidae and Felidae in the United States. **P. canis** Mönnig, 1928 occurs in the stomach of the dog and cat in South Africa and elsewhere. **P. felidis** Ackert, 1936 occurs in the stomach and duodenum of cats in the USA. **P. pseudopraeputialis** Yutuc, 1953 occurs in the stomach and also larynx of cats and coyotes in the Philippines and the USA. **P. tumefasciens** Henry and Blanc, 1912, and **P. dilitata** Rudolphi, 1819 occur in the stomach of simian primates in Asia, Africa and South America. **P. (Abbreviata) caucasia** Linstow, 1902, normally a parasite of the oesophagus, stomach and small intestine of simian primates in south-east Europe,

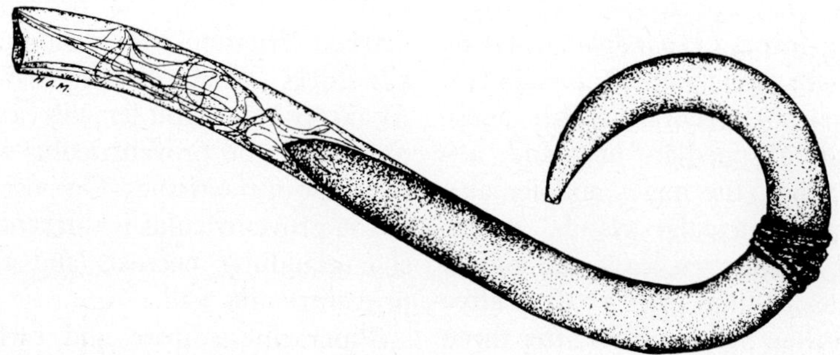

Fig 1.169 Lateral view of the female *Physaloptera praeputialis*.

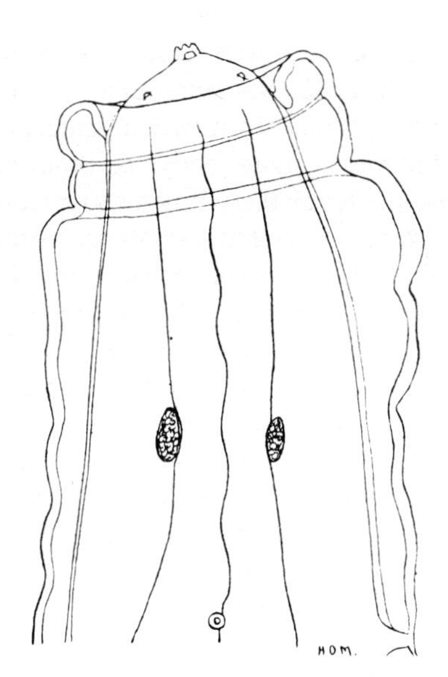

Fig 1.170 Lateral view of anterior end of *Physaloptera canis.*

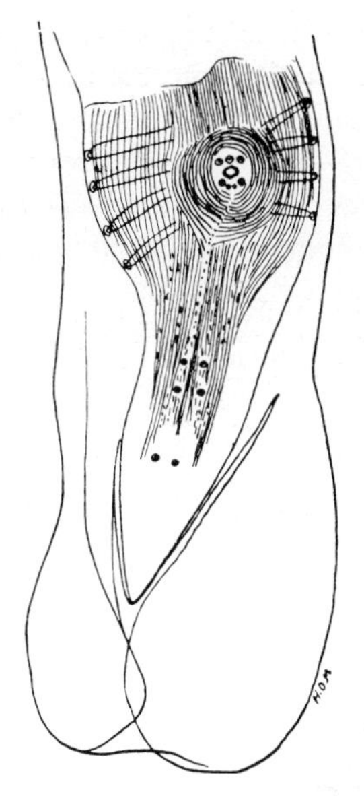

Fig 1.171 Ventral view of hind end of male *Physaloptera canis.*

south-west Asia and Africa, may also infect man. **P. (A.) poicilometra** (Sandground, 1936) occurs in simian primates in Africa. **P. maxillaris** Molin, 1860 is found in the skunk (*Mephitis* spp.) and other animals such as *Procyon* and *Mustela* etc. in North and South America. **P. clausa** Rudolphi, 1819, **P. erinacea** Linstow, 1904 and **P. dispar** Linstow, 1904 occur in hedgehogs in the USSR, Europe and Africa. In birds, **P. alata** Rudolphi, 1891 has been reported from the gizzard and intestine of the dove and other birds in Africa, Asia, South America and Europe (Austria) and **P. gemina** Linstow, 1899 from the fowl (and also the cat) in Africa and Egypt.

Life-cycle. Orthoptera and beetles may be intermediate hosts of *P. rara* and *P. praeputialis.* Petri and Ameel (1950) found that the third-stage larvae of *P. rara* will develop in the cockroach *Blatella germanica*, the field cricket *Gryllus animilis*, the flour beetle *Tribolium confusum* and the ground beetles, *Harpatalus* spp., and that the third-stage larvae of *P. praeputialis* will develop in *Blatella germanica*, *Gryllus animilis* and *Centophilus* spp. *Crotalus viridis* have been found to be naturally infected paratenic hosts and frogs and mice have been experimentally infected. *P. maxillaris* develops in field crickets *Gryllus pennsylvanicus* and the third-stage larvae may overwinter in the skunk before developing to adults (Cawthorn & Anderson 1976). The prepatent period of *P. rara* in the definitive host is 56–83 days and that of *P. maxillaris* is 41–45 days.

Pathogenesis. The parasites, which occur in the stomach, are usually firmly attached to the mucosa, on which they feed, or they may also suck blood. Apparently they occasionally change their site of attachment and leave numerous small oedematous wounds which may continue to bleed. In this way the mucosa becomes eroded and highly inflamed with increased mucus production. There is vomiting and anorexia, the faeces may be dark and tarry and the animals lose weight. A *Physaloptera* sp. in badgers in the USA (Ehlers 1931) has been reported to cause loss of weight, shaggy fur, a quarrelsome nature and tarry faeces. Advanced clinical signs included paresis.

Diagnosis may be made from the clinical signs and the presence of spirurid eggs in the faeces, which must be differentiated from those of *Spirocerca lupi.*

Treatment. Ehlers (1931) was successful in expelling the worms from badgers by the administration of about 0.75 ml carbon disulphide after fasting for 18–24 hours; however, compounds such as dichlorvos and the broad-spectrum benzimidazole anthelmintics should be tried.

Prophylaxis. No definite recommendation can be made but, because the intermediate hosts are the beetles and Orthoptera mentioned above, steps should be taken to prevent the hosts from ingesting these insects.

FAMILY: GNATHOSTOMATIDAE RAILLIET, 1895

The lips are large and trilobed and on their medial surfaces the cuticle forms toothlike ridges which interlock with those of the opposite lip. Genus of importance: *Gnathostoma*.

Genus: Gnathostoma Owen, 1836

Gnathosoma spinigerum Owen, 1836 occurs in the stomach of the cat, dog, mink, polecat and several wild Carnivora and as an erratic parasite under the skin of man. Males are 10–25 mm long and females 9–31 mm. There is a large head-bulb, containing four submedian cavities or 'ballonets' which each communicate with a cervical sac. These cavities contain a fluid and apparently serve to fix the head in the tissues of the host by contraction of the cervical sacs and consequent swelling of the head-bulb. The cuticle of the head-bulb is armed with six to 11 transverse rows of hooks, while the anterior two-thirds of the body bears large, flat, cuticular spines with denticulate posterior edges. The ventral, caudal region of the male bears small spines and four pairs of large pedunculate papillae as well as several smaller sessile ones. The left spicule is 1.1–2.63 mm long and the right 0.4–0.8 mm. The vulva opens 4–8 mm from the posterior end. The eggs are oval with a thin cap at one pole; they have a greenish shell ornamented with fine granulations, are passed in the one-cell or morula stage and measure about 69 by 37 μm.

Life-cycle. Two intermediate hosts are required. The eggs hatch in the water in four days or longer. The larvae are sheathed and motile and die in a few days unless they are ingested by a *Cyclops*. In the latter they grow to 0.372 mm long in seven days or more, developing a definite cephalic bulb with four transverse rows of single-pointed spines, a pair of trilobate lips and two pairs of contractile cervical sacs. When the *Cyclops* is eaten by a fresh-water fish (*Clarias batrachus*, *Ophiocephalus striatus*, *Glossogobius giurus*, *Thereapon argenteus*), or by several species of frogs and reptiles, the parasites grow further to an advanced stage, about 4.5 mm long, and become encysted. The third-stage larvae may also encyst in a variety of mammals such as mice, rats, dogs and primates. Such encysted forms, when fed to cats, become mature in the stomach wall in about six months. Apparently the young worms sometimes migrate in the body of the final host, reaching the liver and other organs. The third-stage and advanced third-stage larvae from fish, etc., can penetrate the skin and prenatal transmission has been recorded.

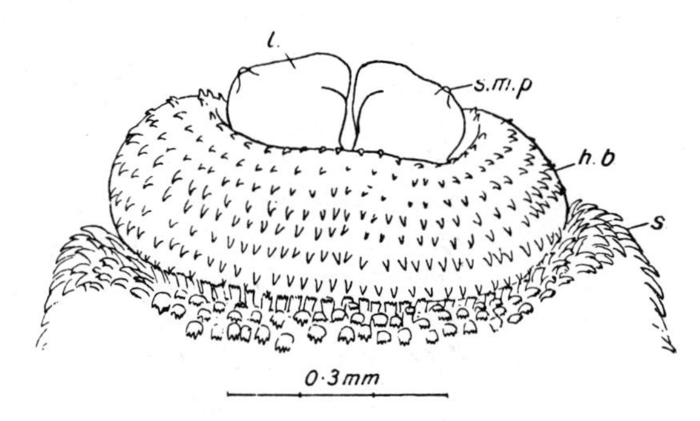

Fig 1.172 Dorsal view of anterior end of *Gnathostoma spinigerum*.
h.b. head bulb
l. lip
s. spines of body
s.m.p. submedian papilla

Pathogenesis. The young worms which pass through the liver cause much destruction of the tissues of this organ, leaving behind them characteristic yellow mosaic markings on the surface and burrows filled with necrotic material in the parenchyma. They also wander through other organs, including the mesentery and diaphragm, and they may enter the pleural cavity. Severe infections may therefore cause various disturbances in these organs. The adult worms penetrate into the wall of the stomach, producing cavities filled with a sanguinopurulent fluid. These cavities later develop into thick-walled cysts, each containing up to nine worms. The connection of the cysts with the stomach lumen may be very small and a similar minute canal may lead into the peritoneal cavity, providing the possibility for the development of peritonitis. Chandler (1925) considers this to be a very pathogenic and fatal parasite of cats, consistently killing its host. Cats with gastric 'tumours' were found only at certain seasons, and they seemed to die out, so that none were found at other times.

Diagnosis. The eggs of the parasites are not always found in the faeces, so that diagnosis may be a difficult matter.

Treatment and prophylaxis. Unknown. The benzimidazole anthelmintics should be tried.

Gnathostoma hispidum Fedtchenko, 1872 occurs in the stomach of the pig in Europe and Asia. Males are 15–25 mm long and females 22–45 mm. The whole body is covered with spines. The left spicule is 0.88–1.29 mm long and the right 0.32–0.4 mm. The eggs measure 72–74 by 39–42 μm.

Life-cycle. The life-cycle of *G. hispidium* has been described by Wang et al. (1976). In water, eggs develop to the second larval stage before hatching. These larvae, when ingested by *Cyclops* spp. and related crustacea, develop to the third stage in 9–10 days. Pigs become infected through ingesting infected *Cyclops*. A second intermediate host is not required in the life-cycle.

The young worms migrate in the abdominal organs of the host, particularly the liver, where they may cause a hepatitis. The adults are found deeply embedded in the gastric mucosa, producing cavities which contain a reddish fluid and are surrounded by an inflamed area. Severe infections may produce a marked gastritis.

Gnathostoma doloresi Tubangui, 1925 occurs in domestic pigs and wild swine in the Far East.

Gnathostoma nipponicum Yamaguti, 1941 has been reported in weasels (*Mustela siberica*) in which it occurs in granulomas in the oesophagus.

GNATHASTOMIASIS IN MAN

Gnathostomiasis in humans is caused by the morphologically, but not sexually, mature stages of *G. spinigerum* and occasionally other species. The parasites do not usually enter the stomach but migrate at random in the body, frequently under the skin but also in the mucous membranes, the eye and the brain where they cause eosinophilic meningitis. In the skin they form abscess pockets or deep cutaneous tunnels. The lesions occur in any peripheral part but the digits and the breasts are frequently affected.

Human infection is most probably acquired by eating inadequately cooked frogs or fish. Surveys of fish sold in markets in the East showed in Thailand that 92% of frogs, 80% of eels and 37% of fish were infected with the larval stages while in Japan 60–100% of *Ophiocephalus argus* were infected (Miyazaki 1954).

REFERENCES

SPIRUROIDEA AND PHYSALOPTEROIDEA

Alicata, J. E. (1935) Early developmental stages of nematodes occurring in swine. *U.S. Dept. Agric. tech. Bull.*, **489**.
Arbuckle, J. B. R. & Khalil, L. F. (1976) *Thelazia* worms in the eyes of British cattle. *Vet. Rec.*, **99**, 376–377.
Bailey, W. S. (1963) Parasites and cancer: sarcoma in dogs associated with *Spirocerca lupi*. *Ann. N.Y. Acad. Sci. U.S.A.*, **108**, 890–932.
Cawthorn, R. J. & Anderson, R. C. (1976) Seasonal population changes of *Physaloptera maxillaris* (Nematoda: Physalopteroidea) in striped skunk (*Mephitis mephitis*). *Can. J. Zool.*, **54**, 522.
Chandler, A. C. (1925) A contribution of the life-history of a Gnathostome. *Parasitology*, **17**, 237.
Cram, E. B. (1927) Bird parasites of the nematode suborders Strongylata, Ascaridata and Spirurata. *U.S. natn. Mus. Bull.*, **140**.

Cram, E. B. (1936) Species of *Capillaria* parasitic in the upper digestive tract of birds. *U.S. Dept. Agric. tech. Bull.*, **516**.
Darne, A. & Webb, J. L. (1964) The treatment of ankylostomiasis and spirocercosis in dogs by the new compound, 2,6-diiodo-4-nitrophenol. *Vet. Rec.*, **76**, 171–172.
Duncan, J. L., McBeath, D. G., Best, J. M. J. & Preston, N. K. (1977) The efficacy of fenbendazole in the control of immature and mature strongyle infections in ponies. *Equine vet. J.*, **9**, 146.
Ehlers, G. H. (1931) The anthelmintic treatment of infestations of the badger with spirurids (*Physaloptera* sp.). *J. Am. vet. med. Ass.*, **31**, 79–87.
Flynn, R. J. (1973) *Parasites of Laboratory Animals*. Ames, Iowa: Iowa State University Press.
Geden, C. J. & Stoffolano, J. C. (1977) *Musca autumnalis* (DeGeer) (Diptera: Muscidae) as vector of *Thelazia* sp. (Bose) (Nematoda: Filaroidea) in Massachusetts. *J. N.Y. ent. Soc.*, **85**, 175.
Halpin, R. B. & Kirkly, W. W. (1962) Experience with an anthelmintic. *Vet. Rec.*, **74**, 495.
Klesov, M. D. (1950) The biology of two nematodes of the genus *Thelazia* Bose, 1819, parasites of the eye of cattle. *Dokl. Akad. Nauk S.S.S.R.*, **75**, 591–594.
Krastin, N. I. (1950) Determination of the cycle of development of *Thelazia gulosa*, parasite of the eye of cattle. *Dokl. Akad. Nauk S.S.S.R.*, **70**, 549–551.
Krastin, N. I. (1952) Determination of the cycle of development of *Thelazia skrjabini* (Erschow, 1928), parasite of the eye of cattle. *Dokl. Akad. Nauk S.S.S.R.*, **82**, 829–831.
Lim, C. W. (1975) The fowl (*Gallus domesticus*) and a lepidopteran (*Setomorpha rutella*) as experimental hosts for *Tetrameres mohtedai* (Nematoda). *Parasitology*, **70**, 143–148.
McGaughey, C. A. (1950) Preliminary note on the treatment of spirocercosis in dogs with a piperazine compound, Caricide (Lederle). *Vet. Rec.*, **62**, 814–815.
Michalski, L. (1976) The efficacy of levamisole and termamisole in the treatment of thelaziasis in cattle. *Medycyna wet.*, **32**, 417.
Miyazaki, I. (1954) Studies on *Gnathostoma* occurring in Japan (Nematoda: *Gnathostomidae*). II. Life history of *Gnathostoma* and morphological comparison of its larval forms. *Kyusu Mem. Med. Sci.*, **5**, 123–140.
Nishiyama, S. (1958) Studies on habronemiasis in horses. *Bull. Fac. Agric. Kagoshima. Univ.*, **7**.
Páv, J. & Zajiček, O. (1972) The epizootiology of the helminths *Ascarops strongylina* (Rud.) and *Physocephalus sexulatus* (Molin) in wild pigs (*Sus scrofa L.*) living in enclosures. *Z. Jadniss*, **18**, 6.
Pence, D. B. & Stone, J. E. (1978) Visceral lesions in wild carnivores naturally infected with *Spirocerca lupi*. *Vet. Path.*, **15**, 322–331.
Petri, L. H. & Ameel, D. J. (1950) Studies on the life cycle of *Physaloptera rara* (Hall and Wigdor, 1918), and *Physaloptera praeputialis* Linstow, 1889. *J. Parasit.*, **36** (Suppl.), 40.
Rao, D. S. P. (1953) Spirocercosis in a dog. *Indian vet. J.*, **29**, 548.
Ruff, M. D. (1978) Nematodes and acanthocephalans. In: *Diseases of Poultry*, M. S. Hofstad, B. W. Calnek, C. F. Helmboldt, W. M. Reid & H. W. Yoder, pp. 705–736. Ames, Iowa: Iowa State University Press.
Vaidyanathan, S. N. (1952) *Spirocera lupi* infection in dogs. A few cases treated with Hetrazan (Lederle). *Indian vet. J.*, **29**, 243–247.
Wang, P., Sun, Y. & Zhao, Y. (1976) On the development of *Gnathostoma hispidum* in the intermediate host, with special reference to its transmission route in pigs. *Acta zool. sinica*, **22**, 45–52.
Weinmann, C. J., Anderson, J. R., Rubtzoff, P., Connolly, G. & Longhurst, W. M. (1974) Eyeworms and face flies in California. *Calif. Agric.*, **28**, 4–5.

SUPERFAMILY: FILARIOIDEA WEINLAND, 1858

These are long and relatively thin worms. As a rule the mouth is small and not surrounded by lips, and there is neither a buccal capsule nor a pharynx. The oesophagus has anterior muscular and posterior glandular portions. The male is frequently much smaller than the female and the spicules are unequal and dissimilar. The vulva is usually situated near the anterior extremity and fully developed larvae are born. The worms live in the body cavities, blood or lymph vessels or connective tissues of their hosts. Families of importance: *Filariidae*, *Setariidae* and *Onchocercidae*.

The larvae are known as microfilariae. They are in some cases enclosed in a thin membrane or sheath which is apparently the very flexible egg-shell. They reach the blood stream or the tissue lymph spaces of the host, whence they may be taken up by species of mosquitoes, fleas etc. In these intermediate hosts the larvae develop to the infective stage and, passing into the body cavity, reach the proboscis of the arthropod. When the latter again sucks blood the larvae break their way out and enter the final host, to complete their development.

The microfilariae of certain species of this group appear in the blood stream of the final host, appearing either only during the day, or only during the night, this phenomenon being known as diurnal or nocturnal periodicity. For instance, in *Wuchereria bancrofti* of man, which is transmitted by mosquitoes, the microfilariae are much more numerous in the peripheral blood vessels at night than during the day and they are said to have a nocturnal periodicity. In *Loa loa*, also a human parasite, which is transmitted by the tabanid fly *Chrysops*, the reverse is the case and the larvae are said to have a diurnal periodicity. Studies by Hawking (1964) with *Wuchereria bancrofti* indicate that periodicity is a function of the microfilariae and not the host. Hawking postulated that the release of microfilariae from the lungs, where they accumulate during the day, is mediated by a 'fixative force' which fixes the microfilariae in the lungs and a 'switch' mechanism which switches the fixative force off at night. The fixative force is increased by raising the oxygen tension of the pulmonary capillaries (e.g. by giving O_2, administration of isoprenaline) and the switch mechanism is postu-

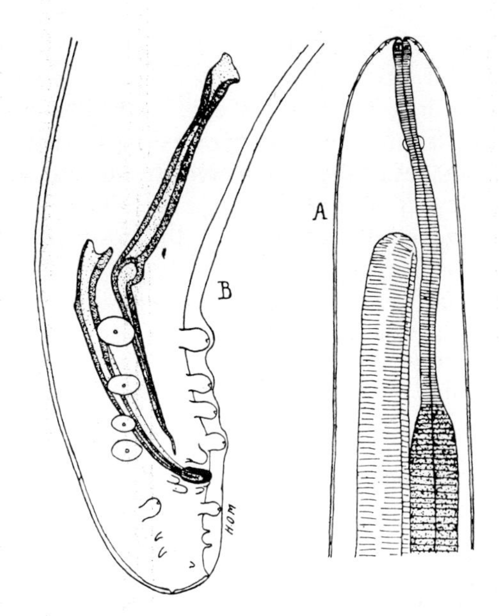

Fig 1.173 *Dirofilaria immitis*. A, Lateral view of anterior end of female. B, Ventrolateral view of hind end of male.

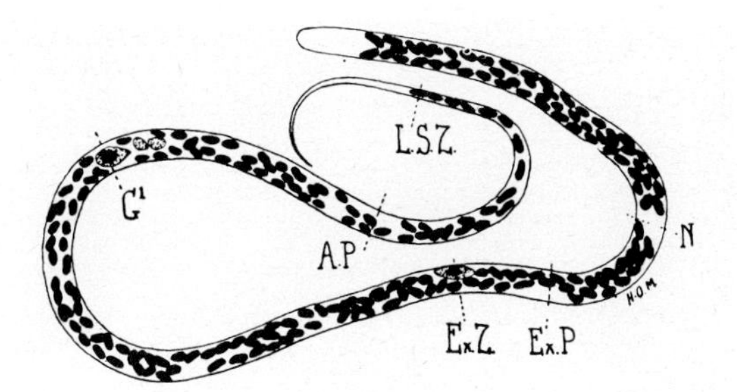

Fig 1.174 The fixed points for measurement of *Dirofilaria immitis microfilariae*.
A.P. anterior pore
Ex.P. excretory pore
Ex.Z. excretory cell
G^1 first genital cell
L.S.Z. last tail cell
N. nerve ring

lated to be an inherent circadian rhythm which is also influenced by similar rhythms of the host.

Microfilariae may survive in the host for several weeks, either after the death of the adult worm or if they are transferred to uninfected animals.

FAMILY: FILARIIDAE CLAUS, 1885

Head of the adult without a peribuccal chitinous ring or other structures. Genera of importance: *Dirofilaria*, *Mansonella*, *Brugia*, *Wuchereria*, *Parafilaria*, *Ornithofilaria*, *Bhalfilaria*, *Elaeophora*, *Suifilaria*, *Wehrdikmansia*.

Genus: Dirofilaria Railliet and Henry, 1911

Dirofilaria immitis (Leidy, 1856) occurs in the dog, cat, fox and wolf in the tropics and subtropics and in some temperate countries. It does not occur in Great Britain. Wild Canidae are infected and may be important reservoir hosts. *D. immitis* also infects man and has been recorded in the horse. California sea lions, harbour seals and black bears have been found infected. Thus, the infection could be important in such animals in zoos, particularly sea lions which are very susceptible, in endemic areas. The worms live mainly in the right ventricle and the pulmonary artery, but have also been found in other parts of the body. Males are 12–16 cm long and females 25–30 cm. The worms are slender and white in colour. The oesophagus is 1.25–1.5 mm long. The hind end of the male is spirally coiled and the tail bears small lateral alae. There are four to six, usually five, pairs of ovoid papillae, of which one pair is postcloacal, two pairs of finger-shaped papillae lateral and posterior to the cloacal opening, and three to four pairs of small conical papillae near the tip of the tail. The left spicule is 0.324–0.375 mm long and pointed, the right is 0.19–0.229 mm long and ends bluntly. The vulva is situated just behind the end of the oesophagus.

Females are ovoviviparous and microfilariae may be found in the blood at all times but there is a tendency towards periodicity. This appears to vary in different countries. Thus in the USA Schnelle and Young (1944) observed minimum microfilaraemia at 11.00 hours and maximum at 16.30 hours; in France, Euzéby and Laine (1951) found the lowest number at 08.00 hours and the greatest at 20.00 hours. With a Chinese strain of *D. immitis*, Webber and Hawking (1955) found minimum parasitaemia at 06.00 hours and maximum at 18.00 hours. Microfilariae have a long,

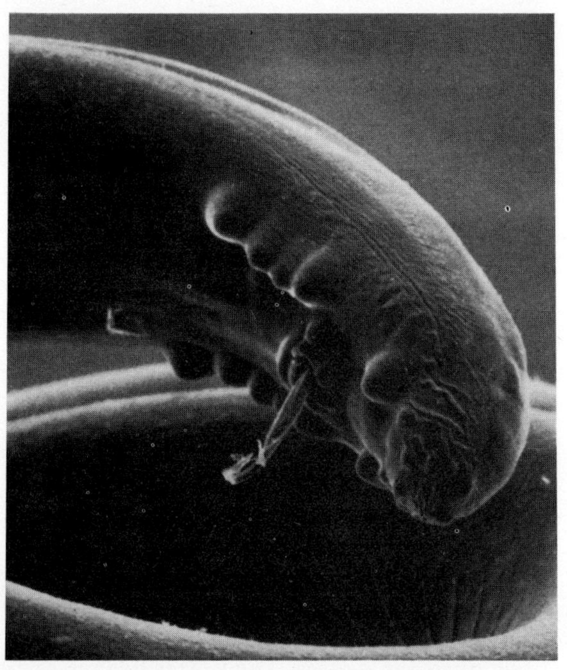

Fig 1.175 Scanning electron micrograph of tail of male *Dirofilaria immitis*. (*Dr Ming Ming Wong, University of California*)

slender tail. Since different methods of fixation cause various degrees of contraction of the microfilariae, it is customary to express the distance of certain fixed points from the anterior extremity as percentages of the total length. In this way accurate descriptions of the larvae are obtained. The fixed points usually used are the position of the nerve ring (N), the excretory pore (Ex.P.), the excretory cell (Ex.Z.), the first genital cell (G.1), second genital cell (G.2), third genital cell (G.3), the anal pore (A.P.) and the last nucleus or last tail cell (L.S.Z.). An elongate 'central body' is seen with some methods of staining. For *D. immitis* microfilariae the mean percentages of the distance of these reference points as given by Newton and Wright (1956) are: nerve ring, 23.8; excretory pore, 32.7; excretory cell, 38.6; first genital cell, 67.9; second genital cell, 74.1; third genital cell, 79.4; anal pore, 82; and last nucleus, 92.7: with a total range in size of 307 to 322 μm and a mean of 313 μm. These figures should be compared with those of other microfilariae of the dog, given in Table 1.2.

Table 1.2 Comparative sizes of microfilariae found in dogs (percentage distance from the anterior end)

Species	NR	EP	EC	G.1	G.2	G.3	G.4	AP	LTC	Range (μm)	Mean (μm)
Dirofilaria immitis	23.8	32.7	38.6	67.9	74.1	75.4	79.4	82.0	92.9	307–322	313
Dipetalonema reconditum	20.8	21.0	34.5	70.1	75.7	77.2	78.6	80.7	89.0	246–293	270.6
Dirofilia repens	23.0	30.0	33.0	62.5	—	—	66.0	70.0	—	—	290 ± 10
Dipetalonema grassii	—	—	—	—	—	—	—	—	—	—	570
Brugia malayi	24.5	35.0	40.0	64.0	—	—	80.0	83.0	—	—	220 ± 20
Brugia patei	Similar to *B. malayi* except cephalic space 4.8 instead of 6.9 μm										
Brugia pahangi	22.0	31.0	33.0	68.0	73.0	74.0	78.0	80.0	—	—	280
Dipetalonema dracunculoides	45*	66–70*	—	—	—	—	—	53–55†	20†	195–230	

* Measurements in μm from anterior end.
† Measurements in μm from posterior end.
NR, nerve ring; EP, excretory pore; EC, excretory cell; G.1, first genital cell; G.2, second genital cell; G.3, third genital cell; G.4, fourth genital cell; AP, anal pore; LTC, last tail cell.

Life-cycle. The intermediate hosts are mosquitoes of the genera *Culex*, *Aedes*, *Anopheles*, *Armigeres*, *Myzorhynchus* and *Taeniorhynchus*. This species does not develop, as has been stated, in fleas. Taylor (1960) has described the development in the mosquito. For the first 24 hours after a blood meal the microfilariae are found in the stomach of the insect: during the next 24 hours they migrate to the malpighian tubules where they develop over the next 15–16 days. For the first six or seven days the developing larvae are found inside the cells of the tubules. By the fourth day after infection of the mosquito the 'sausage stage' larva occurs, this being the second-stage larva; it measures 220–240 μm in length by 20–25 μm in diameter. In this stage the excretory and intestinal cells have increased in number and these ultimately produce their respective organs which are evident in the 'elongated sausage form' which occurs on the ninth day after infection. At this time it measures 500 by 20 μm. This latter stage feeds on the cells of the malpighian tubules and enters the body cavity. From here it migrates through the thorax ending up in the cephalic spaces of the head or in the cavity of the labium. The final, infective, stage is produced in the labium and this measures 800–900 μm.

Development takes about 15–17 days in temperate countries while in tropical areas it may be as short as eight to ten days.

Infection of the dog occurs when the infected mosquito takes a blood meal. Initially, migration in the dog occurs adjacent to the site of infection, larvae being found in the submuscular membranes and a few in the subcutaneous tissue (Kume & Itagaki 1955), From 85 to 120 days after infection developmental stages are found in the heart or pulmonary artery, being 3.2–11 cm in length at this time. In a further two months maturity is reached and microfilariae are shed into the blood. Circulating microfilariae may survive up to two years and transplacental transmission may occur with microfilariae being found in neonatal pups.

Pathogenesis. Many dogs may be infected with *D. immitis* without showing any clinical signs of infection other than microfilariae in the blood. In heavy infections, however, the worms cause circulatory distress, due to mechanical interference and a progressive endarteritis. In large numbers they interfere with the function of the heart valves, masses of worms being found in the right atrium, right ventricle and the upper pulmonary arterial tree. The pathophysiology of heartworm disease is reviewed by Knight (1977). Heartworm disease is primarily a pulmonary vascular disease. There is pulmonary hypertension due to a substantial narrowing of the small peripheral pulmonary arteries. This also occurs in the hepatic veins. Endarteritis takes nine to ten months to become apparent and cause sufficient change to affect the heart (Hennigar & Ferguson 1957). Resultant compensatory hypertrophy of the right ventricle can eventually result in congestive heart failure and ultimately chronic passive congestion manifested by liver enlargement, ascites and occasionally peripheral oedema.

A deep, soft cough is common and haemoptysis may be seen. Affected dogs have decreased stamina and may have a characteristic respiration in that the ribcage remains expended and there is extra inspiratory effect. Emaciation develops gradually. In the vena caval syndrome, or liver failure syndrome, there is anorexia, profound weakness and dark, brownish urine containing haemoglobin and bilirubin.

Radiographic and electrocardiographic changes are described by Knight (1977). In particular, in advanced cases, the main pulmonary artery is dilated and in dorsoventral view a prominent bulge is seen in the 2 o'clock position. The electrocardiogram may be normal at rest but after exercise the T wave may be inverted. Cutaneous manifestations have been described by Vaughan (1952), these consisting of an eczematous dermatitis associated with intense irritation. Similar lesions have been reported from France and Japan. However, since parasites of the genus *Dipetalonema* may also occur along with *D. immitis* it is possible that some of the reported skin lesions may be due to these and not to *D. immitis*.

Diagnosis. Diagnosis is based on the clinical signs and on the demonstration of microfilariae in the blood. Microfilariae may often be readily detected by a simple blood examination; a drop of blood is placed on a slide, covered with a glass and examined directly. The microfilariae will be seen moving actively. Where their numbers are few, a better chance of finding them is by a concentration technique and this is the method of choice. One of the following may be used. In a filter method 1 ml or more of blood in sodium citrate or another anticoagulant is lysed with 0.1% sodium carbonate. The blood is then passed through a filter with a 3 μm porosity. Microfilariae on the filter are stained with 0.1% methylene blue. In Knott's method, 1 ml blood and 9 ml of 2% formalin are mixed and centrifuged for five minutes. The sediment is stained with methylene blue, although 1% cresol blue in 0.8% NaCl and 0.05% Azur II are better stains for somatic structure and the cephalic hook (see below).

It may be important to differentiate microfilariae of *D. immitis* and those of *Dipetalonema reconditum*. In fresh preparations *D. immitis* microfilariae are generally the more motile and usually more numerous; however, these are guides to identification rather than absolute criteria. A method of concentration has been described by Newton and Wright (1956) as follows. Blood is treated by the Knott method (see above) and the concentrated microfilariae are examined with respect to their size, the shape of the head and the tail. *D. immitis* microfilariae are longer (300 μm or more) than those of *Dipetalonema reconditum* and are straight with a straight tail and tapered head; those of *D. reconditum* are curved, with a tail shaped like a buttonhook and a blunt head (see Table 1.2 p. 308, for comparative sizes of microfilariae found in dogs).

A further method of differentiating the two species of microfilariae is given by Sawyer et al. (1965). Thick blood films from the marginal ear vein are dehaemoglobinized in tap water for 10 minutes and transferred without drying to a 1 in 50 dilution of 1% brilliant cresyl blue in 0.8% saline for a further 10 minutes. The slides are then rinsed in saline and mounted in saline. Microfilariae are then examined by both 'high dry' and oil immersion objectives for the presence of a cephalic hook. With this method of staining the cephalic hook is readily seen in *D. reconditum*, while it is absent in *D. immitis*. The method also allows a more satisfactory examination of the internal structure of the microfilariae, the genital cells being especially clear.

Microfilariae may not be detected in 5–20% of infected dogs. Diagnosis must then be made from the history and clinical signs. Thoracic radiography and electrocardiography are recommended aids. The type and severity of angiographic abnormalities are directly related to the parasite burden. Wong and Suter (1979) have assessed the value of serodiagnostic tests for occult dirofilariasis. An indirect fluorescent antibody test combined with thoracic radiography is recommended by Wong and Suter (1979), while Welch et al. (1979) reported high specificity and sensitivity with a cyanogen bromide purified antigen linked to Sepharose 4B beads and used in an indirect fluorescent antibody test and a cell-mediated immunity test using antigens purified by affinity chromatography.

Treatment. This is discussed by the American Heartworm Association (1978) and the American Veterinary Medical Association (1978).

Before treatment the function of the heart, liver, kidneys and lungs must be assessed since these affect the prognosis. Infected dogs are first treated with an adulticide anthelmintic; six weeks later a microfilaricidal anthelmintic is used and the dogs are then placed on prophylactic medication.

As an adulticide, the arsenical thiacetarsamide is given intravenously at a dose of 0.1 ml/0.45 kg twice a day for two or preferably three days. Since thiacetarsamide is a potential nephrotoxin and hepatotoxin, treatment should be stopped if signs of toxicity appear (persistent vomiting, anorexia, icterus). Rest and careful nursing are required for a month after treatment to counteract the effects of emboli of dead worms in the lungs. Melarsoprol 100 mg/kg is also effective against adults (Blair & Campbell 1979). Alternatively surgical treatment may be used to remove the adult worms, via a ventricular puncture and arteriotomy of the main trunk of the pulmonary artery.

To remove the microfilariae, dithiazanine iodide, 2 mg/0.45 kg, is given daily for seven days. If microfilariae are still present the dose may be increased to 5 mg/0.45 kg. Dithiazanine iodide should not be used if hepatorenal disease is present. Levamisole 10 mg/kg may be given orally for up to 15–20 days. The microfilariae are usually eliminated within six to 12 days, at which time treatment should cease. Avermectin B_{1a}, 0.05–0.1 mg/kg, reduced the microfilaraemia by 90% within 24 hours (Blair & Campbell 1979). Fenthion and stibophen may also be used.

Prophylaxis. Control measures are difficult, especially in endemic areas. Insect repellents may have limited benefits. Animals kept indoors in the evenings and during the night have a lower incidence of infection.

Prophylactic medication is effective. In endemic areas thiacetarsamide treatment may be used every six months. Where *D. immitis* is less common dogs should be examined for the presence of infection every six months and treated if necessary.

Diethylcarbamazine (DEC) and styrylpyridinium diethylcarbamazine, 5.5 mg/kg, is an effective prophylactic when given daily to dogs throughout and until two months after the mosquito season. The dogs should be known to be free of infection with *D. immitis* before DEC medication is commenced, since treatment of microfilaraemic dogs may cause gastrointestinal upsets, weakness, collapse and occasionally death.

Mebendazole, 80 mg/kg, given daily for 30 days, beginning two days before experimental infection of dogs, was highly effective and is a possible prophylactic agent (McCall & Crouthamel 1976). Similarly, treatment for five days with avermectin B_{1a} (0.2 mg/kg/day) or melarsoprol (100 mg/kg/day), but not with mebendazole (100 mg/kg/day), was highly effective against 38-day-old immature *D. immitis* in ferrets (Blair & Campbell 1978). These anthelmintics may well be effective prophylactic medications in dogs if given daily or on a monthly basis.

Dirofilaria repens Railliet and Henry, 1911 occurs in the subcutaneous tissue of the dog and cat in Italy, southern France, Sardinia, India, Sri Lanka, the USSR and south-east Asia. Schillhorn (1974) considers it the most common filarid of dogs in Nigeria. Nelson (1959) found it in dogs, cats and genet cats (*Genetta tigrina*) in Kenya and found that its microfilariae develop in *Aedes pembaensis*, *A. aegypti*, *Mansonia uniformis* and *Mansonia africanus*. Microfilariae are not sheathed and are found in the blood and in lymph spaces in the skin. According to the method of fixation they measure 260–360 μm in length (average 290 μm).

D. repens has been found in subcutaneous nodules on various parts of the body of man. Possibly many infections previously ascribed to *D. conjuctivae* (see below) are due to *D. repens*.

Dirofilaria corynodes (Linstow, 1899) (syn. *D. aethiops* Webber, 1955). This species occurs in monkeys (*Cercopithecus aethiops*, *Colobus* spp.) and its vectors are mosquitoes. Nelson (1959) found that its microfilariae develop in *Aedes aegypti* and *A. pembaensis* in Kenya.

Dirofilaria conjuctivae (Addario, 1885) has been reported from humans, being associated with hazelnut-sized nodules on the head, eyelids and other parts of the body. It closely resembles *D. repens* and several workers consider it identical with that species.

Dirofilaria roemeri (Linstow, 1905) (syn. *Dipetalonema roemeri*) is found in the subcutaneous and intramuscular connective tissues and intramuscularly in the pelvic region and hind legs of wallabies and kangaroo. It is transmitted by tabanid flies, particularly *Dasybasis hebes* (Spratt 1975).

Dirofilaria tenuis Chandler, 1942, occurs in the subcutaneous tissues of racoons in the southern part of the USA. Orihel and Beaver (1965) consider this to be the most common species of *Dirofilaria* in the subcutaneous tissues of man in the USA. They also suggest that forms which have previously been reported as *D. conjuctivae* are probably *D. tenuis*.

Dirofilaria ursi Yamaguti, 1941, occurs in the black bears (*Ursus americanus*) of the USA.

DIROFILARIA INFECTIONS IN MAN

Aberrant infections with *D. immitis*, *D. repens* and *D. tenuis* occur in man. Pulmonary lesions are common with *D. immitis* infections. Infection is usually asymptomatic and is seen as small peripheral lesion 1–2 cm in diameter (coin lesions) in the lungs on radiography. Ciferri (1982) reported that 95% of patients had a single nodule, of which 90% contained a single worm. He found that no single laboratory procedure was useful in preoperative diagnosis. *D. tenuis* and *D. repens* occur in subcutaneous nodules, those of *D. repens* occuring particularly round the eye. *D. ursi* and *D. acutuiscula* have also been recorded in man. Welch et al. (1979) have commented that in Australia (Queensland) the occurrence of serum antibody to *D. immitis* in man was proportional to the prevalence of *D. immitis* infections in dogs. In one area where 88% of dogs were infected, 30% of humans were reactive to serodiagnostic tests.

Genus: Suifilaria Ortlepp, 1937

Suifilaria suis Ortlepp, 1937 was described from the pig in southern Africa. The adult parasites live in the subcutaneous and intermuscular connective tissue and are not easily seen. The males are 17–25 mm long, the females 32–40 mm long and 0.15–0.17 mm thick. The hind end of the male is spirally coiled; the spicules are unequal, the right measuring about 0.1 mm, the left 0.655–0.87 mm. The tail of the female ends abruptly and bears on its end a number of small tubercles. Eggs are 51–61 by 28–32 μm.

The life-cycle is unknown. The females appear to lay their eggs in the skin of the pig, because the skin of infected pigs shows numerous small, vesicular eruptions which contain the eggs of the worm.

The worms may produce small, whitish nodules in the connective tissues which may be taken for young *Cysticercus cellulosae* until they are more closely examined. Otherwise they do not affect the health of the pig, except for the fact that the skin vesicles containing the eggs eventually burst and may then become secondarily infected and develop into abscesses.

Genus: Mansonella Faust, 1929

Mansonella ozzardi (Manson, 1897) Faust, 1929 occurs in the peritoneal cavity of man in the northern parts of South America and the Caribbean. The microfilariae are not sheathed; they are found in the skin and in the blood. *Culicoides furens* and also *Simulium amazonicum* can act as intermediate hosts. Morphologically similar species have been seen in neotropical monkeys and in some domesticated animals.

Genus: Brugia Buckey, 1960

All the species of the genus are parasitic in the lymphatic systems of primates, carnivores and insectivores.

Brugia malayi (Brug, 1927) Buckley, 1958. Males measure 2.2 cm, females 4.8 cm, microfilariae 210 by 6 μm. Two strains of *B. malayi* occur: the nocturnally periodic strain affects only man on open plains regions in India, Malaysia and other parts of south-east Asia; the nocturnally subperiodic strain found in swamp/forest areas of south-east Asia (Malaysia, Thailand, Phillipines etc.) is a zoonotic infection affecting man, cats, dogs, monkeys, civet cats and pangolins. In Malaysia leaf monkeys (*Presbytis*) are important reservoir hosts and up to 70% may be infected. The most important vectors are *Mansonia* spp. and some *Anopheles* spp. mosquitoes. In man *B. malayi* may cause enlarged lymph nodes, lymphangitis, lymphoedema and elephantiasis. Only diethylcarbamazine, which has little effect against the adult worm, is used to treat man. Further, mass treatment programmes are complicated by the presence of animals infected with subperiodic strain and suspected reinfection of man from these. The biology and pathology of *B. malayi* are reviewed by Denham and McGreevy (1977).

Brugia timori Partono, Purnomo, Atmosoedjono, Oemijati and Cross, 1977, which affects man, is known only from the Lesser Sunda Islands of south-east Indonesia. The microfilariae are larger than those of *B. malayi* (Purnomo & Partono 1977). This species may cause a more acute lymphatic disease in man than either form of *B. malayi.*

Brugia pahangi (Buckley and Edeson, 1956) Buckley, 1958 affects the lymphatic system of the cat, dog, civet cat (*Arctictus* spp.), tiger, slow loris and leaf monkey in Malaysia. Experimentally it has been transmitted to man. It is transmitted by mosquitoes in the genus *Mansonia* and *Armigeres*. Diethylcarbamazine has been used to treat experimentally infected cats with variable results (Denham & McGreevy 1977). Levamisole is also microfilaricidal. Mebendazole, 50–100 mg/kg/-day for five days, killed nearly all the adults of *B. pahangi* in birds but not in cats (Denham et al. 1978).

Brugia patei (Buckley, Nelson and Heisch, 1958) Buckley, 1958 occurs in the lymph nodes of dogs, cats, the genet cat (*Genetta tigrina*) and the bush baby (*Galago crassicaudatus*) in Africa. Development occurs in *Mansonia uniformis, M. africanus* and *Aedes pembaensis* in Kenya.

Other *Brugia* species include **B. ceylonensis** Jayewardene, 1962, of the dog in Sri Lanka; **B. beaveri**, Ash and Little, 1964, of racoons in the USA: and **B. tupaiae** Orihel, 1966, of the tree shrew (*Tupaia* spp.) in south-east Asia.

Genus: Wuchereria da Silva Aranjo, 1877

This genus contains the well-known species *W. bancrofti* (Cobbold, 1877) Seurat, 1921, which infects the lymphatic system of man in the warmer parts of both the eastern and western hemispheres and causes bancroftian filariasis of man, one feature of which is elephantiasis of the infected tissues. The generic and specific names of this species commemorate the names of its discoverer O. Wucher and the subsequent work of Bancroft on it. The worm and the disease it causes are fully described in textbooks of medical parasitology. As yet there are no records of natural infection in animals although *W. bancrofti* has been experimentally transmitted to macaques.

Genus: Loa Stiles, 1905

Loa loa (Guyot, 1778) occurs in the subcutaneous tissues of man, monkeys and baboons in the rainforest regions of West and Central Africa. Two ecologically distinct strains occur affecting either man or the non-human primates. The intermediate hosts are *Crysops* spp. *L. loa* is the cause of nodules 2–3 cm in diameter which are transient and reappear in different sites.

Genus: Parafilaria York and Maplestone, 1926

Parafilaria multipapillosa (Condamine and Drouilly, 1878) (syn. *Filaria haemorrhagica*) occurs in equines in eastern countries. It has been reported in Great Britain, being imported in horses from eastern Europe. Males are 28 mm long and females 40–70 mm. The anterior end of the body bears a large number of papilliform thickenings. The worms live in the subcutaneous and intermuscular connective tissue and produce subcutaneous nodules which appear suddenly, break open, bleed and then heal up. The condition is seen during the summer and disappears in winter, but may re-appear the next summer. The intermediate host in the USSR is the blood sucking fly *Haematobia atripalpis*. In experimental infections the infective stage was produced in 10–15 days at air temperatures of 20–36°C (Gnedina & Osipov 1960).

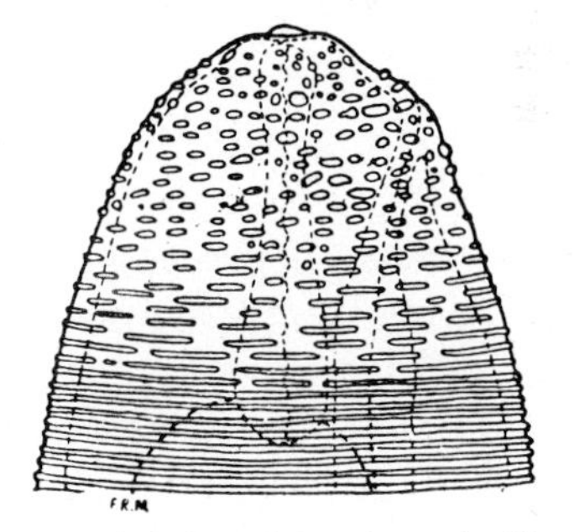

Fig 1.176 Lateral view of anterior end of female *Parafilaria multipapillosa*.

Parafilaria bovicola de Jesus, 1934 causes haemorrhagic nodules on the skin of cattle and buffalo in the Philippines and many parts of Africa, India and Europe as far north as the Scandinavian countries. Adults bear 13 rows of cuticular elevations at the anterior end and the remainder of the cuticle is transversely striated.

Life-cycle. Development occurs in *Musca* spp. such as *M. lusoria* and *M. xanthomelas*, which probably become infected when they feed on the skin lesions. The flies feed primarily on lacrimal secretions and it is likely that it is at this time, or when the flies are feeding on wounds, that the definitive host becomes infected. Adults develop in 167–250 days and bleeding points develop in the skin 242–319 days after infection.

Pathogenesis. The prevalence of infection may be high and 36% of cattle were found to be infected in the northern Transvaal (Carmichael & Koster 1978). Subcutaneous nodules on the body are found primarily on the shoulder region and on the dorsal aspect of the body, loins, withers, neck. The nodules are 12–15 mm in diameter and 5–7 mm in height. The fully gravid female moves into the dermis for oviposition at which time the nodules become enlarged (40 mm in diameter and 10 mm in height) and painful haemorrhage occurs. A fistulous tract leads into acutely inflamed tissue containing the gravid female. In the tissues there is marked oedema and infiltration of lymphocytes and neutrophils (Patnaik & Pande 1963; Pienaar & van den Heever 1964). Most haemorrhage points bleed once, although they may bleed two or three times (Viljoen 1976). At slaughter there are slimy bruise-like lesions on the subcutaneous surfaces of carcasses. The lesions must be trimmed and since exposure of the muscles causes earlier spoilage the carcasses usually are downgraded.

Treatment. Nitroxynil is the most effective anthelmintic and 20 mg/kg repeated 72 hours later reduced lesion area by 95% and visible carcass lesions by 90% (Wellington 1978). High doses of levamisole and fenbendazole given daily for four to five days have also been used, as have the antimonial compounds.

Similar lesions are produced by parasites such as **Parafilaria antipini** Rukhiadev, 1947 in deer and **Indofilaria patabiramani** Alwar, Seneviratna and Gopal, 1959 in elephants.

Genus: Ornithofilaria Gönnert, 1937

Ornithofilaria fallisensis Anderson, 1954. This species occurs in the subcutaneous tissues of anatid birds (see Lapage 1961) and may be transmitted to the domesticated duck. Its life history and transmission by black flies (Simuliidae) have been described by Anderson (1954, 1955).

Genus: Bhalfilaria Bhalerao and Rao, 1944

Bhalfilaria ladamii Bhalerao and Rao, 1944, occurs in the heart of the fowl (Black Minorca Cock) in the Hyderabad area of India. The male is unknown; the female is 34 mm long, with a cuticle with longitudinal striations. Distinct divisions occur in the oesophagus. Microfilariae are spindle-shaped, 200–240 μm long.

Genus: Elaeophora Railliet and Henry, 1912

Elaeophora schneideri Wehr and Dikmans, 1935, is found in the common carotid and internal maxillary arteries of deer, elk and sheep and is wide-spread in the western and south-western USA, while deer in the eastern USA may also be infected. The infection is most common at elevations above 2000 m. Adults may be found in other arteries such as the iliac, tibial and digital arteries.

The adult worms measure 6–12 cm in length. The oesophagus is very long, the tail of the male is tightly coiled and the spicules are unequal, the larger being 1.11 mm long, the smaller 0.4 mm. The worms are frequently found lying together in pairs in the arteries. Microfilariae 270 μm in length by 17 μm in thickness, bluntly rounded anteriorly, tapering posteriorly, occur in the skin.

Life-cycle. Microfilariae are usually found in capillaries on the forehead and face. The intermediate hosts are species of horseflies, *Hybomitra* and *Tabanus*, and up to 43% of horseflies were found to be infected in an endemic area in New Mexico. *H. laticornus* was the most commonly infected species. After two weeks, infective laevae are found in the head and mouthparts of the horsefly. In the definitive host infective larvae begin to develop in the leptomeningeal artieries; three and a half to four weeks after infection they migrate to the carotid arteries and the prepatent period is four and a half months or longer (Hibler & Metzger 1974).

Pathogenesis and clinical signs. The pathology and clinical signs in deer and elk have been reviewed by Hibler and Adcock (1971). White-tailed deer (*Odocoileus virginianus*) and mule deer (*O. hemionus*) appear to be the normal definitive hosts since they usually demonstrate little or no clinical evidence of the infection. In other hosts abnormalities are seen, i.e. filarial dermatitis or 'sorehead' in sheep and blindness and ischaemic necrosis of the brain, muzzle, ears etc. in elk (*Cervus canadensis*).

In sheep filarial dermatitis is seen on the face, usually beginning in the poll region, and also the feet. Lesions are seen in skin supplied by arteries within which the adult parasites are residing. The microfilariae carried in the blood to these sites cause a granulomatous inflamation of the skin with infiltration of eosinophils, lymphocytes and some neutrophils and giant cells. Microscopically vesicles, bullae, ulceration, hyperplasia and hyperkeratosis are seen. There is intense pruritus and self-excoriation and secondary infection may result. Thrombosis of the arteries containing adults may be seen but is not common. The usual lesion is a circumscribed area 5–10 cm in diameter on the poll though this may extend to other parts as stated above. Lesions on the coronary band may produce a club foot appearance (Kemper 1957).

The lesions show periods of quiescence alternating with periods of activity. During the former, the lesions may become encrusted and the crusts may be dislodged during the active period producing a haemorrhagic lesion. The onset of the period of activity is probably due to a new generation of microfilariae reaching the skin.

After prolonged activity the lesions ultimately resolve, healing occurs with growth of new wool and the animal appears normal.

In elk adults are found in the common brachiocephalic arteries and leptomeningeal arteries as well as arteries to the eye. Thrombosis of the artery and inflammation and fibrosis of the arterial wall will cause vascular obstruction, as will granulomatous reactions to dead parasites within the arteries. The microfilariae and the inflammatory reaction to them obstruct arterioles and capillaries. The circulatory impairment results in ischaemic necrosis of the brain, eyes, optic nerves, ears and muzzle. Clinically, bilateral blindness, neurological signs, necrosis of the muzzle, nostrils and ears, abnormal antler growth and emaciation are seen. In elk the infection is frequently fatal and elk cannot be introduced into areas (parks) where the infection is common in deer.

Diagnosis. The lesions are fairly characteristic and microfilariae can be demonstrated in the skin. The most satisfactory method of diagnosis is to macerate a piece of skin in warm saline and examine the material for microfilariae after about two hours. However, in sheep, microfilariae are frequently few in number and may not be found in the skin of infected animals. Similarly, microfilariae have rarely been found in the skin of infected elk. Further, in deer the microfilariae must be differentiated from those such as *Wehrdikmansia cervipedis* and *Setaria yehi*. Post mortem examination may be necessary to confirm the diagnosis.

Treatment. Piperazine hexahydrate at a dose of 180 g per animal or 120 g per gallon of drinking water for three days and Trolene (Dow ET–57) at a dose of 300 mg/kg have been used. Other treatments include Fouadin (4 ml daily until 88 ml given); diethylcarbamazine (100 mg/kg.); sodium antimonyl tartrate in trypan blue with 1% phenol (Trichicide) (15–40 ml for six injections). Unfortunately, treatment and resultant death of large numbers of parasites in the arteries can result in deaths amongst treated animals (Hibler & Adcock 1971).

Elaeophora poeli (Vryburg, 1897) occurs in the aorta of cattle, zebu, buffalo and water buffalo in south-east Asia the Malay Peninsula, Sumatra, Philippines and Central and East Africa. Males are 45–70 mm long and females 40–300 mm. There are no lips and the oesophagus is very long. The tail of the male bears five to seven pairs of papillae, two pairs being pre-cloacal. The spicules measure 0.192–0.25 mm and 0.12–0.132 mm respectively. The uterus has four branches. The embryos are 340–360 μm long, thick anteriorly and tapering posteriorly.

The life-cycle is unknown. The male occurs in nodules in the wall of the aorta, while the female is fixed in the nodules with its anterior extremity, and the rest of its body hangs free in the lumen of the vessel. The site of the parasites is the thoracic portion of the aorta, often along the dorsal aspects near the openings of the vertebral arteries. The affected vessel is diffusely swollen. Its wall is thickened and less elastic than normal on account of the development of connective tissue. The intima is uneven, it contains fibrous tracks, and is raised by the nodules which lie between the intima and the media. The nodules measure 8–13 mm in diameter and contain thrombi in the process of organization and the worms. There is no evidence that the parasites produce clinical disease.

Elaeophora böhmi Supperer, 1953 (*Onchocerca böhmi*, a new combination proposed by Bain et al. 1976) occurs in the medial layer of arteries and veins of the extremities of horses in Austria. Males are 4.5–6 cm long and females 4–20 cm. Microfilariae resemble those of *Onchocerca reticulata*, 230–290 μm. Life-cycle unknown.

The parasites cause thickening of the blood vessel walls and there may be nodules formed containing disintegrated or calcified worms. Usually no clinical signs are associated with the infection.

Genus: Wehrdikmansia Caballero, 1945 (syn. Acanthospiculum, Skrjabin and Schikhobalova, 1946)

Wehrdikmansia cervipedis (Wehr and Dikmans, 1935) occurs in the subcutaneous tissues of deer, frequently of the neck and fore and hind legs. It has been reported in Europe, the USA and the USSR. Males are 5.5–6 cm long and females 18–20 cm. Anterior end rounded, body finely striated, posterior end of male coiled, no caudal alae, spicules unequal. The life-cycle is unknown. A local fibrous lesion may be caused and the parasites are seen as thick, thread-like objects when the animal is skinned. High levels of infection may be seen in red deer (*Cervus elaphus* and *C. nippon*) in eastern Europe.

Wehrdikmansia rugosicauda Böhm and Supperer, 1953 (syn. *Dipetalonema rugosicauda*) occurs in the subcutaneous fascia of the forequarters and back of *Capreolus capreolus* in Austria and surrounding countries. A small form 2.2–2.5 cm in length. The microfilariae concentrate in the nose and to a lesser extent the ears. *Ixodes ricinus* is thought to be the vector (Schulz-Key 1975).

Wehrdikmansia flexuosa (Wedl, 1856) occurs in deer in Eastern Europe and the USSR. Males are 5.4–7.5 cm long and females up to 10 cm. Larvae resembling those in the uterus of the worm are found in the blood sucking insect *Odagmia ornata*.

Cutifilaria wenki Bain and Schulz-Key, 1974, has been described in fallow deer in Europe. Adults occur intradermally and subcutaneously and microfilariae are found in the skin near the site of the adults.

FAMILY: SETARIIDAE SKRJABIN AND SCHIKHOBALOVA, 1945

Head of adult with a peribuccal chitinous ring, with lateral epaulette-like structure or with small teeth. Genera of importance: *Setaria*, *Dipetalonema*, *Stephanofilaria*.

Genus: Setaria Viborg, 1795

The worms of this genus are commonly found in the peritoneal cavity of ungulates. They are several centimetres long, milk-white in colour and taper especially towards the hind end, which is spirally coiled. The mouth is surrounded by a cuticular ring which bears dorsal and ventral and frequently also lateral prominences, giving a characteristic appearance to the worm. The tail of the male bears four pairs of pre-cloacal and usually also four pairs of post-cloacal papillae. The spicules are unequal and dissimilar. The tail of the female may bear spines or a few large, conical projections, while in both sexes there is a pair of small appendages near the tip of the tail. The sheathed microfilariae occur in the blood of the host. Yeh-Liang-Sheng (1959) has provided a revision of the genus, in which there is a deal of confusion. Yeh considers that the generic name *Setaria* should be reserved for forms which occur in equines and the genus *Artionema* applied to those forms which occur in the Bovidae and Cervidae.

Setaria equina (Abildgaard, 1789) is a common parasite of equines in all parts of the world. Males are 40–80 mm long and females 70–150 mm. There are large lateral and smaller, simple, dorsal and ventral peribuccal prominences. The left spicule is 0.63–0.66 mm long and the right 0.14–0.23 mm. The tail of the female ends in a simple point. This worm is found in the peritoneal cavity and sometimes in the scrotum. It has also been recorded from the pleural cavity and the lungs of the horse and from the eye of cattle and horses. Microfilariae are sheathed and 190–256 mm in length. The infection rate may be high and up to 50% of horses may be infected in endemic areas.

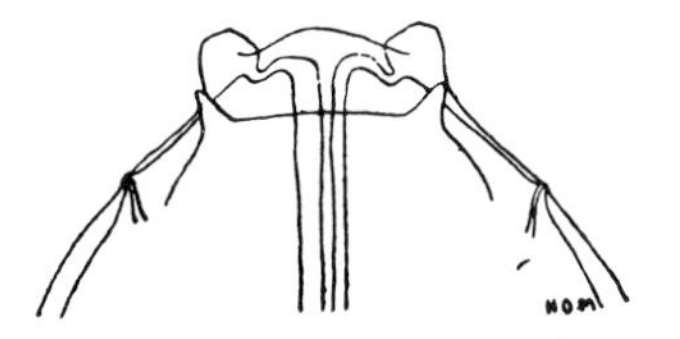

Fig 1.177 Lateral view of anterior end of *Setaria equina*.

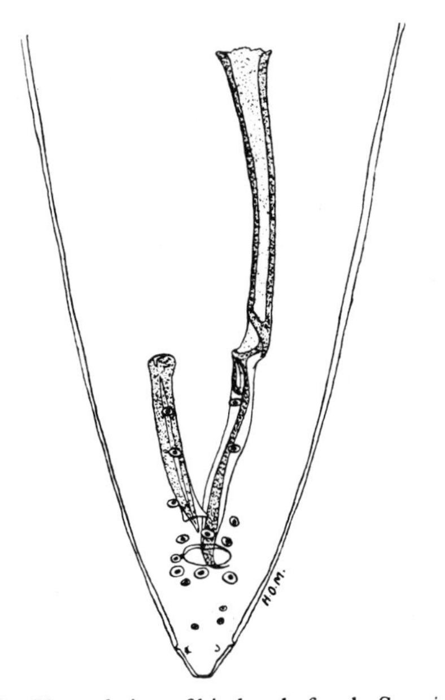

Fig 1.178 Ventral view of hind end of male *Setaria equina*.

Setaria labiato-papillosa (Alissandrini, 1838) occurs in the peritoneal cavity of cattle, deer, giraffe and antelope and is probably cosmopolitan in distribution. This species has frequently been referred to as *S. cervi*; however, there is some confusion in the allocation of specific names in this genus (see Yeh-Liang-Sheng 1959), and Skrjabin and Schikhobalova (1948) give *S. cervi* Maplestone, 1931, as a parasite of *Cervus axis*. Shoho (1958) has proposed that *S. cervi* should be referred to as *S. axis*. Böhm and Supperer (1955) consider *S. labiato-papillosa* and *S. digitata* identical since they were able to find intermediate forms between the two, indicating that a range of characters occur. However, Yeh-Liang-Sheng (1959) reported distinct differences between *S. labiato-papillosa* and *S. digitata* and recommended the species be retained as valid.

Males are 40–60 mm long and females 60–120 mm. Peribuccal ring distinct, the dorsal and ventral prominences are distinct and are 120–150 μm apart. The mouth opening is elongated. The spicules are unequal and measure 120–150 and 300–370 μm. The tail of the female terminates in a marked button which is divided into a number of papillae. The microfilariae are sheathed and measure 240–260 μm.

Setaria digitata von Linstow, 1906 occurs in the peritoneal cavity of cattle, buffalo and zebu in the Far East and Asia. Adults may also be found in lesions in the urinary bladder. The infection rate may be high and in northern Japan 55% of cattle were infected (Yoshikawa et al. 1976). Immature forms are found in the central nervous system of abnormal hosts such as sheep, goats and horses.

Setaria yehi Disset, 1966 occurs in the peritoneal cavity of ungulates of several species such as deer and also moose, caribou and bison in the USA.

Setaria cervi Rudolphi, 1819 is found in the peritoneal cavity of a wide variety of deer in areas such as eastern Europe, the USSR and India.

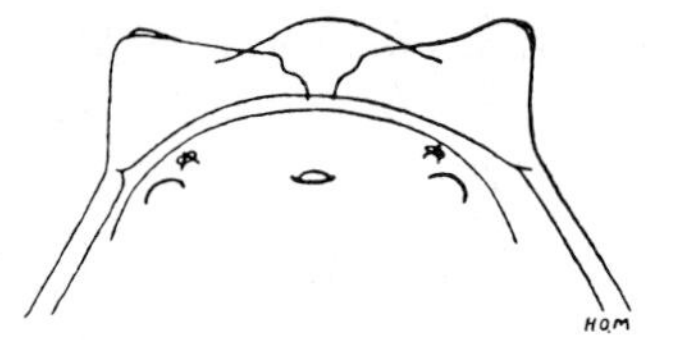

Fig 1.179 Lateral view of anterior end of *Setaria cervi*.

Adults have also been found in the peritoneal cavity of buffalo in India but these may have been *S. digitata*.

Setaria tundrae Issaitschikow and Rajewskaya, 1928, is found in reindeer (*Rangifer tarandus* and *R. asiaticus*).

Several other species of *Setaria* are found in domestic animals, deer and antelope, these include **S. congolensis** Railliet and Henry, 1911, in the peritoneal cavity of pigs in Africa (Zaire), **S. altaica** Rajewskaya, 1928, in *Cervus canadensis asiaticus* in USSR, and **S. cornuta** (von Linstow, 1899), in antelopes in Africa etc.

Life-cycle. Many species show no microfilarial periodicity and only low numbers of microfilariae are present in the blood: in addition, the life-cycle of many species is unknown. However, development of some occurs in the thoracic muscles of mosquitoes such as *Aedes aegypti*, *A. pembaensis* and *Culex* spp. (*Setaria equina*), *Armigeres obturbans*, *Aedes togoi*, *Anopheles*, *Lyrcanus* and *Culex pipiens* (*Setaria digitata*) and *Anopheles gambiae* (*Setaria labiato-papillosa*) and infective larvae are produced in 12–16 days. Development of *S. cervi* has been shown in the USSR to occur in *Haematobia irritans* and *H. stimulans*. In horses and deer the prepatent period is seven to ten months. Prenatal infection has been recorded for a number of *Setaria* species.

Pathogenicity. Adults in the peritoneal cavity are non-pathogenic although they may cause a mild fibrinous peritonitis. However, a moderate to severe peritonitis has been seen in reindeer and *S. cervi* can be a very important parasite on Russian deer farms. *S. digitata* can cause eosinophilic granulomatous lesions in the urinary bladder. On rare occasions adults have migrated erratically and may be found in the eye, particularly of the horse, where they can cause blindness. The major pathogenic effect of *S. digitata* occurs when the immature forms migrate erratically in the central nervous system of abnormal hosts such as sheep, goats and horses. Migration in the CNS has also been recorded for other species of *Setaria* and *S. cervi* has been found in the central nervous system of deer and buffalo and *S. labiato-papillosa* of sheep, goats and horses. Neurological disturbances have also been seen in cattle infected with *Setaria* spp. Enzootic cerebrospinal nematodiasis caused by *S. digitata* occurs in the summer and autumn, when the mosquito intermediate hosts of *S. digitata* are active. The affected animals suffer from acute focal encephalomyelomalacia. The condition has been described in detail by Innes and Shoho (1952, 1953) and the disease has been produced experimentally by Shoho (1960).

The lesions are microscopic and may be overlooked. They are usually single tracks left by migrating young worms and may be found in any part of the central nervous system. Acute malacia occurs in the track of the worm, with disintegration of all tissues at the centre of the lesion and secondary degeneration of the nerve tracts with gigantic swellings of the axis cylinders and eosinophilic infiltration. Occasionally cavities are seen and the lesions are only rarely disseminated. Sugawa et al. (1949) found a malacic lesion with and without the presence of *S. digitata* in horses suffering from Japanese B encephalitis. It is possible that the virus of Japanese B encephalitis may be transmitted to human children and horses by *S. digitata*.

Clinical signs. The signs vary according to the site of the lesion and may be slight if relatively unimportant nervous areas are affected. Clinical signs may vary from muscular weakness and ataxia to paralysis and death.

Treatment. For the treatment of epizootic cerebrospinal nematodiasis Shoho (1952) has recommended 40 mg/kg body weight of diethylcarbamazine given for one to three days. However, a number of compounds including diethylcarbamazine were ineffective against *Setaria* in cattle in India (Manickam & Kathaperumal

1975). Baqui and Ansari (1976) experimentally transferred *S. cervi* from the peritoneal cavity of buffaloes to rats. Treatment of these rats with levamisole at 7.5 and 15 mg/kg/day eliminated the microfilaraemia and killed the adult parasites while diethylcarbamazine, 25–100 mg/kg/day, eliminated the microfilaraemia within three to 13 days but did not kill the adult parasites. Klei et al. (1980) have reported that avermectin given intramuscularly at doses of 0.2 mg–0.5 mg/kg was 80–88% effective against adult *S. equina*.

Genus: Dipetalonema Diesing, 1861 (syn. Acanthocheilonema, Cobbold, 1870)

Dipetalonema dracunculoides (Cobbold, 1870). This species occurs in the peritoneal membranes of dogs and hyaenas in Africa and India. Nelson (1959) found it in these hosts in Kenya and discussed the incidence of this and other species of this genus in Kenya. He stated that its microfilariae are very like those of *Dirofilaria immitis*, but differ in having a short, blunt tail. The male measures 24–30 mm long and 100–200 μm broad and has unequal spicules; the female measures 32–60 mm long by 260–300 μm broad. The microfilariae are not sheathed.

Dipetalonema reconditum (Grassi, 1980). This species has been reported from the body cavity, connective tissues and kidney of dogs in Italy, Africa and the USA. Lindsey (1962) obtained 175 intact adults from 15 of 20 dogs in the USA, the parasites occurring in the subcutaneous tissues. Lindsey recorded the average length of females as 23.4 mm with a range of 17–32 mm. Males had an average length of 13 mm. The spicules are unequal. Microfilariae from gravid uteri examined by Lindsey were 263.9–278.2 μm (mean 270.6 μm) with a width of 4.7–5.8 μm (mean 5.2 μm): microfilariae possessed a button-hook tail. Microfilariae from the gravid uterus of two *D. immitis* worms were 298.6–313.6 μm (mean 308.9 μm) for one worm and 291.2–309.8 μm (mean 298.8 μm) for the other.

Development of *D. reconditum* occurs in fleas, *Ctenocephalides felis*, *C. canis* and *Pulex irritans*, and in ticks *Rhipiciphalus sanguineus* and *Heterodoxus spiniger*. Infective larvae develop in the flea within seven days and the prepatent period in dogs is 61–68 days (Farnell & Faulkner 1978).

No pathogenic effects have been ascribed to the parasite but its microfilariae must be differentiated from those of *D. immitis* in the diagnosis of heartworm infection.

Dipetalonema grassi Noe, 1907, occurs in the subcutaneous tissues of dogs in Italy and Kenya. It is a small form, the female measuring 25 mm in length. The microfilariae are large, 570 μm in length, with a hook-shaped tail. Development occurs in the tick *Rhipicephalus sanguineus*. No pathology is associated with the infection.

Dipetalonema evansi (Lewis, 1882) occurs in the pulmonary and spermatic arteries and lymph nodes of camels in Egypt, the Far East and eastern USSR. This is a fairly large filarid, the male being 75–90 mm and the female 170–215 mm. The microfilariae measure 250–300 μm.

Up to 80% of camels may be infected in the eastern republics of the USSR and it is suspected that the mosquito *Aedes detritus* is an intermediate host. Infected camels suffer arteriosclerosis and heart insufficiency. A parasitic orchitis may occur or aneurysms may be present in the spermatic vessels. The treatment recommended by Soviet workers is fouadin at a dose of 0.5 ml/kg.

Dipetalonema perstans (Manson, 1891) Railliet, Henry and Langeron, 1912, occurs in the peritoneal and pleural cavities of man and anthropoid apes. The infection is fairly widespread in Africa and South America. The intermediate hosts are *Culicoides* spp. such as *C. grahami*, *C. milnei* and *C. austeni*. The worm is not very pathogenic. Its microfilariae are not sheathed and they show no periodicity in the blood. *D. perstans* has been thought to cause cerebral filariasis in man but this is now thought to be due to *Meningonema peruzzi* Orihel and Esslinger 1973 which is a filarial parasite of the CNS of various African monkeys, being found in the subarachnoid spaces of talopin monkeys and *Cereopithecus aethions* (Orihel 1973).

Mebendazole, 400 mg twice a day for 14 days, and mebendazole, 100 mg twice a day plus 50 mg/kg/day of levamisole, have been used to treat *D. perstans* infections. The microfilaraemia has been reduced but apparently only temporarily.

Dipetalonema loxodontis Chabaud, 1952 (syn. *Loxodontofilaria loxodontis*) occurs in the African elephant.

Dipetalonema streptocerca (Macfie & Corson, 1922), the adult and microfilariae of which have been found in the skin of man in Central Africa and in Ghana. They may cause oedema and elephantiasis of the skin. The intermediate hosts are *Culicoides austeni* and probably also *C. grahami*.

Diethylcarbamazine, 50 mg/kg given three times a day for 21 days caused degeneration of adult worms in the subcutaneous tissues within 21–31 days. Some side-effects such as cutaneous papules and mild itching were seen in some treated persons but these resolved by day 31 (Meyers et al. 1978)

Dipetalonema spirocauda Leidy, 1858, is found in the right heart and pulmonary artery of phocids such as the harbour seal (*Phoca vitulina*). and ringed seal (*Cystophora cristata*). Adults cause villous proliferation of the endothelial layer of arteries and verminous emboli may be seen in small branches of the pulmonary artery. Acute vasculitis due to the microfilariae is seen. Also, microfilariae are seen in areas of focal necrosis in the liver and granulomas are present in the spleen (Dunn & Wolke 1976).

Dipetalonema odendhali Perry, 1967 is common in the subcutaneous and intermuscular connective tissue of California sea lions (*Zulophus c. californianus*) in their natural habitat. The microfilaria must be differentiated from those of *Dirofilaria immitis* when sea lions are kept in captivity in endemic areas of *D. immitis*.

Dipetalonema gracile Rudolphi, 1809, **D. marmosetae** Faust, 1935, **D. obtusa** Esslinger, 1966 and **D. tamarinae** Dunn and Lambrecht, 1963 occur in the peritoneal cavity of Central and South American primates and are common in laboratory primates which have been obtained from their natural habitats since the prevalence of infection may be high (Flynn 1973). For instance, *D. gracile* and *D. obtusa* were found in 73% of *Saguinus geoffroyi* and 22.5% of *Cebus capucinus*, respectively. The infections are usually non-pathogenic although heavily infected animals may have peritoneal adhesions.

Genus: Stephanofilaria Ihle and Ihle-Landenberg, 1933

Stephanofilaria dedoesi Ihle and Ihle-Landenberg, 1933, occurs in the skin of cattle in Celebes, Sumatra and Java. Males are 2.3–3.2 mm long and females 6.1–8.5 mm. The oral aperture is surrounded by a protruding cuticular rim which has a denticulate edge. Near to the anterior extremity there is a circular thickening which bears a number of small cuticular spines. Male spicules unequal. Female without an anus.

Stephanofilaria stilesi Chitwood, 1934, is found in the USA and USSR and causes lesions mostly on the underside of the abdomen of cattle. Lesions have been found on 1.5–26.3% of cattle in Uzbek SSR.

Stephanofilaria kaeli Buckley, 1973, causes lesions on the legs of cattle in the Malay Peninsula.

Stephanofilaria assamensis Pande, 1936, causes 'hump sore', a chronic dermatitis, of cattle in the Indian subcontinent (e.g. in Assam, Orissa, West Bengal) and it is widespread in the USSR. The parasite occurs in 4.8 to 37% of cattle in southern and north-eastern Uzbekstan (Dadaev 1978) and in 59% of cattle on the Andaman Islands (Malviya 1972) It has also been reported from lesions of chronic dermatitis of buffaloes and goats in India.

Stephanofilaria zaheeri Singh, 1958, occurs in India and lesions are seen on the inner surface of the pinna of the ear of buffalo. The incidence of infection is high and Agrawal and Dutt (1978) found *S. zaheeri* in 5% of buffalo calves, 20.4% of animals six to 12 months of age, and in up to 95.4% of older buffalo.

Stephanofilaria okinawaensis Ueno and Chibana, 1977 has been described in cattle in Japan where it causes lesions on the muzzle and also on the teat (Ueno et al. 1977). In enzootic areas the infection may be found in 66% of cattle.

Life-cycle. The parasites are transmitted by flies, the adult flies ingesting microfilariae as they feed on the open lesions caused by the worms. Larvae of *S. kaeli* develop in *Musca conducens* and at 26–30°C larvae moult on days 4 and 8 and infective larvae are found in the proboscis of the fly on day 10 (Fadzil 1975). Infective larvae of *S. assamensis* develop to infectivity after 23–25 days in *M. conducens* at 25°C (Patnaik 1973). *S. okinawaensis* is also transmitted by *M. conducens.* The intermediate hosts of *S. stilesi* are *Lyperosia irritans* and *L. titillans. Stomoxys calcitrans* has also been recorded as being an intermediate host.

Pathogenicity and lesions. Bubberman and Kraneveld (1933) describe a verminous dermatitis, commonly known as 'cascado', which is produced by these parasites. Poor condition and high rainfall are predisposing factors and the incidence may be 90% in cattle. A number of small papules develop and coalesce to form a larger lesion covered with crusts. There is ulceration, acanthosis, hyperkeratosis and alopecia and the lesion is rich in blood and lymph which can be squeezed out readily. The lesion extends outwards while the centre becomes hard and covered with a thick, dry crust and may reach 25 cm in diameter. Itching leads to rubbing, and the lesion is aggravated by self-inflicted wounds on the hump by the tip of the horns in Zebu. The worms live in the epithelial layers of the skin and cause an inflammation of the rete Malpighii with proliferation and destruction of epithelial cells, destruction of hair follicles and skin glands and infiltration with small inflammatory cells. Skin lesions contain both microfilariae and adult worms. Secondary bacterial infection may occur. Lesions may partially resolve during the dry, cool, season, but recur at the time of rains. In the buffalo, the lesions are less severe.

Diagnosis. After removal of the crusts deep skin scrapings are made and the microfilariae can be found in this material.

Treatment. A variety of substances, but particularly the organophosphate compounds, have been shown to be effective in the treatment of stephanofilarial dermatitis and are applied topically after the lesions have been cleaned of crusts and necrotic tissues. Trichlorphon applied daily or on alternate days at a 6–10% concentration in petroleum jelly or castor oil cures nearly all the treated animals within seven days (Rahman & Khaleque 1974; Fadzil 1977). However, Baki and Dewan (1975) found that 8% trichlorphon alone was ineffective in the treatment of 'humpsore' while the addition of 4% sulphanilamide to the ointment resulted in healing of the lesions within 20–66 days. Trichlorphon, given parenterally, has also been used.

A common treatment consists of the daily local application of the organophosphate insecticides, Supona 20 and Sumithion. These are used at a 4% concentration and toxic signs are seen if a 6% concentration is used (Das et al. 1977). 2% coumaphos ointment is also effective (Muchlis & Soetijono 1973) and malathion and ambithion are also fairly effective. Broad-spectrum benzimidazoles could be tried (see *Parafilaria bovicola*) and parbendazole as well as levamisole have been reported to be efficacious against *S. okinawaensis* (Ueno & Chibana 1978).

REFERENCES

FILAROIDEA: FILARIIDAE: DIROFILARIA

American Heartworm Association (1978) *Proceedings of Heartworm Symposium 1977.* Bonner Springs, Kansas: Veterinary Medicine Publishing Co.

American Veterinary Medicine Association (1978) Special report. Management of canine heartworm disease. *J. Am. vet. med. Ass.*, 173, 1342–1344.

Blair, L. S. & Campbell, W. C. (1978) Trial of avermectin B_1a, mebendazole and melarsoprol against pre-cardiac *Dirofilaria immitis* in the ferret (*Mustela putonious furo*). *J. Parasit.*, 64, 1032–1034.

Blair, L.S. & Campbell, W. C. (1979) Efficacy of avermectin B_1a against microfilariae of *Dirofilaria immitis*. *Am. J. vet. Res.*, **40**, 1031–1032.
Ciferri, F. (1982) Human pulmonary dirofilariasis in the United States: a critical review. *Am. J. trop. Med.* Hyg., 31, 302–308.
Euzéby, J. & Laine, B. (1951) Sur la periodicite des microfilaires de *Dirofilaria immitis*. Ses variations sous l'influence de divers facteurs. *Revue Vet. Med. Toulouse*, **102**, 231–238.
Hawking, F. (1964) The periodicity of microfilariae. VII. The effect of parasympathetic stimulants upon the distribution of microfilariae. *Trans. R. Soc. trop. Med. Hyg.*, **58**, 178–194.
Hennigar, G. R. & Ferguson, R. W. (1957) Pulmonary vascular sclerosis as a result of *Dirofilaria immitis* infection in dogs. *J. Am. vet. med. Ass.*, **131**, 336–340.
Knight, D. H. (1977) Heartworm heart disease. *Adv. vet. Sci. comp. Med.*, **21**, 107–149.
Kume, S. & Itagaki, S. (1955) On the life-cycle of *Dirofilaria immitis* in the dog as the final host. *Br. vet. J.*, **111**, 16–24.
McCall, J. W. & Crouthamel, H. H. (1976) Prophylactic activity of mebendazole against *Dirofilaria immitis* in dogs. *J. Parasit.*, **62**, 844–845.
Nelson, G. S. (1959) The identification of infective filariae larvae in mosquitoes; with a note on the species found in 'wild' mosquitoes on the Kenya Coast. *J. Helminth.*, **33**, 233–256.
Newton, W. L. & Wright, W. H. (1956) The occurrence of a dog filariid other than *Dirofilaria immitis* in the United States. *J. Parasit.*, **42**, 246–258.
Orihel, T. C. & Beaver, P. C. (1965) Human infection with filariae of animals in the United States. *Am. J. trop. Med. Hyg.*, **14**, 1010–1029.
Sawyer, T. K., Rubin, F. F. & Jackson, R. F. (1965) The cephalic hook in microfilariae of *Dipetalonema reconditum* in the differentiation of canine microfilariae. *Proc. Helminth. Soc. Wash.*, **32**, 15–20.
Schillhorn, T. van Veen (1974) Filariasis in domestic animals in Northern Nigeria and its relation to human health. In: *Parasitic Zoonoses*, ed. E. J. L. Soulsby, pp. 287–293. New York: Academic Press.
Schnelle, G. B. & Young, R. M. (1944) Clinical studies on microfilarial periodicity in war dogs. *Bull. U.S. Army med. Dept.*, **80**, 52.
Spratt, D. M. (1975) Further studies of *Dirofilaria immitis* (Nematoda: Filaroiea) in naturally and experimentally infected miacropodidae. *Int. J. Parasit.*, **5**, 561–564.
Taylor, A. E. R. (1960) Studies on the microfilariae of *Loa loa*, *Wuchereria bancrofti*, *Brugia malayi*, *Dirofilaria immitis*, *D. repens* and *D. aethiops*. *J. Helminth.*, **34**, 13–26.
Vaughan, A. W. (1952) A report on canine filariasis. *Vet. Rec.*, **64**, 454–455.
Webber, W. A. F. & Hawking, F. (1955) Experimental maintenance of *Dirofilaria repens* and *D. immitis* in dogs. *Exp. Parasit.*, **4**, 143–164.
Welch, J. S., Dobson, C. & Freeman, C. (1979) Distribution and diagnosis of dirofilariasis and toxocariasis in Australia. *Aust. vet. J.*, **55**, 265–274.
Wong, M. M. & Suter, P. F. (1979) Indirect fluorescent antibody test in occult dirofilariais. *Am. J. vet. Res.*, **40**, 414–420.

FILARIDAE: MANSONELLA; BRUGIA, PARAFILARIA, ORNITHOFILARIA, ELAEOPHERA, SUIFILARIA, WEHRDIKMANSIA

Anderson, R. C. (1954) The development of *Ornithofilaria fallisensis* Anderson, 1954, in *Simulium venustum* Sag. *J. Parasit.*, **40**, 12.
Anderson, R. C. (1955) Black flies (Simuliidae) as vectors of *Ornithofilaria fallisensis* Anderson, 1954. *J. Parasit.*, **41**, 45.
Bain, O., Muller, R. L., Khamis, Y., Guilhon, J. & Schillhorn van Veen, T. (1976) *Onchocerca raillieti* n. sp. (Filarioidea) in the domestic donkey in Africa. *J. Helminth.*, **50**, 287–293.
Carmichael, I. H. & Koster, S. (1978) Bovine parafilariasis in Southern Africa: a preliminary report. *Onderstepoort J. vet. Res.*, **45**, 213–214.
Denham, D. A. & McGreevy, P. B. (1977) Brugian filariasis: epidemiological and experimental studies. *Adv. Parasit.*, **15**, 243.
Denham, D. A., Suswillo, R. R. & Rogers, R. (1978) Studies with *Brugia pahangi* 19. Anthelmintic effects of mebendazole. *Trans. R. Soc., trop. Med. Hyg.*, **72**, 546–547.
Gnedina, M. P. & Osipov, A. N. (1960) Contribution to the biology of the nematode *Parafilaria multipapillosa* (Gondamine et Drouillu, 1878) parasitic in the horse. *Helminthologia*, 2, 13–16.
Hibler, C. P. & Adcock, J.L. (1971) Elaeophorosis. In: *Parasitic Diseases of Wild Mammals*, ed. J. W. Davis & R. C. Anderson, pp. 263–278. Ames, Iowa: Iowa State University Press.
Hibler, C. P. & Metzger, C. J. (1974) Morphology of the larval stages of *Elaeophora schneideri* in the intermediate and definitive hosts with some observations on their pathogenesis in abnormal definitive hosts. *J. Wildlife Dis.*, **10**, 361–369.
Kemper, H. E. (1957) Filarial dermatosis of sheep. *J. Am. vet. med. Ass.*, **130**, 220–224.
Lapage, G. (1961) A list of the parasitic Protozoa, Helminths and Arthropoda recorded from species of the family Anatidae (ducks, geese and swans). *Parasitology*, **51**, 1–109.
Patnaik, M. M. & Pande, B. P. (1963) A note on parafilariasis in buffalo, *Bos (Bubalus) bubales*. *J. Helminth.*, **37**, 343–348.
Pienaar, J. G. & van den Heever, L. W. (1964) *Parafilaria bovicola* (Tubangui, 1934) in cattle in the Republic of South Africa. *J. S. Afr. vet. med. Ass.*, **35**, 181–184.
Purnomo, D. T. D. & Partono, F. (1977) The microfilaria of *Brugia timori* (Partono *et al.*, 1977 = Timor Microfilaria, David and Edeson, 1964): morphologic description with comparison to *Brugia malayi* of Indonesia. *J. Parasit.*, **63**, 1001–1006.
Schultz-Key, H. (1975) Investigation on the Filariidae of the Cervidae in southern Germany. 3. The Filariidae of the roe deer (*Caprëolus capreolus*) and fallow deer (*Dama dama*). *Tropenmed. Parasit.*, **26**, 494–498.
Soulsby, E. J. L. (1965) *Textbook of Veterinary Clinical Parasitology*, Vol. I, *Helminths*. Oxford: Blackwell Scientific Publications.
Viljoen, J. H. (1976) Studies on *Parafilaria bovicola* (Tubangui, 1934) 1. Clinical observations and chemotherapy. *J. S. Afr. vet. med. Ass.*, **47**, 161–169.
Wellington, A. C. (1978) The effect of nitroxynil in *Parafilaria bovicola* infestations in cattle. *J. S. Afr. vet. med. Ass.*, **49**, 131–132.

SETARIIDAE: SETARIA; DIPETALONEMA, STEPHANOFILARIA

Agrawal, M. C. & Dutt, S. C. (1978) A note on the prevalence of *Stephanofilaria zaheeri* Singh, 1958, infection in buffaloes of different ages. *Indian J. Anim. Sci.*, **48**, 232–234.
Baki, M. A. & Dewan, M. L. (1975) Evaluation of treatment of stephanofilariasis (humpsore) with Neguvon (Bayer) by clinical pathological studies. *Bangladesh vet. J.*, **9**, 1–6.
Baqui, A. & Ansari, J. A. (1976) Comparative studies on the chemotherapy of experimental *Setaria cervi* infection. *Jap. J. Parasit.*, **25**, 409–414.
Böhm, L. K. & Supperer, R. (1955) Untersuchunger über Setarien (Nematoda) bei heimischen Wiederkäuern und deren Beziehung zur 'Epizootischen cerebrospinal Nematodiasis' (Setariosis). *Z. Parasitk.*, **17**, 165–174.
Bubberman, C. & Kraneveld, F. C. (1933) Over een dermatitis squamosa et crustosa circumscripta by het rund in Nederlandsch Indie genamd, cascado, III. Het voorkommen van cascado by de geit. *Ned. Ind. Diergeneesk.*, **46**, 67–73.
Dadaev, S. (1978) The prevalence of the nematode *Stephanofilaria assamensis* Pande, 1936 in cattle in Uzbekistan. *Doklady Akad. Nauk.*, **7**, 70–71.
Das, P. K., Tripathy, S. B. & Mishra, S. K. (1977) Study on the efficacy of some organophosphorus compounds in the treatment of stephanofilarial dermatitis in cattle. *Orissa vet. J.*, **11**, 110–114.
Dunn, J. L. & Wolke, R. E. (1976) *Dipetalonema spirocerca* infection in the Atlantic Harbor Seal (*Phoca vitulina concolor*). *J. Wildl. Dis.*, **12**, 531–538.
Fadzil, M. (1975) The development of *Stephanofilaria kaeli* Buckley, 1937 in *Musca conducens* Walker, 1859. *Kajian Vet.*, **7**, 1–7.
Fadzil, M. (1977) *Stephanofilaria kaeli* infection in cattle in Peninsular Malaysia—prevalence and treatment. *Vet. Med. Rev.*, **1**, 44–52.
Farnell, D. L. & Faulkner, D. R. (1978) Prepatent period of *Dipetalonema reconditum* in experimentally-infected dogs. *J. Parasit.*, **64**, 565–567.
Flynn, R. J. (1973) *Parasites of Laboratories Animals*. Ames, Iowa: Iowa State University Press.
Innes, J. R. M. & Shoho, C. (1952) Nematodes, nervous disease, and neurotropic virus infection. Observations in animal pathology of probable significance in medical neurology. *Br. med. J.*, *x*, 366–368.

Innes, J. R. M. & Shoho, C. (1953) Cerebrospinal nematodiasis. Focal encephalomyelomalacia of animals caused by nematodes (*Setaria digitata*); a disease which may occur in man. *Archs Neurol. Psychiat.*, **70**, 325–349.
Klei, T. R., Torbet, B. J. & Ochoa, R. (1980) Efficacy of Ivermectin (22,23-dihydroavermectinB_1) against adult *Setaria equina* and microfilariae of *Onchocerca cervicalis* in ponies. *J. Parasit.*, **66**, 859–861.
Lindsey, J. R. (1962) Diagnosis of filarial infections in dogs. II. Confirmation of microfilaria identifications. *J. Parasit.*, **48**, 321–326.
Malviya, H. C. (1972) Stephanofilarial infection in cattle and buffalo in Andaman Islands. *Indian J. Helminth.*, **24**, 68–71.
Manickam, R. & Kathaperumal, V. (1975) A note on the treatment trials on setarial microfilariasis in bullocks. *Cherion*, **4**, 68–71.
Meyers, W. M., Moris, R., Neafie, R. C., Conner, D. H. & Bourland, J. (1978) Streptocerciasis: degeneration of adult *Dipetalonema streptocerca* in man following diethylcarbamazine therapy. *Am. J. trop. Med. Hyg.*, **27**, 1137–1149.
Muchlis, A. & Soetijono, P. (1973) A short report on the use of Asuntol ointment in the treatment of Cascado and hoof myiasis. *Vet. Med. Rev.*, **2**, 134.
Nelson, G. S. (1959) The identification of infective filariae larvae mosquitoes; with a note on the species found in 'wild' mosquitoes on the Kenya Coast. *J. Helminth.*, **33**, 233–256.
Orihel, T. C. (1973) Cerebral filariasis in Rhodesia—a zoonotic infection? *Am. J. trop. Med. Hyg.*, **22**, 596–599.
Patnaik, B. (1973) Studies on stephanofilariasis in Orissa. III. Life cycle of *S. assamensis* Pande, 1936. *Z. Tropenmed. Parasit.*, **24**, 457.
Rahman, A. & Khaleque, A. (1974) Treatment of 'humpsore' with Neguvon in local cattle of Bangladesh. *Vet. Med. Rev.*, **4**, 379–382.
Shoho, C. (1952) Further observations on epizootic cerebrospinal nematodiasis. I. Chemotherapeutic control of the disease by 1-diethyl-carbomy-4-methylpoperazine citrate: Preliminary field trial. *Br. vet. J.*, **108**, 134–141.
Shoho, C. (1958) Studies of cerebro-spinal nematodiasis in Ceylon. V. On the identity of *Setaria* spp. from the abdominal cavity of Ceylon spotted deer *Axis axis ceylonensis*. *Ceylon vet. J.*, **6**, 15–20.
Shoho, C. (1960) Studies of cerebrospinal nematodiasis in Ceylon. VII. Experimental production of cerebrospinal nematodiasis by the inoculation of infective larvae of *Setaria digitata* into susceptible goats. *Ceylon vet. J.*, **8**, 2–12.
Skrjabin, D. I. & Schikbobalova, N. I. (1948) *Filariata*. Moscow: Akad. Nauk SSSR.
Sugawa, Y., Mochizuki, H. & Yamamoto, S. (1949) *First Report on Japanese Equine Encephalitis*. Government Experimental Station for Animal Hygiene.
Ueno, H. & Chibana, T. (1978) Stephanofilariasis caused by *S. okinawaensis* of cattle in Japan. *Japan. agric. Res. Q.*, **12**, 152–156.
Ueno, H., Chibana, T. & Yamashiro, E. (1977) Occurence of chronic dermatitis caused by *Stephanofilaria okinawaensis* on the teats of cows in Japan. *Vet. Parasit.*, **3**, 41–48.
Yeh-Liang-Sheng (1959) A revision of the nematode genus *Setaria* Viborg, 1795, its host-parasite relationship speciation and evolution. *J. Helminth.*, **33**, 1–98.
Yoshikawa, T., Oyamada, T. & Yoshikawa, M. (1976) Eosinophilic granulomas caused by adult seterial worms in the bovine urinary bladder. *Jap. J. vet. Sci.*, **38**, 105–116.

FAMILY: ONCHOCERCIDAE CHABAUD AND ANDERSON, 1959

Genus: Onchocerca Diesing, 1841

The species of this genus are elongate, filariform worms. They live in the connective tissue of their hosts, often giving rise to firm nodules in which they lie coiled up. The cuticle is transversely striated and, in addition, bears characteristic spiral thickenings which are usually interrupted in the lateral fields.

Microfilariae are found in the skin in lymph spaces and connective tissue spaces. The intermediate hosts are insects of the families Simuliidae and Ceratopogonidae.

Onchocerca gibsoni Cleland and Johnston, 1910, occurs in cattle and the zebu in Asia, Australasia and southern Africa. The worms are usually found in nodules which may occur especially on the brisket and the external surfaces of the hind-limbs and it is difficult to extricate complete specimens. The male is 30–53 mm long, the female 140–190 mm, but it is stated that she may be 500 mm or more long. The tail of the male is curved ventrad; it bears small lateral alae and six

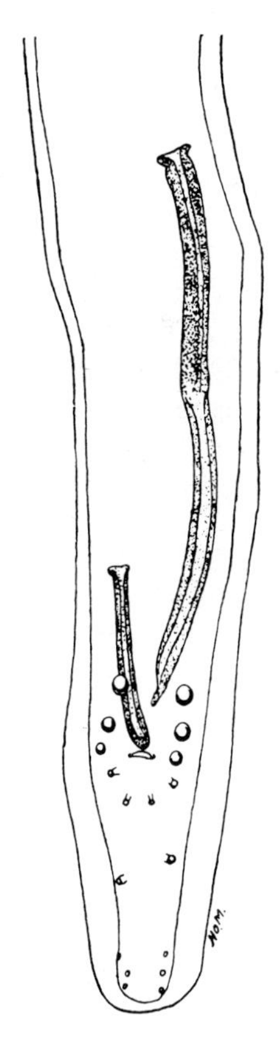

Fig 1.180 Ventral view of hind end of male *Onchocerca gibsoni*.

to nine papillae on either side. The spicules measure 0.14–0.22 mm and 0.047–0.094 mm respectively. The microfilariae are 240–280 μm in length (mean 266 μm); they are not sheathed; the nerve ring is situated about 60–70 μm from the anterior end; the anterior and posterior portions of the larvae which contain no nuclei are respectively 2.5–5 μm and 9–13 μm long; the body cuticle bears rather pronounced transverse striations. Microfilariae are found primarily in the brisket region.

The intermediate host is the midge *Culicoides pungens*.

Onchocerca gutturosa Neumann, 1910 occurs in the ligamentum nuchae and other parts of the body such as on the scapular cartilage and in the hip, stifle and shoulder regions of cattle and buffalo in most parts of the world. Males are 2.9–3.0 cm long and females 60 cm or more Some authors consider *O. gutturosa* to be synonymous with *O. lienalis* (see below).

Bremner (1955) described the morphological differences between *O. gutturosa* and *O. gibsoni*. The microfilariae, 200–230 μm, are found in the skin of the host but the predilection site of the microfilariae appears to be variable. This may be associated with the difficulties in speciation of the *Onchocerca* species in cattle or with differences in the feeding habits of the simuliid intermediate hosts in different geographical areas. For instance Eichler and Nelson (1971) found microfilariae of *O. gutturosa* to be concentrated in the umbilical region of British cattle which is consistent with their transmission by *S. ornatum* (Steward 1937) which preferentially feeds in the midline area. Conversely, Elbihari and Hussein (1978) found microfilariae concentrated in the midline of the hump and back of cattle in the Sudan, while in Australia, the Sudan and Tanzania microfilariae were found primarily in the cephalic region (ear and neck), although they are found in other areas of the body (umbilicus, scrotum, udder) (Mwaiko 1979; Hussein et al. 1975). In Tanzania this was consistent with *O. gutturosa* being transmitted by *S. vorax* which feeds primarily on the ear, although *S. nyasalandicum* which feeds on the ventral abdomen may also be involved (Mwaiko 1979).

The intermediate hosts in the family Simuliidae include *Simulium ornatum*, *S. vorax*, *S. japonicum*, *Odagmia ornata* and *Friesia alajensis*. The parasite develops to the infective stage in about three weeks.

Onchocerca lienalis Stiles, 1892, occurs in the gastrosplenic ligament, on the spleen capsule and above the xiphisternum in cattle in areas such as Australia and the USA. Some authors consider *O. lienalis* to be synonymous with *O. gutturosa* and if this is the case then *O. lienalis* is distributed throughout the world. Microfilariae are more prevalent in the umbilical than the cervical region.

Onchocerca cervicalis Railliet & Henry, 1910 occurs in the ligamentum nuchae of the horse and mule in many parts of the world. The male is 6–7 cm in length and the female up to 30 cm. The microfilariae are 200–240 μm in length and 4–5 μm in diameter, unsheathed and with a short tail. Mellor (1973*a*, *b*) demonstrated that over 95% of the microfilariae were found in the skin over the linea alba, consistent with the fact that the intermediate hosts, *Culicoides nebeculosus* midges, land and feed predominantly in the midline region.

Development occurs in the biting midge *Culicoides nebeculosus*, the larval stages developing in the thoracic muscle of the insect. Infective forms 600–700 μm in length are found in the proboscis of the fly 24–25 days after infection. In addition to *C. nebeculosus*, *C. obsoletus* and *C. parroti* may act as the intermediate host. The mosquitoes *Anopheles maculipennis* and *A. sacharovi* have been reported as intermediate hosts (Yoeli et al. 1948).

Onchocerca reticulata Diesing, 1841 occurs in the connective tissue of the flexor tendons and the suspensory ligament of the fetlock, chiefly the foreleg, of the horse, mule and donkey. The microfilariae occur in the skin and measure 330–370 μm and possess a long, whiplash-like tail. Development is similar to that of *O. cervicalis* in *Culicoides nubeculosus*.

Mellor (1974) considered that there was insufficient evidence for the separation of *O. cervicalis* and *O. reticulata*; however, Lichtenfels (1975) considered the parasites to be separate species.

Onchocerca raillieti Bain, Muller, Khamis, Guilhon and Schilhorn von Veen, 1976 has been reported from the domestic donkey (*Equus asinus*) in Africa. Adults occur in the cervical ligament, in subcutaneous cysts on the penis and in the perimuscular connective tissue.

Onchocerca ochengi Bwangamoi, 1969 is found in subcutaneous and intradermal nodules on the ventral regions and flanks, primarily the udder and scrotum, of cattle in East and West Africa.

Onchocerca dukei Bain, Bussienas and Amegee, 1974. Adults of this species are found in the subcutaneous and perimuscular tissue of cattle in West Africa. The nodules are found particularly in the thorax, abdomen, diaphragm and thighs and may be confused with the metacestodes of *Taenia saginata*.

Onchocerca sweetae Spratt and Morehouse, 1971 has been reported from intradermal nodules in the pectoral region of the buffalo, *Bubalis bubalis*, near Darwin in the Northern Territory of Australia.

Onchocerca cebei Galliard, 1937 occurs in buffalo in the Far East. Adult parasites have the same distribution as *O. gibsoni*.

Onchocerca armillata Railliet and Henry, 1909 occurs in the aorta of cattle, buffalo, sheep and goats and has been reported in donkeys in Africa and Asia. The infection may be common and has been seen in up to 90% of cattle. Males are about 7 cm long and females up to 70 cm. The microfilariae are unsheathed, 285–300 μm by 6 μm and are found primarily in the skin of the hump or the withers, but also on the neck, dewlap and umbilical area (Shastri 1978). The life-cycle is unknown.

Onchocerca synceri Sandground, 1938 has been reported from the subcutaneous tissues of the African buffalo (*Sycerus caffer*).

Onchocerca tarsicola Bain and Schulz-Key, 1974, **O. tubingensis** Bain and Schulz-Key, 1974 and **O. garmsi** Bain and Schulz-Key, 1976 are parasites of deer in Europe (Germany).

Pathogenesis of the onchocerciases

Adult parasites. The incidence of infection with *Onchocerca* species may be high. For instance, up to 100% of cattle may be infected with *O. gutturosa*, *O. lienalis* and *O. gibsoni* in northern Queensland and *O. dukei* has been found in 50–90% of cattle in West Africa. Similarly, the incidence of infection with *O. cervicalis* is high and 22–61% of horses have been found infected in the USA.

Onchocerca gibsoni produces nodules in the brisket and occasionally elsewhere. The nodule, or 'worm nest' contains a coiled-up worm surrounded by fibrous tissue. The 'worm nest' proper is surrounded by a fibrous tissue capsule which increases in thickness as the lesion grows older. The whole nodule may be up to 5 cm in diameter and it is ovoid or flattened in shape. In older nodules degeneration of the tissues and calcification frequently take place, or caseation occurs. Infected animals show no clinical signs except the nodular swellings under the skin. The importance of the parasites is mainly that infected carcasses are not suitable for sale or may have to be trimmed extensively.

Adult parasites of *O. cervicalis*, *O. gutturosa*, *O. linealis* and *O. dukei* in the subcutaneous connective tissues, ligamentum nuchae, etc. are non-pathogenic. The parasites occur in tunnels and nodules in the connective tissue, with usually little reaction to the live parasites. Chuahan and Pande (1972) have described the lesions of *O. gutterosa* in buffaloes and bullocks. Degenerate worms are surrounded by a granulomatous reaction which becomes encapsulated with connective tissue and eventually calcifies. For many years *O. cervicalis* was assumed to be the cause of fistulous withers or poll-evil in horses. This is now known not to be the case. Heavy infections of *O. reticulata* in the flexor tendons and suspensory ligaments of horses have been associated with lameness.

Onchocerca armillata adults are found in tortuous tunnels and nodules in the aorta. Early in infection raised tunnels and a few nodules are seen. In older infections the aortic wall is thickened, the intima shows numerous tortuous tunnels and there are numerous nodules containing yellow caseous or slimy fluid and coiled worms. In chronic infections the nodules become calcified, the aortic wall is thin, with linear, broken, calcified ridges, and aneurysms may be seen (Cheema & Ivoghli 1978). Clinical signs are not usually attributable to the adult parasite although Patnaik (1962) suggested that epileptiform fits, blindness and periodic ophthalmia may be seen in some cattle with high counts of microfilariae. Concomitant infections with *O. armillata* and *Spirocerca lupi* in the aorta of goats have been reported by Chowdhury and Chakraborty (1973).

Microfilariae. In the majority of animals there is no obvious clinical evidence to indicate the presence of microfilariae in the skin. However, a seasonal (summer), sporadic dermatitis given various names such as 'wahi' and 'kasen' in cattle, summer mange, allergic dermatitis etc. in horses is seen in many parts of the world. There is papular or exudative dermatitis with alopecia and severe pruritus. The condition has been associated with the presence of large numbers of microfilariae in the affected dermis (Ishihara 1958) but it has also been observed in animals in which no microfilaria are discernible. Thus, it is considered by many authors that the condition is a result of an allergic reaction to the bites of the ceratopogonid or simuliid insects. However, in man, dermatitis is associated with the presence of microfilariae of *Onchocerca volvulus* in the skin and exacerbation of this dermatitis as the Mazzotti reaction following treatment with diethylcarbamazine may be associated with a reaction to the dead microfilariae although the mechanism of this phenomenon and the resultant dermatitis is unknown (Henson et al. 1979). Thus, it is possible that the aetiology of the dermatitis in animals is two-fold, being associated with an immunological reaction to dead and dying microfilariae in the skin of some animals and with an allergic reaction to the insect bites in other animals.

Ocular onchocerciasis is a well recognized entity in man, in whom *O. volvulus* is the cause of 'river blindness'. *Onchocerca* microfilariae have also been incriminated as causing pathological changes (periodic ophthalmia and blindness) in the eyes of animals, particularly horses. Ocular onchocerciasis in the horse has been reviewed by Cello (1971). Periodic ophthalmia was manifested by recurrent attacks of keratitis, conjunctivitis, kerato-uveitis and anterior uveitis. However, microfilariae may be found in apparently normal eyes and it has not yet been conclusively proved that *Onchocerca* species are involved in periodic ophthalmia in horses and cattle. In a review of the pathology of *O. cervicalis* of horses, Mellor (1973*a*, *b*) concluded that neither the adults in the ligament nor the microfilariae in the skin or in the eyes are of significance.

Diagnosis

Adult *Onchocerca* in the ligamentum nuchae, gastrosplenic ligament etc. can be detected only on post mortem, although nodules of *O. ochengi*, *O. gibsoni* etc. may be palpated in the subcutaneous tissue. Microfilariae will be found in the subcutaneous lymph spaces. A skin biopsy is taken from the site of predeliction of the microfilariae (ear region, umbilicus, brisket, etc.). It is teased apart and suspended in warm normal saline and incubated for at least six hours. Microfilariae, if present, will be found in the deposit. Alternatively, the epidermis may be shaved off a skin site and the fluid expressed examined for microfilariae. In the eye microfilariae may be detected by ophthalmological examination. Cello (1971) found the most constant source of microfilariae to be the stroma of the cornea adjacent to the limbus in the upper quadrant. A snip or biopsy of conjunctiva may be taken from here and treated as above.

Treatment

Diethylcarbamazine is microfilaricidal and is used in both man and animals. 5–8 mg/kg is given daily for 21 days. Systemic corticosteroids will suppress the Mazzotti reaction in the skin and should be applied topically or subconjunctivally to suppress possible inflammation in the eye (Cello 1971). Cello also recommends that horses with uveitis should not be treated until the inflammation has subsided. Suramin has been used to kill adult parasites in man. Metrifonate suppresses microfilaraemia and trichlorophon has been used with success in horses. The avermectins are of value in onchocerciasis. Klei et al. (1980) reported high activity of Ivermectin (22,23 dihydroavermectin B_1) against microfilariae of *O. cervicalis* at doses of 0.2–0.5 mg/kg given by intramuscular injection. No gross or clinical reaction was seen with treatment.

REFERENCES

ONCHOCERCIDAE

Bremner, K. C. (1955) Morphological studies on the microfilariae of *Onchocerca gibsoni* (Cleland & Johnston), and *Onchocerca gutturosa* (Neumann) (Nematoda: Filaroidea). *Aust. J. Zool.*, **3**, 324–330.

Cello, R. M. (1971) Ocular onchocerciasis in the horse. *Equine vet. J.*, **3**, 148–154.

Cheema, A. H. & Ivoghli, B. (1978) Bovine onchocerciasis caused by *Onchocerca armillata* and *O. gutturosa*. *Vet. Path.*, **15**, 495–505.

Chowdhury, N. & Chakraborty, R. L. (1973) On current and early aortic Onchocerciasis and spirocerciasis in domestic goats (*Capra hircus*). *Z. Parasitenkde*, **42**, 207–212.

Chuahan, P. P. S. & Pande, B. P. (1972) Onchocercal lesions in aorta and ligamentum nuchae of buffaloes and bullocks: A histological study. *Indian J. Anim. Sci.*, **42**, 809–813.

Eichler, D. A. & Nelson, G. S. (1971) Studies on *Onchocerca gutturosa* (Neumann, 1910) and its development in *Similium ornatum* (Meigen, 1818). I. Observations on *O. gutturosa* in cattle in south-east England. *J. Helminth.* **65**, 245–258.

Elbihari, S. & Hussein, H. S. (1978) *Onchocerca gutturosa* (Neumann, 1910) in Sudanese cattle. I. The microfilariae. *Revue Élev. Méd. vét Pays trop.*, **31**, 179–182.

Henson, P. M., MacKenzie, C. D. & Spector, W. G. (1979) Inflammatory reactions in onchocerciasis: a report on current knowledge and recommendations for further study. *Bull. Wld Hlth Org.*, **57**, 667–671.

Hussein, M. F., Nur, D. A., Gassouma, M. S. & Nelson, G. S. (1975) *Onchocerca gutturosa* (Neumann, 1910) infection in Sudanese cattle. *Br. vet. J.*, **131**, 76–84.

Ishihara, T. (1958) La filariose chez les animaux domestiques au Japon. II, La gale d'été ('kasan disease') du cheval. III, Le kase ou 'wahi' maladie du betail. *Bull. Off. int. Epizoot.*, **49**, 531–535.

Klei, T. R., Torbert, B. J. & Ochoa, R. (1980) Efficacy of Ivermectin (22,23-dihydroavermectinB_1) against adult *Setaria equina* and microfilariae of *Onchocerca cervicalis* in ponies. *J. Parasit.*, **66**, 859–861.

Lichtenfels, J. R. (1975) Helminths of domestic equids. *Proc. helminth. Soc. Wash.*, **42**, Special issue.

Mellor, P. S. (1973*a*) Studies on *Onchocerca cervicalis* Railliet and Henry 1910: I. *Onchocerca cervicalis* in British horses. *J. Helminth.*, **47**, 97–110.

Mellor, P. S. (1973*b*) Studies on *Onchocerca cervicalis* Railliet and Henry 1910: II. Pathology in the horse. *J. Helminth.*, **47**, 111–118.

Mellor, P. S. (1974) Studies on *Onchocerca cervicalis* Railliet and Henry 1910: III. Morphological and taxanomic studies on *Onchocerca cervicalis* from British horses. *J. Helminth.*, **48**, 145–153.

Mwaiko, G. L. (1979) *Onchocerca gutturosa* in Tanzanian cattle. Its prevalence and distribution in north-eastern Tanzania. *Tanzanian vet. Bull.*, **1**, 8–12.

Patnaik, B. (1962) Onchocerciasis due to *O. armillata* in cattle in Orissa. *J. Helminth.*, **36**, 313–326.

Shastri, U. V. (1978) A note on microfilariae of *Onchocerca armillata*. Indian *vet. J.*, **55**, 741–743.

Steward, J. S. (1937) The occurrence of *Onchocerca gutturosa* Neumann in cattle in England, with an account of its life history and development in *Simulium ornatum* Mg. *Parasitology*, **29**, 212–219.

Yoeli, M., Roden, A. T. & Abbott, J. D. (1948) Smears from the subcutaneous nodules of a mule showing microfilariae *Onchocerca* spp. 2. Developmental forms of same species in *Anopheles sacharovi* and *A. maculipennis*. *Trans. R. Soc. trop. Med. Hyg.*, **41**, 444.

SUPERFAMILY: DRACUNCULOIDEA CAMERON, 1934

Stoma rudimentary; internal circle of cephalic papillae well developed. Family of importance *Dracunculidae*.

FAMILY DRACUNCULIDAE LEIPER, 1912

Anterior end with helmet. Females very much larger than the male. Anus and vulva atrophied in the gravid female. Genera of importance: *Dracunculus*, *Avioserpens*.

Genus: Dracunculus Reichard, 1759

Dracunculus medinensis (Linnaeus, 1758), the 'guinea-worm', is a parasite of man in India, Pakistan, West Africa, north-east Africa and the Middle East. It is also seen in the dog and sometimes horse, cattle and other animals in these areas. The male is 12–29 mm long. The female may be 100–400 cm long and 1.7 mm thick and has no vulva. The adult female lives in the subcutaneous connective tissue, and cutaneous swellings, which later become ulcers, develop around its anterior extremity. When these lesions come in contact with water the uterus prolapses through the anterior end of the worm or through its mouth and ruptures, discharging a mass of larvae which are 500–750 μm long, into the water. These larvae have to develop in a species of *Cyclops* to become infective for the final host. Infection of the latter takes place through drinking water which contains the infected *Cyclops* and the worms develop to maturity in about one year. Shortly before the parasite comes to the surface of the body the host may show signs such as urticaria, itching and a rise of temperature.

The classical method of treatment is to remove the worm, which must be done carefully without breaking it. It is usually tied to a small stick and gradually rolled up in the course of a few days or weeks. Diethylcarbamazine in large doses is stated to kill the adult worms and the developing larval stages. The female parasites may be removed surgically. Niridazole, 25 mg/kg/day given in divided doses for 10 days, and thiabendazole, 50 mg/kg/day for two or three days, promote spontaneous elimination or removal of the worms, suppress the inflammation and promote the healing of ulcers.

Dracunculus insignis (Leidy, 1858) Chandler, 1942 is a form which occurs in wild carnivores and the dog in North America. It may be separated from *D. medinensis* on the basis of length of gubernaculum and number of preanal papillae.

The infection occurs in carnivores throughout North America (Ewing & Hibbs 1966). It may be widespread in racoons (*Procyon lotor*) and mink (*Mustela vision*), of which more than 50% were found to be infected in southern Ontario (Crichton & Beverley-Burton 1974). These appear to be the natural definitive hosts. Intermediate hosts are *Cyclops* species, in which the developmental cycle is similar to *D. medinensis. Rana pipiens* and *R. clamitrans*, but not fish, have been shown experimentally to be suitable paratenic hosts (Crichton & Beverley-Burton 1977). In the definitive host, third- and fourth-stage larvae are found in the tissues of the thorax and abdomen. The female may be found in the extremities as early as 120 days after infection but the prepatent period is 309–410 days (Crichton & Beverley-Burton 1975). The parasite is commonly found in the subcutaneous tissues of the limbs but occasionally it may be found elsewhere, such as the conjunctiva, heart, vertebral column, scrotum, etc. It is associated with swellings on the tiba and tarsus which progress to non-healing ulcers.

Treatment consists of surgical removal of the parasite from the ulcer; alternatively treatment with niridazole or thiabendazole, as for *D. medinensis*, should be tried.

Dracunculus lutrae Crichton and Beverley-Burton, 1973 has been reported in *Lutra canadensis* in Canada and **D. fuelleborni** Travassos, 1934 in the opossum in Brazil and the USA. A number of species have been reported in reptiles. These include **D. globocephalus** Mackin, 1927 in turtles and **D. dahomensis** (Neumann, 1895) Moorthy, 1937, **D. ophidensis** Brackett, 1938 and **D. alii** Deshmukh, 1969 in snakes. However, the taxonomy of these is still uncertain.

Genus: Avioserpens Wehr & Chitwood, 1934

Avioserpens taiwana (Sugimoto, 1919, 1934), is a species found in the subcutaneous tissues of domestic ducks in China, mainly in the dry season (January to April), and in Taiwan, where disease caused by it may also occur in September to October. It affects ducks three weeks to two months old. The male is unknown. The female is up to 25 cm long by 0.8 mm in width. The anterior end is rounded, the mouth being surrounded by a chitinous rim bearing two prominent lateral papillae. There are four smaller papillae further back on the head. The uterus is large and filled with larvae. The vagina and vulva and anus are atrophied. The tail ends in a conical papilla.

The larvae of this species occur in *Cyclops* sp. in Taiwan.

Avioserpens mosgovoyi Supriaga, 1966, is similar to the above and occurs in domestic ducks and wild water fowl in USSR and the Far East. The developmental cycle is probably the same as *A. taiwana*. Fish appear to act as paratenic hosts since the infection in ducklings was stopped when fish caught from a river inhabited by bald coots were removed from the diet (Garkavi & Golub 1974).

Pathogenicity. The worms cause the formation of swellings under the mandible, which are at first soft and movable and after about one month hard and painful. They may reach the size of a large nut. They interfere with swallowing and respiration and may cause death from inanition or asphyxia. Occasionally the swellings occur on the shoulders and legs and interfere with the bird's movements. Numerous microfilariae are found in the blood. Usually the adult worms escape by piercing the swelling and healing occurs, the disease lasting about one and a half months. Sometimes the worms die in the swellings, which may then become abscesses. The general effect on birds that survive is unthriftiness and retardation of growth.

Treatment. Removal of the worms through an incision into the most prominent part of the swelling and antiseptic treatment of the swelling results in cure in a week or so.

Prophylaxis. Arrangements should be made for rearing ducklings in the rainy season and providing them with water free from *Cyclops.* They should not be allowed access to marshland.

REFERENCES

DRACUNCULOIDEA

Crichton, V. F. J. & Beverley-Burton, M. (1974) Distribution and prevalence of *Dracunculus* spp. (Nematoda: Dracunculoidea) in mammals in Ontario. *Can. J. Zool.*, **52**, 163–167.
Crichton, V. F. J. & Beverley-Burton, M. (1975) Migration, growth and morphogenesis of *Dracunculus insignis* (Nematoda: Dracunculoidea). *Can. J. Zool.*, **53**, 105–113.
Crichton, V. F. J. & Beverley-Burton, M. (1977) Observations on the seasonal prevalence, pathology and transmission of *Dracunculus insignis* (Nematoda: Dracunculoidea) in the raccoon (*Procyon lotor*) (L.) in Ontario. *J. Wildl. Dis.*, **13**, 273–280.
Ewing, S. A. & Hibbs, C. M. (1966) *Dracunculus insignis* (Leidy, 1858) in dogs and wild carnivores in the Great Plains. *Am. Mid. Nat.*, **12**, 87–132.
Garkavi, H. L. & Goblub, V. E. (1974) *Avioserpens mosgovoyi* in ducks. *Veterinariya*, **9**, 73–74.

SUBCLASS: ADENOPHOREA CHITWOOD, 1958 (FORMERLY APHASMIDIA)

Nematodes with phasmids absent. Caudal glands present or absent. Caudal alae usually absent.

ORDER: ENOPLIDA SCHURMANS, STEKHOVEN AND DECONINCK, 1933

SUPERFAMILY: TRICHUROIDEA RAILLIET, 1916

In the nematodes of this superfamily the muscular tissue of the oesophagus is much reduced and the oesophageal glands are outside the oesophagus in the form of a single row of cells. Males have one or no spicules. Families of importance are *Trichinellidae, Trichuridae* and *Capillaridae.*

FAMILY: TRICHINELLIDAE WARD, 1907

The adults of the single genus of this family are small, the posterior part of the body being only slightly thicker than the anterior part. The male has neither a copulatory spicule nor a spicule sheath. The female is larviparous.

Genus: Trichinella Railliet, 1895

Trichinella spiralis (Owen, 1833) (syn. *Trichina spiralis*) occurs in the small intestine of man, pig, rat and many other mammals. Even birds have been infected experimentally. It is probably cosmopolitan in distribution; though it has not been reported from some countries it is recognized in most of the northern hemisphere, Africa, Asia and South America.

Strain differences have been demonstrated in *T. spiralis*, including different infectivities for experimental animals and pigs and these differences have been confirmed by interbreeding experiments. As a result it has been proposed that *T. spiralis* should be divided into four species. *T. spiralis* remains the synanthropic–zoonotic species which occurs primarily in rats, pigs and mice. *T. nativa* Britov and Boev, 1972, occurs as a sylvatic cycle in wild carnivores in areas such as Canada and the USSR, mostly north of latitude 38°N. *T. nelsoni* Britov and Boev, 1972, is found in wild carnivores primarily in eastern and southern Africa, but has also been recognized in the southern parts of the USSR, eastern Europe (Bulgaria) and Switzerland; its area of distribution corresponds to the southern part of the Old World up to latitude 47°N. *T. pseudospiralis* Garkavi, 1972 has been isolated from a raccoon in the northern Caucasus; it may also occur in India. It does not produce a muscle cyst and it infects birds as well as mammals. The morphological and embryological characteristics and host specificity of *T. pseudospiralis* are discussed by Bessonov et al. (1978). However, the validity of the new species is questioned by a number of

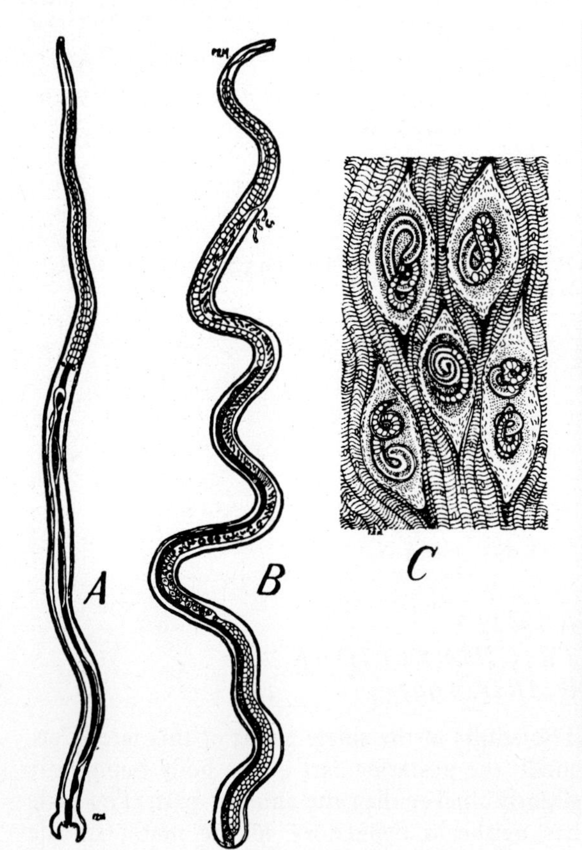

Fig 1.181 *Trichinella spiralis*. A, Adult male. B, Adult female. C, Larvae encapsulated in muscle.

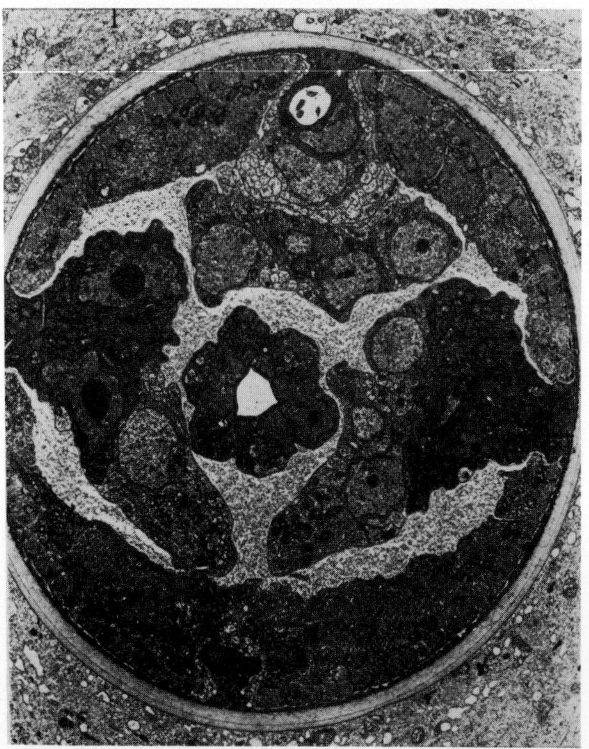

Fig 1.182 Electron micrograph of *Trichinella spiralis* larvae in a muscle fibre ('nurse cell'). Note the close apposition between cell contents and larva. Numerous mitochondria and much rough endoplasmic reticulum occur in the cytoplasm of the nurse cell. (*Dr D. D. Despommier*)

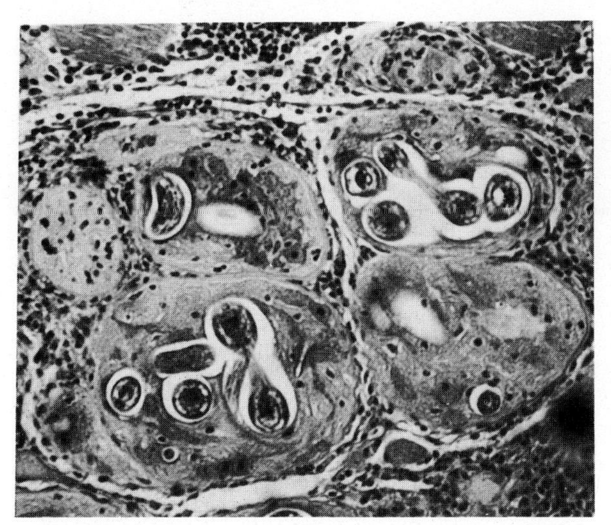

Fig 1.183 *Trichinella spiralis* larvae in striated musculature of infected rat and 30 days after infection. Coiling of the larva has commenced but the dense capsule has not yet formed. (*Dr D. D. Despommier*)

workers and more work is required to verify the separate species status proposed. For example, Pawlowski (1981) considers *T. pseudospiralis* to be a valid species but the others to be intraspecific variants (i.e. *T. spiralis domestica*, *T. s. nelsoni* and *T. s. nativa*).

The male is 1.4–1.6 mm long and the female 3–4 mm. The body is slender and the oesophageal portion is not markedly longer than the posterior part. The hind end of the male bears a pair of lateral flaps on either side of the cloacal opening, with two pairs of papillae behind them. There is neither a spicule nor a sheath. The vulva is situated near the middle of the oesophageal region. The eggs measure 40 by 30 μm and contain fully developed embryos when in the uterus of the female.

Life-cycle. The life-cycle is initiated when encysted muscle larvae are ingested. These are liberated within a few hours by the digestive processes and the first two moults are completed within 26 hours and the fourth moult in less than two days in the small intestine of the host. Development of the adult stage is rapid, being completed in four days. Copulation occurs about 40 hours after infection. After copulation has taken place in the intestine the males die and the females penetrate into the mucosa via Lieberkühn's glands and some may reach the lymph spaces. Here they produce, over a period of several weeks, eggs that hatch inside the uterus of the worm. The longevity of the female worm is not known with certainty, but probably few are left in the bowel after five or six weeks. In experimental infections a marked loss of adults occurs about the twelfth day of infection; however, this depends on the animal and adults may live as long as 16 weeks in man. Some larvae may be passed in the faeces during this stage of adult infection. The loss of adults is greatly accelerated in previously sensitized animals.

The first larvae (newborn larvae), which are about 0.1 mm long, enter the lymph and pass in it by way of the thoracic duct to the left superior vena cava and thus reach the blood, by which they are distributed all over the body. They develop further, especially in the voluntary muscles, especially those of the diaphragm, tongue, larynx, eye and the masticatory and intercostal muscles. They have also been found in the liver, pancreas and kidney.

The larvae enter striated muscle fibres and become surrounded by a capsule formed from the muscle fibre. Following penetration of the muscle cell there is a modulation or redifferentiation in the structure of the muscle cell. It is termed a 'nurse cell' and probably serves in larval nutrition and in the handling of waste products (Teppema et al. 1973; Despommier 1975). In the infected cell there is enlargement of the nuclei and an increase in nucleolar material; there is an increase in the number of mitochondria but these are smaller than those seen in the normal muscle fibre. The myofilaments disappear and there is a marked proliferation of the sarcoplasmic reticulum, particularly rough endoplasmic reticulum; by 10 days of infection the outer plasma membrane is highly hypervoluted and shows a 36-fold increase in the volume of the glycocalyx while a host-derived double membrane completely surrounds the larva. The larvae grow rapidly and after 30 days they measure 800–1000 μm in length and have begun to coil inside the cell. The capsule finally measures 0.4–0.6 by 0.25 mm at about three months; calcification begins after six to nine months but the larvae in them may live for several years. Cases in which the larvae lived for 11 and even 24 years have been recorded. In the cysts the larvae cannot develop further, but must await ingestion of the infected meat by another host. In this other host the larvae are liberated from their cysts in the stomach and grow to adults in the intestine, beginning to deposit larvae within six to seven days. Prenatal infection with this parasite appears to be rare, but it has been produced experimentally following heavy infections.

Epidemiology. Independent sylvatic and synanthropic–zoonotic cycles of infection occur. The sylvatic cycle involves wild carnivores such as foxes, jackals, wild boars, black bears, bushpigs, walrus etc. and these animals maintain the transmission. However, man may become infected following the ingestion of meats from wild boar, bear, walrus etc. Trichinellosis of man associated with the consumption of bear meat is increasingly common. In recent outbreaks in Alaska and California the infection has been associated with wok-prepared meat for exotic dishes (Cinque et al. 1979). Trichinellosis in the Arctic is reviewed by Rausch (1970). The synanthropic–zoonotic cycle occurs primarily in swine and rats; occasionally cats, dogs and man may become infected. Outbreaks of trichinellosis have been associated with horses. Some appear to be due to contamination of horse meat with pork, but, in addition, horses are susceptible to infection with *T. spiralis* and may be infected if processed horse-feed contains remnants of infected meat. The proceedings of the fourth international conference on trichinellosis (Kim & Pawlowski 1978) contain details and reviews of epidemiological aspects of trichinellosis in various countries of the world.

Pathogenesis and clinical signs. The parasite is of principal importance in human medicine and it is unusual for any obvious clinical entity to be associated with the infection in domestic or wild animals. The intestinal forms may produce a certain amount of irritation and cause a marked enteritis in heavy infections. The most important pathogenic effects are produced by the larvae in the muscles. Heavy infections may lead to death, especially through paralysis of the respiratory muscles. The clinical signs which accompany trichinosis are very variable and may simulate those of a variety of other diseases; they include diarrhoea, fever, retroperitoneal pain, stiffness and pain in the affected muscles, dyspnoea, hoarseness, sometimes an oedema of the face and deafness. A marked eosinophilia is usually present. A crisis is usually reached after about four weeks, when the egg-production of the females begins to decline and the larvae become encapsulated.

Epidemics of trichinellosis occasionally occur in human beings when a number of people partake of insufficiently cooked trichinous meat of a pig, bear or other host. For example, Cameron (1962) records serious disease in arctic villages when an infected animal (e.g. walrus, polar bear) is used in a feast.

Diagnosis. In the human a provisional diagnosis may be made on the clinical signs of muscular pain, oedema of the eyelid and face etc., such being pathognomonic when trichinellosis exists in an area. The adult worms may, on occasion, be found in the faeces but this is more accidental than the rule. Larvae may be demonstrated in muscle taken by biopsy, the sample being examined either microscopically when pressed between two pieces of glass or after digestion in an acid pepsin (1%) solution. A circulating eosinophilia may be present, possibly reaching 25% in the more active stages of the disease. It is good supportive evidence for a diagnosis of trichinosis.

A number of immunodiagnostic tests have been used in diagnosis. These include complement-fixation, haemagglutination, flocculation and intradermal techniques. With symptomatic cases of trichinellosis there is usually a good correlation between the immunodiagnostic test and the presence of the infection; however, with asymptomatic cases there may be considerable difficulty in assessing the results. The most satisfactory immunodiagnostic test in man is the bentonite flocculation technique but it does not become positive until three to four weeks after infection. A review of the serodiagnosis of trichinellosis is given by Kagan (1976) and the commercially available reagents are listed by Kagan (1979).

The diagnosis of *T. spiralis* infection in animals depends mainly on the detection of the infection at meat inspection. Various immunodiagnostic tests have been studied, especially in pigs, and though they will adequately detect experimental infections their level of usefulness is much lower in natural infections.

In Germany, pig muscle samples are routinely examined at slaughter houses by 'trichinoscopes'. These consist of a compressorium to press the muscle and the image of the muscle is projected on a screen. A more adequate, but more laborious and time-consuming, method is digestion of pooled muscle samples with an acid pepsin mixture for several hours. This is best carried out in a Baermann apparatus and the resulting sediment is examined for larvae. A further method is to infect experimental animals with the suspected muscle and later to examine the carcase for infection but, obviously, this is impractical on a wide scale. An automated enzyme-linked immunosorbent assay (ELISA) is highly sensitive and specific for detecting antibodies to *T. spiralis* and is being used in some countries for the mass screening of pigs for *Trichinella* infection at slaughter. Similarly, radio-immunoassay techniques are under development for mass screening of pigs at slaughter.

Treatment. Human trichinellosis is usually diagnosed during the muscle phase of the infection. In man, thiabendazole is widely used to treat infections with *T. spiralis.* 25 mg/kg is given twice a day for five to ten days. It is effective against the intestinal infection and against larvae in mice, but its efficacy against muscle larvae in man is unknown. Side effects include nausea, vomiting and fever. A large number of other anthelmintics, particularly the benzimidazole anthelmintics, i.e. mebendazole, flubendazole and cambendazole, are highly effective against muscle larvae in experimental animals and Thienpont (1976) recommends the use of mebendazole or its fluoride derivative, flubendazole, in the treatment of human infections. 200 mg are given three times a day for at least 10 days, the dosage being progressively augmented to 400 or 500 mg given three times a day.

Control and prevention. Human trichinellosis is disseminated chiefly by the pig. This animal is most usually infected from raw garbage which contains scraps of trichinous meat. Grain-fed pigs usually show a low incidence of infection but this is not always true. However, prohibition on the feeding of garbage or regulations requiring garbage to be cooked before being fed to pigs contributes to reducing the incidence of swine trichinellosis.

Infection of various wild animals, including wild boar, polar bear, walrus, seal and certain fur-bearing animals, may be responsible for human infection in certain areas of the world, e.g. Arctic, Central Europe etc.

Man usually acquires the infection mainly from pork and sausages which are eaten raw or partly cooked. Home-cured pork is of particular importance. In Arctic regions the consumption of bear, fox, walrus etc., which are considered delicacies by the Eskimo, frequently leads to trichinellosis. Prophylaxis should therefore aim at the elimination of uncooked garbage in the feed of pigs and, as far as man is concerned, the thorough cooking of all pork products and the meat of wild animals such as wild boar, fox, bear, walrus, etc. The meat should be cooked to a temperature of 58°C (137°F) throughout. The larvae are killed by freezing at −25°C for 10–20 days but salting and other methods of curing are not reliable unless the preserving process is thorough.

Gould et al. (1962) have demonstrated the feasibility of using gamma irradiation to sterilize meat. They maintain the method is simple and rapid and meat so treated still retains its flavour.

FAMILY: TRICHURIDAE RAILLIET, 1915 (TRICHOCEPHALIDAE BAIRD, 1853)

The name of the important genus in this family, *Trichuris*, represents an error, in that it literally means 'hair-tail' and not 'hair-head' as is presumed was intended. Consequently the substitute name *Trichocephalus* was proposed, but under the rules of priority of the International Code of Zoological Nomenclature it is, strictly, not acceptable. It has, however, been adopted by many European and Soviet parasitologists.

Members of the family are medium to large worms, the posterior part of the body being much thicker than the anterior. Genus of importance: *Trichuris*.

Genus: Trichuris Roederer, 1761 (syn. Trichocephalus)

The worms belonging to this genus are generally known as 'whip-worms', since the anterior part of the body is long and slender, while the posterior part is much thicker. The hind end of the male is curled and there is one spicule surrounded by a protrusible sheath which is usually armed with fine cuticular spines. The vulva is situated at the beginning of the wide part of the body.

Trichuris ovis (Abildgaard, 1795) has been recorded from the caecum of the goat, sheep, cattle and many other ruminants. The male of *T. ovis* is 50–80 mm long; the anterior end constitutes three-quarters of the length. The female is 35–70 mm long, of which the anterior end forms two-thirds to four-fifths. The fully evaginated spicule is 5–6 mm long. The sheath bears an oblong swelling a short distance from its distal extremity and is covered with minute spines which decrease in size towards the distal extremity. The eggs are brown, barrel-shaped, with a transparent plug at either pole, and measure 70–80 by 30–42 μm, including the plugs. They contain an unsegmented embryo when laid.

Trichuris discolor (Linstow, 1906) occurs in the caecum and colon of ox, zebra, buffalo, sheep and goat in Europe, South and East Asia and several states of the USA. The females are orange-yellow in colour and eggs measure 60–73 by 25–35 μm.

Trichuris globulosa (Linstow, 1901) occurs in the caecum of the camel, sheep, goats, cattle and other ruminants, and is the common form found in these animals in South Africa and possibly also in some other parts of the world. The male is 40–70 mm long and the female 42–60 mm, the anterior part constituting about two-thirds to three-quarters of the length. The spicule measures 4.2–4.8 mm and its sheath bears a terminal, spherical expansion on which the spines are larger than on the remaining portion. The eggs measure 68 by 36 μm and are similar to those of *T. ovis*.

Trichuris vulpis (Fröhlich, 1789) occurs in the caecum and other parts of the intestine of the dog and fox. The worms are 45–75 mm long, about three-quarters of this being made up by the anterior portion. The spicule is 9–11 mm long and the sheath bears small spines only on the proximal portion. The eggs measure 70–89 μm and have a brown colour. The infection may be common and surveys of dogs, for example, in New York and Detroit revealed infection rates of 31 and 52%, respectively (Kenney & Eveland 1978). Similarly, in Great Britain 32% of greyhounds were infected (Jacobs & Prole 1976).

Trichuris campanula (Linstow, 1889) occurs in the caecum and colon of the cat. It has been reported from South America, the Bahamas Islands, Cuba and the USA. The worms are 21–31 mm long, the spicule is 1.2–1.5 mm long and the eggs measure 70–80 by 30–36 μm.

Trichuris serrata Linstow, 1879 also occurs in cats and closely resembles *T. campanula*. The morphological differences between *T. serrata* and *T. campanula* are given by Ng and Kelly (1975) and Hass and Meisels (1978).

Trichuris suis (Shrank, 1788) occurs in pig, wild pig and wild boar. It is cosmopolitan in distribution. Morphologically it is identical to *T. trichiura* (Linnaeus, 1771) of man and other primates and some workers believe the two to be identical. However, there is no critical evidence to show that the two parasites are interchangeable between the two hosts. The male is 30–50 mm

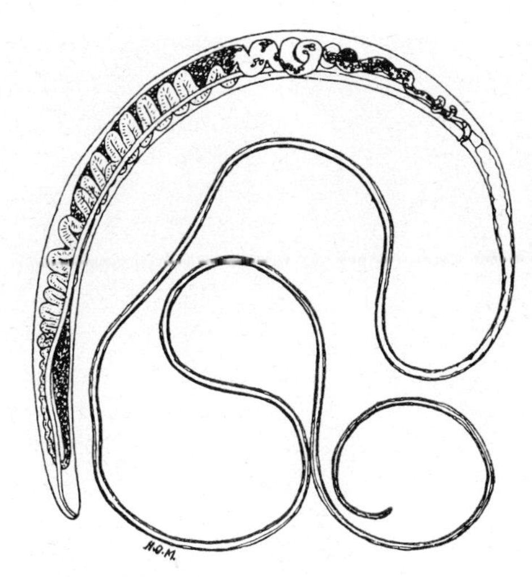

Fig 1.184 Female *Trichuris globosa.*

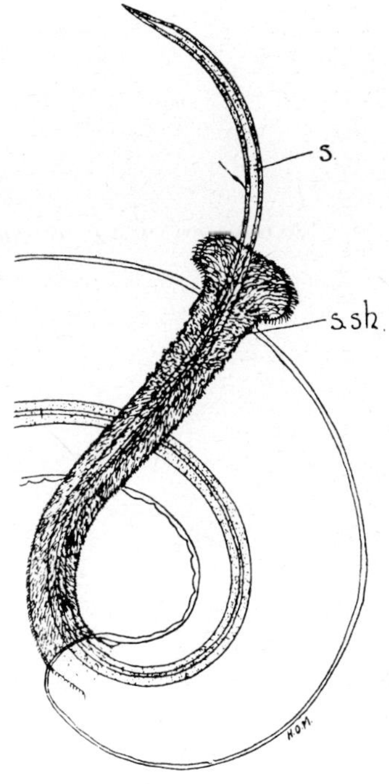

Fig 1.185 Hind end of male *Trichuris globosa.*
s. spicule
s.sh. spicule sheath

long and the female 35–50 mm. The anterior portion forms about two-thirds of the total length. The spicule is 2–3.35 mm long, with a blunt tip, and its sheath is variable in shape and in the extent of its spinous armature. The eggs measure 50–60 by 21–25 μm.

Levels of infection vary but 8–14-week-old pigs are the most heavily infected. Thus, 75.5% of such pigs were found to be infected in Wisconsin while the infection was less common in sows and boars (Powers et al. 1959). Similarly, 20% of pigs on a pig fattening farm near Leipzig and 2.6% of pigs in Berne, Switzerland were infected.

Trichuris trichiura (Linnaeus, 1771) is morphologically similar to *T. suis* but it is biologically distinct and it infects only man and simian primates. Infection rates may be very high; for example 90% of Xhosa children in an urban area near Cape Town were found to be infected although only 9.7% were infected in the Tsolo district (Van Niekerk 1979). Similarly, 84% of four to six-year-old children from squatter kindergardens in the Kuala Lumpur–Petaling Jaya region of Malaysia were infected (Yen et al. 1978). The infection has been seen in 7–27% of rhesus monkeys, 5–58% of cynomolgus monkeys and 29–86% of Japanese and Formosan macques (Flynn 1973).

Other species of the genus include: **T. cameli** Ransom, 1911 of the large intestine of the camel and dromedary; **T. tenuis** Chandler, 1930 and **T. raoi** Alwar and Achuthan, 1960, both of the dromedary; **T. skrjabini** Baskakov, 1924, of the large intestine of camel and dromedary in Turkestan; and **T. leporis** (Froelich, 1789) and **T. sylvilagi** Tiner, 1950 of the caecum and large intestine of cotton-tail rabbits, snowshoe hares, and coypus in Europe and North America.

Life-cycle of Trichuris spp.

The eggs reach the infective stage after about three weeks under favourable conditions; however, development may be much more prolonged at lower temperatures (e.g. 6–20°C), since development is related to soil moisture and temperature. Infective eggs may remain viable for several years. The host acquires the infection by ingesting the eggs, and the larvae penetrate the anterior small intestine for two to ten days before

Fig 1.186 Scanning electron micrograph of *Trichuris suis* in the mucosa of the large intestine of pig. Anterior end of worm occupies 'tunnels' in the mucosa. (*Dr Edward G. Batte, University of North Carolina*)

Fig 1.187 *Trichuris suis* infection of the large intestine of pigs. (*Dr Edward G. Batte, University of North Carolina*)

they move to the caecum where they develop to adults. The prepatent period of *T. ovis* is seven to nine weeks, of *T. vulpis* 11–12 weeks and of *T. suis* six to seven weeks.

Pathogenicity

Opinion about the pathogenicity of these species varies, but there is little doubt that they can produce an acute or chronic inflammation, especially in the caecum of the dog and man.

T. vulpis and probably other *Trichuris* spp. are blood feeders (Burrows & Lillis 1964). *Trichuris* spp. have a mouth stylet, 7–10 μm long, projecting through their mouth opening. The adults tunnel into the intestinal mucosa with their anterior ends and the stylet is used to enter vessels or to lacerate tissues creating pools of blood which the nematodes ingest.

In sheep, cattle and swine naturally acquired infections are not usually severe enough to cause clinical disease and sheep over eight months of age show an age resistance to infection and a resistance to reinfection two to three weeks after a primary infection. However, clinical disease due to *Trichuris* spp. has been recorded in sheep and cattle (Georgi et al. 1972).

Epidemics resulting in mortality may be seen in pigs. Affected pigs show anaemia, dehydration, anorexia, dysentery and loss of weight. The pathology is due to adult *T. suis* tunneling into the mucosa of the large intestine. There is caecitis and colitis, with mucosal necrosis and areas of haemorrhage. The mucosa is oedematous and catarrhal inflammation may be seen (Batte et al. 1977).

Many dogs are infected but the burdens in general seem to have little clinical effect. Nevertheless severe infections, particularly in dogs up to 18 months of age, are not uncommon and several hundred or thousand worms may cause a profuse diarrhoea, loss of weight and unthriftiness. In severe cases the faeces may be markedly haemorrhagic or even frank blood. Then there is evidence of anaemia, sometimes jaundice, the infection terminating in death.

Diagnosis is made by demonstration of the characteristic barrel-shaped eggs in the faeces.

Treatment

Dogs. N-butyl chloride (0.1–10 ml/kg) has been reported effective, especially if given hourly for five hours. The effect, however, may be very variable (0–100%). The compound is well tolerated. *Phthalofyne* (3-methyl-1-pentyn-3-yl sodium phthalate), at a dose of 250 mg/kg either orally or by intravenous injection is generally highly effective and usually without severe side effects. *Glycobiarsol* (p-glycolylamino-phenyl-arsonic-bismuth), as a single dose of 1000 mg/kg or five to ten daily doses of 200 mg/kg, is reported to be quite effective and well tolerated. *Dichlorvos* given orally at a dose rate of 30–40 mg/kg is highly effective and when given in the feed for five days at a rate of 200 ppm dichlorvos showed a 98% efficacy (Roberson et al. 1977). *Mebendazole*, 100 mg twice a day for three to five days, is effective and *fenbendazole*, 150 mg/kg, may also be used. *Diuredosan* was demonstrated to have an efficacy of 88, 85 and 94% when given at dose rates of 25, 50 and 100 mg/kg (Todd & Yates 1976).

Swine. Hygromycin B 80 million units per tonne of feed is highly effective. *Trichlorophon*, 75 mg/kg intramuscularly has been reported highly effective. *Levamisole* given subcutaneously at a dose rate of 7.5 mg/kg shows an efficacy of 96–100% while *tetramisole* may be given in the feed at a dose rate of 15 mg/kg. *Thiophanate* in feed at levels of 0.0225–0.045% may be used (Baines et al. 1979) while a single oral dose of 50 mg/kg removed 96–99% of adult parasites, although the efficacy against immature parasites was lower (Baines et al. 1976). *Chlorophos*, two doses of 100 mg/kg given in the feed at an interval of five days, is said to be 100% effective. *Febantel*, 30 mg/kg, is 98% effective against fourth-stage larvae, while 20 mg/kg was the dose used to eliminate adult parasites (Enigk & Dey-Hazra 1978). *Fenbendazole*, 3–25 mg/kg given as a single oral dose, was 90–97% effective while 3 or 5 mg/kg given daily for three days in the feed showed 99–100% efficacy (Batte 1978). *Oxibendazole* is also effective. *Flubendazole* given in the feed at 30 ppm for five consecutive days controlled mature clinical infections, while the same rate for ten days controlled immature stages (Thienpont et al. 1982).

Sheep and cattle. Methyridine [2(β-methoxyethyl) pyridine] 200 mg/kg orally or subcutaneously is highly effective. *Fenbendazole*, 5–20 mg/kg is also effective (Townsend et al. 1977) and *oxfendazole* at a dose rate of 2.5 mg/kg had an 89–99% efficacy against adult parasites and was 62–100% effective against the immature parasites (Chalmers 1978).

Man. Mebendazole, 100 mg twice a day for five days, has a 95% efficacy and *flubendazole*, at 200 mg/kg/day for three days for children and 300 mg/kg/day for adults, showed an 88% efficacy.

Prophylaxis

Improved hygiene is necessary. Dog kennels and exercise yards etc. should be regularly cleaned. Heavily infected soil should be avoided for several months to allow the natural agencies of sunlight and dryness to kill the eggs.

FAMILY: CAPILLARIIDAE NEVEU-LEMAIRE, 1936

Trichuroidea in which the oesophageal part of the body is shorter than the posterior part and more or less equal in thickness. Genus of importance: *Capillaria*.

Genus: Capillaria Zeder, 1800

The worms of this genus are closely related to *Trichuris*, but they are small and slender and the posterior part of the body is not conspicuously thicker than the anterior part. The life-cycle may be direct or indirect. The eggs are unsegmented when laid and develop to the first larval stage in nine to 14 days; they are then infective for the definitive host if the life-cycle is direct or for earthworms in which the parasite accumulates if the life cycle is indirect. (The life-cycles of *Capil-*

laria hepatica and *Capillaria philippinensis* are unusual.) In comparison to *Trichuris* eggs, the shell is almost colourless, the egg is more barrel-shaped, with the sides nearly parallel, and the bipolar plugs do not project as far.

The genus contains numerous species, among which the following may be encountered by the veterinarian:

SPECIES FOUND IN BIRDS

The taxonomy of these species follows that given by Madsen (1945).

Species found in the intestine

Capillaria caudinflata (Molin, 1858) Wawilowa, 1926 (syn. *C. caudinflatum* (Molin, 1858), *C. longicollis* (Mehlis, 1831)). This species occurs in the duodenum and ileum of the fowl, pigeon and related wild birds. Males are 9–14 mm long and females 14–25 mm. The oesophagus is almost half as long as the body in the male and one-third as long in the female. The tail of the female is cylindrical up to the end. The vulva has a conspicuous projecting appendage. The eggs measure 47–65 by 23 μm, with a thick, finely sculptured shell. The life-history of this species is indirect. The eggs, like those of *C. annulata*, are swallowed by earthworms (*Eisenia foetida*, *Allobophora calignosa*) and birds infect themselves by eating the earthworms. The prepatent period is three weeks.

Capillaria obsignata (Madsen, 1945) (syn. *C. columbae*). This species occurs in the small intestine of the pigeon, fowl, turkey and related wild birds. Males are 9.5–11.5 mm long and females 10.5–14.5 mm. In the male the oesophagus is more than half as long as the body; in the female it is shorter. The tail of the female tapers posteriorly. The eggs measure 48–53 by 24 μm and the shell has a reticulate pattern.The life-cycle is direct; eggs develop in 13 days at 20°C and the prepatent period in birds is 20–21 days.

Capillaria anatis (Schrank, 1790) Travassos, 1915 (syn. *C. retusa, C. collaris, C. anseris, C. mergi*). This species occurs in the caeca of domesticated gallinaceous and anatine birds and the caeca of gallinaceous game birds in Europe and in ducks in North America. Adults are 6–18 mm long. The egg measures 49–65 by 22–35 μm and has a thick rugose outer shell. The life cycle is probably direct.

Heaviest infections are usually seen in young birds, older birds serving as carriers of the infection. In heavily infected birds disease may be apparent during the prepatent period. There is emaciation and diarrhoea or bloody diarrhoea associated with haemorrhagic enteritis. Chronically infected birds have thickened intestinal walls covered with a catarrhal exudate.

Species found in the crop and oesophagus

Capillaria annulata (Molin, 1958). This species occurs in the crop and the oesophagus of the fowl, turkey and related birds. Males are 15–25 mm long and females 37–80 mm. The cuticle of the anterior end forms a characteristic swelling behind the head. The eggs measure 60–65 by 25–28 μm. The life-history of this species resembles that of *C. caudinflata*, earthworms being the intermediate hosts. The earthworms *Eisenia foetida*, *Allolobophora caliginosa* and some of the genera *Lumbricus* and *Dendrobaena* are concerned, the infective larvae being reached in them in 14–21 days (Allen 1949). However, Morehouse (1944) states that no development occurs in the earthworm, rather passage through it to induce hatching of the eggs is the essential requirement. Madsen (1952) regards this species as being synonymous with *C. contorta* (see below).

Capillaria contorta (Creplin, 1839) Travassos, 1915. This species occurs in the crop, oesophagus and mouth of the turkey, duck and many wild birds. The males are 12–17 mm long and the females 27–38 mm. The eggs measure 48–56 by 21–24 μm. The life-history of this species is, so far as is known, direct, but Madsen (1952) concluded that *C. contorta* is synonymous with *C. annulata*, the correct name for both these species being *C. contorta*. He suggests that the life-history of the species described as *C. contorta* may sometimes require the passage of the eggs through earthworms, so that it may be either direct or indirect.

Light infections produce only mild inflammation and thickening of the crop. In heavy infections there is a marked thickening of the crop and oesophageal wall, a marked catarrhal or croupous inflammation and sloughing of the mucosa. The birds are emaciated and weak.

Capillaria spp. can be important parasites of birds kept in straw yards (e.g. *C. caudinflata* in turkeys) and in deep litter houses, where heavy infections may build up in the litter. Those parasites which have an indirect life-cycle are mainly of importance in range birds but they can be responsible for severe losses in pheasants, quail etc. For example, *C. obsignata* is very important in pigeons.

Diagnosis is usually made at post mortem when large numbers of parasites can be found. The eggs may also be detected in large numbers in the faeces.

Several effective compounds are now available for treatment. Coumaphos at levels of 0.003–0.004% in the feed for 10–14 days and hygromycin B, 0.00088–0.00132% are effective. Methyridine given subcutaneously at 150 mg/kg or in the water at a rate of 200–400 mg/100 ml for 24 hours is effective also although slight side effects may be seen (Ruff 1978). Gangadhara Rao (1976) demonstrated that pyrantel tartrate, 75 mg/kg, was 91–99% effective against seven- to 21-day-old experimental infections of *C. obsignata* in chickens; methyridine, 200 mg/kg, and levamisole, 30 mg/kg, showed efficacies of 99–100% and 95–99%, respectively. Fenbendazole in the feed at a dose rate of 8 mg/kg for six days was highly effective against *Capillaria* spp. in pigeons although a single dose of 82.5 mg/kg was inadequate (Vindevogel et al. 1978). Conversely, in another experiment fenbendazole, 20 mg/kg as a single dose, or at a rate of 100 ppm in the feed for three to four days, was completely effective in pigeons (Kirsch et al. 1978).

Prophylaxis is a matter of hygiene, but presents many difficulties. The animals should be fed and watered in such a way that contamination with eggs of the worms is eliminated as far as possible.

SPECIES FOUND IN MAMMALS

Capillaria entomelas (Dujardin, 1845) occurs in the small intestine of the mink, beech marten and polecat, and may cause a haemorrhagic enteritis, particularly in mink. The infection is acquired chiefly by animals kept in boxes and can be prevented only by hygienic measures. The eggs measure 56–63 by 23–27 μm.

Capillaria bovis (Schnyder, 1906) and **C. brevipes** Ransom, 1911, which are considered by some to be synonymous, and **C. longipes** Ransom, 1911 occur throughout the world in the small intestine of cattle, sheep and goats. The males are 8–13 mm in length and the females are 12–20 mm in length. The eggs measure 45–50 by 22–25 μm. The life-cycle is direct. No well-defined pathological entity has been associated with infection by this species.

Capillaria bilobata Bhalerao, 1938 has been reported from the abomasum of zebu in India. Males are 10–16 mm in length and females 14–21 mm. The eggs measure 33–53 by 14–21 μm.

Capillaria megrelica Rodonaja, 1947 is recorded from the abomasum of the goat in the USSR. Males are 12 mm long and females 18–20 mm. Eggs are 55 by 20 μm. The life-cycle is probably similar to that of *C. longipes*.

Capillaria erinacea Rudolphi, 1819 is a parasite of the hedgehog, but adults have been identified in the intestine of dogs, cats and pigs in New Zealand (Collins 1973; McKenna & Buddle 1974).

Capillaria putorii (Rudolphi, 1819) occurs in the small intestine of various mustelids and the cat in Europe and the USSR. It may cause a haemorrhagic enteritis in mink. Males are 5–8 mm and the females 9–15 mm long; the eggs measure 56–72 by 23–32 μm.

Capillaria plica (Rudolphi, 1819) occurs in the urinary bladder and sometimes the pelvis of the kidney of dogs, cats and foxes. It is apparently relatively harmless, although it occasionally causes cystitis and difficulty in urination. No treatment, except sanitation, is known. The male is 13–30 mm long and the female 30–60 mm. It has been stated that the life-history of this species is direct, but Enigk (1950) recorded that it may be indirect, the larvae entering the connective tissue of earthworms and foxes infecting themselves by eating infected earthworms. Adult parasites are found about 60 days after infection. The eggs are passed in the urine and are colourless; they measure 63–68 by 24–27 μm.

Other species of urinary bladder capillarids include **Capillaria felis cati** (Diesing, 1851), which has been recorded from the cat in most parts of the world, and **Capillaria mucronata** (Molin, 1858) in the urinary bladder of mink.

Capillaria hepatica (Bancroft, 1893) (syn. *Hepaticola hepatica*) occurs in the liver of numerous rodents, particularly the rat and mouse, throughout the world. It also infects other animals such as the squirrel, muskrat, rabbit, opossum and rarely dogs, cats and man. The infection has also been recorded in horses (Nation & Dies 1978). Infection rates in rats can be high. Thus, 75% of the Norway rat population in Baltimore Zoo were found to be infected (Farhang-Azad 1977) and in Malaysia 15.5% of rats (17.7% of field rats in agricultural areas, 0.7% of urban rats and 8% of jungle rats) were infected with *C. hepatica* (Sinniah et al. 1979). The worms are very thin and cannot easily be removed from the liver tissue, so that it is difficult to measure their length. Lengths of 4–12 cm have been reported. Their eggs are laid in the liver and the eggs and the granulomatous response to these can be seen in the form of irregular yellow streaks and patches on the surface and in sections.

Life-cycle. The life-cycle is unusual. The unembryonated eggs must be released from the liver by a predator, often called an intercalary host, or by cannibalism, and the eggs are passed in the faeces of the predator or cannibal. Embryonation to the infective stage then takes four weeks at 30°C. Decomposition of dead carcasses is also important in egg dispersal.

Infection takes place by ingestion of food contaminated by eggs shed by the intercalary or cannibalistic host or from a decomposed infected carcase.

Light infections in domestic animals or man may be inapparent but heavy infections may cause acute or subacute hepatitis with splenomegaly, peritonitis, ascites and eosinophilia. In these cases the prognosis is grave. Diagnosis is made by the demonstration of eggs in liver biopsies.

Capillaria aerophila (Creplin, 1839) (syn. *Eucoleus aerophilus*) occurs in the trachea, bronchi and rarely the nasal cavities and frontal sinuses of dogs, foxes and coyotes. It has also been recorded from the cat, pine marten, beech marten, wolf (*Lupus lupus*) and the badger. *C. aerophila* can be important in silver fox colonies and also has been found in 37% of wild foxes in New York State and 38% of coyotes in the Great Plains States of the USA (Zeh et al. 1977; Morrison & Gier 1978). Infection is also seen periodically in man.

The male is 24.5 mm long and the female 32 mm. There is one spicule and a spicule sheath armed with spines. The eggs measure 59–80 by 30–40 μm, including the polar plugs. They have a slight greenish tinge, and the thick shells have a 'netted' surface.

The life-cycle is direct. The eggs are laid in the lungs, coughed up and swallowed, and are therefore passed in the faeces. They develop in the open, reaching the infective stage after five to seven weeks, and may remain viable for over a year under favourable conditions. The infective larva does not hatch out of the egg until the infective egg is swallowed by a suitable host. The eggs then hatch in the intestine and the larvae migrate to the lungs in seven to ten days, reaching maturity 40 days after infection.

Animals with mild infections show no clinical signs. Severe infections cause rhinitis with nasal discharge and chronic tracheitis and bronchitis. The affected animals are susceptible to secondary infections by bacteria which may cause bronchopneumonia. There is a whistling noise while the animal breathes and a deep, wheezing cough, especially at night. The mouth may be held open on account of dyspnoea. The animals become emaciated and anaemic, while the fur grows harshly.

Diagnosis is made by finding the eggs in the faeces, in the sputum or nasal discharge. These should be differentiated from those of *C. plica* and *T. vulpis*.

Levamisole may be effective and three cycles of levamisole, 5 mg/kg given daily for five days at nine-day intervals, has been efficacious (Norsworthy 1975). Fenbendazole may be tried.

The disease occurs chiefly in young foxes up to 18 months old. The parasite is most troublesome where the soil of the pens is shaded or not properly drained. The infection is also acquired in breeding boxes, where the eggs are allowed to accumulate. Dry pens and clean boxes should therefore be the objective of preventive measures.

Capillaria didelphis Butterworth, Beverly-Burton, 1977, is found in the lung parenchyma of the opossum (*Didelphis virginianus* and *D. marsupialis*) in the USA. Well defined granulomata surrounding parasites and eggs are seen in the lung tissues (Prestwood et al. 1977).

Capillaria philippinensis Chitwood, Valsquez and Salazar, 1968 has been reported in the posterior small intestine and anterior large intestine of man in the Philippines and Thailand. Its sudden appearance as an epidemic in 1967–9 (1700 cases, with 120 deaths) suggests that *C. philippinensis* is a newly emerged zoonosis, the normal definitive host of which is not yet known, although many species of animals have been examined. The male is 2–3 mm in length and the female is 2.5–4.3 mm. Parous females may produce typical thick-shelled bipolar eggs, eggs with thin shells or first-stage larvae without shells.

Life-cycle. Cross et al. (1978) have described the unusual life-cycle of these parasites. Normal bipolar eggs passed in the faeces embryonate in water in five to ten days. They are ingested by fresh water fishes such as *Hypselotris bipartita* and *Ambassis commersoni* and develop to infective larvae in the mucosa of the intestines in three weeks. In experimentally infected gerbils these larvae developed to larviparous adults 13–14 days after infection. Larvae laid by these developed to second-generation adults 22–24 days after infection. Most of these second-generation females were oviparous and produced bipolar eggs with a prepatent period of 24–35 days but a few were larviparous. That autoinfection was an integral part of the life-cycle was demonstrated by the fact that gerbils developed infections of 852 to 5353 parasites after receiving only two or three larvae. In addition, the infection can be serially passaged in gerbils by the transfer of intestinal stages.

Clinical signs include diarrhoea, malabsorption, fluid imbalance and a protein-losing enteropathy.

Treatment and prophylaxis. Mebendazole administered at a rate of 400 mg/day for 20 days is effective (Singson et al. 1975). Control involves improved sanitation and raw fish should not be eaten.

Several species of capillarids occur in the gastrointestinal tract of rats and mice and these include: **Capillaria gastrica** (Baylis, 1926) of the stomach of rats, mice and voles in Europe and the Americas; **C. bacielata** (Eberth, 1863) of the oesophagus of Norway rat and mice in Europe; **C. annulosa** (Dujardin, 1845) of the small intestine of the Norway rat and black rat in Europe; **C. intestinalis** Vanni, 1937 of the intestine of rats in Central Europe; and **C. tavernae** Ash 1962 of the intestine of Norway and black rats in Hawaii.

Other species occur in the urinary bladder of rats and these include: **C. papillosa** (Polonio, 1860) in Europe and **C. prashadi** Maplestone and Bhadari, 1942 of the Norway rat in India.

CAPILLARIDS OF FISH

Several species of the genus occur in fish.

Capillaria tomentosa (Dujardin, 1843), **C. tuberculata** Linstow, 1914 and **C. lewaschoffi** Heinze, 1933 require crustacean intermediate hosts and parasitize various freshwater fish in Europe (e.g. *Leucisius*, *Sardinius*, *Blicca* and *Rutilus* spp.

Capillaria cantenata van Cleeve and Mueller, 1932 may cause enteritis in the intestine of bluegills, black bass, catfish etc. in North America and **C. eupomotis** Ghittino. 1961 is found in the liver and may cause intensive liver damage to trout and other fish in Europe (Flynn 1973). **C. catostomi** Pearse, 1924 occurs in the intestine of the white sucker in Wisconsin and **C. petruschewskii** (Shulman, 1948) occurs in the sunfish in Europe.

Genus: Trichosomoides Railliet, 1895

In this genus the male is degenerate and parasitic in the uterus of the female. A single species is recognized.

Trichosomoides crassicauda (Bellingham, 1840) Railliet, 1895, which occurs in the urinary bladder of Norway and black rats throughout the world. The female worms are 10–19 mm long and the males 1–3.5 mm. Eggs are thick-shelled, with a bipolar plug, and 55–80 by 30–48 μm. Infection is by the ingestion of embryonated eggs voided in the urine. The prepatent period is eight to nine weeks. Usually infection is inapparent but granulomatous lesions may occur. Bladder tumours have been associated with this infection but there is no firm evidence for a carcinogenic role for the parasite (Chapman 1964). The infection can be eliminated from rat colonies by caesarean derivation techniques. Infected stock may be treated with nitrofurantoin in the feed (2 g/kg for six weeks) or methyridine (200 mg/kg intraperitoneally in a single dose).

Genus: Anatrichosoma Smith and Chitwood, 1954

The male of this genus is without a spicule and is not parasitic in the uterus of the female. Eggs are operculate and contain an embryo when laid. The members of this species are parasitic in the connective tissues of primates.

Anatrichosoma cutaneum (Swift, Boots and Miller, 1922) occurs in the skin of rhesus monkeys and of man in the USA and Far East. It has also been found in the nasal mucosa of rhesus monkeys in Bethesda, Md, USA. Females are 22–25 mm long; the males are unknown. Eggs are 56–70 by 38–42 μm.

REFERENCES

TRICHINELLIDAE

Bessonov, A. S., Penkova, R. A. & Gumenshshikova, V. P. (1978) *Trichinella pseudospiralis* Garkavi, 1972: Morphological and biological characteristics and host specificity. In *Trichinellosis*. Proc. 4th int. Conf. Trichinellosis. Poznan, Poland, ed. C. W. Kim & Z. S. Pawlowski, pp. 79–99. Hanover, NH: University Press of New England.

Cameron, R. W. M. (1962) Trichinellosis in Canada. In: *Trichinellosis*. Proc. 1st Int. Conf. Trichinellosis, ed. Z. Kozar. Warszawa: Polish Scientific Publishers.

Cinqué, J., Fannin, S., Brodsky, R., Farrell, J. & Woodland, T. L. (1979) Trinchinosis associated with bear meat—Alaska, California. *Mort. Morbid. Weekly Rep.*, **28**, 12.

Despommier, D. D. (1975) Adaptive changes in muscle fibers infected with *Trichinella spiralis*. *Am. J. Path.*, **78**, 477–496.

Gould, S. E., Gomberg, H. J. & Villella, J. B. (1962) Effects of different energies of ionizing radiation on *Trichinella spiralis*. In: *Trichinellosis*. Proc. 1st int. Conf. Trichinellosis, ed. Z. Kozar. Warszawa: Polish Scientific Publishers.

Kagan, I. G. (1976) Serodiagnosis of trichinosis. In: *Immunology of Parasitic Infections*, ed. S. Cohen & E. H. Sadun, pp. 143–151. Oxford: Blackwell Scientific.

Kagan, I. G. (1979) Diagnostic, epidemiologic and experimental parasitology: immunologic aspects. *Am. J. trop. Med. Hyg.*, **28**, 429–439.

Kim, C. W. & Pawlowski, Z. S. (1978) *Trichinellosis*. Proc. 4th int. Conf. Trichinellosis, Posnan, Poland, 1976. Hanover, NH: University Press of New England.

Pawlowski, Z. S. (1981) Control of trichinellosis. In: *Trichinellosis: Proceedings of the Fifth International Conference on Trichinellosis*, ed. J. Ruitenberg and J. S. Teppema, pp. 7–20. Chertsey: Reedbooks.

Rausch, R. L. (1970) Trichinosis in the Arctic. In: *Trichinosis in Man and Animals*, ed. S. E. Gould, pp. 492–511. Springfield, Ill.: Charles C. Thomas.

Teppema, J. S., Robinson, J. E. & Ruitenberg, E. J. (1973) Ultrastructural aspects of capsule formation in *Trichinella spiralis* infection in the rat. *Parasitology*, **66**, 291–296.

Thienpont, D. (1976) Treatment of trichinelliasis. *Nouv. Presse med.*, **5**, 1759–1760.

World Health Organization (1979) Parasitic zoonoses. *Wld Hlth Org. tech. Rep. Ser.*, 637.

TRICHURIDAE: TRICHURIS

Baines, D. M., Dalton, S. E. & Eichler, D. A. (1976) Experimental and field studies with thiophanate in pigs. *Vet. Rec.*, **99**, 119–122.

Baines, D. M., Evans, P., Lake, P. & Frape, D. L. (1979) Field studies with thiophanate given as a 14-day low level in-feed anthelmintic to pigs. *Vet. Rec.*, **105**, 81–82.

Batte, E. J. (1978) Evaluation of fenbendazole as a swine anthelmintic. *Vet. Med. small Anim. Clin.*, **73**, 1183–1186.

Batte, E. G., McLamb, R. D., Muse, K. E., Tally, S. D. & Vestal, T. J. (1977) Pathophysiology of swine trichuriasis. *Am. J. vet. Res.*, **38**, 1075–1079.

Burrows, R. B. & Lillis, W. G. (1964) The whipworm as a blood sucker. *J. Parasit.*, **50**, 675–680.

Chalmers, K. (1978) The efficacy of oxfendazole against natural infections of nematodes of cattle. *N. Z. vet. J.*, **26**, 162–164.
Enigk, K. & Dey-Hazra, A. (1978) The treatment of helminth infestation in pigs with Rintal. *Vet. med. Rev.*, **2**, 134–144.
Flynn, R. J. (1973) *Parasites of Laboratory Animals.* Ames, Iowa: Iowa State University Press.
Georgi, J. R., Whitlock, R. H. & Flinton, J. H. (1972) Fatal *Trichuris discolor* infection in a Holstein–Friesian heifer. Report of a case. *Cornell Vet.*, **62**, 58–60.
Hass, D. K. & Meisels, L. S. (1978) *Trichuris campanula* infection in a domestic cat from Miami, Florida. *Am. J. vet. Res.*, **38**, 1553–1555.
Jacobs, D. E. & Prole, J. H. B. (1976) Helminth infections of British dogs: prevalence in racing greyhounds. *Vet. Parasit.*, **1**, 377–387.
Kenney, M. & Eveland, L. K. (1978) Infection of man with *Trichuris vulpis*, the whipworm of dogs. *Am. J. clin. Path.*, **69**, 199.
Ng, B. K. Y. & Kelly, J. D. (1975) Isolation of *Trichuris campanula* von Linstow, 1889 from Australian cats. *Aust. vet. J.*, **51**, 450–451.
Powers, K. G., Todd, A. C. & Goldsby, M. S. (1959) Swine whipworms in Wisconsin. *Vet. Med.*, **54**, 397–398.
Roberson, E. L., Anderson, W. I. & Hass, D. K. (1977) Anthelmintic drug evaluation: dichlorvos-medicated dry dog feed. *Am. J. vet. Res.*, **38**, 597–600.
Thienpont, D., Vanparijs, O., Hermans, L. & De Roose, P. (1982) Treatment of *Trichuris suis* infections in pigs with flubendazole. *Vet. Rec.*, **110**, 517–520.
Todd, K. S. & Yates, R. L. (1976) Anthelmintic activity of diuredosan in dogs experimentally infected with *Ancylostoma caninum* and *Trichuris vulpis. Am. J. Vet. Res.*, **37**, 1329–1330.
Townsend, R. B., Kelly, J. D., James, R. & Weston, I. (1977) The anthelmintic efficacy of fenbendazole in the control of *Moniezia expansa* and *Trichuris ovis* in sheep. *Res. vet. Sci.*, **23**, 385–386.
Van Niekerk, C. H., Weinberg, E. G., Shore, S. C. L. & Heese, H. de V. (1979) Intestinal parasitic infestation in urban and rural Xhosa children. A comparative study. *S. Afr. med. J.*, **55**, 756–757.
Yen, C. W., Ishak, F., Hee, G. L., Devaraj, J. M., Ismail, K., Jalleh, R. P., Peng, T. L. & Jalil, T. M. A. (1978) The problem of soil transmitted helminths in squatter areas around Kuala Lumpur. *Med. J. Malaysia*, **33**, 34–48.

CAPILLARIDAE: CAPILLARIA, TRICHOSOMOIDES

Allen, R. W. (1949) Studies on the life history of *Capillaria annulata* (Molin, 1858) Cram 1926. *J. Parasit.*, **35** (Suppl.), 35.
Chapman, W. H. (1964) The incidence of a nematode, *Trichosomoides crassicauda*, in the bladder of laboratory rats: Treatment with nitrofuration and preliminary report of their influence on the urinary calculi and experimental bladder tumor. *Invest. Urol.*, **2**, 52–57.
Collins, G. H. (1973) A limited survey of gastro-intestinal helminths of dogs and cats. *N.Z. vet. J.*, **21**, 175–176.
Cross, J. H., Banzon, T. & Singson, C. (1978) Further studies on *Capillaria philippinensis*: development of the parasite in the mongolian gerbil. *J. Parasit.*, **64**, 208–213.
Enigk, K. (1950) Die Biologie von *Capillaria plica* (Trichuroidea, Nematoda). *Z. tropenmed. Parasit.*, **1**, 560–571.
Farhang-Azad, A. (1977) Ecology of *Capillaria hepatica* (Bancroft 1893) (Nematoda). 1. Dynamics of infection among Norway rat populations of the Baltimore Zoo, Baltimore, Maryland. *J. Parasit.*, **63**, 117–122.
Flynn, R. J. (1973) *Parasites of Laboratory Animals.* Ames, Iowa: Iowa State University Press.
Gangadhara Rao, Y. V. B. (1976) Experimental chemotherapy on *Capillaria obsignata* in chickens. *Indian vet. J.*, **53**, 776–777.
Kirsch, R., Petri, J. & Kegenhardt, H. (1978) Treatment of *Capillaria* and *Ascaridia* infections in pigeons with fenbendazole. *Kleintierpraxis*, **23**, 291–298.
McKenna, P. B. & Buddle, R. (1975) *Capillaria* infection in a pig. *N.Z. vet. J.*, **23**, 242–243.
Madsen, H. (1951) Notes on the species of *Capillaria* (Zeder, 1800), known from gallinaceous birds. *J. Parasit.*, **37**, 257–265.
Madsen, H. (1945) The species of *Capillaria* (Nematoda: Trichinelloidea) parasitic in the digestive tract of Danish gallinaceous and anatine game birds, with a revised list of species of *Capillaria* in birds *Dan. Rev. Game Brol.*, **1**, 112.
Morehouse, N. F. (1944) Life cycle of *Capillaria caudinflata*, a nematode parasite of the common fowl. *Iowa St. Coll. agric. Sci.*, **18**, 217–253.
Morrison, E. E. & Gier, H. T. (1978) Lungworms in coyotes on the Great Plains. *J. Wildl. Dis.*, **14**, 314–316.
Nation, P. N. & Dies, K. H. (1978) *Capillaria hepatica* in a horse. *Can. vet. J.*, **19**, 315–316.
Norsworthy, G. D. (1975) Feline lungworm treatment case report. *Feline Pract.*, **5**, 14.
Prestwood, A. K., Nettles, V. F. & Farrell, R. L. (1977) Pathologic manifestations of experimentally and naturally acquired lungworm infections in opossums. *Am. J. vet. Res.*, **38**, 529–532.
Ruff, M. D. (1978) Nematodes and acanthocephalans. In: *Diseases of Poultry*, ed. M. S. Hofstad, B. W. Calnek, C. F. Helmboldt, W. M. Ried & H. W. Yoder, pp. 705–36. Ames, Iowa: Iowa State University Press.
Singson, C. N. Banzon, T. C. & Cross, J. H. (1975) Mebendazole in the treatment of intestinal capillariasis. *Am. J. trop. Med. Hyg.*, **24**, 939–943.
Sinniah, B., Singh, M. & Anuar, K. (1979) Preliminary survey of *Capillaria hepatica* (Bancroft, 1893) in Malaysia. *J. Helminth.*, **53**, 147–152.
Vindevogel, H., Duchatel, J. P. & Fievez, L. (1978) Treatment of capillariasis in pigeons with fenbendazole. *Ann. Med. vet.*, **122**, 109–115.
Zeh, J. B., Stone, W. B. & Roscoe, D. E. (1977) Lungworms in foxes in New York. *N.Y. Fish Game J.*, **24**, 91–93.

SUPERFAMILY: DIOCTOPHYMATOIDEA RAILLIET, 1916

Oesophageal glands multinucleate, caudal glands absent, vagina tubular, reproductive system highly developed. Male with one spicule, muscular caudal sucker present. Families of importance *Dioctophymidae* and *Soboliphymidae*.

FAMILY: DIOCTOPHYMIDAE RAILLIET, 1915

This group contains three genera (*Dioctophyma*, *Hystrichis* and *Eustrongylides*) in which the alimentary canal is attached to the abdominal wall by four longitudinal bands of suspensory muscles. The tail of the male bears a terminal, cup-shaped ' bursa ' without rays and there is a single spicule. The female has a single genital tube and the eggs have thick, pitted shells.

Genus: Dioctophyma Collet–Meygret, 1802 (syn. Dioctophyme)

Dioctophyma renale (Goeze, 1782) (syn. *Eustrongylus gigas*) is the largest nematode known and occurs in various countries in the kidneys and other organs of the dog, fox, otter, beech marten, pine marten, polecat, mink, weasel, other wild

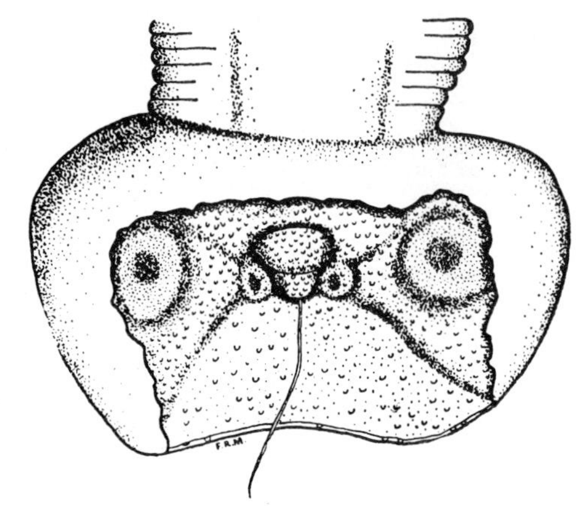

Fig 1.188 Ventral view of posterior end of male *Dioctophyma renale.*

carnivores and the seal, and it has also occasionally been found in the pig, horse, cattle and man. Larvae have also been found in subcutaneous nodules in man. Mink (*Mustela vison*) are the most commonly infected mustelids and are the principal definitive hosts. *D. renale* is an important parasite of mink and large numbers of mink, up to 50%, may be infected (Mace & Anderson 1975). The male measures up to 35 cm by 3–4 mm, and the female up to 103 cm by 5–12 mm. The worms have a blood-red colour. The spicule is 5–6 mm long, but it may reach up to 12–14.5 mm. The eggs are barrel-shaped and brownish yellow, and the shells are pitted except at the poles. They measure 71–84 by 46–52 μm and are not segmented when laid.

The eggs are passed in the urine of the host and develop slowly in water, requiring one to seven months according to the temperature. They do not hatch until they have been swallowed by the intermediate host, but they may remain viable for five years. There has been confusion over the life-cycle but Karmanova (1960) has demonstrated that the free-living oligochaete annelid, *Lumbriculus variegatus*, is the only intermediate host required to complete the life-cycle. A developmental period of over 100 days in the annelid is required. The definitive hosts may become infected by ingesting the infective larvae in annelids. In addition, paratenic hosts may occur in the life cycle. *Idus* spp. in Europe, bullheads (*Ictalurus nebulosus*) in the USA and pike (*Esox lucius*) are important paratenic hosts, becoming infected by the ingestion of infected annelids. Frogs (*Rana clamitans melanota* and *R. pipiens*) are also naturally infected and 0.9–9.6% of frogs have been shown to contain encysted larvae in their stomach wall and in the abdominal muscles in enzootic areas (Mace & Anderson 1975).

In the final host the infective larvae penetrate the bowel wall and initially develop in the body cavity and then penetrate the kidney. The prepatent period in mink is about 138 days. The right kidney appears to be invaded much more frequently than the left. The worms apparently enter the pelvis and destroy the parenchyma. Eventually only the capsule is left as a much-distended bladder, containing one or more worms bathed in an albuminous fluid containing red cells, epithelial cells and eggs. If some of the kidney parenchyma is left, it may be partly calcified. Often a worm projects into the ureter and if the ureter is blocked, uraemia may result. Worms may also wander down into the bladder and pass out through the urethra. In the abdominal cavity the worms may be found free or encapsulated, causing a chronic peritonitis with adhesions in various places. Frequently they lie between the lobes of the liver and destroy the surface of this organ. Small nodules containing eggs of the parasite may be found in the liver and the omentum. The worms have also been found in the pleural cavity.

Frequently no signs are shown, since the normal kidney can serve the needs of the body. In other cases the infected animals grow thin and they may show signs of kidney trouble together with nervous signs. Retention of urine and death from uraemia may occur. In man renal colic, pyuria and haematuria may be seen.

Diagnosis can be made by finding the eggs in the urine if the parasites occur in the kidneys.

The worms may be removed surgically.

Genus: Hystrichis Dujardin, 1845

The species of this genus occur in aquatic birds. The anterior extremity is somewhat swollen and bears several rows of small spines.

Hystrichis tricolor Dujardin, 1845 occurs in the glands of the proventriculus of domestic and wild ducks, where it produces nodules and destruction of the tissues. The male is about 25 mm long and the female up to 40 mm.

The eggs measure 85–88 by 36–40 μm; they have truncated poles and are covered with tubercles. They develop slowly in water, reaching the embryonated stage in about 60 days. Further development occurs in oligochaetes such as *Criodrilis lacuum* (Glossoscolecidae) and *Allolobophora dubiosa* (Lumbricidae) (Karmanova 1956). The first-stage larva hatches in the digestive tract of the annelid and migrates to the supraneural blood vessel where three ecdyses occur producing a fourth-stage larva. Birds became infected by eating infected oligochaetes and the mature parasite is produced in one month, the longevity being about 45–50 days.

Nodules are produced in the wall of the proventriculus: these may perforate to the pleural cavity and cause adhesion of the organs in that area. Suppuration of the nodule and the local area is not uncommon. The parasite is intricately sewn in the lesion and difficult to extract in a complete state.

Genus: Eustrongylides Jägerskiöld, 1909

Eustrongylides tubifex (Nitzsch, 1819) occurs in the intestine of anatine birds in Europe. Head not spiny, mouth small, cuticle markedly annulated, oesophagus very long, without dilatation. Males are up to 34 mm by 2 mm in thickness; the bursal cup is trumpet-shaped. Spicule long and slender. Females are 35–45 mm long, with the vulva near the anus. Eggs are 70 by 44 μm, with an operculum at either pole and markedly pitted.

The developmental cycle is unknown but fish have been incriminated.

Eustrongylides papillosus (Rudolphi, 1802) is similar to *E. tubifex* and is found sewn into the mucosa of the oesophagus and proventriculus of ducks and geese. The species measures up to 30 mm in length, the bursal cup has a fringed margin and the eggs measure 68 by 38 μm.

The intermediate host is unknown but is probably a turbificid oligochaete. However, fish may act as paratenic hosts for many *Eustrongylides* spp. and the larvae are found encysted in capsules in the stomach wall, mesenteries and internal organs such as the gonads. Although larvae encysted in the stomach wall of brown trout (*Salmo trutta*) were not pathogenic (Kennedy & Lie 1976), their presence in the testes and ovary can cause severe damage in infected fishes.

Marked nodule formation occurs in the wall of the anterior digestive tract of infected birds.

FAMILY: SOBOLIPHYMIDAE PETROV, 1930

Possessing a muscular cephalic sucker. Mouth feebly developed. Bursal cup campanulate. Genus of importance *Soboliphyme*.

Genus: Soboliphyme Petrov, 1930

Soboliphyme baturini Petrov, 1930 occurs in the intestines of foxes, sable and cats in the Soviet Union (Siberia) and has been found in the wolverine in North America.

SUPERFAMILY: MERMITHOIDEA WÜLKER, 1934

Species of this superfamily belong to the Class Nematoda. They are not parasitic in domesticated animals, but may be swallowed with drinking water or otherwise found in the neighbourhood of farm stock, and they may be mistaken for parasitic nematodes. They are smooth, hair-like worms, whitish or brownish in colour, and the adults are free-living in soil or water. The pharynx may extend along half the length or more of the body and it is not connected with the intestine, which becomes a double row of cells packed with reserve food materials. The males are much smaller than the females. Parthenogenesis may occur. The larvae are parasitic in terrestrial invertebrates, which they may kill.

FAMILY: MERMITHIDAE BRAUN, 1883

The larvae are economically important because they are parasitic inside the bodies of the nymphs or larvae of insects (grasshoppers, earwigs, ants), whose viscera they destroy, so that these insects may fail to develop or may die. The larvae of *Mermis subnigrescens* are parasitic in grasshoppers and earwigs, those of *Paramermis* in grasshoppers and those of *Allomermis* in ants. *Mermis subnigrescens* has been called the rain-worm, because the adults appear out of the soil after rain in the summer in thundery weather and climb up plants to lay their eggs.

FAMILY: TETRADONEMATIDAE COBB, 1919

The larvae of the relatively few species of this family are parasitic in the larvae of the Diptera *Sepsis* and *Sciara*.

REFERENCES

DIOCTOPHYMATOIDEA

Karmanova, E. M. (1956) An interpretation of the biological cycle of the nematode *Hystrichis tricolor* Dujardin 1845, a parasite of domestic and wild ducks. *Dokl. Akad. Nauk S.S.S.R.*, 111, 245–247.

Karmanova, E. M. (1960) The life cycle of the nematode *Dioctophyme renale* (Goeze, 1782). *Dokl. Akad. Nauk S.S.S.R.*, **127**, 700–702.

Kennedy, C. R. & Lie, S. F. (1976) The distribution and pathogenecity of larvae of *Eustrongylides* (Nematoda) in brown trout *Salmo trutta* L. in Fenworthy Reservoir. *Devon J. Fish Biol.*, 8, 293–302.

Mace, T. F. & Anderson, R. C. (1975) Development of the giant kidney worm, *Diotophyma renale* (Goeze, 1782) (Nematoda: Dioctophymatoidea). *Can. J. Zool.*, 53, 1552–1568.

CLASS: NEMATOMORPHA (GORDIACEA, HAIRWORMS)

Species of this class are not parasitic in domesticated animals. They are cylindrical, relatively very long worms, with a smooth cuticle devoid of rings. The anterior end is clear with a dark ring behind the clear area. The posterior end of the male ends in two broad processes and the posterior end of the female in three similar processes. The nervous system consists of cerebral ganglia and a mid-ventral nerve cord. There is no pharynx and the end of the alimentary canal tends to degenerate. The sexes are in separate individuals and the sexual ducts of both sexes open into the intestine. The adults are not parasitic, but the larval phases are.

ORDER: GORDIOIDEA VON SIEBOLD, 1848 (HAIRWORMS)

The adults live in fresh water and the larvae are parasitic in insects (grasshoppers, crickets, cockroaches, beetles) that live near fresh water, or in centipedes or millipedes.

ORDER: NECTONEMATOIDEA

There is only one genus of this order, *Nectonema*, the species of which are pelagic and marine. They are up to 20 cm long and their larvae are parasitic in the crab and hermit-crab.

Phylum: Acanthocephala Rudolphi, 1808

The Acanthocephala are a group of parasitic worms usually considered as being closely allied to the Nematoda. They are commonly called 'thorny-headed worms'.

Morphology

The body is in most cases cylindrical. The body covering or tegument has five layers and its absorptive surface is considerably expanded, 20–62-fold, by invaginations of the outer plasma membrane into lacunar canals and vesicles (Graeber & Storch 1978).

An alimentary canal is absent. The worms feed like cestodes, by absorbing their nourishment through the body wall. Anteriorly the body bears an evaginable proboscis, which is a hollow, cylindrical or oval structure armed with transverse or longitudinal rows of recurved hooks and lies in a proboscis sac. Next to the proboscis sac a pair of elongate, hollow organs, the 'lemnisci', project into the body cavity. The proboscis is retracted by special muscles which are inserted in the body wall and it is evaginated by the contraction of the proboscis sac. The lemnisci are connected with the proboscis and probably secrete or store the proboscis fluid.

The excretory system is absent or consists of a pair of nephridia which discharge into the genital ducts.

From the proboscis sac a 'suspensory ligament' runs backwards through the body cavity and provides attachment for the genital organs which are enclosed in a ligament sac. The male has two testes which lie in a tandem position. The vasa deferentia unite to form an ejaculatory duct and this opens through a 'penis' which projects into an evaginable sac or 'bursa' at the posterior extremity of the body. A group of cement glands or prostate glands are connected with the ejaculatory duct. They appear to secrete a substance which protects the vulva of the female after copulation. There is a great size disparity since the males are usually much smaller than the females.

In the female the ovary discharges its ova into the body cavity, where they appear to be fertilized and the embryo develops for some time, forming around itself a shell of three layers. The 'uterine bell' is a special organ which swallows the eggs through an anterior opening and passes the mature eggs through into the vagina, while the immature ones are returned to the body cavity through a small pore situated posteriorly. The eggs contain 'acanthor' larvae which are provided with an anterior circlet of hooks. Eggs have three or sometimes four membranes.

Development

As far as the life-cycles are known the eggs require to be ingested by an intermediate host, which is usually an arthropod. Where known the intermediate hosts are insect larvae, beetles and cockroaches for acanthocephalans which are parasitic in land animals and birds, and crustaceans or occasionally molluscs for those parasitic in aquatic vertebrates. The spindle-shaped acanthocephalan eggs resemble diatomes which makes them a favourable food for crustaceans.

The acanthor larva in the egg hatches in the intermediate host and then encysts as a cystacanth in the haemocoel of the arthropod. The cystacanth may require one to three months for further development to the infective stage and it is often orange-red in colour and can be seen through the cuticle of the crustacean. Definitive hosts become infected by ingesting the arthropods and the prepatent period of the infection is usually five to 12 weeks. Cystacanths may re-encyst in vertebrates other than the definitive host following their ingestion. These act as paratenic hosts and may be important epidemiologically acting as a link between the intermediate host and the definitive host. The adult parasites occur chiefly in aquatic vertebrates, mainly fishes and birds.

ORDER: PALAEACANTHOCEPHALA MEYER, 1931

FAMILY: POLYMORPHIDAE MEYER, 1931

Usually small worms with a cylindrical body. Genera of importance: *Polymorphus*, *Filicollis*.

Genus: Polymorphus Lühe, 1911 (syn. Profilicollis)

Polymorphus boschadis (Schrank, 1788) (syn *P. minutus*) occurs in the small intestine of the duck, swan, fowl, goose and various wild aquatic birds throughout the world. Males are about 3 mm long and females up to 10 mm. The worms have an orange colour when fresh. The cuticle is spiny anteriorly, and behind this region the body is constricted. The proboscis has 16 longitudinal rows of seven to ten hooks each, which decrease in size posteriorly. The testes are oval and situated diagonally; the cement glands are elongate.

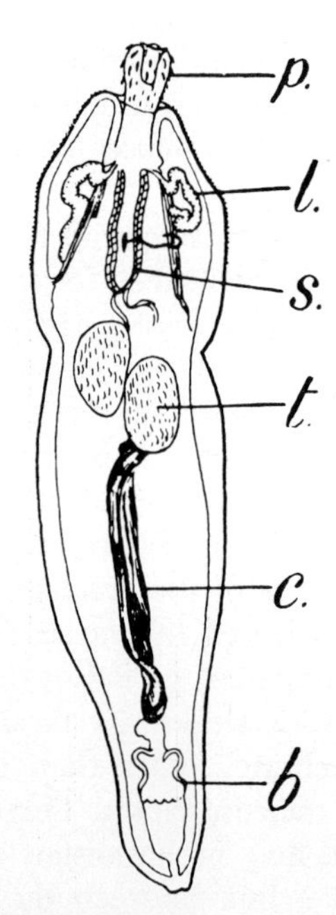

Fig 1.189 Male *Polymorphus boschadis*.
b. bursa copulatrix
s. proboscis sac
p. proboscis
l. lemniscus
c. cement glands
t. testis

The eggs are spindle-shaped and measure about 110 by 19 μm. Their outer shell is thin, while the middle shell is thick and irregularly constricted at the poles. The embryo has a yellowish-red colour.

The cystacanth develops in the 'fresh water shrimp', the amphipod crustacean *Gammarus pulex*, and possibly also in the crayfish *Potamobius astacus*. The final host acquires the infection by ingesting the infected intermediate host.

The worms are located in the posterior part of the small intestine, where they penetrate deeply into the mucosa with their probosces. At the points of attachment small nodules can be seen from the peritoneal surface. The parasites produce cachexia, inflammation of the intestine and anaemia. Heavy infections can cause severe losses in colonies of domestic and wild ducks, geese and swans (Sanford 1978).

Diagnosis can be made by finding the eggs in the faeces.

Little is known about the treatment of *Polymorphus* infections. Petrochenko (1949) has reported that carbon tetrachloride at a dose of 0.5 ml/kg is up to 98% effective.

The birds have to be kept away from water harbouring infected intermediate hosts. Since migrating birds are usually the original source of infection attempts should be made to restrict their entry to breeding grounds.

Several other related species occur in wild water fowl. Thus **Polymorphus botulus** van Cleave, 1916, has been reported to be responsible for seasonal high mortality in the eider duck in Scotland (Rayski & Garden 1961). Development occurs in the common shore crab *Carcinus moenas*. **Polymorphus magnus** Skrjabin, 1913, occurs in the intestines of wild anseriforms in eastern Europe, India and the Soviet Union. Development occurs in *Gammarus lacustris*.

Genus: Filicollis Lühe, 1911

Filicollis anatis (Schrank, 1788) occurs in the small intestine of the duck, goose, swan and wild

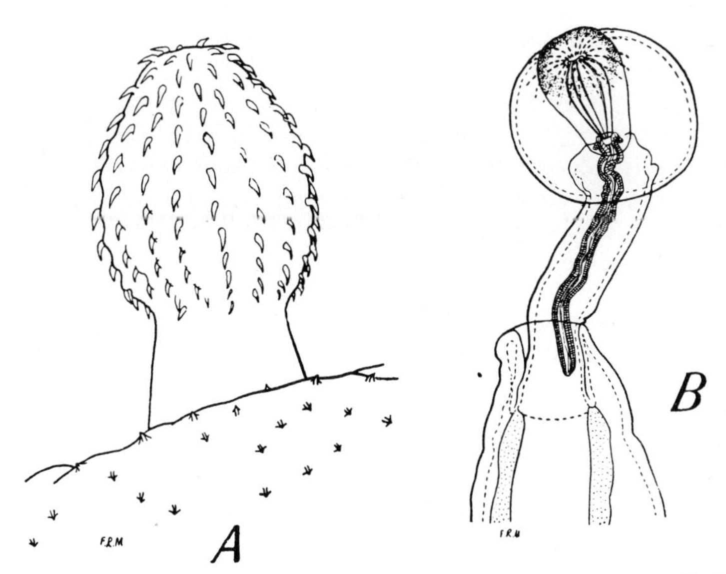

Fig 1.190 *Filicollis anatis*. A, Anterior end of male. B, Anterior end of female (less highly magnified).

aquatic birds. The male is 6–8 mm long and white in colour; the female is 10–25 mm long and yellowish. The male has an ovoid proboscis with 18 longitudinal rows of 10–11 hooks each and the anterior part of the body is armed with small spines. The lemnisci are long in both sexes. The female has a long, slender neck, while the proboscis is globular in shape and about 2–3 mm in diameter. There are 18 rows of 10–11 hooks each arranged in a star-shaped pattern at the apex of the proboscis. The eggs are oval and measure 62–70 by 19–23 μm.

The eggs are passed in the faeces of the host and the larval stage develops in isopods and crustaceans such as the water louse *Asellus aquaticus* (Isopoda). The final host acquires the infection by ingesting such isopods and the worms grow adult in about four weeks.

The parasites are situated in the middle portion of the small intestine and also further back in severe infections. While the male attaches its proboscis in the mucosa, the female worm pierces through the mucous and muscular layers of the wall, so that its proboscis comes to lie directly underneath the peritoneum. In some cases the peritoneum may even be ruptured. The parasites are obviously harmful; they cause emaciation and frequently the death of their hosts.

Diagnosis can be made by finding the eggs in the faeces.

Treatment and prophylaxis are as in the case of *P. boschadis*.

Genus: Corynosoma Lühe, 1904

Corynosoma strumosum (Rudolphi, 1802) and **C. semerme** (Forssell, 1904) are found in the small intestine of animals such as the fox and dog in North America, the Arctic, Asia, Europe and the USSR. Mink are accidental hosts in which the parasite does not reach maturity. The eggs in the faeces measure 79–101 by 16–29 μm.

The intermediate host is believed to be the amphipod, *Pontoporeia affinis*; several species of fish, primarily fresh water fish but also marine teleosts, then act as transport hosts for the juvenile stages.

There have been reports of acute infections causing bloody diarrhoea and anaemia in mink. The parasites are not usually pathogenic in the normal definitive hosts.

Carbon tetrachloride has been used successfully. The benzimidazole–carbamate anthelmintics might be effective.

A number of acanthocephalan parasites occur in the intestines of fresh water and marine fishes. These include **Echinorhynchus salmonis** (Mueller, 1784) seen primarily in salmonids; **Pomphorhynchus laevis** (Zoega in Mueller, 1784) which is common in the intestine of fresh water fishes, particularly chubb and grayling, in Europe and North America, but also occurs in marine fishes such as flatfish and eels off European shores; and **Acanthocephalus** spp. (Kroelreuther, 1771) which can be pathogenic in trout in the wild and in culture.

The life-cycles of these parasites involve crustaceans such as *Gammarus pulex* and *Asellus aquaticus* as intermediate hosts. The adults are not usually very pathogenic parasites of fishes but the proboscis may cause severe local damage to the intestine and occasionally deterioration in the condition of infected fishes is seen. *Acanthocephalus jacksoni* Bullock, 1962 can be highly pathogenic in trout since it transfers its site of attachment and therefore can cause a number of necrotic haemorrhagic ulcers (Needham & Wootten 1978).

ORDER: ARCHIACANTHOCEPHALA MEYER, 1931

FAMILY: OLIGACANTHORHYNCHIDAE MEYER, 1931

Usually large to medium sized worms. Proboscis not retractile. Genera of importance *Macrocanthorhynchus* and *Prosthenorchis*.

Genus: Macrocanthorhynchus Travassos, 1916

Macrocanthorhynchus hirudinaceus (Pallas, 1781) occurs in the small intestine of the domestic pig and wild boars and is present in most countries of the world, although it is absent from western Europe. Adults have been recorded in the ileum of man and they are also seen in animals such as the muskrat, squirrel and peccary. The infection may be fairly common and the pigs on 17–32% of farms in the Belorussian SSR were infected with the infection rate on farms ranging from 0.9–5% and occasionally up to 23%. The worms are usually more or less curved and have a pale reddish colour. The male is up to 10 cm long and the female up to 35 cm or even more and 4–10 mm thick. The cuticle is transversely wrinkled. The proboscis is relatively small and bears about six transverse rows of six hooks each, which decrease in size backwards. The eggs measure 67–110 by 40–65 μm and have four shells, of which the second is dark brown and pitted.

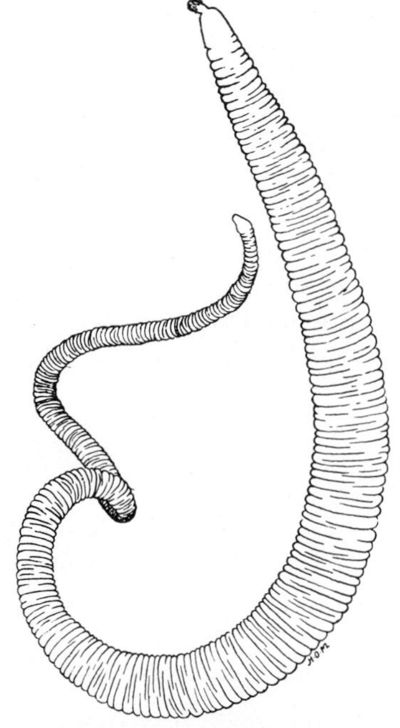

Fig 1.191 *Macracanthorhynchus hirudinaceus.* × $1\frac{1}{2}$

The eggs are passed in the faeces of the pig and are very resistant to cold and dryness, being able to live for several years in the open.

A number of genera in the dung beetle family Scarabaeidae act as intermediate hosts (*Melolontha*, *Cetonia*, *Cotinus*, *Phyllophaga*, *Scarabeus* and water beetles, *Tropisternus*). These become infected when they feed on manure and infected soil and the egg hatches when it is ingested by the larvae of these dung beetles, 'June' beetles and 'May' bugs. The young worm eventually becomes encysted in the body cavity of the insect. Pigs become infected by ingesting either the grubs or the adult beetles which harbour the infective stage of the worm. Development in the pig takes two or three months. A female lays about 260 000 eggs per day and lays for about 10 months.

The parasites penetrate with their probosces deeply into the intestinal wall, where they produce inflammation and a granuloma at the site of attachment. Rarely, perforation and death from peritonitis occur. Apart from this possibility mild infections are not very harmful, but severe infections may cause slow growth or emaciation, which are very important in pigs.

Diagnosis can be made by finding the eggs in the faeces.

Carbon tetrachloride, tetrachlorethylene and nicotine sulphate have been used.

The pigs should be prevented from ingesting the larvae or adults of the intermediate hosts. Where pigs are kept in sties or small runs regular removal and suitable disposal of faeces will assist in reducing the infection.

Other species of this genus, **Macrocanthorhynchus catalinum** Kostylesi, 1927 and **M. ingens** (Linstow, 1879), occur in the small intestine of mammals such as the wolf, badger, fox, domestic dog and skunk, mink, mole, racoon, respectively. The life-cycles, when known, involve scarabaeid beetles and millipeds. The larval stages have been found encysted in the muscles, peritoneal cavity and internal organs of mice, squirrels and woodchuck (*M. catalinum*) and in the peritoneal cavity of *Rana pipiens* (*M. ingens*). These presumably act as paratenic hosts.

Genus: Oncicola Travassos, 1916

Oncicola canis (Kaupp, 1919) is a small parasite which occurs in the intestine of the domestic dog, coyote, domestic cat, lynx and bobcat in North and South America. **O. campanulatus** Diesing, 1851 has been found in the cat in the USA. The male is 6–13 mm long and the female 7–14 mm, both sexes being 2–4 mm thick. The shape of the body is roughly conical, tapering backwards, and it has a dark grey colour. The proboscis bears six transverse rows of six hooks each. The anterior hooks have the shape of a taenioid hook, while the posterior ones resemble rose thorns. The lemnisci are very long and slender. The testes are oval, tandem and situated in the anterior half of the body. The eggs are oval, brown in colour and measure 59–71 by 40–50 μm.

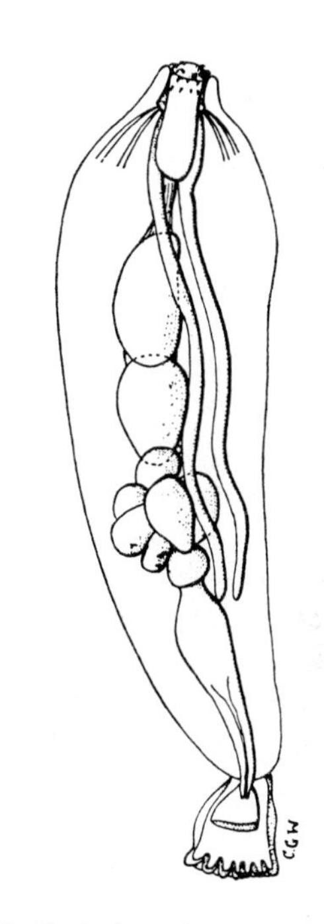

Fig 1.192 Male *Oncicola canis.*

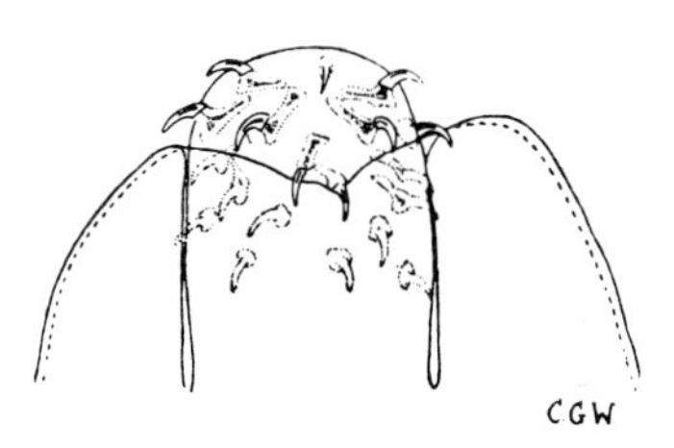

Fig 1.193 Anterior end of *Onicola canis.*

Probably an arthropod acts as an intermediate host. Immature forms, about 4 mm long, which are believed to belong to this species, have been found encysted in the connective tissue and muscles of the armadillo (*Dasypus novemcinctus*) and in the epithelium of the oesophagus of turkeys. These presumably act as paratenic hosts and up to 10% of turkey poults have been found infected in Texas.

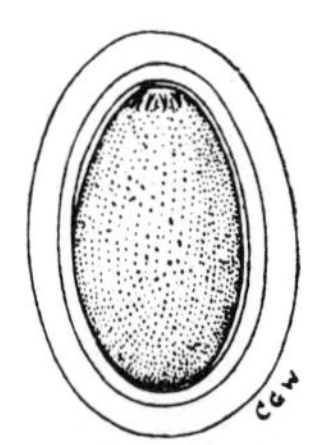

Fig 1.194 Egg of *Oncicola canis*.

Infection may be common and 31% of dogs have been found infected, but the parasite is rarely of clinical importance. The worms lie with their probosces deeply embedded in the intestinal wall, penetrating right through to the peritoneum. The parasite may possibly cause rabiform signs in dogs. Such were observed in a dog which harboured about 300 worms.

Diagnosis can be made by finding the eggs in the faeces.

Treatment and prophylaxis are unknown.

Genus: Prosthenorchis Travassos, 1915

Prosthenorchis elegans (Diesing, 1851) and **P. spicula** (Olfers in Rudolphi, 1819) are very important parasites in the small intestine, particularly in the terminal portion of the ileum but also the caecum and colon, of Central and South American monkeys. *P. elegans* is very common and *P. spicula* is less common. These parasites are now found throughout the world where primates are kept in captivity and where they have introduced the parasites. Adults are 2–5 cm long and the proboscis is globular with five to seven rows of hooks. The eggs measure 65–81 by 42–53 μm.

The intermediate hosts are cockroaches (*Blatella germanica*), which may be common in primate colonies.

In heavy infections there is diarrhoea, anorexia and debilitation; death is not uncommon. On post mortem examination abscesses and granulomatous lesions are seen around the probosces embedded in the intestinal mucosa. The serosal surface of the small intestine appears studded with yellow, pearly nodules 2–6 mm in diameter. The intestine is swollen and obstruction of the ileocaecal valve and perforation of the intestinal wall is not uncommon.

Diagnosis can be made by finding the eggs in the faeces.

Dithiazinine iodide has been effective (Peters & Zwart 1962). Insecticides and good sanitation will control the intermediate hosts.

Phylum: Annelida

CLASS: HIRUDINEA

The Hirudinea or leeches are not usually considered as 'helminths', but are included here as parasitic worms in the wider sense of the term. They are soft-bodied and usually dorsoventrally flattened worms with a true metameric segmentation, but this has become obscured by specialization. The annulations visible externally are more numerous than the metameric segments, the number of which, according to different authorities, is 33 or 34 and these segments are externally subdivided by transverse lines into two to 14 rings each. The anterior region is differentiated to form a head which bears a triangular anteroventral sucker. The posterior seven segments form a large sucker except in one genus, and the anus is situated on the dorsal aspect of the hind end. The ninth, tenth and eleventh segments form the clitellar organ which serves to secrete the cocoon in which the eggs are deposited.

Two orders of leeches are recognized, Gnathobdellidae and Rhynchobdellidae. In the Gnathobdellidae the anterior sucker contains the oral aperture, which is provided with three strong toothed jaws, while the Rhynchobdellidae have a protrusible proboscis, but no jaws. There is a short oesophagus and a stomach (crop), which is usually provided with paired, segmentally arranged, blind sacs for storage of ingested blood, followed by an intestine and a rectum. The body cavity is filled with a lacunar tissue. The excretory organs are segmentally arranged nephridia.

The worms are hermaphrodite. There are several to many testes, which have a common duct on either side, and one pair of ovaries. The median male genital opening is on the tenth segment and the median female genital opening on the eleventh segment. The eggs are laid in cocoons secreted by the clitellum.

The leeches are occasional parasites. They feed on blood of various animals to which they attach themselves and drop off after having engorged. They have pharyngeal glands which secrete an anticoagulatory substance; this is injected into the wound made by the leech, and the bleeding may continue for some time after the leech has dropped off. Leeches are vectors of certain trypanosomes and their relatives are parasitic in marine and freshwater fish.

The species of medical and veterinary interest belong to the Gnathobdellidae.

Genus: Hirudo Linnaeus, 1758

Hirudo medicinalis Linnaeus, 1758 is the leech formerly used for medicinal purposes. It is

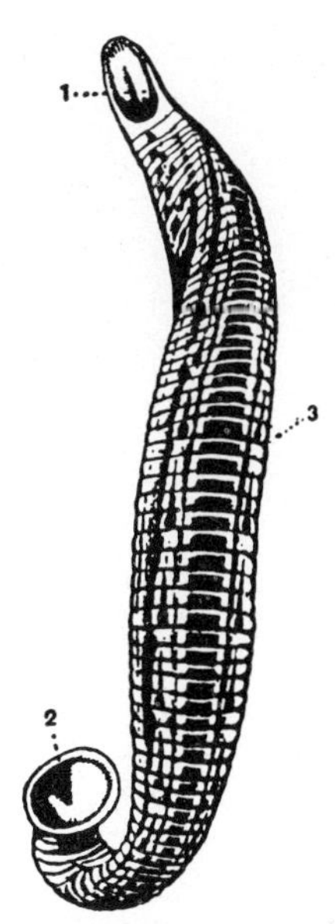

Fig 1.195 *Hirudo medicinalis*. 1, Mouth. 2, Posterior sucker. 3, Sensory papillae on the anterior annulus of each segment. The remaining four annuli which make up each true segment are indicated by the markings on the dorsal surface.

8–12 cm long and 1–2 cm wide. The dorsal surface is greyish-green with six longitudinal reddish bands; the ventral surface is olive-green with a black band on either side. It occurs in Europe and North Africa and lives in marshes, small streams and pools of water. It is of little interest as a possible parasite, but occasionally animals may become infected.

Genus: Limnatis Moquin-Tandon, 1826

Limnatis nilotica (Savigny, 1820) (horse leech) occurs in Europe and North Africa in pools of water which contain plants. It is 8–12 cm long and its body is soft. The dorsal surface is fairly dark brown or greenish and usually has several longitudinal rows of black spots. The ventral surface is darker than the dorsal and there is frequently an orange band on either side. The anterior lip has a longitudinal groove on its inner surface. The adult worms live at the bottom in the mud; the young leeches, however, occur near the surface and are very easily attracted by steps at the water's edge.

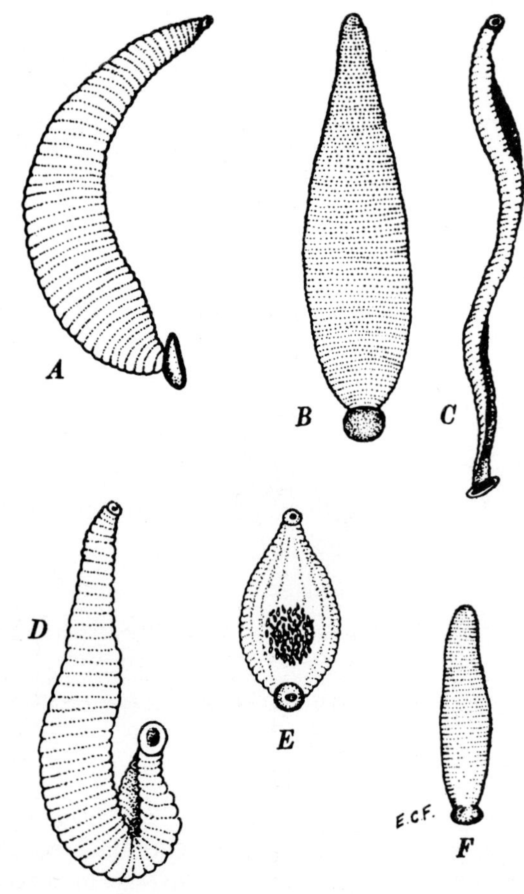

Fig 1.196 Sketches of leeches. A, *Limnatis nilotica*. B, C, Dorsal and lateral views of an extended specimen of *Hirudo medicinalis*. D, *Haemopis sanguisuga*. E. *Placodbella parasitica* (ventral view) with brood of attached young. F, *Haemadipsa zeylandica* (dorsal view).

The young leeches can readily be swallowed by animals drinking in such ponds and infection occurs chiefly in dry years when water is scarce. This leech is very troublesome in central Europe; cattle, buffaloes, equines, sheep, dogs, pigs and man may become infected.

The parasites attach themselves in the pharynx and nasal cavities, where they may stay for days or even weeks. They suck blood, so that in severe infections anaemia and loss of condition may be produced. More serious is the frequent occurrence of oedema in the affected areas. Blood or bloody froth may often be seen exuding from the mouth or nostrils of the animals. There is dyspnoea and in severe cases the neck is extended and the mouth is held open. Oedematous swellings may be seen in the parotid and intermandibular regions. Death may be caused by asphyxia, and it may occur suddenly as a result of oedema of the glottis.

Diagnosis is made from the clinical signs and by finding the parasites in the pharynx.

The injection of chloroform water gives very successful results. An elastic catheter is passed through the inferior nasal meatus and to the free end a 60 ml syringe containing the solution is attached. The solution is injected slowly while the catheter is revolved, the head of the animal being held in a horizontal position. In extreme cases tracheotomy may be necessary.

As prophylaxis the animals should be watered from clean troughs. Leeches in water can be killed by means of copper sulphate as applied for snails.

Limnatis africana occurs in West Africa. It is about 12 cm long when fully extended. It is found attached in the nasal cavities of man, dogs and monkeys and frequently protrudes through the nostril. The leech is taken in while drinking infected water, and it may cause debility and marked local disturbances. It may enter, and cause bleeding from, the vagina or urethra of people bathing.

Dinobdella ferox (Blanchard, 1888) is widespread in southern Asia and is found attached to the pharynx of ruminants and the upper respiratory tract of dogs, monkeys and occasionally man. The leech is 3–6 cm in length, dorsoventrally flattened and dark red in colour when engorged. It is often found in macaques obtained from endemic areas, such as Taiwan.

Genus: Haemadipsa Tennant, 1851

The species of this genus are small leeches, measuring about 2–3 cm in length. They occur in tropical forests of Asia and South America and live on trees, shrubs and rocks, attaching themselves to passing animals and human beings. These leeches are very active and creep in even under tight-fitting clothing. Their bite is not painful, but the wounds may bleed for a long time. When groups of these leeches attach themselves to the ankles of human beings or the legs of horses and other animals, severe irritation and anaemia may result. Death has been attributed to them, but they are not usually fatal.

Aquatic birds of the family Anatidae may harbour in their nasal cavities leeches belonging to various genera, among which are: *Dina parva*, *Protoclepsis tesselata*, *Placobdella rugosa* and *Theromyzon* spp. Roberts (1955) described severe keratoconjunctivitis in domesticated geese in Shropshire, England, caused by numerous individuals of the small species *Theromyzon tessulatum*, normally an inhabitant of the nasal sinuses, which had attacked the eyes of the geese. For species recorded from these and other Anatidae see Lapage (1961).

Checklists of leeches affecting fish and laboratory reptiles and amphibians are given by Flynn (1973). Fish may be treated by immersing them in a solution of neguvon (2–3%) for 15–30 seconds.

REFERENCES

ACANTHOCEPHALA

Graeber, K. & Storch, V. (1978) Electron microscopic and morphometric investigations of the integument of Acanthocephala (Aschelminthes). *Z. Parasitenk.*, **57**, 121–135.
Needham, T. & Wootten, R. (1978) The parasitology of teleosts. In: *Fish Pathology*, ed. R. J. Roberts, pp. 144–182. London: Baillière Tindall.
Peters, J. C. & Zwart, P. (1962) Dithiazinine iodide (Dilombrin) as a treatment against acanthocephalan infection in monkeys. *Nord. VetMed.*, **14**, 284–287.
Petrochenko, V. T. (1949) Cycle of development of *Polymorphus magnus* (Dkthsin 1913), parasite of domestic and wild ducks. *Dokl. Akad. Nauk S.S.S.R.*, **66**, 137–140.
Rayski, C. & Garden, E. A. (1961) Life-cycle of an acanthocephalan parasite of the eider duck. *Nature, Lond.*, **192**, 185–196.
Sanford, S. E. (1978) Mortality in mute swans in southern Ontario associated with infestation with the thorny-headed worms *Polymorphus baschadis*. *Can. vet. J.*, **19**, 234–236.

ANNELIDA

Flynn, R. J. (1973) *Parasites of Laboratory Animals*. Ames, Iowa: Iowa State University Press.
Lapage, G. (1961) A list of the parasitic Protozoa, Helminths and Arthropods recorded from species of the family Anatidae (ducks, geese and swans). *Parasitology*, **51**, 1–109.
Roberts, H. E. (1955) Leech infestation of the eyes in geese. *Vet. Rec.*, **67**, 203–204.

2

Arthropods

Phylum: Arthropoda

The name of this phylum, derived from the Greek words *arthros*, a joint, and *podos*, a foot, refers to the fact that the members of the phylum have jointed limbs. The primitive limb of arthropods was biramous, consisting of an unbranched basal piece, the *protopodite*, which branched into an inner *endopodite* and an outer *exopodite*. Some of the limbs of some species of arthropod are still of this type.

Arthropods have probably descended from ancestors which also gave origin to the soft-skinned annelid worms, an example of which is the earthworm; but the arthropods have developed an outer covering of *chitin*, which forms an *exoskeleton* in which the whole body is enclosed. This chitinous covering is secreted by chitogenous cells beneath it and it not only covers the external surface of the body, but also passes through the mouth into the anterior part of the alimentary canal called the *stomodaeum* and also through the anus into the posterior part of the alimentary canal called the *proctodaeum*, both of which arise as invaginations from the exterior into the body. The exoskeleton is usually present in the form of chitinous plates, called *sclerites*, a typical segment of the body having a dorsal sclerite, called a *tergum*, a ventral sclerite, called a *sternum*, and a lateral plate between the tergum and sternum, which is called a *pleuron*. The tergum, sternum and pleuron of each segment are united by more flexible portions of the chitinous exoskeleton. As the arthropod grows it becomes too big for its chitinous covering and periodically this is cast off and a new exoskeleton is formed. Each casting of the exoskeleton is called an *ecdysis*.

Arthropods are metamerically-segmented animals. The segments of arthropods are associated in groups, the anterior segments forming the *head*, the middle ones the *thorax* and the posterior ones the *abdomen*.

The appendages found on the body of an arthropod are typically paired, one pair usually being found on each segment. The appendages on the head are typically one or two pairs of *sensory antennae* and, behind these, paired appendages modified for feeding. Commonly there is one pair of *mandibles* and behind these two pairs of *maxillae*. Behind these again there may be *maxillipedes*, which are walking legs adapted for feeding. The next group of appendages belongs to the thorax and they are walking legs. Behind them, in aquatic species such as the Crustacea, there are a variable number of abdominal appendages, some or all of which are used for swimming; terrestrial species usually lose these or some of them may become modified to perform other functions.

A dominant feature of the internal anatomy of the Arthropoda is the fact that the general body cavity is not a coelom. It is a space full of blood, which is called the *haemocele*. The blood in it bathes all the organs of the body. The *heart* is an enlarged dorsal blood vessel, which is enclosed in a compartment of the haemocele full of blood called the *pericardium*. As the heart pulsates, it sucks in blood from the pericardium through openings in its walls called *ostia*. It then pumps the blood into the haemocele through short arteries, which are usually the only blood vessels in the body.

The respiratory organs of arthropods are also characteristic of the phylum. They are:

1. *Gills* (branchiae) of various kinds found in larvae, nymphs and adults of species that are aquatic.

2. *Tracheae*, which are fine, elastic tubes, with a thin, chitinous lining, which are held open by rings or spiral thickenings of the chitinous lining;

tracheae branch and ramify among the internal organs, to which they take air that enters them through their external openings or *stigmata*; tracheae are especially characteristic of insects.

Other respiratory structures are *lung-books* and *gill-books* of spiders and crabs respectively. In some forms, e.g. the parasitic mites, respiration is through the cuticle.

The alimentary canal varies in the different classes of arthropods. In all, however, it consists of (*a*) the *stomodaeum* mentioned above, which is lined by chitin and may be divided into a sucking *pharynx*, a *proventriculus* (crop) and a *gizzard*; (*b*) the *proctodaeum* mentioned above, which is also lined by chitin; (*c*) a mid-gut, or *mesenteron*, which connects the proctodaeum with the stomodaeum.

The excretory organs of arthropods vary in the different classes of the phylum. Those of the class Crustacea are a pair of *nephridia* which open on the bases of the second antennae. The excretory organs of the Insecta are tubules, called *malpighian tubules*, which are arranged in a ring round the alimentary canal. Usually they open into the anterior end of the proctodaeum. Arachnida also have malpighian tubules that open into the anterior end of the proctodaeum, but they have, in addition, *coxal glands*, which open on the coxae of the legs. These latter are true nephridia, homologous with the nephridia of the Crustacea.

The nervous system of arthropods consists of cerebral ganglia in the head, united by circumoesophageal commissures to a ventral double nerve cord that runs along the ventral side of the body and has nerve ganglia on it. Typically there is one ganglion in each segment, but fusions of segments carry with them fusions of the ganglia associated with them. Associated with this central nervous system are eyes, sensory setae and other special sense organs, some of which are described below.

The sexes of arthropods are usually separate.

The phylum Arthropoda includes the following classes of importance:

CLASS: CRUSTACEA LARMARK, 1815

This class includes the crayfishes, lobsters, shrimps, crabs, wood-lice and their relatives. Most of the species are aquatic and breathe by means of gills, but some, such as the wood-lice, are terrestrial. Crustacea have two pairs of antennae and numerous pairs of limbs on the thorax and abdomen and these limbs are frequently biramous. The class is divided into two subclasses.

Subclass: Entomostraca Müller, 1785

Species belonging to this subclass are usually small Crustacea with a variable number of body segments. The abdomen often ends in a caudal fork. To this subclass belong some species which are intermediate hosts of parasitic helminths, among which are the species of the genus *Cyclops*, which act as intermediate hosts of the tapeworm *Diphyllobothrium latum* and of the nematode *Dracunculus medinensis*, and *Daphnia*, which act as the intermediate hosts of the spiruroid nematode *Echinuria uncinata*.

The subclass also contains the parasitic copepods, some of which are serious pests of fish culture.

Subclass: Malacostraca Latreille, 1802

Species belonging to this subclass are usually larger than the Entomostraca and possess a constant number of body segments. Typically there are eight segments in the thorax and seven in the abdomen. The character of the appendages clearly marks off the thorax and the abdomen. To this subclass belong the shrimps, lobsters, crayfishes and crabs. This subclass also includes important intermediate hosts of parasitic nematodes, among which are the terrestrial wood-lice and their aquatic relatives belonging to the genus *Asellus* and the 'freshwater shrimp' *Gammarus pulex*.

CLASS: MYRIAPODA LANKESTER, 1904

Species of this class are the centipedes and millipedes. Their bodies consist of a number of segments which are, with the exception of the head, not grouped into definite body areas. There are two orders: (1) the *Diplopoda* (millipedes), which are chiefly vegetarian species and have two pairs of limbs on each segment of the body behind the head; some of them are serious pests of crops; (2) the *Chilopoda* (centipedes), which are chiefly carnivorous and have one pair of limbs on each segment of the body behind the head; some species of them are useful enemies of other pests of garden and other crops and some are poisonous arthropods.

CLASS: INSECTA LINNAEUS, 1758
This class includes all the insects. Their bodies are divided into three parts, namely, a *head*, which bears one pair of antennae, a *thorax*, consisting of three segments, which bears three pairs of legs and, typically, two pairs of wings, and an *abdomen*, consisting of a variable number of segments which either has no appendages or the appendages on it are modified for various special purposes. Insects breathe by means of tracheae.

CLASS: ARACHNIDA LAMARCK, 1815
This class includes the king-crabs, scorpions, spiders, ticks, mites and their relatives. They not only vary much in structure among themselves, but they also differ considerably from other arthropods.

CLASS: PENTASTOMIDA HEYMONS, 1926
These are the tongue worms and are of uncertain systematic position: some consider them as arthropods, while others regard them as annelids; still others assign them to a distinct phylum.

CLASS: INSECTA LINNAEUS, 1758

The class includes about 70% of all the known species of animals of all kinds. With the majority of these the veterinarian is not concerned, but some species are of great veterinary importance.

A brief description of the anatomy of insects is given below. This is intended as a guide to the recognition of the important parts of an insect, especially those which are used in identification. A more extensive account of the morphology and physiology of the Insecta will be found in textbooks of entomology such as Richards and Davies (1977).

Head. The head is an ovoid or globular capsule, composed of a number of plates or sclerites, at the anterior end of the body. Eyes are usually present and are placed laterally, above the cheeks or genae. They are compound eyes and may meet one another in the midline (holoptic) or they may be wide apart (dichoptic). Simple eyes, or ocelli, may be present and are arranged in a triangle on the dorsum or vertex.

Antennae. These are situated between or in front of the compound eyes. Their form varies greatly, some are elongated and many segmented (mosquitoes), some are short and squat (houseflies). They are frequently haired or may carry special bristles (e.g. aristae).

Mouth parts (Fig. 2.1). These consist of the *labrum* or upper lip which forms the upper

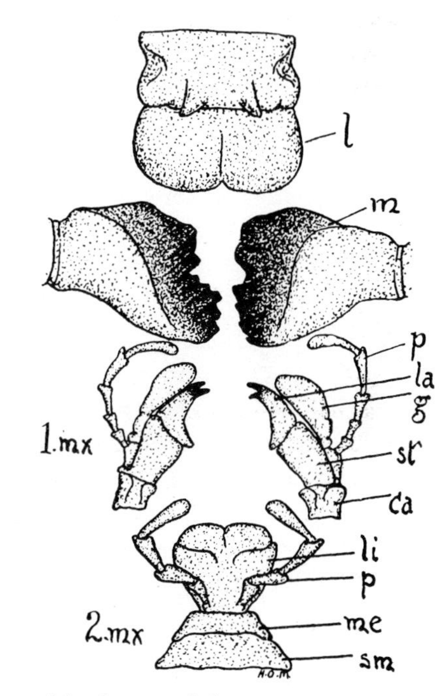

Fig 2.1 Mouth parts of a locust.

ca	cardo
g	galea
l	labrum
la	lacinia
li	ligula
m	mandible
me	mentum
p	palp
sm	submentum
st	stipes
1 mx	first maxilla
2 mx	second maxilla

boundary of the mouth and the *labium* or lower lip which forms the lower boundary of the mouth. Between these two structures are two pairs of biting jaws, an upper pair, the *mandibles*, and a lower pair, the *maxillae*. On the underside of the labrum there is a small membranous structure, the *epipharynx* which bears the organ of taste. These two are frequently fused to form the *labrum–epipharynx*. On the upper surface of the *labium* is a further membranous structure, the *hypopharynx*, which bears the opening of the salivary duct. Both the maxillae and the labium possess jointed *palps* which are sensory in function.

Great modifications occur in this basic mouth part structure. In the chewing insects (e.g. the locust) all the various components can be recognized but in the suctorial forms various structures of the mouth parts may be much modified. Thus the labium may be greatly expanded for imbibing liquid food (house fly) or all parts may be modified into fine piercing stylets (mosquitoes).

Thorax. This consists of three segments, the *prothorax*, *mesothorax* and *metathorax*. These parts may not be distinct in some insects due to fusion. Each segment typically bears a pair of legs and the mesothorax and metathorax typically bear one pair of wings each.

Legs (Fig. 2.2). These consist of a basal coxa by which the leg is attached to the body, this being followed by a trochanter, femur, tibia and tarsus, the latter being composed of a number of joints, usually five. The last tarsal segment is frequently provided with a pair of claws between which is an empodium consisting of a pad, a spine or a bristle. A pair of pads, the pulvilli, occur below the claws.

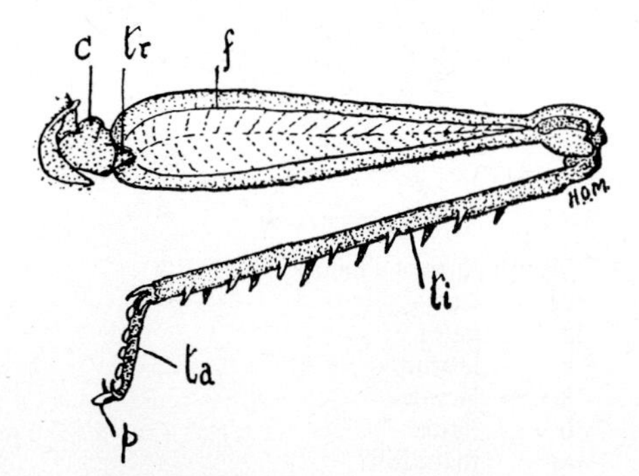

Fig 2.2 Leg of a locust.
c coxa
f femur
p pulvillus
ta tarsus
ti tibia
tr trochanter

Wings. Normally two pairs occur but in the *Diptera* the posterior pair is reduced to a pair of balancers or halteres. Embryonically the wing is sac-like but in the adult insect the two membranes become closely applied to each other. The wings are supported by 'veins' which are breathing tubes or tracheae. The arrangement of the veins is a valuable means of identification in many cases.

Abdomen. Usually clearly segmented and soft and membranous. Various structures may be present on the abdomen such as copulatory claspers, an ovipositor and the external genitalia.

Respiratory system. The respiratory system of insects consists usually of a system of branching tubes, the tracheae, which open through spiracles or stigmata at the sides of the body. The tracheae are composed of a thin layer of chitin strengthened by spiral thickenings secreted by special chitogenous cells. They end in air sacs with very delicate walls. There may be a pair of spiracles to each segment, but they are usually reduced in number and there are none in the head and the prothorax. The spiracles may be bordered by a thick rim of chitin which bears bristles and they open into a vestibule which contains a valve controlled by muscles. The respiratory movements are produced by muscular contractions and elastic distensions of the body wall.

Alimentary canal. The alimentary canal of insects consists of a stomodaeum, a mesenteron and a proctodaeum (Fig. 2.3). Of stomodaeal origin are the buccal cavity with the salivary glands, the epi- and hypopharynx, the pharynx, the proventriculus or gizzard and the crop, or oesophageal diverticulum or food reservoir. The mesenteron forms the mid-gut. At its posterior end is a ring of malphighian tubes, which have an excretory function. The proctodaeum consists of the intestine or hind-gut and the rectum, together with the malpighian tubes and the papillae of the rectum.

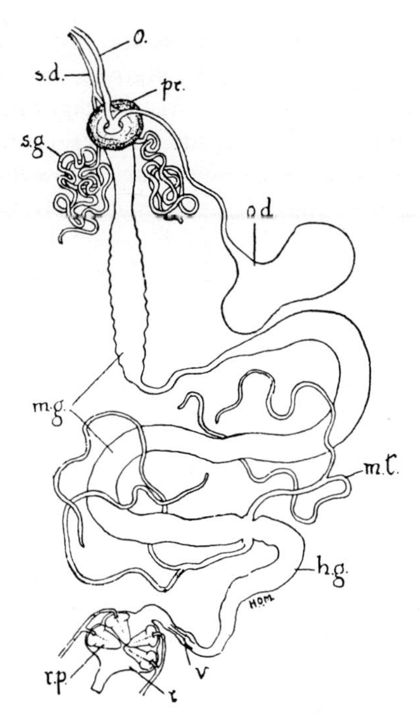

Fig 2.3 Alimentary tract of *Lucilia sericata*.
hg hindgut
mg midgut
mt malpighian tubules
o oesophagus
od oesophageal diverticulum
pr proventriculus
r rectum
rp rectal papilla
sd salivary duct
sg salivary gland
v valve

A crop is present in most *Diptera*, joined to the oesophagus by a narrow tube. In mosquitoes it is represented by three thin bags. The gizzard is present in insects that eat solid food, like the locust, and it contains a complicated set of teeth on its internal surface. The salivary glands are paired and have long ducts, which later join to form a common duct. An oesophageal valve may be present and in the *Cyclorrhapha* the proventriculus is a compact, spherical structure which functions as a valve. In the *Cyclorrhapha* and some other insects there is a delicate, tubular peritrophic membrane within the alimentary canal, extending from the mid-gut to the rectum. Its posterior end is not attached to the intestinal wall. It separates the food from the wall of the intestine, but osmosis takes place through it. In the *Acarina* the mid-gut has several diverticula, which are blind sacs capable of great distension.

Vascular system. The vascular system comprises a dorsally situated heart, an aorta and the general body cavity or haemocele. The heart is a tube surrounded by pericardial cells and its lumen is divided into a number of compartments by valves which allow the blood to pass forward only. Each compartment opens into the haemocele through a pair of ostia. Anteriorly there is an aorta which carries the blood to the head, whence it enters the haemocele and bathes all the organs. The blood is a viscid fluid and contains few cells.

The fat-body consists of numerous cells laden with fat and lies in the body wall, lining the haemocele and surrounding all organs. It is especially large in recently emerged adults, pupae and mature larvae, and may hang into the body cavity in large masses.

Nervous system. The nervous system of insects consists of a circumoesophageal commissure with ganglia and a double ventral chain of ganglia from which nerves are given off. This ventral nerve chain originally had a pair of ganglia to each segment, but usually concentration occurs by fusion of the ganglia, especially in the thorax. In some cases the thoracic and abdominal ganglia may all be fused together.

Reproductive system. The male (Fig. 2.4) has two testes, each with a vas deferens forming distally a vesicula seminalis and then fusing to

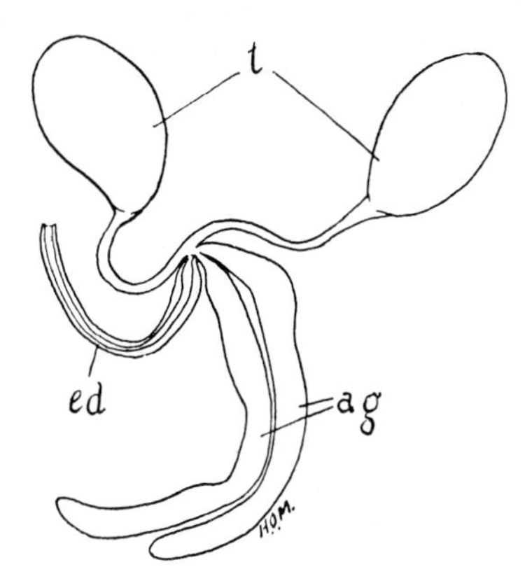

Fig 2.4 Male genital organs of *Lucilia sericata*.
ag acessory glands
ed ejaculatory duct
t testes

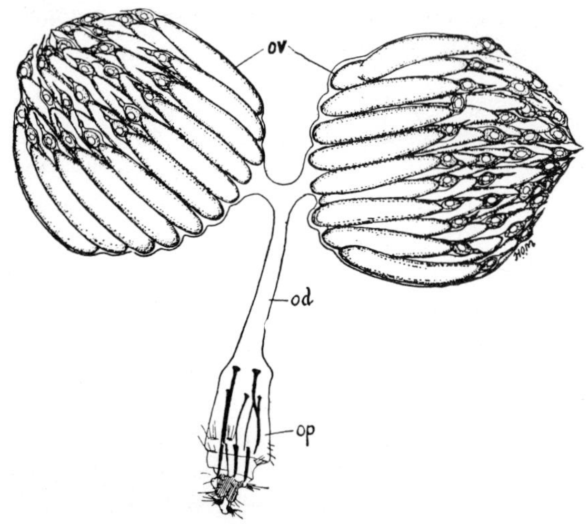

Fig 2.5 Female genital organs of *Lucilia sericata*.
od oviduct
op ovipositor, contracted
ov ovaries

form a common duct. A penis with a sheath and other accessory structures are frequently present.

The female (Fig. 2.5) has two ovaries, which consist of groups of ovarian tubes, all arising from an apical filament. The ducts unite to form a common duct which bears the receptaculum seminis and ends in the ovipositor. Accessory glands are usually present. Some insects are oviparous, others viviparous or larviparous, and some even pupiparous, like *Glossina* and the *Pupipara*, which give birth to larvae that are ready to pupate. In these latter forms the larvae are nursed in the common portion of the oviduct, the uterus, which is provided with 'milk glands' that secrete a milky fluid through a teat to which the mouth of the larva is applied. Only one larva is born at a time. The larva usually lies with its stigmatic plate at the vulva of the fly and is thus able to breathe. Parthenogenesis of various kinds is also found among insects.

Development. As a rule the eggs consist of a large mass of yolk surrounded by the cytoplasm, which undergoes segmentation, and the body of the embryo is formed around the yolk. The rate of development is greatly influenced by the temperature and sometimes the presence of a fair amount of moisture is absolutely essential, else the eggs may remain dormant. In some cases the young arthropod which hatches from the egg resembles the adult; in other cases if differs from the adult only in size and minor features, or there is a very marked difference, and definite stages, termed the larva, pupa and imago, are seen. Ecdyses always occur and the different stages separated by ecdyses are called *stadia*, the form of the insect during each stadium being called an *instar*. During its development the insect undergoes a variable degree of change or *metamorphosis*. The only insects that do not undergo some degree of metamorphosis are the primitive bristle-tails and silver-fish and their relatives, which never develop wings and belong to the subclass *Apterygota*. The other insects all undergo a greater or lesser degree of metamorphosis and this fact is expressed in the classification given below. When metamorphosis is complete, as it is, for instance, in the life-history of a butterfly or house fly, the form that leaves the egg is called the *larva*, which feeds and grows and eventually becomes a quiescent phase, called the *pupa* (chrysalis), inside which the adult insect (*imago*) is formed. When the metamorphosis is simple or slight the form that leaves the egg is more like the adult and it is called the *nymph*, and the nymph merely grows, casting its skin several times, to become the imago. The larvae of species with a complete metamorphosis take various forms. The *polypod larva*, such as the caterpillar of a butterfly, has a well-marked head, a thorax of three segments, each of which bears a pair of clawed legs, and an abdomen of ten segments, which bears five pairs of fleshy hooked legs, called *prolegs*. The *oligopod larvae*, such as those of many beetles, have a well-marked head and three pairs of thoracic legs, but no abdominal legs. The *apodous larva* has no legs either on the thorax or the abdomen and the head is also reduced. The larvae of the house fly and the blow fly and of all Diptera are apodous. Often they are called *maggots*.

The pupae of insects may take three forms. In the most active pupae, such as those of beetles, the wings and legs of the adult insect can be seen externally and they are free from the rest of the body. Pupae of this kind are called *free* or *exarate* pupae. In other pupae, such as those of the butterflies and moths, the horse flies (Brachycera) and the mosquitoes and their relatives (Nematocera), the legs and wings are bound down to the body by moulting fluid, but they can usually be seen externally. Pupae of this kind are called *obtectate* pupae. The pupae of the *Cyclorrhapha* are, however, enclosed in the cast skin of the last larval phase, which is called the *puparium*. Their skin hardens and the insect inside cannot be seen. Pupae of this kind, such as the pupae of the house flies, are called *coarctate* pupae.

CLASSIFICATION OF INSECTA

The classification of insects used in this book is that adopted by Richards and Davies (1977).

SUBCLASS: APTERYGOTA

Wingless insects, the wingless condition being primitive. Metamorphosis absent or very slight. One or more pairs of abdominal appendages present other than genitalia and cerci.

Order: Thysanura (silver fish, bristle tails)
Order: Protura (myrientomata)
Order: Diplura (Campodeids)
Order: Collembola (spring-tails)
(No veterinary importance.)

SUBCLASS: PTERYGOTA

Winged insects, possessing wings in the adult stage or secondarily wingless, having descended from winged forms. Metamorphosis very varied, rarely slight or absent. No abdominal appendages other than genitalia or cerci.

Division: Exopterygota

Wings develop externally as buds. Metamorphosis simple, rarely a pupal stage. The immature stages usually resemble the adults in structure and habit. The division includes the following orders:

Order: Orthoptera (grasshoppers, cockroaches)
Order: Dermaptera (earwigs)
Order: Plecoptera (stone flies)
Order: Isoptera (termites)
Order: Psocoptera (book lice)
Order: Mallophaga (biting lice)
Order: Siphunculata (syn. *Anoplura*) (*sucking lice*)
Order: Odonata (dragon-flies)
Order: Thysanoptera (thrips)
Order: Hemiptera (bugs)

Division: Endopterygota

Wings develop internally. Metamorphosis complete. Pupal stage present. The division includes the following orders:

Order: Coleoptera (beetles)
Order: Hymenoptera (bees, wasps)
Order: Lepidoptera (butterflies, moths)
Order: Neuroptera (lace wings)
Order: Siphonaptera (syn. *Aphaniptera*) (*fleas*)
Order: Diptera (true flies)

SUBCLASS: PTERYGOTA LANG, 1889

DIVISION: EXOPTERYGOTA

ORDER: ORTHOPTERA LATREILLE, 1796

Species of this order have two pairs of wings; the anterior (mesothoracic) pair, which are thickened, act as covers (called *tegmina*) for the hinder (metathoracic) pair, which are membranous. The antennae are usually long and filamentous and many-jointed. The mouth parts are of the type adapted for chewing. Examples of the order are the grasshoppers, crickets, cockroaches, locusts and stick and leaf insects. Grasshoppers belonging to the genus *Melanoplus* are intermediate hosts of the spiruroid nematodes *Tetrameres americana* and *Cheilospirura hamulosa* etc. Among the cockroaches *Blatella germanica* can be experimentally infected with the spiruroid nematode *Gongylonema pulchrum* and *Pycnoscelus surinamensis* is an intermediate host in Australia of the spiruroid nematode *Oxyspirura parvorum*.

COCKROACHES

A cockroach common in many countries is the croton bug, *Blatella* (*Ectobia*) *germanica*. The adults measure about 15 mm in length to the tips of the wings, which are present in both sexes. Their colour is light brown, with two longitudinal dark stripes over the prothorax and wings. Fairly common also is the Oriental roach, *Blatta* (*Periplaneta*) *orientalis*, which is almost black, about 25 mm long, with wings that do not quite reach the tip of the abdomen in the male and are vestigial in the female.

Cockroaches live preferably in warm places and roam about in the dark. At other times they hide in cracks and crevices, behind and along baseboards of walls, around water-pipes and cisterns and in similar places. They feed on starchy or sugary materials, but will eat almost anything if necessary, and are found most frequently in kitchens, bakeries and storerooms in which cereal products are kept. They may be a frequent pest in animal quarters. They are not parasites, but may easily spread disease on account of their habits. They have been proved to carry various fungi and protozoa and to act as intermediate hosts of parasitic nematodes.

The eggs are laid in egg-cases which each contain a number of eggs. They may be carried about for some time and may be seen protruding from the abdomen of the female. The egg-cases are deposited in crevices and the rate of development of the eggs and the young cockroaches depends very much on the temperature and the available food supply.

Control of cockroaches is difficult, because they breed rapidly and new infestations readily occur. The most effective control is through the use of insecticides.

Among the chlorinated hydrocarbon insecticides, dieldrin or lindane sprayed into the hiding places gives effective control, but must be repeated as new infestations occur. Among suggested strengths are 2.5% chlordane emulsion or solution, 0.5% dieldrin solution or 1% dieldrin dust or 2% chlordane. These have been used for the control of *Blatella germanica* in Germany and the United States, but this species has developed resistance to chlordane and lindane and to a less extent to malathion. For strains resistant to chlorinated hydrocarbons organophosphorus insecticides may be tried. Suggested formulae are 1.5% malathion, 1% dipterex, 0.5% diazinon or 1.5% chlorthion. The synthetic pyrethroids are probably of value.

(For further information, see the periodical publications of WHO, the *Tropical Diseases Bulletin* and the *Journal of Economic Entomology*.)

ORDER: MALLOPHAGA NITZSCH, 1890

Species of this order are the biting lice or bird lice. They share affinities with the Psocoptera (booklice, psocids) from which stock Richards and Davies (1977) suggest they may have evolved. However, they also share many features with the sucking lice (Siphunculata) and Clay (1970) prefers to classify the biting and sucking lice in a single order (Phthiraptera) which would then be divided into the suborders Amblycera, Ischnocera, Rhynchophthirina and Anoplura. The present use of two orders, Mallophaga for the biting lice and Siphunculata for the sucking lice, follows the classification proposed by Richards and Davies (1977).

Species of this order are small and wingless and have dorsoventrally flattened bodies. Their antennae are short and are composed of three to five segments. The eyes are vestigial and the segmentation of the thorax is indistinct. The mesothorax and metathorax are fused to form one piece, in front of which the prothorax is a distinct and separate segment. The thoracic spiracles are on the ventral side of the mesothorax. The tarsi consist of one or two segments and each tarsus bears one or two claws. There is one pair of spiracles situated on the mesothorax. Typically there are six pairs of abdominal spiracles, but when fusion of abdominal segments occurs there may be fewer than six pairs. The operculated eggs are cemented, without stalks, to the hairs or feathers. There is little or no metamorphosis. The phase of the life-history that leaves the egg resembles the adult and is called the first nymph. There are three ecdyses, the first nymph becoming the second nymph, which becomes the third nymph and this becomes the adult. The duration of the developmental cycle varies with species of biting louse and the environmental condition. At 37°C eggs of *Columbicola columbae* hatched after four days and each nymphal stage lasted seven days (Martin 1934). The whole life-history is passed on the host. Uninfected hosts are infected by close contact with infected ones, but lice may also be spread by farm equipment and personnel. Thus lice of horses may be spread by brushes, blankets, harness or other stable equipment.

The biting lice are divided into three suborders, the characteristics of which are summarized below.

Suborder: Amblycera
Suborder: Ischnocera
Suborder: Rhynchophthirina

SUBORDER: AMBLYCERA KELLOGG, 1899

In species of this suborder the antennae lie in grooves in the sides of the head and they may not be readily seen. Maxillary palps may, however, be present and these may be visible in mounted specimens and may be confused with the antennae. The antennae, however, may be identified by the fact that usually they consist of four to five segments and the third segment is stalked, being somewhat the shape of an egg-cup that holds the fourth segment. The maxillary palps, when they are present, also have two to five segments, but the third segment is not stalked. The antennae of some species (e.g. those of the genus *Columbicola*) show sexual dimorphism, the antennae of the males being elongate and having a swollen first segment, with an appendage on the third segment. The mandibles bite horizontally. The head is often broader and more rounded anteriorly than that of the *Siphunculata* but this is not a very reliable character. They are parasitic on both mammals and birds. The tarsi of species parasitic on birds have two claws and those parasitic on mammals have one claw.

The following species may be encountered by veterinarians:

AMBLYCERA OF BIRDS

Menopon gallinae (Linnaeus, 1758) (syn. *M. pallidum*), the 'shaft louse' of poultry, is pale yellow in colour. Male 1.71 mm, female 2.04 mm long. The thoracic and abdominal segments have each one dorsal row of bristles. This species occurs on fowls and also on ducks and pigeons. It moves about rapidly. The eggs are laid in clusters on the feathers.

Menopon phaeostomum (Nitzsch, 1818) occurs on the peacock.

Holomenopon leucoxanthum Burmeister, 1838, has been demonstrated to be the cause of 'wet feather' of ducks (Humphreys 1975). Birds show soiled and tattered plumage which they preen continuously. If large areas of the body are affected, the plumage no longer repels water and birds become chilled and may die from pneumonia. A dust of coumaphos (Asuntol) has been used with success for treatment and control.

Menacanthus (Eomenacanthus) stramineus (Nitzsch, 1818) (syn. *Menopon biseriatum*) is the yellow 'body louse' of poultry, occurring on the skin of those parts of the body which are not densely feathered like the breast, thighs and around the anus. It occurs on the fowl, turkey, peacock and Japanese pheasant, and is especially harmful to small chicks. Male 2.8 mm, female 3.3 mm long. The abdominal segments have each two dorsal rows of bristles. The eggs have characteristic filaments on the anterior half of the shell and on the operculum and are laid in clusters on the feathers near the skin (Fig. 2.6).

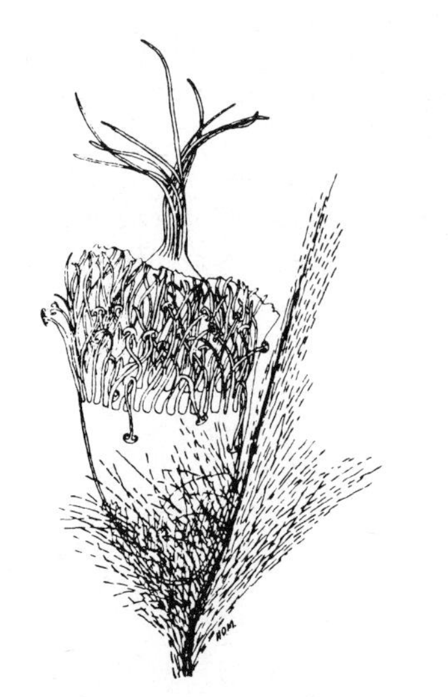

Fig 2.6 Egg of *Menacanthus stramineus* attached to a feather.

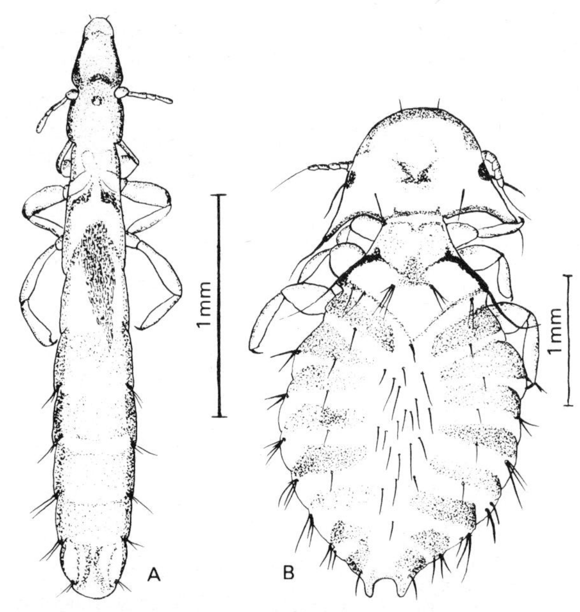

Fig 2.7 A, *Columbicolae columbae*. B, *Chelopistes meleagridis*.

Trinoton anserinum (J. C. Fabricius, 1805) (syn. *Trinoton anseris*) may be found on the duck and the swan.

In addition to the above, a substantial number of genera have been reported from various orders of wild birds. These are listed by Turner (1971*a*).

AMBLYCERA OF MAMMALS

Although most of the Amblycera occur on birds, some species are parasitic on mammals, among which **Gyropus ovalis** Nitzsch, 1818, and **Gliricola porcelli** (Linnaeus, 1758) and **Trimenopon hispidum** (Burmeister, 1838) may all be found on the guinea-pig, and the former two species also occur on other rodents. **Heterodoxus spiniger** (Enderlein, 1909) is common on the dog in warm countries (between latititudes 40°N and 40°S) and **H. longitarsus** (Piaget, 1880) and **H. macropus** Le Souëf and Bullen, 1902 occur on kangaroos and wallabies.

Turner (1971*b*) gives a list of the wild mammals that are parasitized by lice of the suborder Amblycera.

SUBORDER: ISCHNOCERA KELLOGG, 1899

In species of this suborder the antennae are filiform and visible at the sides of the head and they are composed of three to five segments. The head of an ischnoceran louse may therefore look at first sight somewhat like that of a sucking louse, although it is usually broader. There are no maxillary palps, so that these cannot be mistaken, as the palps of Amblycera may be, for the antennae. The mandibles bite vertically. In the abdomen segments one and two, and nine and ten are fused and segment eleven may not be visible. These lice are parasitic on both mammals and birds.

ISCHNOCERA OF BIRDS

Cuclotogaster (Liperus) heterographus Nitzsch, 1886, the 'head louse' of poultry, occurs on the skin and feathers of the head and neck. Male 2.43 mm, female 2.6 mm long. In the male the first segment of the antenna is long and thick, bearing a posterior process. The abdomen is elongate in the male and barrel-shaped in the female, with dark-brown lateral tergal plates. The eggs are laid singly on the feathers. It occurs on fowls and partridges. It is a dangerous parasite of chicks.

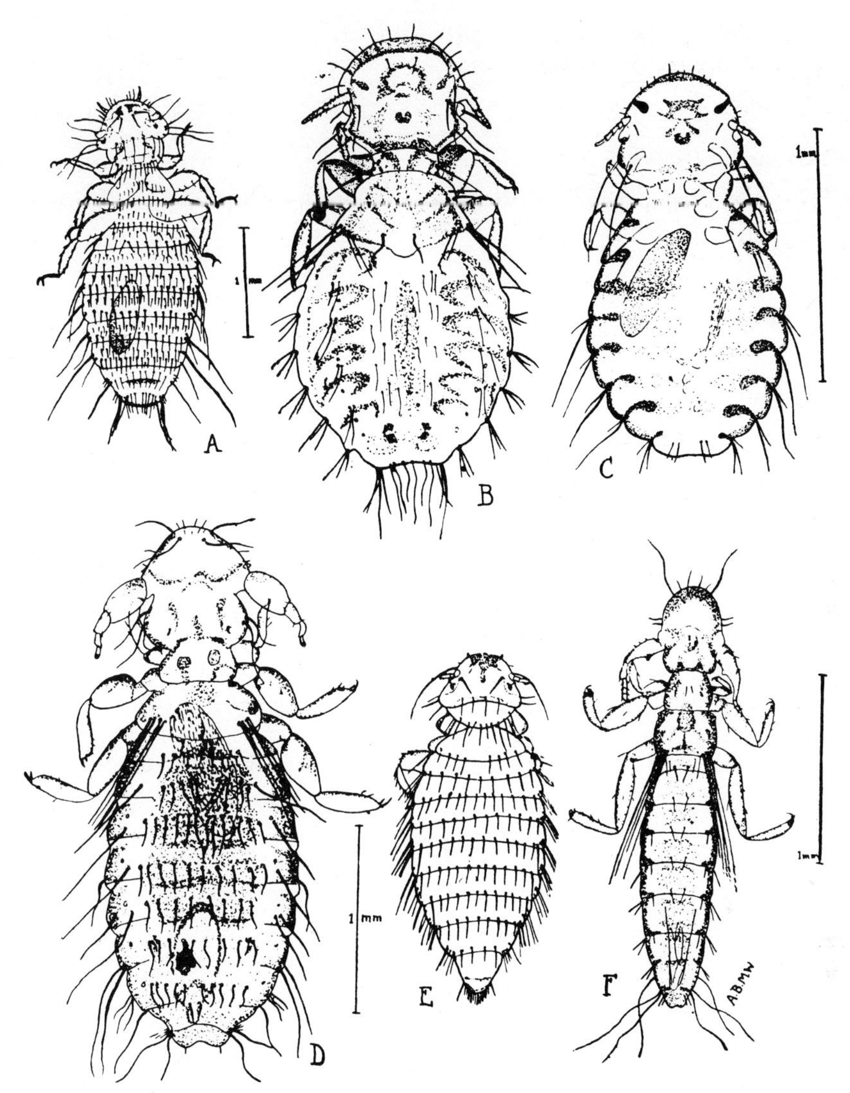

Fig 2.8 Common fowl lice. A, *Menacanthus stramineus*. B, *Goniodes gigas*. C, *Goniocotes gallinae*. D, *Cuclotogaster heterographus*. E, *Menopon gallinae*. F, *Lipeurus caponis*.

Lipeurus caponis Linnaeus, 1758, the 'wing louse', is a slender, elongate louse which occurs on the under-side of the large wing feathers and moves about very little. It occurs in fowls and pheasants.

Goniodes gigas (Taschenberg, 1879) (syn. *Goniocotes gigas*) is a large louse occurring on the body and feathers of the fowl. Male 3.2 mm and female 5 mm long.

Goniocotes gallinae (de Geer, 1778) (syn. *Goniocotes hologaster*; *Goniodes hologaster*), the 'fluff louse', occurs in the fluff at the base of the feathers of fowls, pheasants and pigeons. It is a small louse; the male is 1 mm long and the female 1.6 mm. The body is broad and the head short and wide.

Chelopistes meleagridis (Linnaeus, 1758) (syn. *Goniodes meleagridis*; *Virgula meleagridis*) is a common louse of the turkey.

Columbicola columbae (Linnaeus, 1758) (syn. *Lipeurus baculus*) occurs on domestic and wild pigeons.

Anaticola crassicornis (Scopoli, 1763) and **A. anseris** Linnaeus, 1758 may be found on the duck.

Genera of the Ischnocera which occur on wild birds are listed by Turner (1971*a*).

ISCHNOCERA OF MAMMALS

Damalinia (Bovicola) bovis (Linnaeus, 1758) (syn. *Trichodectes scalaris*) on cattle.

Damalinia (Bovicola) equi (Linnaeus, 1758) (syn. *Trichodectes parumpilosus*) (Fig. 2.9A) on equines.

Damalinia (Bovicola) ovis (Linnaeus, 1758) (syn. *Trichodectes sphaerocephalus*) on sheep.

Damalinia (Bovicola) caprae (Gurlt, 1843) (syn. *Trichodectes climax*) on goats.

Bovicola painei (Kellog & Nakayama, 1914) on goats.

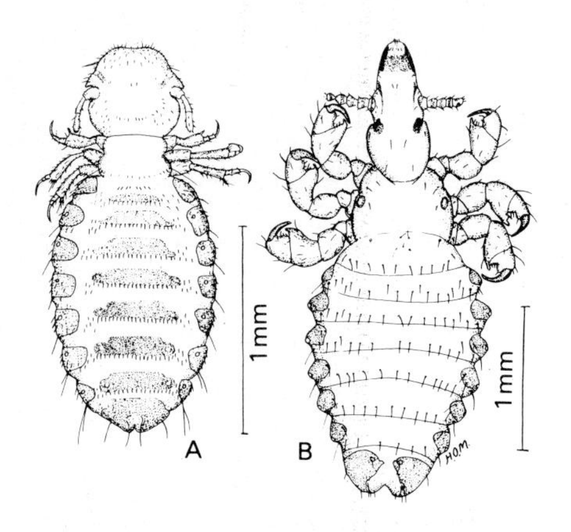

Fig 2.9 Lice of the horse. A, *Damalinia equi*. B, *Haematopinus asini*.

Damalinia limbata (Gervais, 1847) (syn. *Trichodectes limbatus*) on angora goats.

Trichodectes canis (de Geer, 1778) (syn. *T. latus*) (Fig. 2.10A) on dogs.

Felicola subrostratus (Nitzsch, 1838) (syn. *F. subrostrata*) on cats.

The genera of Ischnocera which occur on wild mammals are listed by Turner (1971*b*).

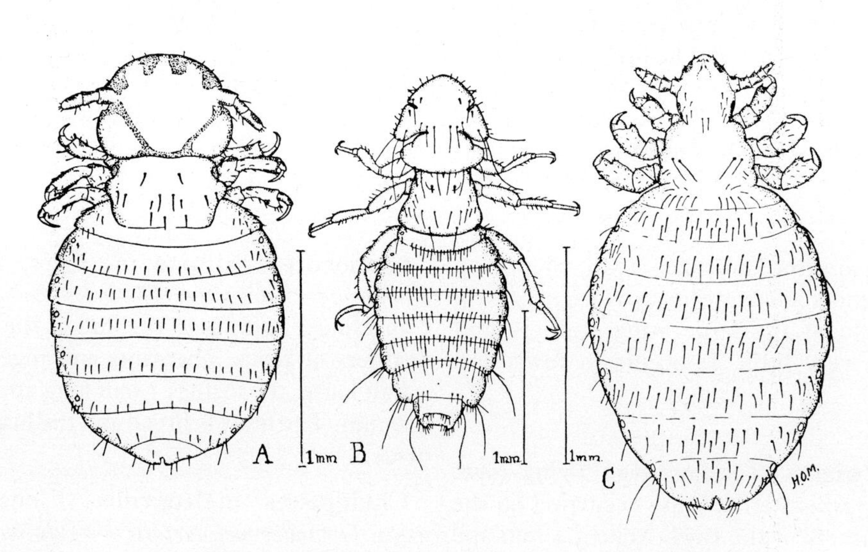

Fig. 2.10 Lice of the dog. A, *Trichodectes canis*. B, *Heterodoxus longitarsus*. C, *Linognathus setosus*.

SUBORDER: RHYNCHOPHTHIRINA HOPKINS AND CLAY, 1952

The one family in this order, Hymatomyzidae, Enderlein, 1904, contains two species which are aberrant forms. The head is prolonged anteriorly in a rostrum, the mandibles are at the apex of the rostrum. The thoracic segments are fused.

Haematomyzus elephantis Piaget, 1869 occurs on Indian and African elephants.

Haematomyzus hopkinsi (Clay, 1963) occurs on wart-hogs.

ORDER: SIPHUNCULATA MEINERT, 1891

Species of this order are the sucking lice. In previous editions they have been classified as a suborder—*Anoplura*—of the order Phthiraptera. Clay (1970) prefers this latter classification but Richards and Davies (1977), whose classification is used in the present edition, prefer to accord ordinal status to the group.

They are wingless insects living as ectoparasites on mammals. The mouth parts are adapted for sucking the tissue fluids and the blood of the host. The two antennae are visible at the sides of the head and are usually composed of five segments. There is no sexual dimorphism. The thorax is small and its three segments are fused together. The abdomen is relatively large, with seven of its nine segments visible; the segments often bear at their sides dark-brown or black areas of thickened chitin, called *paratergal plates*. The eyes are reduced or absent, but are present in the human head louse, *Pediculus humanus*, and on the human pubic louse, *Phthirus pubis*. The first pair of legs is usually smaller with weaker claws; the third pair of legs is usually the largest. The two segments of the tarsus are usually not distinguishable. Each tarsus has only one claw. The hair of the host is held between this claw and a thumb-like process on the ventral apical angle of the tibia. In the Haematopinidae the hold on the hair is helped by a spiny pad, the *tibial pad*, which can be thrust up to lock the grip on the hair. There are thoracic spiracles on the dorsal side of the mesothorax and six pairs of abdominal spiracles. The head is usually more or less pointed anteriorly.

The developmental cycle is similar to that of the Mallophaga; eggs are usually cemented to the hair of the host, with the exception of *Pediculus humanus* (human body louse). The egg hatches in one to three weeks (mean 12 days) and three nymphal stages are undergone (12 days) to produce adults which reach sexual maturity in one to three days.

They are blood-sucking insects and there is considerable host specificity and closely related host species are parasitized by similar sucking lice. Three families of veterinary importance, and two others which contain lice of various rodents and pinnipeds, are dealt with here, namely:

Family: Haematopinidae
Family: Linognathidae
Family: Pediculidae
Family: Hoplopleuridae
Family: Echinophthiriidae

FAMILY: HAEMATOPINIDAE ENDERLEIN, 1904

The eyes are absent, the head has forward prolongations (temporal angles) behind the antennae and the thorax is broad; there are marked paratergal plates and there is one row of spines on each abdominal segment.

To this family belong:

Haematopinus asini (Linnaeus, 1975) (Fig. 2.9B), the sucking louse of equines.

Haematopinus bufali (de Geer, 1788) of buffalo in South Africa.

Haematopinus suis (Linnaeus, 1758), the very large louse of pigs (Fig. 2.11).

Haematopinus eurysternus (Nitzsch, 1818), the 'short-nosed' cattle louse, with a relatively short head and broad thorax and abdomen.

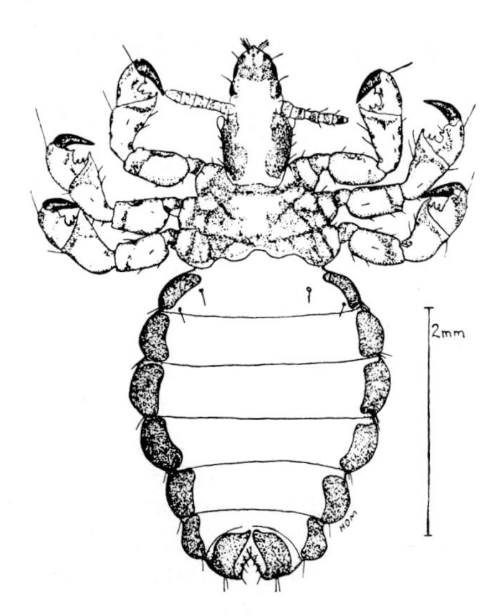

Fig 2.11 Female *Haematopinus suis.*

Haematopinus quadripertusus (Fahrenhotz, 1916). This species occurs on cattle in North America, Queensland, Papua–New Guinea and the Solomon Islands. It was previously thought to be *H. eurysternus.*

Haematopinus tuberculatus (Burmeister, 1839) occurs on buffalo (*B. bubalis*) in Asia and the Pacific area and on the yak. It also occurs on camels and cattle in Australia. Large, 5.5 mm.

FAMILY: LINOGNATHIDAE ENDERLEIN, 1905

The eyes are absent; the abdomen is membranous with numerous hairs on the segments. The first pair of legs are the smallest. Most species are parasitic on ungulates.

To this family belong:

Linognathus ovillus (Neumann, 1907), the body louse or 'blue louse' of sheep, occurs in New Zealand, Australia and Scotland. A better name for it is the 'face louse', because it occurs chiefly on the face. Its head is much longer than wide and also longer than the thorax.

Linognathus vituli (Linnaeus, 1758), the 'long-nosed' cattle louse, which has an elongated head and body.

Linognathus africanus Kellogg & Paine, 1911, the African 'blue louse' of sheep.

Linognathus pedalis (Osborn, 1896), the 'foot louse' of sheep, occurs on the legs and feet of the sheep where there is no wool.

Linognathus stenopsis (Burmeister, 1838) on goats.

Linognathus setosus (v. Olfers, 1816) (syn. *L. piliferus*) on dogs and foxes.

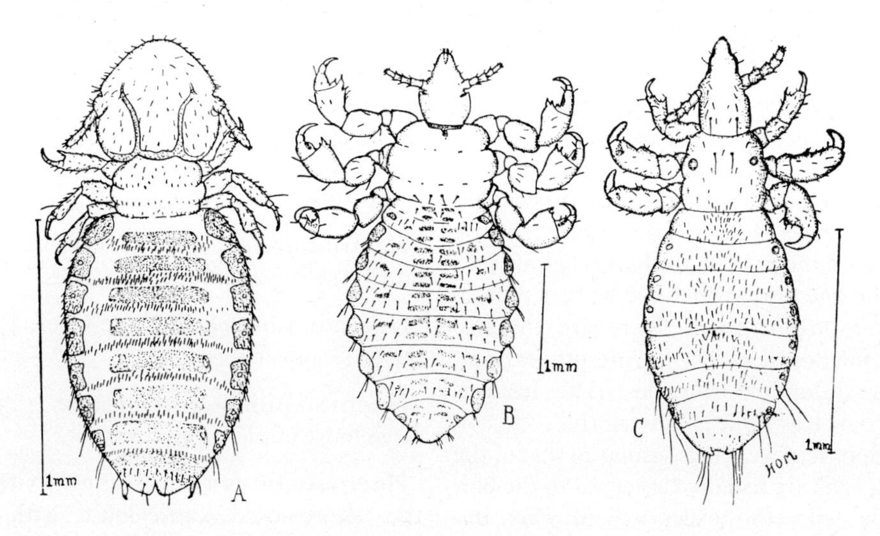

Fig 2.12 Cattle lice. A, *Damalinia bovis.* B, *Haematopinus eurysternus.* C, *Linognathus vituli.*

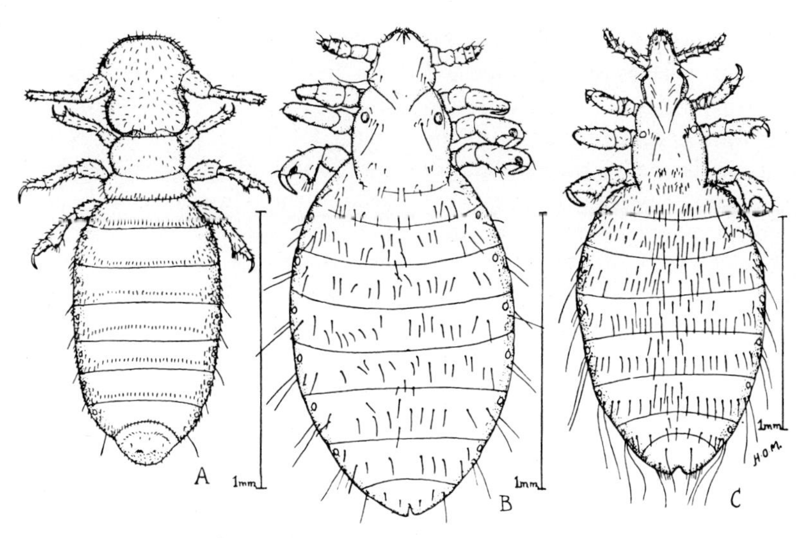

Fig 2.13 Sheep lice. A, *Damalinia ovis*. B, *Linognathus pedalis*. C, *L. africanus*.

Solenopotes capillatus Enderlein, 1904, on cattle in Europe, USA and Australia. This is similar to *L. ovillus* but is distinguished from the genus *Linognathus* by the presence of abdominal spiracles set on slightly sclerotized tubercles which project slightly from the abdomen.

Microthoracius cameli (Linnaeus, 1758) occurs on camel and dromedary in Africa and Asia.

Microthoracius praelongiceps (Neumann, 1909) occurs on the llama in South America, as does **M. mazzai** Werneck, 1932 and **M. minor** Werneck, 1935.

FAMILY: PEDICULIDAE LEACH, 1817

Pigmented eyes are present in species of this family and the abdomen has paratergal plates.

To this family belong the human head and body lice of the species *Pediculus humanus* Linnaeus, 1758, the legs of which are all the same size, the claws being slender; and the human pubic or crab louse, *Phthirus pubis* Linnaeus, 1758, which has a very wide thorax and a small abdomen, the first pair of legs being slender with slender claws.

This family also contains lice in the genus *Pedicinus* which are found on various monkeys. For example, *P. eurygaster* (Burmeister, 1838) occurs on macaques in Asia, *P. obtusus* (Rudow, 1869) on leaf monkeys, green monkeys and baboons of Asia and Africa, *P. patas* (Fahrenholz, 1916) on colobus monkeys and *P. mjobergi* on howler monkeys in South America (Flynn 1973).

FAMILY: HOPLOPLEURIDAE FERRIS, 1951

This is the largest family of sucking lice and its species occur mainly on rodents. The paratergal plates project apically from the body and the tergal and sternal plates are usually distinct. Turner (1971*b*) gives a host list for the various genera in the family. Of these the following may occur on laboratory animals or rodents associated with human habitation (Flynn 1973).

Polyplax serrata Burmeister, 1839 and **P. spinulosa** Burmeister, 1839 (the spined rat louse) are common lice of the laboratory mouse and laboratory rat respectively. These are slender lice 0.6–1.5 mm long, yellowish-brown, with well formed lateral plates on the dorsal and ventral aspects of the abdomen. Both are vectors of various organisms. Thus *P. spinulosa* transmits *Haemobartonella* and *P. serrata* transmits *Eperythrozoon* and *Francisella* species.

Hoplopleura acanthopus Burmeister, 1839 is usually found on various species of wild mice and occasionally laboratory mice. **Hoplopleura captiosa** Johnson, 1960 occurs primarily on the house mouse, but occasionally the laboratory mouse, and **Hoplopleura pacifica** Ewing, 1924 is the tropical rat louse occurring on various species of rats throughout the world, but it is uncommon in laboratory rats. The *Hoplopleura* species are slender forms 1–2 mm in length with large paratergal plates.

Haemodipsus ventricosus Penny, 1842 is the rabbit louse but it is uncommon in laboratory rabbits. It is 1.25–2.5 mm in length with a large oval-shaped abdomen and no dorsal paratergal plates.

FAMILY: ECHINOPHTHIRIIDAE ENDERLEIN, 1905

Species of this family occur on pinniped carnivora. The body is densely clothed with thick setae, sometimes modified into scales. Plates are missing on the abdomen. Two genera are of interest, *Antarctophthirus* Enderlein, 1906 and *Echinophthirius* King, 1964. Dailey and Brownell (1972) give a check list of the species which occur on various pinnipedia. Some species, e.g. *Antarctophthirus microchir* (Trouessart and Neumann, 1888) Enderlein, 1906, are common on sea lions and, in rookery areas, pups with massive infections may be seen, though the adult animals may be free of the parasite (Dailey & Brownell 1972).

EFFECTS OF LICE ON THEIR HOSTS

The chief effects of lice on their hosts are due to the irritation they cause. They are most numerous in the winter, possibly because of longer hair on the host's coat, closer contact of animals and also lack of general vigour. The hosts become restless and do not feed or sleep well and they may injure themselves or damage their feathers, hair or wool by biting and scratching the parts of their bodies irritated by the lice. The egg production of birds and the milk production of cattle may fall. In mammalian hosts scratching may produce wounds or bruises on the animal, while in sheep the wool is damaged and it is also soiled by the faeces of the lice. The coat becomes rough and shaggy and, if the irritation is severe, the hair may become matted. Excessive licking of it by calves may lead to the formation of hair-balls in the stomach. The foot louse of sheep is found most frequently around the dew-claws, and severe infections may produce lameness. In laboratory animals heavy infections may be fatal.

Diagnosis is easily made by finding the lice or by detecting the eggs or 'nits' on hairs or feathers.

CONTROL AND TREATMENT OF LICE

In the use of any of the substances mentioned below, due care should be paid to regulations governing their use, since they may be banned or not yet licensed for use and may require a withdrawal time before animals may be sent for slaughter or before milk may be sold.

The insecticides which are available increase every year. The majority in use are synthetic products and natural products such as nicotine, rotenone and pyrethrins are increasingly discarded in favour of chlorinated hydrocarbon insecticides, organophosphates and the carbamate insecticides. In recent years the synthetic pyrethroids have assumed importance and it is likely that they will be dominant in the next decade or so.

Poultry. Previously the painting of perches with a strong extract of tobacco or nicotine or dusting with sodium fluoride were the limited approaches available. Now control is achieved by dusts or sprays of carbaryl (Sevin) (a carbamate) (5%) or coumaphos (O,O-diethyl O-3 chloro-4-methyl-2-oxo-2H-1-benzopyran-7-yl-phosphorothioate) (Co-Ral, Asuntol, Muscatox) (an organophosphate) 0.06%. Other compounds that may be used include toxaphene, hexachlorocyclohexane (HCH, lindane) and malathion (0.1%).

Cattle. Several insecticides may be used for cattle lice and these include crotoxyphos (dimethyl phosphate of alpha-methyl benzyl 3-hydroxy-cis-crotonate), an organophosphate (Ciodrin), which is used as a 3% dust. Brahman cattle may show idiosincracy to crotoxyphos. Ciovap is a formulation consisting of 1% of crotoxyphos and 0.25% dichlorvos (2,2 dichlorovinyl dimethyl phosphate; DDVP); this is especially useful for *H. eurysternus* infections and is used as a spray.

Coumaphos (as above) used as a 0.06% spray and repeated in seven to ten days gives excellent control. No pre-slaughter interval is required for this compound.

Crufomate (*O*-4-tert-butyl-2-chlorophenyl methyl phosphoramidate) (Ruelene, Dowco 132) an organophosphate may be applied as a dip (35%), a spray (25%) or a 'pour-on' (13.5%).

Famphur (*O*-[*p*-(dimethylsalphamoyl) phenyl] *O*,*O*-dimethyl phosphorothioate) (Cythioate, Famaphos) is an organophosphate which may be used as a 'pour-on' (13.2%) and is especially useful for *L. vituli* infections.

Ronnel (*O*,*O*-dimethyl *O*-(2,4,5-trichlorophenyl) phosphorothioate), an organophosphate (Korlan, Ectoral, Trolene, Nankor, Etrolene, fenchlorphos), is used as a 0.25% spray or as a 1% solution in oil on 'back rubbers'. In a 25% emulsion it may be used as a 'pour-on'.

Methoxychlor (Marlate), a chlorinated hydrocarbon insecticide, may be used as a 0.5% spray or dip or as a 5% in oil solution on 'back rubbers'.

Malathion (*O*,*O*-dimethyl phosphoro-dithioate ester with diethylmercapto-succinate), an organophosphate, is used as a 0.5% spray.

Mitraz (1,4-di-(2,4-dimethylphenyl)-3-methyl-1,3,5-triazapenta-1,4-diene) is a triazapentadiene compound with high acaricidal activity. A 'pour-on' formulation has been shown to be effective against cattle lice (Griffiths 1975).

Cypermethrin is a synthetic pyrethroid which, used in a concentration of 150 ppm as a dip or spray, will be effective in one application.

Ivermectin, an avermectin, given subcutaneously at a dose of 0.2 mg/kg, is 100% effective against *L. vituli* and *H. eurysternus*. Withdrawal times are required for this compound.

Sheep. In sheep, the efficacy of insecticides depends to some extent on the ability of the insecticide to penetrate the fleece. Plunge dipping is generally the most efficient in this respect but is often the least convenient. Jetting, spraying and showering techniques are often employed and the efficacy of showers for the control of ectoparasites of sheep is reviewed by Kirkwood et al. (1978). Showers or sprays are generally satisfactory for the control of lice on sheep but may not be satisfactory for the control of sheep scab (*Psoroptes*), especially during winter with long-fleeced sheep.

Dipping with 0.25% Ronnel or 0.125% Coumaphos (Co-Ral) used not later than 15 days before slaughter provide effective control for lice. Similar concentrations in sprays produce good results.

Application by garden watering-can may be used in small groups of sheep crowded together. Ronnell (24%) or diazinon (*O*, *O*-diethyl *O*-(2-isopropyl-4-methyl-6-pyrimidinyl) phosphorothioate) at a 0.06% concentration are useful for this purpose.

Previously dips containing either 0.1% DDT or 0.007% gamma HCH were recommended. Arsenical dips containing 0.2% arsenic trioxide also gave good results. Resistance of *D. ovis* to HCH and dieldrin has appeared in Great Britain. Such resistant strains have been effectively controlled by organophosphorus insecticides.

The synthetic pyrethroids may be of value in the control of sheep louse (see above).

Pigs. Sprays of coumaphos (0.06%) or crotoxyphos (1%) plus DDVP (0.25%) give good control. They should be repeated if necessary seven days later. Malathion (0.5%) or Ronnel (0.25%) as sprays also give good control.

Horses. Malathion as a 0.5% spray, repeated in two weeks, is effective against horse lice.

Dogs. Coumaphos (Co-Ral) may be used as a 0.5% dust. Ronnel (Ectoral) may be applied topically as a 0.25–1% solution or given systemically at the rate of 110 mg/kg. A combined approach using topical and systemic administration is most useful. Hexachlorocyclohexane (lindane), where its use is not prohibited, applied as a 1% spray or dip, gives effective control, as does a 4% dip of chlordane (Engo). A formulation (Malingo) consisting of 4% chlordane, 1% lindane and 1% malathion is used effectively as a dip for the control of lice of dogs. Carbaryl is available as a shampoo for dogs and cats.

Cats. Because of the extensive self-grooming that cats perform, great care is necessary in the use of insecticides. Generally, the chlorinated hydrocarbons should not be used for cats. Some of the organophosphates may be used. Thus ronnel (as Ectoral) may be used as a 1% solution for the control of ectoparasites and when given systemically (110 mg/kg) is effective also.

Dichlorvos (DDVP) may be incorporated into a 'flea collar' for cats (usually 4.65%). Stronger formulations are prepared for dog 'flea collars' and should not be used on cats. Since DDVP is a cholinesterase inhibitor, cats wearing flea collars should not be exposed to other cholinesterase inhibitors.

Pyrethrum or carbaryl may be used as a shampoo or powder for lice on cats.

Laboratory animals. Generally, the insecticides mentioned above are applicable to laboratory animals. However, various species may show idiosyncracies to different insecticides and it is wise to treat representative members of the infected group of animals before all the group are treated. Treatments for individual species are discussed by Flynn (1973). Where ectoparasite control is necessary with very large numbers of animals, Page (1974) has reported that the use of automatic activators for aerosol dispensers of synergized pyrethrins, e.g. pybuthrin (piperonyl butoxide: pyrethrins), giving a periodic discharge of aerosol (e.g. 100 mg every 15 minutes over a two-month period in a 480 m^3 guinea-pig breeding house) will achieve a high level of control.

Man. Hexachlorocyclohexane (1%) may be applied topically or, alternatively, 0.2% pyrethrins with 2% piperonyl butoxide or 0.03% copper oleate.

Resistance of human lice to the chlorinated hydrocarbon insecticides has been reported from various parts of the world.

ORDER: HEMIPTERA LATREILLE, 1825

These are the true bugs. The mouth parts are adapted for piercing and sucking. Two pairs of wings are usually present, the anterior pair often of harder consistency (hemelytra) than the membranous posterior pair. A large number of plant lice and bugs, of considerable economic importance, occur in this order. Only a small number of species are of medical importance by virtue of their blood-sucking capabilities.

FAMILY: CIMICIDAE LATREILLE, 1825

Genus: Cimex Linnaeus, 1758

Cimex lectularius Linnaeus, 1758, the bed-bug (Fig. 2.14), attacks man and animals to suck blood. The parasites are 4–5 mm long, flat-bodied, elongate oval in shape and yellowish-brown to dark brown in colour. The head bears a pair of long antennae with four joints. The compound eyes project conspicuously at the sides of the head. The prothorax is large and deeply notched anteriorly where the head is inserted in it. The wings are vestigial. The abdomen has

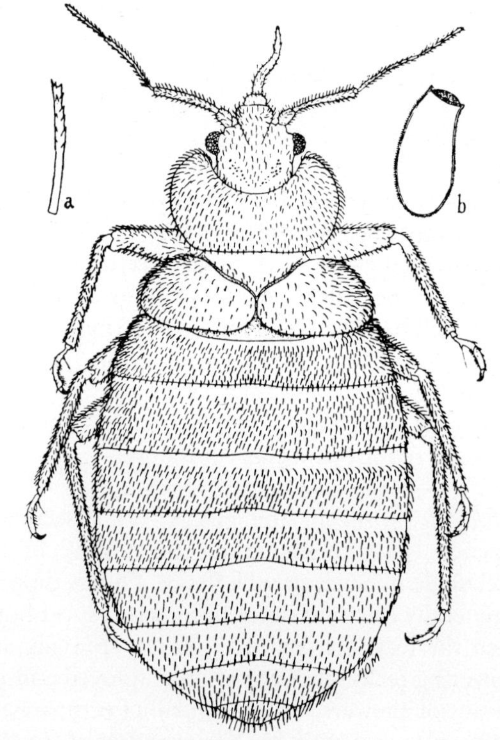

Fig 2.14 Dorsal view of *Cimex lectularius*.
a bristle
b egg

eight visible segments. The whole body is covered with characteristic spinose bristles and some hairs. The tibiae of the legs are long and the tarsi have three joints. The adult has a pair of ventral thoracic stink glands and the young stages have similar dorsal abdominal glands. These glands are responsible for the characteristic odour of the insect. The mouth parts (Fig. 2.15) are modified for piercing and sucking. The labrum is small and immovable. The labium forms a tube with four joints and contains the piercing mandibles and maxillae.

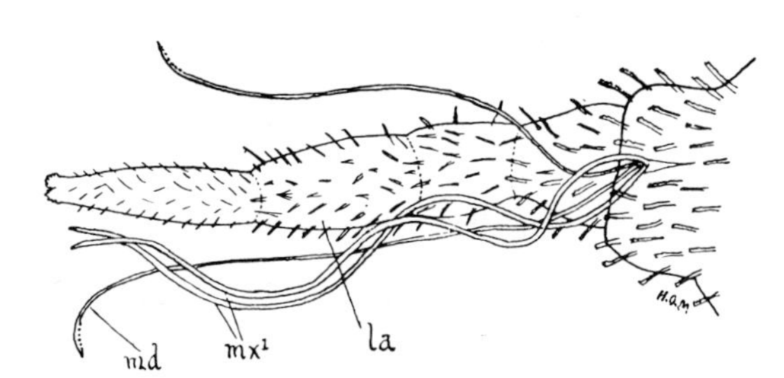

Fig 2.15 Dorsal view of the mouth parts of *Cimex lectularius*.
la labium
md mandible
mx I first maxilla

Life-cycle and habits. The female lays about 150–200 eggs in dark crevices. The egg is creamy white, about 1 mm long and has an operculum. The larva hatches after 3–14 days at 23°C or longer at lower temperatures and resembles the adult. There are five nymphal stages. The rate of development depends greatly on the food supply and the temperature. Under favourable conditions the adult stage is reached in 8–13 weeks after hatching.

The bugs live long and can survive long periods of starvation. The insects live in crevices and cracks of wood near the sleeping-places of their hosts; for instance, in bedsteads or behind picture-rails and skirting boards, or in the nests or perches of poultry. They are mainly nocturnal insects, but will bite in the daytime if hosts are quiet (e.g. sitting hens). After a blood meal the bug defaecates and usually turns round in such a way that its faeces fall on or near the wound, thus providing the possibility of transmission of disease through its faeces. Bugs may travel relatively long distances, passing to adjoining houses from an infected one.

Apart from being very annoying insects in human dwellings, several species of bugs at times cause severe irritation and anaemia in poultry, especially fowls, turkeys and pigeons. Besides *C. lectularius*, other species of this genus include *Cimex hemipterus* (Fabricius, 1803) (syn. *C. rotundatus*) of southern Asia and Africa and *Ornithodorus toledoi* Pinto, 1927 which occurs in Brazil. Both are poultry parasites. *Haematosiphon inodora* Duges, 1892 occurs in Central America and is an important pest of chickens and other poultry.

Control. HCH, chlordane and dieldrin are all powerful killers of bedbugs and can be used as sprays, smokes or powders. The WHO recommends, for the control of *Cimex*, application of a 5% emulsion or a solution of DDT to baseboards, crevices, beds and mattresses. Resistance to the chlorinated hydrocarbon insecticides may occur and the organophosphates, synergized pyrethrins and synthetic pyrethroids are then indicated. The organophosphates may be applied to walls and floors, etc. as a residual spray (e.g. dimethoate 1% or ronnel 1%), as dry baits (e.g. DDVP) or as resin-impregnated strips (e.g. DDVP). Oil-based sprays with pyrethrins plus a synergist provide useful control and the synthetic pyrethroid permethrin, which has shown marked activity against *Musca domestica*, will probably be of value in the control of *Cimex* and the associated genera responsible for ectoparasitic infections of poultry.

FAMILY: REDUVIIDAE FABRICIUS, 1803

Numerous species of this family are hosts of *Trypanosoma cruzi*, the cause of human trypanosomiasis in South America, natural hosts of this trypanosome being the dog, cat, fox, armadillo, monkey and other animals. Reduviidae are larger than Cimicidae and have well-developed wings, a cone-shaped head; the abdomen is less flattened than that of the Cimicidae. They are variously known as cone-nosed bugs, kissing bugs and assassin bugs. Most of the genera are distributed in South and Central America and the genus *Triatoma* is the largest and also the one which contains the most important vector species. Eggs are laid after a blood meal and hatching is dependent on environmental temperature. There are five nymph stages and the duration of these may be from 120 days to two years, though the majority of species take one year to complete their life-cycle.

Because these bugs can fly long distances, control is difficult. Houses may be screened and nets may be used to protect beds. The WHO recommends the application of 0.5 g/m^2 of HCH or 1–1.5 g/m^2 of dieldrin combined with elimination of the breeding places of the bugs. Other insecticides, as for the Cimicidae, are also indicated.

DIVISION: ENDOPTERYGOTA

ORDER: COLEOPTERA LINNAEUS, 1758

Species of this order are the beetles. They have two pairs of wings. The anterior (mesothoracic) pair are thickened to form hard horny or leathery covers for the posterior (metathoracic) pair. They are called the *elytra*. They usually meet in the mid-dorsal line to form a straight suture there. Beneath them the posterior membranous wings are folded. The mouth parts are adapted for chewing. The metamorphosis is complete. The pupa is free.

The beetles are important as carriers of disease-producing organisms. Some species are intermediate hosts of the spiruroid nematodes *Spirocerca lupi*, *Ascarops strongylina*, *Physocephalus sexalatus* and *Gongylonema pulchrum* etc. Scavenger beetles that feed on carcasses have been shown to carry anthrax bacilli and other pathogenic bacteria may also be spread in this way. Dung beetles may possibly spread bacteria occurring in faeces, but they are especially important as intermediate hosts of numerous tapeworms and nematodes.

ORDER: SIPHONAPTERA LATREILLE, 1825

This order was previously referred to as the Aphaniptera. The present classification is that of Hopkins and Rothschild (1953–1971). The fleas are wingless insects with laterally compresed bodies, about 1.5–4 mm long. The chitinous covering is thick and dark brown. Compound eyes are absent, but some species have large or small simple eyes. The abdomen has ten segments. The ninth abdominal segment of both the male and the female bears a dorsal plate called the *sensilium* or *pygidium*, which is covered with sensory setae; its function is not known. The tergum of the ninth abdominal segment of the male is modified to form the claspers. The penis (*aedeagus*) of the male is chitinous and coiled and its structure is complex. These and other anatomical features used for the classification of fleas are shown in Figs 2.16 and 2.17. The legs are long, strong and adapted to leaping. In some species, such as the dog flea, *Ctenocephalides canis* (Curtis, 1826), and the cat flea, *C. felis felis* (Bouché, 1835), there are a number of large spines on the head and the thorax known as 'combs' or *ctenidia*. On the cheek (*gena*) there may be a *genal comb* and on the posterior border of the first thoracic segment a *pronotal* comb. Either or both of these combs are absent from some species. The short, clubbed antennae are sunk in antennal grooves on the sides of the head.

In addition to the above Turner (1971*a,b*) gives a list of families of Siphonaptera and their hosts and a list of bird flea genera and their general host distribution.

Host Distribution

Ctenocephalides felis has four distinct subspecies: *Ctenocephalides f. felis* Bouché 1835 which is cosmopolitan and occurs on cat, dog and occasionally man, mouse, rat and primates, *C.f. strongylus* (Jordan, 1925) which occurs in Africa, *C.f. damarensis* Jordan, 1926 in south-western Africa and *C.f. orientalis* Jordan, 1925 in India,

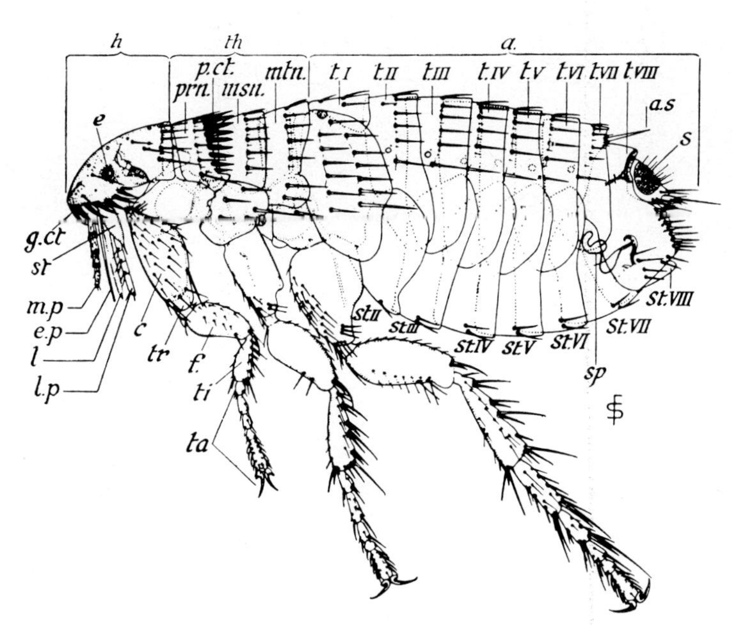

Fig 2.16 *Ctenocephalides felis felis.*

h	head
th	thorax
a	abdomen
e	eye, with antenna behind in antennal groove
p ct	pronotal ctedium
g ct	genal ctenidium
prn	pronotum
msn	mesonotum
mtn	metanotum
t1–t8	terga of abdominal segments
as	antisensilial seta
s	sensilium
st	stipes of first maxilla
l	lacina of first maxilla
mp	palp of first maxilla
lp	palp of second maxilla
c	coxa
tr	trochanter
f	femur
t	tibia
ta	tarsus with claws
st2–8	sterna of abdominal segments

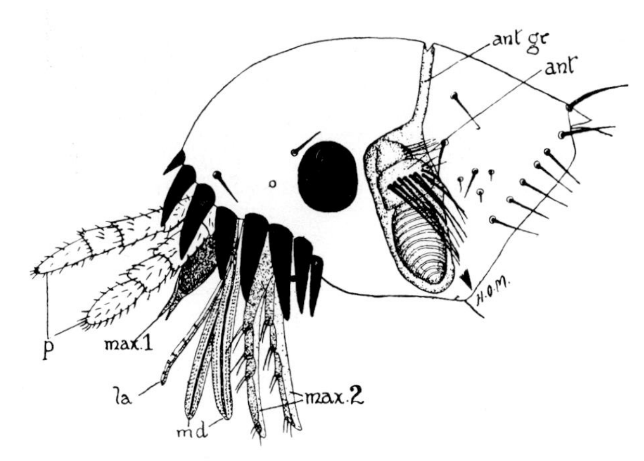

Fig 2.17 Lateral view of the head of *Ctenocephalides canis.*

ant	antenna
ant gr	anterior groove
la	epipharynx
max 1	stipes of first maxilla
max 2	stipes of second maxilla
md	laciniae of first maxilla
p	palp of first maxilla

Sri Lanka and south-east Asia on the cat and other related species. However, the last-named is often regarded as a separate species rather than a subspecies.

Ctenocephalides canis is generally restricted to dogs and related species and is cosmopolitan in distribution.

Archaeopsylla erinacei Bouché, 1835 occurs on hedgehogs in Europe and North America. It may also be found on dogs and cats following accidental exposure.

Spilopsyllus cuniculi Dale, 1878 is restricted to the rabbit and hare and is seldom seen in laboratory-raised animals. *Leptopsylla segnis* (Schöncher, 1811) occurs on the house mouse as well as the rat, field mouse and other wild rodents. It is not common in laboratory-raised animals.

Ceratophyllus (*Nosopsyllus*) *fasciatus* Bosc,

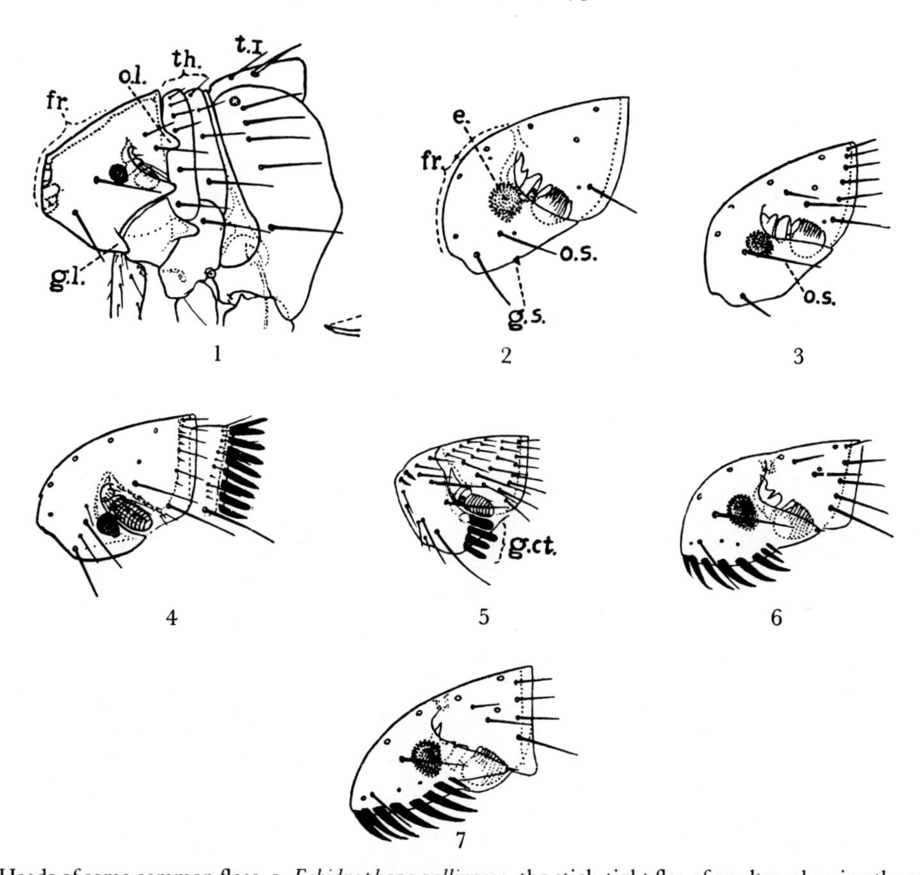

Fig 2.18 Heads of some common fleas. 1, *Echidnophaga gallinacea*, the stick-tight flea of poultry, showing the angulate frons (*fr*); the well-developed occipital lobe (*o.l.*) on the occiput; the genal lobe (*g.l.*) directed backwards, the thorax (*th.*) narrower dorsally than the tergum (*t.i.*) of the first abdominal segment, and the absence of ctenidia (cf. *Pulex irritans* and *Xenopsylla cheopis*). 2, *Pulex irritans* of man and other mammals, especially the pig and badger, showing the absence of ctenidia, the smoothly-rounded frons (*fr.*); the position of the ocular seta (*o.s.*) below the eye (contrast its position in the other fleas figured), and the single small spinelet (*g.s.*) on the genal margin; *e*, eye. 3, *Xenopsylla cheopis*, chiefly of the rat, showing the absence of ctenidia. 4, *Ceratophyllus* (*Nosopsyllus*) *fasciatus* of rodents, chiefly rats, which has a pronotal ctenidium, but no genal ctenidium. Note the antenna in its antennal pit behind the eye. 5, *Leptopsylla segnis*, of the house-mouse, showing the vertical genal ctenidium (g.*ct.*) composed of four spines. This flea has a pronotal ctenidium also, but it is not shown in the figure. 6, *Ctenocephalides canis*, chiefly of Canidae, showing the horizontal genal ctenidium (cf. *C. felis felis*), composed of eight to nine spines, the first spine being only about half as long as the second (contrast *C. felis felis*) and the strongly-rounded head (contrast *C. felis felis*). This flea has a pronotal ctenidium also, but it is not shown in the figure. 7, *Ctenocephalides felis felis* of many mammals, showing the horizontal genal ctenidium, composed of eight to nine spines, the first spine being about as long as the second (contrast *C. canis*) and the elongate head (contrast *C. canis*). This flea has a pronotal ctenidium also, but it is not shown in the figure.

1801 occurs on a range of rats and the house mouse and is the commoner of the two rat fleas. *Xenopsylla cheopis* (Rothschild, 1903), the oriental or black rat flea, is associated with the transmission of plague. It is often present on rodents in ports and occurs on rats, mice and other rodents.

Pulex irritans Linnaeus, 1758 is the human flea but also may occur on the pig and badger as well as the dog, cat and rat. *Tunga penetrans* (Linnaeus, 1758), also of man, is the jigger or chigoe flea and may be found on the pig and baboon.

Ceratophyllus gallinae Schrank, 1803 is a common flea of the chicken; it may occur on other birds and occasionally on rodents. *Ceratophyllus columbae* Gervais, 1844 occurs on the domestic pigeon. *Dasypsyllus gallinulae* (Dale, 1878) is found on various wild birds including the

Table 2.1 Key to the Fleas of Veterinary Importance

1. Thorax reduced, the three thoracic segments together shorter in width than the first abdominal segment (Family: Tungidae)	2
Thorax not reduced, the three thoracic segments together much wider than the first abdominal segment (Family: Pulicidae)	3
2. On poultry, no ctenidia, two bristles on occipital lobe	*Echidnophaga gallinacea* (sticktight flea; poultry)
On man and other mammals, frons sharply angled, tropical	*Tunga penetrans* (jigger, chigoe) (man, pig)
3. Genal and pronotal ctenidia absent	4
Genal and/or pronotal ctenidia present	5
4. Mesopleural rod absent, on human	*Pulex irritans* (man)
Mesopleural rod present, on black rats	*Xenopsylla cheopis* (oriental rat flea)
5. Genal and pronotal ctenidia present	6
Only pronotal ctenidia present	8
6. Genal ctenidium of four elements, vertically placed, on mice	*Leptopsylla segnis* (mouse)
Genal ctenidium of four to six elements, obliquely placed, on rabbits	*Spilipsyllus cuniculi* (rabbit)
Genal ctenidium of eight (sometimes nine) elements, horizontally placed	7
7. Frontal spine of genal ctenidium as long as the second spine, head with low sloping front about two times as long as high	*Ctenocephalides felis* (cat)
Frontal spine of genal ctenidium shorter than second spine, head with rounded front, about one and a half times as long as high	*Ctenocephalides canis* (dog)
8. Some 18–20 spines in pronotal ctenidium, on rodents	*Ceratophyllus* (*Nosopsyllus*) *fasciatus* (northern rat flea) (rats and mice)
More than 24 spines in pronotal ctenidium, on poultry	*Ceratophyllus gallinae* (chicken flea)

grouse, robin and woodcock, while *Ceratophyllus garei* Rothschild, 1902 is found on water fowl, e.g. ducks.

Echidnophaga gallinacea Westwood, 1875 is common on chickens in tropical and subtropical areas of the world. It may also occur on rodents, dogs, cats and occasionally man.

Four species, *Vermipsylla ioffi* Smit, 1953, *V. perplexa* Smit, 1974, *V. alacurt* Schimokovich, 1885, and *V. dorcadia* Rothschild, 1912, occur on sheep, goat, yak, cattle, horse and various hybrids of these in the Kirgizii area of the USSR, the Mongolian People's Republic and northern China.

Life-cycle

The female flea lays up to 20 eggs at a time and some 400–500 during her lifetime. Studies by Mead-Briggs and Rudge (1960) and Rothschild (1965), using the rabbit flea, *Spilopsyllus cuniculi*, have demonstrated that maturation of the ovaries of the female flea occurred only on pregnant rabbits. The oval, glistening eggs are deposited in dust or dirt, or they may be laid on the host, but soon drop off, as they are not sticky. The eggs are about 0.5 mm long, rounded at the poles and pearly-white in colour. The rate of development varies greatly and depends also on the temperature and humidity. The larvae may hatch in two days, or up to 16 days after the eggs have been laid. The egg-shell is broken by means of a chitinous spine, which is present on the head of the first larval instar. The larvae are elongate, slender, maggot-like creatures, consisting of three thoracic and ten abdominal segments, each of which bears a few long hairs. The last abdominal segment bears two hooked processes called the *anal struts,* which are used for holding on to substrata or for locomotion. The larvae are creamy-yellow in colour and very active, hiding from light. They have masticatory mouth parts and feed on dry blood, faeces and other organic matter, but apparently require little food, though in some species the blood which is passed through the body of the adult flea is a necessary part of their nutriment. Some larvae may prod the adult flea to produce faecal blood for them to feed upon (Smit 1973). They are found in crevices in floors, under carpets and in the nest litter or sleeping places of the host animals. A moderate temperature and a high degree of humidity are favourable for development, which lasts seven to ten days or longer. The mature maggot is about 6 mm long. It spins a cocoon which measures about 4 by 2 mm and because of the stickiness of the cocoon it becomes covered with dust and debris. In this the pupal stage is passed, lasting 10–17 days under average conditions, but it may last several months and a low temperature will cause the imago to stay in the cocoon.

Table 2.2. The Longevity (Days) of Various Common Fleas

Flea	*Fed*	*Unfed in air nearly saturated with moisture*
Pulex irritans	125	513
Xenopsylla cheopis	38	100
Ceratophyllus fasciatus	95	106
Ctenocephalides felis	58	234
Ceratophyllus gallinae	127	345

Habits and significance

Fleas are much less permanent parasites than, for example, lice and frequently leave their hosts. Their longevity varies, with different species of fleas, according to whether they are fed or not, and according to the degree of moisture in their surroundings (Table 2.2). Unfed fleas do not live long in dry surroundings, but, in a humid environment, if debris in which they can hide is present, different species may live for one to four months. Specimens of the rat flea *Ceratophyllus fasciatus* have been kept without food for 17 months at a temperature of 15.5°C and 70% humidity.

Fleas are not markedly specific for their host and may feed on other hosts. In the absence of the normal host they may still feed and they then may be able to survive for longer periods than those indicated in Table 2.2. Thus *Pulex irritans* may have a maximum life of about 17 months, *Ctenocephalides canis* of 26 months and *Ceratophyllus gallinae* of 17 months.

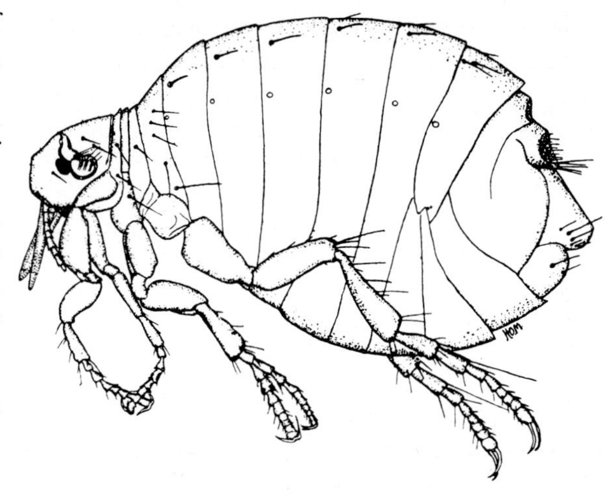

Fig 2.19 *Echidnophaga gallinacea.*

Pathogenesis

Substantial infections of fleas are especially found on animals which are in poor condition or are suffering from a chronic debilitating disease. Such burdens may be seen in old cats. Infected animals become restless, lose condition and spoil their coats by biting and scratching.

The reaction to a flea bite is determined by the stage of sensitization of an animal to the saliva of the flea. A hapten in the saliva becomes immunogenic by fixation to skin collagen (Michaeli et al. 1965). This induces a sequence of reactivity from a stage of no observable reaction (the induction period), to one of delayed dermal

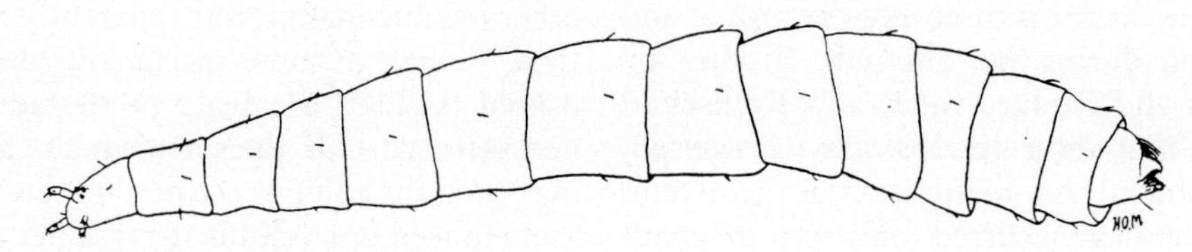

Fig 2.20 Larva of *Echidnophaga gallinacea.*

responsiveness followed by one consisting initially of an immediate-type reaction and then a delayed reaction, then followed by an immediate-type reaction only and finally no reactivity, which is a stage of desensitization (Benjamini et al. 1961). Transition from one stage to the next is not clear cut and the immediate-type reaction is the most usual reaction seen and is characterized by elevated levels of immunoglobulin E and eosinophilia at the site of the reaction.

Animals vary greatly in their sensitivity to flea bites. For example, dogs and cats which do not become hypersensitive to flea saliva show at the most a mild pruritus which often is unnoticed. However, in animals which respond, moderate to severe pruritic reactions occur on the lumbar–sacral area, the abdomen, inside the hind-legs and the neck. Thoday (1979) describes lesions varying from painful areas of moist dermatitis (wet eczema) which may become secondarily infected to less severe manifestations consisting of discrete papules with scaling. Acanthosis, hyperpigmentation and hyperkeratinization are features of long-standing cases.

In cats, many small papules associated with pruritus are a feature of flea hypersensitivity (Thoday 1979), 'miliary dermatitis'. Diffuse alopecia may result from excessive grooming and scratching because of the pruritus.

The 'sticktight' flea of poultry, *Echidnophaga gallinacea*, may in addition attack dogs, cats, horses, rabbits, pigeons and ducks. The female flea attaches herself mainly to the comb, wattles and around the eyes of birds. This flea does not jump away when disturbed and large numbers may be seen clustering together. Young birds are quickly killed by these fleas and even adult birds may succumb to heavy infections. The female burrows into the skin, causing the formation of swellings which may ulcerate. The flea lays its eggs in these ulcers. The eggs hatch there and the larvae fall out to develop like the larvae of other fleas, reaching the adult stage in about four weeks.

In addition to their direct effects, fleas also are important in the transmission of disease. The classical example is bubonic plague (*Yersinia pestis*) which is carried by *Xenopsylla cheopis*. Other species may be infected experimentally but are not regarded as important natural vectors; thus *Xenopsylla astia* and *Xenopsylla braziliensis* are relatively poor transmitters both in nature and artificially. An important factor in plague transmission is the 'blocking' of the chitinous teeth of the proventriculus with blood and plague bacilli. A 'blocked' flea is unable to fill the mid-gut with blood and wanders from host to host attempting to feed. In so doing it contaminates the new hosts with bacilli.

Enzootic plague (sylvatic plague) exists in wild rodents in China, parts of Africa, South America and in the western states of the USA. It is considered that the disease became established in the western USA following the San Francisco earthquake and fire of 1906; rats from the waterfront were driven inland and infected ground squirrels, which are now the major hosts for the infection.

Endemic or murine exanthematous typhus (*Rickettsia mooseri*) is spread by fleas of rats and mice in human dwellings. The rickettsiae multiply in the flea gut and are excreted in the faeces of the flea. Thus there is no active mechanism by which rickettsiae are injected into the animal. Bibikova (1977) comments that this is compensated for by the very high virulence of the organism, such that an infective dose for man is one-fifth of a single excrement of an infected flea.

Tularaemia is spread mechanically by fleas in the Old and New Worlds. The causative organisms, *Francisella tularensis*, do not reproduce in fleas and transmission occurs mechanically in the first five days after contamination.

The myxomatosis virus affects hares and rabbits in Australasia, Europe and South America and is transmitted by *Spilopsyllus cuniculi* or mosquitoes. The virus does not reproduce in fleas and is gradually excreted from the flea. The virus is passed by contaminated mouth parts.

Salmonellosis (*Salmonella enteritidis*) of rats and its transmission by fleas has been studied by Eskey et al. (1949). Organisms reproduce in the stomach of the flea, resulting in contamination by both the oral and the anal route. As with *Yersinia pestis*, blockage of the proventriculus occurs, obstructing its normal function and, *inter alia*, preventing the flow of blood to the anterior digestive tract.

Bibikova (1977) notes that in no case of disease transmission does penetration of the organism occur into the body cavity, salivary glands or gonads of the flea, hence there is no transovarian or trans-stadial transmission of infection.

In adition to the above, *Ctenocephalides canis, C. felis* and *Pulex irritans* serve as intermediate hosts for the tapeworm *Dipylidium caninum* and *C. canis* and *C. felis* as intermediate hosts for the filarial worm of the dog *Dipetalonema reconditum*.

Control

Control includes treatment of an animal to kill fleas, the elimination developmental stages in the hosts environment and the prevention of reinfestation of the environment.

Flea collars are now a very popular means of control. They consist of a plastic strip impregnated with dichlorvos (DDVP) (2,2-dichlorovingl dimethyl phosphate) in the proportions of 9.3% for dogs, 4.65% for cats (18.6% for flea medallions for dogs). Claims for the duration of protection vary, but three months protection from fleas by flea collars is commonly claimed. Animals wearing flea collars should not be exposed to or treated with other cholinesterase inhibitors. Dog flea collars should not be used on cats and whippets and greyhounds are especially susceptible to flea collars. A contact dermatitis may occur from the use of flea collars.

A combination of dichlorvos and fenitrothion (Nuvan Top) is effective for dogs and cats. Carbaryl, formulated as a shampoo, is available for dogs and cats. Hexachlorocyclohexane (0.01% gamma isomer as a dust or wash) is useful in dogs but should be avoided in cats; for the latter pyrethrum or synergized pyrethrins are indicated.

Other organophosphates of use include Coumaphos (Asuntol) as a 0.5% dust for dogs, ronnel as a 1% solution for topical application and a dip of 5% ronnel with 4.65% dichlorvos.

Where additional clinical manifestations have resulted from infestation, corticosteroids may reduce pruritus and symptomatic treatment may need to be applied. Animals with dermatitis should be protected against reinfection.

Premises should be thoroughly cleaned. Cracks and crevices should be thoroughly vacuum cleaned and then an effective insecticide should be blown into likely breeding places. Beds and bedding of cats and dogs should be cleaned or destroyed.

For poultry a 0.1–0.2% dust or 0.06% spray of HCH has proved useful, as has a 4% dust of malathion.

Strains of fleas resistant to chlorinated hydrocarbons have appeared in various parts of the world and for these a 4% dust or 0.5% wash of malathion is recommended. Strains of dog and cat fleas resistant to DDT may be treated with 1% emulsions of diazinon or malathion.

Echidnophaga gallinacea, the sticktight flea, presents a special problem, because the adult fleas are attached to the hosts. DDT at a concentration of 0.1% in dip form is effective, though this may be raised to 1% without harmful effects. Dusts containing 5% DDT kill the fleas but do not give prolonged protection as do the dips. A 5% malathion dust has proved effective.

ORDER: DIPTERA LINNAEUS, 1758

Species of this order have only a single pair of functional membranous wings. These are the mesothoracic pair, the metathoracic pair being modified to form halteres (balancers). The mouth parts are adapted for sucking. Usually they form a proboscis, the labium having at its distal end a pair of fleshy lobes (*labella*). The mandibles are usually absent. The mesothorax, which bears the single pair of wings, is usually large and the prothorax and metathorax are small and fused with the mesothorax. The tarsi usually have five joints. The metamorphosis is complete, the larvae being apodous and often having a reduced head. The pupa may be coarctate and enclosed in the skin (puparium) of the last larval phase or it may be obtectate. The number of veins in the wings is reduced. Many species of this order cause diseases of animals. Examples of them are the mosquitoes and their relations, the house fly, the tsetse fly, the blow fly, the warble fly and the sheep-ked.

The classification of the order Diptera is based on that used by Freeman (1973) and Richards and Davies (1977). The order is divided into the following three suborders:

SUBORDER: NEMATOCERA

The antennae of the adults are longer than the head and thorax. They have more than eight segments and the segments are, with the exception of the first two segments next to the head, all more or less alike. They have no arista. The larvae and pupae of Nematocera described in this book are aquatic. The larvae have a well-developed head and mandibles that bite horizontally. The pupae are obtectate.

The suborder contains several families of veterinary and medical importance, including: family Ceratopogonidae (biting midges); family Simuliidae (black-flies); family Psychodidae (sandflies); and family Culicidae (mosquitoes).

Other families of no veterinary importance are Tipulidae (daddy-long-legs, crane flies), Bibionidae and Chironomidae (non-biting midges).

SUBORDER: BRACHYCERA

The antennae are shorter than the thorax and consist of less than six segments, three only being often present, the last segment being annulated. An arista may be present on the antenna, but, when it is, it is terminal. The maxillary palps are held stiffly forwards (porrect). The abdomen usually has seven visible segments. The larvae have an incomplete and usually retractile head and the mandibles bite vertically. The pupa is obtectate. The only family of veterinary importance in this suborder is the Tabanidae.

SUBORDER: CYCLORRHAPHA

The antennae have only three segments and there is an arista, which is usually on the dorsal side of the antenna. The maxillary palps are usually small and consist of one joint only. The abdomen usually has fewer than seven visible segments. The larva has a vestigial head. The pupa is coarctate.

The head of the *Cyclorrhapha* has a horseshoe-shaped ridge, the *ptilinal suture* or frontal suture, which runs transversely above the antennae and downwards on either side of them. Along this suture the head capsule is invaginated in the form of a much-convoluted membranous sac which

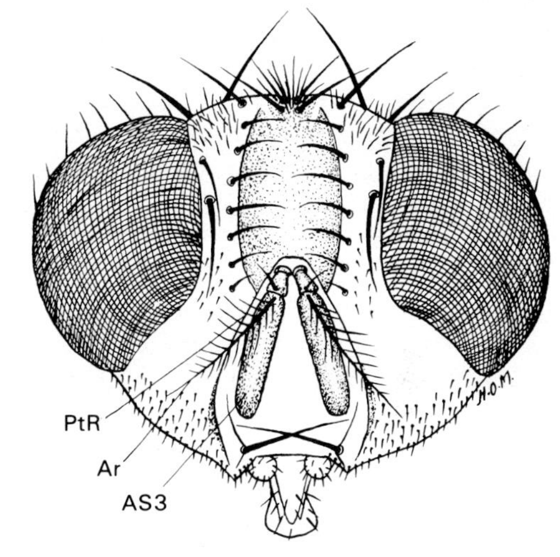

Fig 2.21 Anterior view of the head of *Lucilia sericata*.
Ar arista
PtR ptilinial ridge
AS3 third antennal segment

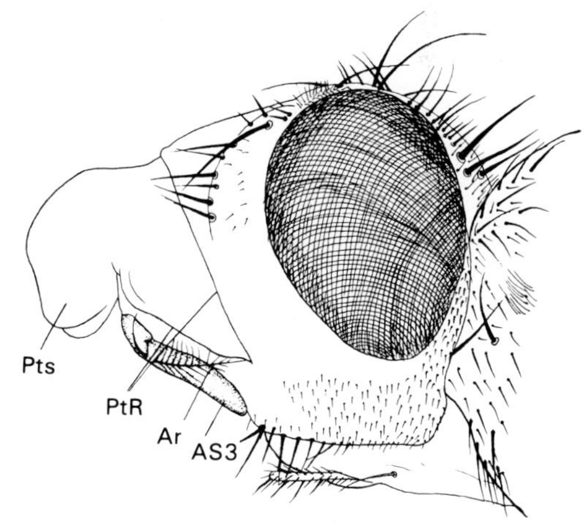

Fig 2.22 Lateral view of head of *Lucilia sericata* just emerged from the puparium, showing inflated ptilinal sac.
PtS ptilinial suture
PtR ptilinial ridge
Ar arista
AS3 third antennal segment

functions only in the young fly. When the latter is ready to emerge from the puparium, it pushes out this *ptilinal sac* by inflating it and thus breaks a circular piece off the anterior end of the pupal case. The sac is then gradually withdrawn and the ptilinal suture closes. Species of the Nematocera and Brachycera have no ptilinal sac and emerge from the pupal case through a T-shaped split on its dorsal surface.

The suborder Cyclorrhapha is divided into the following three series by Freeman (1973). (However, Richards and Davies (1977) refer to these as 'Sections', the Pupipara being included in the Schizophora.)

Series: Aschiza

Frontal or ptilinal suture absent, ptilinum absent. Generally of no veterinary importance except the Syrphidae (hover flies) which may occur in food material.

Series: Schizophora

Frontal or ptilinal suture distinct, ptilinum always present.

Richards and Davies (1977) note that the classification of this group is difficult and prefer not to use the classification based on the size of the squamae and other features. However, Freeman (1973) prefers the use of this older classification and for the purposes of classifying the insects of medical and veterinary importance in this group Freeman's classification will be used.

Section Acalypterae. Squamae small, not concealing the halteres. Thorax without distinct transverse suture. The family Gasterophilidae is of veterinary importance.

Section Calypterae. Squamae well developed, concealing halteres. Thorax with a distinct transverse suture. Families of importance include: Muscidae (house flies, stable flies etc.); Calliphoridae (blow flies), Oestridae (bot flies), Glossinidae, etc.

Series: Pupipara

Head closely united with thorax, dorsoventrally flattened, integument leathery or horny. Adapted for ectoparasitic life. Wings reduced or absent. Viviparous. Pilinum present or absent. One family of veterinary importance: *Hippoboscidae* (louse flies).

SUBORDER: NEMATOCERA LATREILLE, 1825

FAMILY: CULICIDAE STEPHENS, 1829

This family comprises the mosquitoes which are slender Nematocera with small, spherical heads and long legs. The antennae of 14–15 segments are conspicuous and are plumose in the males. The proboscis is long and slender. The abdomen is elongate and the thorax is characteristically wedge-shaped, with the broad end dorsal. The wings are long, narrow and folded flat over the abdomen during rest. They bear elongate, leaf-like scales along the margins and on the veins.

Because of the great importance of mosquitoes as vectors of various infectious agents, they have been subject to detailed examination and the classification of the family is a matter for the specialist. Keys to subfamilies and genera of the Culicidae are given by Mattingly (1973). For the purposes of this volume, division of the family Culicidae into the culicine and anopheline species provides a sufficient basis for consideration of the differences in life-cycle, behaviour and health importance. The major differences between the anopheline and the culicine mosquitoes are summarized in Table 2.3.

Life-cycle

The eggs are laid on water or on floating vegetable matter and each species has its special requirements, which are usually very restrictive. Some species lay their eggs only in fresh rain water, others in stagnant pools or in any vessel that

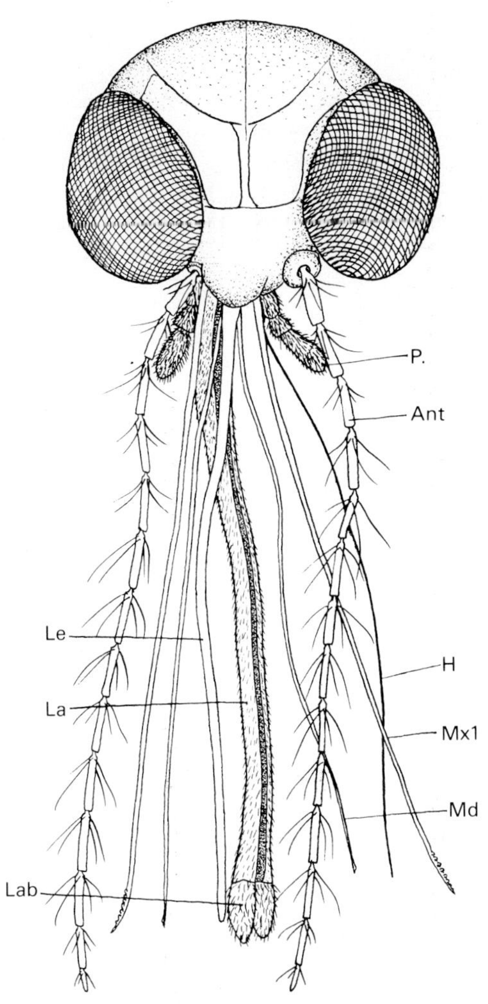

Fig 2.23 Head and mouth-parts of a female *Culex*.

Ant	antenna
H	hypopharynx
La	labium
Le	labrum-epipharynx
Lab	labellum
Md	mandible
Mx1	first maxilla
P	palp

contains water; others lay in quiet pools at the edges of streams, and some even lay in salt water. The temperature of the water, the nature of the microflora in it, the presence or absence of decayed matter and the acidity or alkalinity are all deciding factors. The eggs may be deposited in masses or 'egg-rafts' as in the case of *Culex* (Fig. 2.26) or singly as in the case of *Anopheles* and *Aedes*. In the

Fig 2.24 Wing of a mosquito.

egg-rafts the eggs are arranged vertically with their anterior ends towards the water. The anopheline egg is boat-shaped and is provided with a float on either side and a frilled edge.

The larvae have a well-developed head and a distinct thorax and abdomen. The head bears eyes, antennae and several hairs. The mouth parts are masticatory and are surrounded by brushes that produce currents in order to bring food particles to the mouth. The unsegmented thorax bears feathered hairs and the abdomen, which is segmented and also hairy, is provided, in the anophelines, with certain palmate hairs, by means of which the larva is able to cling to the surface of the water. The stigmata from which tracheae pass through the whole body are situated on the fused eighth and ninth abdominal segments. The tenth segment bears feathered hairs and tracheal gills, which are especially well developed in forms, like *Aedes*, that feed at the bottom of the water. All mosquito larvae, except those of *Anopheles*, are provided with a siphon or tube which arises from the dorsal aspect of the eighth and ninth abdominal segments and surrounds the stigmata. It is closed at its apex by chitinous valves that open when the larva goes to the surface of the water to breathe. The larvae then hang down into the water at an angle, while anopheline larvae lie against the

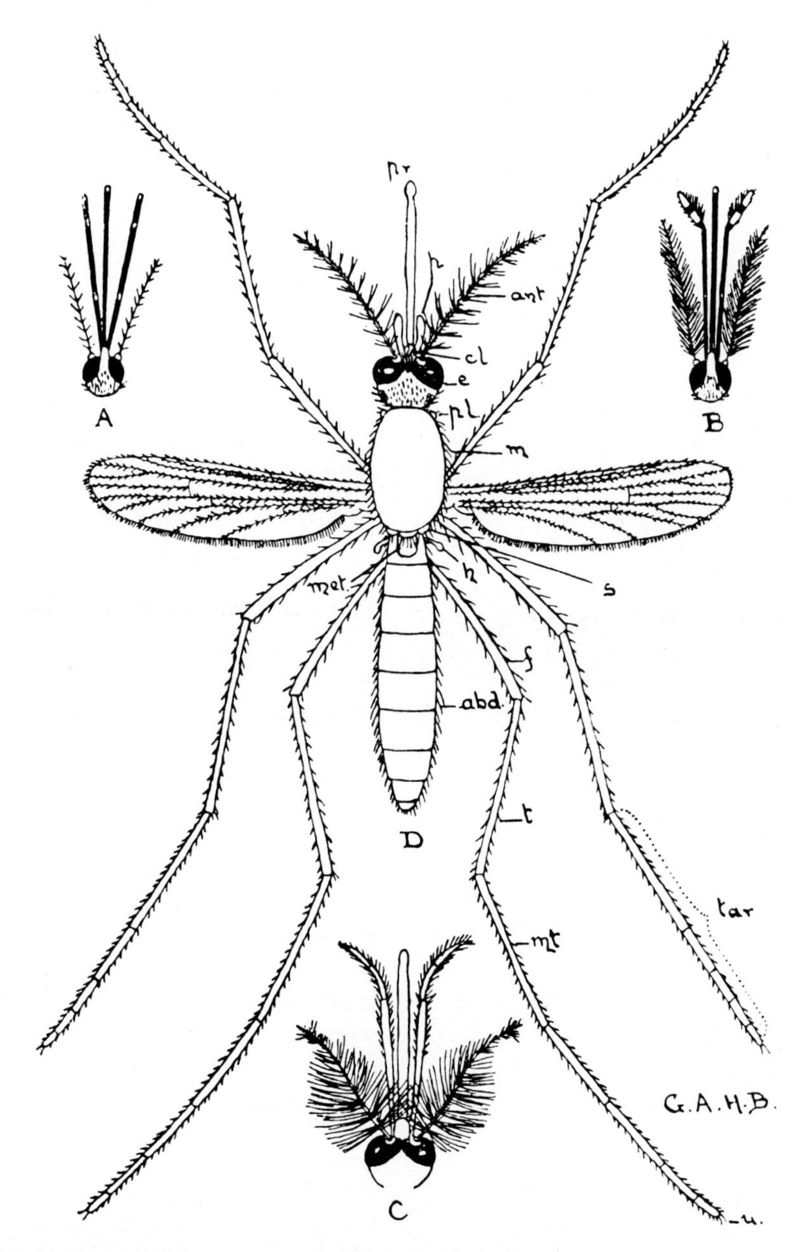

Fig 2.25 External anatomy of mosquitoes. A, Head of female *Anopheles* sp. B, Head of male *Anopheles* sp. C, Head of male *Culex* sp. D, Female *Culex* sp.

abd	abdomen	mt	metatarsus
ant	antenna	p	palp
cl	clypeus	pl	prothoracic lobe
e	eye	pr	proboscis
f	femur	s	scutellum
h	halter	t	tibia
m	mesothorax	tar	tarsus
met	metathorax (post-scutellum)	u	claws

Table 2.3. Distinquishing Characters of Anopheline and Culicine Mosquitoes

	Anopheline	*Culicine*
Eggs	Laid singly, boat-shaped with paired lateral floats	Laid in rafts or singly (*Aedes*). No floats
Larva	No siphon tube Rests parallel to water surface and feeds there Palmate hairs on dorsal surface of abdomen	Well-developed siphon tube Hangs head down from water suface and feeds below it No palmate hairs
Pupa	Breathing trumpet short and broad in lateral view	Breathing trumpets long and narrow in lateral view
Adults	Rest with abdomen directed away from resting surface, i.e. proboscis and abdomen in a straight line Female palps as long as proboscis Male palps as long as proboscis, clubbed Scutellum evenly curved	Rest with abdomen pointing towards resting surface. Proboscis and abdomen at an angle, imparting a humped-back appearance Female palps very short Male palps as long as proboscis, not clubbed Scutellum trilobed
Examples	*Anopheles maculipennis* *Anopheles gambiae*	*Culex pipiens* *Aedes aegypti*

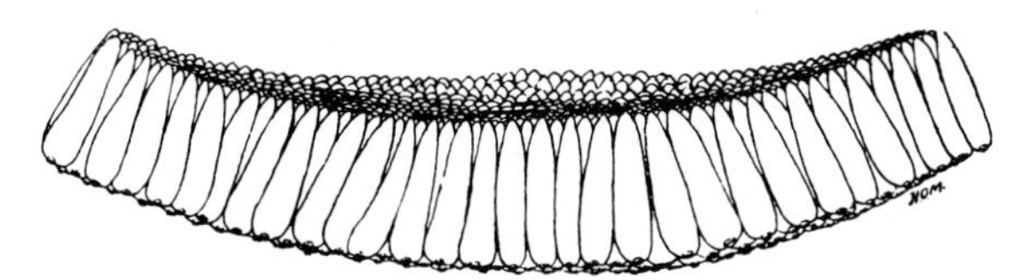

Fig 2.26 *Culex* egg-raft.

surface. The larvae moult four times, the last time at the moment of pupation.

The pupa has a rounded ' body ', which consists of head and thorax, and an elongate abdomen, flattened dorsoventrally and flexed underneath the body. Through the delicate cuticle of the latter the wings and appendages of the adult can be seen. From the dorsal aspect of the thorax, and attached to the lateral stigmata, there arise a pair of tubes or respiratory trumpets, through which the pupa breathes at the surface of the water. The pupae are not quite as active as the larvae.

The time required for development differs according to the species and varies from about seven to 16 days under favourable circumstances, which means especially a sufficiently high temperature. Cold weather may prolong the larval period to several months. The eggs of some species can also resist cold or drying for a considerable time. There are great variations in the seasonal

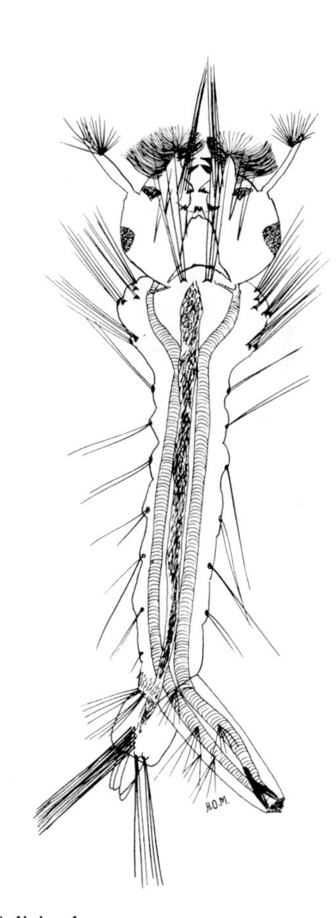

Fig 2.27 Culicine larva.

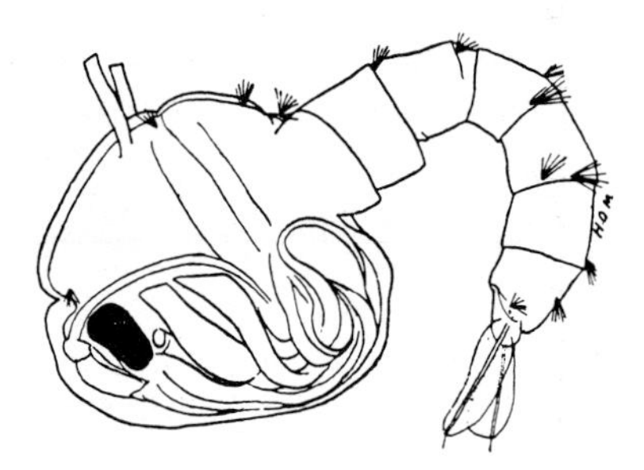

Fig 2.28 Culicine pupa.

prevalence of different species, mainly due to temperature requirements; most species breed in warm weather, but in warm climates some breed only in winter. Females which find it too late to lay may hide away and hibernate, or aestivate in hot climates, until the next season. The hiding-places are usually relatively dark surroundings with an even temperature, like cellars, barns and lofts under thatched roofs.

Apart from the presence of permanent water, rainfall has a marked influence on the number of mosquitoes for several reasons. If there is little or no rain, mosquitoes will be scarce because their numbers are limited by their natural predators in the permanent breeding-places. After a rain the numbers will increase, because the mosquitoes develop rapidly and are not restricted by predators in the new breeding-places. Much rain, on the other hand, washes away the larvae and usually decreases the number of mosquitoes, unless pools remain afterwards. The young stages can be transferred to a new area by water which flows intermittently.

The adults may fly fair distances, especially anophelines, which can travel several kilometres from their breeding-ground to feed and return in the morning. They may be carried much further by winds or cover long distances in successive stages. They can also be transported in all kinds of vehicles. Some species readily enter buildings, while others do not.

Mosquitoes can be fed on fruit juices and sugar water and the males exist normally on such food, but the females are blood-suckers and require a meal of blood in order to lay eggs. The females seem to be attracted by the warmth radiating from the skin of their host. They are active at night and are attracted by light, while during the day they hide in dark corners.

Importance

Although mosquitoes can be a great nuisance and their bites may cause painful reactions (Mellanby 1946), their chief importance lies in the fact that they are the intermediate hosts and vectors of several important parasitic and virus diseases of man and domestic animals. Thus the definitive hosts of species of the protozoan genus *Plasmodium*, which cause human malaria, are various species of *Anopheles* and those of the species of *Plasmodium* which cause bird malaria are species of *Culex*, *Theobaldia* and *Aedes*. Various culicine and anopheline species are intermediate hosts of filariid nematodes. Thus those of *Wuchereria bancrofti* of man are numerous species of *Culex*, *Aedes* and *Mansonia*; those of *Brugia malayi* in Indonesia, south-east Asia and Sri Lanka are species of *Mansonia* and *Anopheles*; and those of *Dirofilaria immitis* of the dog are species of *Anopheles*, *Culex*, *Mansonia* and *Myzorhynchus*. The spirochaete *Borrelia anserina* is transmitted to the fowl by species of *Aedes*.

Mosquitoes are of special importance in the transmission of arboviruses. Mattingly et al. (1973) give a summary of the viruses transmitted by mosquitoes; of these the following are of special interest.

Rift Valley fever of man in southern Africa is transmitted by various genera; yellow fever of man is transmitted chiefly by *Aedes aegypti* and perhaps by several other species of *Aedes*; eastern and western equine encephalitis and St Louis and Japanese B encephalitis of man, and other forms of viral encephalitis in America and elsewhere, are transmitted by species of *Aedes* and *Culex*, as well as by *Dermanyssus gallinae*. Dengue fever is transmitted to man chiefly by *Aedes aegypti* and perhaps by other species of *Aedes*.

Recent studies have demonstrated that with some arboviruses transovarial and venereal transmission occur. For example venereal transmission of La Crosse (Californian encephalitis) virus by males of *Aedes triseriatus* was demonstrated by Thompson and Beaty (1977).

Since transovarially infected males were shown to be capable of venereally infecting females there are thus several routes by which females may transmit the LAC virus. The LAC virus, as with other vector–pathogen systems, is venereally transmitted by male accessory sex gland fluid (Thompson & Beaty 1977).

Control

Measures directed against the larval stage and against the adults are those largely used at present.

Measures against larvae. Anti-larval methods include biological control, the use of larvicidal compounds and removal or reduction of breeding sites. Biological control of larvae (or pupae) by larvivorous fish (mainly *Gambusia* spp.) is applicable in lakes, ponds, pools, rice fields etc., but its effectiveness is somewhat limited. The application of larvicidal compounds is usually effective where there is seasonal breeding due to seasonal rainfall or flooding. The application of larvicides must be repeated periodically during the breeding season, hence it is costly and requires personnel for application and also for supervision and evaluation.

Paris Green mixed with kerosene as a carrier is effective and safe as a larvicide. Pyrethrum, organophosphates (difenphos, fenthion) and chlorinated hydrocarbons such as DDT, dieldrin, chlordane and hexachlorocyclohexane have been used as larvicides. However, the use of chlorinated hydrocarbons as larvicidal compounds is generally discouraged on the ground of environmental pollution and also that their use as larvicides can accelerate the development of resistance to insecticides that may be more effective as residual sprays in houses. Furthermore, the use of larvicides should be regarded as an interim measure until permanent engineering work can abolish the breeding places or limit them to a minimum.

Attention to breeding-places includes drainage, dyking and, where feasible, filling in breeding areas. Long-term programmes may be concerned with reclamation of swamp areas for agricultural and other uses.

Measures against the adult stages. Those which are largely in use at present include residual spraying, especially of the indoor surfaces of houses for malaria control, which is one of the most effective measures. Residual-type insecticides kill by contact on the external surface of insects and subsequent absorption through pores and nerve endings. Some residual insecticides also have a fumigant action but the effective period of this is usually less than the contact effect. Apart from the need for prolonged contact activity, residual insecticides need to have stability in storage, good mixing and application qualities, low cost and low toxicity for man and animals.

DDT (dichlorodiphenyltrichloroethane) was introduced in the 1940s and has been an excellent residual insecticide. It can be formulated in oil or water and its insecticidal effect may persist in certain circumstances for six to 12 months indoors. It is stable at normal temperatures with a shelf-life of years. However, its stability is its disadvantage since it persists in the environment and tends to concentrate in the fat of animals in food-chains. Because of this its use is banned in certain parts of the world. Apart from the food-chain toxicity, mosquitoes may develop a high degree of resistance to DDT.

HCH (hexachlorocyclohexane; benzene hexachloride: Lindane = gamma isomer). Lindane possesses the residual properties of HCH without the musty odour associated with HCH. The residual dosage rate is 0.25–0.30 g/m^2 which gives protection for three months. Lindane is more toxic than DDT and the cost is higher. Its use as a residual spray is as a replacement for DDT where mosquitoes have become resistant to DDT. Because HCH may enter various food-chains its use is banned in certain parts of the world.

Dieldrin, a chlorinated hydrocarbon, was formerly used as a residual spray: it has good persistence and high toxicity to mosquitoes. However, because of the development of resistance and the toxicity to man and animals its use has been largely discontinued in most countries.

Malathion (Cythion, O, *O*-dimethyl phosphorodithioate), an organophosphate, is available as a 25% wettable powder, as a 5% dust or as a concentrated emulsion. It is a broad-spectrum insecticide effective against a range of insects including mosquitoes. For residual spraying a dose on wall surfaces of 2 g/m^2 is satisfactory. This can be used in milking byres and parlours. Persistence is good, for up to six months depending on the surface to be sprayed. It is used where mosquitoes and other insects have become DDT and dieldrin resistant.

Crotoxyphos (dimethyl phosphate of alpha methylbenzyl 3-hydroxy-*cis*-crotonate). Ciodrin is a 3% dust and Ciovap is 1% Ciodrin and 0.25% Vapona (= dichlorvos = 2,2 dichlorovinyl dimethyl phosphate). Crotoxyphos can be used as a residual spray in milking premises or as Ciovap as a spray on the animal (e.g. horses) for the control of mosquitoes: used regularly as a mist spray Ciovap will control biting flies in byres, stables and barns.

Other organophosphates of use as residual insecticides for the control of mosquitoes and other blood-sucking flies include dichlorvos, fenitrothion (2 g/m^2) and pirimiphos-methyl (2 g/m^2) which has an additional vapour toxic effect.

Cabaryl (Sevin), a carbamate insecticide, is effective against a range of biting flies as is Landrin; they are applied at the rate of 2 g/m^2.

Pyrethrins and pyrethroids. Pyrethrins have an immediate knock-down effect. They are usually combined with synergists (e.g. piperionyl butoxide, sulphoxide). They are usually used as sprays or mists for knock-down of biting flies of various kinds. Their high activity, rapid action and low mammalian toxicity make them suitable for enclosed spaces.

The synthetic pyrethroids are of value as space sprays and residual insecticides. The use of these compounds has been reviewed by Elliot et al. (1978). Compounds such as permethrin, cypermethrin, fenvalerate and decamethrin serve as contact insecticides with advantages over DDT and dieldrin in that they are more toxic for insects, more resistant to light and more stable.

It is probable that rapid advances in the development and use of pyrethroids will occur in the coming years. Of the many compounds tested for their repellent action, indalone and dimethyl phthalate seem to be the most effective. The WHO (1959) lists these and other repellents and warns that all of them must be used with care. Some of them damage clothing other than that made of cotton or wool and some repel Simuliidae and other insects as well.

FAMILY: CERATOPOGONIDAE MEIGEN, 1803

Species of this family are the minute insects that are called biting midges, punkies or, in the USA, 'no-see-ums'. In Australia they, and the Simuliidae as well, are often called 'sandflies'. Their mouth parts form a short proboscis adapted for sucking blood, the mandibles acting like scissors. The thorax is humped over the head. The long antennae of Ceratopogonidae are, like those of Culicidae, plumose in the male and pilose in the female. The wings, however, unlike those of the Culicidae, have no scales, but they have hairs. The wings of some Ceratopogonidae are spotted and their anterior veins, which are stouter than the posterior ones, enclose two small areas called the first and second radial cells, and the other veins form two median forks, the branches of which end at the margin of the wing. The wings are folded flat over the abdomen when they are at rest.

The eggs of Ceratopogonidae are laid in water and the worm-like larvae, like those of the non-biting midges, the Chironomidae, are aquatic or semi-aquatic, but, while the larvae of some Chironomidae are coloured red by haemoglobin and are called bloodworms, the larvae of Ceratopogonidae are whitish, with a small head and three thoracic and nine abdominal segments, the terminal segment bearing a few locomotory spines. At the posterior end there are three retractile anal gills, but the larvae also breathe through the skin. The brown, inactive, obtectate pupa has two long, respiratory trumpets on the sides of the mesothorax and nine abdominal segments, which end in two spines with which the pupa anchors itself in, or at the surface of, shallow water or in crevices in manure heaps or rotting vegetation. The whole pupa is covered with spines and tubercles.

Members of the Ceratopogonidae are all predatory in the adult stage and the genus *Culicoides* has veterinary importance. Species of

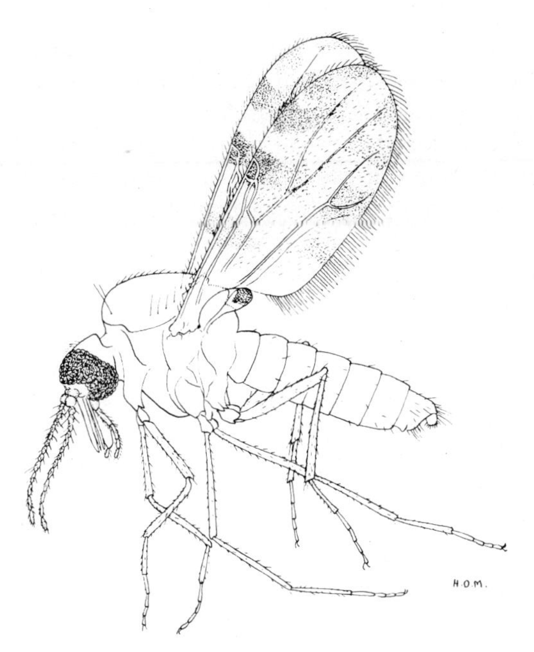

Fig 2.29 *Culicoides.*

this genus are about 1–3 mm long and can pass through ordinary mosquito screens. The females of some species of this genus attack man and animals to suck blood and can cause great annoyance if they occur in large numbers. Their bites cause inflammatory responses which in man and some animals may be severe. Several species are vectors of protozoa and filariid nematodes. They are listed, together with the protozoa and nematodes they transmit, and the literature about them, by Fallis and Bennett (1961). The following species are intermediate hosts of the filariid nematodes named: *C. grahami* and *C. austeni* of *Dipetalonema perstans* in Africa and *C. grahami* of *D. streptocara* in Africa; *C. nubeculosus* and possibly other species of *Onchocerca cervicalis* of horses in Britain; *C. pungens* and possibly three other species also of *O. gibsoni* in Malaya; *C. furens* of *Mansonella ozzardi* in South America. Species of *Culicoides* are also responsible for the transmission of virus diseases, especially bluetongue of sheep in Africa, Asia and North America and horse sickness in Africa. However, Shen et al. (1978) failed to propagate equine infectious anaemia virus in *Culicoides variipenis*.

C. robertsi causes an allergic dermatitis of horses in Queensland, described by Riek (1954) as 'Queensland itch', though the condition occurs in horses throughout the world and is known as 'sweet itch', 'sweat itch', 'summer dermatitis', etc. (McCraig 1973). It is a recurring seasonal dermatitis of individual horses and ponies during the warmer times of the year. There is loss of hair on the dorsal part of the body, e.g. withers and base of the tail. The lesions may be exacerbated by self-inflicted damage due to the pruritus. Stabling of the animals during the flies' active period (16.00 to 07.00 hours) and spraying with DDT (where this is permissible), dichlorvos or other insecticide will reduce the number and severity of cases. Treatment with antihistamines causes regression of the lesions. Applications of benzyl benzoate emulsion to the animal will decrease the number of midges which will attack it.

Control of these species is difficult. Screens treated with repellents and insecticides prevent the entry of the insects. Their larvae may be attacked in their breeding grounds. Smith et al. (1959) found that the adults of *C. furens* can migrate for several kilometres over the salt marshes of eastern Florida. Ditching, dyking and pumping water from these marshes failed to control the insects breeding, but excellent control of the larvae was obtained by the application by aircraft of 5% dieldrin granules; heptachlor was also effective. In 1958 it was found that the larvae had become resistant to dieldrin, heptachlor, chlordane, lindane and endrin, but not to DDT. Parathion, malathion and Bayer L 21/199 were, however, highly effective against strains resistant to chlorinated hydrocarbons. Kettle et al. (1959) found that, in the boglands of Midlothian, Scotland, spraying with dieldrin, chlordane and DDT controlled *C. impunctatus* and that 100% control persisted for three years and seemed to improve with time. HCH was much less effective.

FAMILY: SIMULIIDAE LATREILLE, 1804

The species of this family are often called black flies or buffalo gnats. The thorax is humped over the head and the piercing proboscis is short. The long antennae, which have 11 segments, differ from those of species of the families. Ceratopogonidae and Culicidae in not being plumose or pilose. The wings are broad and they are not spotted. They have no scales and they are not hairy, except for bristles on the thick anterior veins. The body is covered with short golden or silvery hairs.

The eggs are laid on stones or plants just below the surface of the water in running streams. The female inserts her ovipositor into the water to lay, and deposits several hundred eggs at a time. They hatch in four to 12 days, depending on the temperature. The larvae are cylindrical and attach themselves by means of a posterior sucker-like organ which is armed with small hooks, but they are able to move about. Anteriorly are the mouth parts and a pair of brush-like organs; the larvae are carnivorous. Near the anterior extremity the ventral surface bears an arm-like appendage called the *proleg,* which has a circlet of hooks at its free end, and the larva uses this when it moves about. The larvae moult six times; at the last moult the pupa appears. The mature larva spins a triangular cocoon on the surface to which it is attached and in this the pupal stage is passed. The obtectate pupa has one dorsal and one ventral respiratory tube, the branches of which float out of the cocoon.

The simuliids occur in practically all parts of the world, but are troublesome especially in warm countries. They can cause great annoyance and irritation and may even be associated with acute disease characterized by generalized petechial haemorrhages, particularly on areas of fine skin, and oedema of the throat and abdomen. In Central Europe, cattle are particularly affected by massive Simuliid attack in the spring. A *Simulium* toxin is suspected (Gräfner & Heipe 1979). *Simulium erythrocephalum* is a common cause of such lesions. Swarms of these flies may keep cattle from grazing or cause them to stampede. They bite on the legs and abdomen or on the head and ears. The bites give rise to vesicles, which burst, or wart-like papules, which may be very troublesome on the teats of cows and take weeks to heal.

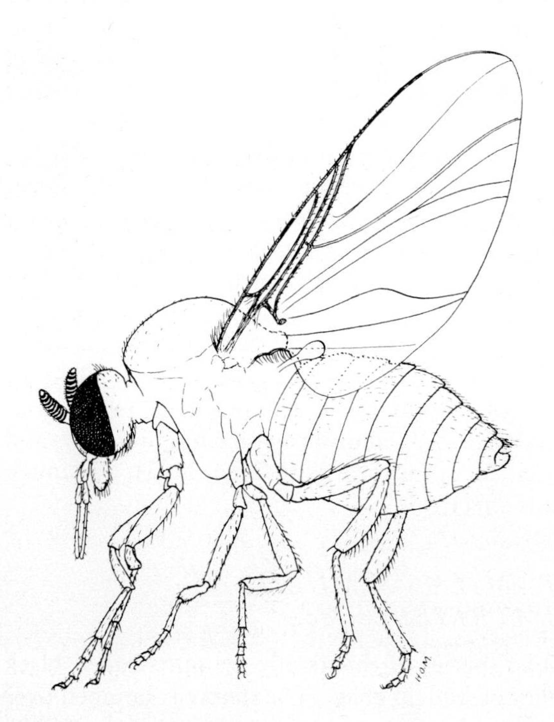

Fig 2.30 *Simulium* imago.

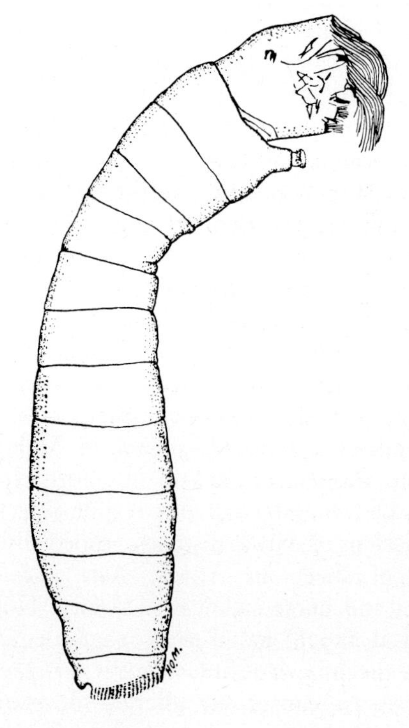

Fig 2.31 *Simulium* larva.

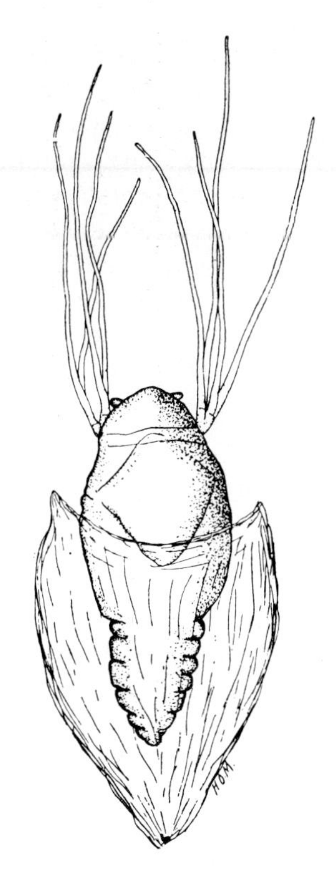

Fig 2.32 *Simulium* pupa.

Poultry are often attacked and may even become anaemic from loss of blood. The flies are active in the morning and evening, resting during the hot part of the day on the under-side of leaves near the ground.

Simulium indicum, the potu fly, is active in the Himalayan mountains. Swarms of *S. columbaczense* on the shores of the River Danube in 1923 are reported to have killed 20 000 horses, cattle, sheep, goats and pigs, as well as deer, foxes, hares and other animals, and in 1934 even larger numbers of animals were killed by it. *S. pecuarum*, the southern buffalo gnat, has caused severe losses of cattle in the Mississippi valley and may especially attack mules. In Canada, in the mid-1940s, *S. arcticum* was responsible for severe livestock losses in milk and beef production until a control programme using DDT was instituted (Steelman 1976). *Simulium* spp. transmit the viruses of eastern equine encephalitis and vesicular stomatitis. They also transmit various protozoa and nematodes. Thus in North America *S. venustum* transmits *Leucocytozoon anatis* of ducks, *L. smithi* of turkeys and *L. caulleryi* of the fowl. *S. rugglesi* may transmit *Haemoproteus nettionis* to ducks and geese. In Canada *S. ornatum* is the intermediate host of the filariid nematode *Onchocerca gutturosa* of cattle. The intermediate hosts of *O. volvulus* of man are *S. damnosum* and *S. neavei* in Africa and, in Guatemala and Mexico, *S. metallicum*, *S. ochraceum* and *S. callidum*. A summary of the organisms transmitted by *Simulium* spp. is given by Crosskey (1973).

Control is difficult because the adults can fly 3–5 km or more. Medicated screens may provide some protection against the adults. WHO points out that the adults prefer to get into clothing before they bite, so that repellents applied to clothing may help. Openings at the neck, wrists, etc. should be closed and light-coloured clothes attract them less. For protecting poultry from these flies a dust bath of 0.5% lindane (20% lindane one part, inert dust 40 parts) is recommended but this may be toxic to young chickens and turkeys. Several workers have shown that 10^{-2} mg/litre of DDT will rid streams of the larvae for several kilometres. WHO states that *S. arcticum* was almost completely eradicated in Canada from 160 km of stream by 10^{-2} mg/litre of DDT applied by aircraft and suggests as a guide the application of 10^{-2} mg/litre of DDT for 15 minutes at intervals of 1.5 km, or 20 g of DDT per hectare of water surface. These amounts will not kill fish, but may harm aquatic arthropods. *S. neavei* was eradicated from the whole of Kenya, excepting one small area, by the application of 10^{-1} or 10^{-2} mg/litre of DDT to the streams in nine cycles, each cycle lasting ten days.

A major control programme against *S. damnosum* in the Volta River basin in West Africa, one of the worst endemic onchocerciasis areas in the world, was started in 1974. This consists of the systematic destruction of blackfly larvae in their breeding sites by the application of abate [*O*,*O*′-(thiodi-4, 1-phenylene)*O*,*O*,*O*′,*O*′-tetramethylphosphorothioate], a biodegradable insecticide, to rivers at weekly intervals using aircraft. This programme will continue for 20 years.

FAMILY: PSYCHODIDAE BIGOT, 1854

The Psychodidae, commonly known as 'sandflies' or 'owl midges', are small, moth-like flies, rarely over 5 mm long. Their bodies and wings are hairy. The legs are long, rarely short. The wings are held rooflike over the abdomen during rest. The mouth parts are short or of medium length. The antennae are long, consisting of 16 segments which often have a beaded appearance, and they are thickly covered with hairs. The palpi are recurved and hairy. Lewis (1973) classifies these nematocerans in the family Phlebotomidae and reserves the family Psychodidae for the moth flies.

The eggs are laid in moist, dark places, e.g. in rock crevices and between stones. The female of *Phlebotomus papatasii* lays about 40–80 eggs at a time. A temperature of over 15°C is required, otherwise the embryo becomes dormant. Under favourable conditions the whole life-cycle can be completed in about six weeks. The larvae resemble small caterpillars and feed on faeces of lizards, bats and other animals and on dried leaves.

The flies are active at night only and hide during the day in dark corners. They are weak fliers and the females of some species are blood-suckers. They can pass through ordinary nets or will otherwise search for an opening to get through. *Phlebotomus* species are especially important in the transmission of leishmaniasis. Thus *P. papatasii, P. sergenti, P. major* and perhaps other species transmit *Leishmania tropica,* the cause of cutaneous leishmaniasis, to man, dogs being reservoir hosts in some areas; *L. tropica* is transmitted from gerbils and ground squirrels to man in Turkestan and Iran by *P. papatasii* and *P. caucasicus*. Various species of *Phlebotomus* transmit *L. braziliensis*, the cause of South American cutaneous leishmaniasis (espundia), to man. *L. donovani*, the cause of kala-azar, is transmitted by *P. argentipes* in India, by *P. chinensis* in China and by several other species of *Phlebotomus* in the Middle East, Africa and the Mediterranean region. Sandfly fever is transmitted in the Mediterranean area by *P. papatasii* and in China possibly by *P. chinensis* and *P. mongolensis*. There is evidence for transovarian transmission of the virus of sandfly fever (Lewis 1973). Other viruses, including the virus of yellow fever, have been isolated from sandflies but the significance of these findings has to be determined fully. *Bartonella bacilliformis*, the cause of Carrion's disease (Oroya fever), is transmitted by *P. verrucarum* and in Colombia possibly by *P. columbianum*.

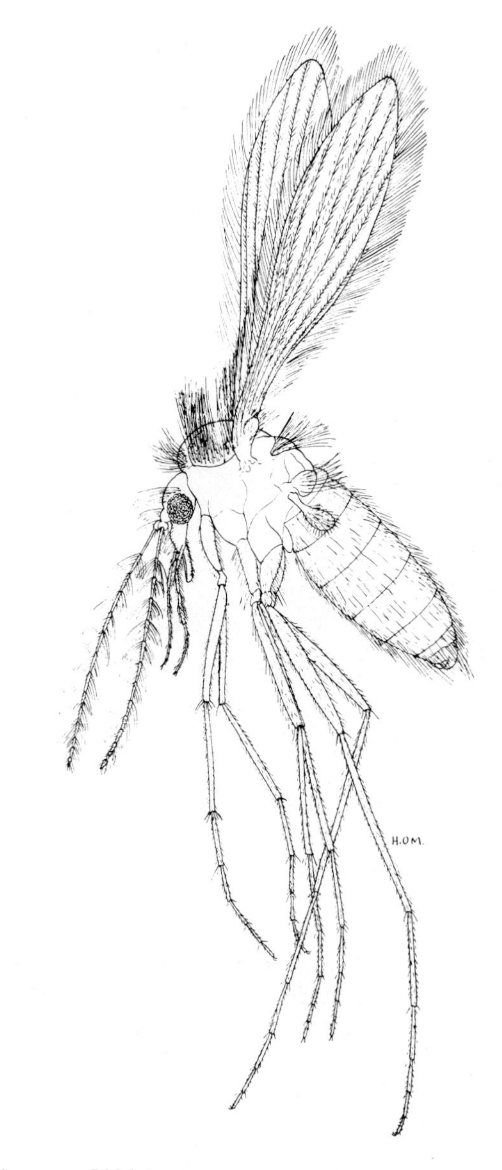

Fig 2.33 *Phlebotomus.*

Control. Removal of dense vegetation discourages the breeding of these flies. WHO states that single residual application of DDT at 1–3 g/m^2 (100–300 mg/ft^2) to the interiors of human and animal dwellings has given good results and protection for one or two years, and that 2.5 mg/m^2 (22 mg/ft^2) of gamma HCH reduced the number of flies for three months. Populations of sandflies have been reduced in many areas as a result of mosquito control for malaria. As malaria eradication is discontinued in many areas, sandfly populations increase.

SUBORDER: BRACHYCERA McQUART, 1834

This suborder includes 16 families, the members of which, in different developmental stages, are predatory upon other insects. The family of major interest in this edition is *Tabanidae*, the members of which are predatory on vertebrates. The family *Rhagionidae* also contains some blood-sucking species.

FAMILY: TABANIDAE LEACH, 1819

These insects, commonly known as 'horse-flies' or 'breeze flies', are large, robust flies with powerful wings and large eyes. The latter are almost contiguous (holoptic) in the males and separated by a narrow space in the females and project posteriorly beyond the lateral margins of the thorax. The antennae have two short basal segments and a third which is large and usually ringed. The wing venation is very characteristic, especially the branching of the fourth longitudinal vein. The proboscis is relatively short in *Tabanus* and *Haematopota*, longer in *Chrysops* and very long in *Pangonia*. These are the most important genera. In species of the genus *Pangonia* the proboscis is very long and projects forwards and the third (terminal) segment of the antenna has six or seven annulations. The wings are held divergent at rest. In species of the genera *Tabanus*, *Chrysops* and *Haematopota* the proboscis is soft and hangs down. These three genera may be distinguished by their antennae. In species of the genus *Chrysops* the first and second segments of the antenna are long, the third (terminal) segment has four annulations, the wings have a dark band passing from the anterior to the posterior border of the wing, the wings are divergent when at rest and the eyes are of a metallic colour. In species of the genus *Haematopota* the first segment of the antenna is large and the second segment narrower, while the terminal segment has three annulations and the wings have a characteristic mottling. In the species of the genus *Tabanus* the first two segments of the antenna are small and the third

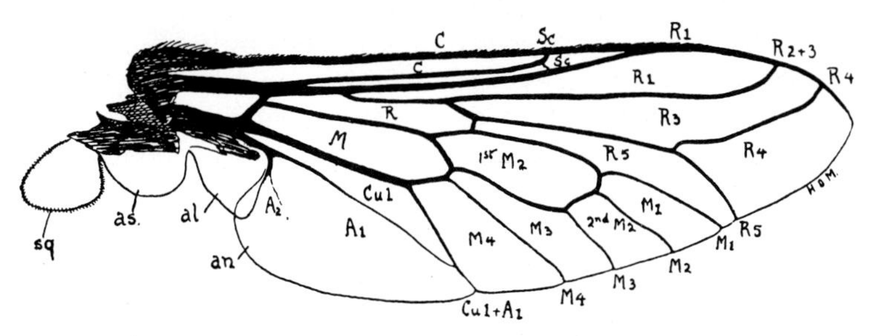

Fig 2.34 Wing of *Tabanus*. Showing veins and cells.

A	anal	Sc	subcosta
C	costa	al	alula
Cu	cubital	an	anal lobe
M	medial	as	antisquama
R	radial	sq	squama

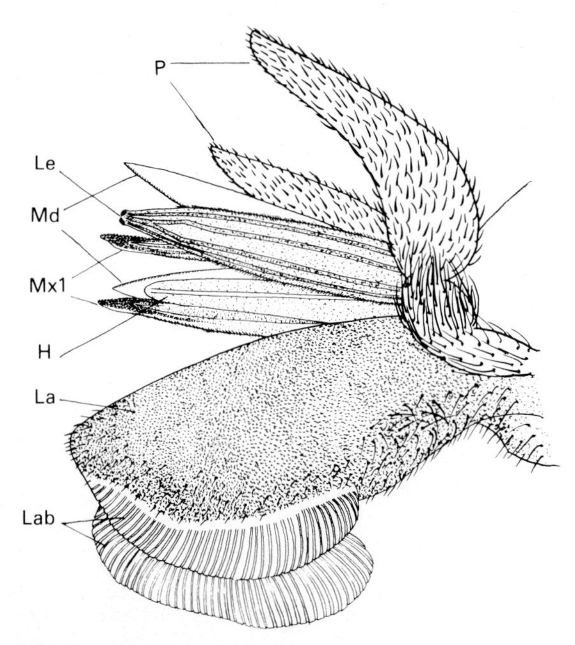

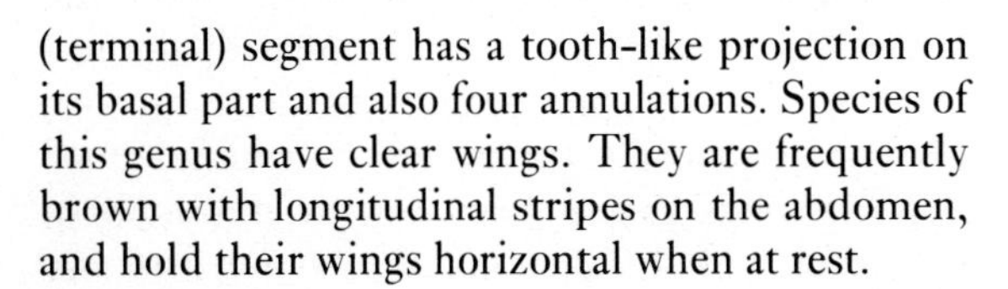

Fig 2.35 Mouth parts of *Tabanus*.

H	hypopharynx
La	labium
Lab	labella with pseudotracheal membrane
Le	labrum-epipharynx
Md	mandibles
Mx1	first maxillae
P	palps

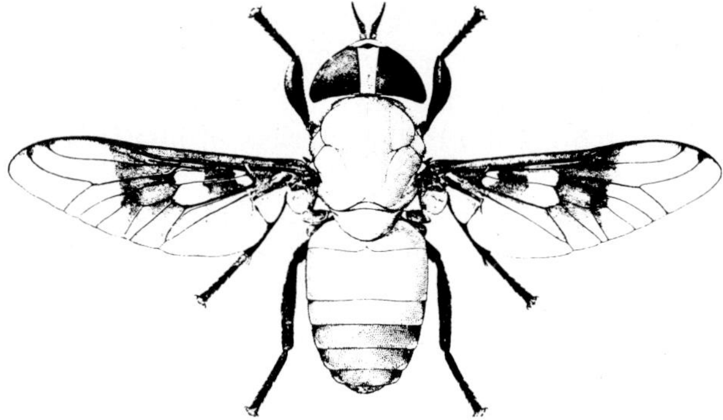

Fig 2.36 *Tabanus latipes*. (*From Smart*)

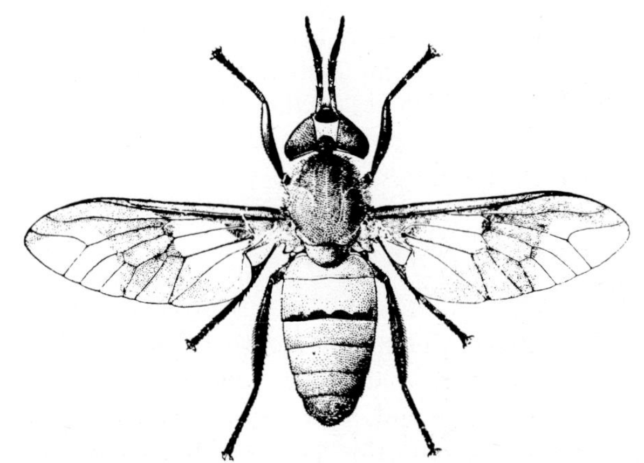

Fig 2.37 *Chrysops fixissimus*. (*From Smart*)

(terminal) segment has a tooth-like projection on its basal part and also four annulations. Species of this genus have clear wings. They are frequently brown with longitudinal stripes on the abdomen, and hold their wings horizontal when at rest.

Life-cycle

The eggs are laid in the vicinity of water, usually on the leaves of plants. The larger species lay 500–600 eggs, the smaller 300. The eggs are about 2 mm long and light in colour, but turn dark after a while. The larvae hatch after four to seven days and drop into the water, or on mud, into which they disappear. They are maggot-like and the body has 11 segments, besides the cephalic portion which is not conspicuous. Each segment has eight fleshy tubercles. The mouth parts are prehensile and masticatory; the larvae are carnivorous. There are three-jointed antennae and the large lateral tracheae open on the penultimate segment, which also bears a retractile siphon tube. The larvae feed on small crustacea, or even on one another, and grow for two to three months, performing several ecdyses. Finally they pass through a quiescent stage and then pupate. The pupa is brown and subcylindrical; the abdominal segments are movable and in the anterior part the appendages of the imago can be distinguished. This stage lasts about 10–14 days. The whole life-cycle takes four to five months under favourable conditions, but low temperatures prolong development and the larvae may hibernate. The eggs are parasitized by certain small Hymenoptera.

Biology of adults

The flies are seen in summer and are very fond of sunlight. They abound especially near their breeding-places and are most active on hot, sultry days. The females are well-known blood-suckers (the males feed on honeydew and on the juice of

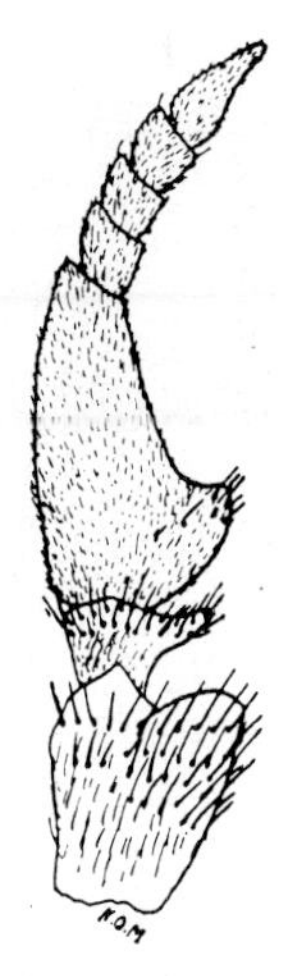

Fig 2.38 Antenna of *Tabanus*.

flowers, as do the females if a suitable host is not available) and feed chiefly on large animals such as horses and cattle. Some feed mainly on the under-side of the abdomen around the navel or on the legs; others bite also on the neck and withers. They bite a number of times in different places before they are replete, and from the wounds made by them small quantities of blood usually continue to escape and are sucked up by non-biting Muscidae. The flies feed about every three days. After feeding they rest for a few hours on the under-side of leaves or on stones or trees.

The bites of the Tabanidae are painful and irritating and may give rise to weals in soft-skinned animals. Horses and cattle are restless when troubled by these flies and may become unmanageable if they are in harness. The flies, as well as the Muscidae that come after them, may act as mechanical transmitters of anthrax, anaplasmosis and the virus of equine infectious anaemia. *Chrysops discalis* is a vector of *Pasteurella tularensis* in the USA. *Chrysops dimidiata*, the Mango fly, and *Chrysops silacea* are intermediate hosts of the filariid nematode, *Loa loa*. Various species are mechanical vectors of *Trypanosoma evansi*, the cause of surra of equines and dogs; *T. equinum*, the cause of mal de Caderas of equines; *T. simiae* of pigs; *T. vivax* and *T. brucei*, which cause nagana of cattle, sheep, equines and other ungulates; and of *T. gambiense* and *T. rhodesiense*, which cause human African trypanosomiasis. *T. theileri* of cattle is transmitted cyclically by species of *Tabanus* and *Haematopota*.

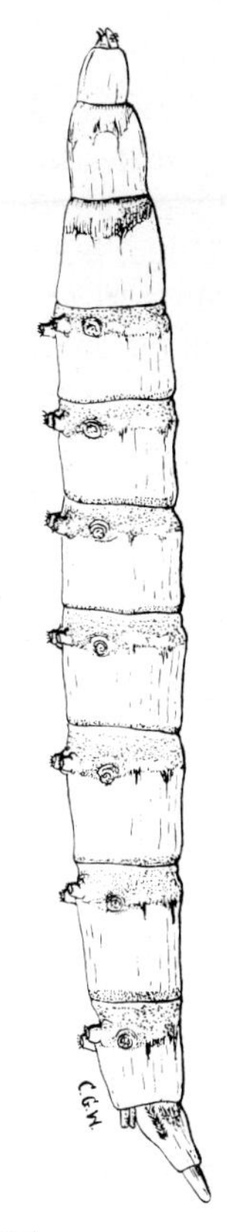

Fig 2.39 Larva of *Tabanus*. × 5.

Pangonia is often seen hovering over flowers on the border of woodlands, but may attack man and animals in various parts of the world. The labium is not adapted for piercing and the proboscis is used for sucking up spilled blood.

Steelman (1976) summarizes the estimates of losses to livestock production caused by tabanids. Thus in the USA losses of $40 million were ascribed to tabanids, directly through a nuisance effect and indirectly through disease transmission. Sustained attack by tabanids may reduce milk production severely (Zumpt 1949) and a study in Illinois, USA, showed an inverse correlation between butterfat production and the abundance of horseflies. Similarly, in beef cattle, tabanid control may lead to significantly greater live-weight gain than when these flies were not controlled.

Control

Control is difficult. Where drainage is possible the breeding-places may be destroyed by this method. The flies have a habit of skimming over water and occasionally dipping their bodies into it; this led to the practice of pouring kerosene onto water, which kills the flies when they dip into it. Animals should be kept away from places where the flies abound during the hot part of the day. Measures similar to those for the control of mosquitoes are applicable. Residual sprays for the inside walls of animal houses include chlorinated hydrocarbon insecticides (where their use is permitted), organophosphates such as malathion and crotoxyphos which can be used as a residual spray or (as Ciovap) as a spray on the animal (see p. 413).

FAMILY: RHAGIONIDAE

Normally these flies (snipe flies) are predatory upon other insects but two genera, *Symphoromyia* (in USA) and *Spaniopsis* (in Tasmania), are recorded as blood-suckers with a painful bite (Smith 1973). They are elongate flies, dully coloured, and breed in damp earth or leaf mould. Disease transmission is not recognized.

SUBORDER: CYCLORRHAPHA BRAUER, 1863

SERIES: ASCHIZA

The insects in this group of Diptera are of no direct veterinary importance. However the *Syrphidae* (or hover flies or drone flies) are frequently found in the summer on dung, rotting wood and decaying animal food. They are brightly coloured flies and superficially resemble wasps. A characteristic feature of the wing venation is a 'vena spuria'. One genus, *Eristalis*, has a larval stage with a long terminal flexible respiratory tube and hence it is referred to as a 'rat-tailed maggot'. It may be found in decaying foodstuffs and may also occur accidentally in the digestive tract of animals (pigs especially) and man leading to 'false myiasis'. The larvae cause no harm.

SERIES: SCHIZOPHORA

SECTION: ACALYPTERAE

Squamae small, not concealing halteres; thorax without distinct transverse suture.

FAMILY: GASTEROPHILIDAE ESSIG, 1925

Many authors consider these flies as a subdivision of the family *Oestridae* but it is now usual to assign family status to the group. Freeman (1973) includes them in the Section *Acalypterae*, whereas the family *Oestridae* is placed in the Section *Calypterae*.

These are hairy flies; their mouthparts are reduced and functionless. The genus *Gasterophilus* occurs in equines and *Cobboldia* in elephants.

Genus: Gasterophilus Leach, 1817

The larvae of several species of this genus are parasites of equines and are known as 'bots'. They are rarely found in dogs, pigs, birds and man. The adult flies are brown in colour and hairy and somewhat resembling bees. *G. intestinalis* (de Geer, 1776) (syn. *G. equi*) is the commonest species. The adults of this species are about 18 mm long and a dark, irregular, transverse band runs across either wing.

The species is most readily recognized by the

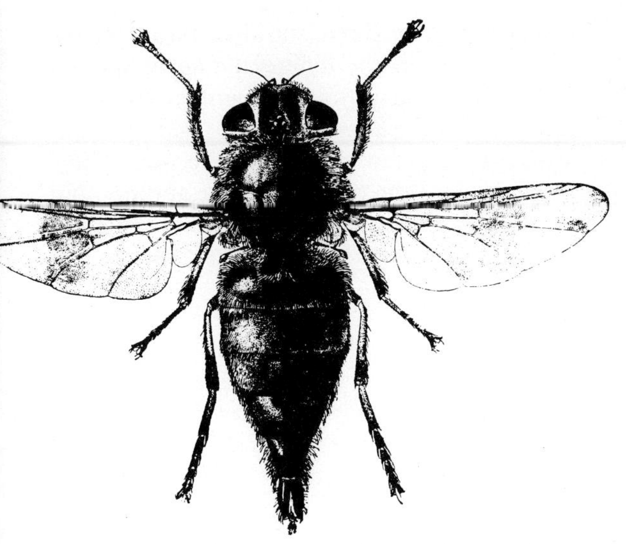

Fig 2.40 Female *Gasterophilus intestinalis*. (*From Smart*)

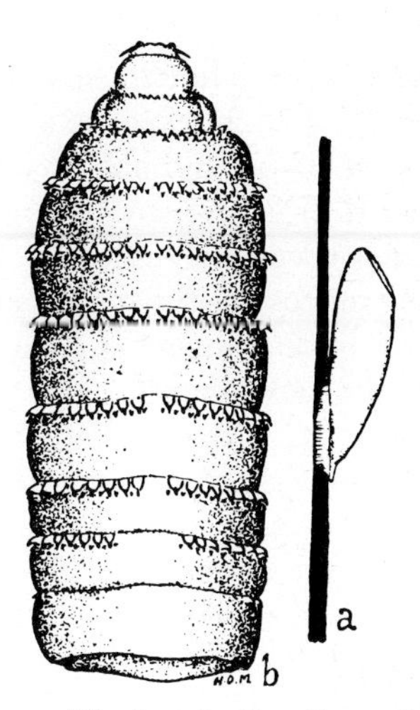

Fig 2.41 *Gasterophilus intestinalis*. a, Egg on hair. b, Dorsal view of larva.

third-stage larvae which can be differentiated as outlined in Table 2.4.

Biology and life-cycle. The adult flies occur during the latter half of the summer and live only a few days, rarely up to three weeks. The female fly hovers about the animal with its ovipositor extended and repeatedly darts at it to glue an egg to a hair. A large number of eggs may be laid in succession. *G. intestinalis* deposits its eggs mainly around the fetlocks of the fore-legs, also higher up the legs and in the scapular region. *G. nasalis* lays on the hairs of the intermandibular region, while *G. haemorrhoidalis* and *G. inermis* deposit their eggs on the hairs around the mouth and on the cheeks. The eggs of *G. pecorum* and *G. haemorrhoidalis* are dark in colour and those of the other species pale yellow; they are elongate, pointed at the attached end and operculate at the other.

The eggs are ready to hatch in five to ten days or more. The eggs laid near the mouth of the horse

Table 2.4. Differentiation of Third-stage *Gasterophilus* larvae

Characters	Species
One row of ventral spines on each segment: third segment always with a row of dorsal spines	*G. nasalis* (Linnaeus, 1758) (syn. *G. veterinus*) (in *Equus* spp.; global)
One row of spines on body segments, segment 3 bare dorsally and ventrally, segment 4 bare dorsally	*G. nigricornis* (Loew, 1863) (in horse, donkey; Spain, Middle East, South Asia, China)
One row of ventral spines on each segment, dorsal spines on segments 5 to 10	*G. meridionalis* (Pillers and Evans, 1926) (in zebra; southern Africa)
Two rows of ventral spines on segments; dorsal spines in complete rows only on segments 2 to 5; spines pointed	*G. pecorum* (Fabricius, 1794) (in horse, donkey; Old World, Africa)
Two complete rows of ventral spines on segment 3; ventral spines interrupted medially on segment 11. Spines pointed	*G. inermis* (Brauer, 1858) (in horse, zebra; Palaearctic region)
Two incomplete rows of ventral spines on segment 3; spines not interrupted medially on segment 11. Spines pointed	*G. haemorrhoidalis* (Linnaeus, 1958) (in horse, donkey, zebra; global)
Two rows of spines dorsally and ventrally to at least 10th segment. Spines blunt-tipped	*G. intestinalis* (De Geer, 1776) (in horse, donkey; global)
Three rows of spines dorsally and ventrally on most body segments	*G. ternicinctus* (Gedoelst, 1912) (in zebra; Central and southern Africa)

After Zumpt, 1965.

hatch spontaneously, while those of *G. intestinalis* and *G. pecorum* hatch in response to an increase in temperature on licking and not, as was previously thought, due to friction or moisture. The larvae are not swallowed directly into the stomach, but penetrate the mucosa of the mouth and gradually wander down in the mucosa at least as far as the pharynx in the next month. The larvae of *G. intestinalis* and *G. haemorrhoidalis* are found chiefly in the mucosa of the tongue, those of *G. pecorum* and *G. inermis* in the mucosa of the cheeks. The larvae of the latter species, perhaps also those of *G. haemorrhoidalis*, even pierce the skin of the face and wander in it to the mouth, leaving conspicuous tracks behind them and this has been known to occur in man. *Gasterophilus*, however, though it rarely infects man, may cause in him a cutaneous swelling at the point at which the first larva penetrates the skin. Beesley (1974*a*) reports infection in a New Zealand farrier. First-stage larvae were recovered from lesions in the axilla and on the forearm and the infection was ascribed to the close association with horses during shoeing. More rarely the larvae reach the human stomach and cause irritation there. In other animals the migratory larvae occasionally go astray and have been found in various thoracic and abdominal organs, the sinuses of the head and even in the brain.

The larvae of *G. intestinalis*, after wandering in the tongue for 21–28 days, become attached in the cardiac portion of the stomach, rarely in the fundus or the pylorus; they have a reddish colour. Those of *G. nasalis* are pale yellow and attach themselves in the pylorus and the duodenum. The larvae of *G. pecorum* are blood-red in colour; the second and sometimes also the third stage is found in the pharynx and the upper part of the oesophagus but the third stage is usually attached in the fundus of the stomach. The young stages of *G. haemorrhoidalis*, which are red, are occasionally found in the pharynx, but later they settle in the stomach.

The larvae remain in the host for 10–12 months and reach the third stage when they measure up to 20 mm in length, are brown in colour and have dense spines on the anterior border of the segment. There are a pair of distinct mouth hooks on the first segment. The shape and distribution of the spines are used in the differentiation of species. The larvae then pass out through the intestine. Those of *G. haemorrhoidalis* again attach themselves for a few days in the rectum. Some larvae may leave the host in the late autumn, but usually they pass out in the spring. They pupate in the ground for three to five weeks and then the flies emerge.

Pathogenesis. The flies which lay their eggs on or near the head annoy the animals and may cause them to panic. The migrating larvae are found in the superficial layers of the buccal epithelium and generally produce no reaction, though Chereshnev (1954) states that *G. pecorum* may cause stomatitis. When later they attach to the pharynx, stomach or duodenum, an inflammatory process produces a ring-like thickening around the larva.

The parasites are very common and various surveys have shown infection rates of, for example, up to 90% in horses in Ireland (Hatch et al. 1976) and similar infection rates have been reported from elsewhere in the world. *Gasterophilus intestinalis* is usually more common than the other species, e.g. *G. nasalis*.

Ulceration of the oesophageal region of the stomach, caused by larvae of *G. intestinalis*, is the most common lesion. Sometimes the larvae are very closely packed together, producing a pallisade effect.

Though dramatic in appearance, opinions differ on the importance of bots in horses. Waddell (1972) found that fewer than 1% of larvae were attached to the glandular part of the stomach and the majority of the infection was located in the non-glandular mucosa, which plays little role in digestion. Possibly this is the reason why there is no strong evidence that pathogenic effects are produced by the infection. Nevertheless, it would seem unreasonable to assume that the at times extensive ulceration of the stomach by a large number of parasites is without a general effect. Steelman (1976) quotes estimates of losses due to bots in horses, which may be considerable in the thoroughbred horse raising areas of the USA. Occasionally more extensive effects are evident, including abscess formation, rupture of the

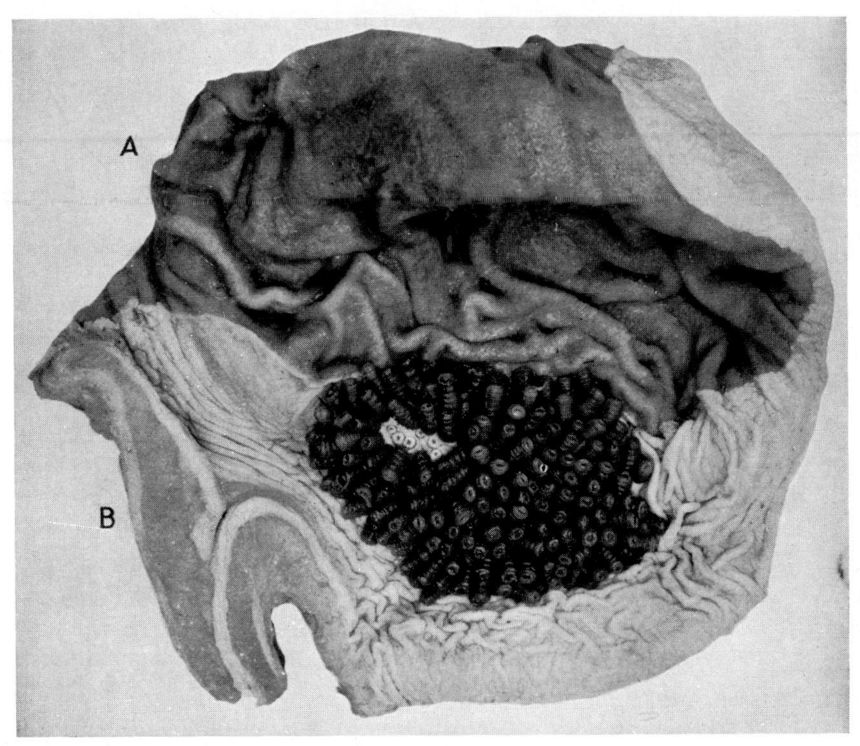

Fig 2.42 Mucous membrane of a horse stomach infected with horse bots. A, Pyloric end of stomach. B, Oesophageal (cardiac) end of stomach. Several bots have been removed to show the circular pits (centre) where the larvae have been attached. (H. Thornton).

stomach with peritonitis or stricture of the pylorus by clusters of larvae and tissue reaction to them in the pyloric region. Larvae which attach temporarily to the rectum after passage through the digestive tract (e.g. *G. haemorrhoidalis*) cause irritation and inflammation.

Despite the lack of firm evidence that the larvae produce ill-effects, horse owners are usually of the opinion that the parasites are of importance and demand treatment for their animals.

Diagnosis. The eggs can be found by examining the sites at which they are deposited and larvae in the pharynx can be seen on direct inspection. There is no way of diagnosing the presence of larval parasites in the stomach.

Treatment. Old treatments included carbon disulphide which is given by stomach tube or in a capsule at a rate of 2–5 ml/kg body weight after an 18-hour fast. A piperazine/carbon disulphide complex (Parvex) or this mixture plus phenothiazine (Parvex Plus) at a dose of 110 mg/kg is effective against *Gasterophilus* larvae and also against ascarids and large and small strongyles of horses. Trichlorofon removes *Gasterophilus*, ascarids and *Oxyuris*. It is given in the feed or in water by stomach tube at a rate of 39.6 mg/kg. Thiabendazole/trichlorfon is effective against intestinal strongyles and *Gasterophilus*. It may be administered in feed or by stomach tube or drench at a rate of 2 g thiabendazole and 18 g trichlorofon per 50 kg body weight. Debilitated animals and pregnant mares should not be treated. Dichlorvos, an organophosphate, is given in the feed at a dose of 26–52 mg/kg or as a paste at 20 mg/kg and is also effective against ascarids and the large and small stongyles. Butonate at a dose rate of 45 mg/kg given by stomach tube is effective against *Gasterophilus* and *Parascaris*.

Treatment against *Gasterophilus* should be carried out once or twice per year. The first dose is given about one month after the first frost, when larvae are in the stomach and no more adult flies remain. The second treatment is given in late winter.

Prophylaxis. Frequent grooming assists in removing eggs before they become infective. Eggs may be hatched by sponging the chest, fore-legs and chin with water at 40–43°C. The water should contain an insecticide to kill the hatched larvae. Although treatment is usually delayed until one month after the first frosts (which kill the adult flies) in order to obtain a maximum kill with a single treatment, eggs on hairs may remain infective for some time despite frosts.

Genus: Cobboldia Brauer, 1887

Members of this genus are found in the stomachs of elephants.

Cobboldia elephantis (Cobbold, 1866) occurs in the stomach of the Indian elephant and **C. loxodontis** Brauer, 1896 has been reported in the stomach of the African elephant and also the rhinocerus.

SECTION: CALYPTERAE

The squamae are well developed, concealing halteres, and the thorax has a distinct transverse suture. Families of importance are: Muscidae, Glossinidae, Calliphoridae, Oestridae and Cuterebridae.

FAMILY: MUSCIDAE LATREILLE, 1802

Hypopleural bristles are absent (see Calliphoridae). The median (M) wing vein is more or less parallel to the radial (R4 + 5) wing vein or curving towards it; there is no definite elbow bend in the radial vein. Genera of importance are *Musca*, *Stomoxys*, *Hydrotaea*, *Haematobia* and *Fannia*.

Genus: Musca Linnaeus, 1758

Musca domestica Linnaeus, 1758, the common house fly, has a cosmopolitan distribution and is important as a mechanical carrier of various infectious agents including viruses, bacteria and protozoa. It also acts as intermediate host for a number of helminths. Other species of this genus bear a close resemblance to *M. domestica* and their identification is a matter for the specialist. The male is 5.8–6.5 mm long and the female 6.5–7.5 mm long. In the wing the M1 + 2 vein curves forward distally and the R5 (first posterior) cell is nearly closed. The thorax is yellowish-grey to dark grey and has four longitudinal dark stripes which are equally wide and extend to the posterior border of the scutum. The abdomen has a yellowish ground colour and a median, black, longitudinal stripe which becomes diffuse on the fourth segment. In addition to this

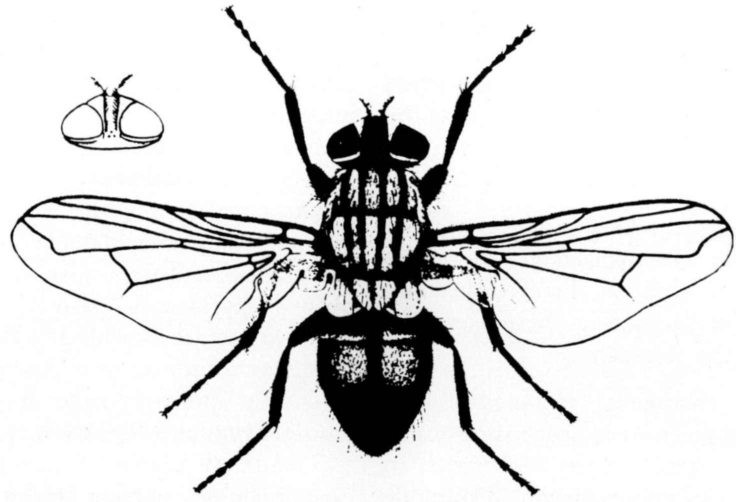

Fig 2.43 Female *Musca domestica*, the house fly. *Inset*, Head of the male. (*From C. G. Hewitt*)

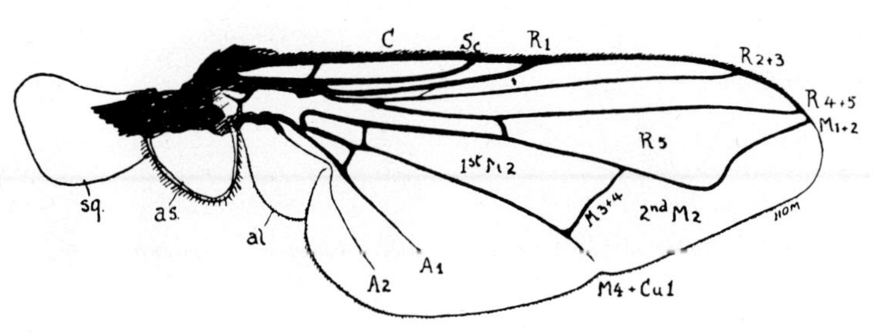

Fig 2.44 Wing of *Musca domestica*.

C	costa	A	anal
Sc	subcosta	al	alula
R	radial	as	antisquama
M	median	sq	squama

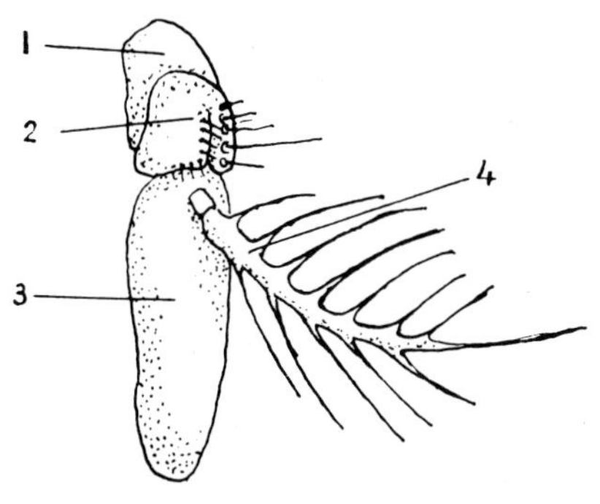

Fig 2.45 Antenna of *Musca domestica*.
1–3 first, second and third segments
4 arista biplumose to its tip

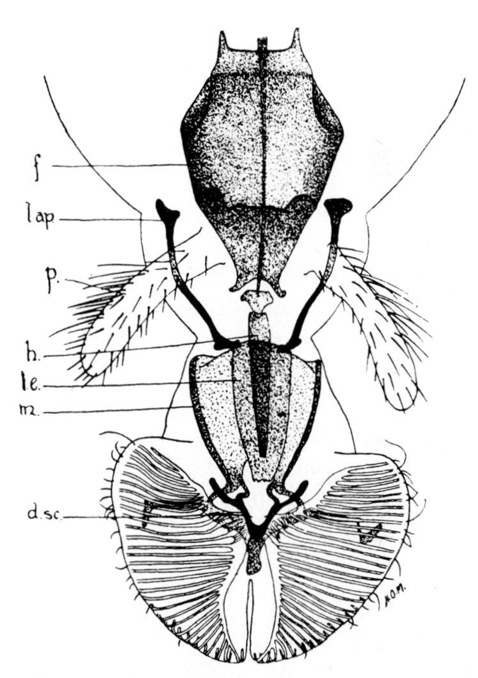

Fig 2.46 Dorsal view of the mouth parts of *Musca domestica*, showing the labium expanded into two labella.

dsc	discal sclerite
f	fulcrum
h	hypopharynx
l ap	labral apodeme
le	labrum-epipharynx
m	mentum
p	palp

stripe the abdomen of the female is marked on either side with a diffuse dark band. The arista is bilaterally plumose up to the tip. The mouth parts are adapted for imbibing liquid food. The labium is expanded distally into two labella. These are capable of marked expansion when the fly is feeding. The median walls are covered by pseudotracheae which bear a system of channels serving to suck up food in fluid form. The labella are also hollow organs and their cavities are connected through that of the labium with the general body cavity or haemocele. Pressure of the haemocele fluid causes the labella to expand and turn their medial surfaces forwards, so that they can be brought in contact with the food. Beginning at the periphery, these grooves converge towards the prestomum, the middle ones opening directly into it, while the dorsal and ventral grooves run first into larger collecting channels. Each groove is strengthened by a series of incomplete rings of chitin, standing closely side by side. The rings are bifid at one end and expanded at the other, and they are so arranged that the bifid and expanded ends alternate on either side of the groove. When

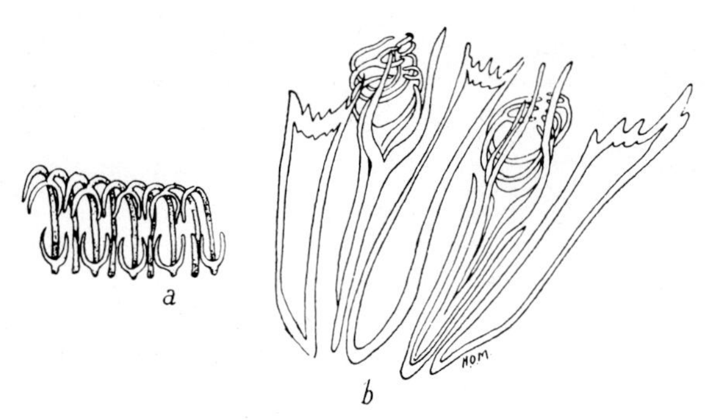

Fig 2.47 Chitinous structures of the pseudotracheal membrane of *Musca domestica*. a, Chitinous supports of food channels. b, Chitinous supports at discal sclerite, with intervening tooth-like blades.

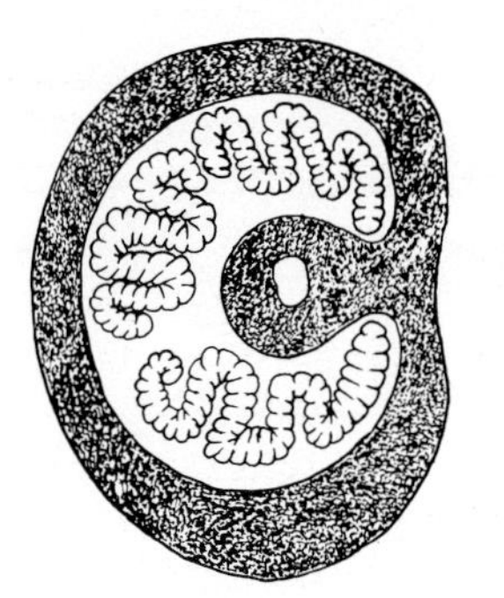

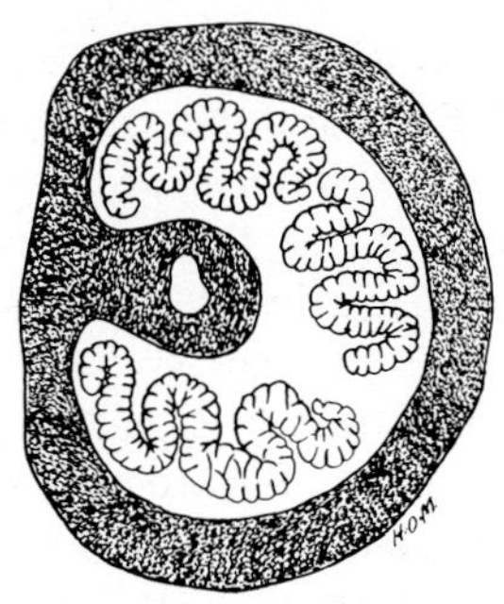

Fig 2.48 Posterior spiracle and stigmatic plates of the third-stage larva of *Musca domestica*.

the fly feeds the edges of the grooves are drawn together and the food is strained through the small openings which remain between the bifid and expanded ends of the rings. Under such conditions the house fly can take in particles of only about 4 μm diameter, but it can swallow larger particles if the grooves are not so tightly shut, or particles of about 45 μm diameter if it separates the labella completely and sucks directly through the prestomum, although it is doubtful whether this is frequently done. Liquefiable solid food, such as sugar, may be made fluid before it is sucked up by ejection on to it of saliva and crop-fluid, these drops of fluid being called *vomit-drops*. These points are important in connection with the disease-transmitting capacity of the fly.

Life-cycle. The house fly lays 100–150 eggs at a time and a total of about 1000. Fresh horse manure is preferred but the fly will also develop in the faeces of other animals and man, as well as in all sorts of decaying organic matter and refuse. The eggs are about 1 mm long, elongate and creamy white in colour and the dorsal surface has two curved, rib-like thickenings. The larva hatches in 12–24 hours and grows into a maggot 10–12 mm long in three to seven days depending on environmental temperatures. The body of the larva is pointed anteriorly and broad at the posterior end, on which the stigmal plates are situated. The distance between the two plates is less than the width of a plate and each bears three winding slits. The second body segment also bears a pair of anterior spiracles which are fan-shaped, consisting of a stalk and five to eight papillae each. Anteriorly the body bears a pair of oral hooks which are connected to an internal cephalopharyngeal skeleton composed of darkly pigmented chitin. Three ecdyses occur during the larval stage and the pupa remains in the last larval skin, which turns brown and becomes rigid to form the puparium. The full-grown larva leaves the material in which it has developed to pupate in the ground. The pupal stage lasts three to 26 days, depending on the temperature. Fertilization and

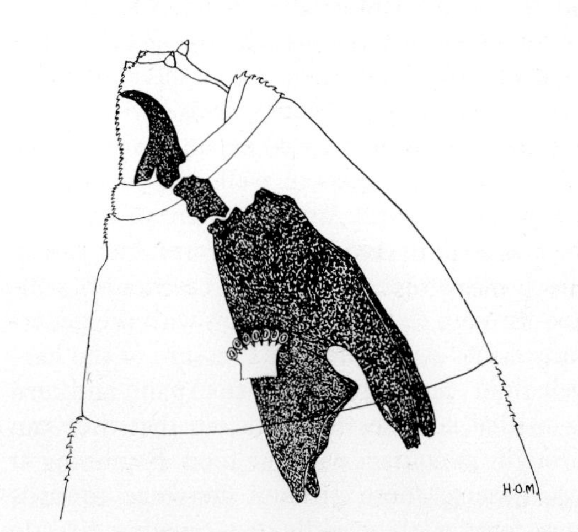

Fig 2.49 Lateral view of the anterior end of the third-stage larva of *Musca domestica*, showing the anterior spiracle and cephalopharyngeal skeleton.

oviposition take place a few days after emergence of the fly and the whole cycle may be completed in about eight days (at 33–35°C), so that a number of generations develop in one summer. The adult flies live a few weeks only in summer, but they live longer in cool weather. They probably rarely hibernate. Development proceeds slowly even in winter, but flies do not emerge in cold regions. The eggs, larvae and pupae can resist a fair degree of cold when they lie protected and are responsible for the new crop of flies in spring.

Habits and importance. The adult fly is a synanthrope and occurs associated with man throughout the world. It enters houses and animal accommodations and will feed on almost anything with a moist surface, especially milk, sugary food, meats, excrement and garbage. Females, particularly, are attracted to protein-containing material, this being necessary for the maturation of ovaries. The fly regurgitates at frequent intervals to aid its feeding (vomit-drop) and it defaecates at random; these activities are responsible for the mechanical transmission of a substantial number of diseases. These include virus infections (e.g. poliomyelitis), bacterial infections (e.g. diarrhoeal and enteric fevers, typhoid, cholera, anthrax, tuberculosis, leprosy, etc.) and protozoa (e.g. *Entamoeba*), as well as the eggs of various helminths (e.g. *Enterobius, Ascaris* etc.). *Musca domestica* is not a blood-feeder habitually, but it may follow other blood-sucking flies and feed on decomposing blood and tissue fluids. *Musca domestica* is the intermediate host of several helminth parasites of domestic animals, for example *Habronema* spp. and *Raillietina* spp.

Control. The veterinarian is concerned with control measures against the breeding of flies mainly in connection with stables, stock-yards, abattoirs, drains, garbage etc.

Manure must be regularly and frequently removed from stables, sties etc. It may be spread on fields directly (e.g. as slurry) and if the application is thin enough it precludes the breeding of flies. If manure is stacked in large compact heaps it ferments and the heat thus produced kills the maggots, as well as the eggs and larvae of internal parasites, in the central portion of the material. The sides of the heap and the surrounding soil may be treated with insecticides to kill larvae and pupae.

Garbage should be collected in cans with tight-fitting lids, or the lid may overhang the top of the can and allow the entrance of flies, which are then caught in a trap placed over a hole in the lid (*Hodge's garbage can trap*). Garbage which is not used should be incinerated.

A range of insecticides is available for the control of houseflies and related species with similar developmental cycles and habits. Many of the insecticides are used for other insect pests (e.g. mosquitoes, biting midges) as well (see p. 391) and only a summary is given here.

Pyrethrins combined with synergists (e.g. piperonyl butoxide) have an immediate knock-down effect. The pyrethroid insecticides are now in common use and can be used in a variety of ways in enclosed spaces, such as volatilization in flares, matches, coils etc. Resistance to pyrethrin and the pyrethroids has developed in house flies in various parts of the world, but there is little or no cross-resistance between pyrethroids and organo-phosphates in houseflies (Elliot et al. 1978).

Residual wall sprays include dimethoate (Cygon) (1%), fenchlorphos (Ronnel) (1%) or rabon (1%), applied as soon as flies become numerous and continued as needed throughout the season.

Organophosphates may be applied as liquid or dry baits. They are formulated with an attractant (e.g. sugar) and placed where flies congregate (e.g. window sills, ledges, gutters etc.) but where livestock cannot get access to them. Compounds such as DDVP, diazinon, malathion and Negu-von are used in this manner.

Resin strips impregnated with DDVP or similar organophosphate are used extensively for housefly control. They are not effective where there is extensive air movement.

For larval stages of *M. domestica* diazinon or dimethoate is applied as a 1–2% solution to the breeding area at weekly intervals.

Musca autumnalis de Geer 1776 resembles *M. domestica*, is slightly larger and is the face fly of cattle and other animals. It is reputed to have been introduced to North America in the 1950s and is now widely spread throughout Canada and the USA. Economic losses ascribed to *M. autumnalis* are high (e.g. $68 million per year) (U.S. Department of Agriculture 1965). It is the intermediate host of the eyeworm *Thelazia rhodesii* and transmits infectious bovine keratitis (pinkeye) (*Moraxella bovis*). The flies gather around the eyes and nostrils, feed on ocular and nasal discharges and cause annoyance and irritation: horses huddle together in the shade, are nervous and interrupt their normal grazing behaviour. In cattle milk yield and growth may be interrupted (Depner 1969).

Eggs of *M. autumnalis* are deposited in fresh cattle faeces and larvae pupate in the soil or in dry dung. The adult fly may overwinter in farm buildings.

Dorsey (1966) assessed various control measures on quarter-horses in West Virginia. He found good to excellent control of face flies was achieved by the use of specially designed halters impregnated with dichlorvos.

Various organophosphate sprays can be used for the control of face flies. These include dichlorvos, crotoxyphos, coumaphos and Ronnel. The use of 'backrubbers' is indicated for face fly control. These consist of sacking wrapped round a wire cable or chain and suspended between two posts at a height allowing cattle to rub their backs and 'self-apply' insecticide. The insecticide is incorporated in in oil (e.g. diesel or mineral oil) and methoxychlor (5%), malathion (2%) and Ronnel (1%) are often used. Insecticidal dust bags, permitting self-application, serve the same purpose and may contain pyrethrins with synergist (0.75%), crotoxyphos (3% dust) or methoxychlor (2% dust).

Other species of the genus *Musca* may constitute livestock pests. They are not bloodsuckers, in that they do not pierce the skin and suck blood, but they feed from wounds and sores and add to the inflammatory response. However *Musca crassirostris* Stein, 1903 of the Mediterranean area is able to fold back the labella and use the prestomal teeth to rasp the skin and draw blood which is then sucked up. *Musca sorbens* Weidemann, 1830 (attracted to eyes and sores), *M. bezzii* Patton and Craig, 1913, *M. lusoria* Weidemann, 1824, *M. vetustissima* Walker, 1857 (Australian bush fly), *M. vitripennis* Meigen, 1826, *M. pattoni* Austen, 1910 and *M. fasciata* Stein, 1910 are species, which, in tropical and subtropical areas, follow the primary bloodsuckers. *Musca sorbens* may be responsible for the transmission of certain eye infections.

Genus: Muscina Robineau-Desvoidy, 1830

In these muscid flies the fourth wing vein curves gradually to the third and the scutellum is yellow at the apex.

Muscina stabulans (Fallén, 1817) is common around stables, byres, etc. The larvae occur in rotting fruits and carrion and may be found as the accidental cause of intestinal myiasis.

Muscina assimilis (Fallén, 1823) and **M. pabulorum** (Fallén, 1817) have similar habits but they seldom enter houses.

Genus: Morellia Robineau-Desvoidy, 1830

Non-metallic muscids, adults of which occur on flowers and vegetation but also are attracted to sweat and mucus of man, cattle and horses. 'Sweat-fly' species include *M. aenescens* Robineau-Desvoidy, 1830, *M. hortorum* (Fallén, 1817) and *M. simplex* Loew, 1857 and they may be very irritating in late summer. Developmental stages occur in dung.

Genus: Fannia Robineau-Desvoidy, 1830

Fannia canicularis (Linnaeus, 1761). This is a common species measuring 4–6 mm in length, greyish to almost black in colour and possessing three dark longitudinal stripes on the dorsal aspect of the thorax. The aristae are bare. The larvae breed in all kinds of decaying vegetable matter and refuse and are readily recognized by the dorsoventral flattening of the body and by the branched fleshy processes which project from the sides of the body. Larvae are often found in deep litter of chicken houses. Occasionally larvae are found in the discharges of the male or female genital organs and may lead to urinogenital myiasis.

Fannia scalaris (Fabricius, 1794), the latrine fly, resembles *F. canicularis* and the larvae have fleshy processes which are more feathered than in *F. canicularis*. This fly breeds in human faeces. It may be concerned with urinogenital myiasis in the same way as *F. canicularis*.

Fannia benjamini Malloch, 1919 is a North American species and may cause annoyance, being attracted by sweat and mucus.

Fannia australis Malloch, 1925, may be a tertiary striker in blow fly strike of sheep in Australia.

Genus Stomoxys Geoffroy, 1762

Stomoxys calcitrans Geoffroy, 1764 is the commonest species of this genus and is known as the 'stable fly'. It occurs all over the world. The flies are about as large as *Musca domestica*. The proboscis is prominent, directed horizontally forwards and has small labella. The M1 + 2 vein curves gently forwards and the R5 cell is open, ending at or behind the apex of the wing. The thorax is grey and has four longitudinal dark stripes, of which the lateral pair are narrow and do not reach the end of the scutum. The abdomen is shorter and broader than that of the house fly and has three dark spots on each of the second and third segments.

Life-cycle. Stomoxys sometimes lays its eggs in horse manure, but prefers decaying vegetable matter like straw and hay, especially when these are contaminated with urine. The material must be moist, otherwise it is unsuitable. A fly lays about 25–50 eggs at a time and may lay a total of 800. The eggs are dirty-white to yellow, about 1 mm long, and bear a longitudinal groove on one side. They hatch in one to four days or longer in cold weather. The larvae feed on the vegetable matter and in warm weather grow mature in 14–24 days. The full-grown larva resembles that of *Musca*, but its stigmal plates are far apart and each has three S-shaped slits.

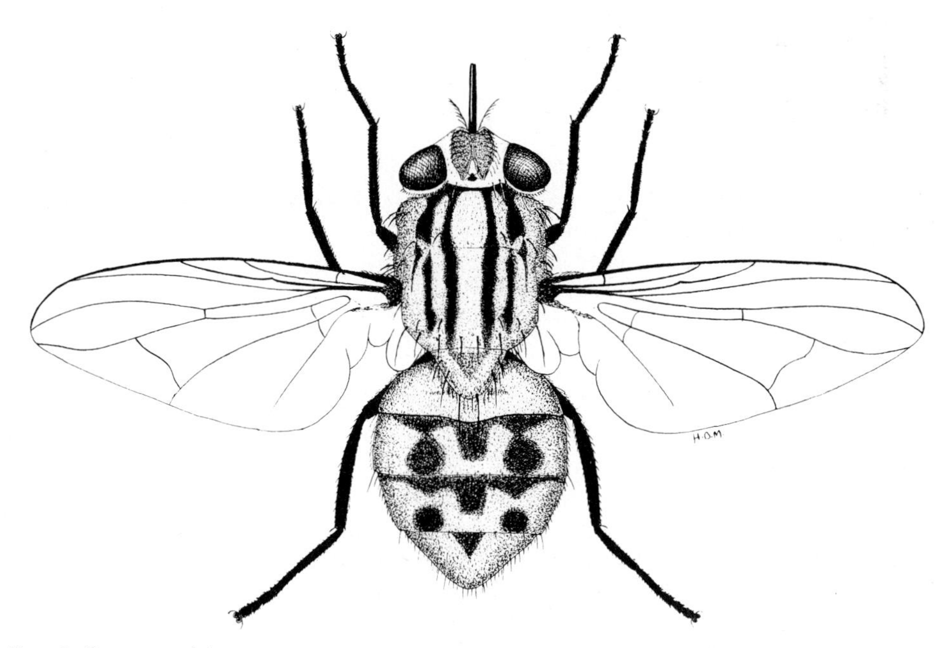

Fig 2.50 Female *Stomoxys calcitrans*.

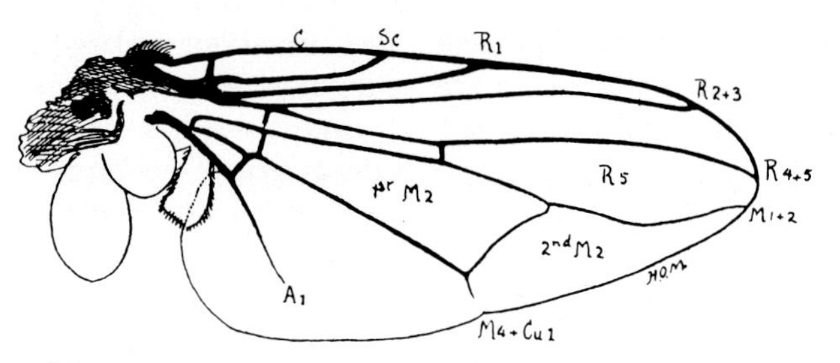

Fig 2.51 Wing of *Stomoxys calcitrans.*

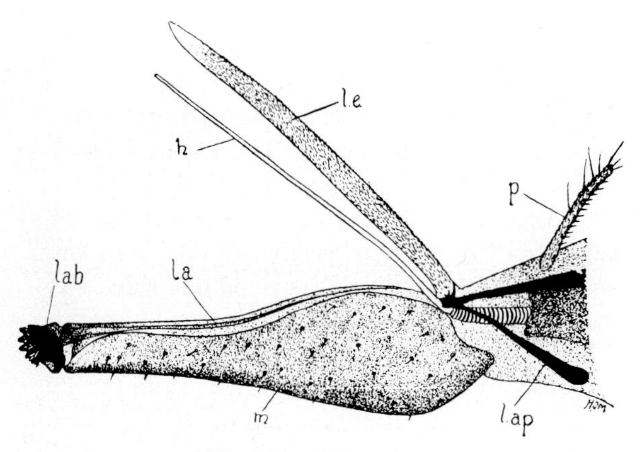

Fig 2.52 Mouth parts of *Stomoxys calcitrans.*

h	hypopharynx
la	labellum
lab	labella with teeth
l ap	labral apodeme
le	labrum-epipharynx
m	mentum
p	palp

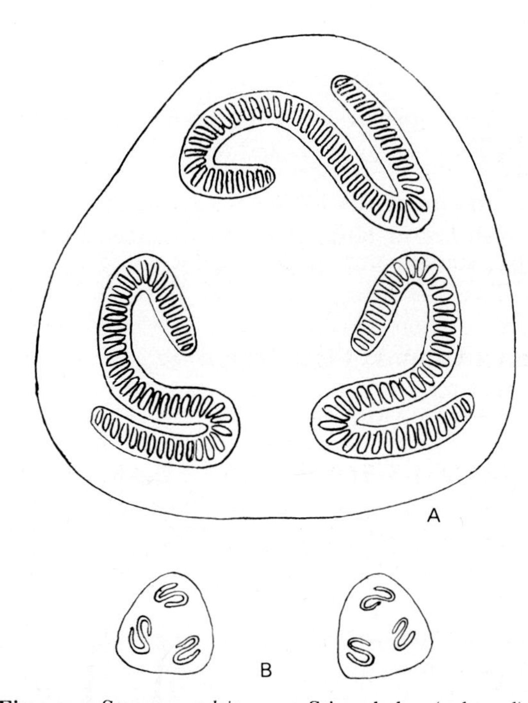

Fig 2.53 *Stomoxys calcitrans.* a, Stigmal plate (enlarged). b, Relative distance between stigmal plates.

Pupation takes place in the drier parts of the breeding material and this stage lasts about six to nine days, or much longer in cold weather. Oviposition begins about nine days after emergence of the fly and after a few meals of blood have been taken. The complete life-cycle may take about 30 days.

Habits and significance. The flies are most abundant in summer and autumn and live about a month under natural conditions. They prefer a fairly strong light and are not seen in dark stables or houses. They enter buildings only in autumn or during rainy weather. They are swift fliers, but do not travel long distances. Both males and females are blood-suckers, attacking man, horses, cattle and other mammals, and even birds and reptiles. About three to four minutes are required for a meal and the fly often changes its position or flies to another animal to continue its feed.

Trypanosoma evansi (surra of equines and dogs) and *T. equinum* (mal de caderas of equines, cattle, sheep and goats) are transmitted mechanically by *Stomoxys.* The species may also mechanically transmit *T. gambiense* and *T. rhodesiense,* the causes of human trypanosomiasis in Africa, and *T. brucei* and *T. vivax*, which cause nagana of cattle, sheep, goats and equines in Africa. It also serves as an intermediate host of the nematode *Habronema majus.*

The role of *S. calcitrans* in the transmission of equine infectious anaemia is still under debate (Steelman 1976). However, the fly is responsible for the mechanical transmission of septicaemic infections such as anthrax. Its nuisance effect has been estimated at $142 million in the USA (Steelman 1976) and Cheng (1958) has estimated that control measures were economical when levels of 25 stable flies per animal per day were reached in beef herds. The importance of biting flies in the transmission of disease and economic loss through 'fly worry' has been reviewed by Stork (1979).

Control. The fly is most troublesome in localities where suitable breeding-places are readily found. Control measures should therefore be directed toward destroying breeding-places by regular removal of moist bedding, hay and faeces from stables and yards, and food wastes from feeding troughs, and by preventing the accumulation of heaps of weeds, grass cuttings and vegetable refuse. Control by insecticides is similar to that for house flies.

Genus: Hydrotaea Robineau-Desvoidy, 1830

Adults of this genus are found in the open on flowers, foliage and decaying matter. Females of the genus may occur in swarms around animals or man. They are attracted by blood and serum from wounds and mucous secretions from the nose, eyes and mouth.

Hydrotaea dentipes Fabricius, 1805 is commonly found on dung and it may enter human or animal housing. Its larvae are parasitic on the larvae of other diptera.

Hydrotaea meteorica Linnaeus, 1758 is a persistent sweat fly of horses and man; its larvae occur in decomposing vegetation. **H. occulta** Meigen, 1826 and **H. albipuncta** (Zetterstedt, 1826) also are troublesome sweat flies in temperate regions.

Hydrotaea irritans Fallen, 1823, the sheep head fly, is a non-biting muscid resembling the housefly, 4–7 mm in length. The thorax is black with grey patches, the abdomen is olive green and the wing bases are orange-yellow. The fly is associated with woodland and plantation and is referred to as the forest or plantation fly. It is widespread in Europe (Great Britain, Denmark, Germany, Sweden). In Great Britain, the fly causes damage to sheep chiefly in the northern counties of England and southern counties of Scotland, though sporadic cases occur in southern England.

Eggs are deposited on decaying vegetation, including manure, and larvae are found in similar material.

The fly is active at the edges of woods and plantations in which the flies shelter. Flies are not active inside woodland and are not found in the open unless animals are present. They are active on warm, calm and sunny days; light intensity may be a limiting factor on activity and this may account for the absence of sheep head flies inside buildings (Hunter 1975). They occur from June to September and generally reach a peak in numbers in July to August.

Flies are attracted to animal movement and wounds. Female flies require protein for ovarian development and seek blood and exudate fluids for this. The flies have well developed prestomal teeth and can penetrate the skin by a scraping action of the proboscis (Tarry 1975). Tarry and Boreham (1977) report that 75% or more of sheep head flies feed on cattle blood whereas only 15% or less feed on sheep blood. Lower feeding levels occurred in flies collected in plantations indicating an extended resting phase, during which a blood meal digestion process of extended duration occurs (Tarry & Boreham 1977).

Sheep head flies cause damage by feeding with other flies on secretions from the eyes and nose of sheep and cattle. Large swarms may occur around an animal and the annoyance and/or irritation associated with this leads to sheep rubbing or knocking against hard objects, enough to cause wounds ('broken heads'). These attract more flies which create additional damage. Secondary bacterial infection may occur and also the lesions may lead to blowfly strike. Sheep of all breeds are attacked but breeds with horns and without wool on the head are most severely affected (Hunter 1975). Lambs are more severely affected than ewes and the number of flies and the disease problem increases with flock size.

As well as attacking sheep, flies often are found on cattle and may attack deer (Hunter 1975). The economic effect of head flies may be marked. Losses in weight gain of 5–10 kg have been associated with interruption of feeding (Hunter 1975). The scarring and disfigurement (e.g. loss of horns) reduce the market value of animals, especially pedigree animals.

The role of *H. irritans* in the transmission of causative agents of summer mastitis was investigated by Wright and Titchener (1977). Adult flies are readily infected with *Corynebacterium pyogenes* and *Streptococcus dysgalactiae* and these bacteria have been recovered from head flies under field conditions. Infected flies can transmit the organisms to blood agar plates, suggesting a role, probably mechanical, for them in summer mastitis. Additional evidence for this was obtained by Tarry et al. (1977) who showed that *C. pyogenes* could be transmitted from a field case of mastitis to normal animals by *H. irritans*. Other flies (e.g. *S. calcitrans* and *Aedes* and *Culex* mosquitoes) were not associated with transmission.

Prevention consists of using canvas head caps, which give good to complete protection (French et al. 1977) or repellants which have had varied efficacies. Crotoxyphos (dimethyl phosphate of alpha-methylbenzyl 3-hydrog-*cis*-crotonate) (Ciodrin, Ciovap, etc.) as a cream (0.05%) has produced good control, a 1% spray of crotoxyphos over the head and body at two-week intervals, a dip of 0.125% 1,1 *bis* (*p*-ethoxyphenyl)-2-nitroprone and a dip or spray of 0.05% to 0.1% permethrin may also be useful in the control of head fly.

As a longer-term measure French et al. (1977) indicate that breeds of sheep with wool on the head are not damaged by head flies and selective breeding for this may reduce the effects of the flies.

Genus: Haematobia Robineau-Desvoidy, 1830

The taxonomic status of species within this genus and the relationship of the genus to the genus *Lyperosia* Rondani, 1856 (syn. *Siphona*) are complex and controversial. In this edition *Lyperosia* is considered a synonym of *Haematobia* though Richards and Davies (1977) recognize both genera as valid.

Several species of this genus may be encountered in various countries and are serious pest of cattle. The flies are some of the smallest of the blood-sucking muscids, measuring about 4 mm in length. The face is silvery-grey, the thorax silvery-grey medially and dark laterally with two well-defined dark stripes. The wing venation is similar to that of *Stomoxys calcitrans*. The palps are yellowish, stout, of uniform thickness and as long as the proboscis. The aristae are haired on the dorsal surface only.

Haematobia exigua (De Meijere, 1903) (syn. *L. exigua*) is the buffalo fly of India, Malaysia, China and Australia, being distributed in the northern part of the latter country. The fly feeds on buffaloes and cattle chiefly and rarely leaves the host except for a brief flight when disturbed, to transfer to another host or to lay eggs. Several thousand flies may occur on cattle, especially bulls. Weight gain and milk production are severely interferred with by heavy infections. This species is capable of transmitting *Trypanosoma evansi*, the cause of surra: it is also an intermediate host for the spirurid *Habronema majus*.

Haematobia minuta Bezzi, 1892, occurs in Africa and has similar habits to the above. Lloyd and Dipeolu (1974) recorded it as a common blood-sucking muscid of cattle in northern Nigeria.

Haematobia stimulans (Meigen, 1824). Smart (1939) states this fly to be generally distributed in the British Isles and its range does not seem to extend beyond Europe.

Haematobia irritans (Linnaeus, 1758) is the 'horn fly'. They may be found in thousands around the base of the horns and also on the back, shoulders and belly of cattle. They occasionally attack horses, sheep and dogs. The fly remains on the animal and leaves it only to pass to another host or to lay eggs when the animal defaecates. It is a blood-sucker and heavy burdens cause injury and irritation due to the constant piercing of the skin.

The animal may develop sores and wounds and these attract the screw-worm fly. Steelman (1976) gives the estimated loss to cattle production in the USA caused by the hornfly as approximately $180 million in 1965. Horn fly control results in significant increases in milk and beef production.

The fly transmits the filarid *Stephanofilaria stilesi*, a parasite of the skin of cattle.

Favourable climatic conditions for horn flies consist of hot, humid weather; hot, dry weather or cold weather is unfavourable. The preferred macroclimate is a temperature of 23–27°C, a relative humidity of 65–90% with scattered light showers and no wind. Within the mantle of the micro-environment of the skin of an animal the flies prefer an air temperature of 29.5°C, a skin temperature of about 36°C and a relative humidity of 65%. Such a micro-environment was most commonly found in the Holstein and there was a significant difference between the number of flies on animals of this breed than on Guernsey or Jersey heifers. The flies prefer the dark-coloured areas of bicoloured cattle during the daylight hours, the black of the Holstein being preferred to the tan of the Guernsey. When the macro-temperature is about 29.5°C many flies are found also on the white skin of the belly and udder areas (Morgan 1964).

Life-cycle. The eggs are laid in the fresh dung of cattle (and buffaloes). They are 1.3–1.5 mm long and hatch in about 20 hours at a temperature of 24–26°C. Lower temperatures retard or arrest development and the eggs are rapidly killed by drying. The larvae burrow into the dung and feed on it, growing mature in about four days at a temperature of 27–29°C. Lower temperatures prolong development considerably. A fair amount of moisture is also required; 68% of free water is optimal for the larvae and development ceases if the moisture falls below 50%. The pupae require about the same temperature as the larvae. The pupal stage lasts six to eight days.

The flies are not inclined to fly about, but remain on their hosts for several days, feeding at intervals or darting down to lay eggs when the animal has defaecated. They are spread, therefore, chiefly by their hosts.

Control. Since horn flies remain for long periods on the animal, control of the adult is relatively easy. Regular spraying with organophosphate insecticides achieves good control. Thus 0.5% methoxychlor (Co-Ral), 0.25% carbaryl or 1% Ciodrin may be used every three weeks; however, the appropriate withdrawal time prior to slaughter should be observed.

Self-application by the use of 'back-rubbers' or 'dust bags' usually achieves good control also. Back-rubber solutions include 5% methoxychlor, 2% malathion, 1% ronnel or oil-based synergized pyrethrins in fuel oil or kerosene. Dust bags are suspended in pens, gangways near feeding troughs etc. Methoxychlor dust (2%) at the rate of one 6 kg bag for every 10–15 animals every four weeks and 3% crotoxyphos in 1.5 kg bags provide good control. Good control of *S. calcitrans* and *H. irritans*, with increased milk production, was achieved in cattle by spray washes containing 0.05–0.1% permethrin (Bailie & Morgan 1980).

Since larvae of the fly breed in fresh cattle dung, the feeding of insecticides and the resulting residuum in the faeces gives good control for horn flies.

Insecticide resistance in horn flies has appeared in recent years and this necessitates the use of alternative insecticides.

FAMILY: GLOSSINIDAE (COCKERELL, 1908) MALLOCH, 1929

Genus: Glossina Wiedemann, 1830

This genus previously has been included in the family *Muscidae*. It is now usual to regard the genus as the sole constituent of the family *Glossinidae*; this has been discussed by Pollock (1971).

The species of the genus *Glossina* or tsetse flies are important blood-sucking flies; they transmit several species of trypanosomes which cause fatal diseases, for example, nagana of domestic animals and sleeping sickness of man. The genus is confined to subtropical and tropical parts of Africa (lat. 5°N to 20°S). *Glossina tachinoides* was

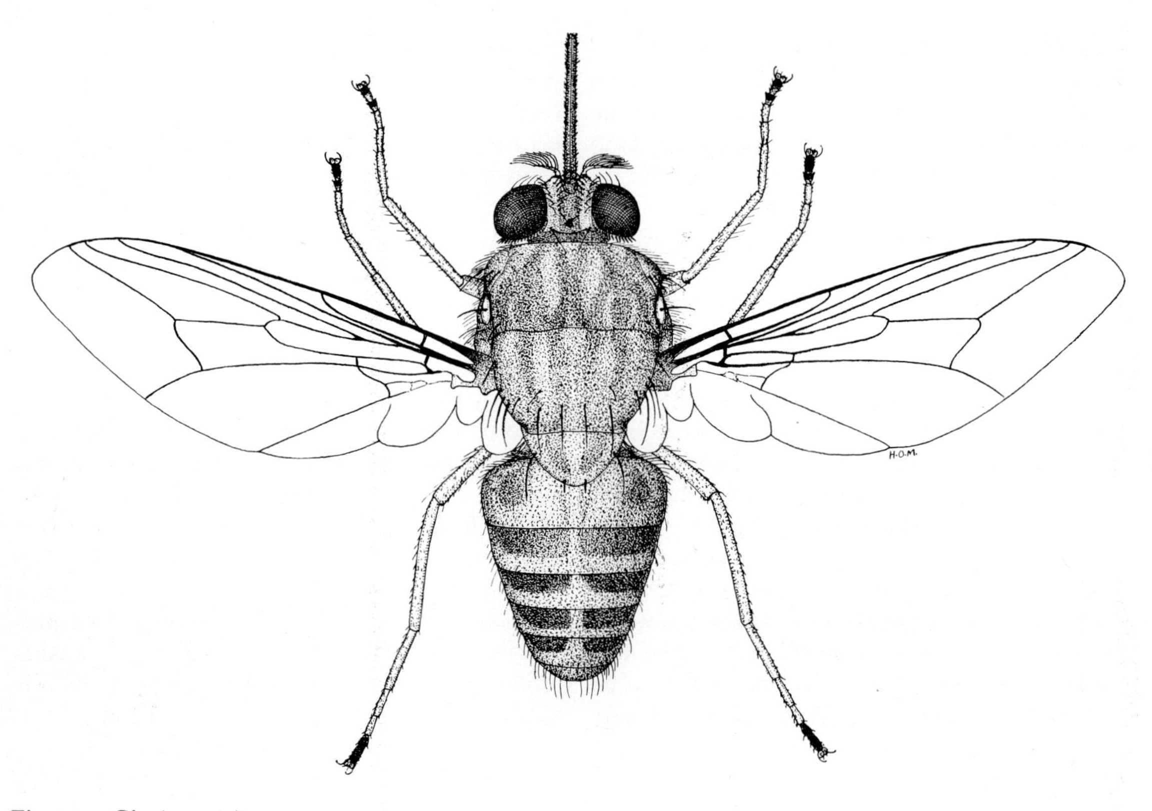

Fig 2.54 *Glossina morsitans.*

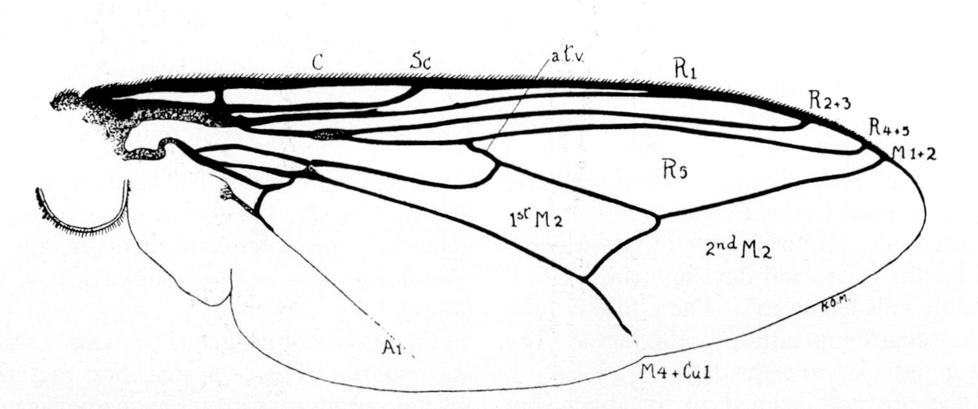

Fig 2.55 Wing of *Glossina pallidipes.*

C	costa	M	median
Sc	subcosta	Cu	cubital
atv	anterior transverse vein	A	anal
R	radial		

recorded in southern Arabia in 1910 but has not been found there again. The flies are narrow-bodied, yellowish to dark brown and 6–13.5 mm long. At rest the wings are held over the back, overlapping almost completely scissor-like. The thorax frequently has a dull greenish ground colour and is marked with inconspicuous stripes or spots. The abdomen is light to dark brown, and six segments are visible from the dorsal aspect. The venation of the wing is very characteristic, especially the course of the M1 + 2 vein which produces the hatchet- or cleaver-shaped discal cell (first M2 cell). The proboscis is long, held horizontally and ensheathed in long palps which are of an even thickness throughout. The antenna has a large, elongate third antennal segment which ends in a blunt, forwardly directed point, and an arista which bears 17–29 dorsal branching hairs.

A description of the different species of *Glossina* goes beyond the scope of this book. They are described by Buxton (1955) and Potts (1973) who provides a key for the identification of species.

Life-cycle. The female fly produces one larva at a time when the latter is full grown and ready to pupate. The larva grows in the uterus of the female, its mouth being attached to a 'teat' from which 'milk' is obtained for nourishment; its posterior extremity, which bears the stigmal plates, lies near the vulva. The gestation period lasts about ten days under suitable conditions, but

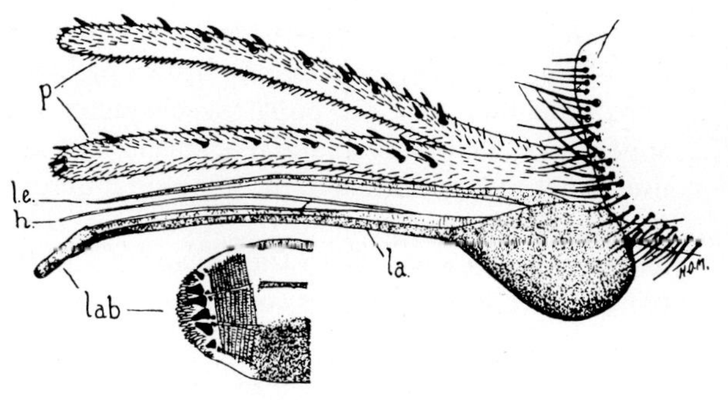

Fig 2.57 Mouth parts of *Glossina pallidipes.*
h hypopharynx
la labium strengthened by mentum
l e labrum-epipharynx
lab labellum with enlarged medial view showing armature
p palps

is prolonged if food is scarce or in cold weather when the flies do not feed readily. It is estimated that a female can produce eight to ten, sometimes up to 12 larvae and one act of mating renders a female fly fertile for life.

The larva is an oval, about 7 mm long. It wriggles into the soil to a depth of about 2 cm and turns into a pupa after 60–90 minutes. The larva has two large respiratory lobes at its posterior end, each perforated by about 500 spiracular openings.

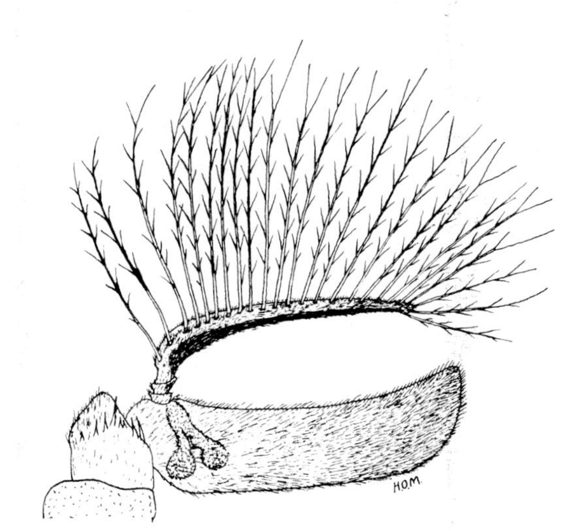

Fig 2.56 Antenna of *Glossina pallidipes.*

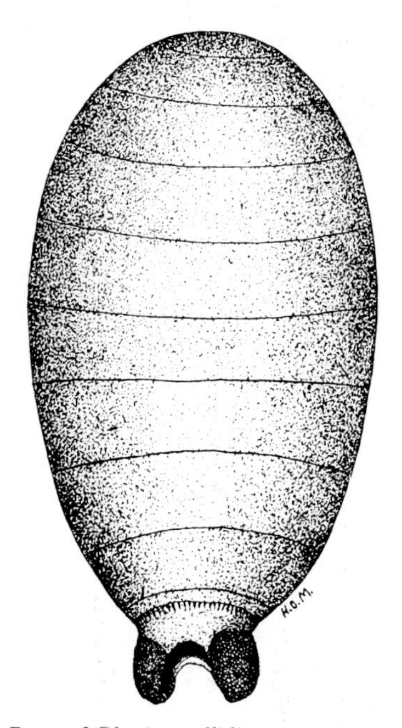

Fig 2.58 Pupa of *Glossina pallidipes.*

These lobes persist with the last larval skin, which forms the puparium covering the brown or black pupa, and they give the pupa the characteristic appearance. They vary in different species and are used for the identification of the pupae of different species. The pupa is about 6–7 mm long. The length of the pupal period varies according to the temperature and the species concerned. It usually lasts about 35 days, with limits of about 17–90 days. For example, the pupal period of *G. pallidipes* varies from 31 days in the summer to 149 days or more, usually about 92 days, in the winter.

Habits. The bionomics of tsetse flies, especially of certain species, have been studied intensively for many years and there is an extensive literature on the bionomics and biology of the fly (see Buxton 1955; Mulligan 1970; Potts 1973).

Tsetse flies are found mainly in the central part of Africa, extending from the southern boundaries of the Sahara to southern Africa (e.g. Zimbabwe). In these regions the flies are confined to definite areas known as 'fly-belts', the limits of which are controlled by various factors such as altitude, moisture, vegetation and the presence of hosts. The different species vary greatly in their adaptation to environment and consequently in their distribution. Potts (1973) gives the distribution of *Glossina* species along with the northern and southern limits in West Africa, Central Africa and East Africa for the various species. For example, *G. palpalis* occurs mainly in the areas drained by the Senegal, Niger and Congo rivers; *G. morsitans* occurs from Senegal to Ethiopia and extends south to Zimbabwe. *G. pallidipes* is essentially an East African species and is found from the Republic of South Africa to Uganda and Kenya. *G. tachinoides* can live at higher temperatures than most other species and is found in hot regions like northern Nigeria. More extensive information on the geographical distribution is given by Ford (1970*a*).

The humidity of the atmosphere, the temperature and the presence of shade have an important bearing on the life of the fly. *G. palpalis* requires an almost saturated atmosphere and much shade and is therefore found near water, especially along the banks of rivers or lakes surrounded by overhanging trees or bushes. It is killed within a short time by direct sunlight and by temperatures over 30°C, especially if the humidity is not high. Its natural range from water is about 30 m, but the fly will follow a host for 300 m or sometimes more away from water. *G. morsitans* and *G. pallidipes* are much less restricted to moisture and shade conditions and are most active in a moderately dry and warm climate. They occur in open 'parkland' type of vegetation. *G. pallidipes* requires a moderate degree of humidity and is less independent of cover than *G. morsitans*. Both these species require trees or scrub for shelter and the former especially does not venture far into open country.

The different species are each particularly associated with certain types of vegetation, an important factor which requires much further study. The vegetation as well as other controlling factors like the presence of water restrict the flies during bad seasons to certain areas known as 'primary fly centres' or permanent haunts, from which they migrate outwards along suitable courses to temporary haunts during favourable seasons. *G. palpalis* for instance, will ascend to the upper limits of rivers in rainy seasons and again descend when dry conditions set in. The physical, vegetational and animal influences, as well as population dynamics, are discussed by Glasgow (1970).

All the species of *Glossina* feed on the blood of vertebrates. The hosts of *Glossina* species are discussed by Weitz (1970). Some species of hosts are more suitable than others and the prevalence of flies is dependent on the number and suitability of hosts. *G. palpalis* thrives best on the blood of warm-blooded animals and is stated to prefer human blood, but it can also feed on cold-blooded animals like crocodiles. It is generally agreed that this species is not dependent on big game for its existence. *G. morsitans* feeds on any large mammal or bird, but is not able to exist permanently where big game or cattle are absent. This fly disappeared from the Transvaal with the big game which was killed out by rinderpest. Occasionally *G. morsitans* has been found where big game was scarce, but since the pupal period is long, such flies may have emerged some time after the main stock of flies left the particular area. *G. palpidipes* is similarly dependent on big game or cattle for its existence, but warthogs are an important source of food for both species.

The flies fly low near the ground and hunt by sight. They are susceptible to the effects of light and shade and are attracted especially by moving objects. As a rule they prefer to bite on dark surfaces, probably being attracted by them just as they are attracted by a shadow. While it is being attracted by or following a host, the fly will leave its natural haunts and fly for variable distances, depending on the species of fly and the prevailing conditions. *G. palpalis* is attracted to boats moving on a river and *G. morsitans* may attack animals or human beings several hundred metres away from its bush cover. In this way the fly may reach new sites after leaving its host or after giving up the chase. When the host stops moving, the fly, especially *G. palpalis*, is apt to leave. The search for food by the tsetse fly has been discussed by Ford (1970*b*).

The flies feed about every three days, depending on the temperature and humidity. In a humid atmosphere longer starvation is possible, while a higher temperature shortens the interval. Most species are active in the forenoon and afternoon, disappearing during the hottest hours of the day. Some, like *G. brevipalpis*, are nocturnal in their habits, feeding especially on moonlight nights, but others may attack hosts around a camp fire at night. Rain or windy weather causes the flies to remain in shelter.

Feeding, digestion and excretion of *Glossina* species are reviewed by Bursell (1970). As the fly engorges, its abdomen becomes much distended. Finally the fly leaves its host and settles on the under-side of a leaf or a log or on the bark of a tree to rest. Soon a drop of dark fluid is voided through the anus, then several drops of clear fluid, and sometimes also a drop of fresh blood, which then fills the gut.

Towards the end of the gestation period the females do not feed but remain in shelter. Unsuitable food or starvation may cause the female to abort; occasionally pupation occurs in the uterus and is invariably fatal to the fly.

The breeding-places of tsetse flies are carefully chosen and are restricted by various factors. *G. palpalis* deposits its larvae usually not more than 25 m from water and about 1 m above water-level in dry coarse sand or in humus around tree trunks, in forks of branches and cracks of bark up to 4 m high above the ground. *G. morsitans* and *G. pallidipes* breed in loose, sandy soil rich in humus on sheltered and well-drained spots, usually near game paths where the flies abound. The pregnant female about to deposit a larva is attracted by objects which provide shelter, like fallen tree trunks or slanting rocks, and deposits the larva underneath them. Shade is essential to the pupae even though they lie covered in the soil. A few hours of sunshine per day rapidly kills them.

Control

Recent developments in techniques for tsetse control are reviewed by Jordan (1978). Almost all present-day control methods depend on the use of insecticides and older methods such as bush-clearing, slaughter of game, etc. are not dealt with here in detail.

Catching and trapping. Catching with hand nets is not practicable except in small areas. It is a method of determining the fly density in a given locality. Numerous types of traps have been designed, mainly on the principle that the fly is attracted by a horizontal shadow and flies in underneath it.

Bush-clearing. In the case of flies like *G. palpalis*, which require much shade, this measure is important and the clearing of bush around settlements, wells, landing-places and fords on rivers and on either side of roads has given excellent results in providing protection from the fly. Wholesale clearing of large areas, however, presents many difficulties, especially the expense incurred and the rapid regrowth of the bush.

Destruction of big game is designed to remove the food of the flies and also to break up their breeding areas. Removal of the big game (antelope) controlled *G. morsitans*, a species which prefers to feed on antelope and other big game, but it failed to control *G. pallidipes*, probably because this species feeds not only on big game, but also on bush pig, bush buck and other smaller hosts.

Fly screens, repellents and similar devices, designed to prevent the flies gaining access to their hosts, may be tried but they can only be, at most, partially successful.

Insecticides. The majority of present-day control programmes depend on the use of insecticides. This is discussed by Jordan (1978). Persistent insecticides, such as DDT and dieldrin, applied from the ground, have provided useful control. In the Sudan and the savanna vegetation zone of Nigeria 2.5% DDT wettable powder applied in the dry season achieved eradication of *G. submorsitans*, *G. tachinoides* and *G. palpalis*. Because of the susceptibility of *Glossina* species to these insecticides, discriminative application and selective spraying are effective methods of use. Thus with *G. tachinoides* spot spraying of riverine vegetation up to 70 cm from the ground is sufficient to give control, while spraying up to 1 m gives control of *G. palpalis*.

Where the seasonal factors vary, application techniques for insecticides also must be modified and are less discriminative than above. However, using the ground spray technique, Nigerian authorities have reclaimed some 180 000 km^2 of tsetse-infested country (Jordan 1978).

The application of insecticide from the air, usually using helicoptors rather than fixed-wing aircraft, has been practiced in East and West Africa with good results. Spraying is limited to the dry season and is carried out when temperature inversion conditions exist. This limits application to a period of one to two hours in early morning or at dusk. The technique of spraying (e.g. speed, height of the helicoptor above the canopy) and the formulation of insecticide, usually dieldrin, is discussed by Jordan (1978).

Non-persistent insecticides applied from the air, often by fixed-wing aircraft, at repeated intervals (five to six), coinciding with the length of the pupal period, have produced useful control. Endosulfan as an ultra-low-volume formulation is used most extensively, while isobenzan has also been used as a non-persistent insecticide.

No resistance to insecticides by *Glossina* species has been reported to date.

Other methods of control. Jordan (1978) considers genetic and physiological approaches in this respect. With the former, the use of sterile insect release has been studied in detail. Jordan concludes that logistic problems associated with the production of sufficient numbers of sterile males for release make it unlikely it will be a technique for tsetse control to be used on its own. However, its success has been demonstrated in several limited areas of Africa and it may well be a useful approach to eliminating small residual populations of *Glossina*. The physiological approach is to interfere with the normal function of the fly. While a number of substances will cause perturbations in the adult fly or pupa when injected into or topically applied to the fly or its pupa, under normal conditions there remains the problem of getting appropriate contact by the substance. Recent studies of the factors determining attractiveness of a host for a fly may assist in this, since attraction of flies to a stationary source would permit exposure to a compound affecting a fly's physiology or to ionizing radiation.

FAMILY: CALLIPHORIDAE (BRAUER AND VON BERGENSTAMM, 1889) TOWNSEND, 1915

This family includes a large number of species whose larvae are saprophagous, flesh feeders or parasites of other arthropods. They are usually bristly and are characterized by the presence of a row of bristles on the hypopleuron—the hypopleural bristles—which are placed like a screen on either side, in front of the metathoracic spiracles.

The family contains two subfamilies of importance, Calliphorinae and Sarcophaginae. The Calliphorinae are the blowflies and are often metallic blue or green in colour. Several species are of veterinary importance. The Sarcophaginae includes the flesh flies which have a grey longitudinal striped thorax with a check-board marked abdomen.

SUBFAMILY: CALLIPHORINAE

Genus: Lucilia Robineau-Desvoidy, 1830 (syn. Phoenicia)

This genus contains the most important blowflies **L. cuprina** Meigen, 1826 and **L. sericata** Weidemann, 1830. The larvae of *L. cuprina* are the chief cause of blowfly strike of sheep in Australia and the Republic of South Africa; those of *L. sericata* are its chief cause in Britain. Larvae of *L. sericata* have also been found in human wounds. Other species of interest are **L. caesar** Linnaeus, 1758 and **L. illustris** Meigen, 1826 which are cosmopolitan in distribution. These flies have bright metallic colours, being bright green or a bronze colour in some kinds of lighting. They are called green-bottle or copper-bottle flies. The eyes are brownish-red. The body is relatively slender, 8–10 mm long. It is difficult to distinguish the two major species, but with *L. sericata* the legs are black whereas in *L. cuprina* the femora of the first pair of legs are bright green.

Genus: Calliphora Robineau-Desvoidy, 1830

Species of this genus are often called blue-bottle flies, the body having a metallic blue sheen.

Calliphora erythrocephala Meigen, 1826 (syn. *C. vicina*) is a large, stoutly-built fly about 12 mm long, which buzzes loudly when it flies. The eyes are red and the genae are red with black hairs. **C. vomitoria** Linnaeus, 1758 is similar, but the genae are black with reddish hairs. Both species strike sheep in England. **C. stygia** Fabricius, 1794, **C. australis** Boisdard, 1825, **C. nociva** Hardy, 1937, **C. augur** Fabricius, 1794 and **C. fallax** Hardy, 1937 strike sheep in Australia.

Microcalliphora varipes (Macquart, 1835) and species of **Sarcophaga** Meigen, 1826 are also sheep blowflies in the Antipodes.

Genus: Phormia Robineau-Desvoidy, 1830

Species of this genus are sometimes called black blowflies.

Phormia regina Meigen, 1826 deposits its eggs in the wool of sheep in the United States. Its thorax is black, with a metallic blue-green sheen, the abdomen blue-green to black, the fly being 6–11 mm long.

Phormia terrae-novae Robineau-Desvoidy, 1836 strikes sheep in Britain and northern Canada, but in Canada *P. regina* is the more important cause of strike.

Genus: Chrysomyia Macquart, 1885

Chrysomyia chloropyga Wiedemann, 1818 and **C. albiceps** Wiedemann, 1819 are sheep blowflies in South Africa; **C. rufifacies** Macquart, 1843, the hairy maggot fly, and **C. micropogon** Bigot, 1860, the steel-blue blowfly, are sheep blow-flies in Australia; **C. bezziana** (Villeneuve, 1914), the Old World screw-worm fly, occurs in Africa and southern Asia. It is a medium-sized stout, bluish-green fly, with four black stripes on the prescutum; the face is orange-yellow. In India it may lay eggs on the skin of man and domesticated animals.

CALLIPHORINE MYIASIS OF SHEEP

This disease, commonly called 'strike', may be caused by the larvae of various species belonging

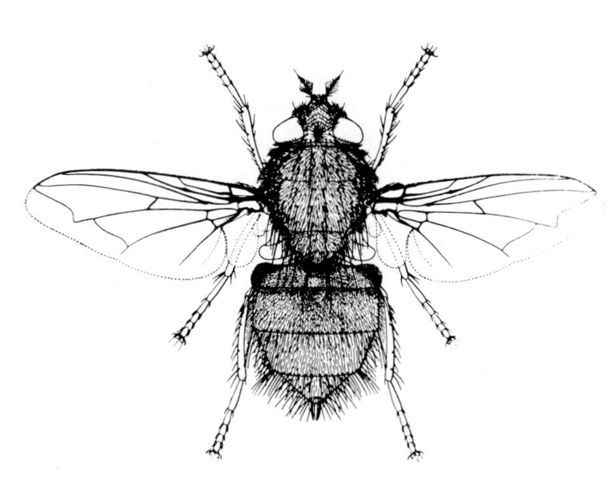

Fig 2.59 *Calliphora erythrocephala.* (*From Smart*)

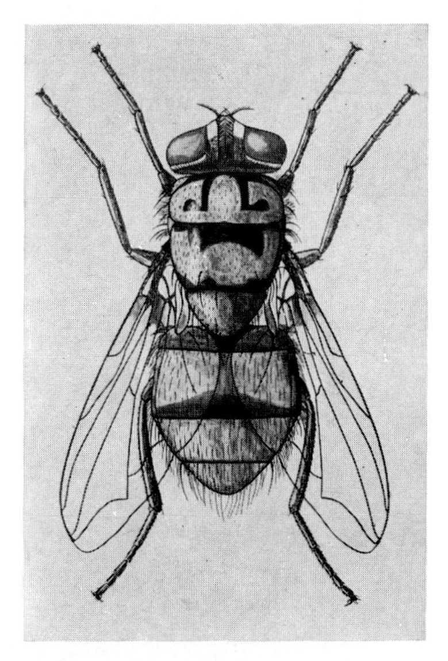

Fig 2.60 *Chrysomyia chloropyga.*

to the genera *Lucilia, Calliphora, Phormia* and *Chrysomyia*. The condition should be distinguished from the form of myiasis caused by the larvae of the screw-worm flies belonging to the genera *Callitroga* and *Chrysomyia*. The latter is described in more detail on p. 426.

Life-cycle of the species causing strike

The flies lay clusters of light yellow eggs in carcasses, wounds or soiled wool, being attracted by the odour of decomposing matter. While the fly is selecting a suitable spot to lay, it feeds on the moist matter present. A female blowfly lays about 1000–3000 eggs altogether and 50–150 in one batch. A meal of protein is required before the ovaries reach full maturity.

The larvae hatch from the eggs in eight hours to three days, depending on the temperature, and begin to feed. They grow rapidly and pass two ecdyses, becoming full-grown maggots in about two to 19 days. The rate of growth depends on the amount and suitability of food, the temperature and the degree of competition with other larvae. The mature larvae roughly resemble those of *Musca*. They are about 10–14 mm long, greyish-white or pale yellow in colour, sometimes with a pink tinge. The anterior extremity bears a pair of oral hooks and on the broad, flattened, posterior end the stigmatic plates are situated. The stigmatic openings consist of three long slender slits, more or less parallel to each other, on the spiracle. The second segment bears a pair of anterior spiracles as in *Musca*.

Two groups of larvae can be recognized: 'smooth' larvae and 'hairy' ones. The 'hairy' larvae bear a number of thorn-like, fleshy projections with small spines at their tips on most of the body segments. The larvae of *Chrysomyia rufifacies, Chrysomyia albiceps* and *Microcalliphora varipes* are 'hairy'; those of *Chrysomyia micropogon* and the other sheep blowflies mentioned above are smooth. The different species of larvae can be further differentiated by means of the structure of their spiracles and the cephalopharyngeal skeletons (see Zumpt 1965).

The mature larvae usually leave the host or the carcass to pupate in the ground, but some may pupate in dry parts of a carcass or even in the wool of the live animal. Before pupating the larva may wander fair distances over or through the soil and most species usually pupate below the surface. If conditions are not favourable, as in cold weather, pupation can be suspended for months and the larvae may hibernate. At pupation the larva loosens its skin, which turns brown and becomes rigid to form the puparium. The pupa therefore bears some resemblance to the last larval stage; the pupae of the 'hairy' maggots retain the projections on their covering. For the rest the

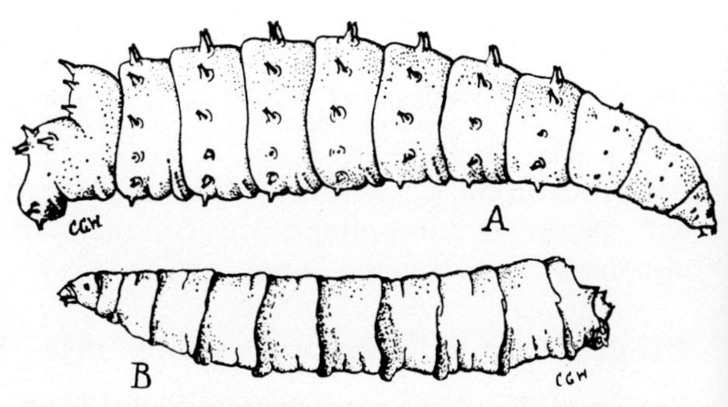

Fig 2.61 A, Larva of *Chrysomyia albiceps*. B, Larva of *Lucilia sericata*. × 3.

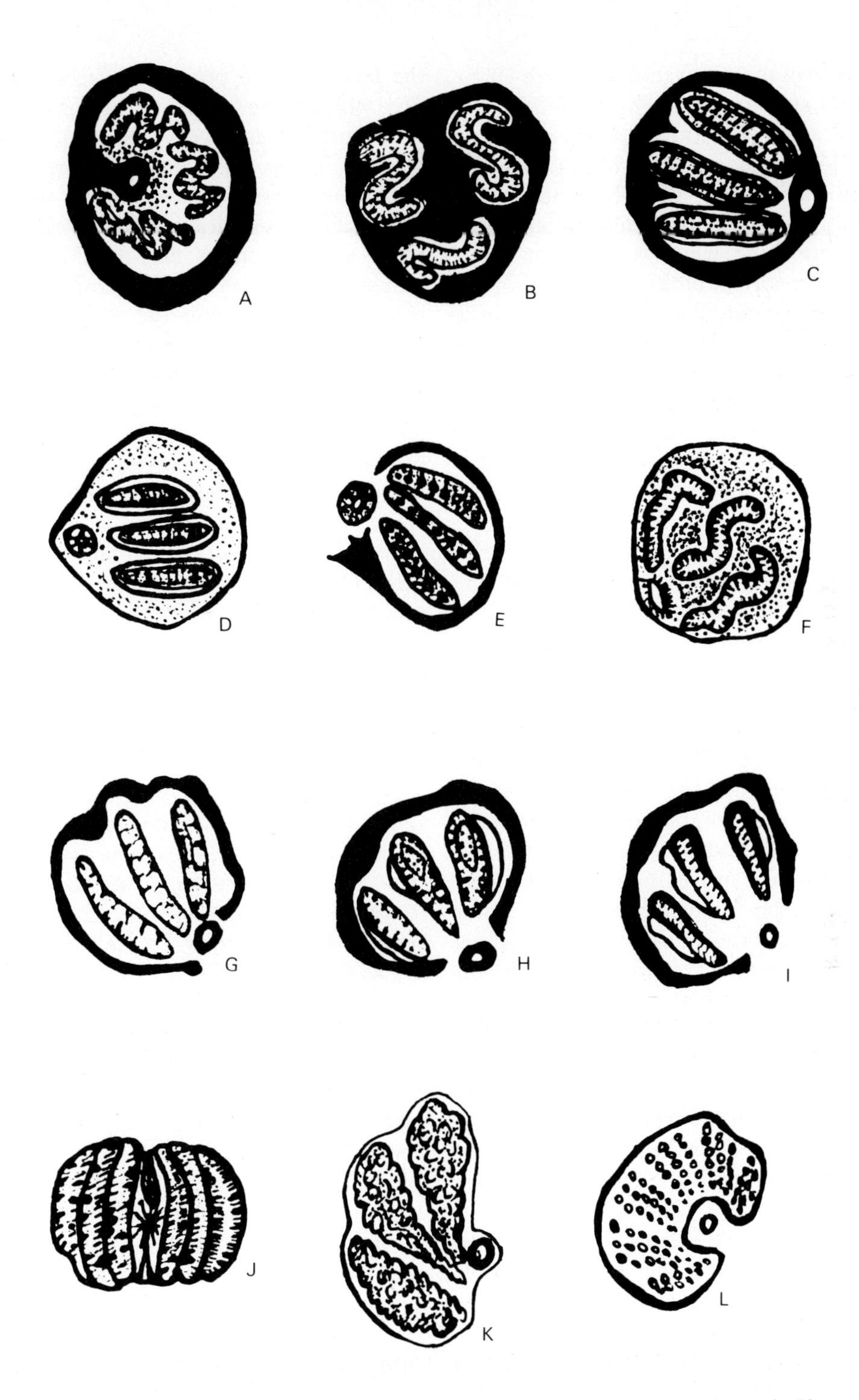

Fig 2.62 Posterior stigmal plates of some myiasis-causing larvae (Redrawn from several authors). A, *Musca domestica*. B, *Stomoxys calcitrans*. C, *Calliphora* spp. D, *Auchmeromyia luteola*. E, *Wohlfahrtia* spp. F, *Cordylobia anthropophaga*. G, *Sarcophaga* spp. H, *Chrysomya* spp. I, *Cochliomyia* spp. J, *Gasterophilus intestinalis*. K, *Cuterebra* spp. L, *Hypoderma bovis*.

pupae are slightly shorter than the mature larvae and their ends are more bluntly rounded.

The pupal stage lasts three to seven days in summer to much longer in winter, hibernation also occurring in this stage. The fly emerges by pushing off the end of the puparium by means of the inflated ptilinal sac, which is further used for progression to the surface of the soil. The shortest time known for completion of the life-cycle is seven days, so that several generations can develop in one year. Nine to ten generations may be completed in a year in certain parts of the Republic of South Africa. The flies live a month or longer and can also hibernate.

Epizootiology

The factors which influence the occurrence of calliphorine myiasis in sheep can be classed into two groups: those controlling the prevalence of flies and those determining the susceptibility of the sheep.

The *prevalence of flies* is seasonal, because the adults are adapted to definite ranges of temperature and to variations of humidity. They are most abundant in late spring and early summer, decreasing in numbers during the hottest part of the year and increasing again in the early autumn. The different species, however, are not all alike in this respect, some preferring lower temperatures than others, and consequently a number of overlapping 'waves' of different species succeed one another during the season.

The abundance and suitability of food for the adults as well as for the larvae is of great importance. The adult flies feed on liquefied protein and on the nectar of certain plants. Since protein food is required for the maturation of the ovaries, its relative abundance greatly influences the fertility of the females. The rate of growth of the larvae and the number that can develop also depend on the same factor. The larvae can obtain this food from either living sheep or dead carcasses of various animals (carrion) on which the adults lay eggs. The flies can be classified into: (*a*) *primary flies*, which initiate a strike by laying eggs on living sheep; (*b*) *secondary flies*, which do not usually do this, but lay their eggs on sheep already struck, the larvae extending the injury done by the larvae of the primary flies; and (*c*) *tertiary flies*, which come last of all, the larvae of which do little further damage. This succession occurs both on living sheep and on carrion, the succession on carrion corresponding to the various stages of decomposition of the carcases. Thus the larvae of the primary flies develop during autolytic and early bacterial decomposition, those of the secondary flies during the succeeding phase of liquefaction, while the tertiary flies follow when the carcass begins to dry out. In Australia, the primary flies occur in two principal waves, being most prevalent in the spring and autumn, and control and treatment are adapted to meet these waves. The waves vary in numbers in different years. The most important primary fly in Australia is *Lucilia cuprina*, other flies concerned being *Calliphora stygia*, *C. augur*, *C. australis*, *C. nociva*, *C. fallax* and *Lucilia sericata*. In New Zealand *Lucilia sericata* causes most strikes while *Calliphora stygia* is second in importance. In Britain the most important primary fly is *L. sericata*. In Australia the secondary flies include *Chrysomyia rufifacies*, *Chrysomyia micropogon*, *Microcalliphora varipes* and carrion flies of the genus *Sarcophaga*. The tertiary flies in Australia are *Musca domestica*, *Fannia australis* and *Peronia rostrata*. The succession of the secondary and tertiary flies is further influenced by competition between them for food. In the battle for this, the larvae of the secondary flies usually overcome those of the primary flies, so that, once the secondary flies have entered the competition, relatively few eggs of *Lucilia* and other primary flies give rise to adult flies. When the carnivorous larvae of *Chrysomyia* arrive, they feed on the larvae of the other species. The importance of this fact was realized in the Republic of South Africa, where it was found that *Chrysomyia marginalis*, which never strikes live sheep but breeds in carcasses only, plays a very important role as a competitor in carcasses. This fly is active only during the warm summer months and then it completely prevents primary sheep blowflies from breeding in carcasses, since it has a strong repellent action on their larvae, as has *C. rufifacies*. During this period the primary sheep blowflies can therefore breed on live animals only, whereas in winter, when *C. marginalis* is inactive, they are able to breed in carcasses and they build up a large population of hibernating larvae and pupae in the soil, from which flies emerge in spring.

Lucilia cuprina, the most important primary fly in Australia and South Africa, prefers living sheep to carrion as is the case with *L. sericata*, the most important primary fly in Britain. Species of this genus are therefore more likely to be derived from living sheep than from carrion.

The *susceptibility of the sheep* depends on inherent factors which can be influenced by selective breeding and temporary factors which can be otherwise controlled.

Sheep are struck most frequently in the breech (*breech strike, crutch strike*) and around the tail (*tail strike*), where the wool is soiled and the skin scalded by diarrhoeic faeces and by urine in the case of ewes. The major predisposing factors lie in the conformation of this region, especially narrowness of breech and wrinkling of the skin, which favour constant soiling by urine and faeces. Rams and wethers, in which the sheath has a narrow opening, soil the wool of this region with urine and become struck there (*pizzle strike*). Rams with deep head folds or with horns lying close to the head develop a 'sweaty' condition of the skin in these parts and attract blowflies, strike in this region being called *poll strike*.

Any other part of the body may become infected if an undressed wound is present, as the result of accidents, dog bites, contacts with barbed wire etc., or due to operations. The term *wound strike* is sometimes given to strikes which occur on wounded areas. They are seen especially on the scrotum after castration, on the tail after docking and on the heads of rams that have been fighting. Myiasis in the dorsal region of the body (*body strike*) is usually due to prolonged wet weather, when the wool becomes soaked with rain and bacterial activity, *fleece rot*, sets in. Areas over the withers are especially susceptible. Length and fineness of wool are important factors in this connection. Sheep with short or coarse wool, which dries rapidly, are less commonly affected in this way than those with longer or finer wool. In Australia fleece rot and dermatitis account for a large proportion of body strike. Additionally foot rot is a major disease predisposing sheep to blowfly strike.

Pathogenesis

Unless a wound is present to attract the flies and provide a suitable substrate for the larvae, bacterial activity appears to be important in preparing favourable conditions. In wool that is kept moist by prolonged wet conditions, the yolk, wool scales and skin scales become pasty and form a suitable medium for bacteria. These then produce, on decomposition, an odour that attracts the flies and probably also an exudative reaction of the skin which provides food for the young larvae until they are able to pierce the skin. Where the skin becomes soiled by urine and faeces, it is directly affected and becomes inflamed, but even here bacteria probably assist to aggravate the condition and to attract flies. The larvae of the primary blowflies initiate the attack and create favourable conditions for those of the secondary flies. They secrete proteolytic enzymes which digest and liquefy the tissues of the host and then feed on this pre-digested material. Only when this stage has been reached are the secondary larvae able to develop in the lesion.

Large wounds are usually produced and the larvae, especially those of the secondary flies, may form deep tunnels in the tissues and underneath the skin. The central portion of the lesion may heal, with the formation of a thick scab, while the action of the larvae extends outwards. The smell emanating from the lesion attracts other flies to deposit eggs in it and further batches of primary and secondary larvae may find suitable conditions for development.

The lesion and the parasites are irritating and the animal does not feed properly, and becomes poor and weak. The immediate cause of death is probably a toxaemia due to absorption of toxic substances from the lesion, or even septicaemia.

The financial losses due to strike result from the decreased value of the fleece, reduced meat and milk production, the death of animals and the labour costs incurred in the surveillance and handling of blowfly strike.

Clinical signs

The affected sheep usually stands with its head down, but does not feed, and presents a characteristic picture, so that it can be readily noticed. It may attempt to bite the affected part. When the lesion is situated around the tail or on the buttocks, the animal will stamp or jerk the hind-legs and wag the tail. Examination shows a patch of discoloured, greyish-brown, moist wool with an evil odour. The maggots may be found in the wool attacking the skin and they crawl away into the surrounding wool when disturbed. In later stages there is an inflamed ragged wound from which a foul-smelling liquid exudes into the wool; the larvae are burrowing into the tissues and only their posterior extremities project. The temperature may be elevated. If the disease progresses, malnutrition and loss of milk occur and death may follow within a few days in bad cases. A condition known as 'lightning strike', in which the affected sheep dies within a few days, occurs in Australia. This is probably due to the introduction of bacteria, which, subsequently, may cause toxaemia or septicaemia.

Diagnosis

Diagnosis is easily made from the clinical signs and by finding the larvae in the wound.

Treatment

Treatment of the lesions aims to kill larvae in the lesions, to promote healing and to prevent re-infestation with more larvae. The extent of the lesion is ascertained by clipping the wool and many larvae can be removed while this is being done. Larvae removed should be killed to prevent them from giving rise to adult flies. It may be difficult to remove larvae in deep pockets, but these also must be killed, especially when the fly concerned is a primary fly such as *L. cuprina* or, in Britain, *L. sericata*, which prefers to feed on living sheep. The dressing should be bland and not toxic to the sheep and should promote healing.

While various dip solutions may be used on maggot-infested wounds, there is a danger that the insecticides will be absorbed rapidly and cause toxicity. Wound dressing compounds are available which contain 2% diazinon or 3% coumaphos.

Prophylaxis

Prophylactic measures may be divided into those which attempt to render the sheep less attractive to the flies and those which are directed against the flies themselves.

Treatment of the sheep to render them less attractive to the flies may be described under the following headings:

1. *Selective breeding.* Narrowness of the breech and folds of skin in this region predispose sheep, especially ewes, to strike. Efforts have been made to eliminate, or to lessen, the influence of these features by breeding sheep with plain, or plainer, breeches and much progress has been made in this respect, although it has been necessarily slow.

2. *Surgical removal of the breech folds.* Mules's operation (Mules 1935) is an older surgical technique for the control of blowfly strike. A crescentic area of skin on either side of the urinogenital area of the Merino sheep is removed and the resultant scarring flattens the skin folding so that the accumulation of excretions is minimized. Richardson (1971) has compared sheep with and without the operation, confirmed its value in controlling strike and reported that it does not affect the rate of weight gain in sheep. Sheep of all ages may be 'mulesed' but the operation is most convenient and most effective when performed on lambs. While 'mulesing' combats breech strike it will have no effect against body strike.

The use of chemicals to produce 'mulesing' has been studied in Australia for the prevention of pizzle strike (Hopkins 1978). A 40% phenol solution is applied along the prepuce of young wethers after shearing or crutching; this produces a dry encrustation which heals leaving a scar which eliminates skin folding in the affected area of the prepuce. Similar treatment can be applied to the breech area.

3. *Docking.* Tests carried out mainly in Australia have shown that docking the tails of lambs behind the fourth, instead of the usual second, caudal vertebra reduces strike

appreciably. The explanation seems to be that the tail, being pressed against the body of the animal, tends to flatten out small skin folds often present on either side of the vulva, which may become soiled when they are prominent. The longer tail is also held well away from the body when the ewe urinates. Short docking tends to produce a stump surrounded by folds—the 'rose tail'—which is often struck.

4. *Crutching* consists of clipping the wool from around the tail and in the breech. This is a very useful measure, tending to promote dryness in this region, and it is effective in preventing strike for four to five weeks under ordinary circumstances. Machine crutching is more effective than hand crutching, because the machine clips the wool shorter than the shears.

5. *Treatment of sheep with insecticides.* During the last two decades there has been extensive development in the use of highly effective insecticides for the protection of sheep against blowfly strike. The once extensively used coal-tar and arsenic dips were replaced by the chlorinated hydrocarbons DDT and HCH (hexachlorocyclohexane) and these were followed by dieldrin with which it was possible to produce a complete season's protection with a single treatment. A common preparation was 0.05% dieldrin with 0.016% HCH, which produced an immediate kill of larvae lasting a few months after dipping and also a delayed effect on pupation and emergence which was evident many months afterward. Dieldrin is also translocated down the hair fibre to the skin where it also travels laterally, permitting low-volume tip spraying techniques to be used (e.g. in Australia) (Beesley 1973). However, its prolonged persistence in the tissues poses a human health hazard due to residues in meat and additionally the persistence has aided the development of blowflies resistant to dieldrin. Resistance has been reported in all the important species of blowflies in the major sheep-producing countries of the world. Dieldrin was replaced by the organophosphate insecticides about 1964 (Beesley 1973) and resistance to these has also developed; however, by alternating insecticides, control problems associated with resistance can be overcome.

The duration of protection generally is proportional to the amount of insecticide deposited on the fleece. Consequently attention to the strength of the dip solution is important and the concentration of insecticide must be maintained by regular replenishment with concentrate. Continuous replenishment to ensure an adequate concentration is preferrable.

The system of application should ensure adequate exposure. In dipping, sheep should be immersed for a minimum of 30 seconds; jet spraying should ensure thorough soaking of the fleece, though tip spraying with concentrated solutions does not demand this since the insecticide is translocated down the wool fibre and into new wool growth.

Occasionally local degradation of insecticide or inhibition of the translocation process occur, for example in the breech region due to soiling with faeces and urine (Yeoman & Bell 1978). The incorporation of insecticide into an aluminium alkoxide gellant avoided degradation and animals sprayed with such formulations were well protected against breech strike.

A number of compounds are available at present for the control of myiasis; it is likely that more will become available in the next few years and it is likely that the synthetic pyrethroids will be used extensively. The marked efficacy of these (e.g. decamethrin) in the control of cattle ticks and protection against biting flies suggests their use for the control of blowflies. A new family of antiparasitic agents, the avermectins, with unprecedented activity against parasitic nematodes, has been ascribed by Egerton et al. (1979). These also have insecticidal activity and Ostlind et al. (1979) report activity against *Cuterebra* larvae in mice and *Lucilia* spp.

The insecticides in general use for blowfly strike include the following: bromophos and bromophos-ethyl (0.05% in dip), carbophenothion (0.042%), chlorfenvinphos (0.05%), coumaphos (0.05%), diazinon (0.04%), dichlofenthion (0.04%), dursban (Chlorpyrifos) (0.0125%) and butacarb (0.05%).

Where the duration of protection against strike has been reduced due to insecticide resistance sheep may be protected (e.g. against *Lucilia cuprina*) by thorough jetting. The insecticides which are most useful in this respect are chlorfenvinphos, diazinon, dichlofenthion and bromophos-ethyl (Shanahan & Hughes 1978).

Improved methods of application include mist, high pressure and low volume techniques. The incorporation of insecticides in acrylic emulsions offers a reservoir of insecticide which is retained in the fleece; the length of the staple does not determine the degree of penetration and the acrylic does not wash off with exposure to rain.

Genetic manipulation approach to control. A strain of blowfly (translocation-male/eye colour) (TM/EC), in which females are blind and cannot survive in the field and males are partially sterile but transmit the genes for blindness to female offspring, offers the possibility that control of blowflies could operate through the majority of blowfly seasons (Foster et al. 1978).

Carcass disposal. The destruction of carcasses is important during those seasons when they are the main breeding grounds of sheep blowflies and there is little or no competition from other species. During such periods—for instance, during the winter in the Republic of South Africa—carcasses should be burnt or treated with insecticides and buried.

Genus: Callitroga Brauer, 1833

This genus includes species the larvae of which cause myiasis of man and other animals and are called screw-worms (see below). They are also known as the genus *Cochliomyia* Townsend, 1915.

Callitroga hominivorax (Coquerel 1858) (syn. *C.* (*Cochliomyia*) *americana*), the American screw-worm fly, occurs in North and South America.

Callitroga (syn. Cochliomyia) macellaria Townsend, 1915 is also an American species; its distribution extends from central Canada to Patagonia. The adults of these two species are 10–15 mm long, the body being bluish-green, with three longitudinal stripes on the thorax, and the face and the eyes orange-brown. The palps of both species are short and thread-like and the antennae of both species are feathered to their tips. It is difficult to distinguish between the adults of these two species, but their larvae differ: those of *Callitroga hominivorax* have pigmented tracheae and large posterior spiracles, while those of *Callitroga macellaria* have non-pigmented tracheae and small posterior spiracles.

SCREW-WORMS OF MAN, CATTLE AND OTHER ANIMALS

The name 'screw-worm' is given to the larvae of *Callitroga hominivorax* and *Callitroga macellaria*, both of which occur in North and South America, and to those of *Chrysomyia bezziana*, the Old World screw-worm fly, which occurs in Africa and southern Asia.

Life-cycle

The female flies deposit clusters of 150–500 eggs at the edge of a wound on the host. These species have become so closely adapted to parasitic life that they breed only in wounds and sores on their hosts and not in carcasses. The larvae hatch in 10–12 hours and grow mature in about three to six days, after which they leave to pupate in the ground. The mature larvae are about 15 mm long and are well armed with bands of spines around the body segments. The pupal period lasts from three days (*C. hominivorax*) or seven days (*C. bezziana*) to several weeks, according to the prevailing temperature. Hibernation occurs most commonly in the pupal stage.

Pathogenesis

Cattle, pigs and equines suffer most frequently, but other animals, including fowls, dogs and man, may also be affected. The flies deposit their eggs in wounds resulting from accidents, castration, dehorning, branding, scalding by dips, tick bites and so forth, as well as around the vulva of cows when there is a bloody discharge, or on the navels of young calves. Rainy weather predisposes to screw-worm infection. The maggots penetrate into the tissues, which they liquefy, and extend the lesion considerably. The wound develops an evil odour and a foul-smelling liquid oozes out. Severe infections are common and death from screw-worm infection is frequent. At one time screw-worm infection was common in the USA and represented a serious pest to livestock in the southern and south-western states. Now, for all practical purposes, it has been eliminated by the use of irradiation-sterilized flies (see below).

Treatment and prophylaxis

The wound should be thoroughly cleaned and the maggots, especially those of *C. bezziana*, destroyed to prevent them from pupating. Dressings used for myiasis in sheep may be applied and further infection guarded against. Prophylaxis requires proper dressing of all wounds and avoiding operations, such as branding, dehorning, earmarking, etc. during the fly season. *C. hominivorax* is especially attracted to newborn animals and it was recommended in the southern USA at one time that breeding should be so controlled as to have no young stock born between 1 May and 15 November, while the fly was most active. Various insecticidal dressings can be applied to wounds to prevent screw-worm infection.

The most dramatic campaign against these flies and their larvae was the one carried out for the eradication of *Callitroga hominivorax* from the USA by the release of male flies sterilized by irradiation. The basis of it was the fact that this fly mates only once a year, so that mating with sterilized males prevents breeding. The campaign is described by Baumhover et al. (1955, 1959), Bushland (1960) and Skerman (1958). For the eradication campaign 50 million irradiated flies a week were required and 36 000 kg of meat a week to feed their larvae. The results of the campaign were that, after preliminary eradication of *C. hominivorax* from the island of Curaçao, it was eradicated from the whole of Florida and also from the Bahama Islands near to the coast of Florida. The campaign was subsequently applied to the whole of the southern part of the USA, with outstanding success. The majority of the cases of screw-worm infections which now occur are traceable to flies which have migrated across the USA–Mexican border, either on the wing or on animals.

Genus: Cordylobia Grünberg, 1903

Cordylobia anthropophaga Grünberg, 1903, the 'tumbu fly' or 'skin-maggot fly', is a stout, compactly built fly, about 9.5 mm long. The general colour is light brown, with diffuse bluish-grey patches on the thorax and a dark grey colour

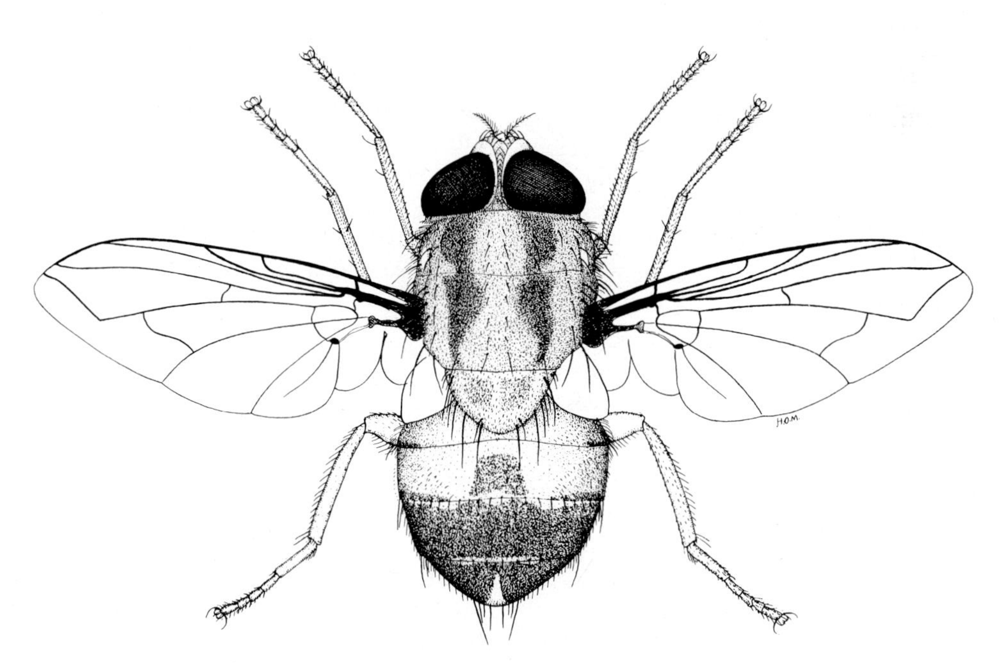

Fig 2.63 *Cordylobia anthropophaga.*

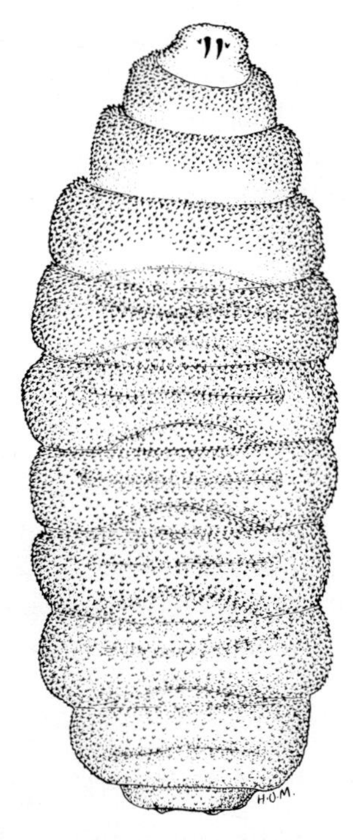

Fig 2.64 Maggot of *Cordylobia anthropophaga*.

on the posterior part of the abdomen. The fly is widely distributed in Africa south of the Sahara. It is a myiasis-producing fly of man, small rodents, monkeys and dogs.

Life-cycle and habits. The fly deposits some 500 eggs in the sleeping-places of man and various animals, on the ground or on straw, sacking, etc., sometimes apparently also on clothing that smells of perspiration or urine. The larvae hatch after two to four days and penetrate into the skin, where they grow mature in 8–15 days. The mature larva is about 12 mm long and is covered with a large number of minute spines. It leaves the host and pupates in the ground for three to four weeks in summer before the fly emerges.

Clinical signs and diagnosis. The larva is situated in a swelling which is about 1 cm in diameter, is rather painful and has a small central opening.

Treatment. The larvae can be easily pressed out and a disinfectant applied.

Prophylaxis. Cleanliness and regular disinfection of sleeping-places is important. In the case of valuable animals, like angora rabbits, which are frequently affected, protection can be afforded by keeping the flies out with gauze wire.

Cordylobia rodhaini Gedoelst, 1911, closely resembles *C. anthropophaga* but is larger, measuring 12.5 mm in length. It is found in tropical Africa, particularly rainforest areas. It is often known as 'Lund's fly' and the larva as 'the larva of Lund'. Its main hosts are antelopes and the giant rat and occasionally man (e.g., Commandant Lund, after whom it was named).

Genus: Booponus Aldrich, 1923

Booponus intonsus Aldrich, 1923 is the 'foot maggot' which attacks cattle, goats and carabao in Celebes and the Philippine Islands. The fly is about as large as a house fly, light yellow with light brown on the anterior dorsal part of the thorax. The eyes are relatively small. The head, adbomen and femora are evenly covered with short, black hairs. The veins of the wings are yellow. The flies are active mainly in the dry season.

The fly lays its eggs on the hairs along the coronet and the posterior part of the pastern. When the larvae hatch they penetrate the skin in these parts and produce wounds in which the hind ends of the larvae can be seen. After about two to three weeks they fall out and pupate in the ground for 10–12 days.

Affected animals are restless and show lameness, which may be serious in severe cases.

Treatment. The maggots should be destroyed and reinfection prevented by applications as in myiasis due to sheep blowflies. A grease containing dieldrin and diazinon is effective.

Genus: Auchmeromyia Schiner, Brauer and Bergenstamm, 1891

Five species of this genus are known, but only one **Auchmeromyia luteola** (Fabricius 1805) is associated with man and animals. The larvae of the fly are blood-sucking, being known as the Congo floor maggot. It attacks people sleeping on the bare floor in tropical Africa. Adult flies feed on faeces and decayed material; the larvae are found attached to the skin which they pierce and suck blood. This is largely a synanthropic species but larvae have been found in association with domestic pigs and the adults at the entrance of burrows of wart-hogs (Zumpt 1965).

Genus: Pollenia Robineau-Desvoidy, 1830

This genus contains the species **Pollenia rudis** Robineau-Desvoidy, 1830, the cluster fly. The larvae of this fly occur in earthworms. The adults congregate in roof spaces of buildings in temperate climates. They are inactive during cold weather but are aroused by a mild spell of weather or by warm air. They constitute a nuisance by beating on windows or falling into water or food.

SUBFAMILY: SARCOPHAGINAE MERDIVENCI, 1966

The 'flesh flies' are medium-sized to large and thick-set, of a light or dark grey colour. The arista is plumose to about its middle and bare in the distal portion. The thorax often has three longitudinal dark stripes and the dorsum of the abdomen has dark spots or is chequered dark and grey. Species of the family are larviparous and may lay their larvae in wounds or sores in which the larvae develop, although some species also lay their larvae in decomposing meat or other decaying matter.

Genus: Sarcophaga Meigen, 1826

Among the species of this genus which may lay their larvae in decomposing flesh, wounds, ulcers, etc. are:

Sarcophaga haemorrhoidalis Fallen, 1810, which occurs in Europe, America, Asia and Africa.

Sarcophaga fusicauda Böttcher, 1913, which occurs in Australia, China, Japan and neighbouring countries.

Sarcophaga carnaria (Linnaeus, 1758), which, like the two species above may lay its larvae in wounds, etc. on the skin of man. This species can drop its larvae from a height of 70 cm through wire-gauze covers put over meat.

Sarcophaga dux Thompson, 1868, the larvae of which have been found in skin lesions on camel, cow and bullock in India.

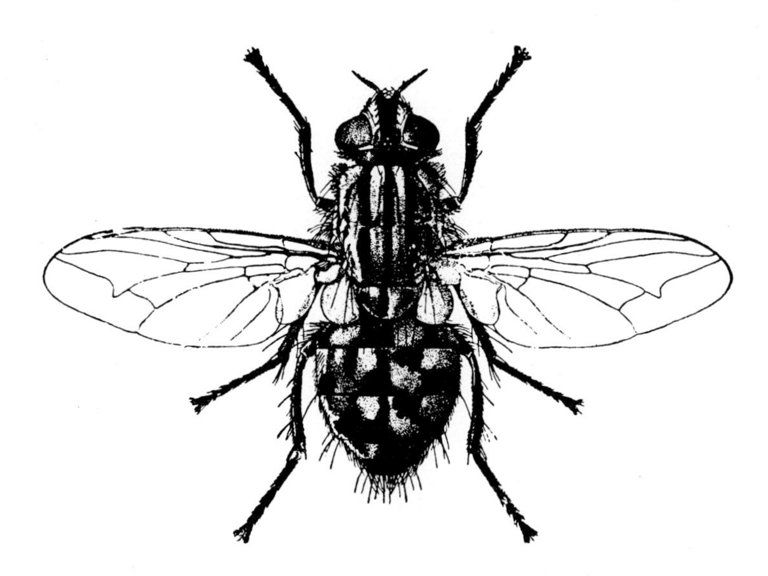

Fig 2.65 Female *Sarcophaga carnaria*. (*From Castellani and Chalmers*)

Genus: Wohlfahrtia Brauer and Bergenstamm, 1889

Species of this genus are also larviparous and have habits similar to those of species of the genus *Sarcophaga*. Important species are:

Wohlfahrtia magnifica (Schiner, 1862), the Old World flesh fly, which occurs in the Mediterranean area, Arabia, Turkey and Russia. It may deposit its larvae in the external ear of man, or in sores around the eyes or elsewhere on the bodies of man and other animals. It is an important pest of sheep in southern Soviet Union.

Wohlfahrtia vigil (Walker, 1949) occurs in Canada and the northern USA.

Wohlfahrtia meigeni (Schiner, 1862), which occurs in the western USA.

These three species have similar habits and the severe disfigurements and suffering that they may cause in man, especially in children, are discussed in textbooks of medical entomology.

Wohlfahrtia nuba (Wiedemann, 1830) occurs in man and animals, especially camels, in Sudan and Ethiopia and eastwards to Karachi.

FAMILY: OESTRIDAE SAMOUELLE, 1819

This family includes the genera *Hypoderma*, *Oestrus* and others, but the genus *Gasterophilus* is now assigned to the family *Gasterophilidae* (Freeman 1973).

The adults are hairy flies which have rudimentary mouth parts and do not feed. They usually lay their eggs on animals. The larvae are parasitic maggots and consist of 12 segments, of which the first two are fused together. Oral hooks are usually present, but there is no head. The posterior stigmata open through semicircular plates which may be retractile. The larvae moult twice during their parasitic life and leave the host when they are full grown to pupate in the ground. They feed on the body fluids of the host or on exudates which surround them.

Genus: Oestrus Linnaeus, 1761

Oestrus ovis Linnaeus, 1761, the 'sheep nasal fly', has a dark grey colour with small black spots which are especially prominent on the thorax and it is covered with light brown hair. The flies hide in warm corners or crevices and in the early morning they can be seen sitting against walls or other objects in the sun. They occur from spring to autumn, particularly in summer, but in warm climates they are active even in winter. The larvae occur in the nasal cavity and the adjoining sinuses in sheep and rarely in goats and have also been found in the blesbock (*Damaliscus albifrons*) and, in Egypt, in the camel. *Oestrus ovis* sometimes also deposits its larvae in the eyes, nostrils and on the lips of man, where they may develop, causing serious trouble. Shepherds are said to be especially susceptible to myiasis by this fly and also persons associated with cheese-making from sheep milk.

The flies deposit their young larvae around the nostrils of the host, whence they crawl upwards. Sometimes they enter cavities which have small openings, like those of the turbinate bones or a branch of the frontal sinus, with the result that they are not able to get out when they have grown fully and so they die there. The rate of development of the first larval instar varies considerably, this instar remaining in the nasal passages for two weeks to nine months during the cold months. The second instar passes into the frontal sinuses and may develop rapidly, leaving

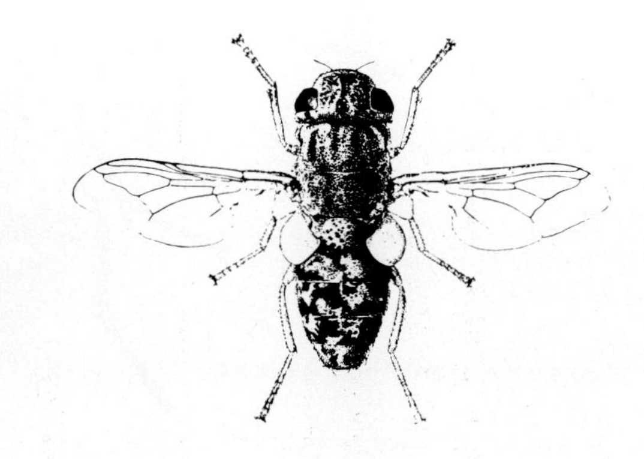

Fig 2.66 Female *Oestrus ovis*. (*From Smart*)

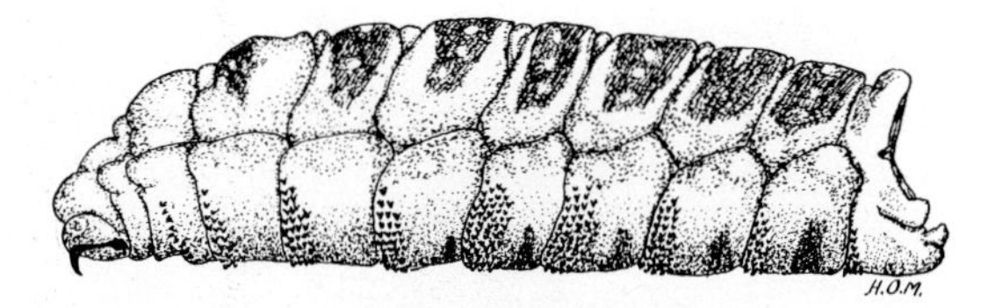

Fig 2.67 Lateral view of the larva of *Oestrus ovis*.

the sheep 25 days after infection or considerably longer. Finally the full-grown larvae crawl out and pupate in the ground for three to six weeks or longer during the cold season, before the fly emerges.

The young larvae are white or slightly yellow; when they become mature, dark transverse bands develop on the dorsal aspects of the segments. The full-grown larva is about 3 cm long, tapering anteriorly and ending with a flat surface posteriorly. There are large, black, oral hooks, connected to an internal cephalopharyngeal skeleton. The ventral surface bears rows of small spines and the black stigmal plates are conspicuous on the posterior surface.

Pathogenesis. The flies cause great annoyance when they attack the sheep to deposit larvae; the animals stop feeding and become restless. They shake their heads or press their noses against the ground or in between other sheep. When the flies are plentiful, they may cause considerable interference with the feeding of the animals. The larvae irritate the mucosa with their oral hooks and spines, causing the secretion of a viscid mucous exudate, on which they feed. Erosion of the bones of the skull may occur and even injury to the brain and then such signs as high-stepping gait and incoordination may suggest infection with *Coenurus cerebralis* (see p. 117). For this reason the infection has been called 'false gid'. Infected sheep have a nasal discharge and sneeze frequently.

Diagnosis can only be made tentatively from the clinical signs, excluding other possible causes like lungworms and chronic bronchial or pulmonary diseases.

Treatment. This is difficult since the larvae are difficult to reach: frequently the openings into the sinuses are narrowed or occluded. At one time direct injections into the frontal sinus were made, using an emulsion of tetrachlorethylene, but this has now been largely discontinued. Instillation of HCH in oil (1–4%) into the nostrils while the sheep is lying on its back has been practised in the Republic of South Africa, with good results.

The use of systemic insecticides, such as the organophosphorus compounds, is a more rational approach. Sheep given 55–88 mg/kg of a mixture of 2 g of Neguvon (Bayer L 13/59) and 0.2 g of Asuntol (Bayer L 21/199) were cleared of the infection (Stampa 1959). Crufomate (Ruelene, Dowco B2) as a dip (35%) or as a 'pour-on' (13.5%) has been useful while rafoxanide as a drench (7.5 mg/kg) greatly reduces infection. Horak and Snijders (1974) compared two large groups of Merino lambs infected with *Oestrus ovis* and one group being treated with rafoxanide (7.5 mg/kg). Over a two-year period the treated group showed a reduction in nasal discharge, an increase in weight gain and virtual freedom from the parasite in the nasal passage and sinuses.

Prophylaxis. This is difficult since the present fly repellents are short lasting. One method is to feed sheep in narrow troughs, the edges of which are smeared with tar. The animals automatically tar themselves and this acts as a repellent.

Genus: Rhinoestrus Brauer, 1886

Adult flies are similar to those of the genus *Oestrus*, the larval stages are obligatory parasites of equines.

Rhinoestrus purpurensis (Brauer, 1858). The larval stages are obligatory parasites of the nasal sinuses and larynx of horses and mules of Europe, Asia and Africa. Occasionally the larvae may cause 'ophthalmomyiasis' of man.

Genus: Cephalopsis Townsend, 1912

Larval stages are obligatory parasites of the nasal cavities and sinuses of camels.

Cephalopsis titillator (Clark, 1816), the camel nasal bot fly. The adult fly occurs in sub-Saharan and other areas where the camel and dromedary are common. The larval stages occur in the nasal cavity, frontal sinus and pharynx of the dromedary and camel. The larva resembles that of *Hypoderma* spp.

Genus: Pharyngobolus Brauer, 1866

Larval stages, which resemble *Oestrus ovis*, are obligatory parasites of elephants.

Pharyngobolus africanus Brauer, 1866, the elephant throat bot fly, occurs in the pharynx of African elephants.

Fig 2.68 *Hypoderma bovis.* (*From Smart*)

Other genera include *Neocuterebra* Grüneberg, 1906 (*Neocuterebra squamosa* Grünberg, 1906), the larvae of which occur in the adipose tissue of the sole of African elephants, and *Cephenemyia* Latreille 1825 (*Cephenemyia trompe* (Modeer, 1786)), the larvae of which occur in the pharynx of the reindeer and other cervids of the Old and New World in the circumpolar region. *Cephenemyia stimulator* Clark, 1915 occurs in the nasal cavities of roe deer, *C. ulrichii* Brauer, 1862 in elk in Europe and *C. auribarbis* Meigen, 1824 in red deer. Other species of this genus include *Cephenemyia phobifer* Clark, 1915 of white-tailed deer in Florida and *C. jellisoni* Townsend, 1914, *C. apicata* Bennett and Sabrosky, 1962 and *C. pratti* Hunter, 1915 in black-tailed and mule deer, and occasionally elk, in western and northern areas of North America (Capelle 1917).

Genus: Hypoderma Latreille, 1818

The larval stages of **H. bovis** (de Geer, 1776) and **H. lineatum** (de Villiers, 1789) (syn. *H. lineata*), the 'ox warbles', are common parasites of cattle, rarely also of man and horses, in many countries in the northern hemisphere. They occur between latitudes 25° and 60°. Apart from infections in imported animals the genus is not established in the southern hemisphere. *H. bovis* is about 15 mm long; *H. lineatum* measures 13 mm. The flies are hairy and have no functioning mouth parts. The hairs on the head and the anterior part of the thorax are yellowish-white in *H. lineatum* and greenish-yellow in *H. bovis*. The abdomen is covered with light yellow hairs anteriorly, followed by a band of dark hairs; the posterior portion bears orange-yellow hairs. *Hypoderma bovis* is referred to as the northern cattle grub in North America and *H. lineatum* as the common cattle grub or the heel fly. This species has also been reported from the American bison (*Bison bison*) where it shares range with cattle (Capelle 1971).

Life-cycle. The flies occur in summer, especially from late May to July. They are most active on warm days, when they attack cattle to lay their eggs. However flies have been observed to oviposit at temperatures as low as 4°C (Andrews 1978). The flies are limited in flight and rarely exceed 5 km though Beesley (1974*b*) has noted flights up to 14 km. *Hypoderma lineatum* tends to appear about one month before *H. bovis*.

The eggs are about 1 mm long and are fixed to the hairs by means of small terminal clasps, especially on the legs, but more rarely on the body as well. *H. bovis* lays its eggs singly, while *H. lineatum* deposits a row of six or more on a hair. The flies are very persistent in approaching the animals and one female may lay 100 or more eggs

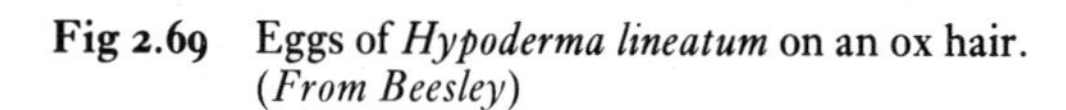

Fig 2.69 Eggs of *Hypoderma lineatum* on an ox hair. (*From Beesley*)

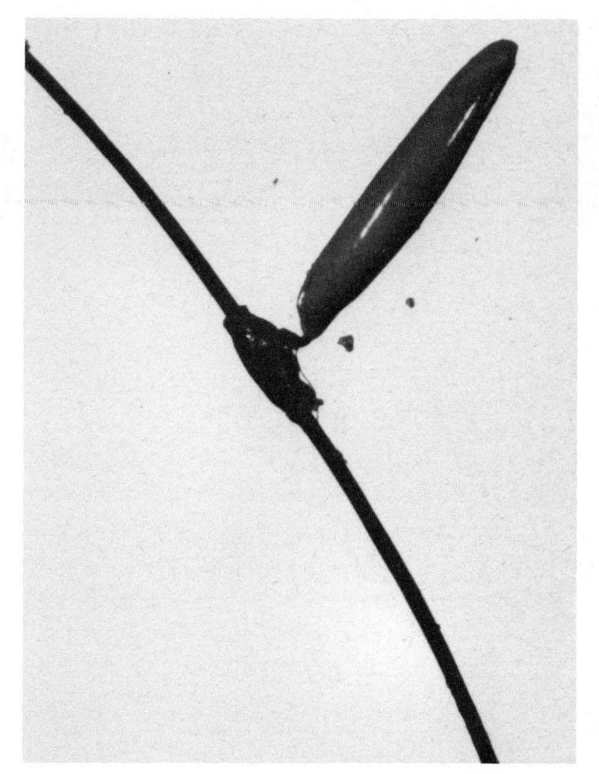

Fig 2.70 Single egg of *Hypoderma bovis*. (*From Beesley*)

on one individual. The larvae hatch in about four days and crawl down the hair to the skin, through which they penetrate. They wander in the subcutaneous connective tissue up the leg and then towards the diaphragm and gradually increase in size, though such first-stage larvae may be found in many different sites of the body, including intermuscular connective tissue, the surface of internal organs etc. Beesley (1974*b*) designates the 'winter resting sites' of first-stage larvae as the submucosal connective tissue of the oesophageal wall in the case of *H. lineatum* and the region of the spinal canal and epidural fat for *H. bovis*. Larvae reach these sites weeks or months after hatching and remain in them for the autumn and winter, growing to about 12–16 mm in length.

Eventually during January and February the now second-stage larvae travel towards the dorsal aspect of the body and reach the subcutaneous tissue of the back where they mature to the third

Fig 2.71 Processed calf skin showing warble holes. (*From Beesley*)

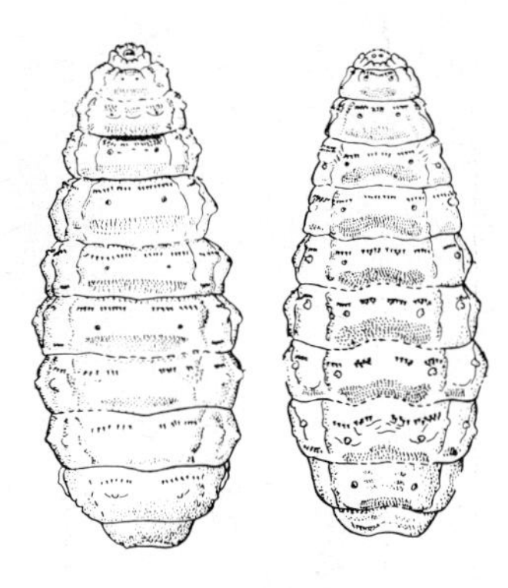

Fig 2.72 Ventral view of third-stage larvae of *H. diana* (left) and *H. lineatum*. (*From Cameron*)

Fig 2.74 Ventral and dorsal surfaces of *H. bovis* larvae. (*From Beesley*)

stage. This is the 'spring resting site' of the larvae (Beesley 1974*b*). When the parasites arrive under the skin of the back, swellings begin to form, measuring about 3 cm in diameter. The skin over each swelling becomes perforated and the larva then lies with its posterior stigmal plate directed towards the pore for the purpose of respiration. This is the 'warble' stage of the infection. This stage lasts about 30 days. The younger larvae are almost white, changing to yellow and then to light brown as they grow older. The full-grown third stage larva of *H. bovis* is 27–28 mm long; that of *H. lineatum* 25 mm. Each segment of the third-stage larva bears a number of flat tubercles and small spines are present on all segments but the last in *H. lineatum* and on all but the last two in *H. bovis*. In spring the mature larva wriggles out of its cyst and falls on the ground into which it penetrates to pupate. The pupal case is black and the fly emerges from it, after 35–36 days, by pushing open an operculum at the anterior end.

Fig 2.73 Processed leather from an adult bovine, showing holes due to warble fly larvae. (*From Beesley*)

Among other hosts on which these two species of *Hypoderma* may lay eggs, especially when cattle are not available, are horses and man, though the larvae do not usually mature in these.

Incidence. Studies in the UK indicate that the overall infection level has increased gradually in the last few years. Regional variations occur and Scotland and northern England show higher infection rates than elsewhere. Based on the percentage of hides with warble larva damage, a lower level occurred in cow and bull hides than from heifer and ox (Andrews 1978). Calves and young cattle are more frequently and more severely infected than older animals. It is possible that cattle develop a degree of immunity to the larvae. However, Rich (1970) did not observe such differences between range cows and their yearling suckler calves in Canada. In other countries the infection rate is lower and some have instituted control programmes which have reduced the incidence to very low levels or have eradicated the parasite (e.g. Cyprus, Norway, Sweden, Malta).

Pathogenesis and economic importance. When adult flies approach an animal to lay eggs, animals become apprehensive and disturbed and attempt to escape the fly by running away, often aimlessly (gadding). Animals may damage themselves on fences or wire or may be killed by gadding into water or falling over heights. Additionally this results in reduced weight gain and milk yield. In the UK in 1978, animal losses due to gadding were estimated at £3 million by the Hides and Allied Trades Improvement Society.

Estimates of the loss in milk production vary from country to country, from 10–15% in Europe to up to 50% in Canada (Andrews 1978). Control by insecticides resulted in a 7–11% increase in milk production. Losses in meat production similarly vary, but all studies indicate a reduction of weight gain associated with the presence of larvae; further a direct relationship between the number of larvae (or warbles) and reduction of weight gain was evident. Several studies have indicated a significant increase in weight gain following the control of infection with organophosphate compounds (Andrews 1978).

Carcass and hide depreciation are important aspects of *Hypoderma* spp. infection. Rich (1970) in Canada associated one to five larvae with the trimming of 0.7 kg of carcass and 11 or more larvae with the necessity to trim 1.2 kg from a carcass. Damage caused includes discoloration of meat along the tracks of larvae, meat may also be gelatinous; the name of 'licked' beef is applied to such lesions. Local abscess formation may occur at the site of a warble, especially if an abortive extraction of an immature larva has been made.

Hide damage is a well recognized consequence of infection and significant differences in monetary value of hides result from holes and other flaws caused by *Hypoderma* larvae. Overall loss were estimated at £13 million in Great Britain in 1978. In the USA, in 1956, losses were estimated at $192 million.

The aberrant migration of larvae in other animals may cause serious results. An acute neurological disease of a horse associated with the intracranial migration of first-stage larvae was reported in western Montana (Hadlow et al. 1977). Olander (1967) has described the lesions caused by the migration of larvae of *Hypoderma lineatum* in the brain of a horse. Warbles may occur in the back of horses, making the wearing of a saddle or harness difficult.

Infection of man rarely occurs. Abdominal pain, subdermal migratory tracts and local warble formation have been reported.

Hypersensitivity responses associated with accidental or deliberate rupture of larvae in the warble may be associated with allergic responses (Hadwen & Bruce 1917) and McDougall (1930) demonstrated the reaction to be anaphylactic in nature. Such response may attend the treatment of infected animals with organophosphate compounds (see below).

Clinical signs. Except for poor growth in bad cases and decreased milk yield the animals show no appreciable signs until the larvae appear along the back, when the swellings can be felt and seen. The larva lies in a cyst which also contains yellow purulent fluid.

Diagnosis. Diagnosis based on the presence of the larvae under the skin of the back. The eggs may also be found on the hairs of the animals in summer. However, as control programmes reach their terminal stages there may be a need for immunodiagnostic tests which will detect animals infected with migrating larvae and indicate those requiring treatment.

Treatment. Treatment may be undertaken in a variety of forms.

1. *Mechanical removal of larvae.* Mature larvae may be squeezed out of the warble swelling. This is less successful when the larvae are not mature. Rupture of the larvae during extraction may lead to a localized inflammation and abscess formation or even a generalized anaphylaxis in a few animals. This method of treatment is the least satisfactory.

2. *Insecticides.* Until the advent of the organophosphorus insecticides, derris, or its active

principle rotenone, was widely used as a larvicide. Derris was usually used as a wash and applied to the back of infected cattle after the scabs covering the warble swellings had been removed by scrubbing with a soapy mixture. This treatment was highly effective but for adequate control repeated treatment of stock was necessary during the warble season. A disadvantage was that the warble fly larvae were killed only when they had migrated through the body and produced damage to flesh and hides. The introduction of the organophosphorus systemic insecticides allowed control of the larvae while they were in the early stages of migration and before they reached the backs of animals. Nevertheless eradication of the parasite was achieved with rotenone in Denmark and Cyprus.

The systemic insecticides are applied once in autumn or winter before larvae appear on the back. Compounds may be administered orally, parenterally or dermally; the last method, using a 'pour-on' formulation, is the most acceptable in most countries though spraying may be used with range cattle. The organophosphate compounds available include: crufomate (Ruelene), famphur, fenthion (Tiguvon), phospmet (Profacte). (As well as being active against *Hypoderma* larvae they also have activity against lice.) These compounds can be used at any time of the year except between December and March inclusive, when larvae may be in the oesophagus or spinal canal. The compounds available will lead to 90% or more control of infection.

Ivermectin at a dose rate of 0.2 mg/kg has been reported to be 99% or more effective against all parasitic stages of *H. bovis* and *H. lineatum*.

A variety of side effects may follow treatment of cattle for warble infection (Khan 1973). These include direct toxicity of the organophosphates which are cholinesterase inhibitors (the antidote is atropine sulphate 0.15 mg/kg or a single dose of 0.3 g). Death of larvae in the wall of the oesophagus may result in oesophagitis with resulting bloat or in the spinal canal may result in temporary or permanent paralysis. Abortion has been claimed as a consequence of the use of organophosphates in cattle, but there is no satisfactory evidence to support this.

Experience with eradication campaigns and toxic side effects have indicated that the prevalence of these is low (e.g. in Ireland one death in 40 000 treated cattle) (Andrews 1978). These effects may be minimized if treatment is restricted to late summer and early autumn when large numbers of larvae have not yet become established in the oesophagus or spinal cord.

Eradication. It is feasible to eradicate the warble fly. Several countries are now free from the parasite and eradication is aimed at in Great Britain in the next five years. This will be achieved through the dermal application of organophosphate insecticides during the period 15 March to 31 July each year.

Hypoderma diana Brauer, 1858. The larvae of this species occur in red deer (*Cervus elaphus*) and roe deer (*Capreolus capreolus*) and occasionally man in Europe. The life-cycle is similar to that of *H. lineatum* and warbles occur on the dorsum in March and April. Infection may be heavy, resulting in severe ill health or death of deer.

Hypoderma aeratum Austen, 1931 is a similar parasite of goats and sheep in Cyprus, Crete and Turkey, while **H. crossi** Patton, 1923 infects goats and sheep in India, especially in the dry, hilly regions of the north-west provinces. The eggs are laid on the long hairs at the sides of the body and the larvae penetrate directly through the skin, remaining there to develop for about seven months. **H. actaeon** Brauer, 1858, occurs in central Europe and **H. silenus** Brauer, 1858, attacks equines and goats in the Balkans, Asia Minor and possibly in North Africa. **H. capreola** Rubtzov, 1939, occurs in roe deer in the USSR and **H. moschiferi** Brauer, 1863 in the musk deer (*Moschus moschiferus*) in Mongolia.

Oedemagena tarandi (Linnaeus, 1758). This species is similar to *Hypoderma* spp. and the larvae occur in cervids in the circum-Arctic and sub-Arctic regions. Thus, it is found in caribou (*Rangifer arcticus*, *R. caribou*), reindeer (*Rangifer tarandus*) and musk-ox (*Ovibos moschatus*).

Adult flies are active from late June to

September and larvae produce warbles on the dorsum in September to October the following year. Fawns and yearlings are most affected by the parasite which produces large oedematous swellings. These may suppurate and provide an attractive focus for the deposition of blowfly eggs. Control of this parasite is difficult in free-ranging wildlife. In reindeer Savelev et al. (1962) have shown that dimethoate is effective against the later larval stages.

Oestromyia leporina (Pallas, 1778) occurs in dermal cysts in voles and muskrats in Europe. The life-cycle is similar to that of the *Hypoderma* spp. Several other species of the genus which occur in rodents are discussed by Zumpt (1965).

FAMILY: CUTEREBRIDAE BRAUER AND VON BERGENSTAMM, 1889

This is a family of New World flies related to Calliphoridae and Oestridae (Oldroyd & Smith 1973). The genus *Dermatobia* affects man while *Cuterebra* is a dermal parasite of rodents.

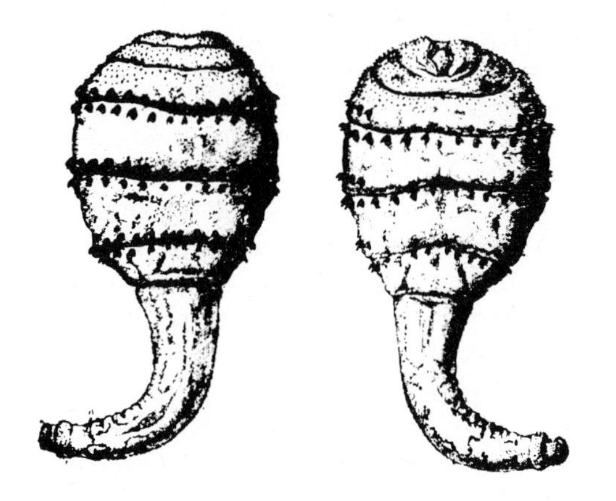

Fig 2.75 Larvae of *Dermatobia hominis* in dorsal (left) and ventral (right) views. (*After Blanchard*)

Genus: Dermatobia Brauer, 1860

Dermatobia hominis (Linnaeus, 1781) (syn. *D. ovaniventris*), also called the 'berne', 'nuche' or 'forcel', occurs in man in tropical America from Mexico to the Argentine. Cattle, dogs, cats, sheep, rabbits and other animals, including man, may become infected and in some areas of Latin America it constitutes an important pest of cattle. It has been estimated to cause losses of $200 million in meat and milk production (Steelman 1976) in this area of the world. The female is about 12 mm long. The thorax is dark blue with a greyish bloom; the abdomen is short and broad and has a brilliant blue colour.

Life-cycle. The adult flies do not feed and nourishment is derived from food stores accumulated during the larval period. When the adult fly is ready to oviposit, she captures a mosquito, or other blood-sucking fly, and glues a batch of eggs to the abdomen of the captive fly. When the transport fly alights on a warm-blooded host the larvae of *D. hominis* hatch from the eggs and penetrate the skin of the host, often using the skin puncture made by the blood-sucking fly. About six days are required for the egg to reach the stage of hatching, but this occurs only when the carrier fly settles on a suitable animal to feed. The most common vectors of *D. hominis* larvae are members of the genus *Psorophora*, though *Culex* and *Stomoxys* are also concerned and non-blood-sucking muscids have also been incriminated. Non-insect transmission may occur when *D. hominis* eggs are deposited on damp clothes or laundry. *D. hominis* breeds in forest country and domestic and wild mammals are commonly parasitized. Man is usually infected when he is associated with domestic animals. As the larva grows under the skin it produces a swelling which has a central opening through which it breathes. These swellings are usually very painful but in addition, in domestic animals, provide an attraction for myiasis-producing muscid flies, including screw-worms. Development in the host requires five to ten weeks, after which the larva escapes and pupates in the ground for an equally long period before the fly emerges. The mature larva is about 25 mm long and has a few rows of strong spines on most of the segments.

Diagnosis. The presence of a superficially situated swelling with a central opening, especially if more than one is present, would lead to the suspicion of myiasis. Specific diagnosis can be made only after extraction of the larva.

Treatment. In man surgical removal of the parasite is usually practised. Topical application of 1% trichlorophon has proved effective. In cattle and other animals the treatments for the *Hypoderma* spp. are indicated.

Genus: Cuterebra Clark, 1815

Large flies (20 mm or more in length), bodies bee-like, mouth parts vestigial. Larvae parasitic, large (25 mm in length), stout and parasitic under the skin of rodents etc. North American forms.

The adults oviposit near the entrance of the burrows of rabbits (*C. buccata* (Fabricius) 1776), *C. americana* (Fabricius) 1775, *C. lepivora* Coquillett, 1898) or of mice and chipmunks (*C. emasculator* Fitch, 1856). Several other species are listed by Capelle (1971). Larvae hatch at intervals and penetrate the skin of the above hosts producing cyst-like subcutaneous lesions in which the larvae mature. Mature larvae are produced in about one month, at which stage they are dark in colour and covered with bands of spines. Younger larvae are lighter in colour. Larvae leave the host to pupate in the soil.

Pathogenesis. A large subcutaneous cyst is produced with associated swelling. *C. emasculator* frequently parasitizes the scrotum destroying the testes and causing parasitic castration.

Though these forms are usually found in wild rodents, occasional infection of cats, dogs and humans may occur. In the cat larvae are frequently found in the neck or submandibular region and owing to scratching the swelling may become secondarily infected. Infection of the cranial cavity of the cat has been reported as well as parasitic orchitis in the dog and cat. Human nasal and dermal infection may also occur.

Treatment. Surgical removal is the most satisfactory. Ostlind et al. (1979) have reported high activity of the avermectins (A_1 being active at a level of 0.078 mg/kg body weight) in mice experimentally infected with larvae of *Cuterebra* spp.

SECTION PUPIPARA MERDIVENCI, 1966

The Pupipara are a group of aberrant Diptera, related to the Muscidae. They are markedly adapted to a parasitic life. All the species, with one exception, live on the blood of mammals or birds. The body is broad and flattened dorsoventrally; the abdomen is indistinctly segmented and usually its wall is soft and leather-like. Wings are present in some species and absent in others. The antenna has one joint and lies in a pit on the forehead. The feet are provided with strong claws, by means of which the parasite clings to the hairs or feathers of its host. As the name Pupipara implies, the females give birth to larvae which are ready to pupate.

FAMILY: HIPPOBOSCIDAE LATREILLE, 1796

This family includes the 'forest flies' which attack horses and cattle, the sheep ked and a variety of species on bats and birds.

Genus: Hippobosca Linnaeus, 1761

Several species of this genus, especially **H. equina** Linnaeus, 1758 (cosmopolitan), **H. rufipes** Olfers, 1816 (Africa) and **H. maculata** Leach, 1817 (tropical and subtropical) are common parasites of horses and cattle. Other animals like dogs and camel may also be attacked. The flies are about 1 cm long to the tip of the abdomen and have a reddish-brown colour with pale yellow spots. There is one pair of wings, the veins of which are crowded together towards the anterior border. The short, thick palpi ensheath the tip of the slender proboscis, the main portion of which is withdrawn into the head during rest.

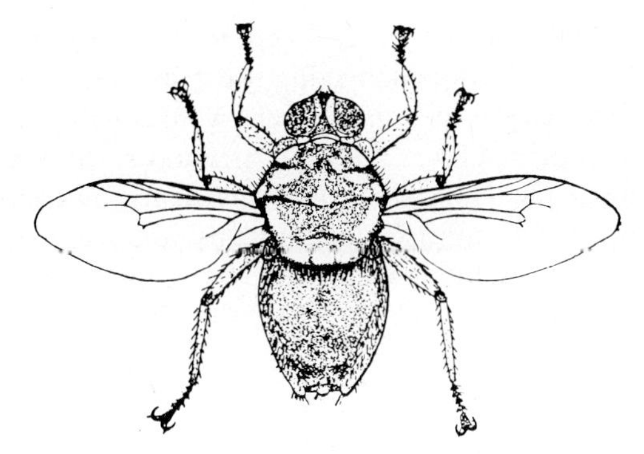

Fig 2.76 *Hippobosca rufipes*. (*After Bedford*)

Life-cycle and habits. The female fly deposits one larva at a time in sheltered spots where there is dry soil or humus. In Wales *H. equina* lays in humus at the roots of certain plants, mainly bracken (*Pteris aquilina*). The larvae pupate almost immediately and gradually turn from yellow to black. The larva is subglobular in shape; it measures about 5 by 4 mm and possesses a dark spot at the posterior pole. The length of the pupal period is greatly influenced by the temperature. The flies are most frequent in summer and attack more particularly in sunny weather. They remain for long periods on their hosts and are not easily disturbed. They cluster in the perineal region and between the hind-legs to the pubic region, but may also bite on other parts of the body. The flies tend to feed mainly on cattle and horses and are not inclined to travel more than a few metres, although they are strong fliers. These flies are a source of great irritation to animals which are not accustomed to them. They transmit the non-pathogenic *Trypanosoma theileri* to cattle and species of *Haemoproteus* to anatid and other birds.

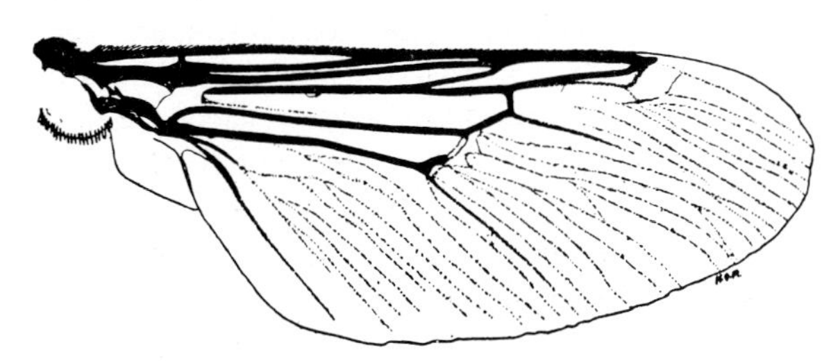

Fig 2.77 Wing of *Hippobosca rufipes*.

Control. The flies can be readily killed by spraying with oil-based insecticides containing pyrethrins or 1% Ciodrin plus 0.25% DDVP.

Hippobosca capensis v. Olfers, 1815 (syn. *H. francilloni*) occurs in Africa and Asia and attacks chiefly dogs and **H. camelina** Leach, 1817, occurs on the camel and horse in northern Africa.

Genus: Melophagus Latreille, 1804

Melophagus ovinus (Linnaeus, 1758), the sheep ked, is found in most parts of the world. It is a wingless, hairy, leathery insect, 4–6 mm long. The head is short, broad and not freely movable; the thorax is brown and the broad abdomen greyish-brown. The legs are strong and armed with stout claws.

Life-cycle and habits. Keds are permanent ectoparasites. The female attaches its larva to the wool of the sheep by means of a sticky substance. Parturition lasts a few minutes. The larva is immobile and soon turns into a chestnut-brown pupa. It is ovoid in shape with broad ends and 3–4 mm long. The pupal stage lasts 19–23 days in summer to 36 days in winter, or longer if the sheep are exposed to very cold conditions. The female ked lives four to five months on a sheep. Copulation occurs three to four days after emergence of the adult and each gestation lasts about 10–12 days. A female may produce 10–15 larvae. Engorged females can live up to eight days off the host. Pupae removed from the sheep, for instance at shearing, can hatch if conditions are favourable, but the emerging adults die very soon if they do not find a sheep to feed on. They usually spread from sheep to sheep by contact and are most numerous in the autumn and winter. A summer decrease in numbers occurs on all sheep of any age or sex. Sheep with dense, long or clotted fleeces are more likely to spread the infection because the keds come to the surface of such fleeces.

Pathogenesis. The parasites live in the wool of the sheep and suck blood. Heavy infections can reduce the condition of the host considerably and even cause anaemia. They produce intense

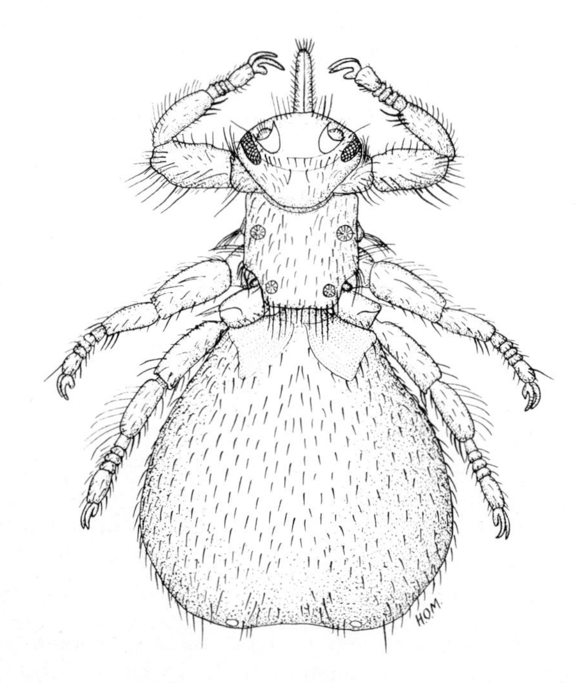

Fig 2.78 Adult *Melophagus ovinus*

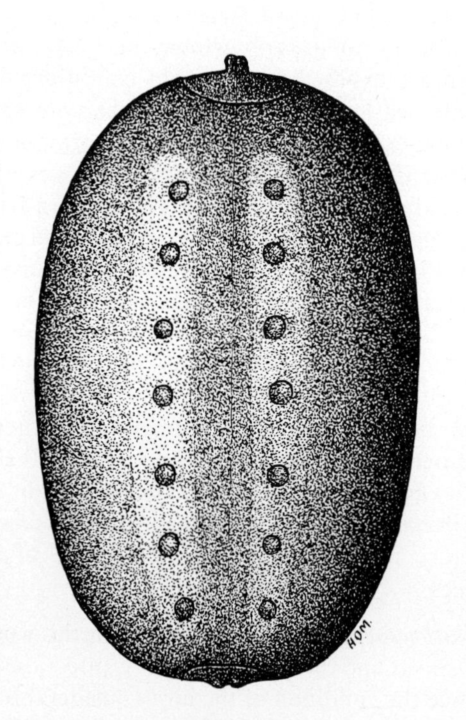

Fig 2.79 Pupa of *Melophagus ovinus.*

irritation, causing the sheep to bite, rub and scratch itself, thus damaging the wool. The faeces of the keds produce stains in the wool which do not wash out readily. The ked transmits the non-pathogenic *Trypanosoma melophagium* to sheep. Poorly fed animals or animals that are not protected against cold weather are most liable to suffer from keds, so that the parasites are particularly troublesome towards the end of winter.

Control. The ked population is markedly reduced by shearing. Where control measures are in operation for lice, blowflies, mange or ticks, these usually result in the control of *M. ovinus.* The insects are very susceptible to organophosphate insecticides (e.g. ronnel, 24% in 4.5 litres of water; diazinon, 0.06%, and Ciodrin, 1% are effective sprays). 'Tip spraying' is an effective control measure with short-fleeced sheep but with the long-woolled breeds a second treatment may be necessary.

Genus: Pseudolynchia Bequaert, 1925

Pseudolynchia canariensis (Macquart, 1840) (syn. *P. maura*) is a dark brown fly, 6 mm long, which resembles the sheep ked, but has a pair of transparent, tapering wings with the venation reduced and concentrated along the anterior border. The claws are strong and spurred. The parasite is widely distributed in warm countries and lives on domestic pigeons and a few wild birds.

Life-cycle and habits. The flies move through the feathers, sucking blood and causing painful wounds, especially on young nestling pigeons two to three weeks old, when the feathers begin to grow and afford protection. They may bite man. The female produces four to five young during her life of 43 days or so. Copulation takes place on the host and the larvae are laid in dark crevices of the pigeon-house in dry dust, in the nests or occasionally on the host, after which they roll off to pupate in cracks and crevices. They are yellow with a dark posterior pole and measure about 3 by 2.5 mm; they turn into black pupae in a few hours.

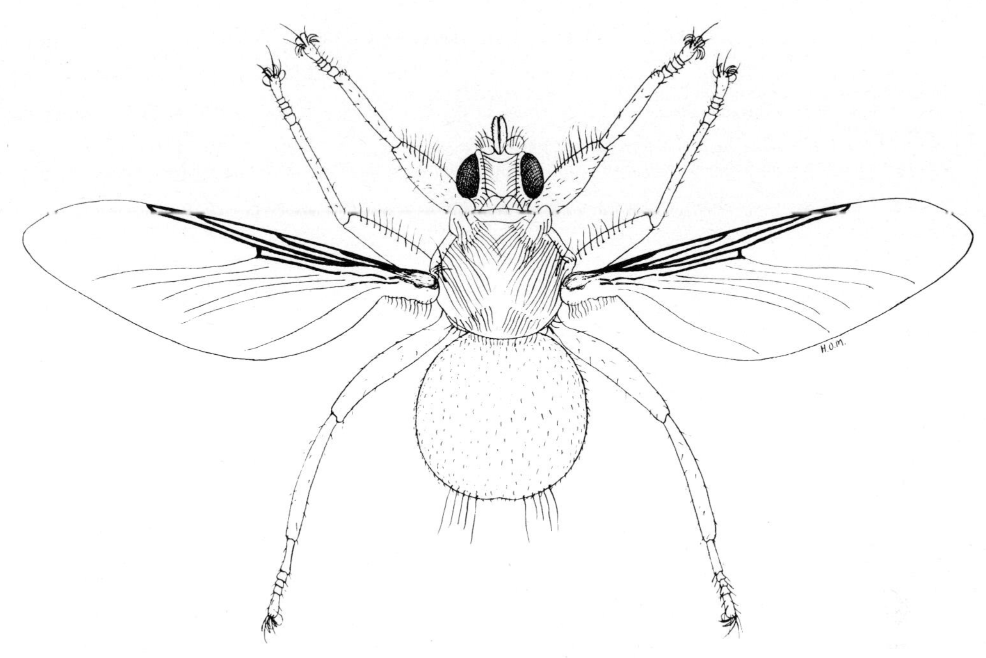

Fig 2.80 *Pseudolynchia canariensis.*

The pupal stage lasts 23–31 days in warm weather. *P. canariensis* and other species (*P. lividicolor* (Bigot, 1885) in Brazil and *P. capensis* (Bigot, 1885) in Spain) transmit *Haemoproteus columbae*, a blood protozoon of pigeons, and the related *H. lophortyx* of the quail.

Control. Organophosphate dusts or sprays containing coumaphos or carbaryl, as used for lice, are effective. Thorough cleaning of the pigeon loft and destruction of the debris is essential in control.

Genus: Lipoptena Nitzsch, 1818

Both sexes of this genus have wings but cast them when they find a host.

Lipoptena cervi (Linnaeus, 1761) is a common parasite of various deer, the wild boar and the badger in Europe. It has been reported to cause skin lesions in man (e.g. porters of carcasses of game animals in Les Halles in Paris). The fly is leathery, brownish in colour and the head sunken into the thorax. The life-cycle is similar to that of *H. equina*.

Lipoptena caprina Austen, 1921 occurs on the goat.

Two families of flies similar to the hippoboscids occur on bats. These are the *Nycteribiidae* which occur on bats in the Old World and *Streblidae* which occur on bats in the tropics and subtropical regions.

REFERENCES

INSECTA—GENERAL

Richards, O. W. & Davies, R. G. (1977) *Imms' General Textbook of Entomology*, 10th ed., vol. 2, pp. 418–1354. London: Chapman and Hall.

MALLOPHAGA AND SIPHUNCULATA

Clay, T. (1970) The Amblycera (Phthiraptera: Insecta) *Bull. Br. Mus. nat. Hist. (Ent.)*, **25**, 73–98.

Dailey, M. D. & Brownell, R. L. (1972) A checklist of marine mammal parasites. In: *Mammals of the Sea, Biology and Medicine*, ed. S. H. Ridgway, pp. 528–589. Springfield, Ill.: Charles C. Thomas.

Flynn, R. J. (1973) *Parasites of Laboratory Animals*, p. 884. Ames: Iowa State University Press.

Griffiths, A. J. (1975) Amitraz. For the control of animal ectoparasites with particular reference to sheep tick (*Ixodes ricinus*) and pig mange (*Sarcoptes scabei*). *Proc. 8th Br. Insecticide Fungicide Conf.*, 557–563.

Humphreys, P. N. (1975) Wet feather associated with *Holomenopon leucoxanthum* in a duck. *Vet. Rec.*, **97**, 96–97.

Kirkwood, A. C., Quick, M. P. & Page, K. W. (1978) The efficacy of showers for control of ectoparasites of sheep. *Vet. Rec.*, **102**, 50–54.

Martin, M. (1934) Life history and habits of the pigeon louse (*Columbicola columbae*). *Can. Entomol.* **66**, 1–16.

Page, K. W. (1974) Automatic control of guinea-pig lice with a synergised pyrethrins aerosol. *Vet. Rec.*, **94**, 254.

Richards, O. W. & Davies, R. G. (1977) *Imm's General Textbook of Entomology*, 10th ed., vol. 2, pp. 418–1354. London: Chapman and Hall.

Turner, E. C. (1971*a*) Ectoparasites. In: *Infectious and Parasitic Diseases of Wild Birds*, ed. J. W. Davis, R. C. Anderson, L. Karstad & D. O. Trainer, pp. 175–184. Ames: Iowa State University Press.

Turner, E. C. (1971*b*) Fleas and lice. In: *Parasitic Diseases of Wild Mammals*, ed. J. W. Davis & R. C. Anderson, pp. 65–77. Ames: Iowa State University Press.

SIPHONAPTERA

Benjamini, E., Feingold, B. F. & Kartman, L. (1961) Skin reactivity in guinea-pigs sensitized to flea bites. The sequence of reactions. *Proc. Soc. exp. Biol. Med.*, **108**, 700–702.

Bibikova, V. A. (1977) Contemporary views on the interrelationships between fleas and the pathogens of human animal disease. *A. Rev. Ent.*, **22**, 23–32.

Eskey, C., Prince, F. & Fuller, F. B. (1949) Transmission of *Salmonella enteritidis* by the rat fleas *Xenopsylla cheopis* and *Nosopsyllus fasciatus*. *Public Hlth Rep., Wash.* **64**, 93–94.

Hopkins, G. H. E. & Rothschild, M. (1953–1971) *An Illustrated Catalogue of the Collection of Fleas*, parts 1–5. London: British Museum of Natural History.

Mead-Briggs, A. R. & Rudge, A. J. B. (1960) Breeding of the rabbit-flea *Spilopsyllus cuniculi* (Dale); requirement of a 'factor' from a pregnant rabbit for ovarian maturation. *Nature, Lond.*, **187**, 1136–1137.

Michaeli, D. E., Benjamini, E., de Buren, F. P., Larrivee, D. H. & Feingold, B. F. (1965) The role of collagen in the induction of flea bite hypersensitivity. *J. Immunol.*, **95**, 162–170.

Rothschild, M. (1965) The rabbit-flea and hormones. *Endeavour*, **24**, 162–168.

Smit, F. G. A. M. (1973) Siphonaptera (fleas). In: *Insects and Other Arthropods of Medical Importance* ed. K. G. V. Smith, pp. 325–371. London: British Museum of Natural History.

Thoday, K. L. (1979) Skin diseases of dogs and cats transmissible to man. *In Practice*, **1**, 5–15.

Turner, E. C. (1971*a*) Ectoparasites. In: *Infectious and Parasitic Diseases of Wild Birds*, ed. J. W. Davis, R. C. Anderson, L. Karstad & D. O. Trainer, pp. 175–184. Ames: Iowa State University Press.

Turner, E. C. (1971*b*) Fleas and lice. In: *Parasitic Diseases of Wild Mammals*, ed. J. W. Davis & R. C. Anderson, pp. 65–77. Ames: Iowa State University Press.

DIPTERA—NEMATOCERA

Crosskey, R. W. (1973) Simuliidae. In: *Insects and Other Arthropods of Medical Importance*, ed. K. G. V. Smith, pp. 109–153. London: British Museum of Natural History.

Elliot, M., Jones, N. F. & Potter, C. (1978) The future of pyrethroids in insect control. *A. Rev. Ent.*, **23**, 443–469.

Fallis, A. M. & Bennett, G. F. (1961) Ceratopagonidae as intermediate hosts for *Haemoproteus* and other parasites. *Mosquito News*, **21**, 21–28.

Freeman, P. (1973) Diptera—Introduction. In: *Insects and Other Arthropods of Medical Importance* ed. K. G. V. Smith, pp. 21–36. London: British Museum of Natural History.

Kettle, D. S., Parish, R. H. & Parish, J. (1959) Further observations on the persistence of larvicides against *Culicoides* and a discussion on the interpretation of population changes in the untreated plots. *Bull. ent. Res.*, **50**, 63–80.

Lewis, D. J. (1973) Phlebotomidae and Psychodidae (sand-flies and moth-flies). In: *Insects and Other Arthropods of Medical Importance*, ed. K. G. V. Smith, pp. 155–177. London: British Museum of Natural History.

McCraig, J. (1973) A survey to establish the incidence of sweet itch in ponies in the U.K. *Vet. Rec.*, **93**, 444–446.

Mattingly, P. F. (1973) Culicidae (mosquitoes). In: *Insects and Other Arthropods of Medical Importance*, ed. K. G. V. Smith, pp. 37–107. London: British Museum of Natural History.

Mattingly, P. F., Crosskey, R. W. & Smith, K. G. V. (1973) Summary of arthropod vectors. In: *Insects and Other Arthropods of Medical Importance*, ed. K. G. V. Smith, pp. 497–532. London: British Museum of Natural History.

Mellanby, K. (1946) Man's reaction to mosquito bites. *Nature, Lond.*, **158**, 554.

Richards, O.W. & Davies, R. G. (1977) *Imms' General Textbook of Entomology*, 10th ed., vol. 2, pp. 418–1354. London: Chapman and Hall.

Riek, R. F. (1954) Studies on allergic dermatitis (Queensland itch) of the horse: The aetiology of the disease. *Aust. J. agric. Res.*, **5**, 109–129.

Shen, D. T., Gorham, J. R., Jones, R. H. & Crawford, T. B. (1978) Failure to propagate Equine Infectious Anaemia virus in mosquitoes and *Culicoides variipenis*. *Am. J. vet. Res.*, **39**, 875–876.

Smith, C. M., Davis, A. N., Weidhma, D. E. & Seabrook, E. L. (1959) Insecticide resistance in the salt marsh sandfly *Culicoides furens*. *J. econ. Entomol.*, **52**, 352–353.

Smith, K. G. V. (1973) Brachycera and Cyclorrhapha Aschiza of minor medical importance. In: *Insects and Other Arthropods of Medical Importance*, ed. K. G. V. Smith, pp. 203–208. London: British Museum of Natural History.

Steelman, C. D. (1976) Effects of external and internal arthropod parasites on domestic livestock production. *A. Rev. Entomol.*, **21**, 155–178.

Thompson, W. H. & Beaty, B. J. (1977) Venereal transmission of La Cross (California Encephalitis) Arbovirus in *Aedes triseriatus* mosquitoes. *Science, N.Y.*, **196**, 530–531.

World Health Organization (1959) Seventh Report: Expert Committee on Malaria. *Tech. Rep. Ser. Wld Hlth Org.*, **162**.

Zumpt, F. (1949) Medical and veterinary importance of horse-flies. *S. Afr. med. J.*, **23**, 359–362.

DIPTERA—CYCLORRHAPHA

Andrews, A. H. (1978) Warble fly: The life cycle, distribution, economic losses and control. *Vet. Rec.*, **103**, 348–353.
Bailie, H. D. & Morgan, D. W. T. (1980) Field trials to assess the efficacy of permethrin for the control of flies on cattle. *Vet. Rec.*, **106**, 124–127.
Baumhover, A. H., Graham, A. J., Bitter, B. A., Hopkins, D. E., New, W. D., Dudley, F. H. & Bushland, R. C. (1955) Screw-worm control through release of sterilized flies. *J. econ. Ent.*, **48**, 462–466.
Baumhover, A. H., Husman, C. N., Skipper, C. C. & New, W. D. (1959) Field observations on the effects of releasing sterile screw-worms in Florida. *J. econ. Ent.*, **52**, 1202–1206.
Beesley, W. N. (1973) Control of arthropods of medical and veterinary importance. *Adv. Parasit.*, **13**, 115–192.
Beesley, W. N. (1974*a*) Arthropods Oestridae, myiasis and acarines. In: *Parasitic Zoonoses*, ed. E. J. L. Soulsby, pp. 349–368. New York: Academic Press.
Beesley, W. N. (1974*b*) Economics and progress of warble fly eradication in Britain. *Vet. Med. Rev.*, **4**, 334–347.
Bursell, E. (1970) Feeding, digestion and excretion. In: *The African Trypanosomiases*, ed. H. W. Mulligan, pp. 305–315. New York: Wiley-Interscience.
Bushland, R. C. (1960) New research results with systemic insecticides. *Proc. 62nd ann. Meet. U.S. live Stock sanit. Ass.*, 192–197.
Buxton, P. A. (1955) *The Natural History of Tsetse Flies: An Account of the Biology of the Genus Glossina (Diptera)*, Memoir No. 10 London School of Hygiene and Tropical Medicine, p. 816. London: H. K. Lewis.
Capelle, K. J. (1971) Myiasis. In: *Parasitic Diseases of Wild Mammals*, ed. J. W. Davis & R. C. Anderson, pp. 279–305. Ames: Iowa State University Press.
Cheng, T. H. (1958) The effect of biting fly control on weight gain in beef cattle. *J. econ. Ent.*, **51**, 275–278.
Chereshnev, N. A. (1954) Stomatitis caused by *Gastrophilus pecorum* in horses. *Trudy Inst. vet. Kazak. Filial. Vses Akad. Sel'skok hoz. Nauk. Alma-Ata*, **6**, 379.
Depner, K. R. (1969) Distribution of the face fly *Musca autumnalis* (Diptera: Muscidae) in western Canada and the relation between its environment and population density. *Can. Ent.* **101**, 97–100.
Dorsey, C. K. (1966) Face-fly control experiments on quarter horses—1926–64. *J. econ. Ent.*, **59**, 86–89.
Elliot, M., Jones, N. F. & Porter, L. (1978) The future of pyrethroids in insect control. *A. Rev. Ent.*, **23**, 443–496.
Egerton, J. R., Ostlend, D. R., Blair, L. S., Eary, C. H., Suhayda, D., Cifelli, S., Rick, R. F. & Campbell, W. C. (1979) Avermectins, new family of potent anthelmintic agents: efficacy of the B1a component. *Antimicrob. Agents Chemother.*, **15**, 372–378.
Ford, J. (1970*a*) The geographic distribution of *Glossina*. In: *The African Trypanosomiases*, ed. H. W. Mulligan, pp. 274–297. New York: Wiley-Interscience.
Ford, J. (1970*b*) The search for food. In: *The African Trypanosomiases*, ed. H. W. Mulligan, pp. 298–304. New York: Wiley-Interscience.
Foster, G. G., Maddern, R. H., Smith, P. H., Vogt, W. G. & Wardlaugh, K. G. (1978) Genetic manipulation of sheep blowfly populations. In: *Aust. Adv. vet. Sci. 1978*, 86–87.
Freeman, P. (1973) Diptera—Introduction. In: *Insects and other Arthropods of Medical Importance*, ed. K. G. V. Smith, pp. 21–36. London: British Museum of Natural History.
French, N., Wright, A. J., Wilson, W. R. & Nichols, D. B. R. (1977) Control of headfly on sheep. *Vet. Rec.*, **100**, 40–43.
Glasgow, J. P. (1970) The *Glossina* community. In: *The African Trypanosomiases* ed. H. W. Mulligan, pp. 348–81. New York: Wiley-Interscience.
Gräfner, G. & Hiepe, T. (1979) Beitrag zum Krankheitsbild und zur Pathogenese des Kriebelmückenbefalles bei Weidetieren. *Mh. VetMed.*, **34**, 538–540.
Hadlow, W. J., Ward, J. K. & Krinsky, W. L. (1977) Intracranial myiasis by *Hypoderma bovis* (Linnaeus) in a horse. *Cornell Vet.*, **67**, 272–281.
Hadwen, S. & Bruce, E. A. (1917) Anaphylaxis in cattle and sheep, produced by the larvae of *Hypoderma bovis, H. lineatum* and *Oestrus ovis*. *J. Am. vet. Med. Ass.*, **51** (N.S. **4**), 15–44.
Hatch, C., McCaughey, W. J. & O'Brien, J. J. (1976) The prevalence of *Gastrophilus intestinalis* and *G. nasalis* in horses in Ireland. *Vet. Rec.*, **98**, 274–276.
Hopkins, P. S. (1978) Blowfly control by chemical mulesing, chemical crutching or mist application of insecticide-acrylic emulsions. *Aust. Adv. vet. Sci. 1978*, 85–86.
Horak, I. G. & Snijders, A. J. (1974) The effect of *Oestrus ovis* infestation on Merino lambs. *Vet. Rec.*, **94**, 12–16.
Hunter, A. R. (1975) Sheep headfly disease in Britain. *Vet. Rec.*, **97**, 95–96.
Jordan, A. M. (1978) Recent developments in techniques for tsetse control. In: *Medical Entomology Centenary Symposium Proceedings*, ed. S. Willmott, pp. 76–84. London: Royal Society of Tropical Medicine and Hygiene.
Khan, M. A. (1973) Toxicity of systemic insecticides: efficacy of dermal and parenteral applications of Crunfomate for system control of *Hypoderma* spp. in cattle. *Vet. Rec.*, **93**, 525–532.
Lloyd, D. H. & Dipeolu, O. O. (1974) Seasonal prevalence of flies feeding on cattle in Northern Nigeria. *Trop. Anim. Hlth Prod.*, **6**, 231–236.
McDougall, R. S. (1930) The warble flies of cattle, *Trans. R. Highld agric. Soc. Scotl.*, **42**, 75–112.
Morgan, N. O. (1964) Autecology of the adult horn fly *Haematobia irritans* (L.) (Diptera: Muscidae). *Ecology*, **45**, 728–736.
Mules, J. H. W. (1935) Crutch strike by blowflies in sheep. A preventive operation. *Queensld. agric. J.*, **44**, 237–241.
Mulligan, W. H. (Ed.) (1970) *The African Trypanosomiases*, p. 950. New York: Wiley-Interscience.
Olander, H. J. (1967) The migration of *Hypoderma lineatum* in the brain of a horse. *Path. Vet.* **4**, 477–483.
Oldroyd, H. & Smith, K. G. V. (1973) Eggs and larvae of flies. In: *Insects and Other Arthropods of Medical Importance*, ed. K. G. V. Smith, pp. 289–323. London: British Museum of Natural History.
Ostlind, D. A., Cifelli, S. & Lang, R. (1979) Insecticidal activity of the antiparasitic avermectins. *Vet. Rec.*, **105**, 168.
Pollock, J. N. (1971) The origin of tsetse-flies. *J. Ent.*, (B) **40**, 101–109.
Potts, W. H. (1973) Glossinidea (Tsetse-flies). In: *Insects and Other Arthropods of Medical Importance*, ed. K. G. V. Smith, pp. 209–249. London: British Musem of Natural History.
Rich, G. B. (1970) The economics of systemic insecticide treatment for reduction of slaughter trim loss caused by cattle grubs. *Can. J. Anim. Sci.*, **50**, 301–310.
Richards, W. O. & Davies, R. G. (1977) *Imm's General Textbook of Entomology*, 10th ed., vol. 2, pp. 418–1354. London: Chapman and Hall.
Richardson, G. (1971) A comparison of mulesed and un-mulesed sheep in the Western District. *J. Agric. Vict. Dept. Agric.*, **61**, 10–11.
Savelev, D. V., Voblikova, N. V., Mezenev, N. P. & Silkov, A. M. (1962) Trials with trichlorophon, fenchlorphos, dichlorvos and dimethoate against the reindeer warble fly. *Veterinariya, Moscow*, **39**, 74.
Shanahan, G. J. & Hughes, P. B. (1978) Larval inplant studies with *Lucilia cuprina*. *Vet. Rec.*, **103**, 582–583.
Skerman, K. D. (1958) The efficiency of new insecticides for control of the body louse of sheep (*Damalina ovis*) by dipping. *Aust. vet. J.*, **35**, 75–79.
Smart, J. (1939) Cychorrhapha. In: *British Blood Sucking Flies*, ed. F. W. Edwards, H. Olroyd & J. Smart, pp. 115–127. London: British Museum.
Stampa, S. (1959) The control of internal parasites of sheep with Neguvon and Asuntol. A preliminary report. *J. S. Afr. vet. med. Ass.*, **30**, 19–26.
Steelman, C. D. (1976) Effects of external and internal arthropod parasites on domestic livestock production. *A. Rev. Ent.*, **21**, 155–178.
Stork, M. G. (1979) The epidemiological and economic importance of fly infestation of meat and milk producing animals in Europe. *Vet. Rec.*, **105**, 341–343.
Tarry, D. W. (1975) The significance of feeding and blood meal identification studies in the planning of sheep headfly control measures. *Proc. 8th Br. Insecticide Fungicide Conf.*, 573–580.
Tarry, D. W. & Boreham, P. F. L. (1977) Studies on the feeding patterns of the sheep headfly *Hydrotaea irritans* (Diptera muscidae) in Great Britain. *Vet. Rec.*, **101**, 456–458.
Tarry, D. W., Wilson, C. D. & Stuart, P. (1977) The headfly *Hydrotaea irritans* and summer mastitis infection. *Vet. Rec.*, **102**, 91.
United States Department of Agriculture (1965) Livestock and poultry losses. In: *Losses in Agriculture*, pp. 72–84, Handbook No. 291. US Department of Agriculture.
Waddell, A. H. (1972) The pathogenicity of *Gasterophilus intestinalis* larvae in the stomach of the horse. *Aust. vet. J.*, **48**, 332–335.
Weitz, B. G. F. (1970) Hosts of *Glossina*. In: *The African Trypanosomiases*, ed. H. W. Mulligan, pp. 317–326. New York: Wiley-Interscience.
Wright, C. L. & Titchener, R. N. (1977) Infection of the sheep headfly, *Hydrotaea irritans*, with bacterial isolates from field cases of summer mastitis. *Vet. Rec.*, **101**, 426.
Yeoman, G. H. & Bell, T. A. (1978) Sheep blowfly breech strike control using aluminium alkoxide gellants. *Vet. Rec.*, **103**, 337.
Zumpt, F. (1965) *Myiasis in Man and Animals in the Old World: A Textbook for Physicians, Veterinarians and Zoologists*, p. 267. London: Butterworth.

CLASS: ARACHNIDA LAMARCK, 1815

This Class includes scorpions, spiders, ticks and mites; and also other species which need not be considered in this book. The arachnids differ fundamentally in structure and function from the insects. Antennae, wings and compound eyes are absent, as is the division of the body into head, thorax and abdomen. Previously, the king-crabs were included in this class but they are now regarded as members of the class Merostomata (Sheals, 1973). The mouth of arachnids is small and they feed chiefly on the tissue fluids of other animals which they suck up by means of a sucking pharynx. They are thus carnivorous animals and many of them possess poison glands and poison claws, with which they paralyse their prey before they suck the juices out of them. The first and second pairs of appendages are modified to help in feeding, the first pair being called the *chelicerae* and the second pair the *pedipalps*. Either of these two pairs of appendages may be pincers and poison glands may be associated with them. The poison glands of the scorpion are, on the other hand, situated on a terminal, postanal segment of the body. The basal joints of the pedipalps of arachnids, and also those of some of the walking legs behind them, may bear teeth which help in chewing the prey. Such basal joints are called *gnathobases*.

The segmentation of the body of arachnids differs from that of other arthropods and different terms are used to describe it. Less confusion arises if the terms thorax and abdomen are abandoned and the term *prosoma* is given to the first six segments of the body and the term *opisthosoma* to the remaining segments. The prosoma bears the chelicerae, pedipalps and four pairs of walking legs. It may be divided into the *gnathosoma*, which bears the chelicerae and pedipalps, and the *podosoma*, which bears the four pairs of walking legs. The opisthosoma corresponds to the abdomen. The podosoma and the opisthosoma are sometimes together called the *idiosoma*. These subdivisions are not, however, evident on the bodies of ticks and mites, because the bodies of these species have lost external signs of segmentation. Most of them show a division into two parts only: (1) an anterior gnathosoma, which bears the chelicerae, and pedipalps, and also the median hypostome developed by these species only (the gnathosoma of ticks and mites is called the *capitulum*) and (2) a posterior single piece, which represents the fused podosoma and opisthosoma and may therefore be called idiosoma. The mouth parts of ticks and mites are modified for sucking the blood or tissue fluids of their hosts and for holding on to the host. They are further described below.

Arachnids breathe by means of the gill-books, lung-books and tracheae described above. *Gill-books* are present on segments nine to thirteen of the aquatic king-crabs. *Lung-books* and tracheae are present in the air-breathing species, which may breathe by means of either tracheae or lung-books or both. Some species, such as some aquatic and other mites, have no special respiratory organs, but absorb oxygen through the cuticle. Various classifications of the Class have been proposed and the following division of the *Euarachnida* is adopted in this volume. This subclass includes the scorpions, spiders, ticks and mites and a number of other related species. The Euarachnida breathe air and their bodies are divided into a prosoma and opisthosoma, except those of the ticks and mites whose bodies consist of one piece only, which is formed by fusion of the thorax and abdomen. This subclass may be divided into the following orders:

SCORPIONIDEA

These are relatively large, terrestrial arachnids which inhabit warm countries. The prosoma consists of a single piece covered by a single dorsal plate. Behind this, the opisthosoma is divided into a portion called the *mesosoma*, consisting of seven segments which are as broad as the segments of the prosoma, and a narrower portion behind this, consisting of five segments, called the *metasoma*, behind which is the terminal, post-anal *sting* containing the poison gland. The chelicerae of Scorpionidea are small, but the pedipalps are large and bear pincers. There are four pairs of walking legs and on the ventral sternal plates of segments

10–13 there are four pairs of stigmata leading into the lung-books. The body ends in a post-anal segment called the sting, which contains a poison gland and is provided with a sharp spine.

PEDIPALPEA

These are predatory species living in warm climates, which prey chiefly on insects. They have large pedipalps and the chelicerae bear claws, which may contain poison glands. For this reason the bites of some species of this order may have severe effects.

ARANEIDEA

These are the spiders, which breathe air. Their bodies are divided into a prosoma (cephalothorax) and an unsegmented soft abdomen, the first segment of which forms a stalk (pedicel) that joins these two parts together. The chelicerae have poison glands and hooks for killing the prey and the pedipalps are relatively small. The bites of some spiders may have serious effects.

PALPIGRADEA

These are microscopic terrestrial species, which look rather like insects, because the prosoma is covered by three separate plates and the opisthosoma is a broader part of the body.

SOLIFUGEA (SOLFUGAE)

These hairy, nocturnal, terrestrial species live in warm countries and breathe by means of tracheae. The first two segments, or according to some experts the first three, form a single piece that bears the very large and powerful pincer-like chelicerae, the leg-like pedipalps and the first pair of legs. This anterior part of the body is jointed to the rest of the body, so that it can be raised. The chelicerae can kill small mammals and birds, but Solifugae feed chiefly on insects.

CHERNETIDEA (CHELIFERAE)

The small species of this order are often called chelifers or pseudo-scorpions. The prosoma is covered by a single plate and the opisthosoma is segmented. The chelicerae are small pincers, but the pedipalps, like those of scorpions, are large and end in pincers. Chelifers feed chiefly on insects and their larvae.

PODOGONEA

This order contains only two small species found in South America and West Africa. The prosoma is a single piece, the opisthosoma is segmented and both the chelicerae and pedipalps bear pincers.

PHALANGIDEA

Species of this order breathe by means of tracheae and are sometimes called harvestmen, but they should not be confused with the larvae of the mites belonging to the acarine genus *Trombicula*, which are often called harvesters. Phalangidea look rather like spiders, but the opisthosoma is segmented and joined to the podosoma across its whole width, so that there is no waist. The chelicerae bear pincers, but the long pedipalps do not. The very long walking legs are brittle.

ACARINA

The species of this order are the hard and soft ticks and their numerous small or minute relatives called mites. Apart from the ticks, most of the species of the order are minute. A few of the numerous species cause the various kinds of mange or inflict other kinds of injury on farm animals. The mouth parts consist of a pair of chelicerae, a pair of pedipalps and, between these, a median toothed structure called the *hypostome*. They are borne on the *gnathosoma*, which consists of a plate called the *capitulum* and the mouth parts just mentioned. The segmentation of the rest of the body is indistinct or absent.

The life-history of Acarina begins with the *egg*, from which emerges a *larva*, which lacks the fourth pair of legs present in the later phases, so that the larva has only six legs. The larva moults to become the *nymph*, which usually resembles the adult, but has no sexual organs. One or more nymphal instars, sometimes called the protonymph, deutonymph and tritonymph, may precede the appearance of the adult phase. The other features of the species of Acarina described in this book are given below.

This order Acarina may be divided into six suborders. Of these the suborders Nostostigmata and Holothyroidea have no economic importance. The remaining suborders, some species of which are further described below, are:

Suborder: Mesostigmata

This suborder includes species of the gamasid mites, examples of which are the red mite of poultry (*Dermanyssus gallinae*), the tropical fowl mite *Ornithonyssus bursa* and their relatives.

Suborder: Ixodoidea (Ixodides)

This suborder includes the hard and soft ticks.

Suborder: Trombidiformes

This suborder includes species which are pests of various fruits and bulbs and also the species of the genus *Demodex*, which cause demodectic mange and those of the genus *Trombicula*, the larvae of which suck the tissue fluids of man and animals and may be important vectors of disease, such as scrub (mite) typhus.

Suborder: Sarcoptiformes

This suborder may be divided into: (*a*) the *Oribatei* (oribatid mites), which are not parasitic, although some species are intermediate hosts of tapeworms belonging to the families Anopolocephalidae and Catenotaeniidae (of rats and mice), and (*b*) the *Acaridae*, a group which includes the mites which cause sarcoptic and other forms of mange and also various species which injure grains, flour and other stored products.

ORDER: ACARINA NITZSCH, 1818

SUBORDER: MESOSTIGMATA CANESTRINI, 1891

Species of this suborder are usually armoured with brown or dark-brown plates. The body, like that of the ticks, to which the species of this suborder are related, is divided into two portions only: an anterior minute gnathosoma, which bears the mouth parts, and a posterior idiosoma. The name of the suborder refers to the fact that the single pair of stigmata are lateral and outside the coxae of the legs. Like the stigmata of ticks, they may be borne on peritremal plates. There are no genital suckers. Among the numerous species of the suborder Mesostigmata the only ones of veterinary importance belong to the group of the suborder called the Gamasides or gamasid mites. Some gamasid mites are not parasitic and live in soil, moss, decaying wood or vegetation or in litter. Others are parasitic on myriapods, beetles and other insects, snakes, birds, bats and other mammals. The following species have veterinary importance.

FAMILY: DERMANYSSIDAE KOLENATI, 1859

Genus: Dermanyssus Dugès, 1834

Dermanyssus gallinae (DeGeer, 1778). This cosmopolitan species (Fig. 2.81) attacks the fowl, pigeon, canary and other cage birds and also many wild birds. It may also feed on man. It is often called the red mite of poultry, but is, like the other mites mentioned below, only red when it has recently fed on its host's blood; otherwise it is whitish, greyish or black. The engorged female adult is about 1 mm long or larger, the other stages being smaller. The dorsal shield does not quite reach the posterior end of the body and its posterior margin is truncated. The setae on it are smaller than those on the skin around the dorsal plate. The anus is on the posterior half of the anal plate, whereas in *O. sylviarum* the anus is on the anterior half of this plate. The chelicerae are long and whiplike.

Life-cycle and habits. The eggs are laid, usually after a blood meal, in cracks and crevices in the walls of the poultry houses or in the nest of the birds, up to seven eggs being laid at a time. The eggs hatch, at outdoor summer temperatures, in 48–72 hours, liberating six-legged larvae which do not feed. These moult in 24–48 hours to become protonymphs, which feed on the host's blood and moult after 24–48 hours to become deutonymphs and these, after a blood meal, moult in 24–48 hours to become the adults. The whole life-cycle can be completed in seven days under optimal conditions. The adults can, under experimental conditions,

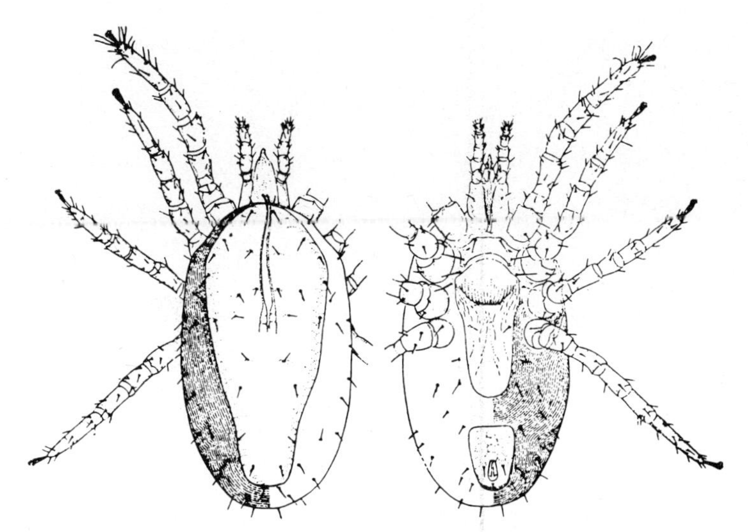

Fig 2.81 *Dermanyssus gallinae*. Left, Dorsal view of the female. Right, Ventral view of the female.

live for four to five months without a meal of blood.

The nymphs and adults periodically visit the hosts to suck blood and hide, during intervals between meals, in cracks and crevices in the quarters of the birds. Under favourable conditions the mites reproduce rapidly and may become a serious pest, causing much irritation and anaemia due to loss of blood. The birds become listless, their egg production may be reduced and loss of blood may cause death. *D. gallinae* may be a frequent parasite of aviaries, especially when these are heated. *D. gallinae* is, in Australia, a vector of *Borrelia anserina*, the cause of spirochaetosis of the fowl. The parasite has been found naturally infected with the viruses of St Louis encephalitis, eastern encephalitis and western encephalitis and consequently it may act as a vector for these infections. *D. gallinae* may occur as a temporary parasite on humans, causing skin lesions. The mites may occur in birds' nests in eaves of houses and the parasites migrate into bedrooms and attack humans in bed.

Diagnosis. The mites can be seen with the unaided eye, especially when they have fed recently on blood and are red. Other causes of anaemia, such as the soft tick, *Argas persicus*, and Simuliidae and Cimicidae, should be eliminated.

Control. The nests of the birds should be removed and well-made houses free from crevices should be provided. Good control can be achieved with 0.25% Sevin (carbaryl) applied to the houses and deep-litter houses and perches, repeated after two to three weeks. The synthetic pyrethrum Amitraz is effective though extended exposure is necessary (Price 1977). Dusts or sprays of malathion give good control and HCH sprays are effective for treatment of the premises, but must not be used to treat the birds or their foods. For the treatment of the premises or litter good results have been obtained with 0.5% lindane, 2% chlordane or 2% malathion dust spread over the litter at the rate of 225 g/m^2. Since cage birds are especially susceptible to the chlorinated hydrocarbon insecticides and some of the organophosphorous compounds, they should be avoided or used with great care for these species. Alternatives are synergized pyrethrins or Sevin (carbaryl) which may be used in the form of aerosol sprays.

Genus: Ornithonyssus Sambon, 1928 (Bdellonyssus Fonseca, 1941; Liponyssus)

Ornithonyssus sylviarum (Canestrini & Fanzago, 1877). This species (Fig. 2.82), often called the northern mite of poultry, is found on the

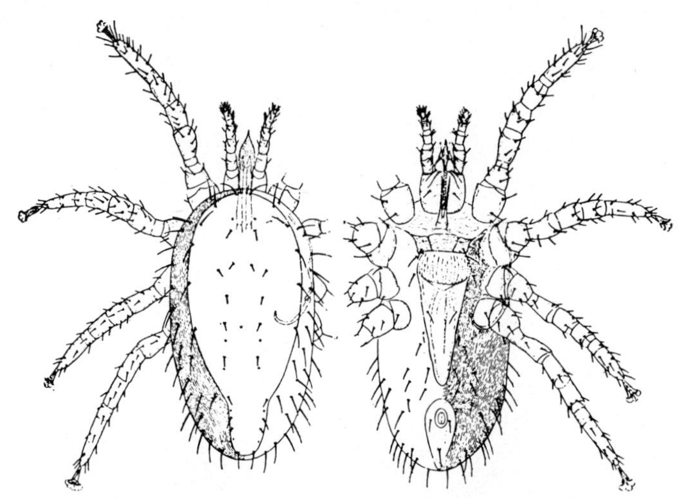

Fig 2.82 *Ornithonyssus sylviarum.* Left, Dorsal view of the female. Right. Ventral view of the female.

fowl and other birds in temperate climates generally and it has been found in Britain and in New Zealand. The elongate to oval adult mites are about 1 mm long. *O. sylviarum* can be distinguished from the other species here described by the shape of its dorsal plate, which is wide for two-thirds of its length and then rather suddenly tapers to form a tongue-like continuation about half as wide for the remainder of its length. The setae on the dorsal plate, like those of *D. gallinae*, are smaller than the setae on the adjacent skin. The ventral plate, like that of *D. gallinae*, bears only two pairs of setae, a third pair being present on the skin immediately behind this plate or almost touching it. The anus is on the anterior half of the anal plate.

Life-cycle and habits. The sticky, whitish eggs are laid largely on the host, one to five eggs being laid after a blood meal. The eggs hatch after one day, or earlier, according to the temperature and humidity. They liberate six-legged larvae, which do not feed, but moult after eight to nine days to become the protonymphs, which feed on blood, needing two blood meals, and then moult to become, after one to three days, the deutonymphs, which do not feed, but moult, after three to four days, to become adults. The whole life-cycle, from the female's blood meal before egg-laying to the adult stage, can occur in five to seven days at optimal conditions, but usually takes longer.

The mites are found on the birds or in their nests or houses and they feed intermittently on the birds, but seldom attack young chickens. They can bite through tender human skin and thus cause pruritus. Heavily infected birds may suffer irritation, loss of weight, reduction of egg production and even, when the loss of blood is great, death. This mite may transmit fowlpox and the viruses of St Louis encephalitis and western equine encephalomyelitis have been found in it.

Control. Effective and safe control has been achieved with Sevin (carbaryl: L-naphthyl methylcarbamate) dust or emulsion and with dusts of fenchlorphos (Dow ET 57), trichlorphon (Bayer L13/59; metrifonate; Dipterex, etc.) or coumaphos (Bayer 21/199; Asuntol; Co-Ral, etc). Good results have been claimed for 4% malathion dust applied to the vents and breasts of the birds or to their litter and nesting boxes and with dusts of 4% nicotine, or 4% malathion, or a mixture of these, which do not, it is claimed, affect the birds or the flavour of their eggs; success is claimed also for a 0.5% emulsion of malathion. Chlordane, toxaphene and lindane have been found toxic to birds.

Ornithonyssus (Bdellonyssus, Liponyssus) bursa (Berlese, 1888). This species (Fig. 2.83), often called the tropical fowl mite, is found on the fowl, pigeon, sparrow and other birds in the

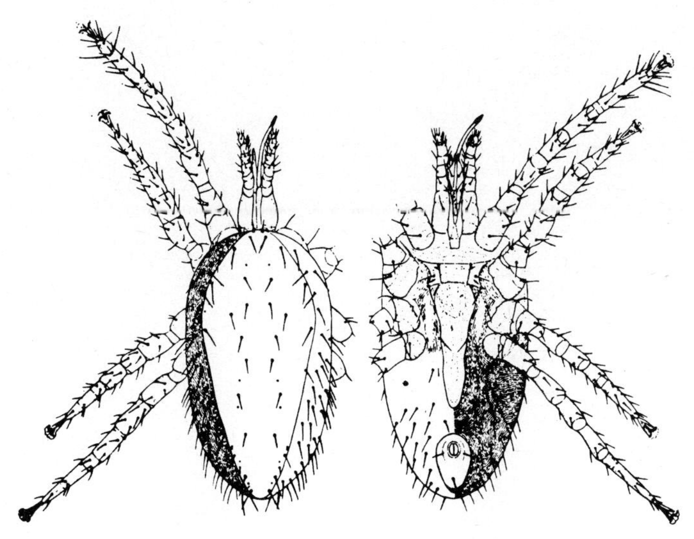

Fig 2.83 *Ornithonyssus bursa*. Left, Dorsal view of the female. Right, Ventral view of the female.

warmer parts of the world, where it replaces *O. sylviarum*, with which it has been confused. It will attack man, causing pruritus, but this is temporary, because this species cannot survive for longer than ten days away from a bird host. It has been found in southern Africa, India, China, Australia, Colombia, Panama and the United States. It can be distinguished from *O. sylviarum* by the shape of its dorsal plate, which gradually tapers to a blunt posterior end, although the setae on this plate are, like those on the dorsal plates of *O. sylviarum* and *D. gallinae*, smaller than those on the adjacent skin. The anus, like that of *O. sylviarum*, is on the anterior half of the anal plate; but in *O. bursa* the ventral plate bears all three pairs of setae, while in *O. sylviarum* and *D. gallinae* only two pairs of these setae are on the ventral plate, the third pair being on the skin behind the plate.

Life-cycle and habits. Baker et al. (1956) state that, in the laboratory, the eggs are usually laid in the litter, not on the host, but in birds in the field, large numbers of eggs may be found in the fluff of the feathers, as well as in the nests of the birds. On sparrows most of the life-cycle takes place in the nests and few mites are found on sparrows flying about. On poultry most of the mites are found on the fluff of the feathers, especially on those around the vent, and they tend to be present on few feathers, large numbers of them giving these feathers a dirty appearance. The distribution of the mites on the birds may be patchy, hundreds being found in small areas. The mites are not usually found on the roosts of the birds.

The eggs hatch in about three days, liberating six-legged larvae, which do not feed, but moult, after about 17 hours, to become protonymphs, which feed on the host's blood and, after one to two days, moult to become deutonymphs, which also feed on the host's blood and become the adults. More details about the life-cycle are required.

Control. The methods for the control of *O. sylviarum* are applicable.

Ornithonyssus (Bdellonyssus, Liponyssus) bacoti (Hirst, 1931). This species (Fig. 2.84), often called the tropical rat mite, is parasitic on rats and man all over the world. The adult female mite is 0.65–1 mm long. Its dorsal plate is narrower than that of other species just described and it tapers gradually to a blunt point; on it are numerous setae, which are the same size as those on the adjacent skin. The chelicerae have no teeth and there is a spur on the distal segment of the pedipalp. The sternal plate bears three pairs of

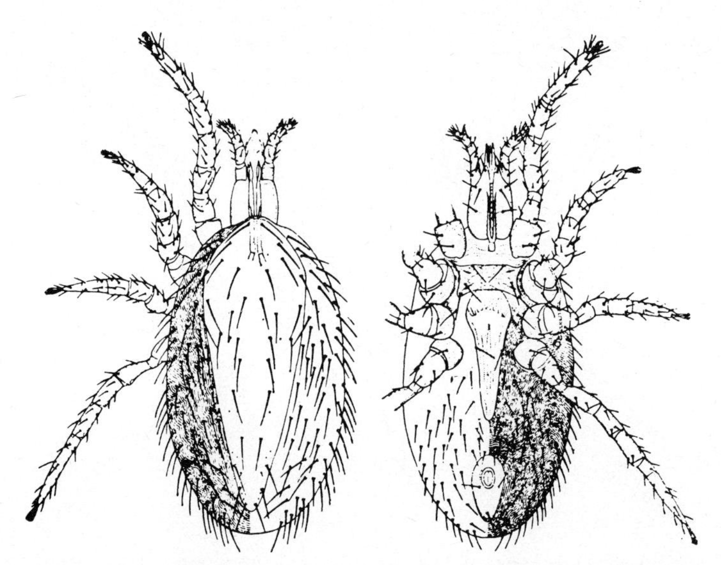

Fig 2.84 *Ornithonyssus bacoti*. Left, Dorsal view of the female. Right, Ventral view of the female.

setae, the anterior pair being on the anterior margin of this plate, the posterior margin of which is concave. The anus is on the anterior half of the anal plate.

Life-cycle and habits. The eggs are laid in the nests and burrows of the rats, not on the hosts. Baker et al. (1956) state that the female mite lays one or two eggs after a blood meal and may lay about 98 eggs during her lifetime, temperature and humidity affecting the egg-laying. The eggs hatch in one to two days, liberating six-legged larvae, which do not feed, but moult after 24 hours to become protonymphs, which take a blood meal and moult to become the deutonymphs, which, like the larvae, do not feed, but moult, after 24–36 hours, to become the adults. Under laboratory conditions 75% of the mites became adult in 11–16 days. Unfertilized eggs may develop parthenogenetically to provide males which can fertilize eggs.

It has been shown that *Yersinia pestis*, the cause of plague, can be experimentally transmitted from rat to rat by the bites of the mites, by eating them or by the injection of infected mites. Murine typhus (*Rickettsia typhi*) and Q fever (*Coxiella burnetii*) are also transmitted by this species. *O. bacoti* is the intermediate host of the filarial nematode, *Litomosoides carinii*, of rodents.

Control. Measures should be taken to control the rats which are the hosts of this mite. Where it is permitted for use, DDT (10% dust), applied to runs and burrows of rats or animal rooms, gives good control. The synthetic pyrethroid 'permethrin' will probably give excellent control.

Allodermanyssus sanguineus (Hirst, 1914). This relatively little-studied blood-sucking mite is called the 'house mouse mite' and occurs on the domestic rat, house mouse and spiny mouse (*Acomys*) in North America, Europe, Africa and Asia. It is generally believed that it may transmit *Rickettsia akari*, the cause of rickettsial pox of man. The mite may be differentiated from the other species just described by its two dorsal plates. The sternal plate bears all three pairs of setae. The life-cycle, which resembles that of the species above, is completed, according to Baker et al. (1956), in 17–23 days.

Control should include control of the mice and the methods indicated for the other mites just described may be tried.

Genus: Pneumonyssus Banks, 1901

Pneumonyssus caninum Chandler and Ruhe, 1940. This species lives in the nasal passages and nasal sinuses of the dogs. It has been found in the

USA, Hawaii, Australia and the Republic of South Africa. Dogs of any age, breed or sex may be affected, but the effects are generally not serious and are usually confined to reddening of the mucosae, sneezing, shaking of the head and rubbing of the nose. The mites are oval and pale yellow and measure 1–1.5 by 0.6–0.9 mm. Their smooth cuticle has scanty setae. The single dorsal plate is irregular in shape; the sternal plate is small and irregular and not well sclerotized; it bears two pairs of setae, the second pair not being on this plate. There are no genital plates and the genital opening is a transverse slit between the fourth coxae Mature females often contain eggs and it has been said that they give birth to larvae, but little is known about the life-history. The mode of transmission is unknown, but is probably by direct contact.

Pneumonyssus simicola Bank, 1901 is a related species parasitic in the bronchi of the rhesus monkey (*Macaca mulatta*). Prevalences of 100% may be found in some batches of monkeys. It is also found in other monkeys (macaques, cynomolgus). Its life-history is little known and attempts to control it have so far failed. Clinical signs of infection are usually absent. However, sneezing and coughing may be a feature of infection and histologically pulmonary lesions vary from a few pale spots or yellowish foci to several hundreds of tubercle-like lesions with associated focal pneumonitis and cellular infiltration (Innes et al. 1954).

A number of dermanyssid mites affect reptiles and amphibians.

Ophionyssus natricis (Gerv̧ais, 1944) is a troublesome blood-sucking mite of captive snakes. It occurs on the skin or under the scales of several species of snakes and lizards in captivity, but is seldom seen in wild specimens. Adult mites measure 0.6–1.3 mm, are yellowish-brown when unengorged and dark red or black when engorged. Two dorsal plates are present; one is lemon-shaped and covers most of the body in an unfed mite, the other is immediately dorsal to the anal plate and is much smaller.

The complete life-cycle takes 13–19 days (at 25°C); there are five developmental stages (egg, larva, protonymph, deutonymph and adult). The larvae are non-feeding but nymphs and adults are blood-feeders, often on the rim of the eye. Heavy infections cause anaemia, listlessness and may lead to death. The mites have been incriminated as mechanical vectors of *Aeromonas hydrophila*, a bacterial pathogen of snakes.

Treatment consists of topical application of malathion (4% dust), pyrethrum applied as a dust, *p*-chlorophenyl phenyl sulphone (Sulphenone) applied as a 10% dust or a 25% emulsion of diazinon. Treatment should be combined with cleaning, disinfection and preferably, sterilization of cages.

Other mites of snakes include the entonyssid mites of the genera *Entonyssus* and *Entophionyssus* which occur in the trachea and lungs of snakes. A number of species are reported but there is no evidence they cause ill-health.

FAMILY: GAMASIDAE LEACH, 1815

The tegument of this family is tough; the dorsal shield does not extend beyond the mouth parts; stigmatal openings occur between the third and fourth coxae; the male genital pore is in front of the sternal plate.

Genus: Raillietia Trouessart, 1902

Raillietia auris (Trouessart, 1902). This is a small species 1 mm in length and occurs on the ears of cattle in North America and Europe.

Raillietia hopkinsi occurs on the ears of antelopes (e.g. *Kobus defassa* in Uganda) (Zumpt 1961). Neither species is associated with clinical disorders.

Other gamasid mites are the laelaptid mites among which are **Echinolaelaps echidninus** (Berlese, 1887), the spiny rat mite, which is the definitive host of the haemogregarine protozoon, *Hepatozoon*, of the rat and hamster; and other species parasitic on rodents described by Baker et al. (1956), who discuss their possible importance as vectors of diseases of man and other animals.

Eulaelaps stabularis (Koch, 1863) is a parasite of small mammals and is found in bedding, poultry houses, grain stores and can cause considerable annoyance to workers in such places. *E. stabularis* is a vector of *Pasteurella tularensis*.

Similar mites are **Haemogamasus pontiger** (Berlese, 1904) (rodents, insectivores and bedding) and **Haemolaelaps casalis** Berlese, 1887 (*Hypoaspis freemani*) of birds and litter and which may attack humans working in association with such animals and/or places.

Mesostigmatid mites of the family *Halarachnidae* occur in the nasal passages of marine mammals. These include the genera *Halarachne* and *Orthohalarachne* which occur on various species of sea lion and seal (Daily & Brownell 1972). Clinical disease is not associated with such infections (Sweatman 1971).

SUBORDER: IXODOIDEA (IXODIDES) LEACH, 1815

This suborder contains the hard and soft ticks and is subdivided into two families, the Argasidae, which includes the fowl ticks and tampans, and the Ixodidae, or true ticks.

FAMILY: ARGASIDAE CANESTRINI, 1890

The integument is leather-like, frequently mammillated, and there is no dorsal shield. In the nymph and imago the capitulum and mouth parts are situated anteriorly on the ventral surface and are not visible from the dorsal aspect. Eyes are absent, or there may be two pairs situated laterally in the supracoxal folds. There is one pair of spiracles situated posterolaterally to the third coxae. Sexual dimorphism not marked.

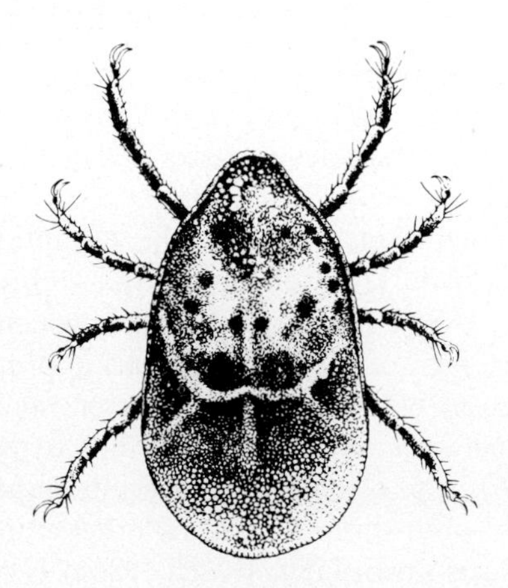

Fig 2.85 Dorsal view of the adult *Argas persicus*.

Genus: Argas Latreille, 1795

Argas persicus (Oken, 1818). The ticks formerly identified as *Argas persicus* represent a complex of species. The true *A. persicus* appears to be confined to the Old World (Europe, Asia, parts of Africa), though a few records of its occurrence exist in the eastern and western USA. **Argas sanchezi** Dugès, 1887 is common in the south-west of the USA and Mexico, **A. radiatus** Railliet, 1893 occurs in central and southern USA and **A. miniatus** Koch, 1884 occurs in Panama and South America. Other species occur on wild birds and bats. *Argas persicus* and its relatives are common parasites in many warm and temperate climates, attacking fowls, turkeys, pigeons, ducks, geese, canaries, ostriches and certain wild birds and it may also bite man. The imago measures 4–10 by 2.5–6 mm and is oval in shape, narrower anteriorly than posteriorly. The edges of the body are sharp. The engorged tick has a slaty-blue colour, while the starved animal is yellowish-brown with the dark intestine showing through. As in other Argasidae there is little difference between the males and the females; the sexes can be distinguished only by the shape of the genital opening, which is situated anteriorly on the ventral surface and is larger in the female than in the male.

Life-cycle and habits. The eggs are laid in cracks and crevices of the fowl-house and under the bark of trees. They are small, spherical, brown in colour and laid in batches of 20–100. The larvae hatch after three weeks or more; they have six relatively long legs and roughly circular bodies, which become spherical after engorging. The larvae attach themselves to the host, frequently under the wings, and engorge in about five days, rarely up to ten days. They then drop off and hide away, to moult after about seven days. There are two nymphal stages, each of which lasts about two weeks and engorges once during this time. The nymphs and the adults hide away in sheltered spots and attack their hosts at night, feeding for about two hours. The adults feed once a month, more or less, and the females lay a batch of eggs after each meal. The larvae can live without food for about three months. The nymphs and the adults survive starving for about five years.

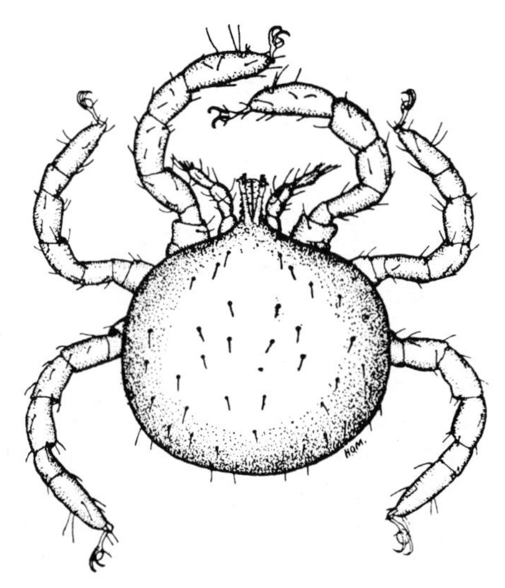

Fig 2.86 Larva of *Argas persicus*.

Pathogenesis. The fowl tick worries the birds at night so that they sleep restlessly and in heavy infections anaemia results from the loss of blood. Egg-laying decreases or may stop completely. *A. persicus* causes tick paralysis of ducks and transmits *Anaplasma marginale* in the USA and *Aegyptianella pullorum* in tropical areas. It is also the vector of *Borrelia anserina*, the cause of fowl spirochaetosis; the spirochaete is passed through the egg to the offspring of an infected female tick and so can be eradicated only with the ticks.

Diagnosis. The parasites will be found in cracks of the woodwork or the walls of the fowl-house.

Control. The classical procedure against fowl ticks in buildings is as follows. The birds must be removed from their run and houses and are placed in wooden crates. The larvae on the birds will drop off within ten days and the fowls can then be returned to their quarters, which have meanwhile been cleaned. All birds subsequently introduced should be treated in the same way. Chicken houses may be thoroughly sprayed with insecticide. Sprays of HCH (0.05% gamma isomer) are effective and an emulsion containing 1.27% gamma isomer of HCH, applied at the rate of 4.5 ml/m to roosts, does not taint the eggs or flesh of the birds. Organophosphorus insecticides are valuable in control as are the synthetic pyrethroids, such as permethrin, decamethrin, etc.

Argas reflexus (Fabricius, 1794) occurs chiefly on pigeons and has been found in Europe, Russia, North and West Africa and in the Far East (India). It may transmit *Borrelia anserina* to poultry.

Argas mianensis, Brumpt, 1921, the Persian miana bug, possibly transmits human relapsing fever there.

Genus: Otobius Banks, 1912

Otobius megnini (Dugès, 1883), the 'spinose ear tick', is found in North and South America, southern Africa and India. Its larval and nymphal stages are most often parasites in the ears of dogs, sheep, horses and cattle, but are sometimes also found in goats, pigs, cats, ostriches and man, and also on rabbits, deer and other wild animals. The engorged larvae are almost spherical. The nymphs are widest at the middle and their skin is mammillated and bears numerous spine-like processes; body colour is bluish-grey, while the legs, mouth parts and spines are pale yellow. Adults, which are not parasitic, have a constriction at the middle, giving the body a fiddle shape.

Life-cycle and habits. The eggs are laid in sheltered spots such as cracks of poles, under food-boxes or stones or in crevices of walls. The infection is therefore mainly associated with sheds, yards and kraals and is hardly ever seen in ranch animals that remain in the open pasture. The larvae hatch in three to eight weeks and may live without food for two to four months. If they find a suitable host, they attach themselves in the ears below the hair-line and engorge in five to ten days. The larvae suck lymph and, when they are engorged, they are 2–3 mm in length and are usually yellowish-white or pink. Their shape is almost spherical and the legs are relatively small. They moult in the ears and the spined eight-legged nymphs feed there and remain in the ears for one to seven months, unless they are accidentally dislodged. The fully-grown, engorged nymph measures 7–10 mm long. They drop off the host and seek dry, protected places in crevices of buildings, fences and trees, where they moult after a few days to become adults, the skins of which are not spiny. The adults do not feed, but the females lay 500–600 eggs. Oviposition may last for as long as six months. When it ends the females die, but unmated females may live longer than a year. The eggs may hatch ten days after they are laid and the larvae that hatch from them are then ready to feed on a host.

Pathogenesis. Masses of these ticks may be present on the host. They suck blood and cause marked irritation, which results in inflammation. Secondary bacterial infection may extend inwards with serious results. The infected animals appear dull; they do not feed well and lose condition. Dogs especially shake the head and scratch the ears. A waxy or oily exudate from the ear is usually present. The ticks, if numerous, can remove much blood.

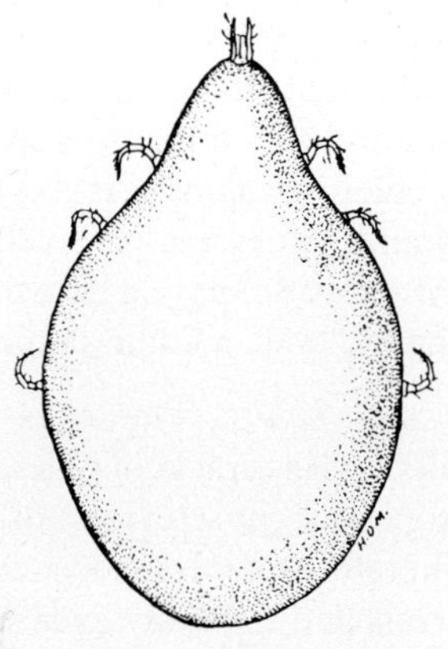

Fig 2.87 Engorged larva of *Otobius megnini.*

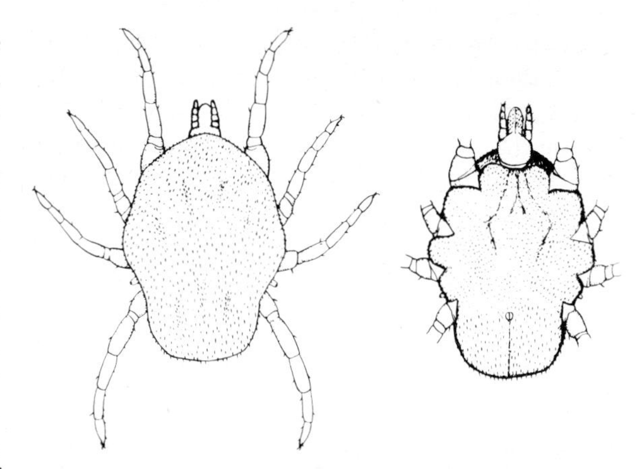

Fig 2.88 Nymph of *Otobius megnini,* dorsal and ventral views.

Diagnosis. Heavy infections when the ear-canals are packed with ticks are readily diagnosed. In other cases the waxy exudate is removed and the ear probed by means of a suitable piece of wire with a small loop at the end, which will assist in dislodging and extracting a few parasites if they cannot be seen by direct inspection.

Treatment and prophylaxis. The United States Department of Agriculture recommends, among other remedies, one part of HCH (15% of the gamma isomer), two parts of xylol and 17 parts of pine oil, all by weight, which penetrates the ear wax, kills the larvae and protects for three to four weeks; or 40% chlordane emulsion one part and pine oil 15 parts may be instilled into the ear. The synthetic pyrethroids used for hard ticks will probably be effective against the soft ticks.

Infected sheds or kraals are most satisfactorily treated by the use of insecticide sprays.

Otobius lagophilus Cooley and Kohls, 1940 is similar to the above, being found on rabbits in the western USA. The adults are not parasitic.

Genus: Ornithodoros Koch, 1844 (emend. Ornithodorus, Agassiz, 1845)

Ornithodoros moubata Murray, 1877, the eyeless tampan of Africa, lives in native huts and in the sand under trees where animals and human beings frequently seek shelter. The life-cycle is described in detail by Hougstraal (1956). The female lays batches of about 100 eggs in the sand. The female 'broods' over the eggs which hatch in about eight days to produce a larva, which after emergence remains quiescent until it has moulted to the nymphal stage. Several nymphal instars are passed through and the nymphs, like the adults, attack their hosts for short periods only to feed. This tampan sucks blood of man and various domestic and wild animals, including birds and even tortoises. In certain localities, for instance in south-western Africa, this parasite causes much trouble by feeding on sheep at their resting-places in the pasture and it is very difficult to combat it under such conditions. It is the only vector, under natural conditions, of *Borrelia* (*Spirochaeta*) *duttoni*, the cause of African relapsing fever of man. This tick constitutes an important reservoir host for the virus of African swine fever, besides acting as an efficient biological vector of the virus among the mammalian reservoir hosts, the wild *Suidae* (*Phacochoerus*, *Potamochoerus*, *Hylochoerus*). It is also cited as a vector of Q fever and may transmit *Borrelia anserina* and *Aegyptianella pullorum* of the fowl. Control may be achieved by spraying the walls of huts, houses and animal buildings with a residual insecticide. HCH gives good control and the organophosphates and the synthetic pyrethroids are likely to be effective. These may be applied to animals directly or, for example with poultry, by providing them with dust baths containing insecticides. Where animal housing is of cheap replaceable materials (e.g. bamboo, palm mat, wattle, etc.) the most effective control is to burn down the housing and rebuild.

Ornithodoros savignyi Audouin, 1827, is a tampan which possesses eyes. Its habits are similar to those of *O. moubata*; its larvae hatch from the eggs, though they do not feed. It occurs in Africa, India and the Near East on the camel, fowl and other domesticated animals, but may bite man.

Ornithodoros turicata (Dugès, 1876) (the relapsing fever tick) occurs in the USA, especially the south-western and western states. It is referred to as the relapsing fever tick since it is responsible for transmission of that disease to man in the south-west. Neitz (1956) cites this as a vector, together with *O. hermsi* Wheeler, Herms and Meyer, 1935, *O. parkeri* Cooley, 1936, *O. erraticus* Lucas, 1849, and *O. gurneyi* Warburton, 1926, of *Coxiella burnetii*, the cause of Q fever. He also cites *O. lahorensis* Neumann, 1908 as a vector to goats of *Theileria ovis* and *Anaplasma*

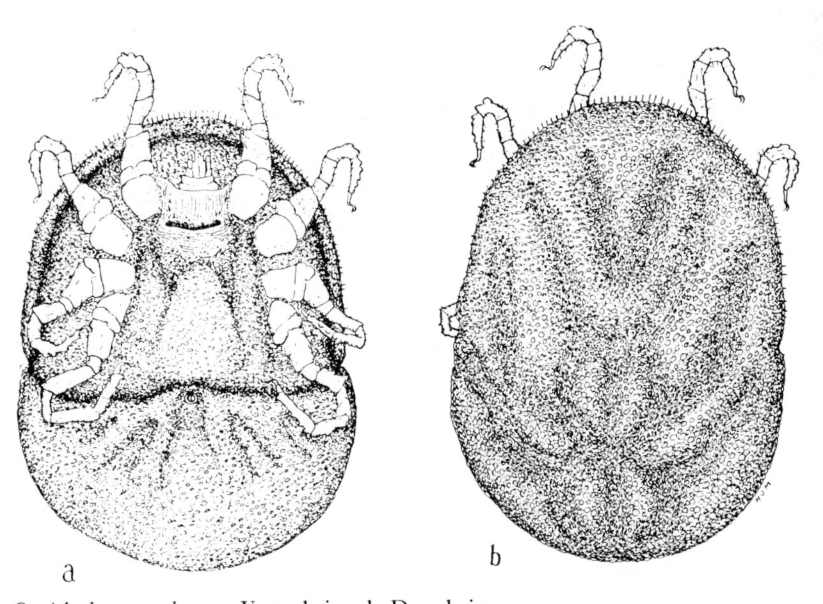

Fig 2.89 Female *Ornithodoros moubata*. a, Ventral view. b, Dorsal view.

ovis and as a cause of tick paralysis of sheep in Central Asia. **O. moubata porcinus** Walton, 1926 (*O. porcinus*) is the form which infests the burrows of warthogs and acts as the reservoir and vector of African swine fever.

Ornithodoros talaje (Guérin-Méneville, 1845) is found in the south-western states of the USA and in Florida.

Ornithodoros coriaceus Koch, 1844, the pajaroello tick, has been found on cattle and deer in California and the Pacific Coast of Mexico. It produces a painful 'bite' but is not known to transmit disease to man, though it is reputed to be a vector of epizootic bovine abortion in Central America.

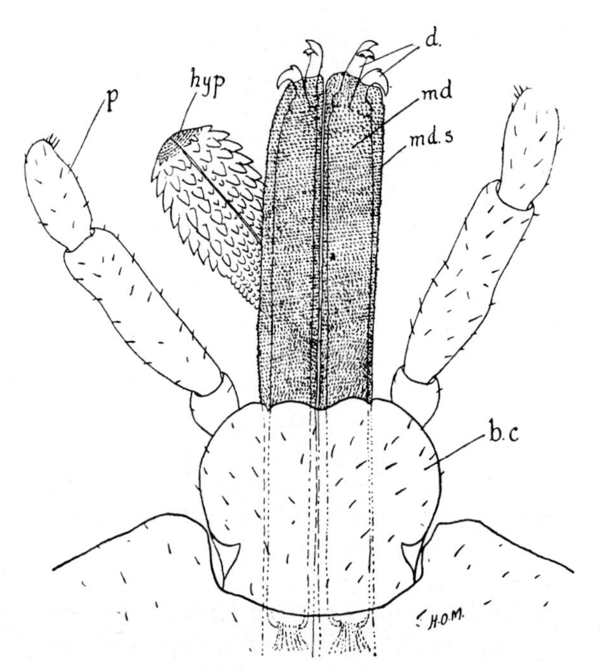

Fig 2.90 Dorsal view of the anterior end of *Amblyomma hebraeum*.

bc	basis capituli
d	digits of mandible
hyp	hypostome, twisted to show ventral surface
md	shaft of mandible
md s	mandibular sheath
p	palp

FAMILY: IXODIDAE MURRAY, 1877

Ticks of this family possess a hard, chitinous shield or scutum which extends over the whole dorsal surface of the male and covers only a small portion behind the head in the larva, nymph and female. The mouth parts are anterior and well visible from the dorsal aspect. Eyes when present consist of one pair situated on the lateral margin of the scutum. The imago has one pair of spiracles situated posterolaterally to the fourth coxae.

The basis capituli or capitulum, which is inserted into the body anteriorly and carries the mouth parts and palps, shows two dorsal porose areas in the female. The scutum has bilateral cervical and lateral grooves, varying in depth and length in different species. The body of the female may have a pair of lateral 'marginal grooves' behind the scutum, while posterolateral and median grooves are usually present on the dorsum in both sexes. The posterior border of the body may be notched, forming the 'festoons', which are generally 11 in number. The genital opening is a ventral transverse slit in front of the middle, the anus being posterior. The male may have ventral plates. *Ornate ticks* have coloured, enamel-like areas on the body, *inornate ticks* have not.

Bionomics

The *Ixodidae* lay their eggs in sheltered spots: under stones and clods of soil or in crevices of walls and cracks of wood near the ground. The eggs are small, spherical, yellowish-brown to dark brown in colour and are laid in large masses. The female lays all her eggs in one batch, up to 18 000 in some species, and then dies. The whole process of subsequent development to the adult stage is greatly influenced by the prevailing temperature, cold weather causing marked prolongation of the different stages, especially hatching of the eggs and the pre-oviposition period of the engorged female.

The newly hatched larvae or 'seed ticks' (Fig. 2.91) climb on to grass and shrubs and wait there till a suitable host passes, to which they attach themselves with their claws.

After having engorged, the larva moults and becomes a nymph. The integument of the latter requires a few days to harden and then the nymph engorges and moults to become an imago. After hardening of the integument, and often also after

Fig 2.91 'Seed ticks' collecting on the tips of grass blades to await an opportunity to attach themselves to passing hosts. (*By kind permission of the Liaison Officer, Agricultural Liaison Section, Commonwealth Scientific and Industrial Research Organisation, Melbourne, Australia*)

copulation, which may take place on the ground or, more usually, on the host, the female engorges, drops off and seeks a sheltered spot to lay her eggs. The males remain much longer on the host than the females, in some cases four months or even longer, and consequently they accumulate on the host. Although it is not known definitely whether the males of all species feed on the host, many of them certainly do so for a few days and then go in search of females. If no males are present on the host, the females may remain attached for much longer periods than under normal conditions.

According to the number of hosts they require during their life-cycle, ticks can be classified into three groups:

One-host ticks. All three instars engorge on the same animal, the two ecdyses also taking place on the host . Examples: *Boophilus decoloratus* and *B. annulatus.*

Two-host ticks. The larva engorges and moults on the host and the nymph drops off after also having engorged; it moults on the ground and the imago seeks a new host. Examples: *Rhipicephalus evertsi* and *R. bursa.*

Three-host ticks. These require a different host for every instar; they drop off each time after having engorged and moult on the ground. Examples: *Ixodes ricinus* and *Rhipicephalus appendiculatus.*

Each species of tick is adapted to certain ranges of temperature and moisture, some occurring only in warm regions with a fair degree of humidity, while others are winter ticks most active in a dry climate. They suck blood and sometimes lymph and are in general not very specific with regard to hosts, although some species, or certain instars of a species, show a particular preference for certain host species, or there may be a definite adaptation to certain hosts. When a tick attaches itself to feed, it buries its mouth parts deeply into the tissues of the host and remains attached until it is engorged. The feeding mechanism of ticks has been reviewed

Table 2.5. Key to the Genera of Ixodidae

1. Anal grooves surrounding the anus anteriorly (Prostriata)	*Ixodes*
Anal grooves surrounding the anus posteriorly (Metastriata) (in *Boophilus* and *Margaropus* the anal groove is faint or obsolete)	2
2. Hypostome and palpi short	3
Hypostome and palpi long	8
3. Eyes absent	*Haemaphysalis*
Eyes present	4
4. Festoons present	5
Festoons absent	7
5. Males with coxae IV much larger than coxae I to III; no plates or shields on ventral surface of male	6
Males with coxae IV not larger than coxae I to III, a pair of adanal shields and usually a pair of accessory adanal shields on ventral surface of the male. Species usually inornate; basis capituli generally hexagonal dorsally	*Rhipicephalus*
6. Species ornate; basis capituli rectangular dorsally	*Dermacentor*
Species inornate; basis capituli hexagonal dorsally with prominent later angles. Coxae IV of male with two long spines	*Rhipicentor*
7. Inornate; coxae I with a small spine. Male with median plate projecting backwards on either side of the anus and with a caudal protrusion when engorged. Fourth pair of legs of male dilated	*Margaropus*
Inornate; coxae I bifid. Male with a pair of adanal and accessory shields and a caudal protrusion. Fourth pair of legs normal	*Boophilus*
8. Eyes present	9
Eyes absent or rudimentary. Species occurring almost exclusively on reptilia	*Aponomma*
9. Festoons absent or present. Males with a pair of adanal shields and two posterior abdominal protrusions. Accessory adanal shields absent or present	*Hyalomma*
Species usually ornate; festoons present. Male without adanal shields, but small plaques may be present on the ventral surface near the festoons	*Amblyomma*

by Tatchell (1969). This is assisted by the host response to the tick and infiltrations of neutrophils lead to destruction of collagen, resulting in a cavity beneath the mouth parts.

The key to the genera of *Ixodidae* shown in Table 2.5 is adapted from Bedford (1932). A key to the Ixodidae of North America is given by Strickland et al. (1976).

Various species of Ixodidae are vectors and reservoirs of important viral, rickettsial and protozoan parasites of man and other animals. The protozoal diseases are discussed in the section on Protozoa. A general consideration of ticks and disease is given by Arthur (1962).

Genus: Ixodes Latreille, 1795

The anal groove surrounds the anus anteriorly. Palpi long. Inornate. Eyes and festoons absent. Ventral surface of male armed with pregenital, median, anal, epimeral and adanal shields, the latter two being paired. Stigmatic plates oval in male, circular in female. Some 200 species of the genus *Ixodes* occur; only a few will be mentioned here. Keys to the species of *Ixodes* in Britain are given by Arthur (1963).

Ixodes ricinus (Linnaeus, 1758), the 'castor-bean tick' or sheep tick, is common in Europe and occurs also in Tunisia and Algeria and limited areas of Asia. It has never been established in North America. It is frequently found on dogs, but also occurs on other domestic and wild mammals, while nymphs and larvae have been recorded from lizards and birds. The tarsi taper away to their ends and are not humped. The postero-internal angle of coxa I bears a spine which is long enough to overlap coxa II.

The following data on the life-cycle of the parasite are compiled from various authors:

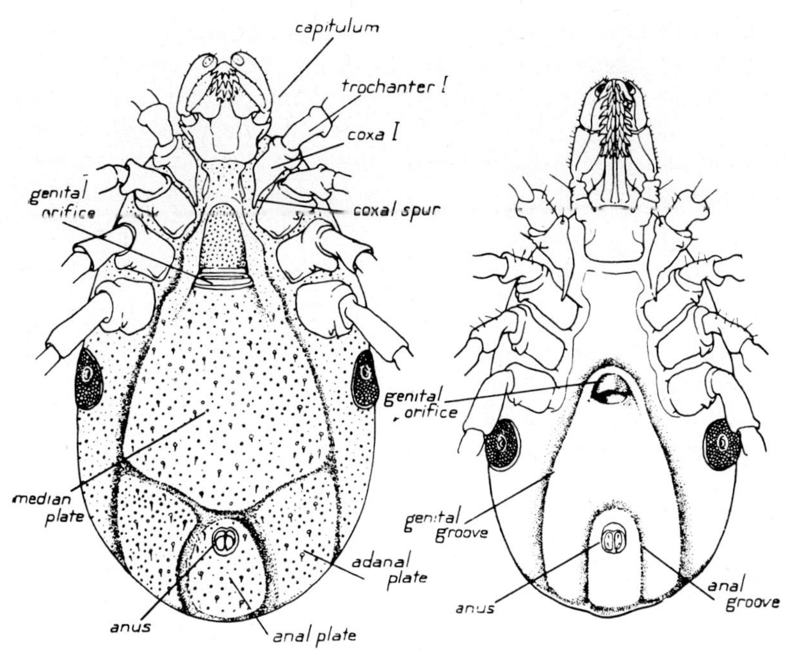

Fig 2.92 *Ixodes ricinus*. Left, Ventral aspect of the male. Right, Ventral aspect of the female. (*From Smart, by permission of the British Museum*)

Pre-oviposition period	7–22 days
Oviposition lasts	about 30 days
Eggs hatch (depending on temperature)	2–36 weeks
Larvae engorge	2–6 days
Larvae moult (depending on temperature)	4–51 weeks
Nymphs engorge	3–7 days
Nymphs moult (depending on temperature)	8–28 weeks
Females engorge	5–14 days
Unfed larvae survive	13–19 months
Unfed nymphs survive	24 months
Unfed adults survive	21–27 months, extreme 31 months

In Britain the life-history of *I. ricinus* extends over three years, the larval stage lasting through the first year, the nymphal stage through the second year, and the adult stage through the third year. In Northumberland and southern Scotland each of these phases feeds only in the spring (March–June) and these ticks are called spring-feeders. In Cumbria, Wales, Ireland and western England and Scotland there are, however, not only spring-feeders but others that feed chiefly in the autumn (August–November). Spring-feeders can be turned in the laboratory into autumn-feeders by raising the temperature to quicken the development, but they then remain autumn-feeders. Wood et al. (1960) state that, in Britain, *Ixodes ricinus* is usually found attached to the face, ears, axillae and inguinal region where the hair is short or the skin bare, that is to say, to parts of the body constantly brushed by vegetation on which the ticks are waiting.

I. ricinus transmits redwater of cattle caused by *Babesia divergens* and *Babesia bovis*, which pass through the egg of the tick, *Anaplasma marginale*, and the viruses of louping-ill and rickettsial 'tick-borne fever' of sheep. In Britain this species is associated with the spread of pyaemia caused by *Staphylococcus aureus* (tick pyaemia), which especially affects lambs two to six weeks old. Neitz

(1956) cites the tick as a cause of tick paralysis in Crete and as a vector of Czechoslovakian encephalitis, Russian spring–summer encephalitis and *Coxiella burnetii* infection in Germany. It also transmits Bukhovinian haemorrhagic fever. (See also Ticks as Parasites, below.)

Ixodes persulcatus (Schulze, 1930) is a Eurasian form and transmits several species of *Babesia*.

Ixodes hexagonus Leach, 1815, sometimes called the hedgehog tick, is found on the hedgehog, dog, otter, ferret and weasel in Britain. The tarsi are humped and the posterointernal angle of coxa I bears a spine that is not long enough to overlap coxa II.

Ixodes canisuga Johnson, 1849, often called the British dog tick, has been found on the dog, sheep, horse and mole. It may be very numerous in dog-kennels and possibly is a parasite of the fox. Its tarsi are humped, but there is no spine on coxa I.

Ixodes pilosus (Koch, 1844), the russet, sour-veld or bush tick, occurs, according to Theiler (1950), at most seasons in most areas of the Republic of South Africa. It is not, as is often stated,

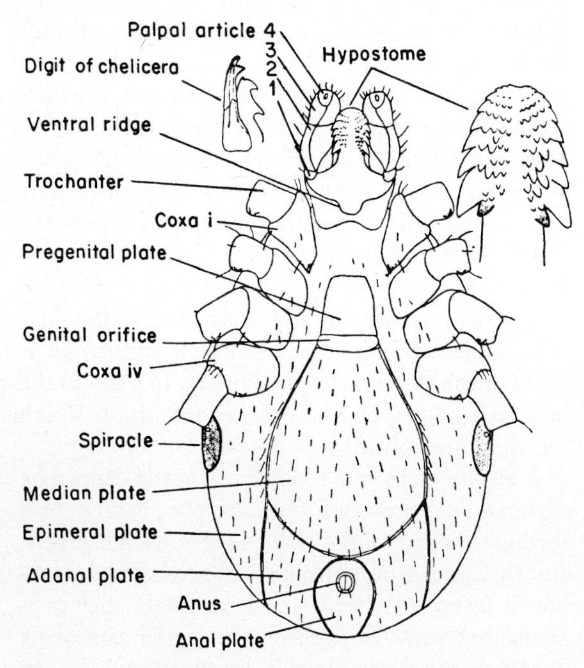

Fig 2.93 Ventral surface of the male *Ixodes pilosus*. (*After Bedford*)

a winter tick and, although it has been accused of causing tick paralysis and has therefore been called the paralysis tick, it does not cause paralysis and statements that it does are probably due to wrong identification. Its hosts are cattle, sheep, goats, horses, dogs, cats and wild ungulates.

The life-cycle may be summarized as follows:

Pre-oviposition and incubation period (summer)	43–93 days
Pre-oviposition and incubation period (winter)	222–309 days
Larvae engorge	$2\frac{3}{4}$ days or longer
Larvae moult (summer)	27 days
Nymphs engorge	4 days
Nymphs moult (autumn)	52 days
Females engorge (males present)	5–6 days

Ixodes rubicundus Neumann, 1904, is the paralysis tick of southern Africa. Theiler (1950) described its distribution there. It is confined to the moister areas of the Karrooveld which are hilly or mountainous. It is active in that country from April–May till July and only exceptionally in the summer. Stampa and du Toit (1958) discussed its biology and life-history. The chief hosts of the adults are domesticated sheep, goats and cattle. Other hosts are wild Artiodactyla, Carnivora and Lagomorpha. The immature stages feed on the red hare, elephant shrew, bush hare and wild Canidae (e.g. the red jackal and dog), but it is never found on the cat or horse or on birds.

Life-history. This requires two years. The engorged females lay eggs for six weeks from the end of August to early September. The eggs are susceptible to dry heat and do not hatch at humidities below 70%; all of them hatch at humidities between 80 and 100%. The larvae are active from March to mid-August, with a peak of activity in April–May, when the peak incidence of the tick paralysis also occurs. After 73–149 days the larvae moult, after feeding, to become nymphs, which are active from July to November, with a peak of activity in September–October. They moult six months after engorgement. The adults feed for four to seven days, attaching on the radial aspects of the legs from the knee or fetlock downwards and on the ventral parts of the host's body. On sheep they prefer woolled to bare areas. The adults are active from February to mid-November of the second year of the cycle, with a peak of activity in April–May, when the tick paralysis also occurs. At high altitudes the ticks are more active in August.

Ixodes holocyclus Neumann, 1899, is the paralysis tick of Australia. It is a coastal species confined to bush and scrub country. It occurs along the northern and east coast of New South Wales and Queensland and in Victoria and Tasmania. It occurs on man, dogs, cats, other domestic animals, long-nosed bandicoots, opossums and spiny anteaters. Its brown legs and long, prominent mouth parts distinguish it from *Boophilus*. Seddon (1951) describes its life-history. It is very susceptible to variations in temperature and humidity. The female lays 2000–3000 eggs and each instar engorges normally in four to seven days.

This species transmits *Coxiella burnetii* in Australia. Seddon (1951) describes the paralysis caused by the adult females, of which one may suffice, or, rarely, by large numbers of nymphs. The paralysis is mainly of importance in dogs. (See Tick Paralysis, below.)

Control measures include weekly applications of derris dust or washes, a bath or spray containing HCH and DDT given weekly and frequent examination of dogs and the removal of ticks.

Ixodes scapularis Say, 1821 is the 'shoulder tick' or 'black-legged tick' occurring on cattle, sheep, horses and dogs and cats in North America. It occurs in the south Atlantic, south central, southern north central and north Atlantic states. It may be responsible for the transmission of anaplasmosis in some areas. *I. scapularis* has been incriminated as a possible vector of piroplasmosis in humans on Nantucket Island, USA, and ticks have been found naturally infected with *Francisella tularensis*, the causal agent of tularaemia.

The life-cycle may be summarized as follows:

Pre-oviposition period	10–19 days
Female lays	approx. 3000 eggs
Eggs hatch	7–19 weeks
Larvae engorge	3–9 days
Larvae moult	3–7 weeks
Nymphs engorge	3–8 days
Nymphs moult	4–8 weeks
Females engorge	8–9 days
Unfed larvae survive	more than 10 weeks
Unfed nymphs survive	more than 9 weeks
Unfed adults survive	undetermined

Ixodes cookei Packard, 1867, occurs on cattle in California and Oregon, horses in Massachusetts, and on dogs and cats in the south-eastern and north-eastern states of the USA and in south-eastern Canada.

Ixodes pacificus Cooley & Kohls, 1945, the 'California black-legged tick', occurs on cattle, sheep, horses and dogs and cats in California, Oregon, Utah and British Columbia.

Several other species of the genus *Ixodes* occur in North America and these include:

Ixodes angustus Neuman, 1899 (Oregon, Washington; dog); **I. kingi** Bishopp, 1911, the 'rotund tick' (Idaho, Montana, Dakotas, Wyoming, Utah and Ontario; dog); **I. muris** Bishopp and Smith, 1937, the 'mouse tick' (Maine; dog); **I. rugosus** Bishopp, 1911 (California, Oregon, Washington; dog); **I. sculptus** Neumann, 1904 (Oregon, Wisconsin; dog); and **I. texanus** Banks, 1909 (Iowa, British Columbia; dog).

Genus: Boophilus Curtice, 1891

Anal groove absent in female, faint in male and surrounding the anus posteriorly. Inornate. Eyes present. Festoons absent. Palps and hypostome short; palps with prominent transverse ridges. Coxa I bifid. Spiracles circular or oval. Males small, provided with adanal or accessory shields and a caudal process; fourth pair of legs ordinary size.

Boophilus annulatus (Say, 1821), the 'North American tick', was widely distributed throughout the southern USA at one time. A campaign, started in 1906, resulted in its eradication from the USA. It is introduced periodically by the illegal movement of cattle from Mexico, where it is still prevalent. It is common in the Sudan and West and Central Africa. It is a one-host tick. It occurs usually on domestic and wild ungulates, but has also been found on other animals and man.

Various authors give the following data on the life-cycle:

Female lays	4500 eggs
Pre-oviposition period	3–25 days
Oviposition lasts	14–59 days
Larvae hatch	23–159 days
Parasitic period on host	15–55 days
Unfed larvae survive	up to 8 months

At one time this tick caused very serious economic loss to the cattle industry in the USA through the transmission of *Babesia bigemina*, the cause of bovine piroplasmosis or Texas fever. In 1907 losses due to the tick were put at $100 million per year. The tick is still important elsewhere than the USA and it transmits *B. bigemina*, the spirochaete *Borrelia theileri* and anaplasmosis.

Boophilus decoloratus (Koch, 1844), the 'blue tick' occurs throughout the Ethiopian region, especially in humid areas. It is parasitic chiefly on cattle and equines, but is also found on sheep and goats, wild ungulates and dogs. The engorged females have a slaty-blue colour and pale yellow legs. This is a one-host tick; the life-cycle is summarized as follows:

Female lays	2500 eggs
Pre-oviposition period	6–9 days
Oviposition and incubation	3–6 weeks
Parasitic period on host	21–25 days
Unfed larvae survive	up to 7 months

This species transmits *B. bigemina* and possibly *B. ovis* also, *Anaplasma marginale* of cattle, spirochaetosis (*Borrelia theileri*) of cattle, horses, goats and sheep and *Babesia trautmanni* of pigs in East Africa. In all instances the infection passes through the eggs of the tick.

Boophilus microplus (Canestini, 1887) (syn. *B. australis* Fuller 1899), the tropical cattle tick, occurs in Australia, West Indies, Mexico, Central America, South America, Asia and the Republic of South Africa. The primary hosts are cattle, but it is also found on horses as well as goats, sheep and deer. It is a one-host tick with the following life-cycle:

Female lays	4400 eggs
Pre-oviposition period	2–39 days
Oviposition period	4–44 days
Larvae hatch	14–146 days
Parasitic period on host	17–52 days
Unfed larvae survive	up to 20 weeks

According to Neitz (1956) it transmits *Babesia bigemina* in Australia, Panama and South America, *B. argentina* in Australia and the Argentine, *Anaplasma marginale* in Australia and South America, *Coxiella burnetii* in Australia and *Borrelia theileri* in Brazil.

Boophilus calcaratus (Birula, 1895) transmits, according to Neitz (1956), *Babesia bigemina* and *B. berbera* in North Africa and *Anaplasma marginale* in the northern Caucasus.

Genus: Margaropus Karsch, 1879

This genus, of which *Boophilus* is regarded as a synonym by some authors, differs from the latter genus in that the males are large, their fourth pair of legs is markedly thickened and they have a median ventral plate which is prolonged into two spines projecting on either side of the anus; coxa I has a small posterior spine.

Margaropus winthemi Karsch, 1879, the 'Argentine tick', is a native of South America which has also been introduced into the Republic of South Africa. It is a parasite of horses and sometimes occurs also on cattle. The engorged female resembles that of *B. decoloratus*, but has dark bands at the joints of the legs.

This is a one-host tick which is especially prevalent in winter. It is not known to transmit any disease.

Margaropus reidi Hoogstraal, 1956, the Sudanese beady-legged tick, occurs on the giraffe in the western Sudan.

Genus: Hyalomma Koch, 1844

Inornate, sometimes ornate. Eyes present. Festoons present or absent. Hypostome and palps long. Male with a pair of adanal shields and sometimes accessory adanal shields; frequently a pair of chitinous protrusions behind the adanal shields. Spiracles comma-shaped in male, triangular in female. The pathogenic agents transmitted by various stages of the species of *Hyalomma* are: *Babesia caballi, B. equi, Theileria parva, T. annulata, T. dispar, Coxiella burnetii, Rickettsia bovis, R. conori* and the causes of haemorrhagic fevers in Russia. Neitz (1956) listed the parasites transmitted by the various species.

There is much confusion in the literature regarding the species of the genus *Hyalomma* and further information about it may be obtained from Hoogstraal (1956). Rousselot (1953) and Kaiser and Hoogstraal (1964) have divided the genus into three subgenera: *Hyalomma*, *Hyalommina* and *Hyalommosta*. Keys are given for these in both publications and full descriptions of species are given by Kaiser and Hoogstraal (1964).

Arthur (1966) has renamed or synonymized various species of the genus as follows:

Hyalomma plumbeum plumbeum Panzer, 1796 (= *H. marginatum*) in southern Europe, the southern USSR and also in the Nile Delta; **H. excavatum** Koch, 1844 (= *H. anatolicum*) in Egypt, Israel, Greece, Asia Minor and extending east through southern USSR to India; **H. detritum scupense** Schulze, 1918 (= *H. volgense*, *H. uralense*) in Transcaucasia. Other species are **H. dromedarii** Koch, 1844 in North Africa, **H. impressum** near **planum** Schultze and Scholttke, 1930 in East Africa and **H. detritum mauretanicum** Senevet, 1922 in North Africa.

Hyalomma spp. are usually two-host ticks, though three hosts may be used in some species. Larvae and nymphs usually feed on small wild mammals and birds, while adults feed on domestic ruminants and horses, including camels in certain areas. A general life-cycle is detailed below:

Pre-ovisposition period	4–12 days
Oviposition lasts	37–59 days
Larvae hatch	34–66 days
Larvae engorge	5–7 days
Larvae moult	2–15 days
Nymphs engorge	7–10 days
Nymphs moult	14–95 days
Larvae and nymphs on host	13–45 days
Females engorge	5–6 days
Unfed larvae survive	12 months
Unfed nymphs survive	3 months
Unfed adults survive	14 months or longer

Genus: Rhipicephalus Koch, 1844

Usually inornate. Eyes and festoons present. Hypostome and palpi short. Basis capituli hexagonal dorsally. Coxae I with two strong spurs. Males with adanal and usually also accessory adanal shields; frequently with a caudal prolongation when engorged. Spiracles comma-shaped, short in the female and long in the male. This genus contains a large number of species which are difficult to differentiate and are important vectors of infectious diseases.

Rhipicephalus appendiculatus Neumann, 1901, the 'brown ear tick', is widely distributed in

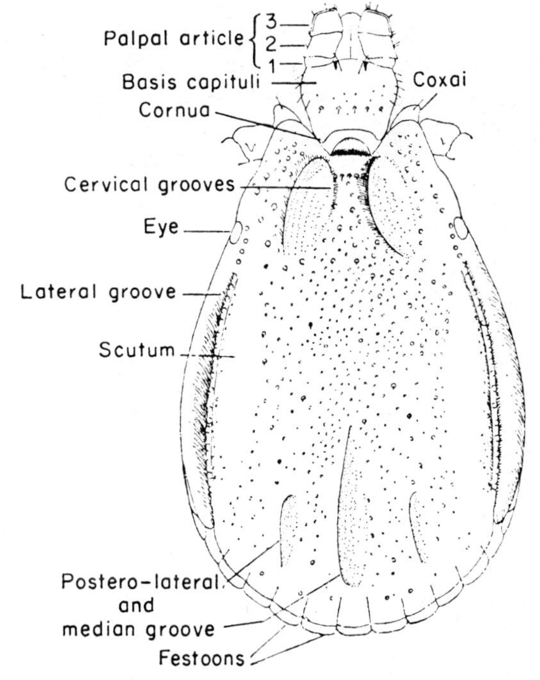

Fig 2.94 Dorsal view of the male *Rhipecephalus appendiculatus*. (*After Bedford*)

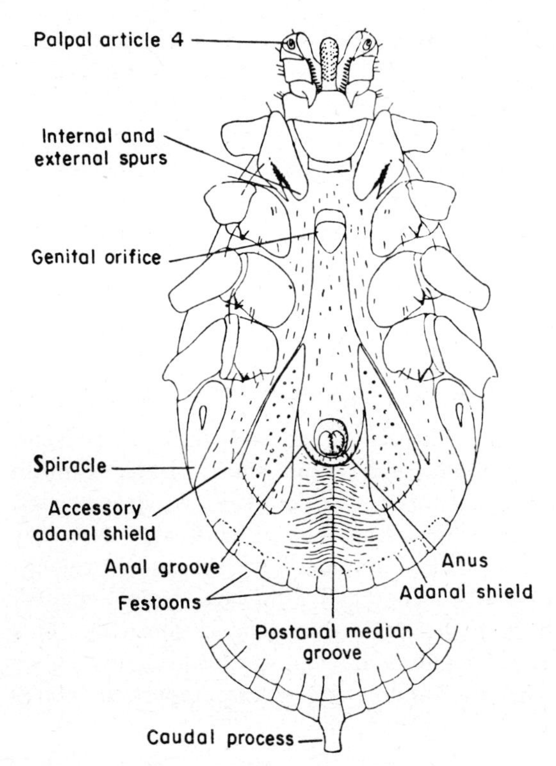

Fig 2.95 Ventral view of the male *Rhipecephalus appendiculatus*. (*After Bedford*)

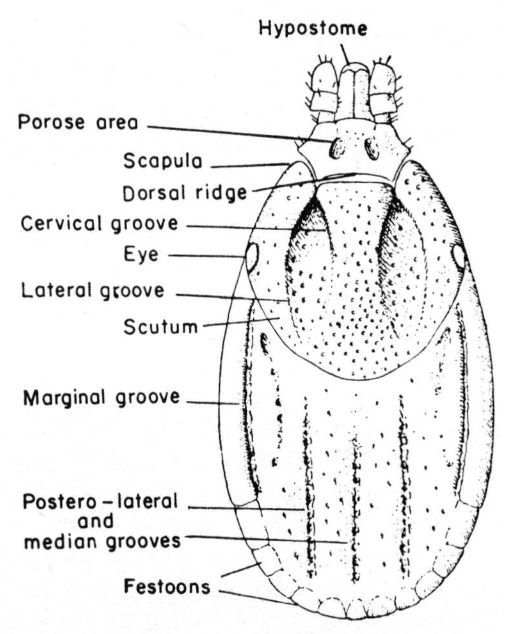

Fig 2.96 Dorsal view of the female *Rhipicephalus appendiculatus*. (*After Bedford*)

southern, central and eastern Africa. It occurs in areas with substantial rainfall and is absent in deserts and areas without shrub cover. It is parasitic on cattle, equines, sheep, goats and wild antelopes and has also been found on the dog and wild rodents. This is a three-host tick.

Female lays	3000–5000 eggs
Pre-oviposition period	5–40 days
Eggs hatch (summer)	28 days
Eggs hatch (winter)	3 months
Larvae engorge	3–7 days
Larvae moult	10–49 days
Nymphs engorge	3–7 days
Nymphs moult	10–61 days
Females engorge	4–10 days
Unfed larvae survive	7 months
Unfed nymphs survive	$6\frac{1}{2}$ months
Unfed adults survive	14 months, sometimes longer and exceptionally over 2 years

This species occurs in a relatively warm climate only. It attaches most frequently under the tail and in the ears, but may also be found on other parts of the body. This tick is the chief vector of East Coast fever (*Theileria parva*) of cattle. It also transmits *Hepatozoon canis* and exanthematic fever of dogs, *Theileria mutans* and *Babesia bigemia* of cattle and *Rickettsia conori* and the viruses of Nairobi sheep disease, Kisenyi sheep disease and louping ill. In none of these cases, except *B. bigemina,* does the infection pass through the egg of the tick. In heavy infections 'tick toxicosis' may occur.

Rhipicephalus capensis Koch, 1844 and **R. simus** Koch, 1844, both three-host ticks of cattle, transmit East Coast fever (*T. parva*), and the last-mentioned also transmits *Anaplasma marginale*.

Rhipicephalus neavei Warburton, 1912, **R. jeanelli** Neumann, 1913 and **R. ayrei** Lewis, 1933 transmit *T. parva* of cattle in Africa and **R. pulchellus** Rondelli, 1926 (Zebra tick) is a vector of Nairobi sheep disease.

Rhipicephalus sanguineus (Latreille, 1806), the 'brown-dog tick', which was probably a native tick of Africa originally, has a more or less cosmopolitan distribution. It is mainly parasitic on dogs and is frequently associated with kennels in the USA, but elsewhere it occurs on a wide variety of mammals and birds (Strickland et al. 1976).

This is a three-host tick.

Female lays	approx. 4000 eggs
Eggs hatch	17–30 days or longer
Larvae engorge	2–4 (–6) days
Larvae moult	5–23 days
Nymphs engorge	4–9 days
Nymphs moult	11–73 days
Females engorge	6–21 days
Unfed larvae survive	up to 8½ months
Unfed nymphs survive	up to 6 months
Unfed adults survive	up to 19 months

This tick transmits canine piroplasmosis (*Babesia canis*) and canine ehrlichiosis (*Ehrlichia canis*), the infection passing through the egg of the tick. There is some confusion in the literature as to the protozoa, rickettsiae, bacteria and viruses that this species can transmit. It seems clear, however, that it transmits *Babesia canis* and *B. vogeli* of dogs, *B. equi* and *B. caballii* of equines, *Anaplasma marginale* in North America, *Hepatozoon canis* of dogs, *Coxiella burnetii*, *Rickettsia conori*, *R. canis*, *R. rickettsii*, *Pasteurella tularensis*, *Borrelia hispanica* and the viruses that cause Nairobi sheep disease and other viral diseases of sheep in Africa. For further information see Neitz (1956), Zumpt (1958) and Arthur (1962). It also causes tick paralysis of the dog. It has also been incriminated as a vector of *B. bigemina* and *Theileria annulata*.

Rhipicephalus evertsi Neumann, 1897, the 'red-legged tick', is common in Africa south of the equator and occurs on many species of domestic and wild mammals. This tick was found in 1960 in game farms in Florida and New York. It was eradicated from the USA in 1962. This species can be distinguished from other members of the genus by its red legs; the shield is black and densely pitted and in the male it leaves a red margin of the body uncovered. This is a two-host tick; the larval and nymphal stages engorge on the same host.

Female lays	5000–7000 eggs
Pre-oviposition period	6–24 days
Eggs hatch	4–10 weeks
Larvae and nymphs on host	10–15 days
Nymphs moult	42–56 days
Females engorge	6–10 days
Unfed larvae survive	7 months
Unfed adults survive	14 months

The larvae and nymphs are usually found in the ears or the inguinal region, the adults mainly under the tail.

This species transmits East Coast fever (*T. parva*), redwater (*B. bigemina*) and *T. mutans* of cattle, *Borrelia theileri* of various animals, and biliary fever (*Babesia equi*) of horses and *R. conori*. In the case of redwater and spirochaetosis, the infections pass through the egg of the tick.

Ripicephalus bursa Canestrini and Fanzago, 1878, is widely distributed in southern Europe, Africa and elsewhere, and transmits *Babesia ovis*, *B. equi*, *B. caballi*, *B. berbera*, *Theileria ovis*, *Anaplasma marginale*, *Rickettsia ovina*, *Coxiella burnetii* and the virus of Nairobi sheep disease.

Genus: Haemaphysalis Koch, 1844

Inornate. Eyes absent. Festoons present. Palps usually short and conical, the second articles having conspicuous lateral projections. The trochanter of the first pair of legs bears a dorsal process. Spiracles in female ovoid or comma-shaped; in males ovoid. Ventral surface of male without chitinous plates. Species usually of small size. A large number of species occur in this genus and further information may be obtained from Hoogstraal (1956).

Haemaphysalis leachi leachi (Audouin, 1826), the 'yellow dog tick', occurs in Africa, Asia and Australia. It is mainly parasitic on domestic and wild carnivora, frequently also on small rodents and rarely on cattle.

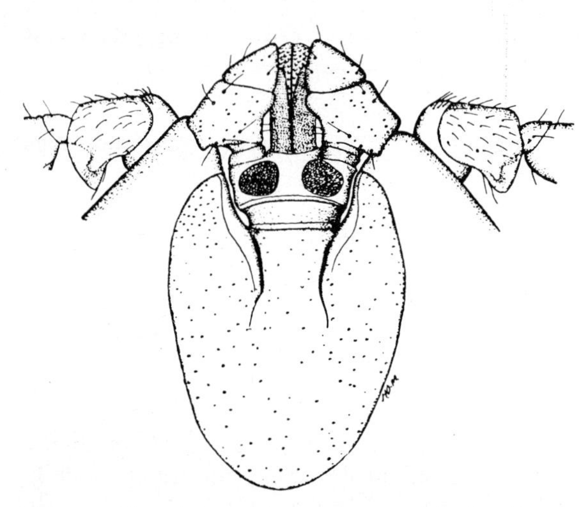

Fig 2.97 Dorsal view of the anterior end of the female *Haemaphysalis leachi leachi*.

H. leachi leachi is a three-host tick.

Female lays	5000 eggs
Pre-oviposition period	3–7 days
Eggs hatch (20°C)	26–37 days
Larvae engorge	2–7 days
Larvae moult	about 30 days
Nymphs engorge	2–7 days
Nymphs moult	10–16 days
Females engorge	8–16 days
Unfed larvae survive	6 months or longer
Unfed nymphs survive	2 months or longer
Unfed adults survive	7 months or longer

The tick lives on the head and body of its host and transmits canine piroplasmosis (*B. canis*), the infection being passed transovarially, and tick-bite fever (*Rickettsia conori*) and *Coxiella burnetii*.

Haemaphysalis leachii muhsami Santos Diaz, 1954, occurs on small carnivores in the Ethiopian region.

Haemaphysalis cinnabarina punctata Canestrini and Fanzago, 1878, occurs in Europe, Japan and North Africa.

While the adults parasitize cattle and other mammals, the larvae and nymphs are also found on reptiles (lizards and snakes). In Europe (including Great Britain) this species has been found on sheep, goats, deer, cattle, horses, rabbits, hedgehogs, the bat *Plecotus auritus* and on the partridge, missel-thrush and curlew. In Asia it has been found on the wolf and bear.

Female lays	3000–5000 eggs
Pre-oviposition period	10 days to 7 months
Oviposition lasts	24–29 days
Larvae hatch (14°C)	38–82 days
Larvae engorge	4–19 days
Larvae moult	14–238 days (winter)
Nymphs engorge	4–33 days
Nymphs moult	7–295 days (winter)
Nymphs engorge	6–30 days
Unfed larvae survive	10 months
Unfed nymphs survive	8½ months
Unfed adults survive	8½ months

This species transmits *Babesia bigemina*, *B. motasi* and *Anaplasma marginale* and *A. centrale* and causes paralysis of sheep and cattle.

Haemaphysalis leporis-palustris Packard, 1867, the 'rabbit tick', is widely distributed in the USA, from Massachusetts to California and it also occurs in South America. Rabbits are preferred hosts but it also feeds on many birds and small mammals may serve as hosts for the immature stages. It transmits Q fever (*Coxiella burnetii*), Rocky Mountain spotted fever and tularemia (*Pasteurella tularensis*) to man. The infection passes through the egg of the tick.

H. leporis-palustris is a three-host tick.

Female lays	2400 eggs
Pre-oviposition period	2–18 days
Oviposition lasts	3–57 days
Larvae hatch	22–61 days
Larvae engorge	4–11 days
Larvae moult	18–134 days
Nymphs engorge	4–11 days
Nymphs moult	14–124 days
Females engorge	19–25 days
Unfed larvae survive	up to 37 weeks
Unfed nymphs survive	up to 1 year
Unfed adults survive	up to 1½ years

Haemaphysalis chordeilis Packard, 1867 (syn. *H. cinnabarina* Koch, 1844), the bird tick, is widely distributed in North and South America.

Birds are preferred hosts for all stages of the parasite. It transmits tularaemia. Deaths in turkeys and wild game birds have been ascribed to heavy infections with this tick.

Haemaphysalis humerosa Warburton and Nuttall, 1909, transmits Q fever (*Coxiella burnetii*) in Australia and is important in maintaining the infection in bandicoots.

Haemaphysalis longicornis Neumann, 1901 is the New Zealand cattle tick or bush tick. It has a wide distribution throughout the East occurring in China, Japan, Australia and New Zealand. In Australia it occurs in south-east Queensland and northern coastal areas of New South Wales.

It occurs on man, cattle, sheep, horse and dog, wild mammals and birds and is a three-host tick.

Pre-oviposition	10–60 days
Oviposition	20–30 days
Incubation	37–90 days
Larvae engorge	3–9 days
Larvae moult	19–22 days
Nymphs engorge	5–7 days
Nymphs moult	23–97 days
Females engorge	7 days

Heavy infections may be seen in cattle and at times dogs, horses and sheep may be severely parasitized. *H. longicornis* is a vector of *Theileria* spp. and *Coxiella burnetii* (Q fever).

Other species of *Haemaphysalis* include: **H. bancrofti** Nuttall and Warburton, 1915 ('wallaby tick'; Queensland; marsupials, cattle); **H. bispinosa** Neumann, 1897, previously confused with *H. longicornis* and occurring in India, Burma, East Africa, Malaysia and Thailand; **H. inermis** Nuttall and Warburton, 1915 (central Europe); and **H. parmata** Neumann, 1905 (antelope, various carnivores; Africa).

Genus: Dermacentor Koch, 1844

Usually ornate. Eyes and festoons present. Hypostome and palps short. Coxa I bifid and coxa IV of male much larger than coxae I to III. No plates on ventral surface of male.

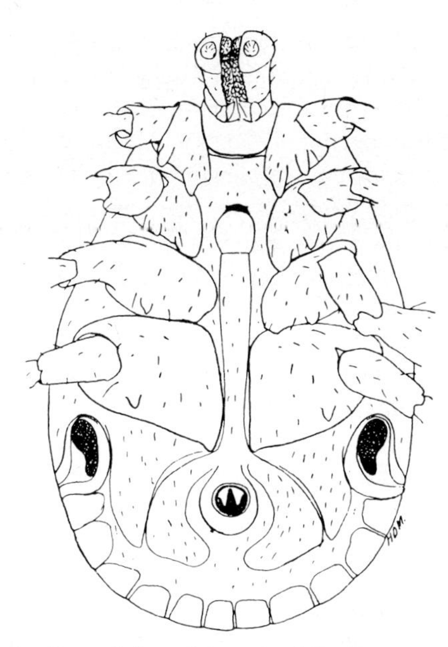

Fig 2.98 Ventral view of the male of *Dermacentor* spp.

Dermacentor reticulatus (Fabricius, 1794) predominates north of latitude 45°. Its microhabitat is associated with a thick layer of plant litter such as occurs in forest or steppe regions (Arthur 1966). It is distributed in the USSR, central Europe, France, Iberian Peninsula and limited areas of south-west England. It is parasitic on many wild and domestic mammals.

Female lays	about 4000 eggs
Larvae hatch	2–3 weeks
Larvae engorge	2 days or more
Larvae moult	about 2 weeks
Nymphs engorge	several days
Nymphs moult	2–3 weeks

This is a three-host tick which transmits *Babesia caballi*, *B. equi* and *B. canis*, the infection passing through the eggs of the tick.

The closely allied form **Dermacentor marginatus** (Sulzer, 1776) Brumpt, 1913 is recognized by Arthur (1966) as a valid species with a distribution more southerly than *D. reticulatus* (e.g. Algeria, Morocco, Tunisia and Canary Islands, as well as Turkey, Iran and Afghanistan). Its development cycle is similar to that of *D. reticulatus*.

Dermacentor andersoni Stiles, 1905 (syn. *D. venustus* Banks, 1908), the Rocky Mountain wood tick, is the primary vector of *Rickettsia rickettsii*, the causal agent of Rocky Mountain spotted fever. It is also an important tick in the production of tick paralysis in man and animals. It occurs in the USA between the Cascade and the Rocky Mountains. The larvae and nymphs occur on practically all available small mammals, especially rodents, while the adults suck blood mainly on horses and cattle but also on other large mammals and man. It is a three-host tick.

Female lays	about 4000 eggs
Pre-oviposition period	7–41 days
Oviposition lasts	about 30 days
Eggs hatch	15–51 days
Larvae engorge	3–8 days
Larvae moult	6–21 days
Nymphs engorge	3–9 days
Nymphs moult	over 3 weeks
Females engorge	8–14 days
Unfed larvae survive	21–117 days
Unfed nymphs survive	300 days or longer
Unfed adults survive	413 days

In addition to Rocky Mountain spotted fever this tick transmits tularaemia (*Pasteurella tularensis*) to man, as well as equine encephalomyelitis (western type). In all cases the infection passes through the egg of the tick. It also transmits *Anaplasma marginale*, *Babesia canis*, *Coxiella burnetii* and *Leptospira pomona*.

Dermacentor variabilis (Say, 1821), the 'American dog tick', which is common on dogs in the USA, transmits Rocky Mountain spotted fever and St Louis encephalitis. This species also transmits *Anaplasma marginale* of cattle and tularaemia to man and causes tick paralysis of dogs in North America. The larvae and nymphae feed on wild rodents, especially the 'short-tailed meadow mouse', *Microtus* spp. This is a three-host tick.

Female lays	4000–6000 eggs
Pre-oviposition period	3–58 days
Oviposition lasts	14–32 days
Eggs hatch	20–57 days
Larvae engorge	2–13 days
Larvae moult	6–247 days
Nymphs engorge	3–12 days
Nymphs moult	16–291 days
Females engorge	5–27 days
Unfed larvae survive	14–540 days
Unfed nymphs survive	29–584 days
Unfed adults survive	up to 1053 days

The seasonal incidence of the adults varies in different parts of the USA, but in general they are most numerous in the spring and early summer; in the south they are found throughout the year and breed slowly through the winter. In colder areas all stages may survive through the winter, except the eggs, which usually hatch before winter comes. The larvae and nymphs are found on mice and other small mammals throughout the winter.

Dermacentor nitens Neumann, 1897, is the 'tropical horse tick' of Mexico, Central and South America, the Caribbean and also occurs in southern Florida and counties in the southern tip of Texas in the USA. Equines are the preferred hosts, though the tick may also occur on sheep, cattle and deer. The preferred site of attachment is the ear though in heavy infections any part of the body may be infected. *D. nitens* is a one-host tick.

Female lays	up to 3400 eggs
Pre-oviposition period	3–15 days
Oviposition lasts	15–37 days
Eggs hatch	19–39 days
Larvae engorge and moult	8–16 days
Nymphs engorge and moult	7–29 days
Females engorge	9–23 days
Unfed larvae survive	71–117 days
Parasitic period on host	26–41 days

Under favourable tropical conditions, several generations are completed each year.

D. nitens is a vector of equine piroplasmosis. Suppuration of the ears may occur in heavy infections and lead to attacks by screw-worm flies.

Dermacentor albipictus (Packard, 1869) is the 'winter tick' or 'moose tick' occurring in the northern part of the USA from Maine to Oregon and through the western states to Texas. It is common in Canada. The preferred host is the moose but it also occurs on elk, horse, cattle, antelope, bear, deer, beaver, bighorn sheep, coyote, mountain goat and mountain sheep. It is most numerous in upland and mountain country. It is a one-host tick.

Female lays	1500–4400 eggs
Pre-oviposition period	7–180 days
Oviposition lasts	19–42 days
Eggs hatch	33–71 days
Larvae engorge and moult	9–20 days
Nymphs engorge and moult	10–76 days
Females engorge	8–30 days
Unfed larvae survive	50–346 days
Parasitic period on host	28–60 days

Under natural conditions the winter tick produces one generation per year.

Heavy infection may occur in the long winter coat of animals during autumn and winter, causing debility, anaemia and at times death of horses, deer, elk and moose, especially when there are food shortages. *D. albipictus* is a vector of anaplasmosis and possibly also of Rocky Mountain spotted fever.

Dermacentor occidentalis (Marx, 1897), the 'Pacific Coast tick', is found in the area between the Sierra Nevada mountains and the Pacific Ocean from Oregon to southern California. Adult ticks occur on deer and also cattle, horse, donkey, rabbit, sheep and man, while the immature forms occur on a variety of rodents. *D. occidentalis* is a three-host tick. It is a vector of anaplasmosis, tick paralysis of livestock, Colorado tick fever, possibly Q fever and tularaemia.

Dermacentor nigrolineatus (Packard, 1869), the 'brown winter tick', is closely related to *D. albipictus* and some authors consider it to be a variety of the latter species. It is widely distributed in the eastern USA. The preferred host is the white-tailed deer but it also occurs on horses and cattle. It is a one-host tick.

Other species of the genus include: **D. marginatus** (Sulzer, 1776) Brumpt, 1913 (Asia, central Europe); **D. nuttalli** Olenov, 1928 (Asia; transmits Siberian tick typhus); **D. silvarum** Olenov, 1931 (Soviet Union; transmits Siberian tick typhus), and **D. halli** McIntosh, 1931 (North America).

Genus: Amblyomma Koch, 1844

Usually ornate. Eyes and festoons present. Hypostome and palpi long. Male without ventral plates, but small chitinous plaques may be present close to the festoons. The species are usually large and broad.

Amblyomma hebraeum Koch, 1844, the 'bont tick', occurs frequently in the warmer parts of South and Central Africa and is parasitic on all domestic and many wild mammals; the young stages also attack birds. This is a three-host tick. This tick has, on occasions, been imported into the USA with game from Africa. It is not established in the USA.

Female lays	up to 20 000 eggs
Pre-oviposition period	6–26 days
Larvae hatch	4–13 weeks
Larvae engorge	4–7 (–20) days
Larvae moult	25–66 days
Nymphs engorge	4–20 days
Nymphs moult	21 days to 3 months
Adults engorge	10–20 days
Unfed larvae survive	7 months
Unfed nymphs survive	6 months
Unfed adults survive	7–20 months

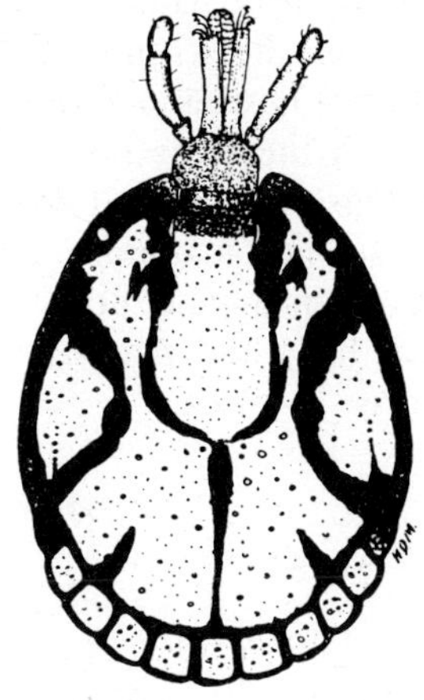

Fig 2.99 Dorsal view of the male *Amblyomma hebraeum* (not showing the legs).

The tick attaches usually in the perineal and genital regions of its host and may produce severe wounds. Zumpt (1958) says that the larvae and nymphs feed on birds, which may carry them over long distances. The species transmits *Rickettsia ruminantium,* the cause of heartwater of cattle, sheep and goats, which is also transmitted by **A. pomposum** Dönitz, 1909 in the Republic of South Africa and in Kenya by **A. gemma** Dönitz, 1909. *A. hebraeum* also transmits *Rickettsia conori* of tick-bite fever.

Amblyomma variegatum (Fabricius, 1794), the 'variegated tick' or 'tropical bont tick', is an African species generally distributed throughout the Ethiopian faunal region, south-western Africa and southern Africa (Hoogstraal 1956). It feeds on many species of mammals and rarely on birds, and transmits heartwater, Nairobi sheep disease and *Coxiella burnetii.*

This is a three-host tick:

Pre-oviposition period	12–18 days
Larvae hatch	53–86 days
Larvae engorge	5–7 days
Larvae moult	14–22 days
Nymphs engorge	5–7 days
Nymphs moult	19–24 days
Adults engorge	10–12 days

Amblyomma americanum (Linnaeus, 1758). This is the 'lone star tick' occurring in the USA from Texas and Missouri eastwards to the Atlantic Coast. It is given the name because of a single large white spot on the scutum of the female. It has a wide range of hosts and although larvae and nymphs occur on the same hosts as the adults, larger populations of the larval stages are found on foxes and smaller mammals.The ear is the favoured site of attachment but in heavy infections the head, belly and flanks are also parasitized. It is a three-host tick.

Female lays	1000–8000 eggs
Pre-oviposition period	5–13 days
Larvae hatch	23–117 days
Larvae engorge	3–9 days
Larvae moult	8–26 days
Nymphs engorge	3–8 days
Nymphs moult	13–46 days
Adults engorge	9–24 days
Unfed larvae survive	48–279 days
Unfed nymphs survive	3–476 days
Unfed adults survive	393–430 days

Apart from the painful bites caused by this tick it transmits Q fever, Rocky Mountain spotted fever and tularaemia. It is associated with tick paralysis in the eastern and southern USA.

Amblyomma cajennense (Fabricius, 1787), the 'cayenne tick', occurs in southern Texas in the USA but is widespread in Mexico, South and Central America and the Caribbean. The horse is chiefly parasitized but other hosts are a wide range of wild animals. It is a three-host tick with a developmental cycle similar to that of *A. americanum.*

This species transmits spotted fever in South America and *Leptospira pomona.* In South America it may cause great damage to cattle, producing 'fever', weakness and, possibly, death.

Amblyomma maculatum (Koch, 1844). This is the 'Gulf Coast tick' occurring in areas of high temperature and humidity on the Atlantic and Gulf of Mexico seaboards of North America. Larvae and nymphs occur mainly on ground-inhabiting birds (e.g. larks) and also occur on rodents. Adults are found on cattle and also horse, sheep, dog and man. It is a three-host tick.

Female lays	4500–18 000 eggs
Pre-oviposition period	3–9 days
Larvae hatch	21–142 days
Larvae engorge	25–100 days
Larvae moult	7–121 days
Nymphs engorge	4–11 days
Nymphs moult	17–71 days
Adults engorge	5–18 days
Unfed larvae survive	up to 179 days
Unfed adults survive	up to 411 days

This species is not known to transmit disease but it causes severe bites and painful swellings and is also associated with tick paralysis.

Genus: Aponomma Neumann, 1899

Eyes vestigial or absent, otherwise resembles *Amblyomma*. The species of this genus occur almost exclusively on reptiles.

Genus: Rhipicentor Nuttall & Warburton, 1908

This genus closely resembles *Dermacentor*, but differs from it especially in the following points: it is inornate, the basis capituli is hexagonal dorsally and has prominent lateral angles, and coxa IV in the male bears two long spurs. Two species in this genus are of interest both of which occur in Africa. **R. bicorinis** Nuttall and Warburton, 1908 occurs on a variety of domestic and wild animals in central and southern Africa, while **R. nuttalli** Cooper and Robinson, 1908 has been recorded from hedgehogs, dogs, hyenas, other carnivores and cattle in southern Africa. The disease relationships are unknown.

TICKS AND DISEASE

Ticks may harm their hosts by: (*a*) injuries done by their bites, which may predispose the hosts to attacks by blowflies, screw-worm flies and biting flies generally; (*b*) sucking blood; (*c*) transmitting the viruses, rickettsiae, bacteria and protozoa mentioned; (*d*) causing tick paralysis. Together, these effects may vary from a situation where it is impossible to raise livestock (either at all or not economically) to one where great expense is incurred in the control of ticks. In tropical and temperate areas, where they pose a problem, ticks are responsible for hundreds of millions of dollars loss per year.

Bite injuries

The harm done by their bites and blood-sucking has been reduced by control measures taken to control the diseases transmitted by ticks, but these two forms of harm are themselves important. Estimates of the amount of blood removed vary according to the species under consideration. A single adult female will remove 0.5–2.0 ml of blood and thus where an animal carries several thousand ticks a substantial blood loss may occur. Such theoretical figures may not apply under natural conditions; for example 20 000 adult *Boophilus* failed to kill a steer whereas 500 did kill a calf (Barnett 1961). Heavy infections of *Dermacentor albipictus* may kill moose and horses especially during winter when food supplies for these hosts are low.

Though very heavy infections do occur in nature, it is more usual for animals to carry a few hundred ticks. These produce a less tangible effect generally known as 'tick worry'. This is probably a combination of several entities including irritation from the tick bites, local skin infection, blood loss and secondary attack by flies. Systematic control of ticks almost always results in improved weight gains and yields.

The secondary effects of tick attack are infection of the local area, producing suppurative lesions (e.g. of the ears, legs etc.) and in lambs the local infection may become pyaemic (e.g. tick pyaemia in lambs in Great Britain). Attacks by blowflies and screw-worm flies are much encouraged by ticks.

A further effect is the damage produced to hides.

Tick paralysis

This is a disease of man and animals characterized by an acute ascending flaccid motor paralysis. The condition may terminate fatally unless the tick(s) are removed before respiratory paralysis occurs. Adult ticks, chiefly females, but sometimes nymphs, are responsible and ticks of the genus *Ixodes* are particularly associated with the condition but other genera, especially *Dermacentor* (*D. andersoni*), are concerned and it has also been ascribed to infections with *Ornithodorus lahorensis* (Mihailov 1957) and *Argus persicus*.

In general the degree of paralysis is proportional to the length of time the tick has been feeding and frequently also on the number of ticks attached. Removal of the ticks is usually followed by recovery provided the heart and respiratory centres have not been affected. Electrophysiological investigations show an almost complete reversibility following removal of ticks (Gothe & Kunze 1974).

In some instances paralysis seems to be produced by ticks only when they are attached near the head or along the spinal column (e.g. *D. andersoni*); however, this situation is the preferred feeding site for this tick and with other species (e.g. *I. ricinus, I. rubicundus, I. holocyclus*) the preferred feeding sites are elsewhere.

The nature of the 'toxin' produced is obscure. It appears to be elaborated in the tick concomitantly with egg production and it has been suggested that the toxin accumulates in the ovaries and passes to the salivary glands in the late stages of engorgement. Toxin, apparently identical with that injected by the engorging tick, has been obtained from the eggs of ticks which induce tick paralysis; however, a similar substance can also be obtained from the eggs of ticks which are not usually associated with tick paralysis (e.g. *Rhipicephalus evertsi, Boophilus decoloratus, Haemaphysalis leachi*, etc.) and consequently there is some doubt that the effective toxin is produced in the ovaries. It would seem rather that the toxin is produced in the salivary glands and some support for this comes from observations that engorging larvae of *I. holocyclus* can induce paralysis (Arthur 1962). Furthermore, unmated females of *D. andersoni* readily produce paralysis (Gregson 1958). Gregson (1959) also found that four to six days of engorgement were required to produce paralysis with *D. andersoni*, but when ticks were transferred from paralysed animals to normal animals signs of paralysis appeared in 12–18 hours. Martin (personal communication) has observed that three different types of glandular acini occur in the salivary glands of female ticks, one of which has the same morphological structure as the poison gland of snakes; it is possible that the toxin is produced in these. Such acini are present in ticks which have fed for several days.

Gothe and Kunze (1974) note that the nature of the toxin is unknown. It acts on motor and sensory nerves and on neuromuscular transmission. The liberation of acetylcholine is diminished and the receptor site is changed in its sensitivity. It is still unclear whether the interaction at the neuromuscular junction is presynaptic or postsynaptic. The toxin must have a very short half-life of activity since paralysis generally disappears if the tick is removed early enough. Its affinity is for peripheral nerve tissue, but nevertheless cerebral and spinal cord effects may occur. Degenerative changes have been observed in the spinal cord and the medulla of paralysed animals. In some cases capillary haemorrhages are present around nerve cells and in the adventitial sheath, accompanied by infiltrations of mononuclear cells (Arthur 1962).

Animals that recover from tick paralysis develop an immunity to the 'toxin' and, in the absence of further infection, may remain refractory to tick paralysis for eight weeks to eight months, the longer period being seen with *I. holocyclus*. Such immunity, however, should not be confused with that directed against the engorgement of the tick in which a marked local cellular reaction at the site of attachment interferes with engorgement and prevents full repletion. Serum from animals exposed to ticks may be used for curative purposes (e.g. *I. holocyclus*).

Eleven ixodid ticks and one argasid tick have been associated with tick paralysis in mammals, while in poultry *Argas persicus* has been recognized as a cause of the condition. A large number of hosts are affected including cattle, sheep, goats, pigs, dogs, various wild ruminants and man. Experimentally the groundhog (*Marmota flaviventris avara*), the hamster and, to a lesser extent, the guinea-pig are susceptible to paralysis. The species which are particularly associated with paralysis are *I. holocyclus*, especially in dogs in Australia, *I. rubicundus*, the cause of Karoo paralysis in South Africa, and *D. andersoni* which causes paralysis in cattle, deer and man in the north-western USA and western Canada. In the eastern and southern USA, *D. variabilis, A. maculatum* and *A. americanum* are implicated (Wilkinson 1965). In western and central Europe *Haemaphysalis inernis, H. punctata* and *I. crenulatus* are responsible.

Tick toxicosis is distinct from tick paralysis and is produced by toxins derived from some species of ticks. Several forms of tick toxicosis occur, the best known being sweating sickness of South, Central and East Africa. It has also been reported from India and Sri Lanka. It affects cattle, sheep, goats and pigs and is at its highest incidence during the summer months. The tick chiefly concerned is

Hyalomma transiens (= *truncatum*), though other species (*H. rufipes rufipes* and *H. rufipes glabrum*) have been incriminated. The adult stage of the tick is responsible for donating the 'toxin' to animals. Neitz (1959) has stated that ticks reared for many generations on unsusceptible animals still retain the ability to transmit the toxin. Animals exhibit profuse moist eczema and hyperaemia of the mucous membranes. It is chiefly a condition of younger animals and mortality may vary greatly, up to 70%.

Another toxicosis in the Republic of South Africa is caused by a leucocytotropic toxin produced by *Rhipicephalus appendiculatus*. Other mild tick toxicoses have been observed in pigs and cattle in Africa.

Virus transmission

The role of ticks in the transmission of viruses, rickettsiae and other organisms has been reviewed by Hoogstraal (1966, 1967, 1968) and Arthur (1962). The role of ticks in the transmission of protozoa is dealt with elsewhere in this volume, but some of the viral and rickettsial diseases transmitted by ticks include Rocky Mountain spotted fever, Siberian tick typhus, Colorado tick fever, Kyasanur Forest disease and Crimea–Congo haemorrhagic fever.

TREATMENT AND CONTROL OF TICK INFECTION

Although ticks are in themselves important parasites, and should be combatted for this reason, control measures are, as a rule, directed against the diseases of which the ticks are the vectors and therefore based on the epizootiology of these diseases as well as on the habits of the ticks.

Because ticks attach to various parts of the bodies of animals, treatment has to be applied to the whole body and may be carried out by dipping the animals in a suitable tank containing the dip in an aqueous solution, suspension or emulsion; however, spray races, showers etc. are replacing conventional plunge dips since they are labour-saving and economical. In some cases (e.g. *Ixodes rubicundus*) ticks attach to the legs and under-sides of the bodies of animals and consequently shallow dips, through which hosts are made to walk, may be sufficient to give control. These modern forms of apparatus drain the dipping fluid and filter it after it has been sprayed on the animals and return it for use again. The various dipping baths and sprays which are available for various animals are discussed in detail by Barnett (1961).

The various stages of ticks may stay on their hosts for only a few days during each year and are often on the hosts only at certain times of the year. Dipping for control of ticks is therefore planned with knowledge of the biology of each species of tick, the duration of each of its stages and of its feeding times and the duration of the whole life-history. An important consideration is whether the tick is a one-host tick, all the stages of which feed on the same individual host, a two-host species, which uses one individual host for the larvae and nymphs and another for the adult, or a three-host tick, each stage of which requires a separate individual host. The one-host tick is obviously much easier to control than the others. Acaricides may act differently on the different stages of the life history. Thus Hitchcock (1953), combatting the one-host tick *Boophilus microplus* in Australia, found that the adults of a strain of this were resistant to HCH, but that the larvae were not and were killed by 500 ppm of HCH; and in the Republic of South Africa it was found that adults of strain of *Boophilus decoloratus* were resistant to toxaphene, while the larvae were killed by dipping every seven days with 0.25% of toxaphene and this gave excellent control of this strain.

In the last two decades control of ticks was achieved by the use of the chlorinated hydrocarbon insecticides (e.g. DDT, HCH, toxaphene etc.), toxaphene having been used extensively against all species of ticks. It has a very good residual effect. With the increased interest in insecticide residues in meat and the development of resistance of *Boophilus* spp. to the chlorinated hydrocarbon insecticides a greater emphasis was placed on the organophosphorus and carbamate insecticides. Several of these are available. Where *Boophilus* spp. have become resistant to all three groups of insecticides it may be necessary to resort to arsenic dips; in some cases, however, the resistance is limited to the adult stage of the tick and an increased frequency of dipping, even with the insecticide to which the adults are resistant, may still result in control. Newer developments in the application of chemical control include the use of insecticide-impregnated ear tags, ear bands and neck bands (Ahrens et al. 1977).

The acaricides of choice in present use are as follows: arsenic (0.16–0.2%); butacarb, a carbamate, formulated as butacarb (20%) with HCH (38%) and phenol (38%) and used as a 0.05% concentration; chlorfenvinphos, formulated to be active as a 0.05% solution, contains chlorfenvinphos (9.8%), HCH (3.2%) and phenol (28%); coumaphos (Asuntol, Co-Ral, Bayer 21/199) as a wettable powder at an effective strength of 0.05%; chlorpyrifos (Dursban) at a concentration of 0.05%; diazinon (0.05%); dioxathion (0.01%); and phosolone (0.04%).

A major problem in tick control by insecticides is the development of resistant strains of ticks. The recent developments in the synthetic pyrethroids (Elliot et al. 1978) suggest that these compounds are more effective and less liable to the development of resistance than the chlorinated hydrocarbons, organophosphates or carbamates. Four synthetic pyrethroids are in use at present for tick control: flumethrin, decamethrin, cypermethrin and cypothrin. These are active against ticks resistant to chlorinated hydrocarbons, organophosphates and amidines. Flumethrin is used in plunge dips and spray races at 30–75 ppm depending on formulation, tick species and country. In addition, concentrations as low as 2–5 ppm prevent oviposition and consequently this compound has an important effect in the reduction of the number of ticks in the local environment. Decamethrin is used at present in South America at 25 ppm in dips and spray races. Cypermethrin, generally used at 150 ppm for dips and spray races, may also be formulated as 2.5% cypermethrin and 13.8% chlorvinphos and used at 100 and 550 ppm. Cypothrin is used in dips and spray races at 150 ppm. Owing to biodegradation, replenishment of dips should be attended to according to the manufacturers' recommendations. Resistance of ticks to insecticides is reviewed by Beesley (1973).

Other measures, which have more limited value, are useful more especially against two- and three-host ticks, which spend relatively long periods of their lives off the host and on the pastures. These methods include:

Burning of pastures. This may kill large numbers of the larvae and other stages, especially if it is correlated with the times of the year when these stages may be expected to be off the hosts.

Cultivation of land. This undoubtedly tends to reduce tick life by controlling the movements of domestic and wild animals, as well as by creating conditions unsuitable for ticks, as, for instance, exposure of eggs to sunlight, or burying them deeply by ploughing. Good drainage helps to reduce the humidity on which the ticks depend. Another practice found effective in Australia, to which the term 'pasture-spelling' has been given, works on the principle of removing cattle from pastures infected with ticks for long enough to ensure that most, if not all, of the larval ticks on the pastures are killed off by starvation or climatic effects. Rotational grazing systems have been devised to apply this method of control and considerable control has been obtained by it (Beesley 1973).

In some parts of the world certain species of grasses are repellent for ticks. Thompson et al. (1978) have shown that species of the genera *Melinis*, *Cynodon* and *Pennisetum* are effective in this respect. Recently, Sutherst et al. (1982) have shown that tropical legumes of the genus *Stylosanthes* immobilize and kill cattle ticks. Larvae of *B. annulatus* are immobilized by a sticky secretion on the hairs of the plant and an unidentified vapour kills the larvae.

Starvation. In general this method is not practicable, because ticks live for long periods without food and wild animals may serve as hosts.

Repellents may be useful in certain circumstances, the most effective ones being indalone and dimethyl phthalate.

Natural enemies. Certain small hymenopterous parasites of ticks, for instance species of *Ixodiphagus* and *Hunterellus*, are known. They parasitize especially the nymphs, in which they lay their eggs and which are literally 'eaten out' by the larvae of the parasite. Certain ants as well as birds (*Bubulcus ibis, Buphagus erythrorhynchus, B. africanus*) destroy a large number of ticks. Specific diseases of the parasites are also known. It is, however, questionable whether biological control by means of parasites and other agents would be able to diminish the number of ticks to a point below which they would no longer be important as disease transmitters. It is, for instance, frequently difficult to find a *Rhipicephalus* on cattle which have for some months been regularly dipped in an East Coast fever area and yet the few remaining ticks may still be sufficient to cause cases of the disease.

Sterile hybrids. Experimental studies have been made with crosses between *B. annulatus* and *B. microplus* (Thompson et al. 1981). Male offspring were sterile and hybrid females produced sterile males for three back-cross generations.

SUBORDER: TROMBIDIFORMES REUTER, 1909

FAMILY: TROMBICULIDAE EWING, 1944

This family contains the mites, whose parasitic larvae are called 'harvest mites', 'chigger mites' and various other names. The term 'chigger' is, however, also given to the flea, *Tunga penetrans.* Trombiculidae usually have a scarlet, red, orange or yellow colour and the adults may be very large. The nymphs and adults are free-living and feed on either invertebrates or plants. Their bodies are covered with dense hairs which give them a velvety appearance. Their bodies are divided into a gnathosoma, a propodosoma bearing the first two pairs of legs and a hysterosoma which bears the third and fourth pairs of legs. The last segment of the large pedipalps opposes a claw on the last segment but one, much as a finger and thumb do. The larvae are parasitic on various animals and man, causing marked irritation and in some cases transmitting important diseases. The larval mites occur most frequently in autumn and attack grazing animals and human beings working in low-lying fields. The natural hosts of the larvae are in most cases small rodents, such as field-mice. The larval mites attach themselves to the host and their salivary secretion hydrolyses the cuticle of the host, forming a tube called the *stylostome,* through which the larva sucks up the host's tissue fluids. When they are ready to moult they drop off and moult to become the non-parasitic nymphal stage.

Genus: Trombicula Berlese, 1905

All legs have seven segments, there are five setae and a pair of sensillae on the scutum; stigmata and tracheae are lacking.

Trombicula (Neotrombicula) autumnalis (Shaw, 1790). This is the 'harvest mite', 'aoutat' or 'lepte automnale' of Europe. The larvae attack man and practically all species of domestic animals, including poultry. The latter may be killed by heavy infections. The larger animals are usually attacked on the head, sometimes the neck and often the extremities. The larvae of this species may cause generalized pruritus and lesions in the interdigital spaces of the dog and be the cause of 'heel-bug' of racehorses. The adults live in the soil and in Britain the parasitic larvae are most numerous in the late summer and autumn. They are commonest on chalky soils and on grassland, cornfields, heathland and scrubby wood-land; they are less common on clay soils. The unfed larvae are about 0.21 mm long and have a deep red colour; when full-fed, they are 0.4 mm long and a pale pink or yellow colour.

Trombicula akamushi (Brumpt, 1910) is found in Japan and New Guinea, **T. delhiensis** Walch, 1922, extends from India to China and Australia, and other species of this genus and also possibly species of **Euschöngastia** Ewing, 1938, may be vectors of *Rickettsia tstusugamushi*, the cause of scrub or mite typhus (tsutsugamushi disease, Japanese river fever) of man. For further information about these and other related species which transmit Korean fever and other diseases see Baker et al. (1956).

Control. For area control of the species which transmit scrub-typhus, the application to the infested ground of 2 kg/hectare of toxaphene or chlordane or 0.5 kg/hectare of lindane as a spray or dust is effective. DDT is relatively ineffective for this purpose.

As a repellent benzyl benzoate may be applied to the clothes, especially to openings in them, either by hand or as a spray, or the repellent may be dissolved in a dry cleaning fluid, or in acetone or a soapy solution, or the clothes may be impregnated with one of these; it remains in the clothes after one or two washes.

Trombicula sarcina (Womersley, 1944) is the cause of blacksoil itch (leg itch) of sheep in Queensland.

This species occurs in areas in Australia which have the true black earths (not 'black soils'), the normal hosts being the kangaroo and wallaroo. The golden-yellow larvae feed on dogs, especially between the toes, and also attack man and sheep and sometimes the horse.

Pathogenesis. The irritation due to the larvae causes stamping and the skin of the coronet, heels and pasterns becomes reddened and abraded. Secondary infection may lead to swelling, thickening of the skin and scab formation and in severe infections similar lesions on the legs may cause them to swell to twice their normal thickness (Seddon 1951).

Treatment. Dressings of kerosene in oil have been used but a foot-bath of 0.1% HCH is more satisfactory. Various organophosphates and synthetic pyrethroids show good promise for the control of these mites.

Baker et al. (1956) describe **T. minor** (Berlese, 1905), the scrub-itch mite of Queensland, which attacks man and rats; and other related species, among which is **Acomatacarus australensis** (Hirst, 1925) which attacks man and dogs in Queensland and New South Wales, for the control of which BHC and dibutyl phthalate are suggested.

Trombicula alfreddugési (Oudemans, 1910) is the common red bug of the USA, though it ranges from Canada to South America and the West Indies and has different local names in different countries. Its larvae are commonest on the borders of forests and in swamps. It attacks many mammals, birds, reptiles and Amphibia and also man. On man the larvae attach themselves on parts of the body constricted by clothing and the lesions produced may be like those due to *Sarcoptes scabiei*. Baker et al. (1956) describe this and related species, among which are **T. spendens** Ewing, 1913 and **T. batatas** (Linnaeus, 1758), which attack man and other hosts in North and South America and Australia.

Neoschöngastia americana (Hirst, 1921) attacks fowls in the southern USA, Mexico, Guatemala and Jamaica and a related variety occurs in Japan and the islands of the Pacific. These attack chickens, quail and turkeys and may cause skin lesions which affect the market value of the birds. For further information about these and related mites, see Baker et al. (1956).

Diagnosis is made by finding the reddish to yellowish larvae in scrapings from the lesions.

FAMILY: PEDICULOIDIDAE BERLESE, 1907

Pediculoides ventricosus (Newport, 1850), the 'grain itch mite', is frequently found as a parasite on insects or their young stages that live in grain or straw. From such infected grain the mites may pass on to human beings or domestic animals, causing a dermatitis known as 'grain itch', which is characterized by intense pruritus.

Treatment and prophylaxis. The mites on animals can be killed by an application of a suitable dip of chlorinated hydrocarbon or organophosphate insecticide. Symptomatic treatment to relieve itching may be necessary; for this purpose benzyl benzoate ointment is recommended. Infected material is best destroyed by burning.

FAMILY: DEMODICIDAE NICOLET, 1855

Genus: Demodex Owen, 1843

This is a very specialized group of parasitic mites which live in the hair follicles and sebaceous glands of various mammals, causing demodectic or follicular mange. The parasites which occur on different species of hosts are usually regarded as distinct species, although it is difficult to distinguish between them morphologically, since the main difference is that of size. Most of the species are called after their hosts; for instance, *D. canis, D. ovis, D. caprae, D. bovis, D. muscardini* of the dormouse, *D. criceti* of the hamster etc., while *D. folliculorum* occurs on man and *D. phylloides* on the pig. Nutting (1976) has reviewed the taxonomy and biology of *Demodex* species of medical and veterinary importance and has constructed a key for their differentiation.

The parasites are elongate, usually about 0.25 mm long; they have a head, a thorax which bears four pairs of stumpy legs and an elongate abdomen which is transversely striated on the dorsal and ventral surfaces. The mouth parts consist of paired palps and chelicerae and an unpaired hypostome. The penis protrudes on the dorsal side of the male thorax and the vulva is ventral in the female. The eggs are spindle-shaped. A detailed description of *Demodex canis* Leydig, 1859 is given by Nutting and Desch (1978).

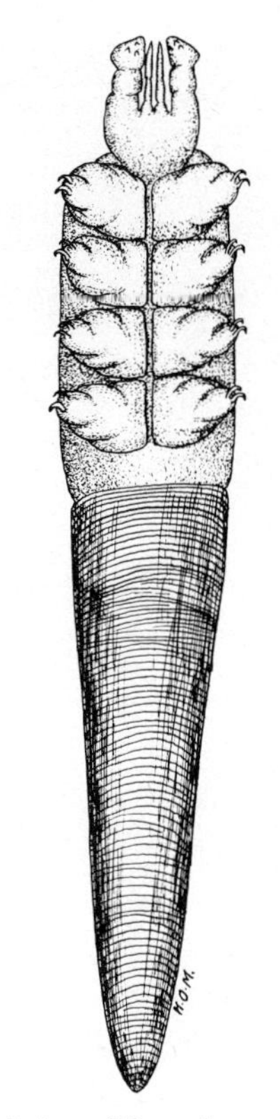

Fig 2.100 Ventral view of *Demodex canis*.

Life-cycle. The entire life-cycle is spent on the host; egg, larva, protonymph, deutonymph and adult stages are recognized (Baker et al. 1956). The morphology has been described by Nutting and Desch (1978). The life-cycle is completed in 18–24 days, in the hair follicles or sebaceous glands according to species. Males occur at or near the skin surface whereas the fertilized female oviposits 20–24 eggs in the hair follicle. Larvae and nymphs are swept by the sebaceous flow to the mouth of the follicle, where they mature and repeat the cycle (Spickett 1961).

Infection is transmitted by direct contact (or, experimentally, by the application of infected skin). Transmission per os or in utero is now discounted. Neonatal contact with an infected mother for as little as one hour may result in infection. But mites are not found in caesarean-derived, orphan or still-born pups (French et al. 1964; Greve 1970). Predisposing factors to clinical disease include age, poor condition, intercurrent infection (especially virus infection) and undue use of alkaline soap or shampoo. However, *Demodex* is very common in normal skins (e.g. 53%; Koutz et al. 1960) and the predisposition to clinical disease may be genetically determined since there is now evidence that demodectic mange of dogs is associated with failure of animals to mount any adequate cell-mediated immune response to infection (Healy 1977).

Pathogenesis. Gross lesions producing clinical disease have been reported for *Demodex canis, D. bovis* Stiles, 1892, *D. caprae* Railliet, 1895, *D. equi* Railliet, 1895, *D. ovis*, Railliet, 1895 and *D. phylloides* Csokor, 1879. The lesions vary from simple squamous epithelial shedding or depilation to small or large papules or nodules. Secondary bacterial infection may occur, leading to suppurative pustule formation.

Canine demodicosis usually becomes evident at the age of three to nine months. The numbers of mites in the skin are proportional to the severity of

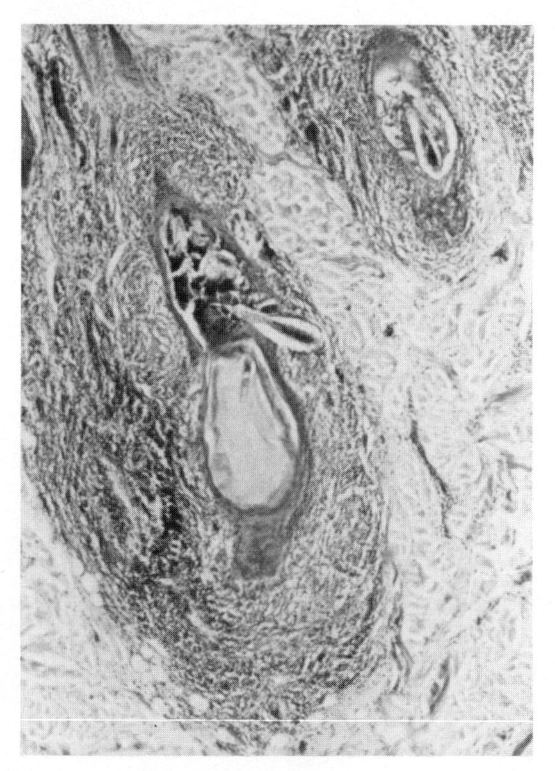

Fig 2.101 Acute phase of demodectic mange in dog skin. Mite in sebaceous gland. ×200

the clinical disease and in a clinical case myriads of mites may be found per cm^2 of skin. The factors which permit the development of such populations are not known, but they are associated with cell-mediated immunity defects (Corbett et al. 1975) which is manifest by a suppressor effect of serum on mitogen-induced transformation of normal lymphocytes or those from dogs with demodicosis (Hirsch et al. 1975). Lesions, initially, occur predominantly on the head and forelimbs and from here may spread over the entire body, though the hairless parts of the body are seldom involved.

Two forms of the disease are usually recognized, a squamous form and a pustular form. In the squamous form the hair follicles are distended with mites and cellular debris, the follicular epithelium is atrophic, hyperkeratosis is progressively evident and pieces of cornified material exfoliate from the surface. The hair may be separated and shed or splintered or disintegrated. Changes in the sebaceous glands vary; some may become atrophied, others may hypertrophy and Nutting (1950) believes that mites feed on hyperplastic cells. Hyperpigmentation occurs giving the skin a coppery-red colour. The condition progresses until large areas of the body are affected and the animal shows alopecia and thickened and wrinkled skin with a 'mousy' odour.

The pustular demodectic mange results from bacterial invasion (*Staphylococcus* spp.) of the dermis. Extensive dermal infiltrations of polymorphonuclear leucocytes and lymphocytes and plasma cells are seen and pustule or abscess formation, with marked inflammation, occurs. This form is usually preceded by the squamous forms.

In extensive forms of canine demodicosis death results from toxaemia or emaciation.

In pigs (*D. phylloides*) large pustules may occur, about the size of a walnut; they may become confluent and rupture.

Demodectic mange of cattle (*D. bovis*) is responsible for considerable financial loss to the tanning industry in Australia (Murray et al. 1976). Holes, craters and scar tissue are produced and these defects are particularly noticeable in split hides.

Demodex equi of horses may produce pruritus, a poor coat with a loss of hair and pustule formation (Bennison 1943).

In goats the pustular form is the most common and may cause economic loss.

In sheep the disease is rare and usually the scaly form develops.

Cats, like dogs, usually develop lesions on the head, but they are rarely affected.

Diagnosis. The mites can be found in deep scrapings and in the contents of pustules and abscesses. Scrapings must be deep enough to assure sampling of the hair follicle and hence should cause capillary oozing. The skin may be softened with a weak solution of potassium hydroxide to assist scraping.

Uncomplicated cases of demodicosis are associated with large numbers of mites in all stages of development (e.g. larva, nymph and adult).

Where only a few adult mites are found, caution is necessary in ascribing the condition to demodicosis.

Treatment. The remedies proposed for demodectic mange are legion. They have varied from the feeding of large amounts of garlic to surgical removal of skin. Many remedies are helpful in resolving demodectic mange if applied in the early stages of the disease, but fail when it is extensive. The recognition that canine demodicosis may be an immunological deficiency disorder provides an explanation for the failure of many treatments in advanced cases.

Rotenone (3%) diluted one in three in spirit is useful for small areas of the body when applied every few days for four to five weeks. An emulsion of benzyl benzoate (20%) or benzyl cresol (0.5%) and BHC (0.25%) may be applied daily, but to small areas only since benzyl benzoate may be toxic if large areas of skin are treated. A satisfactory treatment appears to be the compounds such as fenchlorphos (Ecotoral), trichlorphon or dichlorvos (30 mg/kg at two-week intervals (Hughes & Lang 1973). These are also effective for *Demodex* infections in other animals. More recently Folz et al. (1979) have reported that Mitaban (amitraz: *N*′-(2,4-dimethylphenyl)*N*-[2,e-dimethylphenyl(imino) methyl]-*N*-methylmethanimid amine) (at a rate of 250 ppm given for three to six treatments) achieved a 98% clearance of viable mites in naturally occurring cases of *D. canis*. The anthelminthic closantel has been shown to be effective in localized uncomplicated cases of demodectic mange. A dose of 5 mg/kg for the first subcutaneous injection is followed by 2.5 mg/kg weekly (Losson & Benakhla 1980).

Other species of *Demodex* include **D. odocoilei** Desch and Nutting, 1974 (white-tailed deer) and **D. cervi** Vanselow, 1910 (red deer), which produce nodular lesions; **D. folliculorum** (Simon, 1843) and **D. brevis** Akbulatova, 1963 of man; **D. musculi** Hirst, 1917 (mouse); **D. ratti** Hirst, 1917 (rat); **D. aurati** Nutting, 1961 (hamster); **D. caviae** Bacilgalupo and Roveda, 1954 (guinea-pig); and **D. cuniculi** Pfeiffer, 1903 (rabbit).

FAMILY: CHEYLETIDAE LEACH, 1814

Genus: Psorergates Tyrrell, 1883

The genus is classified by Baker et al. (1956) in this family, but Till (1960) places it in a family called the Psorergatidae.

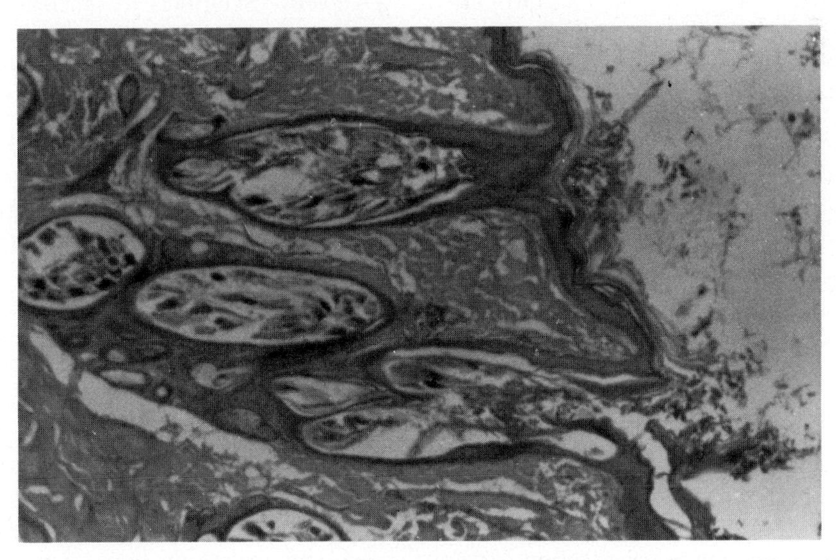

Fig 2.102 Section of skin from an advanced case of demodectic mange. Many organisms in hair follicles, little cellular traction and hyper keratosis. × 200

Psorergates ovis Womersley, 1941. This species is a parasite of the skin of sheep and has been found in Australia, Tasmania, New Zealand and the USA in Ohio, California and New Mexico. It occurs in the Republic of South Africa where the disease is sometimes called 'Australian itch' and it has also been recorded from Argentina. The mites are very small, the female measuring only 189 by 162 μm and the male only 167 by 116 μm. The body is almost spherical, with sides slightly indented between the legs, which are radially arranged and have paired claws. The femur of each leg bears relatively long setae and, on the ventral side, a large, curved spine directed inwards. The chelicerae are very small stylets and the pedipalps are short and conical. There is a single dorsal plate. The female has, on the ventral side, near the posterior end, two pairs of long setae which arise from two lobes. The male has only one pair of ventral setae near the posterior end, which arise from a median lobe. The penis is on the anterior half of the dorsal surface of the body.

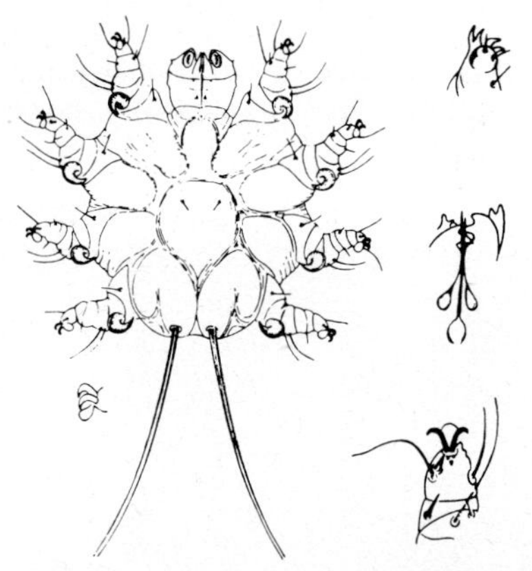

Fig 2.103 Left, Female *Psorergates simplex* (Tyrell). Right, Palps, chelicera and tarsus 1 of *Psorergates ovis* (Wormersley, 1941). (*From Baker et al. 1956*)

Life-cycle and habits. Womersley (1941) described the developmental stages and the life-cycle has been described by Murray (1961). The eggs are delicate structures, 70–90 by 87–117 μm and are produced eight days after infection; larvae develop in them 11 days after infection. Larvae (70–90 μm in diameter) produce protonymphs on day 20; there are three nymphal stages—protonymph, deutonymph and tritonymph—each increasing in size and appearing on days 20, 20 and 27 respectively after infection, adults appearing on day 35. Transmission is by contact. The mites live in the surface layers of the skin and have so far been found in fine-wooled Merino sheep and also in Polwarths, Corriedales, Comebacks and Border Leicesters. They cause a mild but chronic irritation which leads to occasional biting and scratching, so that the wool is disturbed in affected areas, showing a pale tip or even tufts which have been pulled out. The wool growing on affected areas becomes thready, forming pointed tufts. It contains dry scabs and breaks easily. The general appearance is that of a louse-infected sheep. The skin is hardly altered except that it is covered with dry scales. Microscopically there is a hyperkeratosis and marked desquamation; the deeper layers show round-cell infiltration and sometimes eosinophilia in the immediate vicinity of a parasite. The mites apparently do not penetrate into the skin. The infection spreads very slowly over the body, requiring some three to four years to become generalized (Roberts et al. 1965). A detailed account of the infection in sheep is given by Sinclair (1967).

Diagnosis. Scrapings are taken after moistening the skin with a light mineral oil and directly examined in the oil. Mites are more numerous, generally, in winter and spring and the absence of mites in a single examination is not sufficient evidence for a negative diagnosis.

Treatment. The mite is not controlled by the conventional preparations of HCH (which rely on the gamma isomer) and South African workers have suggested that its increase in that country may be due to wide use of DDT and HCH instead of sulphur dips.

Lime–sulphur dips are the most satisfactory, using a concentration of 1% of polysulphide sulphur. The addition of a wetting agent is required to insure good penetration of the fleece. The delta and epsilon isomers of HCH, Bayer L 13/199 and malathion have given promising results. A single dipping with 0.1% delta isomer

HCH or 0.2% malathion cured infestations (Skerman et al. 1960).

Psorergates simplex Tyrrell, 1883. This closely-related species has been found in pustules on the skin of *Mus musculus* in Canada. It has also been found in the multimammate rat (*Mastomys natalensis*) used in laboratories in the Republic of South Africa.

Psorergates oettlei Till, 1960 was found by Till (1960) on *Mastomys natalensis.*

Psorergates bos Johnston, 1964, is an itch mite of cattle in New Mexico and Texas, USA. Roberts et al. (1965) believe the mite to be a widely distributed parasite of cattle in the USA but because of its small size and lack of any marked pathogenic effect it has, hitherto, been undetected. Mites may occur on apparently normal skin and itching or scratching are not features of the infection. In one animal alopecia and desquamation occurred along the sides of the neck, but there appears to be no constant recognizable lesion associated with the infection.

Genus: Syringophilus Heller, 1880

Syringophilus bipectinatus (Heller, 1880) may be found inside the quills of the feathers of the fowl, **S. columbae** (Hirst, 1920) inside the quills of the feathers of the pigeon and **S. (Cheyletoides) uncinata** (Heller, 1880) inside the quills of the feathers of the peacock. These mites have elongated bodies adapted to their life inside the quills of the feathers. Hughes (1959) states that species of this genus found inside the quills of the flight feathers of geese and 'similar birds' may prey on the mite *Syringobia*, although they probably also feed on the material inside the quills.

Genus: Cheyletiella Canestrini, 1886

Cheyletiella parasitivorax (Mégnin, 1878), the rabbit fur mite, has been reported on the fur of rabbits, cats and dogs. In cats and dogs a dry scaly dermatitis has been ascribed to its presence, but it is likely that previous reports have misidentified the species of mite concerned (Smiley 1965).

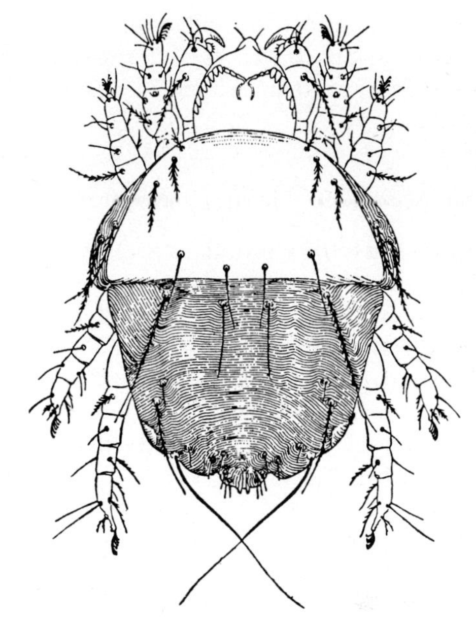

Fig 2.104 Dorsal view of female *Cheyletiella parasitivorax.*

Smiley (1970) has reviewed the genus *Cheyletiella* and considers it consists of five species, namely: **Cheyletiella yasguri** Smiley, 1965 of the dog; **Cheyletiella blakei** Smiley, 1970 of the cat; **Cheyletiella strandtmanni** Smiley, 1970 of the hare; and **Cheyletiella parasitivorax** and **Cheyletiella furmani** Smiley, 1970 of the rabbit.

Differentiation of the dog and cat forms is based on the shape of the sensory organ on genu 1.

Infection of humans with these mites by contact with infected dogs and cats may produce lesions ranging from a mild dermatitis to a more extensive papular eruption (Thoday 1979).

The mites measure 386 by 266 μm and the genus is readily recognized under the microscope by the numerous feathered bristles and the 'combs' on the tarsi.

Satisfactory control is obtained with derris washes or dusts, DDT, HCH or the organophosphorus compounds dichlorvos and fenitrothion.

FAMILY: MYOBIIDAE MÉGNIN, 1877

The small or medium-sized mites of this family are related to Cheyletidae and *Psorergates ovis* has been classified with them. They have a striated, unarmoured skin. Baker et al. (1956) describe species found in rats, in which they cause a mild dermatitis.

Genus: Myobia v. Heyden, 1826

Myobia musculi (Shronk, 1781) is a pelage-inhabiting mite found on laboratory and other mice, associated with loss of hair and dermatitis. The adults are 350–500 μm long, elongate, with transverse striations but no sclerotization. The legs are distinctive in that the first pair are highly modified for clasping hairs. The life-cycle has been described by Haakh (1958) and may be completed in 12–13 days. Transmission is by contact. Control is by acaricidal dips of, for example, 2% malathion or 25% Tetmosol (tetraethylthiuram monosulphide), by exposure to dichlorvos (DDVP) vapour or by dusting with 1% lindane. It is likely that synthetic pyrethrins will be of value.

Radfordia ensifera (Poppé, 1896) is similar to *M. musculi* in morphology and biology, but occurs on wild and laboratory rats. Heavy infections may lead to self-inflicted trauma.

Radfordia affinis Poppé, 1896 (syn. *Myobia affinis*) has been reported from the mouse.

Genus: Sarcopterinus Railliet, 1893

Sarcopterinus (Harpirhynchus) nidulans (Nitzsch, 1818) lives in the feather follicles of pigeon and various other birds. Hughes (1959) states that **S. pilirostris** Berlese and Trouessart, 1889 may be found in the skin on the head of the sparrow.

SUBORDER: SARCOPTIFORMES REUTER, 1909

This suborder includes a large number of species which are mostly free-living and small in size. A few species are important parasites or vectors of disease. The legs are frequently grouped in two pairs on either side in the nymph and adult and they end in suckers, claws or hairs. Some species, in unfavourable environments, produce nymphs adapted to resist adverse conditions. They are called *hypopial nymphs* or *hypopi* and are often provided with groups of suckers and hooks on their legs which they can use to adhere to transport hosts which carry them to more favourable situations. The most important species of the suborder are those that cause mange. These species belong to the families *Sarcoptidae* and *Psoroptidae*.

FAMILY: SARCOPTIDAE TROUESSART, 1892

The genera of importance in this family include *Sarcoptes,* the cause of scabies of man and sarcoptic mange of sheep, goats, cattle, pigs, equines, dogs, foxes, rabbits and other animals; *Notoedres,* the cause of notoedric mange of cats, rabbits and rats; *Cnemidocoptes,* the cause of scaly leg (*C. mutans*) and depluming itch (*C. gallinae*) of poultry.

Species of this family burrow more or less deeply into the skin, causing marked thickening rather than the formation of scabs.

Genus: Sarcoptes Latreille, 1806

The mites belonging to this genus are parasitic on a number of different domestic and wild mammals, causing mange. They are regarded by some authors as belonging to different species and by others as biological or physiological races of the species *S. scabiei,* which are specific to their hosts.

The body of the mite is globose and the striae of the skin are interrupted by scaly and spinose areas. There are dorsal dentate spines with sharp points (also in the genus *Notoedres*). Two vertical setae occur on the dorsum of the propodosoma. The legs are short, the tarsi of the first, second and fourth pair of legs in the male and the first and second pair of legs in the female end in bell-shaped suckers (caruncles) while the third pair of legs in the male and the third and fourth in the female end in bristles. The pedicels which bear these suckers or bristles are not segmented (unlike the pedicels of *Psoroptes* which are three segmented). The anus is terminal and the male has no adanal suckers.

Sarcoptes scabiei (Degeer, 1778) is a minute parasite, roughly circular in outline. The female measures 330–600 μm by 250–400 μm and the male 200–240 μm by 150–200 μm. All the legs of both sexes are short and the third and fourth pairs do not project beyond the margin of the body. On the ventral surface, the epimeres (chitinous extensions of the coxae of the legs) are distinct, those of the first pair of legs are fused into a single rod and those of the third and fourth pair of legs are fused to form a lateral bar. The dorsal surface is covered with fine folds and grooves, mainly transverse in arrangement, and bears a number of small triangular scales. The female bears on either side of the dorsal mid-line anteriorly three short spines and posteriorly six longer spines with bifid tips, in addition to a few hairs.

Life-cycle. The life-cycle of *S. scabiei* in man was described by Mellanby (1952). Probably the life-cycle of the varieties found on other animals is similar. The female burrows into the skin and lays 40–50 eggs in the tunnel it forms. The eggs are laid one or two at a time about three to five being deposited daily. These hatch in three to five days to produce a six-legged larva. Some larvae escape from the breeding tunnels and wander on the skin, but some remain in the parent tunnel or side pockets of it and continue their development there as far as the nymphal stages. Of those that reach the surface many perish; others burrow into the stratum corneum to construct almost invisible moulting pockets, in which they also feed. There are two nymphal stages (protonymph and deutonymph) which may stay in the larval pockets or wander and make new pockets. The nymphs have four pairs of legs, but no genital apertures. Finally, males and females are produced, the development from the time the eggs are laid taking about 17 days. The adult female remains in the moulting pocket until fertilized by a male. She then extends the pocket into a tunnel or makes a new one and, after four to five days, begins to lay eggs, three to five a day. The female probably does not live longer than three to four weeks. Infection is spread mainly by contact by the wandering larvae, nymphs and fertilized young females.

The mites are very susceptible to dryness and cannot live more than a few days off their host. Under optimal laboratory conditions the mites have been kept alive for three weeks.

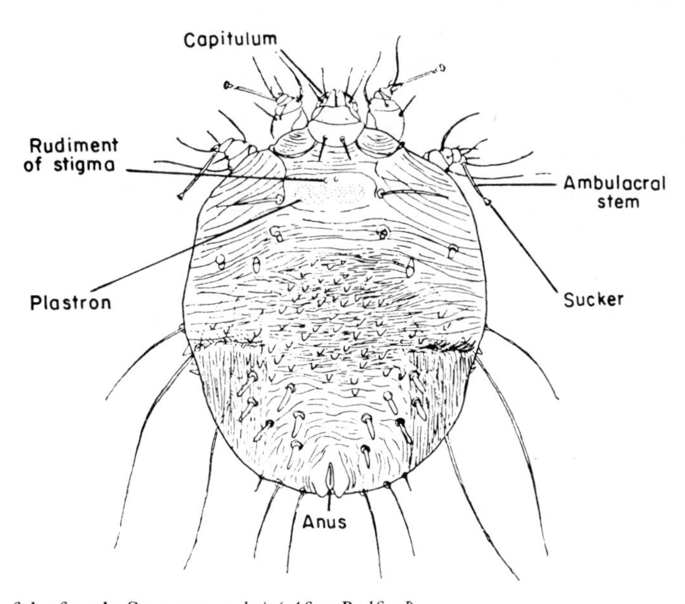

Fig 2.105 Dorsal view of the female *Sarcoptes scabei*. (*After Bedford*)

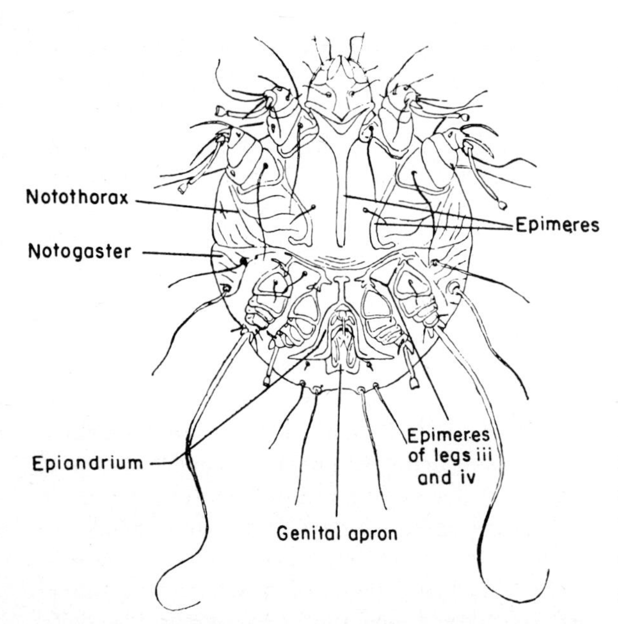

Fig 2.106 Ventral view of the male *Sarcoptes scabei*. (*After Bedford*)

Pathogenesis. The parasites pierce the skin to suck lymph and may also feed on young epidermal cells. Their activities produce a marked irritation which causes intense itching and scratching, which aggravates the condition. The resulting inflammation of the skin is accompanied by an exudate which coagulates and forms crusts on the surface and is further characterized by excessive keratinization and proliferation of connective tissue, with the result that the skin becomes much thickened and wrinkled. There is a concomitant loss of hair which may be very wide-spread. Sensitization of the host to the mite and its products probably plays an important role in the pathogenesis. Usually there is an absence of pruritic signs for the first two or three weeks following infection and in man this period may vary from three to six weeks (Mellanby 1944). Then pruritus appears, being associated with an urticarial response. The duration of the urticarial reaction is variable and often the animals with the most severe urticarial reactions recover more quickly than those with less severe reactions (Mellanby 1944: Sheahan 1974). In some animals, and also in man, severely encrusted lesions containing myriads of mites (crusted scabies, Norwegian scabies) may occur without pruritus.

Sarcoptes tapiri is a species described from the tapir (Kutzer & Gruenberg 1967) and **Trixacarus diversus** Fain, 1968 may cause severe mange in rats, white mice and hamsters. Rats may die from severe infections (Lavoipierre 1960).

Trixacarus (Caviacoptes) caviae (Fain, Hovell and Hyatt, 1972) is responsible for mange in guinea-pigs, characterized by pruritus, keratinization and alopecia (Hovell et al. 1976). Originally described in Europe, it is now also recognized in the USA. Infected guinea-pigs respond well to HCH applied as a shampoo on three occasions at 10-day intervals (Thoday & Beresford-Jones 1977).

SARCOPTIC MANGE IN THE DOG

Initial lesions often occur in the axillary and inguinal region or on the head, around the ears or above the eyes. The initial erythema may be unnoticed, but later a papular dermatitis is evident; and rupture of the papules with encrustation of lymph and pruritus characterizes the early manifestations. Later the skin becomes dry and thickened, there is loss of hair (alopecia) and secondary infection (pyoderma) and self-mutilation may occur (Smith & Catchpole 1967). Cases which progress to a generalized state (and not all do, for reasons which are uncertain) become cachectic and usually die if not treated (Král & Schwartzman 1964).

Diagnosis is based on the clinical signs and the demonstration of the parasite and/or its eggs in deep skin scrapings. These should be made with a scalpel so that capillary oozing occurs. This material is examined in water, glycerine or light mineral oil. In early cases, mites may not be numerous and sometimes only eggs and larvae are seen. Differential diagnosis includes eczema, contact dermatitis, demodectic mange, etc.

SARCOPTIC MANGE IN THE PIG

Sheahan (1974) has described the clinical signs and significance of *S. scabiei* var. *suis* infections in pigs. Up to 35% of pigs may be infected. Light infections may go unnoticed but severe infections consist initially of generalized focal erythema, pruritus and later encrustations, especially in the ears and later of the skin of the back, which may become very thickened, then crack open leaving deep wounds which become secondarily infected. Sheahan (1974) noted that intercurrent disease, especially iron deficiency, exacerbated mange in pigs and other authors have incriminated a variety of entities in this. Despite obvious infection of pigs, Sheahan (1974) was unable to demonstrate a significant influence of mange on the growth rate and feed conversion rate of young pigs similar to the effects reported by other authors who believe the infection has a significant effect on these parameters.

SARCOPTIC MANGE IN THE HORSE

This condition is now rare, being a 'notifiable disease' in many parts of the world. Infection usually starts on the head, neck and shoulders or in the saddle area. Papule formation and pruritus are the early clinical signs. Later the condition progresses to a weeping surface covered with thick crusts and later a scaling condition with papules and scabs, associated with loss of hair, spreads over the whole surface of the body. Marked thickening of the skin may occur, with cracking and secondary infection. In advanced cases emaciation, cachexia and eventually death occur.

Human infection from equine sarcoptic mange is well known and occurs on the fore-arms and back and other places which come into contact with horses. Often the disease takes a mild course and transmission to other humans is not common.

SARCOPTIC MANGE IN OTHER ANIMALS

Sarcoptic mange is rare in cattle; it usually occurs in housed cattle and causes pruritus and bald patches on the head and neck, with marked thickening of the skin. Later the infection spreads to become generalized. In sheep *S. scabiei* var. *ovis* is rare and affects only the non-woolly portions of the body. It usually starts near the mouth (lips, nostrils) and spreads to other parts of the face and then to the carpal and tarsal joints. Along with the more pathogenic *Psoroptes communis* var. *ovis*, it causes a 'notifiable disease' in several countries.

In rabbits, sarcoptic mange first appears on the head and ears and thus becomes generalized, being associated with intense pruritus with loss of hair. The infection is similar to *Notoedres* infection, but more severe. In primates, sarcoptic mange occurs usually on the back, neck and shoulders; scaling, loss of hair and pruritus are features of the infection. *Prosarcoptes pitheci* Katzea and Gruenberg, 1967 is recognized from the white-headed capuchin, the green monkey and baboon (Kutzer & Gruenberg 1967; Philippe 1948) and *P. faini* Lavoipierre, 1970 from baboons.

Sarcoptic mange may be a serious disease of wild carnivores, the fox often being severely affected. Severe infections may be seen in wolves, coyotes, badgers and martins.

Severe, debilitating infections may also occur in wild ungulates such as red and roe deer and chamois in Europe (Kutzer & Onderscheka 1966), llama in South America and a variety of large ungulates in Africa (Sweatman 1971). Sarcoptic mange may be a serious infection of the camel.

TREATMENT

Several compounds are available for the treatment of sarcoptic mange of farm animals, dogs and laboratory animals. The gamma isomer of HCH (lindane) is an effective remedy. It may be applied as a wash, dip or spray at a concentration of 0.016–0.03% and should be repeated on two or three occasions at an interval of 10–14 days. Repeat treatment is necessary to kill larvae hatching from eggs.

Benzyl benzoate is used as an emulsion or mixed with equal parts of soft soap and isopropyl alcohol or methylated spirits. It is painted on and allowed to dry. It is especially useful for primates.

Tetraethylthiuram monosulphide (Tetmosol) is used as a freshly-prepared watery solution, usually 5%. It may be rubbed into the skin by hand after a bath and left on for 24 hours. For prophylaxis a 20% Tetmosol soap may be used. Tetmosol is useful for rabbits. The organophosphates crotoxyphos (Ciodrin) formulated as Ciovap (10% Ciodrin and 2.5% DDVP) in a 0.25% spray is of use for cattle and pigs and coumaphos (30%) along with HCH (30%) appropriately diluted is used for sheep. Bromocyclen may be used as an aerosol spray or dusting powder for sarcoptiform mites on horses, cattle, sheep, pigs, dogs, cats, birds and laboratory animals. As a bath it is used as a 0.07% suspension or up to 0.5% in stubborn infections, or as a 4.25% dusting powder. Carbaryl, formulated as a shampoo, is available for dogs.

The avermectin ivermectin, a broad-spectrum antiparasitic drug, is effective against *Sarcoptes* spp. of cattle at a dose of 200 μg/kg given subcutaneously. It is probably effective against *Sarcoptes* spp. in other hosts at the same dose levels. Given orally at 300–500 μg/kg, ivermectin was 100% effective against *S. scabei* in pigs (Lee et al. 1980).

Where the above remedies are not available a lime–sulphur dip containing 1.5% polysulphide sulphur may be applied as a spray or by means of a brush. Treatment is repeated three to six times, as required, at weekly intervals.

PROPHYLAXIS

General control measures include supportive nutrition, good hygiene in animal quarters. All infected premises should be cleaned out and disinfected by spraying with HCH solution, lime-sulphur dip or organophosphate and they are best left unused for 14–17 days. Utensils such as curry-combs, brushes and harness should also be disinfected and sterilized.

Genus: Notoedres Railliet, 1893

This genus is similar to *Sarcoptes*; the body is globose, the striae are interrupted by scaly or spinose areas and there are two vertical setae on the dorsum of the propodosoma. The legs are short, the pedicels are not segmented and the distribution of the suckers and bristles on the legs of male and female mites is as for *Sarcoptes*. The major point of differentiation is the dorsal anus in *Notoedres* compared with the terminal location in *Sarcoptes*. The major species is *Notoedres cati*, but other species occur on rodents.

Notoedres cati (Hering, 1838) is a minute mite which attacks mainly cats but occasionally rabbits. There are probably different varieties of the parasite on different hosts. It occurs chiefly on the ears and the back of the neck but may extend to the face, foot and hind-paws or even, in young cats, to the whole body. These mites burrow into the skin causing mange-like lesions. The characteristic lesions consist of a yellowing crust in the region of the ears, face or neck; the skin is markedly thickened and wrinkled and the differential diagnosis should include ringworm. The life-history of the parasite is similar to that of *Sarcoptes*.

Since cats are especially susceptible to the toxic effects of the chlorinated hydrocarbon insecticides and also to several of the organophosphate insecticides, special care is necessary in the treatment of *N. cati*. English (1960) has recommended malathion, cats being immersed for one or two seconds in a 0.25–1.25% suspension of the drug. It is stated that the taste of malathion is objectionable and that this prevents cats taking in a toxic dose as a result of licking. Benzyl benzoate and Tetmosol (tetraethylthiuram monosulphide), variously formulated with HCH and antibacterial agents, have been used as local topical medicants. An ointment of 3–10% sulphur has been used successfully.

Notoedres muris Mégnin, 1877 is the cause of ear mange in the rat. It is similar in size and morphology to *N. cati* and may be common in some rat colonies. Lesions occur on the ears, nose and tail and sometimes on the external genitalia. They are often wart-like and horny. Treatment with 0.1–0.25% HCH or 2% aqueous solution of butylphenoxyisopropyl chloroethyl sulphite (aramite-15W) (Flynn 1960) gives good control.

Other species of the genus include **Notoedres oudemansi** Fain, 1965 from *Rattus rattus alexandrinus* and **Notoedres douglasi** Lavoirpierre, 1964 from grey squirrels in California. Sweatman (1971) records severe notedric mange in captive koalas and bandicoots.

Genus: Cnemidocoptes Fürstenberg, 1870

The body of these mites is globose; the striae are interrupted to form scales but there are no spines on the dorsal surface as in *Sarcoptes* spp. Two longitudinal chitinized bars run from the bases of the pedipalps to the level of the legs, where they are united by a transverse bar. The legs and pedicels are short and stumpy. All legs of the male possess suckers, while none of the female do so.

Cnemidocoptes gallinae (Railliet, 1887) is a small mite, resembling *Sarcoptes*, which causes 'depluming itch' in fowls. The mites burrow into the skin alongside the shafts of the feathers and cause an itching, inflammatory condition. The feathers break off readily and they are pulled out by the birds. The lesions are mostly seen on the back and wings, more rarely on the head and neck. The feather mite, *Megninia cubitalis* Mégnin, 1877, may cause a similar disease.

Diagnosis can be made by pulling out a few feathers at the edge of the lesion and searching for the mites on them.

Treatment. The birds are dipped in a mixture of sodium fluoride, sulphur, soap and water. Lindane (0.2%) or pyrethrum may be used and a 10%emulsion of benzyl benzoate applied daily for three days is effective.

Cnemidocoptes mutans (Robin, 1860) causes the condition known as 'scaly leg' in fowls and turkeys. The mites get on to the feet of the birds from the ground, since the lesion usually develops from the toes upwards. The parasites pierce the skin underneath the scales, causing an inflammation with exudate that hardens on the surface and displaces the scales. This process, accompanied by marked keratinization, is responsible for the thickened scaly nature of the skin. The disease may lead to lameness and malformation of the feet. In rare cases the comb and neck may be affected.

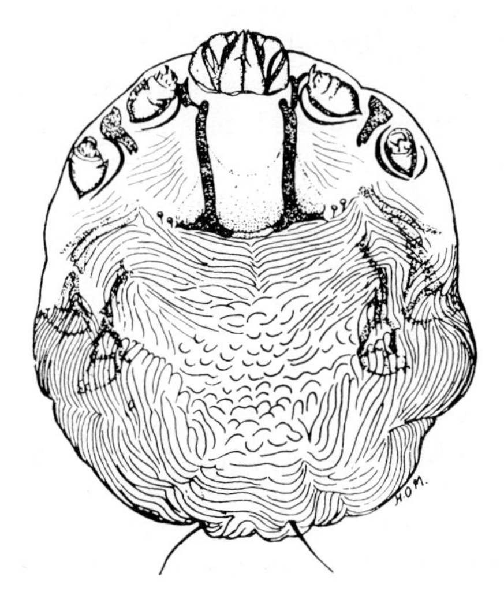

Fig 2.107 Dorsal view of the female *Cnemidocoptes mutans.*

Treatment. Birds may be satisfactorily treated by dipping the legs in a bath containing 0.1% gamma isomer of HCH. Other treatments consist of dipping the affected legs in a 10% sulphur solution which should be repeated several times at ten-day intervals. A further treatment consists of dipping the legs in a 0.5% solution of sodium fluoride. This should be repeated weekly. Dips containing 1% coumaphos or carbaryl (Sevin) are also effective.

Further measures include cleaning poultry houses and spraying perches etc. with creosote or HCH.

Cnemidocolotes pilae. This species was described by Lavoipierre and Griffiths (1951) as the cause of scaly leg in a budgerigar or parakeet (*Melopsittacus undulatus*) in Britain. It affected the shanks and pads of the feet.

Cnemidocolotes pilae may also affect the junction of the beak and feathers (cere) producing a grey-white to yellow crusty mass resembling a honeycomb. Generalized infections may occur (Blackmore 1963; Rickards 1975).

Cnemidocoptes jamaicensis Turk, 1950 is described as the cause of scaly leg of the golden thrush (*Turdus aurantiacus*) in Jamaica and is related to **C. fossar** Ehlers, 1873, found on the Maja finch (*Munia maja*).

Treatment is by topical application of 12–25% benzyl benzoate emulsion or 0.2% of lindane in soft paraffin. Sterilization of cages and shipping containers is indicated.

FAMILY: PSOROPTIDAE CANESTRINI, 1892

The genera of importance in this family include *Psoroptes*, the cause of psoroptic mange of the sheep (sheep-scab), goats, cattle and equines, *Otodectes*, the cause of otodectic mange of dogs, foxes, cats and ferrets and *Chorioptes*, the cause of chorioptic mange of equines and of cattle. The mites do not burrow into the skin, but are parasitic in its surface layers, causing the formation of thick, heavy scabs rather than thickening of the skin. The body is oval, there are no dorsal spines, there are no vertical setae on the dorsum of the propodosoma and the legs are longer and project beyond the margin of the body. There are bell-shaped suckers (caruncles) on stalks (pedicels) on the tarsi of some or all of the legs. The pedicels of the tarsal suckers are long and composed of three segments (in *Psoroptes*) or short and not segmented (*Otodectes* and *Chorioptes*). The anus is terminal and the male has adanal suckers (copulatory discs). The posterior margin of the abdomen of the male is prolonged into two lobes, which are prominent in species of the genera *Psoroptes* and *Chorioptes*, but are not prominent in species of the genus *Otodectes* or in other species of this family. The chitinous epimeres are evident but are not as marked as with species of the Sarcoptidae.

Genus: Psoroptes Gervais, 1841

This genus contains a number of parasites which, unlike the species of the Sarcoptidae, are specific to their hosts, although morphologically they may be difficult to distinguish. They are oval in shape and the tarsal suckers have jointed pedicels. As a rule these parasites live on the skin of parts of the body well covered with hair or wool or in the ears of their hosts.

Sweatman (1958*b*) studied the validity of the so-called species of *Psoroptes* and recognized as valid only the following:

Psoroptes ovis (Hering, 1838) Gervais, 1941, the cosmopolitan body mite of domesticated sheep and possibly cattle.

Psoroptes equi (Hering, 1838) Gervais, 1941, the body mite of the horse and possibly also of the donkey and mule.

Psoroptes natalensis Hunt, 1919, the body mite of domesticated cattle, the zebu and Indian water buffalo in the Republic of South Africa, South America, New Zealand and probably France.

Psoroptes cervinus (cervinae) Ward, 1915, found on American hosts, Bighorn sheep and wapiti (*Cervus canadensis*).

Psoroptes cuniculi (Delafond, 1859) Canestrini and Kramer, 1899, a cosmopolitan species which occurs in the ears of the rabbit, goat, sheep, horse, donkey mule and possibly gazelle (see *P. hippotis*, below).

Other authors also recognize **Psoroptes bovis** (Gerlach, 1857) of cattle; **Psoroptes caprae** Railliet, 1893, of goats; and **Psoroptes hippotis** Railliet and Henry, 1920, of the ears of horses and mules.

Psoroptes ovis is the cause of sheep-scab, an important disease which has been eradicated from Australia. Though absent from Britain for some 20 years, it has recurred there and still occurs in certain states of the USA and elsewhere. The parasite is not normally transmitted to other species of animals. The male, which has a pair of copulatory suckers on the ventral aspect, as well as a pair of large terminal tubercles, each of which bears several hairs, possesses suckers borne on jointed pedicels on the first, second and third pair of legs. Female mites possess suckers in jointed pedicels on the first, second and fourth pair of legs. The pubescent female is provided with a pair of posterior copulatory tubercles which are absent in the ovigerous female, while the latter has a wide genital aperture on the anterior aspect of the ventral surface. The dorsal surface of the body is devoid of scales and spines, but the cuticle shows very fine striations.

Life-cycle. The eggs are laid on the skin at the edges of the lesion and hatch in one to three days. The larvae feed and, two to three days after hatching, moult to the nymphal stage, passing the last 12 hours in a state of lethargy. The nymphal stage lasts three to four days, including a lethargic period of 36 hours before the moult occurs. The smaller nymphs usually become males. The pubescent females appear before the males, sometimes as soon as five days after hatching, while the males do not appear before the sixth day. Copulation begins soon after ecdysis and lasts one day; if there are many more females than males, the period may be shorter. As a rule the proportion of males to females is 1–2:4. The pubescent female moults two days after the commencement of copulation and the ovigerous female begins to lay eggs one day later or nine days after hatching from the egg. Usually, the life-cycle from egg to egg is about 12 days, and even in winter the life-cycle does not vary significantly. Mites may survive for up to two weeks off the host (Wilson et al. 1977).

The female lives 30–40 days and lays up to 90 or more eggs.

Pathogenesis. The mites puncture the epidermis to suck lymph and stimulate a local inflammatory swelling richly infiltrated with serum. The latter exudes on to the surface and coagulates, thus forming a crust. The altered conditions cause the wool to become loose and to fall out, or it is pulled out by the sheep in biting and scratching the lesions, which are markedly pruritic. The bare crusty patches are unsuitable for the mites, which therefore migrate to the margins of the lesion and thus extend the process outwards.

A primary lesion can occur as early as two weeks after contact. This may be evident by a few strands of discoloured wool under which there is a area of moist dermatitis. In winter conditions, this can spread readily to large areas of the body.

Clinical signs. Scab lesions may occur on all parts of the body that are covered with wool or hair, but occur most frequently around the shoulders and along the sides of the body in wooled sheep and along the back, the sternum and the dorsal aspect of the tail in hairy sheep.

In early lesions the wool is disturbed over the lesion by the biting and scratching of the sheep and usually has a lighter colour than the surrounding wool. A lesion two to four days old appears as a small papule, about 5 mm in diameter, with a yellowish colour and a moist surface; the mite will as a rule be found on the affected spot. From about the fifth day onwards the exudate begins to coagulate, forming pale yellow crusts, and the lesion extends outwards as the number of parasites increases. Older lesions are easy to detect on account of the loss of wool and presence of scab, while the mites are producing fresh foci in the surrounding covered parts. In some cases large portions of the body may be affected around an old lesion without showing on the surface. If the wool is opened, it is found to be matted together above the skin by scabs, underneath which numerous parasites are located.

So-called 'latent' cases, in which a small lesion may exist for months, are seen in the form of small, dry lesions on the scrotum, in the perineum, along the sheath, on the sternum, in the ears and infraorbital fossae and at the bases of the horns. Sheep scab is most active in autumn and winter, while latency tends to occur in summer, on account of less active feeding and decreased oviposition by the mites.

Diagnosis. It is relatively simple to dignose the active disease in a scabby flock, but the latent lesions described above make it more difficult to declare a large flock free of scab. In the latter case particular attention should be paid to the infraorbital fossa, the base of the horns and, in rams, the scrotum. Scrapings of suspect lesions should be taken for microscopic examination.

Treatment. Sheep scab (sheep scabies) is a notifiable disease in several countries and consequently treatments are prescribed and legislation regulates the movement of sheep from infected premises. Usually sheep are permitted to be transported for slaughter. In Great Britain preparations must achieve a minimum concentration of 0.016% of gamma HCH initially in the wash. This may vary from country to country, but usually is similar to the British requirement.

All the sheep of a flock should be treated and treatment may be repeated after about two months, especially in mountainous regions etc. where a total 'gather' of sheep may be difficult at any one time. A dip containing 0.03% dieldrin is similarly effective. In countries where the chlorinated hydrocarbons are prohibited because of insecticide residues in the meat the organophosphorous compounds may be used. The latter may also be necessary where *P. ovis* has become resistant to HCH. Premises should be left vacant for a minimum of two weeks and preferably for one month (Wilson et al. 1977).

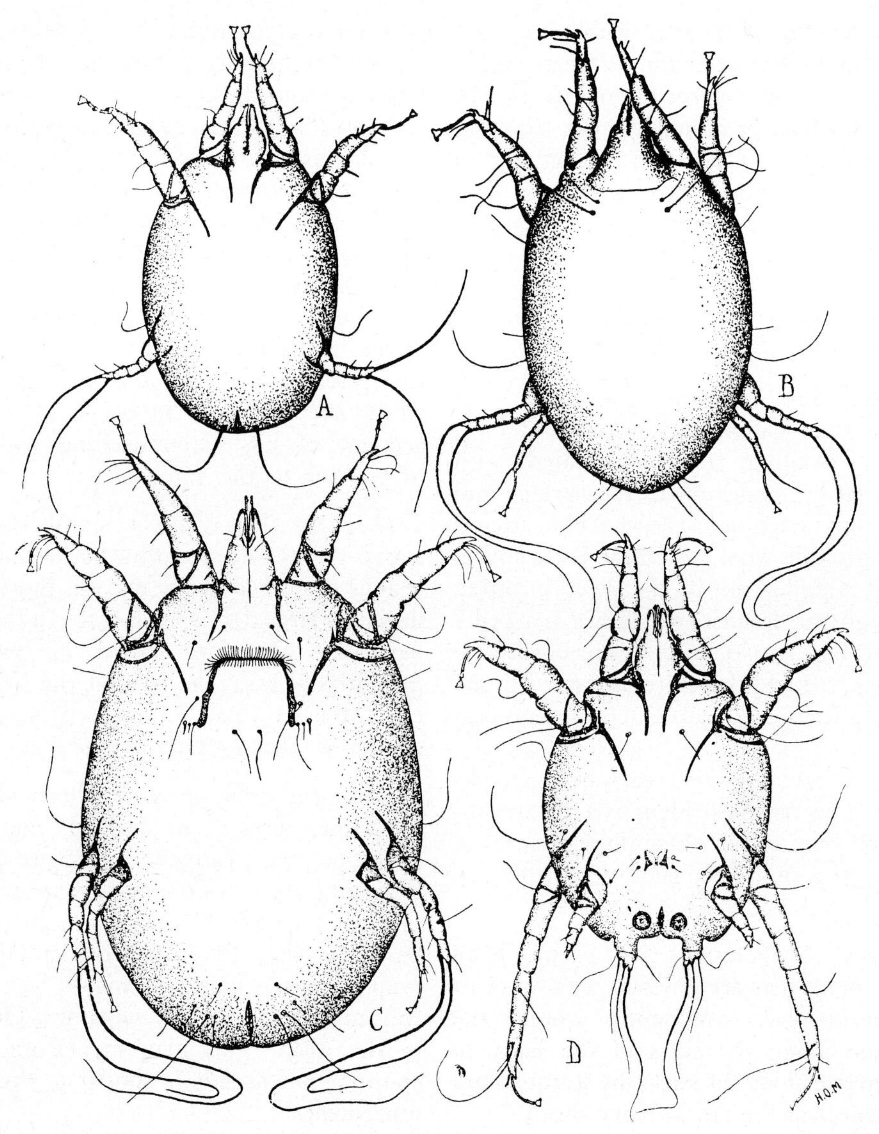

Fig 2.108 *Psoroptes ovis*. A, Ventral view of larva. B, Dorsal view of nymph. C, Ventral view of ovigerous female. D, Ventral view of male.

Psoroptes cervinus and *P. cuniculi* have life-cycles similar to *P. ovis*. *P. cervinus* occurs in the ear canal of bighorn sheep in North America. *P. cuniculi* is more widely found in the ear canal of deer, goats, rabbits and possibly horse, donkey and mule. Both *P. cuniculi* and *P. hippotis* have been associated with 'head-shaking' in horses in Australia and Great Britain (Pascoe 1980). Otacariasis of the goat due to *P. caprae* may extend to the face and neck. Heavy infections may occur and the lesions may spread to other areas of the head and the body. Ear mange, leading to ear canker, is particularly common in laboratory rabbits obtained from commercial sources. Clinical signs include repeated shaking of the head, scratching and crust formation leading to a septic otitis media. Loss of balance may occur and a septic meningitis may ensue.

The chlorinated hydrocarbon and organophosphate insecticides are effective in early and moderate infections: however, where secondary bacterial infection has occurred treatment may be difficult, prolonged and eventually to little avail.

Genus: Chorioptes Gervais, 1859

The different species of this genus live on the skin of several species of domestic mammals, causing chorioptic mange. The parasites resemble *Psoroptes*; the tarsal suckers have unjointed pedicels and suckers occur on the pedicels of all the legs in the male and on the first, second and fourth pair of legs in the female. The male has marked abdominal lobes which bear several hairs which are spatulate at the base. Adanal suckers occur at the base of the abdominal lobes. Copulatory tubercles occur on the posterior dorsal aspect of the pubescent female.

The so-called species of this genus have been named after the hosts on which they occur. Thus the form found on the fetlocks of horses has been called *C. equi*; that on the pasterns of sheep, *C. ovis*; that on goats, *C. caprae*; and that in the ears of rabbits *C. cuniculi*. The life-history is completed in three weeks; it resembles that of *Psoroptes*. Sweatman (1957) concluded that the species called *C. bovis*, *C. equi* and *C. ovis* all belong to the species *C. bovis* (Hering, 1845) Gervais and van Beneden, 1859 and that the only other valid species is *C. texanus* Hirst, 1924, found on a domestic goat in Texas and on Canadian reindeer.

Chorioptes bovis. In horses this species causes the disease called 'foot mange' or 'itchy leg', characterized by itching, scab-like lesions on the fetlocks, especially on those of horses with long hair (feather) in this region, which make the horses rub, stamp, scratch and bite the legs and kick frequently, especially at night. It occurs especially on the pasterns, particularly those of the hind-limbs, causing the formation of papules and, later, scabs. In cattle, camels and wild ruminants the root of the tail is frequently affected and if untreated the condition may spread to the sacral region and other parts of the body. It is most prevalent in the winter. In sheep a scaly condition is seen on the legs and the scrotum of rams. It should be differentiated from the infection due to *P. ovis*.

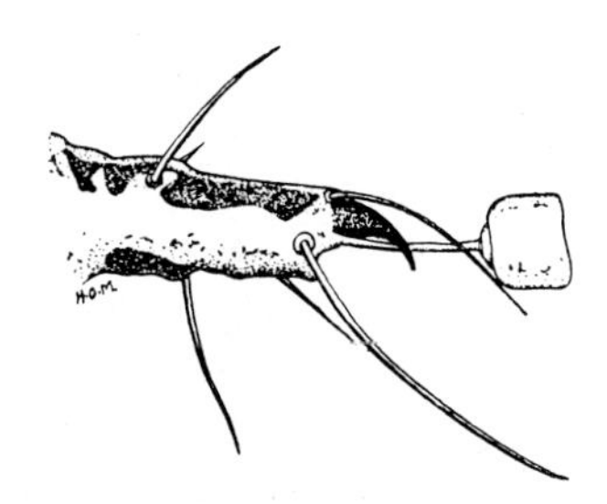

Fig 2.109 Extremity of leg of *Chorioptes*, showing sucker.

Diagnosis of all the members of this genus is made by finding them in skin scrapings.

Treatment. HCH washes are effective for chorioptic mange, as are the organophosphorus insecticides such as fenchlorphos, trichlorphon or diazinon. An older, but effective, remedy is a lime–sulphur wash applied several times at intervals of ten days.

Chorioptes texanus occurs in the ears of reindeer and goats (see above). It is similar morphologically to *C. bovis* (Sweatman 1958*c*).

Genus: Otodectes Canestrini, 1894

Otodectes cynotis (Hering, 1838) occurs in the ears of the dog, cat, fox, red fox, Arctic fox, racoon, ferret and other carnivores, causing ear or otodectic mange. The parasites resemble *Chorioptes*; they have tarsal suckers with unjointed pedicels on the first and second pairs of legs in the female and on all four pairs in the male. The fourth pair of legs in the female is small. In the male the copulatory suckers are not prominent and the abdominal lobes are not marked, though copulatory tubercles occur on them.

Life-cycle and pathogenesis. The eggs are laid singly and the life-history, which resembles that of *Psoroptes* takes three weeks. Sweatman (1958*a*) concluded that there is only one valid species, *Otodectes cynotis* (Hering, 1888), Canestrini, 1894. The mites spend all their lives in the ears of the host, less often on its body and Sweatman concluded that, like *Chorioptes*, they feed on epidermal debris which they chew and do not, as has been stated, pierce the skin to obtain body fluids. The clinical signs are seen in dogs at an earlier stage than in cats and foxes, which usually do not appear to be affected until the disease has reached an advanced stage. The animal shakes its head and scratches the ears in ordinary cases, while in more advanced stages the ears droop, there may be a purulent discharge and in dogs with large ears a haematoma may develop as a result of self-inflicted trauma. There may be torticollis and even epileptiform fits. A purulent inflammation of the external ear frequently sets in and perforation of the tympanic membrane may occur as the result of ulceration, leading to serious affections of the middle and inner ear and the brain.

Diagnosis. In early stages of the infection, mites may be seen when the ear is examined with an auroscope. Confirmation is based on the demonstration of the mites. The mites are readily found in material scraped from the external ear passage. They must be differentiated from *Sarcoptes scabiei* and *Notoedres cati*.

Treatment. The ear should be well cleaned of wax and detritus and an insecticide ointment or cream smeared in the meatus. A concentration of 0.1% HCH is satisfactory and should be repeated after about two weeks to kill mites which have hatched from eggs. Benzyl benzoate emulsion (20%) is also effective. A variety of formulations exist and anti-pruritic and antibacterial agents are frequently added.

In advanced infections, especially when secondary bacterial infection has occurred antibiotics or sulpha drugs are indicated. Surgical intervention may be necessary to insure drainage.

Other species of interest in the family Psoroptidae include:

Choriopsoroptes kenyensis Sweatman, Walker and Bindernagel, 1969 of the African buffalo (*Syncerus caffer*) which causes mange lesions along the back, chest, axillae and groin (Sweatman et al. 1969).

Paracoroptes allenopitheci Fain, 1963 and **Cebalges gaudi** Fain, 1962 from primates in Africa and South America respectively have life-cycles similar to *Chorioptes*. They occur on both the body and the ear.

Caparinia tripilis (Michael, 1889) has been reported from the hedgehog (*Erinaceus europaens*) in which it causes lesions on the head and **Caparinia ictonyctis** Lawrence, 1955 has been reported from the striped polecat and small-spotted genet in the Republic of South Africa. A fuller consideration of these species is given by Sweatman (1971).

FAMILY: ACARIDAE (SYN. TYROGLYPHIDAE) EWING AND NESBITT, 1942

Mites of this family feed on organic material; some are found in flour, grain, cheese, copra and similar

Table 2.6. Leg Suckers in Sarcoptidae and Psoroptidae

Genus	Male				Female			
	1	2	3	4	1	2	3	4
Sarcoptes	S	S	—	S	S	S	—	—
Notoedres	S	S	—	S	S	S	—	—
Cnemidocoptes	S	S	S	S	—	—	—	—
Psoroptes	S	S	S	—	S	S	—	S
Chorioptes	S	S	S	S	S	S	—	S
Otodectes	S	S	S	S	S	S	—	—

S = sucker on tarsus.

products and may become accidental parasites on or in man and animals which handle or feed on such infested materials. At times enormous numbers may be present in old food material such as animal feed, flour etc. They are distinguished from the parasitic mites by the absence of well-marked suckers, the absence of a distinct 'thumb-print' pattern on the body surface; they usually have many more bristles than the parasitic forms and their nymphs may develop hooks and suckers which enable them to cling to other hosts to be transported to a more suitable environment (*hypopi*; hypopial nymphs).

Acarus siro (syn. *Tyroglyphus siro*) (Linnaeus, 1758) is the cheese mite but is also found on other stored foods. It measures 500–550 μm in length and the body is covered with hairs. It is frequently found in food stores and upon accidental ingestion in large numbers may cause gastric or intestinal catarrh with diarrhoea. In food handlers, continued exposure may result in papular eruptions.

Acarus farinae (De Geer, 1778) (syn. *Tyroglyphus farinae*). This is similar to the above, being 330–600 μm in length; there is a distinct line or suture across the body and it is covered with numerous hairs. The mite lives in a variety of stored food products, especially cheeses and grains. In man it may cause digestive upsets if eaten in large numbers and may also cause skin lesions.

Tyrophagus longior (Gervais, 1844), the 'copra mite', is found in grains, seeds and copra and is responsible for 'copra itch'.

Glyciphagus domesticus (de Geer, 1808). This is a very common species, being found in many stored food products. It is the cause of 'grocer's itch' of humans.

Other mites of interest in this family include **Suidasia nesbitti** Hughes, 1948, the cause of wheat pollard itch, and **Carpoglyphus lactis** Linnaeus, 1746, the cause of dried fruit mite dermatitis.

FAMILY: CYTODITIDAE (SYN. CYTOLEICHIDAE) OUDEMANS, 1908

Genus: Cytodites Mégnin, 1879 (syn. Cytoleichus)

Cytodites nudus (Vizioli, 1870), the 'air-sac mite', is a small, oval creamy-coloured mite with suckered legs and practically no hairs, which lives in the respiratory passages and air-sacs and has also been found in other organs of fowls, turkeys and pheasants. Little is known of the life-cycle but larvae and not eggs are produced by the female and two nymphal stages occur (Fain & Bafort 1964).

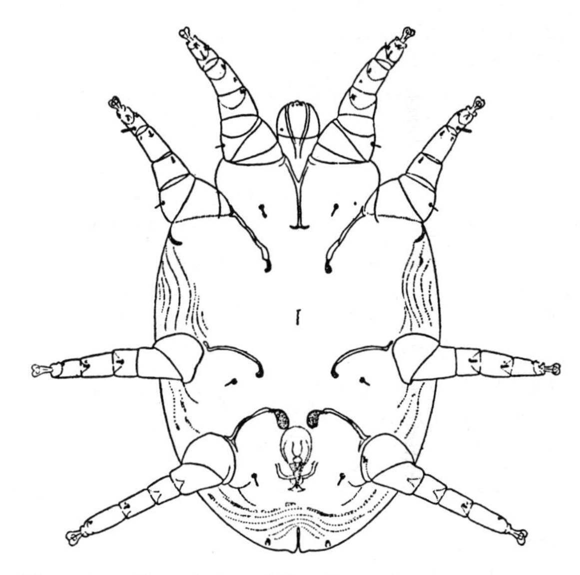

Fig 2.110 Ventral view of *Cytodites nudus*.

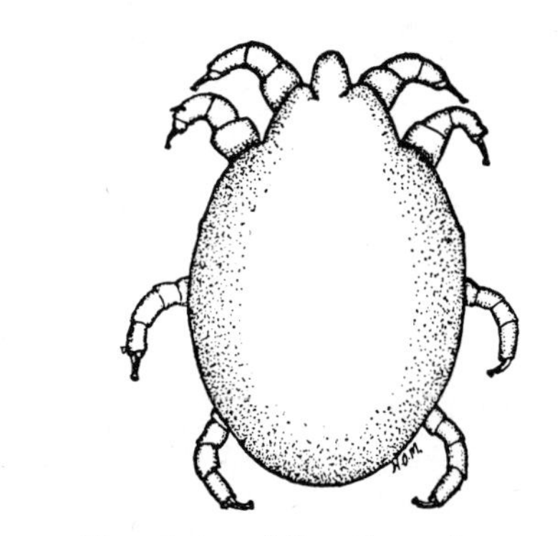

Fig 2.111 Dorsal view of *Cytodites nudus*.

Parasites are probably spread by respiratory mucus.

As a rule the parasites are not pathogenic. In the Republic of South Africa this mite is very common, but it has not been recorded as a cause of disease. It has been stated that heavy infections may predispose to pulmonary disorders.

FAMILY: LAMINOSIOPTIDAE VITZHUM, 1931

Genus: Laminosioptes Mégnin 1880

Laminosioptes cysticola (Vizioli, 1870), the 'subcutaneous mite' of the fowl, is a small oval parasite; the anterior two pairs of legs each end in a claw, and the posterior two pairs each in a claw and a suckerless pedicel.

Life-cycle, treatment and prophylaxis are unknown. The live parasites are rarely seen, but they may be found in the subcutaneous tissues. Usually the presence of the parasites is indicated only by the occurrence of small, flat, oval nodules, which resemble bits of fat in the subcutis. These nodules have a caseous or calcified content and are formed around the dead mites, of which the remains may still be present. The parasites are not pathogenic, but large numbers of these nodules may reduce the value of birds intended for human consumption.

FAMILY: EPIDERMOPTIDAE TROUESSART, 1892

The very small species of this family are not more than 0.4 mm long. They have circular, soft bodies; the males have adanal and copulatory suckers; all the legs bear suckers.

Epidermoptes bilobatus Rivolta, 1876 and **E. (Rivoltasia) bifurcata** Rivolta, 1876 may be found in the skin of the fowl and may, if the mites are numerous, cause ill health.

The genus **Dermatophagoides** (Bogdanoff, 1864) contains species responsible for house-dust allergy of man. The species **D. farinae Hughes**, 1961 in the USA and **D. pteronyssinus** (Trouessart, 1897) in Europe primarily are concerned with the condition. Mites feed on shed epidermal scales and produce an allergen responsible for house-dust allergy (van Bronswijk & Sinha 1971; Voorhorst et al. 1967; Collins-Williams et al. 1976). Lowering the humidity, the use of acaricides or the fungicide (*p*-methyl hydroxybenzoate) are useful measures for control.

FAMILY: LISTROPHORIDAE CANESTRINI, 1892

Species of this family may be found on the hair of small laboratory mammals and on the ferret. Their bodies may be laterally compressed and the pedipalps, legs and sternal region may be modified for clasping the hairs of the host. **Listrophorus gibbus** Pagenstecher, 1861 may occur on the rabbit, **Chirodiscoides caviae** (Hirst, 1917) on the guinea-pig. **Myocoptes musculinus** (Koch, 1844) may occur on laboratory mice and guinea-pigs and may be associated with a mange-like condition. Animals may be treated with 0.1% HCH wash or an organophosphorus insecticide.

Other species of the order Sarcoptiformes, the 'feather mites', are parasitic on the feathers of birds, although some of them may enter the quills of the feathers. They belong to the following families of this order:

FAMILY: ANALGESIDAE TROUESSART, 1915

Species of the family live chiefly between the barbules of the feathers and only rarely enter the quills. The bodies of the adults have an undivided posterior margin and the legs of the males are exceptionally well developed, but cannot be used for walking. Among species of this family are **Analges passerinus** (Linnaeus, 1758), found on the feathers of passerine birds; **Mégninia cubitalis** (Mégnin, 1877), found on those of the fowl; **M. velata** (Mégnin, 1877) on those of the duck; **M. phaisiani** (Mégnin, 1877) on those of the pheasant and peacock; and **M. columbae** (Buchholz, 1870) on those of the pigeon. **Mégninia ginglymura** (Mégnin, 1877) occurs on the fowl in Madras. Lavoipierre (1958) found that *M. cubitalis* was the only detectable cause of a form of depluming itch in a hen, *Cnemidocoptes gallinae* being absent.

FAMILY: DERMOGLYPHIDAE MÉGNIN AND TROUESSART, 1883

Both sexes of species of this family have a propodosomal shield and there may be other dorsal plates. Remarkable modifications of the legs, chelicerae and pedipalps may occur among the males. *Dermoglyphus elongatus* and *D. minor* may enter the feathers, the latter species having an elongated body adapted to this habitat resembling that of the trombidiform species *Syringophilus* mentioned above. *Pterolichus obtusus* lives on the feathers of the fowl, *P. bicaudatus* on those of the South African ostrich, *Freyana chanayi* on those of the turkey, *Pteronyssus striatus* on those of the sparrow, linnet and chaffinch and *Falculifer rostratus* and *F. cornutus* on those of the pigeon. The nymph of the last-named species may enter the subcutaneous tissues or even the peritracheal tissues of the internal organs and formerly, when they were found in these situations, they were given the names *Hypodectes, Cellularia* and *Hypodera*. The nymphs that enter the tissues are *hypopial nymphs*. Control of Dermoglyphidae (chiefly *Pterolichus*) of parakeets and cockatoos may be obtained by scattering malathion or HCH powder on the bare floor of the cage and leaving birds to make contact with it (Cross & Folger 1956).

FAMILY PROCTOPHYLLODIDAE MÉGNIN AND TROUESSART, 1883

Species of this family have several shields on the body and the posterior end of the femal is bilobed and bears leaf-like appendages. Their habits are like those of the species of the two families just mentioned. An example of the family is *Pterophagus strictus*, which lives on the feathers of the pigeon.

REFERENCES

MESOSTIGMATA

Baker, E. W., Evans, T. M., Gould, D. J., Hull, W. B. & Keegan, H. C. (1956) *A Manual of the Mites of Medical or Economic Importance*. New York: National Pest Control Association.

Dailey, M. D. & Brownell, R. I. Jn. (1972) A check list of marine mammal parasites. In: *Mammals of the Sea: Biology and Medicine*, ed. S. H. Ridgway, pp. 528–89. Springfield, Ill.: Charles C. Thomas.

Innes, J. R. M, Colton, M. W., Yevich, P. P. & Smith, C. L. (1954) Pulmonary acariasis as an enzootic disease caused by *Pneumonyssus simicola* in imported monkeys. *Am. J. Path.*, **30**, 813–835.

Price, R. N. (1977) A review of the acaricide Amitraz and a brief report on its activity against the poultry red mite *Dermanissus gallinae*. In: *Perspectives in the Control of Parasitic Disease in Animals in Europe*, ed. D. W. Jolley & J. M. Somerville, pp. 81–90. Surrey: Gresham Press.

Sheals, J. G. (1973) Arachnida (scorpions, spiders, ticks, etc.). In: *Insects and Other Arthropods of Medical Importance*, ed. K. G. V. Smith, pp. 417–72. London: British Museum of Natural History.

Sweatman, G. K. (1971) Mites and pentastomids. In: *Parasitic Diseases of Wild Mammals*, ed. J. W. Davis & R. C. Anderson, pp. 3–64. Ames: Iowa State University Press.

Zumpt, F. (ed.) (1961) *The Arthropod Parasites of Vertebrates in Africa South of the Sahara (Ethiopian Region)*. Johannesburg: South African Institute for Medical Research.

IXODOIDEA

Ahrens, E. H., Gladney, W. J., McWhorter, G. M. & Dean, J. A. (1977) Prevention of screw-worm infestation in cattle by controlling Gulf Coast ticks with slow-release insecticide devices. *J. Econ. Ent.*, **70**, 581–585.

Arthur, D. R. (1963) *British Ticks*, p. 213. London: Butterworth.

Arthur, D. R. (1966) The ecology of ticks with reference to the transmission of Protozoa. In: *The Biology of Parasites*, ed. E. J. L. Soulsby, pp. 61–84. New York. Academic Press.

Barnett, S. F. (1961) The control of ticks on livestock. *F.A.O. agric. Ser.*, No. 54.

Bedford, G. A. H. (1932) A synoptic check-list and host-list of the ectoparasites found on South African Mammalia, Aves, and Reptilia. *18th Rep. Dir. vet. Serv. Anim. Indust. Sth Afr.*, 223–523.

Beesley, W. N. (1973) Control of arthropods of medical and veterinary importance. *Adv. Parasit.*, **11**, 115–192.

Elliot, M., Jones, N. F. & Potter, C. (1978) The future of Pyrethroids in Insect Control. *A. Rev. Ent.*, **23**, 443–469.

Gothe, R. & Kunze, K. (1974) Neuropharmacological investigations on tick paralysis of chickens induced by larvae of *Argas (Persigargas) walkerae*. In: *Parasitic Zoonoses*, ed. E. J. L. Soulsby, pp. 369–82. New York: Academic Press.

Gregson, J. D. (1958) Host susceptibility to paralysis by the tick *Dermacentor andersoni* Stiles. *Can. Ent.*, **90**, 421–424.

Gregson, J. D. (1959) Tick paralysis in groundhogs, guinea-pigs and hamsters. *Can. J. comp. Med.*, **23**, 266–268.

Hitchcock, L. F. (1953) Resistance of the cattle tick *B. microplus* (Canestrini) to benzene hexachloride. *Aust. J. agric. Res.*, **4**, 360–364.

Hoogstraal, H. (1956) *African Ixodoidea*, Vol. 1, *Ticks of the Sudan*. Research Report Naval Medical 005.050.29.07. United States Navy.

Hoogstraal, H. (1966) Ticks in relation to human disease caused by viruses. *A. Rev. Ent.*, **11**, 261–308.

Hoogstraal, H. (1967) Ticks in relation to human disease caused by rickettsia species. *A. Rev. Ent.*, **12**, 377–420.
Hoogstraal, H. (1978) Tickborne diseases of humans—a history of environmental and epidemiological changes. In: *Medical Entomology Centenary*, ed. S. Willmott, pp. 48–55. London: Royal Society Tropical Medicine and Hygiene.
Kaiser, M. N. & Hoogstraal, H. (1964) The Hyalomma ticks (Ixodoidea, Ixodidae) of Pakistan, India, and Ceylon, with keys to subgenera and species. *Acarologia*, **6**, 257–286.
Mihailov, M. (1957) Incidence of *Ornithodorus lahorensis* and tick paralysis in sheep in Bitola, Yugoslavia. *Vet. Glasn.*, **II**, 814–818.
Neitz, W. O. (1956) A consolidation of our knowledge of the transmission of tickborne disease. *Onderstepoort J. vet Res.*, **27**, 115–163.
Neitz, W. O. (1959) Sweating sickness: The present state of our knowledge. *Onderstepoort J. vet. Res.*, **28**, 3–38.
Rousselot, R. (1953) *Notes de Parasitologie Tropicale*, vol. 2, *Ixodes*. Paris: Vigot Freres.
Seddon, H. R. (1951) *Diseases of Animals in Australia*, part 3, *Tick and Mite Infestations*, Publication No. 7. Commonwealth of Australia Department of Health Service.
Sutherst, R. W., Jones, R. J. & Schnitzerling, H. J. (1982) Tropical legumes of the genus *Stylosanthes* immobilize and kill cattle ticks. *Nature, Lond.*, **295**, 320–321.
Stampa, S. & du Toit, A. (1958) Paralysis of stock due to the karoo paralysis tick. *I. rubicundus*. *S. Afr. J. Sci.*, **54**, 241–246
Strickland, R. K., Gerrish, R. R., Hourrigan, J. L. & Schubert, G. O. (1976) *Ticks of Veterinary Importance*, Agriculture Handbook No. 485, p. 122. Washington, D. C.: U.S. Department of Agriculture.
Tatchell, R. J. (1969) Host–Parasite interactions and the feeding of blood-sucking arthropods. *Parasitology*, **59**, 93–104.
Theiler, G. (1950) Zoological survey of the Union of South Africa: tick survey. *Onderstepoort J. vet. Sci.*, **24**, 34–52.
Thompson, G. D., Osburn, R. L., Drummond, R. O. & Price, M. A. (1981) Hybrid sterility in cattle ticks (Acari: Ixodidae). *Experientia*, **37**, 127–128.
Thompson, K. C., Rao, E. J. & Romero, N. R. (1978) Anti-tick grasses as the basis for developing tropical tick control packages. *Trop. Anim. Hlth Prod.*, **10**, 179–182.
Wilkinson, P. R. (1965) A first record of paralysis of a deer by *Dermacentor andersoni* (Stiles) and notes on the 'host potential' of deer in British Columbia. *Proc. ent. Soc. Br. Columbia*, **62**, 28–30.
Wood, J. C., Sparrow, W. B., Page, K. W. & Brown, P. P. M. (1960) The use of Dieldrin, Aldrin and Delnav for the control of the sheep tick *Ixodes ricinus*. *Vet. Rec.*, **72**, 98–101.
Zumpt, F. (1958) A preliminary survey of the distribution and host specificity of ticks (Ixodoidea) in the Bechuanaland Protectorate. *Bull. ent. Res.*, **49**, 201–223.

TROMBIDIFORMES

Baker, W. E., Evans, T. M., Gould, D. J., Hall, W. B. & Keegan, H. C. (1956) *A Manual of the Mites of Medical or Economic Importance*, p. 170. New York: National Pest Control Association, pp. 170.
Bennison, J. C. (1943) Demodicidosis of horses with particular reference to equine members of the genus *Demodex*. *J. R. Army vet. Corps*, **14**, 34–49.
Corbett, R., Banks, K., Hinrichs, D. & Bell, T. (1975) Cellular immune responsiveness in dogs with demodectic mange. *Transplant. Proc.*, **7**, 557–559.
Folz, S. D., Kakuk, T. J., Kratzer, D. D., Conklin, R. D., Nowakowski, L. H. & Rector, D. L. (1979) Evaluation of a new treatment for canine demodicosis. *Abstr. ann. Meet. Am. Ass. vet. Parasitologists, Seattle* 13.
French, F. E., Raun, E. S., Baker, D. L. (1964) Transmission of *Demodex canis* Leydig to pups. *Iowa St. J. Sci.*, **38**, 291–298.
Greve, J. H. (1970) A modern enigma: Canine demodicosis. *Norden News*, Spring, 27–29.
Haakh, U. (1958) Ektoparasitenfreie Laboratorium mäuse; *Myobia*—Räude der weissen Mäuse und ihre Bekämpfung. *Z. Tropenmed. Parasit.*, **1**, 75–87.
Healy, M. C. (1977) Immunobiology of demodectic mange in dogs. *Diss. Abstr.* (1977) 37B, 4919–4920.
Hirsch, D. C., Baker, B. B., Wiger, N., Yaskulski, S. G. & Osburn, B. I. (1975) Suppression of *in vitro* lymphocyte transformation by serum from dogs with generalized demodicosis. *Am. J. vet. Res.*, **36**, 1591–1595.
Hughes, H. C. & Lang, C. M. (1973) Effect of orally administered Dichlorvos on demodectic mange in the dog. *J. Am. vet. med. Ass.*, **163**, 142–143.
Hughes, T. E. (1959) *Mites or the Acari*, p. 225. London: Athlone Press.
Koutz, F. R., Groves, H. F. & Gee, C. M. (1960) A survey of *Demodex canis* in the skin of clinically normal dogs. *Vet. Med.*, **55**, 52–53.
Losson, B. & Benakhla, A. (1980) Efficacé de closantel dans le traitment de la gale démodectique du chien. *Ann. Méd. vét.*, **124**, 521–526.
Murray, M. D. (1961) The life cycle of *Psorergates ovis* Womersley, the itch mite of sheep. *Aust. J. agric. Res.*, **12**, 965–973.
Murray, M. D., Nutting, N. B. & Hewetson, R. W. (1976) Demodectic mange of cattle. *Aust. vet. J.*, **52**, 49.
Nutting, W. B. (1950) Studies on the genus *Demodex* Owen (Acari, Demodicoidea, Demodicidae). Thesis, Cornell University.
Nutting, W. B. (1976) Hair follicle mites (*Demodex* spp.) of medical and veterinary concern. *Cornell Vet.*, **66**, 214–331.
Nutting, W. B. & Desch, C. E. (1978) *Demodex canis* rediscription and re-evaluation. *Cornell Vet.*, **68**, 139–149.
Roberts, I. H., Meleney, W. P. & Colbenson, H. P. (1965) Psorergatic acariasis on a New Mexico range ewe. *J. Am. vet. med. Ass.*, **146**, 24–29.
Seddon, H. R. (1951) *Diseases of Domestic Animals in Australia*. Service Publication 7, Div. Vet. Hygiene, Australian Dept. Public Health.
Sinclair, A. M. (1967) *Studies on the Problem of Sheep Itch Mite (Psorergates ovis Womersley) in Australia and New Zealand*. Sydney: Sunbeam Corporation.
Skerman, K. D., Graham, N. P. H., Sinclair, A. M. & Murray, M. D. (1960) *Psorergates ovis*—the itch mite of sheep. *Aust. vet. J.*, **36**, 317–321.
Smiley, R. L. (1965) Two new Species of the genus *Cheyletiella* (Acarina: Cheyletidae). *Proc. ent. Soc. Wash.*, **67**, 75–79.
Smiley, R. L. (1970) A review of the family Cheyletienidae (Acarina). *A. ent. Soc. Am.*, **63**, 1056–1078.
Spickett, S. G. (1961) Studies on *Demodex folliculorum*, Simon (1842). *Parasitology*, **51**, 181–192.
Thoday, K. L. (1979) Skin diseases of dogs and cats transmissible to man. *In Practice*, **1**, 5–15.
Till, W. M. (1960) *Psorergates oettlei* n. sp., a new mange-causing mite from the multimammate rat (Acarina, Psorergatidae). *Acarologia*, **2**, 75–79.
Womersley, H. (1941) Notes on the Cheyletidae (Acarina, Trombidoidea) of Australia and New Zealand, with descriptions of new species. *Res. South Aust. Mus.*, **7**, 51–64.

SARCOPTIFORMES

Blackmore, D. K. (1963) Some observations on *Cnemidocoptes pilae* together with its effect on the budgerigar. *Vet. Rec.*, **75**, 592–595.
van Bronswijk, F. E. & Sinha, R. N. (1971) Pyroglyphid mites (Acari) and house dust allergy. *J. Allergy*, **47**, 31–52.
Collins-Williams, C., Hung, F. & Bremner, K. (1976) House dust mite and house dust allergy. *Ann. Allergy*, **37**, 12–17.
Cross, R. F. & Folger, G. C. (1956) The use of malathion on cats and birds. *J. Am. vet. med. Ass.*, **129**, 65–66.
English, P. B. (1960) Notoedric mange in cats with observations on treatment with Malathion. *Aust. vet. J.*, **36**, 85–88.
Fain, A. & Bafort, J. (1964) Les acariens de la famille Cytoditidae (Sarcoptiformes) description de sept especes nouvelles. *Acarologia*, **6**, 504–528.
Flynn, R. J. (1960) *Notoedres muris* infestation of rats. *Proc. Anim. Care Panel*, **10**, 69–70.
Hovell, G. J. R., Weston, R. & Fain, A. (1976) Mange in guinea pigs. *Med. Res. Counc. Lab. Anim. Centre Guinea Pig New Letter*, **10**, 8.
Kral, F. & Schwartzman, R. M. (1964) *Veterinary and Comparative Dermatology*, pp. 343–69. Philadelphia: Lippincott.
Kutzer, E. & Gruenberg, W. (1967) Sarcoptesraeude (*Sarcoptes tapiri* nov. spec.) bei Tapiren (*Tapirus terrestris* L.) *Z. Parasitenk.*, **29**, 46–60.
Kutzer, E. and Onderscheka, K. (1966) Die Raeude der Gemse und ihre Bekampfung. *Z. Jagdwiss*, **12**, 63–84.
Lavoipierre, M. M. J. (1958) Some mites responsible for skin disease in birds. *Trans. R. Soc. trop. Med. Hyg.*, **52**, 300.
Lavoipierre, M. M. J. (1960) Notes acarologiques. II. Quelques remarques sur *Trixacarus diversus* Sellnick 1944 (*Sarcoptes anacanthos* Guilhon 1946) et sur trois especes recemment decrites de *Sarcoptes* des singes et des chauves-souris. *Ann. Parasit.*, **35**, 166–170.
Lavoipierre, M. M. J. & Griffiths, R. B. (1951) A preliminary note on a new species of *Cnemidocoptes* (Acarina causing scaly leg in a budgerigar (*Melopsittacus undulatus*) in Great Britain. *Ann. trop. Med. Parasit.*, **42**, 253–254.
Lee, R. P., Dooge, D. J. D. & Preston, J. M. (1980) Efficacy of ivermectin against *Sarcoptes scabei* in pigs. *Vet. Rec.*, **107**, 503–505.

Mellanby, K. (1944) The development of symptoms, parasitic infection and immunity in human scabies. *Parasitology*, 35 197–206.
Mellanby, K. (1952) *Scabies*. Oxford: Oxford University Press.
Pascoe, R. R. (1980) Mites in 'head shaker' horses. *Vet. Rec.*, 107, 234.
Philippe, J. (1948) Note sur les gales du singe. *Bull. Soc. Path. exot.*, 41, 597–600.
Rickards, D. A. (1975) Cnemidocoptic mange in parakeets. *Vet. Med. small Anim. Clin.*, 70, 729–731.
Sheahan, B. J. (1974) Experimental *Sarcoptes scabei* infection in pigs. Clinical signs and significance of infection. *Vet. Rec.*, 94, 202–209.
Smith, E. B. & Catchpole, T. F. (1967) Canine scabies in dogs and humans. *J. Am. med. Ass.*, 199, 95–100.
Sweatman, K. (1957) Life history, non-specificity, and revision of the genus *Chorioptes* a parasitic mite of herbivores. *Can. J. Zool.*, 35, 641–689.
Sweatman, G. K. (1958*a*) Biology of *Otodectes cynotis*, the ear canker mite of carnivores. *Can. J. Zool.*, 36, 905–929.
Sweatman, G. K. (1958*b*) On the life history and validity of the species in *Psoroptes*, a genus of mange mites. *Can. J. Zool.*, 36, 905–929.
Sweatman, G. K. (1958*c*) Redescription of *Chlorioptes texanus* a parasitic mite from the ears of reindeer in the Canadian arctic. *Can. J. Zool.*, 36, 525–528.
Sweatman, G. K. (1971) Mites and Pentastomes. In: *Parasitic Diseases of Wild Mammals*, ed. J. W. Davis & R. C. Anderson, pp. 3–64. Ames: Iowa State University Press.
Sweatman, G. K., Walker, J. B. & Bindernagel, J. A. (1969) Stages in the development of *Choriopsoroptes kenyensis* gen. and sp. n. (Acari: Sarcoptiformes: Psoroptidae) a body mange mite from captive African buffalo, *Syncer caffer* in Kenya. *J. Parasit.*, 55, 1298–1310.
Thoday, K. L. & Beresford-Jones, W. P. (1977) The diagnosis and treatment of mange in the guinea-pig caused by *Trixacarus* (*Caviacoptes*) *caviae* (Fain, Hovell & Hyatt, 1972). *J. small Anim. Pract.*, 18, 591–595.
Voorhorst, R. T. F., Spieksma, M., Varekamp, H., Leupen, M. J. & Lyklema, A. N. (1967) The house dust mite (*Dermatophagoides pteronyssinus*) and the allergens it produces. Identity with the house dust allergen. *J. Allergy clin. Immunol.*, 39, 325–339.
Wilson, G. I., Blachut, K. & Roberts, I. H. (1977) The infectivity of scabies (mange) mites, *Psoroptes ovis* (Acarina: Psoroptidae), to sheep in naturally contaminated enclosures. *Res. vet. Sci.*, 22, 292–297.

CLASS: PENTASTOMIDA HEYMONS, 1926

The taxonomic position of this group of organisms is very unclear. Usually they are considered to be aberrant arthropods, related to the sarcoptiform mites, but some authors suggest they are more related to the annelids (see Self 1969 and Heymons 1935). They occur as internal parasites in the respiratory organs of vertebrates.

The parasites are elongate, often tongue-shaped. The cuticle is transversely striated or sometimes deeply ringed, so that the body may have a beaded appearance in its posterior part. The anterior end is thick and its flattened ventral surface is armed with two pairs of strong, hooked claws, situated on either side of the elongate oral aperture. The small buccal cavity leads into a pharynx which has a suctorial function, and is followed by an oesophagus, an intestine and a rectum. The anus opens posteriorly. The females are usually larger than the males. As a rule the latter have two elongate testes and a cirrus sac with a cirrus or penis is present. The male genital opening is situated near the anus. The female has one elongate ovary, but a double oviduct which leads into a uterus. The vulva is anteriorly or posteriorly situated. The eggs contain a fully-developed embryo when they are laid.

Life-cycle

The egg contains a larva with two or three pairs of rudimentary clawed legs and hatches in the intestine of the intermediate host. The larva bores into the intestinal wall and passes with the blood to the mesenteric glands, liver and lungs, whence it finds its way into a suitable organ, where it becomes encysted. After several ecdyses, during which the legs are lost, a nymphal stage which is infective for the final host appears. The final host acquires the infection by eating the intermediate host containing the young parasites.

FAMILY: LINGUATULIDAE SHIPLEY, 1898

Body flattened and resembling a fluke, segmented and convex on the dorsal aspect. Hooks arranged in an arc at the anteroventral aspect. Digestive tract in the axis of the body and surrounded by the convolutions of the uterovagina.

Genus: Linguatula Fröhlich, 1789

Linguatula serrata Fröhlich, 1779, the 'tongue-worm', is a cosmopolitan parasite and occurs in the nasal and respiratory passages of the dog, fox and wolf, more rarely in man, the horse, goat and sheep. The parasite is tongue-shaped, lightly convex dorsally and flattened ventrally. The cuticle is transversely striated. Male 1.8–2 cm, female 8–13 cm long. The eggs measure about 90 by 70 μm.

Life-cycle. This has been described in detail by Hobmaier and Hobmaier (1940). The eggs are expelled from the respiratory passages of the host

Fig 2.112 Ventral view of female *Linguatula serrata*. (*After Neumann*)

and, when swallowed by a suitable herbivorous animal such as a horse, sheep, goat, bovine, various rodents or rabbit, they hatch in the alimentary canal and the larva reaches the mesenteric lymph glands, in which it develops to the infective nymphal stage. The larval stage is up to 500 μm in size, devoid of annulations and mouth parts, and undergoes a series of moults (six to nine) to produce the nymph which is 4–6 mm in size. It usually lies in a small cyst surrounded by a viscid, turbid fluid. Dogs become infected by eating the infected viscera of animals, especially of sheep and cattle.

Pathogenesis. The parasites attach themselves high up in the nasal passage and, though they usually cause few or no clinical signs, heavy infections may produce a severe irritation which causes the animals to sneeze and cough at intervals. Fits of difficult breathing, uneasiness and restlessness may be observed. A mucous discharge, often blood-stained, may exude from the nostrils. The parasites live about 15 months, after which the animal usually recovers.

Diagnosis is made from the clinical signs and by finding the eggs in the faeces or the nasal discharge.

Treatment is difficult. The parasites may be removed surgically. It is possible that the organophosphate insecticides will be of value, but no information is available on this.

Prophylaxis. The infection can be avoided by preventing dogs and foxes from eating possibly infected material.

FAMILY: POROCEPHALIDAE HEYMANS, 1922

Adults are found in the lungs, trachea and nasal passages of reptiles. The body is cylindrical, segmented and the hooks are arranged in the form of an arc or a trapeze; the internal organs occupy the whole of the abdomen.

Genus: Porocephalus Humboldt, 1811

Species of this genus occur in the respiratory passages of large snakes. The body is deeply ringed and may have a beaded appearance posteriorly. The young stages are found in the mesenteric lymph glands and other organs of various wild and domestic animals, including Herbivora, Carnivora and man.

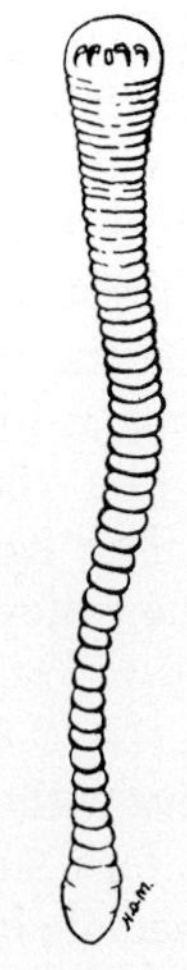

Fig 2.113 Ventral view of *Porocephalus* spp. × 1.5

Porocephalus crotali Homboldt, 1811. Adults are found in the lungs of snakes and nymphs occur in the viscera of various rodents including deer, mice and rats. The nymphs are cylindrical with a smooth annulated body about 8–14 mm in length. Two unequal pairs of hooks are located at the anterior ventral end around the mouth.

Infection is without pathogenic effect generally, either in the snake or the intermediate host, but very heavy infections may lead to death of snakes.

Other species of the genus include **Porocephalus clavatus** Sambon, 1922 and **P. subulifer** (Leukart, 1880) Sambon, 1922 which occur in snakes in South America and tropical Africa respectively.

Genus: Armillifer Sambon, 1922

The body is cylindrical, elongated and segmented, the segmentation being accentuated by a prominent thickening at the anterior end of each segment. Hooks are marked, equal and placed in a straight line. Adult parasites occur in the lungs, trachea and nasal passages of snakes in Africa and the East. Intermediate hosts are a variety of mammals including wild game, rodents and even man.

Armillifer armillatus (Wyman, 1847) is 30–50 mm in length and occurs in the lungs and trachea of various snakes including pythons and venomous snakes. Nymphal stages occur in the inner organs and musculature of many species of primate, including man as well as a number of rodents. The cysts containing the nymphs, which may undergo calcification, are often seen on radiographic examination.

Generally this species is not associated with clinical manifestations.

Other genera of the Pentastomida include **Raillietiella**, which is found in ophidians, lacertilians and birds and may have a direct life-cycle, and **Reighardia**, which is found in the air sacs of gulls and terns.

Nymphs of the genera **Leiperia**, **Sebekia** and **Subtriquetra** occur in various fish in Africa and adults occur in crocodiles.

REFERENCES

PENTASTOMIDA

Heymons, R. (1935) Pentastomida. In: *Bronns' Klassen und Ordungen des Tierreichs*, Leipzig: Akad Verlagsgesel.
Hobmaier, A. & Hobmaier, M. (1940) On the life cycle of *Linguatula rhinaria*. *Am. J. trop. Med. Hyg.*, **20**, 199–210.
Self, J. T. (1969) Biological relationships of the Pentastomida: A bibliography of the Pentastomida. *Exp. Parasit.*, **24**, 63–119.

CLASS: CRUSTACEA PENNANT, 1777

This class includes an array of forms such as the lobsters, crabs, crayfish etc. Most are free-living; a number are also intermediate hosts of various helminths and a few are parasitic. Those of the subclass *Copepoda* are of interest as ectoparasites of fishes and examples of the more important forms are dealt with briefly in this volume. More detailed information on the parasitic crustacea of fish and amphibians may be found in Reichenbach-Klinke and Elkan (1965), Flynn (1973) and Roberts (1978).

The morphology of the parasitic copepods is highly modified as a result of adaptation to the ectoparasitic life. The body is divided into head, thorax and abdomen, these being covered by a rigid or semi-rigid chitinous exoskeleton. The thorax consists of seven segments, the anterior two being fused with the head to form a cephalothorax which is covered by the carapace, a chitinous shield. The lines of division may be fused such that the sutures are not visible. The abdomen typically is composed of four segments, though these may be fused. The appendages are jointed: there are usually five pairs on the head (first and second pairs of antennae, mandibles and first and second pairs of maxillae). The thorax bears ambulatory appendages, some of which are concerned in food acquisition.

The developmental cycle is complex and the body form alters markedly during maturation. Eggs hatch to produce an active larva, the *nauplius*, which moults several times finally acquiring a complement of appendages and becoming a copepodid stage. This moults several times (often five) and becomes the adult stage.

Ergasilus Nordmann, 1832, species. The adult female is cyclops-like (i.e. broad anteriorly, narrow posteriorly), 2–2.5 mm long including egg sacs. The second pair of antennae are modified for clasping the gill filaments of fishes. Each of the thoracic segments bears a pair of swimming legs. The adult male is similar but shorter and more slender than the female. All appendages are reduced.

After copulation the male dies and females attach to a host and become parasitic. Eggs are produced in egg sacs, each containing 18–100 eggs. Eggs hatch in three to six days and the resulting nauplius, copepodid and adult stages require 10–70 days to complete development, depending on temperature.

Ergasilus species (e.g. *E. sieboldi* Nordmann, 1832) are found on the gills of a wide range of

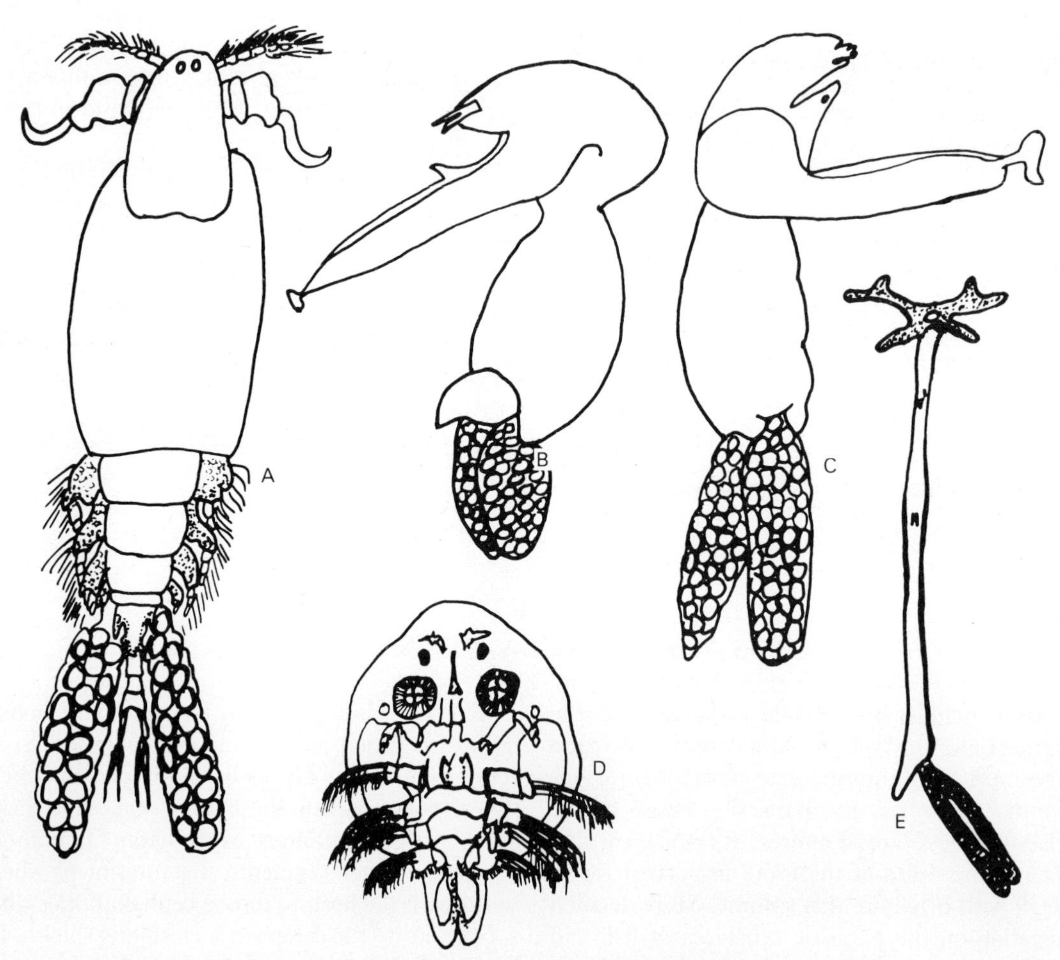

Fig 2.114 Some parasitic crustacea of fish (not to scale). A, Adult female *Ergasilus* sp. B, Adult female *Achtheres* sp. C, Adult female *Salmincola* sp. D, Adult *Argulus* sp. E, Adult female *Lernaea* sp.

fresh-water fish all over the world. Gill filaments are damaged, the parasites feed on blood and epithelium and heavy infections cause impaired respiration and epithelial hyperplasia associated with retarded growth and retarded sexual maturity. Fish also become susceptible to secondary infections, especially fungi. In heavy infections fish become listless and finally die. Peak mortality occurs in summer when water temperature is highest.

Salmincola Wilson, 1915, species. The body (4–7 mm) is divided into a cephalothorax and trunk. Abdomen and abdominal processes are lacking. The head appendages are variously short and peg-like or well developed as with the second maxilla. The adult female produces two egg clusters, each of 60–300 eggs, situated at the posterior aspect of the trunk, twice during a 9–13-week life-span. Eggs hatch to produce copepodids and within two weeks of attachment to the gills of fish five larval moults occur, resulting in maturation and copulation. The species of this genus are found on gills and fins of various salmonids. Clinical signs are similar to infection with *Ergasilus* species.

Achtheres Nordmann, 1832, species. The adult female (2–7 mm) has a small cephalothorax and a small abdomen. The maxillipeds are well developed, elongated and jointed. Egg sacs are produced. Eggs hatch as copepodids and development is complete within two weeks. Species of the genus occur in the gills of various fresh-water fishes. Clinical signs are similar to those of *Ergasilus* species.

Control and prevention of these parasites is achieved by the introduction of parasite-free fish and the use of fish-free water supplies. Filtration of water and pre-filling of ponds at least one week before use are methods to exclude the free-swimming larval stages of the parasites.

Where infection has occurred, infected fish should be destroyed; the density of fish in the pool must be reduced, ponds should be completely dried before restocking and young fish should be treated prophylactically.

Chemicals available for killing the larval stages include calcium chloride (0.85%), copper sulphate (0.2%), magnesium sulphate (1.7%) and the organophosphates, dipterex (*O*,*O*-dimethyl-1-hydroxy-2-trichloromethyl phosphonate) or DDVP at 0.5 ppm.

Lernaea Linnaeus, 1746, species. These are the anchor parasites or anchor worms of fresh-water fish. The genus is ubiquitous and common in hatcheries and fish ponds. *Lernaea elegans* Leigh-Sharpe, 1925 is a common species. The female is 5–22 mm long; the head, which is buried in host tissue, is expanded into large cephalic horns, soft and leathery in texture, and the body is elongated, with degenerate legs. Egg sacs containing about 200 eggs are present at the posterior end.

Eggs hatch into free-swimming and free-living nauplii existing on yolk material. They moult to metanauplii and then through successive copepodid stages before attaching permanently as females to the fish. Females become parasitic after copulation, which occurs during the last copepodid stage; the males never become parasitic. Females develop cephalic horns which serve as anchors; they increase in length, develop egg sacs and complete the life-cycle with the production of eggs.

Heavy mortality may occur in cultured trout, carp, catfish, goldfish etc. The parasite destroys scales and causes ulcerated areas at the site of penetration permitting attack by bacteria, fungi and viruses. Infected fish show retarded growth and swim erratically; heavily infected ones become sluggish and may swim upside down or 'hang' in the water.

Treatment of infected ponds includes the use of Dipterex at 0.5 ppm which kills the larval stages; malathion (0.25 ppm) used weekly for five weeks acts similarly and common salt at a concentration of 0.8 to 1% maintained for three days is effective against larval stages. At a concentration of 3–5% salt can be used as an individual fish dip.

Argulus Müller, 1785, species. These are the fish lice and are common parasites of fresh-water fish of many species. The species is globally distributed.

The mature adult (6–22 mm in length) consists of head, thorax and abdomen. The head is covered by a flattened horseshoe-shaped carapace. There are a series of head appendages including suction caps, antennae, maxillipeds, preoral sting, basal glands, etc. The thorax has four segments, each bearing a pair of swimming legs. The abdomen is a simple bi-lobed segment.

The life-cycle is somewhat complicated. Mature females leave the host and lay eggs on vegetation and various objects in the water. Various developmental stages occur in the egg (nauplius, metanauplius, early copepodid stage) and may hatch as metanauplii or copepodids. A series of moults occurs, during which developmental stages are parasitic, but they leave the host to moult and reproduce. Depending on temperature 40–100 days are required for completion of the life-cycle.

Argulids puncture the skin, feed on blood and inject a cytolytic toxin through the oral sting. This sting may kill larval fish (e.g. eels). Feeding sites become ulcerated and lead to secondary bacterial and fungal infection. *Argulus* spp. transmit spring viraemia of carp.

Clinical signs include erratic swimming, 'flashing' and poor growth. The parasites can be seen with the naked eye.

Treatment includes the use of dipterex (0.25 ppm), malathion (0.25 ppm), pyrethrum (30–100 ppm) or HCH where allowed, for the treatment of ponds. For individual fish, dipping in lysol at 2000 ppm for five to ten seconds is effective. In control, attention should be paid to provision of disease-free fish stock, filtered water, increasing the flow of water to achieve greater dilution and lower pool temperature, removal of objects on which eggs of the parasite might be layed and, when pools are to be restocked, the complete drying of the pools to kill eggs, larvae and adults (Dogiel et al. 1961; Roberts 1978).

REFERENCES

CRUSTACEA

Dogiel, V. A., Petrushevski, G. K. & Polyanski, H. (ed.) (1961) *Parasitology of Fishes,* trans. Z. Kabata, Edinburgh: Oliver and Boyd.
Flynn, R. J. (1973) *Parasites of Laboratory Animals.* Ames: Iowa State University Press.
Reichenbach-Klinke, H. & Elkan, E. (1965) *The Principal Diseases of Lower Vertebrates.* New York and London: Academic Press.
Roberts, R. J. (1978) *Fish Pathology.* London: Baillière Tindall.

INJURIOUS NON-PARASITIC ARTHROPODA

Although the poisonous nature of some arthropods is sometimes much exaggerated, a number of species can be poisonous to man and animals in various ways.

Piercing or biting species inject poisonous or irritating substances into a wound made by their mouth parts or poison claws, e.g. spiders, mosquitoes, centipedes.

Stinging species inject a poison by means of a sting situated at the end of the abdomen, e.g. scorpions. Such a sting is frequently a modified ovipositor, e.g. bees.

Nettling species have hairs or scales which possess irritating properties, e.g. certain caterpillars.

Cryptotoxic species contain irritant or poisonous body fluids, e.g. the blister beetles.

MYRIAPODA

The millipedes (*Diplopoda*) are circular in transverse section and have two pairs of legs on each of most of the segments. They are not poisonous, but it is stated that some of the large tropical species, when irritated, eject poisonous or pungent fluids from cutaneous glands. The centipedes (*Chilopoda*) are flattened dorsoventrally and have a single pair of legs per segment. They have the first pair of legs modified as claws which contain poison glands. These are carnivorous animals and they use the poison claws to stun their prey. The smaller forms can at most cause a painful local reaction in larger animals or man, but the large forms, e.g. the tropical *Scolopendra gigantea*, can cause death.

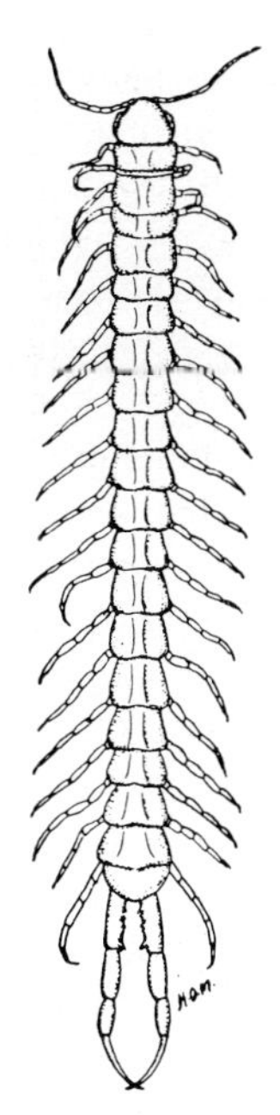

Fig 2.115 Dorsal view of a chilopod.

INSECTA

Piercing or biting species. These and their effects have been described above.

Stinging species. The stinging insects are *Hymenoptera*, the sting being a modified ovipositor found only in the females. In the bee there is a poison reservoir and the poison glands are of two types, one producing an alkaline and the other an acid secretion. When the bee stings, the tip of the abdomen with the whole poison apparatus breaks off and remains in the wound. The muscles of the organ remain active for a considerable time and force poison out of the reservoir into the tissues, so that it is necessary to remove the sting as soon as possible.

The sting produces a local necrosis with infiltration of lymphocytes, hyperaemia and more or less extensive oedema. Some individuals are much more sensitive to bee stings than others and some human beings have a high susceptibility. The stings of a large number of bees may cause death following collapse, shock or suffocation, the latter being due to oedema of the head and neck.

Some species, like bumble-bees, hornets and wasps, sting several times and the sting does not break off so easily. Most ants also possess poison glands and stings, but seldom use them. Some species have no stings, but make a wound by biting and then inject the poison into it.

Nettling species. Many hairy caterpillars produce intense irritation by means of their hairs, which are hollow and bear a poison gland at the base. The hairs break off and remain in the skin of the affected animal or person. The scales of the 'brown-tail moth' (*Porthesia chrysorrhoea*) and of some other moths may cause dermatitis and bronchitis, and the caterpillars of *Porthesia* are also stated to cast off hairs which are very irritating.

Poisonous caterpillars are known to cause stomatitis and even enteritis in animals if they are numerous on a pasture. In cases of dermatitis the application of ammonia, followed by 10% ichthyol ointment, is recommended.

Cryptotoxic species. The well-known Spanish fly or blister beetles (*Meloidae*) possess very irritating body fluids, containing cantharidin as the active principle. Powdered beetles or extracts have long been used for blistering and formerly also as aphrodisiacs. The action in the latter case is due to irritation and the kidneys suffer severely. Cases of poisoning of cattle from ingesting these beetles with their food are known and the flesh of such animals is believed to be poisonous for human beings.

Of other cryptotoxic insects the most important are the caterpillars of the cabbage butterfly (*Pieris brassicae*), which cause stomatitis, colic and paralysis of the hind-legs in animals.

ARACHNIDA

In *scorpions* a post-abdominal segment forms the sting, which contains a pair of well-developed poison glands and ends in a hollow spine. The smaller species may cause painful stings, but these are not fatal. The larger tropical and subtropical species are frequently very poisonous. The poison causes elevation of blood pressure and increased glandular activity, especially lacrimation, salivation and a nasal discharge. There are also muscular spasms and death is due to asphyxia.

Treatment consists of the application of a ligature proximal to the wound—not to be maintained for longer than half an hour—and the injection of a 1% permanganate solution, as in the case of a snake bite. The stings of slightly poisonous forms can be treated by the local injection of procaine and adrenaline, which alleviates pain. For very poisonous forms antivenoms are prepared by hyperimmunization of animals and such a serum, prepared from one species, protects against closely related forms, but not against others. The scorpions themselves appear to contain antitoxins, so that they can resist the stings of other scorpions. For control, the application of 500 mg/m^2 of gamma HCH to walls, attics etc. and to other places in which scorpions hide and the removal of wood stacks, loose bricks and any shelters of these arachnids is recommended.

The Araneidae, or *spiders*, have poison glands which lie in the cephalothorax and open through pores on the tips of the chelicerae. The poison is used to kill small animals, especially insects, for food and most spiders are not able to bite through the skin of larger animals and man, nor are they sufficiently poisonous to do much harm. There are, however, a few other species that are well known to be very dangerous.

The name 'tarantula' is derived from a spider of the family Lycosidae, *Lycos tarantula*, which was believed to be the cause of an hysterical condition and was greatly feared in the Middle Ages. This spider is not poisonous to man or domestic animals. The name 'tarantula' is now usually applied to many large spiders of the Aviculariidae, which may inflict severe bites, but are not fatally poisonous. The wounds they make may become infected with bacteria from their mouth parts.

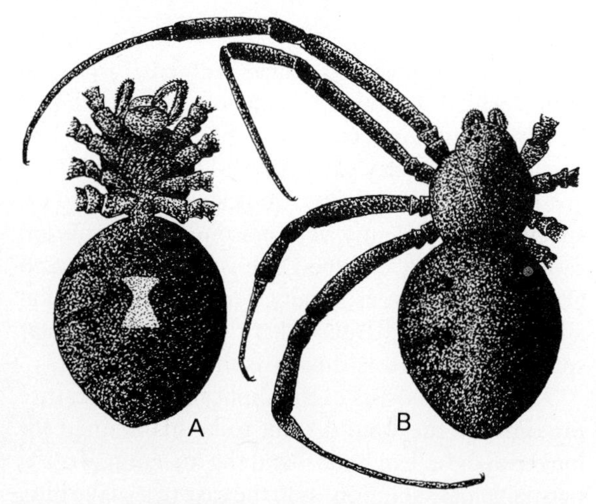

Fig 2.116 A South African species of *Lactrodectus* showing characteristic shape and ornamentation. a, Ventral view. b, Dorsal view. × approx. 1.5

The most poisonous spiders are species of the genus *Latrodectus*, which occur in many tropical and subtropical countries, and are called 'hour-glass spiders' on account of a red hour-glass-shaped spot which is frequently present on the ventral side of the abdomen. *L. mactans*, the 'black widow', is one of the best-known species. The female measures up to 3 cm in length, including the legs. The body is glossy black; on the dorsal side of the abdomen the males and young females have several lighter, spotted stripes which are transversely arranged and run down the sides. On the ventral surface there are brick-red spots which may form the shape of an hour-glass or of a cross. The legs are long and tapering. In the lighter-coloured species the joints of the legs are usually dark. The spiders live in buildings or in shrubs and spin a coarse web; their eggs are laid in spherical masses with an irregular surface. These spiders are very poisonous and frequently also aggressive.

The symptoms are general rather than local, although severe local pain may occur. A general convulsive trembling of medullary origin is stated to be characteristic in guinea-pigs. In man shooting pains in the hips and marked abdominal pains are common and are followed by numbness and paresis of the lower limbs. Dizziness and a general feeling of oppression occur. The blood pressure is elevated; there is marked salivation and cold perspiration, later followed by thirst and sometimes retention of urine. The respiratory system is affected, resulting in dyspnoea and cyanosis. Sometimes vomiting and constipation are seen. The course may be from a few hours to about three days. In cases that die soon, pulmonary oedema is the prominent lesion; in later cases necrotic foci are seen in the liver, kidneys, spleen and adrenals. Recovery is infrequent.

Treatment in general is as for snake-bite. Potassium permanganate is stated to destroy the poison rapidly. Warm baths give relief. Antivenoms are made and used with success, but must be injected very soon after the bite has occurred.

Chiracanthrium nutrix is a European spider which may cause symptoms of poisoning and some of the Epeiridae, or orb-weavers, contain poisonous substances in their body fluids and eggs which may give rise to clinical signs in animals that swallow them accidentally.

3

Protozoa

The protozoa are unicellular animals in which the various activities of metabolism, locomotion, etc. are carried out by organelles of the cell. Comparable forms occur in the plant kingdom (unicellular plants) and, in general, protozoa are differentiated from these by the absence of chlorophyll-containing chromatophores and their mode of nutrition (holozoic). The unicellular plants are frequently bounded by a fairly rigid cell wall made of cellulose and the nuclear material is often dispersed in the cell. The protozoa, on the other hand, have a well-defined nucleus and do not have a rigid cell wall, allowing, at times, a marked variation in size and shape. Nevertheless, these distinctions cannot be rigidly applied to all forms and there is an assemblage of organisms which share the characters of both plants and animals. The term *protista* was introduced for such forms, but this has not been generally adopted.

Since the discovery of protozoa by Antoni van Leeuwenhoek, some 45 000 species have been described. The majority of these are free-living and are found in almost every habitat on land and in water. Although the parasitic protozoa are smaller in numbers, they nevertheless assume an important role as producers of global disease which, apart from producing death or deformity, saps the energy and initiative and decays the moral fibre of mankind in many parts of the world. Of no less importance is the untold loss of livestock and livestock products which is frequently a burden in those communities and areas of the world that can least support it.

STRUCTURE OF PROTOZOA

Nucleus

Protozoa are *eukaryotic* (nucleus enclosed in a membrane) whereas the bacteria are *prokaryotic* (nucleus dispersed in cytoplasm).

Usually only one nucleus is present, although in some forms more than one nucleus may be present in some or all stages of development. The *vesicular* type of nucleus consists of a nuclear membrane which bounds the nucleoplasm in which, lying more or less central, is an intranuclear body, the endosome (or karyosome) or the nucleolus. An endosome is devoid of deoxyribonucleic acid, whereas a nucleolus does possess DNA. Chromatin material (Feulgen-positive for DNA) frequently occurs on the inner surface of the nuclear membrane and may also be seen as strands radiating from the karyosome to the nuclear membrane. The vesicular type of nucleus is seen most commonly in the Mastigophora and the Sarcodina.

The *compact* type of nucleus contains a large amount of chromatin and a small amount of nucleoplasm. This type is found in the ciliates as a macronucleus; it divides amitotically and regulates the cytoplasmic functions of the organism.

Cytoplasm

This is the extranuclear part of the protozoan cell. It may be differentiated into an outer ectoplasm and an inner endoplasm, the former often being homogeneous and hyaline in appearance and the latter frequently containing granules, vacuoles and sometimes pigment. In some forms (e.g. Sarcodina) there is no definite limiting membrane, but usually a pellicle serves as such in the majority of species.

Locomotion

Protozoa may move by gliding or by means of pseudopodia, flagella or cilia.

Gliding is seen in *Toxoplasma*, *Sarcocystis* and other forms, this being achieved without the aid of cilia or flagella.

Pseudopodia are used by the amoeba-like organisms, the structures being temporary locomotor organelles which are formed when required and retracted when not needed.

Flagella are whip-like filamentous structures which arise from a basal granule or blepharoplast in the cytoplasm of the organism. They are composed of a central axial filament, the axoneme, which is surrounded by a contractile cytoplasmic sheath. Ultra-structural studies indicate the axoneme to be composed of two central filaments surrounded by nine peripheral filaments. In some forms the flagellum may be attached to the body of the protozoan by an undulating membrane. Flagella are typically seen in the Mastigophora.

Cilia are fine, short, flagella-like structures originating from a basal granule embedded in the pellicle or ectoplasm. They are the organs of locomotion in the ciliates, but they may also aid in the ingestion of food or serve as tactile structures. Their ultra-structure is similar to that of the flagella and they usually occur in large numbers, arranged in rows over the body of the protozoan.

Organelles for nutrition

In the amoeba-like forms, particulate food material is acquired by means of pseudopodia. An advance on this is a specialized opening called the cytostome through which food particles are engulfed and passed to food vacuoles. In the ciliates the cytostome may be lined with cilia which further assist in the ingestion of food.

Food vacuoles occur in the cytoplasm and contain particulate material in various stages of digestion. Non-digestible material may be extruded from the cell either via a temporary opening or through a permanent cytopyge.

Excretion of waste products may occur directly through the body wall or by means of contractile vacuoles which periodically discharge waste material through the body wall or, in a few instances, through an anal pore.

NUTRITION OF PROTOZOA

Nutrition may be holophytic, holozoic or saprozoic.

The *holophytic* protozoa are forms which possess some characteristics of plants, carbohydrates being synthesized by chlorophyll which is carried in chromatophores or in the bodies of algae or other protophyta which inhabit the cytoplasm of the protozoan. None of these forms is of medical or veterinary importance.

The *holozoic* protozoa utilize preformed food material derived from living animals or plants. Food material is ingested by pseudopodia or through a cytostome and passes to a food vacuole for digestion. Some forms (e.g. *Entamoeba, Balantidium*) ingest the tissue cells of the hosts.

The *saprozoic* protozoa absorb nutrients through the body wall, these being utilized directly by the organisms.

Stored food material may be visible as glycogen granules or chromatoid material.

REPRODUCTION OF PROTOZOA

Reproduction in the protozoan may be either asexual or sexual. *Binary fission* is the commonest form of asexual reproduction. In this two daughter cells result from a 'parent' cell, division being along the longitudinal axis, although in ciliates it is along the transverse axis. The nucleus divides first and cytoplasmic division follows.

In *schizogony*, an asexual form of reproduction, the nucleus divides several times before the cytoplasm does. New progeny are formed along the plasmalemma of the parasite. In some of the sporozoans the nucleus of the parent cell divides mitotically into a large number of nuclear bodies, each of which becomes associated with a portion of cytoplasm and little or nothing of the parent cell remains except the greatly expanded limiting membrane. The dividing form is known as a schizont and the daughter forms are merozoites.

Budding is an asexual reproductive process in which two or many daughter forms are produced by the 'parent' cell. There is usually an unequal fragmentation of the nucleus and cytoplasm, but the budded forms are separated off and then grow to full size.

Endopolyogeny is a form of asexual multiplication (internal budding) whereby new progeny are formed within the parent cell. *Endodyogeny* is a simplified form of endopolyogeny, resulting in two daughter cells. It is seen in forms such as *Toxoplasma* and *Sarcocystis*.

Conjugation is a form of sexual reproduction which occurs in the ciliates. In this, two organisms pair and exchange nuclear material (from the micronucleus). The individuals separate and nuclear reorganization takes place.

Syngamy is sexual reproduction in which two gametes fuse to form a zygote. The male gamete is a microgamete and the female a macrogamete which are produced from microgametocytes (microgamonts) and macrogametocytes (macrogamonts), respectively. The process of gamete formation is gametogony, and the gametes may be similar in size (isogamy) or may markedly differ (anisogamy).

Sporogony normally follows syngamy, and a number of, or very many, sporozoites are formed within the walls of a cyst. This is an asexual process of multiple fission.

Ultra-structural details of morphology, feeding mechanisms, growth and multiplication are presented in a most lucid manner by Aikawa and Sterling (1974).

CLASSIFICATION OF PROTOZOA

The classification of the phylum protozoa adopted in this edition is based on that proposed by the Committee on Systematics and Evolution of the Society of Protozoologists and published by Levine et al. (1980), this being a modification of the classification of Honigberg et al. (1964). The modifications in the classification of the coccidia and piroplasms proposed by Levine (1971, 1973) and the cyst-forming isosporoid coccidia proposed by Frenkel (1977) have been adopted, but the uniform endings of the higher taxa follow those proposed by Honigberg et al. (1964). Only forms of veterinary or related interest are included in this classification.

SUB KINGDOM PROTOZOA, Goldfuss, 1918 emend. von Siebold, 1845

PHYLUM	SARCOMASTIGOPHORA Honigberg and Balamuth, 1963	With flagella, pseudopodia or both; single nucleus; typically no spore formation; sexuality when present, essentially syngamy
Subphylum	MASTIGOPHORA Diesing, 1866	One or more flagella present in trophozoites; asexual reproduction basically binary fission, sexual reproduction unknown in many groups
Class	PHYTOMASTIGOPHOREA Calkins, 1909	Chromatophores present; commonly only one or two emergent flagella; mostly free-living. (Of no veterinary or medical importance)
Class	ZOOMASTIGOPHOREA Calkins, 1909	Chromatophores absent; one to many flagella; amoeboid forms, with or without flagella, in some groups; sexuality in some groups; predominantly parasitic
Order	RHIZOMASTIGIDA Doflein, 1916	Pseudopodia and/or one to four flagella; mostly free-living
Order	KINETOPLASTIDA Honigberg, 1963	One to four flagella, kinetoplast with mitochondrial amnities. Mostly parasitic
Family	TRYPANOSOMATIDAE Doflein, 1901. emend. Grobben, 1905	Leaf-like, may be rounded
Genera	*Trypanosoma* *Leishmania*	
Family	CRYPTOBIIDAE Poche, 1913	Trypanosome-like forms; biflagellate; parasitic in fish, amphibians etc.
Genus	*Cryptobia*	
Order	RETORTAMONADIDA Grassé, 1952	Two or four flagella; one turned posteriorly and associated with ventral cytostomal area
Family	RETORTAMONADIDAE Grassé, 1952	Two or four flagellae
Genus	*Chilomastix*	

Family	COCHLOSOMATIDAE Kotlan, 1923	Six flagella, anterior ventral sucker, parabasal body
Genus	*Cochlosoma*	
Order	DIPLOMONADIDA Wenyon, 1926	Bilaterally symmetrical; with two karyomastigonts, each with four flagella. Mostly parasitic
Family	HEXAMITIDAE Kent, 1880	Bilaterally symmetrical; six or eight flagella; two nuclei
Genera	*Giardia* *Hexamita*	
Order	TRICHOMONADIDA Kirby, 1947	Typically four to six flagella, one recurrent and attached to undulating membrane if present, axostyle present; sexuality unknown; true cysts unknown. Parasitic
Family	MONOCERCOMONADIDAE Kirby, 1947	Three to five anterior flagella, recurrent flagellum usually free
Genera	*Histomonas* *Parahistomonas*	
Family	TRICHOMONADIDAE Chalmers and Pekkola, 1918 emend. Kirby, 1946	Four to six flagella, one recurrent and attached to an undulating membrane
Genera	*Pentatrichomonas* *Trichomonas* *Tritrichomonas*	
Subphylum	SARCODINA Schmarda, 1871	Pseudopodia typically present; flagella when present restricted to developmental stages; asexual reproduction by fission, sexual reproduction if present usually by flagellate gametes. Mostly free-living
Superclass	RHIZOPODA von Siebold, 1845	Locomotion by formation of podia; nutrition phagotrophic
Order	AMOEBIDA Ehrenberg, 1830	Naked; usually uninucleate. Free-living and parasitic
Family	VAHLKAMPFIIDAE Jollos, 1917; Zulueta, 1917	Nuclear division by promitosis
Genus	*Naegleria*	
Family	HARTMANNELLIDAE Volkonsky, 1931	Nuclear division not promitotic
Genera	*Acanthamoeba* *Hartmannella*	

Family	ENDAMOEBIDAE Calkins, 1926	Parasitic in digestive tract
Genera	*Dientamoeba* *Endamoeba* *Endolimax* *Entamoeba* *Iodamoeba*	
PHYLUM	APICOMPLEXA Levine, 1970	Apical complex, including conoid, micronemes, rhoptries etc. present at some stage; single nucleus; cilia and flagella absent (except microgametes); syngamy and cysts often present. All parasitic
Class	SPOROZOEA Leukart, 1879	Apical complex well developed; sexual and asexual reproduction; oocysts present
Subclass	COCCIDIA Leukart, 1879	Mature trophozoites small, typically intracellular; dodyogeny present or absent. Chiefly in vertebrates
Order	EUCOCCIDIIDA Léger and Duboscq, 1910	Asexual and sexual phases in life-cycle; schizogony present; parasites of epithelial cells and blood cells
Suborder	ADELEINA Léger, 1911	Macro- and microgametocytes associated in syzygy
Family	KLOSSIELLIDAE Wenyon, 1926	Zygote inactive; typical oocyst not formed; sporozoites develop in host cell. Often in kidney
Genus	*Klossiella*	
Family	HAEMOGREGARINIDAE Neveu–Lemaire, 1901	Ookinete present; two hosts, one invertebrate; in cells of circulatory system of vertebrates
Genus	*Hepatozoon*	
Suborder	EIMERIINA Léger, 1911	Macro- and microgametocytes develop independently; many microgametocytes produced; zygote non-motile; sporozoites in sporocysts; endodyogeny absent or present
Family	EIMERIIDAE Minchin, 1903	Development in host cell; oocysts with zero to many sporocysts each with one or more sporozoites; schizogony in host, sporogony outside
Genera	*Eimeria* *Isospora* *Tyzzeria* *Wenyonella*	

Family	CRYPTOSPORIDIIDAE Léger, 1911	Development on surface of cell; oocysts without sporocysts
Genus	*Cryptosporidium*	
Family	SARCOCYSTIDAE Poche, 1913	Syzygy absent, endodyogeny present, cysts and/or pseudocysts in cells. Parasites of vertebrates
Subfamily	TOXOPLASMATINAE Biocca, 1956	Pseudocysts present, schizogony and gametogony in intestinal cells; oocysts produced
Genera	*Toxoplasma* *Besnoitia* *Hammondia* *Cystoisospora*	
Subfamily	SARCOCYSTINAE Poche, 1913	Cysts septate, often elongate
Genus	*Sarcocystis* *Frenkelia*	
Suborder	HAEMOSPORINA Danilewsky, 1885	Macro- and microgamates develop independently, zygote motile; schizogony in vertebrate, sporogony in invertebrate; pigment usually formed in host cell
Family	PLASMODIIDAE Mesnil, 1903	Characters of suborder
Genera	*Haemoproteus* *Leucocytozoon* *Plasmodium*	
Subclass	PIROPLASMIA Levine, 1961	Small pyriform, round or pleomorphic; apical complex reduced; reproduction by binary fission or schizogony; pigment not formed in host cell. Parasites of cells of haemopoietic system. Vectors are ticks
Order	PIROPLASMIDA Wenyon, 1926	Characters of class
Family	BABESIIDAE, Poche, 1913	Relatively large pyriform, round or oval parasites; developmental stages usually in erythrocytes; apical complex reduced; binary fission and schizogony occur; no sexual reproduction. Vectors are ticks
Genus	*Babesia*	
Family	THEILERIIDAE du Toit, 1918	Small round, ovoid and other pleomorphic forms; apical complex reduced; schizogony in lymphoid and other cells, then invasion of erythrocytes by forms which may or may not multiply there. Vectors are ticks
Genus	*Theileria*	

PHYLUM	MICROSPORA Sprague, 1977	Spores with one or more polar filaments
Class	MICROSPOREA Delphy, 1963	Spores of unicellular origin, one long tubular polar filament
Genera	*Encephalitozoon* and several in cold-blooded vertebrates	
PHYLUM	MYXOSPORA Grassé, 1970	Amoeboid germinal elements in multicellular spores; trophozoites multicellular showing differentiation of somatic and germinal elements; all species parasitic
Class	MYXOSPOREA Bütschli, 1881	Characteristics of subphylum
Order	MYXOSPORIDA Bütschli, 1881	Parasites of cold-blooded vertebrates. Various genera in fishes
PHYLUM	CILIOPHORA Doflein, 1901	Simple cilia or compound ciliary organelles in at least one stage of life-cycle; usually two types of nucleus, transverse binary fission; sexuality involving conjugation. Mostly free-living
Class	KINETOFRAGMINOPHOREA de Puytorac et al. 1974	Characters of subphylum
Order	TRICHOSTOMATIDA, Bütschli, 1889	
Family	BALANTIDIIDAE Reichenow, 1929	Vestibulum near anterior end of body with cylostome at its base; ciliation uniform. Digestive tract
Genus	*Balantidium*	

REFERENCES

GENERAL

Aikawa, M. & Sterling, C. R. (1974) *Intracellular Parasitic Protozoa*. New York: Academic Press.
Frenkel, J. K. (1977) *Besnoita wallacei* of cats and rodents: with a reclassification of other cyst-forming isosporoid coccidia. *J. Parasit.*, 63, 611–628.
Honigberg, B. M., Balamuth, W., Bovee, E. C., Corliss, J. O., Gojdics, M., Hall, R. P., Kudo, R. R., Levine, N. D., Loeblich, A. R. Jr., Weiser, J. & Wenrich, D. H. (1964) A revised classification of the phylum protozoa. *J. Protozool.*, 11, 7–20.
Levine, N. D. (1971) Taxonomy of the proplasms. *Trans. Am. microsc. Soc.*, 90, 2–33.
Levine, N. D. (1973) *Protozoan Parasites of Domestic Animals and of Man*, 2nd ed. Minneapolis: Burgess.
Levine, N. D., Corliss, J. O., Cox, F. E. G., Deroux, G., Grain, J., Honigberg, B. M., Leedale, G. F., Loeblich, A. R. III, Lan, J., Lynn, D., Merinfeld, E. G., Page, F. C., Poljansky, G., Sprague, V. Vávra, J. & Wallace, F. G. (1980) A newly revised classification of the Protozoa. *J. Protozool.*, 27, 37–58.

Phylum: Sarcomastigophora Honigberg and Balamuth, 1963

Subphylum: Mastigophora Diesing, 1866

CLASS: ZOOMASTIGOPHOREA CALKINS, 1909

The *Zoomastigophorea* are flagellate protozoa possessing one or more thread-like flagella; some also have pseudopodia. In some forms a flagellum may pass along the body, being attached to it by an undulating membrane (e.g. *Trypanosoma, Trichomonas*). The nucleus is usually vesicular and reproduction is generally by longitudinal binary fission. In a few forms encystation may occur.

The neuromotor apparatus consists of a granular blepharoplast or basal granule from which arises the axoneme. Electron micrographs indicate that the axoneme forms the axial structure of the flagellum and consists (in trypanosomes) of two central and nine peripheral fibrils. These are surrounded by a flagellar sheath which extends to the distal end of the axoneme. Closely posterior to the blepharoplast there is a deeply staining granule, the kinetoplast. The kinetoplast contains DNA and is part of a mitochondrion which, in the case of the trypanosomes, runs the whole length of the body.

The *Zoomastigophorea* belong to the subphylum *Mastigophora* which contains also the class *Phytomastigophorea*. The last-named possess chromatophores, which contain chlorophyll, responsible for the synthesis of organic compounds from inorganic materials. Their nutrition is holophytic and they are of no veterinary or medical importance. The *Zoomastigophorea* lack chromatophores and they feed in a holozoic manner; they are classified into several orders, of which the following are of veterinary and medical interest:

Kinetoplastida (e.g. *Trypanosoma*)
Retortamonadida (e.g. *Cochlosoma*)
Diplomonadida (e.g. *Hexamita*)
Trichomonadida (e.g. *Trichomonas*)

ORDER: KINETOPLASTIDA HONIGBERG, 1963

FAMILY: TRYPANOSOMATIDAE DOFLEIN, 1901; EMEND. GROBBEN, 1905

Members of this family, the trypanosomes, are all parasitic and evolved from parasites of the alimentary canal of insects. Now, many are found in the blood and/or tissues of mammals and birds. They are characteristically leaf-like in shape; they have a single flagellum and this is attached to the body of the organism by an undulating membrane. The ultra-structure of the trypanosomes has been summarized by Rudzinska and Vickerman (1968).

During their life-cycle at least one further developmental stage is undergone (with the exception of a few mammalian forms), the various stages being morphologically distinct. The descriptive names for the developmental stages were once also used as the generic names for other members of the family. This source of confusion was removed by the adoption of the terminology of Hoare and Wallace (1966).

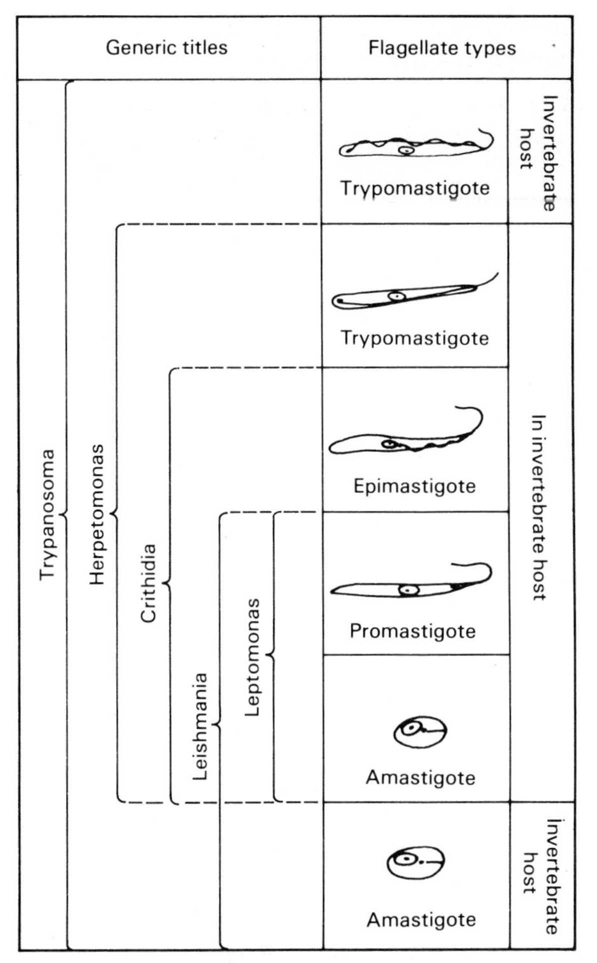

Fig 3.1 Classification of the trypanosomes and allied flagellates. (*From Wenyon 1926*)

DEVELOPMENTAL STAGES (see Fig. 3.1)

Trypomastigote stage (previously, trypanosome stage). A blade-like form with a kinetoplast posterior to the nucleus and usually near the posterior extremity. An undulating membrane is well developed and a free flagellum is often present. This stage is usually found in the vertebrate host but it is also found in arthropods as the infective stage for the vertebrate host.

Epimastigote stage (previously, crithidial stage). The kinetoplast and axoneme lie anterior to the nucleus and the undulating membrane is short. In a few species this stage is found in the vertebrate as part of the vertebrate developmental cycle, but it is principally a stage in arthropods.

Promastigote stage (previously, leptomonad stage). The kinetoplast and axoneme are at the anterior tip of the body. There is no undulating membrane. It is found in arthropods or plants.

Amastigote stage (previously, leishmania stage). The body is rounded; a flagellum is absent or is represented by a short fibril; the kinetoplast is present. The body is found in vertebrates and arthropods.

GENERA

The family includes the following genera:

Trypanosoma, occurs in vertebrates and arthropods; development may include the trypomastigote, epimastigote, promastigote and amastigote stages. In many species, only the trypomastigote stage occurs in the vertebrate host, but in a few more primitive forms, amastigote and epimastigote stages occur in the vertebrate.

Herpetomonas, found in invertebrates. Development may include the trypomastigote, epimastigote, promastigote and amastigote stages.

Crithidia, found in arthropods and other invertebrates. Developmental stages include epimastigote and promastigote forms.

Leptomonas, found in invertebrates. Developmental stages include promastigote and amastigote forms.

Phytomonas, found in plants and arthropods. Developmental stages include promastigote and amastigote forms.

Leishmania, found in vertebrates and arthropods. Developmental stages include amastigote and promastigote forms, the latter in the invertebrate.

Most of the members of the Trypanosomatidae undergo cyclical development. In those genera of veterinary importance (e.g. *Trypanosoma*, *Leishmania*), the life-cycle alternates between the vertebrate and the invertebrate host. When the invertebrate host is infected from the vertebrate host, a cycle of development and multiplication takes place in the invertebrate host, usually involving a developmental stage different from that in the vertebrate host. Thus, in the genus *Trypanosoma*, the trypomastigote form occurs in the vertebrate but it is the epimastigote form which predominates in the arthropod host.

Since the developmental stages of several genera may occur in invertebrates, it is not possible, with certainty, to identify any individual developmental stages as belonging to any specific genus. The stage may be a true parasite of the invertebrate or a developmental stage of a mammalian pathogen in the invertebrate.

Genus: Trypanosoma Gruby, 1843

Members of this genus occur in vertebrates, principally in the blood and tissue fluids, although a few may invade tissue cells. They are transmitted by blood-sucking arthropods in which the developmental stages mentioned above occur. A few species are transmitted mechanically; that is, there is no cyclical development in an arthropod.

Cyclical development may be 'anterior station' or 'posterior station' in type. In the former, developmental stages (e.g. epimastigotes) multiply in the gut of the arthropod, and infective stages, metacyclic trypomastigote forms (which are smaller than the vertebrate forms), accumulate in the mouth parts or salivary glands so that infection is transmitted when the arthropod takes a blood meal. This is the inoculative method of transmission. In the 'posterior station' forms, the metacyclic trypomastigotes accumulate in the hind-gut and are passed in the faeces of the arthropod. Infection of the vertebrate occurs by contamination of the skin or skin wounds.

Any trypanosome can be transmitted mechanically without cyclical changes taking place. Experimentally, this can be done by 'syringe passage' and in nature it is accomplished by blood-sucking insects (and possibly also vampire bats) which feed several times on different animals before repletion. There are a few species, however, where the ability to undergo cyclical development in arthropods has been lost and mechanical transmission is the only means available.

Hoare (1964) has divided the genus into two sections as follows:

Stercoraria (Posterior Station Group; Lewisi Group; Group A)

Morphology. Kinetoplast large and not terminal, posterior extremity tapering, free flagellum present, undulating membrane not well developed.

Biology. Multiplication in vertebrate host discontinuous: may occur in trypomastigote, epimastigote or amastigote forms. Metacyclic trypanosomes in 'posterior station' in arthropod host and transmitted by contamination through faeces. Often non-pathogenic. Blood forms with a high respiratory quotient and low sugar consumption. Cyanide inhibits oxygen consumption.

Four subgenera are recognized by Hoare, namely *Megatrypanum*, *Herpetosoma*, *Schizotrypanum* and *Endotrypanum*. Only the first three contain forms of veterinary or medical importance, these being indicated in Table 3.1.

Salivaria (Anterior Station Group; Group B)

Morphology. Kinetoplast smaller, terminal or subterminal. Posterior extremity blunt, there may be no free flagellum, undulating membrane varying in development.

Biology. Multiplication in vertebrate host continuous in the trypomastigote stage. Metacyclic trypomastigotes in the 'anterior station' of the arthropod host, and transmission is by inoculation. Frequently highly pathogenic. Some species are atypical and are transmitted noncyclically by arthropods or by coitus. Four subgenera are recognized by Hoare (1964), namely *Duttonella*, *Nannomonas*, *Pycnomonas* and *Trypanozoon*. All four contain species of veterinary and medical importance, which are detailed below.

Subgenus: Duttonella Chalmers, 1918

Kinetoplast terminal. Posterior end of body rounded, poorly developed undulating membrane, free flagellum present, shows great motility, monomorphic. Development occurs in the proboscis of *Glossina* spp. only. Blood forms have a

Table 3.1. Stercoraria

Subgenus	*Species*	*Vertebrate hosts*	*Arthropod hosts*	*Stage in which multiplication occurs in vertebrate hosts*
Megatrypanum	*Trypanosoma theileri*	Cattle	Tabanid flies	Epimastigote (binary fission)
	Trypanosoma melophagium	Sheep	Ked (*Melophagus ovinus*)	Uncertain
Herpetosoma	*Trypanosoma lewisi*	Rats	Fleas	Epimastigote (multiple fission)
	Trypanosoma duttoni	Mice	Fleas	Epimastigote (multiple fission)
	Trypanosoma nabiasi	Rabbits	Fleas	Epimastigote (multiple fission)
Schizotrypanum	*Trypanosoma cruzi*	Man, dog, cat, armadillo, opossum, etc.	Bugs	Amastigote (binary fission)
	Trypanosoma rangeli	Dogs, opossum, etc.	Bugs	Trypomastigote (binary fission)
	*Trypanosoma avium**	Rooks, jackdaws, etc.	Hippoboscid flies	? May not multiply, only get larger

* This species and others from birds do not readily lend themselves to classification.

low respiratory quotient and high sugar utilization.

The species include:

Trypanosoma vivax (long forms) in cattle, sheep, goat, antelope.

Trypanosoma uniforme (short forms) in cattle, sheep, goat, antelope.

Subgenus: Nannomonas Hoare, 1964

Kinetoplast marginal. Small size, no free flagellum, undulating membrane moderately developed, monomorphic and polymorphic. Development in mid-gut and then in proboscis of *Glossina* spp. Blood forms have high respiratory quotient and low sugar utilization. Cyanide inhibits oxygen consumption.

The species include:

Trypanosoma congolense, monomorphic (short forms), in cattle, sheep, horses, pigs.

Trypanosoma dimorphon, monomorphic (long forms), in cattle, sheep, horses, pigs.

Trypanosoma simiae, polymorphic in pigs, cattle, horses.

Subgenus: Pycnomonas Hoare, 1964

Monomorphic, stout and short species: kinetoplast small, subterminal, free flagellum short. Development in mid-gut and salivary glands of *Glossina* spp. One species of importance: *Trypanosoma suis*, in pigs.

Subgenus: Trypanozoon Lühe, 1906

Monomorphic or polymorphic. In latter, long, intermediate and stumpy forms occur with long free flagellum, short free flagellum and no flagellum, respectively. Undulating membrane well developed, kinetoplast small and subterminal. Blood forms have low respiratory quotient, low sugar utilization. Cyanide does not inhibit oxygen consumption.

For the purpose of this book the subgenus can be divided into two subgroups depending on the biology:

1. Cyclical development in mid-gut and salivary glands of *Glossina* spp. Species include:

Trypanosoma brucei (polymorphic), in domestic animals and antelopes.

Trypanosoma rhodesiense (polymorphic), principally in man, also in antelopes.

Trypanosoma gambiense (polymorphic), principally in man.

2. No cyclical development in arthropod, transmission mechanical. Species include:

Trypanosoma evansi (polymorphism inconstant), in cattle, camels, equines, dogs, etc.

Trypanosoma equinum (polymorphism inconstant), in equines.

Trypanosoma equiperdum (polymorphism inconstant), in equines.

THE BIOLOGY OF TRYPANOSOMES

In the definitive host. No sexual process has been observed in the life-cycle of trypanosomes and all multiplication is by binary or multiple fission. Division commences at the kinetoplast followed by the nucleus and then the cytoplasm. In the Salivaria, division is chiefly in the trypomastigote stage in the blood or in the lymph glands. In a few instances other developmental stages have been detected; for example, intracellular forms have been found for species of the subgenera *Nannomonas* and *Trypanozoon* (Soltys & Woo 1969; Ormerod & Venkatesan 1971). With the Stercoraria, however, reproduction in the epimastigote and amastigote forms is usual. *Trypanosoma lewisi* undergoes multiple fission in the epimastigote stage, producing rosettes of organisms. With *T. cruzi*, multiplication is intracellular, cells of the reticuloendothelial system and the striated muscle, especially of the heart, being filled with amastigote forms.

As well as the different developmental stages in the several species, polymorphism also occurs and a variation in shape and size is seen. In *T. brucei* the three types (long, intermediate and stumpy forms) have been ascribed an essential role in the biology of the organism. It has been suggested that in cyclically transmitted strains only the slender forms are capable of division and Wijers and Willett (1960) consider that only the stumpy forms are capable of infecting *Glossina* spp.

Some species and strains of trypanosomes can be adapted to abnormal hosts by serial subpassage (Desowitz 1963). This may be a short or prolonged process but is associated with an increasing virulence for the new host. Rat strains of *T. vivax* have been established by supplementing rats with sheep serum and ultimately the dependence on sheep serum has been lost (after 37 subpassages). Such strains produce lethal infections in the rat (Desowitz & Watson 1953). In such cases, the infection has been passaged by mechanical means and it is not known whether the same can be accomplished with forms which are passaged through the arthropod vector. In fact, a loss of ability to infect the arthropod host, and to undergo cyclical development in it, are features of cyclically transmitted forms which are transmitted continuously by mechanical means. The loss of ability to undergo cyclical transmission is unexplained although the associated loss of polymorphism, especially the loss of the stumpy forms (of *T. brucei*), may deplete the system of forms capable of infecting insects. A loss of virulence for the normal host when organisms are passaged in a new host is not an invariable event with all species or strains, although it is a common phenomenon especially when the passage is by mechanical means. Cyclical maintenance in an abnormal host may preserve normal host infectivity for long periods and *T. rhodesiense* has been passaged cyclically in sheep for 18.5 years without losing its infectivity for man (Ashcroft 1959). Similar reports exist of its maintenance in antelope and monkeys for several years.

Changes in trypanosome metabolism accompany morphological changes (and antigenic variation, see below). Vickerman (1971) has suggested a relationship between mitochondrial development and changes in morphology and metabolism.

The blood stream forms of the *T. brucei* group rapidly and incompletely catabolize glucose: most of the hexose is oxidized to pyruvate and excreted. Several enzymes of the Krebs' cycle and the cytochrome pigments are absent. Cyanide or carbon monoxide are without effect on respiration. Vickerman (1962) suggests, from cytochemical observations, that the oxidase system is localized in extramitochondral cytoplasmic granules.

The oxidative pathways of culture forms (see below) and of the developmental forms in the tsetse fly are different and more complex. Under aerobic conditions, glucose is catabolized completely to CO_2 and H_2O. Krebs' cycle enzymes are present, as is a mammalian-type cytochrome system. In such forms, too, the mitochondrion is highly developed, markedly branched and extending the length of the cell.

The pleomorphic blood stream forms also show differences in metabolism and cytology. Thus, Vickerman (1965) has demonstrated NADH-nitrotetrazolium blue reductase activity in the mitochondrion of stumpy forms of *T. brucei* but not in the slender forms. The stumpy forms were able to metabolize α-oxyglutarate, whereas the slender forms were not. The transition from slender to stumpy forms in the blood is associated with a metabolic switch involving the activation of mitochondrial enzymes and Bowman et al. (1970) consider that the stumpy forms contain most of the Krebs' cycle enzymes, whereas the slender forms lack them. The regulatory mechanisms for these metabolic switches are clearly important in infectivity mechanisms and are likely to be important in immunological mechanisms also.

Alterations in drug sensitivity may be marked and may lead to difficulty in drug testing and development. An increase in sensitivity to drugs may be spectacular and this is illustrated with *T. rhodesiense* which develops an increased sensitivity to arsenical compounds on mechanical passage in abnormal hosts. Desowitz (1956) found that the rat-adapted *T. vivax* was 15 times more sensitive to antrycide than the normal sheep strain.

Immunological aspects of trypanosome infection. The immunological response to trypanosomes has occasioned much work in the last decade. The subject has been reviewed by Henson and Noel (1979). The principal feature of infection with salivarian trypanosomes is the ability of these organisms to evade the immune response of the host by the production of relapse strains with different antigenic characteristics. This phenomenon, recognized by Ross and Thompson in 1910, results in successive waves of parasitaemia at intervals of a few days, each wave representing the multiplication of a population of a new antigenic type; the decreasing phase of the wave represents the destruction of this population by antibody (Lumsden 1972). Thus, the persistence of an infection is due to the evasion of the immune response by a repeated change in antigenic character.

There is no satisfactory evidence that antigenic variation occurs in the pathogenic stercorarian trypanosome *T. cruzi*.

The antigens responsible for relapse populations are the 'variant' antigens which are located on the surface of the organism, but may also be found in the plasma of infected animals. In distinction to the 'variant' antigens are the 'stable' antigens, which are probably structural proteins, enzymes, etc. and are shared by the various developmental forms of a trypanosome and even by different species of trypanosomes. The stable antigens are not concerned with the relapse phenomenon and are not thought to be concerned with protective immunity.

Antigenic variation in *Trypanosoma brucei*, and probably in other pathogenic salivarian trypanosomes, can be largely explained by the sequential expression of alternative cell surface glycoproteins (Cross 1977). The number of alternative antigens which may be expressed by a clone is undetermined but it is suggested that several hundred may occur.

The association of variant antigens with the surface coat of trypanosomes was reported by Vickerman (1969) and Vickerman and Luckins (1969) and Cross (1977) considers that in a homogeneous population of *Trypanosoma brucei* the variant glycoprotein is the major, if not the sole, component of the surface coat, 15 nm thick, overlying the three-ply unit plasma membrane of the trypanosome (Vickerman 1969; Wright & Hales 1970). It is absent from forms in the mid-gut of the tsetse fly and in culture, but is reacquired during transformation to infective metacyclic forms in the tsetse salivary glands. The coat is well developed in salivarian trypanosomes which undergo marked antigenic variation; it is less dense in species (e.g. *Trypanosoma lewisi*) where antigenic variation occurs to a minor degree (Lumsden 1972).

Cross (1977) proposes that the surface coat is formed from a monolayer of glycoproteins oriented with their C-terminal ends adjacent to the cell membrane. All the carbohydrate of the molecule appears to be located towards the C-terminal end and hence carbohydrate is absent from the external regions of the surface coat.

Initial characterization of surface antigens by Cross (1977) indicates that the glycoproteins have a molecular weight of 65 000; each has an immunological uniqueness which is reflected in the immense variation in amino acid sequence in that isoelectric points and amino acid compositions differ greatly. Substantial advances are probable in this area in the coming few years since there is a pressing need to develop immunological means for the control of trypanosomiasis.

Variant antigens may be released into ambient fluid (Lumsden 1972) and these probably represent the exoantigen of Weitz (1960). Further, it has been shown that *T. brucei* produces long filamentous structures, 'filopodia' or 'plasmanemes', up to 70 μm in length from the anterior and posterior extremities. The ultra-structure of these is similar to that of the surface coat of the trypanosome. The surface coat may derive from protein secreted into the flagellar pocket being produced by endoplasmic reticulum located between the nucleus and the flagellar pocket (Vickerman 1971). Its role may be to act as a physical barrier to prevent immunoglobulins, or other components of the immune response, from interacting with the underlying cell membrane which contains the invariable antigens. Cross (1977) believes that the variable antigens elicit a strong immune response but, because of the instability of their attachment to the cell membrane, they are shed in an antigen–antibody complex before complement action and membrane damage can occur.

Two main hypotheses have been advanced to explain the phenomenon of antigenic relapse. Vickerman (1971) considers that a switch to a new serotype by organisms possessing a genotypic potential to produce a range of variant antigens is a preferable explanation to genetic mutation. A further hypothesis is proposed by Newton et al. (1973), who envisage that the synthesis of trypanosome antigens may be analogous to that of immunoglobulins in that a 'variable' region would be exposed to the environment and the 'constant' or 'common' region would locate the antigen in the correct configuration at the cellular site.

The predictability of the appearance of antigenic variants has caused debate. Gray (1965) reported that variants appear in a predictable sequence and, furthermore, tend to revert to a constant basic antigen on cyclical transmission. However, Seed and Gam (1966) and McNeillage et al. (1969) produced evidence that variants do not arise in a completely ordered sequence but that certain variants do appear to occur more frequently.

The variant antigens are associated with protective immunity, but only against the homologous antigenic type. Such immunity can be stimulated by homogenates of trypanosomes (Lumsden 1969), irradiated avirulent organisms (Duxbury et al. 1972) and formalin-inactivated organisms (Soltys 1967). An antigen which immunizes mice has been isolated from *T. gambiense* and *T. equiperdum* by Seed (1972). Lumsden (1972) is of the opinion that the wide range of antigen types discourages the idea of protection of animals by polyvalent vaccines. He points to the fact that animals in the field survive repeated infection, not by a sterile immunity but by mechanisms which control parasitaemia at a low, non-pathogenic level.

Suppression of the immune response to a variety of antigens or infections occurs in animals infected with the pathogenic trypanosomes. Goodwin et al. (1972) showed a reduced antibody response to sheep red blood cells (RBC), Urquhart et al. (1972) reported suppression of the immune response in the rat to *Nippostrongylus brasiliensis* infection during trypanosomiasis and Murray et al. (1974) reported suppression of the response of Zebu cattle to a polyvalent clostridial vaccine in animals infected with *Trypanosoma congolense*. However, although profound effects on the immune response occur during trypanosome infection, elimination of the infection by trypanocidal therapy quickly restores the ability of an animal to respond to a variety of antigens (Murray et al. 1974).

In insect vector. With the exception of a few species, the majority of trypanosomes undergo cyclical development in an arthropod vector. When non-cyclical transmission (mechanical) occurs it is usually by the agency of biting-flies such as *Stomoxys* and *Tabanus*. The biting-fly is immediately infective after feeding but it remains so for a short time only and must feed on another animal very soon if the trypanosomes are to be transmitted. *Trypanosoma evansi* and *T. equinum* are solely transmitted in this way, while *T. equiperdum* is transmitted mechanically by coitus. It should be stressed that even the cyclical forms may be transmitted mechanically and even by the arthropod in which they also develop cyclically.

In cyclical development, mammalian blood containing the trypomastigote forms is taken into the intestine of the arthropod and subsequent development depends on whether 'anterior' or 'posterior' station development occurs.

Anterior station development (*Salivarian trypanosomes*). This is exemplified by trypanosomes of the *Trypanozoon* (*brucei*) group. Ingested forms localize in the posterior part of the mid-gut of *Glossina* where they multiply in the trypomastigote stage for the first ten days. It has already been noted that the stumpy forms of a polymorphic population are possibly those which are destined for cyclical development. Initially, the dividing forms in the mid-gut are broad with a kinetoplast midway between the nucleus and the posterior end. By days 10–11, however, long slender forms are produced and these migrate backwards and enter the space around the peritrophic membrane and then penetrate into the proventriculus, being found here 12–20 days after infection. They subsequently migrate anteriorly to the oesophagus and pharynx and then to the hypopharynx and salivary glands. In this latter situation, epimastigote forms are produced and further multiplication takes place. In another two to five days, the metacyclic or infective forms are produced. Metacyclic trypanosomes are small stumpy forms which somewhat resemble the stumpy forms in the blood. These are injected into the host along with saliva when the fly bites, several thousand being injected with each bite.

Infective mechanisms. No satisfactory explanation is available to account for the transition from insect vector form to vertebrate host form. Blood stream forms possess a smooth, electron-dense coat, which appears to contain the variant surface antigens (see above). This coat is lost in the course of cyclical development in *Glossina* spp. (and also in culture) but reappears in the metacyclic trypanosomes in the salivary glands (Vickerman 1972). Other differences between blood and insect forms include changes in the mitochondrion. The long slender blood stages of *T. brucei* have mitochondria with sparse, short, tubular cristae, whereas the stages in the mid-gut of the tsetse fly have mitochondria with numerous plate-like cristae. The mitochondria of the short, stumpy blood forms are intermediate between these two types of mitochondria, this being further evidence that such stumpy forms are infective to the insect vector (Vickerman 1966). These morphological differences are reflected in differences in metabolism. Thus, in the slender blood stream forms, the prime source of energy for adenosine triphosphate (ATP) synthesis is the catabolism of glucose to pyruvate via the glycolytic pathway; pyruvate is then excreted into the blood stream. There are no functional cytochromes and this is reflected in a lack of cyanide sensitivity, and inhibitors of Krebs' cycle enzymes do not affect glucose or oxygen consumption.

On the other hand, with an active mitochondrion, the insect forms catabolize glucose completely to CO_2, pyruvate is further metabolized through the tricarboxylic acid cycle, respiration is inhibited by cyanide, cytochromes are present and Krebs' cycle substrates are used (Trigg & Gutteridge 1977).

It is of interest that the blood stream forms of *Trypanosoma congolense* and *Trypanosoma vivax* possess tubular cristae in the mitochondrion (Vickerman 1971) and there is evidence that glucose is metabolised beyond pyruvate.

The developmental cycle of *T. brucei* in *Glossina morsitans* takes 25 days or more and flies are not infective until the metacyclic forms have been produced. A detailed account of factors which modify this infection rate and the species of flies concerned will be found in Buxton (1955).

If a trypanosome is transmitted by one species of *Glossina* then it is probably transmissible by all species (Hornby 1952). Thus, all species of *Glossina* which have been investigated appear capable of transmitting *T. vivax, T. congolense* and *T. brucei*. The differences in transmission by various species in nature are largely due to factors which determine the prevalence of the species of *Glossina*, e.g. host, feeding, climate, etc. (see Buxton 1955).

Posterior station development (Stercorarian trypanosomes). This type is exemplified by *Trypanosoma lewisi* of the rat. Ingested trypomastigotes enter cells which line the stomach of the rat flea, *Ceratophyllus fasciatus*. Within the cells the trypomastigotes round up to pear-shaped organisms which increase in size, while the nuclei and kinetoplasts divide, ultimately giving a large number of trypomastigote forms. The cells rupture and the liberated trypomastigote stages pass from the stomach to the rectal region. During this migration they change to epimastigote forms, the kinetoplast being displaced anterior to the nucleus. The epimastigote stages attach themselves to the lining cells of the posterior gut, multiply as epimastigotes and then produce metacyclic trypanosomes. The total cycle takes about five days in the rat flea. Metacyclic trypanosomes are passed in the faeces and infection of another rat is by contamination of a flea bite wound with faeces. Alternatively, the infected flea may be ingested and then the metacyclic trypanosomes penetrate the mucous membrane of the digestive tract.

In vitro cultivation. In the past, a major problem in the study of the biology and immunology of trypanosomes has been the inability to maintain infective forms in culture for extended periods. Salivarian trypanosomes rapidly lose infectivity and virulence in the course of in vitro cultivation, and previous reports of maintained or restored virulence and infectivity of culture forms for mice have been shown by Mendez and Honigberg (1972) to be associated with the persistence of blood stream forms in culture. Culture forms are indistinguishable from insect mid-gut stages: they lack the surface coat of the blood stream forms and possess the biochemical and antigenic characteristics of the insect forms (Honigberg et al. 1976). The culture media previously used were complex and usually contained blood. However, semi-defined and defined media for the growth of culture forms of *T. brucei* have been developed by Cross and Manning (1973) and these will now permit a more detailed examination of the biochemistry of the insect stages.

A major contribution has been the propagation of animal infective forms of *T. brucei* in vitro by Hirumi et al. (1977*a*). Of 96 culture conditions tested, the one supporting best growth of parasites consisted of HEPES-buffered RPMI 1640 medium with 20% heat-inactivated fetal bovine serum in the presence of bovine fibroblast-like cells. Culture trypanosomes are morphologically identical to those of blood stream forms and they possess variant surface antigen. They retain the ability to infect mammalian hosts after many passages in culture. Transformation to 'insect mid-gut' forms occurred within 14 days when trypanosomes were cultured at 25°C in the same medium; this was accompanied by a loss of infectivity for mice. Retransformation to infective blood stream forms occurred within 10–14 days after transfer back to 37°C (Hirumi et al. 1977*b*).

Mechanisms of pathogenicity. In an extensive review of the pathology of trypanosomiasis in domestic animals, Losos and Ikede (1972) have emphasized that the animal trypanosomiases are a group of diseases. The disease process of each depends on the species of trypanosome, the strain within the species and the species of host. These authors recognize two groups of tsetse-transmitted pathogenic trypanosomes: the 'haematic group' consisting of *T. congolense* and *T. vivax* which invade and are confined to the plasma of blood vessels; and the 'humoral group' consisting of *T. brucei, T. rhodesiense* and *T. gambiense* which, in addition to occurring in the plasma, are also present in the intercellular tissue and body cavity fluids. Disease produced by the former group is largely due to anaemia while with the latter, although anaemia occurs, it is considered to be of secondary importance to the extensive degenerative, necrotic and inflammatory changes.

The aetiology of the anaemia in trypanosomiasis has been the subject of numerous studies. A depression of erythropoiesis has been frequently stated to be a major factor (e.g. Fiennes 1970) but there is now general agreement that the anaemia is due, at least in part, to immunological mechanisms. Jennings (1976) gives the characteristics of the anaemia of *T. brucei* infections as macrocytosis, reticulocytosis, normoblastic hyperplasia of the bone marrow and spleen, and increased haemosiderin deposits and erythrophagocytosis in the spleen. There is a shortened circulating half-life of ^{51}Cr-labelled RBC, accelerated urinary excretion of ^{51}Cr, but no increase in iron loss as indicated by the retention of injected ^{59}Fe in the body. These are indicative of an anaemia of haemolytic origin. However, there is no evidence of intravascular haemolysis, such as jaundice, etc., and extravascular destruction of RBC is a more likely explanation for the anaemia. Jennings et al. (1974) note that there is no loss of erythropoietic activity in mice infected with *T. brucei*.

Jennings (1976) speculates on the mechanisms by which RBC are made more susceptible to erythrophagocytosis. These include the attachment of trypanosome antigen to RBC, reported by Herbert and Inglis (1973), which may increase the cell susceptibility to erythrophagocytosis, which may be further increased by the union of surface-adsorbed antigen with antibody. Kobayashi et al. (1976) have shown that RBC from cattle infected with *T. congolense* contain immunoglobulin on their surface. The process of opsonization may be completed by the activation of complement and Woodruff et al. (1973) have demonstrated complement components on the RBC of patients with *Trypanosoma rhodesiense* infection. Erythrophagocytosis would be stimulated by the display of various subcomponents of C on the RBC, these being necessary for the attachment to and engulfment by macrophages. Bound complement may stimulate the production of immunoconglutinin—high levels of immunoconglutinin have been detected in experimental animals infected with trypanosomes (Ingram & Soltys 1960) and lead to enhanced haemolysis or erythrophagocytosis. A haemolytic factor, causing direct haemolysis of red cells, may be produced, and Tizard et al. (1977) have demonstrated potent cytotoxic and haemolytic factors after autolysis of *T. congolense*. These factors are free fatty acids of 14–20 carbon atoms and are generated by the action of phospholipase A1 on phosphatidyl choline. Jennings (1976) also suggests that, because of metabolic competition between trypanosomes and RBC, the latter may be made more susceptible to phagocytosis by, for example, their inability to regenerate ATP. An interesting point in this respect is that cells from infected and anaemic animals, when transferred to normal animals, do not behave differently from cells from normal animals. Finally, anaemia may be associated with disorders of coagulation, such as thrombocytopenia and disseminated intravascular coagulation (DIC). These may result in a condition resembling microangiopathic haemolytic anaemia in *T. brucei* infection of rabbits (Boreham 1974), in which microthrombi trap and damage circulating red cells.

Thrombocytopenia may occur in trypanosomiasis (Robins-Brown et al. 1975), and Davis et al. (1974) have claimed that live trypanosomes and extracts of them would cause aggregation of blood platelets. However, Greenwood and Whittle (1976) could not confirm these findings.

The situation with the anaemia in cattle infected with *T. congolense* and *T. vivax* is less clear. A normocytic–normochromic anaemia with an absence of reticulocytes is generally seen (Losos et al. 1973) but Wellde et al. (1974) report anisocytosis and polychromasia as some of the earliest changes. Maxie et al. (1976) have compared calves infected with *T. vivax* or *T. congolense* and showed that either infection resulted in macrocytic normochromic anaemia, leukopenia and thrombocytopenia. Although the anaemia produced was equally severe, the mechanisms may be different for each species. Thus, no immunoglobulin was found on RBC of *T. congolense*-infected cattle, but it was present on cells from *T. vivax*-infected animals; the former would suggest the anaemia is caused by non-immune mechanisms, the latter that immune mechanisms were responsible. An increase in plasma volume is seen in *T. congolense* infections in cattle (Fiennes 1970), *T. vivax* in sheep (Clarkson 1968) and *T. brucei* in rabbits (Boreham 1967) and this may be a factor in the anaemia. Indeed, Holmes and Jennings (1976) conclude that, in rabbits infected with *T. congolense*, haemodilution plays a significant part in the initial development of anaemia. Work by these authors is especially interesting since following treatment the changes in haematological parameters, plasma volume, etc., rapidly return to normal, which suggests that a factor produced by the trypanosomes may play a direct part in this. Tizard et al. (1977) (see above) have reported a potent cytotoxic and haemolytic factor from autolysed *T. congolense* and Nguyen-Huan et al. (1975) have identified a haemolytic factor from *T. brucei*.

Direct effects of metabolic products or toxins of trypanosomes have been incriminated as the cause of injurious effects by several workers. Seed (1972) has identified a substance which disturbs blood sugar levels and glycogen reserves in the liver of infected guinea-pigs. The accumulation of a glucose-6-phosphatase activity-inhibiting factor is postulated to explain these effects.

Additionally, Seed and Hall (1977) have reported that *T. brucei gambiense* can metabolize tryptophane to tryptophol which, in pharmacological doses, induces a lethargic state in animals, changes in body temperature and in the humoral antibody response to heterologous antigens. These authors suggest that enough tryptophol is produced in an infected animal to induce the changes which occur in animals infected with the pathogenic trypanosomes. Further, tryptophol is rapidly eliminated from the body and this is consistent with the rapid recovery of immunological competence and the decrease in lethargy following chemotherapeutic cure.

Other changes in the plasma constituents in trypanosomiasis include an increase in fibrinogen degradation products, suggestive of DIC (Boreham & Facer 1977), an increase in total serum lipid in rabbits infected with *T. brucei* (Goodwin & Guy 1973), but a marked decrease in serum lipid in cattle infected by either *T. congolense* or *T. vivax* (Roberts 1975). There is also a marked increase in macroglobulins. Clarkson (1976) notes that very high IgM concentrations are almost invariably seen in man and animals infected with salivarian trypanosomes. Its origin is in dispute but very likely it is associated with the response to a rapid succession of antigenic variants and also to polyclonal stimulation of B lymphocytes which Urquhart et al. (1973) and Greenwood (1974) ascribe to a 'B' cell mitogen produced by the trypanosome.

The pathophysiological effects of markedly elevated macroglobulin levels include increased plasma viscosity, tending towards stasis of the blood and immunopathological changes.

Earlier studies with the 'humoral' group had ascribed death of the host, infected with *T. brucei* for example, to a progressive and terminally fatal hypoglycaemia, caused directly by the parasite's high rate of carbohydrate consumption. However, summarizing experimental evidence, von Brandt (1966) concluded that this mechanism was inadequate to explain the pathogenicity of such infections.

A major advance in the understanding of the pathogenesis of trypanosome infection was the demonstration that pharmacologically active mediators of inflamation play an important role in the infection. These include the plasma kinins (Boreham 1968*b*, 1970), fibrinogen degradation products (Boreham & Facer 1977), etc. Boreham and Wright (1976) have reviewed the sequence of events resulting in the release of kinins. Trypanosomes contain no free kinin nor kinin-forming enzymes.

It is proposed that immune complexes formed during the infection absorb Hageman factor onto their surfaces, causing its activation. This, in turn, activates kininogenase precursors (e.g. prekallikrein) which then act on kininogens, releasing kinins. Kinins, which are also elevated in immune events such as antigen–antibody reactions (e.g. anaphylactic shock) and thermal injury, cause increased endothelial permeability of vasculature leading to oedema. Goodwin (1970) also suggests that kinins are involved in the more chronic phases of the local inflammatory response. Thus, this recent work has incriminated immunological reactions as the basis of the severe pathophysiological effects. In concert with this, Boreham (1968*a*) has shown that the highest plasma kinin concentrations appear within a few days after each episode of parasitaemia in chronic *T. brucei* infections of cattle. Similarly, increases in the kinin levels were paralleled by decreases in kininogen levels. Increased kinin levels were associated with species-specific antibodies rather than those associated with the variant antigens.

The pathological effect of *T. brucei* on small blood vessels has been studied by Goodwin (1971) in rabbit ear chambers. Increased stickiness of endothelium resulted in accumulations of mononuclear leukocytes, followed by stasis and then disintegration of blood vessels. These changes resembled inflammatory processes produced by other agents, but were widespread rather than localized. Goodwin and Hook (1968) have suggested that the infiltration of mononuclear cells into infected tissues, and the necrosis, may indicate a cell-mediated immunity response in the infection. The mononuclear phagocytes are large and actively engulf parasites, which may suggest activation of such cells by cell-mediated immunity mechanisms.

Parasites of the humoral group only become parasitaemic in the circulation in the terminal stages of the infection. The propensity for extravascular sites has renewed interest in visceral phases of amastigote development, such as those reported in rats by Ormerod and Venkatesan (1971) and in mice by Soltys and Woo (1970).

THE PATHOGENIC TRYPANOSOMES OF DOMESTIC ANIMALS

Section: Salivaria (Anterior Station Forms)

Subgenus: Duttonella Chalmers, 1918 (Vivax Group)

Trypanosoma vivax Zieman, 1905 (syn. *T. cazalboui*; *T. caprae*; *T. angolense*, etc.) (Fig. 3.2). This is a monomorphic species, 20–27 μm (average 22.5 μm) by 3 μm. The posterior part is distinctly broader and bulbous, the kinetoplast is large and terminal and the free flagellum is short, being 3–6 μm in length. The organism is very

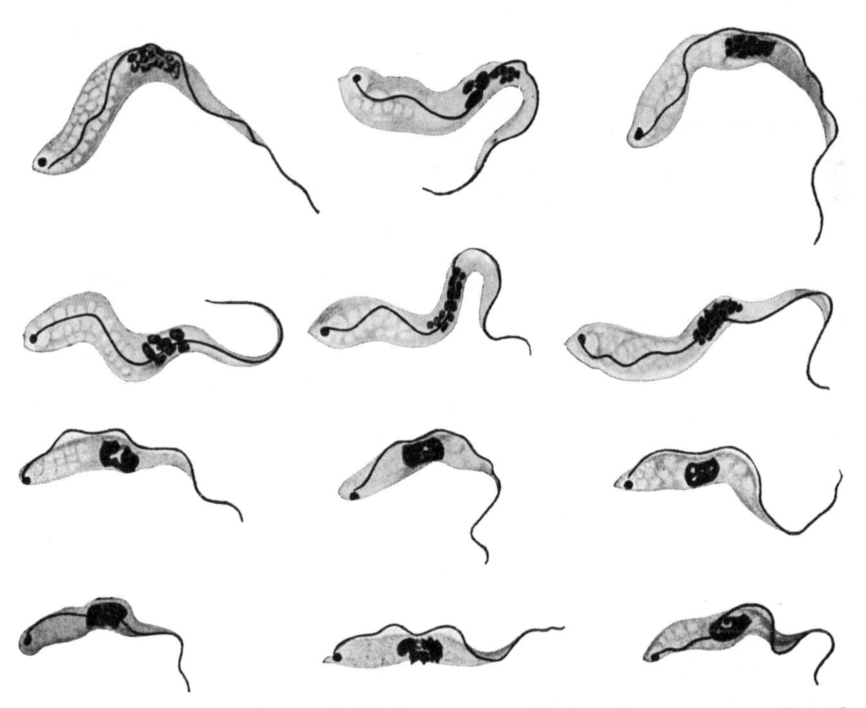

Fig 3.2 Trypanosomes of the subgenus *Duttonella.* Upper two rows: *T. vivax*; lower two rows: *T. uniforme.* (× 2000) (*From Wenyon 1926*)

motile in fresh blood, moving rapidly across the field, pushing red cells aside as it goes.

Its hosts include cattle, water buffalo, sheep, goats, camels, horses; but dogs and pigs are refractory to infection. In Africa, antelopes serve as reservoir hosts and in South America deer act as such. Laboratory rodents are not readily infected, although white rats can be infected if the infection is supplemented with sheep blood (Desowitz 1963).

Geographically it occurs throughout Africa, being transmitted by tsetse flies (especially *Glossina morsitans* and *Glossina tachinoides*), development taking place in the anterior station and in the proboscis only. It is also established in Central and South America, the West Indies and Mauritius where it is transmitted mechanically by biting-flies. Hoare (1967) has suggested that the mechanically transmitted strains of *T. vivax* should be regarded as a separate subspecies, *T. vivax viennei.* Despite the wide geographical locations, the African and South American strains are very similar.

T. vivax causes the most important form of trypanosomiasis of cattle in West Africa (Losos & Ikede 1972). Cattle raised in the Sahel and Sudan zones are relatively free of disease, but because of seasonal climatic conditions and the prospect of better markets, stockmen with their cattle migrate southwards through regions heavily infested with tsetse flies. Morbidity and mortality in cattle are high in such circumstances. In East Africa the disease produced by *T. vivax* is usually mild; in Central and South America it is the most important species of the four that occur in cattle and water buffalo.

T. vivax in cattle may produce an asymptomatic infection or a chronic, acute or peracute disease. Breed susceptibility is recognized and N'Dama cattle rarely develop serious disease, whereas N'Dama–Zebu crosses and pure Zebu may develop severe and fatal infections.

In the peracute infections, a high and persistent parasitaemia occurs, although at the time of death parasitaemia may be very low. Extensive haemorrhages are seen on the mucosal and serosal surfaces of the digestive tract, body cavity, muscles, heart and lymph nodes. Enlargement of spleen and lymph nodes is observed and there is depression of haemoglobin and red blood cell levels. Chronic disease in cattle may last for several months when anaemia, emaciation and severe wasting are evident.

Both acute and chronic disease may occur in goats. Emaciation and splenomegaly are reported, and Losos and Ikede (1972) consider that *T. vivax* causes substantial mortality in goats and sheep in Africa. In horses, *T. vivax* results in a mild chronic disease of low mortality. Emaciation, anaemia and cutaneous urticarial oedematous swellings have been reported.

The course of infection in a water buffalo was reported by Shaw and Lainson (1972) as being characterized by a series of low parasitaemias, and the animal continuously lost weight and eventually died. The cat and dog are not usually susceptible to natural or artificial infection.

Infections are diagnosed by demonstration of organisms in the peripheral blood (Losos & Ikede 1972). Lymph node smears have been regarded as more efficient for diagnosis but Losos and Ikede (1972) consider that evidence from this is inconclusive.

Trypanosoma uniforme Bruce et al., 1911. (See Fig. 3.2) This is similar to *T. vivax* with an average length of 16 μm (12–20 μm) by 1.5–2.5 μm. The free flagellum is shorter than that of *T. vivax*. It infects cattle, sheep, goats, and antelope but is not transmissible to laboratory rodents. It occurs in Uganda and Zaire, being transmitted cyclically by tsetse flies. The disease process is similar to that produced by *T. vivax*, but it is more or less non-pathogenic for goats.

Subgenus: Nannomonas Hoare, 1964 (Congolense Group)

Trypanosoma congolense Broden, 1904 (syn. *T. pecorum; T. nanum; T. montgomeryi* etc.). (Fig. 3.3) This is a small form, 9–18 μm in length, and it is the smallest of the African trypanosomes. The posterior extremity is blunt and the kinetoplast is typically marginal, being some distance from the posterior end. Characteristically, there is no free flagellum although the tapering anterior extremity may, inadvertently, give the appearance of one.

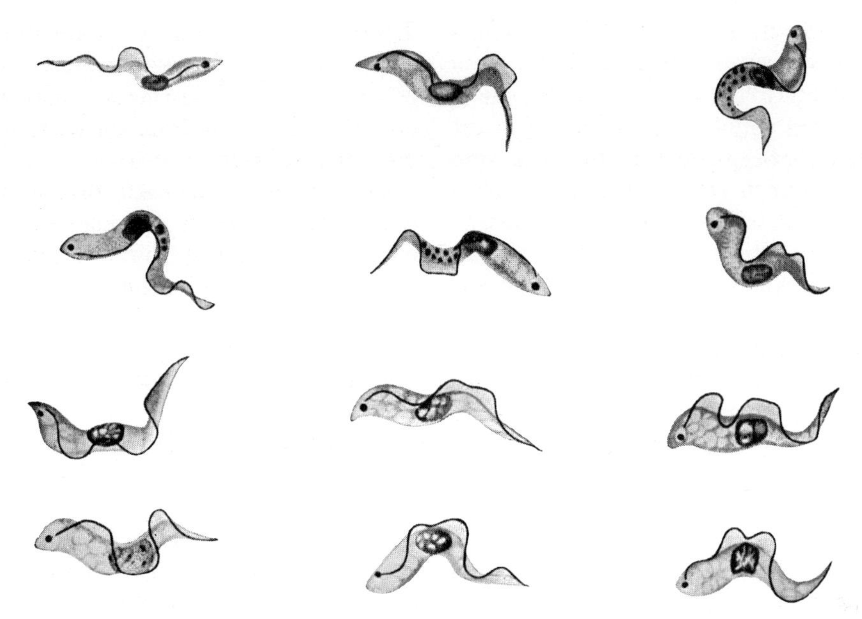

Fig 3.3 Trypanosomes of the subgenus *Nannomonas*. *Trypanosoma congolense*. (× 2000) (*From Wenyon 1926*)

Although the undulating membrane is inconspicuous, the organism is active and it exhibits marked but non-progressive movements in fresh blood.

Its hosts include all domestic animals, and wild game animals—such as antelope, zebra, warthog, and even the elephant—may act as reservoir hosts. The organism is readily transmitted to laboratory rodents. However, its chief association is with cattle and it is the principal cause of the disease 'nagana', this name being derived from a Zulu word meaning 'to be in low or depressed spirits'. It was used by the Zulus to describe trypanosomiasis in cattle, and Bruce, in 1898, brought the word into common usage. The parasite is renowned for the great number of strains which occur; these differ in both virulence and antigenicity.

Geographically, *T. congolense* is widely distributed in tropical Africa and it produces the most important form of animal trypanosomasis in East Africa. It is transmitted cyclically by several *Glossina* spp. Different species of *Glossina* are of importance in different areas. Thus, *G. morsitans* is more effective than *G. tachinoides* and *G. palpalis* in transmission in northern Nigeria, *G. longipalpis* more than *G. palpalis* and *G. submorsitans* in Guinea, and *G. austeni* more than *G. morsitans* in East Africa. Development may occur in any species of *Glossina* if it is present. Development occurs in the anterior station in the mid-gut and the proboscis. Mechanical transmission by biting-flies is common.

T. congolense is regarded as a plasma (haematic) parasite (Losos & Ikede 1972). It may produce peracute, acute or chronic disease. The species has been divided on morphological criteria into the short *congolense* form, the *intermediate* form and the longer *dimorphon* form (Godfrey 1960) and varying pathogenicity may be associated with these. The disease in cattle has been described as consisting of a series of crises (Fiennes 1970). Acute disease and death may coincide with a crisis when there is massive destruction of trypanosomes and red blood cells. The anaemia has been ascribed to the inhibition of haemopoiesis and is normocytic and normochromic with an absence of immature erythrocytes. The severity of anaemia determines the severity of clinical signs; in the peracute disease, which lasts two to four weeks, there is no observable loss of condition but as chronicity supervenes anaemia, emaciation and cachexia are evident. Haemorrhages occur on mucosal and serosal surfaces and in lymph nodes, which may also be enlarged. Splenomegaly and multiple perivascular accumulations of lymphocytes in the liver, kidney and elsewhere, may be seen. Focal polioencephalomalacia of the cerebral cortex has been reported by Losos and Ikede (1972). Fiennes (1952) has described a cryptic form of the infection and, especially in unsuccessfully treated cases, he postulated that trypanosomes accumulated in the capillaries of the myocardium where they are associated with necrosis of myocardial fibres.

A similar disease occurs in sheep and goats with anaemia, weakness and emaciation being prominent clinical signs. In horses the disease is usually chronic, with weakness, anaemia and oedema of the legs and genitalia as common clinical signs. In dogs the disease may be acute, with jaundice, anaemia, enlargement of lymph nodes, keratitis, ulcerative stomatitis, gastroenteritis and subcutaneous oedema (Parkin 1935). A similar disease is seen in cats.

Diagnosis is usually by the demonstration of organisms in the peripheral blood; these are more readily demonstrable in the early stages of infection.

Trypanosoma dimorphon Laveran and Mesnil, 1904. This is now recognized as a valid species (Hoare 1959), it having previously been a synonym of *T. congolense*. It measures 11–24 μm (mean 16.2 μm), is monomorphic, slender, has no free flagellum and the kinetoplast is marginal and subterminal.

It occurs in domestic animals and is distributed similarly to *T. congolense*.

Trypanosoma simiae Bruce et al., 1911 (syn. *T. porci*; *T. ignotum*; *T. rodhaini*). This is a polymorphic form resembling *T. congolense*. The majority are 16–24 μm, being long and stout with a distinct undulating membrane. About 7% of a population are long and slender with an inconspicuous undulating membrane, and a few are short with a poorly developed undulating membrane.

Its natural host is the warthog (*Phacophoerus*) and it is highly pathogenic for pigs and camels. It is distributed in tropical East and Central Africa. Transmission in nature is chiefly by *Glossina morsitans* and *G. brevipalpalis*, development occurring as in *T. congolense*. The *Glossina* flies probably introduce the parasite to domestic pigs but thereafter transmission is mechanical by *Stomoxys* and *Tabanus*. Severe outbreaks of disease may occur in the absence of *Glossina* but with an abundance of biting-flies.

T. simiae generally causes a very acute and fatal form of disease in pigs. A similar entity occurs in camels but only mild infections occur in sheep and goats. In horses, cattle and dogs, although infection does occur, there is no pathogenic effect. van Dijk et al. (1973) have studied the pathology of acute and chronic *T. simiae* infections in pigs. Active proliferation of the reticuloendothelial system occurred throughout the body, which in chronic infections led to proliferative encephalitis, myocarditis and pneumonia. Disseminated intravascular coagulation was a major factor in the pathogenesis of acute infections which, together with emboli of parasites, caused severe effects on the microcirculation resulting in degeneration and necrosis of the liver, spleen, adrenal glands and lymph nodes. Apparent disturbances in the clotting mechanism led to haemorrhages in the heart, lungs and meninges.

Diagnosis is based on the peracute clinical signs and the demonstration of organisms in the peripheral blood.

Subgenus: Pycnomonas Hoare, 1964

Trypanosoma suis Ochmann, 1905. This parasite was once thought to be *T. simiae*, but this is now recognized as a distinct species (Peel & Chardome 1954). It is monomorphic, 14–19 μm in length, stout and with a short free flagellum.

It occurs in Zaire, being transmitted by *G. brevipalpalis*, and cyclical development occurs in the mid-gut and the salivary glands. *T. suis* is pathogenic for pigs. It causes an acute disease with death in two months or less in young swine and a more chronic disease in older animals. Attempts by Peel and Chardome (1954) to transmit it to various other domestic and laboratory animals have failed.

Subgenus: Trypanozoon Lühe, 1906 (Brucei Group)

CYCLICALLY TRANSMITTED FORMS

Trypanosoma brucei Plimmer and Bradford, 1899 (syn. *T. pecaudi*). (Fig. 3.4) This is a polymorphic trypanosome, slender, intermediate and stumpy forms occurring. The undulating membrane is conspicuous in all forms; the kinetoplast is subterminal and the tail pointed. In laboratory animal infections, a fourth type may occur, the 'posterior nuclear' form, in which the nucleus is posterior in position.

The slender forms are 25–35 μm (mean 29 μm) in length, have a long free flagellum and a pointed posterior end. The intermediate forms average 23 μm, the posterior end is more blunt and a free flagellum is present. The stumpy forms average 15 μm, the posterior end is broad, the kinetoplast more terminal and the flagellum is lacking. Movement is snake-like or wriggling, the organism seldom moving out of the field.

All forms may occur in the blood at the same time but usually one form predominates, the peaks in length being significantly different. When the species is transmitted to laboratory animals, polymorphism disappears, as does the ability to undergo cyclical development.

Trypanosoma brucei is widely distributed in tropical Africa between the latitudes of 15° north and 25° south. Transmission is principally by *Glossina morsitans*, *G. tachinoides* and *G. palpalis*, but any of the *Glossina* species may be responsible if they are present. Development occurs in the anterior station, in the mid-gut and the salivary glands of the fly. Transmission may also be mechanical by biting-flies.

T. brucei is classified by Losos and Ikede (1972) as a humoral parasite, being found in the intercellular tissue fluids—such as of connective tissues—and fluids of the body cavity, as well as in the plasma. This has important implications for the pathogenicity of the parasite in various species.

The parasite is most severe in equines. Acute disease lasts two to four weeks, while chronic infection may be of several months' duration. Clinical signs include intermittent fever, rapid emaciation, ocular and nasal discharge, icterus and subcutaneous oedema (Losos & Ikede 1972). In chronic cases paralysis may develop. Reductions in packed cell volume, haemoglobin and red cells are ascribed to a depression of the haemopoietic tissue. (See also mechanisms of pathogenicity, above.) At autopsy, emaciation, anaemia, enlargement of lymph nodes and splenomegaly have been reported. Cerebral lesions, characterized by a diffuse meningoencephalitis with perivascular cuffing of lymphocytes and plasma cells, have been reported (Losos & Ikede 1972).

In cattle, the trypanosomes localize extravascularly and produce inflammatory reactions of the skin, subcutaneous tissue, heart, central nervous system and eye (Ikede & Losos 1972*a*). These authors consider that since natural infections with *T. brucei* are prevalent, often mixed with *T. congolense* and *T. vivax*, they may be more important as a cause of disease than has been thought. In experimental infections, parasitaemia may be detected only occasionally but, nevertheless, severe disease may be present.

In sheep, extensive localization of trypanosomes occurs in the extravascular fluids resulting in mononuclear inflammatory reactions in the heart, skeletal muscle, brain, eye, testis and other organs, and a generalized lymphoid hyperplasia. Clinical signs in experimentally infected animals include fever, anorexia, loss of weight, subcutaneous oedema, nervous disorders and moderate anaemia. No relationship was found between parasitaemia and the severity of disease (Ikede & Losos 1972*b*).

In dogs and cats *T. brucei* produces fever, emaciation, oedema of the eyelids and thorax; often the disease is acute. Ocular and central nervous manifestations may occur.

Of laboratory animals, a fatal fulminating infection occurs in mice, while in rabbits and guinea-pigs a more chronic, but eventually fatal, infection is produced. Losos and Ikede (1972) have presented a detailed account of rodent infections.

Acute infections may be diagnosed by the demonstration of organisms in the peripheral blood, but this is not completely reliable. In chronic infections organisms may be detected in

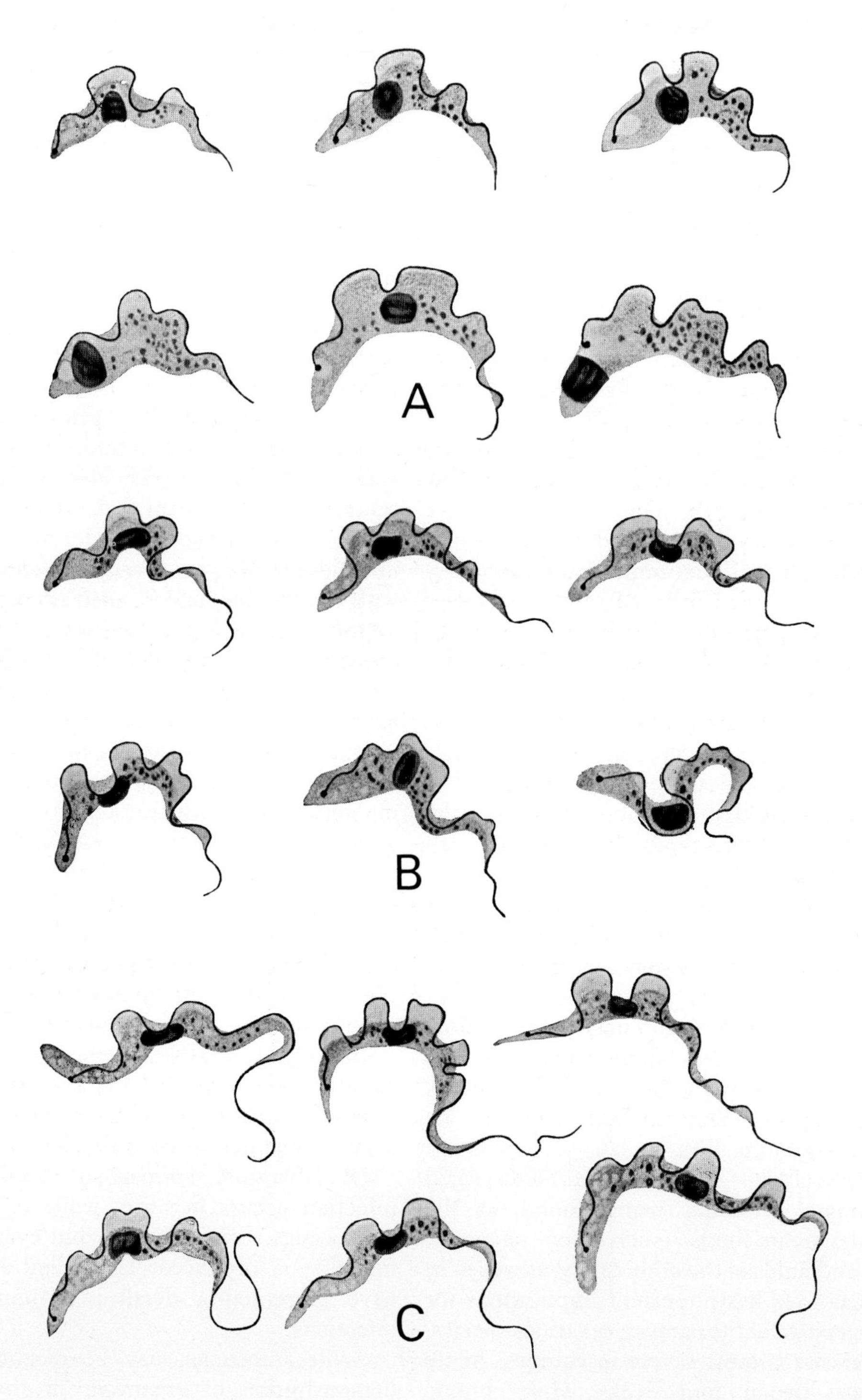

Fig 3.4 *Trypanosoma brucei.* (A) Broad stumpy form; two posterior nuclear forms are shown, one with kinetoplast behind the nucleus, and one with it in front. (B) Intermediate form with short flagellum. (C) Long slender form with flagella. (× 2000) (*From Wenyon 1926*)

lymph node, bone marrow or other organ smears. Mouse inoculation may be necessary in chronic infections.

Trypanosoma gambiense Dutton, 1902 (syn. *T. ugandense*; *T. hominis*; *T. nigeriense*). This is the cause of Gambian sleeping sickness, or trypanosomiasis, of humans. It is morphologically indistinguishable from *T. brucei* of animals, occurring in a polymorphic condition of slender, intermediate, and stumpy forms.

T. gambiense occurs on the west coast of Africa between the latitudes of 15° north and 18° south, with Lake Victoria as an eastern limit. The principal tsetse flies concerned in transmission are the riverine types *Glossina palpalis* and *G. tachinoides*. Development occurs in the anterior station, similar to *T. brucei*, and mechanical transmission due to biting-flies may also occur.

Trypanosoma gambiense infection in cattle, goats, sheep, horses, dogs and cats is discussed by Losos and Ikede (1972). It is generally considered that domestic animals are not readily infected with this species, and the disease produced by successful infections is less severe than that caused by either *T. brucei* or *T. rhodesiense*.

Gambian sleeping sickness is a chronic disease of man. Initially, there is invasion of the blood stream, then of the lymph nodes and finally of the central nervous system, with the production of leptomeningitis and perivascular infiltrations around blood vessels. Diagnosis is based on symptoms, habitat and demonstration of the parasites in the blood, lymph nodes and cerebrospinal fluid.

Trypanosoma rhodesiense Stephens and Fantham, 1910. This is the cause of Rhodesian trypanosomiasis or East African sleeping sickness of humans. The parasite is morphologically indistinguishable from *T. brucei* and *T. gambiense*, and some authors prefer to refer to the three as the *Brucei* subgroup. The organism is polymorphic, showing the three forms, similar to the other species. Geographically it is more restricted than *T. gambiense,* being found in Zambia, Zimbabwe, Tanzania, Botswana, in the southern Sudan and around the shores of Lake Victoria. The principal transmitting flies are the savannah tsetses, chiefly *G. morsitans* and, in some areas, *G. swynnertoni* and *G. pallidipes*. Development occurs in the anterior station, similar to *T. brucei*.

T. rhodesiense is pathogenic for the rat and other laboratory animals.

Trypanosoma rhodesiense is only mildly pathogenic for cattle; experimental infection of goats and sheep has been found to result in keratitis, encephalitis and death within two months of infection; in the horse emaciation, facial oedema and a slight parasitaemia was associated with experimental infection; while in the dog experimental infection produced a disease similar to that seen in man (Losos & Ikede 1972). The disease process produced by *T. gambiense* and *T. rhodesiense* in monkeys is presented in detail by Peruzzi (1928).

The clinical disease produced in man by *T. rhodesiense* is more acute than the Gambian form of the disease. The early stages are similar to the West African form but the climax is reached before any marked involvement of the central nervous system occurs. The marked sleeping sickness stage is not a strong feature of the infection. A fatal termination is common in untreated cases, which may be a matter of months only.

ZOONOTIC ASPECTS OF AFRICAN TRYPANOSOMIASIS

In recent years it has become clear that *T. rhodesiense* infection in man is a zoonosis. The parasite has been identified in wild and domestic animals and, whereas in the past outbreaks of Rhodesian sleeping sickness have been ascribed to the introduction by human carriers into a clean area, it is now clear that animals play an important role in this. Infection may persist for long periods in these species.

The situation with *T. gambiense* is less clear and there is still no firm evidence that infection with this species constitutes a zoonosis. A major problem in such considerations is the identification of the *brucei* subgroup trypanosomes isolated from animals and man. Some progress has been made in this by the usage of the 'blood incubation infectivity test' (Rickman & Robson

1970) which distinguishes between *T. brucei* and *T. rhodesiense*. In this test, organisms isolated from a rodent previously inoculated with the blood of an infected host, are incubated in human blood or saline and then subsequently inoculated into a rodent. If only saline-incubated organisms are viable, then the trypanosome is considered to be non-infective for man, while if both saline- and human blood-incubated organisms are viable, then the trypanosome is considered to be infective for man. A summary of the zoonotic aspects of African trypanosomiasis is given by Ristic and Smith (1974). Detailed considerations of Gambian sleeping sickness and of Rhodesian sleeping sickness are given by Scott (1970) and Apted (1970), respectively.

MECHANICALLY TRANSMITTED FORMS

Trypanosoma evansi (Steel, 1885) Balbiani, 1888 (syn. *T. annamense*; *T. berberum*; *T. cameli*; *T. hippicum*; *T. soudanense*; *T. venezuelense*). (See Fig. 3.5)

This was the first trypanosome shown to be pathogenic for mammals and was identified by Griffiths Evans, a British veterinarian.

In the majority of infections this parasite is monomorphic in character but polymorphism occurs sporadically. The typical form is indistinguishable from the slender form of *T. brucei*, being 15–34 μm in length (mean 24 μm), the kinetoplast is subterminal, the undulating membrane is well developed and there is a substantial free flagellum. Stumpy forms may appear sporadically. Forms which lack a kinetoplast may arise spontaneously, especially after drug treatment (Hoare 1954).

Originally, the distribution of *T. evansi* coincided with that of the camel. Now, it is widespread throughout the Indian subcontinent and the Far East, and it also occurs in the Near East, in North Africa north of latitude 15° north, and in the Philippines, and Central and South America. Transmission is by biting-flies such as *Tabanus, Stomoxys* and *Lyperosia*. No cyclical development occurs in these. An essential factor in

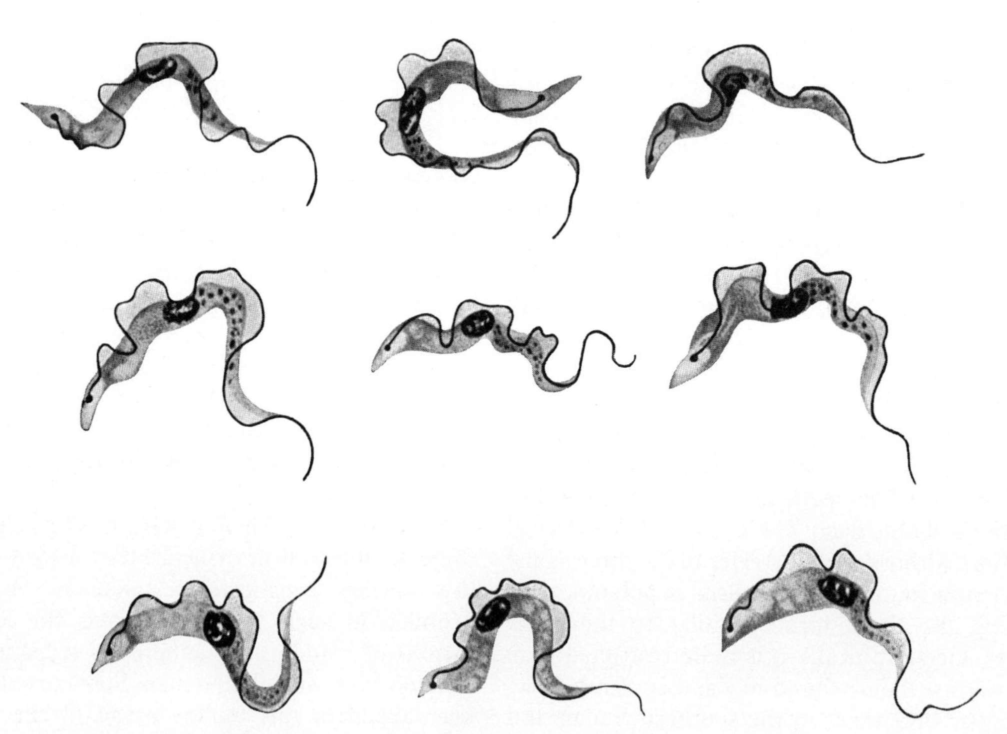

Fig 3.5 *Trypanosoma evansi* from blood of various animals. (×2000) (*From Wenyon 1926*)

mechanical transmission is interrupted feeding on the part of flies which go quickly from one host to the other in order to become replete. Trypanosomes do not survive for more than 10–15 minutes in the proboscis of a fly. In Central and South America the vampire bat may also act as a vector, although the bat may die of infection about a month after an infected blood meal. In the case of dogs, the infection can be transmitted by the ingestion of tissues from infected carcasses.

Trypanosoma evansi affects a wide range of hosts including horse, dog, camel, buffalo, elephant, pig, cat, tapir, capybara and (in Mauritius) deer. Laboratory rodents such as mice, rabbits, rats and guinea-pigs are readily infected. The clinical disease varies according to the strain of the parasite and the species of host. The most severe disease occurs in camels, horses and dogs (Curasson 1943).

The classical disease entity in the Indian subcontinent occurs in horses and is known as surra (a Hindi word meaning rotten). Surra is nearly always fatal to horses if treatment is not applied, death occurring in a few days to a few months depending on the virulence of the strain of organism. Emaciation and oedema are the most common clinical signs, the oedema varying from urticarial plaques on the neck and the flanks to oedema of the legs and lower parts of the body. The plaques may necrose in the centre and haemorrhages occur at the junction of the skin and mucous membranes, especially at the nostrils, eyes and anus. An intermittent fever may be present but in some cases the condition is so acute that this is not obvious. On post mortem there is a marked anaemia, emaciation, enlargement of lymph nodes and splenomegaly. Petechiae occur on the serous surfaces and in the parenchyma of the liver and kidneys.

Camels are highly susceptible to infection, the disease running an acute or chronic course, terminating fatally in untreated cases.

In camels in Asia the disease is chronic. The animals become progressively weaker and emaciated and the infection runs a course of about three years. In African camels the disease is less chronic and in the Sudan, where *T. evansi* is almost entirely restricted to the camel, the infection being known as gufar, deaths may occur within a few months of infection, but a few camels may die in two to three weeks. Clinical symptoms include fever, progressive emaciation and anaemia with oedema of the dependent parts of the body. Because of the differences between this form and that in Asia causing surra, the parasite was formerly referred to as *T. soudanense*. Occasionally, horses in the Sudan show a chronic disease which may end fatally, but donkeys show a mild fever which terminates in a spontaneous cure.

An acute and fatal type of disease is usual in dogs in the East, death possibly occurring in two to four weeks. Oedema is marked, corneal opacity is common and, due to oedema of the larynx, voice changes similar to those which occur in rabies may be noticed.

Cattle and water buffalo are considered to be the main reservoirs of the infection for equines. In these the infection is subclinical in nature; nevertheless, occasional outbreaks of acute disease occur with quick death. This is often associated with the introduction of the parasite or a new strain of it into a new area or to additional stress on animals, such as foot-and-mouth disease vaccination. In the elephant, surra follows a course similar to that seen in camels; animals become emaciated and show marked muscular weakness.

Diagnosis of acute infections is usually relatively easy since the organisms are readily demonstrable in freshly stained blood smears. However, the diagnosis of chronic infections may present difficulty because parasitaemia is usually low. In such cases thick and thin blood smears taken on consecutive days, or lymph node puncture smears, should be attempted. Various indirect tests have been used including the complement fixation test which, although giving a group reaction, is of value when no other trypanosomes occur in the area. Other indirect diagnostic tests depend on disease-induced alteration of serum proteins. A mercuric chloride test consists of adding one drop of serum to 1 ml of a 1 in 30 000 solution of mercuric chloride in distilled water, the tube being shaken gently. A white opalescence, which appears in a few seconds, indicates infection of one month or more standing. The stilbamide test (in India) consists of adding one drop of serum to a 0.3% solution of stilbamide in distilled water. A positive reaction consists of an opalescence of precipitate after one to two minutes. Such tests are to a great extent non-specific and undoubtedly animal inoculation, using rats or guinea-pigs, is more satisfactory.

The South American form of *T. evansi* infection produces similar disease entities as described above. In horses the infection is referred to as murrina (Panama) or derrengadera (Venezuela). A chronic disease associated with high mortality occurs in dogs, untreated animals dying one or two months after infection (Gomez 1956). The capybara, although normally a reservoir host for the horse, may suffer an acute fatal infection similar to that in the horse.

Trypanosoma equinum Voges, 1901. This is a large monomorphic form, 22–24 μm in length, differing from *T. evansi* only in lacking a kinetoplast. The axoneme of the flagellum arises from a small blepharoplast. It occurs in Central and South America, the enzootic areas being Argentina, Bolivia and Paraguay. Transmission is mechanical by biting-flies.

Trypanosoma equinum is chiefly an infection of equines, the horse being most seriously affected, and the clinical condition is referred to as mal de caderas. Mules and donkeys are less susceptible while dogs, cattle, sheep and goats, in that order, can be affected with a mild disease. The capybara (*Hydrochoerus capybara*) is susceptible to infection and may serve as a reservoir host for the equine infection. The parasite is responsible for high mortality of horses in wet areas, particularly swamps, where the incidence of blood-sucking flies is high.

The disease in horses is rarely acute but in such cases death occurs a few weeks after the onset of clinical signs. More usually it runs a chronic course with death occurring two to six months after infection. The incubation period is four to ten days, after which pyrexia and parasitaemia appear. Emaciation commences early in the disease and a marked weakness of the hind-quarters (mal de caderas) results in a staggering gait. This is progressive and the animal finally becomes recumbent. Associated lesions are conjunctivitis, keratitis and oedema of the eyelids. Transient cutaneous plaques occur over the neck and the flanks; these lose their hair and later scab over. On post mortem there is splenomegaly, enlargement of lymph nodes and anaemia. The kidneys show petechial haemorrhages, ascites may be present and there is often an oedematous infiltration in the spinal canal.

Diagnosis in the acute stage of the disease is based on the demonstration of organisms in the peripheral blood. In more advanced infections organisms are difficult to find and inoculation of blood into laboratory rodents (e.g. mice) is the most satisfactory means of diagnosis.

Trypanosoma equiperdum Doflein, 1901. This species is morphologically identical to *T. evansi*. It causes a venereal disease of horses, dourine (Arabic for unclean), that, at present, occurs in North and South Africa, Central and South America, the Middle East and Asiatic Russia. The infection was once widespread in the United States but this country has now been free from the infection for several years. The infection still occurs in Mexico. *Trypanosoma equiperdum* is ordinarily transmitted mechanically by coitus; rarely it is transmitted by biting-flies and by infective discharges contaminating mucous membranes.

The disease is naturally found in horses and donkeys. The jackass may be a symptomless carrier and is therefore especially dangerous. Of the other species, dogs may be infected with some strains of the parasite; cattle are generally resistant to infection, but laboratory mice, rats and rabbits are susceptible, although a second passage may be necessary to establish satisfactory infections. The organism varies in its virulence; some infections may never become clinical whereas others run a more definite clinical course, albeit chronic in character.

The clinical entity of dourine usually progresses through three distinct phases following an incubation period of 2–12 weeks or several months. In some latent cases the disease entity may only be precipitated by other severe disorders. The first phase, the stage of oedema, is initiated by a mucoid vaginal or urethral discharge, a degree of nymphomania and a mild fever with oedema of the genitalia. In the stallion the prepuce and scrotum are swollen and the oedema may extend under the belly as far forward as the chest. In the mare the vaginal mucosa is hyperaemic and ulcers may be present. There is deep pigmentation of circumscribed areas of the vulva and the penis. This phase lasts four to six weeks and in mild cases may go unnoticed, but in severe forms frequent micturition and even abortion in pregnant mares may be evident.

The second phase, the urticarial phase, is characterized by the appearance of oedematous plaques under the skin, especially of the flanks but any part of the body may be affected. The plaques are circular, sharply circumscribed and are 2–10 cm in diameter. They are classically referred to as 'dollar spots' since they appear as if a silver dollar had been inserted under the skin. They may persist for a few hours or three to four days, then disappear, but may appear again. Although they are not an invariable consequence of *T. equiperdum* infection, their presence is almost pathognomonic for the disease and the affected areas may be left depigmented when the swelling has resolved, giving good evidence of their previous presence.

The third phase of infection is that of paralysis. The muscles of the face and nostrils are usually affected first but later paralysis of the muscles is associated with complete paralysis and recumbency, which is followed by death. Mortality varies from 50 to 70%.

At post mortem the carcass is emaciated with marked muscular atrophy, there is oedematous infiltration of the perineal tissues and the abdominal wall and ulcers may occur on the body where the horse has lain. Serous infiltration is present along the large nerve trunks supplying the hind-limbs, and histological examination shows cell infiltrations, oedema and degeneration of these and also the posterior spinal cord.

The clinical disease is typical enough to allow diagnosis in an endemic area. Trypanosomes are not readily detected in the blood but they may be found in the vaginal or preputial discharges or in the serous fluid squeezed from the urticarial plaques. Material may also be inoculated into mice, rats, rabbits, etc., but a second passage may be necessary in order to demonstrate the parasites. The complement fixation test (Watson 1915) is useful in both clinical and latent cases and was used with marked success in the United States' eradication programme. Serum antibodies develop three to four weeks after infection and although a few non-specific reactions may occur, the number of false negatives is small.

Section: Stercoraria (Posterior Station Forms)

Subgenus: Megatrypanum Hoare, 1964

Trypanosoma theileri Laveran, 1902 (syn. *T. americanum*). (See Fig. 3.6) This is a relatively large species, 60–70 μm in length, but forms up to 120 μm in length may be found, especially in chronic infections. *Trypanosoma theileri*-like forms have been reported from animals of the order Artiodactyla (see Wells 1972). The posterior end is long and pointed and the kinetoplast lies some distance from the posterior end. The undulating membrane is well developed and the free flagellum is well defined.

Present evidence indicates that *T. theileri* is transmitted cyclically in the posterior station by tabanid flies such as *Tabanus* and *Haematopota*, infection of the bovine being by contamination. In the bovine both trypomastigote and epimastigote forms occur and multiplication takes place chiefly in the lymph nodes and inner organs. *T. theileri* is readily cultured in a variety of culture media. In those incubated at 27°C only epimastigote forms occur, but in cultures incubated at 37°C both epimastigote and trypomastigote stages develop (Splitter & Soulsby 1967).

Under ordinary circumstances *T. theileri* is a non-pathogenic form and probably occurs in cattle throughout the world. Occasionally, however, it may appear in a parasitaemic form following splenectomy, when stress conditions arise or when concurrent infections with pathogenic agents are present (Wells 1972). For example, it has been incriminated as a cause of death in cattle following Rinderpest vaccination. In some animals the infection was peracute in nature producing an anthrax-like disease. It has also been associated with 'turning sickness' in Uganda (Carmichael

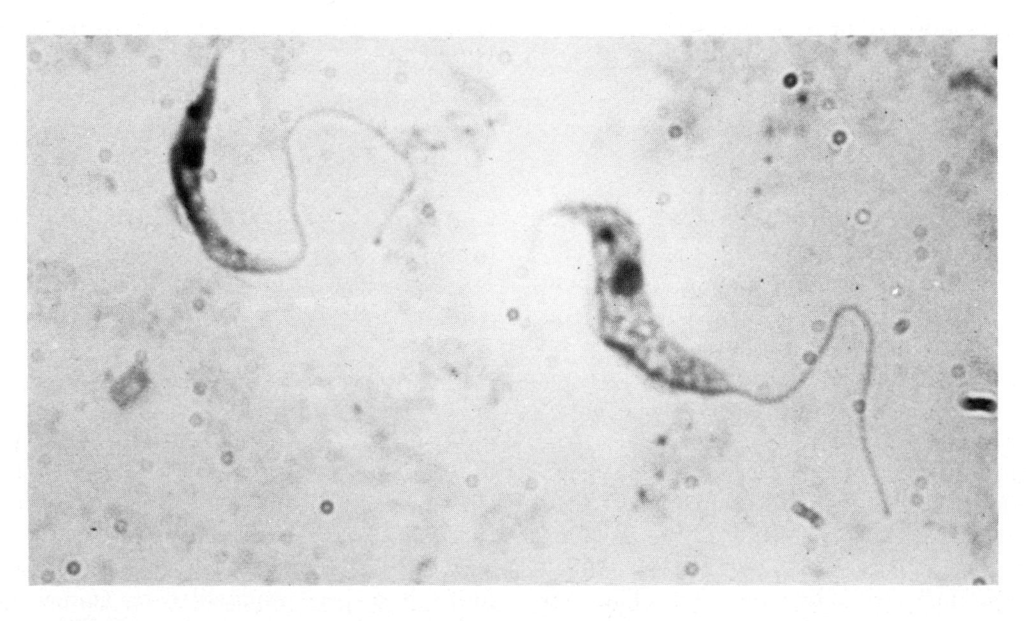

Fig 3.6 *Trypanosoma theileri* of cattle. Giemsa. × 1000 (*From Townsend*)

1939) and with depressed milk production and abortion in cattle (see Wells 1972).

Trypanosoma melophagium (Flu, 1908). This is a non-pathogenic form infecting sheep. It is 50–60 μm in length and generally resembles *T. theileri*. It is transmitted cyclically in the posterior station 'by the sheep ked, *Melophagus ovinus*. Infection is by contamination of the skin and, also, if a ked is eaten the metacyclic trypanosomes may penetrate the buccal mucosa.

The organism is widespread. Usually it is demonstrable only by culture techniques but occasionally it may be demonstrated microscopically in the peripheral blood.

Trypanosoma theodori Hoare, 1931. This is similar to *T. melophagium* and may be a synonym of this species. It is non-pathogenic and occurs in goats in Israel. It is transmitted by the hippoboscid fly *Lipoptena caprina*.

Subgenus: Herpetosoma Doflein, 1901

Members of this subgenus are of little medical or veterinary importance. Some of the species have provided good models for research, especially on the immunological aspect of host–parasite relationships.

Trypanosome lewisi (Kent, 1880) Laveran and Mesnil, 1901. This occurs in rats and is transmitted by the rat flea *Ceratophyllus fasciatus*. It is 26–34 μm in length, has a pointed posterior end and the kinetoplast is some distance from the posterior end. Normally this trypanosome is non-pathogenic to rats but it may cause death in nursling rats.

Trypanosoma nabiasi Railliet, 1895. This occurs in rabbits in Europe and is transmitted by the rabbit flea, *Spilopsylla cuniculi*. It is non-pathogenic.

Trypanosoma rangeli Tejera, 1920 (syn. *T. ariarii*; *T. guatemalense*). This species was first described from the epimastigote stages in the reduviid *Rhodnius prolixus*. Later, it was shown to occur in the blood of man, dog, cat, opossum and monkey and is now recognized as commonly occurring in man, dog and cat, frequently in mixed infections with *T. cruzi*, in South America. The blood forms measure 26–36 μm; the nucleus is anterior to the middle of the body and the kinetoplast is small and subterminal. The undulating membrane is relatively broad and rippled and the free flagellum is short. Reproduction is by longitudinal binary fission.

Development occurs in triatomid bugs, *R. prolixus* being the main vector. Others, such as *Panstrongylus geniculatus, Triatoma dimidiata* and *T. nitida*, have been incriminated. Development occurs in the fore- and hind-gut, a giant stage being produced. Metacyclic trypanosomes may migrate to the salivary glands and infection is either by inoculation or by contamination.

The organism does not appear to be pathogenic for the mammalian host and its principal importance is the differentiation of it from *T. cruzi*.

Trypanosoma species of various primates include *T. minasense* Chagas, 1909 of marmosets and various other monkeys of Central and South America, *T. saimirii* Rodhain, 1941 of squirrel monkeys in Brazil, *T. diasi* Dean and Martins, 1952 of capuchins in Brazil and *T. primatum* Reichenow, 1928 of chimpanzees and gorillas in West and Central Africa.

Subgenus: Schizotrypanum Chagas, 1909 emend. Nöller, 1931

Trypanosoma cruzi Chagas, 1909 (syn. *Schizotrypanum cruzi*; *T. escomeli*). This is the main pathogenic species of the Stercoraria. It is the cause of American human trypanosomiasis or Chagas disease in South America. In the blood it is monomorphic, 16–20 μm in length, and crescent-shaped with a pointed posterior end. The kinetoplast is large and subterminal, and fills the body of the trypanosome at that point. The nucleus is midway along the body and there is a moderately well-developed undulating membrane and free flagellum. No division takes place in the trypomastigote stage and division occurs in the amastigote form. The dividing forms appear as bodies 1.5–4 μm in diameter in muscle and other cells, especially the heart muscle cells.

The geographic distribution of human infection with *T. cruzi* is chiefly South America extending from Argentina northwards, and it also occurs in Central America. A wide variety of animals may be infected and serve as reservoir hosts. Thus dogs, cats, pigs, foxes, ferrets, squirrels, opossums and monkeys may, among others, be hosts and the dog is a frequent host in South America, up to 35% or more being infected (Neghme 1940). Animal infection has been detected in the United States and various reports indicate a substantial prevalence. Thus, the infection has been detected in opossums, raccoons, skunks and foxes in the southwestern states, the southeastern states and as far north as Maryland. Indeed, 14 species of mammals have been found infected in the USA (Levine 1973), including fatal infections in dogs in the southwest United States (Williams et al. 1977).

Trypanosoma cruzi is of greatest importance in man, chiefly affecting young children and infants. The principal reservoir hosts in South America are the armadillo and the opossum; in Central America the opossum acts as such and in the USA woodrats and raccoons may be important in this respect. The close association of man and dog makes the latter a likely source also of human infection. The knowledge that *T. cruzi* is much more widespread in the animal population of the United States than hitherto thought, has posed the question of the importance of this zoonosis to human health in that country. Two autochthonous human cases have been diagnosed in the USA, one in a child in Corpus Christi, Texas (Woody & Woody 1955), and one in a child in Bryan, Texas (Goble 1958).

Trypanosoma sanmartini Garnham and Gonzalez-Mugaburu, 1962 is described from the squirrel monkey in Colombia; it resembles *T. cruzi* and is probably an aberrant strain or a subspecies of *T. cruzi*.

Under natural conditions *T. cruzi* is transmitted by blood-sucking bugs of the family Reduviidae, development occurring in the posterior station. Metacyclic trypanosomes are passed in the faeces of the infected bug 8–10 or more days after initial infection. The human, or mammalian host, is infected when the metacyclic trypanosomes are rubbed into the wound made by the insect, into other skin abrasions or through the mucous membranes. Reduviids (kissing bugs) usually defaecate after feeding and they commonly feed on the thin skin near the eyes and lips. The infected material is therefore easily rubbed into the mucous membranes and the wounds. Animals may be infected by licking the faeces of the bugs or by ingesting infective bugs. Other modes of transmission in man include blood transfusion and transplacental, from mother to fetus. In a large proportion of congenital infections, death of the fetus is a common sequel (Ristic & Smith 1974). Transmission by contact with infected urine has been suggested for opossums.

A large number of species of reduviid bugs have been found naturally infected and several more have been infected experimentally. The most important vectors in South America are *Panstrongylus megistus, Eutriatoma sordida, Panstrongylus geniculatus, Triatoma infestans* and *Rhodnius prolixus*. In the United States at least 15 species have been shown to be infected, the two principal species being *T. protracta* and *T. sanguisuga* (Faust 1949). In view of the high incidence of infection in various species of reduviid bugs, the possibility of inter-triatomid infection has been studied by several investigators (Ristic & Smith 1974).

Following infection of a wound with metacyclic trypanosomes, the organisms enter histiocytes and proliferate in the amastigote form at the local site. There is a local inflammatory response and later encapsulation by fibrous tissue, the whole blocking the local lymphatics and producing oedema of the local area. This is the primary lesion or 'chagoma'. Amastigotes pass from the primary site to local lymph nodes and then by the lymphatic system to the whole body. The liver, lungs, spleen, bone marrow, cardiac muscle and brain cortex are affected. Here, the organisms multiply as amastigotes. Large-scale rupture of host cells releases trypomastigote forms into the blood and this is associated with fever.

The disease in man may be acute or chronic in character, the former usually occurring in infants and young children. Death may occur in 2–4 weeks after the onset of symptoms in acute cases or they may resolve to the chronic form of the infection. The chronic disease is the usual form found in adults. The clinical manifestations depend on the location of the organisms but the cardiac form is common. Megacolon and mega-oesophagus may occur and some workers believe that a toxin is produced by the parasite which affects the nerve ganglia in the area of parasitized tissue. A detailed account of the pathogenesis of Chagas disease is given by Köberle (1974). The pathological changes consist of massive destruction of the reticuloendothelial and muscle cells so that almost every organ of the body may be affected. The amastigote forms are found in cyst-like nests in host cells which ultimately rupture; other cells are then infected and the process continues.

In dogs debility, anaemia and splenomegaly occur, young animals being especially susceptible. Myocardial involvement also occurs in dogs but it is not as extensive as in man. In cats, convulsions and posterior paralysis may be seen. The disease may also be produced in mice and guinea-pigs but the reservoir hosts for the parasite appear to suffer little from the infection.

Diagnosis of the acute stage is based on the demonstration of the trypomastigote stage in thick blood films. However, as the disease progresses the organisms become infrequent in the blood and are present in detectable numbers only during bouts of fever. In such cases blood may be injected into puppies, kittens or guinea-pigs, or spleen or lymph gland biopsies made and the amastigote forms sought for. Xenodiagnosis, that is the feeding of triatomid bugs on the suspect or allowing bugs to feed on the patient's blood through a membrane, may be a valuable technique. Laboratory-reared and *T. cruzi*-free triatomes are used and if the suspected material is positive, metacyclic trypanosomes are found 7–10 days later in the droppings of the bug (Pifano 1954).

A variety of serodiagnostic tests have been developed for Chagas disease. These include the complement fixation (CF) test, an indirect fluorescent antibody (IFA) test and an indirect haemagglutination test (IHA). All three show a high degree of sensitivity and specificity and some investigators recommend that all three should be used to ensure accurate diagnosis. In the majority of acute cases of Chagas disease IFA becomes positive about one month after infection; by two months the CF test is positive. All three tests are positive by the fifth month of infection. A direct agglutination test (DAT), introduced by Vattuone and Yanovsky (1971), utilizes a suspension of trypsinized, formalin-fixed epimastigotes and the test is performed like a haemagglutination test.

The DAT is a very sensitive test for the detection of acute Chagas disease.

Immunity in Chagas disease. There is good evidence that man and other animals develop acquired resistance to acute infections of *T. cruzi.* Both antibody- and cell-mediated immunity mechanisms are concerned in the immune response. The chronic cardiovascular manifestations of Chagas disease are most likely to have a major autoimmune component and an endocardial–vascular–interstitial (EVI) factor, an immunoglobulin which reacts with endocardium and vascular structures, is evidence for this.

Artificial immunization has been attempted by numerous workers. Epimastigotes, killed by various chemical or mechanical procedures have, in general, induced a low degree of immunity to acute infection. However, Neal and Johnson (1977) showed that immunization of mice with killed epimastigotes or trypomastigotes, using saponin as an adjuvant, resulted in an effective protection against homologous and heterologous challenge. Attenuation of culture forms (e.g. γ-irradiation) or by culture with inhibitors of cell reproduction which, however, retain protective immunizing capacity, are recent developments in this field (Hungerer et al. 1976).

TRYPANOSOMES OF BIRDS

Several species of trypanosomes occur in avian hosts but, as far as is known, they are all non-pathogenic. *Trypanosoma avium*, Danilewsky, 1885, occurs in a wide range of birds in Europe and in Canada Geese in North America. *Trypanosoma calmetti*, Mathis and Leger, 1909, occurs in ducklings in South East Asia and *Trypanosoma gallinarum*, Bruce et al., 1911, was reported from fowls in Uganda.

TREATMENT OF TRYPANOSOMIASIS

Reviews of the chemotherapy of trypanosomiasis are given by Goodwin and Rollo (1955), Goodwin (1964) and Whiteside (1962), and of the structure and activity of trypanocidal drugs by Barber and Berg (1962). It is intended here to give a brief account only of the major compounds which are in use at the present time.

Essentially, the trypanocidal drugs available are curative or prophylactic. The latter, in some cases, may give protection against infection for 4–6 months.

Prior to 1938 and the discovery of the phenanthridinium compounds, the only satisfactory drug available was tartar emetic. This is useful in the treatment of *T. congolense* and *T. vivax* in cattle and *T. evansi* in the camel, but it is of little value for *T. brucei* in horses. It is given intravenously in 1 g doses dissolved in 20–35 ml of saline or water and is repeated 6–8 times at weekly intervals. Necrosis may occur if the drug leaks from the vein and it must be used freshly prepared to avoid toxicity. Other compounds of this vintage are Antimosan (40–50 ml of 6.3% solution, weekly for five weeks in cattle); Fuadin (Stibophen) and Surfen C (Congasin). The latter causes marked reactions following intramuscular injection. The compounds in common use for animal trypanosomiasis belong to four chemical groups and are summarized as follows (see also Table 3.2):

Sulphonated naphthylamine—Suramin.
Phenanthridine—homidium bromide or chloride, *iso*-metamidium chloride, pyrithidium bromide.
Diamidine—diminazene aceturate.
6-Aminoquinaldine—quinapyramine di-methyl-sulphate.

Suramin (syn. Antrypol; Bayer 205; Naganol; etc.). This was used for many years as the drug of choice for *T. brucei* and it is similarly useful for *T. evansi, T. equinum* and *T. equiperdum*. In the horse a single dose of 4 g/45 kg body weight is given intravenously, or it can be divided into three parts given over a period of three weeks. For dogs 0.3 g intravenously is repeated for six days. Suramin is the drug of choice for *T. evansi* in camels in which it is well tolerated. It is given intravenously at the rate of 1–2 g/100 kg body weight. In the event of relapse strains of trypanosomes 2 g of quinapyramine dimethyl sulphate/100 kg should be given.

In the phenanthridine series of compounds phenidium chloride, although active against *T. vivax* and *T. congolense*, had the disadvantage of causing muscle necrosis. It was superseded by dimidium bromide and later by homidium bromide. *Homidium bromide* (Ethidium) and *homidium chloride* (Novidium) are given as 1–2.5% solutions at the rate of 1.0 mg/kg by intramuscular injection. They are very effective against *T. vivax* and *T. congolense*, and a single injection will usually sterilize these infections in cattle. *T. brucei* in cattle is less susceptible. The drugs are of value for *T. congolense* infections in horses and dogs. Temporary toxic effects may occur in cattle for up to 30 minutes, but with horses severe symptoms may occur and division of the dose is advisable. The protective effect of homidium bromide may last for only about five weeks (Whiteside 1962). Trypanosomes resistant to homidium should be treated with diminazene or *iso*-metamidium.

Pyrithidium bromide (3-amino-8-(2-amino-6-methyl-4-pyrimidylamino)-6-*p*-aminophenyl-phenanthridine 5,1′ dimethobromide) (Prothidium) (RD 2801). This compound, introduced in 1956, has been widely used in East Africa and is given as a 4% solution at the rate of 2–4 ml (2 mg)/kg body weight, subcutaneously or by deep intramuscular injection. It is active mainly against *T. vivax* and *T. congolense* and, to some extent, against *T. brucei*. Under conditions of low tsetse fly density it will give protection for up to 300 days. Where the prevalence of flies is high, the period of protection is much shorter (Whiteside 1962); however, frequent treatments enable animals to withstand severe challenge. The major disadvantage of the continued use of Prothidium is the ease with which trypanosomes become resistant to it. Furthermore, strains resistant to Prothidium also have an increased resistance to the other phenanthridinium compounds. Prothidium is well tolerated by bovines, ovines and canines, but local reactions may occur with bovines and equines. Strains of organisms resistant to prothidium should be treated with diminazene or *iso*-metamidium.

Iso-metamidium chloride (Samorin, Trypamidium) is given at the rate of 0.5–1 mg/kg body weight by deep intramuscular injection for infections caused by *T. vivax*, *T. congolense* and *T. brucei* in bovines, ovines, caprines, equines and canines; it is well tolerated by all except that local reactions may occur at the point of injection. In pigs with *T. simiae* infection, heavy doses (12.5–35 mg/kg) can be used as a curative treatment. *Iso*-metamidium will afford protection for three to six months. Trypanosomes resistant to this compound should be treated with diminazene.

Diminazene aceturate (N-1,3-diamidino azo-aminobenzene diaceturate tetrahydrate) (Diamidine, Berenil). This compound, given at a dose of 3.5 mg/kg by subcutaneous or intramuscular injection, is very active against *T. congolense* and *T. vivax*, and less active against *T. brucei* and *T. evansi* in bovines, ovines and caprines in which it is well tolerated; it is less well tolerated in equines, with a possibility of local reactions, while in camels and dogs general reactions may occur. Berenil has no prophylactic effect, being rapidly excreted from the body, but it is particularly useful as a 'sanative' drug in that it cures infections which are resistant to other drugs (Whiteside 1962). Resistance to Berenil was previously ascribed to cross-resistance with quinapyramine (Antrycide) but, more recently, resistance to Berenil has been demonstrated in several countries of Africa, particularly by *T. vivax* and also by *T. congolense*. The resistance can be overcome by increasing the dose of Berenil or by the use of *iso*-metamidium (Samorin).

Quinapyramine (4-amino-6-(2-amino-6-methylpyrimidinyl-4-amino)-2-methylquinoline) (Antrycide). Three forms of this compound are in use at present, namely, Antrycide methylsulphate, Antrycide chloride and Antrycide pro-salt (a mixture of the methylsulphate and chloride).

Antrycide methylsulphate (quinapyramine dimethylsulphate) reaches therapeutic levels quickly. It is given as a 10% solution in water, subcutaneously at a rate of 5.0 mg/kg. It is very active against *T. congolense, T. vivax, T. brucei* and *T. evansi*, and is well tolerated by bovines, ovines, caprines and camels. Local and sytemic reactions may occur in horses and the total dose should be divided into two or three parts and given at six-hourly intervals. Local and systemic reactions may also occur in dogs and treatment should be divided as in horses. In pigs infected with *T. simiae*, where the course of the disease is very rapid, repeated treatments with Antrycide methylsulphate may avoid a fatal outcome. However, *T. simiae* infections can be controlled by drug prophylaxis by the use of an Antrycide methylsulphate–Suramin complex or Antrycide chloride (quinapyramine chloride). The former is given at the rate of 40 mg/kg, the latter at 50 mg/kg at three-monthly intervals until the pigs are over nine months of age, and then treatment is reduced to six-monthly intervals (Stephen 1966). Trypanosomes resistant to Antrycide methylsulphate can be controlled by *iso*-metamidium.

Quinapyramine chloride (Antrycide chloride) is slowly absorbed and has been used as a prophylactic, especially for cattle. It can be used for the protection of pigs against *T. simiae* (see above). Usually it is combined with Antrycide methylsulphate (3 parts methyl sulphate: 2 parts chloride) as Antrycide pro-salt. The benefit of the pro-salt is that sufficient methylsulphate is given to serve as a curative dose while depots of the chloride are formed in the subcutaneous tissues from which the drug is slowly absorbed. A dose of 7.4 mg/kg of a solution of 3.5 g in 15 ml of water is given subcutaneously and has good activity against *T. brucei* and *T. evansi*, with protection lasting up to two to three months. It is well tolerated in equines, bovines and camels, although local reactions may occur in horses. Suramin is recommended for the treatment of drug-resistant organisms.

A quinapyramine–Suramin complex (quinapyramine sulphate, 10 g; anhydrous Suramin, 8.9 g; water 200 ml) has been used as a prophylactic for *T. evansi* in horses and will confer protection lasting for six to ten months (Gill & Malhotra 1971). Serious local reactions may occur but these gradually disappear over eight weeks. This complex may also be used for prophylaxis of *T. simiae* infection of pigs (see above). Organisms developing resistance to this complex in pigs can be treated with *iso*-metamidium (12.5–35 mg/kg) and in horses and camels with either *iso*-metamidium or diminazene.

A summary of the indications for chemotherapeutic drugs for trypanosomiasis is given in Table 3.2.

Drug Resistance in Trypanosomes

This has been a continuing problem in the treatment of trypanosomiasis and it is particularly common with *T. congolense*. Bishop (1959, 1962) has discussed the biological aspects of drug resistance while Whiteside (1962) has considered

Table 3.2. **Summary of More Active Chemotherapeutic Drugs for Trypanosomiasis**

	T. congolense			*T. vivax*		*T. brucei*		*T. simiae*	*T. evansi*		
	Ruminants	*Equines*	*Dogs*	*Ruminants*	*Equines*	*Equines*	*Dogs*	*Pigs*	*Equines*	*Camels*	*Treatment of Relapses*
Curatives:											
Homidium	+	+	+	+	+						Diminazene *Iso*-metamidium
Diminazene	++	+		++	+						*Iso*-metamidium
Quinapyramine sulphate			+			++	+		++	++	*Iso*-metamidium
Iso-metamidium	++	++		++	++			+			Diminazene
Suramin									+	+	Quinapyramine
Preventives:											
Iso-metamidium	++	++		++	++		+				Diminazene
Pyrithidium	++	++		++	++						Diminazene *Iso*-metamidium
Quinapyramine sulphate-chloride						++			++	++	Suramin
Suramin + quinapyramine								++	++	++	*Iso*-metamidium (pigs) Diminazine

+ = Active; + + = recommended use.

it from the field aspect. Resistance is developed more readily to prophylactic drugs. Curative drugs are usually rapidly eliminated and the risks of resistance developing after their use are not great, unless treatments have to be repeated frequently, as in areas where the incidence of infection is high. Drug resistance develops in the treated host and naturally occurring drug-resistant trypanosomes have not yet been observed. Resistant strains of trypanosomes can be transmitted by tsetse flies but these disappear after 6–12 months, provided that the cattle source of the resistant organisms is removed.

Resistance can be developed with relative ease by subcurative doses of drugs, which allow a relapse. Usually three to six such treatments are required for this (Whiteside 1962) and, thereafter, the organisms are 40–80 times more resistant to the normal curative doses. Such strains can develop in an individual animal and resistance is retained for at least one year when the infection is syringe-passaged to other cattle. Although the essential factor in the development of drug resistance is underdosing, this can inadvertently arise in several ways, especially with the prophylactic drugs. Thus, where the incidence of infected flies is high, subcurative levels of a drug may exist towards the end of the protection period; animals may be exposed to infection for too long after a single prophylactic treatment, and a further aspect is that a seasonal increase in prevalence of trypanosomiasis may produce a shortening of the protection period which, in turn, leads to a subcurative level of the drug.

Cross-resistance to drugs

With the development of resistance to one compound, trypanosomes may show resistance to compounds of the same series and also to those of other series. Thus, in the phenanthridinium series, resistance to pyrithidium bromide leads to resistance to *iso*-metamidium and homidium. There is also cross-resistance between quinapyramine and the phenanthridinium series. In some cases the resistance developed is not great, and large, but tolerated, doses of other drugs will cure the infection (e.g. *iso*-metamidium). The stage may be reached, however, where the level of resistance exceeds the maximum tolerated dose of other drugs (Whiteside 1962). The degree of cross-resistance depends on the degree of direct resistance developed. A strain of *T. congolense*, resistant to four times the minimal curative dose of quinapyramine, was susceptible to the normal doses of homidium choride and diminazene aceturate; two further exposures to normal doses of quinapyramine resulted in resistance to homidium but not to diminazene; another two exposures to quinapyramine resulted in resistance to diminazene (Whiteside 1962).

Previously, resistance of trypanosomes to diminazene aceturate was ascribed as a consequence of a cross-resistance with quinapyramine. However, it is now recognized that direct resistance to diminazene does occur, particularly with *T. vivax* and *T. congolense*.

Drug resistance poses a potential threat to control measures, but there is evidence that trypanosomes appearing in relapse infections are often of low pathogenicity and may not be readily infective to tsetse flies (Stephen 1962).

No significant differences in metabolism have been found between normal and resistant strains of trypanosomes and Bishop (1962) has suggested that drug resistance develops by mutation. She quotes work in which DNA from resistant strains will transfer resistance to normal trypanosomes. A relationship between drug resistance and antibody resistance has been observed by Soltys (1959). A strain of 'antibody-resistant' *T. brucei* was less sensitive to suramin and quinapyramine than one which had not been exposed to antibody.

The control of drug resistance in the field is discussed by Whiteside (1962). In a curative programme homidium and diminazene are used alternately, since neither produces cross-resistance to the other. Homidium is used over a wide area until evidence of resistance appears; it is then completely withdrawn and diminazene is used instead for at least one year. This drug cures infections whether or not they are resistant to homidium, and it also cures animals reinfected with homidium-resistant forms. After one year the homidium treatment is recommenced.

With prophylactic drugs the situation is more difficult. The two commonly used drugs, quinapyramine and pyrithidium, induce reciprocal cross-resistance. Development of resistance is indicated by a reduction in the length of protection and, when this occurs, diminazene or *iso*-metamidium is introduced and is continued until such time as the resistant strain dies out. Prolonged treatment may be necessary when the resistant strains are introduced into *Glossina*.

The combined use of *iso*-metamidium and diminazene, which do not cause cross-resistance, has been adopted. In Cameroon a large-scale campaign, begun in 1967 (Maikano et al. 1969), has resulted in a decrease in the disease (Eyidi 1971).

CONTROL OF TRYPANOSOMIASIS

Under field conditions, the control of trypanosomiasis is chiefly dependent on chemotherapy. However, many factors militate against complete control by chemotherapy, the development of drug-resistant strains being a major one. The control of *Glossina* offers a much more long-term control measure and this is discussed on p. 417.

The destruction of game, which serves two purposes, that of depriving *Glossina* of its food and of reducing the reservoir of trypanosomes, is a control measure which has aroused much controversy and antagonism. On the one hand, people may rightly fear the demise of a majestic and priceless inheritance, while others fear that in the absence of the large game animals, tsetse flies will feed on more cattle. This subject has been masterfully and thoroughly discussed by Buxton (1955) and Ford (1970). In Zimbabwe, game destruction has resulted in the disappearance of *G. morsitans*, but other species of tsetse, which feed on a variety of small and large game, are little affected by game destruction. In some areas, such as West Africa, game destruction would be pointless since these animals form an unimportant source of the tsetse food supply. Controlled elimination of game by trained hunters may be of value in certain areas to create zones between game areas and those for agricultural development.

The breeding of trypanosome-resistant (trypanotolerant) strains of cattle has been considered on many occasions (Mulligan 1951; Chandler 1958). The trypanotolerant breeds in West Africa belong to *Bos taurus* and are divided into two major groups: the long-horned Hamitic type represented by the N'dama breed, and the short-horned type represented by numerous breeds including the Nigerian Shorthorn in Nigeria, the Lagune in Togo and Dahomey, the Gambian Dwarf, the Bakosi in Cameroon, the Logone in Chad and the Manjaca in Guinea. Studies are in progress in several countries of Africa to develop breeding programmes for these various strains, which can prove economical under the particular conditions of such countries (Payne 1970).

Genetic improvement studies have shown that half-breeds resulting from a Jersey bull and N'dama females show good tolerance to trypanosomiasis, identical with that of the N'dama parents; however, 3/4 Jersey × N'dama show sensitivity to infection. Trypanotolerant breeds produce neutralizing antibodies to infection and the tolerance is associated with the development-acquired resistance since animals reared in non-tsetse areas are susceptible to infection (Desowitz 1959) and may show a short bout of infection when moved to a new area (Stewart 1951). A correlation between trypanotolerance and type A haemoglobin has been noted; N'damas possess only this type of haemoglobin.

Control of *T. evansi* and *T. equinum* is dependant on therapy and the elimination of bloodsucking flies (see p. 400). Since the capybara may act as a reservoir host for *T. equinum*, the destruction of this animal will help control measures.

The control of *T. equiperdum* is achieved by quarantine regulations. The complement fixation test is used to detect infected animals, and reacting animals are destroyed. For Chagas disease, control is more difficult since poor economic conditions play an integral part in the epidemiology of the infection. The elimination of triatomid bugs from dwellings is a practical proposition and the destruction of wild animal reservoirs is also important.

Genus: Leishmania Ross, 1903

Developmental stages of this genus occur in the amastigote form in vertebrates and in the promastigote form in the insect vector and in culture. All the species of the genus are morphologically similar and, previously, species differentiation was based on the pathological entity they produced and the geographical distribution. However, species and subspecies differentiation is now possible using ultra-structural characteristics, serological methods to type excreted factor (EF) released into culture media by promastigotes, buoyant density of nuclear and kinetoplastic DNA and the variation in the electrophoretic mobility of various enzymes (biochemical taxonomy).

Morphology. The amastigote stages in vertebrates are found in the endothelial and macrophage cells of the body. They are circular or oval in outline and 2–4 μm in diameter. When stained with Romanowsky stains, the cytoplasm is blue, the oval nucleus is red and lies to one side, and at right-angles to it is a red- to purple-staining kinetoplast. The parasites multiply in the cytoplasm of cells forming clusters of organisms; dividing forms may be seen in smears.

The ultra-structure of the amastigote has been summarized by Aikawa and Sterling (1974). The organism is surrounded by a unit membrane, below which lie subpellicular microtubules. In the centre of the parasite is a double-membrane-bound nucleus with a centrally located nucleolus. A kinetoplast–mitochondrial complex occurs in a juxtanuclear position along with a short flagellum which is surrounded by an unfolded membrane of the plasmalemma. The kinetoplast–mitochondrial complex contains fibrous DNA which is surrounded by a double membrane to which are connected the cristae of the mitochondrial part of the complex.

Gardener et al. (1977) made morphometric comparisons of ultra-structural features of several species of *Leishmania* amastigotes and demonstrated species-characteristic differences in size and microtubule number.

The reliable identification of *Leishmania* species and strains is important in epidemiological studies. Isolates may be broadly classified on the basis of the buoyant density of nuclear DNA as determined by isopycnic centrifugation in CsCl (Chance et al. 1974) or on the basis of the serotype of the excreted factor produced by promastigotes in culture (Schnur & Zuckerman 1977). A more precise identification of isolates at the species and subspecies level may be obtained by determining the buoyant density of kinetoplastic DNA and the variation of the electrophoretic mobility of various enzymes (Gardener et al. 1974)—enzymes such as malate dehydrogenase (MDH), glucose phosphate isomerase (GPI), glucose 6-phosphate dehydrogenase (G6PDH) and 6-phosphogluconate dehydrogenase (6PGDH). The identifications made by each of these techniques agree closely and when they are used in combination they provide powerful tools for the understanding of the epidemiology of leishmaniasis.

The cycle in the insect. A variety of *Phlebotomus* spp. (*Psychodidae: Nematocera*) have been shown to serve as vectors. During a blood meal the *Phlebotomus* (sand-fly) ingests leucocytes and large mononuclear cells containing amastigotes (Leishman–Donovan bodies). These develop in the mid-gut of the sand-fly and enormous numbers of promastigotes are produced. These pass to the oesophagus and pharynx of the fly and their number may be so great as to block the food canal. When a fly attempts to feed, a plug of organisms may be dislodged and injected into the mammalian hosts. Infection may also occur when infected sand-flies are crushed on the skin.

Speciation of Leishmania. Over the years there has been considerable confusion on the speciation of the leishmaniae. Strains, species and varieties have been designated, in all leading to a jumble in the taxonomic situation. Some authors (e.g. Levine 1973) solve the problem by referring to two species of *Leishmania* in man. However, as the result of extensive experience with these organisms, Lainson and Shaw (1972) and Bray et al., (1973) have proposed new classifications, which are followed in this edition.

VISCERAL LEISHMANIASIS

Leishmania donovani (Laveran & Mesnil, 1903) Ross, 1903 (syn. *Piroplasma donovani*). (Fig. 3.7) This organism is morphologically identical with the other species. It is the cause of kala-azar, dumdum fever or visceral leishmaniasis in humans.

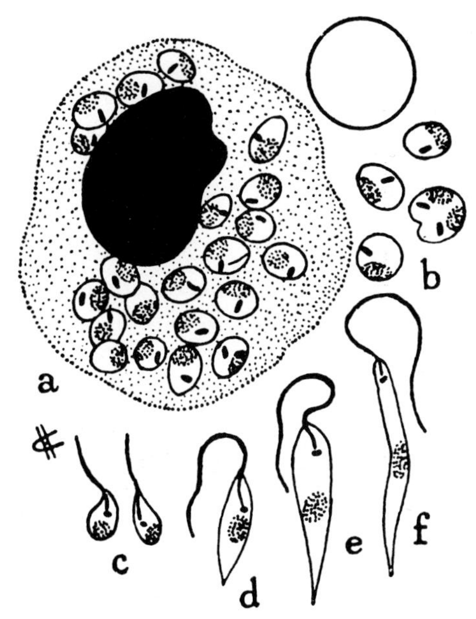

Fig 3.7 *Leishmania donovani* and *L. tropica*. (a) Macrophage containing rounded parasite amastigotes ('Leishman-Donovan bodies'). (b) Parasites outside host cell (one dividing). (c–f) Flagellate (promastigote) forms as seen in sandfly and in cultures. Erythrocyte drawn to scale. (×2000) (*After Hoare 1950*)

Different types of visceral leishmaniasis have been recognized and Biagi (1953) classified them as follows: Indian kala-azar, Chinese kala-azar, Mediterranean (infantile) kala-azar, African kala-azar, Russian kala-azar and American kala-azar. However, recent revision of the classification of the *Leishmania* genus by Lainson and Shaw (1972, 1973) and Bray et al. (1973) assigns species status to the organisms causing Mediterranean (infantile) kala-azar and American kala-azar. Bray (1974), in an evaluation of the zoonotic aspects of the leishmaniae, recognizes ecological forms of the organisms.

Three ecological forms of *Leishmania donovani* are recognized by Bray (1974). These are:

1. *L. donovani donovani* of the Indian subcontinent and Burma, largely Assam and Bengal. This is the cause of Indian kala-azar or dumdum fever. It affects young adults, 60% of infection being in the 10–20-years age group. There is no known reservoir in dogs, although they can be infected experimentally. Transmitted by *Phlebotomus argentipes*.

2. *L. donovani sensu lato* of Kenya, southern Ethiopia, Somalia and Uganda. This is in part the 'African kala-azar' of the Biagi (1953) classification. It resembles Indian kala-azar in that 66% of cases occur in young adults. Bray (1974) considers it is probably an anthroponosis with some involvement of rodents (ground-squirrels *Xerus rutilans*, gerbils *Tatera robusta*) as reservoirs. Naturally infected dogs have not been found in endemic foci. It is transmitted by *P. martini*, and possibly also by *P. vansomerenae* and *P. celiae*.

3. *L. donovani sensu lato* of Sudan, western Ethiopia, Chad, Niger, Republique Centrafricaine and Gabon. It is a zoonosis and carnivores (*Felis serval phillipsi, Genetta genetta senegalensis*) and rodents (*Arvicanthis niloticus luctuosus, Acomys albigena*) serve as reservoir hosts. It is transmitted by *P. langeroni orientalis*.

Leishmania infantum (Nicolle, 1908). Lainson and Shaw (1972) readopted the use of this specific name to differentiate between the Mediterranean and Indian forms of kala-azar. Bray (1974) gives its distribution as north China, south Soviet Asia, Iran, Iraq, Syria, Jordan, Lebanon, Arabia, Turkey, the Mediterranean area and islands, Portugal and North Africa. This species includes the Chinese kala-azar, Mediterranean (infantile) kala-azar and Russian kala-azar forms of Biagi. Reservoir hosts are fox (*Vulpes vulpes*), wolf (*Canis lupus*), jackal (*Canis aureus*) and porcupines (*Histrix hirsutirostris satunini*), with the dog as a domestic reservoir host.

In northern China it causes disease mainly in children with the dog as a reservoir. It is transmitted by *P. chinensis*. In the Mediterranean area, southern Europe and parts of tropical Africa,

80% of cases occur in children under five years of age and 94% in those under ten years. Dogs serve as the main domestic reservoir hosts and in these infection rates may be as high as, or higher than, in the human population.

In Tunisia and Iraq the fox and jackal are important reservoir hosts, the domestic dog serving as the host which brings the infection into the domestic environment. Transmission is chiefly by *P. major*.

The Soviet form is an infection with a zoonotic reservoir in dogs and jackals and corresponds to the Russian kala-azar of Biagi (1953).

In addition to the *Phlebotomus* spp. mentioned above, the following species are concerned in transmission: *P. ariasi, P. syriacus, P. kandelakii, P. simici* and possibly also *P. papatasi, P. perfiliewi, P. perniciosus tobbi, P. chinensis* spp., *P. longicuspis, P. caucasicus, P. mongolensis.*

Leishmania chagasi (Marques da Cunha and Chagas, 1937). Lainson and Shaw (1972) readopted this species for the causal agent of New World visceral leishmaniasis or American kala-azar. Bray (1974) recognizes two ecological forms of the parasite:

1. *Leishmania chagasi* of north-east Brazil. Bray (1974) gives the reservoir host as the 'fox' (*Lycalopex vetelus* and *Cerdocyon thous*), with the dog serving as a domestic reservoir. It is transmitted by *Lu. longipalpis*.
2. *Leishmania chagasi sensu lato* of Venezuela and other South American and Central American states. The reservoir is the crab-eating fox (*Cerdocyon*), possibly others, and Bray (1974) considers the domestic dog may be more important. It is transmitted by *Lu. longipalpis* and possibly other *Lutzomyia* species.

Cases of visceral leishmaniasis have been reported in dogs in the USA (Thorson et al. 1955), the dogs being imported. *Leishmania chagasi* occurs in humans of all ages, but is more common in children.

Pathogenesis of visceral leishmaniasis

Visceral leishmaniasis, especially the Indian form, is frequently a fatal disease in man. Following inoculation, promastigotes proliferate at the local site in macrophages and, after a period of weeks or months, they invade the inner organs to multiply in the spleen, liver, bone marrow and elsewhere, destroying the macrophages in the process. In advanced cases involvement of the digestive tract results in diarrhoea, there is marked emaciation, a distended abdomen and mortality may reach 70–90% in untreated cases, death occurring from a few weeks to a few years after infection.

On post mortem there is emaciation, anaemia, a markedly enlarged spleen which is congested with prominent Malpighian corpuscles, and an enlarged liver shows fatty infiltration. Endothelial and macrophage cells contain masses of amastigotes. The lymph nodes are usually enlarged and the cells parasitized.

In the dog the same general pathology is seen as in humans, thus anaemia, emaciation and ultimately death occur, diarrhoea being a terminal clinical sign. On post mortem there is enlargement of the spleen, liver and lymph glands. Cutaneous lesions may occur, with depilation and ulcerations developing on the lips and eyelids. In chronic cases a chronic eczema may be seen and skin ulceration may be evident.

Immunity in visceral leishmaniasis

Recent reviews of this subject have been given by Turk and Bryceson (1971), Maekelt (1972) and Preston and Dumonde (1976). Spontaneous cure is seldom seen in visceral leishmaniasis but it has been suggested that the massive lymphocyte–macrophage hyperplasia may have the effect of prolonging life. Treated human cases of kala-azar, however, are immune and second infections do not occur in cured patients in endemic areas. Some of the cured cases show a post-kala-azar dermal 'leishmanoid' (PKADL), which contains numerous parasites, but such cases do not show generalized infection. In the dog, however, only temporary clinical improvement may occur after treatment, the infection persisting in the animal. Adler (1964) suggests that it is doubtful whether an effective immunity occurs to the Mediterranean form in the dog.

Genetic aspects of susceptibility and resistance to *L. donovani* have been studied by Bradley (1977) and Bradley and Kirkley (1977). These studies emphasize the spectral nature of leishmaniasis (see below). A single gene control of acute susceptibility to visceral leishmaniasis was demonstrated; this was not linked to the H2 histocompatibility locus.

Diagnosis of visceral leishmaniasis

The only certain method of diagnosis is the demonstration of organisms in spleen pulp, lymph nodes, bone marrow, liver or fixed smears of peripheral blood. The first three may be taken by a biopsy and sternal puncture is increasingly used in diagnosis. In the dog, scrapings should be made from the periphery of skin ulcers or eczematous areas. Cultivation of biopsy or post mortem material on NNN or a similar medium will demonstrate the promastigotes; however, growth may take one to several weeks to become evident.

Animal inoculation, using the golden hamster, may be resorted to for diagnosis, but infection takes some time to develop.

Various immunodiagnostic tests exist for visceral leishmaniasis (see Kagan 1974). These include complement fixation, indirect immunofluorescence and indirect haemagglutination. More recently, a direct agglutination test using promastigote forms of *Leishmania*, treated similarly to the epimastigotes of *Trypanosoma cruzi* (Vattuone & Yanovsky 1971) has given encouraging results.

Non-specific 'serological' tests, such as the formal gel test and the urea Stilbamide test, have been used in India and China for diagnosis. These depend on the marked alterations in the albumin/globulin ratios which occur.

Treatment of visceral leishmaniasis

This has been reviewed by Goodwin and Rollo (1955) and drug resistance in leishmaniasis has been discussed by Peters (1974). An early drug was tartar emetic but many doses had to be given to produce a cure. The pentavalent antimony compounds are widely used for treatment, but Beveridge and Neal (1967) have stressed that there is great variation in the response to antimonials in different geographical areas. Further, 'antimony-resistant' organisms may respond normally to these drugs when they are grown in laboratory animals, e.g. hamsters. Compounds such as the pentavalent sodium stibogluconate (Pentostam) and the trivalent sodium antimonyl gluconate (Triostan, Solustibosan) and others (Glucantime, stibophen) are widely used for cutaneous and visceral leishmaniasis. The diamidine compound, pentamidine isethionate, is very effective for treatment of *L. donovani* infection; however, kala-azar of East Africa, Sudan and Brazil respond somewhat poorly to pentamidine. In the case of Brazil, amphotericin B has been used successfully, but 'drug resistance' ('drug unsusceptibility' is a better term) and its relation to the ability of the host to respond immunologically to the infection is an important aspect in the therapy of leishmaniasis. For example, although infection with *L. donovani* may respond well to diamidines, patients who develop post-kala-azar dermal leishmanoid are often refractory to treatment.

Treatment of canine infection is less effective than that of the human forms, and the Mediterranean type responds poorly to Pentostam; however, Solustibosan is reported to be effective.

Control of visceral leishmaniasis

A major control measure is the routine control of sand-flies. The various chlorinated hydrocarbon and organophosphate insecticides are useful, but measures against the breeding places of flies are also necessary. The latter include the removal of decaying vegetation and the clearing of dense vegetation around the houses. Where kala-azar is a zoonosis involving dogs, these should be treated or destroyed. Control of stray dogs should be instituted.

OLD WORLD CUTANEOUS LEISHMANIASIS

Leishmania tropica (Wright, 1903) Lühe, 1906 (syn. *Helcosoma tropica, Herpetomonas tropica* etc.). The organism is morphologically identical with the other species. It is found in macrophages, endothelial cells of capillaries and the adjacent lymph nodes of the skin; it seldom invades the inner organs. It is the cause of Old World cutaneous leishmaniasis, Oriental sore,

Aleppo button, Delhi boil, etc. The usual host is man. Dogs may suffer from dermal lesions, while in some areas the gerbil and other rodents are important reservoir hosts.

The type locality of this species is Armenia (Bray et al. 1973), and Bray (1974) designates three ecological forms of the genus as follows:

1. A form of the urban Middle East, urban South Soviet Asia, urban Iran, Mediterranean area, urban North India and urban Afghanistan; possibly the dog and the cat serve as domestic reservoirs, other reservoir hosts are not known. Transmitted by *P. papatasi, P. sergenti, P. perfiliewi.*

2. A form of West Africa particularly Senegal, Mali, Upper Volta, Niger, northern Nigeria, Sudan. The reservoir host is the Nile grass rat (*Arvicanthis niloticus*) and it is transmitted possibly by *P. duboscqi* or *Sergentomyia clydei.*

3. A form of rural Israel, Jordan, Syria and Arabia; reservoirs include the sand-rat (*Psammomys*) and a gerbil, and it is transmitted by *P. papatasi.*

Leishmania aethiopica Bray, Ashford and Bray, 1973. The creation of this species was based on geographical and ecological grounds. Its geographical distribution is the highlands of Ethiopia and the slopes of Mt Elgon, Kenya. Reservoir hosts are rock hyraxes and the organism is transmitted by *P. longipes* and *P. pedifer.*

Leishmania major Yakimov, 1915. This organism was raised to specific rank by Bray et al. (1973) and was originally recognized as *L. tropica major*, the cause of rural zoonotic leishmaniasis in the Middle East. With this elevation, Bray et al. (1973) assign other parasites of the Middle East to *L. tropica. Leishmania major* occurs in Turkmenistan, Uzbekistan and Iran, south of Isfahan. Reservoir hosts include the giant gerbil (*Rhombomys opimus*), other gerbils (*Meriones libycus erythrourus, M. meridianus, M. tamariscinus*), and suslik (*Spermophilopsis leptadactylus*) and the hedgehog (*Hemiechinus auritus*). The species is transmitted by *P. caucasicus, P. papatasi, P. mongolensis, P. ansarii,* possibly *S. arpaklensis.*

The classical Oriental sore is found in Middle Eastern and Eastern countries with a hot, dry climate. It occurs in countries bordering the Mediterranean and the Black Sea, in the Sudan, Egypt, Equatorial Africa and West Africa, in the Middle East from the Lebanon to Turkey and thence eastwards to India, Pakistan and Ceylon. In these areas, dogs are commonly infected and probably serve as zoonotic hosts.

Russian authors have distinguished an urban and a rural form of leishmaniasis in Turkmenistan, the former being caused by *Leishmania tropica* var. *minor* and the latter (the rural form) by *L. tropica* var. *major*. The latter organism has been raised to species status by Bray et al. (1973) (see above). The lesions produced by *L. tropica* var. *minor* are similar to the classical Oriental sore and Bray et al. (1973) consider this organism synonymous with *L. tropica. L. major* is found in a variety of rodents of which the gerbil (*Rhombomys opimus*) is the most important. The sand-flies *P. caucasicus* and *P. papatasi* live in the gerbil burrows and maintain the infection in these animals. *Phlebotomus papatasi* probably plays the more important role in the transmission of the gerbil infection to man since it more readily feeds on humans than *P. caucasicus.*

Pathogenesis of Old World cutaneous leishmaniasis

Following introduction of the promastigotes into the skin, they are taken up by macrophages in which they multiply; they then rupture the cell and infect other cells. The first detectable lesion occurs three days to six weeks after the *Phlebotomus* bite, appearing as a reddish papule which gradually develops a crust, forming a shallow ulcer. The ulcer gradually enlarges and may reach several centimetres in diameter: ulcers may coalesce to form larger areas. In an uncomplicated infection the ulcers heal in 2–12 months, leaving a deeply pigmented, depressed scar. The infection is very seldom fatal.

Skin leishmaniasis in the dog is similar to that in the human, ulcers being found on the skin.

In the dry type of infection in Turkmenistan there is a long incubation period, a prolonged clinical phase with papules persisting for several months and these contain large numbers of amastigotes. The moist type has a short incubation period, a short duration and few amastigotes occur in the lesion. Ulcers form rapidly and heal spontaneously.

Immunity to Old World cutaneous leishmaniasis

This is reviewed by Turk and Bryceson (1971), Stauber (1963), Maekelt (1972), Zuckerman (1975) and Preston and Dumonde (1976). In the absence of treatment the local lesion of *L. tropica* runs its full course of reaction and proliferation of local histiocytes occurs with intracellular multiplication of organisms. With the invasion of the lesion by lymphocytes and plasma cells the macrophage proliferation diminishes, as do the organisms, until finally they completely disappear, spontaneous cure having taken place. The process may take up to two years in some cases. Spontaneous cure is followed by immunity which may persist more than 20 years after the disappearance of the lesion.

Turk and Bryceson (1971) view cutaneous leishmaniasis as a spectral disease; one polar form consists of diffuse cutaneous leishmaniasis where there are disseminated nodules, heavily parasitized macrophages, negative delayed skin reactivity to antigen and response to treatment is poor. The other polar form is lupoid or recidiva leishmaniasis, characterized by a local lesion, tuberculoid in nature, which arises when an oriental sore fails to heal completely. Delayed skin reactions are present and the patient is not immune to challenge. The majority of cases of cutaneous leishmaniasis lie in the middle of this spectrum. Various animal models of cutaneous leishmaniasis have been developed, including one in which disseminated lesions are produced in guinea-pigs by infection of a skin site deprived of lymphatic drainage with *L. enrietii* (Kadivar & Soulsby 1975).

The genetic aspects of *L. tropica* infection in mice have been discussed by Preston and Dumonde (1976). Acquired resistance to the parasite in various strains of mice is under the control of more than one gene locus which is in contrast to the situation with *L. donovani* (Bradley 1977).

Immunization of susceptible persons, using material from sores, has been practised for many years in the Middle East and Central Asia. A portion of the body hidden from view is infected, to prevent facial disfigurement from the active infection. A more modern approach to this consists of using living promastigotes from cultures of *L. tropica* (Berberian 1939). Immunity is slow to develop and is not complete until the lesions have healed and the amastigotes are no longer present. The process may take four to six months. If the local lesion is surgically removed before spontaneous cure has taken place the patient remains susceptible to infection.

The immunological relationship between the various biological strains or species of organisms of this genus has yet to be clarified. With the Russian forms it has been stated that *L. minor* (the dry form) does not protect against *L. major* (the wet form) (Manson Bahr 1963). However, Kozhevnikov (1959) was able to immunize against the dry form with the wet form and he also showed that immunization with promastigotes from cultures of *L. major* isolated from gerbils give good protection against both forms. Ansari and Mofidi (1950), in Iran, obtained partial immunity against the wet form, using the dry type infection.

Diagnosis of Old World cutaneous leishmaniasis

Microscopic examination of material from the edge of an ulcer or a local lymph node will show the organisms in the epithelial and mononuclear cells. Culture of materials in NNN medium should also be made.

Treatment of cutaneous leishmaniasis is dealt with in the section on American cutaneous leishmaniasis.

Control measures include treatment and control and/or elimination of dogs. The destruction of gerbils greatly reduces the incidence of the wet form of infection. Vaccination with promastigotes from cultures obtained from experimentally infected animals has been widely practised in the Middle East, but the elimination of sand-flies due to mosquito and malarial control measures has greatly decreased the necessity for immunization in urban areas. It is still applicable in rural communities.

AMERICAN CUTANEOUS LEISHMANIASIS

Lainson and Shaw (1972) consider that the parasites causing cutaneous leishmaniasis in America are indigenous to that area and distinct from *L. tropica* of the Old World. These authors identify strains of two types infecting wild mammals and man in Brazil and elsewhere.

The '*Leishmania mexicana* complex' is only rarely isolated from man and mainly from rodents, less often from opossums, and is transmitted by *Lutzomyia flaviscutellata* and *Lu. olmeca*. Organisms of this complex grow readily in diphasic blood media in vitro, infect hamsters and mice readily and produce large tumour-like histiocytomas containing large numbers of amastigotes without a marked host cellular response.

The '*Leishmania braziliense* complex' is mainly isolated from man, wild animal hosts are poorly known and important vectors are *Lu. intermedia* and *Psychodopygus* groups. Organisms grow poorly in vitro and small, usually non-ulcerative skin lesions containing few parasites and showing a marked cellular response are seen in hamsters.

In the '*Leishmania mexicana* complex' the following parasites are recognized:

Leishmania mexicana mexicana Biagi, 1953. This is a cause of Chiclero ulcer or bay sore of Mexico, Guatemala and Honduras. The primary lesion may be small and will heal within a few weeks. This strain also characteristically attacks the ear, causing a granulomatous lesion which grossly deforms the earlobe. Reservoir hosts include the tree rat (*Ototylomys phyllotis*), the pocket mouse (*Heteromys desmarestianus*), the vesper rat (*Nyctomys sumichrasti*) and the cotton rat (*Sigmodon hispidus*). The organism is transmitted by *Lu. olmeca*.

Leishmania mexicana amazonensis Lainson and Shaw, (1972). This occurs in the Amazon basin and Matto Grosso of Brazil and in rodents in Trinidad. It rarely infects man and is transmitted by *Lu. flaviscutellata*. Reservoir hosts include the cricetid rodents *Orysomys* spp., *Proechmys* spp., *Neacomys spinosus amoenus, Diplomys labilis, Nectomys squamipes, Kannabateomys amblyonyx, Zxgodontomys microtinus,* opossums (*Marmosa* spp.), paca (*Cuniculus paca*) and agouti (*Dasyprocta azarae*).

Leishmania mexicana pifano Medina and Romero, 1959. This causes a rare form of chronic cutaneous leishmaniasis in Venezuela.

In the '*Leishmania braziliensis* complex' the following parasites are recognized:

Leishmania braziliensis braziliensis Vianna, 1911. This causes the classical espundia of the Brazilian rain forest, metastases occurring in the oropharynx in 80% of cases. The skin lesions are chronic and invasion of the mucous membrane sometimes causes great disfigurement by erosion of the soft and cartilaginous tissues. The disease may last for many years and spontaneous recovery is rare. Death may occur due to septicaemia or bronchial pneumonia. It is distributed in Brazil, Venezuela, Columbia and other South American countries. Reservoirs include forest rodents, e.g. *Oryzomys concolor* (Lainson & Shaw 1972). It is transmitted by *Lu. paraensis* and *Lu. wellcomei, Lu. migonei, Lu. whitmani* and by *Lu. anduzei*.

Leishmania braziliensis guyanensis Floch, 1953. Floch (1953) gave this name to the leishmania of the Guyanas which probably causes a similar condition to that in Panama, Costa Rica, Surinam and other parts of northern South America. This type is American forest leishmaniasis, pian bois or buba of northern South America. The skin ulcerations tend to show spontaneous healing and only in about 5% of cases do metastases occur in the nasal mucosa. Dogs may be naturally infected; the important animal reservoirs are unknown. *L. b. guyanensis* is transmitted by *P. anduzei* and possibly other species of phlebotomines.

Leishmania braziliensis panamensis Lainson and Shaw, 1972. This species is found in Panama, Costa Rica and Honduras. Reservoir hosts include sloths (*Choloepus hoffmanni, Bradypus infuscatus*), kinkajous (*Potos flavus*), porcupine (*Coendus rothschildi*) and rodents (*Proechimys semispinous, Hoplomys gymnurus*). It is transmitted by *Lu. trapidoi* and probably also by *Lu. gomezi, Lu. sanguinaria, Lu. panamensis* and *Lu. ylephiletrix*.

Leishmania peruviana Velez, 1913. This is the cause of Uta in the mountains of Peru. It is a benign form of the disease, showing numerous small skin lesions, which generally resembles the Old World form of leishmaniasis in its clinical course. The rodent reservoirs of this species are not known. The dog may serve as a reservoir host. It is probably transmitted by *Lu. verrucarum* and *Lu. peruensis.*

Pathogenesis of American cutaneous leishmaniasis

Although there are a number of different species of organisms, generally the disease process can be divided into the metastasizing form (*L. b. braziliensis*) and the non-metastasizing form (others). In the non-metastasizing form the initial papule gives way to ulceration with induration which heals in 6–18 months. The classical espundia shows metastases which may take several years to develop. The nasal septum, nasopharynx and even the larynx may be involved with appalling disfigurement.

Immunity to American cutaneous leishmaniasis

This is reviewed by Maekelt (1972) and Preston and Dumonde (1976). Recovery is followed by a relatively solid immunity to reinfection in the non-metastasizing form, and it thus resembles the Old World form of cutaneous leishmaniasis. With espundia there is less evidence of spontaneous recovery although the cellular reaction apparently reduces the number of organisms in the lesions to a minimum. Diffuse cutaneous leishmaniasis (DCL) occurs in American cutaneous leishmaniasis, and this is associated with either weak or totally absent cell-mediated immunity responses to the infection.

Cutaneous and mucocutaneous leishmaniases give a delayed skin reaction. This is utilized in the 'Montenegro' test for which organisms obtained from culture serve as antigen. Pelligreno (1951) used a polysaccharide antigen of *L. braziliensis.*

Diagnosis of American cutaneous leishmaniasis

Diagnosis is much the same as for Old World cutaneous leishmaniasis.

An indirect immunofluorescence test has been shown by Walton et al. (1972) to be effective in detecting American cutaneous leishmaniasis and, following successful chemotherapy, antibody titres decreased. The direct agglutination test (DAT) employing trypsinized promastigotes has proved useful in the diagnosis of cutaneous leishmaniasis (Kagan 1974). In long standing infections, where organisms are few, immunodiagnostic tests such as the Montenegro reaction can be employed.

Treatment of American cutaneous leishmaniasis

As with visceral leishmaniasis, there tends to be variation in the effectiveness of drugs in different geographical areas. Most simple cutaneous types of leishmaniasis due to *L. tropica* and *L. mexicana* complex respond satisfactorily to the antimonials and to cycloguanil. The pentavalent antimonial sodium stibogluconate is one of the best for cutaneous infections. However, with the *L. braziliensis* complex, the majority of compounds offer limited effectiveness for the late manifestations of the infection and mucocutaneous metastases. Treatment of disseminated cutaneous leishmaniasis is difficult and organisms rapidly develop a solid resistance to all forms of chemotherapy (Peters 1974). The emergence of drug resistance is associated with the failure of the host immune response and is exemplified in the case of DCL.

Newer compounds for the treatment of cutaneous leishmaniasis include amphotericin B, metronidazole, cycloguanil pamoate and 5-fluorocytosine.

Control of American cutaneous leishmaniasis

American cutaneous leishmaniasis is principally an infection of persons working in forests. Personal protection against sand-flies by using repellents is of immediate but limited value; the long-term control must depend on wide-scale sand-fly control.

OTHER SPECIES OF LEISHMANIA

Leishmania enriettii Muniz and Medina, 1948. This is specific for the guinea-pig. It causes ulcers in the skin of the nose, ears and paws. These usually resolve in 8–10 weeks, to be followed by immunity to reinfection. This parasite has been used extensively for studies of the immune response to cutaneous leishmaniasis.

Nine species of *Leishmania* have been described from lizards (Adler 1964). Of these, *Leishmania adleri* produces infections in both lizards and mammals; in the latter, infections are cryptic and transient. The organism shares antigens with *L. donovani* and *L. tropica* and it may represent a stage in the evolution of mammalian forms of leishmaniae (Adler 1964).

FAMILY: CRYPTOBIIDAE POCHE, 1913

These are biflagellate trypanosome-like forms, one flagellum is free, the other is on the outer margin of an undulating membrane. The kinetoplast is elongated and rod-like. Forms are parasitic in marine and fresh-water fish, various amphibians, snails, leeches, etc.

Genus: Cryptobia (Leidy, 1846)

Of importance in this genus are the parasitic haemoflagellates of fresh-water and marine fish; they are transmitted by leeches.

Cryptobia borreli (Laveran and Mesnil, 1901) occurs in the blood of several fresh-water fish, including rainbow trout, brown trout, coho salmon and king salmon. Serious mortality may occur in salmon hatcheries, pale gills, sunken eyes and emaciation being clinical signs of infection.

Cryptobia brachialis (Leidy, 1846) attaches to gill filaments, causes destruction of epithelium and results in thrombus formation.

Cryptobia cyprini (Plehn, 1903) is associated with anaemia in young carp. It also occurs in goldfish and heavy infections induce loss of weight and listlessness. The infection in goldfish has been termed 'sleeping sickness of goldfish' (Reichenbach-Klinke & Elkan 1965).

Control of *Cryptobia* species is by the elimination of leeches.

REFERENCES

TRYPANOSOMA

Aikawa, M. & Sterling, C. R. (1974) *Intracellular Parasitic Protozoa*. New York: Academic Press.

Apted, F. I. C. (1970) The epidemiology of Rhodesian sleeping sickness. In: *The African Trypanosomiases*, ed. H. W. Mulligan, pp. 645–660. New York: Wiley-Interscience.

Ashcroft, M. T. (1959) The Tinde experiment: a further study of the long-term cyclical transmission of *Trypanosoma rhodesiense*. *Ann. trop. Med. Parasit.*, **53**, 137–146.

Barber, H. J. & Berg, S. S. (1962) Structure and activity of antiprotozoal drugs. In: *Drugs, Parasites and Hosts*, ed. L. G. Goodwin & R. H. Nimmo-Smith, pp. 165–169. London: Churchill.

Bishop, A. (1959) Drug resistance in protozoa. *Biol. Rev.*, **34**, 445–500.

Bishop, A. (1962) Chemotherapy and drug resistance in protozoal infections. In: *Drugs, Parasites and Hosts*, ed. L. G. Goodwin & R. H. Nimmo-Smith, pp. 98–111. London: Churchill.

Boreham, P. F. L. (1967) Possible causes of anaemia in rabbits chronically infected with *Trypanosoma brucei*. *Trans. R. Soc. trop. Med. Hyg.*, **61**, 138.

Boreham, P. F. L. (1968*a*) Immune reactions and kinin formation in 'chronic' trypanosomiasis. *Br. J. Pharm. Chemother.*, **32**, 493–504.

Boreham, P. F. L. (1968*b*) *In vitro* studies on the mechanism of kinin formation by trypanosomes. *Br. J. Pharm. Chemother.*, **34**, 598–603

Boreham, P. F. L (1970) Kinin release and the immune reaction in human trypanosomiasis caused by *Trypanosoma rhodesiense*. *Trans. R. Soc. trop. Med. Hyg.*, **64**, 394–400.

Boreham, P. F. L. (1974) Physiopathological changes in the blood of rabbits infected with *Trypanosoma brucei*. *Rev. Elev. Med. vet. Pays trop.* (Suppl.), **27**, 279–282.

Boreham, P. F. L. & Facer, C. A. (1977) Fibrinogen and fibrinogen/fibrin degradation products in the urine of rabbits infected with *Trypanosoma* (*Trypanozoon*) *brucei*. *Z. Parasitenk.*, **52**, 257–265.

Boreham, P. F. L. & Wright, I. G. (1976) The release of pharamacologically active substances in parasitic infections. In: *Progress in Medicinal Chemistry*, eds G. P. Ellis & G. B. West, vol. 13, pp. 159–204. Amsterdam: North Holland.

Bowman, I. B. R., Flynn, I. W. & Fairlamb, A. M. (1970) Carbohydrate metabolism of pleomorphic strains of *Trypanosoma rhodesiense* and sites of action of arsenical drugs. *J. Parasit.*, **56**, 402–407.

von Brandt, T. (1966) *Biochemistry of Parasites*. New York: Academic Press.

Buxton, P. A. (1955) *The Natural History of Tsetse Flies*. Mem. No. 10, London School Hyg. Trop. Med. London: H. K. Lewis.

Carmichael, J. A. (1939) Turning sickness of cattle and *Trypanosoma theileri*. *Parasitology*, **31**, 498–500.

Chandler, R. L. (1958) Studies on the tolerance of N'dama cattle to trypanosomiasis. *J. comp. Path.*, **68**, 253–260.

Clarkson, M. J. (1968) Blood and plasma volume in sheep infected with *Trypanosoma vivax*. *J. comp. Path.*, **78**, 189–193.

Clarkson, M. J. (1976) Immunoglobulin M in trypanosomiasis. In: *Pathophysiology of Parasitic Infections*, ed. E. J. L. Soulsby. pp. 171–182. New York: Academic Press.
Cross, G. A. M. (1977) Antigenic variation in trypanosomes. *Am. J. trop. Med. Hyg.*, **26** (Suppl.), 240–243.
Cross, G. A. M. & Manning, J. C. (1973) Cultivation of *Trypanosoma brucei* spp. in semi-defined and defined media. *Parasitology*, **67**, 315–331.
Curasson, G. (1943) *Traité de protozoologie vétérinaire et comparée. Tome 1. Trypanosomes*. Paris: Vigot-Frères.
Davis, C. E., Robbins, R. S., Weeler, R. D. & Braude, A. I. (1974) Thrombocytopenia in experimental trypanosomiasis. *J. clin. Invest.*, **53**, 1359–1367.
Desowitz, R. S. (1956) Observations on the metabolism of *Trypanosoma vivax*. *Exp. Parasit.*, **5**, 250–259.
Desowitz, R. S. (1959) Studies on immunity and host–parasite relationships. I. The immunological response of resistant and susceptible breeds of cattle to trypanosomal challenge. *Ann. trop. Med. Parasit.*, **53**, 293–313.
Desowitz, R. S. (1963) Adaptation of trypanosomes to abnormal hosts. *Ann. N.Y. Acad. Sci.*, **113**, 74–87.
Desowitz, R. S. & Watson, H. J. C. (1953) Studies on *Trypanosoma vivax*. IV. The maintenance of a strain in white rats without sheep-serum supplement. *Ann. trop. Med. Parasit.*, **47**, 62–67.
van Dijk, J. E., Zwart, D. & Leeflang, P. (1973) A contribution to the pathology of *Trypanosoma simiae* infection in pigs. *Zbl. vet. Med.*, **20**, 374–391.
Duxbury, R. E., Sadun, E. H. & Anderson, J. S. (1972) Experimental infections with the African trypanosomes. II. Immunization of mice and monkeys with a gamma-irradiated, recently isolated human strain of *Trypanosoma rhodesiense*. *Am. J. trop. Med. Hyg.*, **21**, 885–888.
Eyidi, N. (1971) Controle et prophylaxie des trypanosomiases au Cameroun oriental. *39th Session Committee OIE*.
Faust, E. C. (1949) The etiologic agent of Chagas' disease in the United States. *Boln. Of. Sanit. pan-am.*, **28**, 455–461.
Fiennes, R. N. T.-W. (1952) The cattle trypanosomiases. A cryptic focus of parasites in association with a secondary stage of disease. *Br. vet. J.*, **108**, 298–305.
Fiennes, R. N. T.-W. (1970) Pathogenesis and pathology of animal trypanosomiasis. In: *The Animal Trypanosomiases*, ed. H. W. Mulligan, pp. 729–750. London: George Allen & Unwin.
Ford, J. (1970) Control of populations of Glossina. Control by destruction of the larger fauna. In: *The African Trypanosomiases*, ed. H. W. Mulligan, pp. 557–563. London: George Allen and Unwin.
Gill, B. S. & Malhotra, M. N. (1971) Chemoprophylaxis of *T. evansi* infections in ponies. *Trans. Anim. Hlth Proc.*, **3**, 199–202.
Goble, F. C. (1958) A comparison of strains of *Trypanosoma cruzi* indigenous to the United States with certain strains from South America. *Proc. 6th int. Congr. trop. Med. Malaria*, **3**, 158–166.
Godfrey, D. G. (1960) Types of *Trypanosoma congolense*. I. Morphological differences. *Ann. trop. Med. Parasit.*, **54**, 428–436.
Gomez, R. J. (1956) Estudio do la tripanosomiasis natural del canino (*Canis fam*) en Venezuela. *Rec. Med. vet. Parasit. Caracas*, **15**, 63–105.
Goodwin, L. G. (1964) The chemotherapy of trypanosomiasis. In: *Biochemistry and Physiology of Protozoa*, ed. S. H. Hutner, vol. 3. New York: Academic Press.
Goodwin, L. G. (1970) The pathology of African trypanosomiasis. *Trans. R. Soc. trop. Med. Hyg.*, **64**, 797–817.
Goodwin, L. G. (1971) Pathological effects of *Trypanosoma brucei* on small blood vessels in rabbit ear-chambers. *Trans. R. Soc. trop. Med. Hyg.*, **65**, 82–88.
Goodwin, L. G., Green, D. G., Guy, M. W. & Voller, A. (1972) Immunosuppression during trypanosomiasis. *Br. J. exp. Path.*, **53**, 40–43.
Goodwin, L. G. & Guy, M. W. (1973) Tissue fluid in rabbits infected with *Trypanosoma* (*Trypanozoon*) *brucei*. *Parasitology*, **66**, 499–513.
Goodwin, L. G. & Hook, S. V. M. (1968) Vascular lesions in rabbits infected with *Trypanosoma* (*Trypanozoon*) *brucei*. *Br. J. Pharm. Chemother.*, **32**, 505–513.
Goodwin, L. G. & Rollo, I. M. (1955) The chemotherapy of malaria, piroplasmosis, trypanosomiasis and leishmaniasis. In: *Biochemistry and Physiology of Protozoa*, eds S. H. Hutner & A. Lwoff, vol. 2. New York: Academic Press.
Gray, A. R. (1965) Antigenic variation in a strain of *Trypanosoma brucei* transmitted by *Glossina morsitans* and *G. palpalis*. *J. gen. Microbiol.*, **41**, 195–214.
Greenwood, B. M. (1974) Immunosuppression in malaria and trypanosomiasis. In: *Parasites in the Immunized Host: Mechanism of Survival*, eds Ruth Porter & Julie Knight, CIBA Foundation Symposium 25 (New Series), pp. 137–146. Amsterdam: Elsevier.
Greenwood, B. M. & Whittle, H. C. (1976) Coagulation studies in Gambian trypanosomiasis. *Am. J. trop. Med. Hyg.*, **25**, 390–394.
Henson, J. B. & Noel, J. C. (1979) Immunology and pathogenesis of African animal trypanosomiasis. *Adv. vet. Sci. comp. Med.*, **23**, 161–182.
Herbert, W. J. & Inglis, M. D. (1973) Immunization of mice, against *T. brucei* infection, by the administration of released antigen adsorbed to erythrocytes. *Trans. R. Soc. trop. Med. Hyg.*, 67, 268.
Hirumi, H., Doyle, J. J. & Hirumi, K. (1977*a*) African trypanosomes: cultivation of animal infective *Trypanosoma brucei in vitro*. *Science*, **196**, 992–994.
Hirumi, H., Doyle, J. J. & Hirumi, K. (1977*b*) Propagation and cyclical development of *Trypanosoma brucei in vitro*. *Abstracts 5th Int. Congr. Protozoology, New York, 1977*. Abstract 123.
Hoare, C. A. (1954) The loss of the kinetoplast in trypanosomes with special reference to *Trypanosoma evansi*. *J. Protozool.*, **1**, 28–33.
Hoare, C. A. (1959) Morphological taxonomic studies on mammalian trypanosomes. IX. Revision of *Trypanosoma dimorphon*. *Parasitology*, **49**, 210–231.
Hoare, C. A. (1964) Morphological and taxonomic studies on mammalian trypanosomes. X. Revision of the systematics. *J. Protozool.*, **11**, 200–207.
Hoare, C. A. (1967) Evolutionary trends in mammalian trypanosomes. *Adv. Parasit.*, **5**, 47–91.
Hoare, C. A. & Wallace, F. G. (1966) Developmental stages of trypanosomatid flagellates: a new terminology. *Nature, Lond.*, **212**, 1385–1386.
Holmes, P. H. & Jennings, F. W. (1976) The effect of treatment on the anaemia of African trypanosomiasis. In: *Pathophysiology of Parasitic Infections*, ed. E. J. L. Soulsby, pp. 199–210. New York: Academic Press.
Honigberg, B. M., Balamuth, W., Bovee, E. C., Corliss, J. O., Gojdics, M., Hall, R. P., Kudo, R. R., Levine, N. D., Loeblich, A. R. Jr, Weiser, J. & Wenrich, D. H. (1964) A revised classification of the phylum Protozoa. *J. Protozool.*, **11**, 7–20.
Honigberg, B. M., Cunningham, I., Stanley, H. A., Su-Lin, K.-E. & Luckins, A. G. (1976) *Trypanosoma brucei*: antigenic analysis of bloodstream, vector and culture stages by the quantitative fluorescent antibody methods. *Exp. Parasit.*, **39**, 496–522.
Hornby, H. E. (1952) *African Trypanosomiasis in Eastern Africa*, pp. 39ff. London: HMSO.
Hungerer, K.-D., Enders, B. & Zwisler, O. (1976) On the immunology of infection with *T. cruzi*. 2. The preparation of an apathogenic living vaccine. *Behring Inst. Mitteilungen*, **60**, 84–97.
Ikede, B. O. & Losos, G. T. (1972*a*) Pathological changes in cattle infected with *Trypanosoma brucei*. *Vet. Path.*, **9**, 272–277.
Ikede, B. O. & Losos, G. T. (1972*b*) Pathology of the disease in sheep produced experimentally by *Trypanosoma brucei*. *Vet. Path.*, **9**, 278–289.
Ingram, D. G. & Soltys, M. A. (1960) Immunity in trypanosomiasis. IV. Immunoconglutinin in animals infected with *Trypanosoma brucei*. *Parasitology*, **50**, 231–239.
Jennings, F. W. (1976) The anaemias of parasitic infections. In: *Pathophysiology of Parasitic Infections*, ed. E. J. L. Soulsby, pp. 41–67. New York: Academic Press.
Jennings, F. W., Murray, P. K., Murray, M. & Urquhart, G. M. (1974) Anaemia in trypanosomiasis: studies in rats and mice infected with *Trypanosoma brucei*. *Res. vet. Sci.*, **16**, 70–76.
Kobayashi, A., Tizard, I. R. & Woo, P. I. K. (1976) Studies on the anaemia in experimental trypanosomiasis. II. The pathogenesis of the anaemia in calves infected with *Trypanosoma congolense*. *Am. J. trop. Med. Hyg.*, **25**, 401–406.
Köberle, F. (1974) Pathogenesis of Chagas' disease. In: *Trypanosomiasis and Leishmaniasis with Special Reference to Chagas' Disease*, eds Katherine Elliott, Maeve O'Connor & G. E. W. Wolstenholme. CIBA Foundation Symposium 20 (New Series), pp. 137–158. Amsterdam: Elsevier.
Levine, N. D. (1973) *Protozoan Parasites of Domestic Animals and of Man*. 2nd ed. Minneapolis: Burgess. (First edn published 1971).
Losos, G. J. & Ikede, B. O. (1972) Review of pathology of diseases in domestic and laboratory animals caused by *Trypanosoma congolense, T. vivax, T. brucei, T. rhodesiense* and *T. gambiense*. *Vet. Path.* (Suppl.), **9**, 1–71.
Losos, G. J., Paris, J., Welson, A. J. & Dar, F. K. (1973) Pathology of the disease in cattle caused by *Trypanosoma congolense*. *Bull. epizoot. Dis. Afr.*, **21**, 239–248.
Lumsden, W. H. R. (1969) Some current problems in the seroepidemiology of trypanosomiasis in relation to the epidemiology and control of the disease. *Bull. WHO*, **40**, 871–878.
Lumsden, W. H. R. (1972) Immune response to hemoprotozoa. I. Trypanosomes. In: *Immunity to Animal Parasites*, ed. E. J. L. Soulsby, pp. 287–299. New York: Academic Press.
McNeillage, G. J. C., Herbert, W. J. & Lumsden, W. H. R. (1969) Antigenic types of first relapse variants arising from a strain of *Trypanosoma* (*Trypanozoon*) *brucei*. *Exp. Parasit.*, **25**, 1–7.
Maikano, A., Engueleguele, E. & Ferriot, A. (1969) Prophylaxis trypanosomienne en Adamoua. Premièrs résultats obtenus avec une association Bérénil-Isométamidium *Coll. OCAM élev*. Fort-Lamy. Ed. IEMVT, 1971, 148–151.
Maxie, M. G., Losos, G. J. & Tabel, H. (1976) A comparative study of the hematological aspects of the diseases caused by *Trypanosoma vivax* and *Trypanosoma congolense* in cattle. In: *Pathophysiology of Parasitic Infections*, ed. E. J. L. Soulsby, pp. 183–198. New York: Academic Press.

Mendez, Y. & Honigberg, B. M. (1972) Infectivity of *Trypanosoma brucei*-subgroup flagellates maintained in culture. *J. Parasit.*, **58**, 1122–1136.
Mulligan, H. W. (1951) Tolerance of indigenous West African cattle to trypanosomiasis. *Bull. Bur. perm. Interafr. Tsetse*, **164**, 4.
Murray, P. K., Jennings, F. W., Murray, M. & Urquhart, G. M. (1974) Immunosuppression in trypanosomiasis. In: *Parasitic Zoonoses: Clinical and Experimental Studies*, ed. E. J. L. Soulsby, pp. 133–150. New York: Academic Press.
Neal, R. A. & Johnson, P. (1977) Immunization against *Trypanosoma cruzi* using killed antigens and with saponin as adjuvant. *Acta trop.*, **34**, 87–92.
Neghme, R. A. (1940) La trypanosomosis americana una enfermedas rural en Chile. *Boln Chile méd. Soc.*, **7**, 32–37.
Newton, B. A., Cross, G. A. M. & Baker, J. R. (1973) Differentiation in Trypanosomatidae. In: *23rd Symposium of the Society for General Biology*, eds J. M. Ashworth & J. E. Smith, pp. 339–372. Cambridge: Cambridge University Press.
Nguyen-Huan, Chi, Webb, L., Lambert, P. H. & Miescher, P. A. (1975) Pathogenesis of the anaemia in African trypanosomiasis: characterization and purification of a haemolytic factor. *Schweiz. med. Wschr.*, **105**, 1582–1583.
Ormerod, W. E. & Venkatesan, S. (1971) The occult visceral phase of mammalian trypanosomes with special reference to the life cycle of *Trypanosoma (Trypanozoon) brucei*. *Trans. R. Soc. trop. Med. Hyg.*, **65**, 722–735.
Parkin, B. S. (1935) The symptomatology and treatment of *Trypanosoma congolense* infection of canines. *Onderstepoort J. vet. Sci.*, **4**, 247–250.
Payne, W. J. A. (1970) *Cattle Production in the Tropics*, Vol. 1: *Breeds and Breeding*. London: Longman.
Peel, E. & Chardome, M. (1954) *Trypanosoma suis* Ochmann, 1905-Trypanosome monomorphe pathogène de mammifères évoluant dans les glandes salivaires de *Glossina brevipalpis*, Newst., Mosso (Urundi). *Ann. Soc. belge Méd. trop.*, **34**, 277–295.
Peruzzi, M. R. I. (1928) Pathologic-anatomical and serological observations on trypanosomiases. *Final Report*. League of Nations International Committee on Human Trypanosomiasis, **3**, 245–328.
Pifano, C. F. (1954) Parasitological methods of diagnosing cases of chronic Chagas' disease. *Arch. Venez. Patol. trop. Parasit. méd.*, **2**, 121–156.
Rickman, L. R. & Robson, J. (1970) The blood incubation infectivity test: a simple test which may serve to distinguish *Trypanosoma brucei* from *T. rhodesiense*. *Bull. WHO*, **42**, 650–651.
Ristic, M. & Smith, R. D. (1974) Zoonoses caused by Hemoprotozoa. In: *Parasitic Zoonoses—Clinical and Experimental Studies*, ed. E. J. L. Soulsby, pp. 41–63. New York: Academic Press.
Roberts, C. J. (1975) Ruminant lipid metabolism in trypanosomiasis. *Trans R. Soc. trop. Med. Hyg.*, **69**, 275.
Robins-Brown, R. M., Schneider, J. & Metz, J. (1975) Thrombocytopenia in trypanosomiasis. *Am. J. trop. Med. Hyg.*, **24**, 226–231.
Ross, R. & Thompson, D. (1910) A case of sleeping sickness studied by precise enumerative methods; regular periodic increase of the parasites disclosed. *Proc. R. Soc. Lond. B.*, **82**, 411–415.
Rudzinska, M. & Vickerman, K. (1968) The fine structure. In: *Infectious Blood Diseases of Man and Animals*, ed. D. Weinman & M. Ristic, vol. I. New York: Academic Press.
Scott, D. (1970) The epidemiology of Gambian sleeping sickness. In: *The African Trypanosomiases*, ed. H. W. Mulligan, pp. 614–644. New York: Wiley-Interscience.
Seed, J. R. (1972) *Trypanosoma gambiense* and *T. equiperdum*: characterization of variant specific antigens. *Exp. Parasit.*, **31**, 98–108.
Seed, J. R. & Gam, A. A. (1966) Passive immunity to experimental trypanosomiasis. *J. Parasit.*, **52**, 1134–1140.
Seed, J. R. & Hall, J. E. (1977) The possible role of the trypanosome metabolite indole-3-ethanol in the neuropathology of trypanosomiasis. Abstract 115. In: *Abstracts of Papers read at 5th Int. Congr. Protozoology, New York, 1977*, ed. S. H. Hutner. New York: The Print Shop.
Shaw, J. J. & Lainson, R. (1972) *Trypanosoma vivax* in Brazil. *Ann. trop. Med. Parasit.*, **66**, 25–32.
Soltys, M. A. (1959) Immunity in trypanosomiasis. III. Sensitivity of antibody-resistant strains to chemotherapeutic drugs. *Parasitology*, **49**, 143–152.
Soltys, M. A. (1967) Comparative studies of immunogenic properties of *Trypanosoma brucei* inactivated with β-propiolactone and with some other inactivating agents. *Can. J. Microbiol.*, **13**, 743–747.
Soltys, M. A. & Woo, P. (1969) Multiplication of *Trypanosoma congolense* in vertebrate hosts. *Trans. R. Soc. trop. Med. Hyg.*, **63**, 490–494.
Soltys, M. A. & Woo, P. (1970) Further studies on tissue forms of *Trypanosoma brucei* in a vertebrate host. *Trans. R. Soc. trop. Med. Hyg.*, **64**, 692–694.
Splitter, E. J. & Soulsby, E. J. L. (1967) Isolation and cultivation of *Trypanosoma theileri* in tissue culture media. *Exp. Parasit.*, **21**, 137–148.
Stephen, L. E. (1962) Some observations on the behaviour of trypanosomes occurring in cattle previously treated with prophylactic drugs. *Ann trop. Med. Parasit.*, **56**, 415–421.
Stephen, L. E. (1966) Pig trypanosomiasis in tropical Africa. Commonwealth Bureau of Animal Health. Review Series, No. 8.
Stewart, J. L. (1951) The West African shorthorn cattle. Their value to Africa as trypanosomiasis-resistant animals. *Vet. Rec.*, **63**, 454–457.
Tizard, I. R., Holmes, W., Nielsen, K., Mellors, A. & York, D. (1977) Toxin Production by *Trypanosoma congolense*, Abstract 121. In: *Abstracts of Papers read at 5th Int. Congr. Protozoology, New York, 1977*, ed. S. H. Hutner. New York: The Print Shop.
Trigg, P. I. & Gutteridge, W. E. (1977) Morphological, biochemical and physiological changes occurring during the life cycles of parasitic protozoa. In: *Parasite Invasion*, eds A. E. R. Taylor & R. Miller, pp. 57–81. Oxford: Blackwell Scientific Publications.
Urquhart, G. M., Murray, M. & Jennings, F. W. (1972) The immune response to helminth infection in trypanosome infected animals. *Trans. R. Soc. trop. Med. Hyg.*, **66**, 342–343.
Urquhart, G. M., Murray, M., Murray, P. K., Jennings, F. W. & Bate, E. (1973) Immunosuppression in *Trypanosoma brucei* infections in rats and mice. *Trans. R. Soc. trop. Med. Hyg.*, **67**, 528–535.
Vattuone, N. H. & Yanovsky, J. F. (1971) *Trypanosoma cruzi*: agglutination activity of enzyme treated epimastigotes. *Exp. Parasit.*, **30**, 349–355.
Vickerman, K. (1962) The mechanism of cyclical development in trypanosomes of the *Trypanosoma brucei* subgroup: an hypothesis based on ultrastructural observations. *Trans. R. Soc. trop. Med. Hyg.*, **56**, 487–495.
Vickerman, K. (1965) Polymorphisms and mitochondrial activity in sleeping sickness trypanosomes. *Nature, Lond.*, **208**, 762–766.
Vickerman, K. (1966) Genetic systems in unicellular animals. *Sci. Progr. Oxford*, **54**, 13–26.
Vickerman, K. (1969) On the surface coat and flagellar adhesion in trypanosomes. *J. cell. Sci.*, **5**, 163–193.
Vickerman, K. (1971) Morphological and physiological considerations of extracellular blood protozoa. In: *Ecology and Physiology of Parasites*, ed. A. M. Fallis, pp. 58–91. Toronto: Toronto University Press.
Vickerman, K. (1972) The host–parasite interface of parasitic protozoa. Some problems posed by ultrastructural studies. In: *Functional Aspects of Parasitic Surfaces*, ed. A. E. R. Taylor & R. Muller, vol. 10, pp. 71–91. Oxford: Blackwell Scientific Publications.
Vickerman, K. & Luckins, A. G. (1969) Localization of variable antigens in the surface coat of *Trypanosoma brucei* using ferritin-conjugated antibody. *Nature, Lond.*, **224**, 1125–1127.
Watson, E. A. (1915) Dourine and the complement fixation test. *Parasitology*, 8, 156–182.
Weitz, B. (1960) The properties of some antigens of *Trypanosoma brucei*. *J. gen. Microbiol.*, **23**, 589–600.
Wellde, B., Lotzsch, R., Deindl, G., Sadun, E. H., Williams, J. & Warui, G. (1974) *Trypanosoma congolense*. I. Clinical observations of experimentally infected cattle. *Exp. Parasit.*, **36**, 6–19.
Wells, E. A. (1972) Infections of cattle with trypanosomes of the subgenus Megatrypanum (Hoare, 1964). Review Series, No. 10. Commonwealth Agric. Bureaux. Farnham Royal, England.
Whiteside, E. F. (1962) Interactions between drugs, trypanosomes and cattle in the field. In *Drugs, Parasites and Hosts*, ed. L. G. Goodwin & R. H. Nimmo-Smith. London: Churchill.
Wijers, D. J. B. & Willett, K. C. (1960) Factors that may influence the infection rate of *Glossina palpalis* with *Trypanosoma gambiense*. II. The number and morphology of the trypanosomes present in the blood of the host at the time of the infected feed. *Ann. trop. Med. Parasit.*, **54**, 341–350.
Williams, G. D., Adams, L. G., Yaeger, R. G., McGrath, R. K., Read, W. K. & Bilderback, W. R. (1977) Naturally occurring trypanosomiasis (Chagas' disease) in dogs. *J. Am. vet. med. Ass.*, **171**, 171–177.
Woodruff, A. W., Ziegler, J. L., Hathaway, A. & Gwata, T. (1973) Anaemia in African trypanosomiasis and 'big spleen disease' in Uganda. *Trans. R. Soc. trop. Med. Hyg.*, **67**, 329–337.
Woody, N. C. & Woody, H. B. (1955) American trypanosomiasis (Chagas' disease); first indigenous case in the United States. *J. Am. med. Ass.*, **195**, 676–677.
Wright, K. A. & Hales, H. (1970) Cytochemistry of the pellicle of blood stream forms of *Trypanosoma (Trypanozoon) brucei*. *J. Parasit.*, **56**, 671–683.

LEISHMANIA AND CRYPTOBIA

Adler, S. (1964) Leishmania. *Adv. Parasit.*, **2**, 35–96.
Aikawa, M. & Sterling, C. R. (1974) *Intracellular Parasitic Protozoa*. New York: Academic Press.
Ansari, N. & Mofidi, Ch. (1950) Contribution à l'étude des 'formes humides' de Leishmaniose cutanée. *Bull. Soc. Path. exot.*, **43**, 601–607.
Berberian, D. A. (1939) Vaccination and immunity against oriental sore. *Trans. R. Soc. trop. Med. Hyg.*, **33**, 87–94.
Beveridge, E. & Neal, R. A. (1967) Chemotherapy of cutaneous and visceral leishmaniasis in laboratory animals. *Dermatol. Int.*, **6**, 163–164.
Biagi, F. F. (1953) Alguos comentarios sobre gicos: *Leishmania tropica mexicana*, nueva subespecie. *Med. Rev. Mex.*, **33**, 401–406.
Bradley, D. J. (1977) Regulation of *Leishmania* populations within the host. II. Genetic control of acute susceptibility of mice to *Leishmania donovani* infection. *Clin. exp. Immunol.*, **30**, 130–140.
Bradley, D. J. & Kirkley, J. (1977) Regulation of *Leishmania* populations within the host. I. The variable course of *Leishmania donovani* infections in mice. *Clin. exp. Immunol.*, **30**, 119–129.
Bray, R. S. (1974) Zoonoses in leishmaniasis. In: *Parasitic Zoonoses, Clinical and Experimental Studies*, ed. E. J. L. Soulsby, pp. 65–77. New York: Academic Press.
Bray, R. S., Ashford, R. W. & Bray, M. A. (1973) The parasite causing cutaneous leishmaniasis in Ethiopia. *Trans. R. Soc. trop. Med. Hyg.*, **67**, 345–348.
Chance, M. L., Peters, W. & Shchory, L. (1974) Biochemical taxonomy of *Leishmania*. I. Observations on DNA. *Ann. trop. Med. Parasit.*, **68**, 307–316.
Floch, H. (1953) *Leishmania tropica guyanensis* n. Sp., the agent of tegumentary leishmaniasis in the Guianas and Central America. *Bull. Soc. Path. exot.*, **47**, 784–787.
Gardener, P. J., Chance, M. L. & Peters, W. (1974) Biochemical taxonomy of *Leishmania*. II. Electrophoretic variation of malate dehydrogenase. *Ann. trop. Med. Parasit.*, **68**, 317–325.
Gardener, P. J., Shchory, L. & Chance, M. L. (1977) Species differentiation in the genus *Leishmania* by morphometric studies with the electron microscope. *Ann. trop. Med. Parasit.*, **71**, 147–155.
Goodwin, L. G. & Rollo, I. M. (1955) The chemotherapy of malaria, proplasmosis, trypanosomiasis, and leishmaniasis. In: *Biochemistry and Physiology of Protozoa*, eds S. H. Hutner & A. Lwoff. New York: Academic Press.
Kadivar, D. M. H. & Soulsby, E. J. L. (1975) Model for disseminated cutaneous leishmaniasis. *Science*, **190**, 1198–1200.
Kagan, I. G. (1974) Advances in the immunodiagnosis of parasitic infections. *Z. Parasitenk.*, **45**, 163–195.
Kozhevnikov, P. V. (1959) On cross-immunity between rural and urban skin leishmaniasis. *Med. Parasit. Moscow*, **28**, 695–699.
Lainson, R. & Shaw, J. J. (1972) Leishmaniasis of the New World: taxonomic problems. *Br. med. Bull.*, **28**, 44–48.
Lainson, R. & Shaw, J. J. (1973) Paper presented at the 12th meeting of the PAHO Advisory Committee on Medical Research. *Bull. Pan. Am. Hlth Org.*, **7**, 1–19.
Levine, N. D. (1973) *Protozoan Parasites of Domestic Animals and Man*, 2nd ed. Minneapolis: Burgess.
Maekelt, G. A. (1972) Immune Response to Intracellular Parasites. I. Leishmania. In: *Immunity to Parasites*, ed. E. J. L. Soulsby, pp. 343–363. New York: Academic Press.
Manson Bahr, P. E. C. (1963) Active immunization in leishmaniasis. In: *Immunity to Protozoa*, ed. P. C. C. Garnham, A. E. Pierce & I. Roitt, pp. 235–252. Oxford: Blackwell Scientific Publications.
Pelligreno, J. (1951) Nota preliminar sobre a reação intradérmica feita com a fração Polissacaridea isolada de formas de cultura da *Leishmania braziliensis* en Casos de Leishmaniose Teguimentar Americana. *Hospital (Rio de Janeiro)*, **39**, 859.
Peters, W. (1974) Drug resistance in trypanosomiasis and leishmaniasis. In: *Trypanosomiasis and Leishmaniasis with Special Reference to Chagas' Disease*, eds Katherine Elliott, Maeve O'Connor and G. E. W. Wolstenholme, CIBA Foundation Symposium 20 (New Series), pp. 309–326. Amsterdam: Elsevier.
Preston, P. M. & Dumonde, D. C. (1976) Immunology of clinical and experimental leishmaniasis. In: *Immunology of Parasitic Infections*, eds S. Cohen & E. H. Sadun, pp. 167–202. Oxford: Blackwell Scientific Publications.
Reichenbach-Klinke, H. & Elkan, E. (1965) *The Principal Diseases of Lower Vertebrates. Diseases of Fish*. London: Academic Press.
Schnur, L. F. & Zuckerman, A. (1977) Leishmanial excreted factor (EF) serotypes in Sudan, Kenya and Ethiopia. *Ann. trop. Med. Parasit.*, **71**, 273–294.
Stauber, L. (1963) Immunity to Leishmania. *Ann. N.Y. Acad. Sci.*, **113**, 409–417.
Thorson, R. E., Bailey, W. S., Sherril, W., Hoerlein, A. B. & Siebold, H. R. (1955) A report of a case of imported visceral leishmaniasis of a dog in the United States. *Am. J. trop. Med. Hyg.*, **4**, 18–22.
Turk, J. L. & Bryceson, A. D. M. (1971) Immunological phenomena in leprosy and related diseases. *Adv. Immunol.*, **13**, 209–266.
Vattuone, N. H. & Yanovsky, J. F. (1971) *Trypanosoma cruzi*: agglutination activity of enzyme-treated epimastigotes. *Exp. Parasit.*, **30**, 349–355.
Walton, B. C., Brooks, W. H. & Arjona, I. (1972) Serodiagnosis of American leishmaniasis by indirect fluorescent antibody test. *Am. J. trop. Med. Hyg.*, **21**, 296–299.
Zuckerman, A. (1975) Parasitological review. Current status of the immunology of blood and tissue protozoa. I. Leishmania. *Exp. Parasit.*, **38**, 370–400.

ORDER: TRICHOMONADIDA KIRBY, 1947

Organisms of this order are characterized by possessing four to six flagella, one of which may be a trailing flagellum, frequently attached to an undulating membrane. They may have one or two nuclei and reproduction is asexual, usually by binary fission. Cysts may be produced in some forms.

The majority of the forms are non-pathogenic and a large variety of species is found in the alimentary canal of animals.

From time to time pathological changes have been ascribed to a number of species, since they may be found in large numbers in the faeces or in samples from the digestive tract. However, this is not conclusive evidence of a disease relationship, especially since many species multiply readily in a fluid environment. There are, however, some pathogenic forms, these being found in the genera *Tritrichomonas*, *Trichomonas*, *Giardia* and *Hexamita*.

FAMILY: TRICHOMONADIDAE CHALMERS AND PEKKOLA, 1918 emend. KIRBY, 1946

These are the trichomonads occurring usually in the digestive tract but they may also be found in the reproductive system and elsewhere.

They are pyriform in shape, have a rounded anterior end and a somewhat pointed posterior end. There is a single nucleus in the anterior part of the body and anterior to this is the blepharoplast which is associated with a number of basal granules. Arising from the blepharoplast are the anterior flagella and a posterior flagellum, which runs along the edge of an undulating membrane and often extends posteriorly from the body. A deeply staining costa extends along the base of the undulating membrane and especially characteristic is a rod-like axostyle which runs through the body arising from the blepharoplast and emerging from the posterior end. Several genera occur in the family and the speciation is largely dependent on the number of anterior flagella. The genera of interest include *Tritrichomonas, Trichomonas, Trichomitus, Tetratrichomonas* and *Pentatrichomonas.*

Genus: Tritrichomonas Kofoid, 1920

Members have three anterior flagella and lack a pelta. Details of the many species which occur may be obtained from Grassé (1952), Kudo (1966) and Levine (1973). The following are the species of major interest in domestic animals.

Tritrichomonas foetus (Riedmuller, 1928) Weinrich and Emmerson, 1933 (Fig. 3.8). This is a parasite of cattle. It may also occur in the zebu, pig, horse and deer but pathogenic effects are seen only in the bovine, in which it causes the specific venereal disease, bovine trichomoniasis.

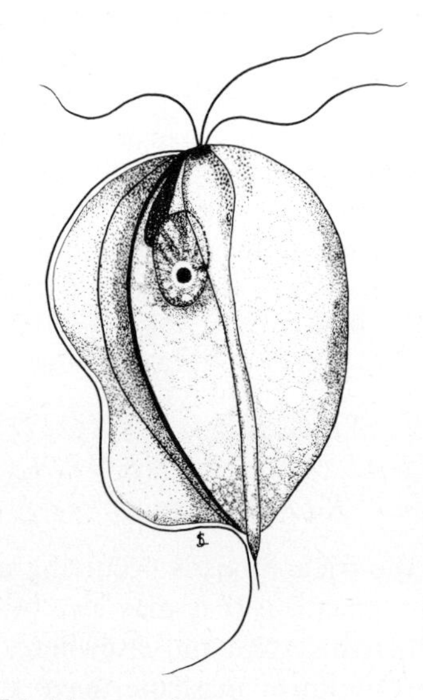

Fig 3.8 *Tritrichomonas foetus.*

The organism is world-wide in distribution and at one time was of major economic importance, especially in dairy herds. Its incidence in cattle is now very much less than formerly, due to the widespread use of artificial insemination and to the decrease in the number of bulls kept on small dairy farms. It is still of importance in beef herds, and in a seven-year survey made in the Rocky Mountain States of the USA Johnson (1964) found it in 62 of 828 beef bulls tested in 34 separate herds, nine herds being infected, giving a herd infection rate of 26%.

Morphology. The organism is roughly pear-shaped, 10–25 μm long by 3–15 μm wide. It moves with a characteristic vigorous jerky movement. The nucleus is anterior, there is a cytostome which is difficult to discern and a sausage-shaped parabasal body. Three anterior flagella are present and the posterior flagellum extends back along the undulating membrane and trails behind the organism. The undulating membrane runs the full length of the body, the costa is prominent, the axostyle is well developed and it emerges from the end of the body through a chromatic ring. The fine structure of *T. foetus* has been studied by Honigberg et al. (1971).

Multiplication is by longitudinal binary fission. No sexual process or encystation has been observed.

The organism can be readily cultured on a variety of media. For immunological work a 'diphasic' glucose–broth–serum medium (Kerr & Robertson 1953) has been used while a trypticase yeast extract, maltose, cysteine and serum medium has been used by Diamond (1957). A detailed discussion of the nutritional requirements of the species is given by Shorb (1964).

Three serologically distinct strains of organism have so far been found; these are the 'Belfast', 'Manley' and 'Brisbane' strains. The Belfast strain is the predominant strain in Europe, Africa and the USA and the 'Brisbane' serotype predominates in Australia (Dennett et al. 1974). The number of outbreaks of disease due to the Manley strain are few. The strains have been maintained in culture for more than 20 years and still retain their immunological distinctness (Kerr 1964). The behaviour and pathogenicity of strains of *T. foetus* in chick liver cell cultures have been studied by Kulda and Honigberg (1969). Three strains of mild, intermediate and high pathogenicity respectively, selected on the basis of a subcutaneous mouse assay, showed distinct differences. A mild strain was readily engulfed and digested by macrophages. The intermediate and highly pathogenic strains caused marked changes in the cell culture, with inhibition of fibroblast division, which were associated with a toxic substance produced by the trichomonads.

Transmission. Under natural conditions the infection is transmitted during coitus. It may also be transmitted by artificial insemination and by gynaecological examination of cows.

Pathogenesis. In the bull, the principal infection site is the preputial cavity. Bartlett (1949) found *T. foetus* only on the surface of the penile and preputial membranes and generally no lesions of diagnostic significance occurred and the infection did not affect the fertility or sexual behaviour of the bull. More recent studies by Parsonson et al. (1974), using culture techniques, confirmed that infection was confined to the preputial cavity. Other workers have indicated that early clinical signs are pain on micturition and a disinclination to serve cows. A mucopurulent discharge may be present. Such signs disappear one to two weeks after infection. Unlike the cow, spontaneous recovery does not occur in the bull and the bull should be regarded as a permanent source of infection.

In the female, natural infection is introduced into the vagina during coitus. The initial lesion is a vaginitis of varying intensity; it may be so slight as to pass unnoticed or it may produce a mucopurulent discharge. The maximum number of organisms are present in the vagina 14–18 days after service (Hammond & Bartlett 1945) and from this site they invade the uterus through the cervix. They may then completely disappear from the vagina, or a low level of infection may remain, producing catarrh and roughness of the vaginal mucosa.

The sequelae to the invasion of the uterus are several. In some cases, despite the infection, the animal may conceive normally, carry the calf to full term, and give birth to a normal healthy calf. The number of infections where this happens is, however, small.

Frequently, a placentitis results with detachment of the placental membranes and death of the fetus. This leads to an abortion, which is characteristically early, usually 8–16 weeks after the infected service. Later abortions (i.e. after six months) are rare and only a few verified cases have been reported (Morgan & Hawkins 1952). If abortion occurs very early (e.g. one to two weeks) the fetus and membranes frequently pass out unnoticed and this may lead to the belief that there has been an irregular oestrous cycle. Following the abortion there may be a uterine discharge and, where this persists, the animal may show a series of irregular heat periods. In some animals the fetus and membranes are not completely eliminated and maceration occurs leading to a chronic catarrhal and sometimes purulent endometritis. This causes more prolonged, and at times permanent, sterility. There is anoestrus, and a persistent uterine discharge occurs, which may be especially noticeable when the animal lies down.

At times the cervix is closed and, in the presence of a retained corpus luteum, a closed pyometra results. Fluid accumulates over several weeks and ultimately the uterus contains a large volume of thin, greyish white, odourless material swarming with trichomonads. There is associated anoestrus and the owner may consider the animal to be pregnant during this period.

Immunity to trichomoniasis in cattle. There is little evidence that bulls become immune to infection. Usually, the organism persists in them for life and the animal is a continuous source of infection for cows.

In the cow, infected non-pregnant animals tend to recover without treatment and persistence of the infection postpartum is rare, although a small proportion of cows may remain carriers for more than a year. Despite the development of immunity to the parasite, permanent sterility may result if there has been extensive involvement of the uterine mucosa.

Reviews of the immunology of *T. foetus* infection have been given by Robertson (1963), Kerr (1964) and Honigberg (1970). Robertson (1963) tabulates the antibodies produced in *T. foetus* infection as:

1. Circulating humoral antibody stimulated by antigen which reaches the systemic circulation from the uterus.
2. Uterine antibody developing in situ.
3. Vaginal antibody developing locally.

These antibodies are most readily detected by an agglutination technique (Kerr & Robertson 1941, 1943). A capillary agglutination test has been described by Feinberg (1952).

A normal agglutinin for *T. foetus* has been detected in cattle, being at a fairly constant titre of 1–48 to 1–96. It is absent from the serum of unsuckled calves at birth and is acquired with other antibodies in colostrum. It persists in a calf for 17–55 days (Kerr & Robertson 1954).

The circulating humoral antibody is developed in field infections only when there is an adequate degree of infection with large numbers of trichomonads in the uterus. This occurs in cases of abortion or pyometra (Pierce 1949*b*). Circulating antibody can also be induced by intramuscular injection of lyophilized organisms, but this does not pass to the lumen of the uterus. Circulating antibody sensitizes the skin of the animal and the development of this can be followed by the intradermal injection of a diethylene glycol extract of lyophilized *T. foetus* (Feinberg & Morgan 1953). The skin reaction is of the immediate type hypersensitivity, producing a local oedema which reaches its maximum in 20–30 minutes and disappears within two to three hours.

Pierce (1953, 1959) demonstrated that vaginal antibody could occur before there was circulating antibody in the serum, and he also found that a high titre of antibody could exist in the vaginal mucus while trichomonads were numerous in the uterine discharge. He concluded that the vaginal mucosa was producing antibody locally, independently of the uterine and circulating antibodies. The protective effects of the vaginal antibody are suggested by Robertson (1963) to be a local control of parasites in the vagina. The uterine antibody is induced by trichomonads in the uterus. It is responsible for the disappearance of *T. foetus* from the uterus in mild infections, especially where pregnancy is not interrupted.

Epidemiology. Under natural conditions, trichomoniasis of the bovine is spread by the bull. Once infected he must be regarded as being a permanent source of infection, whereas in the female the infection is self-limiting and the parasites gradually disappear. Following recovery from infection, a cow will usually conceive and undergo gestation without any danger of abortion. A cow may, however, also serve as a carrier of the infection.

Of the other methods of infection, artificial insemination is the most common, although gynaecological examinations of cows may be responsible for local spread. The use of fresh semen from an infected bull in artificial insemination may lead to widespread outbreaks of disease; however, this danger has been recognized early in the development of artificial insemination and strict examination of bulls in artificial insemination centres avoids this. The development of techniques for preserving bovine semen has also assisted in the control of *T. foetus* infection. A number of investigators have reported, variously, that *T. foetus* does or does not survive when frozen in the presence of glycerol. The organisms are killed by glycerol at 37°C or at ordinary refrigeration temperatures but below this, survival depends on the diluting fluids used to preserve the semen and the stage of the population growth of *T. foetus*. The subject is reviewed by Levine et al. (1958). A detailed study on the survival of *T. foetus* under extended storage conditions in the presence of glycerol has been carried out by Levine et al. (1962). These authors found that the organisms survived much better at −95°C than at −28°C in the presence of 1 M glycerol. At the lower temperature trichomonads remained viable for up to 256 days.

The importance of other animals in the transmission of *T. foetus* to cattle has received attention. Organisms similar to *T. foetus* have been found in the genital tract and aborted fetuses of horses

(Schoop & Oehlikers 1939), and Schoop and Stolz (1939) have found trichomonads resembling *T. foetus* in the genital organs of roe deer in Germany. There is little evidence, however, that these play any part in the epidemiology of the bovine infection. The relationship of *Tritrichomonas foetus* of cattle to the swine form *Tritrichomonas suis* remains to be determined. Switzer (1951) reported that the inoculation of cultures of trichomonads from the nasal and digestive tract of swine into the vagina of cattle produced infections. Fitzgerald et al. (1958) infected a series of heifers with trichomonads from the nose, stomach or caecum of pigs. Infections lasted up to 133 days with the nasal forms, 88 days with the stomach forms, and two infections with caecal trichomonads lasted 84 and 33 days. In one experiment the intrauterine inoculation of swine caecal trichomonads into a four-months' pregnant heifer was followed 20 days later by an abortion and trichomonads were found in the fetal fluids and in the fetus. Such work confirms the idea that swine trichomonads can become established in the reproductive tract of cattle but it is not known whether they play any part in the naturally occurring disease. The spontaneous recovery which is seen with the swine trichomonad infection, especially in the male bovine, would suggest that infection is of a temporary nature. However, Robertson (1960) studied the antigenic relationships between bovine and swine trichomonads and found cross-reactions between them. Two strains of *T. suis* were more closely related to the Belfast strain of *T. foetus* than to the Manley strain, and she concluded that the serological distinction between the organisms did not justify separate speciation. She proposed that they all be called *T. foetus*.

Diagnosis. Presumptive evidence of trichomonads in a herd can be obtained from a history of early abortions, an increased incidence of cows returning to service and a failure of animals to become pregnant except after repeated service. There may also be an increase in the prevalence of vaginal discharge and pyometra in a herd, all of which may be related to the importation of new stock, especially a new bull.

Confirmation of a diagnosis is based on the demonstration of the organisms in the vaginal or uterine discharges or in the fetus, and on serological tests. Trichomonads are found most readily in the stomach of an aborted fetus, in the amniotic and allantoic fluids, and in the uterine discharges after abortion. Where such material is not available uterine discharges may be collected from the cow and this is best done two to three days before the expected time of the next oestrus. Samples may also be obtained by washing the vagina with physiological saline or by irrigation of the uterus with saline. In the bull, preputial washings are the best source of material. In all sampling procedures, however, it is important to avoid contamination with faecal material since this may introduce intestinal protozoa into the sample and these may readily be confused with *T. foetus*. In heavily infected material, especially purulent mucus, the organism may be observed directly, the sample being examined on a slide under a coverslip. In less grossly infected samples the material may be allowed to sediment and be centrifuged before examination. The sample should be kept warm throughout and warm saline added before examination. The main feature of *T. foetus* is its characteristic motility, but old material, or that which has become cold, may show very sluggish organisms and careful observation is required to detect the parasites. In sluggish forms the undulating membrane is seen but care should be taken not to confuse the organism with the many other species of protozoa that may be present in samples, especially if these have been obtained carelessly.

Where organisms are too few to allow an accurate diagnosis, the preparation should be stained or, more satisfactorily, cultures should be prepared from the material. Several media are available for cultivation purposes and have been mentioned previously. These should be examined at 24 and 48 hours and also four days. The cervical mucus agglutination test is the most satisfactory immunodiagnostic test (Pierce 1949*a*). Mucus samples are collected from the vagina of the cow, using a sterile glass tube 50 cm in length and 9 mm in diameter, and bent at an angle of 150° about 9 cm from one end. Mucus is obtained preferably a few days after oestrus, from the anterior end of the vagina by suction. The mucus is mixed with glucose saline and serially diluted or the antibody in the mucus allowed to diffuse into saline which is then diluted. The test organisms consist of a suspension of in vitro cultured *T. foetus*, a density of 100 000 organisms/ml being satisfactory. Although the mucus agglutination test is considerably superior to the serum agglutination test (Pierce 1949*a*), it is essentially a herd test since the status of the sexual cycle of the cow may affect the reaction. Unsatisfactory results are obtained with mucus samples taken during the oestrus period, or shortly after it, and mucus from pregnant animals may give false reactions.

An intradermal test developed by Kerr (1944) gives inconsistent results. Thus, Morgan (1948) obtained negative results with it and other workers have found that animals may be desensitized to the skin test during acute uterine infections with *T. foetus*, or by the injection of antigen intramuscularly.

It is important, with any diagnostic method, to conduct several examinations before an animal is declared uninfected. Even after a satisfactory gestation and the birth of a normal calf an animal should be bred by artificial insemination to avoid the risk of infecting bulls. In the bull, more lengthy testing is necessary before he can be considered free from infection. It is useful to breed the bull to two or more virgin heifers to assess his freedom from infection.

Treatment. A vast number of compounds have been used in the treatment of trichomoniasis in the cow but, since the infection is essentially a self-limiting one and leads to an adequate immunity, management should be aimed at giving the animals a breeding rest and, subsequently, artificial insemination should be used to avoid infecting clean bulls.

Infected bulls are much more difficult to treat since trichomonicidal agents must be introduced to all parts of the preputial cavity. Since treatment is tedious, time-consuming, and must be repeated on several occasions, it is frequently better to slaughter the bull.

In very large herds, for example of beef cattle, artificial insemination may be impractical and the detection of infection in individual animals similarly so. Segregation of young non-infected cattle from older previously infected ones may be impossible. In such circumstances, Bartlett and Dikmans (1949) have recommended ensuring that cows are permitted service only with non-infected bulls and only when at least 90 days have elapsed since the completion of normal pregnancy. This, too, may be impractical in large herds and Clark et al. (1974) described a control system in a breeding herd of 11 000 beef cattle in Australia. Young bulls, aged between one and three years, show a much reduced incidence of infection (this suggests they are refractory to infection) and the disposal of bulls aged four years or more removes the major source of infection. The use of young non-infected bulls aged one, two and three years during a limited breeding season, after which segregation of bulls and cows occurred, resulted in control of trichomoniasis.

In the treatment of the bull, pudendal anaesthesia (Larson 1953) is used to relax the retractor penis muscle. The penis is washed with a weak solution of detergent, dried and flavine ointment introduced into the preputial cavity and massaged in for 15–20 minutes. A solution of acriflavine is also injected into the urethra to kill any organisms which may be there. Fitzgerald et al. (1963) demonstrated that acriflavine ointment was effective in eliminating infection in the majority of treated bulls; however, the use of more than 1% of acriflavine may result in tissue damage. Several treatments may be necessary to eliminate infection but, if the epididymis or testes are affected, treatment is of little use.

Fitzgerald et al. (1963) reported that Berenil (4,4-diamidinodiazoaminobenzene diaceturate tetrahydrate) is also a useful therapeutic agent for bovine trichomoniasis. 100–150 ml of a 1% solution of Berenil is injected into the prepuce and is retained there for 15 minutes while the penis and the prepuce are thoroughly massaged. Five successive daily treatments usually eliminate an infection. McLoughlin (1965) has reported that dimetridazole (1,2-dimethyl-5-nitroimidazole) was effective systemically. Oral administration of 50 mg/kg daily for five days is effective, as is a single intravenous dose of 50 mg/kg or intravenous injections of 10 mg/kg daily for five days. A similar dose was reported to be effective for cows (McLoughlin 1970).

Control. The major control measure for trichomoniasis in cattle is the use of artificial insemination. The practice of using communal bulls is to be discouraged since it leads to the spread of the disease. In an overall policy it is also wise to eliminate cows which have been infected since, although they may breed satisfactorily, their freedom from infection cannot be assumed.

Tritrichomonas suis (Gruby & Delafond, 1843). This is the largest of the pig trichomonads, occurring chiefly in the stomach, but also in the nasal passages, caecum and small intestine. The trichomonad parasites of pigs have been studied in detail by Hibler et al. (1960), who concluded that three forms were present, namely, the large *T. suis* and two small forms, *Trichomitus rotunda* and *Tetratrichomonas buttreyi* (see later).

T. suis is world-wide in distribution. It is elongate or spindle-shaped, 11.19–14.44 μm by 3.3–3.5 μm. Forms as small as 9 μm and up to 16 μm in length may be found. The three anterior flagella are equal in length, each ending in a small spatulate knob. The undulating membrane and the costa run the full length of the body and the axostyle extends 0.6–1.7 μm beyond the body as a cone-shaped projection, narrowing sharply to a short tip.

In the past, the association of *T. suis* with a high percentage of cases of atrophic rhinitis has led to the belief that this organism is responsible for that condition. The disease has been produced in young pigs with nasal washings containing trichomonads but, when *T. suis* is grown axenically, it is no longer able to produce the disease. It is now generally known that atrophic rhinitis is due to another agent. Consequently, *T. suis* is regarded as non-pathogenic.

Tritrichomonas equi (Fantham, 1921). This species has been found in the caecum and colon of horses. It is probably world-wide in distribution, although it has been specifically reported only from South Africa and the United States. It measures 11 μm by 6 μm and has three anterior flagella and an undulating membrane with a slender axostyle.

There is debate regarding the pathogenicity of *T. equi*. Several authors claim it to be the cause of equine protozoan diarrhoea (Wohler 1961) which may vary from a severe peracute enteritis, often fatal, to a chronic intestinal disorder. For example, Bennett and Franco (1969) consider an acute equine diarrhoea of racehorses in Port of Spain, Trinidad, is caused by *T. equi*. They report that 40% of 180 racehorses with the disease have died over a 12-year period. Laufenstein-Duffy (1969) described the condition as a 'functional colitis' with complete or partial loss of intestinal flora. She reports that the onset of clinical disease is associated with a lowering of the resistance of the animal by other infection, e.g. respiratory virus infection, colic, stress, etc. The onset may be sudden with severe diarrhoea and foul-smelling green watery faeces. Iodochlorhydroxyquin (10–20 g/day for five days) is recommended for treatment.

However, other studies have indicated that it is unlikely that *T. equi* causes diarrhoea; the large number of organisms found in the diarrhoeic faeces of horses are a response to, rather than the cause of, the altered environment of the gastro-intestinal tract. Manahan (1970) studied a chronic diarrhoea syndrome of racehorses in Sydney, Australia, and failed to isolate a single causative organism from any case. Trichomonads were seen in some cases and not in others, and large numbers of other organisms (e.g. *Giardia*, *Balantidium*, fungi and bacteria) were also found. Following a survey of *T. equi* infection in horses in Colorado, Damron (1976) concluded that the role of *T. equi* as the aetiological agent of equine diarrhoea was doubtful. Clearly, further studies are necessary to clarify the role of *T. equi* in equine diarrhoea.

Tritrichomonas eberthi (Martin & Robertson, 1911) Kofoid, 1920 (Fig. 3.9b). This is a common trichomonad occurring in the caecum of the chicken, turkey and duck. It is world-wide in distribution and appears to be common, McDowell (1953) finding it in 35% of chickens in

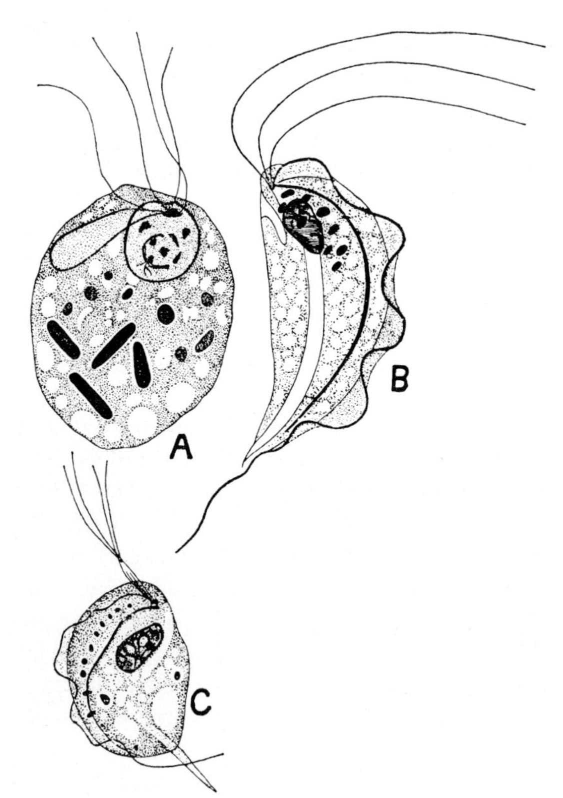

Fig 3.9 Flagellates from the caecum of the fowl. (A) *Chilomastix gallinarum.* (B) *Tritrichomonas eberthi.* (C) *Tetratrichomonas gallinarum.* (× 4000) (*From Wenyon 1926*)

Pennsylvania. It is broadly crescent-shaped, 8–14 μm by 4–7 μm, there are three anterior flagella, the undulating membrane is prominent, extending the full length of the body. The costa and axostyle are distinct.

There is no evidence that this species is pathogenic.

Several other species of *Tritrichomonas* occur in various domestic animals. In domestic animals *Tritrichomonas enteris* (Christl, 1954) has been reported from the caecum and colon of the ox and zebu in Germany and in the East, and an unnamed *Tritrichomonas* species has been reported from the faeces of cattle in the USA. An unnamed species has been found in the faeces of a dog with diarrhoea in Italy (Morganti & Bitti 1968). Neither form is associated with any pathogenic effects.

Tritrichomonas species of rodents, including laboratory rodents, are:

Tritrichomonas muris (Grassi, 1879) in the caecum and colon of rat, mouse, hamster and wild rodents. It measures 16–26 μm by 10–14 μm.

Tritrichomonas minuta (Wenrich, 1924), a small form 4–9 μm by 2–5 μm, occurs in the caecum and colon of rat, house mouse and golden hamster.

Tritrichomonas caviae (Davaine, 1875), a medium-sized form, 10–20 μm by 6–11 μm in the caecum and colon of the guinea-pig.

None of these species is pathogenic.

Genus: Trichomonas Donné, 1837

Members of this genus are pyriform and typically have four free anterior flagella. There is no trailing flagellum. Numerous species exist (Grassé 1952).

Trichomonas gallinae (Rivolta 1878) Stabler, 1938 (syn. *Trichomonas columbae*, *T. hepaticum*, *T. halli*, etc.). *Trichomonas gallinae* is the cause of avian trichomoniasis of the upper digestive tract and is found particularly in pigeons, but the turkey, chicken, hawk, mourning dove, golden eagle, etc. may also be infected (Levine 1973).

The organism is pyriform in shape, usually small, and measures 10 μm by 5 μm; however, it may range in size from 6–19 μm by 2–9 μm. There are four anterior flagella, each up to 13 μm in length. The axostyle is narrow and protrudes a short distance posteriorly from the body, the undulating membrane extends about two-thirds of the total length of the body and does not terminate in a trailing flagellum. A cytosome is present.

T. gallinae is extremely common in the domestic pigeon in which it may cause serious disease. It is also fairly common in the turkey and chicken and it may be responsible for disease in mourning doves (Stabler & Herman 1951). Hawks may be infected when they feed on pigeons.

Pathogenesis. Avian trichomoniasis caused by *T. gallinae* is a disease of young birds. 80–90%, or more, of adult pigeons are infected but show no evidence of disease, but infected pigeon squabs may die within ten days. A difference in the pathogenicity of strains of *T. gallinae* has been reported by Stabler. The most virulent is the

Jones' Barn (JB) strain which has maintained its pathogenicity through 300 serial passages in pigeons (Stabler 1953, 1957). Honigberg (1961) found that the growth rate of three strains of *T. gallinae* was positively correlated to the virulence for pigeons and mice. Thus, the virulent JB strain had a generation time of $3\frac{1}{2}$ hours, a TG strain one of 3 hours 50 minutes, and a less virulent (YG) strain one of 5 hours 50 minutes.

With a virulent strain, the earliest lesions appear as small yellowish circumscribed areas in the mouth cavity, especially the soft palate, 3–14 days after infection. These increase in size and number and extend to the oesophagus, crop and proventriculus; the liver may also be involved and there may be extension to the lungs and the serous surfaces of the intestine, the pancreas and the heart. The lesions of the digestive tract do not extend beyond the proventriculus. Histologically, the lesions consist of inflammation and ulceration of the oropharyngeal mucosa associated with a palisading of the parasites on the mucous surface (Mesa et al. 1961). The organisms invade the pharyngeal glands, penetrate the underlying tissues and reach the liver. It is considered that the abscesses in the liver are the cause of death. Mesa et al. (1961) suggested that the JB strain of *T. gallinae* produces a hepatotoxin.

In the turkey and chicken the lesions most commonly occur in the crop, oesophagus and pharynx and are uncommon in the mouth. Lesions in the mourning dove are similar to those seen in the pigeon.

The pathogenicity of *T. gallinae* in experimental hosts has been studied by various authors (Frost & Honigberg 1962). In studies with mice, only the JB strain was significantly pathogenic, while a mild strain of *T. gallinae* produced a minor effect.

Honigberg et al. (1964), using chicken liver cell cultures, found that the parasites caused degeneration of epithelial and fibroblast cells and stimulated activity of macrophages. Organisms multiplied mainly within macrophages but some invasion of epithelial cells and fibroblasts occurred. Subsequent work by Abraham and Honigberg (1965) demonstrated a reduction of RNA in chick liver cell cultures from the eighth hour after inoculation with the parasites, this effect becoming pronounced with time. The DNA in such cultures remained unchanged.

Clinical signs. Infected pigeon squabs show an initial depression with ruffled feathers and later they become weak and emaciated. There is an accumulation of a greenish fluid, or cheesy material, in the mouth and crop and this may exude from the beak. In turkey poults and chickens, drowsiness and a pendulous crop are usually observed and there is a foul odour from the mouth. On post mortem, yellowish to greyish necrotic lesions are evident in the mouth, crop, and oesophagus with extension to the bones of the skull, the liver and elsewhere. Organisms are usually very numerous in the mouth and crop contents.

Immunity. Recovered birds are immune to reinfection but survive as symptomless carriers. Infection with a non-virulent strain produces immunity against the more virulent strains (Stabler 1951). Certain breeds of birds may be more susceptible than others and Miessner and Hansen (1936) considered that Roller and Tumbler pigeons were the most susceptible. Two major groups of antigens, A and B, were detected by Stepkowski and Honigberg (1972), who examined virulent and avirulent strains of *T. gallinae* by gel precipitation techniques. The results suggested a relationship between antigenic constitution and virulence of strains.

Epidemiology. In pigeons, the organism is transmitted directly from carrier older birds to the newly hatched pigeon squab via the 'pigeon's milk' from the crop. This may occur within a few minutes of hatching. Adult birds may remain infected for a year or more and are a constant source of infection for their young. The mode of infection of turkeys and chickens is most likely to be through drinking water contaminated by trichomonads from the mouths of infected birds (Stabler 1954) but, since the organism does not survive readily, almost direct contamination is necessary. Wild pigeons and other birds probably play an important role in introducing the infection to domestic birds.

Diagnosis. Diagnosis is based on the characteristic lesions in the oral cavity, oesophagus and crop, and usually the organism is readily demonstrated in the greenish fluid material or in the lesions. It is also readily cultured on a variety of media including a glucose–broth–serum medium (Richardson & Kendall 1963) or a trypticase–yeast-extract–maltose–cysteine–serum medium (Diamond 1957).

Treatment. 2-Amino-5-nitrothiazole: Homing pigeons: 30 mg/kg daily for seven days; other birds: 45 mg/kg. Carrier birds may also be treated with this drug (Stabler & Mellentin 1953).

Furazolidone: 25–30 mg/day for seven days, given in gelatin capsules (Stabler 1957).

Control. Control is readily obtained by treatment with 2-amino-5-nitrothiazole. The introduction of infection from wild pigeons and other birds may be prevented by shielding the drinking places.

Trichomonas phasioni Lucas, Toucas, Laroche and Monin, 1956. This causes enteric diseases in pheasant poults in Europe. Birds between three and 12 weeks of age in rearing pens are affected, showing diarrhoea, dehydration and anorexia. Mortality may range from 2 to 40%. Higgins (1980) reports that dimetridazole at 60 g/45 litres of drinking water for five days, followed by half this dose for a further five days, cleared up the diarrhoea and anorexia. Furazolidone has also been recommended at a dose of 1 kg/500 litres of drinking water. The pathology of the infection has been described by Lucas et al. (1956).

Trichomonas vaginalis Donné, 1837. This form occurs in the vagina, prostate and urethra of man. Experimentally, the golden hamster can be infected intravaginally and the mouse subcutaneously.

This is the largest of the human trichomonads, being 7–23 μm by 5–12 μm (mean 17 × 7 μm). There is an oval nucleus at the anterior end; the undulating membrane extends along one-third to two-thirds of the body length; there is no free posterior flagellum and the axostyle is slender and projects from the posterior end as a pointed rod.

Trichomonas vaginalis is world-wide in distribution and various surveys have shown prevalences of from 2% to 90% in various areas. It is more common in females aged 30–49 years; it is uncommon in preadolescent girls, and Faust and Russell (1964) give an overall prevalence in child-bearing women of 25% and in men 4% or more. The infection is venereal in origin, the organism being transmitted during sexual intercourse. In exceptional cases transmission may occur through contaminated towels, clothing and toilet seats.

Pathogenesis. Very frequently *T. vaginalis* infection is asymptomatic, especially so in the male. Definite clinical signs may develop, however, the organism causing degeneration and desquamation of the vaginal epithelium. This is followed by inflammation of the vagina and vulva and a leucocytic discharge may be evident. There may be a tendency for the condition to flare up after each menstrual period. The differences in the intensity of the clinical condition possibly relate to different strains of the organism, since Kott and Adler (1961) demonstrated 19 different serotypes of *T. vaginalis*. Sharma and Honigberg (1969) have studied the cytochemical changes in chick liver cell cultures infected with *T. vaginalis*.

In the male the infection is usually latent and symptomless; occasionally, however, it may be responsible for a urethritis which may be persistent or recurrent. In both males and females bacterial and/or fungal infection may accentuate the local inflammatory reaction. Invasion of the tissues by *T. vaginalis* has been reported by Frost et al. (1961), who found the organism in the cytoplasm of the columnar epithelial cells at the squamocolumnar junction in endocervical biopsy material. The pathogenicity of *T. vaginalis* in experimental hosts such as mice has been studied by Honigberg (1961).

The diagnosis of *T. vaginalis* infection may be made microscopically in the female by the examination of vaginal secretions or urine, and in the male by examination of urine or prostatic secretions. The organism can be readily cultured on several media used for other trichomonads. A specific fluorescent antibody technique was introduced by McEntegart et al. (1958) and this method was assessed by Hayes and Kitcher (1960). These authors found that fluorescent antibody staining gave approximately the same positive diagnosis as the culture method but because it required much less time to perform it was very satisfactory in diagnosis.

Metronidazole is probably the most effective therapy for infection. 250 mg are given twice daily per os for ten days.

Trichomonas tenax (Müller, 1773) Dobell, 1939. This is a non-pathogenic trichomonad of man and monkeys, occurring in the mouth. It is more common in persons with oral disease or dental disorders. It is world-wide in distribution and may be found in up to 26% of persons with dental caries or pyorrhoea and in up to 11% of those with apparently normal, healthy mouths.

The organism is pyriform in outline, 5–12 μm in length, the undulating membrane extends almost to the posterior end of the body but there is no free portion to the posterior flagellum. The axostyle is slender and extends from the posterior part of the body as a pointed rod.

There is no evidence that *T. tenax* is pathogenic. Transmission is probably direct, from droplet spray from the mouth, kissing, or the use of contaminated dishes and drinking water. The organism will survive at normal temperatures in drinking water for several hours.

Although no specific treatment is required, an improvement in oral hygiene will reduce the infection.

Other species of the genus *Trichomonas* which occur in domestic animals are *Trichomonas felistomae*, Hegner and Ratcliffe, 1927, which occurs in the mouth of cats in the United States. It is a non-pathogenic form, 6–11 μm by 3–4 μm. It is possibly the same as *Trichomonas canistomae*, Hegner and Ratcliffe, 1927, which occurs in the mouth of the dog. *Trichomonas equibuccalis*, Simitch, 1939, occurs in the mouth and around the teeth of the horse in Central Europe. It is a non-pathogenic form, 7–10 μm in length. *Trichomonas caballi*, Abraham, 1961, was isolated from the colon of horses in India by Abraham (1961). *Trichomonas macacovaginae*, Hegner and Ratcliffe, 1927, has been found in the vagina of the rhesus monkey.

Genus: Trichomitus Swezy, 1915

Members of this genus have three anterior flagella, a trailing flagellum and a pelta.

Trichomitus rotunda (Hibler, Hammond, Caskey, Johnson and Fitzgerald, 1960) Honigberg, 1963. This is the medium-sized trichomonad of pigs, occurring principally in the caecum and colon. It was recognized as a new species by Hibler et al. (1960) as a result of their extensive study of the trichomonads of swine, it being found in the caecum of 10.5% of approximately 500 pigs in Utah. Up to the present it has been recognized only in North America but it is presumed that it has a similar distribution to that of *T. suis*.

It is broadly pyriform in shape, measuring 8.59 μm by 5.8 μm with a range of 7–11 μm and 5–7 μm. The three anterior flagella are equal in length and terminate in small knob-like structures; the undulating membrane is low, the costa extends along only one-third to two-thirds the length of the body and the posterior free flagellum is shorter than the body. There is no evidence that this organism is pathogenic.

Trichomitus fecalis (Cleveland, 1928) Honigberg, 1963, was isolated from human faeces.

Trichomitus wenyoni (Wenrich, 1946) occurs in the caecum and colon of rats, mice, hamsters and rhesus monkeys.

Genus: Tetratrichomonas Parisi, 1910

Members of this genus have four anterior flagella, a trailing flagellum and a pelta.

Tetratrichomonas buttreyi (Hibler, Hammond, Caskey, Johnson, and Fitzgerald, 1960) Honigberg, 1963. This is the small trichomonad of pigs, occurring in the caecum and colon. So far it has been recognized only in the United States but it is presumed that its distribution is similar to the other two in swine. Hibler et al. (1960) found it in 25% of 496 pigs in Utah. It measures a mean of 5.92 μm by 3.44 μm; there are three or four anterior flagella; the undulating membrane runs the full length of the body and the axostyle is narrow and protrudes 3–6 μm posteriorly from the body. There is no evidence that *T. buttreyi* is pathogenic.

Tetratrichomonas gallinarum (Martin and Robertson, 1911) Honigberg, 1963. (Fig. 3.9*c*) This organism is found in the lower digestive tract and sometimes the liver of turkey, domestic fowl, and guinea-fowl. It has also been found in quail, pheasant and partridge, and forms very similar to it have been found in the Canada goose by Diamond (1957). It is cosmopolitan in distribution.

The organism is pear-shaped, 9–15 μm by 5–9 μm, the axostyle is long and slender and projects from the posterior part of the body. The original form described by Martin and Robertson (1911) had four anterior flagella and one posterior flagellum which extended behind the body. A form with five anterior flagella has been described by Allen (1940), and McDowell (1953) reports that occasionally three or five anterior flagella may occur but four is the most usual number.

Pathogenesis. T. gallinarum has been found in turkeys in liver lesions resembling those due to histomoniasis (Allen 1936, 1941; Walker 1948). These authors consider *T. gallinarum* capable of causing a disease similar to enterohepatitis.

The liver lesions are said to differ from those produced by *Histomonas meleagridis* by having a more irregular outline and by being raised above the liver surface instead of being depressed below it, as in the case of *H. meleagridis*. Allen (1941) reported that the administration to young turkeys of cultures of an organism resembling *T. gallinarum* produced enterohepatitis; however, this finding has not been confirmed and in fact Delappe (1957) was unable to produce any lesions with a strain of *T. gallinarum* isolated from the liver lesions of a turkey with enterohepatitis.

Infection is by the ingestion of trichomonads in contaminated feed or water. McLoughlin (1957) found that *T. gallinarum* could survive for 24 hours in caecal droppings kept at 37°C and for 120 hours when they were kept at 6°C.

Other tetratrichomonads of domestic birds include *Tetratrichomonas anseri* (Hegner, 1929) of the caecum of geese (8 × 4.7 μm) and *Tetratrichomonas anatis* (Kotlan, 1923) from the duck (13–27 μm by 8–18 μm). Neither is known to be pathogenic.

Tetratrichomonas ovis (Robertson, 1932) Honigberg, 1963. This organism was found by Anderson et al. (1962) in the caeca of 12 out of 16 domestic sheep in Illinois and Utah. It is pyriform, 6–9 μm by 4–8 μm, with an anterior nucleus and a slender tapering axostyle which protrudes 5 μm beyond the body. There are four anterior flagella of unequal length and a prominent undulating membrane which extends beyond the body. There is no evidence that this species is pathogenic in sheep.

Tetratrichomonas pavlovi (Levine, 1961) (Syn. *Trichomonas bovis* Pavlov and Dimitrov, 1957) (a new combination created by Levine (1973)) was reported from the faeces of calves with diarrhoea in Bulgaria. The validity of the species and its pathogenesis is open to question.

Tetratrichomonas microti (Wenrich and Saxe, 1950) Honigberg, 1963 is a common form of the caecum of the rat, mouse, hamster, vole and a variety of other rodents. It is 4–9 μm long; it has been transmitted from rodents to the dog and cat but not to man (Simitch et al. 1954).

Genus: Pentatrichomonas Mesnil, 1914

Members of this genus have five anterior flagella and a pelta.

Pentatrichomonas hominis (Davaine, 1860) (syn. *Trichomonas hominis, T. intestinalis* etc.). This is a common intestinal flagellate of man and of other primate species which has been detailed by Flick (1954). It can be transmitted experimentally to a number of laboratory animals including the rat, cat, dog, hamster, guinea-pig and chicken (Levine 1973). Burrows and Lillis (1967) report that it is common in dogs obtained from pounds in the USA.

The body is oval to pear-shaped, 5–20 μm by 3–14 μm; the oval nucleus lies in the anterior half of the body, and there are usually five free anterior flagella, although some forms may possess only three or four; rarely, the number is six or more. The sixth flagellum runs along the undulating membrane and extends behind the body. The undulating membrane extends the full length of the body, a costa is present and the axostyle is thick, hyaline and protrudes outside the body as a pointed process.

Transmission is direct by ingestion, although it has been suggested that flies may serve as mechanical vectors.

There is no evidence that the organism is pathogenic or causes intestinal disturbance. It may be present in large numbers in loose or diarrhoeic faeces.

FAMILY: MONOCERCOMONADIDAE KIRBY, 1947

Organisms of this family possess three to five anterior flagella, with a recurrent flagellum which is usually free.

Genus: Monocercomonas Grassi, 1897

Members of this genus have a pyriform body, three anterior flagella, a trailing flagellum but no undulating membrane. The axostyle is rod-like and protrudes from the posterior end of the body. A list of species is given by Morgan (1944).

Monocercomonas ruminantium (Braune 1914) Levine, 1961, occurs in the rumen of cattle, and *M. gallinarum* (Martin & Robertson, 1911) Morgan and Hawkins, 1948, in the caecum of the chicken. No pathogenic effects are ascribable to these species.

In laboratory animals *Monocercomonas caviae*, *M. pistillum* and *M. minuta* occur in the guinea-pig caecum. The last two occur in North America while *M. caviae* is world-wide in distribution (Nie 1950). *Monocercomonas cuniculi* occurs in the caecum of rabbits. None of these species is pathogenic.

Genus: Histomonas Tyzzer, 1920

Organisms are amoeboid with a single nucleus. A single flagellum arises from a basal granule, close to the nucleus. Previously, this organism was classified in the order *Rhizomastigida* and the family *Mastigamoebidae*.

Histomonas meleagridis (Smith 1895) Tyzzer, 1920. (Figs 3.10 to 3.13) This is found in the caeca and liver and is the cause of histomoniasis, infectious enterohepatitis, or 'blackhead', in the turkey. It may also occur in the chicken, peafowl, guinea-fowl, pheasant, partridge and quail. It is world-wide in distribution and is an important disease entity in turkeys, assuming great economic importance where turkeys are kept in large numbers. It is probably ubiquitous also in chickens, although the incidence of disease in these is low.

Morphology. The organism is somewhat pleomorphic, the morphology depending on the organ location and the stage of the disease. Tyzzer (1919) described three stages of the parasite: an invasive stage of the early caecal and liver lesions; a vegetative stage occurring near the centre of lesions; and a resistance phase. Ultra-structural studies by Lee et al. (1969) have shown that the

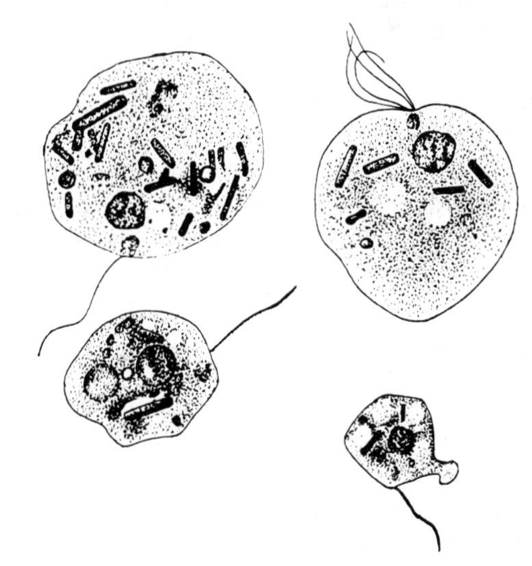

Fig 3.10 Various forms of *Histomonas meleagridis*. (× 2000) (*From Wenyon 1926*)

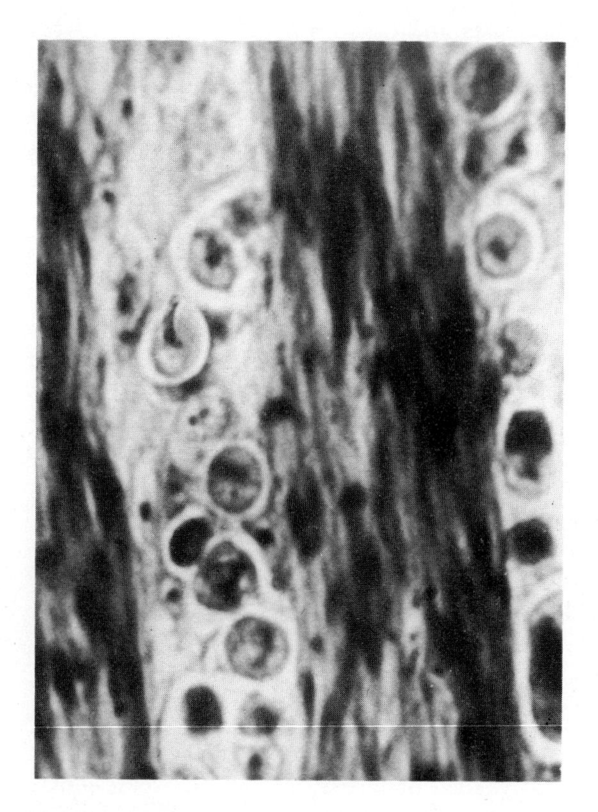

Fig 3.11 *Histomonas meleagridis* in wall of caecum of turkey poult. (× 1200)

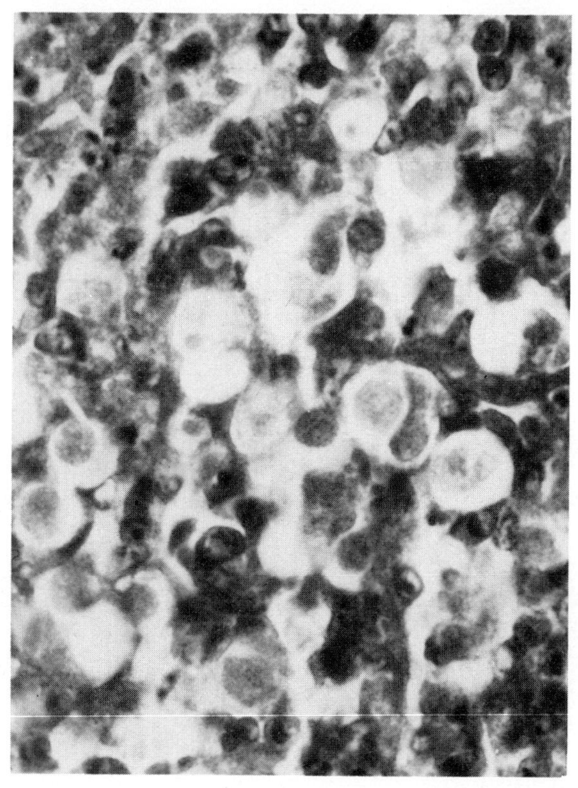

Fig 3.12 Section of turkey liver infected with *Histomonas meleagridis*. (× 750)

organism resembles a trichomonad, which sheds its flagella and becomes amoeboid on invasion of the host tissues.

The stage which occurs in the lumen of the intestine, and in culture, is 5–30 μm in diameter, amoeboid, has a clear ectoplasm, and a granular endoplasm. The latter may contain food vacuoles with bacteria, starch grains, etc. The nucleus is vesicular and a flagellum arises from a small kinetosome near the nucleus. Occasionally, two flagella may be present. A related form, *Parahistomonas wenrichi*, occurs in the pheasant and this possesses four flagella (see below).

In the tissues, the organism is found singly or in clusters and is 8–15 μm in diameter. It is amoeboid, the flagellum being absent.

Although Tyzzer described a 'resistant' stage of the parasite, there is no evidence that any stage is more resistant than another and cysts are not formed. This form is 4–11 μm in diameter and appears to be enclosed in a dense membrane. The cytoplasm is acidophilic and is filled with granules.

Cultivation. In culture the organism feeds on bacteria or red blood cells and it was grown axenically by Lesser (1960) who utilized tissue culture medium 199 supplemented with sterilized cream and a nutrient broth culture of mixed bacterial flora from turkey caecal droppings which had been subjected for at least one hour to the action of high concentrations of penicillin and streptomycin. Histomonads were maintained for more than 40 consecutive subcultures without evidence of bacterial growth. Lesser concluded that a heat-labile intracellular factor, present in turkey caecal bacteria, was required for the successful growth and reproduction of *H. meleagridis*. Dwyer (1970) developed a medium for the agnotobiotic (xenic) culture of *H. meleagridis* and Dwyer and Honigberg (1970) showed that con-

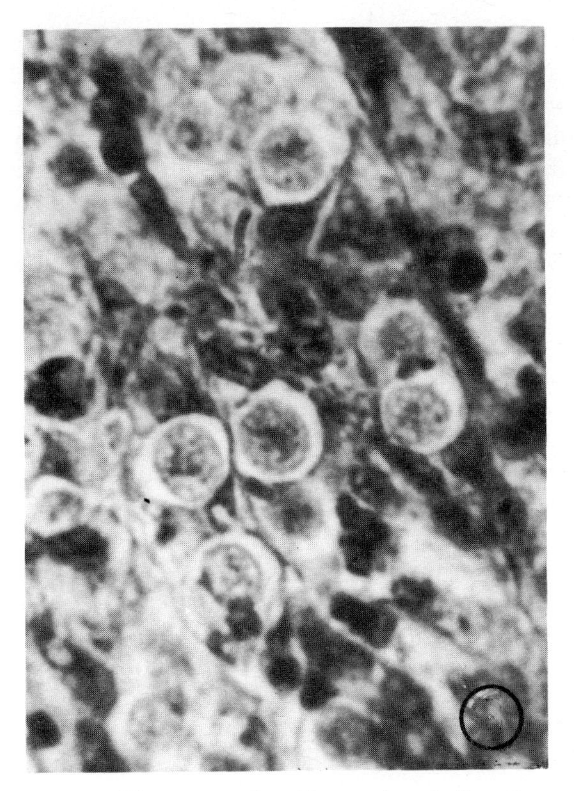

Fig 3.13 *Histomonas meleagridis* organisms in liver tissue of turkey poult. (× 1200)

tinued in vitro cultivation led to attenuation of pathogenicity.

Life-history and transmission. Reproduction of the organism is by binary fission.

Primary infection occurs in the caecum, and under natural conditions the main route of infection is by the ingestion of the embryonated eggs of the caecal worm *Heterakis gallinarum* (Graybill & Smith 1920). This mode of transmission has been fully confirmed by numerous workers but, until recently, the organism has not been demonstrated in the worm. Kendall (1959) observed what he believed to be *H. meleagridis* in the gut cells of *H. gallinarum* obtained from poults killed four days after infection. Lee (1969) demonstrated small 'cells' which probably represent the infective stages of the parasite and these penetrate the developing oocyte of the worm and the organism has been isolated in culture from artifically hatched eggs of *H. gallinarum* (Ruff et al. 1970).

Despite the fact that infection of *H. gallinarum* eggs is widespread, and blackhead infection can be induced with batches of eggs taken from a high percentage of turkeys, or almost any chicken, the rate of infection of eggs is low. Lund and Burtner (1957) reported that less than 0.5% of embryonated eggs of *H. gallinarum* from experimentally infected chickens contained the protozoan.

Under certain circumstances the protozoan discharged from acutely infected turkeys may produce infection if it is ingested at once (Tyzzer & Collier 1925). The acidity of the gizzard is important in such direct infection, and Horton-Smith and Long (1956) found that infection could be created with the free forms when chickens were starved or had received an alkaline diet to raise the pH of the gizzard from a level of 2.9–3.3 to one of 6.2–7.6. Blackhead may also be induced by the rectal injection of cultures of *H. meleagridis* or ground-up infective tissues (Farmer & Stephenson 1949).

Pathogenesis. Histomoniasis is essentially a disease of young turkeys, and, in the absence of treatment, the birds die. The chicken is much more resistant to the pathogenic effects of the infection.

The histomonads are released from the larvae of *H. gallinarum* and they enter the wall of the caecum and multiply. One or both caeca may be involved and the earliest lesions consist of small, raised, pinpoint ulcers in which the organisms are plentiful. These become much enlarged and may involve the whole of the caecal mucosa. The mucosa becomes greatly thickened and its surface necrotic and eventually the caecum contains a hard caseous, adherent core. Macroscopically, the caeca are enlarged and haemorrhagic. The histomonads further invade the mucosa and 'stream' to the deeper layers of the organ. A fall in plasma protein concentration coincident with the development of acute caecal lesions has been reported by Beg and Clarkson (1970). The organisms gain entry to the hepatic portal system and are carried by this to the liver. The liver lesion is a focal necrosis which increases by peripheral extension to produce characteristic circular depressed lesions. These consist of a yellow to yellowish green area of necrosis with a greyish peripheral region with radiating streaks of necrosis. The lesions vary in diameter, being up to 1 cm or more in size; they may be confluent and extend deeply into the liver. They are not encapsulated and in older birds a repair process may be seen with the initiation of a fibrinous and lymphoid tissue reaction.

The detailed pathology of experimentally induced histomoniasis has been studied by Malewitz et al. (1958). Lesions occur in the caeca and liver and also in the spleen, kidney and lungs. Lesions in which the parasites could be readily demonstrated were characterized by hyperaemia, haemorrhage, and infiltration of lymphocytes and macrophages, multinucleated giant cells, necrosis and usually a serous exudate. Areas of hyperaemia and exudate occurred in the lungs, pancreas and heart, but parasites were not demonstrated in such lesions; a lymphocytic infiltration was seen in the lungs.

The organisms disappear from the tissues of birds which recover, and the necrotic lesions are repaired by an invasion of lymphocytic cells and fibroblasts. A firm caecal plug is produced; this becomes contracted and is finally ejected, and then the caeca eventually return to normal size, although much scarring may remain. Extensive scarring may remain in the liver for a very long time.

The relationship between the bacterial flora of the turkey and *H. meleagridis* and the production of lesions in the liver has been investigated by Doll and Franker (1963) who studied the pathogenicity of histomoniasis in gnotobiotic turkeys. Of 12 gnotobiotic turkeys infected with sterile *H. gallinarum* eggs, only one showed signs of histomoniasis and none died of the disease; but of 12 conventional turkeys, 11 developed histomoniasis and died. Subsequent work by Franker and Doll (1964) showed in germ-free turkeys that surface-sterilized embryonated *H. gallinarum* eggs mono-contaminated with either *Escherichia intermedia* or *Streptococcus faecalis* produced lesions of histomoniasis. Similar work by Bradley et al. (1964) demonstrated typical enterohepatitis in gnotobiotic turkeys inoculated with *H. meleagridis* plus *Escherichia coli* or *Clostridium perfringens*.

Under field conditions the most severe disease occurs in young poults between the ages of 3 and 12 weeks, losses ranging from 50% to 100% of the flock. Birds may die within a few days of showing the first clinical signs. In older birds the disease runs a more chronic course and recovery may occur, to be followed by immunity (Kendall 1957). Nevertheless, mature birds may sustain acute infections and die. On a general flock basis the overall mortality may vary from 0% to 90%.

Immunity. Birds which have recovered from infection do not develop any clinical signs on reinfection; they may, however, harbour the organisms in the caeca (Kendall 1957). Early studies by Tyzzer (1933) demonstrated that an attenuated strain of *H. meleagridis* protected against virulent strains which were almost 100% fatal to unprotected birds. Continuous culture of *H. meleagridis* produces initially a loss of virulence (Dwyer & Honigberg 1970) and later a loss of immunizing properties; Tyzzer (1936) was able to demonstrate good protection against newly isolated virulent strains with forms which had lost all virulence in culture. Lund (1959) reported that the administration of a non-pathogenic strain of *Histomonas* by infected *Heterakis* eggs did not protect turkeys against pathogenic strains of the organism. However, immunization by rectal inoculation gave, after three weeks, some protection against a rectal challenge of 10 000 pathogenic histomonads. Although this was almost complete after six weeks, typical enterohepatitis could be produced with *Heterakis* egg-induced infection. Lund believed that the immune barrier was limited to the surface of the caecal mucosa and that the failure to protect against histomonads introduced by *Heterakis* was due to the larval worms penetrating the mucosal barrier to liberate the histomonads into the tissues. A review of the immunology of histomoniasis is given by Cuckler (1970).

Epidemiology. *Heterakis gallinarum* is a common nematode parasite of the domestic turkey and chicken, and usually it is possible to produce histomoniasis in turkeys with any batch of eggs from domestic chickens. Consequently, the chicken is the principal reservoir of infection for turkeys and outbreaks of histomoniasis are likely to occur when turkeys are reared with domestic fowls or when they are placed on ground recently vacated by domestic poultry. The unprotected histomonad dies within a few hours outside the avian host and its survival on pasture is due to the protection afforded by the *Heterakis* egg. Farr (1956) has shown that eggs may retain both helminth and protozoan infectivity for 66 weeks in the soil. In further work, Farr (1961) found that *H. meleagridis* in *Heterakis* eggs were infective as late as week 151 after exposure on soil out of doors.

In addition to the domestic chicken, various wild gallinaceous birds, such as the pheasant, grouse, partridge, quail and wild turkey, may serve as reservoir of infection for domestic turkeys.

Outbreaks of disease vary very much in severity, this probably being an effect of the virulence of the strain. With continued use of infected land, it is likely that the incidence and severity of the disease will increase as time goes by.

Clinical signs of enterohepatitis. The clinical signs of blackhead appear eight or more days after infection. The first evidence of the infection in a flock is a noticeable decline in the feed consumption; this is followed by depression, drooping wings, ruffled feathers and, especially characteristic in the turkey, the appearance of sulphur-yellow coloured droppings. Yellow droppings are not a common sign in the chicken. A few birds may show a darkened or cyanotic discolouration of the skin of the head and wattles (from which the name 'blackhead' arises); however, this is by no means a constant feature of the disease, nor is it a significant clinical sign since it may occur in other disease entities. Occasionally, the disease may be peracute in nature with no marked clinical signs, the birds dying within 24 hours. In older birds a chronic wasting form of the disease is seen; this may be followed by death or recovery.

Diagnosis. In living birds, the characteristic sulphur-yellow droppings are suggestive of enterohepatitis, while at post mortem the necrotic lesions of the liver are pathognomonic. Histological examination of stained sections of the liver shows necrosis and colonies of the organism. The latter appear like punched-out holes in the necrotic tissue, each hole containing a fragment of protoplasm which represents the parasite. The caecal lesions may be confused with those due to *Eimeria tenella* but this infection may readily be eliminated by a microscopic examination of the mucosa when the characteristic developmental stages of the sporozoan are missing.

Treatment. The literature on the chemotherapy of enterohepatitis has been reviewed by Wehr et al. (1958) and by Reid (1967).

2-Amino-5-nitrothiazole [Enheptin-T (USA), Entramin (UK)] can be given in the food or in drinking water. It is recommended in the food at a concentration of 0.05% as a preventive or of 0.1% for curative purposes. It will prevent mortality completely when given over a period of 14 days provided this is begun not later than 72 hours after a single oral infection. When treatment is started as late as 13 days after infection, mortality is reduced and few deaths occur until more than one week after withdrawal of the drug.

2-Acetylamino-5-nitrothiazole [Enheptin-A (USA), Entramin A (UK)] is as efficient as Enheptin-T for prophylaxis. It is less soluble than Enheptin-T and is given in the food at a concentration of 0.025%

Nithiazide [1-ethyl-3-(5-nitro-2-thiazolyl) urea)] is more potent and less toxic than 2-amino-5-nitrothiazole for the treatment of enterohepatitis. It is most effective when given prior to, or 3–7 days after, infection. It is equally effective when administered in the feed or water. Field trials have shown that 0.02% of the drug in drinking water controlled field outbreaks of both histomoniasis and hexamitiasis; 0.025% in the feed controlled outbreaks of histomoniasis in chickens and turkeys, and 0.0125% in the feed provided effective field prophylaxis against blackhead in turkeys.

Furazolidone (NF-180) [N-(5-nitro-2-furfurylidone)-3-amino-2-oxazolidinone], at concentrations of 0.01–0.02% in the food, is almost wholly effective in preventing histomoniasis when continuous medication is commenced before infection takes place. A suitable concentration for preventative medication is 0.015% and for the treatment of established histomoniasis, 0.04% is satisfactory. However, even with the higher doses, the histomonads are only inhibited and they continue to multiply if medication is withdrawn. Consequently, relapses may result if this occurs.

Dimetridazole (1,2-dimethyl-5-nitroimidazole) is effective at concentrations of 0.0125% in the food.

Control of enterohepatitis. Control consists of a combination of good husbandry and preventive medication.

It is essential to avoid contact between turkeys and domestic chickens and where the two are kept on the same farm a separate area should be allocated to each of them. The use of hens for incubating and rearing turkey poults should be avoided and young turkeys should not be placed on land that has carried poultry, unless the land has been rested for at least two years. Regular treatment of all birds with anthelmintics to reduce the incidence of *Heterakis gallinarum* infection is of benefit in reducing the overall incidence of histomoniasis. The eggs of worms passed following therapy are still capable of hatching and releasing histomonads and consequently ground should be regarded as a potential source of infection for some time after anthelmintic treatment.

Where large-scale turkey raising is in operation, young turkeys should be raised on wire floors which are out of contact with the ground. If possible, separate attendants should be used for turkeys and other domestic poultry but, if not, attendants should change footwear and clothing before going from the chicken flock to the turkey flock. Unnecessary visitors should be discouraged. Pens and equipment should be regularly sterilized especially prior to the introduction of new birds.

Where young turkeys are reared on open ground, clean ground should be reserved for this. Light, sandy soil is more satisfactory, thus allowing less survival of *Heterakis* eggs than heavy, damp soil. Low-lying areas and ponds and streams should be avoided and these should be drained or fenced off from poultry.

In large-scale turkey rearing establisments, continuous medication with one or other of the above therapeutic agents is practised.

Genus: Parahistomonas Honigberg, 1969

The single species of this genus is similar to *Histomonas*, but four flagella exist and there is a rod-shaped parabasal body.

Parahistomonas wenrichi (Lund 1963) Honigberg, 1969. This species is a non-pathogenic form of gallinaceous birds, chiefly of pheasant. It is about 1.5 times larger than *H. meleagridis* and has four flagella instead of the one or two of the latter species. It does not multiply in the host tissues, nor does it produce any visible pathological change. *Parahistomonas wenrichi* may occur along with *H. meleagridis* and it is transmitted by the embryonated eggs of *Heterakis gallinarum*. It does not grow in media which support the growth of *H. meleagridis* nor has it been grown in media which supports growth of trichomonads, amoeba or other parasitic protozoa (Lund 1963).

OTHER FLAGELLATES OF THE FAMILY MONOCERCOMONADIDAE

In addition to the genera mentioned above, others occur in the digestive tract of vertebrates. They are largely non-pathogenic, although their numbers may increase during diarrhoeic states. They include the following.

Genus: Chilomitus da Fonseca, 1915. The elongate body is convex on the aboral surface, four anterior flagella occur, there is an anterior nucleus and a rudimentary axostyle. The cytostome is cup-shaped. *Chilomitus caviae* (6–14 μm × 3–5 μm) is common in the caecum of the guinea-pig. *Chilomitus connexus* occurs in the guinea-pig in the USA (Nie 1950).

Genus: Hexamastix Alexeieff, 1912. Pyriform in shape, six flagella one of which is trailing, nucleus and cytostome anterior, axostyle conspicuous, pelta is present. Cysts are not formed. *Hexamastix caviae* and *Hexamastix robustus* are common in the caecum of the guinea-pig. They measure 4–10 μm × 3–5 μm and 7–14 μm × 3–8 μm, respectively. *Hexamastix muris* has been found in the caecum of rat, hamster and other rodents. It is 6–12 μm × 4–10 μm.

Genus: Enteromonas da Fonseca, 1915. Levine (1973) is uncertain of the taxonomic position of this genus. The body is pyriform or spherical, there are three anterior flagella and one trailing flagellum. There is no cytostome, the nucleus is anterior and cysts occur. *Enteromonas hominis* (4–10 μm × 3–6 μm) occurs in the caecum of man, several primates, hamster, rat, rabbit. It is transmissible between the species mentioned above. *Enteromonas suis* was described from the caecum of the pig in India and *Enteromonas caviae* is common in the caecum of the guinea-pig throughout the world (Nie 1950).

Genus: Tetramitus Perty, 1852. This genus belongs to the family TETRAMITIDAE Bütschli, 1878. They are forms with one nucleus and four flagella. The life-cycle involves flagellate and amoeboid forms. Cysts occur. *Tetramitus rostratus*, a coprophilic form, is found in stagnant water and has been found in human and rat faeces. The vegetative stage is 14–18 μm × 7–10 μm, the amoeboid form is 14–48 μm with a single pseudopod, and the cyst is 6–18 μm and is thin-walled and spherical.

Genus: Costia Leclerque, 1890. This is also a member of the family TETRAMITIDAE but is dealt with here for convenience. Organisms are ovoid to pyriform, a shallow depression leads into a cytostome, from which extend two long and two short flagella. Cysts occur. Members of the genus are ectoparasites of fresh-water fish. *Costia necatrix* and *Costia pyriformis* are ectoparasitic flagellates of many species of fish, including the common aquarium species. The organisms are 5–18 μm × 2.5–7.7 μm, and the long flagella are 18 μm in length, the short 9 μm.

The life-cycle is direct, organisms leave the scales of fish, swim about and attach to a new host by means of a flat posterior disc (Schubert 1968). Reproduction is by binary fission. The pathogenic effects are caused by the parasite inserting finger-like processes into the epidermis, resulting in the digestion of the cells. An early effect is a drop in appetite and listlessness; 'flashing' may be evident if the skin is infected but may not occur if only the gills are infected. In heavy infections fish appear to be covered with a grey slime, but fish less than three to four months of age will die before this develops. Organisms are readily demonstrable in scrapings of the skin or gills.

Effective treatments for fish and their holding tanks include: formalin, 200–250 ppm for one hour; pyridylmercuric acetate, 2 ppm for one hour (this may be toxic for some trout); potassium permanganate, 1000 ppm for 30 seconds; and acetic acid, 2000 ppm for one to two minutes (Hoffman 1970). Since the lesions of *Costia* infection may be confused with bacterial gill disease, care in the use of formalin should be exercised when bacterial gill disease is present.

FAMILY: RETORTAMONADIDAE GRASSÉ, 1952

Organisms of this family possess two or four flagella, one of which is turned posteriorly and associated with a vertical cytosomal area.

Genus: Retortamonas Grassi, 1879

Organisms are pyriform, a large cytostome occurs near the anterior end and an anterior flagellum and a trailing flagellum are present. Cysts are pyriform and have one or two nuclei.

Retortamonas intestinalis (Wenyon and O'Connor, 1917) Wenrich, 1932 occurs in man and other primates. Trophozoites are 4–9 μm by 3–4 μm, and cysts are uninucleate. It is not pathogenic.

Retortamonas ovis (Hegner and Schumaker, 1928). This has been found in sheep faeces in the USA. Trophozoites measure 7–9 μm by 2–4 μm and cysts are ovoid, measuring 3–4.5 μm. The organism is not pathogenic.

Other retortamonads include *Retortamonas cuniculi* from the caecum of the rabbit, and *Retortamonas caviae,* a common flagellate of the caecum of the guinea-pig. An unnamed species has been found in the caecum of wild Norway rats in the USA.

Genus: Chilomastix Alexeieff, 1912

Members of this genus are pyriform organisms with a large cytostomal cleft at the anterior end, near the nucleus. There are three anteriorly directed flagella and a short fourth flagellum that lies in the oral groove. Cysts are produced.

Chilomastix mesnili (Wenyon, 1910) Alexeieff, 1912. This is a non-pathogenic flagellate of the human intestine and other primates, and is one of the largest flagellates found in man. It is pear-shaped, 5–20 μm in length (commonly 10–15 μm) and the posterior end of the body is drawn out into a long point. The well-defined cytostomal groove is 6–8 μm in length by 2 μm wide, and the lateral margins of the cystostome are supported by two fibrils. The cysts are pyriform, 7–10 μm in length, contain a single nucleus and the cystome is readily visible, being almost as long as the organisms.

C. mesnili occurs in the caecum and colon and sometimes the small intestine of man, the prevalence of infection varying from 2% to 25% in various parts of the world. It has also been found in monkeys and the pig. Transmission is via the cysts. Generally, *C. mesnili* is considered to be a non-pathogenic commensal; however, Mueller (1959) has reported diarrhoea in children and in himself associated with very large numbers of organisms.

Other species of *Chilomastix* include:

Chilomastix gallinarum Martin and Robertson, 1911 (caecum of chickens and turkey).

Chilomastix caprae da Fonseca, 1915 (rumen of goats in Brazil and India).

Chilomastix cuniculi da Fonseca, 1915 (caecum of rabbits).

Chilomastix equi Abraham, 1961 (intestine of horses in India).

Chilomastix intestinalis Kuezynski, 1914 and **Chilomastix wenrichi** Nie, 1948 (9–28 μm × 7–11 μm and 7.5–12 μm × 4–5 μm, respectively) commonly occur in the caecum of the guinea-pig.

Chilomastix bettencourti da Fonseca, 1915 is similar to *C. mesnili* and has been found in the caecum of mice, rats, hamsters, etc.

FAMILY: COCHLOSOMATIDAE KOTLÁN, 1923

Organisms of this family possess six flagella, there is an anteriorly placed ventral sucker and a parabasal body.

Genus: Cochlosoma Kotlán, 1923

Members of this genus have an oval body which is broad anteriorly and narrow posteriorly. Six flagella of unequal length arise from a complex of blepharoplasts, two of which are trailing flagella. On the anterior ventral surface of the body there is a large sucker.

Cochlosoma anatis Kotlán, 1923. This occurs in the posterior large intestine and caeca of domestic and wild ducks. A similar form has been found in turkeys and Campbell (1945) considers it to be identical with the duck form. It is probably world-wide in distribution. The organism is 6–10 μm in length by 4–6.5 μm; the ventral sucker is about half the body length. Ovoid cysts, with four or more nuclei, occur in the faeces of infected ducks.

The importance of the parasite in ducks is unknown. Campbell (1945) described a disease entity of catarrhal enteritis in turkeys associated with large numbers of *C. anatis*; however, the exact relationship of *C. anatis* to enteritis in turkeys needs further investigation since this organism is usually found associated with *Hexamita meleagridis*, a known pathogen for turkeys.

FAMILY: HEXAMITIDAE KENT, 1880

Organisms of this family are bilaterally symmetrical, with two nuclei and six or eight flagella.

Genus: Hexamita Dujardin, 1838

Members of this genus have a pyriform body with two nuclei (Fig. 3.14). Near the anterior end there arise six anterior and two posterior flagella, and there are two axostyles. Some members of the genus form cysts. A number of free-living species are found in stagnant water whereas others are

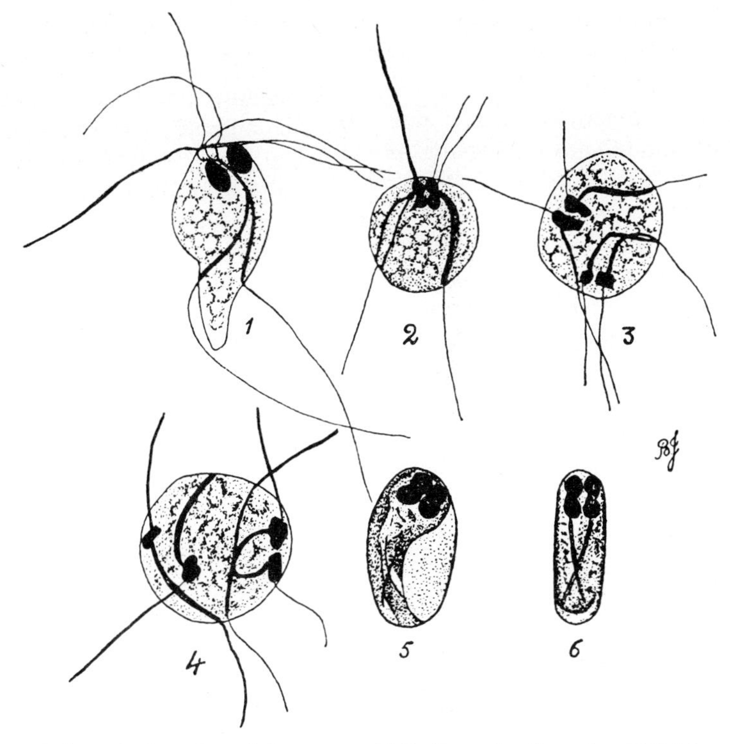

Fig 3.14 *Hexamitus muris* from the intestine of the mouse. (1) Ordinary free form. (2–4) Dividing forms. (5, 6) Encysted forms showing division of nuclei. (× ca 3000) (*From Wenyon 1926*)

parasitic in insects, fish and frogs. The principal species of importance is *Hexamita meleagridis* which occurs in turkeys.

Hexamita meleagridis McNeil, Hinshaw and Kofoid, 1941. This is found principally in the turkey, but organisms morphologically identical with it have been found in quail, partridge, pheasant, etc. *H. meleagridis* has been transmitted experimentally from the turkey to the chicken, quail, domestic duck, etc. (Levine 1973). *H. meleagridis* occurs in the duodenum and small intestine of young turkeys; it has been reported from North and South America, Great Britain, Europe and is probably world-wide in distribution. It is associated with an infectious catarrhal enteritis of turkeys which appears to occur in all the major turkey-producing areas in the Western world.

Morphology. The organism is bilaterally symmetrical, pear-shaped and 6–12 μm in length by 2–5 μm broad. The six anteriorly directed flagella arise in two groups of three, each from distinct blepharoplasts which lie anterior to the two nuclei. The caudal flagella arise on each side from a distinct blepharoplast posterior to the larger anterior blepharoplast and pass posteriorly, emerging at the posterior end of the body.

Life-cycle. Multiplication is by longitudinal binary fission and transmission is through contaminated feed and drinking water. Slavin and Wilson (1954) stated that the life-cycle consisted of the formation of the binucleate cysts and the production of invasive forms which penentrated the reticuloendothelial cells. In this site the authors described the developments of schizonts, merozoites and, eventually mature flagellates. This description was severely criticized by Hoare (1955) who pointed out that insufficient evidence had been produced to support the statements regarding the life-cycle of the organism and further stated that the figures presented were unconvincing and unacceptable. Subsequently, Slavin and Wilson (1960) published a more detailed account of the life-cycle of *H. meleagridis* accompanying this by numerous microphotographs of the life-cycle and the morphology of the flagellate.

Pathogenesis. Young turkey poults up to the age of about two months are the most susceptible. Death may occur within a week of infection in some cases, and the mortality in the flock may reach up to 80%. The essential lesion is a catarrhal enteritis of the upper digestive tract and this produces a marked lack of tone in the duodenum and jejunum so that distended bulbous areas, containing watery contents, appear. The small intestine is inflamed and oedematous; there is congestion of the glandular tissue of the caecum, and myriads of *Hexamita* are found in the bulbous areas.

Clinically, the affected poults become nervous and show a foamy, watery diarrhoea. They usually continue to eat but in the later stages of the disease they become listless, lose weight rapidly and finally die. Recovered birds grow poorly and may act as carriers since large numbers of organisms may persist in the lower jejunum, ileum and bursa of Fabricius.

Epidemiology. The chief sources of infection are carrier adult birds, and Hinshaw and McNeil (1941) found that the prevalence of *H. meleagridis* in adult turkeys may be as high as 32%. Clinical disease may not occur until several batches of turkey poults have been processed through an infected area and this may indicate that the virulence of the organism needs to be increased by passage. Hinshaw (1959) indicated that this was necessary with some strains from carrier turkeys. Studies by McNeil and Hinshaw (1941) showed that *H. meleagridis* could be transmitted to chickens and that the organism could persist in the bursa of Fabricius and the caecal tonsils of these for at least 22 weeks. Clinical disease was not found in infected chickens, but if turkey poults were placed with infected chickens, the turkeys developed the disease. Thus, the chicken must be regarded as a potential carrier of *H. meleagridis.* Quail, pheasant and partridge may also be a source of infection for turkeys. A species in pigeons, *Hexamita columbae*, is not transmissible to turkeys.

Diagnosis. This is most satisfactorily made by the demonstration of the living organisms in a drop of the contents of the small intestine, especially from the bulbous regions. It is important that the preparation should be examined fresh since within an hour or two of death of the bird the organisms are dead and extremely difficult to recognize. Impression smears can be made from cross-sections of the fresh small intestine and stained with Giemsa. The organism has not been cultured in artificial medium; however, Hughes and Zander (1954) were able to obtain axenic cultures in the allantoic cavity of chick embryos with the use of streptomycin and bacitracin.

Treatment. Wilson and Slavin (1955) reported that 2-amino-5-nitrothiazole was 50% effective in artificially infected turkeys. The effectiveness of furazolidone was investigated by Mangrum et al. (1955). A rate of 110 mg/kg, two days prior to experimental infection, reduced the mortality from between 56% and 66% to 10%, and where treatment was begun at the time of the first death after inoculation, the mortality in treated groups was 26%. Briggs (1959) reported that *H. meleagridis* in turkeys could be prevented by the continuous feeding of furazolidone. Fogg (1957) reported that nithiazide [1-ethyl-3-(5-nitro-2-thiazolyl)urea], at a concentration of 0.02% in drinking water, controlled fatal outbreaks due to *Hexamita* and *Histomonas* either singly, or in mixed infections.

Prevention and control. This is based on general hygienic measures. Poults should be reared away from adult turkeys and chickens, overcrowding should be avoided and adequate housing and ventilation should be provided. If possible, separate equipment should be used for the young birds and attendants should disinfect themselves when entering the pens of young poults.

Hexamita columbae (Nöller & Buttgereit, 1923) occurs in the small intestine of pigeons, is a small form, 5–9 μm by 2.5–7 μm, and may be pathogenic for pigeons, causing a catarrhal enteritis (Nöller & Buttgereit 1923). It is not transmissible to turkeys. Treatment is as for *H. meleagridis.*

Hexamita muris (Grassi, 1888) (Fig. 3.11) is found in the posterior intestine and caecum of rats, mice and hamsters. Enteritis in the laboratory mouse has been associated with this species in Israel (Meshorer 1969); an acute to chronic catarrhal duodenitis occurs.

Hexamita pitheci da Cunha and Muniz, 1929 has been reported from the large intestine of rhesus monkeys. It is small (2.5–3 μm × 1.5–2 μm). Other unnamed species occur in primates and Levine (1973) believes they are *H. pitheci.*

Hexamita salmonis Moore 1923 is an intestinal flagellate of fish, e.g. goldfish, salmonids and others in North America.

Hexamita intestinalis Dujardin, 1841 occurs in similar species in Europe (Hoffman 1967). The two species are possibly identical. Organisms are pyriform and are 10–16 μm × 6–8 μm. Cysts are excreted in the faeces of infected fish and infection is created by ingestion of cysts or trophozoites. The pathogenic effects vary from an acute catarrhal enteritis and death to a chronic form associated with poor growth, listlessness and emaciation (Reichenbach-Klinke & Elkan 1965). Treatment included compounds active against *Histomonas* in the turkey, such as Enheptin-T or Enheptin-A (McElwain & Post 1968), dimetridazole (Emtryl) and furazolidone (Neftin) (Richards 1977).

Genus: Giardia Kunstler, 1882

Species of this genus are pyriform to elliptical in outline and are bilaterally symmetrical (Fig. 3.15). The anterior end is broadly rounded and the posterior drawn out and somewhat pointed. The dorsal side is convex and the ventral side concave with a large sucking disc in the anterior half. There are two nuclei, two axostyles, eight flagella arranged in four pairs, and a pair of darkly staining bodies placed medially. Cysts are produced which are oval or elliptical and possess two or four nuclei.

The members of the genus which occur in vertebrates are all morphologically similar, but it is customary to give them different names which indicate the host in which they are found. However, Filice (1952) has suggested that only two species occur in mammals, namely *Giardia muris* and *Giardia duodenalis*, the latter occurring in a wide range of animals including man, ox, dog and cat. The necessary cross-transmission experiments to verify this opinion have not been conducted and for convenience, if nothing else, the more conventional method of describing the various species of the genus will be adopted here.

Giardia lamblia Kofoid and Christiansen, 1915 (syn. *G.* (*Lamblia*) *intestinalis*, etc.). (Fig. 3.16) The organism is found in the duodenum, other parts of the small intestine and occasionally in the colon of man. It has also been found in monkeys, pigs and budgerigars, and experimentally it is transmissible to laboratory rats (*Rattus norvegicus*) but *Rattus rattus* and the laboratory mouse cannot be infected (Haiba 1956). It is world-wide in distribution, the prevalence varying from 2% to 60% or more. It is more common in children and is considered to be the most common flagellate of the human.

It is 9–20 μm (commonly 10–18 μm) by 5–10 μm, and essentially the body of the parasite resembles a pear which has been cut in half longitudinally, the flat side representing the ventral surface of the organism and the convex the dorsal. The cysts are ovoid and refractile, 8–14 μm by 6–10 μm in size. The cyst wall is thin and the organism does not fill the entire cyst. The most prominent structures seen in the cysts are the two, or later four, nuclei and the comma-shaped body.

Reproduction is by binary fission.

Transmission is by the ingestion of cyst-contaminated food and drink. The cyst may remain viable in moist surroundings for up to two weeks.

Pathogenesis. There has been considerable debate over the years concerning the significance of *G. lamblia* in the human. Probably the majority of infections are symptomless, but some, especially in children, in travellers returning from overseas and in persons who have visited sports resorts, are associated with acute, subacute to chronic diarrhoea and duodenal irritation with an excess of mucus production. In a review of the

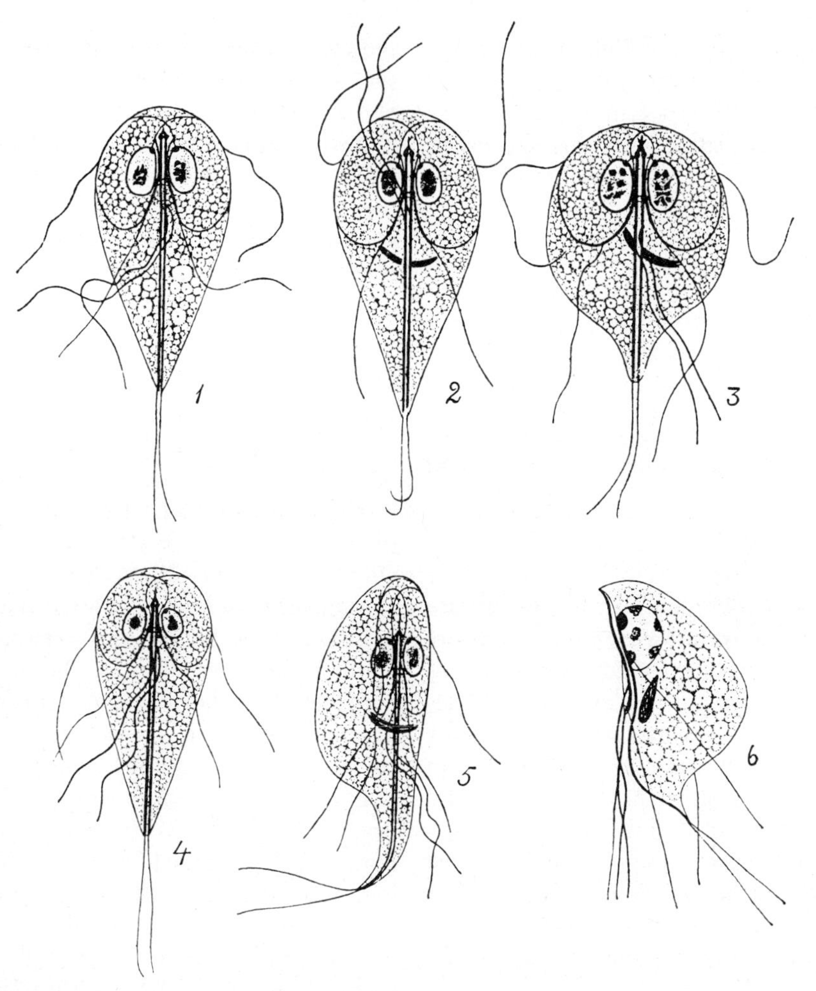

Fig 3.15 *Giardia lamblia* from human intestine. 1–4, Variations in size and shape of body. 5, Partial side view. 6, Side view. (×5000) (*From Wenyon 1926*)

pathogenicity of *G. lamblia*, Pizzi (1957) indicated that an upset fat metabolism may result in a deficiency in fat-soluble vitamins.

There is no evidence that *G. lamblia* has any pathogenic significance in the monkey or in the pig.

Water-borne outbreaks of giardiasis may result in epidemic infections. These are associated with defective water-treatment plants where, for example, filters are defective, chlorination is inadequate and sedimentation is interfered with. The role of wild animals in such outbreaks is unclear. In the study of an outbreak in Camas, Washington, in 1976, Allard et al. (1977) found that giardiasis was not associated with pet ownership but wild animals (three beavers) were found infected near the source of water for the township. Box (1981) found the organism associated with diarrhoea and death in budgerigars and suggested that it may be a public health hazard in these birds.

Diagnosis of *G. lamblia* is based on the microscopic demonstration of the cysts in faecal material. Usually only cysts are passed but in cases of acute diarrhoea, the free flagellates may be found. The cysts are most satisfactorily concentrated by a 33% zinc sulphate flotation technique, and a small amount of iodine solution may be added to aid in the recognition of the cysts.

Giardia infections in humans can be success-

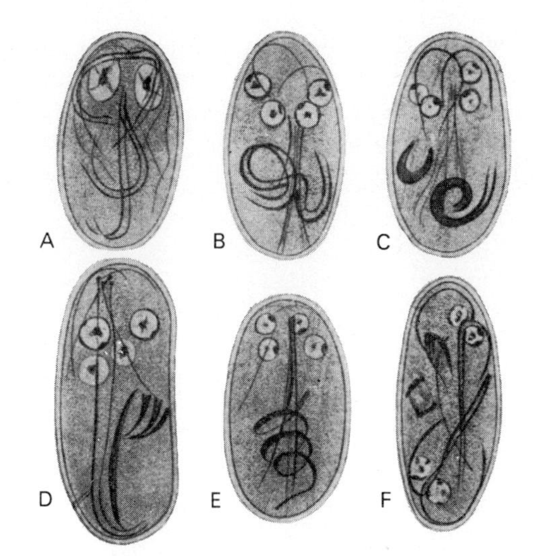

Fig 3.16 *Giardia lamblia*: encysted forms from the human intestine. (A) Form with two nuclei. (B–E) Forms with four terminal nuclei. (F) Form in which two of the nuclei have migrated to the opposite pole and the flagellate is dividing with the cyst. (× 3000) (*From Wenyon 1926*)

fully treated with quinacrine, chloroquine, diodoquin (di-iodohydroxyquinoline) or metronidazole.

Giardia canis Hegner, 1922. This occurs in the duodenum and jejunum of the dog, principally in the United States. The trophozoite form is 12–17 μm long by 7.6–10 μm wide, and the cysts are oval, measuring 9–13 μm by 7–9 μm.

Various surveys on the incidence of infection in dogs have been carried out in the United States. Craige (1948) reported 8.8% of 160 dogs infected in California, and Bemrick (1961) found 7.5% of 2063 dogs infected in the Minnesota area. It is likely that the organism is cosmopolitan in its distribution and, if searched for, a wide prevalence would be evident.

Pathogenesis. The controversy on the significance of *Giardia* in the human is equalled by a similar controversy about the dog. Many clinicians subscribe to the view that the organism has a definite pathological significance, producing diarrhoea and dysentery. This opinion is strengthened by the fact that therapy usually eliminates the organisms and alleviates clinical signs. Craige (1948) found *Giardia* in 17 of 71 dogs with dysentery; however, the organism may be found in large numbers in apparently completely healthy dogs. Duodenal biopsy has demonstrated that *G. canis* can be associated with morphological changes indicative of malabsorption (unpublished data).

Diagnosis of the infection is the same as for the human species. Quinacrine has been found an effective treatment, being given at the rate of 50–100 mg twice daily for two to three days (Craige 1949); chloroquine, diodoquin and metronidazole (up to 250 mg twice daily for ten days) have also been found effective. Control measures consist of an improvement in the hygiene of dog kennels.

Giardia cati Deschiens, 1925 (*Giardia felis*), occurs in the small intestine of the cat in the United States and Europe. It is very similar to *G. canis* with which it may be synonymous. There is no evidence that this form is pathogenic.

Giardia chinchillae Filice, 1952, occurs in the small intestine, especially the duodenum of the chinchilla. The trophozoite stage is 10–20 μm by 6–12 μm.

This organism has been incriminated as the cause of severe blood-stained diarrhoea and death in the chinchilla. The clinical picture is an intense mucoid diarrhoea, and Shelton (1954) has reported a 38% mortality in experimentally induced infections in chinchillas. Severe infections have occurred on chinchilla ranches.

For treatment, 6–9 mg of quinacrine for five to seven days has been recommended (Hagen 1950), while diodoquin diluted in water and sprayed on the hay until the food material is damp, has also been used.

Control consists of the segregation and treatment of sick animals, and an improvement of the hygiene of the ranch.

Giardia bovis Fantham, 1921, occurs in the small intestine of the ox in North America, Europe and South Africa. The trophozoite stage is 11–19 μm by 7–10 μm and the cysts 7–16 μm by 4–10 μm. There is no good evidence that *G. bovis* is pathogenic.

Other species of *Giardia* include *G. caprae* (goat), *G. equi* (horse), *G. duodenalis* (rabbit), *G. muris* (mouse, rat) and *G. caviae* (guinea-pig).

There is no unequivocal evidence that any of these are pathogenic forms.

Giardia muris in the mouse has been developed as an animal model for human giardiasis (Roberts-Thomson et al. 1976). Infection is self-limiting and is followed by resistance to subsequent infection. In part, the immune response is thymus-dependent but a thymus-independent mechanism is also concerned (Stevens et al. 1978).

REFERENCES

TRITRICHOMONAS FOETUS

Bartlett, D. E. (1949) Bovine venereal trichomoniasis: its nature, recognition, intraherd eradication and interherd control. *Proc. 51st ann. Meeting, US Livestock sanit. Ass.*, pp. 170–181.

Bartlett, D. E. & Dikmans, G. (1949) Field studies on bovine venereal trichomoniasis. Effects on birds and efficacy of certain practices of control. *Am. J. vet. Res.*, **10**, 30–39.

Clark, B. L., Parsonson, I. M., White, M. B., Banfield, J. C. & Young, J. S. (1974) Control of trichomoniasis in a large herd of beef cattle. *Aust. vet. J.*, **50**, 424–426.

Dennett, D. P., Reece, R. L., Barasa, J. O. & Johnson, R. H. (1974) Observation on the incidence and distribution of serotypes of *Tritrichomonas foetus* in beef cattle in north-eastern Australia. *Aust. vet. J.*, **50**, 427–431.

Diamond, L. S. (1957) The establishment of various trichomonads of animals and man in axenic cultures. *J. Parasit.*, **43**, 488–490.

Feinberg, J. G. (1952) A capillary agglutination test for *Trichomonas foetus*. *J. path. Bact.*, **64**, 645–647.

Feinberg, J. G. & Morgan, M. T. J. (1953) The isolation of a specific substance, and a glycogen-like polysaccharide from *Trichomonas foetus*. *Br. J. exp. Path.*, **34**, 104–118.

Fitzgerald, P. R., Johnson, A. E. & Hammond, D. M. (1963) Treatment of genital trichomoniasis in bulls. *J. Am. vet. med. Ass.*, **143**, 259–262.

Fitzgerald, P. R., Johnson, A. E., Thorne, J. L. & Hammond, D. M. (1958) Experimental infections of the bovine genital system with trichomonads from the digestive tracts of swine. *Am. J. vet. Res.*, **19**, 775–779.

Grassé, P. P. (1952) *Traité de Zoologie*. I. Fasc. 1. Paris: Masson.

Hammond, D. M. & Bartlett, D. E. (1945) Pattern of fluctuations in numbers of *Trichomonas foetus* occurring in the bovine vagina during initial infections. I. Correlation with time of exposure and with subsequent estrual cycles. *Am. J. vet. Res.*, **6**, 84–90.

Honigberg, B. M. (1970) Immunity to trichomonads. In: *Immunity to Parasitic Animals*, ed. G. J. Jackson, R. Herman & I. Singer, pp. 469–550. New York: Appleton-Century-Crofts.

Honigberg, B. M., Mattern, C. F. & Daniel, W. A. (1971) Fine structure of the mastigote stage in *Tritrichomonas foetus* (Riedmüller). *J. Protozool.*, **18**, 183–198.

Johnson, A. E. (1964) Incidence and diagnosis of trichomoniasis in western beef bulls. *J. Am. vet. med. Ass.*, **145**, 1007–1010.

Kerr, W. R. (1944) The intradermal test in bovine trichomoniasis. *Vet. Rec.*, **56**, 303–307.

Kerr, W. R. (1964) Immobilization and agglutination of *Trichomonas foetus*. In: *Immunological Methods*, ed. J. F. Ackroyd. Oxford: Blackwell Scientific Publications.

Kerr, W. R. & Robertson, M. (1941) An investigation into the infection of cows with *Trichomonas foetus* by means of the agglutination reaction. *Vet. J.*, **97**, 351–365.

Kerr, W. R. & Robertson, M. (1943) A study of the antibody response of cattle to *Trichomonas foetus*. *J. comp. Path.*, **53**, 280–297.

Kerr, W. R. & Robertson, M. (1953) Active and passive sensitization of the uterus of the cow *in vivo* against *Trichomonas foetus* antigen and the evidence for the local production of antibody in that site. *J. Hyg., Camb.*, **51**, 405–415.

Kerr, W. R. & Robertson, M. (1954) Passively and actively acquired antibodies for *Trichomonas foetus* in very young calves. *J. Hyg., Camb.*, **52**, 253–263.

Kudo, R. R. (1966) *Protozoology*, 5th ed. Springfield, Illinois: Charles C. Thomas.

Kulda, J. & Honigberg, B. M. (1969) Behaviour and pathogenicity of *Tritrichomonas foetus* in chick liver cell cultures. *J. Protozool.*, **16**, 479–495.

Larson, L. (1953) The internal pudendal (pudic) nerve block for anesthesia of the penis and relaxation of the retractor penis muscle. *J. Am. med. Ass.*, **123**, 18–27.

Levine, N. D. (1973) *Protozoan Parasites of Domestic Animals and of Man*, 2nd ed. Minneapolis: Burgess.

Levine, N. D., Anderson, F. L., Losch, M. J., Notzold, R. A. & Mehrer, K. N. (1962) Survival of *Trichomonas foetus* stored at −28°C and −95°C after freezing in the presence of glycerol. *J. Protozool.*, **9**, 347–350.

Levine, N. D., Mizell, M. & Houlahan, D. A. (1958) Factors affecting the protective action of glycerol on *Trichomonas foetus* at freezing temperatures. *Exp. Parasit.*, **7**, 236–248.

McLoughlin, D. K. (1965) Dimetridazole, a systemic treatment for bovine venereal trichiomoniasis. I. Oral administration. *J. Parasit.*, **51**, 835–836.

McLoughlin, D. K. (1970) Dimetridazole, a systemic treatment for bovine venereal trichiomoniasis. III. Trials with cows. *J. Parasit.*, **56**, 39–40.

Morgan, B. B. (1948) Studies on the precipitin and skin reactions of *Trichomonas foetus* (Protozoa) in cattle. *J. cell. comp. Physiol.*, **32**, 235–246.

Morgan, B. B. & Hawkins, P. A. (1952) *Veterinary Protozoology*, 2nd ed. Minneapolis: Burgess.

Parsonson, I. M., Clark, B. L. & Dufty, J. (1974) The pathogenesis of *Tritrichomonas foetus* infection in the bull. *Aust. vet. J.*, **50**, 421–423.

Pierce, A. E. (1949*a*) The mucous agglutination test for the diagnosis of bovine trichomoniasis. *Vet. Rec.*, **61**, 347–349.

Pierce, A. E. (1949*b*) The agglutination reaction of bovine serum in the diagnosis of trichomoniasis. *Br. vet. J.*, **105**, 286–294.

Pierce, A. E. (1953) Specific antibodies at mucous surfaces. *Proc. R. Soc. Med.*, **46**, 31–33.

Pierce, A. E. (1959) Specific antibodies at mucous surfaces. *Vet. Rev. Annot.*, **5**, 17–36.

Robertson, M. (1960) The antigens of *Trichomonas foetus* isolated from cows and pigs. *J. Hyg., Camb.*, **58**, 207–213.

Robertson, M. (1963) Antibody response in cattle to infection with *Trichomonas foetus*. In: *Immunity to Protozoa*, ed. P. C. C. Garnham, A. E. Pierce & I. Roitt, pp. 336–345. Oxford: Blackwell Scientific Publications.

Schoop, G. & Oehlikers, H. (1939) Die Züchtung der Rindertrichomonaden in eiweissarmem Nährboden. *Dt. tierärztl. Wschr.*, **47**, 401–403.

Schoop, G. & Stolz, A. (1939) Trichomoniasis bei Rehen. *Dt. tierarztl. Wschr.*, **47**, 113–114.

Shorb, M. S. (1964) The physiology of trichomonads. In: *Biochemistry and Physiology of Protozoa*, ed. S. H. Hutner. New York: Academic Press.

Switzer, W. P. (1951) Atrophic rhinitis and trichomonads. *Vet. Med.*, **46**, 478–481.

OTHER TRITRICHOMONADS, TRICHOMONAS, TETRATRICHOMONAS, PENTATRICHOMONAS, ETC.

Abraham, R. (1961) On *Trichomonas caballi* n. sp. A protozoan parasite from the horse. *Z. Parasitenk.*, **20**, 164–168.

Abraham, R. &. Honigberg, B. M. (1965) Cytochemistry of chick liver cell cultures infected with *Trichomonas gallinae*. *J. Parasit.*, **51**, 823.

Allen, E. A. (1936) A pentatrichomonas associated with certain cases of enterohepatitis or 'blackhead' of poultry. *Trans. Am. microsc. Soc.*, **55**, 315–322.

Allen, E. A. (1940) A redescription of *Trichomonas gallinarum* Martin and Robertson, 1911, from the chicken and turkey. *Proc. helminth. Soc. Wash.*, **7**, 65–68.
Allen, E. A. (1941) Microscopic differentiation of lesions of histomoniasis and trichomonas in turkeys. *Am. J. vet. Res.*, **2**, 214–217.
Anderson, F. L., Levine, N. D. & Hammond, D. M. (1962) The morphology of *Trichomonas ovis* from the cecum of domestic sheep. *J. Parasit.*, **48**, 589–595.
Bennett, S. P. & Franco, D. A. (1969) Equine protozoan diarrhoea (equine intestinal trichomoniasis) at Trinidad racetracks. *J. Am. vet. med. Ass.*, **154**, 58–60.
Burrows, R. B. & Lillis, W. G. (1967) Intestinal protozoan infections in dogs. *J. Am. vet. med. Ass.*, **150**, 880–883.
Damron, G. W. (1976) Gastrointestinal trichomonads in horses: occurrence and identification. *Am. J. vet. Res.*, **37**, 25–28.
Delappe, I. P. (1957) Effect of inoculating the chicken and the turkey with a strain of *Trichomonas gallinarum*. *Exp. Parasit.*, **6**, 412–417.
Diamond, L. S. (1957) The establishment of various trichomonads of animals and man in axenic cultures. *J. Parasit.*, **43**, 488–490.
Faust, E. C. & Russell, P. F. (1964) *Craig and Faust's Clinical Parasitology*, 7th ed. Philadelphia: Lea & Febiger.
Flick, E. W. (1954) Experimental analysis of some factors influencing variation in the flagellar number of *Trichomonas hominis* from man and other primates, and their relationship to nomenclature. *Exp. Parasit.*, **3**, 105–121.
Frost, J. K. & Honigberg, B. M. (1962) Comparative pathogenicity of *Trichomonas vaginalis* and *Trichomonas gallinae* to mice. II. Histopathology of subcutaneous lesions. *J. Parasit.*, **48**, 898–918.
Frost, J. K., Honigberg, B. M. & McLure, M. T. (1961) Intracellular *Trichomonas vaginalis* and *Trichomonas gallinae* in natural and experimental infections. *J. Parasit.*, **47**, 302–303.
Grassé, P. P. (1952) *Traité de Zoologie*, Fasc. 1. Paris: Masson.
Hayes, B. S. & Kitcher, E. (1960) Evaluation of techniques for the demonstration of *Trichomonas vaginalis*. *J. Parasit.*, **46**, 45.
Hibler, C. P., Hammond, D. M., Caskey, F. H., Johnson, A. E. & Fitzgerald, P. R. (1960) The morphology and incidence of the trichomonads of swine, *Trichomonas suis* (Gruby & Delaford) *Trichomonas rotunda* n. sp. and *Trichomonas buttreyi* n. sp. *J. Protozool.*, **7**, 159–171.
Higgins, R. J. (1980) Caecal trichomoniasis of game birds in North Yorkshire. *Vet. Rec.*, **107**, 228.
Honigberg, B. M. (1961) Comparative pathogenicity of *Trichomonas vaginalis* and *Trichomonas gallinae* to mice. I. Gross pathology, quantitative evaluation of virulence, and some factors affecting pathogenicity. *J. Parasit.*, **47**, 545–571.
Honigberg, B. M., Becker, D., Livingston, M. C. & McLure, M. T. (1964) The behavior and pathogenicity of two strains of *Trichomonas gallinae* in cell cultures. *J. Protozool.*, **11**, 447–465.
Kott, H. & Adler, S. (1961) A serological study of *Trichomonas* sp. parasitic in man. *Trans. R. Soc. trop. Med. Hyg.*, **55**, 333–344.
Laufenstein-Duffy, H. (1969) Equine intestinal trichomoniasis. *J. Am. vet. med. Ass.*, **155**, 1835–1840.
Levine, N. D. (1973) *Protozoan Parasites of Domestic Animals and of Man*. Minneapolis: Burgess.
Lucas, A., Toucas, L., Laroche, M., & Monin, L. (1956) Trichomonose intestinale du faisondeau. *Rec. Méd. vét.*, **132**, 613–617.
McDowell, S. (1953) A morphological and taxonomic study of the caecal protozoa of the common fowl, *Gallus gallus*. *J. Morphol.*, **92**, 337–399.
McEntegart, M. G., Chadwick, C. S. & Nairn, R. C. (1958) Fluorescent antisera in the detection of serological varieties of *Trichomonas vaginalis*. *Br. J. Vener. Dis.*, **34**, 1–3.
McLoughlin, D. K. (1957) Age of host and route of administration as factors influencing the susceptibility of turkeys to *Trichomonas gallinarum*. *J. Parasit.*, **43**, 321.
Manahan, F. F. (1970) Diarrhoea in horses with particular reference to a chronic diarrhoea syndrome. *Aust. vet. J.*, **46**, 231–234.
Martin, C. L. & Robertson, M. (1911) Further observations on the caecal parasites of fowls, with some reference to the rectal fauna of other vertebrates. Report. *J. microsc. Sci.*, **57**, 53–81.
Mesa, C. P., Stabler, R. M. & Berthrong, M. (1961) Histopathological changes in the pigeon infected with *Trichomonas gallinae*. *Avian Dis.*, **5**, 48–60.
Miessner, H. & Hansen, K. (1936) Trichomoniasis der Tauben. *Dt. tierärztl. Wschr.*, **44**, 323–330.
Morgan, B. B. (1944) *Bovine Trichomoniasis*, p. 150. Minneapolis: Burgess.
Morganti, L. & Bitti, G. R. (1968) Sul reperimento di due specie di Trichomonas nelle feci di due cani. *Parasitologia*, **10**, 11–15.
Nie, D. (1950) Morphology and taxonomy of the intestinal protozoa of the guinea-pig, *Cavia porcella*. *J. Morphol.*, **86**, 381–493.
Richardson, U. F. & Kendall, S. B. (1963) *Veterinary Protozoology*, p. 311. Edinburgh: Scottish Academic Press.
Sharma, N. N. & Honigberg, B. M. (1969) Cytochemical observations on malic hydrogenase, lipase nonspecific esterase and monoamine oxidase in chick liver cell cultures infected with *Trichomonas vaginalis*. *J. Protozool.*, **16**, 171–181.
Simitch, T., Petrovitch, Z. & Lepech, T. (1954) Contribution à la conaissance de la biologie des Trichomonas. II. Differentiation de *T. wenrichi* Wenrich et Saxe, 1950 et de *T. intestinalis* Leukart, 1879 par leurs caractères biologiques. *Ann. Parasit. Hum. Comp.*, **29**, 199–205.
Stabler, R. M. (1951) Effect of *Trichomonas gallinae* from diseased mourning doves on clean domestic pigeons. *J. Parasit.*, **37**, 437–478.
Stabler, R. M. (1953) Effect of *Trichomonas gallinae* (Protozoa: Mastigophora) on nestling passerine birds. *J. Colo.-Wyo. Acad. Sci.*, **4**, 58.
Stabler, R. M. (1954) *Trichomonas gallinae*: a review. *Exp. Parasit.*, **3**, 368–402.
Stabler, R. M. (1957) The effect of furazolidone on pigeon trichomoniasis due to *Trichomonas gallinae*. *J. Parasit.*, **43**, 280–282.
Stabler, R. M. & Herman, C. M. (1951) Upper digestive tract trichomoniasis in mourning doves and other birds. *Trans. N. Am. Wild Life Conf.*, **16**, 145–163.
Stabler, R. M. & Mellentin, R. W. (1953) Effect of 2-amino-5-nitrothiazole (Enheptin), and other drugs on *Trichomonas gallinae* infection in the domestic pigeon. *J. Parasit.*, **39**, 637–642.
Stepkowski, S. & Honigberg, B. M. (1972) Antigenic analysis and virulent and avirulent strains of *Trichomonas gallinae* by gel diffusion methods. *J. Protozool.*, **19**, 306–315.
Switzer, W. P. (1951) Atrophic rhinitis and trichomonads. *Vet. Med.*, **46**, 478–481.
Walker, R. V. L. (1948) Enterohepatitis (blackhead) in turkeys. I. Pentatrichomonas associated with enterohepatitis and its propagation in developing chick embryos. *Can. J. comp. Med.*, **12**, 43–46.
Wohler, W. H. (1961) Equine protozoal diarrhoea. *Med. vet. Pract.*, **42**, 52.

MONOCERCOMONADIDAE: HISTOMONAS, PARAHISTOMONAS AND OTHER GENERA

Beg, M. K. & Clarkson, M. J. (1970) Effect of histomoniasis on the serum proteins of the fowl. *J. comp. Path.*, **80**, 281–285.
Bradley, R. E., Johnson, J. & Reid, W. M. (1964) Apparent obligate relationship between *Histomonas meleagridis* and *Escherichia coli* in producing disease. *J. Parasit.*, **50**, (Suppl.), 51.
Cuckler, A. C. (1970) Coccidiosis and histomoniasis in avian hosts. In: *Immunity to Parasitic Animals*, ed. G. J. Jackson, H. Herman & I. Singer, vol. 2, pp. 371–397. New York: Appleton-Century-Crofts.
Doll, J. P. & Franker, C. K. (1963) Experimental histomoniasis in gnotobiotic turkeys. I. Infection and histopathology of the bacteria-free host. *J. Parasit.*, **49**, 411–414.
Dwyer, D. M. (1970) An improved method for cultivating *Histomonas meleagridis*. *J. Parasit.*, **56**, 191–192.
Dwyer, D. M. & Honigberg, B. M. (1970) Effect of certain laboratory procedures on the virulence of *Histomonas meleagridis* for turkeys and chickens. *J. Parasit.*, **56**, 694–700.
Farmer, R. K. & Stephenson, J. (1949) Infectious enterohepatitis (blackhead) in turkeys: A comparative study of methods of infection. *J. comp. Path. Ther.*, **59**, 119–126.
Farr, M. M. (1956) Survival of the protozoan parasite, *Histomonas meleagridis*, in feces of infected birds. *Cornell Vet.*, **46**, 178–187.
Farr, M. M. (1961) Further observations on survival of the protozoan parasite, *Histomonas meleagridis*, and eggs of poultry nematodes in feces of infected birds. *Cornell Vet.*, **51**, 3–13.
Franker, C. K. & Doll, J. P. (1964) Experimental histomoniasis in gnotobiotic turkeys. II. Effects of some caecal bacteria on pathogenesis. *J. Parasit.*, **50**, 636–640.
Graybill, H. W. & Smith, T. (1920) Production of fatal blackhead in turkeys by feeding embryonated eggs of *Heterakis papillosa*. *J. exp. Med.*, **31**, 647–655.
Hoffman, G. L. (1970) Control and treatment of parasitic diseases of freshwater fishes. US Bureau Sport Fish Wildlife. Fish Disease Leaflet 28, pp. 7.

Horton-Smith, C. & Long, P. L. (1956) Further observations on the chemotherapy of histomoniasis (blackhead) in turkeys. *J. comp. Path. Ther.*, **66**, 378–388.
Kendall, S. B. (1957) Some factors influencing resistance to histomoniasis in turkeys. *Br. vet. J.*, **113**, 435–439.
Kendall, S. B. (1959) The occurrence of *Histomonas meleagridis* in *Heterakis gallinae*. *Parasitology*, **49**, 169–172.
Lee, D. L. (1969) The structure and development of *Histomonas meleagridis* (Mastigamoebidae: Protozoa) in the female reproductive tract of its intermediate host, *Heterakis gallinarum* (Nematoda). *Parasitology*, **59**, 877–884.
Lee, D. L., Long, P. L., Millard, B. J. & Bradley, J. (1969) The fine structure and method of feeding of the tissue parasitizing stages of *Histomonas meleagridis*. *Parasitology*, **59**, 171–184.
Lesser, E. (1960) Cultivation of *Histomonas meleagridis* in a modified tissue culture medium. *J. Parasit.*, **46**, 686.
Levine, N. D. (1973) *Protozoan Parasites of Domestic Animals and of Man.* 2nd ed. p. 406. Minneapolis: Burgess.
Lund, E. E. (1959) Immunizing action of a nonpathogenic strain of histomonas against blackhead in turkeys. *J. Protozool.*, **6**, 182–185.
Lund, E. E. (1963) *Histomonas wenrichi* n. sp. (Mastigophora: Mastigamoebidae), a nonpathogenic parasite of gallinaceous birds. *J. Protozool.*, **10**, 401–404.
Lund, E. E. & Burtner, R. H. (1957) Infectivity of *Heterakis gallinae* eggs with *Histomonas meleagridis*. *Exp. Parasit.*, **6**, 189–193.
Malewitz, T. D., Runnels, R. A. & Calhoun, M. L. (1958) The pathology of experimentally produced histomoniasis in turkeys. *Am. J. vet. Res.*, **19**, 181–185.
Nie, D. (1950) Morphology and taxonomy of the intestinal protozoa of the guinea-pig, *Cavia porcella*. *J. Morphol.*, **86**, 381–493.
Reid, W. M. (1967) Aetiology and dissemination of the blackhead disease syndrome in turkeys and chickens. *Exp. Parasit.*, **21**, 249–275.
Ruff, M. D., McDongald, L. R. & Hansen, M. F. (1970) Isolation of *Histomonas meleagridis* from embryonated eggs of *Heterakis gallinarum*. *J. Protozool.*, **17**, 10–11.
Schubert, G. (1968) The injurious effects of *Costia necatrix*. *Bull. Off. int. Epizoot.*, **69**, 1171–1178.
Tyzzer, E. E. (1919) Developmental phases of the protozoon of 'blackhead' in turkeys. *J. med. Res.*, **40**, 1–30.
Tyzzer, E. E. (1933) Loss of virulence in the protozoon of 'blackhead', a fatal disease of turkeys, and the immunizing properties of attenuated strains. *Science, N.Y.*, **78**, 522–523.
Tyzzer, E. E. (1936) A study of immunity produced by infection with attenuated culture strains of *Histomonas meleagridis*. *J. comp. Path. Ther.*, **49**, 285–303.
Tyzzer, E. E. & Collier, J. (1925) Induced and natural transmission of blackhead in the absence of Heterakis. *J. infect. Dis.*, **37**, 265–276.
Wehr, E. E., Farr, M. M. & McLoughlin, D. K. (1958) Chemotherapy of blackhead in poultry. *J. Am. vet. med. Ass.*, **132**, 439–445.

RETORTAMONAS, CHILOMASTIX, COCHLOSOMA, HEXAMITA, GIARDIA

Allard, J., Champaign, D. A., Delisle, R., Mires, H. & Lippy, E. (1977) Waterborne giardiasis outbreaks—Washington, New Hampshire. *Morb. Mort. Wkly Rep.*, **26**, 169–170.
Bemrick, U. J. (1961) A note on the incidence of three species of Giardia in Minnesota. *J. Parasit.*, **47**, 87–89.
Box, E. D. (1981) Observations on *Giardia* of budgerigars. *J. Protozool.*, **28**, 491–494.
Briggs, J. E. (1959) Nitrofurans in feeds. *14th Kansas Formula Feed Conf. Kansas State College*, 4–6 Jan., 1959, pp. 15–21.
Campbell, J. G. (1945) An infectious enteritis of young turkeys associated with *Cochlosoma* sp. *Br. vet. J.*, **101**, 255–259.
Craige, J. E. (1948) Differential diagnosis and specific therapy of dysenteries in dogs. *J. Am. vet med. Ass.*, **113**, 343–347.
Craige, J. E. (1949) Intestinal disturbances in dogs: differential diagnosis and specific therapy. *J. Am. vet med. Ass.*, **114**, 425–428.
Filice, F. P. (1952) Studies on the cytology and life history of a giardia from the laboratory rat. *Univ. Calif. Publ. Zool.*, **57**, 53–143.
Fogg, D. E. (1957) *Merck Poultry Nutrition and Health Symposium*, St. Louis, 5 Aug., 1957, pp. 16–18.
Hagen, K. W. (1950) Treatment of giardia in the chinchilla. *Calif. Vet.*, **3**, 11.
Haiba, M. H. (1956) Further study on the susceptibility of murines to human giardiasis. *Z. Parasitenk.*, **17**, 339–345.
Hinshaw, W. R. (1959). In: *Diseases of Poultry*, ed. H. E. Biester & L. H. Schwarte, 4th ed. Ames: Iowa State University Press.
Hinshaw, W. R. & McNeil, E. (1941) Carriers of *Hexamita meleagridis*. *Am. J. vet. Res.*, **2**, 453–458.
Hoare, C. A. (1955) Life cycle of *Hexamita meleagridis*. *Vet. Rec.*, **67**, 324.
Hoffman, G. L. (1967) *Parasites of North American Freshwater Fishes*, p. 486. Berkeley: University of California Press.
Hughes, W. F. & Zander, D. V. (1954) Isolation and culture of *Hexamita* free of bacteria. *Poultry Sci.*, **33**, 810–815.
Levine, N. D. (1973) *Protozoan Parasites of Domestic Animals and of Man*, 2nd ed., p. 406. Minneapolis: Burgess.
McElwain, I. V. & Post, G. (1968) Efficacy of cyzine for trout hexamitiasis. *Progressive Fish-Culturist*, **30**, 84–91.
McNeil, E. & Hinshaw, W. R. (1941) Experimental infection of chicks with *Hexamita meleagridis*. *Cornell Vet.*, **31**, 345–350.
Mangrum, J. F., Ferguson, T. M., Couch, J. R., Willis, F. K. & Delaplane, J. P. (1955) The effectiveness of furazolidone in the control of hexamitiasis in turkey poults. *Poultry Sci.*, **34**, 836–840.
Meshorer, A. (1969) Hexamitiasis in laboratory mice. *Lab. Animal Care*, **19**, 33–37.
Mueller, J. F. (1959) Is chilomastix a pathogen? *J. Parasit.*, **45**, 170.
Nöller, W. & Buttgereit, F. (1923) Über ein neues parasitisches Protozoon der Haustaube (*Octomitus columbae* nov. spec.). *Zbl. Bakt. Orig.*, **75**, 239–240.
Pizzi, T. (1957) Pathogenic role of *Giardia lamblia*. *Boln chil. Parasit.*, **12**, 10–12.
Reichenbach-Klinke, H. & Elkan, E. (1965) *The Principal Diseases of Lower Vertebrates*, p. 600. New York: Academic Press.
Richards, R. (1977) Disease of aquarian fish—3. Disease of the internal organs. *Vet. Rec.*, **101**, 149–150.
Roberts-Thompson, I. C., Stevens, D. P., Mahmoud, A. A. F. & Warren, K. S. (1976) Giardiasis in the mouse: an animal model. *Gastroenterology*, **71**, 57–61.
Shelton, G. C. (1954) Giardiasis in the chinchilla. II. Incidence of the disease and results of experimental infections. *Am. J. vet. Res.*, **15**, 75–78.
Slavin, D. & Wilson, J. E. (1954) *Hexamita meleagridis*. *Nature, Lond.*, **172**, 1179–1181.
Slavin, D. & Wilson, J. E. (1960) A fuller conception of the life cycle of *Hexamita meleagridis*. *Poultry Sci.*, **39**, 1559–1576.
Stevens, D. P., Frank, D. M. & Mahmoud, A. A. F. (1978) Thymus dependency of host resistance to *Giardia muris* infection: studies in nude mice. *J. Immunol.*, **120**, 680–682.
Wilson, J. E. & Slavin, D. (1955) Hexamitiasis of turkeys. *Vet. Rec.*, **67**, 236–242.

SUBPHYLUM: SARCODINA SCHMARDA, 1871

Members of this subphylum do not possess a thick pellicle, they move by means of pseudopodia, and only rarely do they have flagella. The cytoplasm is usually differentiated into ectoplasm and endoplasm but this is not constant. With few exceptions, reproduction is asexual by binary fission. Nutrition is holozoic, the forms being predatory on bacteria, protozoa and small metozoa. Only a few are parasitic and of direct interest in this book. They are classified in the order AMOEBIDA Ehrenberg, 1830 and belong to the families *Endamoebidae*, *Vahlkampfiidae* and *Hartmannellidae*.

FAMILY: ENDAMOEBIDAE CALKINS, 1926

This family contains, exclusively, the parasitic amoebae which occur in the digestive tract of vertebrates and invertebrates. Multiplication is by binary fission and encystment is common. This family should be differentiated from the family *Amoebidae* Bronn, 1859 which contains the free-living amoeba of water, soil, etc. The well-known *Amoeba proteus* occurs in the latter family.

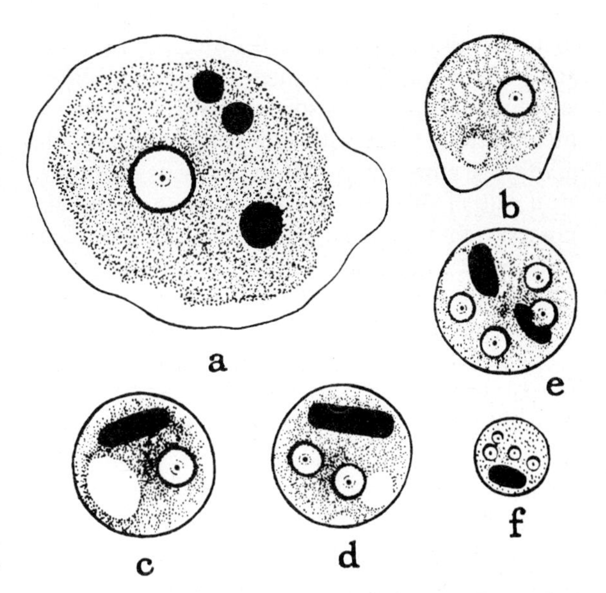

Fig 3.17 *Entamoeba histolytica.* (a) Active (tropic) amoeba with ingested red blood corpuscles. (b) Precystic amoeba. (c–f) Cysts: (c) 1-nucleate with chromatoid body and glycogen vacuole; (d) 2-nucleate with chromatoid body and small glycogen vacuole; (e) 4-nucleate (mature) with two chromatoids and small glycogen vacuole; (f) mature cyst of small race, with chromatoid body. (× 2000) (*From Hoare 1950*)

Genus: Entamoeba Casagrandi and Barbagallo, 1895

Members of this genus have a vesicular nucleus with a comparatively small endosome at or near the centre of the nucleus. A varying number of chromatin granules occur in the periendosomal region and attached to the nuclear membrane. Cysts containing one to eight nuclei are produced.

This genus should be differentiated from the genus *Endamoeba* Leidy, 1897, the latter lacking a defined central endosome. Members of the *Endamoeba* occur in the intestine of invertebrates.

A large number of species of *Entamoeba* occur in a wide range of animals, and a check-list of species has been given by Hoare (1959). The forms can be conveniently grouped according to the trophozoite and cyst morphology and Hoare (1959) recognizes four such groups. Levine (1973) adopted a similar grouping:

1. Entamoebae with eight-nucleate cysts (e.g. *E. coli, E. wenyoni, E. muris,* etc.).
2. Entamoebae with quadrinucleate cysts (e.g. *E. histolytica, E. hartmanni, E. equi,* etc.).
3. Entamoebae with uninucleate cysts (e.g. *E. bovis, E. bubalis, E. suis,* etc.).
4. Entamoebae in which cysts are unknown (e.g. *E. gingivalis, E. canibuccalis, E. equibuccalis,* etc.).

The only species of importance as a pathogen is *E. histolytica.*

Entamoeba histolytica Schaudinn, 1903 (Fig. 3.17). This is the cause of amoebic dysentery of man. It has also been found in many species of monkey and in the dog, cat, rat and pig. Experimentally, the rat, mouse, guinea-pig and rabbit can be infected. Geographically, it is worldwide in distribution, occurring in countries with temperate, subtropical or tropical climates. It is, however, more prevalent in the tropics and subtropics than in cooler climates, although surveys in countries such as Great Britain and the United States have shown a substantial level of infection. The prevalance in various countries is detailed by Hoare (1950), and the results of various other surveys are quoted by Levine (1973).

Morphology. The active trophozoite of *E. histolytica* ranges in size from 10 μm to 60 μm. It has finely granular endoplasm and the ectoplasm is hyaline in appearance and well differentiated from the inner endoplasm. A characteristic of the organism is its active movements, pseudopodia, which are long and finger-like, appearing suddenly and the endoplasm flowing rapidly into them. There is a single spherical nucleus 4–7 μm in diameter and it contains a distinct central endosome about 0.5 μm in diameter. The endosome is surrounded by a clear zone or halo. The nuclear membrane is lined with fine chromatin granules giving the appearance that the nucleus is outlined by a ring of small beads. Active trophozoites also possess food vacuoles which contain red blood cells in the process of digestion. This feature is one which differentiates *E. histolytica* from the non-pathogenic forms.

The cysts of *E. histolytica* are spherical, occasionally ovoid, and measure 5–20 μm in diameter. The cyst wall, visible in living specimens but not seen in stained preparations, is about 0.5 μm in thickness. Initially the cysts are uninucleate but finally a four-nucleate cyst is produced. The nuclei are comparable to those seen in the vegetative form although smaller; moreover, the cysts contain chromatoid bodies and glycogen. The chromatoid bodies appear as refractile rods with rounded ends and stain deeply with chromatin stains. Glycogen vacuoles are seen most clearly in young cysts and stain brown when treated with iodine solution. The glycogen disappears from the cysts when they reach the four-nucleate stage, although the chromatin rods may persist longer. Ultimately, they too disappear, being used up as reserve food supply.

Evidence for viruses in *E. histolytica* was obtained by Diamond et al. (1972), two forms—an icosahedral particle and a filamentous particle —were described (Mattern et al. 1972). Studies suggested that the virus–amoeba relationship may represent a lysogenic system (Diamond & Mattern 1976). All strains of *E. histolytica* studied so far are infected with viruses, although these strains were isolated from clinical cases of amoebiasis and this may indicate that virulent strains are infected with the viruses. *Entamoeba hartmanni* also has an icosahedral viral infection. Using the newborn hamster liver inoculation method (see below) to assess virulence, Mattern et al. (1977) concluded that amoebal viruses probably do not play a role in virulence of the organism.

Differences in the morphology, pathogenicity and culture characteristics of *E. histolytica* have been recognized for many years. A small and large form of the organism are recognized. The small form, or race, is non-pathogenic and has been named *Entamoeba hartmanni* by Brumpt (1925), while the large forms may be divided into an avirulent race and a virulent race (Hoare 1959). However, there is debate on the speciation of the small forms of *E. histolytica*-like organs (see below).

A simple and rapid technique for the assay of virulence of various strains of *E. histolytica* has been developed by Mattern and Keister (1977) and consists of the direct intrahepatic inoculation of newborn hamsters with axenically grown organisms. Virulent strains produced liver abscesses with as few as 20 organisms, while other strains failed to cause infection with 20 000 organisms.

Developmental cycle. (Fig. 3.18) In the trophozoite phase, the organism multiplies by binary fission, this taking place solely in the vertebrate host. The cystic form is passed in the faeces of the host, encystation occurring in the lumen of the bowel. Prior to encystment, the active amoebae divide, producing smaller forms which expel food particles, round up, and cease to feed. The cysts are at first uninucleate but later the nucleus divides into two and each then further divides so that a quadrinucleate cyst is produced. Cysts are passed in the faeces in all stages of development, but apparently only the quadrinucleate forms, which represent the mature stage, remain viable and are capable of inducing new infection.

On subsequent infection of a human, or animal, the mature quadrinucleate cyst excysts in the small or large intestine. The newly released metacystic form undergoes a series of nuclear and cytoplasmic divisions, resulting in the production of eight uninucleate amoebae. These then pass to the large bowel where they grow into the larger forms which may remain in the lumen of the bowel or may invade the tissues.

Pathogenesis. Only the large forms of *E. histolytica* are considered to be pathogenic. The mechanism of tissue invasion by the parasite is by no means clearly understood, and the various factors of the pathogenesis and pathogenic mechanisms of *E. histolytica* have been considered by

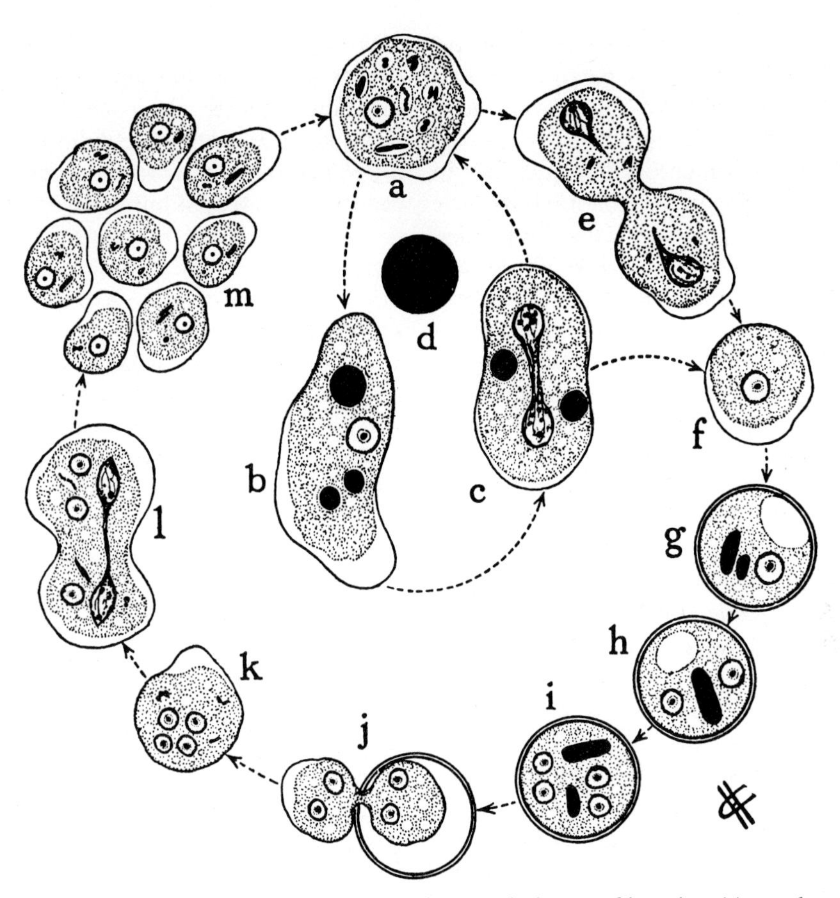

Fig 3.18 Life-cycle of *Entamoeba histolytica*. (a, e–m) Development in lumen of intestine: (a) amoeba with ingested bacteria; (e) division of amoeba; (f) precystic amoeba; (g–i) 1-, 2- and 4-nucleate cysts, with glycogen vacuole and chromatoid bodies; (j) excystation of 4-nucleate amoeba; (k–m) metacystics stages: (k) excysted 4-nucleate amoeba; (l) division of metacystic amoeba; (m) eight 1-nucleate amoebulae, resulting from division of metacystic amoeba. (b, c) Development in intestinal wall: (b) tissue-invading amoeba with ingested erythrocytes; (c) division of tissue-invading amoeba. (d) Normal red blood corpuscle. (×1300) (*From Hoare 1950*)

Maegraith (1963). The penetration of the intestinal epithelium is probably brought about by lysis of the epithelium by proteolytic enzymes, and trypsin and pepsin, but not chymotrypsin, have been found in *E. histolytica*. However, these enzymes have been found in both pathogenic and non-pathogenic strains, and it is not possible at present to differentiate the two on the basis of enzymes. Nevertheless, the presence of the enzymes may be an indication of the mode of invasion. Maegraith (1963) suggests that the ability or otherwise to invade may be determined more by the host than by the parasite. Meerovitch (1977) considers that the initial destruction of mucosal cells results from the action of a potent phosphohydrolase released from the amoebic surface liposomes following contact between cells and parasite.

The relationship between bacteria and amoebae may be of importance since experiments in germ-free guinea-pigs have shown that it was impossible to infect these animals with *E. histolytica* without the addition of intestinal bacteria. Two species of bacteria (*Aerobacter aerogenes* and *Escherichia coli*) permitted the protozoans to become established and invade the tissues (Phillips et al. 1955).

Following invasion of the epithelium the amoebae multiply, forming small colonies, and then penetrate into the deeper tissues and reach the submucosa where they spread laterally. They undermine the mucous membrane and produce a flask-shaped ulcer, which has a narrow canal or neck leading to the lumen of the bowel and a dilated distal part in the submucosa.

The initial lesions, which occur principally in the caecum and the ascending colon, show little cellular reaction or bacterial invasion, the process being exclusively one of lytic necrosis. The lesions may remain confined to the mucosa and repair may keep pace with the disease process, resulting in spontaneous elimination of the organisms. In other cases, the amoebae penetrate more deeply into the intestinal wall, invasion by bacteria occurs and there is associated hyperaemia, inflammation, and an infiltration of neutrophils. Amoebae are found chiefly at the periphery of the ulcer, in contact with healthy tissue into which they gradually penetrate, leaving the cavity of the ulcer filled with necrotic tissue. Maegraith (1963) tested several potentially pathogenic strains of *E. histolytica* for the presence of hyaluronidase but he was unable to equate the presence of this enzyme with pathogenicity. It was absent from non-pathogenic forms and was also absent from two of the potentially pathogenic forms.

Amoebae may pass into the lymphatics or the mesenteric venules and invade other tissues of the body. They have been found in almost all soft tissue of the body but the commonest location is the liver, especially the right lobe. Amoebae are trapped in thrombi in the interlobular veins and here produce lytic necrosis of the walls of the vessels. The lesions increase in size, with little or no cell infiltration, and one or more may enlarge to produce a hepatic amoebic abscess. In rapidly forming abscesses there may be no limiting capsule, but in the more chronic types a fibrous wall is produced. Abscesses may occur elsewhere, such as in the lungs and brain, and rarely they may occur on the skin.

Bray and Harris (1977) have discussed the role of the host's immune response in disease in *E. histolytica* infection. Cell-mediated immunity may be an important determinant of potentially invasive *E. histolytica* infection and the pathology produced by it (Kagan 1974). Since the intestinal and liver lesions show a remarkable absence of cell infiltration, Bray and Harris (1977) suggest this may be indicative of local or central immunodepression, a prerequisite for successful invasion. Further, they believe that the immune response is important in limiting invasion of *E. histolytica* but that amoebic disease must be preceded by a degree of immunodepression.

Pathogenesis of E. histolytica in animals. Dogs, kittens, monkeys, guinea-pigs, hamsters, rats, rabbits and pigs have been infected experimentally. The kitten is especially susceptible and when infected orally with cysts, or per rectum with the trophozoites, an acute amoebic dysentery results with extensive ulceration of the bowel wall. The organisms appear unable to encyst in the cat and natural cat-derived infections are therefore rare.

In the dog, natural infections have been recorded from many parts of the world. These have usually been sporadic infections and probably acquired from human contacts. The majority of cases have been reported from the Far East (Hoare 1959; Levine 1973), but in the United States the organism has been found in association with diarrhoea in a dog in Baltimore (Andrews 1932), and Thorson et al. (1956) recorded a systemic infection in a puppy, organisms being found in large numbers in the lungs, liver, kidneys, and spleen.

A prevalance of 8.4% in dogs in Tennessee detected by cultural methods was reported by Eyles et al. (1954), and Burrows (1968) found the organism in 1% of 835 dogs examined in New Jersey, which may indicate that canine infection with *E. histolytica* is more extensive than has been thought. In the dog, the infection is chiefly localized in the caecum and usually runs a symptomless course. Occasionally, however, the organism invades the tissues and produces clinical signs of acute or chronic amoebiasis.

An outbreak of dysentery in cattle associated with *E. histolytica* has been recorded by Walkiers (1930); this occurred in an area in which there was an epidemic of amoebiasis in Africans. The organism has also been found in the lungs of a zebu cow in Dakar by Thiéry and Morel (1956).

Natural infections with *E. histolytica* are common in a wide range of monkeys, and the human and simian strains are interchangeable as demonstrated by cross-infections. Apparently, the organism is common in Old World monkeys but uncommon in New World monkeys in their natural habitat (Ruch 1959).

New World monkeys are more susceptible than Old World monkeys (Vickers 1969). Usually the infections are symptomless, but chronic mild cholitis, characterized by congestion, petechial haemorrhages and occasionally ulcers, may occur (Bostrom et al. 1968). Organisms isolated from primates can produce typical amoebic dysentry when inoculated into kittens and when cysts of these forms are fed to humans. Fremming et al. (1955) have reported a fatal case of amoebiasis in the chimpanzee.

Monkey infection may be of public health importance, especially since large numbers of these animals are imported into different countries for experimental purposes.

Wild rats may harbour an *Entamoeba* which is indistinguishable from *E. histolytica*, and it is generally assumed that such infections are derived from humans. The pathology of *E. histolytica* infection in the rat varies greatly; the organism may occur as a harmless commensal in the large bowel or it may invade the mucosa, producing typical signs of amoebic dysentery.

Epidemiology. E. histolytica is primarily a parasite of man, infected and carrier humans forming the reservoir of infection. Man is also the reservoir host for animal infections. Infection is by the ingestion of the mature cysts, and trophozoites do not survive long outside the host. The cysts are relatively resistant to adverse conditions; they may remain viable for at least two weeks in a stool sample kept at room temperature and for about two months in a refrigerator. In water, they may remain viable for up to five weeks at room temperature. The thermal death point is 50°C, and desiccation is rapidly fatal. Generally, the cysts are transmitted in food or water, and raw vegetables may also be a source of infection. Various flies (*Musca*, *Lucilia*, etc.) have been shown capable of transmitting cysts in their vomitus (Pipkin 1949). The major outbreaks of amoebiasis in man are usually caused by faulty water supply, and several major outbreaks have been traced to the contamination of drinking water with sewage.

Diagnosis. This is based on the demonstration of the trophozoite or the cyst in the faeces. In a normally formed stool, usually only cysts are found, but in diarrhoeic stools the trophozoite is also seen. Actively motile organisms may be seen in warm, freshly passed, diarrhoeic faeces, but at other times the cysts may be seen more readily if a drop of iodine solution (a saturated solution of iodine in 1% potassium iodide) is added. Cysts may be concentrated by a zinc sulphate flotation solution, but other flotation solutions, such as salt and sugar, should be avoided since these cause undue distortion. A specific diagnosis is more satisfactory, made from a permanent preparation fixed in Schaudinn's solution and stained by the iron haematoxylin method. Cultivation may be helpful if fresh material is available. Various media are available, ranging from the original diphasic coagulated egg slant overlaid with Locke's solution containing serum, to monophasic media in which organisms are grown axenically (Diamond 1968). Most strains of *E. histolytica* will not grow below 30°C but the Laredo strain, isolated from a patient in Laredo, Texas, grows at 20–25°C. At least five strains of *E. histolytica* which grow at 25°C have been found to be non-invasive (Neal & Johnson 1968).

Immunological diagnosis of amoebiasis has been reviewed by Kagan (1974). At present, indirect haemagglutination, indirect immunofluorescent and countercurrent immunoelectrophoresis techniques are widely used. Such tests are very sensitive and specific for patients with invasive amoebiasis. However, the sensitivity of the tests decreases in patients where the tissue invasion is minimal.

Treatment. The therapy of amoebiasis in animals is based on that applicable to humans.

Metronidazole is the drug of choice; for symptomless amoebiasis it is given at the rate of 400 mg three times a day for five days, and in acute cases, at 800 mg three times a day for five days. Other compounds include: diloxanide furoate (500 mg t.i.d., ten days), di-iodohydroxyquinoline (600 mg t.i.d., 21 days), iodochlorhydroxyquin

(250 mg t.i.d., 21 days) for symptomless amoebiasis; and various tetracyclines: chlortetracycline (Aureomycin), oxytetracycline (Terramycin) and tetracycline (Achromycin) at 250 mg t.i.d., for seven to ten days. Fumagillin is useful in primate infections.

Control. The control of *E. histolytica* infection is essentially a question of good sanitation, improved sewage disposal, the avoidance of faecal contamination of food, and an improvement in personal hygiene.

Entamoeba hartmanni von Prowazek, 1912. This form closely resembles the small form of *E. histolytica* and there has been considerable debate as to whether *E. hartmanni* is a valid species. The question has been reviewed by Neal (1966) and Elsdon-Dew (1968) and it is now generally agreed that the '*E. histolytica* complex' consists of: *E. histolytica* trophozoites (20–40 μm), which do not form cysts and are haematogenous; *E. histolytica* trophozoites (7–16 μm), which form cysts (10–15 μm), feed on bacteria and are not haematogenous; and *E. hartmanni* trophozoites (5–11 μm), which form cysts (5–10 μm) and are not haematogenous (Smyth 1976).

Apart from size, other morphological differences occur. Thus, the peripheral nuclear chromatin of *E. hartmanni* is more variable than that of *E. histolytica* and may consist of discrete granules with wide spaces between them, whereas with *E. histolytica* it is usually distributed uniformly on the nuclear membrane.

The prevalence of *E. hartmanni* is unknown, since previously the species has been confused with *E. histolytica*, but it has been reported from the caecum and colon of dog and monkey, as well as man. In a survey of 600 dogs in the USA, Burrows and Lillis (1967) found *E. hartmanni* in one animal.

There is no evidence that *E. hartmanni* is pathogenic.

Entamoeba coli (Grassi, 1879) Casagrandi and Barbagallo, 1895. (See Fig. 3.19) This is a non-pathogenic form in humans, and its importance lies in the fact that it must be distinguished from the pathogenic *E. histolytica*. It is found in the caecum and colon, and up to 30% of some populations may be infected. It is world-wide in distribution, being more common in warm, moist climates.

Morphology. The active trophozoite is 20–30 μm in diameter, with a range of 15–50 μm. The ectoplasm is thin and the cytoplasm possesses food vacuoles containing bacteria, yeast, starch grains and vegetable debris. The nucleus differs from that of *E. histolytica* in having a larger endosome, which is also eccentric in position. The endosome is surrounded by a halo, and the peripheral chromatin of the nucleus is composed of coarser granules than in *E. histolytica*. Chromatin granules also appear on a network between the nuclear membrane and the endosome. The movements of *E. coli* are sluggish, and finger-like pseudopodia are not formed.

The cysts are 10–30 μm in diameter, most frequently 15–20 μm. When newly formed, the cyst is uninucleate but eventually eight nuclei are produced, this representing the mature infective stage. The chromatoid bodies are splinter-like; a glycogen vacuole is present in young cysts but it has generally disappeared by the quadrinucleate stage and is usually absent in the eight-nucleated cyst.

Entamoeba coli may occur in monkeys (Kessel 1928) and the simian strain is transmissible to man. Many simian infections are acquired from humans, and the monkey does not form a major source of human infection. Natural infections with an entamoeba resembling *E. coli* have been found in the dog and in the pig.

Entamoeba gingivalis (Gros, 1849) Brumpt, 1914. This species is commonly found in the mouth of man, occasionally in the gingival tissue around the teeth, especially if inflammation, suppuration, or pyorrhoea is present. It also occurs in primates (Ruch 1959). It is a harmless species, being more common in unhygienic mouths, although it can also be found in those which are well kept.

Only the trophozoite stage is known; encystment apparently does not occur. The trophozoite is 5–35 μm in diameter (usually 10–20 μm), and it has a clear ectoplasm and a granular endoplasm in

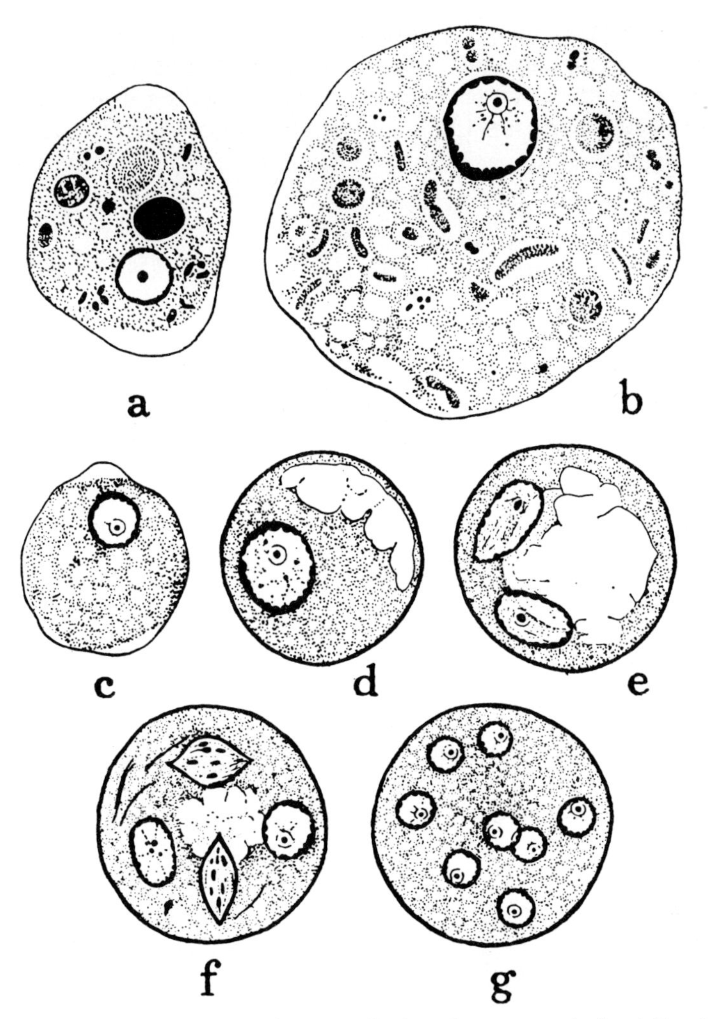

Fig 3.19 Non-pathogenic entamoebae. (a) *Entamoeba gingivalis*, from human mouth. (b–g) *E. coli*, from human intestine: (b) active (tropic) amoeba with ingested bacteria, etc.; (c) precystic amoeba; (d–g) cysts: (d) 1-nucleate with glycogen vacuole; (e) 2-nucleate with large glycogen vacuole; (f) 4-nucleate (two nuclei dividing), with filamentar chromatoids; (g) 8-nucleate (mature). (× 2000) (*From Hoare 1950*)

which food vacuoles containing leucocytes, epithelial debris, bacteria, etc. may be seen. The nucleus is nearly spherical, 2–4 μm in diameter, with a small endosome and a distinct nuclear membrane lined by closely packed chromatin granules.

A form indistinguishable from *E. gingivalis* has been found in the mouths of dogs and cats, and this has been referred to as *Entamoeba canibuccalis* Simitch, 1938. *Entamoeba equibuccalis* Simitch, 1938, and *Entamoeba suigingivalis* Tumka, 1959, occur in horses and pigs, respectively.

OTHER ENTAMOEBA

A large number of other species of the genus occur in domestic and other animals. These have been reviewed by Noble and Noble (1952, 1961), Hoare (1959) and Levine (1973).

Of those which occur, the following are of more interest. All are non-pathogenic.

Entamoeba moshkovskii Chalaya, 1941. In sewage. This resembles *E. histolytica* and may lead to confusion in diagnosis. Attempts to transmit it to kittens, rats and other animals have failed.

Entamoeba bovis Liebetanz, 1905. In cattle. Trophozoites 5–20 μm, cytoplasm filled with vacuoles; nucleus large with a large central endosome made up of a compact mass of granules. Cysts uninucleate, 4–15 μm.

Entamoeba ovis Swellengrebel, 1914. In sheep. Trophozoites 13–14 μm, endosome of the nucleus large and composed of several granules. Cysts uninucleate, 4–13 μm.

Entamoeba gedoelsti Hsiung, 1930. Colon and caecum of horses. Trophozoites 7–13 μm, nucleus resembles that of *E. coli*, endosome is eccentric. No cyst formation.

Entamoeba equi Fantham, 1921. In horse in South America. Trophozoite 40–50 μm by 23–29 μm, nucleus oval. Cysts 15 μm by 24 μm, contain four nuclei.

Entamoeba suis Hartmann, 1913. In swine. Trophozoite 5–25 μm, cysts uninucleate, 4–17 μm.

Entamoeba muris (Grassi, 1879), **Entamoeba caviae** Chatton, 1918, and **Entamoeba cuniculi** Brug, 1918, occur in the caecum and colon of rats and mice, guinea-pigs, and rabbits, respectively.

Entamoeba invadens Rodhain 1934. This species is the cause of amoebiasis of reptiles and an important pathogen of captive snakes and lizards, Morphologically, it resembles *E. histolytica*, trophozoites measuring 10–38 μm by 9–30 μm show active locomotion and feed on leucocytes, cell debris and bacteria and produce cysts 11–20 μm in diameter which contain one to four nuclei, a glycogen vacuole and chromatoid bodies.

Signs of infection consist of anorexia, weight loss and sometimes blood-stained faeces. Lesions occur in the intestinal tract and liver and are similar to the lesions in man caused by *E. histolytica*. Ulcers 1–5 mm in diameter develop in the mucosa of the colon, secondary spread occurs to the liver and spleen, lungs and pancreas may also be affected (Ratcliffe & Gieman 1938; Zwart 1964). Death occurs between two and ten weeks after the onset of clinical signs. Diagnosis is based on the demonstration of trophozoites and cysts in the faeces or in the lesions.

For treatment, diloxanide given orally, 0.5 g/kg body weight or emetine hydrochloride (40 mg/kg) combined with a di-iodohydroxyquin enema has been recommended (Wallach 1969). For prophylaxis, tetracycline at the rate of 400–800 mg/m body length has been recommended (Ratcliffe 1961). Since turtles are reservoirs of infection for snakes, the two should not be housed together (Flynn 1973).

Genus: Endolimax Kuenen and Swellengrebel, 1917

These are small amoebae with a vesicular nucleus and an irregularly shaped, fairly large endosome composed of chromatin granules embedded in an achromatic ground substance. Achromatic threads connect the endosome with the nuclear membrane. Members of the genus are found in the hind-gut of man and other animals. As far as is known, they are all non-pathogenic.

Endolimax nana (Wenyon and O'Conner, 1917) Brug, 1918. This is a non-pathogenic form of man and monkeys. The trophozoite is about 9 μm in diameter, the cytoplasm is pale and vacuolated, and the pseudopodia are short and broad. The nucleus contains a large endosome which is often eccentric in position and may even lie against the nuclear membrane. The cysts are oval, thin-walled and measure 8–10 μm. The mature cyst contains four nuclei. Chromatoid bodies are absent.

Endolimax caviae Hegner, 1926, occurs in the caecum of guinea-pigs; only the trophozoites are known, these being 5–11 μm.

Endolimax ratti Chiang, 1925, occurs in the caecum and colon of laboratory and wild rats. It closely resembles *E. nana* and may be a synonym of it.

Genus: Iodamoeba Dobell, 1919

Amoeba of this genus have a vesicular nucleus with a large endosome rich in chromatin. The cysts are uninucleate and contain a large glycogen vacuole which stains darkly with iodine. The genus occurs in the intestinal tract of man and animals.

Iodamoeba bütschlii (Von Prowazek, 1912) Dobell, 1919. This species occurs in the lower digestive tract of pig, man, a variety of monkeys and baboons. It is world-wide in distribution. The trophozoite is 6–25 μm (average 8–15 μm), the ectoplasm is not clearly differentiated and the endoplasm contains bacteria and yeast cells in food vacuoles. The nucleus has a dense endosome about half the diameter of the nucleus. The cysts are uninucleate, irregular, 6–15 μm in diameter and contain a conspicuous glycogen vacuole which stains deeply with iodine.

Although generally regarded as non-pathogenic, Derrick (1948) has associated the trophozoite with ulceration of the stomach, small intestine, large intestine, etc. of man.

Genus: Dientamoeba Jepps and Dobell, 1918

These are small amoebae frequently showing a binucleate character. The nuclear membrane is delicate, and the endosome consists of several chromatin granules connected to the nuclear membrane by delicate strands.

Dientamoeba fragilis Jepps and Dobell, 1918. This has been found in the caecum and colon of man and monkeys. It is non-pathogenic. The trophozoite is actively amoeboid, 4–18 μm in diameter, and usually there are two nuclei. Encystation has not been observed.

FAMILY: VAHLKAMPFIIDAE JOLLOS, 1917: ZULUETA, 1917

Members of this family are free living, being found in fresh and stagnant water. One genus, *Naegleria*, has been associated with meningoencephalitis in man.

Genus: Naegleria Alexeieff 1912 emend. Calkins, 1913

Members of this genus have an amoeboid stage and a temporary flagellate stage. The nucleus is vesicular, with a large endosome. Cysts are uninucleate.

Members of the genus *Naegleria*, especially *Naegleria fowleri*, have been incriminated as the cause of primary amoebic meningoencephalitis (PAM) in man. This condition is essentially confined to the central nervous system and is almost invariably fatal (Carter 1972; Duma 1972). The disease is world-wide, cases having been reported from Australia, Europe (Czechoslovakia) and North America (Florida, Texas, Virginia). Infection is apparently contracted intranasally from water containing the amoeba, usually during swimming in surface-water polluted pools or in non-chlorinated swimming pools. Thermal pollution of inland waters may increase populations of the amoebae and possibly pose a hazard to children and young adults in contact with such, these groups apparently being the most susceptible. Indeed, Cursons et al. (1976) state that a patient showing PAM symptoms and having swum in thermal water about a week prior to their onset, justifies a tentative diagnosis of PAM and immediate amphotericin B therapy. There are indications that many thermally polluted waters contain pathogenic *Naegleria*. For example, de Jonckheere (1977) examines discharges from 32 thermal polluting factories in Belgium and from seven obtained 24 highly pathogenic strains of *N. fowleri*.

Experimental infections in mice have elucidated the basic feature of the human disease (Culbertson 1971; Martinez et al. 1971, 1973). The natural disease is characterized by an incubation period of five to six days, followed by a short (three days) period of severe complications of the central nervous system leading to coma and death.

Acute inflammatory changes with haemorrhage, oedema and degeneration occur in both the grey and white matter of the brain. Numerous amoebae may be found in many areas of the brain.

Organisms temporarily reside in the nasal mucosa and migrate through submucosal tissues in nerve plexuses and amoebae pass through the cribriform plate into the subarachnoid space, subsequently invading olfactory bulbs and then more widespread invasion of the brain occurs.

Wong et al. (1975) conducted studies on monkeys to determine their susceptibility to pathogenic strains of *N. fowleri*. Intranasal or intravenous injection failed to induce disease, but intrathecal injection produced in 11 of 18 monkeys an acute fatal meningoencephalitis. The authors concluded that the pathogenicity was influenced by strain virulence, cultural conditions and age and immune competence, among other things, of the host.

Diagnosis is usually made at autopsy and serodiagnostic tests are not generally available for this infection. Serological tests are in use for the identification of different strains within the genera (Anderson & Jamieson 1972).

FAMILY: HARTMANNELLIDAE VOLKONSKY, 1931

These are free-living forms, but their association with meningoencephalitis in man and experimentally in monkeys and with other disorders has been reported.

Genus: Hartmannella Alexeieff, 1912

Trophozoite forms are 9–17 μm, the cyst wall is smooth. Saprozoic in soil and fresh water. There is no evidence that *Hartmannella* species are pathogenic although previously, prior to clarification of the distinctions between this genus of the genera *Naegleria* and *Acanthamoeba*, it was thought that *Hartmannella* species could be pathogenic.

Genus: Acanthamoeba Volkonsky, 1931

Small amoebae, with a vesicular nucleus, and large endosome. Hyaline projections occur on the trophozoite. Cyst wall composed of two layers, a stellate endocyst and a closely adherent wrinkled ectocyst (Page 1967; Singh & Das 1970).

Acanthamoeba species are found in soil and water, they have been found as contaminants of monkey kidney cell cultures and have been shown to produce meningoencephalitis when injected intranasally into mice and monkeys (Culbertson et al. 1966).

An organism originally described as *Hartmannella*, but now considered to be *Acanthamoeba* or *Naegleria*, has been reported as the cause of a fatal gangrenous pneumonia in a bull in the Azores (McConnell et al. 1968). Ayers et al. (1972) described *Acanthamoeba* infection in a dog in the Republic of South Vietnam. Multifocal necrohaemorrhagic areas occurred in the heart, lungs, liver and pancreas. The portal of entry of the infection may have been a local wound with haematogenous spread or by the nasal route.

REFERENCES

ENTAMOEBA AND NAEGLERIA

Anderson, K. & Jamieson, A. (1972) Agglutination test for the investigation of the genus *Naegleria*. *Pathology* **4**, 273–278.
Andrews, J. M. (1932) Cysts of the dysentery-producing *Endamoeba histolytica* in a Baltimore dog. *Am. J. trop. Med.* **12**, 401–404.
Ayers, K. M., Billups, L. H. & Garner, F. M., (972) Acanthamoebiasis in a dog. *Vet. Path.*, **93**, 221–226.
Bostrom, R. E., Ferrell, J. F. & Martin, J. E. (1968) Simian amebiasis with lesions simulating human amebic dysentery. Abstract No. 51. *19th Ann. Meeting Am. Ass. Lab. Animal Sci.*, Las Vegas.
Bray, R. S. & Harris, W. G. (1977) The epidemiology of infection with *Entamoeba histolytica* in the Gambia, West Africa. *Trans. R. Soc. trop. Med. Hyg.*, **71**, 401–407.
Brumpt, E. (1925) Etude sommaire de l'*Entamoeba dispar* n. sp. Amibe a Kyster Quadrinuclées parasite de l'homme. *Bull. Acad. Med., Paris*, **94**, 943–952.
Burrows, R. B. (1968) Internal parasites of dogs and cats from Central New Jersey. *Bull. N.J. Acad. Sci.*, **13**, 3–8.
Burrows, R. B. & Lillis, W. G. (1967) Intestinal protozoal infections in dogs. *J. Am. vet. med. Ass.*, **150**, 880–883.
Carter, R. F. (1972) Primary amoebic meningoencephalitis: an appraisal of present knowledge. *Trans. R. Soc. trop. Med. Hyg.*, **66**, 193–213.
Culbertson, C. G. (1971) The pathogenicity of soil amebas. *Ann. Rev. Microbiol.*, **25**, 231–254.
Culbertson, C. G., Ensminger, P. W. & Overton, W. M. (1966) *Hartmannella (Acanthamoeba)*. *Am. J. clin. Path.*, **46**, 305–314.
Cursons, R. T. M., Brown, T. J., Bruns, B. J. & Taylor, D. E. M. (1976) Primary meningoencephalitis contracted in a thermal tributary of the Waikato River—Taupo: a case report. *N.Z. med. J.*, **84**, 479–481.
Derrick, E. H. (1948) A fatal case of generalized amoebiasis due to a protozoon closely resembling, if not identical with, *Iodamoeba bütschlii*. *Trans. R. Soc. trop. Med. Hyg.*, **42**, 191–198.
Diamond, L. S. (1968) Techniques of axenic culture of *Entamoeba histolytica* Schaudinn, 1903 and *E. histolytica*-like amebae. *J. Parasit.*, **54**, 1047–1056.
Diamond, L. S. & Mattern, C. F. T. (1976) Protozoal viruses. *Adv. Virus Res.*, **20**, 87–112.
Diamond, L. S., Mattern, C. F. T. & Bartgis, I. L. (1972) Viruses of *Entamoeba histolytica*. I. Identification of transmissible virus-like agents. *J. Virol.*, **9**, 326–341.

Duma, R. J. (1972) Primary amoebic meningoencephalitis. CRS critical review. *Clin. Lab. Sci.*, **3**, 163–192.
Elsdon-Dew, R. (1968) The epidemiology of amoebiasis. *Adv. Parasit.*, **6**, 1-62.
Eyles, D. E., Jones, F. E., Jumper, J. R. & Drinnon, V. P. (1954) Amebic infection in dogs. *J. Parasit.*, **40**, 163–166.
Flynn, R. J. (1973) *Parasites of Laboratory Animals.* p. 884. Ames: Iowa State University Press.
Fremming, B. D., Vogel, F. S., Benson, R. E. & Young, R. J. (1955) A fatal case of amebiasis with liver abscesses and ulcerative colitis in a chimpanzee. *J. Am. vet. med. Ass.*, **126**, 406–407.
Hoare, C. A. (1950) *Handbook of Medical Protozoology.* London: Baillière, Tindall & Cox.
Hoare, C. A. (1959) Amoebic infections in animals. *Vet. Rev. Annot.*, **5**, 91–102.
de Jonckheere, J. F. (1977) Pathogenic and non-pathogenic *Naegleria fowleri* in the environment. In: *Abstracts of Papers, 5th int. Congr. Protozool.*, Abstract 411.
Kagan, I. G. (1974) Advances in the immunodiagnosis of parasitic infections. *Z. Parasitenk.*, **54**, 163–195.
Kessel, J. F. (1928) Intestinal protozoa of monkeys. *Univ. Calif. (Berkeley) Publ. Zool.*, **31**, 275–306.
Levine, N. D. (1973) *Protozoan Parasites of Domestic Animals and of Man,* 2nd ed., p. 406. Minneapolis: Burgess.
McConnell, E. E., Garner, F. M. & Kirk, J. H. (1968) Hartmannellosis in a bull. *Path. Vet.*, **5**, 1–6.
Maegraith, B. G. (1963) Pathogenesis and pathogenic mechanisms in protozoal diseases with special reference to amoebiasis and malaria. In: *Immunity to Protozoa*, ed. P. C. C. Garnham, A. E. Pierce & I. Roitt. Oxford: Blackwell Scientific Publications.
Martinez, A. J., Nelson, E. C. & Duma, R. J. (1973) Animal model: primary amebic (*Naegleria*) meningoencephalitis in mice. *Am. J. Path.*, **73**, 545–548.
Martinez, A. J., Nelson, E. C., Jones, M. M., Duma, R. J. & Rosenblum, W. I. (1971) Experimental *Naegleria* meningoencephalitis in mice: an electron microscope study. *Lab. Invest.*, **25**, 465–475.
Mattern, C. F. T., Diamond, L. S. & Daniel, W. A. (1972) Viruses of *Entamoeba histolytica*. II. Morphogenesis of the polyhedral particle ($ABRM_2 \rightarrow$ HK-9) $\rightarrow$ HB-301 and the filamentous agent $(ABRM)_2 \rightarrow$ HK-9. *J. Virol.*, **9**, 342–358.
Mattern, C. F. T. & Keister, D. B. (1977) Experimental amebiasis. II. Hepatic amebiases in the newborn hamster. *Am. J. trop. Med. Hyg.*, **26**, 402–411.
Mattern, C. F. T., Keister, D. B. & Diamond, L. S. (1977) Variation in virulence of *Entamoeba histolytica* subpopulations. In: *Abstracts of Papers, 5th Int. Congr. Protozool.*, Abstract 118.
Meerovitch, E. (1977) Pathogenesis of amoebiasis. In: *Abstracts of Papers, 5th int. Congr. Protozool.*, Abstract 117.
Neal, R. A. (1966) Experimental studies on *Entamoeba* with reference to speciation. *Adv. Parasit.*, **4**, 1–51.
Neal, R. A. & Johnson, P. (1968) The virulence to rats of five *Entamoeba histolytica*-like strains capable of growth at 25°C and attempts to discover similar strains. *Parasitology*, **58**, 599–603.
Noble, E. R. & Noble, G. A. (1961) *Parasitology. The Biology of Animal Parasites.* London: Kimpton.
Noble, G. A. & Noble, E. R. (1952) Entamoebae in farm mammals. *J. Parasit.*, **38**, 571–595.
Page, F. C. (1967) Re-definition of the genus *Acanthamoeba* with descriptions of three species. *J. Protozool.*, **14**, 709–724.
Phillips, B. P., Wolfe, P. A., Rees, C. W., Gordon, H. A., Wright, W. H. & Reyniers, J. A. (1955) Studies on the ameba-bacteria relationship in amebiasis. Comparative results of the intracecal inoculation of germ-free, monocontaminated and conventional guinea-pigs with *Entamoeba histolytica. Am. J. trop. Med. Hyg.*, **4**, 675–692.
Pipkin, A. C. (1949) Experimental studies on the role of filth flies in the transmission of *Endamoeba histolytica. Am. J. Hyg.*, **49**, 255–275.
Ratcliffe, H. L. (1961) *Report of the Penrose Research Laboratory of the Zoological Society of Philadelphia*, p. 18.
Ratcliffe, H. L. & Gieman, Q. M. (1938) Spontaneous and experimental amebic infection in reptiles. *Archs Path.*, **25**, 160–184.
Ruch, T. C. (1959) *Diseases of Laboratory Primates.* p. 600. Philadelphia: Saunders.
Singh, B. N. & Das, S. R. (1970) Studies on pathogenic and nonpathogenic small free-living amoeba and the bearing of nuclear division on the classification of the order Amoebida. *Phil. Trans. R. Soc.*, **259**, 435–476.
Smyth, J. D. (1976) *Introduction to Animal Parasitology*, 2nd ed., pp. 466. London: Hodder & Stoughton.
Thiéry, G. & Morel, P. (1956) Amibiase pulmonaire chez un zebu. *Rev. Elev.*, **9**, 343–350.
Thorson, R. E., Seibold, H. R. & Bailey, W. S. (1956) Systemic amebiasis with distemper in a dog. *J. Am. vet. med. Ass.*, **129**, 335–337.
Vickers, J. H. (1969) Diseases of primates affecting the choice of species for toxicologic studies. *Ann. N.Y. Acad. Sci.*, **162**, 659–672.
Walkiers, J. (1930) Un cas d'amibiase intestinale chez un bovidé. *Ann. Soc. Belge Méd. trop.*, **10**, 379–380.
Wallach, J. D. (1969) Medical care of reptiles. *J. Am. vet. med. Ass.*, **155**, 1017–1038.
Wong, M. M., Karr, S. L. & Balamuth, W. B. (1975) Experimental infections with pathogenic free-living amebae in laboratory primate hosts: 1 (A). A study on susceptibility to *Naegleria fowleri. J. Parasit.*, **61**, 199–208.
Zwart, P. (1964) Studies on renal pathology in reptiles. *Path. Vet.*, **1**, 542–556.

Phylum: Apicomplexa Levine, 1970

CLASS: SPOROZOEA LEUCKART, 1879

Members of the *Sporozoea* are parasitic and produce spores. They possess no organs of locomotion, such as cilia or flagella, except in the gamete stage. Reproduction is asexual by binary or multiple fission (schizogony) or sexual (gametogony). Gametogony leads to the formation of a zygote which in turn initiates the process of sporogony or spore formation. The general ultra-structure of the class has been summarized by Aikawa and Sterling (1974) and Scholtyseck (1973).

The classification of the sporozoan organisms has undergone many revisions in the last decade. Knowledge from ultra-structural studies has revealed a closer relationship of organisms than previously supposed. The demonstration that organisms of the genera *Toxoplasma* and *Sarcocystis* possess coccidian life-cycles has necessitated even more extensive revision of the systematic position of several organisms. The classification adopted for the sporozoan organisms in this edition is that proposed by Levine (1975) and Frenkel (1977). It departs from previous classifications in several respects, e.g. placing the sporozoans in the phylum APICOMPLEXA instead of the subphylum SPOROZOA, including the genera *Toxoplasma* and *Sarcocystis* of the family Sarcocystidae along with the family Eimeriidae as members of the suborder Eimeriina and assigning the genera *Haemoproteus, Leucocytozoon* and *Plasmodium* to the family Plasmodiidae.

SUBCLASS: COCCIDIA LEUCKART, 1879

Organisms of this subclass are typically intracellular. They occur chiefly in vertebrates. The majority of medical or veterinary importance belong to the families *Eimeriidae* and *Sarcocystidae*.

ORDER: EUCOCCIDIIDAE LÉGER AND DUBOSCQ, 1910

SUBORDER: EIMERIINA LÉGER, 1911

FAMILY: EIMERIIDAE MINCHIN, 1903

These organisms are, with a few exceptions intracellular parasites of the epithelial cells of the intestine. They have a single host in which they undergo asexual (schizogony, merogony) and sexual (gametogony) multiplication. The macro- and microgamonts develop independently, the latter producing many gametes. A zygote results from the union of these, and by a process of *sporogony*, a variable number of spores (*sporocysts*) containing one or more *sporozoites* are formed. Sporogony occurs outside the host. Homoxenous forms in which tissue cysts are absent or unknown. At present, 25 genera are recognized in the family *Eimeriidae*, and these are illustrated diagrammatically by Levine (1973) and in tabular form by Pellérdy (1965). The following genera are of importance in this book:

Tyzzeria Allen, 1936; no sporocysts, eight sporozoites in the oocyst.

Isospora Schneider, 1881; two sporocysts each containing four sporozoites.

Eimeria Schneider, 1881; four sporocysts each containing two sporozoites.

Wenyonella Hoare, 1933; four sporocysts each with four sporozoites.

Cryptosporidium Tyzzer, 1907, possessing four sporozoites, not enclosed in a sporocyst, and which is not parasitic within the host cell and is a member of the family *Cryptosporidiidae* Léger, 1911, is included, for convenience, in the following account of coccidial infections of domesticated animals.

The majority of the coccidia of importance in domestic animals belong to the genus *Eimeria*. The following account of the morphology and life-cycle of the coccidia of veterinary importance is based principally on this genus.

Life-cycle and morphological stages of coccidia

Oocyst. The oocyst, which contains a zygote, is extruded from the host tissues and passed to the exterior in the faeces. This is the resistant stage of the life-cycle, and under appropriate conditions it forms the mature infective oocyst.

The most common shapes for oocysts are spherical, sub-spherical, ovoid or ellipsoidal, and they vary in size according to species. The oocyst wall is composed of two layers and is generally clear and transparent with a well-defined double outline; in some species, however, it may be

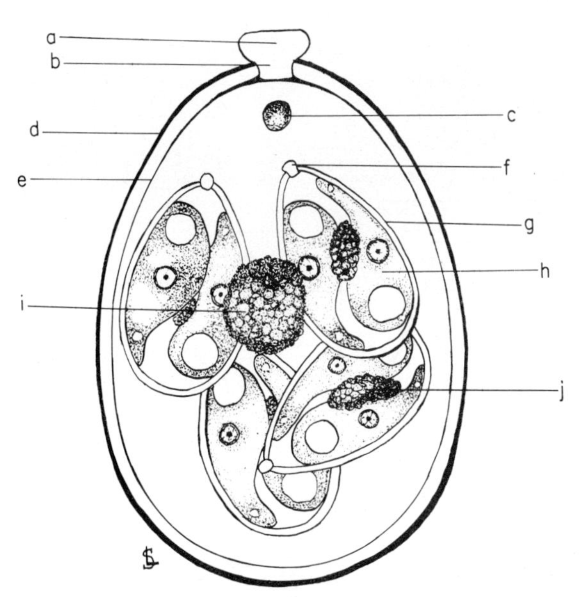

Fig 3.20 Sporulated oocyst of *Eimeria*. (a) Polar cap. (b) Micropyle. (c) Polar granule. (d) Outer layer of oocyst wall. (e) Inner layer of oocyst wall. (f) Stieda body. (g) Sporocyst wall. (h) Sporozoite. (i) Oocyst residual body. (j) Sporocyst residual body.

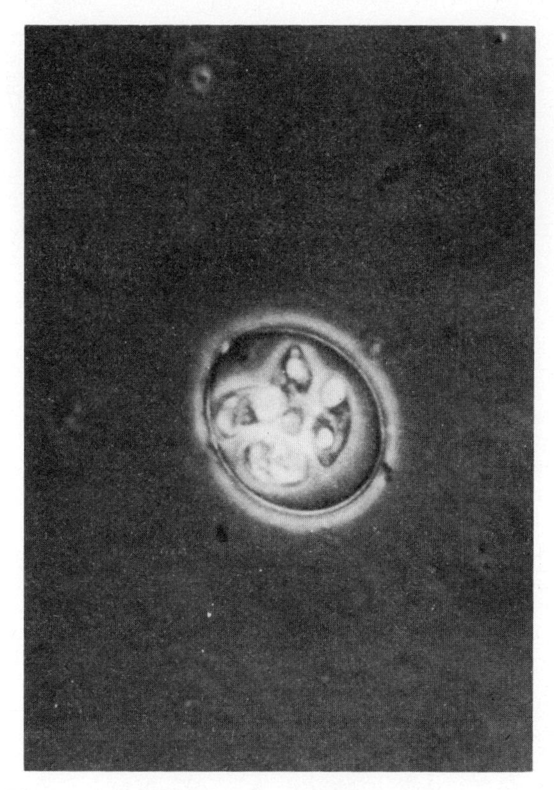

Fig 3.21 Sporulated oocyst of *Eimeria* sp. Phase contrast. (× 600)

yellowish or even green in colour. Other species possess striations or punctations. Several species of coccidia possess a micropyle at one extremity, this usually being the pointed end. The micropyle may be covered by a micropyle cap, and occasionally there may be a dome-shaped projection of the cyst wall to the exterior in the form of a polar cap.

In the sporulated oocyst (Figs 3.20 and 3.21) there are, according to genus, four sporocysts (*Eimeria*) or two sporocysts (*Isospora*). The sporocysts in *Eimeria* are more or less elongate ovoid forms with one end more pointed than the other. At the more pointed end is the Stieda body, and in some forms a micropyle may occur at the same place. An oocystic residual body and a polar granule may also be present in the oocyst. Each sporocyst contains two sporozoites, each having a granular cytoplasm and a distinctly placed central nucleus. Typically, the sporozoites are bent and comma-shaped and contain a round homogenous vacuole at one end. A secondary, or sporocystic, residual body may be present. Detailed accounts of the ultra-structure and the cytochemistry, physiology and biochemistry of the coccidia are given by Scholtyseck (1973) and Ryley (1973), respectively.

The parasitic life-cycle of the coccidia (Fig. 3.22) is initiated when the infective oocyst is ingested by the appropriate host. Excystation releases the contained sporozoites.

Two separate stimuli are necessary for excystation (Jackson 1962; Nyberg et al. 1968). The first is provided by CO_2 and the second by trypsin and bile. Using sheep coccidia, Jackson found that exposure to at least 15% CO_2 in the gas phase was necessary to prepare oocysts for the second stimulus; however, the concentration of CO_2 and the exposure time varied from species to species. With chicken coccidia, pretreatment with CO_2 results in up to 90% excystation compared to less than 10% in air, nitrogen, oxygen or helium (Nyberg et al. 1968). Apparently, CO_2 stimulates the activation or production of an enzyme or an

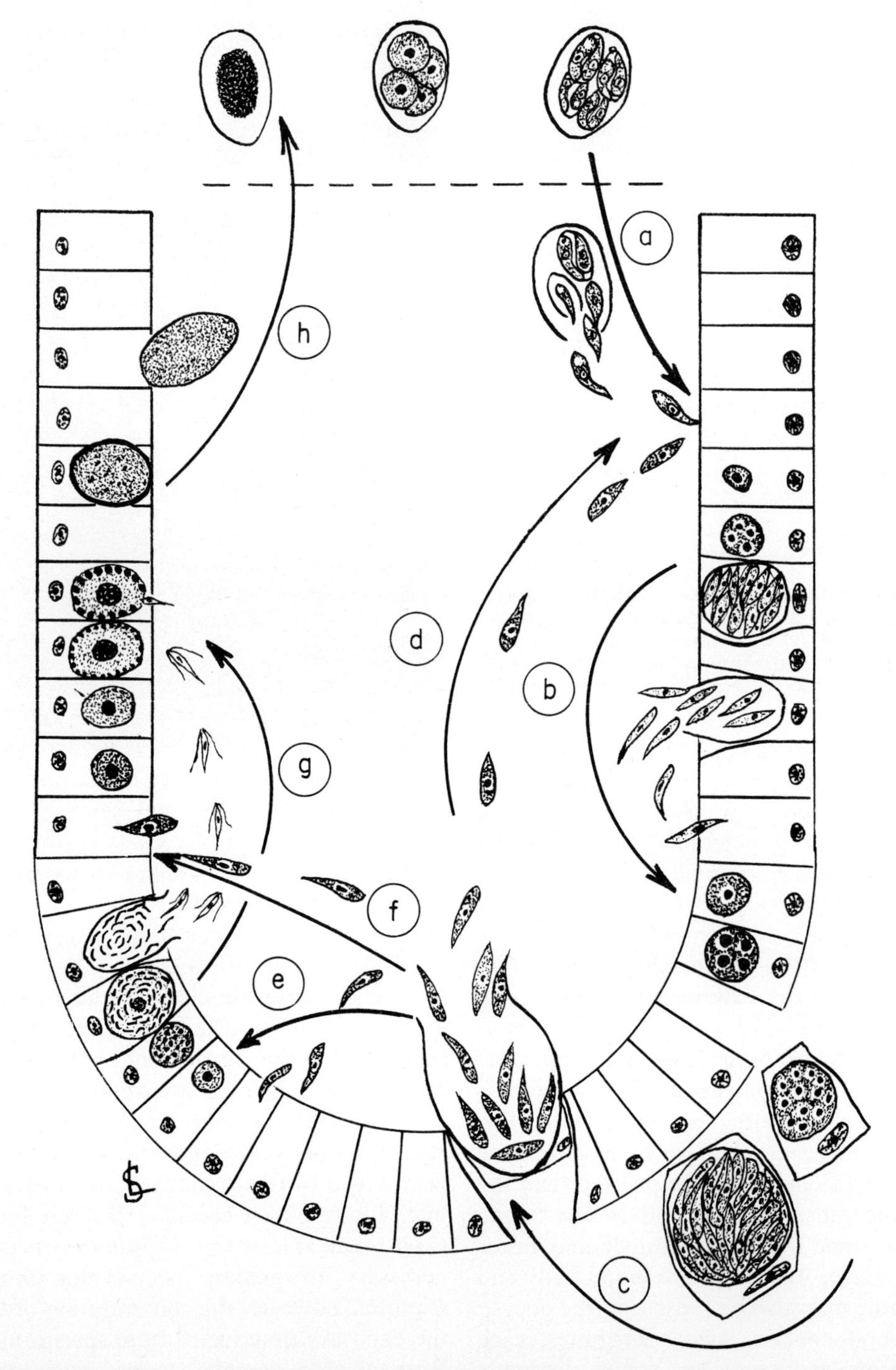

Fig 3.22 Life-cycle of *Eimeria tenella*. (a) Ingestion of sporulated oocyst and liberation of sporozoites. (b) First generation of schizogony. (c) Second generation of schizogony with migration of schizonts to subepithelial tissues. (d) Second generation merozoites initiate third generation of schizogony. (e) Second generation merozoites initiate macrogametocyte formation. (f) Second generation merozoites initiate macrogametocyte formation. (g) Microgametes fertilize macrogametes with formation of zygote. (h) Oocyst is shed from cell and passes to exterior to undergo sporogony.

enzymic rate-limiting step which increases the permeability of the micropyle. The second stage of excystation is pH-dependent and effects the escape of the sporozoites. Bile facilitates the entry of trypsin through the altered micropyle, which then digests the sporocystic plug permitting escape of the motile sporozoites. Doran (1966) considers that the sporozoite may also secrete enzymes which attack the plug.

Liberated sporozoites measure 10 μm by 1.5 μm and are transparent, fusiform organisms which show contraction and elongation and rapid gliding movements. The conoid probably serves as an organ of penetration into host cells. The penetration process is quick and completed within a few seconds (Hammond 1973). In the case of *Eimeria necatrix* van Doorninck and Becker (1957) have shown that, initially, the sporozoites invade the intestinal epithelium at the tips of the villi and are there engulfed by macrophages and carried by them through the lamina propria of the villi to reach the epithelium in the depths of the glands of Leiberkühn. Here, they leave the macrophages and enter the epithelial cells to undergo further development. A similar form of development has been demonstrated for *Eimeria tenella* (Challey & Burns 1959; Pattillo 1959) and for *E. meleagrimitis* (Clarkson 1959).

Asexual reproduction or schizogony. This process is initiated when the sporozoite enters the epithelial cell and becomes rounded up. In many forms development occurs above the nucleus of the epithelial cell, in a few below it, and in one bovine species, within the nucleus. The rounded-up sporozoite is, at this stage, known as a *trophozoite*, and within a few days the nucleus of the trophozoite divides by schizogony to become a schizont. This is the first generation of schizogony or the first generation schizont. The nuclear division in schizogony is considered to be of the mitotic type (Pellérdy 1965).

Initially the cytoplasm is undivided but later the daughter nuclei are each surrounded by a clear zone of cytoplasm, and eventually a number of elongate fusiform organisms are produced: the first generation *merozoites*. These, according to species, measure approximately 5–10 μm by 1.5 μm. They have a granular cytoplasm with a centrally-placed round nucleus. The mature schizont is surrounded by a distinct wall and the parasitized host cell is generally enlarged and distorted and protrudes into the lumen of the gut, etc.

The number of merozoites which are formed in the first generation schizont varies according to species. In some of the large forms, e.g. *Eimeria bovis*, more than 100 000 first generation merozoites may occur, but in *Isospora bigemina* only 16 may be formed.

When the schizont is mature the first generation merozoites are released, and they then enter other epithelial cells in the area and continue the cycle of asexual development. In some species this results in 'colonies' of second generation schizonts, but in others the second generation forms are spread widely in the tissues. In the new host cell the merozoite first rounds up to become a trophozoite and then undergoes multiple fission as before. In some species the second generation schizont is much larger than the first, whereas in others it is much smaller. The number of merozoites produced also varies according to the species.

The second generation merozoites may proceed to a third or more generations of asexual reproduction, or they may differentiate into sexual or gametogonous forms.

Sexual reproduction or gametogony. The factors responsible for the initiation of the gametogonous cycle are not fully understood. Although generally considered to be genetically determined, host responses may play a role, through phenotypic determination, in terminating schizogony (Haberkorn 1970; Long & Rose 1970). In some coccidia the merozoites destined for the sexual process may show sexual dimorphism. Thus, Rutherford (1943) found with rabbit *Eimeria* species that two distinct types of schizonts and merozoites of both the first and second generation were formed in areas of the intestine where microgamonts or macrogamonts developed.

In general, the number of microgametes (male forms) greatly exceeds the number of macrogametes (female forms), the former being very much smaller than the latter. However, the macrogamonts greatly outnumber the microgamonts.

The young macrogamont is initially morphologically indistinguishable from the asexual trophozoite. Later, however, it is readily distinguished from it since the nucleus of the macrogamont does not divide. It is round and roughly equivalent to the size of the oocyst which will result from it. The nucleus is large and clearly seen, and in stained preparations a nucleolus is visible. Horton-Smith and Long (1963) reported a weak Feulgen reaction in the macrogamont nucleus of *Eimeria maxima*.

In the young macrogamete small granules are initially found in the vicinity of the nucleus; later, these enlarge and become scattered over the cytoplasm, the larger granules being found on the periphery of the cell. These are 'plastic granules' or 'wall-forming' granules, which form the wall of the oocyst following fertilization of the macrogamete. A detailed consideration of the histochemistry of the plastic granules and also of the macrogamont is given by Horton-Smith and Long (1963). Fertilization by the microgamete may occur at any point on the surface of the macrogamete: a zygote is formed and the oocyst wall is laid down around it. When the cyst wall is complete the oocyst is extruded from the tissues and passed to the exterior.

The microgamont arises in the same way as the macrogamont, but, as it enlarges, the nucleus undergoes multiple division with the production of a large number of microgametes. Initially, the nuclei are scattered over the cytoplasm of the microgamont, but later they assume a comma shape and accumulate on the periphery of the cell, leaving a residual mass of cytoplasm in the cell. The microgametes are slender, slightly bent, and the anterior end is pointed with two flagella for locomotion. They measure about 5 μm in length, stain intensely with haematoxylin and give a strong reaction for DNA by the Feulgen reaction.

Rupture of the microgamont liberates the microgametes which fertilize the macrogametes.

Sporogony. With few exceptions, sporulation does not occur until the oocyst is shed to the exterior of the body. Initially, the zygote almost fills the oocyst cavity, but within a few hours outside the host the protoplasm contracts from the wall of the oocyst to form a *sporont* and leaves a clear space between it and the wall. The sporont divides into four sporoblasts, any remaining cytoplasm being left as an oocystic residual body. The sporoblasts are, initially, more or less spherical, but later they elongate into ovoid or ellipsoid bodies which then become sporocysts by the laying down of a wall of refractile material around each sporoblast. The protoplasm inside each sporocyst further divides to form two sporozoites. Protoplasm remaining from the division is left as a sporocystic residual body.

The time required for sporulation to the infective stage is a specific feature of each species of coccidium and is used as a characteristic in identification. Oxygen and adequate moisture are necessary for the sporulation, and at constant temperatures an increasing percentage of oocysts are killed as relative humidity decreases. Temperature also has an important influence on sporulation. The optimum temperature for sporulation is about 30°C. In general, sporulated oocysts are more resistant to dessication and cold and may survive for up to two weeks at temperatures of −12°C to −20°C. Unsporulated forms are killed in 96 hours at these temperatures. The survival of oocysts under natural conditions has received substantial consideration from various workers over the years. Factors such as soil types, exposure to direct sunlight or otherwise, the amount of humus in soil, moisture, etc. are important in the longevity of oocysts.

General. Coccidial infections are self-limiting, and asexual reproduction does not continue indefinitely. In the absence of reinfection, therefore, only one cycle of development can take place. Under natural conditions, however, repeated infection usually occurs. As a result of repeated infection, the host may develop immunity and with certain species of coccidia immunity may be marked following a single infection. One of the effects of immunity is to reduce the biotic potential of the coccidium. Whereas an initial infection may produce a maximum number of oocysts, as immunity supervenes the life-cycle is progressively inhibited, so that at one level only a few oocysts may be produced, and at another the infective sporozoites may fail to get further than the initial penetration of the host cell.

COCCIDIA OF SHEEP AND GOATS

It has been thought for many years that the coccidia of sheep and goats are interchangeable. It is clear that this is not the case for all the species, and further clarification of this is necessary.

The following species are considered in this edition:

Eimeria ahsata Honess, 1942
Eimeria arkhari Yakimoff and Matschoulsky, 1937
Eimeria arloingi (Marotel, 1905) Martin, 1909
Eimeria christenseni Levine, Ivens and Fritz, 1962
Eimeria crandallis Honess, 1942
Eimeria danielle Dida, Acsinte and Purchera, 1972
Eimeria faurei (Moussu and Marotel, 1902) Martin, 1909
Eimeria gilruthi (Chatton, 1910) Reichenow and Carini, 1937
Eimeria gonzalezi Bazalar and Guerro, 1968
Eimeria granulosa Christensen, 1938
Eimeria hawkinsi Ray, 1952
Eimeria intricata Spiegl, 1925
Eimeria marsica Restani, 1971
Eimeria ninakohlyakimovae Yakimoff and Rastegaieff, 1930
Eimeria ovina Levine and Ivens, 1970
Eimeria pallida Christensen, 1938
Eimeria parva Kotlan, Mocsy, and Vajda, 1929
Eimeria punctata Landers, 1955
Eimeria weybridgensis Norton, Joyner and Catchpole, 1974
Cryptosporidium agni Barker and Carbonell, 1974

Eimeria ahsata. *Hosts:* domestic sheep, Rocky Mountain big-horn sheep, Moufflon (*Ovis musimon*), Siberian ibex (*Capra ibex siberica*). At one time this species was considered to be a form of *E. arloingi* (Lotze 1953); however, Smith et al. (1960) confirmed its validity. *Oocysts*: ellipsoidal, wall smooth, pinkish yellow, a dome-shaped polar cap over the micropyle, 32.7 μm by 23.7 μm in forms from the big-horn sheep, 33.4 μm by 22.6 μm in forms from domestic sheep (Pellérdy 1974). Prepatent period, 18–20 days (Smith et al. 1960). Davis et al. (1963) reported that this species underwent globidial schizogonic development in the small intestine, sometimes measuring up to 265 μm by 162 μm at 15 days after infection. Davis and Bowman (1970) found intranuclear development of this parasite in experimental infections. Smith et al. (1960) consider *E. ahsata* to be the most pathogenic coccidium in sheep. Lambs, infected with from 100 000 to 800 000 oocysts, showed severe disease; four out of nine lambs, one to three months of age, were killed by the lower dose. Clinical signs consisted of diarrhoea and loss of body weight. At autopsy the wall of the ileum was thickened, especially in the anterior part, and there was inflammation of Peyer's patches.

Eimeria arkhari. *Hosts:* wild sheep in the Soviet Union. *Oocysts:* ellipsoidal to oval, 22.4 μm by 17.4 μm, double contoured oocyst wall with a yellowish tint, micropyle absent.

Eimeria arloingi (Fig. 3.23). *Hosts:* domestic goat, Ibex, angora (also in chamois, red deer and roe deer, although Pellérdy (1974) doubts that these are true records). *Oocysts:* 27 μm by 18 μm, but a wide range occurs, 17–42 μm by 13–31 μm (Christensen 1938*a*). Levine et al. (1962) give a range of 22–31 μm by 17–22 μm with a mean of 28 μm by 20 μm. Oocysts predominantly ellipsoidal, ovoid ones may occur, micropyle 2–3 μm in width with a distinct polar cap. Sporulation time 48–72 hours.

Eimeria arloingi is probably the most common coccidium of goats.

Previously, this species was considered to be one of the most common species of sheep. However, Levine and Ivens (1970) established that the species in sheep, which closely resembles it, is *E. ovina*. These authors showed that *E. arloingi* from goats could not be transmitted to sheep, nor could *E. ovina* of sheep be transmitted to goats.

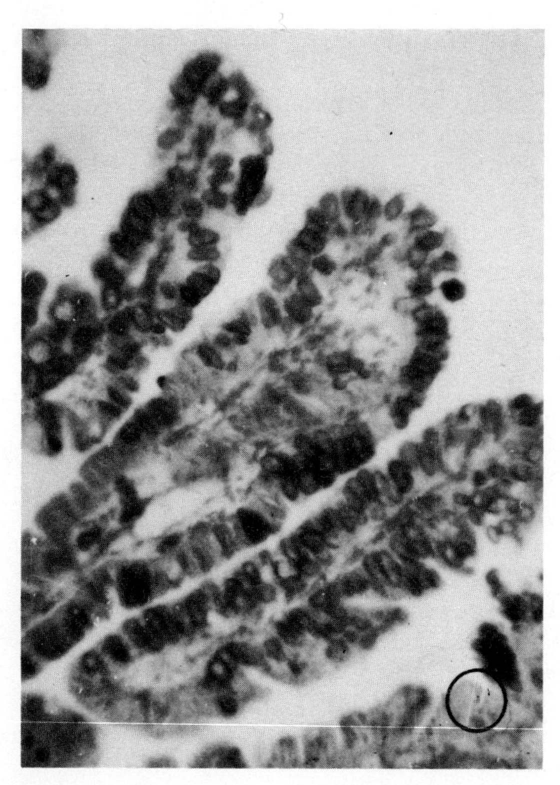

Fig 3.23 Small intestinal mucosa of sheep infected with *Eimeria ovina*. Gametogonous stages are visible in the epithelial cells of the villi, which have undergone marked hypertrophy. (× 300)

Developmental cycle. Not known in detail. Levine et al. (1962) suggest that large schizonts occur in the epithelial cells of lacteals. The prepatent period is unknown.

Pathogenicity. Uncertain. Polyp formation of the villi and hyperplasia of the small intestine of young goats has been reported by Deiana and Delitala (1953) and other authors have described oedema of the intestine, local haemorrhage and desquamation of the epithelium.

Eimeria christenseni. *Hosts:* domestic goat (*Capra hircus*). *Oocysts:* ovoid and slightly flattened at one end, 38 μm by 25 μm, range 34–41 μm by 23–28 μm, micropyle covered by a prominent dome-shaped micropylar cap. Not known to be pathogenic.

Eimeria crandallis. *Hosts:* Rocky Mountain big-horn sheep, domestic sheep, Moufflon, Siberian ibex. *Oocysts:* spherical to broadly ellipsoidal, 23 μm by 19 μm, range 20–27 μm by 17–20 μm, visible micropylar cap. *Distribution:* North America, Soviet Union. Life-cycle and pathogenicity unknown.

Eimeria danielle. *Host:* domestic sheep. Roumania. Full description not available. Didă et al. (1972) reported this species in low numbers in 20% of lambs examined. It is not pathogenic.

Eimeria faurei. (Fig. 3.24f) *Hosts:* sheep, goat, Rocky Mountain big-horn sheep, chamois, ibex, Siberian ibex, Moufflon, aoudad or barbary sheep (*Ammotragus lervia* = *Ovis tragelapus*) and various wild sheep. World-wide in distribution and relatively common. Levine and Ivens (1970) question whether this species is a parasite of both sheep and goats. *Oocysts:* ovoidal, micropyle distinct, no polar cap, wall transparent, brownish yellow to salmon pink, 28.9 μm by 21 μm, range 25–33 μm by 18–24 μm (Christensen 1938*a*). Sporulation time one to two days.

Developmental cycle. Not known in detail. Lotze (1953) described schizonts which measured up to 100 μm in diameter and contained thousands of merozoites.

Pathogenicity. At the most, the species is only mildly pathogenic. Lotze (1954) reported that infection of three-month-old lambs with five million oocysts produced only a slight softening of the faeces, and infections of 50 million oocysts failed to cause death.

Eimeria gilruthi. *Hosts:* sheep and goat. World-wide in distribution. Occurs principally in the abomasum, rarely seen in the small intestine. Only the schizonts have been described, and these occur in the abomasal wall. They are large (giant schizonts), up to 700 μm in length, and are readily visible to the naked eye; many thousands of merozoites fill the schizont.

This species was originally described by Maske (1893) from the abomasum of a sheep. It was later found by Gilruth (1910) and studied in detail by Chatton (1910), who named it *Gastrocystis gilruthi*. Since that time it has been found in various parts of the world; Alicata (1930) found it in up to 11%

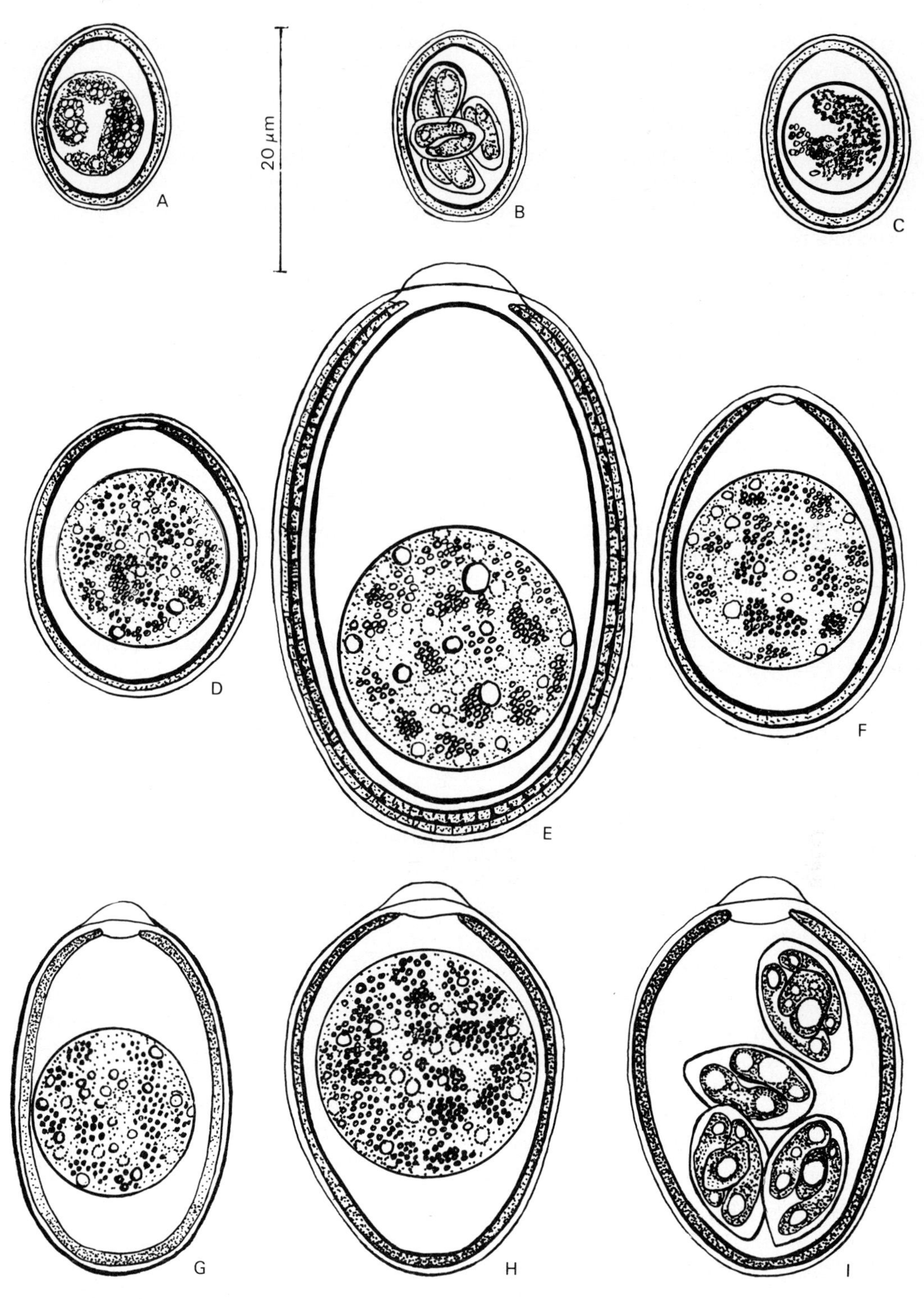

Fig 3.24 Coccidia of sheep. (A) *Eimeria pallida*. (B) *E. pallida* (sporulated). (C) *E. parva*. (D) *E. ninakohlyakimovae*. (E) *E. intricata*. (F) *E. faurei*. (G) *E. arloingi*. (H) *E. granulosa*. (I) *E. granulosa* (sporulated). (*Redrawn with permission from Christensen 1938*)

of sheep in the United States; Sarwar (1951) found it in 32% of sheep and in 40% of goats in the Sudan.

Virtually nothing is known of its endogenous developmental cycle. Levine (1973) considers that this form is a schizontal stage of one of the species of sheep coccidia. Several species of goat and sheep coccidia have giant schizonts in their life-cycle (e.g. *E. arloingi*, *E. parva*), but until further evidence is available about it, it would seem wise to follow the suggestion of Levine (1973) that, as a temporary measure, it be called *Eimeria gilruthi*.

Eimeria gonzalezi. *Host:* domestic sheep. South America (Peru) and Europe (Poland). Location in the body unknown. *Oocysts:* ellipsoidal or ovoid 26–38 μm by 20–26 μm (mean, 30.5 μm by 22.4 μm). Wall of oocyst smooth with transparent yellow outer layer; prominent micropyle covered by a micropylar cap. Sporulation time five to six days. Developmental cycle and pathogenesis unknown. This species has been reported in 17% of 240 sheep in Peru and 3% of 222 sheep in Poland (Levine 1973).

Eimeria granulosa. (Fig. 3.24h,i) *Hosts:* domestic sheep, Rocky Mountain sheep. North America and Germany. Location inside the host is not known, only the oocysts having been found in the faeces. *Oocysts:* urn-shaped, distinct micropyle, 3–5 μm in diameter, with a micropylar cap. Oocyst wall transparent, brownish to yellowish in colour. Mean size 29.4 μm by 20.9 μm, range 22–35 μm by 17–25 μm. Sporulation time three to four days.

Eimeria hawkinsi. *Hosts:* domestic sheep and goat. India. Known only from oocysts in the faeces. *Oocysts:* sub-spherical 20–25 μm by 15–23 μm, micropyle present, triangular polar cap, sporocyst residual body present. Sporulation time five to six days at 21–23°C. Life-cycle and pathogenicity unknown.

Eimeria intricata. (Fig. 3.24e) *Hosts:* domestic sheep, Rocky Mountain big-horn sheep, other members of the genus *Ovis*. Relatively common and world-wide in distribution. *Oocysts:* largest of the species of *Eimeria* in sheep, 47 μm by 32 μm, range 39–53 μm by 27–34 μm, ellipsoidal, well developed micropyle, 6–10 μm in diameter, with a distinct light-coloured polar cap. Sporulation time three to five days. The endogenous development has been studied by Davis and Bowman (1965), Pande et al. (1966) and Lotze and Leek (1970). Schizonts occur in the lower small intestine, in the cells of the intestinal crypts. Gamonts and oocysts occur from the mid-small intestine to the rectum, mainly in the caecum. The prepatent period is 20–27 days. This appears to be a mildly pathogenic species producing mucoid faeces, anorexia and decreased weight gain.

Eimeria marsica. *Host:* domestic sheep. Italy, England and Wales. Location in host unknown. *Prevalence:* probably low, oocysts have been observed on a few occasions in surveys in Britain (Catchpole et al. 1975) and in Italy (Restani 1966). *Oocysts:* ellipsoidal 15.4–22.3 μm by 11.5–14.6 μm (mean 19.1 μm by 13.1 μm). Wall two-layered, smooth, colourless to pale yellow. Micropyle inconspicuous, shallow dome-shaped micropylar cap. Oocyst polar granule present. Sporocyst residuum composed of small scattered granules. Sporulation time 72 hours at 25°C.

The prepatent period is 14–16 days and the pathogenicity is uncertain, but is probably slight. Restani (1971) observed no pathogenic effects in two lambs being given 50 000 oocysts, and Norton and Catchpole (1976) did not produce clinical effects in a lamb with 10 000 oocysts.

Eimeria ninakohlyakimovae. (Fig. 3.24d) *Hosts:* domestic sheep, domestic goat, Rocky Mountain big-horn sheep, Barbary sheep, Siberian ibex, Moufflon, other wild sheep, and also in Persian gazelle, roe deer and red deer. World-wide in distribution. Occurs in posterior part of small intestine and caecum and colon. *Prevalence:* 52% of sheep and 31% of goats were found infected in Kazakhstan (Svanbaev 1957); 5% of sheep in Germany (Jacob 1943); 3% of sheep in the USA (Christensen 1938*a*). *Oocysts:* ellipsoidal, sometimes ovoid; 23.1 μm by 18.3 μm, range 20–28 μm by 15–22 μm (Christensen 1938*a*). Variations in the size of oocysts from other host species have been recorded; thus, those from the moufflon measure an average 25.2 μm by 19.5 μm (Yakimoff et al. 1933). Generally no micropyle, no polar cap. Cyst wall thin, smooth and transparent, slightly brownish yellow. Sporulation one to two days.

The endogenous development of this species was described by Lotze (1954). Sporozoites entered the epithelial cells at the base of the villi in the crypts of Lieberkühn of the small intestine, where they developed to giant schizonts (up to 300 μm in diameter) containing many thousands of merozoites. Wache et al. (1971) report that a second schizogonic development appears in the large intestine 10–11 days after infection and takes one to two days to develop. Following rupture of the schizonts, the gametogonous stages are found in the epithelial cells of the ileum, caecum and upper part of the large intestine. The prepatent period is 15 days (Shumard 1957).

Lotze (1954) considered this species to be the most pathogenic of sheep coccidia. As few as 50 000 oocysts caused diarrhoea in a three-month-old lamb, half a million causing death. Profuse diarrhoea occurred in a two-year-old sheep given a million oocysts. Faeces became soft 12–17 days after infection, remaining in this state for a week or more: in heavily infected animals the faeces were mixed with blood. Haemorrhages were found in the posterior part of the small intestine of severely affected animals by the fifteenth day, and the wall of the intestine was thickened and inflamed. During the gametogonous part of the cycle the caecum and colon were thickened, oedamatous and haemorrhagic by the nineteenth day, and large areas of the small intestine were denuded of epithelium. Chapman (1974) induced severe clinical signs and death in 50% of lambs infected with 100 000 oocysts. Marked reductions in weight gain occurred along with significant increases in haematoglobin and haematocrit values. In contrast to these findings, Shumard (1957), using an obviously less pathogenic strain, found that only slight diarrhoea and lowered food consumption occurred with an infection of seven million oocysts.

Eimeria ovina. (Fig. 3.24g) *Hosts:* domestic sheep, big-horn sheep, argali (*Ovis ammon*) and moufflon. World-wide in distribution. Probably the commonest species of sheep. Development occurs in the small intestine. Levine and Ivens (1970) concluded that *Eimeria arloingi*, previously considered as a coccidium of sheep and goats, was a parasite of goats. Therefore, they named the sheep species *Eimeria ovina*. Interhost transmission of *E. arloingi* and *E. ovina* has not proved possible. In sheep in Great Britain, Pout et al. (1973) recognized large and small forms of *E. arloingi*, designated as types 'A' and 'B'. The type 'B' is now renamed *Eimeria weybridgensis* and the type 'A' is *E. ovina*. *Oocysts:* ovoid or ellipsoidal 23–36 μm by 16–24 μm (mean 27 μm by 20 μm). Oocyst wall two-layered with outer layer smooth, yellowish; micropyle present covered with a micropylar cap. Sporulation time is two to four days (44 hours at 27°C).

Developmental cycle. The liberated sporozoites may be found for several days free in the lumen of the intestine; later they penetrate the epithelial layer and enter the endothelial cells lining the central lacteals of the villi (Lotze 1953). Only one generation of schizonts occurs, and schizonts became mature 13–21 days after infection. They reach a size of 122–146 μm in diameter and contain several million merozoites, each about 9 μm in length. Merozoites are released from the giant schizonts about 19 days after infection, and they enter the epithelial cells of the small intestine. In heavy infections large numbers of epithelial cells are parasitized and the affected villus may be greatly enlarged to form a papillomatous growth. The macrogamonts are markedly granular and there is a preponderance of these forms in the infection. The prepatent period is 22–29 days, mean 24 days (Norton et al. 1974).

Pathogenesis. Lotze (1952) found, in three-month-old lambs, that no clinical signs were seen in lambs receiving less than one million oocysts. With higher doses clinical signs of soft, watery faeces appeared on the thirteenth day. The majority of animals had returned to normal by the twentieth day. Slight haemorrhages occurred in the small intestine up to the thirteenth day, and by the thirteenth and nineteenth days the small intestine was oedematous and thickened, the villi distended with schizonts, and there was a loss of epithelium. In pure infections in coccidia-free lambs, Norton et al. (1974) reported a short patent period, with low oocyst output and an absence of clinical signs. The polyps on the mucosa of the small intestine, which are usually associated with this species (Michael & Probert 1970), are difficult to produce experimentally.

Eimeria pallida. (Fig. 3.24a, b) *Hosts:* domestic sheep and goat in North America. Site of development in the host is not known. *Oocysts:* ellipsoidal, 14.2 μm by 10 μm, range 12–20 μm by 8–15 μm (Christensen 1938*a*). Micropyle imperceptible, no polar cap, oocyst wall thin, pale yellow to yellowish-green and appearing fragile. Sporulation time 24 hours.

The developmental cycle of this species is unknown and no pathogenesis is ascribed to it.

Not all authors agree that this is a valid species, for example Kotlán et al. (1951) and Pellérdy (1974) consider it a synonym of *Eimeria parva*.

Eimeria parva. (Fig. 3.24c) *Hosts:* domestic sheep and goat; also Siberian ibex, Rocky Mountain big-horn sheep, Barbary sheep. World-wide in distribution. Asexual stages in the small intestine, gametogonous stages in the caecum, colon and small intestine. *Prevalence:* 50% of sheep infected in USA (Christensen 1938*a*), 52% of sheep and 9% of goats infected in Germany (Jacob 1943). *Oocysts:* sub-spherical to spherical, 16.5 μm by 14.1 μm, range 12–22 μm and 10–18 μm (Christensen 1938*a*). Oocyst wall smooth with a uniform thickness, no visible micropyle, no polar cap; pale yellow to yellowish green. Sporulation time, one to two days.

The endogenous development has been described by Kotlán et al. (1951). Giant schizonts of the globidial type occur in the small intestine. Two types of schizonts were recognized, smaller ones measuring 60 μm by 40 μm and large sub-spherical forms 185 μm by 179 μm; the latter sometimes became elongate, reaching lengths of up to 256 μm. The larger schizonts were visible as whitish bodies in the mucosa, being found over the entire length of the small intestine. The schizonts reach maturity in 12–14 days, and there appears to be only one asexual generation. The gamonts occur in epithelial cells of the mucous membrane of the caecum and colon. Prepatent period, 16–17 days.

This species is not markedly pathogenic. Massive infections produce a catarrhal inflammation with loss of epithelium; faeces are dark and mixed with blood. The large intestine may become necrotic and infiltrated with leucocytes and neutrophils.

Eimeria punctata. *Hosts:* domestic sheep in North America (Wyoming) and Europe (Germany). Site of development in the host is unknown. *Oocysts:* sub-spherical to spherical, 21.2 μm by 17.7 μm, range 17.8–25.1 μm by 16.2–21.1 μm (Landers 1952*a*, *b*); micropyle present, small polar cap present. Wall of oocyst covered by an even distribution of cone-shaped pits about 0.5 μm in depth, inner part of the wall with a greenish tint. Sporulation time 36–48 hours. The developmental cycle is unknown, and no pathogenesis has been ascribed to the parasite.

Eimeria weybridgensis. *Hosts:* domestic sheep in Great Britain. This species was previously recognized as the small form, or type 'B', of *Eimeria arloingi* (Pout et al. 1973). Development in the small intestine, chiefly jejunum. *Oocysts:* ellipsoidal to sub-spherical 17.1–30 μm by 14.4–19 μm (mean 24.4 μm by 16.9 μm). Oocyst wall two-layered, outer layer smooth, colourless or pale yellow, inner dark. Micropyle present, dome-shaped micropylar cap. Oocyst polar granule present. Sporocyst residuum composed of a number of small granules. Sporulation time is 45 hours at 27°C.

Endogenous development is unknown. The prepatent period varies between 23 and 33 days (mean 26 days). The pathogenicity of this species is difficult to assess. Laboratory infections are well tolerated, patency is short and animals rapidly become immune. In mixed infections anorexia, diarrhoea and weight loss occurred and *E. weybridgensis* oocytes predominated at the onset of diarrhoea (Pout & Catchpole 1974).

Cryptosporidium agni. *Hosts:* domestic sheep in Australia. Organisms of this genus parasitize the microvillous border of epithelial cells in the gastrointestinal tract. The oocysts contain four naked sporozoites. This species was reported in young lambs which suffered concomitant infection with *Salmonella typhimurium* (Barker & Carbonell 1974). Trophozoites 1.5–3.0 μm in diameter were found in the microvillous border, schizonts, 3–4 μm in diameter, containing up to seven microzoites were seen. Bodies thought to be macrogamonts were rarely seen. Cryptosporidia were attached to the luminal border of epithelial cells by a zone of attachment (Vetterling et al. 1971) and resembled *C. wrairi* of the guinea-pig (Vetterling et al. 1971). The mode of transmission is unknown.

Villous atrophy, low surface epithelium, dilated intestinal crypts and diffuse infiltration of the lamina propria with leucocytes were associated with the infection (Barker & Carbonell 1974), although an associated *Salmonella* infection may have confused the picture. For treatment, antibiotics have been used, but doubt exists as to their value.

COCCIDIOSIS IN SHEEP AND GOATS

Coccidiosis in sheep and goats is chiefly confined to young animals up to four to six months of age. Mixed infections are usual and the species of clinical importance are *E. ovina, E. ahsata, E. arloingi, E. parva* and *E. ninakohlyakimovae*.

In the United States coccidiosis is primarily a disease of feed lot lambs, outbreaks of disease occurring when range-reared lambs are moved to intensive conditions of husbandry. The disease frequently appears 2–3 weeks after the lambs are placed in the feed lot. Although the feed lot may become relatively free of infection during the time it is left vacant, infection is introduced by the range lambs, and as a result of crowding and poor hygienic conditions the output of oocysts in the faeces rises markedly. Oocyst output reaches a peak about one month after lambs enter the feed lot; it remains at a high level for a further one to three weeks and then falls quickly so that at the end of the feeding period only a few oocysts are passed in the faeces (Christensen 1940, 1941*a*).

A mortality of 10% due to *E. arloingi* (= *E. ovina*) and *E. ninakohlyakimovae* has been reported in lambs in Georgia by Becklund (1957); a similar outbreak was recorded by Davis et al. (1957*b*) in a flock which had been corralled nightly for three weeks to guard against attacks of dogs. Coccidiosis has been reported as a cause of morbidity and mortality in lambs in other countries. Salisbury and Whitten (1953) considered coccidiosis the cause of unthriftiness and death in lambs aged four to six months in New Zealand; Favati and Guerrieri (1961) reported the disease to be common in Italy, and Rao and Hiregaudar (1954) stated that coccidiosis was a severe parasitic disease among sheep and goats in India.

In Great Britain, infections of coccidia with a high morbidity and a low mortality have been observed by the author in young lambs, aged two to three months, in flocks where nursing ewes were fed grain and concentrates in open troughs in fields. The same area was used continuously for feeding and the environment of the troughs was heavily contaminated with sporulated oocysts. The pathogenic process was largely due to *E. ovina*. Coccidiosis of sheep in Britain has been reviewed by Pout (1976).

The seasonal incidence of coccidiosis is determined by the availability of young animals for the development of the parasite (older animals being immune), by the survival of oocysts from one season to the next and the increased output of oocysts by ewes during the postparturient period. Freezing conditions either kill oocysts or prevent them from sporulating; consequently, pastures, or feed lots, are usually relatively free of coccidia at the time of the new season's crop of lambs.

Clinical signs and pathology

These consist of a brownish to yellowish green diarrhoea, which may be streaked with blood, especially when caused by those species which undergo gametogony in the large intestine. Abdominal pain, some anaemia, inappetence, weakness and loss of weight may occur. The diarrhoea may continue for up to two weeks and, although lambs may die from it or the consequent dehydration, the majority recover and it is unusual to see a mortality exceeding about 10%. Nevertheless, the setback in growth and production may be economically important. Shumard (1957) found that 80 lambs, experimentally infected with a mixture of *E. ninakohlyakimovae* and *E. ovina*, lost an average of 0.205 lb/lb of feed consumed over a 24-day period, as compared with a gain of 0.062 lb/lb of feed consumed in a group of control lambs. The severity of the clinical entity is dependent on the size of the initial infection. Where this is low, clinical disease may never appear; rather, the animal develops immunity but it may continue to shed small numbers of oocysts for the rest of its life.

The pathological changes vary with the species concerned. With *E. ovina*, lesions occur in the posterior part of the small intestine, the giant schizonts and the gametogenous stages producing enlargement of the villi, so much so that these may be visible to the naked eye. Where groups of villi have enlarged, polyp-like growth may be readily visible on the mucosa of the small intestine. In more acute cases the wall of the intestine is thickened, oedematous and it may be haemorrhagic. Scrapings of the mucosa show large numbers of schizonts and oocysts. With *E. parva* infection, the mucosae of the caecum and colon are thickened, oedematous and haemorrhagic. Necrotic areas may occur on the mucosa, and the contents of the bowel are fluid and dark brown to haemorrhagic in colour. With *E. ninakohlyakimovae* infection, the intestinal mucosa is covered with petechial haemorrhages, and the wall is thickened and inflamed.

Pout (1974) has described villous atrophy with a 'flat mucosa' syndrome in ovine coccidiosis. This occurs in the area of the intestine with a high density of parasites. Malabsorption may be associated with these lesions, although compensatory increases in absorption in the lower bowel may make malabsorption less obvious.

Diagnosis

This is based on the history of the outbreak (there is usually evidence of poor hygiene either in feed lots or in grazing management), the lesions seen at post mortem and by an examination of the faeces. Usually very large numbers of oocysts are found in the faeces (several thousand oocysts/g of faeces). However, in peracute infections clinical signs may arise before oocysts are shed. It should be recognized that coccidia are probably present in all sheep, and the mere presence of oocysts in the faeces is not grounds for a diagnosis of coccidiosis. It is advisable, therefore, to conduct a post mortem examination on a representative member of the flock before a definite diagnosis is reached.

Treatment

Sulphonamides. Foster et al. (1941) reported that sulphaguanidine, at a dose of 2 g/day for six days, suppressed oocyst output in subclinical infections and prevented the acquisition of natural infections. In field studies in New Zealand, Whitten (1953) reported no growth response following treatment with sodium sulphadimidine.

Nitrofurazone. This has been shown to be effective against *E. faurei* at a dose of 7–10 mg/kg daily for seven days (Tarlatzis et al. 1955). It may be given in the feed at a level of 0.0165% (Shumard 1959*a*). Shumard (1959*b*) also demonstrated that 0.008%, 0.01% and 0.0133% of nitrofurazone in the drinking water prevented mortality and reduced morbidity resulting from experimental infections produced with ten million oocysts consisting of a mixture of *E. ninakohlyakimovae*, *E. ovina*, *E. intricata*, *E. parva*, *E. faurei* and *E. pallida*.

Amprolium. Recent reports indicate this compound is effective in sheep and goat coccidiosis. Intakes of 50–62.5 mg/kg body weight given in drinking water or feed for sheep, or 100 mg/kg for goats, for four days or longer, results in a rapid reduction in oocyst output and clinical recovery (Ross 1968; Horak et al. 1969).

Ajayi and Todd (1977) reported that feed supplemented with an aureomycin–sulphamethazine premix (equal parts of the two drugs) at levels of 100 mg and 500 mg/lamb/day, inhibited development of a mixed population of coccidia in lambs. However, the higher dose, which interfered with parasite development, prevented the development of acquired resistance and lambs died after withdrawal of the drug. Lambs given 100 mg/day survived when the premix was withdrawn.

Prevention and control

Feed lots should be kept dry and clean, the feed troughs should be constructed so that there is no wastage from them and so that they are not contaminated by faeces. Proper drainage of the feed lot is necessary.

Where nursing ewes are fed concentrates at pasture, the feeding area should be changed regularly. If inclement weather necessitates that lambs and ewes be kept in yards or barns, then the bedding should be changed regularly to avoid an accumulation of large numbers of sporulated oocysts.

COCCIDIA OF CATTLE

Bovine coccidiosis occurs in all parts of the world and serious outbreaks may occur in dairy herds where young stock are kept in large numbers. The following species have been described from cattle:

Eimeria alabamensis Christensen, 1941
Eimeria auburnensis Christensen and Porter, 1939
Eimeria bombayansis Rao and Hiregaudar, 1954
Eimeria bovis (Züblin, 1908) Fiebiger, 1912
Eimeria brasiliensis Torres and Ramos, 1939
Eimeria bukidnonensis Tubangui, 1931
Eimeria canadensis Bruce, 1921
Eimeria cylindrica Wilson, 1931
Eimeria ellipsoidalis Becker and Frye, 1929
Eimeria illinoisensis Levine and Ivens, 1967
Eimeria mundaragi Hiregaudar, 1956
Eimeria pellita Supperer, 1952
Eimeria subspherica Christensen, 1941
Eimeria wyomingensis Huizinga and Winger, 1942
Eimeria zuernii (Rivolta, 1878) Martin, 1909
Cryptosporidium bovis Barker and Carbonell, 1974

Eimeria alabamensis. (Fig. 3.25e) *Hosts:* domestic cattle, zebu, North America. Davis et al. (1955) found this species in 93% of 102 dairy calves examined and subsequent reports indicate that it is a common coccidium of cattle in the United States. The developmental stages occur in the posterior part of the ileum and may extend to the caecum and colon. *Oocysts:* predominantly pear-shaped, some ellipsoidal, subcylindrical or asymmetrical; 18.9 μm by 13.4 μm, range 13–24 μm by 11–16 μm. Oocyst wall thin, homogeneous, transparent and generally colourless, no micropyle. Sporulation time, 96–120 hours.

Life-cycle. The endogenous developmental cycle has been described by Davis et al. (1957). It is unusual in that the developmental stages occur in the nucleus of the epithelial cell. Sporozoites penetrate the intestinal cells as early as the second day after infection, and schizonts are visible in the nucleus two to eight days after infection. The cells usually parasitized are those at the tips of the villi. Multiple invasion of the nucleus may occur, and schizonts may reach a size of 12.6 μm by 9.7 μm at six days. They produce 15–32 merozoites, and free merozoites may be found in the small intestine as early as four days after infection. Davis et al. (1957) considered that more than one asexual generation occurs.

Gamonts are found in the posterior third of the small intestine, but they may also occur in the mucous membrane of the caecum and colon in heavy infections. Oocysts may be seen in the tissues of the lower part of the ileum as early as six days after infection, but the average prepatent period is 8.6 days with a range of 6–11 days. The duration of oocyst production from a single infection is one to ten days with an average of 4.6 days for low-grade infections, and 1–13 days with an average of 7.2 days for heavy infections.

Pathogenesis. It is low under field conditions and is generally considered to be unimportant in clinical bovine coccidiosis. Disease may be produced if large numbers of oocysts are given to young calves. Thus, Broughton (1945) fed 200 million oocysts to young calves and produced severe diarrhoea; two of five animals died. Davis et al. (1955) found that 140 million oocysts produced a yellowish green diarrhoea, admixed with blood, in 14-month-old animals. The small intestine showed hyperaemia, destruction of the epithelium, a leucocytic infiltration and oedema.

Eimeria auburnensis. (Fig. 3.25i, j) *Hosts:* domestic cattle, zebu, water buffalo. *E. auburnensis* is one of the most common coccidia in cattle in North America; reports indicate that it is widespread elsewhere in the world. *Oocysts:* ovoidal, varying from ellipsoidal to tapering, 38.4 μm by 23.1 μm, range 32–46 μm by 20–25 μm (Christensen & Porter 1939; Christensen 1941*b*). Oocyst wall smooth and homogeneous, transparent,

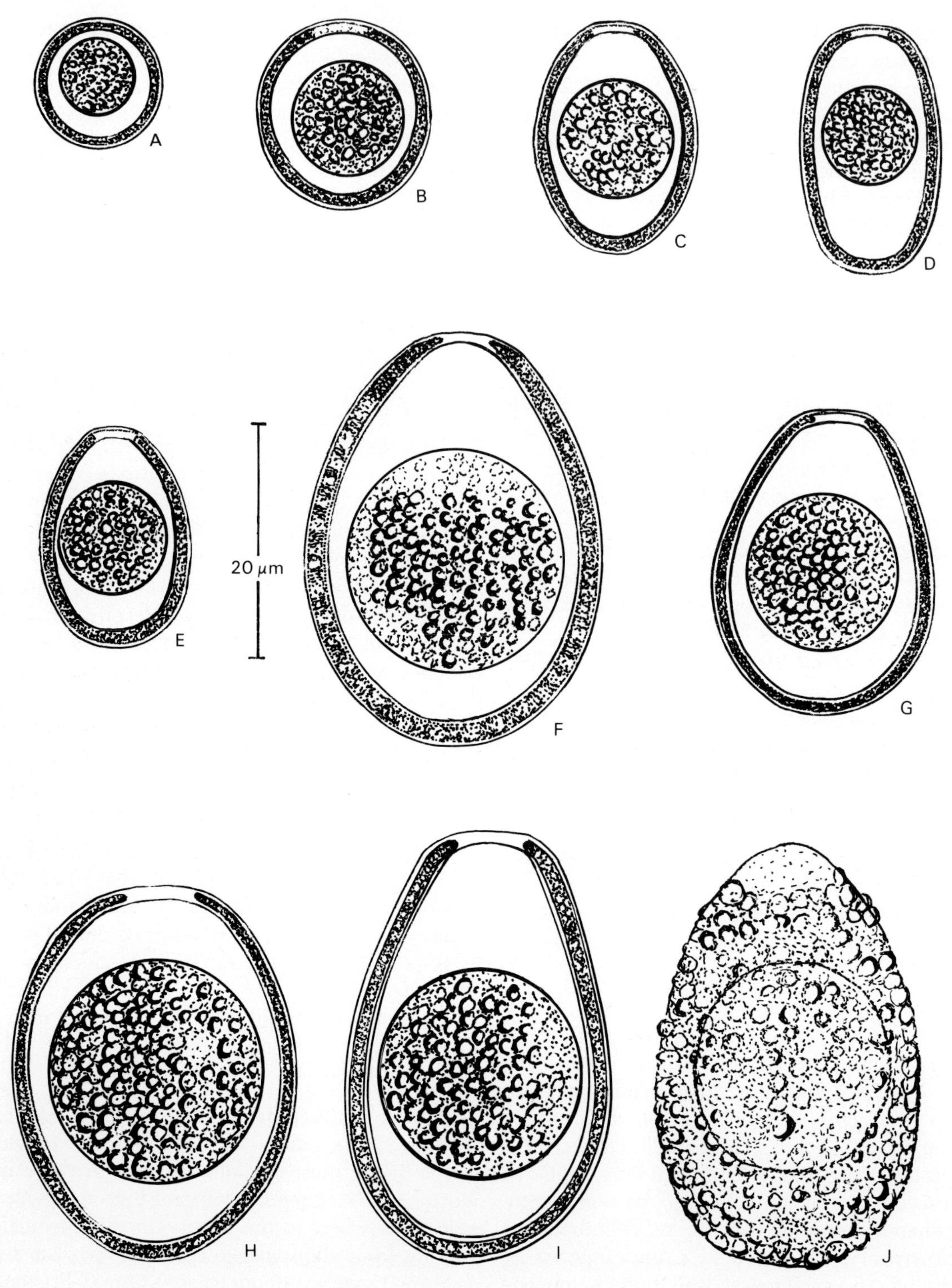

Fig 3.25 Coccidia of cattle. (A) *Eimeria subspherica*. (B) *E. zurnii*. (C) *E. ellipsoidalis*. (D) *E. cylindrica*. (E) *E. alabamensis*. (F) *E. bukidnonensis*. (G) *E. bovis*. (H) *E. canadensis*. (I) *E. auburnensis* (with homogeneous wall). (J) *E. auburnensis* (with mammillated wall). (*Redrawn with permission from Christensen 1938*)

yellowish-brown in colour; occasionally, it may have a coarsely granulated surface and at times be heavily mammillated. Micropyle appears as a pale area at narrow end. Sporulation time, 48–72 hours.

Life-cycle. The endogenous developmental cycle of *E. auburnensis* was investigated by Hammond et al. (1961), Davis and Bowman (1962) Chobotar (1968) and Chobotar and Hammond (1967).

Two generations of schizonts occur. The first is a giant (globidial) schizont in the epithelial cells of the crypts of Lieberkühn in the jejunum and ileum, or in cells of the connective tissue in the lamina propria. Several thousand merozoites are produced and these develop into second generation schizonts in cells of the lamina propria of the small intestine 12–14 days after infection.

The sexual stages occur in cells of mesodermal origin, especially those lying beneath the epithelium of the villi of the small intestine. The microgamonts are unusually large and may be seen with the naked eye, each producing several thousand microgametes. More than one gamont may be seen in an individual cell in heavy infections, and both male and female forms may occur in the same cell. Oocyst production is complete in about 18 days after infection; however, the epithelium needs to be broken before the oocysts can be shed. The prepatent period is 24 days, oocysts being discharged in maximal numbers for three days (Christensen & Porter 1939).

Pathogenesis. This is low under field conditions. A profuse green diarrhoea was produced in a two-week-old calf given 8000 oocysts by Christensen and Porter (1939), clinical signs appearing between the ninth and fifteenth days of infection. Davis and Bowman (1952) reported the passage of blood and mucus, with straining, in artificial infections with large numbers of oocysts and in natural outbreaks.

Eimeria bombayansis. *Hosts:* zebu (India). This species has been recognized from oocysts in faeces, and Rao and Hiregaudar (1954) indicated it was common in dairy calves near Bombay. *Oocysts:* ellipsoidal, 37 μm by 22.4 μm, range 32–40 μm by 20–25 μm. Oocyst wall smooth, transparent and homogeneous. Pellérdy (1974) considers this species to be identical with *E. auburnensis.*

Eimeria bovis. (Fig. 3.25g) *Hosts:* cattle, zebu, water buffalo. Together with *Eimeria zuernii*, this species is most commonly involved in clinical coccidiosis of cattle. World-wide in distribution. The asexual stages occur in the small intestine, the gametogonous stages in the terminal part of the ileum, and in the caecum and colon. *Oocysts:* ovoidal, blunted across the narrow end; in massive infections a variation in shape of oocysts may occur: 27.7 μm by 20.3 μm, range 23–34 μm by 17–23 μm. Oocyst wall smooth, homogeneous, transparent, greenish-brown in colour, micropyle present appearing as a lighter area of the wall. Sporulation time, 48–72 hours at room temperature.

Life-cycle. The endogenous developmental cycle was investigated by Hammond et al. (1946). The sporozoites invade the endothelial cells of the central lacteals of villi in the posterior half of the small intestine. Schizonts appear as early as five days after infection, and swollen cells may become detached from the lacteal cavity. These forms grow to mature giant (globidial) schizonts, and by 14–18 days after infection they may measure up to 400 μm in diameter, the mean size being 281 μm by 303 μm. Each schizont contains an average of 120 000 merozoites, each about 11 μm in length. A second generation schizont has been reported by Hammond et al. (1963*a*), this being smaller than the first (about 100 μm) and occurring in the epithelial cells of the caecum and colon. The second schizont produces 30–36 merozoites and takes 1.5–2 days to develop. Merozoites from the second generation schizonts invade the epithelial cells chiefly of the caecum and colon, although in heavy infections they may be found also in the terminal metre of the small intestine. The gametogonous cycle is initiated about 15–16 days after infection (Hammond et al. 1960), and the gametes are mature three days later. Oocysts are formed in a minimum of 18 days and peak numbers of oocysts are discharged 19–22 days after infection. Small numbers of oocysts continue to be discharged for two to three weeks after the initial major discharge.

Pathogenesis. Eimeria bovis is one of the most common coccidia of cattle. 40% or more of cattle have been reported infected in various parts of the world; Boughton (1945) found it in 41% of more than 2000 bovine faecal samples in the south-eastern USA; similar figures have been reported by Hasche and Todd (1959) in the USA, by Supperer (1952) in Austria and Rao and Hiregaudar (1954) in India.

Although the later stages of the development of the first stage schizont cause distortion of the villi and disruption when the merozoites escape (as is also the case with the second schizonts), it is the gamonts which cause the greatest pathogenic effect. Hammond et al. (1946) estimated that if the full potential of *E. bovis* were to be realized, 1000 oocysts could result in the destruction of 24 billion intestinal cells. In experimental infections, Hammond et al. (1946) found that an infective dose of 125 000 oocysts or more caused marked signs of illness with diarrhoea occurring on the eighteenth day when the faeces were streaked with blood. A calf given 120 000 oocysts became moribund, and with higher doses animals became moribund or died within three to four weeks of infection.

In severe infections the majority of the crypts of the large intestine, and sometimes of the terminal part of the small intestine, are destroyed, the epithelial layer denuded and the lumen of the intestine filled with blood. The mucosa is necrotic and sloughed and this damage may extend to the submucosa; the wall of the intestine is congested and oedematous, and large numbers of gamonts and oocysts are visible microscopically.

Immunity. A resistance to reinfection with *E. bovis,* lasting at least 3–6 months and possibly longer, was demonstrated by Senger et al. (1959). Infections of 10 to 100 000 oocysts induced a rapid immunity, and animals became resistant 14 days after initial infection. Schizonts and merozoites of the first generation, second generation schizonts and merozoites and the gamonts were affected by the immune reaction, the immune response affecting the numbers but not the timing of the various stages of the life-cycle (Hammond et al. 1963*a*). In other studies, Hammond et al. (1964) reported that the invasion of the epithelial cells by first generation merozoites of *E. bovis* was inhibited in immune calves, and merozoites were lysed when placed in immune serum (Anderson et al. 1965). A similar effect was seen with normal serum but not at the dilutions which caused the effect with immune serum.

Klesius and Fudenberg (1977) have demonstrated delayed dermal hypersensitivity in cattle to *E. bovis* antigen, and transfer of such sensitivity to recipient cattle was achieved with a dialysable transfer factor (TFd). Further, transfer of immunity to *E. bovis* in calves was reported using TFd (Klesius & Kristensen 1977). Similar transfer of immunity, using transfer factor, has been achieved in rats infected with *Eimeria nieschulzi* (Liburd et al. 1972).

Eimeria brasiliensis. *Hosts:* cattle, zebu, water buffalo, in North and South America, Europe (Austria), Nigeria and the Soviet Union. Marquardt (1959), following a comparison of *E. brasiliensis* with *Eimeria böhmi* Supperer, 1952, considered the two were synonymous. This species is known from the oocysts only. *Oocysts:* oval, 37.5 μm by 27.1 μm, range 34.2–42.7 μm by 24.2–29.9 μm (Torres & Ramos 1939). Oocyst wall, colourless to yellow, smooth, with a distinct micropyle and a polar cap. Sporulation time, six to seven days (Lee & Armour 1958).

The endogenous developmental cycle is unknown, and no pathogenesis is ascribed to it. Prevalence generally low, although Lee and Armour (1958) reported it a frequent parasite in cattle in Nigeria.

Eimeria bukidnonensis. (Fig. 3.24f) *Hosts:* domestic cattle, buffalo, and zebu. Found originally in the Philippines but subsequently in North and South America, Africa and the Soviet Union. This species is known from the oocysts only (Christensen 1938*b*). *Oocysts:* pear-shaped to oval, yellowish brown to dark brown in colour, 44 μm by 31.1 μm (Lee 1954), micropyle at the narrower end, oocyst wall shows radial striations. Sporulation time, about 17 days (Lee 1954).

The endogenous developmental cycle is unknown. Baker (1939) stated the prepatent period to be about ten days and observed diarrhoea in an experimentally infected calf.

Eimeria canadensis. *Hosts:* domestic cattle, zebu, European bison and water buffalo in North America and Soviet Union. This species is known only from oocysts in the faeces. *Oocysts:* ellipsoidal, occasionally cylindrical, 32.5 μm by 23.4 μm, range 28–37 μm by 20–27 μm (Christensen 1941*b*). Oocyst wall, smooth, transparent, slightly yellowish brown in colour, micropyle present. Sporulation time, 72–96 hours.

The developmental cycle is unknown, as is the pathogenicity. The species is relatively common in the United States, 35% of cattle being found infected in Wisconsin (Hasche & Todd 1959).

Eimeria cylindrica (Fig. 3.24d). *Hosts:* domestic cattle, zebu, water buffalo; North America and India. It is known only from its oocyst. *Oocyst:* cylindrical, some may be narrow cylinders, 23.3 μm by 13.3 μm, range 16–27 μm by 12–15 μm. Oocyst wall, thin, colourless, smooth, no micropyle. Sporulation time, two days. The endogenous developmental cycle is not known. The prevalence of this species is quite high, Hasche and Todd (1959) finding it in 20% of cattle in Wisconsin. Rao and Hiregaudar (1954) found it quite prevalent in zebu calves in Bombay and considered it pathogenic in such.

Eimeria ellipsoidalis (Fig. 3.24c). *Hosts:* domestic cattle, zebu, European bison and water buffalo; North America, Europe and Soviet Union. Developmental stages occur in the small intestine. *Oocysts:* ellipsoidal, occasionally spherical or cylindrical, 16.9 μm by 13 μm, range 12–27 μm by 10–18 μm. Oocyst wall, thin, homogeneous, and transparent, no micropyle. Sporulation time, 48–72 hours.

The endogenous developmental cycle has been studied by Hammond et al. (1963*b*). Schizonts occur in the epithelial cells of the crypts of the ileum and colon. They are fairly small, 10.6 μm by 9.4 μm and contain 24–26 merozoites, 8 μ–11 μm in length. The number of generations of the asexual cycle is not known. Gamonts and oocysts occur in the lower part of the small intestine, mature oocysts being formed ten days after infection.

Boughton (1945) reported that *E. ellipsoidalis* may cause diarrhoea in young calves three months of age, and Hammond et al. (1963*a*), in addition to demonstrating pathogenesis, showed that a varying degree of immunity developed after infection with 50 000 to one million oocysts.

This species is common in cattle, Boughton (1945) finding it in 45% of more than 2000 bovine faecal samples in the south-eastern USA. Christensen (1941*b*) reported it to be more frequent than any other species in healthy cattle in Alabama, and similar incidences have been found in other countries.

Eimeria illinoisensis. *Host:* domestic cattle, USA. This species is known only from the oocysts. *Oocysts:* ellipsoidal or ovoid 24–29 μm by 19–22 μm, mean 26 μm by 21 μm. Oocyst wall, smooth, colourless. Sporulation time, developmental cycle and pathogenesis unknown.

Eimeria pellita. *Hosts:* bovine, Austria (Supperer 1952). It is known only from the oocysts. *Oocysts:* egg-shaped, 39.48 μm by 28.4 μm, range 36.1–40.9 μm by 26.5–30.2 μm. Oocyst wall, relatively thick, dark brown, uniformly placed protuberances give it a velvety appearance, micropyle present. Sporulation time, 10–12 days. The developmental cycle is unknown, as is its pathogenicity.

Eimeria mundaragi. *Hosts:* bovine and zebu, India. Known only from the oocysts. *Oocysts:* oval to egg-shaped, 36–38 μm, by 25–28 μm. Oocyst wall, thin, smooth, transparent, and yellowish in colour, distinct micropyle. Sporulation time, one to two days. Developmental cycle and pathogenicity unknown. Patnaik (1965) considers this a synonym of *E. auburnensis.*

Eimeria subspherica (Fig. 3.24a). *Hosts:* domestic cattle, zebu, water buffalo, world-wide. Known only from the oocysts. *Oocysts:* smallest of all of the bovine *Eimeria* spp., 11 μm by 10.4 μm, range 9–11 μm by 8–12 μm, ellipsoidal to subspherical. Oocyst wall, uniformly thin, smooth and transparent, no visible micropyle. Sporulation time, four to five days. The developmental cycle is unknown; the organism is not known to be pathogenic.

Eimeria wyomingensis. *Hosts:* domestic cattle, zebu, water buffalo, world-wide. This species is similar to *E. bukidnonensis* but the oocysts are slightly smaller. *Oocysts:* ovoidal, 40.3 μm by 28.1 μm, range 37–44.9 μm by 26.4–30.8 μm. Oocyst wall, yellowish-brown to greenish-brown, slightly speckled, micropyle present. Sporulation time, five to seven days. The developmental cycle and the pathogenicity are unknown.

Eimeria zuernii (Fig. 3.24b). *Hosts:* cattle, zebu, and water buffalo, world-wide in distribution. This species is the most common and also the most pathogenic of the bovine coccidia. Boughton (1945) found it in 42% of more than 2000 bovine faecal samples in the south-eastern USA; Hasche and Todd (1959) reported a prevalence of 26% in cattle in Wisconsin, and figures of 10–30% prevalence have been reported in cattle and buffaloes from other parts of the world. Developmental stages occur in the small intestine, caecum, colon and rectum. *Oocysts:* spherical, sub-spherical to ellipsoidal, 17.8 μm by 15.6 μm, range 15–22 μm by 13–18 μm (Christensen 1941*b*). Oocyst wall, thin, homogeneous, transparent, colourless to pale yellow, no visible micropyle. Sporulation time, nine to ten days at 12°C, three days at 20°C, 23–24 hours at 30–32.5°C (Marquardt et al. 1960).

Life-cycle. The endogenous developmental cycle has been described by Davis and Bowman (1957). By the second and third days of infection, trophozoites are found in the mucosa, some penetrating as far as the muscularis mucosa. By the sixth day schizonts are found in the epithelial cells of the upper and lower parts of the small intestine, the parasites lying distally to the cell nucleus. Schizonts may still be present up to the nineteenth day, and by this time occur throughout the small intestine and also in the caecum and colon. Mature schizonts measure up to 7–9.8 μm and produce 24–26 merozoites. Davis and Bowman (1957) suggested that more than one asexual generation occurred. Stockdale (1977) reported two generations of schizogony, the first occurring in the lower ileum and the second in the colon and caecum. The earliest sexual stage is the macrogamont, first seen 12 days after infection in the epithelial cells of the villi of the lower small intestine and in the caecum, colon and rectum. The microgamonts are seen first on the fifteenth day, being found in the lower colon and rectum. They are much fewer than the macrogamonts. Oocysts may be found in the tissues of the caecum and colon as early as 12 days after infection, but oocyst production is highest 19–20 days after infection.

Pathogenesis. Eimeria zuernii is the major pathogenic coccidium of cattle. In Europe it is the most frequent cause of bovine coccidiosis. The acute disease is characterized by haemorrhagic diarrhoea, and the condition may become so intense that the faeces are frank blood. Tenesmus is marked, there is anaemia, weakness and emaciation. In severe infections death may occur as early as seven days after the onset of clinical signs. At post mortem the major lesions occur in the large intestine, although general catarrhal enteritis may be present in both the small and large intestines. In severe cases the caecum and colon may be filled with semi-fluid haemorrhagic material or even frank blood with fibrinous clots. The epithelium may slough away, leaving large denuded areas which are infiltrated with lymphocytes and leucocytes. In less acute cases the mucous membrane is roughened and spotted with petechial haemorrhages. Smears from the mucosa show very large numbers of developmental stages and oocysts.

Immunity. Wilson and Morley (1933) reported that resistance to reinfection developed in two calves which they studied. However, Davis and Bowman (1954) were able to reinfect animals and induce transient clinical signs, indicating that immunity was not complete.

Cryptosporidium bovis. *Hosts:* domestic cattle, Australia, but probably world-wide. This species was described by Barker and Carbonell (1974) from a two-week-old calf with a fatal enteritis which was unresponsive to sulphonamide therapy. Developmental stages similar to those of *C. agni* in sheep were seen on the microvillous border. Macrogametes up to 7 μm in diameter were identified and oocysts undergoing sporogony were seen, in which developing sporozoites and an oocyst residuum were evident.

Organisms similar to *C. bovis* have been seen in calves with diarrhoea by Panciera et al. (1971) and by Kelly and Soulsby (1974, unpublished data). The mode of transmission is not known. It is likely to occur via the faeces but whether by oocysts (which have not been demonstrated in the faeces of infected animals) or by sporocysts is unclear.

The lesions associated with the infection include villous atrophy, loss of microvilli and cellular infiltration into the lamina propria.

COCCIDIOSIS IN CATTLE

Foster (1949) has estimated that the annual loss in the United States from coccidiosis is 10 million dollars. It is probable that this could be increased several-fold to reflect present-day economics. The two major pathogenic species are *E. zuernii* and *E. bovis*, the former being the principal parasite in Europe. Other species, such as *E. auburnensis*, may, at times, contribute to the general clinical picture. In general, the infection occurs in animals three weeks to six months of age, but occasionally clinical disease occurs in yearlings and even adults, especially if massive infections are acquired. The most serious losses are seen in dairy herds where large numbers of calves are kept; nevertheless, appreciable losses can also occur in range cattle, and occasionally severe outbreaks may occur in stabled or yarded animals, especially with *E. zuernii*. The latter entity is frequently referred to as 'winter coccidiosis', and it is assumed that the bedding provides enough warmth and moisture for sporulation of oocysts even in sub-zero temperatures. In other parts of the world, also, infection may be severe, and Biswal (1948) considered bovine coccidiosis to be the primary disease in buffalo herds in India.

Older cattle are carriers of coccidia and, although immune, continue to pass oocysts in the faeces. Successive passage of the parasites in young animals results in a build-up of infection in yards, barns and on pasture, so that severe and fatal coccidiosis may result when a new batch of calves is placed on a pasture or in a yard which hitherto has appeared perfectly safe. Outbreaks of coccidiosis sometimes reach epidemic proportions among calves in the autumn and winter months in the USA. In Great Britain, coccidiosis is an important disease in the late summer and autumn in the south-west of England and in Northern Ireland. In the latter area, it is seen particularly in animals at pasture which graze, following a dry summer, around pools, ponds or damp areas, such sites being grossly contaminated with sporulated oocysts.

Diagnosis

Diagnosis of coccidiosis in cattle is based on the clinical signs, especially the haemorrhagic diarrhoea in acute cases and the demonstration of large numbers of oocysts in the faeces. Boughton (1945) reported that 5000 to 10 000 oocysts/g of faeces may occur in clinical cases. In peracute cases, especially those due to *E. zuernii*, oocysts may be few, the marked pathogenic effects being produced by the developmental stages prior to the shedding of the oocysts. It is important to distinguish this condition from the intestinal form of anthrax.

Where a post mortem examination is possible, scrapings of the small and large intestinal mucosae should be examined for the developmental stages of *Eimeria* spp.

Treatment

The clinical signs of coccidiosis appear usually only when the life-cycle of the parasite is advanced and marked destruction of the mucosa may already have taken place. Consequently, treatment of clinically affected animals cannot be expected to induce a radical cure. Nevertheless, coccidiosis is usually a herd problem, and treatment of the whole group of animals, including those which are not showing clinicals signs, will be beneficial.

The most effective drug at present is amprolium. At doses of 20–25 mg/kg daily given in the feed for four to five days, amprolium has proved active against *E. bovis* and *E. zuernii* infections (Farizy et al. 1970; Slater et al. 1970). lincomycin hydrochloride, at a dose of 1 g/calf given in the drinking water for 21 days, is effective against experimental infection with *E. bovis* (Arawaka & Todd 1968). Earlier compounds used for bovine coccidiosis include various sulphonamides, but because of their limited activity and their toxicity when given in high doses their use is not indicated.

Prevention and control

The prevention of bovine coccidiosis is based on treatment and good sanitation. In feed lots the feeding troughs and water containers should be high enough to prevent faecal contamination and wastage of fodder. The feed lot should also be kept dry and well drained and be cleaned out regularly. Where dairy calves are reared in yards, the bedding should be kept dry, well drained, and should be cleaned out regularly. When outbreaks occur at pasture, water holes and ditches should be fenced off and young calves denied access to them. Bedding and soil may be sterilized by 1.25% sodium hypochlorite, 0.5% cresol or phenol or by fumigation with formaldehyde.

COCCIDIA OF BUFFALOES

A number of species of coccidia have been reported from the water buffalo (*Bubalis bubalis*). Some of these also occur in cattle and other ruminants (see above) but a number have been reported only from the water buffalo. The validity of some of the latter requires clarification. In many cases the description is based on oocysts and no information is available on the developmental cycle, relationship to other ruminant coccidia or the pathogenesis of these species. In addition to the species given under 'Coccidia of cattle', the following have been recorded for water buffaloes.

Eimeria ankarensis Sayin, 1969. Turkey. *Oocysts:* 39.2 μm by 26.4 μm. Sporulation time, three to four days.

Eimeria azerbaidschanica Yakimoff, 1933. Azerbaidzhan. *Oocysts:* 4.5 μm by 22 μm.

Eimeria bareillyi Gill, Chabra and Lall, 1963. India. *Oocysts:* pyriform, 26–35 μm by 19–25 μm (mean 30.8 μm by 21.6 μm). Sporulation time, three to four days. Lesions similar to *E. ovina* in sheep are reported by Pande et al. (1971).

Eimeria gokaki Rao and Bhatavdekar, 1959. Mysore and Agra, India. *Oocysts:* 22–31 μm by 18–25 μm. Sporulation time, less than seven days.

Eimeria ovoidalis Ray and Mandal, 1962. West Bengal. *Oocysts:* 32–40 μm by 20–28 μm (mean 35.5 μm by 23.1 μm). Pinkish orange. Sporulation time, 90–120 hours at 29°C.

Eimeria thianethi Gwéléssiany, 1935. Georgian SSR, Agra, India. *Oocysts:* 34–49 μm by 26–34 μm (mean 42.6 μm by 28.6 μm). Sporulation time, five days.

For a fuller consideration of the coccidia of buffaloes, see Pellérdy (1974) and Levine (1973).

COCCIDIA OF CAMELS

The following species have been reported from the bactrian or two-humped camel (*Camelus bactrianus*) and the dromedary or one-humped camel (*Camelus dromedarius*).

Eimeria bactriani (Nöller, 1933) Levine and Ivens, 1970. USSR, bactrian and dromedary. *Oocysts:* 32 μm by 25–27 μm; light yellowish to yellowish brown, spherical to ellipsoidal, micropyle. Sporulation time, ten days. Development in the small intestine (Pellérdy 1974).

Eimeria cameli (Henry and Masson, 1932) Reichenow, 1953. Bactrian and dromedary; world-wide. *Oocysts:* truncate, 81–100 μm by 63–94 μm. Sporulation time, 10–15 days. Developmental stages occur in the small intestine. Giant (globidial) schizonts up to 350 μm in diameter occur (Henry & Masson 1932; Enigk 1934). Inflammatory lesions of the small intestine have been associated with this species (Henry & Masson 1932).

Eimeria dromedarii Yakimoff and Matschoulsky, 1939. Bactrian and dromedary; world-wide. *Oocysts:* 23–33 μm by 20–25 μm (mean 27.7 μm by 23.2 μm). Polar cap present. Sporulation time, 15–17 days at 10–12°C. Developmental cycle unknown.

Eimeria pellerdyi Prasad, 1960. Bactrian; London Zoo. *Oocysts:* mean 23.2 μm by 12.6 μm. Sporulation time, five days.

Eimeria rajasthani Dubey and Pande, 1963. Dromedary; India. *Oocysts:* ellipsoidal, 34–39 μm by 25–27 μm (mean 36 μm by 25 μm). Sporulation time, one week.

Isospora orlovi Tsygankov, 1950. *Camelus* spp. USSR. *Oocysts:* 30.4 μm by 19.2 μm. Pellérdy (1974) suggests this is an avian form, accidentally ingested.

Further consideration of the coccidia of camels is given by Pellérdy (1974) and Levine and Ivens (1970).

COCCIDIA OF LLAMAS AND ALPACAS

The following species have been recorded from the llama (*Lama glama*) and the alpaca (*Lama pacos*).

Eimeria alpacae Guerrero, 1967. Alpaca; Peru. *Oocysts:* ellipsoidal 22–26 μm by 18–21 μm; wall thick, slightly greenish blue; micropyle.

Eimeria lamae Guerrero, 1967. Alpaca; Peru. *Oocysts:* ovoid to ellipsoidal, 30–40 μm by 21–30 μm; micropyle. Developmental cycle and pathogenesis unknown.

Eimeria macusaniensis Hernández, Bazalar and Alva, 1971. Alpaca; Peru. *Oocysts:* ovoid 81–107 μm by 61–80 μm (mean 93.6 μm by 67.4 μm); micropyle. Developmental cycle and pathogenesis unknown.

Eimeria peruviana Yakimoff, 1934. Llama; USSR. *Oocysts:* ovoid 28–37 μm by 18–22 μm; no micropyle.

Eimeria punoensis Guerrero, 1967. Alpaca; Peru. *Oocysts:* ellipsoidal or oval, 17–22 μm by 14–18 μm; wall thick, bluish; micropyle.

Other unidentified members of the genus in alpacas and vicunas are given by Pellérdy (1974), and an account of the coccidia in alpacas is given by Guerrero (1967).

COCCIDIA OF SWINE

Low-grade infections of coccidia are common in pigs. However, the prevalence of clinical disease attributable to coccidiosis is low, and it is possible that many of the disease entities ascribed to coccidiosis have been based on an uncritical assessment of the illness and the presence of oocysts in faeces of pigs.

A recent evaluation of the coccidia of swine in the United States has been made by Vetterling (1965). He considers that the various species can be placed in two major groups: the first is the *debliecki* group in which the oocysts have a smooth, colourless oocyst wall, no distinct micropyle, and range in size from 12 μm to 40 μm; the second group is the rough-walled group, being a heterogeneous group with a rough, yellow to brownish oocyst wall. The following list of coccidia of swine follows Vetterling's re-evaluations but also includes species mentioned by Pellérdy (1974):

Eimeria cerdonis Vetterling, 1965
Eimeria debliecki, Douwes, 1921
Eimeria guevarai Romero, Rodriguez and Lizcano Herrera, 1971
Eimeria neodebliecki Vetterling, 1965
Eimeria perminuta Henry, 1931
Eimeria polita Pellérdy, 1949
Eimeria porci Vetterling, 1965
Eimeria scabra Henry, 1931
Eimeria scrofae Galli-Valerio, 1935
Eimeria spinosa Henry, 1931
Eimeria suis Nöller, 1921
Isospora almaataensis Paichuk, 1951
Isospora suis Biester and Murray, 1934

Eimeria cerdonis (Fig. 3.26b). *Host:* pig. North America, India. Location in the host unknown. *Oocysts:* ellipsoidal 26–32 μm by 20–23 μm (mean 29 μm by 21 μm); wall rough, yellowish to colourless; no micropyle. The prepatent period is eight days and the period of patency six days. Rommel (1970*b*) considers this species as a synonym of *E. polita*.

Eimeria debliecki (Fig. 3.26h). *Host:* pig. World-wide in distribution. Small intestine and occasionally the large intestine. Probably the most common species in the pig. *Oocysts:* ovoid to subspherical, 21.8–28.8 μm by 12.8–19.2 μm. Oocyst wall; smooth, no visible micropyle. Sporulation time, four to nine days.

Life-cycle. Wiesenhütter (1962*b*) found schizogony stages in the jejunum and ileum. Vetterling (1966) reported schizonts in the distal parts of epithelial cells in the jejunum. Second generation schizonts occur in the jejunum and ileum, as do the gamonts. The prepatent period is approximately seven days.

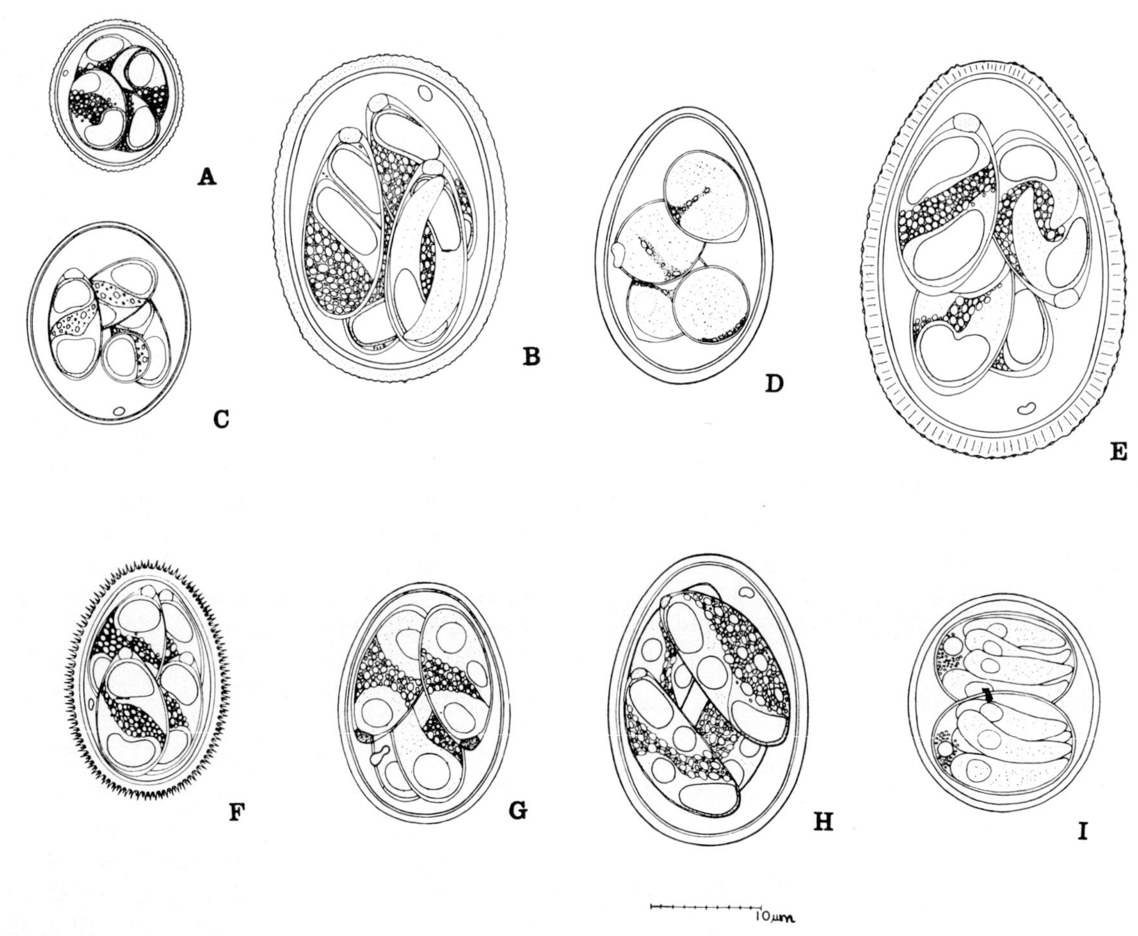

Fig 3.26 Coccidia of swine. (A) *Eimeria perminuta*. (B) *E. cerdonis*. (C) *E. suis*. (D) *E. porci*. (E) *E. scabra*. (F) *E. spinosa*. (G) *E. neodebliecki*. (H) *E. debliecki*. (I) *Isospora suis*. (*From Vetterling 1965*)

Pathogenesis. Pathogenic effects are confined to young pigs; older animals are seldom, if ever, clinically affected. Vetterling (1966) reported that pure infections were only slightly pathogenic. Biester and Murray (1934) observed diarrhoea, loss of appetite, emaciation and stunting of growth when young pigs were fed large numbers of sporulated oocysts. Some pigs died from the infection. Using mixed infections, Alicata and Willet (1946) confirmed that heavy infections (20–30 million oocysts) produced diarrhoea and anorexia on the seventh day after infection, this lasting 2–15 days. More recent work by Wiesenhütter (1962*b*) and Boch and Weisenhütter (1963) demonstrated that infections of 10 000 oocysts or more of *E. debliecki* produced diarrhoea and emaciation in young pigs.

The pathological changes consist of catarrhal inflammation of the small and large intestine in association with large numbers of oocysts. The wall of the large intestine may be greatly thickened, a mucofibrinous exudate may be adherent to the wall, and a marked necrotic enteritis may occur. However, in view of the preponderance of enteritic conditions in the young pig, it would seem wise to exercise caution in attributing pathological changes of the intestine in pigs to coccidiosis, unless there is overwhelming evidence of massive numbers of organisms in the intestinal mucosa.

Eimeria guevarai. *Hosts:* pig. Spain. This species was described on the basis of oocysts in the faeces of domestic pigs. *Oocysts:* pyriform, 26–32 μm by 15–19 μm; no micropyle. Sporulation time, longer than ten days at 20°C. Prepatent period, nine to ten days.

Eimeria neodebliecki (Fig. 3.26g). *Hosts:* domestic pig, probably also wild boar. North America (common), India. Location in the host unknown. *Oocysts:* ellipsoidal 17–26 μm by 13–20 μm (mean 21.2 μm by 15.8 μm); no micropyle. Sporulation time, 13 days. Prepatent period, ten days; patency, six days.

Eimeria perminuta (Fig. 3.26a). *Hosts:* pig, probably world-wide in distribution. This species is known only by the oocysts in the faeces. *Oocysts:* ovoid, occasionally spherical, 11.2–16 μm by 9.6–12.8 μm. Oocyst wall, rough, frequently yellowish in colour; no visible micropyle. Sporulation time, 11 days. No information is available on the endogenous developmental cycle or the pathogenesis.

Eimeria polita. *Hosts:* domestic pig, wild boar. Hungary and USA (Alabama). This species is known only from oocysts in the faeces. It is considered as a synonym of *E. debliecki* by Levine (1973), and Rommel (1970*a*) considers *E. cerdonis* to be a synonym of *E. polita*. Further discussion on this point is given by Pellérdy (1974), who also details the developmental cycle. *Oocysts:* ellipsoidal, occasionally broad to oval, 23.8 μm by 17.9 μm, range 23–27 μm by 10–27 μm (Pellérdy 1949). (Boch et al. 1961 state a range of 24.5–40 μm by 20–26.8 μm). Oocyst wall, generally smooth and yellowish brown to pinkish brown in colour, occasionally roughened; no micropyle present. Pellérdy (1974) states that oocysts of *E. polita* may be confused with those of *E. scabra* and *E. debliecki*. They may be distinguished from *E. scabra* by the smooth wall and from *E. debliecki* by size differences. Sporulation time, eight to nine days.

Eimeria porci (Fig. 3.26d). *Host:* domestic pig. North America and India. This species is known only by the oocysts. *Oocysts:* ovoid 18–27 μm by 13–18 μm, smooth, colourless; indistinct micropyle. Prepatent period is seven days and patency lasts six days. The pathogenesis of this species is unknown. Vetterling (1965) considers *E. polita* as a synonym of *E. porci*.

Eimeria scabra (Fig. 3.26e). *Hosts:* domestic pig and wild boar. Probably world-wide in distribution. Developmental stages occur in the epithelial cells of the large intestine. *Oocysts:* ellipsoidal to ovoid, 25–35.5 μm by 16.8–25.5 μm (Kutzer 1960). Oocyst wall, yellowish brown and rough, micropyle is present at the narrower end. Sporulation time, 9–12 days.

Life-cycle. The endogenous developmental cycle has been described by Rommel and Ipezynski (1967). Three schizogonic generations occur, all in epithelial cells of villi of the small intestine. The first occurs in the jejunum, producing schizonts 12–19 μm by 11–17 μm by the third day. Schizonts of the second and third generation mature on the fifth and seventh day of infection in the ileum, developing between the brush border and the nucleus of the epithelial cells. Second generation schizonts measure 10–20 μm by 9–15 μm producing 14–22 merozoites, while those of the third generation measure 16–27 μm by 13–27 μm and produce 14–28 merozoites. Gamonts appear eight days after infection. The microgametes mature on the ninth day. The macrogametes measure 14–23 μm by 9–16 μm and have a large nucleus. The prepatent period is nine days and patency lasts for four to five days.

Pathogenesis. Alicata and Willett (1946) infected pigs experimentally with 20–30 million mixed oocysts (*E. scabra* and *E. debliecki*) and produced profuse diarrhoea which lasted 2–15 days. Pellérdy (1965) reported on the death of a single piglet weighing 50–60 kg in which large numbers of developmental stages of *E. scabra* were found in the intestinal mucosa.

Rommel and Ipenzynski (1967) found that as few as 220 oocysts caused diarrhoea with a haemorrhagic enteritis in the posterior small intestine. Villi were denuded of epithelial cells in the area where schizonts developed and the mucosa was infiltrated with leucocytes and eosinophils.

Rommel (1970*b*) has also investigated the immune response to *E. scabra*. Antilymphocyte serum prevented the acquisition of immunity to infection, but did not reduce an established immune response. However, paramethazone acetate and dexamethazone were immunosuppressive.

Eimeria scrofae *Hosts:* domestic pig, Lausanne, Switzerland. This species is known only from oocysts in the faeces. *Oocysts:* cylindrical, one end flattened, 24 μm by 15 μm; distinct micropyle. The endogenous developmental cycle and pathogenesis of this species are unknown. Pellérdy (1949, 1974) suggests that *E. scrofae* may be a form of *E. debliecki*.

Eimeria spinosa (Fig. 3.26f) *Hosts:* domestic pig, USA, Hawaii and Soviet Union. Uncommon. *Oocysts:* ellipsoidal to ovoid, 16–22.4 μm by 12.8–16 μm. Oocyst wall, brown, opaque, and its entire surface is covered with spines approximately 1 μm in height and 1 μm apart; no micropyle. Sporulation time, 15 days (Kutzer 1960).

Life-cycle. The endogenous developmental cycle has been described by Wiesenhütter (1962*a*), who reported a prepatent period of seven days. The developmental stages were found in the epithelial cells of the small intestine. Mature schizonts, 8–10 μm; macrogamonts, 7–9 μm; microgamonts, 6–8 μm.

Pathogenesis. Wiesenhütter (1962*a*) reported that experimental infections of 12 000 oocysts caused severe diarrhoea and slight fever in young pigs. One eight-week-old pig died of coccidiosis on the eleventh day. Lesions consisted of a marked inflammation of the small intestine with destruction and desquamation of the epithelium. Andrews and Spindler (1952) found this species was only slightly pathogenic. No clinical signs were observed in a pig which passed as many as seven million oocysts/g of faeces.

Eimeria suis (Fig. 3.26c). *Hosts:* domestic pig; probably world-wide. Vetterling (1965) reinstated this as a valid species and differentiated it from *E. debliecki*. *Oocysts:* ellipsoidal to sub-spherical 13–20 μm by 11–15 μm (mean 17 μm by 13 μm); wall smooth, colourless; no micropyle. The prepatent period is ten days and the period of patency six days (Vetterling 1965).

Isospora almaataensis. *Hosts:* domestic pig, Kazakhstan, USSR. This species is known only by the oocysts in the faeces. *Oocysts:* large, oval to spherical, dark grey in colour, 27.9 μm by 25.9 μm, range 24.6–31.9 μm by 23.2–29 μm. Sporulation time, five days. The endogenous developmental cycle and the pathogenicity of this species are unknown.

Isospora suis (Fig. 3.26i). *Hosts:* domestic pig, USA (Iowa) and Soviet Union (Kazakhstan). Developmental stages in the small intestine. *Oocysts:* sub-spherical, 20–24 μm by 18–21 μm. Oocyst wall, light yellow in colour, micropyle absent. Sporulation time, four days. Two ellipsoidal sporocysts produced, 16–18 μm by 10–12 μm, each containing four sporozoites.

Pellérdy (1974) has suggested that *Isospora suis* is an *Isospora* species from a wild bird, e.g. the sparrow, since these birds are frequent cohabitants of pig pens and pig feeding areas. However, Biester and Murray (1934) described the endogenous developmental cycle of this species, indicating the prepatent period to be 6–8 days; Vetterling (1965) described it as five days. Developmental stages occurred in the small intestine, especially the jejunum and ileum, the parasitized epithelial cells migrating subepithelially. A catarrhal enteritis was produced on infection, with a diarrhoea which lasted three to four days.

COCCIDIOSIS OF PIGS

There is relatively little information on the clinical entity in swine. The disease is primarily a disease of the young animal, older pigs being carriers. *Eimeria debliecki* and *E. scabra* are probably the most pathogenic species, but *I. suis* has also been incriminated in disease. Swanson and Kates (1940) briefly reported details of an outbreak of coccidiosis in a litter of 4½-month-old pigs in which up to 145 000 oocysts/g of faeces were found, the condition being one of a profuse diarrhoea. In Zaire, Deom and Mortelmans

(1954) reported that considerable losses may occur in young pigs due to *E. debliecki*. The clinical signs consisted of diarrhoea, emaciation and constipation in the final stages. Repeated infection leads to immunity (Rommel 1970*b*). Piglets reared in an infected herd are usually resistant to natural infection (e.g. by *I. suis*) for the first three weeks of life but are susceptible thereafter (O'Neill & Parfitt 1976).

Diagnosis

It is based on the clinical signs and, preferably, the demonstration of large numbers of endogenous developmental stages in the intestine. Since coccidia and enteritic conditions are common in swine, diagnosis by a faecal examination alone, unless oocysts are present in very large numbers, is unsound.

Treatment

Deom and Mortelmans (1954) found that 0.44% of nitrofurazone in the feed for seven days was an effective treatment in two pigs infected with *E. debliecki*. Pellérdy (1974) states that amprolium, at a dose of 25–65 mg/kg once or twice daily, is effective.

Control

This is based on improved hygiene; overcrowding of young piglets should be avoided, pens cleaned out regularly and feeding facilities improved.

COCCIDIA OF HORSES

Little information is available on the coccidia of equines, and this group of organisms in horses is in need of re-examination. The species which have been reported are as follows:

Eimeria leuckarti (Flesch, 1883) Reichenow, 1940
Eimeria solipedum Gousseff, 1934
Eimeria uniungulati Gousseff, 1934
Klossiella equi Baumann, 1946

Eimeria leuckarti. Known also as *Globidium leuckarti*, this species is found in the small intestine of horses and asses in Europe, North America and the Indian subcontinent. Its prevalence is unknown. *Oocysts:* some of the largest in the genus *Eimeria*, 80–87.5 μm by 55–59 μm, oval, flattened at the narrower end, oocyst wall thick, 6.5–7 μm, dark brown; distinct micropyle. Sporulation time, prolonged, 20–22 days at 20°C.

Life-cycle. Various stages in the endogenous developmental cycle have been described from natural cases of parasitism but there is disagreement as to what these represent. Hemmert-Halswick (1943) reported sexual stages beneath the epithelium in the villi of the small intestine, but Pellérdy (1974) maintains that Hemmert-Halswick's illustrations are of schizonts. Consequently, *E. leuckarti* possesses the 'giant schizont' type of development.

Barker and Remmler (1970) reported that experimentally infected ponies discharged oocysts 15–33 days after infection and the patent period lasted for 12–32 days.

Pathogenesis. Hemmert-Halswick (1943) described marked inflammatory changes in the intestine. Barker and Remmler (1970) produced no clinical manifestations of disease with up to 200 000 oocysts. However, Sheahan (1976) and Wheeldon and Greig (1977) describe acute and chronic diarrhoea as a result of infection with *Globidium*. Numerous parasite developmental stages were present in the mucosa, villi were clubbed, but cellular infiltration was moderate. In diagnosis of the infection, it is to be noted that the oocysts are heavy and do not rise to the surface of normal flotation solutions. They must be looked for by sedimentation techniques.

Eimeria solipedum. *Hosts:* domestic horse, mule and donkey, Soviet Union. It is known only from the oocysts in the faeces. *Oocysts:* spherical, double-contoured wall, 15–28 μm. Distinctively orange-red or yellowish-brown in colour with no micropyle. No information is available regarding endogenous developmental cycle or pathogenicity.

Eimeria uniungulati. *Hosts:* domestic horse and mule, Soviet Union. It is known only by oocysts found in the faeces. *Oocysts:* oval to ellipsoidal, bright orange in colour, 15.5–24 μm by 12.4–17 μm. The developmental cycle and pathogenesis are unknown.

Klossiella equi. *Hosts:* horse, ass, donkey and zebra. North America, Europe, Turkey, Australia. The prevalence is unknown and its presence is usually detected incidentally during histopathological examination. 'Oocysts' occur in the epithelial cells of the loop of Henle; however, Vetterling and Thompson (1972) believe that a typical coccidian oocyst is not formed. These authors propose the following life-cycle for *K. equi* in the kidney, although the complete life-cycle is unknown. Schizogony occurs in the endothelial cells of Bowman's capsule, merozoites pass down to the proximal convoluted tubules and penetrate epithelial cells to produce another generation of schizonts. Merozoites from these pass to the thick limb of Henle's loop where they penetrate epithelial cells. Here, some develop into microgametocytes, which form eight to ten microgamonts, which form eight to ten microgametes, and others form macrogametes. Syzygy does not occur. After fertilization, sporogony occurs with the formation of about 40 sporoblasts each of which produce 10–15 sporozoites. Sporocysts probably pass with the urine but this has not been observed.

Lee and Ross (1977) observed swelling and desquamation of tubular epithelium, but in general there is no marked inflammatory response to this infection in the horse.

COCCIDIA OF DOGS AND CATS

Previously, it was usual to consider the coccidia of dogs and cats together since it was thought that species, especially of the genus *Isospora*, were interchangeable between hosts. It is now clear this is not so. In fact, a remarkable host specificity exists for individual species. In the last few years there has been a dramatic increase in the study of the role of dogs and cats in the predator–prey life-cycles of a number of coccidial organisms. These include the genera *Isospora, Sarcocystis, Toxoplasma, Besnoitia* and *Hammondia*.

Relationship of the various genera of isosporoid coccidia found in dogs and cats

Frenkel (1977) has reclassified the cyst-forming isosporoid coccidia. This reclassification is based on the behaviour of various stages of development. These stages are *meronts* (merozoites) produced by a process of *merogony* (schizogony), meronts produce *tachyzoites* (a rapidly multiplying part of the precystic stage), *metrocysts* (non-infectious cystic stages) or *bradyzoites* (infectious intracystic stages). The recognized divisional stages are schizogony (merogony), endodyogeny, endopolygeny and splitting (Dubey & Frenkel 1972). The several genera can be classified according to transmission route (homoxenous—requiring one host; or heteroxenous—requiring two hosts); the absence of a propagative cycle in the gut of the final host; whether the propagative cycle has shifted to the cyst of the intermediate host which contains metrocysts and bradyzoites, the latter being equivalent to gamonts (Ruiz & Frankel 1976); the presence of an asexual propagative cycle in the gut of the final host, with only one morphological type of organism in the cyst, the bradyzoite.

The relationships of these genera are illustrated in Fig. 3.27. With the exception of *Toxoplasma*, feline and canine coccidia are probably non-pathogenic to their definitive hosts and their major importance lies in the transmission of infection to intermediate hosts such as cattle, sheep, swine, horses, man and other animals.

The disease entities associated with the various parasites in their intermediate hosts will be dealt with separately under the appropriate genus of *Toxoplasma*, *Sarcocystis* etc.

Table 3.1. Relationship of Isosporid Coccidia of Dogs and Cats (Frenkel 1977)

Family	EIMERIIDAE Minchin, 1903	Merogony (schizogony) and gametogony intracellular, oocyst with zero to four or many sporocysts each with one or more sporozoites. Sporogony outside the host, homoxenous, tissue cysts absent or unknown

Genus	ISOSPORA Schneider, 1881	Oocysts with two sporocysts, each with four sporozoites
Family	SARCOCYSTIDAE Poche 1913	Facultative or obligatory heteroxenous forms. Life-cycle proliferative in intermediate host leading to cyst formation. Propagative cycle may be present or absent in final host. Oocyst with two sporocysts, each with four sporozoites
Subfamily	SARCOCYSTINAE Poche, 1913	Proliferative cysts only in intermediate host containing non-infectious metrocysts and later bradyzoites. Bradyzoites develop directly into gametes in the gut of final host. Obligatorily heteroxenous. Tachyzoites multiply by endo-dyogeny or schizogony and precede cyst development
Genus	SARCOCYSTIS Lankester, 1882	Cysts in muscle, elongate, cell wall variable
Genus	FRENKELIA Biocca 1968	Cysts in brain and spinal cord, sub-spherical occupying whole of host cell
Subfamily	TOXOPLASMATINAE Biocca, 1956	Proliferative cysts contain only bradyzoites. Propagative cycle precedes gametogony in the gut of final host
Genus	TOXOPLASMA Nicolle and Manceaux, 1909	Cysts in many cell types, cyst wall develops from limiting membrane of parasitophorous vacuole. Cysts formed in intermediate and definitive host. Sporozoites, bradyzoites and tachyzoites infective for intermediate and definitive host
Genus	BESNOITIA Henry, 1913	Cysts in fibroblasts, possibly other cells. Bradyzoites in parasitophorous vacuole; cyst wall formed around cell, nucleus of which undergoes hypertrophy. Infectivity of tachyzoites and bradyzoites varies with species
Genus	HAMMONDIA Frenkel and Dubey, 1975	Cysts in skeletal muscle, occasionally heart and brain. Cyst wall develops from membrane of parasitophorous vacuole of cell. Cysts infect only final host, propagative cycle in intestinal of final host. Tachyzoites not transmitted
Genus	CYSTOISOSPORA Frenkel, 1977	A new genus created by Frenkel (1977) to accommodate certain species of the genus *Isospora*. Monozoic cyst in many tissues, especially mesenteric lymph nodes. Cyst wall develops from membrane of parasitophorous vacuole of parasitized cell. Cyst forms only in intermediate host and infects only definitive host. Sporulated oocyst infects intermediate and definitive host. Schizogony in the intestine of final host. Tachyzoites not transmitted

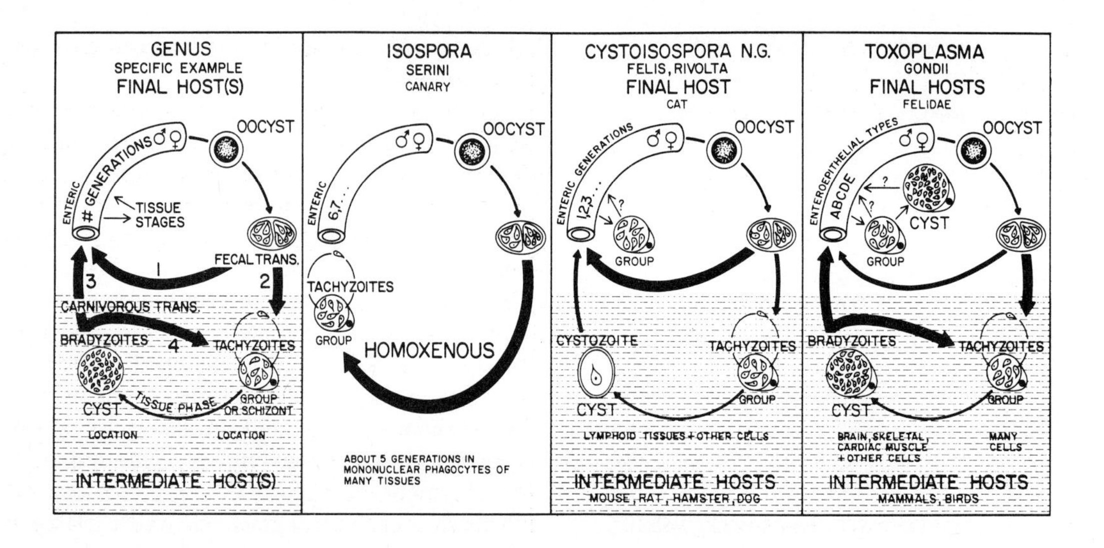

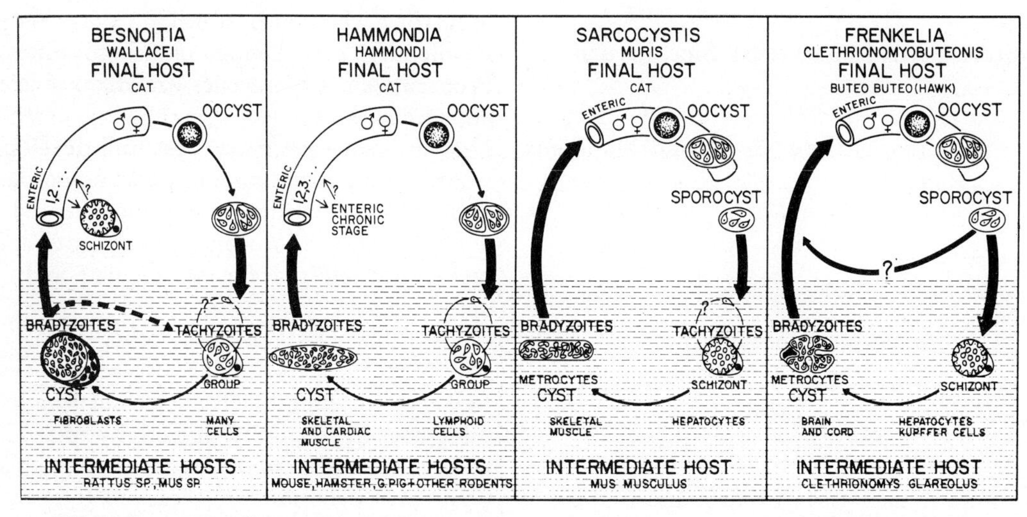

Fig 3.27 Life-cycle patterns of isosporoid coccidia. The first panel serves as a legend and identifies the four major transmission routes indicated by heavy arrows. These are: 1—oocyst to same host (homoxenous, faecal); 2—oocyst to intermediate host (heteroxenous, faecal); 3—'cyst' to final host (heteroxenous, carnivorous); 4—'cyst' to intermediate host (homoxenous, carnivorous). Light arrows indicate less important routes of transmission. *Isospora serini* of the canary represents a homoxenous form with tissue phase but no cyst formation. *Cystoisospora* has route 1 as the common form of transmission and routes 2 and 3 as less common. *Toxoplasma* has routes 2, 3 and 4 as most effective and route 1 as least effective in transmission. *Besnoitia* uses routes 2 and 3; route 1 does not occur. *Hammondia* uses routes 2 and 3 only. *Sarcocystis* uses routes 2 and 3 and the pregametocytic development occurs in the intermediate host and not the gut of the final host. *Frenkelia* is similar to *Sarcocystis* but route 1 transmission may occur. (*From Frenkel 1977*)

Species of the genus *Eimeria* have been reported from dogs (and cats) but it is now considered that these are spurious infections resulting from the ingestion of carcasses or faeces of other animals. Attempts to infect coccidia-free dogs and cats with *Eimeria* species found in dog and cat faeces have failed (Christie et al. 1976; Streitel & Dubey 1976).

The following list includes forms from the families Eimeriidae and Sarcocystidae. The proposed new nomenclature for the sarcosporidia (Heydorn et al. 1975) is used in part, although the validity of this has been questioned (Levine 1977).

COCCIDIA OF THE DOG (Fig. 3.28)

Isospora bahiensis Levine and Ivens, 1965
Isospora burrowsi Trayser and Todd, 1978
Isospora canis Nemeséri, 1959
Isospora heydorni Tadros and Laarman, 1976
Isospora ohioensis Dubey, 1975
Isospora wallacei Dubey, 1976
Sarcocystis bertrami Doflein, 1901
Sarcocystis bigemina (Stiles, 1891) Levine, 1977
Sarcocystis cruzi (Hasselman, 1926) Levine, 1977
Sarcocystis fayeri Dubey, Streitel, Stromberg and Toussant, 1977
Sarcocystis miescheriana (Kühn, 1865) Lankester, 1882
Sarcocystis ovicanis Heydorn, Gestrich, Mehlhorn and Rommel, 1975
Sarcocystis hemionilatrantis Kistner and Hudkins-Vivion, 1977
Hoaresporidium pellerdyi Pande, Bhatia and Chauhan, 1972

Isospora bahiensis. A rare form of the small intestine of dogs. *Oocysts:* sub-spherical, pale with smooth wall, 12–14 μm by 10–12 μm with a micropyle. Sporocysts broad and ellipsoidal, sporocyst residuum present. Developmental forms occur in the epithelial cells of the small intestine. May be pathogenic, causing diarrhoea (Levine 1978).

Isospora burrowsi. *Host:* dog. *Oocyst:* 17–22 μm by 16–19 μm and sporocysts 12–16 μm by 8–11 μm. Levine (1978) states that it may be confused with *I. ohioensis*.

Isospora canis. *Host:* dog. Probably world-wide in distribution. For many years this species in the dog has been referred to as *Isospora felis*, which it was thought occurred in both cat and dog. Nemeséri (1960) was unable to transmit the canine form to the cat and concluded that *I. canis* was a valid species. *Oocysts:* broadly ovoid, 34–40 μm by 28–32 μm. Oocyst wall, colourless, micropyle absent. This is the largest of the oocysts found in the faeces of dogs. Sporulation time, four days. Dogs can acquire infection by ingesting sporulated oocysts or infected mice (Lepp & Todd 1974; Dubey 1975), this representing a facultative heteroxenous developmental cycle.

Three asexual generations of development occurred in the lower third of the small intestine, directly beneath the epithelium of the distal portion of the villi. First generation schizogony lasted up to seven days, second generation was present at six to seven days after infection, and third generation schizogony was evident six to eight days after infection. The prepatent period was 9–11 days (Lepp & Todd 1974). Patency is four weeks.

The pathogenicity of *I. canis* under natural conditions is not known. Dubey (1976) was unable to produce clinical signs of disease in parasite-free dogs given 100 000 sporulated oocysts. However, Nemeséri (1960) reported inflammation and haemorrhage of the small intestine in experimental infections.

Because of the facultative heteroxenous developmental cycle, Frenkel (1977) has proposed that this species be placed in a new genus *Cystoisospora*.

Isospora heydorni. The taxonomic position of this species is uncertain. Dubey (1977) considers it not to be a true *Isospora* because oocysts do not induce oocyst formation in dogs. Its name was proposed by Tadros and Laarman (1976) for the small race of the canine *I. bigemina* and hence it is synonymous with *Isospora wallacei* (below), but because of the uncertainty of the life-cycle Dubey (1977) has suggested the combination *Hammondia heydorni*. *Host:* domestic dog. *Oocysts:* 11 by 13 μm. Sporulated oocysts will induce tissue stages in dogs but not oocysts. However, infected dog tissue will induce oocyst production. Oocysts are also produced upon feeding naturally infected bovine musculature, but the role of the bovine is not clear. The infection is otherwise non-pathogenic for the dog. (See also *I. wallacei*.)

Isospora ohioensis. *Hosts:* domestic dog, also dingo (*Canis dingo*), and probably occurs in a number of other *Canidae*. World-wide in distribution. Prevalence rates of 13–70% have been recorded from various areas. This species was previously known as *Isospora rivolta* in the dog but Mahrt (1966) was unable to transmit dog-derived

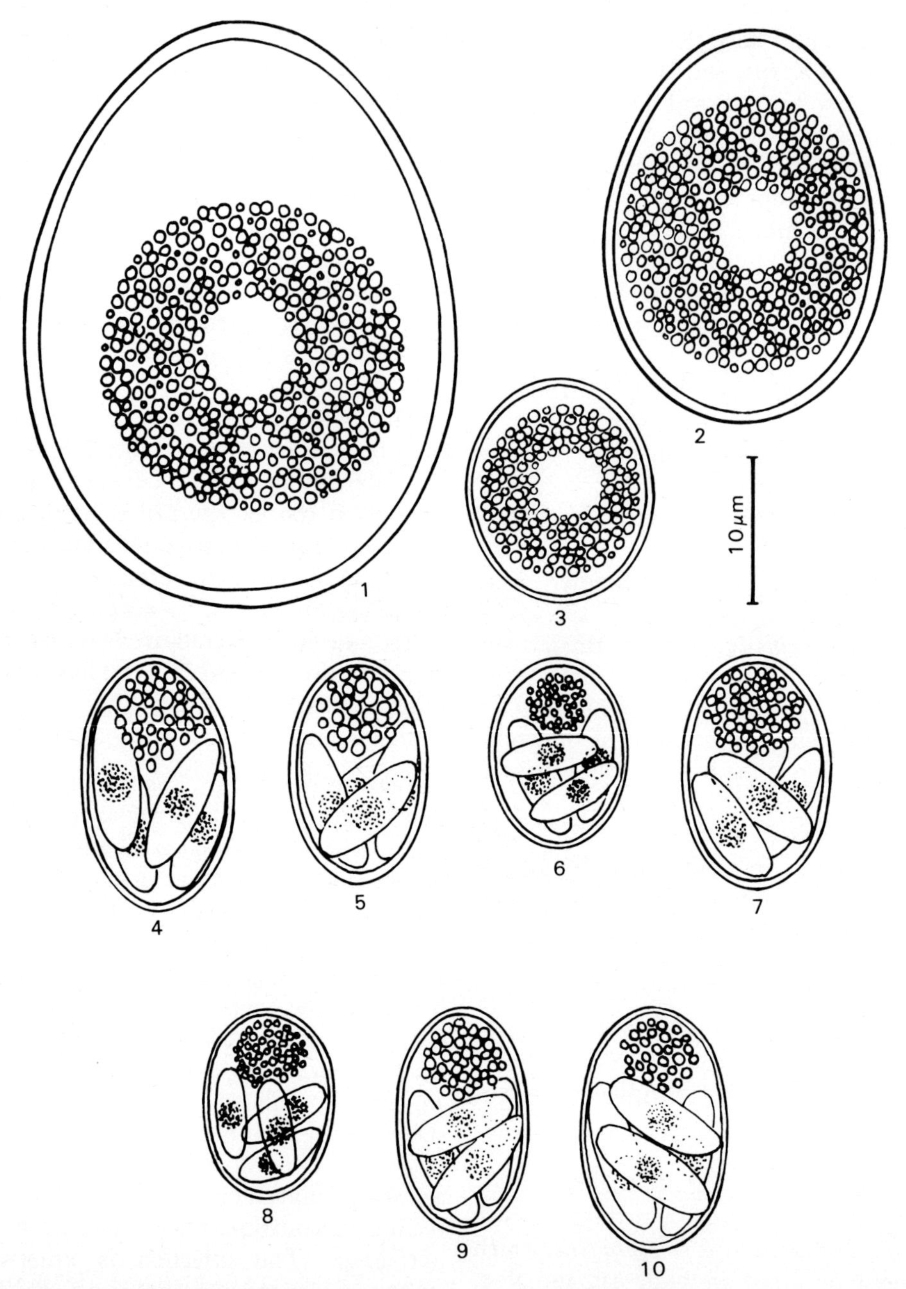

Fig 3.28 Coccidia of the dog. Oocysts in fresh faeces. (1) *Isospora canis*. (2) *I. ohioensis*. (3) *I. wallacei*. Sporocysts in fresh faeces. (4) *Sarcocystis cruzi*. (5) *S. ovicanis*. (6) *S. miescheriana*. (7) *S. bertrami*. (8) *S. fayeri*. (9) *S. hemionilatrantis*. (10) *Sarcocystis* sp. from Grant's gazelle (*From Dubey 1976*)

I. rivolta to cats or cat-derived *I. rivolta* to dogs. Developmental stages occur in the small intestine. *Oocysts:* ellipsoidal, 20–27 μm by 15–24 μm; micropyle absent. Sporulation time, four days.

Life-cycle. Two generations of schizonts in the epithelium and lamina propria of the small and large intestine. Gametogony occurs in a similar situation. The prepatent period is four to six days (Mahrt 1966; Dubey 1975) and patency lasts 13–23 days.

Isospora wallacei. This species was originally the small race of the canine *Isospora bigemina*. The large race of *I. bigemina* has been renamed *Sarcocystis bigemina*. *Oocysts:* 10–14 μm by 7.5–9 μm. Shed unsporulated. Sporulation time, 12–48 hours at 23–37°C. Sporulated oocysts are infectious to dogs but do not induce oocyst formation in dogs; however, ingestion of the tissues of dogs fed oocysts led to the shedding of oocysts. Prepatent periods were 7–15 days and patency one to three days. Sporulated oocysts from naturally infected dogs were not infective to cattle, cats or mice but structurally identical oocysts were shed by two dogs fed hearts of naturally infected cattle (Dubey & Fayer 1976).

There is no evidence that *I. wallacei* is pathogenic. However, in dogs infected with '*I. bigemina*' Lee (1934) has reported clinical disease. Younger animals are most seriously affected, older dogs serving as carriers. In experimental infections, diarrhoea is seen on the third day and blood is found in the faeces between the fourth and sixth days. In severe natural cases, a catarrhal enteritis leads to a haemorrhagic enteritis in which the faeces may be frank blood. This is associated with dehydration, general anaemia, emaciation, weakness and, ultimately, death. If the animal survives the acute phase, the bloody diarrhoea gives place to a gelatinous mucous discharge, the dog showing signs of recovery seven to ten days after the onset of clinical signs. At post mortem there is an haemorrhagic enteritis throughout the small intestine, especially in the lower small intestine. In light infections petechiae are present; there may also be ulcers and the mucosa is thickened.

Sarcocystis bertrami (*Syn: Sarcocystis equicanis*). The intermediate host for this species is the horse in which sarcocysts are found in the heart and striated muscle. Gametogony occurs in the intestine of dog. Sporocysts measure 15 μm by 10 μm; they are sporulated when shed in the faeces. The prepatent period is eight days (Rommel & Geisel 1975; Dubey 1976). (See also *Sarcocystis* in horses, p. 684.)

Sarcocystis bigemina. Previously this was the large race of the dog *Isospora bigemina*. It developed in the lamina propria of the intestine of dogs producing sporocysts 13–16 μm by 8.5–11 μm, which were shed sporulated. The intermediate host is unknown.

Sarcocystis cruzi (*Syn: Sarcocystis fusiformis, S. blanchardi, S. bovicanis*). This is the most pathogenic species of *Sarcocystis* in cattle. The definitive hosts are dogs, wolves, coyotes, racoons and foxes and the parasite is probably world-wide in distribution. *Sporocysts:* 14.3–17.4 μm by 8.7–13.3 μm (mean 16.3 μm by 10.8 μm); sporulated when shed in the faeces. Prepatent period after feeding infected beef is nine to ten days. Macrogamonts develop in the lamina propria of the small intestine of dogs two to six days after ingestion of infected musculature. They measure 5–6 μm by 8.5–11.5 μm. By the seventh day unsporulated oocysts are visible and sporulated oocysts or sporocysts are shed on the ninth day.

On the ingestion by cattle, sporozoites are released in the small intestine. Two or more generations of schizonts develop in extraintestinal organs, primarily in vascular endothelial cells (tachyzoite formation). Young cysts with metrocysts develop in cardiac and striated muscle, and are evident within one month but are common only after two months. Fully formed cysts containing bradyzoites develop within 76 days (Fayer & Johnson 1973). Acute disease occurs in cattle infected with *S. cruzi*. (See Sarcocystosis of cattle, p. 682.)

Sarcocystis fayeri. The intermediate host is the horse, cysts occurring in the cardiac and striated muscles. *Sporocysts:* 12 μm by 18 μm are passed in the faeces of dog. Developmental cycle unknown.

Sarcocystis miescheriana. The intermediate host is the pig. Infection is common in swine. The definitive host is the dog which sheds sporulated sporocysts (13 μm × 10 μm) in the faeces. This species is not known to be pathogenic.

Sarcocystis ovicanis. The intermediate host is the sheep. The parasite is common in sheep throughout the world and *S. ovicanis* is highly pathogenic for lambs in which it produces cysts microscopic in size. Sporulated sporocysts shed in the faeces of the dog are 13.1–16.1 μm by 8.5–10.8 μm (mean 14.8 μm by 9.9 μm). The prepatent period is eight to nine days. In sheep schizonts (tachyzoites) are found in endothelial cells of almost all organs. Cyst wall is 2–5 μm and radially striated. (See Sarcocystosis of sheep, p. 684.)

Sarcocystis hemionilatrantis. The intermediate host is the mule deer, in which the parasite may cause a fatal disease. Definitive hosts are dogs and coyote which shed sporulated sporocysts in their faeces 11 days after ingestion of tissues of infected mule deer. Sporocysts measure 13.8–16.1 μm by 9.2–11.5 μm (mean 14.4 μm by 9.3 μm). In mule deer, when fawns were fed sporocysts, schizonts were found in various tissues, especially around blood vessels in skeletal muscles.

Hoaresporidium pellerdyi. *Hosts:* Dog, India. *Oocysts:* 12–17 μm by 8.5–10.5 μm (mean 14.8 μm by 9.1 μm), ellipsoidal; no micropyle. Sporulation occurs in the intestine, four sporozoites developing without a sporocyst. A mild inflammation is seen in affected tissues.

COCCIDIA OF THE CAT (Fig. 3.29)

Isospora felis Wenyon, 1926
Isospora rivolta (Grassi, 1879) Wenyon, 1923
Besnoitia besnoiti (Marotel, 1912) Henry, 1913
Besnoitia darlingi Smith and Frenkel, 1977
Besnoitia wallacei Tadros and Laarman, 1976
Hammondia hammondi Frenkel and Dubey, 1975
Sarcocystis bovifelis Heydorn, Gestrich, Melhorn and Rommel, 1975
Sarcocystic gigantea (Railliet, 1886)
Sarcocystis hirsuta Moulé, 1888
Sarcocystis muris Blanchard, 1885
Sarcocystis ovifelis Heydorn et al., 1975
Sarcocystis porcifelis Dubey, 1976
Sarcocystis tenella Railliet, 1886
Toxoplasma gondii Nicolle and Manceaux, 1908

Recent consideration of the confused taxonomy of the Sarcocystidae by Frenkel et al. (1979) indicates that a number of *nomina dubia* occur (e.g. *S. hirsuta*, *S. miescheriana*, *S. tenella*, *S. cruzi*, *S. bertrami*, *Isospora bigemina* (*S. bigemina*), *I. hominis* (*S. hominis*)). Consequently, the above list of coccidia of cats may need revision if the proposals by Frenkel et al. (1979) are accepted.

Isospora felis. *Hosts:* cat, also lion, lynx and tiger; possibly other *Felidae.* World-wide in distribution. Developmental stages occur in the small intestine and occasionally the large intestine.

For many years it has been accepted that *I. felis* occurs in both cat and dog. Work by Nemeséri (1960) indicated that the dog form of *I. felis* could not be transmitted to the cat and consequently he suggests it be renamed *I. canis* (see above). Levine and Ivens (1965) concur with this view. *Oocysts:* ovoidal, largest of the *Isospora* species in cats. 38–51 μm by 27–29 μm. Oocyst wall, smooth, possibly pinkish in colour; no micropyle. Sporocysts 20–26 μm by 17–22 μm. Sporulation time, 72 hours.

Cats and non-feline hosts (rodents, dogs, birds) become infected by ingesting sporulated oocysts. Thus, generalized infection occurs in the mouse (transport host). There is only limited multiplication. Cysts occur chiefly in the mesenteric lymph nodes, they contain only one bradyzoite and may remain viable for at least 23 months (Frenkel & Dubey 1972).

Ingestion of sporulated oocysts, or cysts in a transport host, by a cat, leads to three generations of schizonts and gamonts in the intestinal wall (Shah 1970), although extraintestinal forms may occur in the cat. The first generation schizonts are 11–30 μm by 10–23 μm and produce 16–17 banana-shaped merozoites. Each form second generation schizonts which produce two to ten

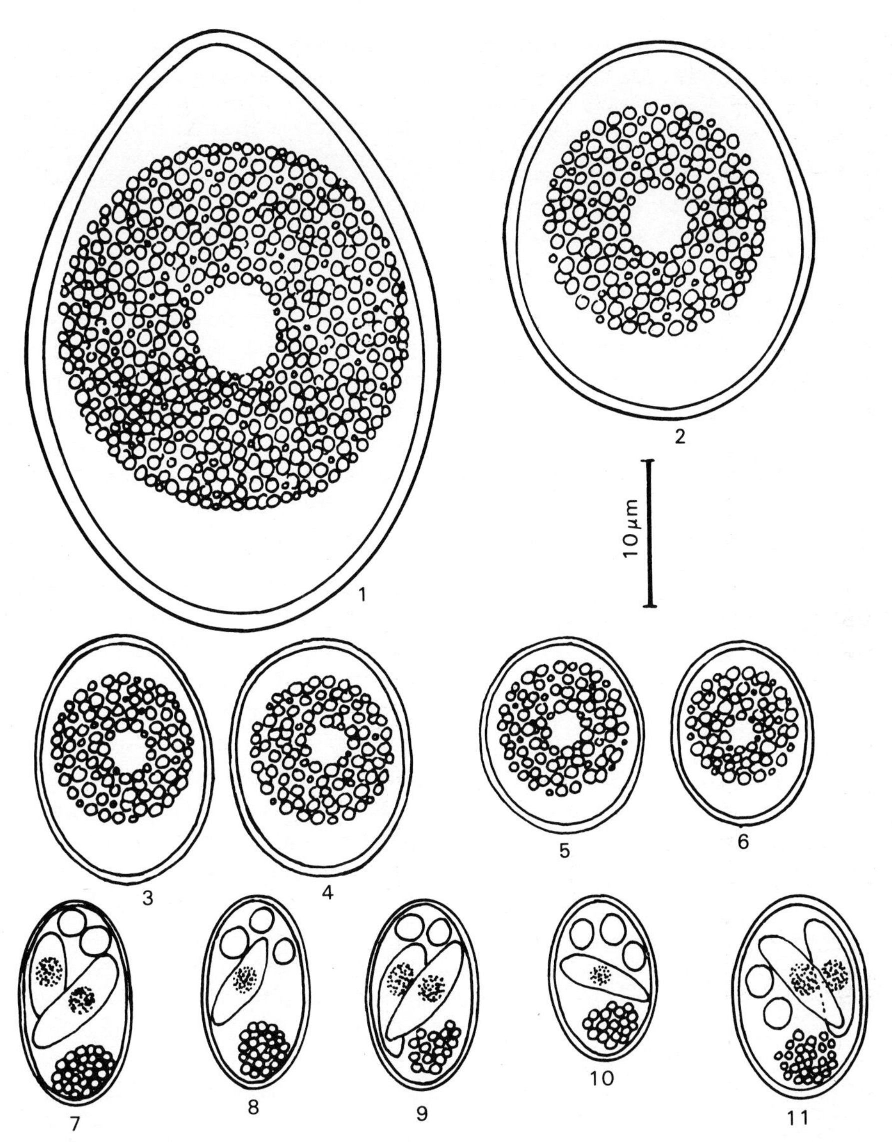

Fig 3.29 Coccidia of the cat. Oocysts in fresh faeces. (1) *Isospora felis.* (2) *I. rivolta.* (3) *Besnoitia besnoiti.* (4) *Besnoitia* sp. (probably *B. wallacei*). (5) *Hammondia hammondi.* (6) *Toxoplasma gondii.* Sporocysts in fresh faeces. (7) *Sarcocystis porcifelis.* (8) *S. hirsutal.* (9) *S. tenella.* (10) *S. muris.* (11) *Sarcocystis* sp. from Grant's gazelle (*From Dubey 1976*)

merozoites and third generation schizonts form while within the second generation schizont. Sexual stages occur on the seventh to eighth days of infection, mainly in the small intestine but some are also found in the caecum. The prepatent period is seven to eight days (Hitchcock 1955; Nemeséri 1960) and patency lasts 10–11 days.

Pathogenesis. Isospora felis appears to be fairly benign under natural circumstances. Andrews (1926) was able to kill young cats by feeding 100 000 sporulated oocysts, Hitchcock (1955) was unable to produce severe clinical signs in four- to nine-week-old kittens with 100 000 oocysts and, similarly, Dubey and Streitel (1976) were unable to produce clinical disease in parasite-free cats fed 100 000 sporulated oocysts. Dubey and Streitel (1976) infected cats with either 1–100 000 oocysts or with the tissues of mice which had been fed 1–100 000 oocysts. Cats remained normal. However, Tomimura (1957) reported diarrhoea, anorexia, emaciation and death in experimental cats. The pathological changes consist of a catarrhal enteritis in mild cases, to a haemorrhagic enteritis with heavy infections.

Isospora rivolta. This occurs in the small intestine of the cat. It is not transmissible to the dog. The oocyst is the second largest in the cat, it is ellipsoidal, 21–28 μm by 18–23 μm (mean 25 μm by 20 μm). A micropyle is present. Sporulation time is four days. The developmental cycle is similar to that of *I. felis*. Schizonts and gamonts occur in the epithelial cells of the small intestine, but may be found in subepithelial tissues. Prepatent period, seven days.

Mice, rats, dogs and chickens can act as experimental intermediate hosts (Dubey 1977) but their role in the normal life-cycle has yet to be clarified.

Under natural conditions, this species is nonpathogenic.

Besnoitia besnoiti. The intermediate hosts of this species are cattle, producing besnoitiosis or olifantvel (elephant hide). It is an important entity in Africa, Asia, southern Europe, USSR and South America. It has not been reported in the USA. The oocysts are passed in the faeces of cats, and are 14–16 μm by 12–14 μm.

Intermediate hosts are infected by ingestion of sporulated oocysts, but also by blood-sucking flies and by the parenteral inoculation of blood during the acute stage of infection. In the bovine, cysts are found in the skin, especially in fibroblasts; they are large and contain numerous bradyzoites. (See Besnoitiosis, p. 686.) Extraintestinal infection in the cat has not been observed.

Besnoitia darlingi. This species was isolated from the opossum by Smith and Frenkel (1977). Tissue cysts fed to cats produced oocysts 11.9 by 12.3 μm, which sporulated in 48–72 hours. The infection appears to be of low pathogenicity for cats, though acute lethal infections may occur in mice injected with tissue stages from opossums.

Besnoitia wallacei. Intermediate hosts are rats and mice. Tissue cysts, occurring in fibroblasts, reach a diameter of up to 200 μm in three months. Final host is the cat in which schizogony, gametogony and oocyst development occur in the intestine. *Oocysts:* 12 μm by 17 μm, shed unsporulated and at sporulation contain two sporocysts, each of which contains four sporozoites. Prepatent period is 12–15 days and patency lasts for 5–12 days. Only one generation of schizonts appears to occur in the cat and gametocytes develop in the goblet cells of the small intestine. (See Besnoitiosis.)

Hammondia hammondi. *Host:* cat. The experimental intermediate hosts are mice, rats, hamsters, guinea-pigs and dogs; the natural intermediate host is not known. In mice, tachyzoites multiply in cells of the lamina propria muscles and Peyer's patches, and during the second week of infection cysts appear in the skeletal muscle. *Oocysts:* 11–13 μm by 10–12 μm, unsporulated when shed by cats. The prepatent period is five to ten days and patency lasts one to two weeks, although infection may persist in the intestine of cats for up to 85 days (Dubey 1976). Extraintestinal forms do not occur in the cat.

There is no evidence that *H. hammondi* is pathogenic in the cat.

Levine (1977) has suggested that *H. hammondi* is a species of the genus *Toxoplasma* and has renamed it *Toxoplasma hammondi.*

Sarcocystis bovifelis (syn. *S. hirsuta*) is the sarcocyst of cattle with a cattle–cat cycle (Frenkel et al. 1979), corresponding to *S. bovicanis* (cattle–dog) and *S. bovihominis* (cattle–man). *Hosts:* domestic and feral cat. Sporocysts: 12.5 by 7.8 μm. The prepatent period is seven to nine days. It is not pathogenic for the cat and only slightly so for cattle. Sarcocysts occur in striated muscle and appear radially striated.

Sarcocystis gigantae. Frenkel et al. (1979) consider this a synonym of *S. tenella* and in turn of *S. ovifelis.* This is in accord with Levine (1977).

Sarcocystis hirsuta. This is considered to be a synonym of *S. bovifelis* (see above).

Sarcocystis muris. The intermediate host is the house mouse and other rodents in which elongated muscle cysts, up to several millimetres in length, occur. Sporocysts 12 μm by 7.5–9 μm are passed in the faeces of cat. There is no evidence that this species is pathogenic to either the definitive or the intermediate hosts.

Sarcocystis ovifelis is recognized by Frenkel et al. (1979) as the sarcocyst with the sheep-cattle cycle. It is synonymous with *S. tenella*, which Levine (1978) considers pathogenic for lambs.

Sarcocystis porcifelis. The intermediate host is the pig in which infection is common (Levine 1973) and probably pathogenic (Dubey 1976). This species was created by Dubey (1976) who separated three species (*S. porcifelis*, *S. porcihominis*, *S. miescheriana*) from the previously recognized simple species *S. miescheriani* of the pig. Sporocysts 13 μm by 8 μm are passed in the faeces of the cat, and the prepatent period is five to ten days.

Sarcocystis tenella. Dubey (1976) states that *S. ovifelis* is a synonym of this, as does Levine (1977). However, Levine (1977) also synonymizes this species with *Balbiania gigantea* which has become *S. gigantea*, the species with the sheep-cat cycle. *Sarcocystis gigantea* would then be a synonym of *S. ovifelis*. *Sarcocystis tenella* is recognized by Levine (1977) as a parasite of the cat with the sheep as intermediate host, in which it causes microscopic sarcocysts. Levine (1978) considers it pathogenic for lambs. In cats, sporocysts 13–16 μm by 8.5–11 μm are passed in the faeces.

Toxoplasma gondii (Fig. 3.30). Over 200 species of intermediate hosts are known and these

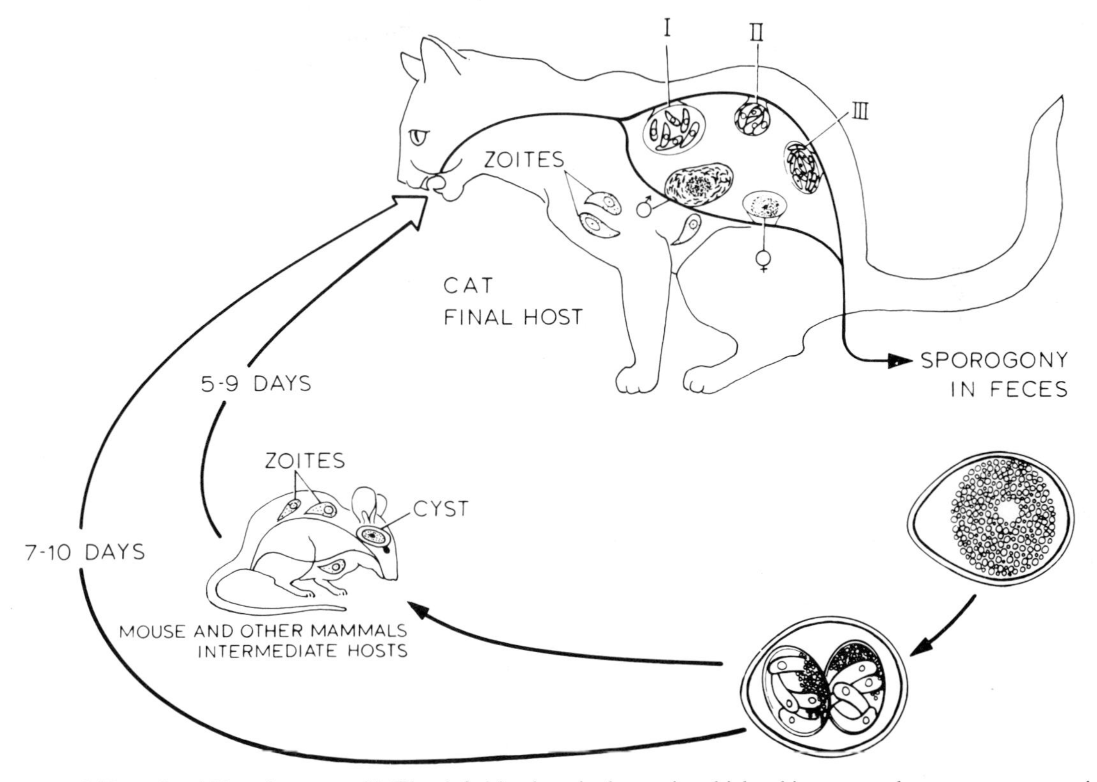

Fig 3.30 Life-cycle of *Toxoplasma gondii*. The definitive host is the cat in which schizogony and gametogony occur in the intestine. Cats may be infected by ingestion of sporulated oocysts, by ingestion of tachyzoites (in rodent) or by ingestion of bradyzoites in cyst. Infection in animals other than cats may be transmitted by ingestion of meat infected with tachyzoites or bradyzoites or transplacentally. (*From Dubey 1976*)

include mammals and birds. Tachyzoites, bradyzoites and sporulated oocysts (sporocysts) are infective. Pathogenicity for intermediate host varies from nil to severe. *Definitive hosts:* domestic cat and various wild *Felidae*. *Oocysts:* 11–13 μm by 9–11 μm. Sporulation occurs outside the host in 2–3 days. The prepatent period in the cat varies according to the stage ingested: three to five days after ingestion of bradyzoites from cysts; five to ten days after ingestion of tachyzoites; and 20–24 days after ingestion of oocysts.

This organism causes severe disease in intermediate hosts; more details of 'Toxoplasmosis' are given in the section on that subject (p. 670).

COCCIDIOSIS OF THE DOG AND CAT

Although coccidia can be readily detected in dogs and cats, there is no general agreement as to the pathogenicity of naturally occurring infections. Experimental evidence would suggest that none of the coccidia of dogs and cats are pathogenic for their final (definitive) hosts. However, there are clinical entities which strongly suggest that coccidiosis is an important cause of intestinal upset; these are seen especially in breeding or boarding kennels where sanitation is poor and animals are crowded together. Sometimes, frank blood may be passed in the faeces and large numbers of oocysts accompany this, or other diarrhoeic syndromes associated with coccidia.

Diagnosis

Diagnosis of coccidiosis is based on the clinical signs and the presence of very large numbers of oocysts in the faeces. Since young puppies and kittens are prone to various enteritides, care should be taken in the diagnosis of a condition as 'coccidiosis', especially since an enteritis due to other causes may release oocysts or sporocysts from the subepithelial tissues. Undoubtedly, a post mortem examination is the most adequate diagnostic method at which large numbers of developmental stages can be demonstrated in the mucosa.

Treatment

There has been little critical work in the treatment of canine and feline coccidiosis. The activity of the various sulphonamides against the genus *Isospora* is low, although McGee (1950) used sulphamethazine, and Brumpt (1943) found mepacrine at a dose of 0.01 g/kg effective for cats. Altman (1951) reported satisfactory results with aureomycin, and Parkin (1943) used an enema of 1% sodium sulphanilyl sulphanilate at the rate of 10 ml/kg repeated after 24 hours. Smith and Edmonds (1959) used nitrofurazone at the rate of 15.4 mg/kg body weight three times a day for ten days. Duberman (1960) gave three equally divided daily doses of 8.8–22 mg/kg body weight of nitrofurazone for an average of 9.3 days to 20 dogs, these being compared with 20 other dogs treated with a combination of sulphonamides consisting of 0.166 g of sulphamezathine, 0.166 g of sulphathiazole, and 0.166 g of sulphamethazine per lb. Although nitrofurazone was effective for *I. felis* (*sic*), it did not show any significant advantage over the combined sulphonamides. To some extent both these treatments may well be considered ineffective since some dogs took as long as 25 days to recover from the infection.

Dubey (1976) considers that there are no satisfactory drugs for either dog or cat coccidia at present.

COCCIDIA OF THE DOMESTIC FOWL

The coccidia of the domestic fowl are responsible for substantial losses to the poultry industry in various countries of the world. The following species have been reported from the domestic chicken:

- *Eimeria acervulina* Tyzzer, 1929
- *Eimeria brunetti* Levine, 1942
- *Eimeria hagani* Levine, 1938
- *Eimeria maxima* Tyzzer, 1929
- *Eimeria mitis* Tyzzer, 1929
- *Eimeria mivati* Edgar and Siebold, 1964
- *Eimeria necatrix* Johnson, 1930
- *Eimeria praecox* Johnson, 1930
- *Eimeria tenella* (Railliet and Lucet, 1891) Fantham, 1909
- *Cryptosporidium tyzzeri* Levine, 1961
- *Wenyonella gallinae* Ray, 1945

Eimeria tenella and *E. necatrix* are the most pathogenic and important species in domestic poultry. However, in recent years the practice of feeding premixed coccidiostats to poultry in large poultry-raising and production establishments has reduced the significance of these species and emphasized the importance of other species. *Eimeria acervulina, E. maxima* and *E. mivati* are common and slightly to moderately pathogenic; *E. brunetti* is uncommon, although markedly pathogenic when it does occur; and both *E. mitis* and *E. praecox* are relatively non-pathogenic and common. *Eimeria hagani* is only slightly pathogenic and is rare.

Eimeria tenella (Fig. 3.31). *Host:* One of the most common and pathogenic coccidia of domestic poultry. World-wide in distribution. Developmental stages occur in the caecum. *Oocysts:* broadly ovoidal, 22.9 μm by 19.16 μm, range 14.2–31.2 μm by 9.5–24.8 μm. Oocyst wall, smooth, no micropyle. Sporulation time, 18 hours at 29°C; 21 hours, 26–28°C; 24 hours at 20–24°C; 24–48 hours at room temperature; no sporulation below 8°C (Edgar 1955).

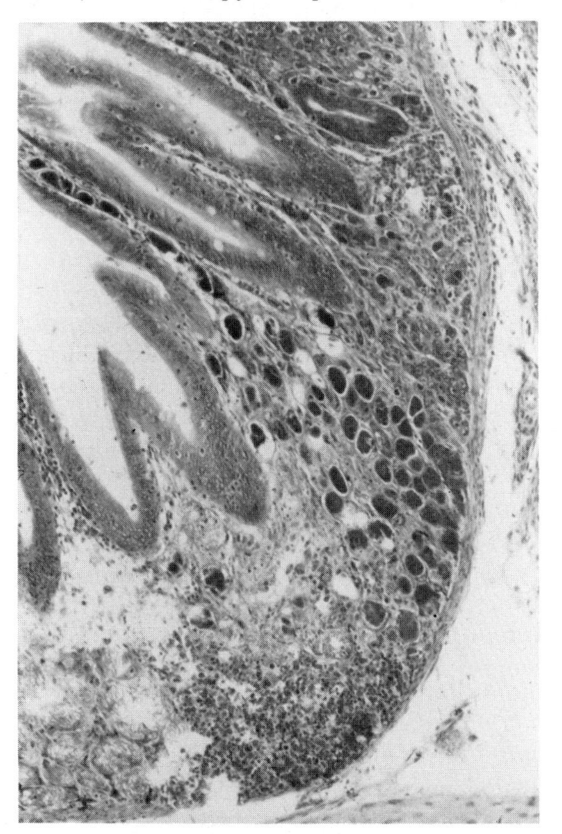

Fig 3.31 Caecal mucosa of chicken infected with *Eimeria tenella*. A colony of second generation schizonts is visible in a subepithelial position. (× 300)

Life-cycle. Excystation of sporulated oocysts has been discussed by Ryley (1973). Two separate stimuli are necessary. The first is caused by carbon dioxide (which may be replaced by mechanical rupture by the gizzard), and the second is provided by trypsin and bile in the small intestine. The liberated sporozoites invade the surface epithelium of the caecum, Pattillo (1959) having observed 'penetration tubes' in the striated border of the epithelium through which the sporozoites passed. They are engulfed by macrophages in the lamina propria and are transported in these to the glands of Lieberkühn. Here, they leave the macrophages and enter the epithelial cells lining the glands, developmental forms being found distal to the host cell nucleus (Challey & Burns 1959).

In the epithelial cell the sporozoite rounds up and becomes a trophozoite. Subsequent development has been described by Tyzzer (1929), and numerous authors have confirmed his original description of the life-cycle.

Mature first generation schizonts are found at the bottom of the crypts of the caecal glands; they measure 24 μm by 17 μm and the host cell is hypertrophied to several times its normal size so that it bulges into the lumen of the gland. Approximately 900 first generation merozoites, each 2–4 μm in length by 1–1.5 μm, are produced. The first generation schizonts rupture into the lumen of the gland about 60–72 hours after infection, and the merozoites penetrate other epithelial cells, round up and form the young second generation schizonts. These are found proximal to the host cell nucleus. The parasitized cell increases in size, breaks loose from its epithelial position and migrates to the subepithelial tissues where mature second generation schizonts are produced. Colonies of second generation schizonts are first apparent by 72 hours, and by 96 hours mature schizonts, up to 50 μm in diameter, but commonly 21–31 μm, are found. These contain 200–350 second generation merozoites, 16 μm by 2 μm. Disruption of the second generation schizonts and the overlying epithelium releases the merozoites into the lumen of the caecum, and when large numbers of second generation schizonts do this, a massive haemorrhage into the caecal lumen may be evident at about the ninety-sixth hour of infection.

Second generation merozoites penetrate new epithelial cells and initiate either a third generation of schizonts or the gametogonous cycle; the majority undertaking the latter process. Third generation schizonts lie distally to the cell nucleus; they are smaller than the previous stages, 9 μm by 7.6 μm, and produce 4–30 merozoites, each 6.8 μm by 1 μm.

The gametogonous stages appear initially as rounded trophozoites. Repeated nuclear division indicates the formation of the microgamonts which lie distal to the cell nucleus, as do the macrogametocytes. The prepatent period is seven days, oocyst production rising to a peak by the tenth day and then rapidly decreasing.

The ultra-structural details of the developmental cycle have been presented by Scholtyseck (1973) and have been summarized by Pellérdy (1974).

In vitro cultivation of *E. tenella* and other coccidia has been reviewed by Doran (1973). *Eimeria tenella* has been cultured in a variety of cell types to various stages of development from the sporozoite stage.

Pathogenesis. Caecal coccidiosis due to *E. tenella* most frequently occurs in young birds, especially those aged four weeks. Gardiner (1955) reported that chicks of one to two weeks of age were more resistant. Nevertheless, it is possible to infect day-old chicks. Older birds are generally immune as a result of previous infection. In general, clinical caecal coccidiosis is produced only when heavy infections are acquired over a relatively short period of time, not exceeding 72 hours (Davies et al. 1963). Resistance to infection may be demonstrable by the ninety-sixth hour following infection, and in the absence of a pathogenic burden in the early stages, an adequate resistance may be acquired to prevent fatal effects (Kendall & McCullough 1952). The number of oocysts required to produce clinical disease has been investigated by Gardiner (1955) and others. In chickens aged one to two weeks, 200 000 oocysts were required to produce mortality, while 50 000–100 000 produced mortality in birds a few weeks older.

On a flock basis, coccidiosis first becomes noticeable at about 72 hours after infection. Chickens droop, cease feeding, huddle to keep warm, and by 96 hours blood appears in the droppings. The greatest haemorrhage occurs on the fifth or sixth day of infection, and by the eighth or ninth day the bird is either dead or on the way to recovery. Mortality is highest between the fourth and sixth days, death sometimes occurring unexpectedly due to excessive loss of blood. In birds recovered from the acute disease, a chronic illness may develop as a result of a persistent caecal core; however, this is usually expelled about 14 days after infection.

The pathological changes which have been discussed by Long (1973) are mainly due to the second generation schizonts. Petechial haemorrhages occur during the first three days, and noticeable lesions consisting of marked haemorrhagic spots appear on the fourth day. By the fifth or sixth day the caeca are dilated, the contents containing unclotted and partly clotted blood, schizonts and merozoites; from the seventh day onwards gametogonous stages are found in the mucosa. By this time, too, the caecal contents have become more consolidated and caseous and adherent to the mucous membrane, and by the eighth day the consolidated caseous plug completely fills the lumen of the caecum. The caecal core detaches from the mucous membrane by eight to ten days and may be shed in the faeces. At this time the caecal wall is still thickened but it has lost its intense haemorrhagic appearance and, following shedding of the core, regeneration of the mucosa occurs and the wall contracts, although a degree of fibrosis may remain for some time.

Immunity. The genetic aspects of resistance to caecal coccidiosis have been investigated by Champion (1954) and Rosenberg et al. (1954). Experimental matings involving resistant and susceptible F1 and F2 individuals showed that selective breeding was effective in establishing lines of chickens resistant or susceptible to caecal coccidiosis. Sex linkage, maternal effect or cytoplasmic inheritance did not play a significant role in this, and it was concluded that resistance or susceptibility to caecal coccidiosis is controlled in a large part by multiple genetic factors which do not exhibit dominance and presumably act in an additive manner. However, Long (1973) warns that if true genetic differences are to be examined, then embryos or very young chickens may not be suitable subjects, since maternally transferred antibodies may occur in chicks and, further, protective amounts of anticoccidial drugs may occur in eggs laid by treated birds.

A striking feature of the acquired immunity to *E. tenella* is its specificity; birds immune to *E. tenella* are completely susceptible to other *Eimeria* spp. The subject of acquired immunity to *E. tenella* and other coccidia has been reviewed in a most comprehensive and effective manner by Rose (1972, 1973). The following is a brief summary from her reviews. The second generation schizonts are the principal stages responsible for the induction of the immunity. Various techniques have been used to determine this including abbreviated infections using therapeutic agents and the manipulation of the developmental cycle. Thus, the introduction of sporozoites into the rectum to produce second generation schizonts and the sexual cycle induced a high degree of resistance, but the introduction of second generation merozoites to produce sexual stages and oocysts induced a much lower level of immunity.

The sporozoite is affected by the immune response very shortly after penetration of the epithelium and few or no first generation schizonts develop from a challenge infection in resistant birds. Protective substances are not present in the lumen of the caecum in sufficient concentration to affect the free sporozoites. A similar immune mechanism probably operates against the second generation merozoites as against the sporozoites. Merozoites may be found deep in the glands, but they fail to develop to gamonts.

Infectious bursal disease (IBD) is an immunosuppressive viral infection of chickens and infection with IBDV frequently renders chickens more susceptible to various infections. In chickens infected with coccidia, a significantly higher mortality occurred in those infected with IBDV compared to chickens without infection (Giambrone et al. 1977).

Epidemiology. Caecal coccidiosis is primarily a disease of young chickens; older birds, although immune, act as carriers. Under ordinary farm or poultry rearing conditions, it is likely that all birds are exposed to infection, but the severity of disease is dependent, very largely, on the number of oocysts ingested. Clinical disease arises only when heavy infection, in relation to previous experience, is acquired over a period of time not exceeding 72 hours (Davies et al. 1963). Infections acquired more slowly than this tend to result in resistance rather than clinical disease. Consequently, the majority of a flock is likely to acquire a clinical infection over a short period of time, and Davies and Kendall (1954*a*) estimate this period to be ten days.

Diagnosis, treatment and control of *E. tenella* coccidiosis will be dealt with under the heading 'Coccidiosis of poultry'.

Eimeria necatrix. *Host:* domestic fowl. World-wide in distribution, extremely common. Asexual development in the small intestine, gametogony cycle in the caecum. This species is one of the most important pathogens of the small intestine of poultry. *Oocysts:* similar to those of *E. tenella*, ovoidal, 16.7 μm by 14.2 μm, range 13.2–22.7 μm by 11.3–18.3 μm. (Davies, 1956, gives a mean of 20.5 μm by 16.8 μm, range 15.5–25.3 μm by 13.6–20.4 μm.) Oocyst wall, smooth, colourless, no micropyle. Sporulation time, two days, but may be 18 hours at 29°C (Edgar 1955).

Life-cycle. The initial behaviour of sporozoites is similar to that of *E. tenella*. Van Doorninck and Becker (1957) observed that sporozoites passed through the epithelium of the tips of villi into the lamina propria and migrated towards the muscularis mucosa. During this migration most of the sporozoites were engulfed by macrophages and were carried in them to the epithelial cells of the fundus of the crypts of Lieberkühn. The first generation schizonts occur proximal to the host cell nucleus; merozoites appear in the gland lumen two to three days after infection, and they enter adjacent epithelial cells and develop into second generation schizonts. This stage of *E. necatrix* is relatively large, and the epithelial cells containing the developing schizont leave their epithelial position and migrate into the subepithelial tissues and sometimes into the submucosa. They are evident here from the fourth day of infection onwards, developing in a colonial manner (Tyzzer et al. 1932). The second generation schizonts of *E. necatrix* are relatively large, 63 μm by 49 μm, this serving to differentiate them from the schizonts of the other species of coccidia which occur in the small intestine. The second generation merozoites are liberated on the fifth to eighth days of infection, although a few may be released for several days thereafter. They are then carried by the peristaltic action of the small intestine to the caecum. Here, they penetrate the epithelial cells and may undergo a further generation of schizogony or differentiate to the gametogonous cycle. The third generation schizonts are small, and multiple infection of cells may occur, three to four third generation schizonts being found in a single host cell. Gametogonous stages may arise from either second or third generation merozoites. They lie distal to the host cell nucleus, displacing it and they may also grossly distort it.

The prepatent period, according to Tyzzer et al. (1932), is seven days, and to Davies (1956), six days. The peak of oocyst production occurs from the eighth to the tenth days.

Pathogenesis. Next to *E. tenella*, *E. necatrix* is considered to be the most common pathogenic species of *Eimeria* in domestic poultry. It tends to cause a more chronic disease than the former and affects older birds; however, disease can be produced in young chickens. The age at which chickens suffer coccidiosis due to *E. necatrix* depends on the biotic potential of the parasite (it is a poor producer of oocysts, and a longer time is necessary for a high level of environmental contamination to occur) and on the degree of immunity which results from light infections.

The principal lesions are found in the middle third of the small intestine. In acute cases a severe submucosal haemorrhage occurs on the fifth and sixth days, the wall of the small intestine is markedly swollen, haemorrhagic, and the contents filled with unclotted blood. The haemorrhages are associated with the large, deeply seated, second generation schizonts, and at times these may be seen as white, opaque foci surrounded by a zone of haemorrhage. Where there has been excessive haemorrhage, blood may also be found in the caeca so that the condition may be confused with *E. tenella* infection, although dual infections may be seen in natural outbreaks.

In mild infections, scattered white spots indicating the colonies of schizonts are surrounded by a zone of petechial haemorrhage, but there is no evidence of gross haemorrhage into the lumen. In contrast to *E. tenella* infection, birds which recover may remain emaciated for several weeks or months afterwards. The chronic form of the infection is a marked contrast to the acute type of disease seen in caecal coccidiosis.

Immunity. Immunity to *E. necatrix* has been reviewed by Rose (1972, 1973). Chickens surviving a severe infection with *E. necatrix* become sufficiently resistant to withstand a new fatal infection; however, this immunity is not as strong as that seen in *E. tenella* infection. Experimentally, it is possible, by repeated exposure, to immunize birds to the level where parasitism is prevented. The developmental stages associated with the induction of immunity are the asexual phases in the small intestine, the gametogonous stages having little immunizing power. The intrarectal inoculation of second generation merozoites causes no diminution in oocyst production with successive inoculations, and when such birds were challenged with oocysts, little or no immunity was demonstrated.

Eimeria acervulina (Fig. 3.32). *Hosts:* domestic poultry, quail; world-wide in distribution. Very common. It is less pathogenic than the two previous species and is responsible for subacute or chronic intestinal coccidiosis in older birds and chickens at the point of lay. Developmental stages occur in the anterior part of the small intestine. *Oocysts:* ovoid, 19.5 μm by 14.3 μm, range 17.7–22.2 μm by 13.7–16.3 μm (Tyzzer 1929).

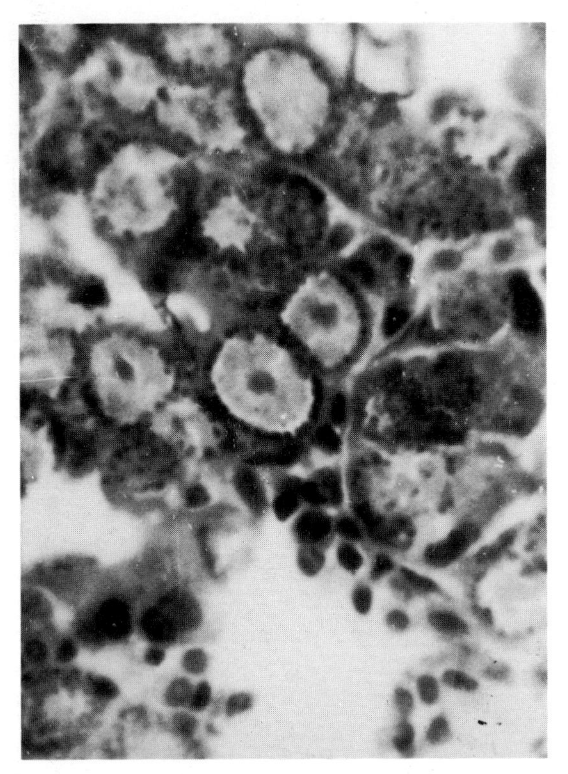

Fig 3.32 Small intestinal mucosa of chicken infected with *Eimeria acervulina*. Macrogametocytes are visible, central field. (×800)

Becker (1956) recorded a mean size of 16.4 μm by 12.7 μm. Oocyst wall, smooth, thinner at narrow end, inconspicuous micropyle present. Sporulation time, 25 hours at room temperature, 17 hours at 28°C (Edgar 1955).

Life-cycle. The asexual stages of *E. acervulina* are found proximal to the nucleus in the epithelial cells of the anterior small intestine. More than one parasite may be found within a single cell. Mature schizonts containing merozoites are seen by the third day of infection. Such schizonts produce 16–32 merozoites, each 6 μm by 0.8 μm, and these initiate another asexual cycle of development. More asexual cycles may occur before the gametogonous stages are produced. Gametogony occurs in the anterior small intestine from the fourth day of infection onwards. The prepatent period is four days, and oocyst production continues for a longer period than for the other chicken coccidia.

Pathogenesis. Tyzzer (1929) originally regarded this species as a producer of chronic inflammation of the small intestine; later (1932), however, he concluded that it was of little or no pathogenic significance. Bracket and Bliznick (1950) were unable to kill chickens with any level of infection, but found 500 000 oocysts reduced weight gains of two-week-old chickens; Morehouse and McGuire (1958), using chickens of four to ten weeks, found that single or multiple doses of five million or more sporulated oocysts produced a 75% mortality, and this was confirmed by Horton-Smith and Long (1959), who produced death in eight-day-old chickens with a total of 10 million oocysts in three doses over three successive days. Within recent years, especially in large poultry establishments, the significance of *E. acervulina* as a pathogen has increased steadily. The biotic potential of this species is great, the prepatent and sporulation periods are short, and very large numbers of oocysts may accumulate in the environment.

The clinical signs consist of weight loss and a watery, whitish diarrhoea. At post mortem greyish white, pin-point foci or transversely elongated areas are visible from the serous surface of the duodenum; these consist of dense foci of oocysts and gametogonous stages and, when examined microscopically, excessive numbers of sexual stages and unsporulated oocysts are found to be present. The intestinal wall and mucosa are thickened and may be covered with catarrhal exudate, but haemorrhage is rare except when excessive numbers (several million) of oocysts are administered.

Eimeria maxima. *Hosts:* domestic poultry. World-wide in distribution. Common. Developmental stages in the small intestine. *Oocysts:* large ovoidal, 29 μm by 23 μm, range 21.4–42.5 μm by 16.5–29.8 μm. Oocyst wall, slightly yellow, some may be roughened, micropyle absent. Sporulation time, two days.

Life-cycle. Accounts of the developmental cycle have been given by Tyzzer (1929), Long (1959) and Scholtyseck (1959, 1963). Sporozoites enter the epithelial cells of the tips of the villi in the duodenum and schizonts develop proximal to the nucleus. There are two generations of schizonts: both types are small, 10 μm by 8 μm, and the second generation schizonts produce 8–16 merozoites by the fourth day. Gametogonous stages are found distal to the epithelial cell nucleus; they are larger than the asexual stages and, as they enlarge, the parasitized cell may move to a subepithelial position, some reaching the muscularis mucosa. The first oocysts may appear 120 to 121 hours after infection, but large numbers usually occur between 123 and 136 hours of infection (Long 1959). Oocyst output lasts for a few days.

Pathogenesis. This species is moderately pathogenic, the most serious effects being due to the sexual stages. Few marked changes occur in the small intestine until the fifth day after infection, and then in severe infections numerous petechial haemorrhages occur on the intestinal wall. There is a marked production of mucus, the mucosa is thickened, there may be a loss of tone and the intestine becomes flaccid and dilated. The mucosal surface is inflamed and the intestinal contents consist of a pinkish mucoid exudate. Microscopic examination of scrapings of the mucosa reveals large numbers of oocysts.

Immunity. The onset of immunity in *E. maxima* infection is quick, resulting in a rapid termination of the infection. Of the poultry species studied by Rose and Long (1962), *E. maxima* possessed the major immunizing power. No information is available on developmental stages which are responsible for the immunizing effect.

Eimeria mivati. *Hosts:* domestic fowl. USA and Canada but probably world-wide. Developmental stages in small intestine but may extend from duodenum to rectum. *Oocysts:* ellipsoidal to broadly ovoidal, 15.6 μm by 13.4 μm, range 10.7–20.0 μm by 10.1–15.3 μm. Oocyst wall, colourless, smooth, micropyle present. Sporulation time, 11–12 hours at 29°C.

Life-cycle. The developmental cycle was described by Edgar and Siebold (1964) and by Long (1967*a*). Four asexual generations are undergone. Sporozoites develop in the epithelial cells at the bases of villi, particularly of the duodenum. They are located just below the surface and well above the nucleus of the host cell. Mature first generation schizonts, measuring 10.4 μm by 10.1 μm, are produced about 36 hours after infection; 10–30 merozoites are produced. Second generation schizonts develop in a similar situation and are mature between 55 and 65 hours after infection. These are 9.2 μm by 7.2 μm and produce 16–20 merozoites. The second generation merozoites parasitize cells of the jejunum, ileum, caeca and rectum in addition to those of the anterior small intestine. Mature third generation schizonts are produced about 80 hours after infection. Some third generation merozoites develop into sexual stages, but others may develop into fourth generation schizonts. All fourth generation merozoites, along with some of the third generation merozoites, enter epithelial cells anywhere from the duodenum to the rectum, infection being most marked in the anterior segments. Mature oocysts appear 93–96 hours after infection. Peak oocyst production occurs during the fifth to seventh days after infection.

Pathogenesis. Marked changes of the anterior small intestine appear by the fourth day when the chickens show listlessness, anorexia and a watery diarrhoea. The affected area of the intestine is swollen, oedematous and shows scattered petechiae. Death, if it occurs, takes place on the sixth or seventh day. Depression of growth, a drop in egg production and impaired food conversion are seen in growing and laying birds.

Immunity. A strong immunity is developed with *E. mivati* which is species-specific. This can persist for at least three months.

Eimeria mitis. *Hosts:* domestic poultry. World-wide distribution. Common. Developmental stages occur in the anterior small intestine, occasionally also in the posterior small intestine and caecum. *Oocysts:* sub-spherical, slightly tapering, 15.8 μm by 13.83 μm, range 11.5–20.7 μm by 10.35–18.4 μm (Joyner 1958) (Tyzzer, 1929, gives 16.2 μm by 15.5 μm). Sporulation time, two days at room temperature, 18 hours at 29 °C (Edgar 1955).

Life-cycle. This has been studied by Tyzzer (1929) and Joyner (1958). The early stages in the life-cycle have yet to be determined. Schizonts are demonstrable 67 hours after infection, being superficially situated in the epithelial cells throughout the length of the small intestine. Merozoites are found in the intestinal contents about four days after infection.

Gamont formation is increasingly present from the fifth day of infection onwards, and sexual stages are numerous by the eighth day. Both the asexual and the sexual stages of the cycle occur together. The prepatent period is about 100 hours.

Pathogenesis. Under normal conditions, the species is, at the most, only a mild pathogen. Tyzzer (1929) was unable to produce clinical signs or lesions in chickens with large doses of sporulated oocysts. However, Joyner (1958) produced 38% mortality in six-day-old chicks following the administration of 2.5 million oocysts. Half a million oocysts reduced weight gains of chickens aged 6–26 days. The pathological changes associated with *E. mitis* are minimal and are not characterized by any visible haemorrhage.

The immunity developed to this parasite is of a low order. Both Tyzzer (1929) and Joyner (1958) noted that chickens could be infected several times before a decrease in the susceptibility of the host occurred.

Eimeria brunetti. *Hosts:* domestic poultry. USA, Europe, New Zealand, probably worldwide. Sporadic. Developmental stages in the small intestine, caecum and cloaca. *Oocysts:* ovoid, 26.8 μm by 21.7 μm, range 20.7–30.3 μm by 18.1–24.2 μm (Levine 1942) (Becker et al. 1955, record 23.4 μm by 19.7 μm, range 13.8–33.7 μm by 12.4–28.3 μm). Oocyst wall, smooth, micropyle absent. Sporulation time one to two days at room temperature, 18 hours at 24°C (Edgar 1955).

Life-cycle. The endogenous development has been described by Boles and Becker (1954), Pellérdy (1960) and Davies (1963). First generation schizonts develop in the epithelial cells of the villi, especially at points in contact with, or close to, the basement membrane. The upper small intestine is the major site of parasitism, but stages may also occur in the lower part and even in the caeca. Mature first generation schizonts, 30 μm by 20 μm, containing approximately 200 merozoites, are found 50–76 hours after infection. By the fourth day, second generation schizonts occur in the posterior part of the small intestine, the rectum, the caeca and the cloaca. Two types of second generation schizonts have been described: large ones, 29.6 μm by 16.2 μm, containing 50–60 merozoites, are usually found at the tips of the villi at 95 hours, and a smaller type, 9.8 μm by 8.8 μm, containing 12 merozoites, may also be seen. Boles and Becker (1954) suggest that the size difference may be caused by crowding; alternatively, it may indicate a third generation of schizonts or sexual dimorphism. Schizonts may enter the subepithelial tissues in heavy infections, but usually they are confined to the epithelium of the villus. Gamonts occur in the epithelial cells of the lower intestine, caecum, rectum and cloaca. The prepatent period is five days.

Pathogenesis. Eimeria brunetti may cause severe disease in chickens between four and nine weeks of age. The lesions are characteristically confined to the posterior part of the small intestine, between the yolk stalk and the caeca, and the condition is typically a rectal coccidiosis. In severe infections the gut wall is thickened, there is a haemorrhagic catarrhal exudate, which appears four to five days after experimental infection, and the droppings are fluid and may be blood-stained. The haemorrhagic condition may become necrotic, due principally to asexual stages developing in the subepithelial tissues.

Allen et al. (1973) have studied some of the pathophysiological changes associated with this infection in chickens. During the acute phase of infection of 16-week-old chickens infected with 3.2×10^5 or 1.28×10^6 oocysts, a dose-dependent decline in total protein, sodium and chloride concentrations occurred, potassium levels increased, while the packed cell volume increased initially and then decreased. Mortality was 10% and 45%, respectively in the two dosage schedules. A progressive alteration in plasma proteins, electrolytes and the packed cell volume were attributed to the severe enteritis produced by the infection. The return of plasma electrolytes to normal concentrations corresponded to the time of resolution of intestinal lesions. Rapid and severe dehydration appeared to be a significant factor in the loss of body weight, while rapid rehydration contributed to the rapid recovery of weight. Complete inappetence, demonstrated by severely affected chickens, indicated that reduced food consumption is an important factor in weight loss in coccidiosis.

Eimeria hagani. *Hosts:* domestic poultry. North America, Europe, India. This species was differentiated from the other species of the chicken by cross immunity tests. Developmental stages in the small intestine. Generally regarded as of little or no pathogenecity. *Oocysts:* broadly ovoid, 19 μm by 18 μm, range 16–21 μm by 14–19 μm (Levine 1938). (Edgar, 1955, reported a mean of 18.1 μm by 16.5 μm.) No micropyle. Sporulation time, one to two days (Edgar 1955).

Life-cycle. The endogenous developmental cycle is not known in detail but occurs in the anterior part of the small intestine. The developing stages cause small petechial haemorrhages which are visible from the serous surface. The prepatent period is six days (Edgar 1955). The patent period is eight days.

Levine (1942) fed large numbers of *E. hagani* to ten-week-old chickens and reported a catarrhal inflammation of the duodenum on the sixth day of infection.

Eimeria praecox. *Hosts:* domestic poultry. World-wide. Developmental stages in the upper part of the small intestine. *Oocysts:* ovoid, 21.2 μm by 17 μm, range 19.7–24.7 μm by 15.6–19.7 μm. Oocyst wall, smooth, colourless, no micropyle. Sporulation time, two days at room temperature.

Life-cycle. The endogenous developmental cycle has been studied by Tyzzer et al. (1932). Long (1967*b*) and Lee and Millard (1971). At least three generations of schizonts occur. First generation schizonts give rise to 16 merozoites, while those of the second generation produce 16–32 merozoites. Gametogony commences about 80 hours after infection but schizogony is still in progress at this time and asexual and sexual stages are seen together. The prepatent period is four days, and the infection remains patent for about four days.

The pathogenicity of this species is low, it being regarded as a non-pathogenic form (Long 1968). Tyzzer et al. (1932) were unable to produce gross changes in the intestines with heavy infections of oocysts. Nevertheless, despite the lack of marked pathogenesis, immunity to infection develops quickly.

Cryptosporidium tyzzeri. *Hosts:* domestic chicken. North America (Massachusetts). Rare. All developmental stages occur extracellularly on the microvilli of the epithelial cells of the tubular part of the caecum. There is no obvious pathogenecity associated with this form. *Oocysts:* ovoid, small 4–5 μm by 3 μm. Contain four naked sporozoites. During the life-cycle minute schizonts 3–5 μm in diameter are attached to the surface of the cell; these produce eight merozoites which then produce small micro- and macrogamonts.

Wenyonella gallinae. *Hosts:* domestic poultry. India. This species was reported by Ray (1945) from the intestine and caecal contents of four- to six-week-old chickens which had died of an outbreak of coccidiosis in India. Developmental stages occur in the terminal part of the intestine. *Oocysts:* oval, 29.4–33.5 μm by 19.8–22.7 μm. Oocyst wall, thick, the outer layer having a rough coat. Sporulation time, four to six days.

Life-cycle. The endogenous developmental cycle is unknown. In experimental infections Ray (1945) found the prepatent period to be seven to eight days, oocyst production continuing for three days.

Infected chickens passed blackish green semi-fluid faeces, but no marked lesions were noted except in natural outbreaks.

PATHOPHYSIOLOGICAL CHANGES IN AVIAN COCCIDIOSIS

In addition to the pathogenic effects mentioned for the individual species of coccidia (see above), general effects include changes in the cellular kinetics and morphology of the villi, changes in the intestinal pH and, associated with this, decreased absorption of vitamin A and carotenes resulting in lowered blood carotenoid levels.

Pout (1968) observed a time relationship between host reaction and life-cycle of the parasite in that the most severe changes in villus height to total mucosal thickness ratio (VH/TMT), surface area, morphology and enzyme activity occurred when maximum host–parasite cell interaction was evident. Infection with *E. acervulina* resulted in no change of VH/TMT during the first generation of schizogony, but subsequent developmental stages did cause a reduced ratio. An overall reduction in surface area, a reduction in goblet cells and a reduction in alkaline phosphatase were features of the infection.

A generalized increase in intestinal acidity has been reported in chickens infected with various species of coccidia. Ruff and Reid (1975) showed that the greatest and most consistent reduction in pH occurred in the region of the intestine where the particular species characteristically produces the severest infection. Kouwenhoven and van der Horst (1972) have reported that lowered pH decreases absorption of vitamin A and carotenes by the intestine (at least for *E. acervulina*), and Ruff et al. (1974) have demonstrated that with any one of the six major species of avian coccidia infection led to lower levels of blood carotenoids than in control birds. Carotenoids decreased as early as four days after infection but returned to normal by 14 days. Maximum decreases occurred seven to eight days after infection, being 49% with *E. tenella* and 62–74% for the intestinal species. The magnitude of the decrease was related to the number of oocysts given.

COCCIDIOSIS OF POULTRY

Infections with a single species of coccidium are rare in natural conditions, mixed infections being the rule; nevertheless, in many outbreaks the clinical entity can be ascribed principally to one species, or occasionally a combination of two or three. *Eimeria tenella* is the most pathogenic and important species, followed by *Eimeria necatrix*. Many coccidiostatic drugs have been directed against *E. tenella*, with the result that other species are increasingly incriminated as a cause of poultry coccidiosis. *Eimeria brunetti* may be markedly pathogenic but is uncommon. *E. maxima* and *E. acervulina* are increasingly common but are of moderate pathogenicity, and *E. mitis* and *E. praecox* are considered to be less pathogenic than the others.

Coccidiosis should be regarded as ubiquitous in poultry management, since even under the extreme conditions of experimental work, it is difficult to avoid infection completely for any length of time. Essentially, the clinical disease entity is dependent on the number of oocysts ingested by individual birds. If the environmental hygiene is poor, this number may be very large and this is particularly so with *E. tenella* which has a high biotic potential. Where young birds are placed on heavily contaminated litter, deaths may occur within a few days and up to 100% of the flock of chickens may die. In order to produce severe and fatal caecal coccidiosis, the intake of the pathogenic level of oocysts must take place within 72 hours, otherwise a rapidly developing immune response will protect against a fatal infection. Such conditions frequently obtain in poorly maintained litter houses and in broiler systems. However, caecal coccidiosis may also occur in 'free-range' birds, especially if they frequent shaded and damp areas to feed.

The environment is being contaminated continuously, even from immune birds, and the initiation of an outbreak depends upon factors which allow oocysts to sporulate and remain viable. For sporulation, oocysts require moisture and warmth and survive best in shaded, moist conditions. Poorly maintained litter houses may well supply such needs, and excessive numbers of sporulated oocysts may be found in poorly kept quarters.

The resistance which is developed from a previous infection will protect birds from subsequent exposure to that species, but it is a specific resistance and does not induce immunity to another species. Thus, for example, an attack of caecal coccidiosis does not preclude subsequent disease caused by *E. necatrix* or any of the other species.

Various measures can be adopted to avoid the intake of large numbers of oocysts by susceptible poultry. In litter houses the removal of the litter may be timed so that the majority of oocysts are removed before they have sporulated (e.g. every two days); however, under modern conditions of poultry husbandry, this is completely uneconomical and impossible. Steps may be taken to keep litter dry so that oocysts do not sporulate, and litter may be redistributed frequently to avoid concentrations of oocysts at places such as feeding or watering troughs, etc. It is important that watering appliances should be effective and do not allow localized areas of dampness to occur. When broiler houses are emptied for a new batch of chickens, the litter should be stacked so that the heat evolved is sufficient to kill oocysts. Heaped litter, left for 12 hours or more, will normally generate a temperature of about 51°C, which is sufficient to destroy the oocysts. Poultry in outside pens should be moved regularly to other pens and the contaminated pen left vacant or, better still, brought under cultivation.

Diagnosis

Diagnosis of coccidiosis in chickens is best accomplished by a post mortem examination of a representative number of birds. Diagnosis by faecal examination may lead to quite erroneous results. In some instances the major pathology is produced before oocysts are shed in the faeces (e.g. *E. tenella*) and, conversely, the presence of large numbers of oocysts may not necessarily indicate a serious pathogenic condition. Thus, with *E. acervulina*, which has a high biotic potential, comparatively larger numbers of oocysts are shed per oocyst given than, for example, with *E. necatrix*. Furthermore, the accurate identification of the oocysts of various poultry coccidia is not easy.

All this may be avoided by a post mortem examination. The location of the major lesions gives a good indication of the species of coccidia concerned (Fig. 3.33). Thus, haemorrhagic lesions in the central part of the small intestine would suggest *E. necatrix*; those in the caecum, *E. tenella*; those in the rectum, *E. brunetti*. It is not sufficient to look for oocysts only, since these may be found with regularity in the small intestine or caecum of chickens; rather, it is necessary to determine if large numbers of schizonts are present in the subepithelial tissues for the major pathogens, and in an epithelial position for the other species.

Treatment and control

A formidable and extensive literature exists on the treatment of coccidiosis in poultry: it is possible only to give a brief outline of it here. Reid (1975) has reviewed progress in the control of coccidiosis with anticoccidial drugs. He estimates that new compounds continue to be introduced to the poultry on an average of about one every two years. In general, mortality due to coccidiosis can be eliminated by any of the 25 or more approved anticoccidial drugs if they are used properly. Anticoccidials are used usually in starter rations for meat-type birds raised under floor-pen management. Protection is more important with these fast-growing birds than with egg-producing types, where immunity and caging alter the demands for anticoccidial drugs. Selection of an anticoccidial is based on the ability of the drug to improve weight and feed conversion and to suppress the development of lesions (Reid 1975).

The emergence of drug-resistant strains of coccidia presents a major problem. Methods used to avoid the development of drug resistance include switching around 13 classes of drugs and the 'shuttle programme', which is a planned switch of drug in the middle of the growing period of birds (Reid 1975). The speed of emergence of resistant strains of coccidia in the field is given by Reid (1975) as follows: (1) glycomide—very rapid; (2) quinolines (buquinolate, deconquinate, nequinate)—rapid; (3) clopidol—less rapid; (4) sulphonamides, nitrofurans, robenidine—moderate; (5) amprolium, zoalene, nitromide—slow; (6) nicarbazin—very slow; (7) monensin—absent or very slow.

The subject of drug resistance has been discussed by Cuckler et al. (1969).

With the anticoccidial drugs dealt with below, various regulations apply to the premarket withdrawal of the drug, although some compounds are classed as 'no-withdrawal' and are often substituted for a 'withdrawal' anticoccidial during the final days before marketing.

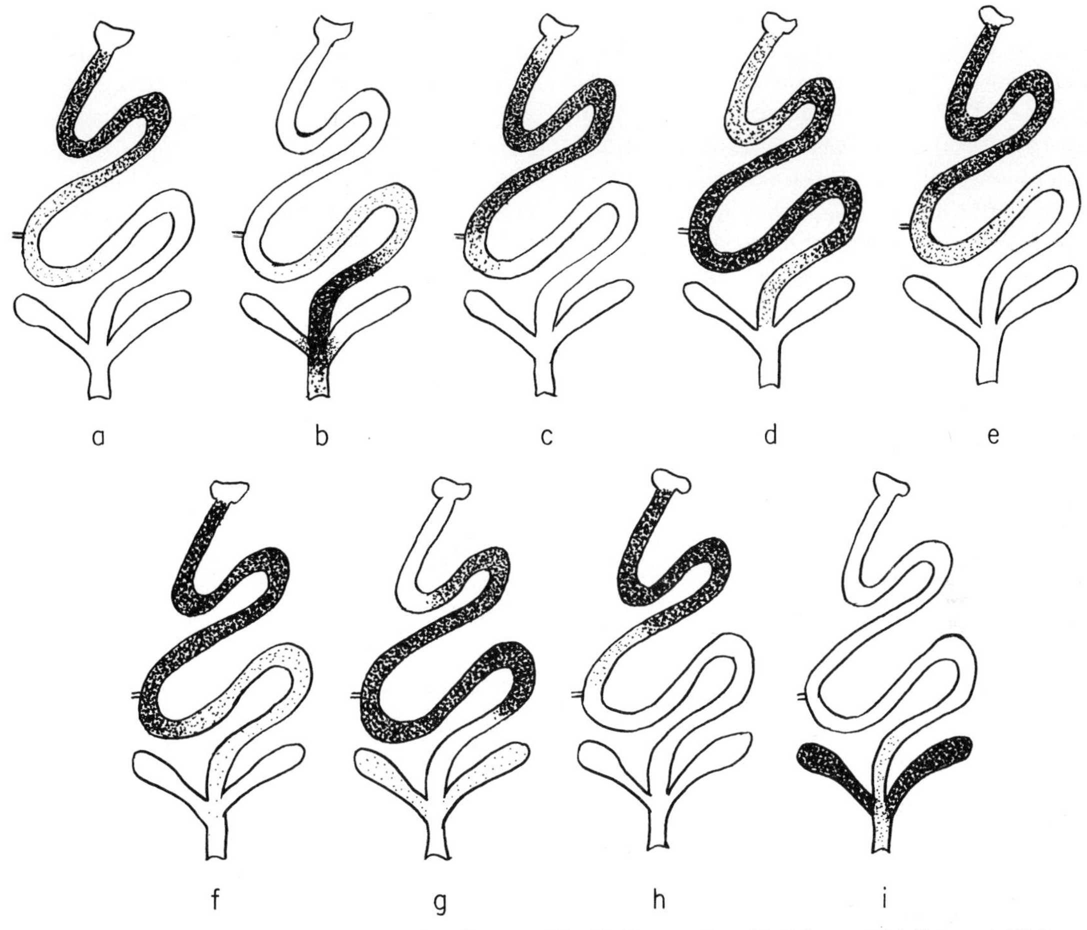

Fig 3·33 Location of lesions for nine species of poultry coccidia. (a) *E. acervulina*. (b) *E. brunetti*. (c) *E. hagani*. (d) *E. maxima*. (e) *E. mivati*. (f) *E. mitis*. (g) *E. necatrix*. (h) *E. praecox*. (i) *E. tenella*. (*Adapted from Redi 1964*)

Curative treatment should be instituted immediately after a diagnosis of coccidiosis is made. An interrupted form of treatment is more satisfactory with the sulpha drugs than continuous treatment, the aim of which is to avoid undue concentrations of the compounds which inhibit the earlier developmental stages of the parasite and thus interfere with the acquisition of immunity. To avoid this, Davies and Kendall (1954*a*, *b*) have suggested that sodium sulphadimidine be given at a concentration of 0.2% in drinking water for two periods of three days, separated by two days without treatment. Sodium sulphaquinoxaline is given in the feed at the rate of 0.5% Nitrofurazone, with furazolidone, to a final concentration of 0.0126%, is given over a seven-day treatment, and this may be repeated after a five-day interval.

The sulphonamides have a coccidiostatic action rather than a coccidiocidal effect. Consequently, they have no direct curative effect, but rather their value lies in halting the onset of disease in other members of the flock. They are active against the schizontal stages and especially the second generation schizonts of *E. tenella* and *E. necatrix*. Increased concentrations affect the first generation schizonts of these species, but much higher doses are required to affect the gamonts.

Preventive medication consists of prolonged or continuous use of coccidiostatic compounds in the feed or water. These include: amprolium (0.0125% in the feed), buquinolate (0.0055%), decoquinate (0.003%), clopidol (0.0125%), monensin (0.0121%), robenidine (0.003–0.006%), zoalene (0.0125%); nicarbazin (0.0125%), furazolidone (0.0055%), nitrofurazone (0.005–0.01%), sulphaquinoxaline (0.0125%), methyl benzoquate (0.001–0.002%), lasalocid (0.005–0.0075%) and salinomycin (0.006–0.01%). Other compounds are under development and it is likely this list will be extended in the next few years.

Details of the actions, indications, modes of use and toxicities of these and other compounds are dealt with below.

Sulphonamides

Sulphadimidine (sulphadimethylpyrimidine: 4, 6-dimethyl-2-sulphanilamidopyrimidine: 'sulfamethazine,' 'sulphamezathine'). This compound is still used as a curative drug in certain parts of the world, but its use has largely been discontinued in Western Europe and North America where it has been replaced by other compounds. It is given in the food at the rate of 0.4% or in the drinking water as a 0.2% solution of the sodium salt. It has been used most satisfactorily in the control of clinical outbreaks of coccidiosis. Active against *E. tenella, E. necatrix* and the other species of coccidia.

Toxicity. Prolongation of blood coagulation time, probably due to an interference with the vitamin K synthesis in the intestine. Male birds, given excessive doses, show hyperplasia of the seminiferous tubules of the testicles; reduced egg production in laying hens occurs.

Sulphaquinoxaline (2-sulphanilamidoquinoxaline). An important, effective and commonly used coccidiostat, in general use throughout the world. For preventive medication, doses ranging from 0.025% to 0.033% may be given over fairly long periods. For therapeutic purposes a dose of 0.043% in the water given for two treatments each for two days with a three- to five-day interval between them, is satisfactory. Sulphaquinoxaline has been used primarily for *E. tenella* and *E. necatrix* infection, but it is also active against *E. acervulina*. It exerts a marked inhibitory effect on schizogony, and Cuckler and Ott (1947) stated that a level of 0.1% in the ratio inhibited invasion by the sporozoites.

Toxicity of sulphaquinoxaline has been demonstrated. A level of 0.1% in the ration for a few days has no adverse effect on chickens, but when this concentration is fed for a longer time (e.g. 30 days) or when lower percentages are incorporated in the ration over a long period, toxic manifestations may be seen. These consist of a haemorrhagic syndrome characterized by multiple haemorrhages in many of the organs, sometimes accompanied by necrotic lesions in the spleen. There is also hypoplasia of the bone marrow and an agranulocytosis (Sadek et al. 1955). Toxicity is associated with an interference with vitamin K metabolism. In the USA there is a ten-day withdrawal premarketing requirement for this compound.

Thiamine analogues

Amprolium [1-(4-amino-2-*n*-propyl-5-pyrimidinylmethyl)-2-picolinium chloride hydrochloride]. This quarternized derivative of pyrimidine, which is a thiamine-antagonist, is highly active against *E. tenella, E. necatrix* and *E. acervulina*.

It is available as a premix and is given prophylactically to birds in a final concentration of 0.0125%. A combination of amprolium and sulphaquinoxaline at levels of 0.006% of each in the food is more effective against poultry coccidia than either of the two drugs used alone (Long 1963). There is no premarketing withdrawal requirement for this compound.

Nitrobenzamides

Zoalene (2-methyl-3,5-dinitrobenzamine). This is given prophylactically at a level of 0.01–0.015% in the food, and is active against both the caecal and intestinal forms of coccidiosis (Arundel 1960; Hymas & Stevenson 1960). Peterson (1960) conducted a comparative study of various anticoccidial drugs against infection with *E. tenella* and *E. necatrix* and found that zoalene, at the recommended level, was completely protective against *E. tenella* and was even more active against *E. necatrix*; full protection was achieved at a level of 0.005% of the active ingredient in the feed. Zoalene inhibits the development of the second generation schizont but does not inhibit the development of immunity under field conditions. Zoalene is coccidiostatic and will not cure the disease once signs have appeared.

Nitromide, marketed as Unistat, is a mixture consisting of 30% N_4-acetyl-N_1-*p*-nitrophenylsulphanilamide, 25% of 3,5-dinitrobenzamide and 5% of 4-hydroxy-3-nitrophenylarsonic acid. At a level of 0.1% in the feed it prevents death and allows more or less normal weight gains with heavy infections of *E. tenella*, *E. necatrix* and *E. acervulina*.

There is no premarketing withdrawal requirement for these compounds.

Nitrofurans

Nitrofurazone (5-nitro-2-furfuraldehydesemicarbazone) (Furacin, Furazol). As well as being a coccidiostat, it also possesses bacteriostatic properties, being active against both Gram-positive and Gram-negative organisms. For preventive medication, nitrofurazone is recommended at levels of 0.0055–0.0056% in the food. Harwood and Stunz (1949) suggest that newly hatched chickens be placed on such a concentration continuously until they are marketed at about 12 weeks of age. For therapeutic purposes it is used at a concentration of 0.022%, but if this level is continued for more than ten days, toxic effects may be seen, these being manifest by nervous signs. When 0.0055% of nitrofurazone is combined with 0.0008% of furazolidone (Bifuran), there is a marked coccidiostatic effect against *E. tenella* and *E. necatrix*.

Furazolidone [N-(5-nitro-furfurylidene)3-amino-2-oxazolidinone]. This compound has been used for infections due to the enteric bacteria, and it is also of value against *E. tenella* at a dose of 0.011% or 0.0055% in the feed. However, it is usually employed in combination with nitrofurazone (Bifuran) at the doses mentioned above, and is chiefly used against *E. tenella* infections. Horton-Smith and Long (1959) found it effective against *E. necatrix* if fed at twice the recommended level.

In the USA, a five-day withdrawal premarketing requirement exists for these drugs.

Substituted carbanilides

Nicarbazin (aryl derivative having an equimolecular complex of 4,4-dinitrocarbanilide and 2-hydroxy-4,6-dimethylpyrimidine). This compound is used principally as a prophylactic; the therapeutic dose lies near the toxic dose. Nicarbazin is usually available as a 22.5% premix, and it is incorporated into the food to give a final concentration of 0.0125%. It is effective against *E. tenella*, *E. necatrix* and *E. acervulina* and does not interfere with the acquisition of immunity (Horton-Smith & Long 1959). McLoughlin and Wehr (1960) confirmed that it has a marked inhibitory effect on the second generation schizonts, and they also reported a moderate action on the asexual stages prior to this.

The drug is suitable for administration to broiler flocks, and it is usually given for the first 12 weeks of the chicken's life. It is unsuitable for laying or breeding stock because of the effect on egg colour and hatchability. A four-day withdrawal premarketing requirement exists for this compound in the USA.

Toxicity may be seen at the level of 0.03% or above in the ration. There is interruption of egg laying, eggs are depigmented, and yolk mottled, and the hatchability is reduced. Ataxia has been seen at feed levels of 0.05–0.1% for three weeks (Newberne & Buck 1956). Where death is due to toxicity, there is degeneration of the epithelium of the renal tubules and liver cells.

Nitrophenide (3,3-di-nitrodiphenyldisulphide). This is used under field conditions at the level of 0.025% in the feed. Horton-Smith and Long (1959) reported good coccidiostatic effect against both *E. tenella* and *E. necatrix*. Its maximum effect occurs at 49–96 hours after infection, which suggests that the drug inhibits the development of second generation schizonts. The difference between the toxic and therapeutic dose is small, and death may occur with a level of 0.16% in the feed. At levels of 0.04% in the feed, administered continuously for 4–12 weeks, there was no effect on growth or egg production, nor on egg hatchability (Waletsky et al. 1949).

Trithiadol. A formulation of bithionol [2,2-dihydroxy-3,3,5,5-tetrachlorodiphenyl sulphide] and methiotriazamine [4,6-diamino-1-(4-methyl-mercaptophenyl) - 1,2 - dihydro - 2,2 - dimethyl-1,3,5-trizaine] in the proportion of five parts bithionol to one part the latter. It is active against *E. tenella, E. necatrix, E. maxima* and *E. acervulina* at the level of 0.06–0.09% in the diet. Good immunity develops to coccidia at these levels. It is not harmful to growing chicks and has no effect on egg production, colour or quality, but hatchability may be affected, and hence it is not recommended for breeding birds.

Hydroxyquinolines

Buquinolate (ethyl 4-hydroxy-6,7-di-isobutoxy-3-quinoline carboxylate). This is given at the level of 0.00825% in the feed and possesses a broad spectrum of activity against all chicken coccidia. The drug arrests sporozoite development, but does not kill these forms and, if withdrawn too early, the inhibited stages may recommence development. Buquinolate favourably influences food conversion rates, toxicity is low and it is quickly eliminated from the tissues following withdrawal of medicated feed. (See Pellérdy, 1974, for references regarding this compound.)

Methyl benzoquate (7-benzyloxy-6-*n*-butyl-3-methoxycarbonyl-quinol-4-one). This is a broad-spectrum coccidiostat with action on sporozoites. It is given as a premix in the feed at a concentration of 0.001–0.002% Toxicity is very low, egg production, quality and hatchability are not affected and weight gains are improved with its use (Pellérdy 1974).

Decoquinate (ethyl-6-*n*-decyloxy-7-ethoxy-4-hydroxyquinoline-3-carboxylate). This is also a broad-spectrum coccidiostat and is used at a concentration of 0.003% in the feed. It inhibits sporozoite development.

There is no premarketing withdrawal requirement for these compounds.

Pyridinoles

Metrichlorpindol (3,5-dichloro-2,6-dimethyl-4-pyrimidol) (Coyden, Clopidol). This is a broad-spectrum coccidiostat affecting sporozoite development. It is given as a premix in the feed at a level of 0.0125% (Pellérdy 1974). There is no premarketing withdrawal requirement.

Guanidines

Robenzidene (1,3-bis(*p*-chlorbenzylide amino) guanidine hydrochloride). This is a broad-spectrum coccidiostat, active as a 0.0066% mixture in the feed. The drug influences feed conversion in a favourable manner. At the 66 ppm level an unpleasant taste is detected by some people from low residues in muscle and liver, but reduction to 0.0033% in the feed with a five-day premarket withdrawal overcomes this (Reid 1975).

Ionophorous antibiotics

Monensin. This is a fermentation product of *Streptomyces albus* which gives protection against mortality for all species at 0.01–0.0121% in the feed. Good weight gains and feed conversion occur, and in some cases suppression of necrotic enteritis also occurs (Reid 1975).

Delays in the development of immunity to coccidia during feeding of monensin-medicated feed were reported by Reid et al. (1977): at levels of 0.012% there was retardation of the development of immunity, and this was greatest with *E. tenella*. Immunity was progressively less delayed as the monensin level was decreased to 0.01% or 0.006%.

There is a three-day premarketing withdrawal requirement for this compound in the USA.

Lasalocid, another fermentation product, is effective at 0.005–0.0075% and permits increased weight gains, feed conversion, and reduces lesion score under severe coccidiosis exposure (Reid et al. 1975).

Salinomycin, a further fermentation product of *Streptomyces albus*. At 0.01% in the feed this compound showed significant anticoccidial activity and it was as effective as 0.0121% monensin in controlling coccidiosis (Danforth et al. 1977).

Other compounds

Polystat. A formulation active against *E. tenella* and *E. necatrix*, given at the rate of 0.02% in the feed. When so presented, it contains 0.03% of acetyl-(*p*-nitrophenyl)-sulphanilamide, 0.002% of dibutylin dilaurate, 0.02% of dinitrodiphenylsulphonylethylene diamine and 0.0075% of 4-hydroxy-3-nitro phenylarsonic acid.

Antibiotics. Ball (1959) re-examined several antibiotics and found that aureomycin, chloramphenicol, erythromycin, spiromycin and terramycin were all active against *E. tenella* infections, spiromycin giving the most satisfactory effect.

Vaccination against coccidia

The high degree of immunity which occurs with several species of poultry coccidia would suggest that artificial immunization would be an advantageous method of controlling these infections. An application of this principle was developed by Edgar (1958). Small doses of oocysts are given to young birds which are then allowed to run on litter but the infection is controlled by a low level of prophylactic coccidiostat. A commercial vaccine, Coccivac, consists of eight *Eimeria* species of the chicken. Birds are 'innoculated' at three days of age, and low-level medication is started at about 13 days of age and continued until the birds are $5\frac{1}{2}$–6 weeks of age. Edgar et al. (1956) have reported that inoculated chickens usually develop a solid immunity to three or five types of coccidiosis by four weeks of age, even when fed continuous prophylactic levels of certain coccidiostatic drugs. The choice of drug is important, and those which are satisfactory and compatible with immunization include 0.0125% nitrophenide, 0.0125% sulphaquinoxaline, 0.000325% 2-sulphanilamido-6-chloropyrazine, 0.0055% nitrofurazone, 0.5 lb/ton of feed of bifuran, and 2 lb/ton of feed of trithiadol or polystat. Nicarbazin, at a level of 0.0125%, was not compatible with the development of immunity to three or more species of coccidia, birds being susceptible at 12–16 weeks of age.

The use of such a vaccine has been criticized by several people. For example, it is difficult to control the vaccine and, more especially, the factors on which the vaccine depends for immunization; thus, an unsatisfactory environment for the sporulation of oocysts from the initial infection may provide the chickens with little or no immunizing infection from the litter. Careful management of litter moisture is necessary to permit satisfactory oocyst sporulation. Nevertheless, the method has been used extensively in several parts of the United States and, despite the theoretical objections to it, it would appear to be a practical method of control in many areas.

Methods of attenuating sporulated oocysts so that a non-pathogenic but immunizing infection may be produced in chickens, have been investigated with special reference to the use of X-rays. Waxler (1941) reported that the exposure of oocysts to X-rays at the rate of 450/minute to provide a total exposure of 4500, 9000 and 13 500 r produced a progressive attenuation, which was indicated by a decrease in the severity of infection. More recently, Hein (1963) has demonstrated that the attenuation of oocysts of *E. tenella* can result in the production of a useful vaccine against this species.

COCCIDIA OF THE TURKEY

Seven species of coccidia occur in the turkey, but only two of these, *Eimeria adenoeides* and *Eimeria meleagrimitis*, are considered to be of pathogenic significance. The species which have been described are as follows:

Eimeria adenoeides Moore and Brown, 1951
Eimeria dispersa Tyzzer, 1929
Eimeria gallopavonis Hawkins, 1950
Eimeria innocua Moore and Brown, 1952
Eimeria meleagridis Tyzzer, 1927
Eimeria meleagrimitis Tyzzer, 1929
Eimeria subrotunda Moore, Brown and Carter, 1954
Cryptosporidium meleagridis Slavin, 1955

Eimeria adenoeides. *Hosts:* domestic turkey. North America, Great Britain, Eastern Europe, USSR. Developmental stages occur in the small intestine, large intestine and caecum. *Oocysts:* ellipsoidal, but show a wide variation in shape and size; 25.6 by 16.25 μm, range 21.5–30 μm by 13.5–19.8 μm. Oocyst wall, double-contoured with a smooth surface; a micropyle may or may not be present. Sporulation time, 24 hours at room temperature (Edgar 1955).

Life-cycle. The endogenous developmental cycle has been investigated in great detail by Clarkson (1956, 1958). First generation schizonts occur in the epithelial cells of the tubular part of the caeca and the terminal part of the ileum as early as six hours. These mature about 60 hours after infection and at this stage measure 30 μm by 18 μm, and contain approximately 700 merozoites. Second generation schizonts are distributed proximal to the nucleus in the epithelial cells throughout the small intestine and rectum. They are mature at 96–108 hours after infection, measure 10 μm by 10 μm, and produce 12–24 merozoites. Sexual stages appear by the fifth day of infection, and the prepatent period is 114–132 hours. Patency lasts for about two weeks.

Pathogenesis. Eimeria adenoeides is one of the most pathogenic species of coccidia in turkeys; young poults are particularly susceptible, but even older turkeys may be seriously affected if they have not had immunizing infections in earlier life. Moore and Brown (1951) produced 100% mortality in poults up to five weeks of age with large doses of sporulated oocysts. Clarkson (1956) found that 200 000 oocysts invariably resulted in death of three-week-old poults, whereas 100 000 oocysts caused a mortality of 45%. Lower doses of 30 000 and 10 000 oocysts failed to produce death, although in poults less than five weeks old they caused severe disease associated with anorexia and blood in the droppings. On the other hand, six-week-old turkeys, given one million oocysts, suffered a mortality of 33% and 11-week-old poults given three million oocysts showed no mortality (Clarkson 1958; Clarkson & Gentles 1958).

The clincial signs commence on the fourth day of infection, consisting of loss of appetite, droopiness and ruffled feathers. Deaths begin on the fifth or sixth day when the droppings are liquid and contain mucus and blood. The main lesions are confined to the posterior small intestine, caecum and rectum, the walls of these being swollen and oedematous; petechial haemorrhages are visible, and the contents of the lower digestive tract are white or grey in colour and contain mucus or blood at a later stage.

Immunity. Immunity induced by *E. adenoeides* is specific, and Clarkson (1959*a*,*b*) found no cross-immunity to *E. meleagridis*.

Eimeria dispersa. *Hosts:* domestic turkey; possibly also Bobwhite quail, Hungarian partridge, pheasant. North America. The developmental stages occur in the small intestine. *Oocysts:* broadly ovoid, 26.1 μm by 21.04 μm, range 21.8–31.1 μm by 17.7–23.9 μm (Hawkins 1952). Oocyst wall, no double-contoured appearance, no micropyle. Sporulation time, two days at room temperature. (Differences in sizes of oocysts from different hosts have been reported by Tyzzer, 1929. Oocysts from the turkey were smaller than those from the quail, the latter measuring 22.7 μm by 18.8 μm, range 17.6–26.4 μm by 15.4–22.4 μm.)

Life-cycle. The parasites develop in the epithelial cells of the tips of the villi. Two types of first generation schizonts have been described (Tyzzer 1929; Hawkins 1952). One form appears at about 55 hours after infection and reaches a size of 6 μm and produces 15 merozoites; the other is larger, measuring 24 μm by 18 μm and producing 50 merozoites. First generation schizonts are mature by 48 hours, and second generation forms are mature four days after infection. The latter measure 11–13 μm and produce 18–23 merozoites. The gamonts are observed from the fourth day onwards, and oocysts appear in the faeces from the fifth to the sixth days after infection.

Pathogenesis. This species is only mildly pathogenic for the turkey. Hawkins (1952), with experimental infections, produced only a slight tendency to liquid faeces and a mild depression in weight gains. He also reported that infection with a few thousand oocysts daily would immunize birds so that massive infections are resisted after the second week.

Eimeria gallopavonis. *Hosts:* domestic turkey. North America, India, USSR. Developmental stages in the lower small intestine, rectum and caecum. *Oocysts:* ellipsoidal, 27.1 μm by 17.2 μm, range 22.2–32.7 μm by 15.2–19.4 μm, difficult to distinguish from *E. meleagridis*. Oocyst wall, double-contoured, no micropyle. Sporulation time, 24 hours.

Life-cycle. The endogenous developmental cycle has been studied by Hawkins (1952) and Farr

(1964). Schizonts occur proximal to the nucleus in the epithelial cells of the tips of the villi of the ileum and the rectum three days after infection. Second generation schizonts are found in a similar situation. Two types of second generation schizont occur: a small form producing 10–12 large merozoites and a large form producing many small merozoites. Farr (1964) reported that large schizonts occurred first, while the smaller schizonts were derived from the second generation of schizogony. A few scattered third generation schizonts may occur, but the majority of second generation merozoites give rise to the sexual cycle. This is first apparent on the fourth day, epithelial cells of the ileum, caecum and, mainly, the rectum being affected. The prepatent period is six days.

Pathogenesis. Generally thought to be non-pathogenic. Hawkins (1952) reported oedema, sloughing of the mucosa and lymphocytic infiltration of the intestine in experimental infections. Hawkins (1952) also reported a strong, specific immunity. Turkeys immune to *E. meleagridis, E. meleagrimitis* and *E. dispersa* could be readily infected with oocysts of *E. gallopavonis.*

Eimeria innocua. *Hosts:* domestic turkey, USA. Probably world-wide. As its name suggests, it is non-pathogenic. *Oocysts:* sub-spherical, 22.4 μm by 10.9 μm, range 18.6–25.9 μm by 17.3–24.5 μm. No micropyle. Sporulation time, 24–48 hours.

The endogenous developmental cycle takes place in the epithelial cells of the tips of the villi of the small intestine. The prepatent period is five days, and patency lasts for no longer than 14 days.

Moore and Brown (1952) found this species to be non-pathogenic, and they could observe no microscopic lesions even in heavy experimental infections.

Eimeria meleagridis. *Hosts:* domestic and wild turkey. Probably cosmopolitan. At one time it was thought that this species was pathogenic for young turkeys; however, it is probable that earlier reports of its pathogenicity should rightly be ascribed to *E. adenoeides.* It is now generally considered that *E. meleagridis* is, at the most, only a mild pathogen. The developmental stages occur initially in the small intestine and later in the caecum and rectum. *Oocysts:* ellipsoidal, 23.8 μm by 17.3 μm (Tyzzer 1927) (Clarkson, 1959*b*, gives 22.5 μm by 16.25 μm). Oocyst wall, smooth, micropyle absent. Sporulation time, 24 hours.

Life-cycle. First generation schizonts develop in the epithelial cells of the small intestine in the region of the rudimentary yolk stalk; merozoites from this generation are found at the fifty-fourth hour of infection (Clarkson 1959*b*). Second generation schizonts and gamonts occur in the caeca, but some gametogonous stages may be found in the lower ileum and the rectum. Second generation schizonts are found in their greatest number 84 hours after infection; they measure 9 μm in diameter and contain 8–16 merozoites. A third generation of schizogony has been reported by Hawkins (1952), but Clarkson makes no mention of this in his description of the life-cycle. Gamonts are found from 91 hours after infection onwards, and the average prepatent period is 110 hours.

Pathogenesis. This is minimal. Clarkson (1959*b*) failed to produce disease in two-week-old turkey poults with doses of up to one million oocysts, and Hawkins (1952) observed only a minor loss in poults infected with ½–1 million oocysts.

Eimeria meleagrimitis. *Hosts:* domestic turkey. Probably world-wide distribution. Common. Developmental stages occur in the small intestine. *Oocysts:* sub-spherical, 18.1 μm by 15.3 μm, range 16.2–20.5 μm by 13.2–17.2 μm (Tyzzer 1929). (Hawkins, 1952, states 19.2 μm by 16.3 μm and Clarkson, 1959*a*, 20.1 μm by 17.3 μm, range 16–25.5 μm by 13.7–22 μm.) Oocyst wall, double-contoured, smooth surface, no micropyle. Sporulation time, 24 hours.

Life-cycle. The endogenous developmental cycle has been studied by Hawkins (1952), Clarkson (1959*a*) and Horton-Smith and Long (1961). Hawkins described two generations of schizonts, Clarkson described three, the latter being confirmed by Horton-Smith and Long. The earliest trophozoites appear in the epithelial cells lining the glands of the small intestine 24 hours after infection. First generation schizonts are mature at 48 hours, reaching 17 μm by 13 μm and containing 80–100 merozoites. Second generation schizonts occur in the epithelial cells of the glands in the same vicinity and form further colonies of schizonts. These are mature at 66 hours of infection, are smaller than the first generation forms, being 8 μm by 7 μm and contain 8–16 merozoites. A third generation schizont occurs prior to the gametogonous cycle. Forms of this generation occur as early as 72 hours after infection, and they reach maturity by 96 hours. They are small, 8 μm by 7 μm and produce 8–16 merozoites. Gametogonous stages are seen in the epithelial cells 4–5 days after infection, occurring in the jejunum, mainly in the epithelial cells at the tips of the villi. The prepatent period is 114–118 hours.

Pathogenesis. Peterson (1949) reported 70–90% mortality in turkey poults aged four to five weeks due to this parasite. Clarkson (1959*a*) and Clarkson and Gentles (1958) produced heavy mortality in poults aged 1.5–3 weeks by feeding 100 000 oocysts. Poults aged four weeks, fed 400 000 oocysts, suffered 100% mortality. Birds aged five to ten weeks showed no mortality when fed 200 000 to two million oocysts.

Clinical signs appear on the fifth day, birds cease to eat, huddle together, and the droppings become fluid and may be brown or blood-tinged. Deaths commence about the sixth day in severe infections and continue for the next few days.

Thickening and dilatation of the intestinal wall is seen on the fourth day, and from the fifth day onwards the duodenal mucosa is greyish white, the blood vessels engorged and the contents fluid, containing caseous mucous threads. Sometimes the contents are reddish due to haemorrhage, although excessive haemorrhage is not a feature of the condition. Sometimes the duodenum may be plugged with a core of reddish brown material which is firmly adherent to the mucosa, later it becomes detached, recedes and is eventually discharged.

Eimeria subrotunda. *Hosts:* domestic turkey. USA. This species, which was isolated by Moore et al. (1954) on the basis of immunological studies and cross-transmission, is essentially non-pathogenic. The developmental stages occur in the small intestine. *Oocysts:* very similar to those of *E. innocua*, sub-spherical, 21.8 μm by 19.8 μm, range 16.5–24.4 μm by 14.2–22.4 μm. Oocyst wall, smooth, no micropyle. Sporulation time, 48 hours.

Life-cycle. The endogenous developmental cycle is unknown, although material from two infected turkey poults has been examined by Moore et al. (1954), and they described developmental stages in the epithelial cells of the tips of the villi, especially in the duodenum, the jejunum and the upper part of the ileum. The prepatent period is 96 hours, and patency continues for 12–13 days.

Pathogenesis. Moore et al. (1954) regarded *E. subrotunda* as a harmless parasite, and even massive infections of poults aged five weeks failed to induce any departure from the normal. Recovery from *E. subrotunda* infection results in immunity. This is specific.

Cryptosporidium meleagridis. *Hosts:* domestic turkey. Great Britain. This organism has been reported by Slavin (1955) as a cause of diarrhoea and moderate mortality in turkey poults in the first two weeks of life in Scotland. As far as can be ascertained, it is the only report of a species of this genus causing trouble in turkeys. In the outbreak reported by Slavin, six different parasites were identified in the turkeys; however, he considered that *C. meleagridis* was responsible for the diarrhoea and the mortality. The parasitic forms are found on the epithelium of the intestine and do not invade the tissues. *Oocysts:* small, oval, 4.5 μm by 4 μm, cytoplasm foamy in consistency, nucleus stains poorly, located eccentrically. Sporulated oocysts and sporozoites have not been described.

The endogenous developmental cycle includes an asexual generation with the production of schizonts and merozoites. Young schizonts are attached to the epithelium of the villi, sometimes in enormous numbers. Slavin (1955) described an organ of attachment which penetrated distal to the microvilli of the epithelial cells; developmental forms were also seen in goblet cells and between the epithelial cells as far down as the basement membrane. Mature schizonts measure 5 μm by 4 μm and contain eight merozoites, although sometimes only two or four occur. Microgamonts are about 4 μm in diameter and produce 16 rod-shaped microgametes; the macrogametes measure 4.5–5 μm by 3.5–4 μm. Slavin suggested that the microgametes were not motile, and the process of fertilization consisted of the male and female forms attaching themselves closely to one another. Oocysts may be found in the faeces in large numbers, but sporulated oocysts were not detected (Slavin 1955).

The pathogenic entity produced by this parasite consisted of diarrhoea and moderate mortality in turkey poults 10–14 days of age. The terminal third of the small intestine was especially affected; however, since the organisms occur on the surface of the epithelial cells, gross microscopic or histopathological lesions were absent.

COCCIDIOSIS OF TURKEYS

Coccidiosis may be responsible for severe disease and heavy economic loss in the turkey industry, the organisms of special importance being *E. adenoeides* and *E. meleagrimitis*. It is primarily a disease of young turkey poults, although older birds serve as carriers, and it has been suggested by Skamser (1947) that turkeys beyond the age of eight weeks are not seriously affected. The increased importance of coccidiosis in turkeys has derived in part from intensive production systems which are now used to rear turkeys. Where overcrowding and poor sanitation are present, the prevalence of coccidiosis may be very high. This is especially so when turkeys are kept in open yards since these are often damp and provide ideal conditions for the sporulation of oocysts.

Diagnosis

This is similar to that for coccidiosis of the chicken. The finding of large numbers of oocysts in the faeces is not an infallible guide, and diagnosis is more satisfactorily done at post mortem examination.

Treatment

Essentially the same compounds are used for turkeys as for chickens, with the exception that the sulphonamides are generally more effective against the turkey coccidia. Thus, Boyler and Brown (1953) found that sulphamezathine and sulphaquinoxaline in the feed, or 0.05–0.1% sulphamezathine in the water, was effective against *E. adenoeides, E. gallopavonis, E. meleagridis, E. innocua, E. subtrotunda, E. dispersa* and *E. meleagrimitis*. This was confirmed by Horton-Smith and Long (1959). Warren and Ball (1963) have demonstrated that both sulphaquinoxaline and amprolium, at the rate of 0.0125% and 0.008% in the food, controlled *E. adenoeides* and *E. meleagrimitis* in turkeys. Some evidence that amprolium may interfere with the acquisition of immunity was produced in that when poults were challenged with a severe infection 14 days after cessation of the drug, those treated with sulphaquinoxaline were less susceptible than those treated with amprolium. Some doubt exists as to the effectiveness of nicarbazin against *M. meleagrimitis*. Cuckler et al. (1955) reported it to be ineffective; in addition, they found nitrofurazone and glycarbylamide also ineffective against this pathogenic species. Raines (1968) reported that the chicken coccidiostat buquinolate was less active against turkey coccidia than the more commonly used preparations.

COCCIDIA OF GEESE AND DUCKS

Eight species of coccidia have been reported from domestic geese and ducks, and some of these have been associated with severe disease and death. A much larger number of species has been recorded from wild ducks and geese; a brief list of these is given on p. 653.

The species of importance in domestic geese and ducks are:

Eimeria anatis Scholtyseck, 1955
Eimeria anseris Kotlán, 1932
Eimeria battakhi Dubey and Pande, 1963
Eimeria kotláni Gräfner and Graubmann, 1964
Eimeria nocens Kotlán, 1933
Eimeria parvula Kotlán, 1933
Eimeria saitamae Inoue, 1967
Eimeria schachdagica Musaev, Surkova, Jelchiev and Alieva, 1966
Eimeria stigmosa Klimeš, 1933

Eimeria truncata (Railliet and Lucet, 1891) Wasielewski, 1904
Tyzzeria anseris Nieschulz, 1947
Tyzzeria perniciosa Allen, 1936
Wenyonella anatis Pande, Bhatia and Srivastava, 1965
Wenyonella gagari Sarkar and Ray, 1968
Wenyonella philiplevinei Leibovitz, 1968

Eimeria anatis. *Hosts:* mallard, possibly domestic duck. Europe (Germany and United Kingdom). This species is principally a parasite of the mallard (*Anas platyrhynchos platyrhynchos*); however, a species of *Eimeria* with oocysts similar to those of *E. anatis* was found by Davies (1957) in domestic ducks in Britain, and he has shown that this species will develop both in the domestic duck and in the mallard. In all probability, the infection of domestic ducks is derived from wild ducks which may frequent the pasture on which the domestic ducks graze. *Oocysts:* ovoid, 16.8 μm by 14.1 μm, range 14.1–19.2 μm by 10.8–15.6 μm. Oocyst wall, smooth, uniform thickness, distinct micropyle at the more pointed end and a distinct ring-shaped projection of the wall forming shoulders around the micropyle. Sporulation time, four days at 20°C (Scholtyseck 1955).

The endogenous developmental cycle is unknown, but it is assumed that both schizogony and gametogony occur in the small intestine.

As far as is known, this species is non-pathogenic. Davies (1957) was unable to induce pathogenic effects in experimental studies.

Eimeria anseris. *Hosts:* domestic goose, blue and lesser snow goose and Richardson's Canada goose. Europe and North America. Prevalence in these geese is generally low. *Oocysts:* oval to pear-shaped, 21.7 μm by 17.2 μm, range 20–24 μm by 16–19 μm. Oocyst wall, narrow part has a distinct truncated cone, micropyle present, smooth and colourless, slightly thickened in the area of the micropyle and incised sharply to form a plate or shelf across the micropyle. Sporulation time, 24–48 hours.

Life-cycle. The endogenous developmental cycle has been described by Kotlán (1933) and Klimeš (1963). Only one schizogonous cycle occurs, developmental stages being found in compact clumps in the epithelial cells of the villi of the posterior part of the small intestine. Some stages may penetrate to near the muscularis mucosa. Mature schizonts measure 12–20 μm and produce 15–25 sickle-shaped merozoites. Gametogenous stages occur in the subepithelial tissues of the villi in heavy infections, but in milder infection they often occur superficially. The prepatent period is seven days, and patency lasts from two to eight days.

Pathogenesis. Kotlán (1933) considered the pathogenicity low; however, Klimeš (1963) and Pellérdy (1956) indicated that this species may produce a haemorrhagic enteritis which may terminate fatally in geese as old as nine months. Klimeš (1963) also observed that immunity developed rapidly after infection.

Eimeria battakhi. *Hosts:* domestic duck. India. This species is known only from oocysts found in the faeces. *Oocysts:* subspherical or ovoid 19–24 μm by 16–21 μm (mean 21 μm by 18 μm). Oocyst wall smooth, bilayered, 1–2 μm thick, no micropyle, polar granule present. Sporulation time, 24 hours. Elongated ovoid sporocysts 12 μm by 7 μm, no oocyst residual body but a sporocyst residual body occurs.

Eimeria kotláni. *Host:* domestic goose. Central Europe. *Oocysts:* oval, truncated at anterior pole, 29–33 μm by 23–25 μm. Bilayered oocyst wall 2 μm in thickness. Broad micropyle present, surrounded by a thickening of the internal layer of the oocyst wall to produce a lip. Sporulation time, 14 days; sporocysts elongate and oval; no oocyst residual body, but a sporocyst residual body occurs.

Life-cycle. Not fully known. Schizonts and gamonts occur in the large intestine (Levine 1973). The macrogamonts occur in the epithelial and subepithelial tissues.

Pathogenesis. Pathological changes similar to those produced by *E. brunetti* in the chicken were reported by Gräfner and Graubmann (1964). Thus oedema, inflammation, haemorrhage and diphtheritic membrane formation occurred. In severe infection these lesions extend into the small intestine. Histologically, there was marked destruction of the large intestinal mucosa, and infiltration by lymphocytes and leucocytes. The schizonts were the most pathogenic stages of development. Clinically, the infection causes mild to profuse diarrhoea and may result in death of geese.

Eimeria nocens. *Hosts:* domestic goose, blue and lesser snow goose (*Anser coerulescens coerulescens*). Europe and North America. Uncommon. *Oocysts:* ovoidal to ellipsoidal, flattened at the micropyle end, 31 μm by 21.6 μm, range 29–33 μm by 10–24 μm (Hanson et al. 1957). (Kotlán 1933, quotes a range of 25–33 μm by 17–24 μm.) Oocyst wall, smooth, green to pale yellow, micropyle present, but covered by the outer layer of the oocyst wall. Sporulation time, unknown.

Life-cycle. Developmental stages occur in the epithelial cells of the tips of the villi in the posterior part of the small intestine. The nucleus of the host cell may be displaced or destroyed by the developing forms, and they may also pass to the subepithelial tissues. Mature schizonts measure 15–30 μm and produce 15–35 merozoites. No information is available on the time sequence of these developmental stages.

Eimeria nocens is infrequently found in domestic geese. Kotlán (1933) described two outbreaks of intestinal coccidiosis in goslings in Hungary, but the infection consisted of *E. nocens* and *E. anseris.* Klimeš (1963) considers it affects mainly young geese and that it may be pathogenic in these.

Eimeria parvula. *Host:* domestic geese. Hungary. Developmental stages occur in the small intestine. *Oocysts:* spherical to subglobular, 13 μm by 10 μm, range 10–14 μm. Oocyst wall, smooth, colourless, no micropyle. Klimeš (1963) has suggested that the species described by Kotlán (1933) is, in fact, a member of the genus *Tyzzeria*, and proposed it be called *Tyzzeria parvula.* Klimeš noted that sporulation took 24 hours and resulted in the formation of eight sporozoites but no sporocysts.

The endogenous developmental cycle occurs in the epithelial cells of the villi in the posterior part of the small intestine. The prepatent period is about five days, and the output of oocysts may continue for several months. Pellérdy (1974) states that this species causes extensive epithelial damage in areas when large numbers of organisms occur, chiefly during gametogony. However, Kotlán (1933) considered it to be a relatively non-pathogenic species.

Eimeria saitamae. *Hosts:* domestic duck. Japan. *Oocysts:* colourless 17–21 μm by 13–15 μm. Sporulation time, one to two days. Second generation schizonts occur in the epithelial cells of the intestine on the second day. Gametogony occurs 96 hours after infection and oocysts are produced on the fourth day (Inoue 1967).

Eimeria schachdagica. *Hosts:* domestic duck. Azerbaidzhan, USSR. This species is known only from oocysts. *Oocysts:* colourless, egg-shaped, 16–26 μm by 12–20 μm; no micropyle. Sporulation time 72–96 hours.

Eimeria stigmosa. *Hosts:* domestic goose. Czechoslavakia. A non-pathogenic coccidium of young geese. *Oocysts:* oval, 23 μm by 16.7 μm, thickening at the poles. Oocyst wall, markedly striated transversely, dark brown in colour, micropyle present. Sporulation time, 48 hours at room temperature.

Klimeš (1963) found the prepatent period to be five days, the developmental stages occurring in the epithelial cells of the villi in the anterior part of the small intestine.

Eimeria truncata. *Hosts:* domestic goose, Graylag goose (*Anser cinereus*), Ross's goose (*Anser rossi*), Canada goose (*Branta canadensis*), etc. World-wide in distribution. Developmental stages occur in the kidney tubules. *Oocysts:* ovoid to ellipsoidal, truncated narrower end, 14.3–23.5 μm by 11.7–16.3 μm (Becker 1934). Oocyst wall, smooth, micropyle present. Sporulation time, one to five days.

Life-cycle. Schizonts occur in the epithelial cells of the kidney tubules. When fully grown, they measure approximately 13 μm in diameter and contain 20-30 merozoites. The infected epithelial cells are destroyed, and the adjacent cells show pressure atrophy and destruction. No information is available at present on the route which sporozoites take to reach the epithelial cells of the kidney tubules. Macrogamonts stain intensely blue and reach a size of 15–17 μm, while the microgamonts are smaller, reaching a size of 7 μm by 13 μm. The prepatent period is five days.

Pathogenesis. This species is a highly pathogenic form for goslings, occasionally causing 100% mortality within a few days of the onset of clinical conditions. However, under natural conditions, and especially in wild geese, there may be a high rate of infection with apparently low mortality. Affected birds show marked weakness and emaciation, and the condition may be so acute as to kill goslings within a day or so. Birds drink water copiously and may show muscular incoordination and a staggering gait.

The pathological changes consist of markedly enlarged kidneys, these being light in colour, and show numerous small, white nodules, streaks and lines on the surface and in the substance (Stubbs 1957). There is destruction of the epithelial cells of the kidney tubules, and the affected tubules are packed with urates, oocysts and gametogonous stages in various stages of development. A marked and general infiltration of round cells occurs.

Eimeria truncata occurs as a sporadic parasite in domestic geese, and it is most likely to occur when geese are kept crowded together in unsanitary surroundings. Contact with wild geese may be responsible for the introduction of the infection. The organism has been associated with losses in the Canada goose in its winter quarters at Pea Island, North Carolina (Critcher 1950); however, the true incidence of the species in wild geese is not known.

Tyzzeria anseris. *Hosts:* domestic goose, white-fronted goose, snow goose, Ross's goose, Canada goose and several other forms. North America and Europe. Developmental stages in the small intestine. *Oocysts:* ellipsoidal, 13 μm by 11 μm, range 10–16 μm by 9–12 μm (Levine 1952), no micropyle. When sporulated, the oocysts contain eight banana-shaped sporozoites, there being no sporocysts, this being characteristic of the genus. The endogenous developmental cycle is unknown.

Tyzzeria perniciosa. *Hosts:* domestic duck. USA, Britain, Holland. Developmental stages occur in the small intestine. *Oocysts:* ellipsoidal, 10–13.3 μm by 9–10.8 μm. Oocyst wall, colourless, no micropyle. When passed in the faeces, the oocyst space is completely filled with granular material. Sporulation time, 24 hours, eight sporozoites being formed without any sporocysts.

Life-cycle. The endogenous developmental cycle has been described by Allen (1936) and Versényi (1967). Sporozoites invade the mucosa and submucosa throughout the whole length of the small intestine; first generation schizonts appear 24 hours after infection, these measure 11.6 μm by 9.3 μm and contain four small merozoites. At least three asexual generations have been reported by Allen (1936), but Versényi (1967) states there are only two. According to the latter, first generation merozoites penetrate mesenchymal cells of the tunica propria and develop there to second generation schizonts. These are evident 60 hours after infection and they cause severe tissue damage. Gamonts appear about 48 hours after infection, and oocysts first appear in the faeces six days after infection.

Pathogenesis. This species is fairly pathogenic, especially in young ducks, and mortality may reach 10% in acute outbreaks. The clinical signs consist of anorexia, there is marked loss of weight, difficulty in standing, and it is reported that baby ducklings cry continuously. The lesions consist of a markedly thickened intestinal wall with haemorrhagic spots and areas showing greyish-white nodules. In severe cases there may be sloughing of the mucosa and plugging of the lumen with a haemorrhagic or cheesy exudate. The parasitic stages may penetrate as far as the muscular layer and here produce massive destruction of tissue; Allen (1936), on occasion, noticed penetration of the muscular layer.

Wenyonella anatis. *Hosts:* domestic duck. India. *Oocysts:* oval, with a punctate surface 11–17 μm by 7–10 μm. Micropyle present. Sporulation time, 48 hours producing four sporocysts, each containing four sporozoites, as is characteristic for the genus. Developmental cycle and pathogenicity unknown.

Wenyonella gagari. *Hosts:* domestic duck. India. *Oocysts:* pitcher-shaped 23–26 μm by 17–19 μm; large micropyle present. Sporulation time, 24–48 hours. Developmental cycle and pathogenicity unknown.

Wenyonella philiplevinei. *Hosts:* domestic duck. USA (New Jersey). *Oocysts:* ovoid, slightly asymmetrical, 15–21 μm by 12–16 μm (mean, 18.7 μm by 14.4 μm). Oocyst wall trilayered, small micropyle present. Sporulation time, 24–33 hours. Oocyst residual body absent; sporocysts 9.4 μm by 6.1 μm, sporocyst residual body present.

Life-cycle. Experimental infections (Leibovitz 1968) demonstrated that sporozoites invaded epithelial cells posterior to Meckel's diverticulum and in the rectum. First generation schizonts appear 24 hours after infection, and second generation schizonts occur 49 hours after infection, being 14 μm by 10 μm in size. Third generation schizonts develop by 74 hours after infection, measure 15 μm by 13 μm and each produce 12 banana-shaped merozoites. Gametogony commences after the ninety-third hour of infection. Oocyst discharge occurs at 120 hours after infection.

Pathogenesis. Leibovitz (1968) reported that the late schizogonic and gamont stages caused severe changes in the mucosae of the ileum and rectum. These included diffuse inflammation, petechial haemorrhages, oedema and necrosis of the surface of the rectum.

COCCIDIOSIS OF GEESE AND DUCKS

There is scant information about the clinical importance of coccidia in ducks and geese and still less about the economic importance of the disease. With the exception of *E. truncata,* which may cause 100% mortality in acute cases of renal coccidiosis, the other species appear to be important only in sporadic outbreaks, presumably where the environment is overcrowded and unhygienic. However, Randall and Norton (1973) have recorded acute disease caused by *E. anseris* and *E. nocens* in geese in Great Britain.

Treatment of coccidiosis of geese and ducks has been little studied. McGregor (1952) found that sodium sulphamezathine was satisfactory for *E. truncata* infections, and Davies et al. (1963) reported that a 0.1% solution of sodium sulphamezathine in the water, given in the form of an interrupted treatment for two periods of three days separated by two days, was successful for acute coccidiosis in ducks.

Prevention and control depends on the same considerations as have been discussed with regard to the chicken and turkey coccidioses.

COCCIDIA OF WILD DUCKS AND GEESE

A large number of species are found in wild ducks and geese, and the list, presented below, indicates some of the more common species.

Eimeria abramovi Svanbaev and Rakhmatullina, 1967 (Mallard, *Anus platyrhynchos platyrhynchos*) (Kazakhstan, USSR).

Eimeria boschadis Waldén, 1961 (Mallard, *Anas platyrhynchos platyrhynchos*) (Sweden). Found in the kidney.

Eimeria brantae Levine, 1953 (Hutchins goose, *Branta canadensis hutchensi;* Lesser Canada goose, *Branta canadensis parvipes*) (North America).

Eimeria bucephalae Christiansen and Madsen, 1948 (Golden Eye, *Bucephala clangula clangula*) (Denmark).

Eimeria christianseni Waldén, 1961 (Mute swan, *Cygnus olor*) (Sweden).

Eimeria clarkei Hanson, Levine and Ivens, 1957 (Lesser snow goose, *Anser coerulescens coerulescens*) (North America).

Eimeria danailovi Gräfner, Graubmann and Betke, 1965 (Mallard, and has also been transmitted experimentally to the domestic goose) (see Pellérdy 1974) (Central Europe).

Eimeria farri Hanson, Levine and Ivens, 1957 (white fronted goose, *Anser alibifrons frontalis*) (North America).

Eimeria fulva Farr, 1953 (Lesser snow goose; Canada goose, *Branta canadensis canadensis*) (USA).

Eimeria hermani Farr, 1953 (Lesser snow goose; Canada goose and cackling goose, *Branta canadensis minima*) (North America).

Eimeria koganae Svanbaev and Rakhmatullina, 1967. (Teal, Mallard etc.) (Kazakhstan, USSR).

Eimeria magnalabia Levine, 1951 (White fronted goose; lesser snow goose; Hutchins goose; Canada goose and cackling goose) (North America).

Eimeria somateriae Christiansen, 1952 (Long-tailed duck, *Clangula hymenalis*; eider duck, *Somateria mollissima mollissima*) (Developmental stages in kidney) (Europe).

Eimeria striata Farr, 1953 (Canada goose) (Hanson et al. 1957 considered this synonymous with *Eimeria magnalabia*) (North America).

Tyzzeria alleni Chakravarty and Basu, 1947 (Cotton teal, *Chenicus coromandelianus*) (India).

Tyzzeria pellérdyi Bhatia and Pande, 1966 (*Anas strepera*, gadwall; *Nyroca nyroca*, white-eyed pochard) (India).

COCCIDIA OF GUINEA-FOWL

Three species occur in this host, namely:

Eimeria gorakhpuri Bhatia and Pandi, 1967
Eimeria grenieri Yvoré and Aycardi, 1967
Eimeria numida Pellérdy, 1962

Eimeria gorakhpuri. *Hosts:* guinea-fowl. India. *Oocysts:* ovoid to ellipsoidal 16–24 μm by 13–17 μm, no micropyle. Sporulation time, 48 hours. Developmental stages occur in epithelial cells along the whole length of the intestine.

Eimeria grenieri. *Hosts:* guinea-fowl. France. *Oocysts:* ovoid 24 μm by 16 μm; micropyle present. Sporulation time, 12–24 hours. Developmental stages occur in epithelial cells along the whole length of the intestine. Pathogenicity is low, since Yvoré and Aycardi (1967) failed to produce serious disease by the administration of 400 000 oocysts to guinea-fowls.

Eimeria numidae. *Hosts:* guinea-fowl. Europe. *Oocysts:* elliptical, 19 μm by 15 μm, range 15–21 μm by 12–17 μm. Oocyst wall, smooth, no visible micropyle. Sporulation time, one to two days.

The endogenous developmental cycle occurs in the small and large intestines. First generation schizonts are found in the epithelial cells of the duodenum and jejunum from the second day onwards; they measure 4–5 μm and produce two to ten merozoites. Second generation schizonts are larger, 12–14 μm, producing 6–14 merozoites and occur in the jejunum, ileum and rectum. The prepatent period is five days.

Severe clinical signs appear on the fourth or fifth day of infection, a dose of 50 000 oocysts being sufficient to kill young birds (Pellérdy 1974). This is a parasite specific for the guinea-fowl since Pellérdy (1956) reported failure to infect the chicken with this species, or the guinea-fowl with chicken coccidia.

For therapy of guinea-fowl coccidiosis Yvoré (1969) reported that the several modern coccidiostats for domestic chickens were effective.

COCCIDIA OF PIGEONS

Coccidiosis of the pigeon may occasionally be seen in young squabs, especially where these are reared intensively and when conditions of hygiene are poor. Older birds serve as carriers and remain apparently healthy. Four species have been reported to be of importance in the pigeon:

Eimeria columbae Mitra and DasGupta, 1937
Eimeria columbarum Nieschulz, 1935
Eimeria labbeana (Labbé, 1896) Pinto, 1928
Eimeria tropicalis Malhotra and Ray, 1961

Eimeria columbae. This occurs in the pigeon (*Columba livia intermedia*) in India. *Oocysts:* 16.4 μm by 14.3 μm, having a thin wall. Sporulation time four to five days at room temperature.

Eimeria columbarum. This has been found in the rock dove (*Columba livia livia*); however, Levine (1973) considers this species a synonym of *Eimeria labbeana*. As described by Nieschulz (1935) the oocysts are spherical, colourless, 20 μm by 18.7 μm and have no micropyle.

Eimeria labbeana. *Hosts:* domestic pigeon (*Columba domestica*); rock dove; ring dove (*Columba palumbus*) and the turtle dove (*Streptopelia turtur*). World-wide in distribution. *Oocysts:*

spherical or sub-spherical, 16.7 μm by 15.3 μm, range 15–18 μm by 14–16 μm (Nieschulz 1935). (Levine, 1973, gives a size of 13–24 μm by 12–23 μm.) Oocyst wall, an inner dark and outer lighter layer, no micropyle. Sporulation time, 24 hours at room temperature.

The endogenous developmental cycle has been studied by Nieschulz (1925). Developmental stages occur in the epithelial cells of the intestine from the anterior region down to the rectum. Mature first generation schizonts are produced in about three days, and second generation forms may penetrate into the deeper tissues. The prepatent period is six to seven days. Oocyst production may show periodicity, most oocysts being passed from 9 A.M to 3 P.M.

Pathogenesis. In young pigeons severe clinical signs commence four to five days after artificial infection. Infected squabs show inappetence, diarrhoea and thirst; the droppings may be markedly green and may even be blood-tinged. In severe infections there is a high mortality, and the major lesion is a marked inflammation of the intestinal mucosa, the lumen being filled with a haemorrhagic exudate.

For treatment, Morini (1950) reported that sulphaguanidine was satisfactory, while Hauser (1959) commented that nitrofurazone was effective for the infection. Probably the modern coccidiostats available for domestic poultry would be useful for pigeon coccidiosis.

Prevention consists of an improvement in hygiene of the pigeon loft and, in general, of following the procedures for the control of coccidiosis in the domestic chicken.

Eimeria tropicalis. *Hosts:* pigeon. India. *Oocysts:* spherical to sub-spherical, 19–24 μm by 18–23 μm. Sporulation time, 40–48 hours. Development occurs in the duodenum and high doses of oocysts cause death in pigeon squabs (Pellérdy 1974).

COCCIDIA AND COCCIDIOSIS OF PHEASANTS

Coccidiosis may be a problem in the young pheasant, especially in rearing establishments, resulting in serious disease and high mortality.

The following coccidia have been reported from the pheasant:

Eimeria colchici Norton, 1967
Eimeria dispersa Tyzzer, 1929
Eimeria duodenalis Norton, 1967
Eimeria langeroni Yakimoff and Matschoulsky, 1937
Eimeria megalostromata Ormsbee, 1939
Eimeria pacifica Ormsbee, 1939
Eimeria phasiani Tyzzer, 1929

Eimeria colchici. Reported from the pheasant in Great Britain. *Oocysts:* ellipsoidal, 19–33 μm by 13–21 μm. Developmental stages occur in the epithelial cells of the small intestine. It may be associated with heavy losses in ducks.

Eimeria dispersa. This species is primarily a parasite of the turkey, but Tyzzer believed that a pheasant-adapted strain existed which could pass from the pheasant to the bobwhite quail. More recent work has indicated that a turkey strain transmitted to the bobwhite may not subsequently infect the pheasant (Moore 1954).

Eimeria duodenalis. Occurs in the ring-necked pheasant in Great Britain. *Oocysts:* subspherical, 18–24 μm by 15–21 μm. No micropyle. Development in the anterior small intestine. Prepatent period five days. An oedematous mucoid enteritis is produced on experimental infection.

Eimeria langeroni. Occurs in the pheasants *Phasianus colchicus chrysomelas* and *P. colchicus tschardynensis. Oocysts:* 32.5 μm by 18.4 μm, range 30–36 μm by 16–20 μm. Oocyst wall, double-contoured, pinkish-yellow with no micropyle. No information is available on the endogenous life cycle or pathogenicity.

Eimeria megalostromata. Occurs in the ring-necked pheasant (*P. colchicus torquatus*). *Oocysts:* ovoid, 24 μm by 19 μm, range 21–29 μm by 16–22 μm. Oocyst wall, smooth, bright yellow to brown, micropyle present. Sporulation time, 48 hours. No information is available regarding developmental time or pathogenicity.

Eimeria pacifica. Occurs in the ring-necked pheasant. *Oocyst:* ovoid, 23 μm by 18 μm, range 17–26 μm by 14–20 μm. Oocyst wall, double-contoured, bright yellow in colour, no micropyle. Sporúlation time, 48 hours. Development occurs in the anterior part of the small intestine, schizonts developing proximal to the cell nucleus. Apparently of low pathogenicity.

Eimeria phasiani. Occurs in the ring-necked pheasant. *Oocysts:* ellipsoidal, 23 μm by 16 μm, range 19.8–26.4 μm by 13.2–17.8 μm. Oocyst wall, brownish yellow, no micropyle. Sporulation time, 24 hours. Endogenous development occurs in the epithelial cells of the villi of the small intestine, developmental stages occurring distal to the cell nucleus. Prepatent period, five days. Tyzzer (1929) considered that severe disease could be produced in young pheasants by this species.

Trigg (1967) found that the modern coccidiostats used for chicken coccidiosis, e.g. Clopidol, were, in general, effective for pheasant coccidiosis (see also Norton & Wise, 1982).

COCCIDIA OF GAME BIRDS

A few studies have indicated that infection rates may be high in these hosts. Madsen (1941) indicated an incidence of 48% in adults and 62% in young birds in Denmark, and Herman et al. (1942) reported a high infection rate in both wild and domesticated birds.

COCCIDIA OF PARTRIDGE

Eimeria kofoidi Yakimoff and Matikaschwili, 1936
Eimeria lyruri Galli-Valerio, 1927
Eimeria procera Haase, 1939

Eimeria kofoidi. Occurs in stone partridge and grey partridge (*Perdix perdix perdix*). *Oocysts:* 20 μm by 17.6 μm.

Eimeria lyruri. This species occurs in the red-legged partridge (*Perdix ruber*), the black grouse (*Lyrurus tetrix tetrix*) and the capercaillie (*Tetrao urogallus aquitanicus*). *Oocysts:* 24–27 μm by 15 μm, cylindrical in outline.

Eimeria procera. This occurs in the grey partridge. *Oocysts:* 28.8–31.2 μm by 16.4–17.2 μm, no micropyle, oocysts elliptical.

COCCIDIA OF GROUSE

Eimeria augusta Allen, 1934
Eimeria bonasae Allen, 1934
Eimeria nadsoni Yakimoff and Gouseff, 1936
Eimeria tetricis Haase, 1936

Eimeria augusta. This occurs in the spruce grouse (*Canachites canadensis*), sharp-tailed grouse (*Pediocetes phasianellus campestris*) and the ruffed grouse (*Bonsar umbellus*) in the United States. *Oocysts:* 29.6 μm by 18.8 μm, range 25–33.9 μm by 16–27.1 μm. Developmental stages occur in the caecum.

Eimeria bonasae. Occurs in the spruce grouse, willow grouse or ptarmigan (*Lagopus lagopus*), a sharp-tailed grouse and hazel hen in Canada. *Oocysts:* 21 μm in diameter and spherical. No micropyle. Parasite of caecum.

Eimeria nadsoni. Occurs in the black grouse. *Oocysts:* 24.9 μm by 21.3 μm, spherical; no micropyle.

Eimeria tetricis. Occurs in black grouse. *Oocysts:* 29.8–31.4 μm by 14.2–15.4 μm; micropyle present, wall smooth.

COCCIDIA OF PTARMIGAN

Eimeria brinkmanni Levine, 1953
Eimeria fanthami Levine, 1953
Eimeria lagopodi Galli-Valerio, 1929

Eimeria brinkmanni. Occurs in rock ptarmigan (*Lagopus mutus rupestris*) in Canada. *Oocysts:* 28.6 μm by 18.8 μm, ellipsoidal; micropyle present. Oocyst wall, brownish yellow, slightly roughened.

Eimeria fanthami. Occurs in rock ptarmigan in Canada. *Oocysts:* 28.3 μm by 18.8 μm, ellipsoidal; no micropyle.

Eimeria lagopodi. Occurs in ptarmigan. *Oocysts:* 24 μm by 15 μm, subcylindrical; micropyle present.

COCCIDIA OF QUAIL

Eimeria coturnicus Chakravarty and Kar, 1947
Eimeria dispersa Tyzzer, 1929

Eimeria coturnicus. Occurs in grey quail (*Coturnix coturnix coturnix*) in India. *Oocysts:* 26.4–38.8 μm by 19.8–26.4 μm; no micropyle.

Eimeria dispersa. The turkey form of this species has been transmitted to the bobwhite quail (*Colinus virginianus*).

COCCIDIA OF PEAFOWL

Eimeria mandali Banik and Ray, 1964

Eimeria mayurai Bhatia and Pande, 1966
Eimeria pavonina Banik and Ray, 1961
Eimeria pavonis Mandal, 1965

Eimeria mandali. *Host:* peafowl. India. *Oocysts:* spherical to subspherical 14–20 μm by 14–18 μm, micropyle present. Development unknown.

Eimeria mayurai. *Host:* peafowl. India. Developmental stages in the small intestine. *Oocysts:* ellipsoidal, smooth 18–27 μm by 13–20 μm; no micropyle. Sporulation time two days. Bhatia and Pande (1968) describe the development in the peachick. This occurs beneath the host cell nucleus of the epithelial cells of the villi and crypts of Lieberkühn of the anterior small intestine. Second generation schizonts are 9–15 μm by 6–10 μm and prepatency is seven days.

In peachicks infected with 100 000 oocysts, diarrhoea, weakness and ruffled feathers were evident and a mucous exudate mixed with denuded cells was present in the duodenum.

Eimeria pavonina. *Host:* peafowl. India. *Oocysts:* ovoid, without micropyle. Wall said to have two layers, the outer one is brilliant blue (Levine 1973). Development unknown.

Eimeria pavonis. *Host:* peafowl. India. *Oocyst:* ovoid 20–25 μm by 18 μm; micropyle present. Sporulation time 65–70 hours at room temperature. Life-cycle and pathogenesis are unknown.

COCCIDIA OF THE RABBIT

Coccidial infection in the rabbit may, at times, cause serious disease and death, especially in young rabbits in intensive breeding establishments for fur, flesh or experimental purposes. One of the most severe is *Eimeria stiedai*, the cause of hepatic coccidiosis. In addition, the intestinal species may be responsible for severe ill health and death. At one time it was thought that the coccidia of rabbits were similar to those of hares (*Lepus* species); however, after extensive studies, Pellérdy (1974) concluded that the majority of the species from domestic and cottontail rabbits were specific for these animals and were not found in hares or jackrabbits.

Nevertheless, for convenience, the species of coccidia of rabbits and hares are listed below.

Eimeria coecicola Cheissin, 1947
Eimeria elongata Marotel and Guilhon, 1941
Eimeria exigua Yakimoff, 1934
Eimeria intestinalis Cheissin, 1948
Eimeria irresidua Kessel and Jankiewicz, 1931
Eimeria magna Pérard, 1925
Eimeria matsubayashii Tsunoda, 1952
Eimeria media Kessel, 1929
Eimeria nagpurensis Gill and Ray, 1961
Eimeria neoleporis Carvalho, 1942
Eimeria perforans (Leuckart, 1897) Sluiter and Swellengrebel, 1912
Eimeria piriformis Kotlán and Pospesch, 1934
Eimeria stiedai (Lindemann, 1865) Kisskalt and Hartmann, 1907

Eimeria coecicola. *Hosts:* domestic and wild rabbit. Hungary, Soviet Union. Development occurs in the posterior ileum and caecum. *Oocysts:* ovoid, 25–40 μm by 15–21 μm. Oocyst wall, smooth, light yellow in colour; micropyle present. Sporulation time, three days at room temperature. The developmental cycle has been described by Cheissin (1947). Asexual development occurs in the epithelial cells of the villi of the posterior small intestine, and the gametogonous stages are found in the cells of the crypts of the caecum. The prepatent period is nine days. This species has negligible pathogenicity. Pellérdy (1974) maintains that this species is synonymous with *Eimeria neoleporis*.

Eimeria elongata. *Host:* Domestic rabbit. France. Levine (1973) considers this the same as *E. neoleporis*. *Oocysts:* ellipsoidal, elongate, greyish in colour; micropyle broad, readily visible. Sporocysts also elongate. Sporulation time, four days. Life-cycle and pathogenesis unknown.

Eimeria exigua. *Hosts:* tame rabbit, Greenland hare and the cottontail rabbit. *Oocysts:* small, more or less sub-spherical, 14.5 μm by 12.7 μm. Oocyst wall, smooth, no visible micropyle. No information is available on the endogenous developmental cycle, and as far as is known, there is no pathogenicity associated with this species.

Eimeria intestinalis. *Hosts:* domestic rabbit (*Oryctolagus cuniculus*). India, Hungary and the Soviet Union. Relatively uncommon. Developmental stages occur in the small intestine. *Oocysts:*

pyriform, 27 μm by 18 μm, range 23–30 μm by 15–20 μm. Oocyst wall, yellowish, smooth; micropyle present. Sporulation time, one to two days at room temperature.

Life-cycle. The developmental cycle has been summarized by Pellérdy (1974). First generation schizonts are found in epithelial cells of the distal portion of the ileum, developing in a colonial manner at the base of the villi. Up to three generations of schizonts have been reported. Gamonts are visible as early as seven to eight days after infection, and the prepatent period is ten days.

Artificial infection of young rabbits produced moderate to severe intestinal inflammation, associated with diarrhoea and, at times, death.

Eimeria irresidua. *Hosts:* domestic rabbit; cottontail rabbit (*Sylvilagus floridanus*); California jackrabbit (*Lepus ruficaudatus*) and the white-tailed jackrabbit (*Lepus townsendii*). World-wide in distribution. *Oocysts:* ovoid, 38.3 μm by 25.6 μm, range 31–43 μm by 22–27 μm. Oocyst wall, smooth, light yellow in colour; distinct micropyle. Sporulation time, 50 hours at room temperature.

The endogenous development of *E. irresidua* has been studied by Rutherford (1943). Developmental stages occur in the epithelium of the villi of the whole of the small intestine. Two types of schizonts occur, a smaller form which produces two to ten merozoites and a larger type which produces 36–48 merozoites. Gamonts are apparent from the eighth day of infection, and the prepatent period is nine to ten days (Rutherford 1943). However, Cheissin (1946) reported that the endogenous development required 7–7½ days, and maintained that sexual stages developed from the third and fourth generation merozoites.

Pathogenesis. Heavy infections produce destruction of large numbers of epithelial cells with associated inflammation of the mucosa. In severe infections there may be haemorrhage into the intestinal lumen with marked diarrhoea and high mortality.

Eimeria magna. *Hosts:* domestic rabbit, California jackrabbit, cottontail, also in two species of hare (*Lepus timidus*, *Lepus europaeus*). World-wide in distribution. Common. Developmental stages in jejunum and ileum. *Oocysts:* broadly ovoidal, 35 μm by 24 μm, range 31–40 μm by 22–26 μm. Oocyst wall, yellow to yellowish-brown, area around the micropyle especially prominent, with a shoulder-like or collar-like protrusion formed by the outer oocyst wall. The outer wall may become detached, especially during sporulation, and then the distinct collar-like protrusion is absent. Sporulation time, two to three days.

Developmental stages occur distal to the nuclei of the epithelial cells, and parasitized epithelial cells migrate into the submucosal tissues. Rutherford (1943) recognized two types of schizonts which reached up to 10–20 μm in diameter. The larger schizonts produced fewer merozoites and appeared earlier than the second schizontal stages. Gamonts appear from the fifth day after infection onwards, and these, too, may become subepithelial in position. The prepatent period is seven to eight days, and patency persists for 15–19 days.

Pathogenesis. Rutherford (1943) considered *E. magna* as a marked pathogen for the domestic rabbit, probably the subepithelial location of developmental stages contributing to this. Lund (1949) observed that 300 000 oocysts of some strains may lead to death of young rabbits, whereas other strains were less pathogenic and up to one million oocysts failed to cause death. Clinical signs consist of progressive emaciation and diarrhoea with mucoid faeces.

Eimeria matsubayashii. *Hosts:* domestic rabbit, Japan. *Oocysts:* broadly ovoidal, 24.8 μm by 18.2 μm, range 22–29 μm by 16–22 μm. Oocyst wall, smooth, light yellow, micropyle present. Sporulation time, 32–40 hours.

The endogenous developmental cycle is not known in detail. Tsumoda (1952) reported it to be similar to *E. magna*, developmental stages occurring in the epithelial cells of the ileum. The prepatent period is seven days.

No detailed information is available on the pathogenesis of this species, although following heavy infection a diphtheritic enteritis may occur in the ileum with the clinical signs of diarrhoea.

Eimeria media. *Hosts:* domestic rabbit, cottontail, jackrabbit. World-wide in distribition. *Oocysts:* ellipsoidal, 31.2 μm by 18.5 μm, range 27–36 μm by 15–22 μm. Oocyst wall, smooth, light pinkish in colour, well-defined micropyle. Sporulation time, two days.

Life-cycle. The endogenous developmental cycle has been studied by Rutherford (1943) and Pellérdy and Babos (1953). Early developmental stages occur in the epithelial cells of the villi but later are found in a subepithelial position. First generation schizonts are mature by the fourth day. Two forms have been reported: a type'A' schizont which produces two to ten merozoites, and a type 'B', smaller than the first, which produces 12–36 merozoites. Second generation schizonts appear on the sixth day of infection, and again two types of schizonts are seen. Gamonts may be present five to six days after infection, and the prepatent period is six to seven days (Rutherford 1943).

Pathogenesis. Pellérdy and Babos (1953) reported that 50 000 oocysts were fatal to young rabbits, these producing a severe enteritis with excessive destruction of the intestinal epithelium. On post mortem, lesions occur in the small intestine and frequently extend into the large intestine. The wall of the caecum may be markedly thickened and greyish white in colour due to the accumulation of large numbers of developmental stages.

Eimeria nagpurensis. *Host:* domestic rabbit. India, Iran (laboratory rabbit). Known only from the oocysts which are barrel-shaped, 20–27 μm by 10–15 μm, thin-walled and without micropyle or residuum. Sporocysts have sharply pointed ends and are oat-shaped. Life-cycle and pathogenesis unknown.

Eimeria neoleporis. *Hosts:* cottontail, domestic rabbit (experimentally). USA, Europe, USSR. *Oocysts:* subcylindrical to ellipsoidal, 38.8 μm by 19.8 μm, range 32.8–44.3 μm by 15.7–22.8 μm. Oocyst wall, smooth, yellowish, distinct micropyle. Sporulation time, two to three days.

Life-cycle. The endogenous developmental cycle occurs in the posterior part of the small intestine and the caecum. Carvalho (1943) described four types of schizonts resulting in an endogenous developmental cycle of 12 days. The first schizonts develop in the epithelial cells lining the crypts of Lieberkühn and produce 40–48 merozoites which are released about the fifth day. The second generation schizonts are produced on the seventh day, and the contain 60–70 merozoites. The third generation schizont is mature by the ninth day; two types occur: a small form producing 14 merozoites and a larger form producing 60–86 merozoites. Gametogony is detectable on the tenth day, and the prepatent period is approximately 12 days. Patency continues for a further ten days.

Pathogenesis. This species is moderately pathogenic. It is more pathogenic in the cottontail than the domestic rabbit. Pellérdy (1954) found that doses of 50 000 to 100 000 oocysts produced death in young rabbits by the tenth day of infection. Pathologic changes occurred mainly in the region of the ileocaecal valve and in the vermiform appendix; there was thickening of the intestinal wall, which was whitish grey in colour due to the large number of developmental stages. Severe infections produced necrosis of the superficial mucosa.

Eimeria perforans. *Hosts:* domestic rabbit, jackrabbit, cottontail, also reported from hare (*Lepus europeaus*), Greenland hare (*L. articus groenlandicus*). World-wide in distribution. Developmental stages occur in duodenum and ileum. *Oocysts:* ovoid to ellipsoidal, 22.7 μm by 14.2 μm, range 15–29 μm by 11–7 μm. Oocyst wall, smooth, colourless to lightish pink, micropyle not readily distinguishable. Sporulation time, 30-56 hours at room temperature.

Life-cycle. The endogenous developmental cycle has been described by Rutherford (1943). Two types of first generation schizonts have been observed: type 'A' forms produce four to eight merozoites and type 'B' up to 24 merozoites. Second generation schizonts occur on the fifth day of infection, and gamonts are first evident on the fourth to fifth days. The prepatent period is five to six days, this being the shortest endogenous cycle of the rabbit coccidia. An extensive account of the developmental stages of this species is given by Pellérdy (1974).

Eimeria perforans is of low pathogenicity. In young rabbits it may cause mild to moderate diarrhoea. On post mortem the wall of the anterior small intestine is thickened and whitish in colour due to the accumulation of developmental stages.

Eimeria piriformis. *Hosts:* domestic rabbit; France and Hungary. *Oocysts:* pyriform, 29 μm by 18 μm, range 26–32 μm by 17–21 μm. Oocyst wall, smooth, double contoured, yellowish-brown in colour, prominent micropyle visible at the tapering end. Sporulation time, 24–48 hours.

Life-cycle. This has been described in detail by Pellérdy (1974).

Parasites develop in the jejunum and ileum proximal to the nucleus of the epithelial cell. Two types of schizont are reported, type 'A' producing up to 12 merozoites and type 'B' producing a large number of slender forms. The majority of the first generation merozoites appear to be of the type 'A' form. In the second generation schizogony, the type 'B' merozoites (long, slender) predominate. Gamonts are seen on the seventh day, and the prepatent period is nine to ten days.

An infection of 30 000 oocysts may produce death in animals, irrespective of age. The pathological lesions consist of a catarrhal inflammation of the small intestine.

Eimeria stiedai. (Fig. 3.34). *Hosts:* domestic rabbit, cottontail, various hares (*L. europaeus, L. americanus, L. timidus*). Developmental stages occur in the liver. *Oocysts:* ovoidal to ellipsoidal, 36.9 μm by 19.9 μm, range 28–40 μm by 16–25 μm. Oocyst wall, smooth, yellowish-orange or salmon coloured, distinct micropyle. Sporulation time, three days at room temperature, 58 hours at 22°C.

Life-cycle. The extensive information on the developmental cycle of *E. stiedai* has been summarized by Pellérdy (1974). Smetana (1933*a*, *b*, *c*) provided the first extensive account of the life-cycle. Excystation occurs in the small intestine, and sporozoites penetrate the intestinal mucosa and pass via the hepatic-portal blood system to the liver and enter the epithelial cells of the bile ducts.

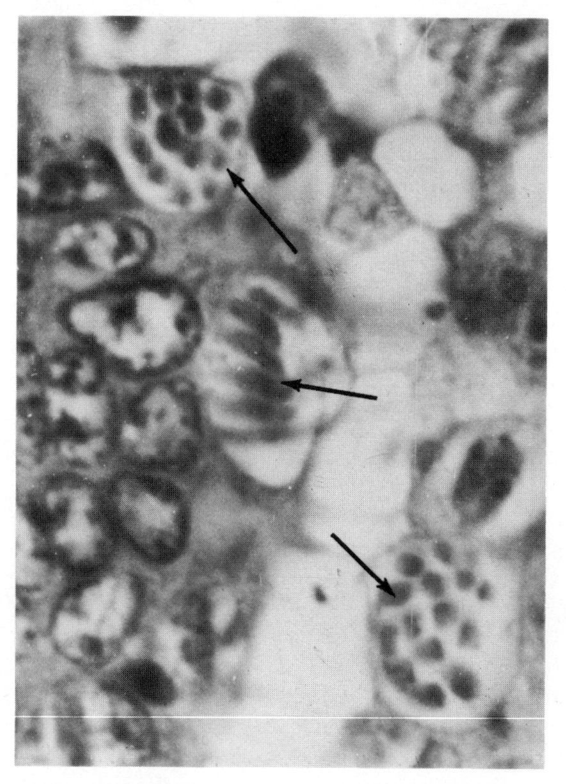

Fig 3.34 Bile duct epithelium of rabbit infected with *Eimeria stiedai*. Three schizonts are visible. (× 1200)

Developmental stages are normally found here five to six days after infection, but in heavy infections stages may be seen at 72 hours. Development occurs proximal to the nucleus of the epithelial cell. Mature schizonts measure 15–18 μm and there is no definite information concerning the number of asexual generations which take place. Gametogony is observed as early as the eleventh day after infection, but large numbers of gamonts are not evident until several days later. Both schizogony and gametogony can be observed in the later stages of infection. The prepatent period is 18 days, excessive numbers of oocysts are shed about day 23 and patency lasts until about day 37 after infection.

Pathogenesis. In mild infections little or no clinical signs are evident, but in heavy infections severe liver involvement produces progressive emaciation, an enormous enlargement of the liver, and death. When several hundred oocysts are given to young rabbits, death may be expected from three weeks onwards. Animals lose appetite, there is progressive emaciation and diarrhoea, and there is an enormous distention of the abdomen in which the enlarged liver can be felt; this is also associated with ascites.

In fatal cases the liver may be five to ten times its normal size. It is pale and shows a mass of yellow hepatic lesions filled with pus-like material. Petechial haemorrhages occur on the liver, in the kidneys and elsewhere in the body; oedema is seen in the peritoneal cavity, and the whole body may be oedematous.

The liver lesions are due essentially to an extensive enlargement of the bile ducts caused by a proliferation of the bile duct epithelium, this being thrown into a very large number of folds. The epithelial covering is filled with developmental stages of the parasite. The pus-like material in the bile ducts consists of desquamated epithelial cells, gametogonous stages and oocysts. There is a massive infiltraton of lymphocytes, plasma cells, eosinophils and a smaller number of neutrophils. If the animal survives, an invasion of the lesions by fibrous tissue occurs, new bile ducts proliferate, and large areas of the liver are transformed into fibrous tissue.

COCCIDIOSIS IN RABBITS

Coccidiosis is essentially a disease of the young rabbit and is seen especially in breeding and rearing establishments where sanitation is poor. However, outbreaks of disease are not uncommon under natural conditions, especially where the warren type of habitat is found. The important species are *E. stiedai* of the liver and *E. irresidua* and *E. magna* of the intestine.

A diagnosis of coccidiosis may be made on the demonstration of very large numbers of oocysts in the faeces; however, since coccidia are frequently present in rabbits and may be present in large numbers without any serious clinical signs, the most satisfactory diagnosis is made at a post mortem examination. With liver coccidiosis only massive infection is associated with ill health, and there is no evidence that a-small number of lesions cause ill health.

The most effective compounds for treatment are the sulphonamides. Davies et al. (1963) state that sodium sulphamezathine in the drinking water, at a concentration of 0.2%, is highly effective for hepatic coccidiosis and can be given for a long time without danger of toxicity. Other compounds for hepatic coccidiosis are sulphaguanidine, succinyl sulphathiazole at the rate of 0.5% in the feed (Horton-Smith 1947) and sulphaquinoxaline at the rate of 0.03% in the feed (Lund 1954). The nitrofurans are unsatisfactory for hepatic coccidiosis. Nitrofurazone has been recommended for intestinal coccidiosis by Boch (1957) at a dose of 0.5–1 g/kg. The author found that zoalene and amprolium had no protective action when they were fed at various levels in the food to rabbits experimentally infected with *E. stiedai*. Erka Z 6000, a pelleted food containing sulphadimethoxine (2,6-dimethoxy-4-sulphanylamido-pyrimidine) in a 3 : 1 ratio with diaveridine (2,4-diamino-5(3,4-dimethoxybenzyl)-pyrimidine) given at a concentration of 0.125% for three days is a reliable preventative medication for hepatic coccidiosis (Lämmler & Dürr 1969).

Control of rabbit coccidiosis is based on the improvement of hygiene in breeding and rearing establishments. Cages, hutches or pens should be cleaned out regularly, preferably every day; proper facilities should be made for feeding animals, and food should not be thrown on the floor for the animals. Feeding and watering troughs should be so placed that they are not contaminated with droppings. In large establishments a periodic post mortem examination on a few animals is a valuable check on the status of the infection.

COCCIDIA OF THE RAT

The more important coccidia of the rat, including the laboratory rat, are given below. While caesarean-derived, barrier-sustained colonies of laboratory rats do not have coccidial infections, others may do so and such infections may complicate the interpretation of experimental data. One species, *Eimeria nieschultzi*, has been used for studies of the immunology of coccidial infections.

Eimeria miyairii Ohira, 1912
Eimeria nieschultzi Dieben, 1924
Eimeria separata Becker and Hall, 1931

Eimeria miyairii. *Hosts:* Norway brown rat (*Rattus norvegicus*) and house rat (*Rattus rattus*) throughout the world. Not common and prevalence in laboratory rats is unknown. *Oocysts:* spherical to sub-spherical, 16.8–29 μm by 16.1–26 μm. Wall brownish yellow, thick, coarse, radially striated. Micropyle absent. Sporulation time 96–126 hours. Endogenous development was described by Roundabush (1937) and occurs in the superficial cells of the mucosa of the small intestine. Three generations of schizonts occur which are completed in four days. Pathogenicity is unknown.

Eimeria nieschultzi. *Hosts:* house rat, Norway brown rat; uncommon in laboratory rats. World-wide in distribution. *Oocysts:* ellipsoidal to ovoid, tapering at both ends, wall smooth, yellowish, 16–26 μm by 13–21 μm. No micropyle. Sporulation time 65–72 hours. Ovoid sporocysts have a small Stieda body and a residuum. The developmental cycle has been described by Roundabush (1937). The prepatent period is seven days, four generations of schizonts are produced, the fourth being evident in the small intestine on the fourth day after infection. The period of patency is four to five days and this is followed by immunity to reinfection. Pathogenic effects are seen mainly in young rats; diarrhoea, weakness and emaciation have been described (Pérard 1926). Lesions consist of a catarrhal enteritis in the distal third of the small intestine.

Eimeria separata. *Hosts:* Norway rat and other rats, world-wide. *Oocysts:* ellipsoid to ovoid 10–19 μm by 10–17 μm, colourless to pale yellow, no micropyle. The endogenous development has been described by Roundabush (1937) and occurs in the caecum and colon. The prepatent period is five to six days. Pathogenesis is mild.

Other species of coccidia in rats include *Eimeria nochti* Yakimoff and Gousseff, 1936 (*R. rattus. Oocysts:* 17 μm by 14 μm), *Eimeria hasei* Yakimoff and Gousseff, 1936 (*R. rattus. Oocysts:* ovoid 12 μm by 24 μm), *Eimeria ratti* Yakimoff and Gousseff, 1936 (*R. rattus. Oocysts:* cylindrical ovoid 16–28 μm by 14–16 μm), *Eimeria carinii* Pinto, 1926 (*R. norvegicus. Oocysts:* round, 22–24 μm, coarse, yellowish brown with a striated shell. While this species is similar to *E. miyairii*, animals immune to *E. carinii* are susceptible to *E. miyairii*). *Isospora ratti* Levine and Ivens, 1965 (Norway rat. *Oocysts:* sub-spherical, smooth tan-coloured, 22–24 μm by 20–21 μm).

COCCIDIA OF THE MOUSE

Eimeria falciformis (Eimer, 1870) Schneider, 1875
Eimeria ferrisi Levine and Ivens, 1965
Eimeria hindlei Yakimoff anf Gousseff, 1938
Eimeria keilini Yakimoff and Gousseff, 1938
Eimeria krijgsmanni Yakimoff and Gousseff, 1938
Eimeria musculi Yakimoff and Gousseff, 1938
Eimeria schueffneri Yakimoff and Gousseff, 1938
Cryptosporidium muris Tyzzer, 1910
Cryptosporidium parvum Tyzzer, 1912

Eimeria falciformis. *Host:* house mouse (*Mus musculus*). Common and world-wide in distribution. Occurs in the small and large intestine. *Oocysts:* oval to spherical, colourless, no micropyle 14–26 μm by 11–24 μm. Sporocysts ovoid. Endogenous development occurs in the epithelium of the large and small intestine. Schizonts are visible at four to five days after infection and oocyst discharge commences on the fifth day after infection. Moderate to severe enteritis may occur during this infection (Owen 1968).

The other species of *Eimeria* have been found in the house mouse (details of the oocysts are given by Pellérdy 1974) but their association with pathogenic signs is doubtful. A full account of coccidia of the mouse is given by Levine and Ivens (1965).

The two species of the genus *Cryptosporidium*, *C. muris* and *C. parvum* (Fig. 3.35), are extracellular forms, developmental stages occurring on the glandular epithelium of the stomach and small intestine, respectively. Both are common in the house mouse.

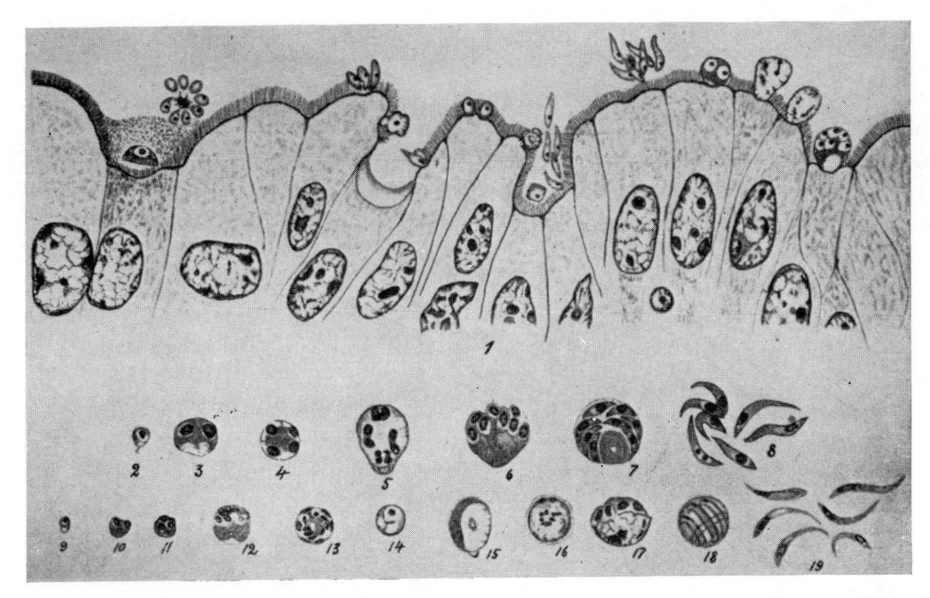

Fig 3.35 *Cryptosporidium parvum* from the small intestine of mice. (1) Various stages of parasite on the surface of the epithelium. (2–8) Schizogony cycle. (9–13) Development of microgametocyte and formation of microgametes. (14, 15) Development of macrogametocyte. (16–18) Development of oocyst and formation of sporozoites. (19) Escape of sporozoites from oocysts. (× ca 2000) (*From Wenyon 1926*)

COCCIDIA OF THE GUINEA-PIG

Eimeria caviae Sheather, 1924

Eimeria dolichotis Morini, Boero and Rodriguez, 1955

Cryptosporidium wrairi Vetterling, Jervis, Merrill and Sprinz, 1971

Klossiella cobayae Seidelin, 1914

Eimeria caviae. *Hosts:* guinea-pig (*Cavia aperea*—wild guinea-pig; *Cavia cobaya*—domestic guinea-pig). Common and world-wide. *Oocysts:* oval to ellipsoidal 17-25 μm by 13–18 μm (mean: 19 μm by 16 μm). Development occurs in the mucosa of the colon, schizogonic stages occurring seven to eight days after infection (Lapage 1940). Prepatent period is 7–12 days. Lapage (1940) noted that the infection was so common that almost every animal was subjected to mild or severe exposure sooner or later. Although usually non-pathogenic, *E. caviae* may cause diarrhoea and, ultimately, death. Mortality may reach 40% in severe outbreaks of disease.

Pathological lesions consist of distended colon, hyperaemia of the wall, pin-point haemorrhages and greyish white nodules of developmental stages.

Control of the infection is achieved by improved hygiene in a guinea-pig colony, treatment with succinylsulphathiazole (0.1% in the drinking water) is effective (Kleeberg & Steenken 1963).

Eimeria dolichotis. *Host:* occurs in the Patagonian cavey (*Dolichotis patagonica*) or 'marra'. *Oocysts:* sub-spherical, 22–26 μm by 18–19 μm, oocyst wall thin, no micropyle. Although this species resembles *E. caviae,* the two appear to be distinct since Zwart and Strik (1961) failed to infect guinea-pigs with the parasite of the 'marra'.

Cryptosporidium wrairi. *Host:* guinea-pig. Reported from the guinea-pig colony of the Walter Reed Army Institute of Research, Washington, DC, USA, where the overall infection rate was 30–40%. No oocysts have been reported. Schizonts develop from trophozoites and measure 3.4–4.4 μm when mature. Microgamonts resemble schizonts and produce 12–16 microgametes. Macrogametes measure 4–7 μm and have many polysaccharide granules. The 'attachment' organ of Tyzzer (1910) is found on all stages; however, it is considered a feeding structure rather than a holdfast structure. All forms are seen near the epithelial surface within the striated border of epithelial cells (Vetterling et al. 1971*a*). Vetterling et al. (1971*b*) have shown that although the parasites appear extracellular, they are actually within the host cell.

Pathological changes include shortening and broadening of villi, sometimes they become flattened. Irregularities appear in microvilli. An increase in monocytes and eosinophils occurs in the lamina propria of the anterior ileum. In more severe infections the terminal ileum and upper jejunum may be affected.

Clinical signs include weight loss but diarrhoea is not a usual sign.

Control. Infections can be eliminated by the addition of 0.2% sulphaquinoxaline or sulphamethazine to the water.

Klossiella cobayae. *Host:* guinea-pig. Probably cosmopolitan in distribution. Occurs in the kidney. It is doubtful whether a true oocyst occurs in the life-cycle. First generation schizonts occur in the endothelial cells of glomerular capillaries. They reach 2–7 μm in diameter, produce 8–12 merozoites, some of which enter other endothelial cells and others initiate schizogony in cells of the proximal convoluted tubule. These are large, contain 100 merozoites which, after release, migrate to the thick limb of Henle's loop and enter epithelial cells where gamont formation is initiated. Syzygy has been reported (developing gamonts occurring together) (Stevenson, in Wenyon 1926), but since this does not occur in *Klossiella equi* (Vetterling & Thompson 1972), its occurrence in *K. cobayae* probably needs re-evaluation. The fertilized zygote undergoes sporogony, growing to 30–40 μm and producing 30 sporoblasts, each of which produces about 30 sporozoites. What has been considered to be an oocyst wall is the host cell membrane. On rupture of the host cell, sporulated oocysts pass out in the urine. It is assumed infection is by ingestion and sporozoites enter capillaries, invade endothelial cells and repeat the cycle.

Pathological changes include accumulation of inflammatory and fibroblastic cells and obstruction of tubules by masses of sporocysts and degenerative cells (Cossel 1958).

The parasite does not appear to produce clinical signs of disease.

COCCIDIA OF FISH

Coccidia are not uncommon in fish. They occur mainly in the intestinal tract but some species parasitize the liver, kidney, swim bladder or other organs. Only those that may be encountered in ornamental, laboratory or fresh-water cultured fish are mentioned here.

Eimeria aurata Hoffman, 1965
Eimeria carpelli Léger and Stankovitch, 1921
Eimeria cyprini Plehn, 1924
Eimeria subepithelialis Moroff and Fiebiger, 1905
Eimeria truttae Léger and Hesse, 1919

Eimeria aurata. *Host:* goldfish. *Oocysts:* 16–24 μm by 14–17 μm. Sporulation time two to five days. This species has been associated with enteritis and lethargy in goldfish (Hoffman 1965).

Eimeria carpelli. *Host:* carp (*Cyprinus carpio*). *Oocysts:* small, spherical 13–14 μm, thin-walled. Schizonts, 6–7 μm in diameter, occur in the epithelial cells of the intestine. Prepatent period is 12–19 days. This species is primarily responsible for intestinal coccidiosis of carp. It causes clinical disease when fish are raised under crowded conditions and chiefly affects the fry of carp, producing coccidiosis in spring-time. As many as 80% of yearling carp may die from it (Pellérdy 1974). Autopsy changes include yellowish intestinal contents and faeces, reddish discoloration of the intestinal wall, yellow nodules which contain myriads of coccidial oocysts.

Eimeria cyprini. *Host:* carp and tench (*Tinca tinca*). *Oocysts:* small, spherical, 9 μm in diameter. Schäperclaus (1954) considers this species identical with *E. carpelli*. Developmental stages may occur subepithelially in the intestinal wall. The species produces a similar disease to *E. carpelli* in carp fry and if fish are overcrowded mortality may be high.

Eimeria subepithelialis. *Host:* carp. *Oocysts:* spherical 18–21 μm. Sporulation completed within the host cell. Pea-sized cluster of developmental stages are seen as raised projections in the intestine covered by epithelium. The species is the cause of 'granular coccidiosis' of carp. Fish become emaciated, increasingly susceptible to other fish diseases and heavy infection leads to increased mortality in fish rearing ponds.

Eimeria truttae. *Host:* salmon. *Oocysts:* spherical, 10–12 μm, colourless wall; passed in sporulated condition. A parasite of the small intestine and the epithelium lining the pyloric sac.

Coccidiosis in fish in fish ponds tends to occur in spring-time. Adult carriers shed oocysts which, during the winter, sink to the bottom of the pond where they accumulate. With the advent of spring, increased crowdedness and greater activity, oocysts are disturbed with the mud from the bottom of the pond, leading to infection by ingestion.

Control is by sanitation. Furazolidone mixed in the food has been used for treatment (Musselius & Strelkov 1968).

REFERENCES

COCCIDIA OF CATTLE, SHEEP, BUFFALOES, PIGS, HORSES ETC.

Aikawa, M. & Sterling, C. R. (1974) *Intracellular Parasitic Protozoa*, p. 76. New York: Academic Press.

Ajayi, J. A. & Todd, A. C. (1977) Relationships between two levels of aureomycin-sulphamethazine supplementation and acquisition of resistance to ovine coccidiosis. *Br. vet. J.*, **133**, 166–174.

Alicata, J. E. (1930) Globidium in the abomasum of American sheep. *J. Parasit.*, **16**, 162–163.

Alicata, J. E. & Willett, E. L. (1946) Observation on the prophylactic and curative value of sulphaguanidine in swine coccidiosis. *Am. J. vet. Res.*, **7**, 94–100.

Anderson, F. L., Lowder, L. J. & Hammond, D. M. (1965) Antibody production in experimental *Eimeria bovis* infection in calves. *Exp. Parasit.*, **16**, 23–25.

Andrews, J. S. & Spindler, L. A. (1952) *Eimeria spinosa* recovered from swine raised in Maryland and Georgia. *Proc. helminth. Soc. Wash.*, **19**, 64.

Arawaka, A. & Todd, A. C. (1968) Cellular responses of calves to first generation schizonts of *Eimeria bovis* after treatment of calves with sulphamezathine and linomycin hydrochloride. *Am. J. vet. Res.*, **29**, 1549–1559.

Baker, D. W. (1939) Species of eimerian coccidia found in New York State cattle. *Rep. N.Y. State vet. Coll.*, 1937–1938, pp. 160–166.

Barker, I. K. & Carbonell, P. L. (1974) *Cryptosporidium agni* sp.n. from lambs and *Cryptosporidium bovis* sp.n. from a calf, with observations on the oocyst. *Z. Parasitenk.*, **44**, 289–298.

Barker, I. K. & Remmler, O. (1970) Experimental *Eimeria leuckarti* infections in ponies. *Vet. Rec.*, **86**, 448-449.

Becklund, W. W. (1957) An epizootic of coccidiosis in western feeder lambs in Georgia. *N. Am. Vet.*, **38**, 262–264.

Biester, H. E. & Murray, C. (1934) Studies on infectious enteritis of swine. *J. Am. vet. med. Ass.*, **85**, 207–219.

Biswal, G. (1948) Coccidiosis in buffalo calves. *Indian vet. J.*, **25**, 36–38.

Boch, J., Pezenburg, E. & Rosenfeld, V. (1961) Ein Beitrag zur Kenntnis der Kokzidien der Schweine. *Berl. Münch. tierärztl. Wschr.*, **74**, 449–451.

Boch, J. & Wiesenhütter, E. (1963) Beitrag zur Klärung der Pathogenität der Schweinekokzdien. *Tierarztl. Umschau*, **18**, 223–225.

Boughton, D. C. (1945) Bovine coccidiosis: from carrier to clinical cases. *N. Am. Vet.*, **26**, 147–153.

Catchpole, J., Norton, C. C. & Joyner, L. P. (1975) The occurrence of *Eimeria weybridgensis* and other species of coccidia in lambs in England and Wales. *Br. vet. J.*, **131**, 392–401.

Challey, J. R. & Burns, W. C. (1959) The invasion of the cecal mucosa by *Eimeria tenella* sporozoites and their transport by macrophages. *J. Protozool.*, **6**, 238–241.

Chapman, H. D. (1974) The effects of natural and artificially acquired infections of coccidia in lambs. *Res. vet. Sci.*, **16**, 1–6.

Chatton, E. (1910) La kyste de Gilruth dans la muqueuse stomacle des ovides. *Arch. Zool. exp. gén.*, **5**, 114.

Chobotar, B. (1968) The asexual and sexual stages of *Eimeria auburnensis* in calves. Dissertation, University of Utah, Logan.

Chobotar, B. & Hammond, D. M. (1967) The sexual endogenous stages of *Eimeria auburnensis. J. Protozool.*, **14** (Suppl.), 21.

Christensen, J. F. (1938) Occurrence of the coccidian *Eimeria bukidnonensis* in American cattle. *Proc. helminth. Soc. Wash.*, **5**, 24.

Christensen, J. F. (1940) The source and availability of inefective oocysts in an outbreak of coccidiosis in lambs in Nebraska feedlots. *Am. J. vet. Res.*, **1**, 27-35.

Christensen, J. F. (1941*a*) Experimental production of coccidiosis in silage-fed feeder lambs with observations on oocyst discharge. *N. Am. Vet.*, **22**, 608–610.

Christensen, J. F. (1941*b*) The oocysts of coccidia from domestic cattle in Alabama. *J. Parasit.*, **27**, 203–220.

Christensen, J. F. & Porter, D. A. (1939) A new species of coccidium from cattle with observations on its life history. *Proc. helminth. Soc. Wash.*, **6**, 45–48.

Clarkson, M. J. (1959) The life history and pathogenicity of *Eimeria meleagrimitis*, Tyzzer 1929, in the turkey poult. *Parasitology*, **49**, 70–82.

Davis, L. R., Boughton, D. C. & Bowman, G. W. (1955) Biology and pathogenicity of *Eimeria alabamensis* Christensen, 1941, an intranuclear coccidium of cattle. *Am. J. vet. Res.*, **16**, 274–281.

Davis, L. R. & Bowman, G. W. (1952) Coccidiosis in cattle. *Proc. US Livestock Sanit. Ass., 58th Ann. Meeting*, pp. 39–50.

Davis, L. R. & Bowman, G. W. (1954) The use of sulphamethazine in experimental coccidiosis of dairy calves. *Cornell Vet.*, **44**, 71–79.

Davis, L. R. & Bowman, G. W. (1957) The endogenous development of *Eimeria zürnii*, a pathogenic coccidium of cattle. *Am. J. vet. Res.*, **18**, 569–574.

Davis, L. R. & Bowman, G. W. (1962) Schizonts and microgametocytes of *Eimeria auburnensis* Christensen and Porter, 1939, in calves. *J. Protozool.*, **9**, 424–427.

Davis, L. R. & Bowman, G. W. (1965) The life history of *Eimeria intricata* Spiegal, 1925, in domestic sheep. 2nd. int. Congr. Protozool. London. In *Progress in Protozoology*, p. 160. New York: Academic Press.

Davis, L. R. & Bowman, G. W. (1970) Intranuclear stages and second generation schizonts of *Eimeria ahsata* in domestic sheep. *J. Parasit.*, **68**, 122.

Davis, L. R., Bowman, G. W. & Boughton, D. C. (1957*a*) The endogenous development of *Eimeria alabamensis* Christensen, 1941, an intranuclear coccidium of cattle. *J. Protozool.*, **4**, 219–225.

Davis, L. R., Bowman, G. W. & Smith, W. N. (1963) Observations on the endogenous cycle of *Eimeria ahsata* Honess, 1942, in domestic sheep. *J. Protozool.*, **10** (Suppl.), 18.

Davis, L. R., Herlich, H. & Rohrbacher, G. H. (1957*b*) Observations on an outbreak of coccidiosis in a flock of western lambs in Alabama. *J. Protozool.*, **4** (Suppl.), 10.

Deiana, S. & Delitala, G. (1953) La coccidiosis dei piccoli rumunanti. *R. Parassit.*, **14**, 165–170; 201–212.

Deom, J. & Mortelmans, J. (1954) Observations sur la coccidiose du porc a *Eimeria debliecki* en congo Belge. *Ann. Soc. belge Méd. trop.*, **34**, 43.

Didă, I., Acsinte, N. & Purchera, V. (1972) Contributii la studiul coccidiozei rumegătoarelor mici. *Rev. Zootech. Med. Vet. Bucaresti*, **10**, 59–63.
van Doorninck, W. M. & Becker, E. F. (1957) Transport of sporozoites of *Eimeria necatrix* in macrophages. *J. Parasit.*, **43**, 40–43.
Doran, D. J. (1966) Pancreatic enzymes initiating excystation of *Eimeria acervulina* sporozoites. *Proc. helminth. Soc. Wash.*, **33**, 42–43.
Enigk, K. (1934) Zur kenntnis des *Globidium cameli* und der *Eimeria cameli*. *Arch. Protistenk.*, **83**, 371–380.
Farizy, P., Gomy, J. L. & Taranchon, P. (1970) Coccidiosis of cattle in France. *Rev. Med. vet.*, **121**, 1137–1144.
Favati, V. & Guerrieri, E. (1961) Ovine coccidiosis in Tuscany, Italy. *Ann. Fac. Med. vet. Univ. Pisa*, **14**, 305–313.
Foster, A. O. (1949) The economic losses due to coccidiosis. *Ann. N.Y. Acad. Sci.*, **52**, 434–442.
Foster, A. O., Christensen, J. F. & Habermann, R. T. (1941) Treatment of coccidial infections of lambs with sulfaguanidine. *Proc. helminth. Soc. Wash.*, **8**, 33–38.
Frenkel, J. K. (1977) *Besnoitia wallacei* of cats and rodents: with a reclassification of other cyst-forming isosporoid coccidia. *J. Parasit.*, **63**, 611–628.
Gilruth, J. A. (1910) Notes on a protozoan parasite found in the mucous membrane of the abomasum of a sheep. *Bull. Soc. Path. exot.*, **3**, 297–299.
Guerrero, C. A. (1967) Coccidia (Protozoa; Eimeriidae) of the alpaca, *Lama pacos*. *J. Protozool.*, **14**, 613–616.
Haberkorn, A. (1970) Die Entwicklung von *Eimeria falciformis* (Eimer 1870) in der weissen Maus (*Mus musculus*). *Z. Parasitenk.*, **34**, 49–67.
Hammond, D. M. (1973) Life cycles and development of coccidia. In: *The Coccidia*, ed. D. M. Hammond & P. L. Long, pp. 44–79. Baltimore: University Park Press.
Hammond, D. M., Anderson, F. L. & Miner, M. L. (1963*a*) The site of the immune reaction against *Eimeria bovis* in calves. *J. Parasit.*, **49**, 415–424.
Hammond, D. M., Anderson, F. L. & Miner, M. L. (1964) Response of immunized and non-immunized calves to cecal inoculation of first generation merozoites of *E. bovis*. *J. Parasit.*, **50**, 209–213.
Hammond, D. M., Bowman, G. W., Davis, L. R. & Simms, B. T. (1946) The endogenous phase of the life cycle of *Eimeria*. *J. Parasit.*, **32**, 409–427.
Hammond, D. M., Clark, W. N. & Miner, M. L. (1961) Endogenous phase of the life cycle of *Eimeria auburnensis* in calves. *J. Parasit.*, **47**, 591–596.
Hammond, D. M., Miner, M. L. & Anderson, F. L. (1960) The timing of the merozoite stage in the life cycle of *Eimeria bovis*. *J. Protozool.*, **7** (Suppl.), 11.
Hammond, D. M., Sayin, F. & Miner, M. L. (1963*b*) Developmental cycle and pathogenicity of *Eimeria ellipsoidalis* in calves. *Berl. Münch. tierärztl. Wschr.*, **76**, 331–333.
Hasche, M. R. & Todd, A. C. (1959) *Eimeria brasiliensis* Torres and Ramos, 1939, in Wisconsin. *J. Parasit.*, **45**, 202.
Hemmert-Halswick, A. (1943) Infektion mit *Globidium leuckerti* beim Pferd. *Z. Vetkde.* **55**, 192–199.
Henry, A. & Masson, G. (1932) La coccidiose du dromedaire. *Rec. Med. vet. exot.*, **5**, 185–193.
Horak, J. G., Raymund, S. M. & Louw, J. P. (1969) The use of amprolium in the treatment of coccidiosis in domestic ruminants. *J. S. Afric. vet. med. Ass.*, **40**, 293–299.
Horton-Smith, C. & Long, P. L. (1963) Coccidia and coccidiosis in the domestic fowl and turkey. *Adv. Parasit.*, **1**, 67–107.
Jackson, A. R. B. (1962) Excystation of *Eimeria arloingi* (Marotel, 1905). Stimuli from the host sheep. *Nature. Lond.*, **194**, 847–849.
Jacob, E. (1943) Zur Verbreitung der Kokzidienarten bei Schafen, Ziegen und Rehen. *Berl. Münch. tierärztl. Wschr.*, **31–32**, 258–260.
Klesius, P. H. & Fundenberg, H. H. (1977) Bovine transfer factor: *in vivo* transfer of cell-mediated immunity to cattle with alcohol precipitates. *Clin. Immunol. Immunopath.*, **8**, 238–246.
Klesius, P. H. & Kristensen, F. (1977) Bovine transfer factor: effect on bovine and rabbit coccidiosis. *Clin. Immunol. Immunopath.*, **7**, 240–252.
Kotlan, S., Pellerdy, L. & Versenyr, L. (1951) Experimentelle Studien über die Kokzidiose der Schafe. I. Die endogene Entwicklung von *Eimeria parva*. *Acat. Vet. Budapest*, **1**, 317–331.
Kutzer, E. (1960) Über die Kokzidien des Schweines in Österreich. *Z. Parasitkde*, **19**, 541–547.
Landers, E. J. (1952*a*) A new species of coccidia from domestic sheep. *J. Parasit.*, **38**, 569–570.
Landers, E. J. (1952*b*) The effect of low temperatures upon the viability of unsporulated oocysts of ovine coccidia. *J. Parasit.*, **39**, 547–552.
Lee, C. G. & Ross, A. D. (1977) Renal coccidiosis of the horse associated with *Klossiella equi*. *Aust. vet. J.*, **53**, 287–288.
Lee, R. P. (1954) The occurrence of the coccidian *Eimeria bukidnonensis* Tubangui, 1931, in Nigerian cattle. *J. Parasit.*, **40**, 464–466.
Lee, R. P. & Armour, J. (1958) A note on *Eimeria braziliensis* Torres and Ramos, 1939 and its relationship to *Eimeria böhmi* Supperer, 1952. *J. Parasit.*, **44**, 302–304.
Levine, N. D. (1973) *Protozoan Parasites of Domestic animals and of man*, 2nd ed., p. 406. Minneapolis: Burgess.
Levine, N. D. & Ivens, V. (1970) *The Coccidian Parasites (Protozoa, Sporozoa) of Ruminants*. Ill. Biol. Monogr. Series, No. 44. Urbana: University of Illinois Press.
Levine, N. D., Ivens, V. & Fritz, T. E. (1962) *Eimeria christenseni* sp. n. and other coccidia (Protozoa: Eimeriidae) of the goat. *J. Parasit.*, **48**, 255–269.
Liburd, M., Pabst, H. F. & Armstrong, W. D. (1972) Transfer factor in rat coccidiosis. *Cell. Immunol.*, **5**, 487–489.
Long, P. L. & Rose, M. E. (1970) Extended schizogony of *Eimeria mivati* in betamethasone-treated chickens. *Parasitology*, **60**, 147–155.
Lotze, J. C. (1952) The pathogenicity of the coccidian parasite, *Eimeria arloingi*, in domestic sheep. *Cornell Vet.*, **42**, 510–517.
Lotze, J. C. (1953) Life history of the coccidian parasite, *Eimeria arloingi* in domestic sheep. *Am. J. vet. Res.*, **14**, 86–95.
Lotze, J. C. (1954) The pathogenicity of the coccidian parasite *Eimeria ninakohlyakimovae*, Yakimov and Rastegaeva, 1930, in domestic sheep. *Proc. Am. vet. med. Ass.*, 1953, pp. 141–146.
Lotze, J. C. & Leek, R. G. (1970) Failure of development of the sexual phase of *Eimeria intricata* in heavily inoculated sheep. *J. Protozool.*, **17**, 414–417.
Marquardt, W. C. (1959) The morphology and sporulation of the oocyst of *Eimeria brasiliensis*, Torres and Ildefonso Ramos 1939, of cattle. *Am. J. vet. Res.*, **20**, 742–746.
Marquardt, W. C., Senger, C. M. & Seghetti, L. (1960) The effect of physical and clinical agents on the oocyst of *Eimeria zurnii* (protozoa, coccidia). *J. Protozool.*, **7**, 186–189.
Maske, H. (1893) Gregarinen im Labmagen des Schafes. Vorläufige Mittheilung. *Z. Fleisch. Milchhyg.*, **4**, 28–29.
Michael, E. & Probert, A. J. (1970) Histopathological observations on some coccidial lesions in natural infections of sheep. *Res vet. Sci.*, **11**, 441–446.
Norton, C. C. & Catchpole, J. (1976) The occurrence of *Eimeria marsica* in the domestic sheep in England and Wales. *Parasitology*, **72**, 111–114.
Norton, C. C., Joyner, L. P. & Catchpole, J. (1974) *Eimeria weybridgensis* sp. nov. and *Eimeria ovina* from the domestic sheep. *Parasitology*, **69**, 87–95.
Nyberg, P. A., Bauer, D. H. & Knapp, S. E. (1968) Carbon dioxide as the initial stimulus for excystment of *Eimeria tenella* oocysts. *J. Protozool.*, **15**, 144–148.
O'Neill, P. A. & Parfitt, J. W. (1976) Observations on *Isospora suis* infection in a minimal disease pig herd. *Vet. Rec.*, **98**, 321–323.
Panciera, R. J., Thomasson, R. W. & Garner, F. M. (1971) Cryptosporidial infection in a calf. *Vet. Path.*, **8**, 479–484.
Pande, B. P., Bhatia, B. B. & Chauhan, P. P. S. (1966) Endogenous stages of *Eimeria intricata* (Sporozoa: Eimeriidae) in Indian sheep. *Ind. J. Microbiol.*, **6**, 35–40.
Pande, B. P., Bhatia, B. B. & Chauhan, P. P. S. (1971) Sexual stages and associated lesions in *Eimeria bareillyi* in buffalo calves. *Ind. J. Anim. Sci.*, **41**, 151–154.
Patnaik, M. M. (1965) On the coccidian infections of buffalo calves. A study of the oocysts. *Agra Univ. J. Res. Sci.*, **13**, 239–256.
Pattillo, W. H. (1959) Invasion of the cecal mucosa of the chicken by sporozoites of *Eimeria tenella*. *J. Parasit.*, **45**, 253–258.
Pellérdy, L. P. (1949) Studies on coccidia occurring in the domestic pig, with the description of a new *Eimeria* species (*Eimeria polita* sp. N.) of that host. *Acta vet. hung.*, **1**, 101–109.
Pellérdy, L. P. (1965) *Coccidia and Coccidiosis*. Budapest: Akad Kiado.
Pellérdy, L. P. (1974) *Coccidia and Coccidiosis*, 2nd ed. Budapest: Akademiai Kiado, and Berlin and Hamburg: Paul Parey.
Pout, D. D. (1974) Coccidiosis of lambs. III. The reaction of the small intestinal mucosa to experimental infections with *E. arloingi* 'B' and *E. crandallis*. *Br. vet. J.*, **130**, 45–53.
Pout, D. D. (1976) Coccidiosis of sheep: a review. *Vet. Rec.*, **98**, 340–341.
Pout, D. D. & Catchpole, J. (1974) Coccidiosis of lambs. V. The clinical response to long-term infection with a mixture of different species of coccidia. *Br. vet. J.*, **130**, 388–399.
Pout, D. D., Norton, C. C. & Catchpole, J. (1973) Coccidiosis in lambs. 2. The production of faecal oocyst burdens in laboratory animals. *Br. vet. J.*, **129**, 568–582.
Rao, S. R. & Hiregaudar, L. S. (1954) Coccidial fauna of cattle in Bombay State with particular reference to a recent outbreak at Aery Milk Colony, together with a description of two species, *Eimeria bombayensis* and *Eimeria khurodensis*. *Bombay vet. Coll. Mag.*, **4**, 24–28.
Restani, R. (1966) Richerche sulla diffusione della coccidiosa nei greggi ovini della Marcia. *Att. Soc. Ital. Sci. vet.*, **20**, 719–723.
Restani, R. (1971) *Eimeria marsica* Nov. sp. (Protozoa: Eimeriidae) parassita di *Ovis aries* L. *Parassitologia*, **13**, 309–312.
Rommel, M. (1970*a*) Verlauf der *Eimeria scabra* und *E. polita*-Infektion in vollempfänglischen Ferkels und Lauferschweinen. *Berl. Münch. tierärztl. Wschr.*, **83**, 181–186.

Rommel, M. (1970*b*) Ausbildung and Dauer der Immunität gegen *Eimeria scabra* (Henry, 1931) und *E. polita* (Pellérdy, 1949). *Berl. Münch. tierärztl. Wschr.*, **83**, 236–240.
Rommel, M. & Ipezynski, V. (1967) Der Lebenszyklus des Schweinekokzids *Eimeria scabra* (Henry, 1931). *Berl. Münch. tierärztl. Wschr.*, **80**, 65–70.
Ross, D. B. (1968) Successful treatment of coccidiosis in lambs. *Vet. Rec.*, **83**, 189–190.
Rutherford, R. L. (1943) The life cycle of four intestinal coccidia of the domestic rabbit. *J. Parasit.*, **29**, 10–32.
Ryley, J. F. (1973) Cytochemistry, Physiology and Biochemistry. In: *The Coccidia*, eds D. M. Hammond & P. L. Long, pp. 145–181. Baltimore: University Park Press.
Salisbury, R. M. & Whitten, L. K. (1953) Coccidiosis in sheep. A review. *N.Z. vet. J.*, **1**, 69–72.
Sarwar, M. M. (1951) Occurrence of *Globidium gilruthi*, a protozoon parasite of sheep and goats from the Indo-Pakistan sub-continent. *Parasitology*, **41**, 282.
Scholtyseck, E. (1973) Ultrastructure. In: *The Coccidia*, ed. D. M. Hammond & P. L. Long, pp. 81–144. Baltimore: University Park Press.
Senger, C. M., Hammond, D. M., Thorne, J. L., Johnson, A. E. & Wells, M. (1959) Resistance of calves to reinfection with *Eimeria bovis*. *J. Protozool.*, **6**, 51–58.
Sheahan, B. J. (1976) *Eimeria leuckarti* infection in a thoroughbred foal. *Vet. Rec.*, **99**, 213–214.
Shumard, R. F. (1957) Ovine coccidiosis—incidence, possible endotoxin, and treatment. *J. Am. vet. med. Ass.*, **131**, 559–561.
Shumard, R. F. (1959*a*) Experimentally induced ovine coccidiosis. I. Use of nitrofurazone in the feed. *Vet. Med.*, **54**, 421–425.
Shumard, R. F. (1959*b*) Experimentally induced ovine coccidiosis. II. Use of water-soluble nitrofurazone as a therapeutic. *Vet. Med.*, **54**, 477–479.
Slater, R. L., Hammond, D. M. & Miner, M. L. (1970) *Eimeria bovis*: development in calves treated with thiamine antagonist (amprolium) in feed. *Trans. Am. microsc. Soc.*, **89**, 55–65.
Smith, W. N., Davis, L. R. & Bowman, G. W. (1960) The pathogenicity of *Eimeria ahsata*, a coccidium of sheep. *J. Protozool.*, **7** (Suppl.), 8.
Stockdale, P. H. G. (1977) Proposed life cycle of *Eimeria zuernii*. *Br. vet. J.*, **133**, 471–473.
Supperer, R. (1952) Die Coccidien des Rindes in Oesterreich. *Öst. Zool.*, **3**, 591–601.
Svanbaev, S. K. (1957) Sur la question de la faune et de la morphologie des coccidia des moutons et des chèvres de l'ouest du Kazakhstan. *Trudy. Inst. Zool., Alma-Ata*, **7**, 252–257.
Swanson, L. E. & Kates, K. C. (1940) Coccidiosis in a litter of pigs. *Proc. helminth. Soc. Wash.*, **7**, 29–30.
Tarlatzis, C., Panetsos, A. & Dragonas, P. (1955) Furacin in the treatment of ovine and caprine coccidiosis. *J. Am. vet. med. Ass.*, **126**, 391–392.
Torres, S. & Ramos, J. I. (1939) Eimerias dos bovinos em Pernambuco, *E. idlefonsoi* e *E. braziliensis* spcs. ns. *Archos Inst. Pesq. agron., Pernambuco*, **2**, 79–96.
Vetterling, J. M. (1965) Coccida (Protozoa: Eimeriidae) of swine. *J. Parasit.*, **51**, 897–912.
Vetterling, J. M. (1966) Endogenous cycle of the swine coccidium *Eimeria debliecki* Douwes, 1921. *J. Protozool.*, **13**, 290–300.
Vetterling, J. M., Takeuchi, A. & Madden, P. A. (1971) Ultrastructure of *Cryptosporidium wrairi* from the guinea-pig. *J. Protozool.*, **18**, 248–260.
Vetterling, J. M. & Thompson, D. E. (1972) *Klossiella equi* Baumann, 1946 (Sporozoa: Euccocidia: Adelina) from equids. *J. Parasit.*, **58**, 589–594.
Wacha, R. S., Hammond, D. M. & Miner, M. L. (1971) The development of the endogenous stages of *Eimeria ninakohlyakimovi* (Yakimoff and Rastagieff, 1930) in domestic sheep. *Proc. Helminth. Soc. Wash.*, **38**, 167–180.
Wheeldon, E. B. & Grieg, W. A. (1977) *Globidium leuckarti* infection in a horse with diarrhoea. *Vet. Rec.*, **100**, 102–103.
Whitten, L. K. (1953) A preliminary field experiment on the treatment of coccidiosis in lambs. *N. Z. vet. J.*, **1**, 78–80.
Wiesenhütter, E. (1962*a*) Ein Beitrag zur Kenntnis der endogen Entwicklung von *Eimeria spinosa* des Schweines. *Berl. Münch tierärztl. Wschr.*, **75**, 72–173.
Wiesenhütter, E. (1962*b*) Experimentelle Studien über die Entwicklung von *Eimeria debliecki* und *Eimeria spinosa* des Schweines. Inaug. Diss., Berlin.
Wilson, I. D. & Morley, L. C. (1933) A study of bovine coccidiosis. II. *J. Am. vet. med. Ass.*, **82**, 826–850.
Yakimoff, W. L., Gousseff, W. F. & Rastegaieff, E. F. (1933) Die Coccidiose der wilden kleinen Wiederkauer. *Z. Parasitkde*, **5**, 85–93.

COCCIDIA OF POULTRY, DOGS, CATS, RABBITS, ETC.

Allen, E. A. (1936) *Tyzzeria perniciosa* gen. et sp. Nov., a new coccidium from the small intestine of the pekin duck, *Anas domesticus L. Arch. Protistk.*, **87**, 262–267.
Allen, W. M., Berrett, S., Hein, H. & Herbert, C. N. (1973) Some physio-pathological changes associated with experimental *Eimeria brunetti* infection in the chicken. *J. comp. Path.*, **83**, 369–375.
Altman, J. E. (1951) Treatment of canine coccidiosis (*Isopora bigemina*) with aureomycin. *J. Am. vet. med. Ass.*, **119**, 207–209.
Andrews, J. M. (1926) A factor in host-parasite specificity of coccidiosis. *Anat. Rec.*, **34**, 154.
Arundel, J. H. (1960) Chemotherapy of caecal coccidiosis in chickens: 3:5 dinitro-*o*-tolumide. *Aust. vet. J.*, **36**, 49–53.
Ball, S. J. (1959) Chemotherapy of caecal coccidiosis in chickens: the activity of nicarbazin. *Vet. Rec.*, **71**, 86–91.
Becker, E. R. (1934) *Coccidia and Coccidiosis of Domesticated, Game and Laboratory Animals and of Man*. Ames: Iowa State University Press.
Becker, E. R. (1956) Catalogue of Eimeriidae in genera occuring in vertebrates and not requiring intermediate hosts. *Iowa State Coll. J. Sci.*, **31**, 85–139.
Becker, E. R., Zimmerman, W. J. & Pattillo, W. (1955) A biometrical study of the oocysts of *Eimeria brunetti*, a parasite of the common fowl. *J. Protozool.*, **2**, 145–150.
Bhatia, B. B. & Pande, B. P. (1968) On the endogenous development of *Eimeria mayurai* Bhatia and Pande, 1966 in a baby peafowl (*Pavo cristatus L.*). *Ind. J. Anim. Hlth*, **7**, 105–107.
Boch, J. (1957) Versuche zur Behandung der Kaninchenkokzidiose mit Nitrofurazon (Furzcin-W). *Berl. Münch. tierärztl. Wschr.*, **70**, 264–267.
Boles, J. I. & Becker, E. R. (1954) The development of *Eimeria brunetti* Levine, in the digestive tract of chickens. *Iowa State Coll. J. Sci.*, **29**, 1–26.
Boyler, C. I. & Brown, J. A. (1953) The comparative coccidiostatic activity of some drugs against turkey coccidia. *Proc. 90th Meet. Am. vet. med. Ass.*, 20–30 July, 1955, pp. 328–336.
Brackett, S. & Bliznick, A. (1950) The occurrence and economic importance of coccidiosis in chickens. *Bull. Lederle Lab.*, **77**.
Brumpt, L. C. (1943) Le traitment des coccidioses des animaux domestiques par quinacrine ou atebrine. *Ann. Parasit. hum. comp.*, **19**, 95–115.
Carvalho, J. C. M. (1943) The coccidia of wild rabbits of Iowa. I. Taxonomy and host specificity. *Iowa State Coll. J. Sci.*, **18**, 103–134.
Challey, R. & Burns, W. C. (1959) The invasion of the cecal mucosa by *Eimeria tenella* sporozoites and their transport by macrophages. *J. Protozool.*, **6**, 238–241.
Champion, L. R. (1954) The inheritance of resistance to caecal coccidiosis in the domestic fowl. *Poultry Sci.*, **33**, 670–681.
Cheissin, E. M. (1946) Duration and life cycle in rabbit coccidia. *C.r. Acad. Sci. USSR.* **52**, 557–570.
Cheissin, E. M. (1947) A new species of rabbit coccidia (*Eimeria coecicola* N. sp.). *C.r. Acad. Sci. USSR*, **55**, 177–179.
Christie, E., Dubey, J. P. & Pappas, P. W. (1976) Prevalence of *Sarcocystis* infection and other intestinal parasitisms in cats from a humane shelter in Ohio. *J. Am. vet. med. Ass.*, **168**, 421–422.
Clarkson, M. J. (1956) Experimental infection of turkey poults with *Eimeria adenoeides* (Moore and Brown, 1951) isolated from a natural case in Great Britain. *Nature*, **178**, 196–197.
Clarkson, M. J. (1958) Life history and pathogenicity of *Eimeria adenoeides*, Moore and Brown, 1951 in the turkey poult. *Parasitology*, **48**, 70–88.
Clarkson, M. J. (1959*a*) The life history and pathogenicity of *Eimeria meleagrimitis* Tyzzer 1929 in the turkey poult. *Parasitology*, **49**, 70–82.
Clarkson, M. J. (1959*b*) The life history and pathogenicity of *Eimeria meleagridis* Tyzzer, 1927 in the turkey poult. *Parasitology*, **49**, 519–528.
Clarkson, M. J. & Gentles, M. A. (1958) Coccidiosis in turkeys. *Vet. Rec.*, **70**, 211–214.
Cossel, L. (1958) Nierenbefunde beim Meerschweinchen bei Klossielleninfektion (*Klossiella cobayae*). *Schweiz. Z. allg. Path. Bakt.*, **21**, 62–73.
Critcher, S. (1950) Renal coccidiosis in Pea Island Canada geese. *Wildl. in N. Carolina*, **14**, 14–15.
Cuckler, A. C., Malanga, C. M. Basso, A. J. & O'Neill, R. C. (1955) The antiparasitic activity of substituted complexes. *Science, N.Y.*, **122**, 244–245.
Cuckler, A. C., McManus, E. C. & Campbell, W. C. (1969) Development of resistance in coccidia. *Acta vet.* (Brno), **38**, 87–90.
Cuckler, A. C. & Ott, W. H. (1947) The effect of sulfaquinoxaline on the developmental stages of *Eimeria tenella*. *J. Parasit.*, **33** (Suppl.), 10–11.
Danforth, H. D., Ruff, M. D., Reid, W. M. & Johnson, J. (1977) Anticoccidial activity of salinomycin in floor-pen experiments with broilers. *Poultry Sci.*, **56**, 933–938.
Davies, S. F. M. (1956) Intestinal coccidiosis in chickens caused by *Eimeria necatrix*. *Vet. Rec.*, **68**, 853–857.

Davies, S. F. M. (1957) An outbreak of duck coccidiosis in Britain. *Vet. Rec.*, **69**, 1051–1052.

Davies, S. F. M. (1963) *Eimeria brunetti*, an additional cause of intestinal coccidiosis in the domestic fowl in Britain. *Vet. Rec.*, **75**, 1–4.

Davies, S. F. M., Joyner, L. P. & Kendall, S. B. (1963) *Coccidiosis*. Edinburgh: Scottish Academic Press.

Davies, S. F. M. & Kendall, S. B. (1954*a*) The principal application of sulphamezathine therapy for caecal coccidiosis. *Vet. Rec.*, **66**, 19.

Davies, S. F. M. & Kendall, S. B. (1954*b*) The effect of sodium sulphaquinoxaline and sodium sulphamezathene in interrupted schedules of treatment on the development of *Eimeria tenella*. *J. comp. Path. Ther.*, **64**, 87–93.

Van Doorninck, W. M. & Becker, E. R. (1957) Transport of sporozoites of *Eimeria necatrix* in macrophages. *J. Parasit*, **43**, 40–43.

Doran, D. J. (1973) Cultivation of coccidia in avian embryos and cell culture. In: *The Coccidia*, eds D. M. Hammond & P. L. Long, pp. 183–252. Baltimore: University Park Press.

Duberman, D. (1960) Treatment of canine coccidiosis using nitrofurazone and sulphonamides. *J. Am. vet. med. Ass.*, **136**, 29–30.

Dubey, J. P. (1975) *Isospora ohioensis* sp.n. proposed for *I. rivolta* of the dog. *J. Parasit.*, **61**, 462–465.

Dubey, J. P. (1976) A review of *Sarcocystis* of domestic animals and of other coccidia of cats and dogs. *J. Am. vet. Med. Ass.*, **169**, 1061–1078.

Dubey, J. P. (1977) *Toxoplasma*, *Hammondia*, *Besnoitia*, *Sarcocystis* and other tissue cyst-forming coccidia of man and animals. In: *Parasitic Protozoa*, ed. J. P. Kreier, pp. 101–237. New York: Academic Press.

Dubey, J. P. & Fayer, R. (1976) Development of *Isospora bigemina* in dogs and other mammals. *Parasitology*, **73**, 371–380.

Dubey, J. P. & Frenkel, J. K. (1972) Cyst-induced toxoplasmosis in cats. *J. Protozool.*, **19**, 155–177.

Dubey, J. P., Miller, N.I. & Frenkel, J. K. (1970) The *Toxoplasma gondii* oocyst from cat feces. *J. exp. Med.*, **132**, 636–662.

Dubey, J. P. & Streitel, R. H. (1976) *Isospora felis* and *I. rivolta* infections in cats induced by mouse tissue or oocysts. *Br. vet. J.*, **132**, 649–651.

Edgar, S. A. (1955) Sporulation of oocysts at specific temperatures and notes on the prepatent period of several species of avian coccidia. *J. Parasit.*, **41**, 214–216.

Edgar, S. A., (1958) Control of coccidiosis of chickens and turkeys by immunization. *Poultry Sci.*, **37**, 1200.

Edgar, S. A., Flanagan, C. & Hwang, J. (1956) Breeding and immunizing chickens for resistance to coccidiosis. I. Immunization phase. *66th and 67th Ann. Rep. Agric. exp. Stn, Alabama Polytech. Inst.*, 1955–1956, pp. 46–47.

Edgar, S. A. & Siebold, C. T. (1964) A new coccidium of chickens. *Eimeria mivati* sp. N (Protozoa: Eimeriidae) with details of its life history. *J. Parasit.*, **50**, 193–204.

Farr, M. M. (1964) Life cycle of *Eimeria gallopavonis* Hawkins in the turkey. *J. Parasit.*, **50** (Suppl.), 50.

Fayer, R. & Johnson, A. J. (1973) Development of *Sarcocystis fusiformis* in calves infected with sporocysts from dogs. *J. Parasit.*, **59**, 1135–1137.

Frenkel, J. K. (1977) *Besnoitia wallacei* of cats and rodents: with a reclassification of other cyst-forming isosporoid coccidia. *J. Parasit.*, **63**, 611–628.

Frenkel, J. K. & Dubey, J. P. (1972) Rodents as vectors for feline coccidia *Isospora felis* and *Isospora rivolta*. *J. infect. Dis.*, **125**, 69–72.

Frenkel, J. K., Heydorn, A. O., Melhorn, H. & Rommel, M. (1979) Sarcocystinae: *Nomina dubia* and available names. *Z. Parasitenk.*, **58**, 115–139.

Gardiner, J. L. (1955) The severity of cecal coccidiosis infection in chickens as related to the age of the host and the number of oocysts ingested. *Poultry Sci.*, **34**, 415–420.

Giambrone, J. J., Anderson, W. I., Reid, W. A. & Eidson, C. S. (1977) Effects of infectious bursal disease on the severity of *Eimeria tenella* infections in broiler chicks. *Poultry Sci.*, **56**, 243–246.

Gräfner, G. & Graubmann, H. D. (1964) *Eimeria kotlani* n.s.p., eine neue pathogene Kokzidienart bei Gansen. *Mh. vet. Med.*, **19**, 819–821.

Hanson, H. C., Levine, N. D. & Ivens, V. (1957) Coccidia (Protozoa: Eimeriidae) of North American wild geese and swans. *Can. J. Zool.*, **35**, 715–733.

Harwood, P. D. & Stunz, D. (1949) Nitrofurazone in the medication of avian coccidiosis. *J. Parasit.*, **35**, 175–182.

Hauser, K. W. (1959) Erfahrungen bei der Bekämpfung der Kokzidiose der Tauben. *Berl. Münch. tierärztl. Wschr.*, **72**, 481–483.

Hawkins, P. A. (1952) Coccidiosis in turkeys. *Mich. State Coll. Technol. Bull.*, **206**, 1–87.

Hein, H. (1963) Vaccination against infection with *Eimeria tenella* in broiler chickens. *Proc. 17th vet. Congr. Hannover*, **1**, 1443–1452.

Herman, C. M., Jankiewicz, H. A. & Saarni, R. W. (1942) Coccidiosis in California quail. *Condor*, **44**, 168.

Heydorn, A-O., Gestrich, R., Mehlhorn, H. & Rommel, M. (1975) Proposal for a new nomenclature of the *Sarcosporidia*. *Z. Parasitenk.*, **48**, 73–82.

Heydorn, A. O. & Rommel, M. (1972) Beiträge zum Lebenszyklus der Sarkosporidien. II. Hund und Katze als Überträger der Sarksporidien des Rindes. *Berl. Münch. tierärztl. Wschr.*, **85**, 121–123.

Hitchcock, D. J. (1955) The life cycle of *Isospora felis* in the kitten. *J. Parasit.*, **41**, 383–397.

Hoffman, G. L. (1965) *Eimeria aurati* n. sp. (Protozoa: Eimeriidae) from goldfish (*Carassius auratus*) in North America. *J. Protozool.*, **12**, 273–275.

Horton-Smith, C. (1947) Treatment of hepatic coccidiosis in rabbits. *Vet. J.*, **103**, 207–213.

Horton-Smith, C. & Long, P. L. (1959) The effects of different anticoccidial agents on the intestinal coccidioses of the fowl. *J. comp. Path.*, **69**, 192–207.

Horton-Smith, C. & Long, P. L. (1961) The effect of sulphonamide medication on the life cycle of *Eimeria meleagrimitis* in turkeys. *Exp. Parasit.*, **11**, 93–101.

Hymas, T. A. & Stevenson, G. T. (1960) A study of the action of zoalene on *Eimeria tenella* and *Eimeria necatrix* when administered in the diet or in the drinking water. *Poultry Sci.*, **39**, 1291–1262.

Inoue, J. (1967) *Eimeria saitamea* n.sp. a new cause of coccidiosis in domestic ducks. (*Anas platyrhynchos* var. *domestica*). *Jap. J. vet. Res.*, **29**, 209–215.

Joyner, L. P. (1958) Experimental *Eimeria mitis* infections in chickens. *Parasitology*, **48**, 101–112.

Kendall, S. B. & McCullough, F. S. (1952) Relationships between sulphamezathine therapy and the acquisition of immunity to *Eimeria tenella*. *J. comp. Path.*, **62**, 116–124.

Kleeberg, H. H. & Steenken, W. (1963) Severe coccidiosis in guinea-pigs. *J. S. Afr. vet. med. Ass.*, **34**, 49–52.

Klimeš, B. (1963) Coccidia of the domestic goose (*Anser anser dom.*). *Zbl. vet.* Med., **10**, 427–448.

Kotlán, S. (1933) Zur Kenntnis der Kohzidiose des Wassergeflugels. Die Kokzidiose der Hausgans. *Zbl.* **129**, 11–21.

Kouwenhoven, B. & van der Horst, C. J. G. (1972) Disturbed intestinal absorption of vitamin A and carotenes and the effect of a low pH during *Eimeria acervulina* infection in the domestic fowl (*Gallus domesticus*). *Z. Parasitenk.* **38**, 152–161.

Laarman, J. J. (1963) *Isospora hominis* (Railliet and Lucet, 1891) in the Netherlands. In: *Progress in Protozoology*, eds J. Ludvik, J. Lom & J. Vavra, pp. 445–447. New York: Academic Press.

Lämmler, G. & Dürr, V. (1969) Zur Chemotherapie der Leberkokzidoise des Kaninchens mit Sulfadimethoxyn-Diaveridin. *Berl. Münch. tierärztl. Wschr.*, **82**, 480–484.

Lapage, G. (1940) The study of coccidiosis (*Eimeria caviae* Sheather, 1924) in the guinea-pig. *Br. vet. J.*, **96**, 144–154; 190–202; 242–254; 280–295.

Lee, C. D. (1934) The pathology of coccidiosis in the dog. *J. Am. vet med. Ass.*, **85**, 760–781.

Lee, D. L. & Millard, B. J. (1971) Fine structure of the schizonts of *Eimeria praecox*. *Int. J. Parasit.*, **1**, 37–41.

Leibovitz, L. (1968) *Wenyonella philiplevinei* n.sp. a coccidial organism of the white Pekin Duck. *Avian Dis.*, **12**, 670–681.

Lepp, D. L. & Todd, K. S., Jr. (1974) Life cycle of *Isospora canis* Nemeséri 1959 in the dog. *J. Protozool.*, **21**, 199–206.

Levine, N. D. (1952) *Eimeria magnalabia* and *Tyzzeria* spp. (Protozoa:Eimeriidae) from the common Canada goose. *Cornell Vet.*, **42**, 247–252.

Levine, N. D. (1973) *Protozoan Parasites of Domestic Animals and of Man*, 2nd ed. Minneapolis: Burgess.

Levine, N. D. (1977) Nomenclature of *Sarcosystis* in the ox and sheep and of fecal coccidia of the dog and cat. *J. Parasit.*, **63**, 36–51.

Levine, N. D. (1978) *Textbook of Veterinary Parasitology*. p. 236. Minneapolis: Burgess.

Levine, N. D. & Ivens, V. (1965*a*) The coccidian parasites (Protozoa; Sporozoa) of rodents. *Ill. Biol. Monogr.* Series, No. 33, p. 365. Urbana: University of Illinois Press.

Levine, N. D. & Ivens, V. (1965*b*) *Isospora* species in the dog. *J. Parasit.*, **51**, 859–864.

Levine, P. P. (1938) *Eimeria hagani* N. sp. (Protozoa: Eimeriidae) a new coccidium of the chicken. *Cornell Vet.*, **28**, 263–266.

Levine, P. P. (1939) The effect of sulphanilamide on the course of experimental avian coccidiosis. *Cornell Vet.*, **29**, 309.

Levine, P. P. (1942) A new coccidium pathogenic for chickens, *Eimeria brunetti* N. sp. (Protozoa:Eimeriidae). *Cornell Vet.*, **32**, 430–439.

Long, P. L. (1959) A study of *Eimeria maxima* Tyzzer, 1929. A coccidium of the fowl (*Gallus gallus*). *Ann. trop. Med. Parasit.*, **53**, 325–333.

Long, P. L. (1963) The effect of a combination of sulphaquinoxaline and amprolium against different species of *Eimeria* in chickens. *Vet. Rec.*, **75**, 645–650.

Long, P. L. (1967*a*) Studies on *Eimeria mivati* in chickens and a comparison with *Eimeria acervulina*. *J. comp. Path.*, **77**, 313–325.

Long, P. L. (1967*b*) Studies on *Eimeria praecox* Johnson, 1930 in the chicken. *Parasitology*, **37**, 351–361.

Long, P. L. (1968) The pathogenic effect of *Eimeria praecox* and *E. acervulina* in the chicken. *Parasitology,* **58**, 691–700.
Long, P. L. (1973) Pathology and pathogenicity of coccidial infections. In: *The Coccidia,* eds D. M. Hammond & P. L. Long pp. 183–252. Baltimore: University Park Press.
Lund, E. E. (1949) Considerations on the practical control of intestinal coccidiosis of domestic rabbits. *Ann. N.Y. Acad. Sci.,* **52**, 611–620.
Lund, E. E. (1954) The effect of sulphaquinoxaline on the course of *Eimeria stiedae* infections in the domestic rabbit. *Exp. Parasit.,* **3**, 497–503.
McGee, H. L. (1950) Coccidiosis in the dog. Clinical observation. *J. Am. vet. med. Ass.,* **117**, 227–228.
McGregor, J. K. (1952) Renal coccidiosis in geese. *J. Am. vet. med. Ass.,* **121**, 452–453.
McLoughlin, D. K. & Wehr, E. E. (1960) Stages in the life of *Eimeria tenella* affected by nicarbazin. *Poultry Sci.,* **39**, 534–538.
Madsen, H. (1941) The occurrence of helminths and coccidia in partridges and pheasants in Denmark. *J. Parasit.,* **27**, 29–34.
Mahrt, J. L. (1966) Ph.D. Thesis. University of Illinois, Urbana.
Moore, E. N. (1954) Species of coccidia affecting turkeys. *Proc. 91st ann. Meeting. Am. vet. med. Ass.,* pp. 300–304.
Moore, E. N. & Brown, J. A. (1951) A new coccidium pathogenic for turkeys, *Eimeria adenoeides* N. sp. (Protozoa: Emeriidae). *Cornell Vet.,* **31**, 124–125.
Moore, E. N. & Brown, J. A. (1952) A new coccidium of turkeys, *Eimeria innocua* N. sp. (Protozoa: Eimeriidae). *Cornell Vet.,* **42**, 395–409.
Moore, E. N., Brown, J. A. & Carter, R. D. (1954) A new coccidium of turkeys: *Eimeria subrotunda* N. sp. (Protozoa: Emeriidae). *Poultry Sci.,* **33**, 925–929.
Moorehouse, N. F. & McGuire, W. C. (1958) The pathogenicity of *Eimeria acervulina. Poultry Sci.,* **37**, 665–672.
Morini, E. G. (1950) Coccidiosis in pigeons. *Rev. Med. vet. B. Aires,* **32**, 207–226.
Musselius, V. A. & Strelkov. J. A. (1968) Diseases and control measures for fishes of Far East Complex in farms of the USSR. *Bull. Off. Int. Epizootiol.,* **69**, 1603–1611.
Nemeséri, L. (1960) Beiträge zur Atiologie der Coccidiose der Hunde. I. *Isospora canis* sp. Nov. *Acta vet. hung.,* **10**, 95–99.
Newberne, P. M. & Buck, W. B. (1956) Studies on drug toxicity in chicks. *Poultry Sci.,* **35**, 1259–1264.
Nieschulz, O. (1925) Über die Entwicklung des Taubencoccids *Eimeria pfeifferi* (Labbe, 1896). *Arch. Protistenk.,* **51**, 479–494.
Nieschulz, O. (1935) Über Kokzidien der Haustaube. *Z. Bakt. T. Orig.,* **134**, 390–393.
Norton, C. C. & Wise, D. R. (1982) Efficacy of clopidol as an anticoccidial for pheasants. *Vet. Rec.,* **110**, 406.
Owen, D. (1968) Investigations: B. Parasitological studies. *Lab. Animal Care News Letter,* **35**, 7–9.
Parkin, B. S. (1943) Treatment of canine coccidiosis. *J. S. Afr. vet. med. Ass.,* **14**, 73–76.
Patillo, W. H. (1959) Invasion of the cecal mucosa of the chicken by sporozoites of *Eimeria tenella. J. Parasit.,* **45**, 253–258.
Pellérdy, L. P. (1954) Beiträge zur Spezifizität der Coccidien des Hasen und Kaninchens. *Acta vet. hung.,* **4**, 481–487.
Pellérdy, L. P. (1956) On the status of the *Eimeria* species of *Lepus europaeus* and related species. *Acta vet. hung.,* **6**, 453–467.
Pellérdy, L. P. (1960) Adatok a maxima-coccidiosis ismeretchez es hazai eloforduIasahoz magy. *Allatorv. Lap.,* **15**, 307–309.
Pellérdy. L. P. (1974) *Coccidia and Coccidiosis.* 2nd. ed. Berlin & Hamburg: Paul Parey.
Pellérdy, L. P. & Babos, S. (1953) Untersuchungen über die endogene Entwicklung sowie pathologische Bedeutung von *Eimeria media. Acta vet. hung.,* **3**, 173–178.
Pérard, C. (1926) Sur la coccidiose du rat. *Acad. Vet. France Bull.,* **102**, 120–124.
Peterson, E. H. (1949) Sulfonamides in the control of experimental coccidiosis in the turkey. *Vet. Med.,* **44**, 126–128.
Peterson, E. H. (1960) A study of anti-coccidial drugs against experimental infections with *Eimeria tenella* and *necatrix. Poultry Sci.,* **39**, 739–745.
Pout, D. D. (1968) The reaction of the small intestine of the chicken to infection with *Eimeria* spp. In: *Reaction of the Host to Parasitism,* ed. E. J. L. Soulsby. pp. 28–38. Marburg: Elwert.
Raines, T. V. (1968) Buquinolate—a review. *Poultry Sci.,* **47**, 1425–1432.
Randall, C. J. & Norton, C. C. (1973) Acute intestinal coccidiosis in geese. *Vet. Rec.,* **91**, 46–47.
Ray, H. N. (1945) On a new coccidium *Wenyonella gallinae* N. Sp. from the gut of the domestic fowl, *Gallus gallus domesticus. Linn. Curr. Sci.,* **14**, 275.
Reid, W. M. (1975) Progress in the control of coccidiosis with anticoccidials and planned immunization. *Am. J. vet. Res.,* **36**, 593–596.
Reid, W. M., Dick, J., Rice, J. & Stino, F. (1977) Effects of monensin-feeding regimens on flock immunity to coccidiosis. *Poultry Sci.* **56**, 66–71.
Reid, W. M., Johnson, J. & Dick, J. (1975) Anticoccidial activity of lasalacid in control of moderate and severe coccidiosis. *Avian Dis.,* **19**, 12–18.
Rommel, M. & Geisel, O. (1975) Untersuchungen über die Verbreitung und den Lebenszyklus einer Sarksporidienart des Pferds (*Sarcocystis equinacnis* n. spec). *Berl. Münch. tierärztl. Wschr.* **88**, 468–471.
Rommel, M., Heydorn, A. O. & Gruber, F. (1972) Beitrage zur Lebenszyklus der Sarkosporidien. I. Die Sporozyste von *S. tenella* in den Fazes der Katze. *Berl. Münch. tierärztl. Wschr.,* **85**, 101–105.
Rose, M. E. (1972) Immune response to intracellular parasites: coccidia. In *Immunity to Animals Parasites,* ed. E. J. L. Soulsby, pp. 295–341. New York: Academic Press.
Rose, M. E. (1973) Immunity. In: *The Coccidia,* ed. D. M. Hammond & P. L. Long, pp. 295–341. Baltimore: University Park Press.
Rose, M. E. & Long, P. L. (1962) Immunity to four species of *Eimeria* in fowls. *Immunology,* **5**, 79–92.
Rosenberg, M. M., Alicata, J. E. & Palafox, A. L. (1954) Further evidence of hereditary resistance and susceptibility to cecal coccidiosis in chickens. *Poultry Sci.,* **33**, 972–980.
Roundabush, R. L. (1937) The endogenous phases of the life cycles of *Eimeria nieschultzi, Eimeria separata* and *Eimeria miyairii* coccidian parasites of the rat. *Iowa State J. Sci.,* **11**, 135–163.
Ruff, M. D. & Reid, W. M. (1975) Coccidiosis and itnestinal pH in chickens. *Avian Dis.,* **19**, 52–58.
Ruff, M. D., Reid, W. M. & Johnson, J. K. (1974) Lowered blood carotenoid levels in chickens infected with coccidia. *Poultry Sci.,* **53**, 1801–1809.
Ruiz, A. & Frenkel, J. K. (1976) Recognition of cyclic transmission of *Sarcocystis muris* by cats. *J. infect. Dis.,* **133**, 409–418.
Rutherford, R. L. (1943) The life cycle of four intestinal coccidia of the domestic rabbit. *J. Parasit.,* **29**, 10–32.
Ryley, J. F. (1973) Cytochemistry, physiology and biochemistry. In: *The Coccidia,* eds D. M. Hammond & P. L. Long, pp. 145–181. Baltimore: University Park Press.
Sadek, S. E., Hanson, L. E. & Alberts, J. (1955) Suspected drug-induced anemias in the chicken. *J. Am. vet. med. Ass.,* **127**, 201–203.
Schäperclaus, W. (1954) *Fischkrankheiten,* 3rd ed., p. 708. Berlin: Akademie Verlag.
Scholtyseck, E. (1955) *Eimeria anatis* N. sp. ein neues Coccid aus der Stockente (*Anas platyrhynchos*). *Arch. Protistenk.,* **100**, 431–434.
Scholtyseck, E. (1959) Zur Pathologie der *Eimeria maxima*-Coccidiose. *Zbl. Bakt. Abt. I. Orig.,* **175**, 305–317.
Scholtyseck, E. (1963) Vergleichende Untersuchungen uber die Kennverhaltnisse und das Wachstum bei Coccidiomorphen unter besonderer Berucksichtigung von *Eimeria maxima. Z. Parasitkde,* **22**, 428–474.
Scholtyseck, E. (1973) Ultrastructure. In: *The Coccidia,* eds D. M. Hammond & P. L. Long, pp. 81–144. Baltimore: University Park Press.
Shah, H. L. (1970) *Isospora* species of the cat and attempted transmission of *I. felis* Wenyon, 1923 from the cat to the dog. *J. Protozool.,* **17**, 603–609.
Skamser, L. M. (1947) Coccidiosis in poults. *Turkey Wld.,* 3.
Slavin, D. (1955) *Cryposporidium meleagridis* sp. Nov. *J. comp. Path. Ther.,* **65**, 262–266.
Smetana, H. (1933*a*) Coccidiosis of the liver of rabbits. I. Experimental study on the excystation of oocysts of *Eimeria stiedae. Archs Path.,* **15**, 175–192.
Smetana, H. (1933*b*) Coccidiosis of the liver of rabbits. II. Experimental study of the mode of infection of the liver of sporozoites of *Eimeria stiedae. Archs Path.,* **15**, 175–192.
Smetana, H. (1933*c*) Coccidiosis of the liver of rabbits. III. Experimental study of the histogenesis of coccidiosis of the liver. *Archs Path.,* **15**, 516–536.
Smith, D. D. & Frenkel, J. K. (1977) *Besnoitia darlingi* (Protozoa: Toxoplasmatinae): cyclic transmission by cats. *J. Parasit.,* **63**, 1066–1071.
Smith, M. J. & Edmonds, R. S. (1959) Use of nitrofurazone in canine coccidiosis. *Mod. vet. Pract.,* **40**, 31–32.
Streitel, R. H. & Dubey, J. P. (1976) Prevalence of *Sarcocystis* infection and other intestinal parasitisms in dogs from a humane shelter in Ohio. *J. Am. vet. med. Ass.,* **168**, 423–424.
Stubbs, E. L. (1957) Case report—renal coccidiosis in geese. *Avian Dis.,* **1**, 349.
Tadros, N. & Laarman, J. J. (1976) *Sarcocystis* and related coccidian parasites: a brief general review together with a discussion on some biological aspects of their life cycles and a new proposal for their classification. *Acta leiden.,* **44**, 1–107.
Tomimura, T. (1957) Experimental studies on coccidiosis in dogs and cats. I. The morphology of oocysts and sporogony of *Isospora felis* and its artificial infection in cats. *Riseichugaku Zasshi,* **6**, 12–14.

Trigg, P. J. (1967) *Eimeria phasiani* Tyzzer, 1929—a coccidium from the pheasant (*Phasianus colchicus.*) I. The life cycle. II Pathogenicity and drug action. *Parasitology*, 57, 135–145; 147–155.
Tsumoda, K. (1952) *Eimeria matsubayashii* sp. nov., a new species of rabbit coccidium. *Exp. Rep. Govt Exp. Sta. Anim. Hyg. Tokyo*, 25, 109–119.
Tyzzer, E. E. (1910) An extracellular coccidium, *Cryptosporidium muris* (gen. et sp. nov.) of the gastric glands of the common mouse. *J. med. Res.*, 23, 487–509.
Tyzzer, E. E. (1927) species and strains of coccidia in poultry. *J. Parasit.*, 13, 215.
Tyzzer, E. E. (1929) Coccidiosis in gallinaceous birds. *Am. J. Hyg.*, 10, 269–382.
Tyzzer, E. E. (1932) Criteria and methods in the investigation of avian coccidiosis. *J. Am. vet. med. Ass.*, 80, 474.
Tyzzer, E. E., Theiler, W. & Jones, E. E. (1932) Coccidiosis of gallinaceous birds. II. A comparative study of species of *Eimeria* of the chicken. *Am. J. Hyg.*, 15, 319–393.
Versényi. L. (1967) Adatok a *Tyzzeria perniciosa* (Allen 1935) endogen fejlodesi ciklusahoz. *Magy. Allatory. Lap.*, 22, 299–303.
Vetterling, J. M., Jervis, H. R., Merrill, T. G. & Sprinz, H. (1971*a*) *Cryptosporidium wrairi* sp. n. from the guinea-pig *Cavia porcellus* with an amendation of the genus. *J. Ptotozool.*, 18, 243–247.
Vetterling, J. M., Takeuchi, A. & Madden, P. A. (1971*b*) Ultrastructure of *Cryptosporidium wrairi* from the guinea-pig. *J. Protozool.*, 18, 248–260.
Vetterling, J. M. & Thompson, D. E. (1972) *Klossiella equi* Beaumann, 1946. (Sporozoa: Encoccidia: Adeleina) from equids. *J. Parasit.*, 58, 589–594.
Waletsky, E., Hughes, C. O. & Brandt, M. C. (1949) The anticoccidial activity of nitrophenide. *Ann. N.Y. Acad. Sci.*, 52, 543.
Warren, E. W. & Ball, S. J. (1963) The effect of sulphoquinoxaline and amprolium in the life cycle of *Eimeria adenoeides* Moore and Brown, 1951, in turkey poults. *Parasitology*, 53, 653–662.
Waxler, S. H. (1941) Immunization against cecal coccidiosis in chickens by the use of X-ray attenuated oocysts. *J. Am. vet. med. Ass.*, 99, 481–485.
Wenyon, C. M. (1926) *Protozoology*, vols 1 and 2, pp. 1563. London: Baillière Tindall and Cox.
Yvoré, P. (1969) Control of *Eimeria grenieri* coccidiosis in guinea fowl. (*Numida meleagridis*) by coccidiostats. *Acta vet.* (*Brno*), 38, 129–136.
Yvoré, P. & Aycardi, J. (1967) Une nouvelle coccidie *Eimeria grenieri* N. sp. (Protozoa: Eimeriidae) parasite de la pintade *Numida meleagridis. Cr. Acad. Sci.* (*Paris*), 264, 73–76 (Ser. D).
Zwart, P. & Strik, W. J. (1961) Farther observations on *Eimeria dolichotis* a coccidium of the Patagonian cavey (*Dolichotis patagonica*). *J. Protozool.*, 8, 58–59.

FAMILY: SARCOCYSTIDAE POCHE, 1913

Levine (1973*b*) has classified the genera *Toxoplasma* and *Sarcocystis* in the family Sarcocystidae, previously (Levine 1973*a*) having placed them in separate families, namely Toxoplasmatidae and Sarcocystidae. Subsequently, Frenkel (1974) reinstated the two families (Toxoplasmatidae Biocca, 1956 and Sarcocystidae Poche, 1913) but later (Frenkel 1977) he reclassified the cyst-forming isosporoid coccidia in the family Sarcocystidae in which he recognized two subfamilies Sarcocystinae Poche, 1913 and Toxoplasmatinae Biocca, 1956. This classification will be followed for the present edition. Nevertheless, it is likely that further revision of the systematics will occur in the coming years. Indeed, Levine (1977) places all the genera of isosporoid coccidia in the family Eimeriidae Minchin 1903.

Genus: Toxoplasma Nicolle and Manceaux, 1908

Oocysts with two sporocysts, each with four sporozoites; facultatively or obligatorily heteroxenous. Definitive host a felid. Merogony in both intermediate and definitive hosts and can cause infection in intermediate and definitive hosts. Metrocytes not formed. Schizonts (meronts) and gametocytes (gamonts) in the enteric cells of felidae. Sporogony outside host.

A single species is recognized: *Toxoplasma gondii* Nicolle and Manceaux, 1908, although Levine (1977) has proposed that *Hammondia hammondi* be regarded as a separated, additional species.

Toxoplasma gondii (Fig. 3.36). *Definitive hosts:* domestic cat (*Felis catus*), jaguarundi (*F. yagouaroundi*), ocelot (*F. paradalis*), mountain lion (*F. concolor*), leopard cat (*F. bengalensis*), bobcat (*Lynx rufus*). *Intermediate hosts:* there is little host-specificity and almost every warm-blooded animal, including man, can be infected. World-wide in distribution. Prevalence varies according to climatic region, presence of cats, etc. *Oocysts:* Described previously (p. 629). Spherical to sub-spherical 11–13 μm by 9–11 μm (mean 12 μm by 10 μm). Sporulation time two to three days at 24°C. Sporulated oocysts 12–15 μm by 10–13 μm (mean 13 μm by 12 μm). Sporocysts ellipsoidal 8.5 μm by 6 μm each with four sporozoites 8 μm by 2 μm (Fig. 3.37).

The following account of *Toxoplasma gondii* is drawn largely from the reviews by Frenkel (1973, 1974). Additional information on the ultrastructure of *T. gondii* is reviewed by Scholtyseck (1973), Aikawa and Sterling (1974) and the clinical manifestations in man and the zoonotic aspects of the infection are reviewed by Beverley (1974).

The classification, developmental cycle and zoonotic importance of *T. gondii* have remained obscure for years. The discovery of a coccidian-type developmental cycle in the cat has clarified

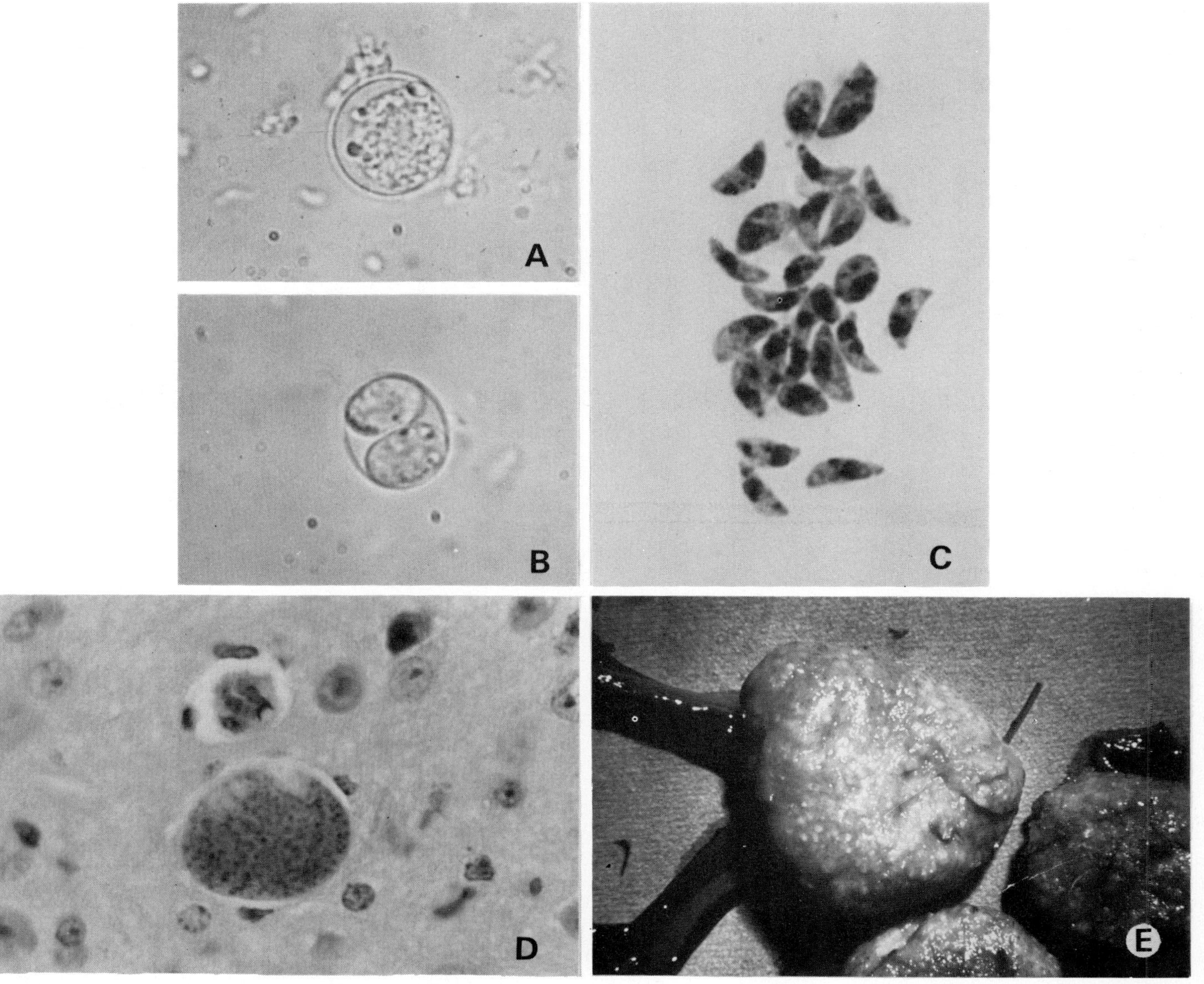

Fig 3.36 *Toxoplasma gondii*. Unsporulated (A) and sporulated (B) forms from cat faeces. C, Tachyzoites from the peritoneal cavity of the mouse. Giemsa. D, Cyst containing bradyzoites in mouse brain. E, Cotyledon of infected sheep showing focal areas of necrosis. A, B, C, × 1000; D, × 500. (A, B, *courtesy of S. Lloyd*; D, E, *courtesy of A. A. Watson*)

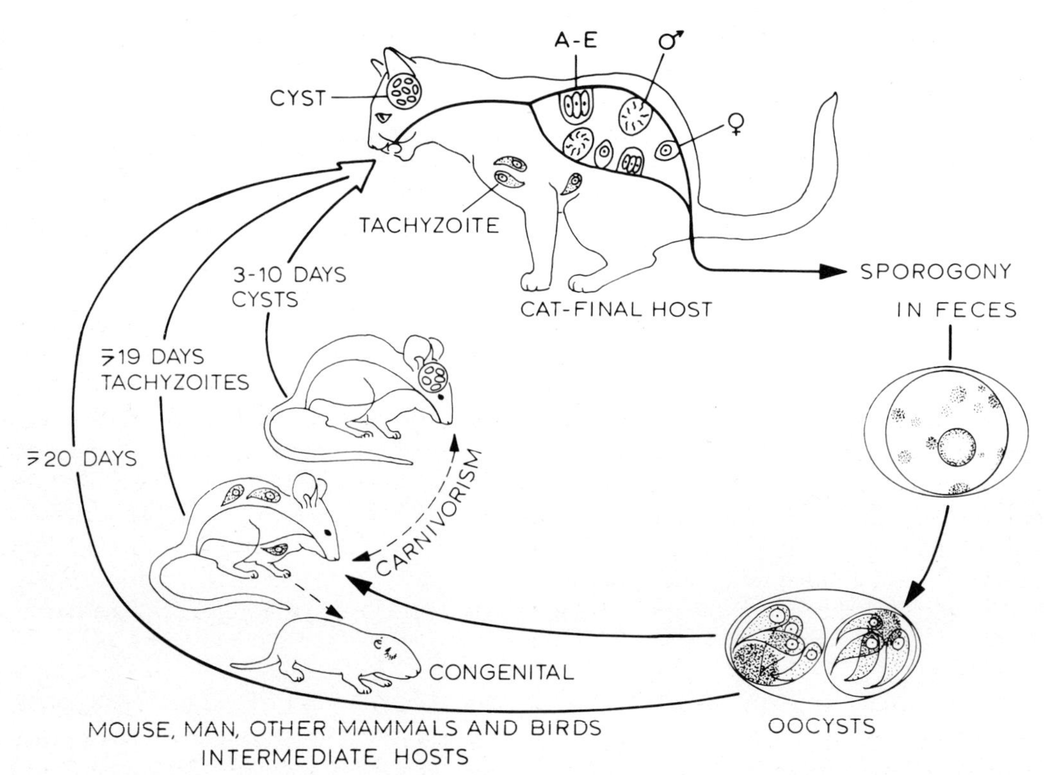

Fig 3.37 Life-cycle of *Toxoplasma gondii*. Definitive host is the cat. Following an enteroepithelial development in the intestine (involving asexual and sexual multiplication) oocysts are shed in the faeces and, after sporulation to reach the infective stage, may infect a variety of hosts (300 or so). In these, an extraintestinal cycle of development occurs producing tachyzoites or bradyzoites (cysts). Infection of the final host (cat) may occur through ingestion of sporulated oocysts, tachyzoites or bradyzoites. Transplacental transmission is important in man and sheep. (*From Dubey 1976*)

much of the confusion which existed about this parasite. The historical aspects have been referred to by Frenkel (1973, 1974) and his reviews should be consulted for details.

It is likely that revision of the developmental cycle will be necessary in the future and consequently, by the time this account is published, it will probably be somewhat out of date.

Developmental stages of Toxoplasma

The developmental cycle is summarized in Fig. 3.37. Two cycles of development, an 'enteroepithelial' cycle and an 'extraintestinal' cycle are recognized, involving five developmental stages. The enteroepithelial cycle occurs in cats and is similar to that of other coccidia consisting of enteroepithelial multiplicative stages and gamonts resulting in oocyst production, with sporogony. The extraintestinal stages occur in the extraintestinal tissues of cats and other mammalian and avian hosts and Frenkel (1973) has designated these as tachyzoites (rapidly multiplying stages) and bradyzoites (slowly multiplying stages).

Enteroepithelial cycle. This has been studied extensively in kittens infected with cysts (see below) containing bradyzoites derived from mice. Bradyzoites enter intestinal epithelial cells and a number of morphological types of multiplicative stages occur.

Multiplicative stages. Frenkel (1973) has designated these as types A, B, C, D and E. Type A appears 12–18 hours after infection, it is the smallest of the multiplicative types and is evident as collections of two or three organisms in the jejunum. Division is by endodyogeny (formation of daughter cells by internal budding). Type B occurs 12–54 hours after infection. It has a centrally located nucleus, a prominent nucleolus. It divides by endodyogeny and endopolygeny (internal budding resulting in many daughter forms). Type C occurs 24–54 hours after infection and divides by schizogony (merogony). They are elongate and have a subterminal nucleus. Type D forms occur from 32 hours to 15 days after infection and according to Frenkel (1973) account for over 90% of all *Toxoplasma* found in the small intestine at this time. The type D forms are smaller than type C and divide by endodyogeny, schizogony and by separation of single merozoites from the nuclear mass. Frenkel (1973) states that it is unclear whether this stage represents a sequential group since the three modes of division occur simultaneously. Type E divides by schizogony; it occurs 3–15 days after infection and resembles the type D.

Gamonts. Gamonts occur throughout the small intestine and are common in the ileum 3–15 days after infection.

Oocyst stages. Oocyst formation occurs in the epithelial cells of the small intestine. Initially, their development is identified by the occurrence of plastic granules in the cytoplasm of macrogametes; later, they are surrounded by argyrophilic membranes. Oocysts are discharged from the epithelial cells and shed in the faeces.

The above developmental cycle in the intestine of the cat relates to infection induced by feeding cysts containing bradyzoites, derived from the brain of mice. In this case the prepatent period is three to five days and peak oocyst production occurs between five and eight days with a patent period from 7 to 20 days (Dubey & Frenkel 1972). After feeding sporulated oocysts the prepatent period in cats is 21–24 days and after feeding tissue containing tachyzoites it is 9–11 days (Frenkel et al. 1970).

Extraintestinal cycle. These stages are the only forms in the life-cycle which occur in non-felines. However, they may also occur in the cat and the extraintestinal cycle may start almost simultaneously with the enteroepithelial cycle of development in that animal (Frenkel 1973).

Several descriptive terms have been used for the stages of the extraintestinal development. Originally, two morphological forms were described: a proliferative form (trophozoite) and a cyst-like form (pseudocyst). These terms have now, more or less, been discarded. Other terms have been introduced, including 'zoites', 'merozoites', and 'proliferating form' and 'terminal colony' to describe variously the trophozoite or the pseudocyst stages. However, Frenkel (1973) considers these terms, and others, are insufficient to describe the stages which are known at present and has introduced the term 'tachyzoite' for the rapidly multiplying forms (trophozoites) of acute infections, and 'bradyzoites' for the slowly multiplying encysted form seen in chronic infection. This terminology will be adopted in the following account. (Alternative terms for tachyzoites and bradyzoites are endozoites and cystozoites, respectively.)

Tachyzoite formation. Tachyzoite development is seen especially in acute visceral infections. In cats tachyzoite development occurs in the lamina propria, mesenteric lymph nodes and distant organs, coexistently with the enteroepithelial cycle. In other animals tachyzoites are the first stages found following ingestion of sporulated oocysts. Tachyzoites develop in a vacuole in a variety of cell types, including fibroblasts, hepatocytes, reticular cells and myocardial cells. Organisms multiply by endodyogeny. Eventually, 8–16 or more organisms accumulate in a host cell, which then disintegrates and new cells are infected. Frenkel (1973) indicates that the accumulation of tachyzoites has been termed 'terminal colonies', 'aggregates' and 'pseudocysts' but considers that these terms are not sufficiently descriptive of these forms.

Bradyzoite formation. Bradyzoites contained in cysts are characteristic of chronic infection and occur mainly in the brain, heart and skeletal muscle. Bradzoites multiply slowly, mainly by intracellular endodyogeny. The cysts containing thousands of these forms may persist for months or years after infection. Cysts may measure up to 100 μm in diameter and contain up to 60 000 organisms. Bradyzoites in the cysts are closely packed together, somewhat lancet-shaped and possess a terminal nucleus. Cyst formation usually coincides with the development of immunity. If immunity wanes, bradyzoites are capable of initiating renewed tachyzoite proliferation and additional cysts containing bradyzoites may be formed from these tachyzoites if immunity returns. However, bradyzoite formation may occur in the absence of immunity, for example, in cell cultures and it has been suggested that cysts are produced whenever multiplication is retarded, e.g. in older cell cultures.

Bradyzoites resist peptic and tryptic digestion. The cyst in which they are contained is surrounded by an argyrophilic cyst wall and usually this does not provoke an inflammatory reaction around it in the host tissue.

Comparative infectivity. When given by the oral route, oocysts are more infective than tachyzoites and bradyzoites in cysts. Similarly, oocysts are more infective than tachyzoites and bradyzoites in cysts when given subcutaneously or intraperitoneally.

Summary of extraintestinal infection. Following ingestion of oocysts, bradyzoites in cysts or trachyzoites, enteric infection spreads to regional lymph nodes and thence via the portal circulation to the liver, or via the thoracic duct to the lungs. Subsequently, a variable number of organisms are disseminated systemically to other tissues. During acute infection, *T. gondii* can be isolated from the blood, often in high titres but in chronic infections parasitaemia is sporadic and titres are low (Frenkel 1973).

Role of the cat in toxoplasmosis

The discovery that feline species excrete a highly resistant coccidian oocyst indicates that the cat is essential to the life-cycle of the parasite. In the limited number of investigations to date, infection with *Toxoplasma* has been absent, or almost so, in man and animals in areas where cats do not occur (see Wallace 1973*a*, *b*). The epidemiology of *Toxoplasma* infection in relation to the biology of the cat has been discussed by Wallace (1973*a*, *b*). Thus, seasonal breeding trends may determine the predominance of excretion of oocysts in an area.

Dubey (1973) has emphasized that the mode of infection of cats in nature is important. Although toxoplasmosis can be transmitted congenitally in man (see below) and by the ingestion of infected meat, these modes of infection do not account for the widespread toxoplasma infection in herbivorous animals. In these cases transmission by oocysts would seem to be the major mode of infection. In a study of domiciled and stray cats in the USA, all recently weaned kittens ages $4\frac{1}{2}$–10 weeks, had only low titres of passively transferred antibody which decreased to negligible levels three months after birth. This was interpreted to indicate that kittens in this age group were not infected with *T. gondii* which, if true, indicates either that congenital infection rarely occurs or that congenitally infected kittens die early in life (Dubey 1973). Recent studies by Dubey (1977*a*, *b*) and Dubey and Hoover (1977) indicate that congenital infection in the cat is unlikely.

Dubey's (1973) data suggest that cats acquire infection by carnivorism (e.g. cats older than 11 weeks) and this would account for the finding that the prevalence of toxoplasma infection was higher in stray cats than in domiciled cats. The epidemiological evidence suggests that predatory behaviour facilitates transmission, and this is in line with the experimental evidence that bradyzoites contained in cysts provide the most satisfactory source of infection for cats.

Whether all infections with *Toxoplasma,* which cannot be explained by congenital infection or by the consumption of infected meat, are attributable to contact with oocysts in cat faeces remains to be determined. Epidemiological patterns do not suggest a ready and easy route of infection from the cat to other animals. Thus, various surveys have shown a relatively low prevalence of cats shedding *T. gondii* oocysts (e.g. 12 of 1604 stray cats on Oahu, Hawaii; Wallace 1973); cats excrete oocysts for a relatively short time and those that have excreted oocysts once usually do not do so on reinfection (Dubey 1973). Dubey (1976*a*, *b*) has reported reshedding of *Toxoplasma* oocysts in chronically infected cats following feeding of *Isosporan* oocysts (e.g. *I. rivolta* and *I. felis*). He attributes this to an interference with local bowel immunity by the isosporan infections.

Despite this, probably only 1% or less of infected cats shed oocysts at any given time. Wild Felidae may replace the domestic cat in epidemiological situations where the latter do not occur. Nevertheless, one must postulate that the oocyst of *T. gondii* is very resistant and widely dispersed to explain *T. gondii* infections in the wide variety of hosts in which it occurs.

It is possible that the infection can be maintained in nature by mammals and birds whose habits include cannibalism and carrion feeding (Wallace 1973). It is also possible that invertebrates, such as coprophagous insects and molluscs, play a role in dissemination (Wallace 1973). Nevertheless, despite these alternatives there are several unsolved problems in the epidemiology of the disease.

Pathogenesis of toxoplasmosis: general characteristics of the infection

Most infections with *T. gondii* are probably asymptomatic. Numerous reports attest the widespread prevalence of toxoplasmosis in domestic and wild animals and in man. Despite the 'sea of *Toxoplasma* infection around us' (Jacobs 1956), relatively few clinical cases occur. Asymptomatic acute toxoplasmosis plays an important role in congenital infection in man and sheep, and asymptomatic chronic human infection may be important where immunosuppressive therapy is used for other medical disorders.

In most acute infections the route of infection is the intestinal tract (Frenkel 1973). Organisms are disseminated by the lymphatics and portal blood with invasion of various organisms and tissues.

The parasites multiply in the tachyzoite form, producing areas of necrosis; parasitism reaches high levels, and the animals may succumb during this period. During the height of this phase, organisms may appear in secretions and excretions such as urine, faeces, milk, conjunctival fluid and even in the saliva, but this is rare. These forms are unable to survive for any length of time outside the host. There is little or no spread of toxoplasmosis from one animal to another in the acute phase, even when the animals are confined in a close space (e.g. mice in the same jar).

The subacute form of the disease is characterized by the appearance of antibodies which clear the blood and tissues of the tachyzoites. The brain is cleared of organisms very late, followed by clearing of the heart, while the liver, spleen and lungs are cleared of organisms relatively quickly. Persistence of the bradyzoites in cysts is characteristic of the chronic infection. This phase can live for some considerable time; in dogs for up to ten months, and in rats, mice and pigeons they have been found for as long as three years after infection.

Virulence

A variation in strains of the organism is seen. The main criteria for differentiating strains are virulence and the characteristics of the disease produced when they are injected into laboratory animals. The most virulent strains are those which are highly pathogenic for mice and also produce severe disease in other laboratory animals. With strains of low virulence there is, in general, a lower parasitaemia, less tissue invasion and shorter persistence of the parasite. Organisms isolated from animals which have been sick or dying of the infection are usually more virulent than those that have been obtained from an animal which shows no clinical evidence of the disease. In nature the organism seems very well adapted to its hosts, and the majority of infections are avirulent or, at most, subclinical in character. The factors which lead to the organism adopting a more virulent behaviour are unknown.

Immunology of toxoplasma infection

This topic has been reviewed by Remington and Krakenbuhl (1976), McLeod and Remington (1977) and Jones et al. (1977). A feature of *Toxoplasma* is its ability to survive in macrophages which would otherwise kill extracellular organisms. *Toxoplasma* not only has the ability to induce phagocytosis in cells not ordinarily phagocytic but also it is able to block the delivery of liposomal constituents into the phagocytic vacuole in which the organism lives.

Cellular factors play a major role in resistance to reinfection. The adoptive transfer of lymphocytes leads to protection against virulent *Toxoplasma* challenge whereas passive transfer of serum yields only slight protection. Immune lymphocytes from *Toxoplasma*-infected animals are able to activate macrophages which then have an enhanced capacity to kill *Toxoplasma*, and also other intracellular organisms. Jones et al. (1977) have described an inhibitory factor (IF) (a lymphokine) which is released from immune lymphocytes after interaction with *Toxoplasma* antigen. *Toxoplasma* IF interacts with a glycoprotein on the surface of the macrophage. Cyclic AMP is increased, cGMP is decreased, protein synthesis occurs and inhibition of *Toxoplasma* multiplication results.

TOXOPLASMOSIS IN MAN

An extensive account of toxoplasmosis in man is given by Beverley (1974). Infection may be acquired or congenital. Most of the cases of toxoplasmosis in children are congenital in origin, the mother usually showing a mild infection or no evidence of infection. Congenital infection occurs only when a woman has a primary infection during pregnancy. The time in pregnancy during which toxoplasmosis is acquired is a major determinant of the outcome of pregnancy. Maternal to fetal transmission of parasites occurred less commonly early in pregnancy than later, yet the severity of the damage to the fetus was greater with early infection than later (Desmonts 1977). It has been variously estimated that from 1/500 to 1/20 000 births, according to country, are associated with congenital toxoplasmosis (Beverley 1974).

The outcome of congenital infection varies according to fetal damage which, in turn, depends on the virulence of the strain of organism and the time of infection during pregnancy. In severe infections, acquired early in pregnancy, abortion is a common sequel. *Toxoplasma* organisms are readily demonstrable in the products of abortion. In less severe infections pathological changes are found more frequently in the central nervous system than in the viscera and somatic tissues. The lesions are marked and the symptoms characteristic, namely cerebral calcification, choroidoretinitis, hydrocephalus or microcephaly and psychomotor disturbances. The child may be born either alive or dead, and if born alive may suffer serious mental retardation within a few weeks of birth. Generalized infection may be present at birth, and fever, adenopathy, and enlargement of the spleen and liver occur shortly afterward. Mild cases of congenital toxoplasmosis occur, and these are more difficult to diagnose than the clinical form.

Acquired toxoplasmosis (i.e. non-congenital) should be suspected when lymphadenopathy, lassitude accompanied by fever, lymphocytosis, meningoencephalitis, eye lesions of doubtful origin or myocarditis are observed. Probably hundreds or thousands of cases of human toxoplasmosis go unrecognized since only a mild illness characterized by slight fever and slight enlargement of the lymph glands is presented.

The relative roles of infection by ingestion of oocysts or by ingestion of undercooked infected meat is not known. The role of infected meat in transmission would seem minor since in one study *T. gondii* was not isolated from the tissues of 60 cattle, but was found in 4 of 84 sheep and in 8 of 50 pigs obtained at an abattoir (Jacobs et al. 1960).

TOXOPLASMOSIS IN DOGS

Mello (1910) first described toxoplasmosis in a dog. The disease in this case was characterized by fever, anaemia, respiratory distress and haemorrhagic diarrhoea. On autopsy the animal had a serosanguinous exudate in body cavities, small nodules in the lungs and numerous small ulcers in the small intestine. Intracellular and extracellular parasites were seen in smears and sections. Clinical manifestations may vary greatly. Beverley (1957) has stated that in approximately half the cases of canine toxoplasmosis, respiratory signs are in evidence, alimentary disturbances occur in another quarter and neurological signs in the remainder. No sex difference was apparent, and all types of dog were infected. The onset of illness is marked by an insidious development of fever with lassitude, anorexia and diarrhoea; occasionally, it may be sudden with vomiting followed by fits and paralysis and other neurological manifestations (Beverley 1957). Infection in the dog often occurs in conjunction with canine distemper (Campbell 1956; Roberts 1966). Indeed, Beverley (1974) notes that canine toxoplasmosis is often the explanation for the supposed breakdown of distemper vaccination. In most cases illness starts within four to six weeks of vaccination.

Dogs are frequently infected in nature. Miller and Feldman (1953), using a dye test, found that 59% of 51 dogs in New York were infected, Siim et al. (1963) found 18.5% of a group of dogs in Copenhagen had a high dye test titre, and Lainson (1956) found 42.5% of sera from 113 dogs in London positive by the complement fixation test.

The disease is characterized by necrosis, and the cellular infiltration is predominantly mononuclear. In the brain gliosis may develop alongside perivascular infiltrations which sometimes include plasma cells. A leptomeningitis may be present, and foci of necrosis occur in the grey matter just beneath the ependyma. The pseudocyst stage is demonstrable in the brain (Beverley 1957).

In the lungs necrotic nodules may be found in the parenchymatous tissue, and a pleural exudate may be present. Associated glands are swollen, and the organisms can be easily found in the cells lining the alveoli, trachea or bronchi.

The spleen and liver are usually enlarged, and organisms can be found in the liver cells, the epithelium of the biliary tubules and in the reticuloendothelial cells of the spleen. Ulceration of the intestinal mucosa is common in canine toxoplasmosis. The ulcers of the digestive tract are usually deep and occur in the duodenum or the rectum; organisms are found in the adjacent mucosa or lying underneath the muscle layers.

TOXOPLASMOSIS IN CATS

Because of the important role of the cat in the epidemiology of toxoplasmosis, numerous surveys of feline infection, based on the presence of oocysts of *Toxoplasma* in the faeces and/or antibodies in the serum, have been carried out. These show that cats are commonly infected in nature (Dubey 1973) and produce millions of oocysts. However, rarely does the infection result in clinical disease in the cat. Dubey (1968) has summarized the clinical and pathological manifestations of spontaneous disease in cats. These are somewhat protean but include enteritis, with ulceration, enlargement of mesenteric lymph nodes, pneumonia, perivascular and degenerative changes in the central nervous system, encephalitis, chronic interstitial nephritis, etc. There is little doubt that these require re-evaluation in the light of modern knowledge of toxoplasmosis. For example, Frenkel (1973) states that ulceration rarely occurs in cats, although infected young kittens may develop diarrhoea. In such animals lesions of the lymph nodes and liver are usually prominent and at death widely disseminated lesions are seen in lymph nodes, heart and brain (see Frenkel 1973).

TOXOPLASMOSIS IN CATTLE

Reports are scarce concerning toxoplasmosis in cattle. Sanger et al. (1953) recorded an outbreak of illness in a herd of Brown Swiss cows in which 45 of 78 calves, born over a period of a year, died at various ages. Clinical signs were dyspnoea, coughing, sneezing, a nasal discharge, trembling and shaking of the head. An elevated temperature was seen in several cases, but in others the temperature was more or less normal. At times death was sudden without any previous evidence of illness, or in other cases the disease lasted for several months. Autopsy of a four-week-old calf showed a fibrinous deposit in the peritoneal cavity, enlargement of submaxillary and bronchial lymph glands, a haemorrhagic tracheitis and pneumonia with consolidation. *Toxoplasma* were found in the brain, lungs and lymph nodes. Koestner and Cole (1961) have studied the neuropathology of toxoplasmosis in experimentally infected cattle and sheep, the lesions in both species being similar. In the early stages lesions consist of damage to the vascular walls producing endothelial swelling, perivascular oedema and proliferation of the adventitia cells. The process spreads to adjacent nervous tissue producing foci of necrosis which contain numerous *Toxoplasma* organisms. Glial nodules consisting of pleomorphic microglia and oligodendroglia, astrocytes and monocytes are formed, and at this stage pseudocysts are in evidence. Healing of these lesions with scar formation occurs, and a particular feature in the chronic form is calcification of the blood vessel walls.

TOXOPLASMOSIS IN SHEEP

An early report of toxoplasmosis in sheep was made by Olafson and Monlux (1942), who found the organism in sheep affected with nervous signs. Wickham and Carne (1950) recorded a case of 'circling disease' in Australian sheep, toxoplasma being found in the brain along with congestion and perivascular cuffing.

The association of *Toxoplasma* with ovine perinatal mortality was recorded by Hartley and Marshall (1957) in New Zealand. These authors demonstrated the association of abortion with *Toxoplasma gondii*, and in Great Britain, Beverley and Watson (1961) recorded in Yorkshire that *Toxoplasma* infection was associated with ovine abortion. Beverley (1974) estimates that annual lamb losses in Britain due to toxoplasmosis are 80 000–100 000. Serological surveys showed toxoplasma infection to be widespread in Yorkshire flocks, and antibodies to *Toxoplasma* were found in high titres in ewes which had aborted from hitherto unknown causes. In New Zealand toxoplasmosis is probably the most widespread and most important cause of infectious ovine perinatal mortality (Hartley & Marshall 1957). The severity of congenital infection depends on the duration of gestation at the time of infection. Infection early in gestation (e.g. 45–55 days) causes death of the fetus and expulsion such that it is probably seldom observed. Infection at a later time of gestation (e.g. 90 days) results in fetal death and expulsion and organisms are readily demonstrable in fetal tissues. Infection at 120 days of gestation causes fetal infection but not death and lambs usually survive (see Beverley 1974).

Histological examination of the fetal membranes, particularly of those where the infection is still active, shows oedema of the mesenchyme of the fetal villi with a moderately diffuse invasion of mononuclear cells. Focal areas of epithelial swelling and necrosis, with desquamation, are evident, and where large areas of the trophoblast are shed necrotic nodules may result. Cotyledons show intra- and extracellular toxoplasms (Hartley et al. 1954; Hartley & Marshall 1957).

Ewes having aborted one year usually lamb normally in subsequent years. Beverley (1974) considers this indicative of the persistence of an effective immunity.

It is interesting to note that cattle grazing the same pasture as sheep acquire infection less frequently than sheep (Dubey, personal communication).

The neuropathological lesions in experimental toxoplasmosis in sheep are similar to those found in cattle (Koestner & Cole 1961) and occur in 75% of cases. The principal lesion is a focal necrosis in the acute form, while in the more chronic form glial nodules are much in evidence, and toxoplasma cysts are found associated with these.

TOXOPLASMOSIS IN PIGS

Toxoplasmosis in pigs was first recorded by Farrell et al. (1952). Since then pigs in North America, Norway and Japan have been found infected, and Harding et al. (1961) reported the infection in British pigs. Newly born animals and those up to three weeks of age tend to be affected, and usually the disease is manifest by excessive losses of young pigs at farrowing time. The clinical signs include fever, shivering, weakness, coughing, incoordination, relaxation of abdominal muscles and diarrhoea. Pulmonary signs may be a common feature of acute toxoplasmosis in the young pig, and frequently this is associated with subclinical pneumonia in the herd.

Cole et al. (1954) state that the most common post mortem signs are pneumonia, focal necrosis of the liver, hydrothorax, ascites, lymphadenitis and enteritis.

TOXOPLASMOSIS IN HORSES

Serological studies in various countries indicate that toxoplasmosis occurs in horses but clinical disease attributable to the infection is relatively uncommon. For example, Vanderwagen et al. (1974) found positive tests in 14 of 105 horses in California, Seeman (1959) reported 129 of 389 horses in Eastern Europe seropositive, Engster and Joyce (1976) noted high titres (IFA) in 68 of 200 horses in Texas, and von Seyerl (1970) found 100 of 561 horses positive.

The clinical entities ascribed to toxoplasmosis in horses are associated with '*Toxoplasma*-like' developmental stages in tissues, especially the central nervous system. An encephalomyelitis with '*Toxoplasma*-like' organisms has been reported by Cusick et al. (1974), Dubey et al. (1974)

and Beech (1974). However, Dubey (1976*a*) believes that these clinical disease entities are attributable to *Sarcocyst*-like organisms rather than to *Toxoplasma*.

Von Alton et al. (1977) infected seronegative ponies with one million *Toxoplasma* oocysts, and while infection was produced as assessed by serodiagnostic studies, little clinical effect was evident.

TOXOPLASMOSIS IN BIRDS

A large number of records have been made concerning the prevalence of *Toxoplasma* infection in birds. However, many of these records have been made upon the histological appearance of an organism resembling *Toxoplasma*, and no direct serological or mouse inoculation evidence exists to support many of the reports.

In domestic poultry, a disease ascribed to toxoplasmosis has been reported by Ericksen and Harboe (1953) in Norway. Fowls were found dead without any evidence of previous illness, and others showed signs of anorexia, emaciation and pallor for periods of up to a month. Diarrhoea and blindness were features in some of the poultry. Histologically, there was a pericarditis, focal or diffuse myocarditis, focal encephalitis, a necrotic hepatitis and ulcers of the gastrointestinal tract. *Toxoplasma* were found in various organs.

Beverley (1974) describes the reactive changes in birds as follicular hyperplasia in the spleen and lymphoid deposits in various organs which develop into follicles with pale histiocytes. Active follicles develop in the bursa of Fabricius.

Jacobs et al. (1952) carried out a survey in Washington of the prevalence of *Toxoplasma* in pigeons. A prevalence of 12.5% was found, and strains isolated from them were morphologically and serologically identical with the virulent RH (human) strain.

TOXOPLASMOSIS IN OTHER ANIMALS

Christiansen (1948) and Christiansen and Siim (1951) recorded outbreaks of toxoplasmosis in hares in the winter months in Denmark. In nearly all the cases the infections were of the acute, fatal type and showed general systemic infections. A marked increase in the size of the spleen was evident, and the livers were enlarged, pale and contained scattered submiliary foci. The mesenteric lymph nodes were swollen, there was oedema of the lungs and a serosanguinous fluid in the body cavities. *Toxoplasma* were present in large numbers in most of the organs, being particularly associated with the changes in the lungs, liver and spleen.

Toxoplasmosis in wild rabbits was studied by Beverley et al. (1954), who found in Great Britain that 34% showed serological evidence of infection. Lainson (1955) demonstrated that the organism may be found in the brain of rabbits, even in those which had been bred under conditions which excluded them from contact with the wild rabbit. Strains recovered from such rabbits failed to produce disease in mice.

The prevalence of *Toxoplasma* in rats has not been widely investigated. Perrin et al. (1943) found 8.7% of Savannah rats infected with *Toxoplasma* in the brain, and Eyles (1952) found a prevalence of 3.2% in rats in the Memphis area. Spontaneous disease in the rat, due to toxoplasmosis, is rare, and it is also difficult to induce disease artificially in rats, even with large doses of the organism.

Toxoplasmosis in laboratory animals may cause a reduced life-span, complicate interpretation of data or, on occasions, decimate a colony. The role of the disease in laboratory animal care is discussed by Flynn (1973).

Public health significance of toxoplasmosis

Beverley (1974) has reviewed the epidemiology of this zoonotic infection. Transmission by the cat is clearly an important factor in the epidemiology of the infection. The role of non-cat transmission (e.g. by meat) has yet to be fully assessed. However, apart from transmission there is the unanswered question of why *Toxoplasma* infection is common, but clinical disease associated with it is relatively rare.

Westphal and Bauer (1952) have stressed that toxoplasmosis is 'not normally a disease', rather it appears to be a condition of symbiosis between the host and the parasite, and only rarely is the balance tipped in favour of the parasite. Beverley et al. (1954), in a study of toxoplasmosis in the Sheffield area of Great Britain, ranked various occupational groups according to their susceptibility to toxoplasmosis. The general prevalence of toxoplasmosis, judged by serological tests, was 25% of the adult population. The prevalence increased up to the age of 20 years, after which it remained steady. There was no significant difference between the prevalence in men or in women, but significantly higher antibody titres were found in the sera of veterinarians and abattoir workers. A still higher prevalence was noted in individuals who handled rabbits, and highest of all was the group who were concerned with the trapping of rabbits.

Veterinarians are frequently called upon to advise on the potential danger(s) of cats and *Toxoplasma* infection to children and pregnant women. Dubey (1976*a*) has summarized the situation as follows and this provides a balanced view of the dangers as assessed by present knowledge.

As many as 64% of cats in the USA and elsewhere (Dubey 1968) are seropositive for toxoplasmosis; only 1% of them shed oocysts at any time. Natural transmission of toxoplasmosis between cats is probably by eating *Toxoplasma*-infected animals or tissues and not by the ingestion of oocysts (even though oocyst transmission is probably the most important form from feline to non-feline). Infected cats start shedding oocysts within one week of ingesting bradyzoites in mice (19 days for tachyzoites in tissues and 20 days or so for sporulated oocysts) and they shed oocysts for one to two weeks. Congenital toxoplasmosis in cats probably does not occur (Dubey & Hoover 1977). Clinical disease due to toxoplasmosis in the cat is rare and although infected cats may produce millions of oocysts, Dubey (1976*a*) has rarely seen any digestive disorders resulting from the infection.

Following shedding of oocysts, cats become carriers. Developmental stages persist in the intestine and other tissues (Dubey 1977*a,b*) and reshedding of oocysts can occur after superinfection with *Isospora felis*, which also produces an asymptomatic infection (Dubey 1976*b*). Reshedding can also be induced by treatment with corticosteroids (Dubey & Frenkel 1974). A further consideration is the possible role of virus infections as immunodepressive infections in the reshedding of oocysts of *Toxoplasma*.

The importance of antibody titres in the diagnosis of feline toxoplasmosis requires further evaluation. Dubey (1976*a*) states that a four-fold rise in antibody titre (IHA) within two weeks in paired serum samples is indicative of acute infection but not necessarily clinical disease.

Preventive aspects for pregnant women include changing cat litter daily, washing of hands before eating; gloves should be worn when gardening; sand boxes should be covered when not in use and uncooked meat should not be fed to cats. In this respect, there has been a general failure to isolate *Toxoplasma* from beef in the USA and Western Europe, although elsewhere the success of isolation has been greater (Dubey 1976*c*).

Diagnosis of toxoplasmosis

Diagnosis of toxoplasmosis on clinical grounds is usually difficult, and recourse must be made to the demonstration of either the organism or antibodies against it. The most convincing diagnosis is the isolation of the parasite by inoculation of suspect material into mice. Mice are probably the most useful animals for this since they are highly susceptible and rarely suffer from spontaneous infection. A highly virulent strain produces an acute and generalized fatal infection 1–14 days after the intraperitoneal route of injection, and a few days earlier if the intracerebral route has been used. Ascites develops after intraperitoneal injection, and abundant proliferative stages can be found in films prepared from peritoneal or pleural fluids or in smears from the cut surfaces of lung, liver, spleen and brain.

With strains of lower virulence a transient disease is produced in a few mice at about the third week after inoculation, but mostly the disease is asymptomatic. Infected mice can be detected by serological tests about three weeks after inoculation and the infection confirmed by examination of the brain for pseudocysts. This is done by emulsifying the brain in saline and examining a drop of this under the low power of a microscope. The number of cysts found in a smear may vary from 1 to 100 or more.

Where diagnosis is uncertain it is sometimes desirable to carry out blind passage of material from the first set of mice to a second or further set of mice. Several passages may be necessary before the organism appears in a form virulent enough to cause acute peritonitis. Alternatively, cortisone can be given to enable strains of low virulence to produce the acute fatal type of infection.

Various strains of *Toxoplasma* exist. Some are highly virulent, as illustrated by the RH (human) strain, which will cause an acute fulminating and fatal disease in mice (the RH strain of *Toxoplasma* has been serially passaged in mice since its original isolation from a human case of encephalitis); but on the other hand apparently completely avirulent toxoplasms appear.

Serological tests. Fulton (1963), Kagan (1974) and Jacobs (1976) have reviewed the serological procedures available for the diagnosis of toxoplasmosis. Of many tests available the complement fixation test, the dye test, the indirect immunofluorescent test and the haemagglutination test are in more general use.

Dye test. The dye test depends on the principle that antibody and an accessory factor (a complement-like serum factor, probably properdin) modify living *Toxoplasma* so that these fail to stain with methylene blue at pH 11. Proliferative forms of *Toxoplasma* which have not been modified by antibody stain readily, and the test is quantitated by finding the highest dilution of serum which will modify 50% of the toxoplasms in a standard suspension. Constant practice is necessary in order to perform and interpret the dye test, and it is best carried out in a large laboratory which is routinely doing serological surveys. Because of the exacting value of the test and, at times, the difficulty in obtaining sera containing accessory factor, many diagnostic laboratories have stopped using it. Additionally, Jacobs (1976) notes that it may be a dangerous test to use since it requires live parasites.

Complement fixation test. This has been widely used as a diagnostic test. The titres, though lower, generally follow those delineated by the dye test. Complement-fixing antibodies usually appear later and disappear sooner than those detected by the other tests, and in most cases antibodies disappear following the disappearance of clinical signs.

Haemagglutination test. The passive haemagglutination test has gained wide acceptance and now serves as a routine serological test in many laboratories. Various modifications of it have increased its usefulness; thus, formalinized cells provide a stable supply of standardized cells, and human group O cells avoid heterophile reactions which may occur with sheep cells.

Indirect fluorescent antibody test (IFA). This test, which uses killed stable antigen, is replacing, for example, the dye test. It has been shown to be sensitive, specific and reproducible. In association with heavy-chain specific antisera to human immunoglobulins it is used to differentiate between maternally transferred antibody and fetal antibody in congenital toxoplasmosis in man (Jacobs 1976).

Treatment

No completely satisfactory treatment for toxoplasmosis is known. Daraprim (pyrimethamine) (2,4-diamino-5-*p*-chlorophenyl-6-ethyl-pyrimidine) has been found effective in monkeys and humans at adequate blood levels and is effective against the proliferative forms of the organisms. It is of little value against the cyst forms.

Pyrimethamine, with triple sulpha drugs, has given good results in many cases, especially in ocular cases of toxoplasmosis. There is synergy between the sulphonamides and the diamino pyrimidines, this producing sequential blocks in the metabolic pathways involving *p*-aminobenzoic acid, folic acid and folinic acid (Eyles 1956).

Frenkel (1975) has reported that oocyst shedding in *Toxoplasma*-infected cats was reduced with a combination of sulphadiazine (120 mg/kg) and pyrimethamine (1 mg/kg) and Sheffield and Melton (1976) stated that intramuscular injection of 2 mg/kg pyrimethamine with 100 mg/kg sulphadiazine inhibited oocyst shedding. However, Dubey and Yeary (1977) found pyrimethamine in conjunction with sulphadiazine unpalatable for cats. Oocyst shedding was much reduced following administration of 2-sulphamoyl-1-4,4-diaminodiphenylsulphone (SDDS) (160–1000 mg/kg) three to four days after feeding *Toxoplasma*, but oocyst shedding was not completely eliminated.

Clindamycin, an antibiotic, has shown activity against acute and chronic toxoplasmosis in mice (Aranjo & Remington 1974) and Dubey and Yeary (1977) state it reduces the output of oocysts in infected cats. Oocysts from cats given any of the above drugs sporulated normally and were infective for mice.

Genus: Sarcocystis Lankester, 1882

The general developmental cycles of members of this genus in the final or definitive hosts have been discussed above in relation to the coccidia of cats and dogs. This account is concerned with the effects of *Sarcocystis* organisms in the intermediate host. To recapitulate, infective sporocysts (derived from the faeces of the final host: cat, dog, man, etc.) are ingested, sporozoites are released in the intestine and they invade many tissues. Schizogony occurs in the endothelial cells of blood vessels in most organs (Johnson et al. 1975), preceding the development of typical cysts in the striated musculature.

While cysts generally occur in the muscles, they are occasionally found in the brain. Cysts vary in size, depending on the species, from a few millimetres to several centimetres in length. Two distinct regions are recognizable within the cyst (Fig. 3.38). The peripheral region contains globular forms called metrocytes, which by internal budding (endodyogeny) produce two daughter cells and these, after several further replications, give rise to banana-shaped bradyzoites. Bradyzoites have many micronemes and resemble ultrastructurally coccidian merozoites.

Infection of the final host is by ingestion of muscle cysts containing bradyzoites. Schizonts and metrocysts are not infective for the definitive host (Ruiz & Frenkel 1976).

SARCOCYSTIS IN CATTLE

Three species of sarcocysts occur in cattle:

Sarcocystis bovifelis (syn. *S. hirsuta*) Moulé, 1888

Sarcocystis cruzi (syn. *S. fusiformis*, *S. bovicanis*) (Hasselman, 1926) Levine 1977

Sarcocystis hominis (syn. *S. bovihominis*) Railliet and Lucet, 1891; Levine 1977

Sarcocystis bovifelis has the cat and feral cat (*Felis silvestris*) as definitive hosts. The prepatent period in the cat is seven to nine days and sporocysts 12 μm by 1 μm are produced. This species is considered innocuous to cattle (Dubey 1976*b*).

Sarcocystis cruzi is the most pathogenic species in cattle. Dog, wolves, coyotes, racoons, foxes and hyenas serve as definitive hosts and shed sporulated oocysts or sporocysts in their faeces (Fayer 1974). Following ingestion of sporulated forms by cattle, two or more generations of schizonts occur in the vascular endothelial cells, metrocytes are found in the stri-

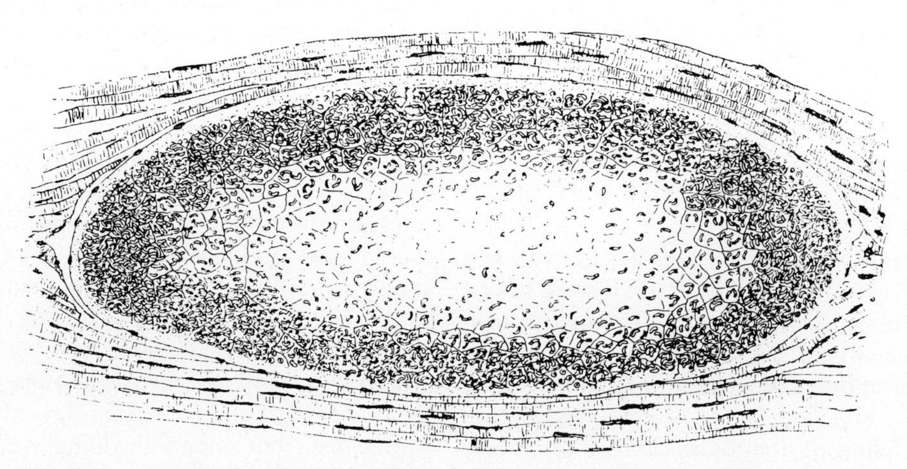

Fig 3.38 Longitudinal section of a sarcocyst in the muscles of the ox. (× ca 500) (*From Wenyon 1926*)

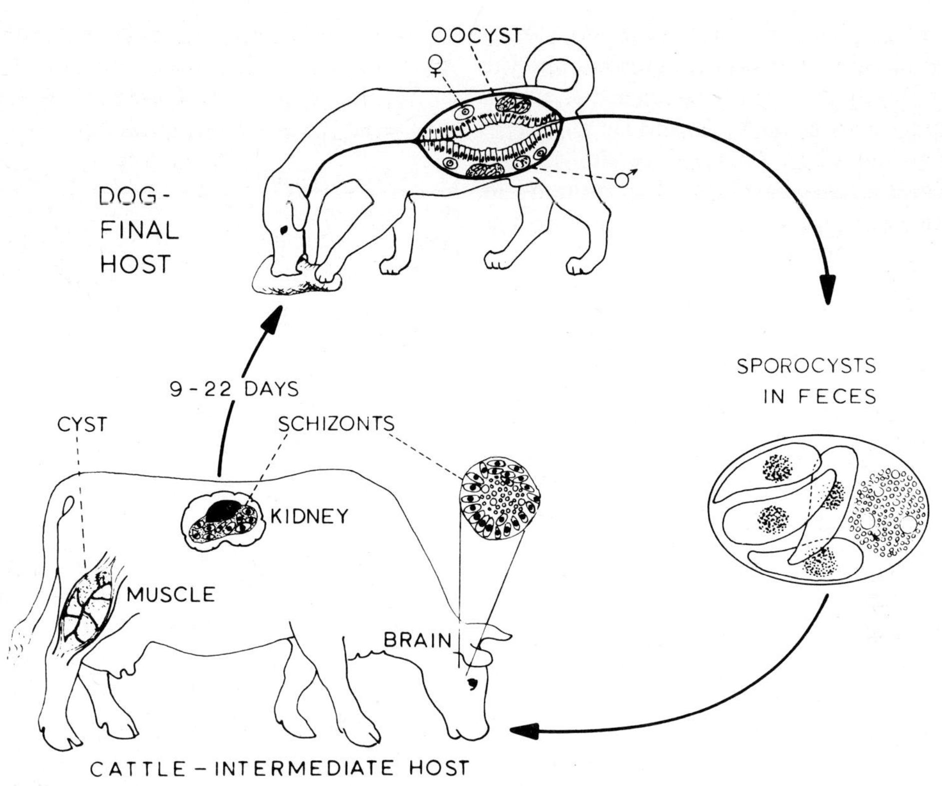

Fig 3.39 Life-cycle of *Sarcocystis cruzi* of cattle. Dogs, wolves, etc. shed sporulated sporozoites in the faeces. Schizonts (tachyzoites) occur in the endothelial cells of blood vessels, especially the kidney and brain. Later cysts (metrocytes and bradyzoites) are formed in the muscles. (*From Dubey 1976*)

ated musculature about one month after infection and fully formed muscle cysts are evident within $2\frac{1}{2}$–3 months (see Fig. 3.39 for life-cycle) (Fayer & Johnson 1973, 1974).

Acute disease occurred in calves experimentally infected with material from dog faeces (Johnson et al. 1975). Clinical signs included anorexia, pyrexia, anaemia and loss of weight. Mortality was produced in 33 days by feeding 10^5–10^6 sporocysts. At post mortem a generalized lymphadenopathy was evident.

The recent information on the developmental cycle of *Sarcocystis* spp. suggests that previously unexplained disease conditions may be due to the parasite. Markus et al. (1974) reviewed the evidence that Dalmeny disease in Canadian cattle, reported by Corner et al. (1963), may have been caused by schizogony of *Sarcocystis*. In this infection an unidentified protozoan was seen in the endothelial cells of blood vessels of many tissues.

Clinical signs included intermittent pyrexia, reduced milk yield, loss of condition and dyspnoea: of 17 pregnant animals, 10 aborted. Chronic illness was characterized by emaciation, submandibulary oedema and exophthalmia. At autopsy ecchymotic haemorrhages occurred in various organs including the myocardium. Schizonts were found in the endothelial cells of almost all organs. A similar disease has been produced experimentally by feeding sporocysts to cattle (Johnson et al. 1975). Abortion is a common clinical eventuality in cows experimentally infected with *Sarcocystis* (Fayer et al. 1977). A further outbreak of Dalmeny disease was reported in Canada (Meads 1976) and Clegg et al. (1978) reported a clinical entity in cattle in England which resembled Dalmeny disease. A field outbreak of *Sarcocystis* in cattle in the USA was reported by Frelier et al. (1977) and an unidentified protozoan was found in the tissues of a calf which died in England. This organism closely resembled that found in Dalmeny disease and the schizonts were like those of *Sarcocystis* (Lainson 1972; Markus et al. 1974). Fayer and Lunde (1977) and Lunde and Fayer (1977) have shown high titres of indirect haemagglutinating antibody in affected animals. Uninfected animals had titres of 1 : 500 or less, while infected animals had titres in the thousands.

Sarcocystis hominis has man, rhesus monkey, baboon and possibly the chimpanzee as definitive hosts and the ox (*Bos taurus*) as the intermediate host. In man, the prepatent period of infection is nine to ten days and sporocysts measuring 15 μm by 9 μm are passed in the faeces. *S. hominis* apparently is non-pathogenic for cattle.

SARCOCYSTIS IN SHEEP

Two species of sarcocysts occur in sheep:

Sarcocystis ovicanis Heydorn, Gestrich, Melhorn and Rommel, 1975
Sarcocystis tenella (Syn. *S. ovifelis*) Railliet, 1866

Infection with *Sarcocystis* in sheep is common, Dubey (1976*a*) reporting, for example, that 73% of adult sheep in an abattoir in Michigan were infected.

Sarcocystis ovicanis is highly pathogenic for lambs which become anorectic, weak and may die. Leek and Fayer (1978) report that 8 of 11 pregnant ewes aborted when infected, became anaemic and lost weight. At necropsy of acutely ill animals, the heart was the most severely affected organ, schizonts being found in epithelial cells. However, stages of *S. ovicanis* have not been found in the fetus, placenta or uterine tissue and this indicates that intrauterine infection is rare.

Encephalomyelitis, probably due to *Toxoplasma*, has been reported from the USA, Europe and Australia (Hartley & Blakemore 1974; McErlean 1974). Affected sheep have myelomalacia and schizonts were present in astrocytes.

Sarcocystis tenella is non-pathogenic.

SARCOCYSTIS IN SWINE

Three species of sarcocysts occur in swine:

Sarcocystis miescheriana (Kühn 1865) Lankester, 1882
Sarcocystis porcifelis Dubey, 1976
Sarcocystis porcihominis Dubey, 1976

Sarcocystis infection is quite common in pigs, for example, 12.7% of swine aged more than one year were infected in a survey of abattoir animals in Detroit, Michigan (Seneviratana et al. 1975), although animals under one year of age were not infected.

Sarcocystis miescheriana has a dog–pig cycle, is world-wide in distribution and has little or no pathogenic effect on the pig. Cysts occur in the striated and heart muscles.

Sarcocystis porcihominis has a human–pig cycle. In man, after infection with cyst material, the prepatent period is 10–17 days, sporocysts passed in the faeces are 13 μm by 9 μm (Rommel et al. 1974). This species was previously confused with *S. miescheriana* in that Rommel and Heydorn (1972) produced human infection by feeding raw pork containing cysts of '*Sarcocystis miescheriana*' and raw beef containing cysts of '*S. fusiformis*'. At that time it was believed that a single parasite of man, originally recognized as *Isospora hominis*, was capable of infecting both cattle and swine. It is now recognized that two species of sarcocyst occur in man, namely *S. hominis* (intermediate host: bovine) and *S. porcihominis* (intermediate host: pig).

This species is not known to be pathogenic.

Sarcocystis porcifelis has a cat–pig cycle and is reported to be pathogenic for swine (Golubkovan & Kisliakova 1974). Sporocysts in cat faeces are 13 μm by 8 μm and the prepatent period is five to ten days.

SARCOCYSTIS IN HORSES

Two species of sarcocysts occur in horses:

Sarcocystis bertrami (syn. *S. equicanis*) Doflein, 1901.
Sarcocystis fayeri Dubey, Streitel, Stromberg and Toussant, 1977

Sarcocystis bertrami (Syn. *S. equicanis*) occurs in horses, asses and mules, in the heart, diaphragm and other muscles. World-wide in distribution and common. Muscle cysts are up to 10 mm in length and are compartmented. It has a dog–horse cycle, with a prepatent period in the dog of eight days which lead to the shedding of sporocysts 15 μm by 10 μm in size. It is not known to be pathogenic.

Sarcocystis fayeri was described by Dubey et al. (1977); it has a dog–horse cycle and is differentiated from *S. bertrami* by its shorter sporocysts (12 μm by 10 μm compared to 15 μm by 10 μm) and longer prepatent period (12–15 days compared to 8 days for *S. bertrami*). This species appears to be the only sarcocyst of importance in horses in the USA. It is not known to be pathogenic.

SARCOCYSTIS OF MAN

Three species of *Sarcocystis* occur in man:

Sarcocystis hominis (Railliet and Lucet, 1891) Levine, 1977
Sarcocystis porcihominis Dubey, 1976
Sarcocystis lindemanni (Rivolta, 1878)

Man is the final host for *S. hominis* (Syn. *S. bovihominis*) and *S. porcihominis*, and these have been discussed earlier in relation to their intermediate hosts. Infection with them in man has been associated with anorexia, nausea and diarrhoea (Jarpa 1966).

Man is the intermediate host for *S. lindemanni*. Cysts vary considerably in size, some being up to 5 cm in length, but they are usually smaller. Jeffrey (1974) has reviewed *Sarcocystis* infection in man and concludes that of the 28 reported cases 16 can be accepted as valid infections. Clinical manifestations occur in a minority of cases. These include fever, malaise and muscular and subcutaneous swellings.

The final host of *S. lindemanni* is unknown.

SARCOCYSTIS IN GAME ANIMALS

In a survey of the incidence of *Sarcocystis* infection of deer, moufflons and wild boars in Hungary, Kavai and Sugar (1976) reported that 88 of 99 roe deer, 46 of 59 red deer, 39 of 42 fallow deer, 12 of 19 moufflons and 12 of 25 wild boar were infected. Infection occurred in the heart, tongue, diaphragm, oesophagus and skeletal muscles. There was no preference for male or female, but animals under one year of age showed very reduced infection rates. The species of these sarcocysts and their definitive hosts were not determined.

In North America heavy infections of the fawns of mule deer were reported (Hudkins-Vivion et al. 1976). Dogs and coyotes shed sporocysts after being fed tissues of infected mule deer, and the sporocysts measure 13.8–16.1 μm by 9.2–11.5 μm (mean 14.4 μm by 9.3 μm); fawns fed sporocysts from coyote faeces died of acute sarcocystosis between 27 and 63 days after infection (Hudkins-Vivion & Kistner 1977). This species is named *Sarcocystis hemionilatrantis* (Hudkins-Vivion & Kistner 1977).

Sarcocystis cervi Destombes, 1957. This species was described from a deer in Vietnam. The final host is unknown.

SARCOCYSTIS IN OTHER ANIMALS

Sarcocystis cuniculi Brumpt, 1913 occurs in the striated and heart muscles of the domestic rabbit and cottontail. it is world-wide in distribution and apparently is non-pathogenic.

Sarcocystis muris (Blanchard, 1885) Labbé, 1899 occurs in the cat and the intermediate hosts are the house mouse, and various species of rat and voles. Cysts measure up to several millimetres in length but are apparently non-pathogenic for the intermediate hosts.

Sarcocystis kortei Castellani and Chambers, 1909 is found in the striated muscle of the rhesus monkey. Cysts measure 0.4–0.8 mm by 0.1–0.15 mm and occur in striated muscles. No pathogenic effects of the infection are known and the definitive host also is not known.

Sarcocystis nesbitti Mandour, 1969 was described from striated muscles of rhesus monkeys. There is no information about the pathogenicity of this species.

Other unnamed species of *Sarcocystis* have been reported from baboons (Strong et al. 1965) and other primates (see Flynn 1973).

Sarcocystis rileyi (Stiles, 1893) occurs in the heart and striated muscles of a wide variety of domestic and wild ducks but in addition it has been reported from the domestic chicken, grouse and a number of other wild birds. Prevalence is world-wide and there is no evidence of pathogenicity.

Wenyon (1926) gives a list of species recorded from various birds.

Genus: Besnoitia Henry, 1913

This genus is characterized by having cysts containing bradyzoites in fibroblasts and possibly other cells. The cyst wall occurs around the infected cell and bradyzoites occur in a parasitophorous vacuole. The host cell nucleus within the cyst undergoes hyperplasia and hypertrophy (Frenkel 1977). Sporulated oocysts of the definitive host infect only the intermediate host.

Previously, the name *Globidium* has been used, but it is now clear that the intestinal globidia represent macroschizont development of one or more species of *Eimeria* (Pellérdy 1974).

Cutaneous besnoitiasis is a serious skin condition of cattle and horses, especially in the Republic of South Africa. It is characterized by painful swellings, thickening of the skin, loss of hair and necrosis.

The following species are of interest:

Besnoitia besnoiti (Marotel, 1912) Henry, 1913
Besnoitia bennetti Babudieri, 1932
Besnoitia darlingi (Brumpt, 1913) Smith and Frenkel, 1977
Besnoitia jellisoni Frenkel, 1955
Besnoitia tarandi (Hadwen, 1922) Levine, 1961
Besnoitia wallacei Tadros and Laarman, 1976

Besnoitia besnoiti *Hosts*: definitive host, cat; intermediate hosts cattle, rabbit experimentally. The parasite is endemic in the Republic of South Africa and also occurs in Zaire, Sudan, Angola, southern Europe, USSR, Asia and South America. In the cat, oocysts 14–16 μm by 12–14 μm (mean 15 μm by 13 μm) are shed in the unsporulated state. Peteshev et al. (1974) demonstrated that this species was transmitted by cats.

In the intermediate host—the bovine—the parasite is found in the dermis, subcutaneous tissues and fascia and in the laryngeal, nasal and other mucosae.

The *Besnoitia* cyst may reach up to 600 μm in diameter. It is usually spherical, and when mature it is packed with crescentic trophozoites (bradyzoites) each 2–7 μm in length. Animals may be infected artificially by the intraperitoneal, intravenous or subcutaneous injection of blood from an infected animal.

The morphology of the developmental stages has been followed by Pols (1960) in artificially infected rabbits. Trophozoites (bradyzoites) appear 16–18 days after inoculation and occur extracellularly or in monocytes in the blood. The early forms are mainly crescentic, 5–9 μm by 2–5 μm, with a blue granular cytoplasm and a more or less central nucleus. Invasion of histiocytes occurs and organisms multiply in a parasitophorous vacuole. As the *Besnoitia* grow the host cell nucleus divides to form a multinucleate cell, and as the cyst becomes larger the cytoplasm of the host cell is flattened to form the inner coat of the pseudocyst. As the parasite becomes larger, collagen is laid down around the host cell membrane to form the hyaline capsule of the cyst.

The natural mode of transmission requires further study. For example, *B. besnoiti* from Kazakhstan is transmitted by cats, but Rommel (1975) was unable to transmit *B. besnoiti* from Uganda by means of cats. In the Republic of South Africa, Hofmeyer (1945) has suggested that contaminated watering troughs may be an important source of infection. Bigalke (1960) has reported that the *Besnoitia besnoiti* may be transmitted mechanically by *Glossina palpalis*.

Pathogenesis. The mortality of infection is usually below 10%, although animals may lose condition markedly, pregnant animals may abort, bulls may become sterile and the hides are of little value for leather-making purposes. Animals of all ages, from six months upwards, may be infected, and after an incubation period of six to ten days there is a temperature rise which lasts two to ten days. Cyst formation in the skin occurs one to four weeks after the start of the temperature rise. The febrile stage of the infection is accompanied by photophobia, enlargement of the lymph nodes and oedematous swellings on the limbs and the lower parts of the body. These are tender and warm, the animals are reluctant to move and are anorexic, respiration is rapid, diarrhoea may be present and abortion may occur at this stage. The initial stage may be mild with little obvious clinical alteration, and such animals usually recover. However, the acute febrile stage may progress to the second, or depilatory, stage in which the skin becomes markedly thickened; it loses its elasticity, and it may crack open and ooze a serosanguinous fluid. The hair falls out over the swollen parts, and the skin is left very much wrinkled 'olifantvel'. Animals may start to feed again, but the loss of condition is very marked. Death may occur during the second stage, but if the animal survives then the third, or seborrhoea sicca, stage supervenes. The animals remain emaciated for several months, there is a further loss of hair over the areas which were oedematous, and a markedly thickened and folded hide results. This skin is also scurfy, and the whole appearance is one of extensive mange. The hairless condition may remain for several months if not for the life of the animal, although in milder cases a certain amount of regrowth of hair may occur giving checkered markings on the animal. There is no satisfactory treatment for the condition.

Besnoitia bennetti. *Hosts*: horse, ass. Africa, Sudan, southern France, Mexico. Two cases have been found in imported burros in the United States (Terrell & Stookey 1973). The species is less common than *B. besnoiti* of cattle. The parasite was originally described in both horses and cattle in the Sudan, but Pols (1960) has separated the parasites of the horse and those of the cow.

The morphology is essentially the same as for *B. besnoiti*, and the life-cycle is presumed to be the same, although the definitive host is unknown.

Pathogenesis. The pathogenesis of *B. bennetti* was described by Bennett (1933). The disease has a chronic course, animals being ill for seven to eight months. There is marked weakness, a general thickening of the skin, loss of hair and a scurfy appearance. Occasionally the disease may run a more acute course, ending in death.

Other species of *Besnoitia* include:

Besnoitia jellisoni. *Hosts*: deer mice (*Peromyscus maniculatus*), kangaroo rat (*Dipudomys* spp.), opossum in the USA and *Microxus torques* in Peru. It has been transmitted experimentally to a wide range of rodents and to chick embryos. Definitive host unknown. Cysts occur in the subcutaneous tissues, on the serosal membranes of many visceral organs such as the liver, spleen, heart, intestines and in the venous sinuses of the dura mater. Acute or chronic disease may be produced. The acute infection is characterized by yellowish masses in the dural venous sinuses. In the chronic disease cysts, up to 1 mm in diameter, may cause pits on the surface of bones and also occur elsewhere; however, in general, the infection does not produce clinical disease (Jellison 1971).

Besnoitia tarandi. Reindeer and caribou in Alaska. Cysts occur in the periosteum, on the surface of tendons and in the fibrous connective tissue. Cysts are spherical, 0.1–0.45 μm in diameter, with thick walls. The infection produces cornmeal disease of reindeer, since the lesions on the periosteum and tendons are granular. The definitive host is not known.

Besnoitia wallacei (Fig. 3.40). *Hosts*: definitive host, cat (*Felis catus*). Intermediate hosts, *Rattus norvegicus, R. exulans* and *Mus musculus. Oocysts*: 10–13 μm by 16–19 μm (mean 12 μm by 17 μm) passed in the faeces unsporulated. Prepatent period 12–15 days. In rodents tissue cysts probably occur in fibroblasts (Frenkel 1977). Cysts reach up to 200 μm in size. Bradyzoites not transmissible to other intermediate hosts or only poorly. This species has a distinct host range compared to other species of *Besnoitia* and, for example, is not transmissible to *R. rattus* or hamsters.

Besnoitia darlingi. *Hosts*: intermediate hosts, lizards, Panama. Opossum. Definitive host unknown, but the cat has been shown to be an experimental host (see p. 628). It can be transmitted to laboratory mice (Schneider 1967; Conti-Diaz et al. 1970) and bats. *Besnoitia panamensis* is a synonym.

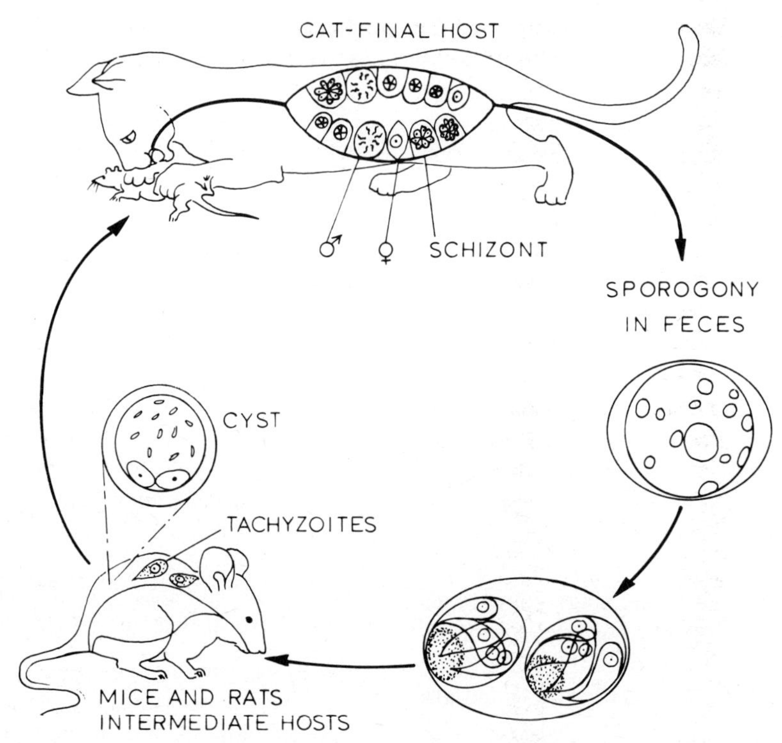

Fig 3.40 Life-cycle of *Besnoitia wallacei* of murines. Cats shed unsporulated oocysts in faeces. After sporulation oocysts are infective for rats and mice and tachyzoites develop in various organs of the intermediate hosts. Cats become infected by ingestion of cysts in the intermediate host; schizonts, gametocytes and oocysts are formed in the intestine. Extraintestinal forms do not occur in the cat. (*From Dubey 1976*)

Genus: Hammondia Frenkel and Dubey, 1975

The definitive host is the cat. Intermediate hosts are rodents. Cysts occur in skeletal muscle, occasionally heart muscle and the brain (Frenkel 1977). The cyst wall develops from the limiting membrane of the parasitophorous vacuole in which bradyzoites occur. Host cell nuclei present outside the cyst. Cyst stages are infective only for the definitive host and sporulated oocysts from the definitive host only infect the intermediate host (Frenkel 1977). (See Fig. 3.41.)

Only one species of this genus is known:

Hammondia hammondi Frenkel and Dubey, 1975

This organism is similar to *Toxoplasma* and Levine (1977) considers it a synonym of *Toxoplasma hammondi*. The definitive host is the cat and the intermediate hosts are mice, guinea-pigs, rats, dogs and hamsters, which are infected by the ingestion of sporulated oocysts.The genus is of low virulence for mice. Tachyzoites multiply in the lamina propria of the intestine, Peyer's patches and the musculature. Necrosis of infected cells occurs and cysts appear after two weeks of infection (Dubey 1976*a*).

Genus: Frenkelia Biocca, 1968

In this genus cysts occur in the brain and spinal cord: they occupy the host cell completely, the nucleus undergoing hypertrophy. The definitive host is a bird of prey and the intermediate host(s) consist of the quarry of the predators.

Frenkelia clethrionomyobuteonis Rommel and Krampitz, 1975. *Hosts*: *Buteo* and *Microtus* spp. (Rommel & Krampitz 1975).

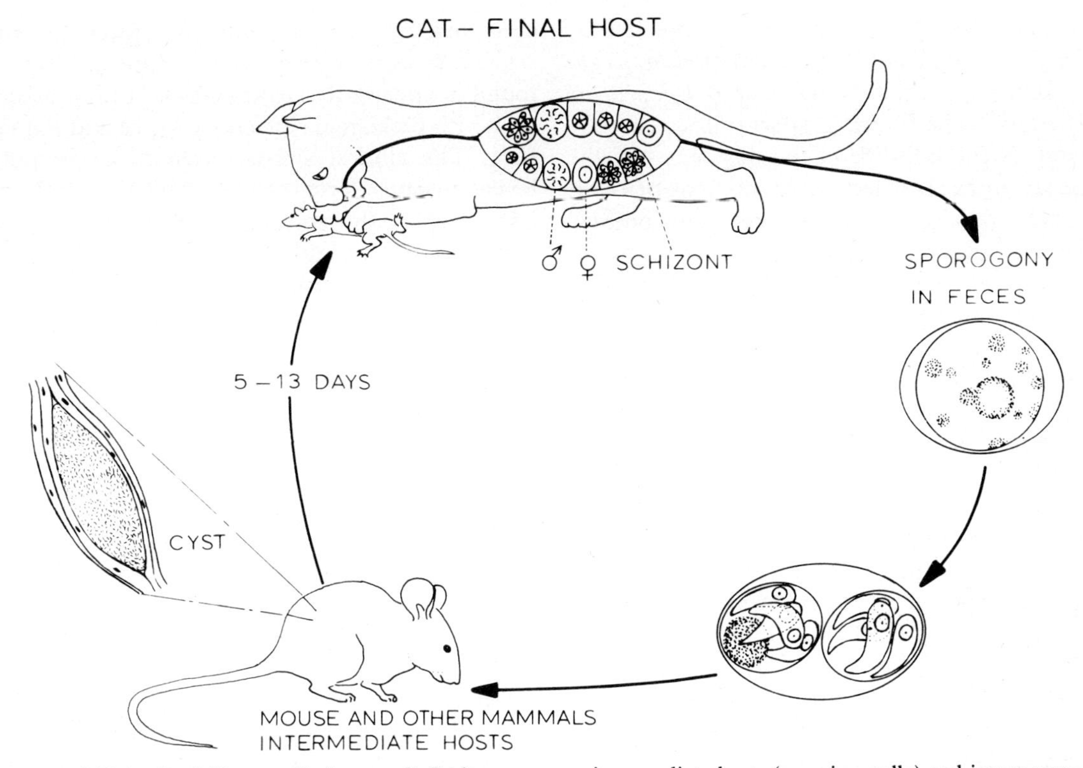

Fig 3.41 Life-cycle of *Hammondia hammondi*. Rodents serve as intermediate hosts (experimentally) and ingest sporulated oocysts. Muscle cysts develop in the rodent host. (*From Dubey 1976*)

FAMILY: HAEMOGREGARINIDAE NEVEU-LEMAIRE, 1901

Organisms of this family belong to the suborder Adeleidea Léger, 1911 and are similar to the *Eimeridae*, but micro- and macrogamonts become attached to each other in pairs (syzygy) during development into gametes; the zygote becomes an oocyst producing numerous sporoblasts, each of which develops into a spore containing two or four sporozoites. Parasites of the above family occur in the cells of the circulatory system of vertebrates, and the genus *Hepatozoon* Miller, 1908, is the only group of interest.

Genus: Hepatozoon Miller, 1908

Schizogony occurs in the endothelial cells of the liver, and the gametocytes are found in leucocytes or erythrocytes according to the species of organism. Sporogony occurs in various blood-sucking arthropods.

Hepatozoon canis (James, 1905). *Hosts*: dog, cat, jackal, hyena. Far East, Central and North Africa, Middle East, Italy.

Schizonts occur in the endothelial cells of the spleen, bone marrow and liver as round or oval bodies, more or less filling the host cell and containing 30–40 nuclei.

Various types of schizonts have been described; one produces a small number of large merozoites (usually three) which Wenyon (1926) considered became schizonts, and another produces a large number of small merozoites which are the forms that enter leucocytes.

The blood forms, gamonts, occur in the leucocytes and are elongate, rectangular bodies measuring 8–12 μm by 3–6 μm (mean 6 μm by 3 μm). They are surrounded by a delicate capsule, stain pale blue with a dark reddish purple nucleus and have a number of pink granules in the cytoplasm. In citrated blood these forms may be found free in the plasma.

Developmental cycle. The dog is infected by the ingestion of the infected vector tick, *Rhipicephalus sanguineus*, which contains sporocysts in its body cavity. The liberated sporozoites penetrate the wall of the intestine of the dog, pass via the blood stream to the spleen, liver and bone marrow, and here they enter tissue cells and become schizonts. Several generations of schizonts occur, but ultimately merozoites enter the circulating leucocytes and become gametocytes or gamonts. These show no sexual dimorphism and undergo no futher change until ingested by the tick. Gamonts leave the host leucocyte in the alimentary canal of the tick, become associated in pairs, and the microgamont produces two non-flagellate microgametes, one of which fertilizes the macrogamete to produce a zygote. This is motile (ookinete), and it penetrates the intestinal wall to enter the haemocoel of the tick where it grows to become an oocyst which, when mature, is about 100 μm in length. Sporoblasts (30–50) and then sporocysts are formed, each of which produces about 16 sporozoites. On ingestion of the tick, the oocysts and sporocysts rupture to release the sporozoites.

Pathogenesis. Schizonts of a *Hepatozoon*-like parasite have been found in the myocardium of domestic cats in Israel by Klopfer et al. (1973). The schizonts appeared to be located in the capillary lumen and no parasites were seen in the peripheral blood, spleen or lymph nodes. Two forms of schizonts were seen, one containing peripheral rosettes of merozoite formations and the other filled with merozoites, showing a cyst-like structure. Schizonts were found in 30% of 50 cats submitted for autopsy and 42% of 50 apparently healthy cats. Inflammatory reactions with schizonts were scarce. The organism may be found in apparently healthy dogs, but it is associated with pathogenic effects in Africa and the Far East. The clinical signs consist of an irregular fever, anaemia, progressive emaciation with enlargement of the spleen. Lumbar paralysis has been described. Death occurs four to eight weeks after the onset of clinical signs.

Hepatozoon canis infection is diagnosed by the demonstration of the gametocytes in stained blood smears, or the schizonts in the spleen or bone marrow.

There is no known treatment, and control is based on tick control.

Other species of *Hepatozoon* are:

Hepatozoon muris (Balfour, 1905). Occurs in the brown rat (*Rattus norvegicus*) and the black rat (*Rattus rattus*) throughout the world. The schizogony cycle takes place in the parenchymal cells of the liver, the gametocytes being found in the monocytes of the blood. Development occurs in the rat mite, *Echinolaelaps echidninus*, rats being infected by ingestion of the infected mite. The infection is usually considered to be non-pathogenic, although anaemia and emaciation, with splenomegaly, have been reported in rats with severe infections (Miller 1908).

Hepatozoon musculi (Porter, 1907). Is found in the white mouse in England.

Hepatozoon cuniculi (Sangiorgi, 1914). Occurs in the rabbit in Europe.

Hepatozoon griseisciuri (Clarke, 1958). Is found in the grey squirrel in the United States.

REFERENCES

TOXOPLASMA, SARCOCYSTIS, BESNOITIA, HAMMONDIA AND HEPATOZOON

Aikawa, M. & Sterling, C. R. (1974) *Intracellular Parasitic Protozoa.* New York: Academic Press.
von Alton, Y., Heydorn, A. O. & Janitschke, K. (1977) Zur Infektiosität von Toxoplasma-oozystem für das Pferd. *Berl. Münch. tierärztl. Wschr.*, **90**, 433–435.
Aranjo, F. G. & Remington, J. S. (1974) Effect of clindamycin on acute and chronic toxoplasmosis in mice. *Antimicrob. Ag. Chemother.*, **5**, 647–651.
Beech, J. (1974) Equine protozoan encephalomyelitis. *Vet. Med. small Anim. Clin.*, **69**, 1562–1566.
Bennett, S. C. J. (1933) Globidium infections in the Sudan. *J. comp. Path. Ther.*, **46**, 1–14.
Beverley, J. K. A. (1957) Toxoplasmosis. *Vet. Rec.*, **69**, 337–341.

Beverley, J. K. A. (1974) Some aspects of toxoplasmosis, a world wide zoonosis. In: *Parasitic Zoonosis: Clinical and Experimental Studies*, ed. E. J. L. Soulsby, pp. 1–26. New York: Academic Press.

Beverley, J. K. A., Beattie, C. P. & Roseman, C. (1954) Human toxoplasma infection. *J. Hyg., Camb.*, **52**, 37–46.

Beverley, J. K. A. & Watson, W. A. (1961) Ovine abortion and toxoplasmosis in Yorkshire. *Vet. Rec.*, **73**, 6–10.

Bigalke, R. D. (1960) Preliminary observations on the mechanical transmission of cyst organisms of *Besnoitia besnoiti* (Marotel, 1912) from a chronically infected bull to rabbits by *Glossina brevipalpis* Newstead, 1910. *J. S. Afr. vet med. Ass.*, **31**, 37–44.

Campbell, R. S. F. (1956) Canine toxoplasmosis. *Vet. Rec.*, **68**, 591–592.

Christiansen, M. (1948) Toxoplasmose Nos Hare i Damark. *Medlemsbl. Danske Dyrlaegeforen*, **31**, 93–104.

Christiansen, M. & Siim, J. C. (1951) Toxoplasmosis in hares in Denmark. Serological identity of human and hare strains of Toxoplasma. *Lancet*, **260**, 1201–1203.

Clegg, F. G., Beverley, J. K. A. & Markson, L. M. (1978) Clinical disease in cattle in England resembling Dalmeny disease associated with suspected sarcocystis infection. *J. comp. Path.*, **88**, 105–114.

Cole, C. R., Sanger, V. L., Farrell, R. L. & Kirnder, J. D. (1954) The present status of toxoplasmosis in veterinary medicine. *N. Am. Vet.*, **53**, 265–270.

Conti-Diaz, I. A., Turner, C., Tweeddal, D. T. & Furcolou, M. L. (1970) Besnoitiasis in the opossum (*Didelphis marsupialis*). *J. Parasit.*, **56**, 457–460.

Corner, A. H., Mitchell, D., Meads, E. B. & Taylor, P. A. (1963) Dalmeny disease. An infection of cattle presumed to be caused by an unidentified protozoan. *Can. vet. J.*, **4**, 252–264.

Cusick, P. K., Sells, D. M., Hamilton, D. P. & Hardenbrook, H. J. (1974) Toxoplasmosis in two horses. *J. Am. vet. med. Ass.*, **164**, 77–80.

Desmonts, G. (1977) Toxoplasma, mother and child. Abstracts, *1st Int. Symp. Parasit. Soc. Chilean Parasit*, Santiago, Oct. 1977, p. 19.

Dubey, J. P. (1968) Feline toxoplasmosis and its nematode transmission. *Vet. Bull.*, **38**, 495–499.

Dubey, J. P. (1973) Feline toxoplasmosis and coccidiosis: a survey of domiciled and stray cats. *J. Am. vet. med. Ass.*, **162**, 873–877.

Dubey, J. P. (1976*a*) A review of *Sarcocystis* of domestic animals and other coccidia of cats and dogs. *J. Am. vet. med. Ass.*, **169**, 1061–1078.

Dubey, J. P. (1976*b*) Reshedding of *Toxoplasma* oocysts by chronically infected cats. *Nature*, **262**, 213–214.

Dubey, J. P. (1976*c*) Prevalence of *Toxoplasma* infection in cattle slaughtered at an Ohio abattoir. *J. Am. vet. med. Ass.*, **169**, 1197–1199.

Dubey, J. P. (1977*a*) Attempted transmission of feline coccidia from chronically infected queens to their kittens. *J. Am. vet. med. Ass.*, **170**, 541–543.

Dubey, J. P. (1977*b*) Persistence of *Toxoplasma gondii* in the tissues of chronically infected cats. *J. Parasit.*, **63**, 156–157.

Dubey, J. P., Davis, G. W., Koestner, A. & Kiryu, K. (1974) Equine encephalomyelitis due to a protozoan parasite resembling *Toxoplasma gondii*. *J. Am. vet. med. Ass.*, **165**, 249–255.

Dubey, J. P. & Frenkel, J. K. (1972) Cyst-induced toxoplasmosis in cats. *J. Protozool.*, **19**, 155–177.

Dubey, J. P. & Frenkel, J. K. (1974) Immunity to feline toxoplasmosis: modification by administration of corticosteroids. *Vet. Path.*, **11**, 350–379.

Dubey, J. P. & Hoover, E. A. (1977) Attempted transmission of *Toxoplasma gondii* infection from pregnant cats to their kittens. *J. Am. vet. med. Ass.*, **170**, 538–540.

Dubey, J. P., Streitel, R. H., Stromberg, P. C. & Toussant, M. J. (1977) *Sarcocystis fayeri* sp. n. from the horse. *J. Parasit.*, **63**, 443–447.

Dubey, J. P. & Yeary, R. A. (1977) Anticoccidial activity of 2-sulfamoyl-4,4-diaminodiphenylsulfone, sulfadiazine pyrimethamine and clindamycin in cats infected with *Toxoplasma gondii*. *Can. vet. J.*, **18**, 51–57.

Engster, A. R. & Joyce, J. R. (1976) Prevalence and diagnostic significances of *Toxoplasma gondii* antibodies. *Vet. Med. small Anim. Clin.*, **71**, 1469–1473.

Ericksen, S. & Harboe, A. (1953) Toxoplasmosis in chickens. I. An epidemic outbreak of toxoplasmosis in a chicken flock in south-eastern Norway. *Acta path. microbiol. scand.*, **33**, 56–71.

Eyles, D. E. (1952) Incidence of *Trypanosoma lewisi* and *Hepatozoon muris* in the Norway rat. *J. Parasit.*, **38**, 222–225.

Eyles, D. E. (1956) Newer knowledge of the chemotherapy of toxoplasmosis. *Ann. N.Y. Acad. Sci.*, **64**, 252–267.

Farrell, R. L., Docton, F. L., Chamberlain, D. M. & Cole, C. R. (1952) Toxoplasmosis. I. Toxoplasma isolated from swine. *Am. J. vet. Res.*, **13**, 181–185.

Fayer, R. (1974) The development of *Sarcocystis fusiformis* in the small intestine of dogs. *J. Parasit.*, **60**, 660–665.

Fayer, R. & Johnson, A. J. (1973) Development of *Sarcocystis fusiformis* in calves infected with sporocysts from dogs. *J. Parasit.*, **59**, 1135–1137.

Fayer, R. & Johnson, A. J. (1974) *Sarcocystis fusiformis*: development of cysts in calves infected with sporocysts from dogs. *Proc. helminth. Soc. Wash.*, **41**, 105–108.

Fayer, R., Johnson, A. J. & Lunde, M. N. (1977) Abortion and other signs of disease in cows experimentally infected with *Sarcocystis* from dogs. *J. infect. Dis.*, **134**, 624–628.

Fayer, R. & Lunde, M. N. (1977) Changes in serum and plasma proteins and in IgG and IgM antibodies in calves experimentally infected with *Sarcocystis* from dogs. *J. Parisit.*, **63**, 438–442.

Flynn, R. J. (1973) *Parasites of Laboratory Animals*, pp. 884, Ames: Iowa State University Press.

Frelier, P., Mayhew, I.G., Fayer, R. & Lunde, M. N. (1977) Sarcocystosis: a clinical outbreak in dairy calves. *Science, N.Y.*, **195**, 1341–1342.

Frenkel, J. K. (1973) Toxoplasmosis: parasite life cycle, pathology and immunology. In: *The Coccidia*, ed. D. A. Hammond & P. L. Long, pp. 343–410. Baltimore: University Park Press.

Frenkel, J. K. (1974) Advances in the biology of sporozoa. *Z. Parasitenk.*, **45**, 125–162.

Frenkel, J. K. (1975) Toxoplasmosis in cats and mice. *Feline Pract.*, **5**, 28–41.

Frenkel, J. K. (1977) *Besnoitia wallacei* of cats and rodents: with a reclassification of other cyst-forming isosporoid coccidia. *J. Parasit.*, **63**, 611–628.

Frenkel, J. K., Dubey, J. P. & Miller, N. L. (1970) *Toxoplasma gondii*: fecal stages identified as coccidial oocysts. *Science*, **167**, 893–896.

Fulton, J. D. (1963) Serological tests in toxoplasmosis. In: *Immunity in Protozoa*, ed. P. C. C. Garnham, A. E. Pierce & I. Roitt, pp. 259–272. Oxford: Blackwell Scientific Publications.

Golubkovan, D. I. & Kisliakova, Z. I. (1974) The sources of infection for swine *Sarcocystis* [in Russian]. *Veterinariia*, **11**, 85–86.

Harding, J. D. J., Beverley, J. K. A., Shaw, I. G., Edwards, B. L.& Bennett, G. H. (1961) Toxoplasma in English pigs. *Vet. Rec.*, **73**, 3–6.

Hartley, W. J. & Blakemore, W. F. (1974) An unidentified sporozoan encephalomyelitis in sheep. *Vet. Path.*, **11**, 1–12.

Hartley, W. J., Jebson, J. L. & McFarlane, D. (1954) New Zealand type II abortion in ewes. *Aust. vet. J.*, **30**, 216–218.

Hartley, W. J. & Marshall, S. C. (1957) Toxoplasmosis as a cause of ovine perinatal mortality. *N. Z. vet. J.*, **5**, 119–124.

Hofmeyer, C. F. B. (1945) Globidiosis in cattle. *J. S. Afr. vet. med. Ass.*, **16**, 102–109.

Hudkins-Vivion, G. & Kistner, T. P. (1977) *Sarcocystis hemionilatrantis* (sp. nov.); life cycle in mule deer and coyote. *J. Wildl. Dis.*, **13**, 80–84.

Hudkins-Vivion, G., Kistner, T. P. & Fayer, R. (1976) Possible species differences between *Sarcocystis* from mule deer and cattle. *J. wildl. Dis.*, **12**, 86–87.

Hutchinson, W. M. (1965) Experimental transmission of *Toxoplasma gondii*. *Nature*, **206**, 961–962.

Jacobs, L. (1956) Propagation, morphology and biology of toxoplasmosis. *Ann. N.Y. Acad. Sci.*, **64**, 154–179.

Jacobs, L. (1976) Serodiagnosis of toxoplasmosis. In: *Immunology of Parasitic Infections*, eds S. Cohen & E. H. Sadun, pp. 94–106. Oxford: Blackwell Scientific Publications.

Jacobs, L., Melton, M. L. & Jones, F. E. (1952) The prevalence of toxoplasmosis in wild pigeons. *J. Parasit.*, **38**, 457–461.

Jacobs, L., Remington, J. S. & Melton, M. L. (1960) A survey of meat samples from swine, cattle and sheep for the presence of encysted *Toxoplasma*. *J. Parasit.*, **46**, 23–38.

Jarpa, G. A. (1966) Coccidiosis humana. *Biologica (Santiago)*, **39**, 3–26.

Jeffrey, H. C. (1974) Sarcosporidiosis in man. *Trans. R. Soc. trop. Med. Hyg.*, **68**, 17–29.

Jellison, W. L. (1971) Besnoitiosis. In: *Parasitic Disease of Wild Mammals*, ed. J. W. Davis & R. D. Anderson, pp. 354–357. Ames: Iowa State University Press.

Johnson, A. J., Hildebrandt, P. K. & Fayer, R. (1975) Experimentally induced *Sarcocystis* infection in calves: pathology. *Am. J. vet. Res.*, **36**, 995–999.

Jones, T. C., Masur, H., Len, L. & Tzu Lin Tom Fu (1977) Lymphocyte–macrophage interaction during control of intracellular parasitism. *Am. J. trop. Med. Hyg.*, **26** (Suppl.), No. 6, 187–193.

Kagan, I. G. (1974) Advances in the immunodiagnosis of parasitic infections. *Z. Parasitenk.*, **45**, 163–195.

Kavai, A. & Sugar, L. (1976) On the incidence of Sarcosporidia in the game of Hungary [in Hungarian]. *Parasit. Hung.*, **9**, 17–19.

Klopfer, V., Nobel, T. A. & Neumann, F. (1973) Hepatozoon-like parasite (schizonts) in the myocardium of the domestic cat. *Vet. Path.*, **10**, 185–190.

Koestner, A. & Cole, C. R. (1961) Neuropathology of ovine and bovine toxoplasmosis. *Am. J. vet. Res.*, **22**, 53–66.

Lainson, R. (1955) Toxoplasmosis in England. I. Variation factors in the pathogenesis of Toxoplasma infections; the sudden increase in virulence of a strain after passage in multimammate rats and canaries. *Ann. trop. Med. Parasit.*, **49**, 397–416.

Lainson, R. (1956) Toxoplasmosis in England. II. Toxoplasmosis infections in dogs: the incidence of complement-fixing antibodies among dogs in London. *Ann. trop. Med. Parasit.*, **50**, 172–186.
Lainson, R. (1972) A note on sporozoa of undetermined taxonomic position in an armadillo and a heifer calf. *J. Protozool.*, **19**, 582–586.
Leek, R. G. & Fayer, R. (1978) Sheep experimentally infected with *Sarcocystis* from dogs. II. Abortion and disease in ewes. *Cornell Vet.*, **68**, 108–123.
Levine, N. D. (1973*a*) *Protozoan Parasites of Domestic Animals and of Man*, 2nd ed. Minneapolis: Burgess.
Levine, N. D. (1973*b*) Introduction, history and taxonomy. In: *The Coccidia*, ed. D. M. Hammond & P. L. Long, pp. 1–22. Baltimore: University Park Press.
Levine, N. D. (1977) Nomenclature of *Sarcocystis* in the ox and sheep and of fecal coccidia of the dog and cat. *J. Parasit.*, **63**, 36–51.
Lunde, M. N. & Fayer, R. (1977) Serologic tests for the detection of antibody to *Sarocystis* in cattle. *J. Parasit.*, **63**, 222–225.
McErlean, B. A. (1974) Ovine paralysis associated with spinal lesions of toxoplasmosis. *Vet. Rec.*, **94**, 264–266.
McLeod, R. & Remington, J. S. (1977) Influence of infection with *Toxoplasma* on macrophage function, and role of macrophages in resistance to *Toxoplasma*. *Am. J. trop. Med. Hyg.*, **26**, (Suppl.), No. 6, 170–186.
Markus, M. B., Killick-Kenrick, R. & Garnham, P. C. C. (1974) The coccidial nature and life cycle of *Sarcocystis*. *J. trop. Med. Hyg.*, **77**, 248–259.
Meads, E. B. (1976) Dalmeny disease—another outbreak—probably sarcocystosis. *Can. vet. J.*, **17**, 271.
Mello, U. (1910) Un cas de toxoplasmose du chien observe a Turin. *Bull. Soc. Path. exot.*, **3**, 359–363.
Miller, L. T. & Feldman, H. A. (1953) Incidence of antibodies for Toxoplasma among various animal species. *J. infect. Dis.*, **92**, 118–120.
Miller, W. W. (1908) *Hepatozoon perniciosum* (n.g. n.sp.)—a haemorgregarine pathogenic for white rats; with a description of the sexual cycle in the intermediate host, a mite (*Laelaps echidninus*). *US Treasury Dept. Hyg. Lab. Bull.*, **46**, pp. 51.
Olafson, P. & Monlux, W. S. (1942) Toxoplasma infections in animals. *Cornell Vet.*, **32**, 176–190.
Pellérdy, L. (1974) *Coccidia and Coccidiosis*, 2nd edn. Budapest: Akad. Kiado.
Perrin, T., Brigham, C. & Kajahn, E. (1943) Toxoplasmosis in wild rats. *J. infect. Dis.*, **72**, 91–96.
Peteshev, V. M., Galuzo, I. G. & Polomoshnov, A. P. (1974) Cats, definitive hosts of *Besnoitia* (*Besnoitia besnoiti*) [in Russian]. *Izv. Akad. Nauk. Kazak SSR.*, Ser. Biol. 1974 (1), 33–38.
Pols, J. W. (1960) Studies on bovine besnoitiosis with special reference to the aetiology. *Onderstepoort J. vet. Res.*, **28**, 265–356.
Remington, J. S. & Krahlenbuhl, J. L. (1976) Immunology of toxoplasma infection. In: *Immunology of Parasitic Infections*, eds S. Cohen & E. H. Sadun, pp. 236–267. Oxford: Blackwell Scientific Publications.
Roberts, R. M. (1966) Encephalitis. In: *Current Veterinary Therapy 1966–1967: Small Animal Practice*, ed. R. W. Kirk, pp. 219–268. Philadelphia: Saunders.
Rommel, M. (1975) Neue Erkenntinisse zur Biologie der Kokzidien, Toxoplasmen, Sarkosporidien und Besnoitien. *Berl. Münch. tierärztl. Wschr.*, **88**, 112–117.
Rommel, M. & Heydorn, A.-O. (1972) Beiträge zum Lebenszyklus der Sarkosporidien. III. Isospora lominis (Raillet und Lucet, 1891) Weynon, 1923, eine Danerform der Sarcosporidien des Rindes und des Schweins. *Berl. Münch. tierärztl. Wschr.*, **85**, 143–145.
Rommel, M., Heydorn, A.-O., Fischle, B. & Gestrich, R. (1974) Beiträge zum Lebenszyklus der Sarkosporidien. V. Weitere Endwirte der Sarkosporidien von Rind, Schaf und Schwein und die Bedeutung des Zwischenwirten für dieser Parasitose. *Berl. Münch. tierärtzl. Wschr.*, **87**, 392–396.
Rommel, M. & Krampitz, H. E. (1975) Beiträge zum Lebenzyklus der Frenkelien. I. Die Identität von *Isospora buteonis* aus dem Mäusebussard mit einer Frenkelienart (*F. clethrionomyobuteonis* spec. n.) aus der Rötelmaus. *Berl. Münch. tierärztl. Wschr.*, **88**, 338–340.
Ruiz, A. (1974) Fase sexual del llamado Sarcocystis muris en el gato. *3rd Int. Congr. Parasit.*, München, 1974.
Ruiz, A. & Frenkel, J. K. (1976) Recognition of cyclic transmission of *Sarcocystis muris* by cats. *J. infect. Dis.*, **133**, 409–418.
Sanger, V. L., Chamberlain, D. N., Chamberlain, R. W., Cole, C. R. & Farrel, R. S. (1953) Toxoplasmosis. V. Isolation of Toxoplasma from cattle. *J. Am. vet. med. Ass.*, **123**, 87–91.
Schneider, C. R. (1967) *Besnoitia darlingi* (Brumpt, 1913) in Panama. *J. Protozool.*, **14**, 78–82.
Scholtyseck, E. (1973) Ultrastructure. In: *The Coccidia*, ed. D. M. Hammond & P. L. Long, pp. 81–144. Baltimore: University Park Press.
Seeman, J. (1959) Serological fundings of toxoplasmosis in horses and other domestic animals. *Ces. epidem. Mikrobiol. Immunol.*, 8, 228–234.
Senviratana, P., Edward, A. G. & DeGuisti, D. L. (1975) Frequency of *Sarcocystis* spp. in Detroit, metropolitan area, Michigan. *Am. J. vet. Res.*, **36**, 337–339.
von Seyerl, F. (1970) Untersuchungen über die Häufigheit der Infektion mit *Toxoplasma gondii* bei Equiden. *Tierärztl. Umschau*, **25**, 447–449.
Sheffield, H. G. & Melton, M. L. (1976) Effect of pyrimethamine and sulfadiazine on the intestinal development of *Toxoplasma gondii* in cats. *Am. J. trop. Med. Hyg.*, **25**, 379–383.
Siim, J. C., Biering-Sørensen, V. & Møller, T. (1963) Toxoplasmosis in domestic animals. *Adv. vet. Sci.*, **8**, 335–429.
Spindler, L. A., Zimmerman, H. E. & Jaquette, D. S. (1946) Transmission of sarcocystis to swine. *Proc. helminth. Soc. Wash.*, **13**, 1–11.
Strong, J. P., Miller, J. H. & McGill, H. C., Jr. (1965) Naturally occurring parasitic lesions in baboons. In: *The Baboon in Medical Research*, ed. H. Vagtborg, pp. 503–512. Austin: Texas University Press.
Terrell, T. G. & Stookey, J. L. (1973) *Besnoitia bennetti* in two Mexican burros. *Vet. Path.*, **10**, 177–184.
Vanderwagen, L. C., Behymer, D. E., Riemann, H. P. & Franti, C. E. (1974) A survey for Toxoplasma in northern California livestock and dogs. *J. Am. vet. med. Ass.*, **164**, 1034–1037.
Wallace, G. D. (1973) The role of the cat in the natural history of *Toxoplasma gondii*. *Am. J. trop. Med. Hyg.*, **22**, 313–322.
Wenyon, C. M. (1926) *Protozoology*, 2 vols, pp. 1563. London: Baillière, Tindall & Cox.
Westphal, A. & Bauer, F. (1952) Weitere Untersuchungen und Betrachtungen zur Taxoplasmose-Komplementbindungsreaktion nach Westphal. *Z. Tropenmed. parasit.*, **3**, 326–329.
Wickham, N. & Carne, H. R. (1950) Toxoplasmosis in domestic animals in Australia. *Aust. vet. J.*, **26**, 1–3.

SUBORDER: HAEMOSPORINA DANILESKY, 1885

The development of the Haemosporina is similar to that of the Coccidia; however, in the former the life-cycle is shared by two hosts, schizogony occurring in vertebrates, and gametogony and sporogony occurring in blood-sucking invertebrates.

This suborder contains one family of medical and veterinary importance: Plasmodiidae. Three genera of importance are assigned to this family: *Plasmodium*, *Haemoproteus* and *Leucocytozoon*. Previously these genera were assigned to the families Plasmodiidae, Haemoproteidae and Leucocytozoidae, respectively. However, Levine (1973) considers that, on the basis of present knowledge of these organisms, a division into families is not warranted. Levine's classification will be used in this revision.

FAMILY: PLASMODIIDAE MESNIL, 1903

Macro- and microgamonts develop independently, the zygote is motile, schizogony occurs in vertebrates and sporogony in invertebrates and pigment is usually formed in the host cell. Three genera are of importance: *Plasmodium Haemoproteus* and *Leukocytozoon.*

Genus: Plasmodium Marchiafava and Celli, 1885

This genus contains the malarial organisms of man, other mammals and vertebrates. Schizogony occurs in the red blood cells and also in endothelial cells of inner organs, while the sexual phase of the cycle occurs in blood-sucking insects; for mammalian forms these are anopheline mosquitoes and for the avian forms, culicine mosquitoes.

Extensive accounts of the malarial organisms have been given by numerous authors, and a detailed account of malaria and malariology with a consideration of the taxonomy, morphology, life-cycle, vector and other aspects may be found in the textbook *Practical Malariology* by Russell et al. (1963).

Although mammalian malaria is of no immediate concern to the veterinarian, it is, nevertheless, of considerable importance as a global disease, and no text of protozoology would be complete without reference to it. The major consideration is devoted to avian malaria.

AVIAN MALARIA

An extensive, general review of avian malaria is given by Hewitt (1940), and a catalogue of the plasmodia of birds is given by Coatney and Roundabush (1949). Levine and Hanson (1953) list the species which occur in waterfowl, and Levine and Kantor (1959) list those which are found in columbiform birds. The more recent information on avian malaria is reviewed by Huff (1963).

The more common species of avian malarial organisms are listed below (after Russel et al. 1963):

Species with round or irregular gamonts which displace the nucleus of the host cell

Plasmodium cathemerium Hartman, 1927
Plasmodium gallinaceum Brumpt, 1935
Plasmodium juxtanucleare Versiani and Gomes, 1941
Plasmodium relictum (Grassi & Feletti, 1891)
Plasmodium griffithsi Garnham, 1966

Species with elongate gamonts which do not usually displace the host cell nucleus

Plasmodium circumflexum Kikuth, 1931
Plasmodium durae Herman, 1941
Plasmodium elongatum Hugg, 1930
Plasmodium fallax Schwetz, 1930
Plasmodium hexamerium Huff, 1935
Plasmodium lophurae Coggeshall, 1938
Plasmodium polare Manwell, 1935
Plasmodium rouxi Sergent and Catanei, 1928
Plasmodium vaughani Novy and MacNeal, 1904.

Developmental cycle

A major advance in the understanding of the life-cycle of the malarial organism (Fig. 3.42) was made by the demonstration that the infective sporozoites did not enter erythrocytes directly, but rather developed as exoerythrocytic forms in cells of the reticuloendothelial system prior to invasion of the erythrocytes.

Following the introduction of the sporozoites from infected culicine mosquitoes, numerous pre-erythrocytic schizonts are found in the macrophages and fibroblasts of the skin near the point of entry. These are referred to as cryptozoites. Merozoites from this first generation of pre-erythrocytic schizonts form a second generation of pre-erythrocytic schizonts: the metacryptozoites. Merozoites from the metacryptozoites enter erythrocytes and other cells of the body and in the latter form exoerythrocytic schizonts. In the case of *P. gallinaceum, P. relictum* and *P. cathemerium,* these other cells are endothelial cells, but in the case of *P. elongatum* and *P. vaughani* they are cells of the haemopoietic system. In some species of avian plasmodia, e.g. *P. gallinaceum* and *P. elongatum,* the exoerythrocytic developmental stages may be added to by forms which are derived from the erythrocytic cycle. These are known as phanerozoites, being derived from the merozoites of the schizonts in the erythrocytic cycle.

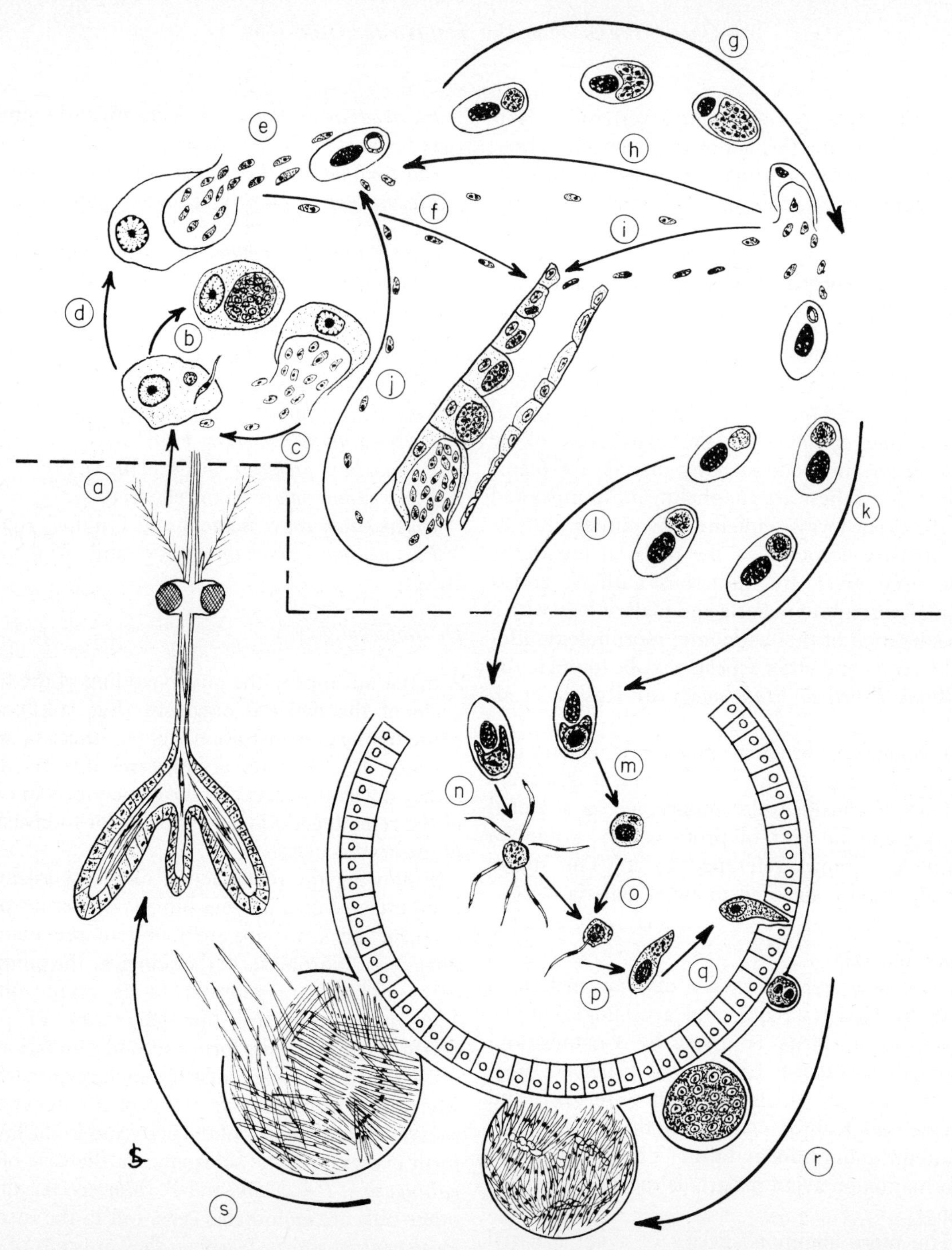

Fig 3.42 Life-cycle of *Plasmodium gallinaceum*. (a) Introduction of sporozoites by infected culicine mosquito. (b) Development of preerythrocytic schizonts (cryptozoites) in skin macrophages. (c) Release of merozoites from cryptozoites which enter other macrophages of skin. (d) Formation of metacryptozoites. (e) Entry of merozoites from metacryptozoites into erythrocytes. (f) Entry of merozoites from metacryptozoites into endothelial cells to form exoerythrocytic schizonts. (g) Erythrocytic schizogony. (h) Erythrocytic merozoites initiate further erythrocytic schizogony. (i) Erythrocytic merozoites initiate further exoerthyrocytic schizonts (phanerozoites). (j) Merozoites from exoerythrocytic schizonts initiate further erythrocytic stages. (k) Development of macrogametocytes. (l) Development of microgametocytes. (m) Maturation of macrogamete in mosquito mid-gut. (n) Maturation and exflagellation of microgamete in mosquito mid-gut. (o) Syngamy. (p) Formation of motile zygote (oökinete). (q) Penetration of oökinete to outside wall of mid-gut. (r) Sporogony. (s) Rupture of oocyst and migration of sporozytes to salivary glands of mosquito.

The erythrocytic cycle is initiated seven to ten days after infection by merozoites from metacryptozoites and at other times by merozoites from exoerythrocytic schizonts located, according to species, in the endothelial or haemopoietic cells. On entering the red blood cell, the merozoite rounds up to form a trophozoite. This is a small rounded form containing a large vacuole which displaces the cytoplasm of the parasite to the periphery of the cell. The nucleus is situated at one of the poles, giving the young form a 'signet ring' appearance when stained by the Romanowsky stains. The early trophozoites undergo schizogony to produce merozoites, the number produced depending on the species of parasite. During the process of schizogony, the parasite takes in host cell cytoplasm by invagination, haemoglobin is digested and the residual haematin pigment is deposited in granules within the food vacuoles. Apparently, schizogony may continue indefinitely, the length of each cycle of schizogony depending on the species of parasite. The release of merozoites from the schizonts occurs synchronously in the host, and in human malaria this is associated with a paroxysm of fever. Fever does not appear to be a significant part of the syndrome in avian hosts (Russell et al. 1963).

After a number of asexual generations has occured, some merozoites undergo sexual development with the formation of microgamonts and macrogamonts. Levine (1973) claims that the female forms should be referred to as macrogametes since they possess a haploid number of chromosomes. The haploid nature is maintained throughout the whole of the life-cycle of the malarial parasite, except that a diploid state is found following fertilization and zygote formation. The female forms are generally more numerous than the male forms, and they stain more intensely blue with Romanowsky stains than do the male forms. In addition, of course, the nucleus of the microgamont is more diffuse than in the female cell. Further development of the gamonts can take place only when the blood is ingested by a suitable mosquito.

Development in the mosquito is rapid. Within 10–15 minutes the nucleus of the microgamonts divides, and through a process of exflagellation, six to eight long, thin, flagella-like microgametes are extruded from the parent cell. These remain attached to the parent cell for a few minutes, lashing actively; they then become detached and swim away to find, and fertilize, the macrogamete. The zygote resulting from fertilization is motile and is called an oökinete. This oökinete penetrates the mid-gut mucosa and comes to lie on the outer surface of the stomach, forming an early oocyst about 50–60 μm in diameter. The nucleus of the oocyst divides repeatedly to produce a very large number of sporozoites. These are about 15 μm in length with a central nucleus. Maturation of the oocyst takes a variable period of time depending on the species of parasite, temperature and the species of mosquito; but in general, it is 10–20 days. When mature, the oocyst ruptures, liberating the sporozoites into the body cavity of the mosquito, and these then migrate all over the body of the mosquito but eventually reach the salivary glands. Here they may lie intracellularly, extracellularly or in the ducts of the salivary glands. They are now infective to a new host, infection occurring when the mosquito takes a blood meal. A mosquito remains infected for its life-span, transmitting malarial parasites every time it takes a blood meal.

More detailed information on the exoerythrocytic and erythrocytic cycles of malarial organisms may be found in Bray (1957), Bishop (1955) and in the account of the life-cycle of *P. gallinaceum* given by Huff and Coulston (1944).

Plasmodium cathemerium. *Hosts:* common in passerine birds, e.g. English sparrow, red-winged blackbird, etc. It has been extensively used for experimental work in canaries, in which it causes an acute fatal disease. Gamonts are rounded, with rod-shaped rather coarse pigment granules, and displace the nucleus of the cell. Gamonts and shizonts, about 7–8 μm in diameter; schizogony, 24-hour cycle with a peak of segmentation about 6–10 P.M.; 6–24 merozoites produced. Transmitted by several species of *Culex* and *Aedes*. Developmental cycle is comparable to that already described and has been investigated by Bray (1975).

Splenic infarcts occur in experimentally infected canaries, and there is marked enlargement of the spleen and liver with anaemia and subcutaneous haemorrhages.

Plasmodium gallinaceum. *Hosts:* primarily a parasite of the domestic fowl in India. Other birds, pheasant, goose, partridge and peacock, can be infected experimentally; however, the canary, duck, guinea-fowl, pigeon and English sparrow are resistant to infection. Gamonts, round and possessing pigment granules of relatively large size and few in number; schizonts, round to irregular, produce 8–30 merozoites. Both developmental stages cause displacement of the host cell nucleus. Schizogony, 36-hour cycle, peak of segmentation noon and midnight, alternately.

The developmental cycle is comparable to that described above (Fig. 3.42). Exoerythrocytic stages occur in the endothelial cells, and the reticuloendothelial cells of the spleen, brain and liver.

The natural vectors have not been identified. Experimentally, the parasite will develop in species of the genus *Aedes, Armigeres,* etc.

Pathogenesis. The chicken is particularly susceptible, and even adult birds may suffer a mortality of up to 80% in some areas. Birds become progressively emaciated as the disease progresses, there is anaemia and spleen and liver enlargement. Paralysis may occur due to massive numbers of exoerythrocytic forms in the endothelial cells of the brain capillaries.

Plasmodium juxtanucleare. *Hosts:* domestic chicken, South and Central America. Turkeys have been infected experimentally but not ducks, guinea-fowl, pigeon or canary. Gamonts, round to irregular, relatively small, parasites tend to be in contact with the host cell nucleus, host erythrocyte is often distorted. Schizogony, 24-hour cycle; three to seven, usually four, merozoites produced. Developmental cycle poorly known.

The species is highly pathogenic, Brazilian strains appearing to be especially so. Chickens become listless, weak with anaemia, and occasionally central nervous system involvement may be seen (Al-Dabagh 1961).

Plasmodium relictum. *Hosts:* pigeon, mourning dove, a number of Anatidae and various passerines. Experimentally transmissible to canary, duck, chicken and other birds. Worldwide in distribution. Gamonts, round to irregular, displace the host cell nucleus, which may be expelled from the erythrocyte, pigment fine and pinpoint. Schizogony, from 12-hour to 36-hour cycle; merozoites produced, 8–32, depending on the strain.

The development cycle of this species has been studied extensively by Bray (1957), who demonstrated exoerythrocytic stages in the endothelial cells. Several species of *Culex, Anopheles, Aedes,* etc. serve as vectors.

This parasite is highly pathogenic for pigeons. Young birds become weak and anaemic, the anaemia being the principal cause of death. On post mortem there is a markedly enlarged spleen and pigmented liver.

Plasmodium griffithsi. *Host:* turkey. Burma. Rare. This species was found in turkeys dying of malaria on a farm near Rangoon.

Plasmodium circumflexum. *Hosts:* wide variety of hosts, e.g. passerine birds, ruffed grouse and Canada goose (Levine & Hanson 1953). A large species, both the schizonts and gamonts are elongate and tend to encircle the erythrocyte nucleus, although they are not in contact with it; nor do they displace it. Schizogony, 48-hour cycle, 12–30 merozoites produced.

The life-cycle is similar to the above, exoerythrocytic stages occurring in the endothelial cells.

Plasmodium durae. *Hosts* turkey, Africa. Transmissible to ducks. Gamonts, elongate, nucleus of the host cell frequently displaced; pigment granules usually large and stain intensely. Schizogony, 24-hour cycle, 6–14 merozoites produced.

Exoerythrocytic stages of *P. durae* have been found in the endothelial cells of liver, spleen, lungs and brain of turkeys. Cerebral involvement may occur, and other clinical signs are emaciation and oedema. In young turkeys the disease takes an acute course (Purchase 1942). At post mortem there is congestion of the liver, spleen, kidneys and the capillaries of the brain and meninges.

Plasmodium elongatum. *Hosts:* English sparrow, canaries and ducks experimentally. Gamonts, elongate.

Plasmodium fallax. *Hosts:* owl, African guinea-fowl; chickens and pigeons experimentally. Gamonts, large, resembling those of *Haemoproteus*, tending to surround the nucleus of the host cell without displacing it.

Plasmodium hexamerium. *Hosts:* passerine birds. Gametocytes and schizonts, elongate. Four to eight merozoites produced.

Plasmodium lophurae. *Hosts:* originally found in fire-back pheasant (*Lophura igniti*); chickens and ducklings are susceptible experimentally, but canaries are not. Gamonts, large, elongate, host cell nucleus not displaced. 8–18 merozoites produced. Schizogony, 24-hour cycle.

Plasmodium rouxi. *Hosts:* sparrows, finches, Near East. Gamonts, elongate, host cell nucleus not displaced. The schizonts have a cycle of 24 hours and produce four merozoites.

Plasmodium vaughani. *Hosts:* American robbin (*Turdus migratorius migratorius*), starling (*Sturnus vulgaris vulgaris*), also other birds. Gamont, elongate, host cell nucleus not displaced. Schizogony, 24-hour cycle, four merozoites produced.

Therapy of avian malaria

Many of the compounds which are used for the treatment of human malaria have been developed, initially, against avian species. Reviews of the compounds which may be used, their mode of action and the development of resistant forms are given by Rollo (1964) and Bishop (1962). Chloroquine at the rate of 5 mg/kg, paludrine at 7.5 mg/kg and pyrimethamine at 0.3 mg/kg are all effective against *P. gallinaceum*.

MALARIAL PARASITES OF MAN

The species of the Plasmodia of man are:

Plasmodium falciparum Welch, 1897

Plasmodium malariae (Laveran, 1881) Grassi and Felleti, 1890

Plasmodium ovale Stephens, 1922

Plasmodium vivax (Grassi & Felletti, 1890) Labbé, 1899

These four species are those which are considered to be natural parasites of man. However, recent evidence has indicated that under certain circumstances man may be infected also with simian malaria. Accidental laboratory infection of two humans with *Plasmodium cynomolgi* has been reported by Eyles et al. (1960) and Boyle et al. (1961). It has also been realized that *Plasmodium malariae* is apparently common to man and chimpanzee, previously being known in the latter as *Plasmodium rodhaini* (Manwell 1963). *Plasmodium knowlesi* of *Macaca irus* is experimentally transmissible to man by blood inoculation, but mosquito-induced infections are not yet known. *Plasmodium vivax* can produce mild infections in chimpanzees when human infected blood is inoculated, but mosquito infections fail since sporozoites seem unable to develop in chimpanzees. Jeffery (1961) failed to find any pre-erythrocytic forms in monkeys inoculated with sporozoites of *P. vivax*.

Endogenous developmental cycle of human malaria

Detailed accounts of the life-cycle of each species may be found in Russell et al. (1963) and Faust and Russell (1964).

Following the bite of the infected mosquito, the sporozoites remain in the blood for a short time, but after an hour the blood is no longer infective for another host. Sporozoites enter the parenchyma cells of the liver and here develop into pre-erythrocytic schizonts (cryptozoites). The exo-erythrocytic forms are confined to the liver in mammalian malaria and, unlike the avian forms, are sparse. The hepatic form grows to become a large schizont, the time for this and the size of the mature form depending on the species of parasite. Thus, in *P. falciparum* pre-erythrocytic growth is rapid, taking 5–6 days, and schizonts measure 60 μm when mature and contain about 40 000 merozoites. There appears to be only a single generation of cryptozoites in *P. falciparum* and a one-way passage of merozoites from the exoerythrocytic cycle. With *Plasmodium cynomolgi* pre-erythrocytic schizogony is observed from the second day onwards, and by the eighth day schizonts measure 38 μm in diameter and contain about 10 000 merozoites. The majority of the merozoites from this schizogony enter the erythrocytic cycle, but some may return to the liver parenchyma cells and continue exoerythrocytic schizogony, these being a constant source of infection for the erythrocytes. Details of the exoerythrocytic development of other species may be found in Faust and Russell (1964).

Following exoerythrocytic development, merozoites invade the erythrocytes, and the cycle of development in the blood and subsequently in the mosquito is comparable to that described previously for avian malaria.

Plasmodium falciparum. The cause of malignant tertian malaria, falciparum malaria or subtertian malaria. There is a tendency for infected erythrocytes to clump, and consequently the schizonts and merozoites are found almost exclusively in the capillaries of the inner organs. 8–18 (8–32) merozoites are produced per schizont, the pigment granules are dark brown or black and usually occur in a compact mass. The gamonts are sausage- or crescent-shaped and appear in the peripheral blood. Macrogametes stain blue and the pigment granules are grouped around the nucleus, whereas the microgametes stain bluish to reddish and have scattered pigment granules. This parasite differs from the other species of human Plasmodia in that normally only the ring forms of trophozoites and gametocytes are seen in the peripheral blood.

The parasite is widely distributed in the tropics and is relatively uncommon in temperate zones. It is generally regarded as the most malignant form of malaria in the human.

Plasmodium malariae. The cause of quartan malaria. The schizonts appear in the cirulating blood and frequently assume a band form across the erythrocytes; when mature they almost fill the host cell. Six to twelve merozoites are produced per schizont, and these may be arranged around a mass of pigment granules. The gamonts are round, the macrogametes stain more deeply than the male forms and they have a smaller, more deeply staining nucleus and coarse pigment granules. Microgamonts possess a larger, lightly staining nucleus and finer and more numerous pigment granules.

This species is less common than the other three species and occurs in tropical and subtropical areas. Though widely distributed, it has a low prevalence.

Plasmodium ovale. The cause of mild tertian malaria. A distinctive feature of the parasitized red cell is the appearance of Schuffner's dots, and the host cell is often fimbriated. Schizonts produce 8–10 merozoites. The gametocytes resemble those of *P. malariae*, and the host cells are markedly affected with Schuffner's dots and slightly enlarged. Pigment is evenly distributed. This species has a limited distribution, being confined to Africa, the Philippines and India.

Plasmodium vivax. The cause of benign tertian or vivax malaria. The developing schizonts are irregular, amoeboid and active and extend over the cell which is enlarged, pale and contains Schuffner's dots. Mature schizonts almost fill the host cell and produce 8–24 merozoites (usually 12–18). The macrogametes are 9–10 μm in diameter, more or less fill the host cell, stain deeply and contain a compact nucleus with evenly distributed pigment granules. The microgamonts are a little smaller, have a paler blue cytoplasm with a larger nucleus that stains less deeply than the female form.

Benign tertian malaria is the commonest and the most widely distributed malarial infection in the world; it extends into the temperate zones and occurs in North America and the more northern parts of Europe.

IMPORTANT MALARIA PARASITES OF NON-HUMAN PRIMATES

Non-human primate malaria is used extensively in the study of the biology and immunology of *Plasmodium* species. For example, studies of the mechanism of host specificity have revealed the need for a specific receptor on the surface of the erythrocyte (e.g. Duffy blood group factor for *Plasmodium knowlesi*; Miller et al. 1977) and specific immunity can be induced with isolated merozoites of *P. knowlesi* (Cohen et al. 1977).

Quotidian malaria

In this type the schizogony cycle takes one day.

Plasmodium knowlesi Sinton and Mulligan, 1932. *Hosts:* cynomolgus monkey, pigtail macaque and other macaques. A natural case has been found in man (Chin et al. 1965). Malay Peninsula, Philippines, Taiwan. Developmental cycle similar to other species, transmitted naturally by *Anopheles hackeri*. Disease produced is mild in natural hosts, but acute and fatal in experimentally infected rhesus, velvet and patas monkeys.

Tertian malaria

In this type the schizogony cycle takes two days.

Plasmodium schwetzi Brumpt, 1939. *Hosts:* chimpanzee, gorilla, Africa. Experimentally, in man. It causes a mild tertian malaria in chimpanzees in endemic areas.

Plasmodium reichenowi (Sluiter, Swillengrebel and Ihle, 1922). *Hosts:* chimpanzee, gorilla. Africa. Causes a mild tertian malaria.

Plasmodium eylesi Warren, Bennett, Sandosham and Coatney, 1965. *Host:* whitehanded gibbon in Malaya. Probably non-pathogenic for the gibbon.

Plasmodium coatneyi Eyles, Fong, Warren, Guinn, Sandosham and Wharton, 1962. *Host:* cynomolgus monkey, Malay Peninsula and Philippines. Mildly pathogenic in natural host. Transferrable to rhesus, macaque and leaf monkeys.

Plasmodium cynomolgi Mayer, 1907. *Hosts:* cynomolgus monkey, macaques, bonnet and leaf monkey. Asia. Transmissible to man. Causes tertian type malaria in monkeys and man.

Plasmodium gonderi *Hosts:* mandrills, mangabeys. West Central Africa. Readily transmissible to the rhesus monkey (Bray 1959). Causes a mild tertian disease in natural hosts.

Plasmodium simium Fonseca, 1951. *Host:* howler monkey, Brazil. Has been reported in man (Deane et al. 1966). Levine (1973) considers this species is either *P. ovali* or *P. vivax* gone wild in New World monkeys.

Quartan malaria

Schizogony cycle takes three days.

Plasmodium inui (Halberstädter and von Prowazek, 1907). *Hosts:* cynomolgus monkey, pigtail and other macaques in South Asia, Indonesia, Philippines, Taiwan. Common. Readily transmitted to the rhesus monkey. Experimentally transmitted to man (Coatney et al. 1966). Causes a mild quartan type of infection in monkeys. Vectors include *A. hackeri* and *A. leucosphryus*.

Plasmodium brasilianum Gonder and von Berenberg-Gossler, 1908. *Hosts:* various South American monkeys including howler, spider and squirrel monkeys. Experimentally, man and marmosets. Causes a quartan-type infection which may be fatal in spider and howler monkeys but is more usually fatal in experimentally infected night monkeys and marmosets. Dunn (1965) has suggested that this species is *P. malariae* gone wild in monkeys, having been introduced by early explorers.

Plasmodium malariae as mentioned above (p. 698) is common to man and chimpanzee.

Hepatocystis kochi (Laveran 1899) (Syn. *Plasmodium kochi*). This is the commonest malarial parasite of the green monkey, other guenons, baboons, mangakeys in Central Africa. Prevalences up to 50–70% have been reported in primates in endemic areas. The parasite rarely causes clinical disease but may confuse research data (Vickers 1966). The genus *Hepatocystis* Levaditi and Schoen, 1932 is characterized by the presence of gamonts but no schizonts in the erythrocytes. Schizogony occurs in the hepatocytes where large cysts (merocysts or megaloschizonts) are formed, which may measure up to 4 mm in diameter in baboons. These contain a mass of merozoites which, when released, invade erythrocytes. The vector is the midge *Culicoides* (Garnham et al. 1961) in which development is similar to *Plasmodium*.

The pathological changes in *H. kochi* infection include white to greyish nodular foci on the surface of the liver (Vickers 1966)

Non-human primate malaria as a zoonosis

This topic has been reviewed by Coatney (1971) and Ristic and Smith (1974). Infection may be accidental, from laboratory infections, or from field exposure. Coatney (1971) proposed that *P. cynomolgi, P. knowlesi, P. inui* and *P. schwetzi* were true zoonotic species. *Plasmodium schwetzi* may be an anthropozoonotic (human form infecting animals) form of *P. ovale*. It has been mentioned above that *P. simium* and *P. brasilianum* may be anthropozoonotic forms of *P. vivax* and *P. malariae*, respectively.

MALARIA PARASITES OF RODENTS

Rodent malaria has been used extensively for research purposes. The species in common use include:

Plasmodium berghei Vincke and Lips, 1948. Occurs naturally in the tree rat in Central Africa. It is transmissible to rat, mouse, hamster but not guinea-pig or rabbit.

Plasmodium vinckei Rodhain, 1952. Occurs in various rats of Africa and experimentally causes severe disease in the mouse. It is not transmissible to the rat or hamster.

Plasmodium chabaudi Landau, 1965. Occurs in the tree rat in Central Africa. It can be transmitted to mouse but not rat.

Genus: Haemoproteus Kruse, 1890

Previously, this genus was classified in the family Haemoproteidae Doflein, 1916. Organisms are transmitted by blood-sucking insects in which developmental stages, comparable to those of the genus *Plasmodium*, occur.

Gametocytes occur in the erythrocytes and possess a halter-shaped appearance encircling the nucleus (synonym *Halteridium*). Pigment granules also occur. Schizogony occurs in the endothelial cells of the blood vessels of inner organs, especially in the lungs. The parasites are transmitted by hippoboscid flies and in some cases by members of the genus *Culicoides*. The genus is widespread in birds and also occurs in reptiles. The following species have been recorded from birds:

Haemoproteus canachites Fallis and Bennet, 1960
Haemoproteus columbae Kruse, 1890
Haemoproteus danilewskii Kruse, 1890
Haemoproteus lophortyx O'Roke, 1930
Haemoproteus meleagridis Levine, 1961
Haemoproteus nettionis (Johnston and Cleland, 1909) Coatney, 1936
Haemoproteus sacharovi Novi and MacNeal, 1904

A check-list and a host list of the species of the genus *Haemoproteus* has been prepared by Coatney (1936). Herman (1944) has prepared a check-list of the species in North American birds and Cook (1971) has reviewed the distribution, transmission and pathogenesis of the genus.

Haemoproteus columbae. *Hosts:* domestic and wild pigeons; also in mourning doves, turtle doves and a number of other wild birds. World-wide in distribution.

The only forms which occur in erythrocytes are the gamonts. These may range from tiny forms to elongate, crescent-shaped gamonts which partially encircle the nucleus of the host cell in the form of a halter. The nucleus may be displaced but not to the edge of the cell. Macrogamonts stain dark blue with Romanowsky stains, the nucleus is compact, staining dark purple to red, and the pigment granules are dispersed throughout the cytoplasm. Microgametocytes stain pale blue to pinkish, the nucleus is pale pink and diffuse, and pigment granules are collected into a spherical mass.

Life-cycle (Fig. 3.43). The endogenous developmental cycle, which has been described by Aragão (1908) and Huff (1942), is initiated when sporozoites are injected by an infected hippoboscid fly. Sporozoites enter the blood stream, penetrate endothelial cells of blood vessels and here develop into early schizonts. The early stages are minute cytoplasmic bodies with a single nucleus, but by growth and nuclear division 15 or more small, unpigmented masses, or cytomeres, each with a single nucleus, are produced. Each cytomere continues to grow, and its nucleus undergoes repeated division until the now greatly enlarged endothelial cell is filled with a large number of

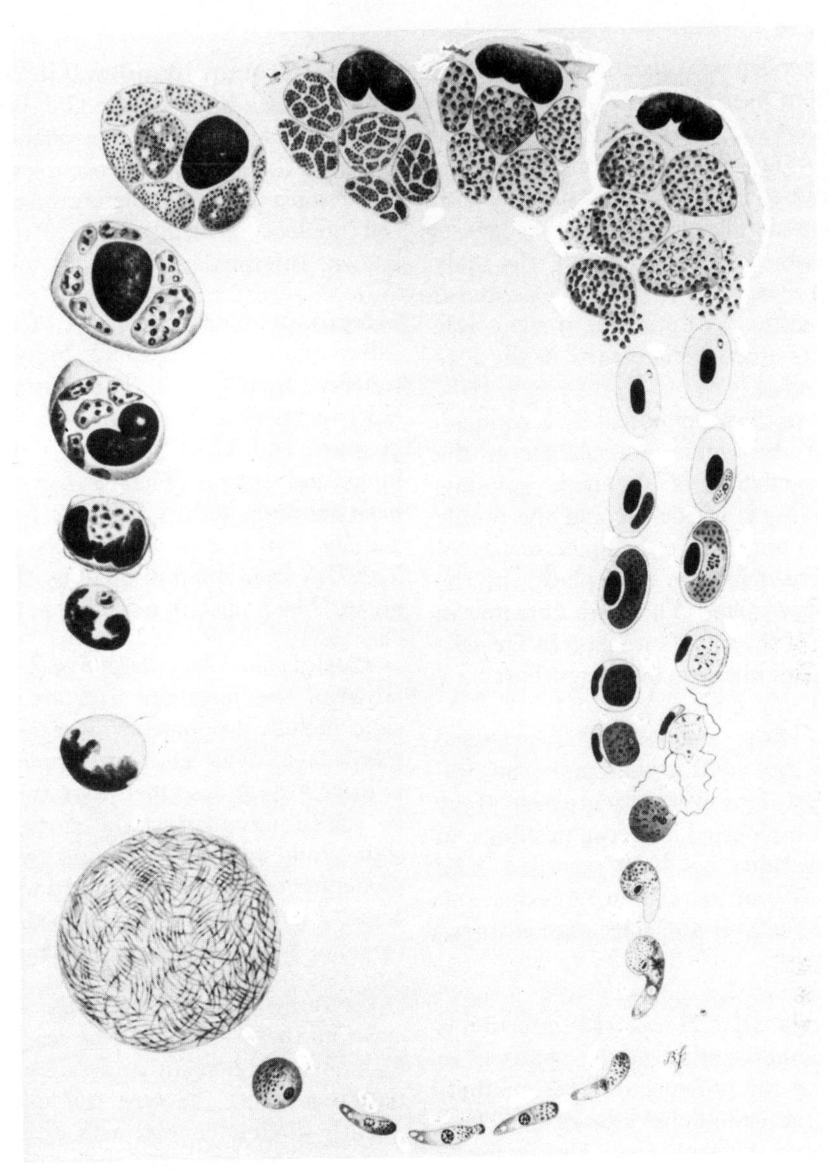

Fig 3.43 Life-cycle of *Haemoproteus columbae*. (*From Wenyon 1926*)

multinucleate bodies, or cytomeres, surrounded by a fine cyst wall. Each cytomere produces an enormous number of minute merozoites. Subsequently, the endothelial cell breaks down and the cytomeres are liberated; these accumulate in the capillaries which they may block, but soon after liberation, the cytomeres rupture and the merozoites escape into the blood stream. The development to this stage takes about four weeks.

The merozoites enter red blood cells and become gamonts, although it is probable that others enter further endothelial cells and repeat the asexual cycle, this being carried on for several generations. The young gamonts first appear in the blood about 30 days after infection. Although multiple infections of erythrocytes with trophozoites may occur, it is rare for more than one mature gamont to exist in a cell.

Subsequent development occurs in hippoboscid flies. The only proven vector is *Pseudolynchia canariensis*. Baker (1957), in England, showed that *Ornithomyia avicularia* could support sporogony, but he was unable to infect domestic pigeons with infected hippoboscid flies. Levine (1973) suggests that hippoboscids are probably not the only vectors for *H. columbae,* and since *Haemoproteus nettionis* is transmitted by *Culicoides,* it is possible that these insects are also concerned in the life-cycle of *H. columbae*.

Development in the hippoboscid fly is comparable to that of the genus *Plasmodium* in the mosquito. Exflagellation of the male gamonts occurs in the mid-gut of the fly, and the motile zygote (oökinete) migrates to the outer surface of the mid-gut. Here sporogony takes place with the production of sporozoites. These are liberated in the body cavity of the insect and pass to the salivary glands to await injection into a new host.

Pathogenesis. The pathogenicity of *H. columbae* is generally low, and adult birds usually show no evidence of disease. However, an acute form of the infection has been reported in pigeon nestlings, in which heavy mortality has been recorded. The clinical signs consist of anorexia and anaemia, and on post mortem the liver and spleen are enlarged and dark in colour.

Diagnosis. Diagnosis of *H. columbae* infection is based on the demonstration of the gamonts in blood smears and the presence of large numbers of schizonts in the endothelial cells of the blood vessels of the lungs. Schizonts may also be found in the liver, spleen and kidneys.

Little is known regarding treatment. Coatney (1935) has indicated that quinacrine is effective against the young gamonts, but no information is available regarding treatment against the schizontal stages. Prevention and control is dependent upon the control of the insect vector.

Haemoproteus canachites. *Host:* spruce grouse (*Canachites canadensis*), Canada. Experimentally, ruffed grouse. Transmitted by *Culicoides sphagnumensis* (Fallis & Bennet 1960). Prepatent period is 14 days. Pathogenesis, unknown.

Haemoproteus danilewskii. *Hosts:* hooded crow, various birds of the Old World and New World (see Cook 1971). The characteristic of this species is that the parasite occupies the entire host cell cytoplasm: it completely surrounds the host cell nucleus or displaces it. Pathogenesis, unknown. Intermediate host, unknown.

Haemoproteus lophortyx. *Hosts:* California valley quail, Gambel quail, bobwhite quail and Catalina Island quail. The mature gamonts are halter-shaped and contain numerous pigment granules, and the schizonts are found in the liver, lungs and spleen. The fly *Lynchia hirsuta* has been incriminated as a vector for this parasite. Tarshis (1955) has successfully transmitted *H. lophortyx* from quail to quail by *Stilbometopa impressa*. The prepatent period is 21 days.

Pathogenesis. An enlarged spleen and pigmentation of the lungs and liver are found. Clinical signs include droopiness, anorexia and death may follow ten days after infection (Cook 1971). O'Roke (1932) described four types of infection in California quail: mild chronic, no clinical signs; mild acute, anorexia for two to four days; moderate, chronic, anaemia and weakness; heavy, acute, a late spring or summer relapse, anorexia, droopiness, death—this form is rare.

Haemoproteus meleagridis. *Hosts:* domestic and wild turkeys, North America. Relatively rare. Gamonts are elongate and sausage-shaped, partially encircling the host cell nucleus, and frequently in close contact with it. The life-cycle is unknown. No pathogenesis has been ascribed to it.

Haemoproteus nettionis. *Hosts:* domestic duck and goose, and other wild ducks, geese and swans. Lists of species which may be infected have been given by Herman (1954) and Fallis and Wood (1957). World-wide in distribution and reasonably common. A survey by Bennett and Fallis (1960) of 3000 birds in Algonquin Park, Canada, showed *Haemoproteus* in 26% of them.

Gamonts are elongate and sausage-shaped, partially, and in some cases almost completely, encircling the host cell nucleus. The cell nucleus may also be displaced. Pigment granules are usually coarse and tend to be grouped at the poles of the parasite.

Life-cycle. Fallis and Wood (1957) demonstrated that *Culicoides* (possibly *piliferus*) could transmit the infection. Oökinetes, oocysts and sporozoites were found 14–21 days later. Fallis and Bennet (1960) also demonstrated that *Haemoproteus canachites* of the spruce grouse developed in the biting midge *Culicoides spagnumensis*.

Haemoproteus nettionis is only slightly, if at all, pathogenic.

Haemoproteus sacharovi. *Hosts:* domestic pigeon, mourning dove and turtle dove; North America and Europe.

Gamonts are distinctive in that they completely fill the host cell when they are mature; they distort it and push the host cell nucleus to one side. They also possess little pigment compared with the other species of *Haemoproteus*. The natural vectors of this species are not known. It has been transmitted by *Pseudolynchia canarensis*, although Levine (1973) suggests that a species of *Culicoides* may also be concerned.

The pathogenic effect of *H. sacharovi* is low, although it is reported to cause hepatomegaly in pigeon squabs.

Genus: Leucocytozoon Danilewsky, 1890

Parasites of this genus undergo schizogony in the endothelial and parenchymatous cells of the liver, heart, kidney and other organs of avian hosts. Large schizonts are produced. The gametogonous stages occur in the circulating blood, and the infected host cells become grossly distorted and assume a spindle shape. No pigment is produced. The vectors are black flies of the genus *Simulium*. A check-list of species and their host has been given by Coatney (1937), while Herman (1944) has listed the North American species. The more common forms which occur are:

Leucocytozoon simondi Mathis and Léger, 1910
Leucocytozoon smithi Laveran and Lucet, 1905
Leucocytozoon caulleryi Mathis and Léger, 1909

and in addition:

Leucocytozoon bonasae Clark, 1945
Leucocytozoon mansoni Sambon, 1908
Leucocytozoon marchouxi Mathis and Léger, 1910
Leucocytozoon sakharoffi Sambon, 1908

Leucocytozoon simondi. *Hosts:* domestic and wild ducks and geese; North America, Europe and the East. Relatively common.

Life-cycle. The endogenous developmental cycle has been described by Huff (1942) and Fallis et al. (1951). Sporozoites injected by the *Simulium* fly are carried by the blood stream to various cells of the body, two types of schizonts being produced. The first asexual generation takes place in the Kupffer cells of the liver. The schizonts are small, and they produce merozoites, some of which may enter blood cells to become gamonts, while others initiate hepatic schizonts and megaloschizonts. The hepatic schizonts occur in the parenchyma liver cells, and they produce a number of cytomeres, which, by multiple fission, form a large number of small merozoites. The megaloschizonts are more numerous than the hepatic forms and apparently develop in lymphoid cells or macrophages. They are found in the brain, liver, lungs, kidney, intestinal tissue and lymphoid tissues four to six days after infection. As their name suggests, they are large, 60–160 μm in diameter, and they contain a large number of cytomeres which in turn produce a much larger number of merozoites. With rupture of the hepatic schizonts and the megaloschizonts, merozoites are released into the blood and these appear as gamonts in the peripheral circulation six to seven days after infection. The majority of merozoites probably develop into gamonts, but it is presumed that some may initiate further asexual reproduction. The identity of the host cell for the gamonts has been debated for a number of years. One of the problems in this is that the cells containing the mature gamonts are so distorted that their origin is difficult to determine. Huff (1942) considered the cells to be lymphocytes, or lymphocytes which have been stimulated to undergo transformation by the presence of the parasite.

Other studies by Fallis et al. (1951) indicated that young gamonts may occur in both lymphocytes and erythrocytes, and a similar observation was made by Cook (1954) and Desser (1967).

Mature gametocytes are elongate, oval bodies, usually 14–15 μm in length but may reach up to 22 μm, by 4.5–5.5 μm. The infected host cell is grossly distended and elongated, and it may measure up to 48 μm in length. The nucleus of the host cell is elongate and forms a long, thin, dark crescent along one side of the parasitized cell. Occasionally, round forms of the parasite occur in undistorted host cells, and there is no evidence that these differ functionally from the elongate forms. The macrogametes stain dark blue with Romanowsky stains, the nucleus is compact, and several vacuoles may occur in the darkly stained cytoplasm. The microgamonts are slightly smaller than the macrogametes, the cytoplasm stains less deeply, usually a pale blue colour, and the nucleus is diffuse and stains pale pink.

The vectors of *L. simondi* are members of the genus *Simulium*. *Simulium venustum* has been recognized as such for some time, and *S. croxtoni*, *S. euryadminiculum* and *S. rugglesi* were added to the list by Fallis et al. (1956). Development in the insect vector is essentially the same as *Plasmodium* in the mosquito.

Pathogenesis. *Leucocytozoon simondi* is markedly pathogenic for young ducks and geese. The clinical signs of leucocytozoonosis are sudden in onset, and death may occur within a day or so. Ducklings are listless, anorexic, show rapid breathing (due to the large number of megaloschizonts in the capillaries of the lung) and may show nervous derangements prior to death. The disease in older birds is less acute and develops more slowly. Birds become emaciated and listless but seldom die in less than four days from the onset of the disease.

Diagnosis. Diagnosis of *L. simondi* in ducks is based on the demonstration of the gamonts in blood smears or, and more satisfactorily, the megaloschizonts in smears of the lung. There is no known effective treatment for *L. simondi* infection, and control consists of controlling *Simulium* flies. Young ducklings should be isolated from the older birds since these may serve as carriers. Total freedom from the disease is only achieved when ducklings and goslings are raised in regions where black flies do not occur.

Leucocytozoon smithi. *Hosts:* domestic and wild turkeys; North America, Europe. This is a markedly pathogenic species for young turkeys; heavy losses due to it have been recorded in North America. The general prevalence ranges from about 20% to more than 80% in domestic and wild turkeys, especially in areas where conditions are good for the breeding of *Simulium* flies.

Morphologically, this parasite resembles *L. simondi*. The mature gamonts are elongate, approximately 20–22 μm by 6 μm, and the host cell is greatly distorted and elongated and may measure up to 45 μm by 14 μm. The nucleus of the host cell is elongated, forming a dark band on one side of the parasites, and this is frequently split to form a dark band on each side of the parasite.

The endogenous developmental cycle is comparable to that of *L. simondi*, but megaloschizonts have not been observed. Hepatic schizonts (10–20 μm in diameter) occur in the cells of the liver parenchyma and contain cytomeres which produce large numbers of merozoites. The prepatent period of infection is approximately nine days. The vectors of *L. smithi* are *Simulium occidentale*, *S. nigroparvum* and *S. slossonae* in which the same developmental cycle occurs as in *L. simondi*.

Pathogenesis. This parasite may be extremely pathogenic for young turkeys; very heavy losses having been reported from various parts of the world, the death rate reaching up to 90% in some flocks. The clinical signs consist of anorexia, emaciation, debility, while leg weakness and incoordination occur in the later stages of the disease. In fulminating outbreaks of the disease, birds usually die within two or three days after the onset of clinical signs but, if not, some undergo a chronic type of infection where there is a persistent cough and moist bronchitis, while others may appear completely recovered. In all cases recovered birds remain as carriers.

On post mortem there is jaundice, congestion of the duodenum, splenic enlargement and congestion of the lungs and kidneys.

Diagnosis. Diagnosis of turkey leucocytozoonosis is based on similar evidence as that for *L. smithi.* For treatment Bierer (1950) has recommended sulphaquinoxaline; however, this needs confirmation as to its efficacy. Prevention and control are the same as for *L. simondi.*

Leucocytozoon caulleryi. *Hosts:* domestic chicken, Far East, south-east Asia and South Carolina (Levine 1973).

The mature gamonts of this species are round, approximately 15 μm in diameter; the host cell is not distorted as in the other species but is enlarged, measuring up to 20 μm in diameter. The host cell nucleus is compressed to one side of the cell, forming a band extending about one-third of the way around the parasite.

Disease due to this parasite has been reported in Japan by Akiba (1960). He demonstrated that *Culicoides arakawae* was probably the natural vector of *L. caulleryi* in that oökinetes, oocysts and sporozoites were demonstrable in specimens of this fly which had fed upon infected chickens. He was also able to demonstrate gamonts in chickens 14 days after the injection of a suspension of sporozoites. An outbreak of leucocytozoonosis in chickens in Taiwan in which 20% of the chickens died was reported by Liu (1958). Post mortem changes included anaemia, haemorrhages in the lungs, liver and kidneys, enlargement of the spleen and white spots on the heart muscle. Megaloschizonts were found in the kidney, lung and heart, but hepatic schizonts were not.

For treatment, sulphamonomethoxine sodium (4-methoxy-6-sulphanilamidopyrimidine monohydrate; Daimeton Sodium) at a rate of 1 g/litre of drinking water has been shown to be effective. For prophylaxis, continuous medication at the level of 1 g/20 litres has been effective.

The other species of *Leucocytozoon* occur as follows:

Leucocytozoon bonasae, ruffed grouse and species of grouse, ptarmigan, USA and Canada.

Leucocytozoon mansoni, capercaillie, black grouse, hazel grouse, Sweden.

Leucocytozoon marchouxi, doves and pigeons. Cosmopolitan.

Leucocytozoon sakharoffi, crow, blue jay, raven, rook, jackdaw, Europe, North America.

REFERENCES

HAEMOSPORINA (PLASMODIUM)

Al-Dabagh, M. A. (1961) Symptomatic partial paralysis in chickens infected with *Plasmodium juxtanucleare. J. comp. Path.*, **71**, 217–221.

Bishop, A. (1955) Problems concerned with gametogenesis in haemosporidea with particular reference to the genus *Plasmodium. Parasitology*, **45**, 163–185.

Bishop, A. (1962) Chemotherapy and drug resistance in protozoal infections. In: *Drugs, Parasites and Hosts*, eds L. G. Goodwin & R. H. Smith, Biological Council Symposium, pp. 98–111. London: Churchill.

Boyle, H. K., Getz, M. E., Coatney, G. R., Elder, H. A. & Eyles, D. E. (1961) Simian malaria in man. *Am. J. trop. Med. Hyg.*, **10**, 311.

Bray, R. S. (1957) Studies on the exoerythrocytic cycle in the genus Plasmodium. *Lond. School Hyg. trop. Med.*, **12**, 192.

Bray, R. S. (1959) Pre-erythrocytic stages of human malaria parasites: *Plasmodium malariae. Br. med. J.*, **2**, 679–680.

Chin, W., Contacos, R. G., Coatney, G. R. & Kimball, H. L. (1965) A naturally acquired quotidian-type malaria in man transferable to monkeys. *Science, N.Y.*, **149**, 865.

Coatney, G. R. (1971) The Simian malarias: zoonoses, anthroponoses, or both. *Am. J. trop. Med. Hyg.*, **20**, 795–803.

Coatney, G. R., Chin, W., Contacos, P. G. & King, H. K. (1966) *Plasmodium inui*, a quartan type malaria parasite of Old World monkeys transmissible to man. *J. Parasit.*, **52**, 660–663.

Coatney, G. R. & Roundabush, R. L. (1949) A catalogue of the species of the genus *Plasmodium* and index of their hosts. In: *Malariology*, ed. M. F. Boyd. Philadelphia: Saunders.

Cohen, S., Butcher, G. A., Mitchell, G. H., Deans, J. A. & Langhorne, J. (1977) Acquired immunity of vaccination in malaria. *Am. J. trop. Med. Hyg.*, **26**, (6) Suppl., 223–229.

Deane, L. M., Deane, M. P. & Neto, J. F. (1966) Studies on transmission of simian malaria and on a natural infection of man with *Plasmodium simium* in Brazil. *Bull. WHO*, **35**, 805–808.

Dunn, F. L. (1965) On the antiquity of malaria in the Western Hemisphere. *Human Biol.*, **37**, 385–393.

Eyles, D. E., Coatney, G. R. & Getz, M. E. (1960) Vivax-type malaria parasite of Macaques transmissible to man. *Science, N.Y.*, **131**, 1812–1813.

Faust, E. C. & Russell, P. F. (1964) *Craig and Faust's Clinical Parasitology*, 7th ed. Philadelphia: Lea & Febiger.

Garnham, P. C. C., Heisch, R. B. & Minter, D. M. (1961) The vector of *Hepatocystis* (*Plasmodium*) *kochi*; the successful conclusion of observations in many parts of tropical Africa. *Trans. R. Soc. trop. Med. Hyg.*, **55**, 497–502.

Hewitt, R. I. (1940) Bird Malaria. *Am. J. Hyg. Monogr. Series No.*, **15**, Baltimore: John Hopkins Press.

Huff, C. G. (1963) Experimental research, on avarian malaria. *Adv. Parasit.*, **1**, 1–65.

Huff, C. G. & Coulston, F. (1944) The development of *Plasmodium gallinaceum* from sporozoite to erythrocytic trophozoite. *J. infect. Dis.*, **75**, 231–239.

Jeffery, G. M. Inoculation of human malaria into a simian host, *Macaca mulatta. J. Parasit.*, **47**, 90.

Levine, N. D. (1973) *Protozoan Parasites of Domestic Animals and of Man*, 2nd ed. p. 406. Minneapolis: Burgess.

Levine, N. D. & Hanson, H. C. (1953) Blood parasites of the Canada goose, *Branta canadensis interior. J. wildl. Mgemt*, **17**, 185–196.

Levine, N. D. & Kantor, S. (1959) Check-list of blood parasites of birds of the order Columbiformes. *Wildl. Dis.*, **1**, 1–38.

Manwell, R. D. (1963) Factors making for host-specificity, with special emphasis on the blood protozoa. *Ann. N.Y. Acad. Sci.*, **113**, 332–342.
Miller, L. H., McAuliffe, F. M. & Mason, S. J. (1977) Erythrocyte receptors for malaria merozoites. *Am. J. trop. Med. Hyg.*, **26**, (6) Suppl., 204–208.
Purchase, H. S. (1942) Turkey malaria. *Parasitology*, **34**, 278–283.
Ristic, M. & Smith, R. D. (1974) Zoonoses caused by Hemoprotozoa. In: *Parasitic Zoonoses Clinical and Experimental Studies*, ed. E. J. L. Soulsby, pp. 41–63. New York: Academic Press.
Rollo, I. M. (1964) The chemotherapy of malaria. In: *Biochemistry and Physiology of Protozoa*, ed. S. H. Hunter, vol. III. New York: Academic Press.
Russell, P. F., West, L. S., Manwell, R. D. & MacDonald, G. (1963) *Practical Malariology*, 2nd ed. London: Oxford University Press.
Vickers, J. H. (1966) *Hepatocystis kochi* in *Cercopithecus* monkeys. *J. Am. vet. med. Ass.*, **149**, 906–908.

HAEMOSPORINA (HAEMOPROTEUS, LEUCOCYTOZOON)

Akiba, S. K. (1960) Studies on the Leucocytozoon found in the chicken in Japan. II. On the transmission of *L. caulleryi* by *Culicoides arakawae*. *Jap. J. vet. Sci.*, **22**, 309–317.
Aragão, H. B. (1908) Über den Entwicklungsgang und die Übertragung von *Haemoproteus columbae*. *Arch. Protistenk.*, **12**, 154–167.
Baker, J. R. (1957) A new vector of *Haemoproteus columbae* in England. *J. Protozool.*, **4**, 204–208.
Bennet, G. F. & Fallis, A. M. (1960) Blood parasites of birds in Algonquin Park, Canada, and a discussion of their transmission. *Can. J. Zool.*, **38**, 261–273.
Bierer, B. W. (1950) Leucocytozoon infection in turkeys. *Vet. Med.*, **45**, 87–88.
Coatney, G. R. (1935) The effect of atebrin and plasmochin on the Haemoproteus infection of the pigeon. *Am. J. Hyg.*, **21**, 249–259.
Coatney, G. R. (1936) A check-list and host-index of the genus Haemoproteus. *J. Parasit.*, **22**, 88–105.
Coatney, G. R. (1937) A catalog and host-index of the genus Leucocytozoon. *J. Parasit.*, **23**, 202–212.
Cook, A. R. (1954) The gametocyte development of *Leucocytozoon simondi*. *Proc. helminth. Soc. Wash.*, **21**, 1–9.
Cook, R. S. (1971) *Haemoproteus* Kruse, 1890. In: *Infectious and Parasitic Diseases of Wild Birds*, eds J. W. Davis, R. C. Anderson, L. Karstad & D. O. Trainer, pp. 300–308. Ames: Iowa State University Press.
Desser, S. S. (1967) Schizogony and gametogony of *Leucocytozoon simondi* and associated reactions in the avian host. *J. Protozool.*, **14**, 244–254.
Fallis, A. M., Anderson, R. C. & Bennett, G. F. (1956) Further observations on the transmission and development of *Leucocytozoon simondi*. *Can. J. Zool.*, **34**, 389–404.
Fallis, A. M. & Bennett, G. F. (1958) Transmission of *Leucocytozoon bonasae* Clarke to furred grouse (*Bonasa umbellus L.*) by the black flies *Simulium laptipes* Mg. and *Simulium aureum* Fries. *Can. J. Zool.*, **36**, 533–539.
Fallis, A. M. & Bennett, G. F. (1960) Description of *Haemoproteus canachites* n. sp. (Sporozoa: Haemoproteidae) and sporogony in Culicoides (Diptera: Ceratopogonidae). *Can. J. Zool.*, **38**, 455–464.
Fallis, A. M., Davies, D. M. & Vickers, M. A. (1951) Life history of *Leucocytozoon simondi* Mathis and Leger in natural and experimental infections and blood changes produced in the avian host. *Can. J. Zool.*, **29**, 305–328.
Fallis, A. M. & Wood, D. M. (1957) Biting midges (Diptera: Ceratopogonidae) as intermediate host for *Haemoproteus* of ducks. *Can. J. Zool.*, **35**, 425–435.
Herman, C. M. (1944) The blood protozoa of North American birds. *Bird-Banding*, **15**, 89–112.
Herman, C. M. (1954) Haemoproteus infections in waterfowl. *Proc. helminth. Soc. Wash.*, **21**, 37–42.
Huff, C. G. (1942) Schizogony and gametocyte development in *Leucocytozoon simondi* and comparisons with Plasmodium and Haemoproteus. *J. infect. Dis.*, **71**, 18–32.
Levine, N. D. (1973) *Protozoan Parasites of Domestic Animals and of Man*, 2nd ed. Minneapolis: Burgess.
Liu, S. K. (1958) The pathology of Leucocytozoon disease in chicks. *Mem. Coll. Agric. natn. Taiwan Univ.*, **5**, 74–80.
O'Roke, E. C. (1932) Parasitism of the Californian valley quail by *Haemoproteus lophortyx*, a protozoan blood parasite. *Calif. Fish Game*, **18**, 223.
Tarshis, I. B. (1955) Transmission of *Haemoproteus lophortyx* O'Roke of the California quail by hippoboscid flies of the species *Stilbometropa impressa* (Bigot) and *Lynchia hirsuta* Ferris. *Exp. Parasit.*, **4**, 464–492.

SUBCLASS: PIROPLASMIA LEVINE, 1961

Blood parasites of vertebrates. Small, round or pleomorphic. The apical complex is reduced, reproduction is by binary fusion or schizogony. Parasites of cells of the haemopoetic system. Vectors are ticks.

ORDER: PIROPLASMIDA WENYON, 1926

With characters of the subclass. Two families are of importance in the order: *Babesiidae* and *Theileriidae*.

FAMILY: BABESIIDAE POCHE, 1913

Relatively large, pyriform organisms. Developmental stages occur in erythrocytes. Various opinions exist regarding the placing of the babesias in a separate family.

Kudo (1966) considers this family to be one of the three of the order Haemosporidia, containing blood protozoa which lack pigment granules and are minute parasites of erythrocytes. He includes the two genera *Babesia* and *Theileria* in the family. Neitz (1956) has proposed that the genera *Babesia* and *Theileria* be assigned separately to two families, Babesiidae and Theileriidae, these being families of a suborder Piroplasmidea Wenyon, 1926. The two-family scheme is adopted here.

Organisms of the family Babesiidae are round to

pyriform, amoeboid, forms occurring in the erythrocytes. They multiply by binary fission, or schizogony, in the red blood cells. The vectors are ixodid ticks.

Genus: Babesia Starcovici, 1893

Organisms multiply in the erythrocytes by asexual division, producing two, four or more non-pigmented amoeboid parasites. When stained with Romanowsky stains they show a blue cytoplasm and a red chromatin mass, usually at one pole. A string of chromatin granules may extend from the larger mass. Characteristically, they are pear-shaped forms lying at an angle with the narrow ends in apposition.

A number of species of *Babesia* have been recognized, and in general they fall into two major groups: large forms with an average length of more than 3 μm and small forms which have an average length of less than 2.5 μm.

Life-cycle of Babesia

Multiplication of *Babesia* organisms in the vertebrate host occurs in the erythrocytes by a budding process (schizogony) to form two, four or more trophozoites. These are liberated from the erythrocyte and invade other cells, the process being repeated until a large percentage of red blood cells are parasitized. Occasionally, a cell shows multiple infection with a large number of trophozoites, but it is considered that this represents a series of binary fissions rather than a multiple invasion of the cell. The blood forms are readily transmissible by mechanical means to another animal, and these then initiate a further cycle of asexual reproduction.

Preferential invasion of young erythrocy,tes by merozoites occurs in acute *Babesia bigemina* infections (Wright & Kerr 1974) and probably in all babesial infections. A mechanism for the penetration of host erythrocytes by merozoites has been described by Chapman and Ward (1977). Penetration of RBCs depends on factors of the alternative pathway of complement (C) activation (properdin and factor B) as well as C3 and C5. With *Babesia rodhaini*, cells of various species are invaded when appropriate C components are provided. The fixation of C3 to the RBC membrane appears critical since sheep RBC conjugated as EAC 143 will permit penetration in the absence of exogenous C, whereas other conjugations (EA, EAC1 or EAC14) will not. Penetration commences with the indentation of the erythrocyte membrane by the blunt end of the merozoite, which is followed by rapid penetration of the cell by the parasite.

Under natural conditions the *Babesia* species are transmitted by ticks, the first demonstration of this being by Smith and Kilborne (1893) for the causal agent (*Babesia bigemina*) of Texas fever. This observation was a major landmark in the history of arthropod-borne diseases. The development of *Babesia* in ticks has occasioned much work and speculation. Dennis (1932) maintained that sexual reproduction occurred with *Babesia bigemina* in the tick *Boophilus annulatus*, but since his work the idea of a phase of sexual reproduction has been discounted although in recent years the possibility of a sexual cycle of development in the tick has been reintroduced, especially since a sexual developmental cycle has been demonstrated for *Theileria* spp. (see below).

Essentially, development and transmission of *Babesia* spp. in ticks is either by transovarian transmission or by stage to stage transmission. The former is the only mode of transmission for one host ticks, since following the attachment of larvae the rest of the tick developmental stages occur on the same animal. With two or three host ticks, stage to stage transmission becomes of importance, adult stages transmitting infection which they acquired as nymphs, or nymphs doing the same with infection acquired as larvae.

An account of the life-cycle of *Babesia bigemina* in the tick *Boophilus microplus* is given by Riek (1964, 1966); that of *B. caballi* in *Dermacentor nitens* is given by Holbrook et al. (1968*a*); and that of *B. ovis* in *Rhipicephalus bursa* is given by Friedhoff (1969). Electron microscope studies of the developmental stages in larvae, nymphs and non-replete females of *Boophilus decoloratus* have been reported by Potgieter and Els (1977). Although the descriptions differ in detail, there is good general agreement on the developmental cycle of the genus.

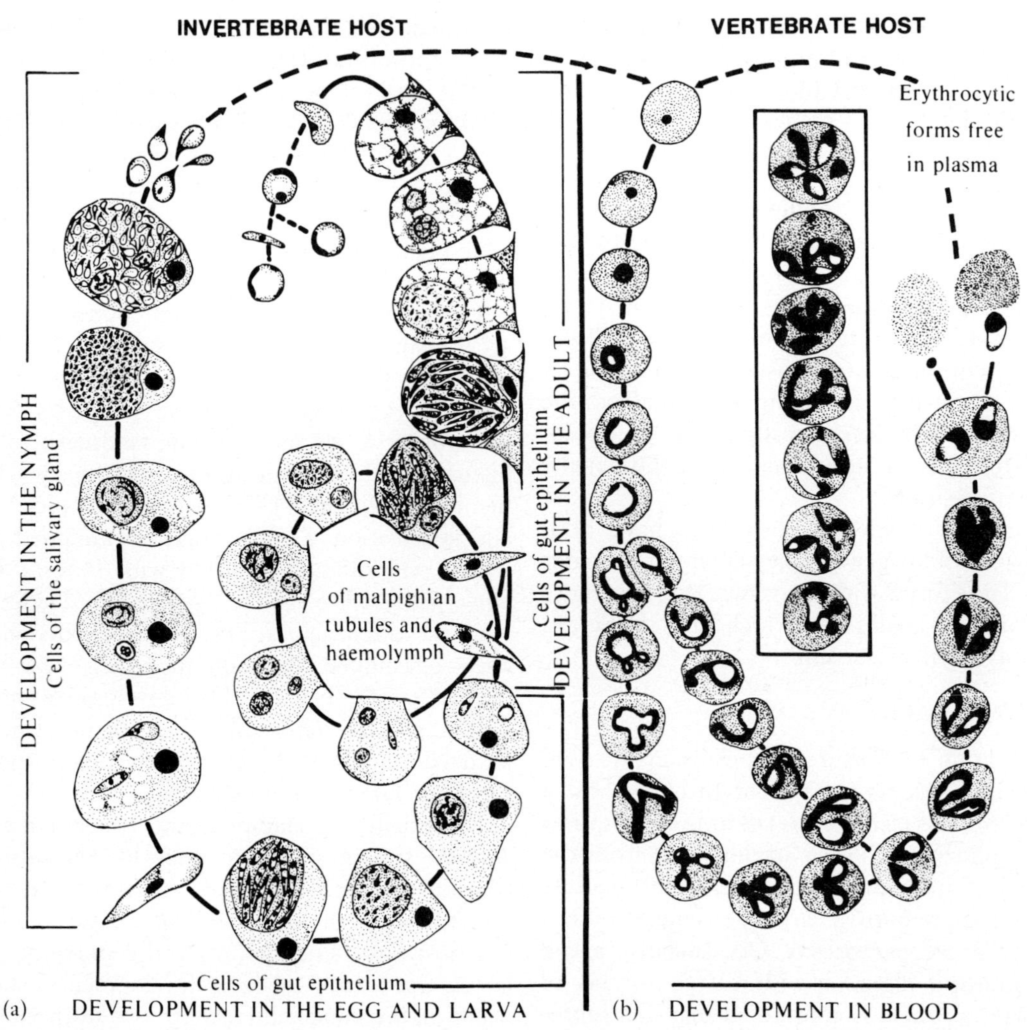

Fig 3.44 (a) Life-cycle of a large Babesia (*B. bigemina*) in the one-host tick *Boophilus microplus* (*from Reik 1964*). Development of vermicules in the adult tick and in the larva and nymph. (b) Life-cycle of a large Babesia (*B. caballi*) in erythrocytes of the vertebrate host (horse) (from Holbrook *et al.* 1968). Inset shows atypical forms that arise only from muliple infection of cells or from multiple division (*From Mahoney 1972*)

Transovarian transmission of Babesia

The rate of development in the tick is dependent on the environmental temperature; it is much more rapid in ticks held at 28°C. Riek (1964) considered that parasites in red blood cells ingested early in the engorgement of a tick either were destroyed or their development was retarded until the tick was replete. Immediately on repletion erythrocytic forms of the parasite were seen lying free in large numbers in the gut contents. Many were irregular in shape with long rays or pseudopodia and frequently were clumped together. Koch (1906) described sexual union between forms similar to this, but his observations have not been confirmed. A further stage in development consisted of a spherical body, 3–5 μm in diameter, with a large vacuole surrounded by a thin layer of cytoplasm containing a single peripheral nuclear mass. Two other forms of spherical bodies were observed. The first, infrequent in appearance, showed three to four separate chromatin dots on the periphery around which the cytoplasm concentrated leaving a central vacuole. Riek considered that this form could divide into discrete elongated bodies, 4–6.9 μm by 1–2.1 μm, with a central chromatin dot. A second spherical form possessed two nuclei, one elongated and extending around the periphery of the vacuolated organism and the other a round mass at the opposite pole. Riek was unable to observe division of the second spherical form into elongate bodies. He suggested that these early stages might represent part of a gametogenous cycle and syngamy with the production of a zygote. He based this opinion on several considerations, including the fact that a large number of parasite stages are taken in by the tick vector but only a minority undergo further development. These are almost exclusively oval or spherical forms. Riek suggests that the spherical form with two nuclei may result from a union of an elongated form with the spherical form with a single nuclear mass. The result of this union may be a blunt, curved, cigar-shaped body, 8–10 μm long by 3.5–4.5 μm wide. This form was found in the gut contents at the end of the first 24 hours of development. The chromatin of the cigar-shaped body was discrete and stained intensely red while the cytoplasm was intensely blue. It assumed an ovoid shape as it grew, and subsequently it became a spherical body with a vacuolated central cytoplasm with chromatin spread on the periphery.

Between 24 and 48 hours after repletion, the first evidence of development in epithelial cells of gut was found, being indicated by an irregular, spindle-shaped body with a more or less centrally placed single chromatin mass. Following invasion of the epithelial cell the parasitic stage appeared to undergo multiple fission. Development proceeded rapidly, and the chromatin was distributed in the cell in a number of small dots; each dot later collected a ring of cytoplasm around it to produce a number of separate, oval or globular elements 3.2–6.5 μm in diameter. Between 40 and 48 hours after repletion, large numbers of these spherical forms, 'fission bodies', were seen in smear preparations from the gut of ticks. The 'fission body' grew within the epithelial cell, and ultimately the cytoplasm and chromatin became separated into vacuolated bodies, these being contained within the limiting membrane of the parasite. The mature 'fission body' contained up to 200 forms which, when liberated in making smears, measured 5–8 μm by 3–4 μm. The mature 'fision body' ruptured from the epithelial cell and liberated club-shaped bodies or 'vermicules' into the lumen of the gut. Initially, the vermicule had a homogeneous cytoplasm, and the nucleus lay to the broader end to give a cap-like appearance. Vermicules migrated through the gut wall to the haemolymph and, as they aged, the anterior red-staining area disappeared, the cytoplasm became vacuolated. At this stage they measured 9–13 μm by 2–2.9 μm. By 72 hours after repletion, such forms were present in the haemolymph, the ovary and other tissues of the body, while numerous fission bodies in various stages of development were still found in the epithelial cells of the gut. Vermicules occurred in mature ova from 72 hours onwards. In other studies Riek (1964, 1966) found that transmission of infection to cattle by larvae, developed from eggs of infected ticks, occurred only with eggs laid later than 96 hours after repletion.

A second cycle of development was initiated about 96 hours after infection. This involved the entry of vermicules into cells of the Malpighian tubules and the haemolymph, and here they underwent further multiple fission. The cycle undertaken appeared to be essentially the same as that occurring in the epithelial cells of the gut except that these cells were not involved.

In the initial development in the egg of the tick, the vermicules were present in yolk material. For further development vermicules entered the epithelial cells of the gut of the larvae, and here the same sequence of development occurred as was seen in the Malpighian tubule cells of the parent female. Further vermicules were produced by this phase of development, and these were liberated into the gut lumen or haemolymph of the larva.

Development of the *Babesia* to the infective stage for the bovine is dependent on a moult of the tick from the larval to the nymphal stage, and Riek was unable to observe transmission of *B. bigemina* to the bovine before eight to ten days after larval attachment of the tick *Boophilus microplus*. After the moult to the nymphal stage, vermicules were found in the cells of the salivary gland, these being similar to those in the haemolymph. The vermicules assumed a spherical form, they enlarged and their chromatin became scattered in the cell and broken up into a large number of small dots, many hundreds being found in a single cell. Cytoplasm was organized around the chromatin dots to produce spherical or pyriform organisms, which Riek regarded as mature, infective forms ready for liberation from the salivary gland. Such forms, too, were comparable in appearance to the pyriform stages found in the bovine erythrocyte.

The transovarian mode of development and transmission described by Regendanz and Reichenow (1933) for *Babesia canis* of the dog in *Dermacentor reticulatus* and for *B. canis* in *Rhipicephalus sanguineus* described by Shortt (1973) is essentially the same as the above. Following entry of the vermiform stage into the egg, it divides a few times to form some very small, round individuals, and they do not undergo further development until the larval tick has hatched and undergone a moult. Then the small forms enter the nymphal salivary glands, and by a series of binary fissions give rise to thousands of minute vermiform infective parasites. As well as occurring in the nymphal stage, this process also takes place in the adult tick.

Stage to stage transmission of Babesia

In a study of the stage to stage transmission of *Babesia canis* by *Rhipicephalus sanguineus*, Shortt (1936, 1973) found no evidence of sexual reproduction. Multiplication of developmental forms was seen in phagocytes which lay contiguous to the hypodermis in the body cavity of the tick. 'Pseudocysts' of organisms occurred about seven days after the nymph had dropped off the host, and by 11–15 days club-shaped organisms, 9 μm by 2 μm, were present in the cysts. The club-shaped forms were liberated from the host cell and migrated to the muscle sheaths of the nymphal tick where they penetrated muscle cells, rounded up and divided repeatedly to form a large number of small, ovoid forms about 1.2 μm in length. Subsequent development occurred when the recently metamorphosed adult fed on a dog; the parasites migrated to the salivary glands, entered the cells of the acini and underwent repeated binary fission to form the large number of small, ovoid, infective stages.

The question of sexual reproduction by *Babesia* spp. in ticks has yet to be clarified. Originally, Dennis (1932) described the formation of isogametes with fusion of these to form a motile zygote (oökinete) which passed through the intestinal wall and invaded the ovary and then the ova of the tick. Comparable developmental stages were observed by Petrov (1941) in *Ixodes ricinus* infected with *Babesia bovis*, but Muratov and Kheisin (1959) failed to find it with *Babesia bigemina* in two species of ticks, Polyanskii and Kheisin (1959) failed to find it in *Ixodes ricinus* infected with *Babesia bovis* and Shortt (1936) failed to observe it for *Babesia canis* in *Dermacentor reticulatus* or *Rhipicephalus sanguineus*. However, Riek (1964) (see above) suggests it does occur and it is very likely that in the next few years it will be demonstrated that sexual development occurs in the *Babesia* spp. infections in general.

Host specificity of Babesia

Rosenbusch (1927) suggested that *Babesia* organisms were, in reality, parasites of ticks and did not necessarily require mammals as alternative hosts. Support for this idea came from the fact that *Babesia* infection of ticks can be maintained when ticks are fed over a long period on hosts refractory to the parasite. Thus, ticks reared for five generations on hedgehogs may still be capable of infecting dogs with *Babesia canis*. Enigk (1944) found that equine piroplasms could be retained for several generations in ticks fed on hosts other than horses, and Markov and Abramov (1966) reported that *Babesia ovis* (of sheep) was still present 32 generations later in ticks maintained on rabbits. However, there is increasing evidence that the babesias are not as strictly host-specific as was previously thought. Callow (Riek 1965) has observed inapparent infections with *Babesia bigemina* in sheep, goat and horse, and the progeny of female ticks fed on such animals were infective for cattle. Enigk and Friedhoff (1962*a*) were able to transmit *Babesia divergens* of cattle to splenectomized red deer, fallow deer, roe deer and moufflons, and *Babesia divergens* has been found in chimpanzees following the injection of heparinized blood containing this parasite (Garnham & Bray 1959).

Pathogenesis of babesial infections

The release of pharmacologically active substances and the destruction of erythrocytes play a major role in the pathogenesis of *Babesia* infection. The proportionate role of each varies with the individual species of *Babesia*, e.g. the disease caused by *B. bigemina* resembles a haemolytic anaemia while with *B. bovis* infection, kinin production is the more important.

Plasma kallikrein levels have received detailed attention but those of kinin and kininogen require further study. Kallikrein activity has been assessed by its esterolytic activity using TAME (N-α-tosyl-L-arginine methyl ester) as a substrate.

Plasma kallikrein levels rise markedly three days after infection and then fall to subnormal levels terminally. In calves infected by the transfer of infected cells, the rise occurs too soon to be explicable on the basis of antigen–antibody reactions or as the byproduct of tissue damage (Mahoney 1977). Activation of prekallikrein to kallikrein occurs one to two days before parasites are detectable in the peripheral blood and this continues until prekallikrein levels are less than 10% of normal values, 11–12 days after infection (Boreham & Wright 1976). Small amounts of *B. bovis* extract activate the kallikrein system when injected intravenously into normal calves (Mahoney 1977) and an enzyme extracted from such material activates plasma kallikrein in vitro (Wright 1975). Consequently, the parasite produces an in vitro activator of kallikrein. These studies suggest that in acute *Babesia* infections (*B. argentina*) there is massive mobilization and activation of kallikrein. Kallikrein produces increased vascular permeability and vasodilatation leading to circulatory stasis and shock. The initial fall in packed cell volume (PCV) in *B. bovis* infection is largely attributable to this disturbance rather than erythrocyte destruction (Wright & Kerr 1975). Kallikrein also triggers intravascular coagulation and this is reflected in the changes of coagulation parameters in *B. bovis* and *B. caballi* infections. However, Mahoney (1977) warns that the kinin system may not be activated to the extent described above by all species of *Babesia*. For example, he notes that disease caused by *B. bigemina* resembles an uncomplicated haemolytic anaemia.

The topic of the release of pharmacologically active substances in parasitic infections has been reviewed by Boreham and Wright (1976).

The anaemia is associated with the emerging parasites from red cells; often, however, erythrocyte loss exceeds that attributable to the mechanical rupture of cells by parasites, although there have been no detailed studies of this in domestic animals (Mahoney 1977). Mahoney (1977) states there is evidence for direct removal of non-infected erythrocytes by phagocytosis and suggests that the osmotic fragility of non-infective RBC in *B. bovis* and *B. argentia* infections may predispose them to spontaneous lysis. Other explanations for excessive erythrocyte loss include the adsorption of circulating antigen–antibody complexes to the surface of RBC leading to RBC removal by phagocytosis (Sibinovič et al. 1969), although Mahoney (1977) considers this unlikely in domestic animals and believes that previous studies on this were complicated by the presence of filterable replicating agents capable of causing anaemia, circulating antigens closely resembling blood group antigens and extracts of *Babesia* organisms causing shock.

Glomerulonephritis may be associated with *Babesia* infection and has been demonstrated experimentally in *Babesia rodhaini* infection in the rat (Iturri & Cox 1969; Annable & Ward 1974). An acute to moderate proliferative glomerulitis was seen and this was associated with glomerular deposits of immunoglobulin G and the third component of complement (C3). Coincident with the parasitaemia in the rat there is hypocomplementaemia, glomerular deposits of IgG and C3 and severe anaemia (Annable & Ward 1974). Further study is needed to assess the importance of glomerular changes in domestic animals.

Central nervous system damage is a feature of some babesial infections (e.g. *B. bovis, B. canis*). Selective concentration of parasitized cells occurs in brain capillaries leading to obstruction of the blood flow (Mahoney 1977). Infected cells stick to one another and to the vessel endothelium, and the increased stickiness has been ascribed to a parasite enzyme (Wright & Goodger 1973) or antigen which alters the surface charge (Mahoney 1977).

BABESIA OF CATTLE

Six species of *Babesia* have been reported from cattle; these are:

Babesia argentina (Lignieres, 1909)
Babesia berbera (Sergeant, Donatien, Parrot, Lestoquard, Plantureux & Rougebief, 1924)
Babesia bigemina (Smith & Kilborne, 1893)
Babesia bovis (Babes, 1888)
Babesia divergens (M'Fadyean & Stockman 1911)
Babesia major Seargeant, Donatien, Parrot, Lestoquard & Plantureux, 1926

However, Mahoney (1977) believes that *B. argentina* and *B. berberi* are synonyms of *B. bovis*. His arguments for this are convincing and based on his opinion four species of *Babesia* will be recognized in cattle, namely: *B. bigemina*, *B. bovis*, *B. divergens* and *B. major*.

Babesia bigemina. (Fig. 3.45) This organism is the cause of cattle tick fever, red water fever, piroplasmosis and, formerly in North America, Texas fever. *Hosts:* principally the bovine, also in zebu, water buffalo. (Levine, 1973, also gives deer (*Mazama americana reperticia*) and white-tailed deer (*Odocoelius virginianus chiriquensis*) as hosts.) Occurs throughout the tropics and subtropical areas such as Central and South America, North and South Africa, Australia and southern Europe.

Morphology. *Babesia bigemina* is a large piroplasm, 4–5 μm in length by about 2 μm wide, round forms 2–3 μm in diameter. The organisms are characteristically pear-shaped and lie in pairs forming an acute angle in the red blood corpuscle. Round, oval or irregularly shaped forms may occur, depending on the stage of the development of the parasite in the red cell.

Developmental cycle. Multiplication in the vertebrate host has been discussed previously (p. 707), and the developmental cycle in the tick has been described by Riek (1964) (see above and Fig. 3.44).

The following ticks have been incriminated as vectors: one-host ticks: *Boophilus annalatus*, North America; *Boophilus calcaratus*, North Africa; *Boophilus decoloratus*, South Africa; *Boophilus microplus*, Australia, Panama, South America; two-host ticks: *Rhipicephalus evertsi*, South Africa; *Rhipicephalus bursa*, South Africa; three-host ticks: *Haemaphysalis punctata*, Europe and Eurasia; *Rhipicephalus appendiculatus*, South Africa.

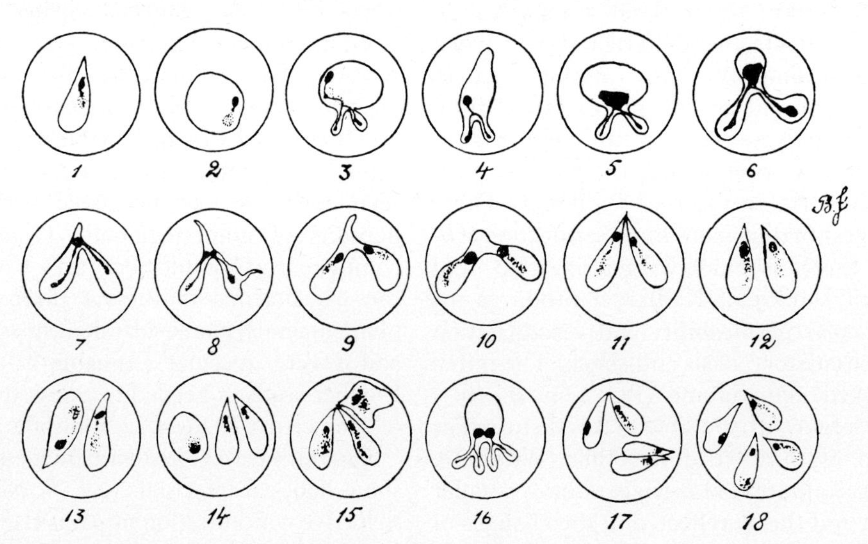

Fig 3.45 *Babesia bigemina* of cattle. Method of multiplication in red blood corpuscles. (× ca 3000) (*From Wenyon 1926*)

There is no evidence that *B. bigemina* can be transmitted mechanically by blood-sucking arthropods. On rare occasions it seems that intrauterine transmission may occur from mother to fetus, but this mode of infection does not seem to be of any significant importance in the general epidemiology of babesiosis.

Pathogenesis. The general aspects of the pathogenesis of babesial infections have been discussed above (p. 711) and a detailed account is given by Mahoney (1977). In young animals the infection is frequently symptomless and associated with a low parasite density. There are a few reports of fatal infection in newborn calves (Smith & Kilborne 1893), and Hall (1960) has reported the development of babesiosis in calves, born of non-immune parents, when they were inoculated with infective blood or exposed to infected ticks at 12–55 days of age. The natural resistance of the young calf to infection usually disappears at 9–12 months of age.

The incubation period after infection, or exposure to infected ticks, is one to two weeks, and the first evidence of the disease is a spectacular rise in body temperature to 41–42°C. The high fever lasts from two to seven days or more, and during this period a profound anaemia frequently develops. There is haemoglobinuria and cardiac palpitation. Initially, there is a profuse diarrhoea, and this is later followed by marked constipation. At the height of fever, up to 75% of red blood cells may be destroyed, and the mortality may be very high in acute cases, death occurring four to eight days after the onset of clinical signs. Animals which survive the acute phase go into a chronic disease syndrome which may extend over several weeks with an irregular course and intermittent temperature rises, at times reaching 40–40.6°C. Animals become thin and emaciated, but there is usually no marked haemoglobinuria in this stage, and finally the animal recovers.

A cerebral form of *B. bigemina* infection has been reported in cattle in Tanzania by Zlotnik (1953). The onset of disease is sudden, the body temperature reaching 41.7°C in a few hours, and death may occur within 12–36 hours after the onset of clinical signs. The parasites appear to accumulate, and probably multiply, in the cerebral capillaries since organisms are only rarely seen in blood smears.

At post mortem the lesions consist of subcutaneous and intramuscular oedema with icterus, the fat is yellow and gelatinous and the blood thin and watery. On sedimentation, the blood plasma contains haemoglobulin, and the urine in the bladder is frequently red or dark brown. The spleen is markedly enlarged with a soft, dark splenic pulp; the liver is enlarged, pale and yellowish, and the gall bladder is distended with thick, dark bile. In the cerebral form there is perivascular, perineuronal and interstitial oedema throughout the brain and spinal cord.

Microscopically, there is centrilobular necrosis in the liver, deposits of haemosiderin in the Kupffer cells and congestion in a variety of organs, such as lungs, heart, spleen and kidney. Degeneration of the tubular epithelium and cast formation is seen in the kidneys, along with deposits of haemosiderin in various cells of this organ. Depletion of the germinal centres of the spleen and lymph nodes occurs, with hyperplasia of the reticular tissue and large numbers of macrophages containing haemosiderin.

Immunology. An inverse age susceptibility occurs in *Babesia* infections, young animals being naturally resistant while older animals are fully susceptible. Calves in an enzootic area are free of clinical signs and have a very low parasite density. The passive transfer of maternal antibodies via the colostrum is probably responsible in part for this resistance (Hall et al. 1968); however, Riek (1963) suggests that additional explanations are necessary since animals of several months of age develop only a low parasitaemia when they are introduced from a clean to an infected area.

Breed of animal. It has been suggested that some races of cattle are more resistant to *B. bigemina* than others (*Bos indicus* has been suggested to be more resistant than *Bos taurus*) (Johnston 1967); however, Arnold (1948) reported no difference in the severity of *Babesia* infections when either Brahman or European breed animals were introduced into an infected area in Jamaica. Daly and Hall (1955) observed similar reactions in all breeds tested with blood-transmitted infections of *B. bigemina*.

Acquired immunity. Clinical recovery from babesial infection is followed by apparent removal of parasites from the peripheral blood. However, latent infection may persist for some considerable time (up to several years) and be demonstrable by the injection of blood into splenectomized animals. Possibly because of such latent infections, it has been thought that acquired immunity in babesial infection is contingent on their presence, i.e. a premunition. The concept of premunity was reviewed by Sergent (1963). However, there is now good evidence that acquired immunity can persist for several years in the absence of demonstrable organisms (Mahoney et al. 1973) and the termination of a subclinical infection by drug treatment resulted in immunity to reinfection for at least six months (Callow et al. 1974). Mahoney (1977) concludes that one infection with *B. bigemina* (and *B. bovis*) confers species-specific protection for life.

However, immunological differences between geographical strains of *B. bigemina* occur (Curnow 1968) and clinical disease may be produced in an immune animal by infection with a strain of *B. bigemina* from another area.

Antigenic variation in *Babesia* spp. is recognized and has been studied experimentally in *B. rodhaini* infection in rodents (Thoongsuwan & Cox 1973) where three serological types were reported, whereas in cattle infected with *B. bovis*, Ross and Mahoney (1974) estimated, by computer simulation techniques, that the parasite had the potential to produce over 100 different antigenic types.

The spleen plays an important role in maintaining the immune state to *Babesia* infection, since immunity may be broken down by removal of the spleen. With *B. bigemina* a fatal infection may be precipitated by this procedure. Hence, antibodies play a central role in immunity, protection being transferable by serum or colostrum and it is likely that antibodies to the variant antigens are the protective elements.

Antibodies are detectable 7–21 days after tick-transmitted infection, and they persist for more than ten months. Antibodies may be present with no detectable parasites in the blood, but such blood may be shown to be infective by injection into splenectomized calves. At a later stage, antibodies occur when blood will not create infection in susceptible animals.

Immunization against *B. bigemina* infection is practised by the inoculation of infected blood and, if necessary, the subsequent treatment of animals with a babesiacidal drug to prevent severe or fatal disease. The use of a depot-forming babesiacidal drug (e.g. imidocarb, see below) has alleviated the problems associated with variations in virulence in the 'vaccine' and imidocarb allowed effective immunization when given up to a month before infection (Callow & McGregor 1970). Care must be exercised that infection is not terminated too early, otherwise effective immunity would not be induced. Vaccination against bovine babesiasis has been reviewed by Callow (1977). A detailed account of the vaccine in current use in Australia for *B. argentina* infection is presented.

Experimental vaccines include the injection of infected blood previously exposed to 20–50 kr of ionizing radiation (Brocklesby et al. 1972) and dead vaccines prepared from infected RBCs or from the plasma of infected animals. This is discussed by Mahoney (1977).

Diagnosis. Diagnosis is based on the clinical signs and confirmed by the detection of parasites in the peripheral blood. Both thick and thin blood smears may be employed, being stained by one of the Romanowsky stains; however, the organisms may not always be apparent, and it may be necessary to examine a number of smears to establish their presence. In the cerebral form, examination of cerebral capillaries is necessary. In endemic areas high fever associated with haemoglobinuria and anaemia is suggestive of *Babesia* infection, and frequently animals are treated without recourse to blood examination.

Immunodiagnostic tests are increasingly used to detect infection, especially in the subclinical situation when organisms are not demonstrable in the blood. The provision of antigen free from host protein contamination presents problems, but Goldman and Bukovsky (1975) prepared a soluble antigen from parasite-stroma suspensions of *B.*

bigemina for use in immunodiagnostic tests. The serodiagnostic tests available include the complement fixation test (Mahoney 1962), the indirect fluorescent antibody test (IFA) (Ross & Lohr 1968), the indirect haemagglutination test (IHA) (which is approximately 80% effective in field infections; Mahoney 1977) and a rapid agglutination test which uses a suspension of stained parasites (Curnow 1973). The last-named test showed a degree of strain-specificity which precluded its general use in field diagnosis (Mahoney 1977).

Treatment. Older compounds used in treatment include trypan blue (100 ml of 1–2% solution in normal saline given intravenously) and acriflavine (20 ml of 5% aqueous solution intravenously). These have now largely been replaced by quinuronium sulphate and the diamidines.

Pirevan (Acapron, Babesan, Piroparv, Acaprin, Piroplasmin) (quinuronium sulphate) [6 : 6′-di(*N*-methylquinolyl) urea dimethosulphate]. Dose: 1 ml of a 5% solution subcutaneously per 50 kg body weight. Intravenous injections are contraindicated. The toxic reactions consist of vasodilatation with sweating, salivation, diarrhoea, urination and sometimes collapse and death. The drug has a specific action on the parasympathetic nervous system. Pirevan has been used extensively for the treatment of babesiosis caused by all species, but it does not eliminate the infection.

Phenamidine (4,4-diamidinodiphenyl ether). Dose, 12 mg/kg subcutaneously in a 40% aqueous solution. This compound eliminates all the parasites from the animal, and treated animals are no longer premune. Occasional cases of toxicity may result from its use, these being anaphylactic in type, but usually the animal recovers.

Berenil (4,4-diamidinodiazoaminobenzene aceturate). Diminazene. Dose: 2–3.5 mg/kg by deep intramuscular injection. It is used to treat all species of *Babesia*.

Diampron (3,3′-diamidinocarbanilide diisethionate). Amicarbilide. Introduced for the treatment of *B. divergens* infection, it is also effective against *B. bigemina* (Shone et al. 1961) and will eliminate infection with this species. Dose: 10 mg/kg intramuscularly or subcutaneously, preferably by deep intramuscular injection.

Imidocarb [3,3-*bis*(2-imidazolin-2-yl)] carbanilide. Imidazole. This compound has been used as a therapeutic and prophylactic drug against *B. bigemina* and *B. argentina* (= *B. bovis*) (Callow & McGregor 1970). Dose: 0.5–1 mg/kg given subcutaneously.

Dalgliesh and Stewart (1977) have reported tolerance to imidocarb by *B. argentina* (=*B. bovis*) but it is not yet known whether *B. bigemina* behaves similarly following repeated exposure to the drug. The residual effect of imidocarb will provide short-term protection of susceptible animals on their introduction to an endemic area (Kuttler et al. 1975).

Control. Since the natural transmission of *B. bigemina* is dependent on certain species of ticks, infection can be prevented by adequate tick control measures which keep animals free from tick infestation. This can be done by the regular dipping of cattle, this method having resulted in the elimination of *B. bigemina* from the United States. The methods and frequency of dipping are factors which must be determined for the local area and are considered on page 473.

The inverse age resistance of cattle to *B. bigemina* has been exploited for control by immunization. Young animals are inoculated, preferably with a mild strain of *B. bigemina*, and the subsequent infection, if necessary, is controlled by drug treatment. This method is especially useful when cattle are to be shipped to endemic areas. Control procedures are discussed by Mahoney (1977) and Callow (1977).

Babesia bovis. (Syn. *B. argentina*; *B. berbera*) *Hosts:* cattle, also roe deer and stag. Mahoney (1977) considers there are insufficient differences to retain *B. argentina* and *B. berbera* as separate species and his opinion is accepted. Hence, the distribution of *B. bovis* is southern Europe, Africa, Asia and Central and South America and Australia. At one time it was considered to extend into northern Europe, including Great Britain; however, investigations by Simitch et al. (1955) and Davies et al. (1958) showed that the northern European form was *Babesia divergens*.

Morphology. A small piroplasm, 2.4 μm by 1.5 μm and slightly larger than *B. divergens*. In further contrast to *B. divergens* there are usually no divergent forms lying superficially in the red blood corpuscle, and vacuolated signet ring forms are particularly common, consisting of a centrally placed vacuole with a nuclear mass at one pole.

Developmental cycle. The vectors of *B. bovis* include *Ixodes ricinus*, a ubiquitous tick in Europe, *Ixodes persulcatus*, which has a more northerly and easterly distribution, *Boophilus calcaratus*, *B. microplus* and *Rhipicephalus bursa*.

The pathogenesis of *B. bovis* infection has been discussed in detail by Mahoney (1977). The important events in pathogenesis are the activation of the kinin system rather than the production of anaemia, although the latter occurs in *B. bovis* infections (see p. 711).

In Australia *B. bovis* (= *B. argentina*) is responsible for the majority of field outbreaks of babesiosis (Riek 1963). It is relatively uncommon in animals under one year of age, and this age group is predominantly infected by *B. bigemina*. In animals over two years of age, however, *B. bovis* (= *B. argentina*) is the main infection. Under Australian conditions *B. bovis* (= *B. argentina*) is the more pathogenic species (Pierce 1956), and cattle inoculated with the smaller form showed twice the mortality of those infected with *B. bigemina*.

Clinical signs are similar to those for *B. divergens*. High fever is evident about a week to ten days after infection, and shortly afterwards haemoglobinuria occurs. A cerebral form of disease due to *B. bovis* (= *B. argentina*) has been recorded by Callow and McGavin (1963). Post mortem changes consisted of congestion of the grey and white matter of the brain and generalized dilatation of capillaries by red blood cells, the majority of which were infected with *B. bovis* (= *B. argentina*). In addition, perivascular, perineuronal and interstitial oedema occurred throughout the brain and cord. Clinical signs varied, consisting of convulsions, incoordination and coma. Other studies by Callow and Johnston (1963) suggested that the brain capillaries may be a predilection site for *B. bovis* (= *B. argentina*) in healthy animals.

Immunology. Recovery from the acute disease is usual and this leads to resistance to reinfection. Subsequently organisms may persist for two to four years; nevertheless, animals are resistant to reinfection and do not need artificial vaccination during this period (Mahoney et al. 1973).

Attenuation of the virulence of *B. bovis* occurs by repeated passage through splenectomized calves. Using strains so attenuated, Callow and Mellors (1966) were able to standardize a vaccination dose (10^7 organisms) which caused a mild reaction, not necessitating treatment, and which led to good immunity. A detailed account of the preparation and use of this is given by Callow (1977).

Treatment. The chemotherapeutic compounds dealt with under 'Treatment of *B. bigemina*' are generally applicable to *B. bovis*. Imidazole is the only compound which eliminates *B. bovis* from carrier cattle, although Dalgliesh and Stewart (1977) have reported tolerance to imidocarb in *B. bovis* (= *B. argentina*) infection.

Babesia divergens. *Hosts:* cattle. It is essentially a northern European form, and Simitch et al. (1955) and Davies et al. (1958) have differentiated it from the more southern European and Danube Basin form, *B. bovis*. Enigk and Friedhoff (1962*a*) have demonstrated the transmission of *B. divergens* to splenectomized moufflon, fallow deer, red deer and roe deer. Organisms were demonstrated microscopically in these animals, and infections were subsequently transmitted back to calves. The infection persisted in red deer for eight months, but sheep and goats were not susceptible to the infection. These authors considered that the wild game may serve as natural reservoirs for *B. divergens*.

Enigk and Friedhoff (1962*b*) described a species of *Babesia* of the roe deer, *Babesia capreoli*. This was transmissible to other roe deer, but splenectomized cattle and sheep were refractory to the infection.

Morphology (Fig. 3.46). *B. divergens* is smaller than *B. bovis*, 1.5 μm by 0.4 μm, and commonly appears as paired, divergent forms lying superficially on the red blood cell. Other forms may be

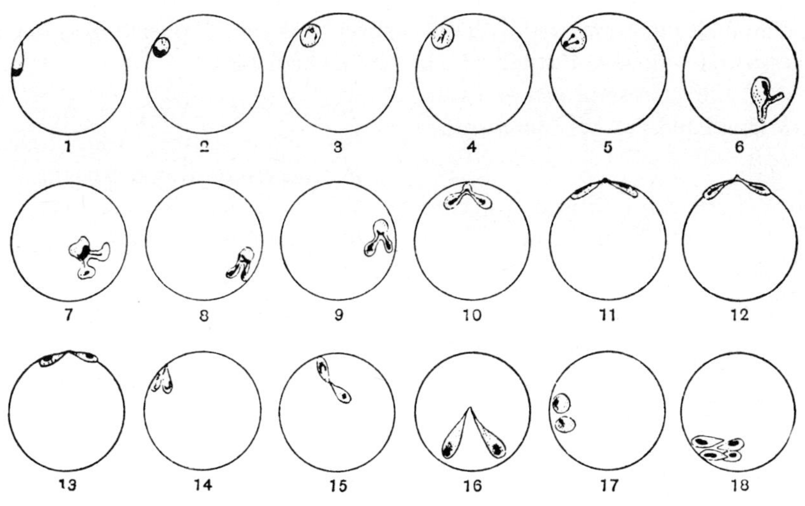

Fig 3.46 *Babesia divergens* of cattle. (× ca 3000) (*From Wenyon 1926*)

stout and pyriform, measuring 2 μm by 1 μm, some may be circular, and a few are vacuolated and up to 2 μm in diameter. Garnham (1962) has suggested that *Babesia pitheci* of African monkeys may be a form of *B. divergens* of cattle.

Developmental cycle. This has yet to be studied in detail. It is transmitted by *Ixodes ricinus*, and Joyner et al. (1963) have demonstrated trans-ovarian transmission of *B. divergens* to larvae of *I. ricinus*. Arthur (1966) has implicated *Dermacentor reticulatus* as a host tick.

Cases of piroplasmosis are associated with the activity of the tick. For example, in Great Britain, where *I. ricinus* may show a biannual activity, cases are to be expected in late spring and autumn. Serious disease may occur in animals that have been moved into an endemic area. As with *B. bigemina*, animals up to one year of age seldom show clinical signs of the infection, although they serve as carriers for the infection and as a source of infection for ticks.

Pathogenesis. The disease entity produced by *B. divergens* is less severe than that caused by *B. bigemena*. The incubation period is four to ten days, and the first clinical sign is a marked elevation of body temperature, up to 41°C, which persists for two to three days. Within 48 hours of the peak of fever, haemoglobinuria is observed, followed by jaundice of the mucous membrane and palpitations of the heart. In severe infections death may occur. If an animal survives the acute syndrome, gastrointestinal upsets become evident by a thin and watery diarrhoea which is followed by a marked constipation, with the faeces being bile-stained. The animal progressively loses flesh, appetite is reduced, and death may occur if this phase is prolonged, although recovery from this clinical entity is not uncommon.

Diagnosis. This is based on the clinical signs, the history of the area and the demonstration of the parasites in blood smears. These are not readily found but are to be expected in their greatest number during the period of high fever. Recovered animals are immune, and Davies et al. (1958) reported that two heifers which had lost their infection with *B. divergens* remained immune to reinfection despite the lack of organisms.

Treatment. The compounds used for *B. bigemina* are effective.

Control. This is essentially the same as for *B. bigemina* and consists of tick control and, where indicated, the immunization of younger animals with blood from infected animals. Vaccination against *Babesia* is regularly carried out in Sweden, and Bodin and Hildar (1963) have reported that of 78 000 inoculated animals, 0.72% showed clinical reactions and 0.007% died. In a study of vaccinated animals at pasture, a morbidity rate of 12%

in non-vaccinated animals was compared with one of 0.12% in vaccinated animals. Purnell et al. (1981) have reported the successful use of a vaccine consisting of irradiated (28 kr) blood infected with *B. divergens*.

Babesia major. *Hosts:* cattle. South America, North and West Africa, southern Europe, Great Britain, Soviet Union.

This species resembles *B. bigemina* except that it is smaller and lies in the centre of the erythrocyte. The pyriform bodies are 2.6 μm by 1.5 μm, and the angle formed by the organisms is less than 90°. Round forms about 1.8 μm in diameter may occur.

Developmental cycle. This is comparable to that of *B. bovis. Boophilus calcaratus* is the tick vector in the Soviet Union, and in Holland, Bool et al. (1961) suggested that *Haemaphysalis punctata* may be a vector.

Babesia major is considered less pathogenic than *B. bovis*; temperature elevation is not so marked, and haemoglobinuria and anaemia are mild. An indirect fluorescent antibody test for the diagnosis of *B. major* infection has been described by Morzaria et al. (1977). IFA titres persist at a significant level for up to 11 months after recovery from infection.

For treatment the compounds mentioned above are satisfactory. An experimental vaccine using irradiated piroplasms has been described by Purnell et al. (1979).

BABESIA OF SHEEP AND GOATS

Four species of *Babesia* have been reported from sheep and goats, consisting of one large form and three small forms:

Babesia motasi Wenyon, 1926
Babesia ovis (Babes, 1892) Starcovici, 1893
Babesia foliata Ray & Rhaghavachari, 1941
Babesia taylori (Sarwar 1935)
Babesia capreoli Enigk & Friedhoff, 1962 (in roe and red deer)
Babesia odocoilei Emerson & Wright, 1968 (in white-tailed deer)

Babesia motasi (Fig. 3.47). *Hosts:* sheep and goats. Southern Europe, Middle East, Soviet Union, South-East Asia, also Africa and other parts of the tropics.

This is a large form of measuring 2.5–4 μm by 2 μm, the pyriform stages resembling those of *B. bigemina*, the angle at which they meet being acute. They may occur singly or in pairs.

Developmental cycle. This is similar to that of *B. bigemina*. The tick vectors include *Dermacentor silvarum, Haemaphysalis punctata* and *Rhipicephalus bursa*. Both transovarian and stage to stage transmission have been demonstrated for this parasite in *Rh. bursa*. Markov and Abramov (1957) described clavate parasites in the ovary and eggs of the tick, and Li (1958) described similar forms in other developmental stages.

Pathogenesis. The disease caused by *B. motasi* may be acute or chronic. In the former it follows a course comparable to *B. bigemina*, being characterized by high fever, haemoglobinuria and marked anaemia with prostration. Death is not uncommon. In the chronic form there are no

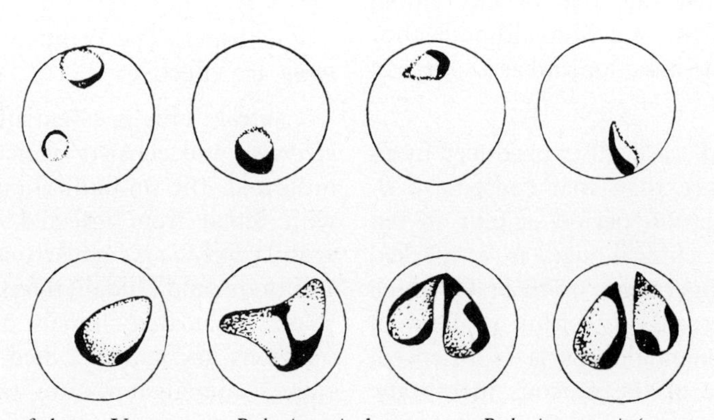

Fig 3.47 *Babesia* species of sheep. Upper row: *Babesia ovis*; lower row: *Babesia motasi*. (× 3000) (*From Wenyon 1926*)

characteristic signs and death is unusual. Recovered animals are immune to the infection but are still susceptible to *Babesia ovis*.

Diagnosis. This is based on clinical signs and the demonstration of the parasites in the peripheral blood, these being most numerous at the time of the acute fever.

Treatment. Trypan blue is effective, being given intravenously. The dose for sheep is 10–25 ml of a 1% solution in normal saline. It is probable that compounds effective against *B. bigemina* are useful for *B. motasi* and would include the diamidines and quinuronium sulphate (p. 715).

Babesia ovis. *Hosts:* sheep and goats. Distributed throughout tropical and subtropical areas, also in southern Europe and the Soviet Union.

Babesia ovis is much smaller than *B. motasi*, being 1–2.5 μm in length. The majority of organisms are round, occurring at the margin of the red cell. Pyriform organisms are comparatively rare, and when they occur in pairs the angle between them is obtuse, the organisms usually lying at the margin of the erythrocyte.

The developmental cycle of *B. ovis* in *Rhipicephalus bursa* has been described by Friedhoff (1969). He suggested the erythrocytic forms transformed to clavate forms, rather than going through the process of fertilization and the production of an oökinete as was described by Riek (1964) for *B. bigemina*. *R. bursa* is a two-host tick, and transovarian transmission and stage to stage transmission have been reported in it (Markov & Abramov 1957).

Pathogenesis. The effects are usually less severe than those of *B. motasi*, although an acute phase characterized by fever, jaundice, haemoglobinuria and anaemia may be seen. In the chronic form of the disease about 1% of the erythrocytes are infected.

Recovered animals are immune to infection, and there is no cross-immunity with *B. motasi*.

Treatment. Trypan blue is ineffective. Pirevan (quinuronium sulphate) has been used for *B. ovis* at a dose of 2 ml of a 0.5% solution per 10 kg. Simitch et al. (1956) used berenil at the rate of 3 mg/kg for *B. ovis* infection in sheep. A rapid recovery occurred in 161 out of 169 infected sheep given a single intramuscular injection.

Babesia foliata. This species has been recorded from sheep in India. It resembles *B. ovis* but lies more centrally in the erythrocyte. The vector has not yet been identified. It may be a synonym of *B. ovis*.

Babesia taylori. *Host:* goat. India. This is a small form which may reach up to 1.5–2 μm in length, but usually it is ovoid to round, about 1 μm in diameter, and appears to undergo several fissions to produce 8 or even 16 parasites per erythrocyte. The host red cell is often enlarged, and dividing forms of the organism may be seen in the plasma. The vector of this species has yet to be identified.

Pathogenesis. This is low, haemoglobinuria not being in evidence.

The control of *Babesia* spp. in sheep is similar to that in cattle and depends on tick control, the therapeutic use of drugs and immunization of young animals with blood from older infected animals.

Babesia capreoli occurs in roe and red deer in Europe, being transmitted by *Ixodes ricinus*. Müller and Rapp (1971) report *B. capreoli* as a cause of death in red deer. **B. odocoilei** does not appear to cause serious ill-health in white-tailed deer.

BABESIA OF HORSES

Two species of *Babesia* occur in the horse:

Babesia caballi (Nuttall, 1910)
Babesia equi (Laveran, 1901)

Babesia caballi. *Hosts:* horse, also donkey and mule. Southern Europe, Asia, Soviet Union, Africa, Panama Zone and USA. Two species of *Babesia* occur in equines in Florida, *B. caballi* representing the majority of the parasite population, but a second form typical of *Babesia equi* is also present.

Morphology (Fig. 3.48). *B. caballi* is a large species resembling *B. bigemina*. Parasites commonly occur as pairs, are pyriform and measure

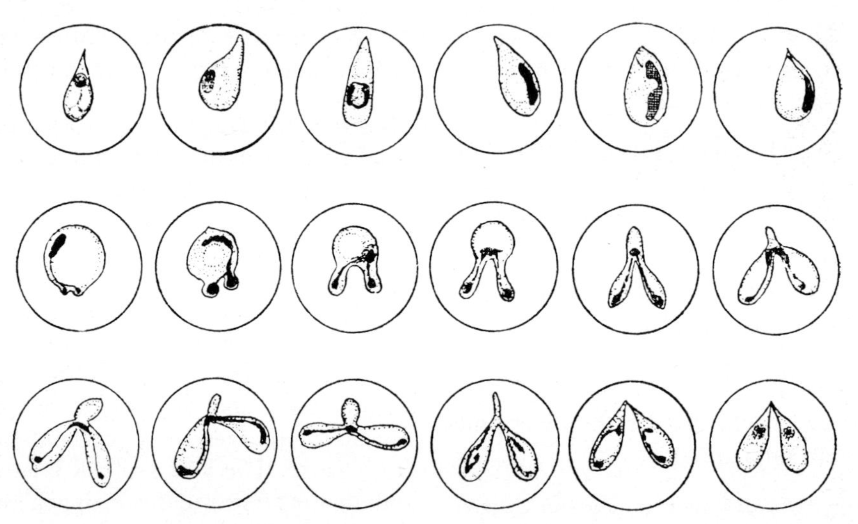

Fig 3.48 *Babesia caballi* of the horse. Various forms seen during multiplication in the red cells. (× ca 3000) (*From Wenyon 1926*)

2.5–4 μm in length; the angle formed by the organisms is acute. Round or oval forms, 1.5–3 μm in diameter, may also occur.

Developmental cycle. The tick vectors of the *Babesia* species of horses are considered in detail by Enigk (1943, 1944, 1951). These are as follows: *Dermacentor marginatus* (southern and eastern Soviet Union, Germany), *D. reticulatus* and *D. silvarum* (European Soviet Union), *D. nitens* (Florida, Panama), *Hyalomma excavatum* and *H. dromedarii* (North Africa), *H. scupense* (Ukraine), *Rhipicephalus bursa* (Bulgaria), *Rh. sanguineus* (Greece), etc. The developmental stages have been described by Holbrook et al. (1968*a*) in *D. nitens* and in the mammalian erythrocyte (Holbrook et al. 1968*b*).

Pathogenesis. There is a great variation in the clinical manifestations of *B. caballi* infection. The course may be acute or chronic, mild or severe, and in some cases it may end in death. Changes in the coagulation parameters may indicate activation of kallikrein during infection (Mahoney 1977). Persistent fever and anaemia with icterus commonly occur, but haemoglobinuria is rare and is not characteristic of the infection. In acute cases death may occur from one to four weeks after the onset of clinical signs. Disturbances of the central nervous system are common and may result in posterior paralysis. Malherbe (1956) records incoordination in foals aged four to five months during midsummer in South Africa. Clinical signs of this consisted of restlessness, nervousness and walking in circles with incoordination.

All breeds of horses are equally susceptible to *B. caballi* infection, although the disease is more marked in older horses, and the inverted age resistance is comparable to that seen in *B. bigemina*. Following recovery the animal is premune, and in general horses are resusceptible to piroplasmosis one to two years after recovery in the absence of reinfection.

Diagnosis. Diagnosis is based on the clinical signs, the presence of tick vectors, history of the area and the demonstration of the organism in the peripheral blood. Several samples may need to be examined since those taken at the onset of disease may show only a small number of parasites. The most satisfactory site to obtain the blood sample is the skin of the ear, and frequently the first drop of blood contains the greatest number of organisms. A reduction in haemoglobin amount and erythrocytes and an increase in erythrocyte sedimentation rate supply supportive evidence.

Serological tests have been used for the diagnosis of *Babesia* infection in horses. Hirato et al. (1945) prepared an antigen from the stromata of blood cells from an acutely infected horse and demonstrated the persistence of complement fixation antibodies for at least 100 days after infection. Ristic et al. (1964) have used several serological tests for diagnosis. Using a soluble antigen prepared by protamine sulphate precipitation of sonically lysed infected erythrocytes, Ristic and Sibinovic (1964) were able to detect precipitins in the serum of horses convalescing or recovered from *Babesia* infection. The specificity of the test was shown by the absence of the reactions with sera of horses with various other infections, including viral infectious anaemia.

The protamine sulphate precipitating antigen is a mucoprotein in nature (Sibinovic 1965), and at least one component of the antigen can withstand boiling at 90°C for 30 minutes. This component has the characteristics of polysaccharide, and it can be adsorbed onto sheep erythrocytes so that they can be used for passive haemagglutination tests. Ristic (1966) reports a good correlation between the results of precipitation in gels and the haemagglutination tests. Soluble antigens from acute phase serum of infected animals can be prepared and adsorbed onto bentonite particles or tanned erythrocytes and used in agglutination procedures for the detection of antibodies to *Babesia* (Ristic 1966).

Treatment. Trypan blue is effective and is given intravenously as a 1% solution at the rate of 50–100 ml according to the size of the horse. It is well tolerated and may be repeated 24 hours later.

Pirevan (Acapron, Babesan, Piroparv), quinuronium sulphate, is effective and superior to trypan blue. It is given subcutaneously as a 5% solution at the rate of 1.2 ml/100 kg body weight.

In the Soviet Union, haemosporidin [(*N,N'*-di-4-dimethyl-aminophenyl) urea methylmethosulphate] is given subcutaneously as a 2% solution, 5–6 ml being given per horse. The therapeutic dose is based on the level of 0.2 mg/kg (Ershov 1956). Amicarbilide is effective against horse *Babesia* and can be used to eliminate *B. caballi* (Taylor et al. 1969). Imidocarb, given subcutaneously in doses of 0.5–1 mg/kg, is effective.

Symptomatic treatment consists of housing animals if possible, appropriate nursing and a balanced diet. In cases of severe anaemia, blood transfusions may be indicated.

Prophylaxis of *B. caballi* infection in horses is comparable to that for *B. bigemina* in cattle.

Babesia equi. *Hosts:* horse, mule, donkey, certain zebra. Asia, Africa, Europe, South and East Africa, South America and Soviet Union. It occurs in the United States (Florida).

This parasite is readily distinguished from *B. caballi*. It is smaller, about 2 μm in length, and characteristically divides into four daughter organisms which frequently form a Maltese cross (Fig. 3.49). (Some authors prefer to refer this species to the genus *Nuttalia*.) Less usual forms in the erythrocytes are rounded or amoeboid stages. Recent studies by Schein et al. (1981) indicate that this species may belong to the genus *Theileria* rather than *Babesia*.

Developmental cycle. The tick vectors are considered by Enigk (1943, 1944). These include: *Dermacentor reticulatus* (European Soviet Union), *D. marginatus* (eastern Europe), *Hyalomma excavatum, H. plumbeum* (Greece, Central Asia), *H. dromedarii* (North Africa), *Rhipicephalus bursa, Rh. turanicus* (Soviet Union), *Rh. evertsi* (South Africa) and *Rh. sanguineus* (Central Asia, North Africa). The vector in the USA is unknown and *D. nitens* cannot be infected.

Exoerythrocytic schizogony has been demonstrated in vivo and in vitro with the development of macro- and microschizonts and the invasion of erythrocytes by merozoites (Schein et al. 1981). Equine lymphoblastoid cell lines infected and transformed by *B. equi* sporozoites (as has been described for *Theileria* spp. in cattle) have been established by Rehbein et al. (1982).

Pathogenesis. In general, this species is more pathogenic than *B. caballi*, but mixed infections of *B. caballi* and *B. equi* may occur. Ershov (1956) considers that simultaneous primary infections occur rarely, and he suggests an antagonism between the two, *B. equi* infection having a tendency to develop more frequently than *B. caballi*. Following infection, the incubation period is eight to ten days, the first clinical sign being a marked increase in body temperature, which may reach 41.7°C, and this coincides with the appearance of the organisms in the circulating blood. In acute cases the disease process lasts eight to ten days, and if recovery is to occur there is a fever crisis about the tenth day; thereafter, the body temperature falls to normal, and the

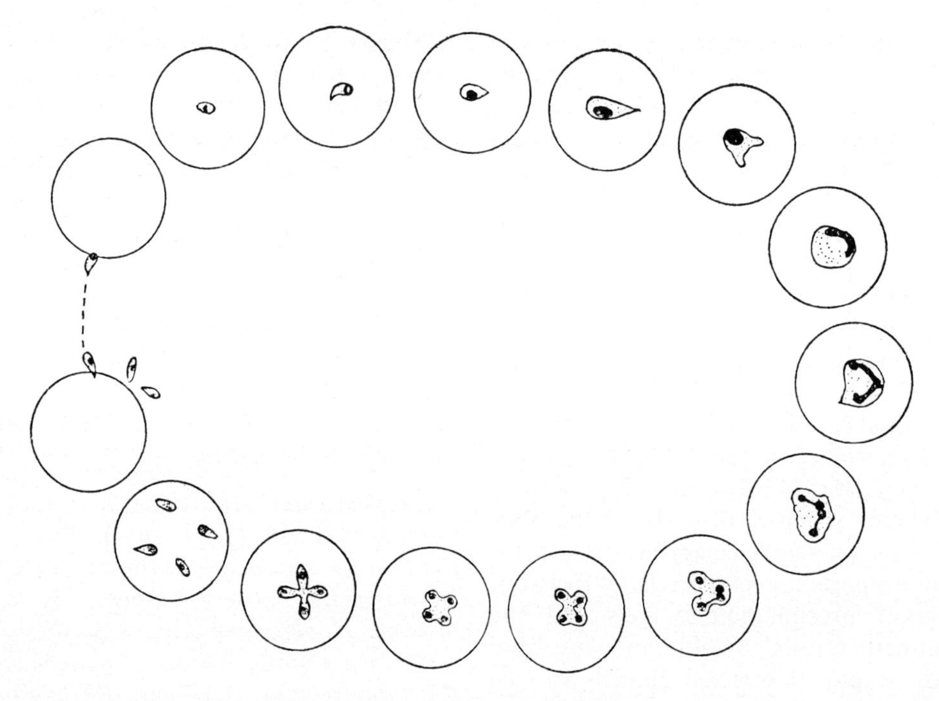

Fig 3.49 *Babesia equi* of the horse. Various forms seen during multiplication in the red cells. (× ca 3000) (*From Wenyon 1926*)

animal rapidly recovers and becomes a carrier. In peracute cases death may occur in one to two days after the onset of clinical signs. Anaemia and haemoglobinuria may be marked, and there is listlessness, depression and inappetence. Oedema of the dependent parts of the body and the head may occur. There may also be gastrointestinal upsets and hard faeces covered with yellowish mucus. Posterior paralysis, common in *B. caballi* infections, is not usually seen. The subacute infection develops more slowly and is more prolonged. Recovery may take several weeks or months.

The pathological changes are most readily seen in acute infections and consist of general jaundice, petechial haemorrhages, enlargement of the spleen and liver, and the kidneys are flabby and may show petechial haemorrhages. In severe cases oedema of the lungs and sometimes terminal pneumonia may be found.

Animals recovered from infection are immune and the persistence of immunity is comparable to that with the piroplasms of cattle. Abramov (1952) has observed different strains which varied in virulence, and Ershov (1956) records that a transcaucasian strain is more virulent and pathogenic than one from the temperate zones of the Soviet Union. Consequently, superimposed infection of the former on infections induced by the latter may lead to clinical disease. There is no cross-immunity between *B. equi* and *B. caballi*.

Diagnosis. This is made on the clinical signs and the identification of parasites in stained blood smears. Blood examination is best made during the period of fever since, subsequently, organisms become scarce in the blood. The Maltese cross formation is typical of the species, and these may be readily seen if a fluorescent antibody technique is used (Ristic et al. 1964). The various serodiagnostic tests used for the detection of *B. bigemina* in cattle are also applicable for the equine *Babesia*.

Treatment. Since *B. equi* is a small piroplasm, it is not affected by trypan blue. Pentamidine (lomidine 4,4′-diamidino 1,5-diphenoxy pentane) has been used widely by French workers in North Africa (Carmichael 1956), and haemosporidin has been used for the infection of horses in the Soviet Union. The diamidines may be used and imidocarb given at the rate of 4 mg/kg on four occasions at 72-hour intervals eliminated *B. equi* from horses. This dose was lethal to donkeys (Frerichs et al. 1973).

The control measures appropriate to *B. equi* infection are comparable to those for the other *Babesia*.

BABESIA OF SWINE

Two species have been recorded:

Babesia trautmanni (Knuth and Du Toit, 1918)
Babesia perroncitoi (Cerruti, 1939)

Babesia trautmanni. *Hosts:* pig; wart hog and bush pig may act as carriers (Neitz 1956). Soviet Union, southern Europe (Italy, Bulgaria), Zaire, Tanzania.

Morphology. Organism, 2.5–4 μm long by 1.5–2 μm wide, characteristically long and narrow. Organisms frequently occur in pairs, but the cell may contain up to four organisms and sometimes five or six. Oval, amoeboid and ring forms may occur, and occasional parasites may be seen in the plasma.

The developmental cycle in the vertebrate host is probably the same as for the other species of *Babesia*. Tick vectors include *Rhipicephalus turanicus* (Soviet Union), *Boophilus decoloratus* (Tanzania) and *Rh. sanguineus* and *Dermacentor reticulatus* (Europe) (Neitz 1956).

Pathogenesis. Babesiosis of swine is a seasonal disease occurring during the spring and reaching a peak incidence during May and June. Piglets two to four months of age and adult swine are equally susceptible, and in the Soviet Union the wild boar may be infected and thereby serve as a natural reservoir for the disease (Ershov 1956). In the acute disease there is fever with anaemia, haemoglobinuria, jaundice, oedema of the dependent parts and incoordination. Pregnant sows may abort and mortality may reach 50%. The incubation period is 12–25 days.

Diagnosis is made on clinical signs, especially the haemoglobinuria and the icterus, and the demonstration of the organisms in blood smears.

Treatment. Trypan blue is effective being given intravenously as a 1% solution at the rate of 10–25 ml/animal. In Italy, Puccini et al. (1958) have reported that berenil is effective against *B. trautmanni* being given as a 7% solution at the rate of 3.5 ml/kg intravenously. Lawrence and Shone (1955) found that phenamidine given subcutaneously at the rate of 1.5 ml of a 40% solution per 45 kg body weight was effective. The diamidines and quinuronium sulphate are probably effective.

Babesia perroncitoi. *Hosts:* pig. Sardinia, Sudan.

This is a small rounded vacuolated form 0.7–2 μm in diameter. Oval to pyriform forms may occur, 1.2–2.6 μm long by 0.7–1.9 μm wide.

The tick vectors have yet to be established experimentally; however, Cerruti (1939) has suggested that *Rhipicephalus sanguineus* and *Dermacentor reticulatus* may be the vectors in Italy.

The pathogenesis of the infection is comparable to that caused by *B. trautmanni*. Puccini et al. (1958) found that berenil was effective for treatment.

BABESIA OF DOGS AND CATS

Canine babesiosis is widespread throughout the world; four species have been described:

Babesia canis (Piana and Galli-Valerio, 1895)
Babesia gibsoni (Patton, 1910)
Babesia vogeli Reichenow, 1937
Babesia felis Davis, 1929

Of the above, *Babesia canis* is the species of major importance.

Babesia canis. *Host:* domestic dog. Asia, Africa, southern Europe, United States, Puerto Rico, Central and South America. Naturally infected wolves, striped jackals and black-backed jackals have been found in Turkestan, East Africa and South Africa, respectively. The red fox and silver fox have been artificially infected in Germany.

Morphology (Fig. 3.50). This is a large piroplasm, pyriform in shape, 4–5 μm in length, pointed at one end and round at the other. Frequently there is a vacuole in the cytoplasm. The pyriform forms may lie at an angle to one another, but pleomorphism of shape may be seen,

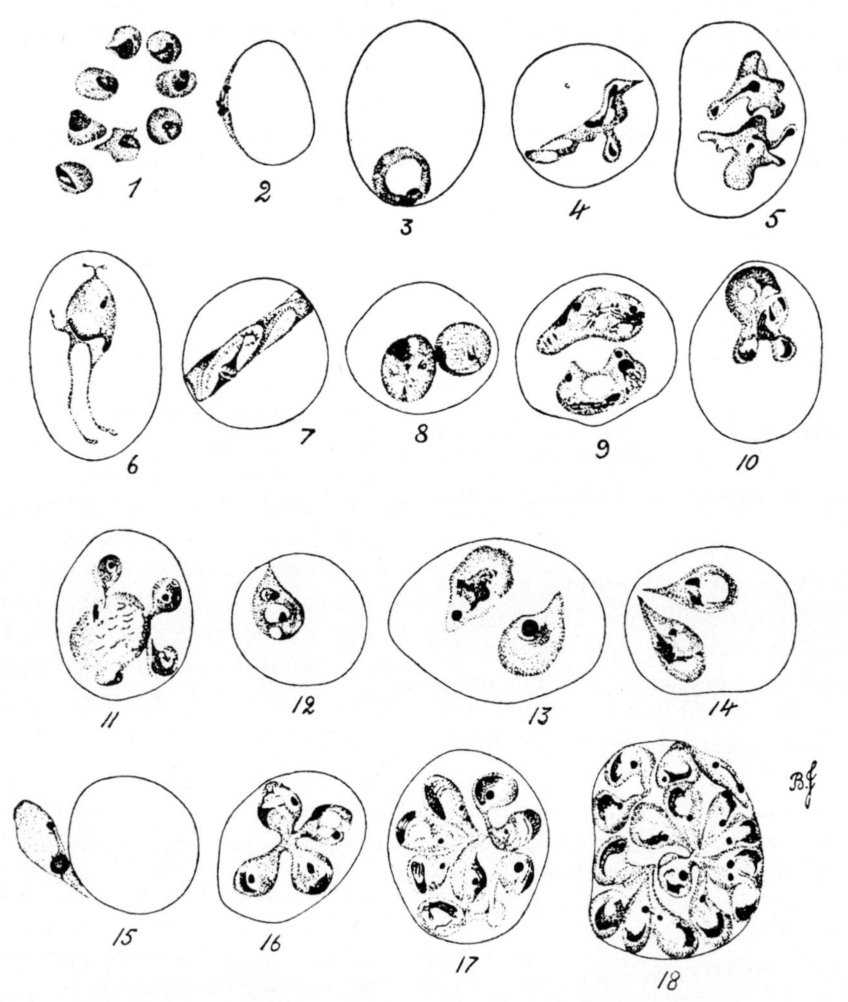

Fig 3.50 *Babesia canis* in the blood of the dog. (a) Group of free forms, probably resulting from rupture of a cell with multiple infection. (b) Marginal form. (c–i) Various types of parasite. (j) Form producing two buds. (k) Form producing four buds. (l–n) Pear-shaped individuals. (o) Free pear-shaped form. (p) Form producing four buds. (q, r) Cells containing several budding parasites. (×ca 2250) (*From Wenyon 1926*)

organisms varying from amoeboid to ring forms. Multiple infection of erythrocytes may be seen, up to, and sometimes more than, 16 organisms being found in a single red blood cell. Organisms may also be found in endothelial cells of the lungs and liver and also in macrophages, this probably being due to erythrophagocytosis.

Developmental cycle. The life-cycle of *B. canis* in *Rhipicephalus sanguineus* and *Haemaphysalis leachi* has been described and magnificently illustrated by Shortt (1936, 1973). Shortt (1973) summarizes the life-cycle in the tick host as follows: Following ingestion of infected blood by an adult tick, the majority of the parasites in the red blood cells are quickly destroyed. Those which survive leave the red cells and become motile, penetrate the walls of the diverticula and enter the coelomic cavity and make their way in the haemolymph to the ovaries and invade the ova. Multiplication occurs in the ova and these forms are the source of transovarian transmission when ova hatch into larvae. Infected larvae can transmit the infection but the parasites can also survive through subsequent instars of ticks, retaining the ability to infect through several generations of ticks even when these are fed on refractory hosts.

When a nymphal stage ingests infective blood there is a similar destruction of parasites and the survivors undergo a complicated developmental cycle in cells of the subcuticular layer and produce club-shaped forms which become free, are motile and are found in the haemolymph. On development of the nymph to an adult these forms enter muscles, round up and undergo marked multiplication and then become static until the adult feeds, when they again become motile, move to the salivary glands and there undergo marked multiplication in the acinar cells. These are the infective forms which pass into the host when the tick feeds.

When larval stages ingest infected blood, the same destruction of parasites occurs and the surviving forms appear as uninucleate bodies which are incorporated into the nymphal tissues as ecdysis occurs.

Comparison of the illustrations made by Shortt (1973) of developmental forms of *B. canis* in the tick with those made by Riek (1966) of the developmental forms of *B. bigemina* and *B. argentina* (= *B. bovis*) in the tick *Boophilus microplus* reveals a close similarity of many developmental stages.

The principal vector of *B. canis* is *Rhipicephalus sanguineus* which occurs throughout the world; it has been specifically demonstrated as a vector in India, Germany, France, South Africa, United States and Brazil. *Dermacentor marginatus* and *D. reticulatus* have been incriminated in France, Germany and the Soviet Union. *Dermacentor venustus* (*D. andersoni*) is a possible vector in France, *Haemaphysalis leachi* in South Africa and *Hyalomma plumbeum plumbeum* in the Soviet Union.

Pathogenesis. Unlike other animals, the puppy may show clinical disease as severe as that seen in adult dogs. Shortt (1973) states that the younger the dog the more susceptible it is, that there is no difference in the susceptibility of dogs from Europe, Africa or Asia and that the general physical condition of the dog is not related to susceptibility.

Artificial infection of susceptible dogs with infective blood was studied by Ewing (1965). Initially, there was a transient parasitaemia which lasted three to four days, after which the organisms disappeared from the peripheral blood for about ten days. A second parasitaemia developed about two weeks after exposure, and the increase in the number of organisms in the red blood cells resulted from binary fission within the cells; cells which harboured multiple trophozoites contained either two or exponential multiples of two.

The severity of infections with *B. canis* varies markedly according to the strain of the parasite and this may be the most important factor in determining the outcome of the infection (Shortt 1973). Ewing and Buckner (1965), in the USA, found that although anaemia developed in uncomplicated infections, there was also enough active haemopoiesis to allow recovery in many cases. On the other hand, Maegraith et al. (1957) and Tella and Maegraith (1965) reported a strain of *B. canis* obtained from Elberfeld (Germany) in 1938, which had been maintained by subinoculation through dogs and puppies, to be highly virulent for young puppies, killing them in 4–5 days. The inoculation of 1–2 ml of blood either intravenously or intraperitoneally resulted, almost invariably, in death a few days after the onset of clinical signs. In fully grown dogs the infection was not always fatal although severe haemolysis often developed. The role of plasma kinins (see page 711) in the pathogenesis of canine babesiosis is not known, but warrants investigation.

Under natural conditions and in endemic areas, a wide variety of clinical manifestations of the disease may be seen. Malherbe (1956) states, 'there is almost no guise under which the disease does not masquerade at some time or another'. The common clinical signs seen in the majority of cases are as follows: An incubation period of 10–21 days is followed by a fever of 38.9–40.6°C, malaise and listlessness. There is depression, disinclination to move, anorexia, the mucous membranes become pale, and jaundice develops in advanced or neglected cases. Haemoglobinuria is usually associated with peracute cases where the loss of red blood cells is marked. Splenic enlargement is present, the faeces are markedly yellow (except in very early or peracute cases), and usually there is a lot of bilirubin in the urine. Progressive debility is seen, and emaciation may become extreme; however, the animal frequently dies before this stage is reached. In the majority of cases organisms are readily demonstrable in smears from the peripheral blood (Malherbe 1956).

In chronic infections there may be an irregular temperature, a capricious appetite and a marked loss of condition.

In a study of severe and fatal infections, Maegraith et al. (1957) found no direct relationship between the clinical condition and the degree of parasitaemia. Thus, the degree of anaemia may not be correlated with a high parasite density, and in fatal infection Tella and Maegraith (1965) found the average parasite count in puppies dying on the fifth day of infection to be 6.6% in contrast to a PCV of 29.3% and an RBC count of 24.6% at that time. Maegraith et al. (1957) noted that the fall of erythrocyte numbers and haemoglobin concentration may be accompanied by intravascular haemolysis, but frequently anaemia developed without obvious haemoglobinaemia and haemoglobinuria. Maegraith et al. (1957) consider that very active phagocytosis of both parasitized and unparasitized RBC by macrophages in the circulating blood and in the spleen, bone marrow and liver, is largely responsible for the loss of cells in cases where haemoglobinuria is not present.

Erythropoiesis is active even in profound cases of anaemia, and reticulocytes appear at an early stage of the infection and continue to do so throughout the disease process.

The mode of death in *B. canis* infection depends on the length of illness. In rapid, fulminating infections, which cause death in four to five days, the animal remains conscious to the end with a strong heart action, and death is associated with acute respiratory failure, often with extensor spasm. In infections which do not kill so quickly, the animal becomes weaker and finally unconscious; it is completely relaxed, there is always a profound anaemia, the extremities are cold, respirations are shallow and rapid, the heart beat rapid and feeble, and the animal dies from circulatory failure associated with pulmonary oedema (Maegraith et al. 1957).

The atypical manifestations vary from a simple catarrhal bronchitic condition to pneumonia, both of which show a remarkable response to specific therapy. Subcutaneous oedema, ascites and purpura have been described by Malherbe and Parkin (1951). In the ascites cases, characteristically there is marked abdominal distension usually, but not invariably, associated with emaciation. This is seen usually in half-grown puppies or young dogs under a year of age; animals have very pale mucous membranes, the temperature may be normal or subnormal, and blood smears may or may not show the *Babesia* organisms. In some cases organisms are undetected even after extensive examination. Such cases show a striking response to specific treatment, the peritoneal exudate being rapidly resorbed, and by 7–12 days the oedema fluid has disappeared. Purpura haemorrhagica may occur in a few cases. Petechial haemorrhages or ecchymoses occur on the iris, the mucous membranes of the mouth and lips and the skin of the abdominal wall and the inside of the thighs. Some cases may pass red urine or, occasionally, blood clots, and blood may be observed in the faeces indicating haemorrhage in the posterior bowel.

Central nervous system involvement is less common than the other manifestations. Piercey (1947) has described a case of acute cerebral canine babesiosis characterized by sudden death, and Purchase (1947) reported a case in which parasites were scarce in the peripheral blood but abundant in brain smears and in histological sections. Maegraith et al. (1957), in their acute experimental infections, failed to observe neurological signs other than weakness of the limbs and coma, which developed as a terminal effect. Peripheral nerve lesions have been reported by Malherbe and Parkin (1951). These may be manifest by rheumatoid muscular pains, chiefly of the legs, causing lameness and even paraplegia. Dogs may scream with pain if their heads are touched or their mouths opened.

At post mortem there is enlargement of the liver and spleen. Centrilobular degeneration or necrosis occurs in the liver, and in a few cases this may extend almost to the periphery of the lobule. The kidneys show medullary congestion in fatal cases, and there are degenerative changes in the tubular epithelium in the cortical region. Other post mortem changes include oedema in the pleural and the peritoneal cavities and petechial haemorrhages on various organs and the mucous membranes. The latter are also icteric and there is a profound anaemia.

The effect of combined infection with *Babesia canis* and *Ehrlichia canis* has been discussed by Ewing and Buckner (1965). They reported grave illness accompanied by severe anaemia of the normocytic, normochromic type which was due to destruction of mature erythrocytes and an impairment of erythropoiesis. These workers considered that the *Babesia* were responsible for the former and *Ehrlichia* for the latter. Dogs infected with either organism alone did not succumb to the infection. Dual infections in young dogs were often fatal, death being expected 19–45 days after exposure. It remains to be seen how commonly *Ehrlichia* is a component of the disease entity which hitherto has been attributed to *Babesia*.

The genus *Ehrlichia* comprises a group of rickettsial organisms which parasitize circulating leucocytes. *Ehrlichia canis* is the cause of tropical canine pancytopenia (Huxsoll et al. 1969). Smith and Ristic (1977) have reviewed the genus *Ehrlichia* (see p. 757).

Diagnosis. In areas where the infection is endemic, any dog with a high fever and clinical signs of anaemia and jaundice is a suspect for babesiosis. Frequently, dogs are treated without a blood examination being made. Even in the absence of demonstrable organisms in the peripheral blood, there may be sufficient justification for treatment since demonstration of the organisms in blood smears may not be easy. Organisms are most readily found in the first drop of capillary blood from an ear puncture.

When atypical forms of the disease occur diagnosis may need to be based on the response to specific therapy (Malherbe 1956), and usually this is rapid and more or less complete. Alternatively, immunodiagnostic tests such as complement fixation, IFA, and IHA (p. 714) may be useful.

Supportive evidence for a diagnosis includes splenomegaly, an increased bleeding time, anaemia, accelerated erythrocyte sedimentation and the presence of increased amounts of bilirubin in the serum.

Treatment. Trypan blue is effective against *B. canis.* A single intravenous injection of 4–5 ml of a 1% solution is usually effective for an average-sized dog (16 kg). Care should be taken that the drug is given intravenously and not subcutaneously.

Phenamidine (4,4′-diamino diphenylether) has given excellent results at the rate of 10 ml of a 5% solution per kg subcutaneously. A single dose is usually effective but it may be repeated 24 hours later.

Pirevan (Acapron, Babesan, Piroparv), quinuronium sulphate, is given subcutaneously as a 0.5% solution at the rate of 0.05 ml/kg. The drug is generally well tolerated, although in some dogs minor convulsions may occur, but these are transitory. It may be repeated in 24 hours with safety.

Prevention and control. This is essentially the same as that for other *Babesia*. Since *Rhipicephalus sanguineus* may occur in dog kennels, and even in human habitation, a regular programme of tick control should include periodic cleaning and fumigation of kennels.

Recovered animals are immune to *B. canis.* Kobalskii et al. (1963) have shown that the blood of recovered dogs may remain infective for 16 months; however, after $1\frac{1}{2}$–2 years the blood ceased to be infective for clean animals and the dogs were no longer immune.

Babesia gibsoni. *Hosts:* domestic dog. India, Sri Lanka, parts of China, Turkestan, possibly parts of North Africa. In addition, the jackal (*Canis aureus*) in India, the wolf (*Canis lupus*) in Turkestan and the fox (*Vulpes vulpes dorsalis*) in Sudan are naturally infected (Neitz 1956).

B. gibsoni is a small form, is pleomorphic and lacks the usual pyriform-shaped trophozoites. Characteristically the trophozoites are annular or oval; signet ring forms may occur and, rarely large ovoid to circular blue forms, about half the diameter of the host cell, or elongate forms stretching across the cell, may be seen.

The developmental cycle is similar to that of *Babesia canis.* The transmitting ticks include *Haemaphysalis bispinosa* and *Rhipicephalus sanguineus* in India, both of which are three-host ticks.

The disease produced by *B. gibsoni* is more chronic than that caused by *B. canis*. There are periodic exacerbations of fever and progressive anaemia and haemoglobinuria. Death may take place after several weeks, or even several months, of illness. There is a marked enlargement of the spleen and the liver, but jaundice is not a frequent clinical sign.

Diminazene (Berenil) is effective against *B. gibsoni* and phenamidine has shown activity against it (Groves & Vanniasingham 1970).

Babesia vogeli. *Host:* Dog. North Africa and southern Asia.

Morphologically it is similar to *B. canis* but larger. It may be a synonym of *B. canis*, and the species was originally established because dogs infected with it failed to show immunity to a strain of *B. canis* transmitted by *Dermacentor*. It is transmitted by *Rhipicephalus sanguineus*, and the clinical entity is comparable to that produced by *B. canis*.

Babesia felis. *Hosts:* domestic cat. Sudan, South Africa. Also in wildcat (Sudan), Sudanese lion (Sudan), Indian leopard (*Panthera pardus fusca*) (India), American puma (*Felis concolor*) (Zoological Gardens, Cairo), American lynx (*Lynx rufus*) (Zoological Gardens, London).

This is a small species, the majority of the forms being round or oval, 1.5–2 μm in diameter; pyriform stages are uncommon. Division is into four, forming a Maltese cross arrangement; however, binary fission is also seen.

The developmental cycle is not known.

The pathogenic effects of *B. felis* consist of anaemia and icterus. Prolonged cases lead to emaciation, splenomegaly and occasionally haemoglobinuria.

The infection is effectively treated by trypan blue and acaprin (quinuronium sulphate) and possibly other babesiacidal drugs would be effective but are untried.

BABESIA SPP. AS ZOONOTIC AGENTS

Ristic and Smith (1974) and Ristic and Lewis (1977) have reviewed the evidence for human babesiosis being derived from animal infections. The initial cases of human babesiosis were reported in splenectomized individuals (Škrabalo & Deanović 1957; Fitzpatrick et al. 1968) but later infections were recorded in persons with a spleen (Western et al. 1970). The species of organism responsible for the human infection varies; in Yugoslavia *B. bovis* was incriminated (Škrabalo & Deanović 1957), in Ireland *B. divergens*, and in persons with a spleen in the north-eastern USA (Nantucket Island) *Babesia microti* of rodents was the causal agent (Healy et al. 1976). Walter (1982) has demonstrated development of *B. microti* in the nymphs of several species of ixodid ticks.

Inapparent or latent infections are probably more common than anticipated. For example, Orsono (1975) found that approximately 38% of individuals in a rural endemic area of animal babesiosis in Mexico showed serological evidence of infection.

A vast number of species of *Babesia* have been reported from wild animals (see Ristic & Lewis 1977) and potentially these may present sources of infection for man. The factors concerned in this are discussed by Ristic and Lewis (1977).

Various species occur in primates and rodents in the wild (e.g. *B. pitheci* in monkeys, *B. microti* and *B. rodhaini* in rodents) but they do not cause disease in laboratory animals.

PIROPLASMOSIS OF POULTRY

Previously, the genus *Aegyptianella*, which contains the species *A. pullorum* and *A. moshkovskii* which occur in poultry, was considered a protozoan and a member of the family Babesiidae. It is now recognized as a form with similarities of the rickettsiae and chlamydiae (Gothe & Kreier 1977) and is not a protozoan. A brief account of the disease caused by these agents is given later (see p. 756).

FAMILY: THEILERIIDAE DU TOIT, 1918

Members of this family are round, ovoid, rod-like or irregular forms, found in lymphocytes, histiocytes and erythrocytes. They do not produce pigment and are transmitted by ixodid ticks. They occur in cattle, sheep and goats and other ruminants.

Three genera are recognized in the family Theileriidae, namely: *Theileria, Haematoxenus* and *Cytauxzoon*. Some authors synonomize *Cytauxzoon* with *Theileria* (e.g. Levine 1973) but Barnett (1977) considers the genera to be separate.

Genus: Theileria Bettencourt, Franca and Borges, 1907

Organisms multiply by schizogony in lymphocytes and finally invade erythrocytes. Transmitted by ticks.

THEILERIA SPECIES OF CATTLE

Theileria parva (Theiler, 1904) Bettencourt, Franca and Borges, 1907

Theileria lawrenci (Neitz, 1955)

Theileria annulata Dschunkowsky and Luhs, 1904

Theileria mutans (Theiler, 1906) Franca, 1909

(Some authors, e.g. Mpangala et al. 1976, assign two subspecies to *T. parva: T. parva parva* and *T. parva lawrenci*).

Theileria parva. *Hosts:* Cattle (*Bos taurus*), East, Central and South Africa; African buffalo (*Syncerus caffer*), East Africa; Indian water buffalo (*Bubalis bubalis*), East and South Africa.

Classically, *T. parva* causes the disease East Coast fever, or bovine theileriasis, which is responsible for high mortality among susceptible and imported stock. The zebu (*Bos indicus*) in endemic areas has a high natural resistance to *T. parva*; however, animals imported into endemic areas are highly susceptible.

Under natural conditions, the distribution of *T. parva* is limited essentially to the distribution of the tick *Rhipicephalus appendiculatus*. This is found in Central, East and southern Africa; it is a three-host tick, and stage to stage transmission of infection occurs. Several other species of tick have been demonstrated as capable of transmitting the infection, and these include the *Rhipicephalus* species: *R. ayrei, R. capensis, R. evertsi, R. jeanelli, R. neavei, R. simus*, and the *Hyalomma* ticks: *H. anatolicum* (syn. *H. excavatum*), *H. dromedarii* and *H. truncatum*. Transmission of the parasite in all is on a stage to stage basis, and the organism does not survive in the ticks for more than one moult.

Morphology (Fig. 3.51). The forms in red blood cells are mainly rod-shaped, 1.5–2 μm by 0.5–1 μm; however, round, oval, comma- and ring-shaped forms may also occur. With Romanowsky stains they show a blue cytoplasm with a red chromatin granule at one end. Several parasites may occur in individual erythrocytes, but there is no evidence of multiplication in the red cells.

The actively multiplying forms of the parasite occur chiefly in the cytoplasm of lymphocytes and occasionally in the endothelial cells, especially of the lymphatic glands and the spleen. These are schizonts (Koch's blue bodies), being circular or irregularly shaped structures about 8 μm in diameter, but they may vary from 2 to 12 μm or more. With Romanowsky stains they show a blue cytoplasm and a varied number of red chromatin granules. Two forms of schizonts are recognized. Those which contain large chromatin granules, 0.4–2 μm in diameter (mean 1.0 μm), are referred to as macroschizonts and produce macromerozoites, 2–2.5 μm in diameter. The other forms

Fig 3.51 *Theileria parva* of cattle. Piroplasm forms in erythrocytes. (×ca 2500) (*From Wenyon 1926*)

contain smaller chromatin granules, 0.3–0.8 μm in diameter (mean 0.5 μm), and are referred to as microschizonts and produce micromerozoites. The latter invade the red blood cells and may represent sexual stages of the parasite (see below).

Developmental cycle in the bovine. Cattle are infected when vector ticks which possess large numbers of uninucleate infective particles (sporozoites) engorge on an animal. Transmission to the animal does not occur immediately on attachment, but sporozoites develop in the salivary gland during the first two to four days of engorgement, of nymph or adult stages. Sporozoites are 1.5 μm in diameter in the tick but they have not been identified in the tissues of the bovine during the first five days after infection. The first visible stages occur in the local lymph nodes five to eight days after infection. Since the tick *R. appendiculatus* frequently attaches to the ear region, the lymph nodes of the neck and prescapular are those that show developmental stages first. The first visible stages are small rounded bodies, 2 μm in diameter, with a round nucleus in a pale-staining cytoplasm (Barnett 1977). They are found in lymphocytes and lymphoblasts and the infected lymph node shows an increase in mitotic figures. Apparently the organism stimulates cells into mitosis. Within the next several days parasites are increasingly apparent in the local nodes and elsewhere in the lymphoid and reticuloendothelial tissues. They appear as multinucleate macroschizonts (Koch's blue bodies). Macroschizonts contain an average of eight nuclei, each approximately 1 μm in size. Barnett (1977) notes that while there is a progressive increase in macroschizonts during the course of infection, the release of merozoites from macroschizonts has not been observed. There is evidence from in vitro culture studies that the replication of the parasite is dependent on replication of the host cell (Malmquist & Brown 1974) but whether this is the situation in vivo is undetermined. Jarrett et al. (1969) studied the kinetics of replication of macroschizonts and concluded that the growth rate was logarithmic and a ten-fold increase occurred every three days ($T_{10} = 3$) regardless of the infecting dose. However, Radley et al. (1974) noted that the T_{10} was dose-dependent. Following macroschizont replication, microschizont formation commences. Jarrett et al. (1969) considered this time-dependent since it occurred about the same number of days after infection regardless of the dose of infective particles or the number of macroschizonts present, and this suggests a switch to microschizont production after a fixed number of replications by macroschizonts. The studies by Radley et al. (1974) confirmed that microschizont production is time-related.

Initially, the microschizont is a compact body in the cytoplasm of lymphocytes or in reticular cells or macrophages (Barnett 1977). They contain many (50–120) nuclear particles and these appear to be derived by budding from a large nuclear mass which may be the denser macroschizont nucleus. The appearance of microschizonts coincides with the appearance of piroplasms in the erythrocytes and morphological similarities between micromerozoites and piroplasms are such that it is concluded the piroplasms are derived from microschizonts.

Hulliger et al. (1966) have obtained microschizontal development in vitro when lymphocyte cultures were incubated at around 42°C. With this temperature of incubation, little multiplication of lymphocytes occurred but the theilerial particles continued to multiply without inhibition, producing stages containing many nuclear particles. Hulliger et al. (1966) comment that although the significance of the high temperature requirements for microschizontal development in culture is not clear, the disease in the animal is characterized by high temperatures of up to 42.4°C during the period of microschizont development.

It is generally considered that the piroplasm stage of *T. parva* in the erythrocyte is a non-dividing form and is the end-form of the infection in the mammalian host. However, Barnett (1977) raises the question of whether *T. parva* can be transmitted by blood inoculation and also whether such inoculated piroplasms can revert to schizonts.

Throughout the years there has been debate whether a sexual cycle of development occurs in *Theileria* spp. This has now been demonstrated (see below) and studies by Schein et al. (1977*a*) have demonstrated with *Theileria annulata* that two different forms were found in the erythrocytes. A comma form and an ovoid form were recognized in ultra-structural studies and there is a strong suggestion that these represent micro- and macrogamonts.

Developmental cycle in the tick. Studies of the life-cycle of *T. parva* in the tick vector date from 1906 when Koch (1906) described developmental stages in ticks. Work on the life-cycle of *T. parva* by Gonder (1910, 1911) and later by Cowdry and Ham (1932) suggested sexual union of parasites and the formation of a zygote, oökinete, sporoblasts and sporozoites. Such work was criticized by Reichenow (1940), who regarded the 'zygotes' as symbionts and the other stages as degenerated tissue cells. Reichenow's criticism delayed the elucidation of the cycle in the tick for more than two decades until Schein et al. (1977*a*) described the sexual development of *T. parva* in *R. appendiculatus*. In summary: after intraerythrocytic stages are ingested by ticks, lysis of erythrocytes occurs and merozoites are liberated which differentiate into sexual stages. In the lumen of the gut of infected nymphs spindle-shaped microgamonts develop from ring forms. These break up into several thread-like microgametes after nuclear division and the development of thread-like cytoplasmic projections (Schein et al. 1977*b*). Ring forms also develop into round forms 3–4 μm in diameter which are considered to be macrogametes. Six days after repletion, zygotes appear in the epithelial cells of the gut. There is an increase in size and progressively denser cytoplasm occurs up to the third day after moulting to adult ticks. By the fifth day after moulting a club-shaped, motile oökinete is produced (Schein et al. 1977*b*). In the case of *T. annulata*, oökinetes pass to the salivary gland and round up into sporonts from which sporogony, with the production of infective particles, results (Schein 1975). Parasites are found in the cytoplasm of the granular cells of the type III acinar cells of the tick salivary gland. Martin et al. (1964) have shown that three distinct types of acinar cells were found in the salivary glands of *R. appendiculatus*, and developmental stages of *T. parva* were found only in the granular secretory cells. An infected cell is often recognized by its hypertrophied nucleus. Although occasionally stages may be found in the unfed tick, preinfective forms are generally seen in the salivary glands only after the infected nymph or adult tick has fed for 24 hours or more. Development at this stage is by multiple fission. Cytomere formation occurs during the second to fourth day of engorgement, and from the fourth day onwards mature sporozoites distend the acinar cells. Sporozoites show a nucleus, the cytoplasm contains rhoptries and probably micronemes (Barnett 1977).

Cultivation. Successful cultivation of macroschizonts of *T. parva* has been achieved (Malmquist & Brown 1974). These culture methods have also been applied to *T. lawrenci* and were developed from cultivation techniques used for *T. annulata* (Tsur-Tchernomoretz & Adler 1965). Once established, infected lymphoblasts replicate readily in suspension and can be grown in large volume for immunodiagnostic and vaccine purposes. Aspects of in vitro culture of *T. parva* are dealt with by Purnell (1977). Culture forms of *T. parva*, when injected into irradiated mice, produce tumour-like masses and Irvin et al. (1975) adapted bovine infected cells in irradiated athymic mice. However, in these cases there was no evidence that the organism had been adapted to mouse lymphoid cells.

Pathogenesis. East Coast fever is a serious disease with high mortality in susceptible stock, being characterized by lymphoid hyperplasia, followed by exhaustion of the lymphoid tissues and leucopenia. In recently imported stock the mortality may reach 90–100%. The zebu (*Bos indicus*) has a high level of natural resistance, and in enzootic areas a calfhood mortality of 5–10% may be expected (Barnett 1963). Where zebu are introduced from non-enzootic areas the mortality may be very high. There is little difference in mortality between calves and adults in highly susceptible European cattle, but where the disease is endemic young cattle are relatively more resistant than older cattle.

The incubation period following exposure is 10–25 days (mean, 13 days), and the acute form of the disease is the most common, lasting 10–23 days. This commences with fever, the temperature rising to 40–41.7°C, which is maintained until death or recovery occurs. A few days after the onset of fever animals cease to eat, there is swelling of the superficial lymph nodes, and there may be a nasal discharge, lacrimation and swelling of the eyelids and ears. The heart beat is rapid, diarrhoea with blood and mucus in the faeces may occur and there is marked emaciation. Lung oedema occurs in the acute form, and this is probably the immediate cause of death.

Post mortem findings in the acute form consist of a marked enlargement of the spleen and the liver, which is also yellowish-brown in colour, friable and shows degeneration. The lymph nodes are markedly swollen and hyperaemic, the kidneys show haemorrhagic or greyish-white 'infarcts', the lungs are congested and oedematous, and there may be fluid in the thorax, pericardium and underneath the kidney capsule. Ulceration of the abomasum and small and large intestines occurs, the ulcers consisting of a central necrotic area surrounded by a haemorrhagic zone.

A subacute form of the disease frequently occurs in calves. The clinical signs are similar to the acute form but not as pronounced, and recovery from this form is more common than with the acute disease.

A mild form of the disease is seen in calves born of immune dams in endemic areas. In this form the temperature rise is mild, lasting a few days to a week, there is swelling of the superficial lymph nodes, and recovery is more common.

In East and Central Africa, the condition of 'turning sickness' has been associated with *T. parva* and also *T. mutans*. Affected animals make circling movements and show abduction of the hind-limbs. On post mortem there is an increase in cerebrospinal fluid, extravasations of blood in various areas of the cortex, and localized necrotic areas occur in the brain in which forms resembling the schizonts of *Theileria* are found. Barnett (1963) considers the condition to be a local or generalized breakdown in immunity. With *T. parva* the condition sometimes occurs when animals are moved from one enzootic area to another in which there is severe reinfection challenge.

Artificial transmission. Normally *T. parva* is not transmissible to other animals by blood inoculation. This is in distinction to the other species of *Theileria* where blood inoculation readily creates an infection. However, it is usually fairly readily transmitted by spleen or lymph node suspensions from infected animals.

Immunology. The immunity which results in a recovered animal is solid, specific, and does not depend upon premunity. The immunity is stable and in an endemic area and for a given strain of organism it is effective for the life of the animal. In the absence of reinfection, immunity wanes slowly, although occasionally animals may lose their immunity within a few months and be susceptible to a second infection. The level of immunity is not dependent on the degree of response of the first infection. However, strains of organisms exist which will cause clinical disease in recovered animals. There is no immunity to reinfection between any of the other *Theileria* species with the exception of *T. lawrenci* which induces a high level of protection against *T. parva*. The immunity is not influenced by splenectomy, and Barnett (1963) considered that the increased resistance of calves to *T. parva* is not due to immunity transferred passively via the colostrum.

Active immunization of animals against *T. parva* is based on the observation by Theiler (1911) that artificially induced infections have a lower mortality rate and a higher recovery rate than tick-induced infections. Immunization of cattle by the injection of spleen or lymph node suspensions was practised for a period in South Africa; however, the level of mortality (25%) was high and, moreover, not all animals infected were immune, only 60–70% being so on exposure to tick infection.

Purnell (1977) has summarized the studies on immunization against *T. parva*. The approach of 'infection and treatment' employs the injection of a standard dose of infective particles (obtained from the supernatant of ground-up ticks after feeding) and concomitant treatment with an oxytetracycline drug. By this method, immunity to homologous challenge was excellent but significant reactions occurred to other strains of *T. parva*. By using a 'cocktail' of strains followed by treatment, Radley et al. (1975) demonstrated that immunized animals could withstand a very severe challenge from infected ticks.

Immunization with lymphoblast cultures of *T. parva* macroschizonts has been investigated, but results are disappointing in that animals responded in a varying manner to the injection of infected lymphoblasts, this being traceable to a relationship of the lymphocyte grouping of the recipient animal to that of the donor animal (Purnell 1977).

Diagnosis. Under general field conditions, the most satisfactory diagnosis is made by the demonstration of the schizonts in material obtained from superficial lymph nodes or by spleen puncture. The forms in the erythrocytes may be difficult to see at times, and in the early part of the infection they may be very few. Differential diagnosis between *T. parva* and other *Theileria* species is not always easy, this being based on the enzootic area, the pathology, the epidemiology and, preferably, cross-immunity tests. Purnell (1977) has reviewed the advances in the immunodiagnosis of East Coast fever. Several tests are available and include complement fixation, indirect fluorescent antibody (IFA), indirect haemagglutination (IHA) and capillary-tube agglutination (CA). The IFA test has been the most effective.

Treatment. Oxytetracycline and chlortetracycline will arrest macro- and microschizont formation, but they must be given prior to infection or at the time infection takes place; if delayed until clinical signs are evident, they are useless. Most clinical cases have passed the point of effective therapy before a diagnosis can be made. However, the recent demonstration that Menoctone (2-hydroxy-3-(8-cyclohexyloctyl)1, 4-naphthoquinone) is active against *T. parva* may indicate that an effective therapeutic agent is available (McHardy et al. 1976). The drug is active against the parasite in vitro and it halted macroschizont production in animals when injected from the first day of fever onwards. Infected animals treated with menoctone were immune to subsequent challenge with the homologous strain. Schein and Voigt (1979) have reported that the anticoccidial substance Halofuginone, at a dose of 1.2 mg/kg orally, was highly effective against theileriasis. Normally lethal experimental infections were treated successfully, even during the advanced stage of the disease.

Control of *T. parva* infection is dependent on tick control and, in some areas, quarantine measures. Transmission of the infection is from stage to stage, and the aim should be to prevent infected tick larvae or nymphs from transmitting the infection to a clean animal. In some parts of East Africa animals are dipped at intervals of three to four days for this purpose. If an infected tick attaches to an animal which is not susceptible to the infection, it then loses its infectivity; however, in the absence of a suitable host an adult tick may remain viable and infective for as long as 18 months.

Theileria lawrenci. *Hosts:* buffalo (*Syncerus caffer*), cattle and water buffalo: East and Central Africa, Angola. Transmitted by *R. appendiculatus* and possibly *R. duttoni*. This organism produces severe and fatal disease in cattle and water buffalo but a mild disease in *S. caffer*. There is cross-immunity between *T. parva* and *T. lawrenci* but not with other species of *Theilera*.

A fatal disease of cattle, 'corridor disease', is encountered in an area called 'the corridor', a stretch of country 100 square miles in extent lying between the Hluhluwe and Umfolozi game reserves in Zululand. Corridor disease is highly pathogenic for cattle, mortality reaching 80% or more, but the African buffalo is highly resistant and serves as a reservoir for the infection. Corridor disease is due to *T. lawrenci*, although Barnett and Brocklesby (1966*a*, *b*, *c*) and Brocklesby and Barnett (1966*a*, *b*) concluded that *T. lawrenci* was a strain of *T. parva*. For the purposes of this volume, it is regarded as a separate species.

Kenya strains of *T. lawrenci* on passage in ticks come to resemble *T. parva*. However, the strain of *T. lawrenci* in Zimbabwe largely retains its characteristics on tick transmission.

Theileria annulata. *Hosts:* bovine, zebu, water buffalo: North Africa, Middle and Far East, Soviet Union and southern Europe. A parasite which occurs in Siberia, Korea and Japan has been considered a form of *T. annulata*, but has also been referred to as *T. sergenti* (Yakimoff and Dekhtereff, 1930). Uilenberg (1981) considers *T. sergenti* a valid species.

The organism produces a highly fatal disease of cattle in North Africa and is transmitted by ticks of the genus *Hyalomma*. Elsewhere, it produces a moderate disease in cattle and a mild infection in the buffalo. In Siberia '*T. sergenti*' produces mild to moderate clinical disease.

Morphology. The piroplasm forms in the red blood cells are more or less indistinguishable from those of *T. parva* but more commonly occur as round, oval- or ring-shaped (0.5–1.5 μm) forms. Rod shapes, commas (1.6 μm) and anaplasma-like organisms may also be found, the latter measuring 0.5 μm. The erythrocytic forms undergo binary fission with the formation of two daughter individuals. Division into four to produce a cross may also occur. Macroschizonts and microschizonts are found in the lymphocytes of the spleen and lymph nodes, being similar to those of *T. parva*. *Theileria annulata* is readily transmissible by blood passage, and schizonts are fairly numerous in the circulating blood.

Developmental cycle. The developmental cycle in the vertebrate host is comparable to that of *T. parva*. Recent work on the growth of *T. annulata* in tissue culture systems has shown that the organism may be propagated for several serial passages in lymphoid cells (Hulliger et al. 1965; Tsur-Tchernomoretz & Adler 1965).

The tick vectors of *T. annulata* are all members of the genus *Hyalomma*. These include: *H. detritum* (North Africa, Soviet Union), *H. dromedarii* (Central Asia), *H. excavatum* and *H. turanicum* (Middle East), *H. savignyi* (syn. *H. marginatum*) (Middle East, India), *H. plumbeum plumbeum* and *H. scupense* (Soviet Union).

Extensive studies of the developmental cycle of *T. annulata* in ticks have been carried out by Schein and his colleagues. They (Schein et al. 1977*a*) have demonstrated that two forms of erythrocytic stages occur which probably represent gamonts from which gamete stages are formed when ingested by the tick. In the tick (*Hyalomma anatolicum excavatum*) gamete formation occurs and zygote formation is deduced, even though syngamy has not been observed. Zygotes increase in size to reach a diameter of approximately 10 μm at 12 days after repletion. At 13 days after repletion invagination occurs and a club-shaped form is produced which is considered to be the oökinete of *T. annulata*. These oökinetes pass to the acinar cells of the salivary glands (types II and III), round up into sporonts and commence sporogony (Schein 1975; Mehlhorn & Schein 1977).

Pathogenesis. In general, the disease entity is comparable to East Coast fever. Mortality varies considerably, being 10% in some areas and up to 90% in others. For this reason there has been a tendency to ascribe Algerian theileriasis to a mild strain, referred to as *Theileria dispar* by French workers. The more pathogenic strains occur in the Soviet Union, Israel, Iran and India.

The disease entity may be acute, subacute or chronic. The acute disease occurs in all breeds and all ages of cattle as well as buffalo and zebu, but the latter two recover more readily from the disease. The incubation period is 9–25 days, and the disease may last as little as three to four days in the acute form or may be prolonged for about 20 days. A marked rise in body temperature, reaching 40–41.5°C, is followed by depression, lacrimation, nasal discharge and swelling of the superficial lymph nodes. Emaciation is rapid and haemoglobinuria may occur. The severity of the disease does not necessarily correspond to the extent of the parasitaemia. An animal may be seriously ill when less than 25% of the blood cells are infected with piroplasm stages, and less severe reactions may occur where 45% of the red blood cells are infected.

The post mortem findings consist of a markedly enlarged spleen and liver, there are 'infarcts' in the kidneys, the lungs are usually oedematous, and the lymph nodes may be swollen, especially in the acute form of the disease. The mucous membranes show icterus and often petechiae, and those of the abomasum and small intestine are swollen, reddened and show characteristic ulcers 2–12 mm in diameter and surrounded by a zone of inflammation. Necrotic infarcts of the brain may occur in a small number of cases (Barboni 1942), and cutaneous lesions have been reported by Tsur-Tchernomoretz et al. (1960), schizonts being found in the dermis.

Immunology. Recovery from *T. annulata* infection leads to the development of premunity, and infection can be transmitted from such animals by blood inoculation. There is no cross-immunity between *T. annulata, T. mutans* or *T. parva*. Practical vaccination procedures, based upon the difference in virulence of various strains, have been practised in Algeria and in Israel. Field strains of low virulence are used, but since this may not protect all calves from virulent field strains, it is often necessary to boost immunity by a virulent strain 1–2 months after the use of the mild strain. Advantages of this vaccination include the fact that the low pathogenic strain can be maintained in calves by blood or tissue inoculation and that serial passage results in the loss of the erythrocyte phase after 10–13 passages. Thus, vaccinated animals are not infective for ticks.

The duration of immunity to *T. annulata* is less than that seen in *T. parva*, and Barnett (1963) estimates that 10% of animals may be completely susceptible 17 months after the first infection; however, Sturman (1959) believed protection induced by vaccination lasted two to three years under conditions of field exposure. Immunity to *T. annulata* is less stable than that to *T. parva*. Relapse and death may occur after a considerable period of premunity.

Diagnosis and treatment. Diagnosis is based on the demonstration of parasites in the red blood cells or in smears of material obtained from lymph nodes or spleen. Differentiation between *T. annulata* and *T. parva* is not easy, and diagnosis is based on the evaluation of the enzootic parasitic conditions in the area. It is not uncommon for *T. annulata* to occur along with *Babesia* or *Anaplasma*, and the disease entity may be a combination of two or all of these.

Chlortetracyclines and oxytetracyclines given throughout the incubation period reduce the clinical reaction by decreasing the degree of parasitaemia. These drugs have little effect once clinical disease is apparent. Recently Menoctone (McHardy et al. 1979) and Halofuginone (Schein & Voigt 1979) have been shown to be active against clinical infections with *T. annulata*.

The control of *T. annulata* is based on tick control measures. Immunization is practised in Algeria and Israel, and in some areas this has greatly reduced the incidence of the disease.

Theileria mutans. *Hosts:* cattle. Africa, Asia, Australia, Soviet Union. Splitter (1950) found the organism in a splenectomized calf in the USA, Kreier et al. (1962) have found a *Theileria* species in splenectomized deer in the USA, and Hignett (1953) reported the parasite in England in cattle which had been artificially infected with *Babesia divergens* but it is now widely distributed in the British Isles.

It is the cause of benign bovine theileriasis and it is almost always non-fatal.

Morphologically, this parasite is indistinguishable from the other species of *Theileria*. The forms in the erythrocytes are round, oval, pyriform or anaplasma-like, and measure 1–2 μm in diameter, two or four parasites occurring in a single red blood cell. Schizonts are not readily detectable but when they are found they occur in the lymphocytes of the spleen and lymph nodes and measure about 8 μm in diameter, but they may be up to 20 μm in diameter. They resemble the macroschizonts of the more pathogenic species.

Developmental cycle. Little work has been done on the endogenous cycle in the mammalian host, and it is assumed to be comparable to that in the other species.

The ticks responsible for the transmission of *T. mutans* are listed by Neitz (1956) and include: *Rhipicephalus appendiculatus* and *R. evertsi* (South Africa), *Haemaphysalis bispinosa* (Australia) and *H. punctata* (Soviet Union). In Great Britain *Haemaphysalis punctata* is the transmitting tick. *Boophilus annulatus* and *B. microplus*, both one-host ticks, have been incriminated as vectors (experimentally), but since the infection is transmitted from stage to stage only the role of these ticks in natural transmission requires clarification.

The developmental stages of *T. mutans* in the tick *H. bispinosa* were studied by Riek (1966). Parasites similar to those in the erythrocytes were found free in the gut contents of larval ticks on repletion, but 24 hours later only small numbers were found despite a parasite density of more than 10% in blood. 24–48 hours after repletion small numbers of organisms were seen inside the epithelial cells of the gut wall, but no development was detected. By ten days, spherical bodies, up to 15 μm in diameter, with a homogeneous cytoplasm and a nucleus about 3 μm in diameter were seen in several squash preparations. In such preparations these were extracellular, but Riek considered they had been liberated from the epithelial cells when the preparation was made. The next recognizable developmental forms were found in the salivary glands of nymphs during the 24 hours after attachment. Thereafter, development was rapid, and by 48 hours to four to five days of multiple fission produced bodies, 40 μm by 30 μm, which contained large numbers of infective forms. Of the three types of acini which had been found in the salivary glands of *H. bispinosa*, developmental stages occurred only in the granular secretory cells, only a few acini were infected, and only one or two cells in these were parasitized.

An indirect fluorescent antibody test for *T. mutans* has been described by Morzaria et al. (1977). The test is specific and does not cross-react with *Babesia major*, another parasite of cattle in Great Britain.

The pathogenic effects of *T. mutans* are minimal, although subclinical anaemia may occur in experimental infections.

THEILERIA SPECIES OF SHEEP

Two species of *Theileria* have been described from sheep and goats:

Theileria hirci Dschunkowsky and Urodschevich, 1924
Theileria ovis Rodhain, 1916

Theileria hirci. *Hosts:* sheep, goats; North and East Africa, Iraq, Turkey, southern Soviet Union and Greece.

The disease may be highly fatal, mortality ranging from 50 to 100%.

Morphology. Erythrocytic piroplasms are round to oval in the majority of cases, about 18% are rod-shaped and a small percentage are anaplasma-like. Round forms measure 0.6–2 μm in diameter, they may be found in pairs or in fours and multiplication takes place in the erythrocytes. Schizonts occur in the lymphocytes of the spleen and lymph nodes. They range in size from 4 to 10 μm (mean 8 μm) and contain up to 80 chromatin granules, 1–2 μm in diameter. Both macroschizonts and microschizonts occur.

The tick vector of *T. hirci* has yet to be established, but in enzootic areas it is likely to be *Rhipicephalus bursa* (Neitz 1956).

Pathogenesis. The disease is highly pathogenic for sheep and goats, mortalities up to 100% having been reported in endemic areas. The infection is mild in young lambs and kids, possibly due to maternal immunity. An acute form of the disease is more usual, but subacute and chronic forms have been observed. In general, it resembles East Coast fever; there is high fever associated with listlessness, a nasal discharge, jaundice, petechial haemorrhages in submucous, subserous and subcutaneous tissues, marked enlargement of the spleen and lymph nodes, the kidneys are enlarged and pale and show infarcts, and there may be a transitory haemoglobinuria. Animals which recover are premune, and there is no cross-immunity with *Theileria ovis.*

Diagnosis of the infection is based on the detection of piroplasms in blood smears or schizonts in lymph node and spleen smears.

There is no known treatment to date, but Menoctone and Halofuginone might be of value (see *T. parva*). Control depends on tick control.

Theileria ovis. *Hosts:* sheep, goats. Africa, Asia, India, Soviet Union, parts of Europe.

It is much more widely distributed than *T. hirci* and causes a benign disease.

Morphologically, the organism resembles *T. hirci*, but the blood forms are relatively scarce, as are the schizonts which occur in the lymph nodes.

The ticks responsible for transmission include *Rhipicephalus bursa* (Soviet Union), *R. evertsi* (South Africa), *Dermacentor sylvarum, Haemaphysalis sulcata* and nymphs of *Ornithodorous lahorensis* (Soviet Union) (Bitukov 1953).

The pathogenic entity is mild, and there is seldom mortality or any distinct clinical signs.

THEILERIDAE OF OTHER RUMINANTS

Several species of *Theileria* have been reported from various African ruminants. Neitz (1959) has given an account of these.

Howe (1971) provides an extensive list of the artiodactyla in which erythrocytic stages of unidentified *Theileria* species have been observed.

The following species are of special interest:

Theileria camelensis Yakimoff, 1917 (Camel)
Theileria cervi Bettencourt, Franca and Borges, 1907 (White-tailed Deer)
Theileria tarandi Kertzelli, 1909 (Reindeer)

Theileria camelensis. *Host:* camel, Egypt, Somalia and Turkestan SSR. Transmission is presumed to be by *Hyalomma dromedarii*. Schizonts have not been described and Wenyon (1926) suggests that the organism found by Yakimoff (1917) was *Babesia equi*. Barnett (1977) notes that a fatal infection with this parasite was reported in a llama in an Egyptian zoo.

Theileria cervi. This organism was found in the splenectomized white-tailed deer (*Dama virginiana*) by Kreier et al. (1962) and was subsequently identified as *T. cervi* by Schaeffler (1962). It cannot be transmitted to the ox or the sheep. It is non-pathogenic. The erythrocytic stages are pleomorphic, occurring as bipolar, comma or signet ring forms in addition to the usual rod- and oval-shaped forms. Maltese-cross forms occur as the result of multiplication of piroplasms in the erythrocyte.

Theileria tarandi. *Host:* Reindeer, northern Soviet Union. The organism is probably transmitted by *Ixodes persulcatus* (Barnett 1977). Schizonts have not been described but piroplasms are numerous in the acute disease, which has been described by Kertzelli (1909).

Genus: Haematoxenus Uilenberg, 1964

This genus resembles *Theileria* but the erythrocytic stages have a rectangular 'veil' extending from their side. This veil stains violet with Giemsa stains. Two species are recognized in this genus:

Haematoxenus veliferus Uilenberg, 1964
Haematoxenus separatus Uilenberg and Andreasen, 1974

Haematoxenus veliferus. *Hosts:* ox, zebu; Madagascar, East Africa, Central African Republic, Chad, Nigeria. As far as is known the organism is non-pathogenic, but is very common in the above areas and has been found in splenectomized animals and those exposed to natural tick infection (Mpangala et al. 1976). Transmission is by *Amblyomma variegatum*. Mpangala et al. (1976) demonstrated that infection with *H. veliferus* did not induce antibodies which cross-reacted with either *T. parva* or *T. mutans*.

Haematoxenus separatus. This has been reported from sheep in Tanzania. It is non-pathogenic and is transmitted by *Rhipicephalus evertsi* (Uilenberg & Andreasen 1974).

Genus: Cytauxzoon Neitz and Thomas, 1948

This genus was created by Neitz and Thomas (1948) to differentiate those *Theileria* which underwent schizogony in histiocytes rather than lymphocytes and multiplied by fission in the erythrocytes. The genus has not been recorded from domestic livestock. Species of interest include:

Cytauxzoon sylvicaprae Neitz and Thomas, 1948
Cytauxzoon strepsicerosi Neitz and de Lange, 1956
Cytauxzoon taurotragi Martin and Brocklesby, 1960
Cytauxzoon sp. Wagner, 1976

Cytauxzoon sylvicaprae. *Host:* duiker; South Africa. Organisms indistinguishable from the erythrocytic stages of the *Theileria* of cattle. Anaemia, icterus, pulmonary congestion, greyish liver and an enlarged spleen occur (Neitz & Thomas 1948).

Cytauxzoon strepsicerosi. *Host:* kudu; Transvaal. Affected animals show fever, anorexia, icterus, anaemia, swelling of lymph nodes.

Cytauxzoon taurotragi. This organism was originally described by Martin and Brocklesby (1960) and Brocklesby (1962) from a fatal infection in a yearling eland (*Taurotragus oryx pattersonianus*). Brocklesby (1962) reported schizonts in sections of the liver, lung and lymph node but not in the spleen or kidney. Erythrocytic stages of the parasite were indistinguishable from those of *T. parva*, and the schizonts of the liver were of the cytauxzoon type.

Cytauxzoon sp. *Host:* cat. A fatal cytauxzoon-like disease of cats in the USA was reported by Wagner (1976). The early cases were reported from south-western Missouri, but cases have now been discovered in Texas, Georgia and Arkansas. Clinical signs include lethargy, pale icteric mucous membranes and fever. The disease occurs in cats which are allowed to roam rural wooded areas and hence tick transmission of the infection is probable. Because the disease is so acute in the cat, Wagner (1976) suggests that another animal species serves as the primary host for the parasite.

Large numbers of schizonts characteristic of *Cytauxzoon* sp. are present in the liver, lung, spleen and lymph nodes of cats and they occur in the reticuloendothelial cells. Erythrocytic piroplasms occur as ring forms.

Wightman et al. (1977) believe that the erythrocytic phase of the infection causes haemolytic anaemia while the tissue phase produces severe circulatory impairment. Indeed, massive accummulations of cytauxzoon schizonts parasitize the reticuloendothelial cells lining blood vessels and partially or completely occlude major blood channels of the spleen, liver and lungs.

The course of the disease lasts for about one week. Prognosis is poor and treatment has been to no avail.

REFERENCES

PIROPLASMIDA (BABESIIDAE)

Abramov, I. V. (1952) Summary of the 36th Plenary Session of the USSR Leningrad Academy Veterinary Section, on Protozoan Diseases. *Veterinariya Moscow*, **9**, 55–57. (Abstr. *Vet. Bull.* (1953), 23, 140.)

Annable, C. R. & Ward, P. A. (1974) Immunopathology of the renal complications of babesiosis. *J. Immunol.*, **112**, 1–8.

Arnold, R. M. (1948) Resistance to tick-borne disease. *Vet. Rec.*, **60**, 426.

Arthur, D. R. (1966) The ecology of ticks with reference to the transmission of protozoa. In: *The Biology of Parasites*, ed. E. J. L. Soulsby, pp. 61–84. New York: Academic Press.

Bodin, S. & Hildar, G. (1963) Immunization of Swedish cattle against piroplasmosis. *Proc. 9th Nordic vet. Congr. Copenhagen*, **1**, 328–333.

Bool, P. H., Goedbloed, E. & Keidel, H. J. W. (1961) The bovine babesia species in the Netherlands: *Babesia divergens* and *Babesia major*. *Tijdschr. Diergeneesk.*, **86**, 28–37.

Boreham, P. F. L. & Wright, I. G. (1976) The release of pharmacologically active substances in parasitic infections. In: *Progress in Medicinal Chemistry*, eds G. P. Willis & G. B. West, vol. II, pp. 160–204. Amsterdam: North Holland Publishing.

Brocklesby, D. W., Purnell, R. E. & Selwood, S. A. (1972) The effect of irradiation on intra-erythrocytic stages of *Babesia major*. *Br. vet. J.*, **128**, iii–v.

Callow, L. L. (1955) *Babesia bigemina* in ticks grown on non-bovine hosts and its transmission to these hosts. *Parasitology*, **55**, 275–381.

Callow, L. L. (1977) Vaccination against bovine babesiosis. In: *Immunity to Blood Parasites of Animals and Man*, ed. L. H. Miller, J. A. Pino, & J. J. McKelvey jun., pp. 121–149. New York and London: Plenum.

Callow, L. L. (1974) Epizootiology, diagnosis and control of babesiosis and anaplasmosis. *Anim. Quarantine*, **3**, 6–12.

Callow, L. L. & Johnston, L. A. Y. (1963) *Babesia* spp. in the brains of clinically normal cattle and their detection by a brain smear technique. *Aust. vet. J.*, **39**, 25–31.

Callow, L. L. & McGavin, M. D. (1963) Cerebral babesiosis due to *Babesia argentina*. *Aust. vet. J.*, **39**, 15–20.

Callow, L. L. & McGregor, W. (1970) The effect of imidocarb against *Babesia argentina* and *Babesia bigemina* infections of cattle. *Aust. vet. J.*, **46**, 195–200.

Callow, L. L., McGregor, W., Parker, R. J. & Dalgliesh, R. J. (1974) The immunity of cattle to *Babesia argentina* after drug sterilization of infections of varying duration. *Aust. vet. J.*, **50**, 6–11.

Callow, L. L. & Mellors, L. T. (1966) A new vaccine for *Babesia argentina* infection prepared in splenectomized calves. *Aust. vet. J.*, **42**, 464–465.

Carmichael, J. (1956) Treatment and control of babesiosis. *Ann. N.Y. Acad. Sci.*, **64**, 147–151.

Cerruti, C. G. (1939) Recherches sur les piroplasmoses du porc. *Ann. Parasit.*, **17**, 114–136.

Chapman, W. E. & Ward, P. A. (1977) *Babesia rodhaini*: requirement of complement for penetration of human erythrocytes. *Science, N.Y.*, **196**, 67.

Curnow, J. A. (1968) *In vitro* agglutination of bovine erythrocytes infected with *Babesia argentina*. *Nature, Lond.*, **217**, 267–268.

Curnow, J. A. (1973) Studies on the epizootiology of bovine babesiosis in north eastern New South Wales. *Aust. vet. J.*, **49**, 284–289.

Dalgliesh, R. J. & Stewart, N. D. (1977) Tolerance to imidocarb induced experimentally in tick-transmitted *Babesia argentina*. *Aust. vet. J.*, **53**, 176–180.

Daly, G. D. & Hall, W. T. K. (1955) A note on the susceptibility of British and some Zebu-type cattle to tick fever (Babesiosis). *Aust. vet. J.*, **31**, 152.

Davies, S. F. M., Joyner, L. P. & Kendall, S. B. (1958) Studies on *Babesia divergens* (McFadyean and Stockman, 1911). *Ann. trop. Med. Parasit.*, **52**, 206–215.

Dennis. E. W. (1932) The life-cycle of *Babesia bigemina* (Smith and Kilborne) of Texas cattle-fever in the tick *Margarpus annulatus* (Say) with notes on embryology of *Margaropus*. *Univ. Calif. (Berkeley) Publ. Zool.*, **36**, 263–298.

Enigk, K. (1943) Die Überträger der Pferdepiroplasmose, ihre Verbreitung und Biologie. *Arch. wiss. prakt. Tierheilk.*, **78**, 209–240.

Enigk, K. (1944) Weitere Untersuchungen zur Überträgerfrage der Pferdepiroplasmose. *Arch. wiss. prakt. Tierheilk.*, **79**, 58–80.

Enigk, K. (1951) Der Einfluss des Klimas auf das Auftreten der Pferdepiroplasmosen. *A. Tropenmed. Parasit.*, **2**, 401–410.

Enigk, K. & Friedhoff, K. (1962*a*) Zur Wirtsspezifität von *Babesia divergens* (Piroplasmidea). *Z. Parasitkde*, **21**, 238–256.

Enigk, K. & Friedhoff, K. (1962*b*) *Babesia capreoli* n. sp. beim Reh (*Capreolus capreolus* L). *Z. Tropenmed. Parasit.*, **13**, 8–20.

Ershov, V. S. (1956) *Parasitology and Parasitic Diseases of Livestock*. Moscow: State Publishing House for Agricultural Literature. Engl. ed.—Washington DC: National Science Foundation and Department of Agriculture, USA.

Ewing, S. A. (1965) Method of reproduction of *Babesia canis* in erythrocytes. *Am. J. vet. Res.*, **26**, 727–733.

Ewing, S. A. & Buckner, R. G. (1965) Manifestations of babesiosis, ehrlichiosis and combined, infections in the dog. *Am. J. vet. Res.*, **26**, 815–828.

Fitzpatrick, J. E. P., Kennedy, C. C., McGeown, M. G., Oreopouldous, D. G., Robertson, J. H. & Soyannwo, M. A. (1968) Human case of piroplasmosis (babesiosis). *Nature, Lond.*, **217**, 861–862.
Frerichs, W. M., Allen, P. C. & Holbrook, A. A. (1973) Equine piroplasmosis (*Babesia equi*): therapeutic trials of imidocarb dihydrochloride in horses and donkeys. *Vet. Rec.*, **93**. 73–75.
Friedhoff, K. T. (1969) Lichtmikroskopische Untersuchungen über die Entwicklung von *Babesia ovis* (Piroplasmidea) in *Rhipicephalus bursa* (Ixodoidea). I. Die Entwicklung in weiblichen Zecken nach der Repletion. *Z. Parasitenk.*, **32**, 191–219.
Garnham, P. C. C. (1962) Discussion in: *Aspects of Disease Transmission by Ticks*, ed. D. R. Arthur, Symposium No. 6., pp. 257–258. London: Zoological Society.
Garnham, P. C. C. & Bray, R. S. (1959) The susceptibility to piroplasms. *J. Protozool.*, **6**, 352–355.
Goldman, M. & Bukovsky, E. (1975) Extraction and preliminary use for diagnosis of soluble precipitating antigen from *Babesian bigemina*. *J. Protozool.*, **22**, 262–264.
Gothe, R. & Kreier, J. P. (1977) Aegyptianella, Eperythrozoon and Haemobartonella. In: *Parasitic Protozoa*, ed. J. P. Kreier, vol. IV, pp. 251–294. New York: Academic Press.
Groves, M. G. & Vanniasingham, J. A. (1970) Treatment of *Babesia gibsoni* infections with phenamidine isethionate. *Vet. Rec.*, **86**, 8–10.
Hall, W. T. K. (1960) The immunity of calves to *Babesia argentina* infection. *Aust. vet. J.*, **36**, 361–366.
Hall, W. T. K., Tammemagi, L. & Johnston, L. A. Y. (1968) Bovine babesiosis: the immunity of calves to *Babesia bigemina* infection. *Aust. vet. J.*, **44**, 259–264.
Healy, G. R., Spielman, A. & Gleason, N. (1976) Human babesiosis: reservoir of infection on Nantucket Island. *Science, N.Y.*, **192**, 479–480.
Hirato, K., Ninomiya, Y., Uwano, Y. & Kuth, T. (1945) Studies on the complement fixation reaction for equine piroplasmosis. *Jap. J. vet. Sci.*, **7**, 197–205.
Holbrook, A. A., Anthony, D. W. & Johnson, A. J. (1968*a*) Observations on the development of *Babesia caballi* (Nuttall) in the tropical horse tick *Dermacentor nitens* Neumann. *J. Protozool.*, **15**, 391–396.
Holbrook, A. A., Johnson, A. J. & Madden, P. A. (1968*b*) Equine piroplasmosis: intraerythrocytic development of *Babesia caballi* (Nuttall) and *Babesia equi* (Laveran). *Am. J. vet. Res.*, **29**, 297–303.
Huxsoll, D. L., Hildebrandt, P. K., Nims, R. M., Ferguson, J. A. & Walker, J. S. (1969) *Ehrlichia canis*—the causative agent of a haemorrhagic disease of dogs. *Vet. Rec.*, **85**, 587.
Iturri, G. M. & Cox, H. W. (1969) Glomerulo-nephritis associated with actue hemosporidian infection. *Milit. Med.* (Special issue), **134**, 1119–1128.
Johnston, L. A. Y. (1967) Epidemiology of bovine babesiosis in northern Queensland. *Aust. vet. J.*, **43**, 427–431.
Joyner, L. P., Davies, S. F. M. & Kendall, S. B. (1963) The experimental transmission of *Babesia divergens* by *Ixodes ricinus*. *Exp. Parasit.*, **14**, 367–373.
Koch, R. (1906) Beiträge zur Entwicklungsgeschichte der Piroplasmen. *Z. Hyg. Infektkr.*, **54**, 1–9.
Kobalskii, N. A., Gaidukov, A. K. & Tarverdyan, T. N. (1963) Duration of immunity and carriage of piroplasms in animals recovered from piroplasmosis. *Trudy Vsesoyuz Inst. eksp. Vet.*, **28**, 170–176. (Abstr. *Vet. Bull.*, **33**, 674.)
Kudo, R. R. (1966) *Protozoology*, 5th ed. Springfield, Ill.: Charles C. Thomas.
Kuttler, K. L., Graham, O. H. & Trevino, J. L. (1975) The effect of imidocarb treatment on *Babesia* in the bovine and the tick (*Boophilus microplus*). *Res. vet. Sci.*, **18**, 198–200.
Lawrence, D. A. & Shone, D. K. (1955) Porcine piroplasmosis. *Babesia trautmanni* infection in Southern Rhodesia. *J. S. Afr. vet. med. Ass.*, **26**, 89–93.
Levine, N. D. (1973) *Protozoan Parasites of Domestic Animals and of Man*, 2nd ed., p. 406. Minneapolis: Burgess.
Li, P. N. (1958) Developmental forms of *Babesiella ovis* in the larvae and nymphae of *Rhipicephalus bursa*. *Nauch. Trud. Ukrainsk. Inst. exp. Vet.*, **24**, 283–287.
Maegraith, B. G., Gilles, H. M. & Dekavul, K. (1957) Pathological processes in *Babesia canis* infections. *Z. Tropenmed. Parasit.*, **8**, 485–514.
Mahoney, D. F. (1962) Bovine babesiosis: diagnosis of infection by a complement fixation test. *Aust. vet. J.*, **38**, 48–52.
Mahoney, D. F. (1973) Babesiosis of cattle. *Aust. Meat Res. Comm. Rev.*, No. **12**, pp. 1–21.
Mahoney, D. F. (1977) Babesia of domestic animals. In: *Parasitic Protozoa*, ed. J. P. Kreier, vol. IV, pp. 1–52. New York: Academic Press.
Mahoney, D. F., Wright, I. G. & Mirre, G. B. (1973) Bovine babesiosis: The persistence of immunity to *Babesia argentina* and *B. bigemina* in calves (*Bos taurus*) after naturally acquired infection. *Ann. trop. Med. Parasit.*, **67**, 197–203.
Malherbe, W. D. (1956) The manifestations and diagnosis of babesia infections. *Ann. N.Y. Acad. Sci.*, **64**, 128–146.
Malherbe, W. D. & Parkin, B. S. (1951) A typical symptomatology in *Babesia canis* infection. *J. S. Afr. vet. med. Ass.*, **22**, 25–36.
Markov, A. A. & Abramov, I. V. (1957) Peculiarities of circulation of *Babesiella ovis* (Babes, 1892) in the tick *Rhipicephalus bursa* Can. et Fanz. 1877. *Veterinariya Moscow*, **34**, 27–30.
Markov, A. A. & Abramov, I. V. (1966) Reciprocal adaptations of certain blood parasites and their hosts. *Proc. 1st Int. Congr. Parasitology* (Rome), 1964, **1**, 268–269.
Morzaria, S. P., Brocklesby, D. N. & Harradine, D. L. (1977) Evaluation of the indirect fluorescent antibody test for *Babesia major* and *Theileria mutans* in Britain. *Vet. Rec.*, **100**, 484–487.
Muratov, E. A. & Kheisin, E. M. (1959) Development of *Piroplasma bigeminum* in the tick *Boophilus calcaratus*. *Zool. Zh.*, **38**, 970–986.
Müller, B. & Rapp, J. (1971) Babesiose als Todesursache bei einem Rehkitz. Tierärtz. Umsch., **26**, 314–315.
Neitz, W. O. (1956) Classification, transmission and biology of piroplasms of domestic animals. *Ann. N.Y. Acad. Sci.*, **64**, 56–111.
Orsono, M. B. (1975) Public health importance of babesiosis. (Quoted by Ristic and Lewis, 1977.)
Petrov, V. G. (1941) Development of *Babesiella bovis* in the tick *Ixodes ricinus* L. and a method of investigating the ability of ticks to transmit viruses. *Vestnik. Sel. skokhoz. Nauk. Vet.*, **3**, 136.
Pierce, A. E. (1956) Protozoan diseases transmitted by the cattle tick. *Aust. vet. J.*, **32**, 210–215.
Piercey, S. E. (1947) Hyper-acute canine babesia (tick fever). *Vet. Rec.*, **59**, 612–613.
Polyanskii, Y. I. & Kheisin, E. M. (1959) Some data on development of *Babesiella bovis* in tick-vectors. *Trudy. Karel. Filiala. Akad. Nauk. SSSR*, **14**, 5–13.
Potgieter, F. T. & Els, H. J. (1977) Light and electron microscope observations on the development of *Babesia bigemina* in larvae, nymphae and non-replete females of *Boophilus decoloratus*. *Onderstepoort J. vet. Res.*, **44**, 213–232.
Puccini, V., Muzio, F. & Gianubilo, G. (1958) Efficacia del 'Berenil' nella cura della Piroplasmosi suina da *Piroplasma trautmanni* e da *Babesiella perroncitoi*. *Vet. Ital.*, **9**, 611–616.
Purchase, H. S. (1947) Cerebral babesiosis in dogs. *Vet. Rec.*, **59**, 269–270.
Purnell, R. E., Lewis, D., Brabazol, A., Francis, L. M. A., Young, E. R. and Grist, C. (1981) Field use of an irradiated blood vaccine to protect cattle against redwater (*Babesia divergens* infection) on a farm in Dorset. *Vet. Rec.*, **108**, 28–31.
Purnell, R. E., Lewis, D. & Brocklesby, D. W. (1979) *Babesia major*: protection of intact calves against homologous challenge by the injection of irradiated piroplasms. *Int. J. Parasit.*, **8**, 69–71.
Purnell, R. E., Lewis, D. & Brocklesby, D. W. (1980) Bovine babesiosis: protection of cattle by the inoculation of irradiated piroplasms. In: *Isotope and Radiation Research on Animal Diseases and Their Vectors*, pp. 77–94. SM 240/1). Geneva: International Atomic Energy Agency.
Regendanz, P. & Reichenow, E. (1933) Die Entwicklung der Piroplasmen. *Z. Bakt. Parasit.*, Abt. I. orig., **135**, 108–119.
Rehbein, G., Zweygarth, E., Voigt, N. P. & Schein, E. (1982) Establishment of *Babesia equi*-infected lymphoblastoid cell lines. *Z. Parasitenk.*, **67**, 125–127.
Riek, R. F. (1963) Immunity to babesiosis. In: *Immunity to Protozoa*, eds P. C. C. Garnham, A. E. Pierce & I. Roitt, pp. 160–79. Oxford: Blackwell Scientific Publications.
Riek, R. F. (1963) Immunity to babesiosis. In: *Immunity to Protozoa*, eds P. C. C. Garnham, A. E. Pierce & I. Roitt. Oxford: Blackwell Scientific Publications.
Riek, R. F. (1964) The life cycle of *Babesia begemma* (Smith & Kilborne, 1893) in the tick vector *Boophilus microplus* (Canastrini). *Aust. J. agric. Res.*, **15**, 802–821.
Riek, R. F. (1966) The development of *Babesia* spp. and *Theileria* spp. in ticks with special reference to those occurring in cattle. In: *Biology of Parasites*, ed. E. J. L. Soulsby, pp. 15–32. New York: Academic Press.
Ristic, M. (1966) The vertebrate developmental cycle of *Babesia* and *Theileria*. In: *Biology of Parasites*, ed. E. J. L. Soulsby, pp. 128–41. New York: Academic Press.
Ristic, M. & Lewis, G. E., Jr. (1977) Babesia in man and wild and laboratory-adapted mammals. In: *Parasitic Protozoa*, ed. J. P. Kreier, vol. IV, pp. 53–76. New York: Academic Press.
Ristic, M., Oppermann, J., Sibinovic, S. & Phillips, T. N. (1964) Equine piroplasmosis: a mixed strain of *Piroplasma caballi* and *Piroplasma equi* isolated in Florida and studied by the fluorescent-antibody technique. *Am. J. vet. Res.*, **25**, 15–23.

Ristic, M. & Sibinovic, K. H. (1964) Equine babesiosis. Diagnosis by a precipitation in gel and by a one-step fluorescent antibody inhibition test. *Am. J. vet. Res.*, **25**, 1519–1526.
Ristic, M. & Smith, R. D. (1974) Zoonoses caused by Haemoprotozoa. In: *Parasitic Zoonoses*, ed. E. J. L. Soulsby, pp. 41–63. New York: Academic Press.
Rosenbusch, F. (1927) Study of Tristeza (Piroplasmosis): development of *Piroplasma bigeminum* in the tick (*Boophilus microplus* Can. Lah.). *Rev. Univ. Buenos Aires*, **5**, 863–867.
Ross, D. R. & Mahoney, D. F. (1974) Bovine Babesiosis: computer simulation of *Babesia argentina* parasite rates in *Bos taurus* cattle. *Ann. trop. Med. Parasit.*, **68**, 385–392.
Ross, J. P. J. & Lohr, K. F. (1968) Serological diagnosis of *Babesia bigemina* infection in cattle by the indirect fluorescent antibody test. *Res. vet. Sci.*, **9**, 557–562.
Sergent, E. (1963) Latent infections and premunition. Some definitions of microbiology and immunology. In: *Immunity to Protozoa*, eds P. C. C. Garnham, A. E. Pierce & I. Roitt, pp. 39–47. Oxford: Blackwell Scientific Publications.
Schein, E., Rehbein, G., Voigt, N. P. & Zweygarth, E. (1981) *Babesia equi* (Laveran, 1901). 1. Development in horses and in lymphocyte culture. *Tropenmed. Parasit.*, **32**, 223–227.
Shone, D. K., Wells, G. E. & Walter, F. J. A. (1961) The activity of Amicarbalide against *Babesia bigemina*. *Vet. Rec.*, **73**, 736.
Shortt, H. E. (1936) Life-history and morphology of *Babesia canis* in the dog-tick *Rhipicephalus sanguineus*. *Ind. J. med. Res.*, **23**, 885–920.
Shortt, H. E. (1973) *Babesia canis*: the life cycle and laboratory maintenance in its arthropod and mammalian hosts. *Int. J. Parasit.*, **3**, 119–148.
Sibinovic, K. H. (1965) *Serological Activity and Biologic Properties of a Soluble Antigen of Babesia caballi*. M.Sc. Thesis. University of Illinois, Urbana.
Sibinovic, K. H., Milar, R., Ristic, M. & Cox, H. W. (1969) *In vivo* and *in vitro* effects of serum antigens of babesial infection and their antibodies on parasitized and normal erythrocytes. *Ann. trop. Med. Parasit.*, **63**, 327–336.
Simitch, C. P., Nevenic, V. & Sibalic, S. (1956) Le traitment de la piroplasmose ovine et la piroplasmose bovine par 'berenil'. *Acta vet. Belgrad*, **6**, 3–13.
Şimitch, T., Petrovitch, Z. & Rakovec, R. (1955) Les espèces de *Babesiella* du boeuf d'Europe. *Arch. Inst., Pasteur Alger.*, **33**, 310–314.
Škrabalo, Z. & Deanović, Z. (1957) Piroplasmosis in man. Report on a case. *Doc. Med. Geogr. Trop.*, **9**, 11–16.
Smith, R. D. & Ristic, M. (1977) Ehrlichiae. In: *Parasitic Protozoa*, ed. Kreier, J. P., Vol. IV, pp. 295–328. New York: Academic Press.
Smith, T. & Kilborne, F. L. (1893) Investigations into the nature causation and prevention of Texas or southern cattle fever. *US Dept Agric. Bur. Ann. Ind. Bull.*, **1**, 1–301.
Taylor, W. M., Bryant, J. E., Anderson, J. B. & Williers, K. H. (1969) Equine piroplasmosis in the United States: a review. *J. Am. vet. med. Ass.*, **155**, 915–919.
Tella, A. & Maegraith, B. G. (1965) Physiopathological changes in primary acute blood transmitted malaria and *Babesia* infections. I. Observations on parasites and blood cells in rhesus monkeys, mice, rats and puppies. *Ann. trop. Med. Parasit.*, **59**, 135–146.
Thoongsuwan, S. & Cox, H. W. (1973) Antigenic variants of the haemosporidian parasite *Babesia rodhaini*, selected by *in vitro* treatment with immune globulin. *Ann. trop. Med. Parasit.*, **67**, 373–385.
Walter, G. (1982) Versuche zur Übertragung von *Babesia microti* durch nymphen von *Dermacentor marginatus*, *D. reticulatus*, *Haemaphysalis punctata*, *Rhipicephalus sanguineus* und *Ixodes hexagonus*. *Z. Parasitenk.*, **66**, 353–354.
Western, K. A., Benson, G. D., Gleason, N. N., Healy, G. R. & Schultz, M. G. (1970) Babesiosis in a Massachusetts resident. *N. Engl. J. Med.*, **283**, 854–856.
Wright, I. G. (1975) The probable role of *Babesia argentina* esterase in the *in vitro* activation of plasma prekallikrein. *Vet. Parasit.*, **1**, 91–96.
Wright, I. G. & Goodger, B. V. (1973) Proteolytic enzyme activity in the intra-erythrocyte parasites *Babesia argentina* and *Babesia bigemina*. *Z. Parasitenk.*, **42**, 213–220.
Wright, I. G. & Kerr, J. D. (1974) The preferential invasion of young erythrocytes in acute *Babesia bigemina* infection of splenectomized calves. *Z. Parasitenk.*, **43**, 63–69.
Wright, I. G. & Kerr, J. D. (1975) Effects of trasylol on packed cell volume and plasma kallikrein activation in acute *Babesia argentina* infection of splenectomized calves. *Z. Parasitenk.*, **46**, 189–194.
Zlotnik, I. (1953) Cerebral piroplasmosis in cattle. *Vet. Rec.*, **65**, 642–643.

PIROPLASMIA (THEILERIIDAE)

Barboni, E. (1942) Multiple brain haemorrhage in cattle with *T. annulata* infection. *Nuova Vet.*, **21**, 11–15.
Barnett, S. F. (1963) The biological races of the bovine *Theileria* and their host–parasite relationship. In: *Immunity to Protozoa*, eds P. C. C. Garnham, A. E. Pierce & Roitt, pp. 180–95. Oxford: Blackwell Scientific Publications.
Barnett, S. F. (1977) Theileria. In: *Parasite Protozoa*, ed. J. P. Kreier, vol. IV, pp. 77–113. New York: Academic Press.
Barnett, S. F. & Brocklesby, D. E. (1966*a*) A mild form of East Coast fever (*Theileria parva*) with persistence of infection. *Br. vet. J.*, **112**, 361–370.
Barnett, S. F. & Brocklesby, D. W. (1966*b*) Susceptibility of the African buffalo (*Syncerus caffer*) to infection with *Theileria parva* (Theiler, 1904). *Br. vet. J.*, **112**, 379–386.
Barnett, S. F. & Brocklesby, D. W. (1966*c*) The passage of *Theileria lawrencie* (Kenya) through cattle. *Br. vet. J.*, **112**, 396–409.
Bitukov, P. A. (1953) Experiments on the transmission of ovine theileriasis and anaplasmosis by the ticks *Ornithodorus lahorensis* and *Haemaphysalis sulcata*. *Trudy Akad. Nauk. Kazahkstan. SSR Inst. Zool.*, **1**, 30–36.
Brocklesby, D. W. (1962) *Cytauxzoon taurotragi* Martin and Brocklesby, 1960, a piroplasm of the eland (*Taurotragus oryx pattersonianus* Lydekker, 1906). *Res. vet. Sci.*, **3**, 334–344.
Brocklesby, D. W. & Barnett, S. F. (1966*a*) A review of the literature concerning *Theileridae* of the African buffalo (*Syncerus caffer*). *Br. vet. J.*, **112**, 371–378.
Brocklesby, D. W. & Barnett, S. F. (1966*b*) The isolation of *Theileria lawrenci* (Kenya) from a wild buffalo (*Syncerus caffer*) and its passage through captive buffaloes. *Br. vet. J.*, **112**, 387–395.
Cowdry, E. V. & Ham, A. W. (1932) Studies on East Coast fever. I. Life cycle of the parasite in ticks. *Parasitology*, **24**, 1–49.
Gonder, R. (1910) The life cycle of *Theileria parva*: The cause of East Coast fever of cattle in South Africa. *J. comp. Path.*, **23**, 328–335.
Gonder, R. (1911) Die Entwicklung von *Theileria parva*, dem Erreger des Küstenfiebers der Rinder in Afrika. 2. Teil. *Arch. Protistenk.*, **22**, 170–178.
Hignett, P. G. (1953) *Theileria mutans* detected in British cattle. *Vet. Rec.*, **65**, 893–894.
Howe, D. L. (1971) Theileriosis. In: *Parasitic Diseases of Wild Mammals*, eds J. W. Davis & R. C. Anderson, pp. 343–353. Ames: Iowa State University Press.
Hulliger, L., Brown, C. G. D. & Wilde, J. K. H. (1965) Theileriosis (*T. parva*) immune mechanism investigated *in vitro*. In: *Progress in Protozoology*, vol. II, p. 37. International Congress Protozoology, Int. Conf. Series, No. 91. Amsterdam: Excerpta Medica.
Hulliger, L., Brown, C. G. D. & Wilde, J. K. H. (1966) Transition of developmental stages of *Theileria parva in vitro* at high temperature. *Nature*, **211**, 328–329.
Hulliger, L., Brown, C. G. D. & Wilde, J. K. H. (1966) Transition of developmental stages of *Theileria parva in vitro* at high temperature. *Nature, Lond.*, **211**, 328–329.
Irvin, A. D., Stagg, D. A., Kanhai, G. K. & Brown, C. G. D. (1975) Heterotransplantation of *Theileria parva*-infected cells to athymice mice. *Nature, Lond.*, **253**, 549–550.
Jarrett, W. F. H., Crighton, G. W. & Pirie, H. M. (1969) *Theileria parva*: kinetics of replication. *Exp. Parasit.*, **24**, 9–25.
Kertzelli, S. (1909) Piroplasmosis *Taranda rangiferis*. *Arch. Vet. Nauk.*, **5**, 549.
Koch, R. (1906) Beiträge zur Entwicklungsgeschichte der Piroplasmen. *Z. Hyg. Infecktkr.*, **54**, 1–9.
Kreier, J. P., Ristic, M. & Watrach, A. M. (1962) *Theileria* sp. in a deer in the United States. *Am. J. vet. Res.*, **23**, 657–662.
Levine, N. D. (1973) *Protozoan Parasites of Domestic Animals and of Man*, 2nd ed., p. 406. Minneapolis: Burgess.
McHardy, M., Haigh, A. J. B. & Dolan, T. T. (1976) Chemotherapy of *Theileria parva* infection. *Nature, Lond.*, 261, 698–699.
Malmquist, W. A. & Brown, C. G. D. (1974) Establishment of *Theileria parva* infected lymphoblastoid cell lines using homologous feeder layers. *Res. vet. Sci.*, **16**, 134–135.
Martin, H. M., Barnett, S. F. & Vidler, B. O. (1964) Cyclical development and longevity of *Theileria parva* in the tick *Rhipicephalus appendiculatus*. *Exp. Parasit.*, **15**, 527–555.

Martin, H. M. & Brocklesby, D. W. (1960) A new parasite of the eland. *Vet. Rec.*, **72**, 331–332.
Mehlhorn, H. & Schein, E. (1977) Electron microscopic studies of the development of kinetes of *Theileria annulata* Dschunkowsky and Luhs, 1904 (Sporozoa, Piroplasmea). *J. Protozool.*, **24**, 249–257.
Morzaria, S. P., Brocklesby, D. W. & Harradine, D. L. (1977) Evaluation of the indirect fluorescent antibody test for *Babesia major* and *Theileria mutans* in Britain. *Vet. Rec.*, **100**, 484–487.
Mpangala, C., Uilenberg, G. & Schreuder, B. E. C. (1976) Studies on Theileria (Sporozoa) in Tanzania. II. Serological characterization of *Haematoxenus veliferus. Tropenmed. Parasit.*, **27**, 192–196.
Neitz, W. O. (1956) Classification, transmission and biology of piroplasms of domestic animals. *Ann. N.Y. Acad. Sci.*, **64**, 56–111.
Neitz, W. O. & Thomas, A. D. (1948) *Cytauxzoon sylvicaprae* gen. nov., spec. nov., a protozoan responsible for a hitherto undescribed disease in the duiker (*Sylvicapra grimmia*). *Onderstepoort. J. vet. Res.*, **23**, 63–76.
Purnell, R. E. (1977) East Coast fever: some recent research in East Africa. *Adv. Parasit.*, **15**, 83–132.
Radley, D. E., Brown, C. G. D., Burridge, M. J., Cunningham, M. P., Pierce, M. A. & Purnell, R. E. (1974) East Coast fever: quantitative studies of *Theileria parva* in cattle. *Exp. Parasit.*, **36**, 278–287.
Radley, D. E., Brown, C. G. D., Cunningham, M. P., Kimber, C. D., Purnell, R. E., Stagg, S. M. & Punyua, D. K. (1975) East Coast fever: homologous challenge of immunized cattle in an infested paddock. *Vet. Rec.*, **96**, 525–527.
Reichenow, E. (1940) Der Entwicklungsgang des Kustenfieber-regers im Rinde und in der übertragenden Zecke. *Arch. Protistenk.*, **94**, 1–56.
Riek, R. F. (1966) The development of *Babesia* spp. and *Theileria* spp. in ticks with special reference to those occurring in cattle. In: *The Biology of Parasites*, ed. E. J. L. Soulsby, pp. 15–22. New York: Academic Press.
Schaeffler, W. L. (1962) *Theileria cervi* infection in white-tailed deer (*Dama virginiana*) in the United States. Ph.D. Diss. University of Illinois, Urbana. (*Diss. Abstr.*, **23**, 389–391.)
Schein, E. (1975) On the life cycle of *Theileria annulata* (Dschunkowsky and Luhs, 1904) in the midgut and hemolymph of *Hyalomma anatolicum excavatum* (Koch 1844). *Z. Parasitenk.*, **47**, 165–167.
Schein, E., Mehlhorn, H. & Warnecke, M. (1977*a*) Zur Feinstruktur der erythrocytaren Stadien von *Theileria annulata* (Dschunkowsky, Luhs, 1904). *Trop. Med. Parasit.*, **28**, 349–360.
Schein, E. & Voigt, W. P. (1979) Chemotherapy of bovine theileriosis with Halofuginone. *Acta trop.*, **36**, 391–394.
Schein, E., Warnecke, M. & Kirmse, P. (1977*b*) Development of *Theileria parva* (Theiler 1904) in the gut of *Rhipicephalus appendiculatus* (Neumann, 1901). *Parasitology*, **75**, 309–316.
Splitter, E. J. (1950) *Theileria mutans* associated with bovine anaplasmosis in the United States. *J. Am. vet. med. Ass.*, **117**, 134–135.
Sturman, M. (1959) Tick fever in Israel. *Symposium for Veterinarians. Rehovah.*
Theiler, A. (1911) Progress report on the possibility of vaccinating cattle against East Coast fever. *1st Rep. Dir. vet. Res. Pretoria*, 1911, 47–207. Dept Agric. South Africa.
Tsur-Tchernomoretz, I. & Adler, S. H. (1965) The cultivation of lymphoid cells and *Theileria annulata* schizonts from infected bovine blood. *Refuah. Vet.*, **22**, 60–62.
Tsur-Tchernomoretz, I., Davidson, M. & Weissenberg, I. (1960) Two cases of bovine Theileriasis (*Th. annulata*) with cutaneous lesions. *Refuah. Vet.*, **17**, 100–199.
Uilenberg, G. (1981) Theilerial species of domestic livestock. In: *Advances in the Control of Theileriosis*, ed. A. D. Irwin, M. P. Cunningham & A. S. Young, pp. 4–37.
Uilenberg, G. & Andreasĕn, M. P. (1974) *Haematoxenus separatus* sp. n. (Sporozoa, Theileriidae), a new blood parasite of domestic sheep in Tanzania. *Rev. Elev. Méd. vét. Pays trop.*, **27**, 459–465.
Wagner, J. E. (1976) A fatal cytauxzoonosis-like disease in cats. *J. Am. vet. med. Ass.*, **168**, 585–588.
Wenyon, C. M. (1926) *Protozoology*, 2 vols. London: Baillière, Tindall & Cox.
Wightman, S. R., Kier, A. B. & Wagner, J. E. (1977) Feline cytauxzoonosis: clinical features of a newly described blood parasite disease. *Feline Practice*, May 1977, 23–26.
Yakimoff, W. L. (1917) Maladies animales du Turkestan russe à parasités endoglobulaires. A. Piroplasmoses. III. Piroplasmoses des moutons. B. Theilérioses. III. Theilériose des Chameaux. C. Nuttallioses. II. Nuttalioses des ânes. D. Anaplasmoses. III. Anaplasmoses des Chiens. *Bull. Soc. Path. exot.*, **10**, 302.

Phylum: Microspora Sprague, 1977

Members of this phylum are all obligate intracellular parasites; they produce spores, have a vesicular nucleus and have a polar filament. Levine et al. (1980) include the class Microsporea in this phylum. A detailed consideration of the classification is given by Levine et al. (1980) in the new revision of the classification of protozoa.

CLASS: MICROSPOREA DELPHY, 1963

With characteristics of the phylum. Spores possess a polar filament originating in a polar sac and coiled in the peripheral layers of the cytoplasm. The polar filament serves to conduct the sporoplasm out of the spore, through the hollow tube of the filament, into the host cell. Asexual division is by binary and multiple fission. Parasites of vertebrates and invertebrates. *Nosema bombycis* Nageli, 1857 is probably the best-known member of the class, being the cause of silkworm disease (pébrine) which caused devastation of the silk industry in Europe in the nineteenth century. Canning (1977*a*, *b*) has reviewed the morphology, taxonomy and host–parasite interactions of the microsporidia.

Genus: Encephalitozoon Levaditi, Nicolau and Schoen, 1923

Obligate intracellular parasites in which sporogony is by binary fission. Parasites of warm-blooded vertebrates.

One species, *Encephalitozoon cuniculi* Levaditi, Nicolau and Schoen, 1923, is of interest.

Encephalitozoon cuniculi (Fig. 3.52). *Hosts:* mouse, rat, rabbit, guinea-pig, hamster, dog and man. World-wide in distribution. It was originally found in the brain and the kidney of a rabbit affected with motor paralysis.

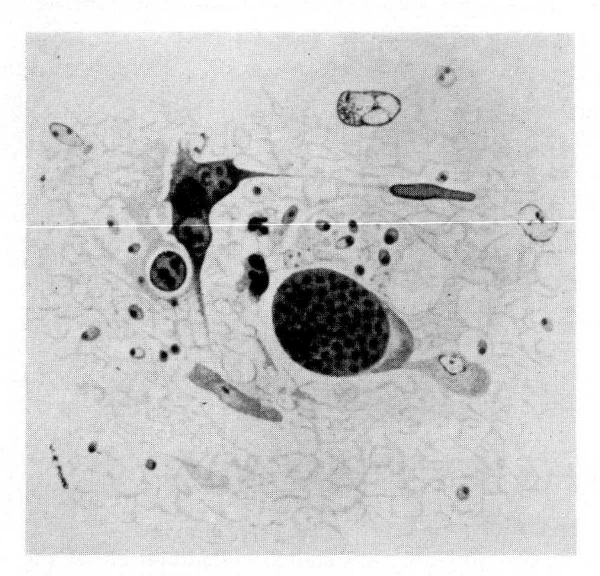

Fig 3.52 *Encephalitozoon cuniculi* in atrophying nerve cell and scattered through the brain substance. (× 1200) (*From Wenyon 1926*)

This organism was first described by Levaditi et al. (1923) and was named *Encephalitozoon cuniculi*. Subsequently, Lainson et al. (1964) demonstrated the organisms were microsporidial in nature and reassigned them to the genus *Nosema*. However, Canning (1977*b*) states that *E. cuniculi* is the correct name. Sporogony and schizogony occur within one host (monoxenous). The spore, which is passed in the urine of its host into the environment, is the infective stage. Spores are ellipsoidal to pyriform and the posterior half contains the spirals of a polar filament, which can be extruded under various stimuli. At the free end is a small body, the sporoplasm, which multiplies by schizogony within the host cell to produce further spores. The organisms resemble *Toxoplasma* and in the past have been confused with this genus.

The cytoplasm of *Encephalitozoon* stains uniformly light blue with Giemsa stain, whereas that of *Toxoplasma* is granulated. *Encephalitozoon* is Gram-positive and *Toxoplasma* Gram-negative; *Encephalitozoon* stains poorly with haematoxylin and eosin, whereas *Toxoplasma* stains well; with iron haematoxylin *Encephalitozoon* stains black while *Toxoplasma* does not. *Toxoplasma* grows readily in tissue culture of monkey kidney monolayers, but *Encephalitozoon* fails to do so; *Toxoplasma* is usually fatal to mice, especially on serial passage, but *Encephalitozoon* is not and its virulence is not enhanced by serial passage.

In section, the trophozoites of *E. cuniculi* appear as straight or slightly curved, uninucleate rods with rounded ends, one end being a little larger than the other. Extracellular forms are 2.4–3.4 μm by 1.8–2.8 μm and intracellular forms are 1.5–3 μm by 1.4–2.8 μm. Parasites may be found in compact pseudocyst accumulations containing 100 or more trophozoites in nerve cells, macrophages or other tissue cells.

The developmental cycle has been studied by Barker (1975) and Pakes et al. (1975) and is as follows. Spores are ingested, but the hatching of these and their migration to viscera has not been followed. Asexual division by binary fission occurs in parasitophorous vacuoles in peritoneal macrophages. Following schizogony, sporonts are produced which give rise to pairs of sporoblasts which then develop into spores. Macrophages may become packed with spores as a result of repeated sporogony. In chronic infections the parasite is found primarily in the kidney and the brain. Spores are passed in the urine. The organisms can be transmitted to mice, rats and rabbits by intracerebral and intraperitoneal inoculation of material from brain, liver, spleen or peritoneal exudate. Congenital infection may occur also (Plowright 1952).

Pathogenesis. In many instances the infection is not apparent and, at the most, is indicated by a mild temperature increase. In the rabbit the infection is usually chronic, and motor paralysis may occur with death. Lesions occur in the brain, and fatal or acute cases show necrotic areas with lymphocytic cuffing about the blood vessels of the cerebrum. Granulomatous lesions occur in the kidneys and other organs, and necrotic lesions have been described in the heart and kidneys.

Disease due to *E. cuniculi* has been reported in dogs by Plowright (1952) and Plowright and Yeoman (1952). Foxhound puppies were affected with posterior weakness and incoordination, the animals fatigued quickly, and there was loss of condition and ocular changes. The animals died between 6 weeks and 15 months of age. In both outbreaks the clinical signs resembled those of rabies. Animals became vicious, attempted to bite people, and some puppies had epileptiform fits or uncontrolled spasms. The major lesions consisted of encephalitis and nephritis, the lesions being similar to those seen in the rabbit.

Nordstoga and Westbye (1976) have reported polyarteritis nodosa in blue foxes associated with infection with the parasite in Norway.

There have been several reports of *E. cuniculi* infection in primates (Canning 1977*b*). Human infection has been reported by Matsubayashi et al. (1959) and Brown et al. (1973) reported the infection in squirrel monkeys (*Saimiri sciureus*).

Diagnosis. E. cuniculi infection is based on the demonstration of the lesions and organisms in sections. Differentiation from *Toxoplasma* is necessary, and this is done on serological evidence and on staining reactions.

No treatment is known. Improvement in sanitation is indicated when the infection occurs.

Genus: Nosema Nageli, 1857

Sporogony occurs by binary fission and diffuse infiltrations of parasites occur in host tissues. Previously, *Encephalitozoon cuniculi* was assigned to this genus. Organisms are chiefly parasites of invertebrates (see Kudo 1966), and some occur in fish. The latter include *Nosema lophi* (Doflein, 1898) which produces tumours of the ganglia of cranial and spinal nerves of angler fish (*Lophius piscatorius* and *L. americanus*) and *Nosema branchialis* Nemeczek 1911 which produces small white tumours on the gills of haddock.

Nosema connori Sprague, 1974 was found in a four-month old immunodeficient child (Margileth et al. 1973). Spores were diffusely distributed in the smooth musculature of the intestinal tract and blood vessels but also in the myocardium and diaphragm. Spores are oval, 4 μm by 2 μm and the polar filament has 11 coils (Sprague 1974).

Genus: Plistophora Gurley, 1893

Spores are ovoid and uninucleate, produced in variable numbers within a pansporoblast membrane (an outer covering for the sporont and sporoblasts, isolating these forms from the host cytoplasm). Species of importance are parasites of fish.

Plistophora hyphessobryconis (Schäperclaus, 1941) occurs in the skeletal muscles of the neon tetra (*Hyphessobrycon innesi*) and many other tropical fish kept in aquaria. Muscle bundles of the fish become packed with spherical pansporoblasts, each containing numerous spores.

The condition is named 'neon fish disease'. The disease is progressive and causes congestion and muscular paralysis with loss of balance and fin degeneration (Reichenbach-Klinke & Elkan 1965). The muscles become pale and transparent and cause the surface of the fish to appear spotty and patchy. There is no known treatment and control is based on elimination of infected fish and sanitation.

Plistophora macrozoarcidis (Nigrelli, 1946) occurs in the skeletal muscle of the Ocean pout (*Macrozoarces americanus*). It causes small whitish cysts in musculature which contain schizonts while brown cysts contain mature spores. Heavy infections cause hyalinization and destruction of tissue. Large tumours, several centimeters in size, may occur.

Plistophora cepedianae Putz, Hoffman and Dunbar, 1965 was associated with the death of gizzard shad (*Dorosoma cepedianum*) in Ohio,

USA. This species produces large cysts in the visceral cavity. The cysts may protrude from the body wall (Putz et al. 1965).

Plistophora salmonae Putz, Hoffman and Dunbar, 1965 occurs in trout (*Salmo gairdneri, Oncorhynchus nerka* and *Coltus* species). Major losses due to this parasite have been reported by Wales and Wolf (1955) in British Columbia. The lesions occur as cysts on the gill lamellae and are surrounded by a thin layer of epithelium. Epithelial hyperplasia, with fusion of lamellae resulting in clubbing of gill filaments, is evident.

Awakura and Kurahashi (1967) reported acquired resistance of salmonids to *Plistophora* following infection, this immunity lasting up to one year. Amprolium fed at 0.06% of body weight is able to eliminate schizogony from the heart musculature.

Plistophora ovariae Summerfelt, 1964 infects the ovaries of golden shiners (*Notemigonous crysoleucas*) in the USA. Although not associated with high mortality, the infection reduces fecundity in minnows which are used as bait.

Plistophora anguillarum Hoshina, 1951 occurs in the muscle of eels (*Anguilla japonica*) in culture ponds in Japan. Serious epizootics may occur associated with body deformities due to the lysis of muscles (Hoshina 1951).

Other species of the genus which occur in fish are listed by Putz et al. (1965).

Genus: Glugea Thelohan, 1891

Sporonts arise from a multinucleate plasmodium and each sporont develops into two spores. The host cells become enormously hypertrophied, forming 'Glugea-cysts' or xenomas.

Glugea anomala (Moniez, 1889) (Gorley, 1893). This is a common parasite of the three-spined stickleback (*Gasterosteus aculeatus*). It occurs in Europe, Soviet Union and North America. Xenomas (tumours) are found in the connective tissue in many parts of the body; they may reach up to 8 mm in diameter and the superficial lesions cause marked deformity.

Other *Glugea* spp. of interest include *Glugea hertwigi* Weissenberg, 1911 which occurs in the viscera and body wall of smelt (*Osmerus* spp.) Heavy mortality has been recorded in American smelt taken from the Great Lakes and the north-east coast (Haley 1953; Légault & Delisle 1967). The intestinal wall is mainly affected, hundreds of cysts up to 8 mm in diameter are produced and these occlude the intestinal lumen (Canning 1977*b*). *Glugea stephani* (Hagenmuller 1899) Woodcock, 1904 occurs in the connective tissue of the gut of flat fish such as plaice (*Pleuronectes platessa*), flounder (*P. flesus*) and winter flounder (*Pseudopleuronectes americanus*). Highest prevalence is in small fish and fish heavily infected in their first year do not survive into the second (Stunkard & Lux 1965).

Genus: Thelohania Henneguy, 1892

Each sporont develops into eight spores. The sporont membrane may degenerate at different times of development.

This genus contains species parasitic in mammals and fish.

Thelohania apodemi Doby, Jeannes and Rault, 1963, occurs in the brain of field mice (*Apodemus sylvaticus*) producing colonies without any tissue reaction around them. Division occurs within a pansporoblast membrane, eight spores being produced. Colonies measure 30–100 μm in diameter. The spores have a strong natural fluorescence when exposed to light of wavelength of about 400 nm (Canning 1977*b*).

Thelohania baueri Voronin, 1974 has been reported from the ovary of the stickleback.

Genus: Ichthyosporidium Caullery and Mesnil, 1905

In this genus sporogony is by binary fission, development occurs in a cyst composed of proliferated host cytoplasm devoid of nuclei. One species is of interest: *Ichthyosporidium giganteum*

(Thelohan, 1895), which is parasitic in the connective tissue of the body wall of fish (corkwing—*Crenilabrus melops*; and spot—*Leiostomus xanthurus*). The lesions cause marked bulging from the ventral surface in the region of the pectoral fins. The tumours contain numerous cysts and proliferating host cytoplasm and they lack a thick wall of the type seen with *Glugea* spp.

Phylum: Myxozoa Grassé, 1970 emend.

Organisms with amoeboid germinal elements (sporoplasms) in multicellular spores; trophozoites are multicellular, showing differentiation of somatic and germinal elements (Mitchell 1977).

The taxonomic position of the myxosporids has been under debate. Multicellularity occurs and some authors (e.g. Lom 1973) do not consider the myxosporids true protozoa.

CLASS: MYXOSPOREA BÜTSCHLI, 1881

With characteristics of the subphylum.

The order BIVALVULIDA Schulman, 1959 is of interest in this volume. The spores of these organisms are formed with two valves, each containing one to six cnidocysts. Usually an extracellular parasite, in contrast to the microsporids which are intracellular. Parasites of this order are responsible for serious disease in free-living fish but are rarely found in aquarium fish.

A key to the genera of the myxosporidia is given by Hoffman (1967).

Genus: Myxosoma Thelohan, 1892

The spore of this genus is ovoid in front view, lenticular in profile. Two pyriform capsules occur at the anterior end. Sporoplasm without iodinophilic vacuole. Histozoic forms in fresh-water and marine fish. Several species occur; these are listed by Hoffman (1967).

Myxosoma cerebralis (Hofer) (Plehn, 1905) Kudo, 1933. This is the cause of 'whirling disease' or 'twist disease' of salmonid fish. The parasite affects the cartilage and perichondrium and young fish are especially affected. Whirling disease is one of the most widespread diseases of rainbow trout.

The developmental cycle of the parasite is not fully known but is presumed to be direct (Hoffman et al. 1965) by ingestion of spores released from disintegrated infected tissue. Fish-eating birds have been suspected to play a role in transmission and Taylor and Lott (1978) have demonstrated transmission by mallard ducks and black-crested night heron which had fed on trout infected with *M. cerebralis*. The need for maturation of spores in mud (Hoffman et al. 1969) indicates that cement-lined pools are needed in hatcheries where the disease occurs.

Following infection, parasites invade and erode skeletal cartilage by about 40–60 days after infection. Spores are found in the cartilage, chiefly of the cranium; they are lemon-drop-shaped with two polar capsules.

The major disease manifestations result from invasion of the cartilage supporting the CNS. Invasion of the cartilaginous capsule of the semicircular canals produces loss of equilibrium, tail-chasing, particularly when fish are startled.

Damage to the nerves controlling melanophores results in excessive pigmentation of the tail (black tail). Whirling subsides with age, but the fish are left mis-shapen with sunken heads and twisted spines.

The world-wide distribution of *M. cerebralis* is ascribed to intercontinental shipment of infected trout. Freezing does not kill the spores (Mitchell 1977).

Control of the disease is by raising trout in spore-free water. Various chemicals are available for application to drained pond floors, filtration may be used and ultra-violet irradiation of flowing hatchery water is practised in the USA (Mitchell 1977).

Myxosoma cartilaginis Hoffman, Putz and Dunbar, 1965, localizes in the cartilaginous parts of the head of the bluegill, sunfish and black bass, but it does not cause deformation or impaired movement, although small cysts occur in the cartilage.

Myxosoma dujardini Thélohan, 1892 localizes in the gills of carp and other fishes in Europe and Asia. It causes white to yellowish 1-mm cysts on the gills, leading to dyspnoea and death.

Genus: Ceratomyxa Thélohan, 1892

Spores of this genus are arched: shell valves conical and hollow.

Ceratomyxa shasta Noble 1950, occurs in salmonids in the western United States. The spore of *C. shasta* are small and elongate and are found in large numbers in the gut wall, liver, kidney, spleen and musculature. Severe losses in fingerling trout due to this parasite have been reported (Wales & Wolf 1953). It also causes losses in rainbow trout, steelhead, coho and chinook salmon, especially before spawning.

Infection is direct and is contracted as adult salmon enter infected fresh water. Prespore development stages are found in the intestine 18 days after infection and spores occur 20–30 days after infection, but development is temperature-dependent.

Marked swelling occurs in the area of the vent, the abdomen is distended with fluid and large boils may develop and protrude from the surface of the body. In fingerlings, death usually occurs about 40 days after infection.

No treatment is known and control depends on hygiene, provision of parasite-free water and avoidance of transfer of infected fish to new water.

Genus: Henneguya (Thélohan, 1892) Davis, 1944

Spores are ovoid with two polar capsules at the anterior end. The posterior shell valves are prolonged into extended processes. Usually forming cysts. Numerous species occur and have been listed by Hoffman (1967).

Henneguya exilis Kudo, 1920 occurs in the channel catfish. It is found in cysts in the capillaries and interlamellar spaces of the gills. The interlamellar forms may occlude the space between the gill lamella and cause suffocation. White to pinkish cysts are seen in the gills, often in large numbers. Mortality may reach 95% in very young fish.

Henneguya zschokkei (Gurley, 1894) causes boil disease in a number of salmonid fishes. This parasite occurs in the rivers which drain into the Arctic Ocean, the Baltic and North Seas and also in lakes in Switzerland.

Genus: Thelohanellus Kudo, 1934

Spores are pyriform, each with one polar capsule. Parasites of freshwater fish.

Thelohanellus piriformis (Thélohan, 1892) Kudo, 1933 causes boil disease in cyprinid and coregonid fishes in the rivers of central and eastern Europe and Siberia.

Other genera of Myxosporea occur but are less important than those listed above. A check-list of these various genera is given by Hoffman (1967).

REFERENCES

MICROSPOREA AND MYXOSPOREA

Awakura, T. & Kurahashi, S. (1967) Studies of the Pleistophora disease of salmonid fishes. III. On prevention and control of the disease. Scientific Reports of the Hokkaido Fish Hatchery, No. 22, pp. 51–68. Hokkaido Fish Hatchery, Nakanoshima, Sapporo, Japan [in Japanese].

Barker, R. J. (1975) Ultrastructural observations on *Encephalitozoon cuniculi* Levaditi, Nicolau et Schoen, 1923, from mouse peritoneal macrophages. *Folia. Parasitol.* (*Prague*), **22**, 1–9.

Brown, R. J., Hinkle, D. K., Trevethan, W. P., Kupper, J. L., & McKee, A. E. (1973) Nosematosis in a squirrel monkey (*Saimiri sciureus*). *J. med. Primatol.*, **2**, 114–123.

Canning, E. F. (1977*a*) New concepts of Microsporida and their potential in biological control. In: *Parasites, Their World and Ours*, ed. A. M. Fallis, pp. 101–140. Ottowa: Royal Society of Canada.

Canning, E. F. (1977*b*) Microsporida. In: *Parasitic Protozoa*, ed. J. P. Kreier, vol. IV, pp. 155–196. New York: Academic Press.

Haley, J. A. (1953) Microsporidian parasite *Glugea hertwigi* in American smelt from the Great Bay region, New Hampshire. *Trans. Am. Fish. Soc.*, **83**, 84–90.

Hoffman, G. L. (1967) *Parasites of North American Freshwater Fishes*, p. 486. Berkeley: University of California Press.

Hoffman, G. L., Dunbar, C. E. & Bradford, A. (1969) Whirling disease of trouts caused by *Myxosoma cerebralis* (Protozoa: Myxosporida) in the United States. *US Bur. Sport Fish. Wildlife. Spec. Sci. Rep. Fish.*, No. **427**, pp. 15.

Hoffman, G. L., Putz, R. E. & Dunbar, C. E. (1965) Studies on *Myxosoma cartilaginis* n.sp. (Protozoa: Myxosporidea) of centrarchid fish and a synopsis of the Myxosoma of North American freshwater fishes. *J. Protozool.*, **12**, 319–332.

Hoshina, T. (1951) On a new microsporidian *Plistophora anguillarum* n.sp. from the muscle of the eel *Anguila japonica*. *J. Tokyo Univ. Fish.*, **38**, 1.

Kudo, R. R. (1966) *Protozoology*, 5th ed. Springfield, Ill.: Charles C. Thomas.

Lainson, R., Garnham, P. C. C., Killick-Kendrick, R. & Bird, R. G. (1964) Nosematosis, a microsporidial infection of rodents and other animals including man. *Br. med. J.*, **22**, 470–472.

Légault, R.-O & Delisle, C. (1967) Acute infection by *Glugea hertwigi* Weissenberg in young-of-the-year rainbow smelt *Osmerus eperlanus mordax* (Mitchell). *Can. J. Zool.*, **45**, 1291–1294.

Levaditi, C., Nicolau, S. & Schoen, R. (1923) L'étiologie de l'encéphalite. *C.R. Hebd. Seances. Acad. Sci.*, **177**, 985–988.

Levine, N. D., Corliss, J. O., Cox, F. E. G., Deroux, G., Grain, J., Honigberg, B. M., Leedale, G. F., Loeblich, A. R., Lom, J., Lynn, D., Merinfeld, E. G., Page, F. C., Poljansky, G., Sprague, V., Vavra, J. & Wallace, F. G. (1980) A newly revised classification of the Protozoa. *J. Protozool.*, **27**, 37–58.

Lom, J. (1973) Current status of myxo- and microsporidia. In: *Progress in Protozoology*, 4th Int. Congr. Protozoology, 1973, p. 254.

Margileth, A. M., Strano, A. J., Chandra, R., Neafie, R., Blum, M. & McCully, R. M. (1973) Disseminated nosematosis in an immunologically comprised infant. *Archs Path.*, **95**, 145–150.

Matsubayashi, H., Korke, T., Mikata, I., Takei, H. & Hagiwara, S. (1959) A case of Encephalitozoon-like body in man. *Archs Path.*, **67**, 181–187.

Mitchell, L. G. (1977) Myxosporida. In: *Parasitic Protozoa*, ed. J. P. Kreier, vol. IV, pp. 115–154. New York: Academic Press.

Nordstoga, K. & Westbye, Kr. (1976) Polyarteritis nodosa associated with nosematosis in blue foxes. *Acta path. microb. scand.*, **84**, 291–296.

Pakes, S. P., Shadduck, J. A. & Cali, A. (1975) Fine structure of *Encephalitozoon cuniculi* from rabbits, mice and hamsters. *J. Protozool.*, **22**, 481–488.

Plowright, W. (1952) An encephalitis-nephritis syndrome in the dog probably due to congenital Encephalitozoon infection. *J. comp. Path. Ther.*, **62**, 83–92.

Plowright, W. & Yeoman, G. (1952) Probable *Encephalitozoon* infection of the dog. *Vet. Rec.*, **64**, 381–383.

Putz, R. E., Hoffman, G. L. & Dunbar, C. E. (1965) Two new species of *Pleistophora* microsporidea from North American fish with a synopsis of microsporidea of freshwater and Euryhaline fishes. *J. Protozool.*, **12**, 228–236.

Reichbach-Klinke, H. H. & Elkan, E. (1965) *The Principal Diseases of Lower Vertebrates*, p. 205. New York: AcademicPress.

Sprague, V. (1974) *Nosema connori*, n.sp. a microsporidian parasite of man. *Trans. Am. microsc. Soc.*, **93**, 400–403.

Stunkard, H. W. & Lux, F. E. (1965) A microsporidian infection of the digestive tract of the winter flounder, *Pseudopleuronectes americanus*. *Biol. Bull.*, **129**, 371–387.

Taylor, R. L., & Lott, M. (1978) Transmission of salmonid whirling disease by birds fed trout infected with *Myxosoma cerebralis*. *J. Protozool.*, **25**, 105–106.

Wales, J. H. & Wolf, H., (1955) Three protozoan diseases of trout in California. *Calif. Fish and Game*, **41**, 183–187.

Phylum: Ciliophora Doflein 1901

CLASS: KINETOFRAGMINOPHOREA DE PUYTORAC ET AL., 1974

Organisms of this class possess cilia for locomotion. They are highly organized forms possessing two nuclei, a macronucleus which is large and massive and responsible for the cytoplasmic activities of the organism, and a micronucleus, which is vesicular and is concerned with the reproductive process. Reproduction is asexual by transverse binary fision or, in the sexual phase, by conjugation.

A large number of ciliates exist, the majority of which are free living, inhabiting water of all kinds. The whole group has been reviewed by Corliss (1974, 1975) and recently reclassified by Levine et al. (1980).

In his revision of the ciliates, Corliss (1974) designated the Ciliophora as a phylum in its own right. In the phylum Ciliophora, Doflein, 1901, three classes are recognized: Kinetofragmophora, de Puytorac et al. 1974; Oligohymenophora, de Puytorac et al. 1974; and Polymenophora Jankowski, 1967. This has been adopted by Levine et al. (1980).

Only one genus, *Balantidium*, is associated with disease in mammals. It is a parasite of the large intestine of man, pig, monkey and possibly other animals. In addition to this pathogen, however, a large number of ciliates occur in the rumen of ruminants and in the large intestine of equines. These are not parasitic in the sense that they produce disease, rather, they are concerned in digestive processes, but their exact role in this has yet to be determined. No attempt will be made to discuss these forms, and further information on them may be obtained from Lubinsky (1957) and Levine (1973).

In addition to the above, a number of ciliates occur in fish, a notable example being *Ichthyophyhirius multifiliis* ('Ich'), a serious pathogen of aquarium fish.

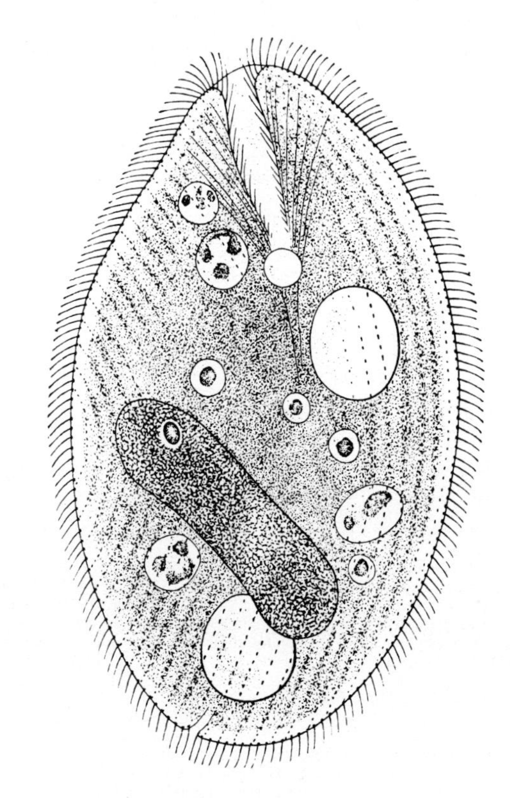

Fig 3.53 *Balantidium coli* from the human intestine. (× 1200) (*From Wenyon 1926*)

Genus: Balantidium Claparède and Lachmann, 1858

Organisms of this genus are oval to ellipsoidal in outline; there is a distinct macronucleus and a small micronucleus. The cilia are arranged in longitudinal rows over the whole of the body; a mouth, or peristome, is situated near the anterior end of the organism, and there is a weakly developed cytopharynx.

A large number of species of *Balantidium* have been described, the description of them being based largely on the host in which they have been found and on the size of the body and the macronucleus. The synonymy of the genus has not been investigated in detail, but it is likely that many of the forms are synonyms.

Balantidium coli (Malmsten, 1857) Stein, 1862 (Fig. 3.53). The vegetative forms average 50–60 μm in length, but larger forms are not uncommon, and some may measure up to 150 μm in length. The body surface is covered with slightly oblique longitudinal rows of cilia, the peristome is subterminal and at the narrower end, the macronucleus is kidney-shaped, and the micronucleus lies in the notch of the macronucleus. One contractile vacuole occurs near the posterior end of the body, another near the centre, and the cytoplasm contains numerous food vacuoles. The organism is actively motile and moves quickly over the microscopic field.

Cysts are produced; these are ovoid to spherical, and measure 40–60 μm. They are faintly yellowish green in colour, and the organism can be recognized within the cyst by the macronucleus.

Reproduction is by transverse binary fission, but conjugation may also take place. Transmission to other hosts is by the cysts.

Balantidium coli is widespread in swine, and it is likely that it will be found in any pig if an adequate examination is made. The prevalence of the infection in man is much lower, and prevalences of 0.6–1% have been recorded. The organism also occurs in other higher primates, and it may be a troublesome infection in zoological gardens. Occasional infections of other animals with *B. coli* have been reported; thus, Bailey and Williams (1949) reported a clinical infection in a dog in the USA.

Pathogenesis. The pig appears to be the primary host, and in it *B. coli* is generally regarded as a commensal, since under normal conditions it is found in the lumen of the large intestine and is associated with no change in the mucosa. Occasionally, and for undetermined reasons, it may invade the mucosa and cause superficial and even deep ulcerations, these being associated with a mild to severe enteritis. Almejew (1963) has reported acute and at times fatal infections in pigs characterized by dysentery with haemorrhage. Organisms were found as deep as the muscular layer of the caecum and colon, being associated with lymphocytic and leucocytic infiltrations. Almejew suggested that *B. coli* had a plasmolytic effect and produced damage to the nuclei of the epithelial cells of the mucosa. Tempelis and Lysenko (1957) have shown that *B. coli* produces hyaluronidase, and this may assist the organism to enter the tissues.

In man, *B. coli* produces superficial to deep ulcers associated with dysentery. The early lesions of *B. coli* infection resemble those produced by *Entamoeba histolytica,* but they do not show the same tendency to enlarge and spread. Only rarely does *B. coli* invade other tissues, such as the liver.

Human infection is a zoonosis and is usually acquired from swine through the contamination of foodstuffs, fingers, etc. with pig faeces. Normally, the cysts are the source of infection, and these remain viable for days or weeks in moist pig faeces. Under conditions of gross contamination, trophozoites may initiate human infection, but these are very much less resistant than the cysts, and they die within 15–30 minutes at temperatures above 40°C, although in a moist environment they may survive one to three days at room temperature. *Balantidium* infection in the higher primates in zoological gardens is normally maintained by the animals themselves.

Diagnosis of *B. coli* infection is based on the clinical signs, post mortem evidence of an ulcerative condition of the large intestine and the presence of very large numbers of *B. coli.* Because of the common occurrence of the organism in apparently normal animals, it is unwise to ascribe intestinal upsets to its presence without additional evidence of its invasive properties.

Treatment. Acute infections may be treated with the tetracycline antibiotics, and in man carbarsone and tetracyclines have been used. Carbarsone is also of value in captive primates, a dose of 250 mg being given daily for ten days.

Genus: Ichthyophthirius Fourquet, 1876

Members of this genus have an oval to round body which is very plastic, 50–1000 μm in diameter. Ciliation is uniform, the pellicle is longitudinally striated, a peristome is present, the macronucleus is horseshoe-shaped and can be seen unstained. Parasites of the skin and gills of fish. No division of the organism occurs while it is in the skin of fish but multiplication occurs within a cyst, which is formed after the parasite leaves the fish, to produce up to 1000 juvenile forms, tomites, which measure 30–45 μm; these are oval, ciliated and have an anterior knob (perforatorium). Infection is by the tomites which rupture from the cyst, attach to the fish and penetrate the skin where they grow to mature trophozoites.

Ichthyophthirius multifiliis Fourquet, 1976 (Fig. 3.54). The cause of ichthyophthiriasis, 'Ich' or white spot disease, and may occur in all freshwater fish throughout the world. It is especially common in aquarium-maintained fish, especially goldfish, but it may also occur in trout hatcheries, etc.

The life-cycle is temperature-dependent (Hoffman 1967) and at an optimal temperature of 24–26°C multiplication within the cyst is complete in seven to eight hours; unattached tomites survive for about 48 hours and the entire life-cycle can be completed, including the development in the fish, in about four days. Infection can be transferred to clean tanks by plants and other objects.

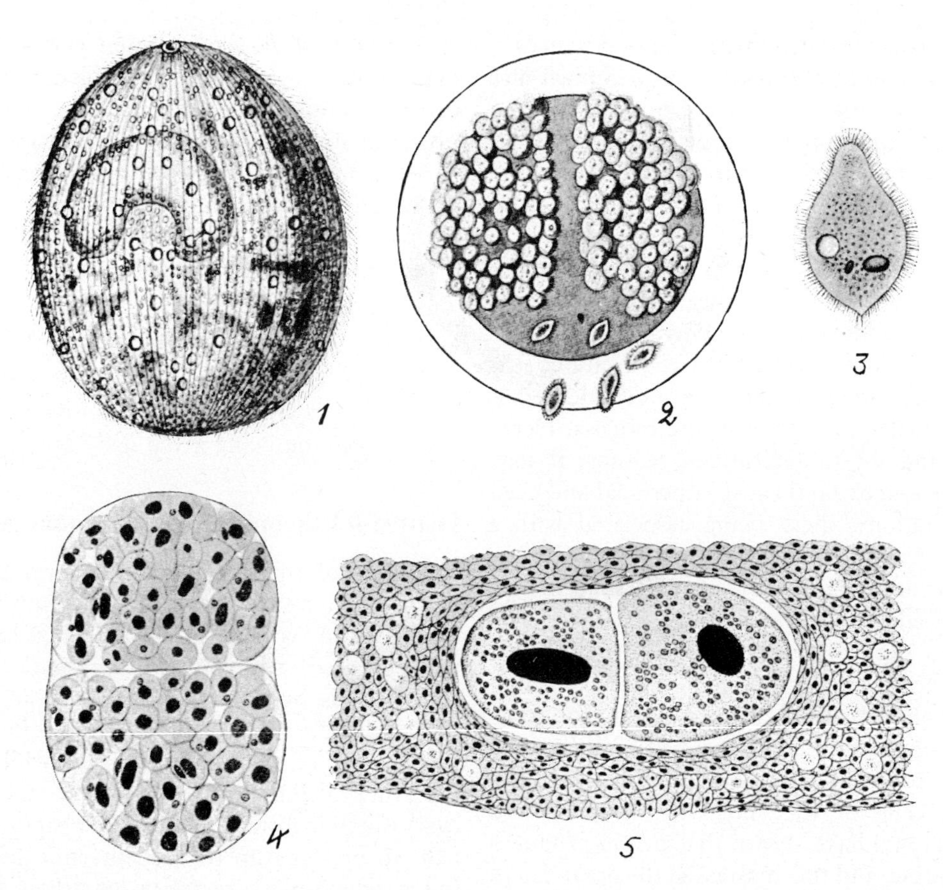

Fig 3.54 *Ichthyophthirius multifiliis* from the skin of fish. 1. Mature trophozoite (× 75). 2. Mature cyst filled with juvenile forms (tomites) (×75). 3. Juvenile form, or tomite, a ciliated form with an anterior knob. 4. Section through a cyst. 5. Section of skin of carp, showing two ciliates in a vacuole. (*From Wenyon 1926*)

Penetration by the tomite causes severe irritation accompanied by mucous secretion and hyperplasia of the epithelium. White pustules are produced ('white spot'). In severe infections the whole body surface and gills are affected, sloughing of the epithelium occurs and death is common in such cases (Bauer 1959; Hoffman 1973).

Fish which survive a number of 'Ich' attacks acquire a temporary immunity, which, however, is lost if attacks cease (Reichenback-Klinke & Elkan 1965).

Diagnosis is based on the lesions and the demonstration of the large ciliate which rotates slowly in a wet preparation.

There is no treatment for infected fish, which should be destroyed.

Control is based on treatment of infected tanks or aquaria and quarantine of fish. Since tomites will survive for only 48 hours at 24–26°C, removal of fish from tanks for 72 hours will free the tank from infection. Where fish cannot be removed treatment is aimed at destruction of free-living forms. There is no treatment that will kill the parasite in the fish without killing the fish (van Duijn 1967). Water may be treated with formalin (15–25 ppm daily in aquaria and ponds), malachite green (2 ppm for 30 min daily or 0.1 ppm in ponds and aquaria) or methylene blue (3 ppm in ponds and aquaria) (Hoffman 1973).

New fish to be added to a clean colony should be held in quarantine at 20–26°C for one week (van Duijn 1967).

Cryptocaryon irritans is the salt-water counterpart of *I. multifiliis* causing 'salt water Ich'.

Genus: Chilodonella Strand, 1926

(Syn: *Chilodon* Ehrenberg, 1833). These ciliates are ovoid, dorsoventrally flattened, the dorsal surface convex, ventral flattened. Central surface with ciliary rows, anteriorly flattened dorsal surface with a cross-row of bristles. Macronucleus rounded. Occur in fish and amphipods. Many species occur.

Chilodonella cyprini, Moroff, 1902, is oval, 40–60 μm with an indentation in the posterior part of the body. A parasite of the gills and skin of trout and many other species of fish. The life-cycle is direct, by contamination of water. Binary fission occurs on the skin. The parasite is an ectoparasite and does not penetrate the epidermis. Usually infection is inapparent but it may be severe in debilitated fish when kept at 5–10°C. In heavy infections there is epithelial cell hyperplasia, excess mucus production, the gills may be more severely damaged than the rest of the body and death may occur in heavy infections when the skin may fall away in strips.

Diagnosis is by microscopic examination to differentiate the condition from others, e.g. Ich. Control is similar to that for ichthyophthiriasis.

Chilodonella hexasticha Kiernik, 1909, has been reported from *Tinca tinca*.

Genus: Trichodina, Ehrenberg, 1833

These are members of the subclass Peritricha. They possess an oral ciliature with three ciliary girdles. They have a chitinoid attaching disc with radially arranged hooked teeth. Flat, disc-like organisms, with a sausage-shaped macronucleus. Several species occur on the gills and skin of various fish. *Trichodina domerguei* is common on the skin and gills of goldfish and other aquarium fish in Europe and Asia; *Trichodina fultoni* occurs on chub, shiners, black bass, brook trout, etc. in North America; *Trichodina megamicronucleatum* occurs on carp in the Soviet Union; and *Trichodina truttae* occurs on the gills of salmon, also in the Soviet Union.

Reproduction is by binary fission and sexual reproduction by conjugation. Transmission is by direct contact.

Light infections are harmless, but heavy infection, which occurs under conditions of poor hygiene, leads to dermal lesions such as white blotches on the head and back, frayed fins, epithelial hyperplasia and excess mucus production. Death may occur in heavy infections. On fish farms in Europe and the Soviet Union, serious infections occur mainly in July and August (Bauer 1958).

Control is similar to that for *Ichthyophthirius multifiliis*.

Genus: Epistylis Ehrenberg, 1833

Organisms have an inverted bell form, usually on branched non-contractile stalks forming colonies. Juvenile forms are free swimming and have an additional circle of cilia around the posterior end of the body as well as the zone of cilia at the adoral end, which occurs in adults. Macronucleus is ribbon-like. Reproduction is by binary fission and sexual reproduction is by conjugation. Transmission is by direct contact.

This is a common parasite of the skin of various fresh-water fish, trout, black bass, etc. It is usually non-pathogenic, but heavy infections may cause dermal lesions (Hoffman 1973). Control measures are similar to those for *I. multifiliis*.

Genus: Ambiphyra Raabe, 1952

(Syn: *Scyphidia*) Cylindrical forms; at the posterior end is a holdfast (scopula) body, cross-striated. In addition to the adoral circle of cilia there is a collar of cilia one-third of the distance from the posterior end. Reproduction is by binary fission and conjugation. Transmission is direct by young free-swimming forms.

Heavy infections cause increased mortality in young fish. Control is as for *I. multifiliis*.

REFERENCES

CILIOPHORA

Almejew, C. (1963) Propagation of balantidium in the porcine intestine. *Mh. vet. Med.*, **18**, 250.
Bailey, W. S. & Williams, A. G. (1949) Balantidium infection in the dog. *J. Am. vet. med. Ass.*, **114**, 238.
Bauer, O. N. (1958) Parasitic diseases of cultured fishes and methods of their prevention and treatment. In: *Parasitology of Fishes*, eds V. A. Dogiel, G. K. Petrashevski & Yu. I. Polyanski, pp. 265–298. Leningrad University Press, 1958. Translated by Z. Kabata, 1961. Edinburgh: Scottish Academic Press.
Bauer, O. N. (1959) The ecology of parasites of fresh water fish [in Russian]. Translated ed. 1962. US Dept Commerce. Office Tech. Serv, TT61-31056, pp. 236.
Corliss, J. O. (1974) The changing world of ciliate systematics: historical analysis of past efforts and a newly proposed phylogenetic scheme of classification for the protistan phylum ciliophora. *System. Zool.*, **23**, 91–138.
Corliss, J. O. (1975) Taxonomic characterization of the subfamilial groups in a revision of recently proposed schemes of classification for the phylum Ciliophora. *Trans. Am. microsc. Soc.*, **94**, 224–267.
van Duijn, C. (1967) *Disease of Fishes*, 2nd ed, p. 309. London: Iliffe Books.
Hoffman, G. L. (1967) *Parasites of North American Freshwater Fishes*, p. 486. Berkeley: University of California Press.
Hoffman, G. L. (1973) Parasites of Laboratory Fishes. In: *Parasites of Laboratory Animals*, ed. Flynn, R. J., pp. 645–768. Ames: Iowa State University Press.
Levine, N. D. (1973) *Protozoan Parasites of Domestic Animals and Man*, 2nd ed., Minneapolis: Burgess.
Lubinsky, G. (1957) Studies on the evolution of the Ophryoscolecidae (Ciliata: Oligotricha). I. A new species of Entodinium with 'caudatium', 'lobosospino-sum', and 'dubardi' forms, and some evolutionary trends in the genus Entodinium. II. On the origin of the higher Ophryoscolecidae. III. Phylogeny of the Ophyroscolecidae based on their comparative morphology. *Can. J. Zool.*, **35**, 111–159.
Reichenbach-Klinke, H. & Elkan, E. (1965) *Principal Diseases of Lower Vertebrates. I. Diseases of Fishes*, p. 205. Hong-Kong: T.F.H. Publications.
Tempelis, C. H. & Lysenko, M. G. (1957) The production of hyaluronidase by *Balantidium coli*. *Exp. Parasit.*, **6**, 31–36.

ORDER: RICKETTSIALES GIESZCZKIEWICZ, 1939

Certain rickettsia have previously been considered to be protozoa; these include organisms in the genera *Anaplasma, Eperythrozoon, Haemobartonella, Aegyptianella, Grahamella,* etc. and, although there is little doubt as to their non-protozoal nature, they are still frequently dealt with in courses in protozoology, at conferences on parasitology and in textbooks dealing with these subjects. For the convenience of persons who may inadvertently turn to a textbook of veterinary parasitology for information on these forms, the following general outline is presented.

Genus: Anaplasma Theiler, 1910

With the light microscope, *Anaplasma* appear as small, spherical bodies, red to dark red in colour when stained with Romanowsky stains, inside the red blood cells of cattle, deer, sheep and goats. They are 0.2–0.5 μm in diameter, with no cytoplasm, but a faint halo may appear around them. Sometimes two organisms may lie close to each other and occasionally multiple invasion of a cell may occur.

Ultra-structural studies (Ristic & Watrach 1963) have shown *Anaplasma* to consist of an initial body which enters an erythrocyte by invagination of the cytoplasmic membrane leading to a vacuole formation, after which the initial body multiplies by binary fission to form an inclusion body consisting of four to eight initial bodies. Organisms are structurally similar to the members of the psittacosis lymphogranuloma group of organisms.

Three species of *Anaplasma* are of interest:

Anaplasma marginale Theiler, 1910 (Cattle)
Anaplasma centrale Theiler, 1911 (Cattle)
Anaplasma ovis Lestoquard, 1924 (Sheep and goats)

Anaplasma marginale (Fig. 3.55). This organism is widely distributed throughout tropical and subtropical areas of the world, as well as some temperate areas. It is common in Africa, Middle East, southern Europe, Far East, Central and South America and the United States. Cattle are the major hosts but infections also occur in zebra, water buffalo, bison, various African antelopes, American deer (southern black-tailed, Virginia white-tailed, mule deer), elk and camel. Sheep and goats may develop inapparent infections and the African buffalo is refractory to infection. It is transmitted by ticks of various species and mechanical transmission by blood-sucking flies is important in some areas.

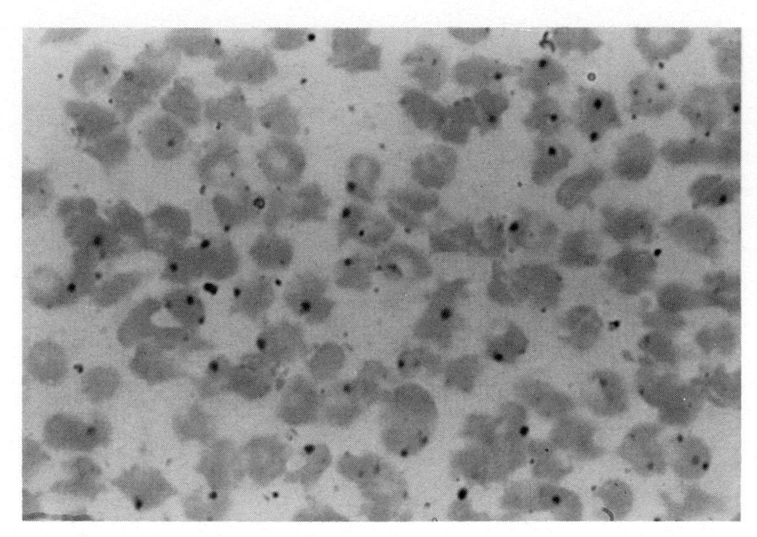

Fig 3.55 *Anaplasma marginale.*

Espana and Espana (1963) distinguished two morphological types of *A. marginale*: a normal, rounded form and a filamented form, both of which occurred in the majority of infected animals, but in some cases only the rounded form of organism was seen. *Paranaplasma caudatum* has been proposed for the filamented or tailed form which was initially found in cattle in Oregon (Kreier & Ristic 1963*a,b*). However, Carson et al. (1974) have demonstrated that the appendage on *P. caudatum* occurs in bovine erythrocytes and not in deer erythrocytes. Fluorescent antibody and cross-immunity studies indicate some antigenic distinctness between *A. marginale* and *P. caudatum*. A third species, *Paranaplasma discoides*, with ovoid or disc-like structures at each pole has been described by Kreier and Ristic (1963*c*).

Transmission of Anaplasma. Some 20 species of ticks have been shown to transmit *Anaplasma*, but detailed field evidence of such transmission is lacking. Organisms have been demonstrated in various tick tissues including gut contents and the Malphigian tubules; little is known of the developmental cycle in the tick (Ristic 1977). Transovarian transmission occurs.

Transmission by blood-sucking flies is well recognized and tabanids, deer flies, stable flies and mosquitoes are the insects chiefly concerned. Insect transmission is particularly common in the southern United States. Direct transfer of infected blood must take place for insect transmission and this must occur within a few minutes after feeding on an infected animal.

Carrier cattle play an important role in the epidemiology of infection, although deer have also been demonstrated to serve as carriers for cattle, and deer to deer transmission may occur in the absence of cattle (Ristic 1968).

Mechanical transmission of anaplasmosis is well known, and major and minor operations in cattle husbandry such as dehorning, castration, vaccination, blood sampling, etc. may be responsible for the transmission of anaplasmosis both in and out of season.

Clinical manifestations of anaplasmosis. Anaplasmosis is essentially a disease of adult cattle and, in general, severe clinical infections do not occur until an animal is about 18 months of age. Younger animals are susceptible to the infection but exhibit little detectable reaction, although they can be made clinical cases by splenectomy. In mature cows the incubation period is 15–36 days with an average of 26 days. There is an increase in body temperature, and in mature animals the infection may be fatal during the fever period. Organisms appear in the red blood cells several days before the fever; at first only a few are found, but by the time fever is initiated, 30–48% of cells may be parasitized and, as the fever progresses, the number of parasitized red cells increases. Anorexia develops and animals show severe anaemia which is especially noticeable at the time of, or shortly after, the fever crisis.

Mortality, especially in susceptible imported cattle, may be as high as 80%, but in an enzootic area the seasonal death rate may be of the order of 10%.

In more chronic cases there is a severe anaemia, and recovery is slow, the animal being susceptible to numerous other conditions which may affect it; in Africa, for example, malnutrition, virus diseases, etc. may terminate the life of the animal.

Tetracyclines (chlortetracycline, tetracycline and oxytetracycline) are used for treatment, being given at the rate of 6–10 mg/kg body weight. Other compounds, such as Imidocarb, are chemotherapeutic and will also eliminate parasites from carrier animals (Roby & Mazzola 1972).

Control by vaccination has been attempted by several means for many years. Thus, Theiler (1910) used vaccination with *A. centrale* to reduce the severity of an *A. marginale* infection and this approach is still used in several countries. Premunization with small doses of *A. marginale*-infected blood with or without subsequent control of the infection with drugs has been used, although this leads to a carrier state in immunized animals. Attenuation of *A. marginale* by irradiation or by passage in deer and sheep has been employed in the preparation of vaccines in various countries where the disease is endemic.

Anaplasma centrale. This organism is morphologically similar to and possibly a variant of *A. marginale*. As its name suggests, it is centrally placed in the erythrocyte. *A. centrale* infections are comparable to those of *A. marginale* except that they are milder.

Anaplasma ovis. This organism is referred to in the literature, but there is some doubt as to the validity of the species. *Anaplasma marginale* may be maintained by subpassage through sheep and goats in which it is usually non-pathogenic, and during passage the virulence of the organism for cattle is reduced or lost. This organism resembles *A. marginale*, and it may, under natural conditions, be a form which has become adapted to sheep. Infections in sheep have been reported from North America, North and South Africa, Israel and the Soviet Union.

Genus: Eperythrozoon Schilling, 1928

Organisms of this genus are minute prokaryotic forms occurring on the surface of erythrocytes and in the plasma; they are usually minute rings or coccoid-shaped granular bodies, 0.5–3 μm in diameter, and stain reddish-purple with Romanowsky stains. Electron microscope studies show them to be oblong, rod-shaped or pleomorphic forms, surrounded by a single limiting membrane with a cytoplasm containing an accumulation of electron-dense material near each end.

Differentiation from *Haemobartonella* may present difficulty. *Haemobartonella* rarely occur as ring forms and are more closely associated with the cell surface.

Several species of *Erythrozoon* have been reported from various parts of the world; the following are of immediate interest:

Eperythrozoon coccoides Shilling 1928 (mice, rats, rabbits, hamsters)

Eperythrozoon ovis Neitz, Alexander and du Toit 1934 (sheep, goats)

Eperythrozoon parvum Splitter 1950 (pigs)

Eperythrozoon suis Splitter, 1950 (pigs)

Eperythrozoon wenyoni Adler and Ellenbogen 1934 (cattle)

(*Eperythrozoon felis* Clark 1942 of cats is probably a misidentification of *Haemobartonella felis*)

The developmental cycle of *Eperythrozoon* is not fully known. It is believed the parasites multiply in the blood and have an affinity with erythrocytes. The infection is readily transmissible by blood inoculation, but the natural mode of transmission has not been fully investigated. In some cases arthropods may be responsible for transmission.

Organisms are susceptible to arsenicals and tetracyclines but not the penicillins, streptomycin or sulphonamides. Gothe and Kreier (1977) give a partial list of drugs tested for activity against *Eperythrozoon*.

Eperythrozoon suis and E. parvum occur in pigs. *E. suis* is the larger form, occurring as rings 2–3 μm in diameter. It is also the more pathogenic, severe infections causing icteroanaemia (Splitter 1950), which is an anaplasmosis-like disease of swine. The smaller form, *E. parvum*, is 0.5–0.8 μm in diameter, and it is generally non-pathogenic, although clinical signs may be seen in splenectomized pigs. The organisms are transmitted by all parenteral routes and Jansen (1952) has reported that the pig louse (*Haematopinus suis*) may transmit the infections.

The pathogenecity of *E. suis* has been described by Splitter (1950). Following an incubation period of about nine days, the organisms appear in the blood, and this coincides with the elevation of body temperature, which may reach 41.7°C. Severe infections cause anaemia, anorexia, and jaundice. The highest morbidity and mortality occur in suckling pigs, and in acute infections pigs may die in less than five days. In older animals, such as weanlings, death may be delayed, or the animal may recover. Although the infection is widespread it is, in general, not responsible for significant losses.

The pathological changes consist of anaemia, jaundice (yellow belly), the liver is brownish in colour and the bile yellowish green and viscid. The spleen is enlarged and hyperplastic, and there is hyperplasia of the bone marrow. Blood smears show large numbers of organisms on the red blood cells and in the plasma.

Diagnosis of eperythrozoonosis in pigs is based on the clinical signs and the demonstration of organisms in blood smears.

The organisms are susceptible to the tetracycline antibiotics, and oxytetracycline (Terramycin) given at the rate of 50–100 mg/kg orally, or intramuscularly or intravenously at the rate of 4 mg/kg daily, is effective.

Eperythrozoon ovis. This infection occurs as an occasional pathogen of sheep in South Africa, Australia, USA and Europe. The organisms are small pleomorphic forms, assuming ovals, rods or rings, on and between the red blood cells. The infection is usually benign in nature, but occasionally it produces an anaemia with icterus, an irregular fever, progressive emaciation and poor condition. Transmission by horse flies has been reported (Øverås 1959).

Eperythrozoon wenyoni. Cattle infected with this organism usually show little or no clinical manifestation. Occasionally, in severe infections, acute fever occurs, to be followed by emaciation and some jaundice. At this time parasites are very common in the blood. Distribution of the organism is world-wide.

Eperythrozoon coccoides. This is a common infection of laboratory mice. Parasitaemia may be very heavy but, despite this, anaemia may be very mild or non-existent. However, when mixed infections with other micro-organisms occur clinical or fatal disease may result. Thus, fatal mouse hepatitis results from combined infection with *E. coccoides* and mouse hepatitis virus (Gledhill 1956), possibly through the depression by *E. coccoides* of interferon production (Glasgow et al. 1974). However, mixed infections of *E. coccoides* and malaria were similar to malaria infection alone (Finerty et al. 1973). Murine eperythrozoonosis may be a complicating infection in haematological and immunological studies. Infection is transmitted by the parenteral and the oral routes and the mouse louse (*Polyplax serrata*) may transmit the infection. Arsenicals and tetracyclines may be used to eliminate infections.

Genus: Haemobartonella Tyzzer and Weinman, 1939

Obligate prokaryotic parasites which occur on or within erythrocytes of many species. As seen by the light microscope they are bacilli-form, or coccoid forms in chains; they occur singly or in groups in indentations on the erythrocyte surface. They rarely occur in the plasma. They stain intensely with Romanowsky stains, such that on first appearance the blood smear appears to be very poorly stained. Ultrastructurally they possess a single or a double limiting membrane with no distinct nuclear structures (Tanaka et al 1965).

Cross-serological reactions with *Haemobartonella* are seen with *Anaplasma* and *Eperythrozoon*.

As mentioned previously, *Haemobartonella* and *Eperythrozoon* may be difficult to differentiate. The major criterion is the close attachment of *Haemobartonella* to the erythrocyte compared to *Eperythrozoon* which is readily dislodged and frequently seen free in the plasma.

The growth of *Haemobartonella* is inhibited by arsenicals and tetracyclines but not by penicillin or streptomycin.

A number of species occur and are listed by Gothe and Kreier (1977); of these the following are of immediate importance:

Haemobartonella canis (Kikuth) Tyzzer and Weinman 1939 (dog)
Haemobartonella felis (Cork) Flint and McKelvie 1956 (cat)
Haemobartonella muris (Mayer) Tyzzer and Weinman 1939 (rat, mouse, hamster)

Haemobartonella canis. This occurs in the dog and on occasion may be responsible for producing anaemia, emaciation and anorexia. Young puppies are the most susceptible to the infection. Splenectomy exacerbates the clinical signs. The dog tick, *Rhipicephalus sanguineus*, has been incriminated as a vector (Seneviratna et al. 1973).

Haemobartonella felis. Previously, this organism was described as *Eperythrozoon felis*. It is responsible for a severe and often fatal anaemia in cats. Acute, subacute and chronic forms of the infection may occur. In the acute form there is intermittent fever with a progressive anaemia, corresponding to the level of organisms in the circulating blood. The infection is most common in young cats; if not diagnosed and treated, it results in extended illness with anaemia and perhaps death. The tetracycline antibiotics are effective when given orally for 18–21 days at the rate of 100 mg/kg three times a day for chloramphenicol (Chloromycetin), or 50 mg/kg daily orally or 5 mg/kg daily intravenously or intramuscularly for Terramycin.

Haemobartonella muris. This is world-wide in distribution and the rat louse *Polyplax spinulosa* is an important vector. Infection is usually inapparent but may become clinical on splenectomy or when the animal is otherwise compromised immunologically.

Genus: Aegyptianella Carpano, 1929

Previously, this genus was thought to be related to *Babesia* but produced 4–16 trophozoites instead of two; indeed, Laird and Lari (1957) thought it unnecessary to assign a separate genus to the organisms and retained them in the genus *Babesia*.

Four species have been described, chiefly in birds, but also in the tortoise and snakes. Two species are of interest:

Aegyptianella pullorum Carpano, 1929 (poultry)
Aegyptianella moshkovskii (Schurenkova, 1938) Poisson 1953 (poultry)

Aegyptianella pullorum. *Hosts:* domestic chicken, goose, also duck and turkey. (Experimentally in doves, pigeons, quail, canaries and other birds.) Sudan, North and South Africa, South-east Asia, India, south-eastern Europe and Soviet Union.

Morphology. The early trophozoites or initial bodies occur in erythrocytes and are small, 0.5–1.0 μm, round to oval and consist of a chromatin granule with a small ring of cytoplasm. Spherical bodies up to 4 μm may occur, containing up to 26 small granules. Electron microscopic studies reveal parasites surrounded by a double membrane enclosing electron-dense aggregates of granular material. The parasites occur in a vacuole and are separated from the erythrocyte cytoplasm by a limiting membrane. A detailed description is given by Gothe and Kreier (1977).

The developmental cycle in the avian host consists of the formation of initial bodies, developmental forms and marginal bodies and these have been considered in detail by Gothe (1971) who reports their occurrence in erythrocytes, other cells and in the plasma of infected birds.

Transmission is through the fowl tick, *Argas persicus*, and the developmental stages in the tick have been summarized by Gothe and Kreier (1977). Transmission in the tick is by neither the stage to stage nor the transovarian routes. Following feeding by an adult tick on an infected fowl, 25 days or more are required before the organism is transmissible to another bird.

Pathogenesis. The disease may be acute, subacute or chronic. Indigenous poultry rarely suffer the acute disease, but freshly introduced stock may die within a few days of the onset of the clinical entity. The incubation period is 12–15 days after which there is fever, diarrhoea, anorexia and jaundice. At post mortem there is anaemia, enlargement of the spleen, degeneration of the kidneys. The clinical condition is often complicated by fowl spirochaetosis (*Borrelia*), which is also transmitted by *A. persicus*.

Treatment. The tetracyclines and dithiosemicarbazone can be used. A single parenteral dose of 25–50 mg/kg of oxytetracycline or chlortetracycline is effective.

Aegyptianella moshkovskii. *Hosts:* chicken, possibly also turkey, pheasant, house crow and other birds. Indian subcontinent, South-east Asia, Egypt and eastern parts of the Soviet Union. The early initial forms are 0.2–0.6 μm in diameter, and later there are larger ring forms 2.1 by 1.4 μm and large oval or irregular forms 0.9–5.3 μm in diameter. The developmental cycle and the pathogenesis of this form are unknown.

Genus: Grahamella Brumpt, 1911

These are intraerythrocytic organisms of small mammals, and are world-wide in distribution. They are long or short rod-shaped, some may be dumb-bell-shaped. They stain intensely with Giemsa stain and are 0.5–1.0 μm long and 0.2 μm wide. They are bacteria, related to the *Bartonella* and may be confused with *Haemobartonella*. Blood-sucking ectoparasites are concerned in transmission and fleas are important in this respect.

Many species have been described and these are listed by Weinman and Kreier (1977). Infections in rodents may complicate immunological or haematological data.

Genus: Ehrlichia Moshkovski, 1945

Small pleomorphic, coccoid to ellipsoidal intracytoplasmic forms in circulating leucocytes of various mammals. Organisms may occur singly or in compact colonies as a 'morula' which is the characteristic form of the organism. Organisms are transmitted by Ixodid ticks, transtadial transmission occurs but transovarian transmission does not (Smith & Ristic 1977).

Organisms are susceptible to the tetracyclines. The following species are of importance:

Ehrlichia bovis Moshkovski, 1945 (cattle, sheep)
Ehrlichia canis Moshkovski, 1945 (dogs)
Ehrlichia equi Stannard, Gribble and Smith, 1969 (horse, donkey)
Ehrlichia ovina Moskovski, 1945 (sheep)
Ehrlichia phagocytophilia Philip, 1962 (cattle, sheep, goat)

Ehrlichia bovis occurs in mononuclear cells of cattle in North and Central Africa, the Middle East and Ceylon and is transmitted by *Hyalomma* species of ticks. The disease is known as 'Nopi' or 'Nofel' in West Africa and may be acute, subacute or chronic. Anorexia, fever, incoordination and enlargement of lymph nodes occur (Smith & Ristic 1977).

Ehrlichia canis is the cause of canine ehrlichiosis or tropical canine pancytopenia. Organisms occur in monocytes and the disease may be mild (Ewing 1969) but often is an acute febrile entity characterized by pancytopenia and, particularly, thrombocytopenia. The disease is often fatal. The organism is transmitted by *Rhipicephalus appendiculatus* and occurs globally.

Ehrlichia equi has been reported to cause equine ehrlichiosis in the Sacramento Valley in California. Clinical signs include fever, anorexia, incoordination and oedema of the legs. The disease is rarely fatal. Organisms occur in neutrophils and eosinophils.

Ehrlichia ovina occurs in mononuclear cells of sheep in North and Central Africa. It is transmitted by *Rhipicephalus bursa* and usually causes only a mild disease.

Ehrlichia phagocytophilia is the cause of tick-borne fever of sheep (and cattle) in Great Britain and Europe. It occurs in neutrophils and eosinophils and is transmitted by *Ixodes ricinus*. A febrile disease with relatively low mortality is produced; abortion may occur in pregnant animals.

Species of Uncertain Classification

Pneumocystis carinii Delanoë, 1912 is the aetiological agent of an interstitial pneumonia of man and other animals characterized by a massive mononuclear cell, predominantly plasma cell, infiltration. Latent infection is common and clinical disease occurs in the very young and the old, especially when debilitative factors occur.

Pneumocystis carinii shows associations with the protozoa but also with the fungi. Some authors (e.g. Faust et al. 1975) place it in the *sporozoa* with relationships to *Toxoplasma*. Electron microscopic studies have shown it to be a eukaryotic organism, but it lacks the ultra-structural affinities of the spore and cyst forming protozoa (Seed & Aikawa 1977).

Four morphological forms are recognized during the life-cycle: trophozoites, precystic forms, cystic forms and intracystic bodies (Seed & Aikawa 1977). Development occurs in the alveoli of infected lungs. Initially, extracellular trophozoites increase in size, resulting in either cell division or spore formation similar to ascospore formation of yeasts. Development of the intracystic stage is described by Seed and Aikawa (1977).

The organism is widely distributed, occurs in a number of animals and is considered to be zoonotic. It is an air-borne infection, and latent infections are common. Clinical disease is associated with entities which comprise cell-mediated immunity (e.g. leukaemia, multiple myeloma, etc.) or with the use of immunosuppressive drugs. In recent years pneumocystis infection has increased in patients with tissue and organ transplants.

The role of animals in human infection has yet to be clarified. The organism has been demonstrated in a wide range of animals, particularly rodents, although the dog is considered to be an important reservoir host. In man, infection appears to be transmitted from infant to infant or infant to adult; however, in institutions (e.g. orphanages) where the infection is endemic, mice have been found to have the parasites in their lungs.

Treatment of *P. carinii* infection has been reviewed by Steck (1971). Two groups of compounds have shown efficacy: the antifolates and the stilbamidines. Thus, a combination of sulphadiazone and pyrimethamine, stilbamidine or pentamidine (Steck 1971) are the drugs of choice in the treatment of *Pneumocystis* infection.

REFERENCES

RICKETTSIALES (INCLUDING *PNEUMOCYSTIS*)

Carson, C. A. Weisinger, R., Ristic, M., Thwimon, J. C. & Nelson, D. R. (1974) Appendage related antigen production by *Paranaplasma candatum* in deer erythrocytes. *Am. J. vet. Res.*, **35**, 1529–1531.

Espana, E. & Espana, C. (1963) *Anaplasma marginale*. II. Further studies of morphologic features with phase contrast and light microscopy. *Am. J. vet. Res.*, **24**, 713–722.

Ewing, S. A. (1969) Canine ehrlichiosis. *Adv. vet. Sci. comp. Med.*, **13**, 331–353.

Faust, E. C., Beaver, P. C. & Jemg, R. C. (1975) *Animal Agents and Vectors of Human Disease*, 4th ed., Philadelphia: Lea and Febiger.

Finerty, J. F., Evans, C. B. & Hyde, C. L. (1973) *Plasmodium berghe*; and *Eperythrozoon coccoides*. Antibody and immunoglobulin synthesis in germ-free and conventional mice simultaneously infected. *Exp. Parasit.*, **34**, 76–84.

Glasgow, L. A., Murrer, A. T. & Lumbardi, P. S. (1974) *Eperythrozoon coccoides*. II. Effect on interferon production and role of humoral antibody in host resistance in mice. *Infect. Immun.*, **9**, 266–272.

Gledhill, A. W. (1956) Quantitative aspects of the enhancing of eperythrozoa on the pathogenicity of mouse hepatitis virus. *J. gen. Microbiol.*, **15**, 292–304.

Gothe, R. (1971) Wirt-Parasit-Verhältnis von *Aegyptianella pullorum* Carpano, 1928 im biologischen Überträger *Argas (Persicargas) persicus* (Oken 1818) und im Wirbeltier-wirt *Gallus gallus domesticus* L. *Z. vet. Med. Bh.*, **16**, 144.

Gothe, R. & Kreier, J. P. (1977) *Aegyptianella, Eperythrozoon*, and *Haemobartonella*. In: *Parasitic Protozoa*, ed. J. P. Kreier, vol. IV, pp. 251–294. New York: Academic Press.

Jansen, B. C. (1952) The occurrence of *Eperythrozoon parvum* Splitter, 1950 in South African swine. *Onderstepoort. J. vet. Res.*, **25**, 5–6.

Kreier, J. P. & Ristic, M. (1963*a*) Anaplasmosis. X. Morphologic characteristics of the parasites present in the blood of calves infected with the Oregon strain of *Anaplasma marginale*. *Am. J. vet. Res.*, **24**, 672–676.
Kreier, J. P. & Ristic, M. (1963*b*) Anaplasmosis. XI. Immunoserologic characteristics of the parasites present in the blood of calves infected with the Oregon strain of *Anaplasma marginale*. *Am. J. vet. Res.*, **24**, 688–696.
Kreier, J. P. & Ristic, M. (1963*c*) Anaplasmosis. XII. The growth and survival in deer and sheep of the parasites present in the blood of calves infected with the Oregon strain of *Anaplasma marginale*. *Am. J. vet. Res.*, **24**, 697–702.
Laird, M. & Lari, F. A. (1957) The avian blood parasite *Babesia Moshkovskii* (Schurenkova 1938) with a record from *Corvus spendens* Vieillot in Pakistan. *Can. J. Zool.*, **35**, 783–795.
Øverås, J. (1959) *Eperythrozoon ovis*, a new blood parasite in sheep in Norway. *Nord. vet. Med.*, **11**, 791–800.
Ristic, M. (1968) Anaplasmosis. In: *Infectious Blood Diseases of Man and Animals*, eds D. Weinman & M. Ristic, pp. 478–542. New York: Academic Press.
Ristic, M. (1977) Bovine anaplasmosis. In: *Parasitic Protozoa*, ed. J. P. Kreier, vol. IV, pp. 235–249. New York: Academic Press.
Ristic, M. & Watrach, A. M. (1963) Anaplasmosis. VI. Studies and a hypothesis concerning the cycle of development of the causative agent. *Am. J. vet. Res.*, **24**, 267–277.
Roby, T. O. & Mazzola, V. (1972) Elimination of the carrier state of bovine anaplasmosis with imidocarb. *Am. J. vet. Res.*, **33**, 1931–1933.
Seed, T. M. & Aikawa, M. (1977) Pneumocystis. In: *Parasitic Protozoa*, ed. J. P. Kreier, vol. IV, pp. 329–357. New York: Academic Press.
Seneviratna, P., Weerasinghe, N. & Ariyadasa, S. (1973) Transmission of *Haemobartonella* canis by the dog tick *Rhipicephalus sanguineus*. *Res. vet. Sci.*, **14**, 112–114.
Smith, R. D. & Ristic, M. (1977) Ehrlichiae. In: *Parasitic Protozoa*, ed. J. P. Kreier, vol. IV, pp. 295–328. New York: Academic Press.
Splitter, E. J. (1950) Ictero-anaemia of Swine. *Proc. 54th Ann. Meeting US Livestock Sanit. Ass.*, Phoenix 1950, pp. 279–386.
Steck, E. (1971) *The Chemotherapy of Protozoan Diseases*, p. 26. 1–27.13. Washington DC: Division of Medicinal Chemistry, Walter Reed Army Institute of Research.
Tanaka, H., Hall, N., Scheffield, J. & Moore, D. (1965) Fine structure of *Haemobartonella muris* as compared with *Eperythrozoon coccoides* and *Mycoplasma pulmonis*. *J. Bacteriol.*, **90**, 1735–1749.
Theiler, A. (1910) *Anaplasma marginale*. The marginal points in the blood of cattle suffering from specific disease. *Sov. vet. Bacteriol. Transvaal. South Africa*, pp. 6–64.
Weinman, D. & Kreier, J. P. (1977) *Bartonella* and *Grahamella*. In: *Parasitic Protozoa*, ed. J. P. Kreier, vol. IV, pp. 197–233. New York: Academic Press.

4

Technique

COLLECTION AND PRESERVATION OF HELMINTHS

The collection of parasites from an animal should be carried out as systematically and completely as possible. For this purpose it is wise to follow a definite scheme and examine each organ as it is presented at autopsy, beginning with the outside of the body, then the subcutaneous tissues, the body cavities and so forth.

Filarioid worms found under the skin and in the body cavities and blood vessels are best placed immediately in a 10% solution of formalin, without washing in saline, unless they are soiled with blood.

Large nematodes should be collected and washed by shaking in 0.9% saline and immediately dropped into hot 70% alcohol or 5% formol saline, which causes them to be fixed in an extended state. They can be stored in the fixative.

It may be difficult to remove nematode larvae or small adult nematodes situated in tissues and it may be necessary to digest tissue with an artificial digestion mixture (see below). However they may be freed of tissue if the material is teased apart in warm physiological saline. The worms should be removed and fixed as soon as they are free.

Cestodes must be collected with scolices, as these are of great importance for identification. The worms are placed into a dish of warm water at about 40°C, and if their heads are attached to the intestinal wall, a piece of the latter is cut out and also placed in the dish. The worms will die fully extended usually in about an hour and the heads will be free or may have to be dissected out. The worms are fixed, for permanent preparations, in 5% formol saline, in cold 70% alcohol with 5% glycerin, or equal parts of 70% alcohol, glycerin and distilled water, or Zenker's fluid, drawing large forms a few times over the hand or the edge of the vessel and small ones rapidly through the fluid, in order to get them fully extended. When fixing in Zenker's fluid the worms must be removed after 24 hours and thoroughly washed in running water.

The segments can be stained alive or fixed in a freshly made mixture of 97 parts of a saturated solution of carmine in 45% acetic acid and three parts of a saturated solution of ferric acetate in glacial acetic acid, staining for 5–30 minutes and then mounting in lactophenol.

Trematodes are treated like cestodes, but if greater extension is required, they should be placed between glass slides held together by rubber bands. Trematodes may be cleaned by shaking them in a 1% salt solution or in cold or lukewarm water. Much of their anatomy may be seen before they are fixed. Fix by pouring off the salt solution and adding 10% formalin, replacing this with 3% formalin when the flukes are fixed. Flukes may be fixed in Bouin's, Müller's or Helly's fixative, the corrosive sublimate in the latter two fixatives being removed with iodine in 70% alcohol.

Acanthocephala should be pressed between glass slides, in order to get the proboscis extended, and so fixed in 70% alcohol.

Small worms of all kinds are usually of great importance and should not be missed. After the larger worms have been removed the digestive tract is cut into convenient lengths, and each is vigorously shaken in warm water, collecting the worms that are liberated.

Faeces containing eggs are fixed by mixing them with an equal quantity of hot 10% formalin. Cold formalin can be used, but the eggs of ascarids, for example, may continue to develop in this fixative. The material should be transferred gradually to 70% alcohol if it is to be kept for an indefinite period.

COLLECTION AND PRESERVATION OF ARTHROPOD PARASITES

Specimens for dry mounting, such as adult *Diptera*, are killed in a cyanide killing bottle, made by placing into a wide-mouthed bottle a freshly mixed paste of potassium cyanide and plaster of Paris, which is covered with a few layers of blotting paper. Strips of paper are placed into the bottle to prevent the insects from clinging together and damaging their appendages. Specimens are set out on a mounting board in the ordinary way and then pinned down in boxes, the floors of which are covered with a mixture of wax and naphthalin. Some material may also be collected in 70% alcohol for making mounts of important parts of the arthropods.

Dipterous larvae, lice, fleas, mites, etc, are fixed and stored in 70% alcohol or 10% formalin. Care must be exercised not to miss the small parasites, and in the case of skin mites it may be necessary to make deep skin scrapings of affected areas in order to find them. Fleas should be put in small tubes containing 70–80% alcohol, which will preserve them indefinitely. A drop of glycerin may be added if they are to be preserved for a long time. Formalin should not be used for fleas.

Ticks can be preserved in their natural colours by dropping them alive into a solution of chloroform in 10% formalin. The formalin should be made up with distilled water. Chloroform is added in slight excess; the mixture is shaken and left for a few minutes to settle, and then the solution is poured into a bottle with a well-fitting glass stopper. The live ticks are dropped in and the bottle is left closed for about a month.

MAKING PERMANENT PREPARATIONS

Canada balsam is an excellent mounting medium, but requires careful preparation of the material by passing it through increasing concentrations of alcohol to draw out all water, during which shrinkage, especially of nematodes, is difficult to avoid. Glycerin gelatin is simpler, but also frequently causes shrinkage. It is very useful for eggs and small nematodes that tend to become too transparent in other media.

For all unstained material, such as nematodes, tapeworm heads, acanthocephala and arthropods, a suitable medium is the following mixture: gum arabic 60, glycerin 40, chloral hydrate 100, and thymol 1. The specimens are transferred from 70% alcohol into 50% glycerin and then mounted in the gum arabic medium. For small insects and mites Berlese's fluid is useful. It is made by dissolving 15 g gum arabic in 20 ml distilled water and adding 10 ml glucose syrup and 5 ml acetic acid, the whole being then saturated with chloral hydrate (up to 100 g). This should set in one or two weeks and the slide may then be sealed by ringing the edges of the cover-slip. Mites may also be mounted in a jelly made of 20 parts gelatin, 100 parts glycerin, 120 parts carbolic acid and 2 parts distilled water, the slide being ringed after mounting.

Arthropod material, such as heads of flies, whole fleas etc., may first be boiled in 10% potassium hydroxide solution to remove the soft internal parts, and it may be necessary to puncture the body for this purpose. The material is then washed in water and transferred to 70% alcohol etc.

While nematodes are usually examined in lactophenol, into which they can be placed from 70% alcohol, cestodes and trematodes have to be stained so that the internal organs can be seen, and the preparation is mounted in balsam in the ordinary way. Various stains are employed, e.g. Delafield's haematoxylin, Ehrlich's acid haematoxylin, paracarmine, acid alum carmine, lithium carmine and haemalum with 2% acetic acid. It is usually advisable to dilute the stain considerably with distilled water and to stain for a prolonged period, rather than to use the concentrated stain. The specimens should be slightly overstained and differentiated in a 0.1% acid alcohol or alum solution.

CLINICAL DIAGNOSTIC METHODS

Examination of the Outside of the Body

The body is searched for external parasites or their eggs (bots, oxyurids), not only on the surface but also in the ears and in the conjunctival sac (eyeworms), and the skin should be palpated to determine the presence of subcutaneous larvae. If mange-like lesions are present, the hair round the affected area should be clipped and scrapings made with a scalpel, the blade being held at such an angle that the material scraped away falls onto a piece of card or paper or a microscope slide held underneath. A little oil on the blade used will cause the material to adhere to the blade, so that it is not lost. Scraping should continue until a little blood appears, especially when sarcoptic mange is suspected. The lesions should then be dressed and the material examined for the presence of mites or of fragments of them. Some material may be examined directly, either in water or saline or in light oil, e.g. clove oil. If it is too dense for direct examination, it should be brought just to the boil in 10% caustic soda or caustic potash to break it up. It may then be examined in the hydroxide used, but it is better to centrifuge it lightly and examine the sediment. It may be possible to find mites in the external auditory canal by rotating a cotton-wool swab in this canal. A little oil on the swab will help to capture the mites. Examination with an illuminated auriscope may be useful.

Examination of Excretions

Excretions of the body may contain parasite eggs or larvae. The nasal discharge and the sputum may, therefore, aid in the diagnosis of parasites in the air-passages, the vomit may bear evidence of parasites in the stomach, and the urine may contain eggs which can be concentrated by centrifuging. The faeces are by far the most important, as the eggs or larvae of gastro-intestinal parasites and many others leave the body in the faeces. It should be remembered, however, that no eggs may be found if the worms are still immature or if only males are present. In some cases abnormally shaped eggs may be found.

Faeces. Faeces are examined in the first place for adult parasites, larval stages of insects (e.g. bots) or segments of tapeworms and the student should be familiar with their appearance, otherwise he may be at a loss to recognize these bodies. If a tapeworm infection is suspected, a purgative may be given to cause the expulsion of segments in case they are not readily found.

Birds that have caeca, as the domestic fowl for example, pass two kinds of faeces, those from the small intestine being relatively coarse and loose with particles of varying colour, while those from the caeca are of a fine, pasty nature with a homogeneous brown or brownish-green colour. The eggs of small-intestinal worms will be found in both types of faeces, while those of caecal worms are found only in the caecal faeces.

Egg counts. For the counting of the number of eggs in faeces there are various methods. The following are some of the more commonly used techniques:

1. *Direct smear.* A small quantity of faeces is placed on a slide, mixed with a drop of water, spread out covered with a slip and examined directly. At least three slides from different parts of the faecal sample should be examined. This method is suitable for a very rapid examination, but will usually fail to detect low-grade infections and it is not a quantitative method.

2. *Concentration methods.* The purpose of these methods is to detect light infections as well as others, to save time by concentrating the eggs in a small volume and to eliminate the trouble caused by large faecal particles. Advantage is taken of the low specific gravity of most helminth eggs to separate them from the faeces, as described below.

In the *Willis technique* about 1 ml of mixed faecal specimen is diluted with 10–20 ml of saturated common salt solution in a suitable narrow cylinder, which is filled to the top with the liquid. A clean slide or cover glass is slid sideways over the top of the cylinder so that it is in contact with the liquid, care being taken that there are no air bubbles between the slide or cover glass and the salt solution. After about 10–20 minutes the slide or cover glass is quickly removed and examined under a low power of the microscope. This method is not suitable for eggs of trematodes or most cestodes, but is suitable for the majority of nematode eggs. A saturated sugar solution may be used instead of salt solution, one advantage of it being that it is sticky so that the eggs are less likely to be washed off the slide or cover glass when it is removed.

By *centrifugal flotation* methods more time is saved and greater accuracy obtained. A sample of faeces, 1–5 g, is well mixed with water (about 30–50 ml) and strained through a sieve (1 mm mesh) to remove coarse faecal material. The mixture is sedimented for 10–15 minutes on the bench, or by light centrifugation for two or three occasions, until the supernatant is clear. The sediment is then mixed with a saturated solution of sugar, salt, or zinc sulphate in a centrifuge tube

(15–50 ml volume) and centrifuged for one or two minutes at 500 g. The eggs will float to the surface and may be removed by touching the surface with the end of a square-cut glass rod and transferred to a slide; alternatively the surface may be touched with a coverslip. If a 33% solution of zinc sulphate is used some eggs, e.g. those of *Fasciola hepatica*, may be distorted and, in addition, the higher the specific gravity of the flotation solution, the more debris is produced in the preparation. The specific gravities of various common solutions used in diagnosis are: saturated sodium chloride 1.200; sugar (sucrose) 1.12–1.30; zinc sulphate 33% 1.18; magnesium sulphate 35% 1.28; sodium nitrate 1.36; and potassium mercury iodide (111 g potassium iodide, 150 g mercury iodide; water 399 ml) 1.44.

The faeces may also contain larvae, for instance those of lungworms. Such larvae will be found by the flotation methods used for detecting the presence of eggs. A simple method for detecting larvae in sheep faeces is to place a few pellets in a Petri dish containing a small quantity of water, when the larvae will rapidly leave the pellets and can be found in the water. In the case of soft faeces the material is spread out in a Petri dish and several small holes are made by means of a glass rod going down to the bottom. The holes are filled with water and the larvae migrate into them from the faeces.

The eggs of many worms can be recognized by their shape and size and the student should become familiar with the eggs which may be found in the faeces of different species of domestic animals, as well as with the other objects which may appear in faeces and which may thus be mistaken for eggs, such as pollen grains, plant cells or hairs, fungus spores and cysts of protozoa. Figs 4.1 to 4.8 illustrate the eggs of worm parasites of various hosts and the infective larvae of some nematodes of sheep.

Egg counting techniques. To obtain accurate information with regard to the severity of an infection, egg-counting methods have been devised in order to determine the number of eggs per gram of faeces. All these methods, however, involve unavoidable sources of error, so that the results given by them must be regarded as only approximate and as supplemental to a thorough clinical examination of the individual animal or group.

Stoll's dilution method for counting nematode eggs uses 3 g of faeces weighed into a test-tube graduated to 45 ml. The tube is then filled to the 45 ml mark with decinormal caustic soda solution and 10 or 12 glass beads are added. The tube is then closed with a rubber stopper and is shaken to give a homogeneous suspension of the faecal material. After shaking, 0.15 ml of the well mixed suspension is drawn off with a pipette graduated to show this amount and placed on a slide. The total number of eggs in the 0.15 ml sample is then counted and this number, multiplied by 100, gives the number of eggs in 1 gram of faeces. Coarse material, such as horse faeces, can be sieved to remove fibres which may block the pipette. For field work a heavy flask graduated at 56 and 60 ml may be used. The flask is filled to the 56 ml mark with decinormal caustic soda solution and faeces are added until the fluid reaches the 60 ml mark. The 0.15 ml is withdrawn and the number of eggs in it is multiplied by 100 to give the number of eggs per gram of faeces. This method gives reliable counts of trematode as well as nematode eggs.

In the *McMaster egg-counting technique* the eggs are floated up in a counting chamber which dispenses with the need for a graduated pipette. The counting chamber is made of two glass slides, separated by three or four narrow, transversely placed strips of glass 1.5 mm thick, so that two or three spaces of 1.5 mm depth are obtained between the two slides. On the under-surface of the upper slide an area of 1 cm^2 is ruled over each space. The volume underneath this ruled area will therefore be 0.15 ml. From the faeces 2 g are weighed and soaked in 30 ml of water until they are sufficiently soft. To this is added 30 ml of saturated salt solution. After thorough shaking a sample is withdrawn by means of a wide (8 mm) pipette and run into the counting chamber, filling all the spaces. The eggs float to the under-surface of the upper slide, so that they are all in focus against the ruled area. The number of eggs within each ruled area, multiplied by 200, represents the

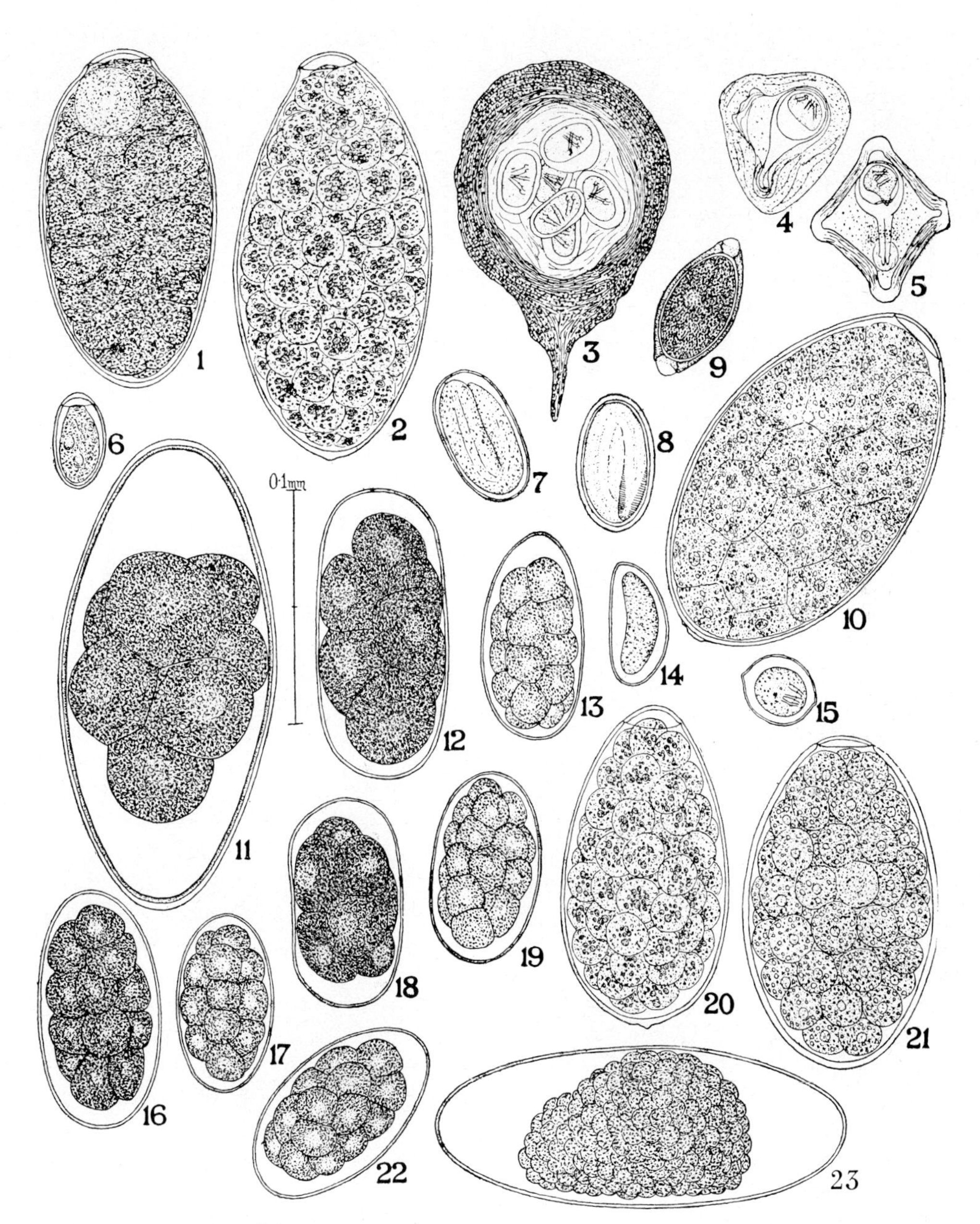

Fig 4.1 Eggs of worm parasites of sheep.

1 *Fasciola hepatica*
2 *Paramphistomum cervi*
3 *Thysaniezia giardi*
4 *Moniezia expansa*
5 *Moniezia benedeni*
6 *Dicrocoelium dendriticum*
7 *Stronglyloides papillosus*
8 *Gongylonema pulchrum*
9 *Trichuris globulosa*
10 *Fasciola gigantica*
11 *Nematodirus spathiger*
12 *Gaigeria pachyscelis*
13 *Trichostrongylus* spp.
14 *Skrjabinema ovis*
15 *Avitellina centripunctata*
16 *Chabertia ovina*
17 *Haemonchus contortus*
18 *Bunostomum trigonocephalum*
19 *Oesophagostomum columbianum*
20 *Cotylophoron cotylophorum*
21 *Fascioloides magna*
22 *Ostertagia circumcincta*
23 *Marshallagia marshalli*

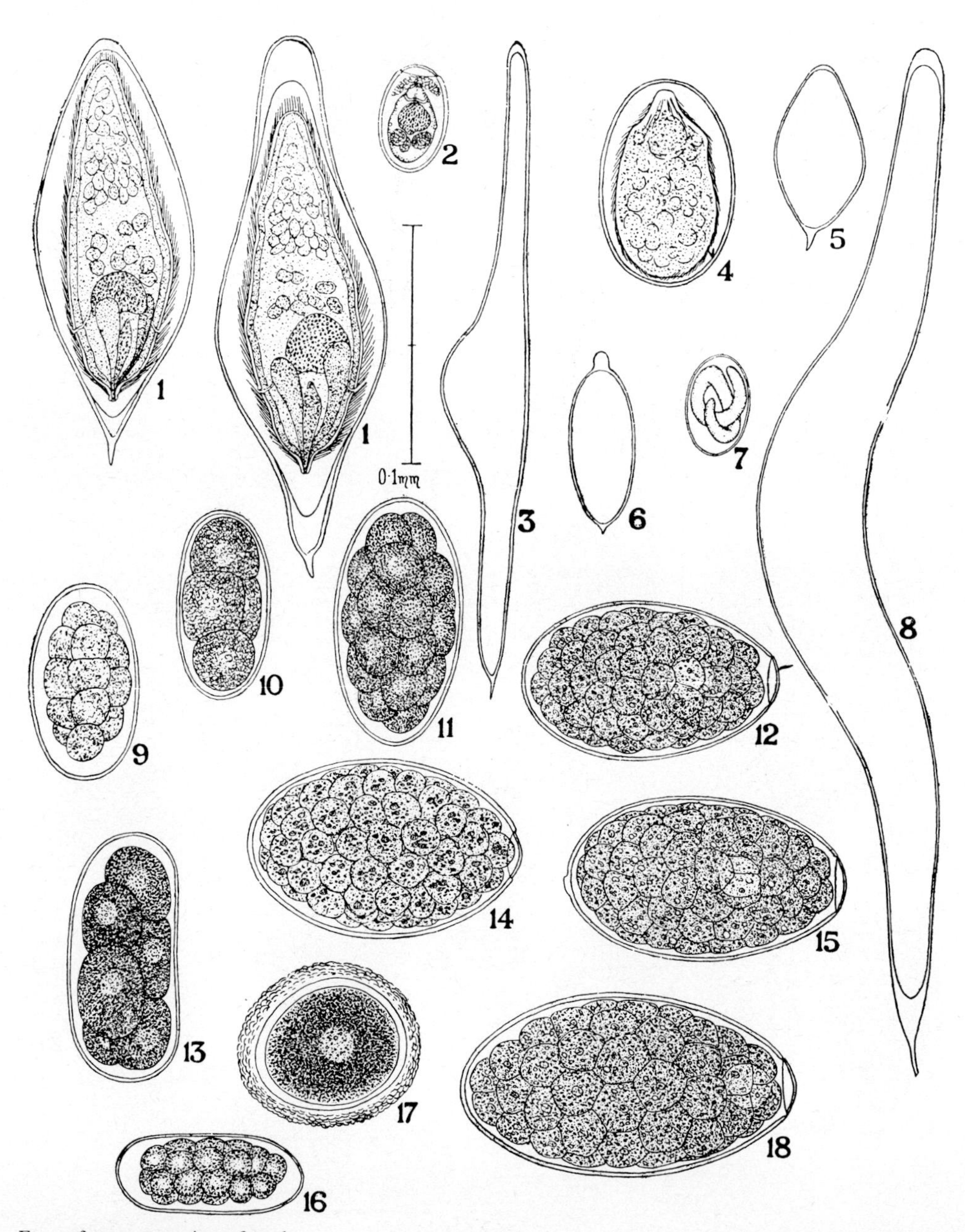

Fig 4.2 Eggs of worm parasites of cattle.

1 *Schistosoma bovis*
2 *Eurytrema pancreaticum*
3 *Schistosoma spindalis*
4 *Schistosoma japonicum*
5 *Schistosoma indicum*
6 *Ornithobilharzia turkestanicum*
7 *Thelazia rhodesii*
8 *Schistosoma nasalis*
9 *Oesophagostomum radiatum*
10 *Syngamus laryngeus*
11 *Mecistocirrus digitatus*
12 *Fischoederius cobboldi*
13 *Bunostomum phlebotomum*
14 *Carmyerius spatiosus*
15 *Gastrothylax crumenifer*
16 *Cooperia pectinita*
17 *Toxocara vitulorum*
18 *Fischoederius elongatus*

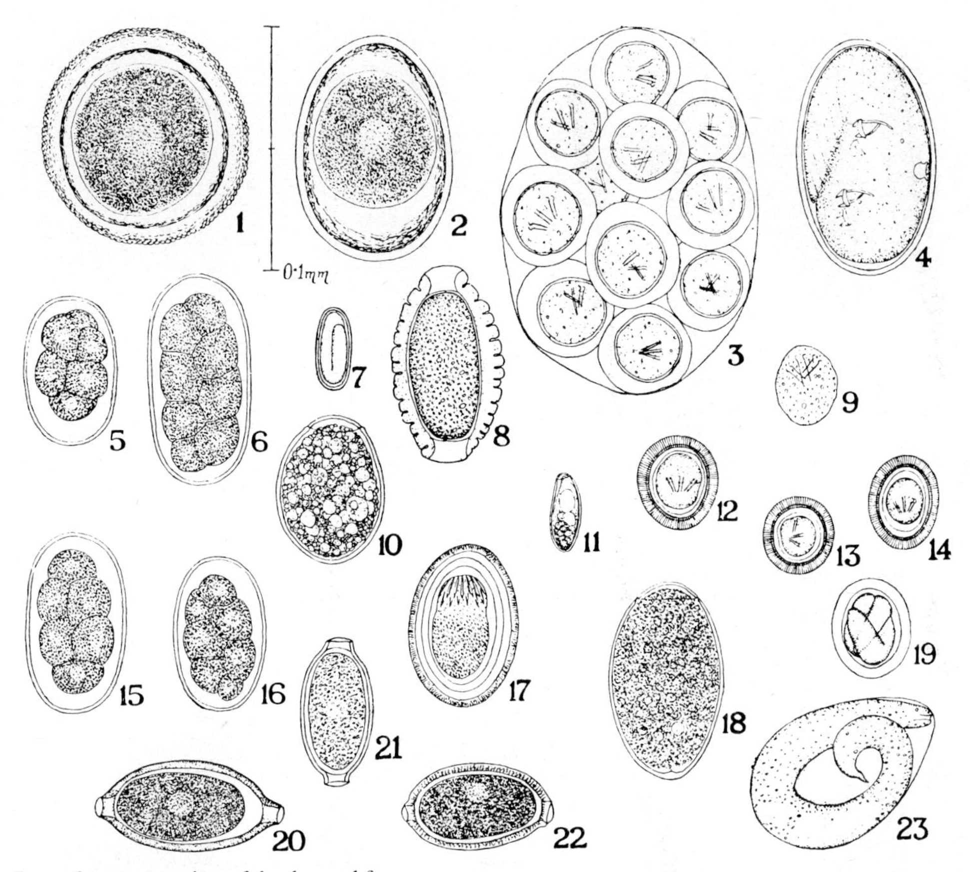

Fig 4.3 Eggs of worm parasites of the dog and fox.

1 *Toxocara canis*
2 *Toxascaris leonina*
3 *Dipylidium caninum*
4 *Linguatula serrata*
5 *Ancylostoma caninum*
6 *Ancylostoma braziliense*
7 *Spirocerca lupi*
8 *Dioctophyma renale*
9 *Mesocestoides lineatus*
10 *Diphyllobothrium latum*
11 *Euryhelmis squamula*
12 *Echinococcus granulosus*
13 *Taenia hydatigena*
14 *Taenia ovis*
15 *Uncinaria stenocephala*
16 *Necator americanus*
17 *Oncicola canis*
18 *Troglotrema salmincolo*
19 *Physaloptera canis*
20 *Trichuris vulpis*
21 *Capillaria plica*
22 *Capillaria aerophila*
23 *Filaroides osleri*

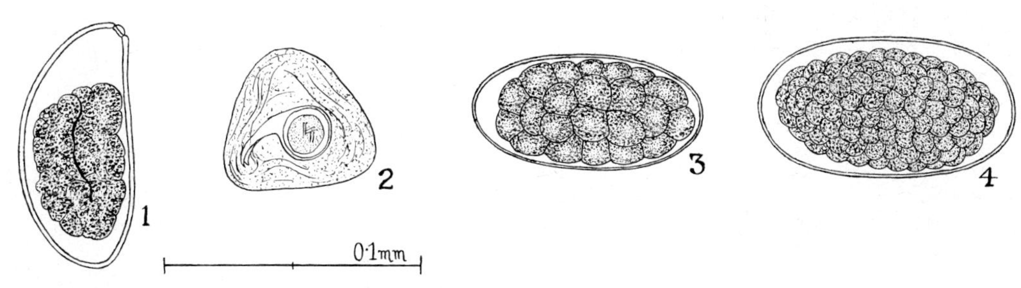

Fig 4.4 Eggs of worm parasites of the rabbit.

1 *Passalurus ambiguus*
2 *Cittotaenia ctenoides*
3 *Trichostrongylus retortaeformis*
4 *Graphidium strigosum*

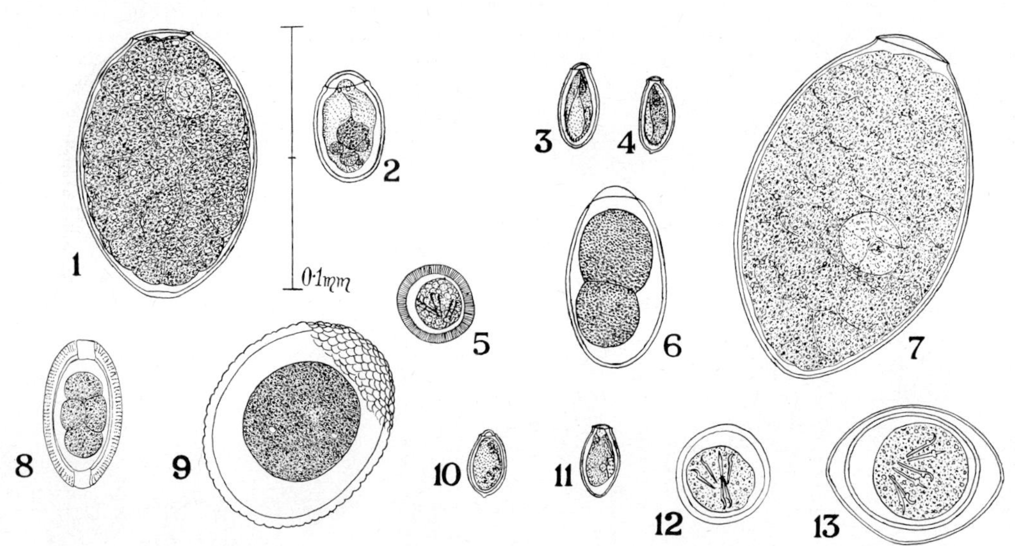

Fig 4.5 Eggs of worm parasites of the cat.
1 *Echinochasmus perfoliatus*
2 *Platynosomum concinnum*
3 *Opisthorchis sinensis*
4 *Opisthorchis tenuicollis*
5 *Taenia taeniaeformis*
6 *Gnathostoma spinigerum*
7 *Euparyphium melis*
8 *Capillaria hepatica*
9 *Toxocara mystax*
10 *Heterophyes heterophyes*
11 *Metagonimus yokogawai*
12 *Diplopylidium zchokkei*
13 *Joyeuxiella furhmanni*

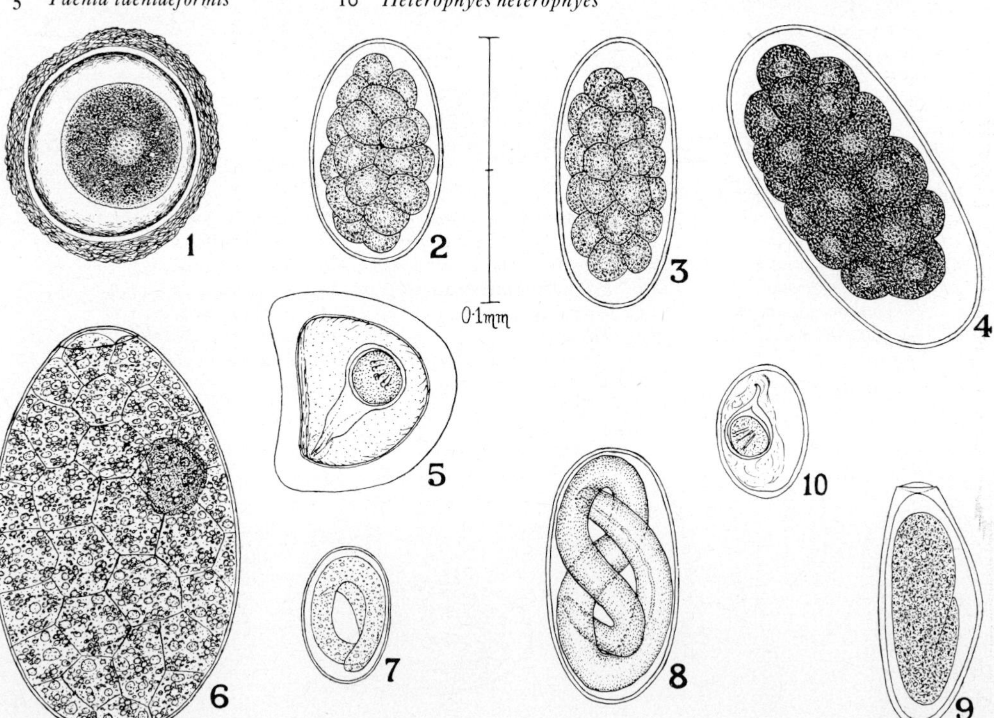

Fig 4.6 Eggs of worm parasites of equines.
1 *Ascaris equorum*
2 *Strongylus* spp.
3 *Trichomena* spp.
4 *Triodontophorus tenuicollis*
5 *Anoplocephala* spp.
6 *Gastrodiscus aegyptiacus*
7 *Strongyloides westeri*
8 *Dictyocaulus arnfieldi*
9 *Oxyuris equi*
10 *Paranoplocephala mamillana*

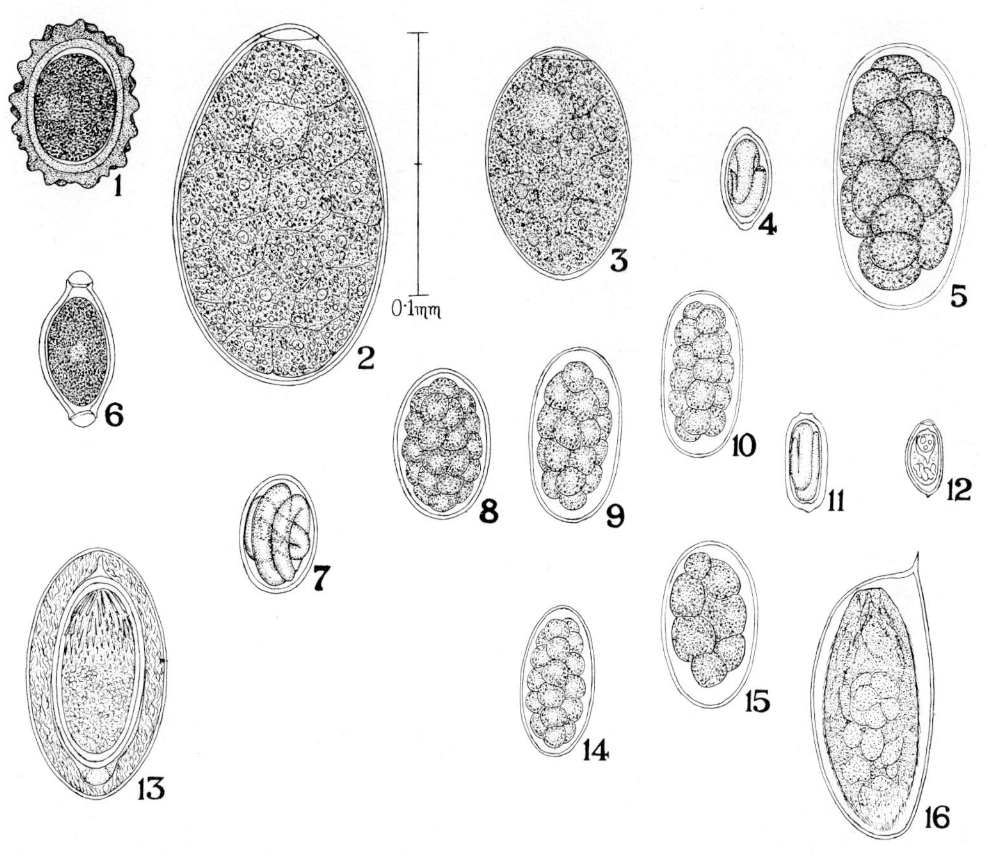

Fig 4.7 Eggs of worm parasites of pigs.

1 *Ascaris lumbricoides*
2 *Fasciolopsis buski*
3 *Paragonimus westermonii*
4 *Ascarops strongylina*
5 *Stephanurus dentatus*
6 *Trichuris trichura*
7 *Metastrongylus apri*
8 *Bourgelatia diducta*
9 *Oesophagostomum dentatum*
10 *Hyostrongylus rubidus*
11 *Physocephalus sexalatus*
12 *Brachylaemus suis*
13 *Macracanthorhynchus hirudinaceus*
14 *Globocephalus connorfilii*
15 *Necator* sp.
16 *Schistosoma suis*

number per gram of the original sample. According to the number of ruled areas examined and the amount of faeces used, an estimate of the faecal egg count within the limits of 10 epg may be obtained. There have been a number of modifications of this technique, many designed to remove the debris in the faecal sample. These include, for example, sieving the material through a metal coffee strainer or sieving through fine nylon mesh using two polystyrene cups, the bottom of one being cut off to form a plunger. A similar method is to weigh 2 g faeces into a mortar, grind it up in 10 ml of water and add 50 ml of saturated salt solution; enough of this mixture is taken, while it is being stirred, to fill the counting chambers. When large numbers of samples are being counted an electrically operated stirrer may be used.

Trematode eggs. Trematode eggs are often heavier than those of other helminths and the flotation methods described above, with the exception of zinc sulphate, will fail to detect them. The Stoll technique is useful when a large number of eggs are present, but often it is important to detect infections when there are few eggs in the faeces, e.g. during early patency or when anthelmintics have been administered.

For counting the eggs of *Fasciola hepatica* the following methods are available.

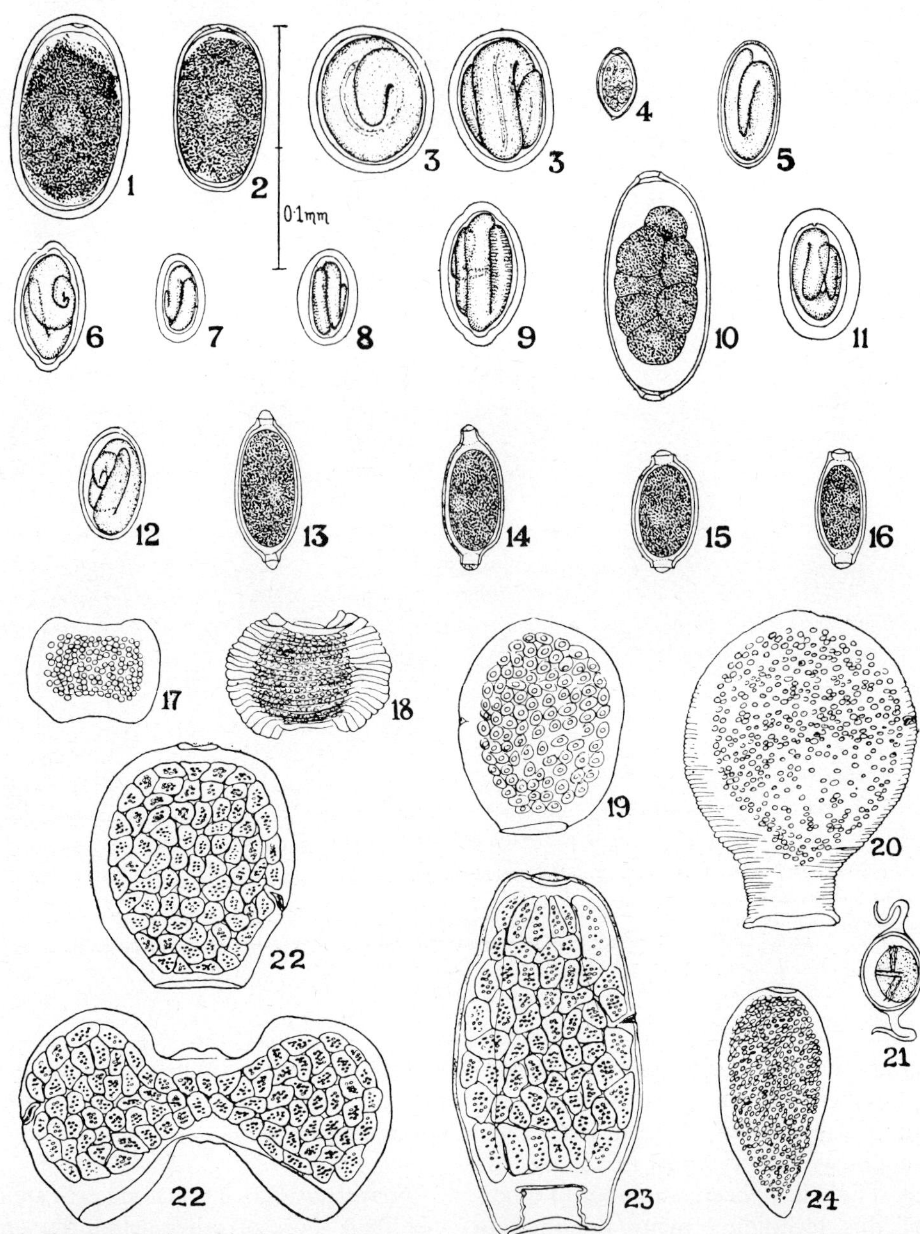

Fig 4.8 Eggs of worm parasites of fowls.

1 *Ascaridia galli*
2 *Heterakis galliae*
3 *Subulura brumpti*
4 *Prosthogonimus* sp.
5 *Strongyloides avium*
6 *Tetrameres americana*
7 *Acuaria spiralis*
8 *Acuaria hamulosa*
9 *Gongylonema ingluvicola*
10 *Syngamus trachea*
11 *Hartertia gallinarum*
12 *Oxyspirura mansoni*
13 *Capillaria annulata*
14 *Capillaria retusa*
15 *Capillaria columbae*
16 *Capillaria longicollis*. Ripe segments of tapeworms (not drawn to scale)
17 *Amaebotaenia sphenoides*
18 *Hymenolepis carioca*
19 *Raillietina cesticillus*
20 *Choanotaenia infundibulum*
21 single egg of *C. infundibulum*
22 *Raillietina echinobothrida*
23 *Raillietina tetragona*
24 *Davainea proglottina*

For *sheep faeces*, 10 g of faeces are weighed into an Erlenmeyer flask, which is calibrated to the 300 ml mark and partially filled with 0.4 N sodium hydroxide solution. The flask is stoppered and allowed to stand over night. Next morning it is vigorously shaken to break up the faeces and is then filled to the 300 ml mark with 0.4 N sodium hydroxide solution and again shaken. Immediately after this shaking 7.5 ml of the mixture in the flask is transferred to a 15 ml centrifuge tube and centrifuged. The supernatant fluid is then drawn off and replaced by saturated sodium chloride solution with a specific gravity of 1.2. This is again centrifuged and the supernatant fluid drawn off. Centrifugation is continued with repeated changes of sodium chloride solution until the supernatant fluid is clear. The final sediment is then passed through an 80-mesh screen into a counting chamber made by marking a 70 mm Petri dish with parallel lines 3 mm apart. The eggs in this are then counted under a dissecting microscope. Since the 7.5 ml sample contained 0.25 g of faeces, the number of eggs counted is multiplied by four to give the number of eggs per gram.

For cattle faeces a similar method is used, except that 30 g of cattle faeces are taken and 5 ml samples are taken for centrifugation. Since these contain 0.5 g of faeces, the number of eggs counted in the final sediment is multiplied by two to give the number of eggs per gram. The addition of a drop of methylene blue to the microscope slide will assist in examination, the fluke eggs showing up as yellowish-brown objects against a blue background.

Factors influencing egg counts. All egg counts will be influenced by the consistency of the faeces, wet faeces being much heavier than drier specimens, and by the total amount of faeces passed per day, because in dry faeces the eggs are concentrated in a smaller output. The first source of error may be corrected by taking 2 g of normal formed faeces, 2.5 g of soft faeces, 3 g of a medium-soft stool not formed into pellets, 5 g of a pultaceous stool and 7 g of a watery stool. Egg counts may vary widely in individual animals in a flock or herd and it is wise to examine a number of samples from both clinically ill and apparently normal animals. It is advantageous if samples taken on different days can be examined. The interpretation of results requires experience and in all cases should be related to clinical findings.

The following figures can be taken as a guide, but clinical signs and other circumstances should also be taken into account.

1. *Nematodes.* In equines 500 epg suggests a mild infection, 800–100 a moderate infection and 1500–2000 a severe infection. In lambs, much depends on the species of worm present, but 2000–6000 eggs per gram indicates a severe infection and treatment is advisable when 1000 eggs per gram or more are found. In cattle, 300–600 eggs per gram indicates the advisability of treatment.

2. *Trematodes.* The presence of 100–200 eggs of *Fasciola hepatica* per gram of faeces of cattle, or 300–600 eggs per gram of faeces of sheep, indicate an infection likely to be pathogenic.

For experimental work, an estimate of the egg output per day may be required. For this the count per gram is multiplied by the total weight of the faeces passed in 24 hours. If this is done for three consecutive days, the average obtained is a very fair measure of the infection and can be used for accurate experimental studies, for instance, the testing of anthelmintics. It is important, however, to remember that the egg production of the female worms of different species is influenced by various factors. These include the immunity of the host, the age of the parasites, the consistency of the faeces, diet etc.

Faeces cultures. Because a specific diagnosis cannot always be made from the eggs, it may be necessary to cultivate the larvae from those eggs that hatch in the free state. For this purpose the faeces are broken up and placed in a glass jar, which is closed and kept at a temperature of about 26°C for a suitable time, usually seven days. The faeces must be neither too dry nor too wet; the consistency of normal sheep and horse faeces is the correct one. If they are too dry, the faeces are moistened with water; if too wet, sterilized and dried sheep faeces are added or a good grade of animal charcoal can be used instead. After incubation the jar is placed in a dull light and many species of larvae will migrate up the walls of the jar, from which they can be removed by means of a needle or soft brush for examination in a drop of water on a slide. It is advisable to kill the larvae by gentle heat until movement ceases and the worms lie fully extended.

Larvae that do not migrate can be separated by means of the Baermann technique as follows: a glass funnel 20 cm wide at the top and closed at the bottom with a clamped piece of rubber tubing is fixed in a stand and a circular piece of wire gauze, about 9 cm in diameter, is placed in the funnel and on it a larger piece of linen. The incubated faeces are spread two cm deep on the linen and water at 40°C is poured down the side of the funnel until it covers the faeces. The apparatus is left for a few hours, when the larvae will have migrated from the faeces into the water and sedimented into the neck of the funnel and will be found in a few ml of water drawn off at the bottom. This method was devised for the separation of larvae from soil and is very useful for that purpose as well.

Another culture method which gives good results, for instance with dog faeces, is to mix equal parts of faeces and charcoal, placing the mixture in a Petri dish and in the lid of the dish a piece of moist blotting paper. The larvae will later be found on the paper.

Counts of immature and mature worms. These may be required in the diagnosis of parasitism in a flock or herd by post mortem of an animal which has died or one killed as a representative of the sick animals. In general the procedure is to ligature off the various parts of the alimentary canal and to pass their contents through a series of graded screens to remove solid debris. The worms are then washed off the screens and are counted and identified. Total counts may be made or aliquots of well mixed material may be taken. Some workers prefer to avoid sieving or screening and rely on sedimentation techniques. In this case, the contents of the organ are sedimented in water several times until the supernatant is free of debris, two to three hours sedimentation being allowed each time. Aliquots of the sediment are then examined. If necessary the contents of portions of the alimentary canal can be preserved in formalin and dealt with when time is available. The mucosa of the alimentary tract may be digested with acid pepsin (10 g pepsin, 30 ml HCl, 1000 ml of water), at 37°C for several hours. The material is then sieved through a mesh of 200 apertures to the linear inch, which retains the immature worms. These are preserved in 5% formol saline. Alternatively, the material may be sedimented several times. On examination, parasites may be more easily recognized if they are stained with a drop or so of Lugol's iodine.

Some workers prefer not to use a digestion mixture for immature worms in the mucosa. Thus for *O. ostertagi* in the bovine abomasum, Williams et al. (1977) found that as many immature parasites migrated from the abomasal mucosa when it was soaked in normal saline as when a digestive mixture was used. The opened abomasum, cleaned of its contents, is placed in a 9 litre (2 gall) plastic bucket and covered with normal saline at 40°C and held at this temperature for four or five hours. After this the organ is dipped several times in a second lot of saline (9 litres) at 40°C. The contents of both containers are sieved through a screen with apertures of 30 μm and the residue collected and examined.

BLOOD EXAMINATION FOR LARVAE

The microfilariae of most filarial worms are found in the blood and the diagnosis of filariasis is made by finding them. Several methods may be employed and these are described in detail on p. 310 along with the criteria for the identification of microfilariae.

1. A drop of fresh blood is placed on a slide, covered with a cover-slip and examined immediately. The microfilariae will be seen moving about. This method can be carried out as a preliminary, but is not suitable for a specific identification.

2. If the larvae are abundant, a thin smear can be made; if they are rare, a thick film is made and

this gives better results in the majority of cases. The films are completely air-dried as quickly as possible and the thick film is then placed into a vessel of distilled water in a slanting position and facing downwards, until it has been completely dehaemoglobinised. It is then air-dried again, fixed in methyl alcohol for ten minutes and then stained (see p. 310).

3. When the microfilariae are rare concentration techniques such as the Knott technique (p. 310) are applicable.

ESTIMATION OF HERBAGE INFECTIVE LARVAE

Apart from the importance of obtaining an estimate of the population of infective larvae herbage for experimental purposes, pasture herbage counts of infective larvae are used increasingly in the diagnosis and prognosis of parasitic disease in farm animals.

Various techniques are available for the sampling of herbage, the recovery of larvae from the herbage and the estimation of the pasture burden. The various techniques have been described by Taylor (1939), Lancaster (1970), Sievers Prekehr (1973) and have been evaluated by Bürger (1981) and Raynaud and Gruner (1981).

Herbage sampling. Taylor (1939) recommended a W-shaped route across a pasture; 100 samples are collected by hand by pinching the grass off at soil level. Grass may be cut at soil level with shears. A double W route on which 300–400 samples are taken has been used. Alternatively, random strips may be sampled and ten areas each of 100 × 10 cm (total 1 m^3) sampled. Sampling around faecal pats may be done by hand and four samples are collected within a distance of 10 cm from each of 50 randomly selected faecal pats on the pasture.

Soil sampling. Soil cores, plus the herbage growing on them may be collected with the cylinder-like garden tool used for planting flower bulbs (height 7 cm, diameter 5 cm). Sixty soil cores are collected at random, this representing an area of 0.12 m^2 (Raynaud & Gruner 1981).

Treatment of samples. 1. *Soaking.* Herbage samples may be treated individually or bulked. They are soaked in buckets of water (sometimes with the addition of a small amount of detergent) for a day and the material is then sieved through screens of 200–250 μm and 20–25 μm and the retained material examined.

2. *Washing.* Herbage samples are placed in a domestic washing machine or in a commercial cement mixer, water is added and washed for 30 minutes. The contents are poured over two screens with apertures of 200–250 μm and 20–25 μm respectively. Herbage samples may be washed three times by the above procedure, to ensure full recovery of larvae. Larvae obtained from the sieves may be examined directly or may be concentrated by centrifugation in saturated magnesium sulphate solution (SG 1.28). Estimates of larvae based on dry weight of the herbage (larvae per kg) may be obtained by drying the soaked or washed herbage to a constant weight at 70°C and adjusting larval counts accordingly. *Soil samples:* Cores of soil are placed herbage side down on a large screen in a bucket filled with water.

Herbage sampling is subject to a number of variables which depend on the time of the year, climatic conditions and the time of the day when samples are taken. As much standardization as possible should be introduced into the procedures.

EVALUATION OF ANTHELMINTICS

Registration authorities demand extensive information on several aspects of antiparasitic compounds; this includes efficacy and for this the controlled or critical tests are usually used.

Critical test (Gibson 1963). The critical test is applicable to laboratory studies and studies using *Dictyocaulus viviparus* and *Fasciola hepatica.* An animal infected with helminths is treated with the anthelmintic under study. The parasites expelled in the faeces, tracheal exudate, etc., are counted and identified. Five to seven days later the animal is killed and the remaining worm burden is counted and identified. The time of autopsy is important and depends on the speed of action of a drug. Efficacy is calculated as follows:

$$\frac{x}{x + y} \times 100 = \% \text{ efficacy}$$

where x = no. of parasites expelled
y = no. of parasites remaining at autopsy

Each animal serves as its own control and hence individual variation in worm burden is allowed for. The critical test is of little or no value for immature stages of parasites and errors may occur when small gastrointestinal nematodes, which may be digested after treatment, are under consideration.

Controlled test (Moskey & Harwood 1941). Matched groups of parasitized animals either naturally or artificially infected are established. One group is treated and one serves as an untreated control group. All animals are killed, usually five to seven days after treatment, and the parasites counted and identified in each animal. Assuming equal numbers in each group the efficacy is calculated as follows:

$$\frac{a - b}{a} \times 100 = \% \text{ efficacy}$$

where a = no. of parasites in control animals
b = no. of parasites in treated animals

IDENTIFICATION OF HELMINTHS IN TISSUE SECTIONS

A detailed account with numerous keys for the identification of parasitic metazoa in tissue sections is given by Chitwood and Lichtenfels (1972).

DIAGNOSTIC METHODS IN PROTOZOOLOGY

Faeces or Intestinal Contents

Where motile organisms are to be searched for (e.g. *Hexamita*, *Trichomonas* etc.), material should be examined as soon as possible since the organisms lose their motility in cold material. The preparation should be kept warm until examination and a simple smear is made, mixing the specimen with warm saline. Direct examination is useful for the detection of motile organisms, and the use of a phase contrast or dark field microscope greatly facilitates this.

With less delicate forms (e.g. cysts of *Entamoeba*, *Balantidium* or coccidia, etc.), concentration can be employed, the faeces being passed through a screen to remove coarse debris and then repeatedly sedimented or concentrated by a flotation technique described for helminth eggs. The sample from the concentration may be stained to facilitate examination. Thus, an aqueous solution of eosin (1%) stains faecal debris a bright red, but living protozoa and cysts fail to stain and are recognized as colourless bodies in the sample. An aqueous solution of iodine (1%) is useful to facilitate detection of protozoal cysts.

Where it is necessary to sporulate oocysts for accurate identification, these may be obtained by concentration techniques (e.g. the top layer of a salt or sugar flotation is washed in water and the oocysts collected by sedimentation or light centrifugation), or, if oocysts are numerous in a faecal sample, the sample may be used in an unconcentrated form. For sporulation the material is mixed with an excess of 2% potassium dichromate solution and poured into Petri dishes; the depth of fluid should not exceed 1 cm. The dishes are then incubated at room temperature or 27°C for the appropriate time to allow sporulation.

Intestinal contents or faeces may need to be stained for more accurate identification of intestinal protozoa. Cover glass preparations are the most suitable, and these should be coated with albumen fixative to facilitate adherence of faeces etc. to the glass. A thin layer of faeces or intestinal contents is spread on the cover glass, and this is fixed before it is allowed to dry. The most satisfactory fixative is Schaudinn's fluid (saturated solution of mercuric chloride, two parts; absolute alcohol, one part; glacial acetate acid, few drops), and cover glasses should be gently dropped into this or floated on it, smear down, and left for about 10–15 minutes. Fixed smears are then washed in 70% alcohol and can be stored in this. It is useful to add a little iodine (to give a port wine colour) to remove excess mercuric chloride. Smears are most satisfactorily stained with Heidenheim's iron haematoxylin. For best results, overnight staining should be used (e.g. stain overnight in a mixture of one part 3% iron alum in distilled water to 9 parts 50% alcohol: wash in 70% alcohol: stain overnight in one part haematoxylin and 9 parts 70% alcohol: differentiate in dilute iron alum: wash in 70% alcohol: dehydrate, mount); however, for rapid diagnostic purposes 2% iron alum in distilled water for one hour, 0.1% aqueous haematoxylin for two hours and destaining with either saturated aqueous picric acid or iron alum can be used to produce an adequate preparation.

Other staining methods include Mayer's acid haemalum and Mallory's phosphomolybdic-acid haematoxylin. A quick diagnosis may often be made if the smear is fixed in absolute methanol and stained with Giemsa, though this stain is not recommended for faecal smears when the finer structures of intestinal protozoa are to be examined.

Blood or Tissue Fluids

Wet blood smears can be examined for living trypanosomes, the search for organisms being greatly facilitated by phase contrast or dark field illumination.

Blood smears should be stained by one of the Romanowsky stains (methylene blue-eosin combination). Where organisms are plentiful, a thin blood smear is satisfactory and allows a more satisfactory examination. Slides should be absolutely clean and the blood smear should be spread evenly and thinly. Thick blood smears allow more blood cells to be examined but require more care in interpretation. Thick blood smears cannot be used with avian or camel blood because of the nucleated erythrocytes. Thick blood smears need to be dehaemoglobinized before staining and this may be done by placing them in distilled water until the colour has disappeared. Thin smears or dehaemoglobinized smears are fixed in absolute methanol and may then be stained by Leishman, Giemsa, Wright's or Field's stain. Wright's stain is the most useful for quick staining, but added detail is produced if smears are stained overnight in dilute Giemsa. For best results it is important to adjust the pH of the stain to pH 7.2 for mammalian blood and pH 6.7 for avian blood.

Tissues

Usually, the most satisfactory method of examination of tissues is for sections to be cut and stained with haematoxylin and eosin, Giemsa or other appropriate stain. Frozen sections provide a rapid means of examination. A diagnosis may often be made by mixing a scraping, or a small sample, of the tissue with a little saline and examining the preparation in the fresh state (e.g. schizonts of coccidia, toxoplasma, pseudocysts, sarcosporidia etc.). Such preparations may also be stained to achieve a more critical examination.

Cultural Techniques

A variety of these exist for the isolation of protozoa (e.g. for trichomonads, amoebae, leishmaniae, etc.). Reference to these may be found in the appropriate section dealing with the parasite and information is also available in Levine (1973).

Immunodiagnostic Techniques

A wide range of techniques exists; these are reviewed by various authors in Cohen and Sadun (1976). In addition the appropriate section of this volume should be consulted.

REFERENCES

Bürger, H.-J. (1981) Experiences with our techniques for the recovery of nematode larvae from herbage. In: *The Epidemiology and Control of Nematodiasis in Cattle: Current Topics in Veterinary Medicine and Animal Science 9*, ed. P. Nansen & R. J. Jorgensen, & E. J. L. Soulsby, pp. 25–30. The Hague: Martinus Nijhoff.

Chitwood, M. & Lichtenfels, J. R. (1972) Identification of parasitic metazoa in tissue sections. *Exp. Parasit.*, **32**, 407–519.

Cohen, S. & Sadun, E. H. (1976) *Immunology of Parasitic Infections.* Oxford: Blackwell Scientific.

Gibson, T. E. (1963) The use of the critical and controlled test for the evaluation of anthelmintics against gastro-intestinal worms. In: *The Evaluation of Anthelmintics.* Proc. 1st Int. Cong. WAAVP (Ed. E. J. L. Soulsby) Merck, Sharp & Dohm. Int. New York

Lancaster, M. B. (1970) The recovery of infective nematode larvae from herbage samples. *J. Helminth.*, 44, 219–230.
Levine, N. D. (1973) *Protozoan Parasites of Domestic Animals and of Man*, 2nd ed. Minneapolis: Burgess.
Moskey, H. E. & Harwood, P. D. (1941) Methods of evaluating the efficacy of anthelmintics. *Am. J. vet. Res.*, 2, 55–59.
Raynaud, J.-P. & Gruner, L. (1981) Comparisons of techniques for assessment of the contamination of pasture herbage with infective nematode larvae. In: *The Epidemiology and Control of Nematodiasis in Cattle: Current Topics in Veterinary Medicine and Animal Science* 9, ed. P. Nansen, R. J. Jorgensen & E. J. L. Soulsby, pp. 51–68. The Hague: Martinus Nijhoff.
Sievers Prekehr, G. H. (1973) Methode zur Gewinnung, III. Strongyliden larven aud den Weidengras. Vet. Med. Diss. Hannover.
Taylor, E. L. (1939) Technique for estimation of pasture infestation by strongyle larvae. *Parasitology*, 31, 473–478.
Williams, J. C., Knox, J. W., Sheehan, D. & Fuselier, R. H. (1977) Efficacy of albendazole against inhibited early fourth stage larvae of *Ostertagia ostertagi*. *Vet. Rec.*, 101, 484–492.

Host–Parasite List

This list includes only the parasites mentioned in this volume and it is therefore not necessarily a complete check list of parasites of domesticated animals. Occasional parasites and temporary parasites, such as the blood-sucking Diptera, are not included. Ticks are listed only under those hosts which are the most important or to which they transmit diseases. *In all cases the main text should be consulted to ascertain the full extent of the host range of a parasite.*

MAN AND OTHER PRIMATES

DIGESTIVE TRACT

Trematodes
Opisthorchis sinensis
Fasciolopsis buski
Echinostoma revolutum
Echinostoma ilocanum
Echinostoma hortense
Heterophyes heterophyes
Metagonimus yokogawi
Nanophyetus schikhobalowi
Gastrodiscoides hominis

Cestodes
Bertiella mucronata
Bertiella studeri
Inermicapsifer cubensis
Inermicapsifer madagascariensis
Diphyllobothrium latum
Diphyllobothrium dendriticus
Diphyllobothrium dalliae
Diphyllobothrium pacificum
Diphyllobothrium strictum
Diphyllobothrium minus
Mesocestoides spp.
Dipylidium caninum
Diplogonoporus grandis
Taenia solium
Taenia saginata
Hymenolepis nana

Nematodes
Ascaris lumbricoides
Ascaris suum
Strongyloides stercoralis
Strongyloides fülleborni
Enterobius vermicularis
Ternidens deminatus
Oesophagostomum aculeatum
Oesophagostomum bifurcatum
Oesophagostomum stephanostomum
Trichinella spiralis
Trichuris trichiura
Capillaria philippinensis
Ancylostoma caninum
Ancylostoma braziliense
Ancylostoma duodenale
Ancylostoma ceylanicum
Necator americanus
Trichostrongylus colubriformis
Trichostrongylus vitrinus
Trichostrongylus probolurus
Trichostrongylus axei
Trichostrongylus orientalis
Mecistocirrus digitatus
Angiostrongylus costaricensis

Nematodes (cont.)
Streptopharagus armatus
Physaloptera tumefasciens
Physaloptera dilitata
Physaloptera caucasia
Physaloptera poicilometra

Acanthocephala
Prosthenorchis elegans
Prosthenorchis spicula

Protozoa
Trichomonas tenax
Trichomonas fecalis
Pentatrichomonas hominis
Enteromonas hominis
Chilomastix mesnili
Giardia lamblia
Retortaemonas intestinalis
Entamoeba coli
Entamoeba histolytica
Entamoeba hartmanni
Entamoeba gingivalis
Endolimax nana
Iodamoeba butschlii
Dientamoeba fragilis
Balantidium coli
Sarcocystis hominis
Sarcocystis porcihominis

LIVER

Trematodes
Athesmia foxi
Dicrocoelium dendriticum
Opisthorchis tenuicollis
Clonorchis sinensis
Concinnum brumpti
Pseudamphistomum truncatum
Fasciola hepatica
Schistosoma spp.

Cestodes
Hydatid cyst

Nematodes
Toxocara spp. (larvae)

CIRCULATORY SYSTEM

Trematodes
Schistosoma mansoni
Schistosoma japonicum
Schistosoma haematobium
Schistosoma bovis
Schistosoma mattheei

Nematodes
Dirofilaria immitis
Depetalonema perstans
Dipetalonema streptocerca
Mansonella ozzardi
Brugia malayi
Brugia timori
Wuchereria bancrofti
Toxocara spp. (larvae)

Protozoa
Trypanosoma gambiense
Trypanosoma rhodesiense
Trypanosoma rangeli
Trypanosoma cruzi
Trypanosoma minasense
Trypanosoma saimirii
Trypanosoma diasi
Trypanosoma primatum
Trypanosoma sanmartini
Toxoplasma gondii
Babesia divergens
Babesia microti
Babesia pitheci
Babesia spp.
Plasmodium vivax
Plasmodium falciparum
Plasmodium ovale
Plasmodium malariae
Plasmodium cynomolgi
Plasmodium rodhaini
Plasmodium schwetzi
Plasmodium reichenowi
Plasmodium eylesi
Plasmodium kochi
Plasmodium coatneyi
Plasmodium gonderi
Plasmodium simium
Plasmodium knowlesi
Plasmodium inui
Plasmodium brasilianum
Leishmania donovani
Leishmania infantum
Leishmania tropica
Leishmania mexicana mexicana
Leishmania mexicana amazonensis
Leishmania mexicana pifano
Leishmania braziliensis braziliensis
Leishmania braziliensis guyanensis
Leishmania braziliensis panamensis
Leishmania peruviana
Pneumocystis carinii

RESPIRATORY SYSTEM

Trematodes
Paragonimus westermanii

Cestodes
Hydatid cyst

Nematodes
Filaroides cebus
Filaroides gordius
Metastrongylus apri
Metathelazia ascaroides

Leeches
Limnatis africana
Dinobdella ferox

Protozoa
Toxoplasma gondii

Arthropods
Pneumonyssus simicola

URINARY SYSTEM

Nematodes
Dioctophyma renale

Protozoa
Trichomonas vaginalis
Trichomonas macacovaginae

SKIN AND SUBCUTANEOUS TISSUE

Cestodes
Sparganum mansoni

Nematodes
Dracunculus medinensis
Gnathostoma spinigerum
Dirofilaria corynodes
Loa loa
Anatrichosoma cutaneum

Leeches
Limnatis africana
Hirudo medicinalis

Arthropods
Auchmeromyia luteola (larva)
Gastrophilis intestinalis (larva)
Hypoderma bovis (larva)
Hypoderma lineata (larva)
Dermatobia hominis (larva)
Callitroga hominivorax (larva)
Cordylobia anthropophaga (larva)
Pulex irritans
Tunga penetrans
Ceratophyllus gallinae
Pediculus humanus

Arthropods (cont.)
Pedicinus erygaster
Pedicinus obtusus
Pedicinus patas
Pedicinus mjobergi
Phthirus pubis
Argas persicus
Otobius megnini
Ornithodoros moubata
Ornithodoros turicata
Haemaphysalis leporispalustris
Ixodes spp.
Dermacentor spp.
Sarcoptes scabiei
Ornithonyssus bacoti
Ornithonyssus sylvarium
Ornithonyssus bursa
Trombicula spp.

Arthropods (cont.)
Demodex folliculorum
Prosarcoptes pitheci
Prosarcoptes faini
Paracoroptes allenopitheci
Cebalges gaudi

MUSCLES, TENDONS, ETC

Cestodes
Cysticercus cellulosae

Nematodes
Trichinella spiralis

Protozoa
Sarcocystis lindemanni
Sarcocystis kortei
Sarcocystis nesbitti

EYE

Nematodes
Thelazia callipaeda
Toxocara spp. (larva)
Dirofilaria conjunctivae

Arthropods
Oestrus ovis (larva)

Protozoa
Toxoplasma gondii

CENTRAL NERVOUS SYSTEM

Cestodes
Taenia solium (*Cysticercus cellulosae*)

Nematodes
Angiostrongylus cantonensis

Protozoa
Naegleria fowleri
Hydatid cyst

SEROUS CAVITIES

Nematodes
Mansonella ozzardi
Dipetalonema perstans
Dipetalonema gracile
Dipetalonema marmosetae
Dipetalonema obtusa
Dipetalonema tamarinae

HORSE, MULE AND DONKEY

DIGESTIVE TRACT

Trematodes
Gastrodiscus aegyptiacus
Gastrodiscus secundus
Pseudodiscus collinsi

Cestodes
Anoplocephala magna
Anoplocephala perfoliata
Paranoplocephala mamillana

Nematodes
Parascaris equorum
Habronema muscae
Habronema majus
Draschia megastoma
Gongylonema pulchrum
Rhabditis gingivalis
Strongyloides westeri
Strongylus equinus
Strongylus edentatus
Strongylus vulgaris
Strongylus asini
Triodontophorus serratus
Triodontophorus brevicauda
Triodontophorus minor
Triodontophorus tenuicollis
Oesophagodontus robustus
Craterostomum acuticaudatum
Craterostomum tenuicauda
Gyalocephalus capitatus
Cyathostomum spp.
Cylicocyclus spp.
Cylicodontophorus spp.
Cylicostephanus spp.
Poteriostomum spp.
Trichonema spp.
Trichostrongylus axei
Cooperia oncophora
Probstmayria vivipara
Oxyuris equi
Oxyuris poculum
Oxyuris tenuicauda

Arthropods
Gasterophilus intestinalis (larva)
Gasterophilus inermis (larva)

Arthropods (cont.)
Gasterophilus nasalis (larva)
Gasterophilus pecorum (larva)
Gasterophilus haemorrhoidalis (larva)
Gasterophilus nigricornis
Gasterophilus meridionalis
Gasterophilus ternicinctus

Protozoa
Tritichomonas equi
Trichomonas equibuccalis
Trichomonas caballi
Chilomastix equi
Giardia equi
Entamoeba equi
Entamoeba equibuccalis
Entamoeba gedoelsti
Eimeria leuckarti
Eimeria solipedum
Eimeria uniungulati

LIVER

Trematodes
Dicrocoelium dendriticum
Fasciola gigantica
Fasciola hepatica
Fascioloides magna

Cestodes
Taenia hydatigena (*Cysticercus tenuicollis*)
Hydatid cyst

CIRCULATORY SYSTEM

Trematodes
Schistosoma bovis
Schistosoma indicum
Schistosoma mattheei
Schistosoma intercalatum
Schistosoma nasalis
Schistosoma margrebowiei
Schistosoma japonicum
Ornithobilharzia turkestanicum

Nematodes
Elaeophora böhmi
Strongylus vulgaris (larva)

Protozoa
Babesia caballi
Babesia equi
Trypanosoma brucei
Trypanosoma congolense
Trypanosoma dimorphon
Trypanosoma equinum
Trypanosoma equiperdum
Trypanosoma evansi
Trypanosoma vivax
Ehrlichia equi

UROGENITAL SYSTEM

Nematodes
Dioctophyma renale

Protozoa
Trypanosoma equiperdum
Klossiella equi

RESPIRATORY SYSTEM

Trematodes
Schistosoma nasalis

Cestodes
Hydatid cyst

Nematodes
Dictyocaulus arnfieldi
Habronema spp. (larva)

Leeches
Limnatis nilotica

Arthropods
Rhinoestrus purparensis (larva)

SKIN AND SUBCUTANEOUS TISSUE

Nematodes
Dracunculus medinensis
Habronema spp. (larva)
Parafilaria multipapillosa

Leeches
Limnatis nilotica

Arthropods
Hypoderma bovis (larva)
Hypoderma lineata (larva)
Callitroga hominivorax (larva)
Chrysomyia bezziana (larva)
Hippobosca equina
Hippobosca maculata
Hippobosca rufipes
Damalinia equi
Haematopinus asini
Vermipsylla ioffi
Vermipsylla perplexa
Vermipsylla alacurt
Vermipsylla dorcadia
Otobius megnini
Ixodes ricinus
Ixodes canisuga
Ixodes pacificus
Ixodes cookei
Ixodes scapularis
Boophilus annulatus
Boophilus decoloratus
Boophilus microplus
Margaropus winthemi
Rhipicephalus appendiculatus
Rhipicephalus evertsi
Rhipicephalus bursa
Dermacentor albipictus
Dermacentor nigrolineatus
Dermacentor venustus
Dermacentor nitens
Dermacentor occidentalis
Dermacentor reticulatus
Amblyomma hebraeum
Amblyomma americanus
Amblyomma cajennense
Amblyomma imitator
Amblyomma maculatum
Sarcoptes scabiei
Psoroptes equi
Psoroptes hippotis
Chorioptes equi
Trombicula spp.
Demodex equi

Protozoa
Besnoitia bennetti

MUSCLES, TENDONS, ETC.

Nematodes
Onchocerca cervicalis
Onchocerca reticulata
Onchocerca raillieti

Protozoa
Sarcocystis bertrami
Sarcocystis fayeri

EYE

Nematodes
Thelazia lacrymalis
Setaria equina (*microfilariae*)

CENTRAL NERVOUS SYSTEM

Cestodes
Hydatid cyst
Taenia multiceps (*Coenurus cerebralis*)

Nematodes
Setaria digitata
Micronema deletrix

SEROUS CAVITIES

Cestodes
Taenia hydatigena (*Cysticercus tenuicollis*)

Nematodes
Setaria equina

SHEEP, GOAT, DEER AND ANTELOPE

DIGESTIVE TRACT

Trematodes
Cymbiforma indica
Skrjabinotrema ovis
Paramphistomum cervi
Paramphistomum ichikawai
Paramphistomum microbothrium
Paramphistomum microbothrioides
Cotylophoron cotylophorum
Calicophoron calicophorum
Calicophoron raja
Calicophoron cauliorchis
Ceylonocotyle streptocoelium
Ceylonocotyle scoliocoelium
Gastrothylax crumenifer
Platynosomum ariestis
Ogmocotyle indica

Cestodes
Moniezia expansa
Moniezia benedeni
Avitellina centripunctata
Avitellina chalmersi
Avitellina goughi
Avitellina tatia
Stilesia globipunctata
Thysanosoma actinoides
Thysaniezia giardi

Nematodes
Gongylonema pulchrum
Gongylonema verrucosum
Gongylonema mönnigi
Gaigeria pachyscelis
Haemonchus similis
Haemonchus contortus
Haemonchus placei
Haemonchus bedfordi
Haemonchus dinniki
Haemonchus krugeri
Haemonchus larwrenci
Haemonchus mitchelli
Haemonchus vegliai
Mecistocirrus digitatus
Marshallagia marshalli
Marshallagia orientalis
Marshallagia mongolica
Marshallagia schikhobalovi
Marshallagia dentispicularis
Trichostrongylus axei
Trichostrongylus colubriformis
Trichostrongylus falculatus
Trichostrongylus vitrinus

Nematodes (cont.)
Trichostrongylus capricola
Trichostrongylus probolurus
Trichostrongylus rugatus
Trichostrongylus longispicularis
Trichostrongylus drepanoformis
Trichostrongylus hamatis
Trichostrongylus skrjabini
Trichostrongylus orientalis
Trichostrongylus retortaeformis
Trichostrongylus affinis
Ostertagia ostertagi
Ostertagia circumcincta
Ostertagia trifurcata
Ostertagia lyrata
Ostertagia leptospicularis
Ostertagia pinnata
Ostertagia orloffi
Ostertagia hamata
Ostertagia podjapolskyi
Pseudostertagia bullosa
Skrjabinagia popovi
Skrjabinagia dagestanica
Skrjabinagia kolchida
Spiculopteragia spiculoptera
Spiculopteragia böhmi
Longistrongylus albifrontis
Longistrongylus meyeri
Teladorsagia davtiani
Cooperia curticei
Cooperia punctata
Cooperia pectinata
Cooperia oncophora
Cooperia surnabada
Cooperia spatulata
Nematodirus spathiger
Nematodirus battus
Nematodirus filicollis
Nematodirus helvetianus
Nematodirus abnormalis
Nematodirus tarandi
Nematodirus odocoelei
Nematodirus andreevi
Nematodirus hsuei
Nematodirella longispiculata
Oesophagostomum columbianum
Oesophagostomum venulosum
Oesophagostomum aspersum
Oesophagostomum multifoliatum
Oesophagostomum okapi
Oesophagostomum walkeri
Chabertia ovina
Bunostomum trigonocephalum
Skrjabinema ovis
Skrjabinema alata
Skrjabinema africana

Nematodes (cont.)
Skrjabinema caprae
Skrjabinema tarandi
Strongyloides papillosus
Agriostomum cursoni
Agriostomum equidentatum
Agriostomum gungunis
Trichuris ovis
Trichuris globulosa
Trichuris discolor
Capillaria longipes

Protozoa
Tetratrichomonas ovis
Retortamonas ovis
Chilomastix caprae
Giardia caprae
Entamoeba ovis
Eimeria absata
Eimeria arkhari
Eimeria arloingi
Eimeria christenseni
Eimeria crandallis
Eimeria danielle
Eimeria faurei
Eimeria gilruthi
Eimeria gonzalezi
Eimeria granulosa
Eimeria hawkinsi
Eimeria intricata
Eimeria marsica
Eimeria ninakohlyakimovae
Eimeria ovina
Eimeria pallida
Eimeria parva
Eimeria punctata
Eimeria weybridgensis
Cryptosporidium agni

LIVER

Trematodes
Dicrocoelium dendriticum
Eurytrema pancreaticum
Fasciola hepatica
Fasciola gigantica
Fascioloides magna
Cymbiforma indica
Parafasciolopsis fasciolaemorpha

Cestodes
Stilesia hepatica
Stilesia globipunctata
Thysanosoma actinioides
Taenia hydatigena (*Cysticercus tenuicollis*)
Hydatid cyst

CIRCULATORY SYSTEM

Trematodes
Schistosoma mattheei
Schistosoma bovis
Schistosoma spindale
Schistosoma indicum
Schistosoma japonicum
Schistosoma intercalatum
Schistosoma nasalis
Schistosoma margrebowiei
Ornithobilharzia turkestanicum

Nematodes
Elaeophora schneideri
Parelaphostrongylus odocoilei

Protozoa
Trypanosoma brucei
Trypanosoma vivax
Trypanosoma congolense
Trypanosoma dimorphon
Trypanosoma evansi
Trypanosoma melophagium
Trypanosoma theodori
Babesia motasi
Babesia ovis
Babesia foliata
Babesia taylori
Trypanosoma melophagium
Trypanosoma uniforme
Trypanosoma vivax
Theileria hirci
Theileria ovis
Theileria cervi
Theileria tarandi
Haematoxenus separatus
Cytauxzoon sylvicaprae
Cytauxzoon strepsicerosi
Cytauxzoon taurotragi
Anaplasma ovis
Eperythrozoon ovis
Ehrlichia ovina
Ehrlichia phagocytophilia
Toxoplasma gondii

RESPIRATORY TRACT

Trematodes
Schistosoma nasalis
Paragonimus westermanii
Paragonimus kellicotti

Cestodes
Hydatid cyst

Nematodes
Mammomonogamus nasicola
Mammomonogamus laryngeus

Nematodes (cont.)
Dictyocaulus filaria
Protostrongylus rufescens
Protostrongylus skrjabini
Protostrongylus kochi
Protostrongylus stilesi
Protostrongylus rushi
Protostrongylus hobmaieri
Protostrongylus davtiani
Protostrongylus brevispiculum
Cystocaulus nigrescens
Cystocaulus ocreatus
Muellerius capillaris
Spiculocaulus leuckarti
Spiculocaulus austriacus
Spiculocaulus kwongi
Spiculocaulus orloffi
Bicaulus schulzi
Bicaulus sagittatus
Neostrongylus linearis

Arthropods
Oestrus ovis (larva)
Cephenemyia trompe (larva)
Cephenemyia stimulator (larva)
Cephenemyia alrichii (larva)
Cephenemyia auribarbis (larva)
Cephenemyia phobifer (larva)
Cephenemyia jellisoni (larva)
Cephenemyia opicata (larva)
Cephenemyia pratti (larva)

Protozoa
Toxoplasma gondii

SKIN AND SUBCUTANEOUS TISSUE

Nematodes
Parafilaria antipini
Elaeophora schneideri
Wehrdikmansia cervipedis
Wehrdikmansia rugosicauda
Wehrdikmansia flexuosa
Cutifilaria wenki

Arthropods
Hypoderma aeratum (larva)
Hypoderma crossi (larva)
Hypoderma diana (larva)
Hypoderma silenus (larva)
Hypoderma actaeon (larva)
Hypoderma capreola (larva)
Hypoderma moschiferi (larva)
Oedemagena tarandi (larva)
Dermatobia hominis (larva)
Hydrotaea irritans
Lucilia sericata (larva)
Lucilia caesar (larva)
Lucilia caeson (larva)
Lucilia illustris (larva)
Calliphora albifrontalis (larva)
Calliphora augur (larva)
Calliphora australis (larva)
Calliphora nociva (larva)
Phormia regina (larva)
Phormia terraenovae (larva)
Calliphora erythrocephala (larva)
Calliphora fallax (larva)
Calliphora stygia (larva)
Calliphora vomitoria (larva)
Microcalliphora varipes (larva)
Callitroga hominivorax (larva)
Chrysomyia bezziana (larva)
Chrysomyia chloropyga (larva)
Chrysomyia albiceps (larva)
Chrysomyia rufifacies (larva)
Chrysomyia micropogon (larva)
Melophagus ovinus
Damalinia ovis
Damalinia caprae
Damalinia limbata
Bovicola painei
Linognathus ovillus
Linognathus africanus
Linognathus pedalis
Linognathus stenopsis
Vermipsylla ioffi
Vermipsylla perplexa
Vermipsylla alacurt

Arthropods (cont.)
Vermipsylla dorcadia
Otobius megnini
Ornithodoros moubata
Ornithodoros lahorensis
Ixodes ricinus
Ixodes pilosus
Ixodes rubicundus
Ixodes pacificus
Ixodes persulcatus
Ixodes scapularis
Boophilus decoloratus
Boophilus microplus
Haemaphysalis cinnabarina punctata
Haemaphysalis leporispalustris
Rhipicephalus appendiculatus
Rhipicephalus evertsi
Rhipicephalus sanguineus
Dermacentor venustus
Dermacentor variabilis
Dermacentor nitens
Dermacentor occidentalis
Amblyomma hebraeum
Amblyomma variegatum
Amblyomma americanum
Amblyomma cajennense
Amblyomma maculatum
Sarcoptes scabiei
Chorioptes texanus
Chorioptes ovis
Psoroptes ovis
Psoroptes cervinus
Psoroptes caprae
Psorergates ovis
Trombicula sarcina
Demodex ovis

MUSCLES AND TENDONS

Cestodes
Taenia ovis (*cysticercus ovis*)

Nematodes
Parelaphostrongylus odocoilei
Parelaphostrongylus andersoni
Onchocerca tarsicola
Onchocerca tubingensis
Onchocerca garmsi

Protozoa
Sarcocystis tenella
Sarcocystis ovicanis
Sarcocystis cervi
Sarcocystis hemionlatrantis
Besnoitia tarandi

EYE

Nematodes
Thelazia rhodesii
Thelazia californiensis

CENTRAL NERVOUS SYSTEM

Cestodes
Taenia multiceps (*Coenurus cerebralis*)

Nematodes
Elaphostrongylus cervi
Elaphostrongylus panticola
Elaphostrongylus rangiferi
Parelaphostrongylus tenuis
Setaria digitata (immature)

Protozoa
Toxoplasma gondii

SEROUS CAVITIES

Nematodes
Setaria cervi
Setaria tundrae
Setaria altaica
Setaria cornuta

BOVINE (INCLUDING ZEBU)

DIGESTIVE TRACT

Trematodes
Ogmocotyle indica
Paramphistomum cervi
Paramphistomum scotiae
Paramphistomum hiberniae
Paramphistomum ichikawi
Paramphistomum gotoi
Paramphistomum liorchis
Paramphistomum microbothrioides
Calicophoron calicophorum
Calicophoron raja
Calicophoron cauliorchis
Cotylophoron cotylophorum
Gastrothylax crumenifer
Fischoederius cobboldi
Fischoederius elongatus
Carmyerius spatiosus
Carmyerius gregarius
Ceylonocotyle streptocoelium
Cymbiforma indica

Cestodes
Moniezia expansa
Moniezia benedeni
Avitellina spp.
Thysanosoma actinioides
Thysaniezia giardi

Nematodes
Toxocara vitulorum
Gongylonema pulchrum
Gongylonema verrucosum
Mecistocirrus digitatus
Haemonchus contortus
Haemonchus placei
Haemonchus similis
Trichostrongylus axei
Trichostrongylus colubriformis
Trichostrongylus longipicularis
Ostertagia ostertagi
Ostertagia trufurcata
Ostertagia lyrata

Nematodes (cont.)
Ostertagia leptospicularis
Ostertagia bisonis
Ostertagia orloffi
Ostertagia podjapolskyi
Skrabinagia boevi
Cooperia punctata
Cooperia pectinata
Cooperia oncophora
Cooperia surnabada
Cooperia spatulata
Nematodirus filicollis
Nematodirus helvetianus
Nematodirus spathiger
Agriostomum vryburgi
Bunostomum phlebotomum
Strongyloides papillosus
Trichuris ovis
Trichuris globulosa
Trichuris discolor
Capillaria bovis
Capillaria longipes

Nematodes (cont.)
Capillaria bilobata
Oesophagostomum radiatum
Chabertia ovina

Acanthocephala
Macracanthorhychus hirudinaceus

Protozoa
Giardia bovis
Entamoeba bovis
Tritrichomonas enteris
Tetratrichomonas pavlovi
Monocercomonas ruminantium
Eimeria alabamensis
Eimeria auburnensis
Eimeria bombayensis
Eimeria bovis
Eimeria brasiliensis
Eimeria bukidnonensis
Eimeria canadensis
Eimeria cylindrica

Protozoa (cont.)
- *Eimeria ellipsoidalis*
- *Eimeria illinoisensis*
- *Eimeria mundaragi*
- *Eimeria pellita*
- *Eimeria subspherica*
- *Eimeria wyomingensis*
- *Eimeria zuernii*
- *Crytosporidium bovis*

LIVER

Trematodes
- *Dicrocoelium dendriticum*
- *Dicrocoelium hospes*
- *Eurytrema pancreaticum*
- *Eurytrema coelomaticum*
- *Fasciola hepatica*
- *Fasciola gigantica*
- *Fascioloides magna*
- *Gigantocotyle explanatum*

Cestodes
- Hydatid cyst
- *Taenia hydatigena* (*Cysticercus tenuicollis*)
- *Stilesia hepatica*
- *Stilesia globipunctata*
- *Thysanosoma actinioides*

CIRCULATORY SYSTEM

Trematodes
- *Schistosoma japonicum*
- *Schistosoma bovis*
- *Schistosoma mattheei*
- *Schistosoma spindale*
- *Schistosoma mansoni*
- *Schistosoma nasalis*
- *Schistosoma margrebowiei*
- *Schistosoma indicum*
- *Ornithobilharzia turkestanicum*
- *Ornithobilharzia bomfordi*

Nematodes
- *Elaeophora poeli*
- *Onchocerca armillata*

Protozoa
- *Trypanosoma brucei*
- *Trypanosoma congolense*
- *Trypanosoma dimorphon*
- *Trypanosoma evansi*
- *Trypanosoma theileri*
- *Trypanosoma uniforme*
- *Trypanosoma vivax*
- *Babesia bigemina*
- *Babesia bovis*

Protozoa (cont.)
- *Babesia divergens*
- *Babesia major*
- *Theileria annulata*
- *Theileria mutans*
- *Theileria lawrenci*
- *Theileria sergenti*
- *Theileria parva*
- *Haematoxenus veliferus*
- *Toxoplasma gondii*
- *Anaplasma marginale*
- *Anaplasma centrale*
- *Eperythrozoon wenyoni*
- *Ehrlichia bovis*
- *Ehrlichia phagocytophilia*

UROGENITAL SYSTEM

Nematodes
- *Stephanurus dentatus*
- *Dioctophyma renale*

Protozoa
- *Tritrichomonas foetus*

RESPIRATORY SYSTEM

Trematodes
- *Fasciola hepatica*
- *Schistosoma nasalis*

Cestodes
- Hydatid cyst

Nematodes
- *Dictyocaulus viviparus*
- *Mammomonogamus laryngeus*
- *Mammomonogamus nasicola*

Leeches
- *Limnatis africana*
- *Dinobdella ferox*

Protozoa
- *Toxoplasma gondii*

SKIN AND SUBCUTANEOUS TISSUE

Nematodes
- *Rhabditis bovis*
- *Stephanofilaria dedoesi*
- *Stephanofilaria stilesi*
- *Stephanofilaria kaeli*
- *Stephanofilaria assamensis*
- *Stephanofilaria okinawaensis*
- *Parafilaria bovicola*
- *Onchocerca dukei*
- *Dracunculus medinensis*

Arthropods
- *Hypoderma bovis* (larva)
- *Hypoderma lineata* (larva)
- *Dermatobia hominis* (larva)
- *Chrysomyia bezziana* (larva)
- *Callitroga hominovorax* (larva)
- *Callitroga macellaria* (larva)
- *Haematobia exigua*
- *Haematobia irritans*
- *Haematobia minuta*
- *Haematobia stimulans*
- *Hippobosca equina*
- *Hippobosca maculata*
- *Hippobosca rufipes*
- *Damalinia bovis*
- *Haematopinus eurysternus*
- *Haematopinus quadripertasus*
- *Linognathus vituli*
- *Solenopotes capillatus*
- *Vermipsylla ioffi*
- *Vermipsylla perplexa*
- *Vermipsylla alacurt*
- *Vermipsylla dorcadia*
- *Raillietia auris*
- *Raillietia hopkinsi*
- *Otobius megnini*
- *Ixodes ricinus*
- *Ixodes pilosus*
- *Ixodes rubicundus*
- *Ixodes cookei*
- *Ixodes pacificus*
- *Ixodes persulcatus*
- *Ixodes scapularis*
- *Boophilus annulatus*
- *Boophilus decoloratus*
- *Boophilus microplus*
- *Boophilus calcaratus*
- *Margaropus winthem*
- *Hyalomma plumbeum plumbeum*
- *Hyalomma detritum scupence*
- *Hyalomma impressum*
- *Hyalomma detritum mauretanicum*
- *Rhipicephalus appendiculatus*
- *Rhipicephalus capensis*
- *Rhipicephalus simus*
- *Rhipicephalus neavei*
- *Rhipicephalus jeanelli*
- *Rhipicephalus ayrei*
- *Rhipicephalus pulchellus*
- *Rhipicephalus sanguineus*
- *Rhipicephalus simus*
- *Rhipicephalus evertsi*
- *Haemaphysalis cinnabarina punctata*
- *Haemaphysalis bispinosa*
- *Haemaphysalis bancrofti*
- *Haemaphysalis longicornis*
- *Dermacentor albipictus*
- *Dermacentor reticulatus*
- *Dermacentor andersoni*

Arthropods (cont.)
- *Dermacentor nitens*
- *Dermacentor occidentalis*
- *Dermacentor variabilis*
- *Dermacentor venustus*
- *Dermacentor nigrolineatus*
- *Amblyomma americanum*
- *Amblyomma cajennense*
- *Amblyomma imitator*
- *Amblyomma hebraeum*
- *Amblyomma variegatum*
- *Amblyomma maculatum*
- *Rhipicentor bicornis*
- *Sarcoptes scabiei*
- *Psoroptes bovis*
- *Psoroptes natalensis*
- *Chorioptes bovis*
- *Psorergates bos*
- *Trombicula* spp.
- *Demodex bovis*
- *Raillietia auris*

Protozoa
- *Besnoitia besnoiti*

MUSCLES AND TENDONS

Cestodes
- *Taenia saginata* (*Cysticercus bovis*)

Nematodes
- *Onchocerca gibsoni*
- *Onchocerca gutturosa*
- *Onchocerca lienalis*

Protozoa
- *Sarcocystis bovifelis*
- *Sarcocystis cruzi*
- *Sarcocystis hominis*

EYE

Nematodes
- *Thelazia rhodesii*
- *Thelazia gulosa*
- *Thelazia alfortensis*
- *Thelazia skrjabini*
- *Setaria* spp. (immature)

CENTRAL NERVOUS SYSTEM

Cestodes
- *Coenurus cerebralis*

Protozoa
- *Toxoplasma gondii*

SEROUS CAVITIES

Nematodes
- *Setaria labiato-papillosa*
- *Setaria digitata*
- *Setaria yehi*

BUFFALO

DIGESTIVE TRACT

Trematodes
- *Eurytrema pancreaticum*
- *Paramphistomum cervi*
- *Paramphistomum gotoi*
- *Paramphistomum microbothrium*

Trematodes (cont.)
- *Cotylophoron cotylophorum*
- *Calicophoron calicophorum*
- *Carmyerius spatiosus*
- *Carmyerius gregarius*
- *Fischoederius elongatus*

Trematodes (cont.)
- *Fischoederius cobboldi*
- *Gigantocotyle explanatum*
- *Gastrothylax crumenifer*
- *Ceylonocotyle scoliocoelium*
- *Ceylonocotyle streptocoelium*

Cestodes
- *Moniezia expansa*
- *Moniezia benedeni*
- *Avitellina centripunctata*
- *Thysaniezia giardi*

Nematodes
Toxocara vitulorum
Paracooperia nodulosa
Mecistocirrus digitatus
Gongylonema pulchrum
Gongylonema verrucosum
Haemonchus contortus
Haemonchus similis
Haemonchus bedfordi
Mecistocirrus digitatus
Oesophagostomum radiatum
Chabertia ovina
Strongyloides papillosa
Ostertagia ostertagi
Ostertagia circumcincta
Skrjbinagiar boevi
Longistrongylus meyeri
Cooperia oncophora
Cooperia punctata
Cooperia pectinata
Trichostrongylus axei
Bunostomum phlebotomum
Gaigeria pachyscelis
Agriostomum vryburgi
Trichuris discolor
Trichuris ovis
Capillaria bovis

Protozoa
Eimeria ankarensis
Eimeria azerbaidschanica
Eimeria bareillyi
Eimeria bovis
Eimeria bukidnonensis
Eimeria canadensis
Eimeria cylindrica
Eimeria gokaki
Eimeria ellipsoidalis
Eimeria zuernii
Eimeria auburnensis
Eimeria thianethi

Protozoa (cont.)
Eimeria braziliensis
Eimeria subspherica
Eimeria wyomingensis
Eimeria ovoidalis

LIVER

Trematodes
Eurytrema pancreaticum
Gigantocotyle explanatum
Fasciola hepatica
Fasciola gigantica
Dicrocoelium dendriticum

Cestodes
Cysticercus spp.
Hydatid cysts

CIRCULATORY SYSTEM

Trematodes
Schistosoma bovis
Schistosoma spindale
Schistosoma nasalis
Schistosoma indicum
Schistosoma japonicum
Ornithobilharzia turkestanicum

Nematodes
Onchocerca armillata
Elaeophora poeli

Protozoa
Babesia bigemina
Theileria parva
Theileria lawrenci
Theileria annulata
Trypanosoma vivax
Trypanosoma congolense
Trypanosoma theileri
Anaplasma marginale

RESPIRATORY SYSTEM

Cestodes
Hydatid cysts

Nematodes
Mammomonogamus laryngeus
Dictyocaulus viviparus

Leeches
Limnatus nilotica
Dinobdella ferox

EYES

Nematodes
Thelazia rhodesii
Thelazia gulosa
Thelazia bubalis

PERITONEAL CAVITY

Cestodes
Taenia hydatigena (*Cysticercus tenuicollis*)

Nematodes
Setaria digitata
Setaria labiato-papillosa

MUSCLES, TENDONS AND LIGAMENTS

Cestodes
Taenia saginata (*Cysticercus bovis*)

Nematodes
Onchocerca gutturosa

Protozoa
Sarcocystis fusiformis
Sarcocystis tenella

EARS

Nematodes
Stephanofilaria kaeli
Stephanofilaria zaheeri

SKIN AND CONNECTIVE TISSUE

Nematodes
Stephanofilaria kaeli
Stephanofilaria assamensis
Onchocerca gibsoni
Onchocerca sweetae
Onchocerca synceri
Onchocerca cebei
Onchocerca armillata (*microfilariae*)
Parafilaria bovicola

Arthropods
Lyperosia exigua
Lyperosia minuta
Lyperosia irritans
Psoroptes natalensis
Hyalomma detritum
Hyalomma excavatum
Hyalomma dromedarii
Sarcoptes scabiei
Demodex spp.
Dermatophagiodes spp.
Haemaphysalis bispinosa
Haematopinus tuberculatus
Boophilus decoloratus
Boophilus annulatus
Rhipicephalus sanguineus
Rhipicephalus appendiculatus
Amblyomma variegatum
Haematopinus bufali
Haematopinus eurysternus
Chorioptsoroptes kenyensis
Hypoderma bovis (larva)
Chrysomyia bezziana (larva)

CAMEL AND LLAMA

DIGESTIVE TRACT

Trematodes
Eurytrema pancreaticum

Cestodes
Moniezia expansa
Stilesia globipunctata

Nematodes
Oesophagostomum venulosum
Trichostrongylus orientalis
Trichostrongylus colubriformis
Trichostrongylus vitrinus
Trichostrongylus probolurus
Trichuris globulosa
Trichuris cameli
Trichuris skrjabini
Trichuris tenuis
Trichuris raoi
Ostertagia ostertagi
Ostertagia circumcincta
Ostertagia occidentalis
Ostertagia trifurcata
Camelostrongylus mentulatus

Nematodes (cont.)
Spiculopteragia peruviana
Nematodirus abnormalis
Nematodirus helvetianus
Nematodirus oiratianus
Nematodirus spathiger
Nematodirus lamae
Nematodirella cameli
Nematodirella dromedarii
Haemonchus longistipes
Haemonchus contortus
Gongylonema pulchrum
Strongyloides papillosus
Chabertia ovina
Cooperia oncophora
Marshallagia marshalli
Marshallagia mongolica
Physocephalus cristatus

Protozoa
Eimeria bactriani
Eimeria cameli
Eimeria dromedarii
Isospora orlovi

Protozoa (cont.)
Eimeria pellerdyi
Eimeria rajasthani
Eimeria alpacae
Eimeria lamae
Eimeria macusaniensis
Eimeria peruviana
Eimeria punoensis

LIVER

Trematodes
Eurytrema pancreaticum
Fasciola hepatica
Fasciola gigantica
Dicrocoelium dendriticum

Cestodes
Hydatid cysts
Stilesia hepatica

CIRCULATORY SYSTEM

Trematodes
Schistosoma bovis

Trematodes (cont.)
Schistosoma indicum
Schistosoma nasalis
Ornithobilharzia turkestanicum

Nematodes
Dipetalonema evansi
Onchocerca gibsoni

Protozoa
Trypanosoma evansi
Trypanosoma brucei
Trypanosoma vivax
Trypanosoma congolense
Trypanosoma dimorphon
Trypanosoma simiae
Theileria camelensis
Anaplasma marginale

RESPIRATORY SYSTEM

Cestodes
Hydatid cysts

Nematodes
Dictyocaulus filaria
Dictyocaulus viviparus
Dictyocaulus cameli

Arthropods
Oestrus ovis (larva)
Cephalopsis titillator (larva)

MUSCLES, TENDONS AND LIGAMENTS

Cestodes
Taenia solium (*Cysticercus cellulosae*)
Hydatid cyst

EYES

Nematodes
Thelazia rhodesii
Thelazia leesei

CENTRAL NERVOUS SYSTEM

Cestodes
Taenia multiceps (*Coenurus cerebralis*)

SKIN AND CONNECTIVE TISSUE

Nematodes
Onchocerca fasciata

Nematodes (cont.)
Onchocerca gibsoni

Arthropods
Hippobosca camelina
Sarcophaga dux (larva)
Ornithodoros savignyi
Amblyomma variegatum
Boophilus decoloratus
Hyalomma detritum
Hyalomma dromedarii
Hyalomma impressum
Hyalomma plumbeum
Rhipicephalus appendiculatus
Rhipicephalus sanguineus
Microthoracius cameli

Arthropods (cont.)
Microthoracius praelongiceps
Microthoracius mazzi
Microthoracius minor
Vermipsylla ioffi
Vermipsylla perplexa
Vermipsylla alacurt
Vermipsylla dorcadia

PIG

DIGESTIVE TRACT

Trematodes
Fasciolopsis buski
Echinochasmus perfoliatus
Postharmostomum suis
Gastrodiscoides hominis
Gastrodiscus aegyptiacus
Brachylaemus suis
Metagonimus yokogawai

Cestodes
Diphyllobothrium latum

Nematodes
Ascaris suum
Ascarops strongylina
Ascarops dentata
Physocephalus sexalatus
Simondsia paradoxa
Gongylonema pulchrum
Gnathostoma hispidum
Gnathostoma doloresi
Strongyloides westeri
Strongyloides ransomi
Trichinella spiralis
Trichuris suis
Bourgelatia diducta
Oesophagostomum dendatum
Oesophagostomum brevicaudum
Oesophagostomum quadrispinulatum
Oesophagostomum granatensis
Oesophagostomum georgianum
Ancylostoma duodenale
Necator americanus
Globocephalus urosubulatus
Globocephalus samoensis
Globocephalus longemucronatus
Globocephalus versteri
Trichostrongylus colubriformis
Trichostrongylus axei
Hyostrongylus rubidus
Ollulanus tricuspis
Mecistocirrus digitatus

Acanthocephala
Macracanthorhynchus hirudinaceus

Arthropods
Gastrophilus haemorrhoidalis (larva)
Gastrophilus intestinalis (larva)

Protozoa
Chilomasti mesnili
Giardia lamblia
Tetratrichomonas buttreyi
Tritrichomonas suis
Trichomitus rotunda
Enteromonas suis
Entamoeba suis
Iodamoeba butschlii
Eimeria cerdonis
Eimeria debleicki
Eimeria guevarai
Eimeria neodebliecki
Eimeria perminuta
Eimeria polita
Eimeria porci
Eimeria scabra
Eimeria scrofae
Eimeria spinosa
Eimeria suis
Isopora suis
Isopora almaataensis
Balantidium coli

LIVER

Trematodes
Dicrocoelium dendriticum
Opisthorchis tenuicollis
Clonorchis sinensis
Fasciola hepatica
Eurytrema pancreatum (pancreas)

Cestodes
Taenia hydatigena (*Cysticercus tenuicollis*)
Hydatid cyst

Nematodes
Ascaris suum (aberrant)

CIRCULATORY SYSTEM

Trematodes
Schistosoma japonicum
Schistosoma suis

Protozoa
Trypanosoma brucei
Trypanosoma congolense
Trypanosoma cruzi
Trypanosoma dimorphon
Trypanosoma evansi
Trypanosoma simiae
Trypanosoma suis
Babesia perroncitoi
Babesia trautmanni
Toxoplasma gondii
Eperythrozoon parvum
Eperythrozoon suis

UROGENITAL SYSTEM

Nematodes
Stephanurus dentatus
Dioctophyma renale

RESPIRATORY SYSTEM

Trematodes
Paragonimus westermannii
Paragonimus kellicotti

Nematodes
Metastrongylus elongatus
Metastrongylus pudendotectus
Metastrongylus salmi
Metastrongylus madagascariensis

SKIN AND SUBCUTANEOUS TISSUE

Nematodes
Suifilaria suis

Arthropods
Callitroga hominivorax (larva)
Cuterebra spp. (larva)
Haematopinus suis
Haematomyzus hopkinsi
Pulex irritans
Tunga penetrans
Ornithodorus porcinus
Boophilus decoloratus
Amblyomma americanum
Amblyomma maculatum
Dermacentor nitens
Dermacentor variabilis
Dermacentor venustus
Ixodes scapularis
Sarcoptes scabiei
Demodex phylloides

MUSCLES AND TENDONS

Cestodes
Taenia solium (*Cysticercus cellulosae*)
Spargana

Nematodes
Trichinella spiralis

Protozoa
Toxoplasma gondii
Sarcocystis miescheriana
Sarcocystis porcifelis
Sarcocystis porcihominis

SEROUS CAVITIES

Nematodes
Setaria congolensis

EYE

Nematodes
Thelazia erschowi

DOG, CAT, FOX, MUSTALIDS, PROCYONIDS ETC.

DIGESTIVE TRACT

Trematodes
Alaria alata
Alaria canis
Alaria americana
Alaria michiganensis
Alaria mustelae
Alaria arisaemoides
Alaria marcianae
Apophallus mühlingi
Apophallus donicum
Cryptocotyle lingua
Cryptocotlye concava
Cryptocotyle jejuna
Echinochasmus perfoliatus
Isthmiophora melis
Euparyphium melis
Euparyphium ilocanum
Euryhelmis squamula
Euryhelmis monorchis
Metagonimus yokogawai
Heterophyes heterophyes
Rossicotrema donicum
Plagiorchis lutrae
Nanophyetus salmincola

Cestodes
Atriotaenia procyonis
Oschmarenia pedunculata
Oschmarenia wallacei
Oschmarenia oklahomensis
Spirometra mansoni
Spirometra mansonoides
Spirometra erinacei
Diphyllobothrium latum
Mesocestoides lineatus
Mesocestoides corti
Mesocestoides variabilis
Dipylidium caninum
Dipylidium sexcoronatum
Taenia hydatigena
Taenia pisiformis
Taenia ovis
Taenia krabbei
Taenia taeniaeformis
Taenia bubesi
Taenia crocutae
Taenia brauni
Taenia erythraea
Taenia gongamai
Taenia hlosei
Taenia hyaenae
Taenia laticodis
Taenia lycaontis
Taenia macrocystis
Taenia martis
Taenia mustela
Taenia omissa
Taenia parva
Taenia polyacantha
Taenia regis
Taenia rileyi
Taenia twitchelli
Taenia multiceps
Taenia serialis
Echinococcus granulosus
Echinococcus multilocularis
Echinococcus oligarthus
Echinococcus vogeli
Joyeuxiella spp.
Diplopylidium spp.

Nematodes
Toxascaris leonina
Toxascaris transfuga
Toxocara canis
Toxocara cati
Strongyloides stercoralis
Strongyloides cati
Strongyloides tumefaciens
Strongyloides procyonis
Ancylostoma caninum
Ancylostoma tubaeforme
Ancylostoma braziliense
Ancylostoma ceylanicum
Ancylostoma duodenale
Ancylostoma kusimaensis
Ancylostoma paraduodenale
Uncinaria stenocephala
Uncinaria criniformis
Necator americanus
Spirura rytipleurites
Protospirura numidia
Protospirura bestianum
Trichinella spiralis
Trichuris vulpis
Trichuris campanula
Trichuris serrata
Capillaria entomelas
Capillaria erinacea
Capillaria putorii
Physaloptera praeputialis
Physaloptera rara
Physaloptera canis
Physaloptera felidis
Gnathostoma spinigerum
Gnathostomum nipponicum
Ollulanus tricuspis
Spirocerca lupi
Spirocerca artica
Soboliphyme baturini

Acanthocephala
Corynosoma strumosum
Corynosoma semerme
Macrocanthorhynchus catalinum
Macrocanthorhynchus ingens
Oncicola canis

Protozoa
Giardia canis
Giardia cati
Trichomonas canistomae
Trichomonas felistomae
Entamoeba histolytica
Hammondia hammondi
Isospora canis
Isospora bahiensis
Isospora burrowsi
Isospora heydorni
Isospora ohioensis
Isospora wallacei
Sarcocystis bertrami
Sarcocystis bigemina
Sarcocystis cruzi
Sarcocystis fayeri
Sarcocystis miescheriana
Sarcocystis ovicanis
Sarcocystis hemionilatrantis
Sarcocystis bovifelis
Sarcocystis gigantea
Sarcocystis hiruta
Sarcocystis muris

Protozoa (cont.)
Sarcocystis ovifelis
Sarcocystis porcifelis
Sarcocystis tenella
Isospora rivolta
Isospora felis
Besnoitia besnoiti
Besnoitia darlingi
Besnoitia wallacei
Toxoplasma gondii
Hammondia hammondi
Hoaresporidium pellerdyi

LIVER

Trematodes
Dicrocoelium dendriticum
Concinnum procyonis
Concinnum ten
Opisthorchis tenuicollis
Opisthorchis viverrini
Clonorchis sinensis
Pseudamphistomum truncatum
Fasciola hepatica
Fasciolopsis buski
Platynosomum fastosum
Eurytrema procyonis
Metorchis albidus
Metorchis conjunctus
Parametorchis complexus

Cestodes
Mesogyna hepatica

Protozoa
Hepatozoon canis
Leishmania donovani

CIRCULATORY SYSTEM

Trematodes
Schistosoma japonicum
Schistosoma rodhaini
Schistosoma spindale
Schistosoma mekongi
Schistosoma incognitum
Ornithobilharzia turkestanicum
Heterobilharzia americanum

Nematodes
Dirofilaria immitis
Angiostrongylus vasorum
Brugia patei
Brugia pahangi
Brugia ceylonensis
Brugia beaveri
Gurltia paralysans

Protozoa
Trypanosoma vivax
Trypanosoma brucei
Trypanosoma congolense
Trypanosoma cruzi
Trypanosoma dimorphon
Trypanosoma evansi
Trypanosoma rangeli
Leishmania donovani
Leishmania infantum
Leishmania chagasi
Babesia canis
Babesia felis
Babesia gibsoni

Protozoa (cont.)
Babesia vogeli
Cytauxzoon spp.
Eperythrozoon felis
Haemobartonella canis
Haemobartonella felis
Ehrlichia canis

UROGENITAL SYSTEM

Nematodes
Dioctophyma renale
Capillaria plica
Capillaria felis cati
Capillaria mucronata

RESPIRATORY SYSTEM

Trematodes
Paragonimus westermanii
Paragonimus kellicotti
Paragonimus ohirai
Paragonimus iloktsuensis
Paragonimus africanus
Paragonimus uterobilateralis
Paragonimus caliensis
Paragonimus peruvianus
Paragonimus mexicanus
Troglotrema acutum

Nematodes
Mammomonogamus auris
Mammomonogamus ierei
Mammomonogamus megaughei
Aelurostrongylus abstrusus
Angiostrongylus vasorum
Filaroides bronchialis
Filaroides osleri
Filaroides hirthi
Filaroides milski
Filaroides martis
Anafilaroides rostratus
Perostrongylus pridhami
Perostrongylus falciformis
Broncostrongylus subcrenatus
Vogeloides massinoi
Metathelazia californica
Metathelazia felis
Metathelazia multipapillata
Skrjabingylus nasicola
Skrjabingylus chitwoodorum
Skrjabingylus petrowi
Skrjabingylus magnus
Pneumospiruria capsulata
Capillaria aerophila
Capillaria didelphis
Crenosoma vulpis
Crenosoma petrowi
Crenosoma mephiditis
Troglostrongylus subcrenatus

Leeches
Limnatis africana
Dinobdella ferox

Arthropods
Pneumonyssus caninum
Linguatula serrata

Protozoa
Toxoplasma gondii

SKIN AND SUBCUTANEOUS TISSUE

Nematodes
Pelodera strongyloides
Rhabditis macrocerca
Rhabditis clavopapillata
Dirofilaria repens
Dirofilaria tenuis
Dirofilaria ursi
Dipetalonema reconditum
Dipetalonema grassi
Dracunculus medinensis
Dracunculus insignis
Dracunculus lutrae
Dracunculus fuelleborni

Arthropods
Dermatobia hominis (larva)
Cordylobia arthropophaga (larva)
Cuterebra americana (larva)
Trichodectes canis
Felicola subrostratus
Heterodoxus spiniger

Arthropods (cont.)
Heterodoxus longitarsus
Linognathus setosus
Archaeopsylla erinacei
Ctenocephalides canis
Ctenocephalides felis
Echidnophaga gallinacea
Otobius megnini
Amblyomma americanum
Amblyomma cajennense
Amblyomma maculatum
Ixodes ricinus
Ixodes holocyclus
Ixodes hexagonus
Ixodes canisuga
Ixodes cookei
Ixodes kingi
Ixodes muris
Ixodes pacificus
Ixodes rubicundus
Ixodes persulcatus
Ixodes scapularis
Ixodes angustus

Arthropods (cont.)
Ixodes rugosus
Ixodes sculptus
Ixodes texanus
Boophilus decoloratus
Rhipicephalus simus
Rhipicephalus sanguineus
Rhipicephalus bursa
Haemaphysalis leachi
Haemaphysalis bispinosum
Haemaphysalis leporispalustris
Haemaphysalis parmata
Dermacentor variabilis
Dermacentor reticulatus
Dermacentor venustus
Rhipicentor nutalli
Sarcoptes scabiei
Notoedres cati
Otodectes cynotis
Demodex canis
Cheyletiella yasguri
Cheyletiella blakei
Trombicula spp.

Arthropods (cont.)
Ornithonyssus bacoti

MUSCLES, TENDONS ETC.

Nematodes
Trichinella spiralis

EYE

Nematodes
Thelazia californiensis
Thelazia callipaeda

CENTRAL NERVOUS SYSTEM

Protozoa
Toxoplasma gondii
Encephalitozoon cuniculi

SEROUS CAVITIES

Nematodes
Dipetalonema dracunculoides
Dipetalonema reconditum

MARSUPIALS (KANGAROO, WALLABY)

SKIN AND SUBCUTANEOUS TISSUE

Arthropods
Heterodoxus longitarsus
Heterodoxus macropus
Haemaphysalis bancrofti

Nematodes
Dirofilaria roemeri

MUSCULATURE

Protozoa
Besnoitia darlingi

LIVER

Trematodes
Fasciola hepatica

RESPIRATORY SYSTEM

Nematodes
Filaroides pilbarensis

ELEPHANT

DIGESTIVE TRACT

Nematodes
Leiperenia leiperi
Leiperenia galebi
Choniangium epistomum
Choniangium magnostomum
Decrusia additicta
Equinurbia sipunculiformis
Khalilia pileata
Khalilia buta
Khalilia sameera
Murshidia murshida
Quilonia spp.
Bathmostomum sangeri

Arthropods
Cobboldia elephantis (larva)
Cobboldia loxodontis (larva)

LIVER

Trematodes
Fasciola hepatica
Fasciola jacksoni

Nematodes
Grammocephalus clathratus
Grammocephalus intermedius
Grammocephalus hybridatus
Grammocephalus varedatus

RESPIRATORY SYSTEM

Nematodes
Mammomonogamus indicus
Mammomonogamus loxodontus

Arthropods
Pharyngobolus africanus (larvae)

SKIN AND SUBCUTANEOUS TISSUE

Arthropods
Haematomyzus elephantis

Nematodes
Indofilaria patabiramani
Dipetalonema loxodontis

MUSCLES, TENDONS ETC.

Arthropods
Neocuterebra squamosa (larva)

CIRCULATORY SYSTEM

Trematodes
Bivitellobilharzia nairi
Bivitellobilharzia loxodontae

Protozoa
Trypanosoma congolense

RODENTS AND LAGOMORPHS

DIGESTIVE TRACT

Trematodes
Echinostoma ilocanum
Echinostoma hortense
Hasstilesia tricolor

Cestodes
Cittotaenia ctenoides
Cittotaenia denticulata

Cestodes (cont.)
Cittotaenia pectinata
Inermicapsifer spp.
Hymenolepis nana
Hymenolepis diminuta
Hymenolepis microstoma

Nematodes
Passalurus ambiguus
Syphacia obvelata

Nematodes (cont.)
Aspicularis tetraptera
Dermatoxys veligera
Heterakis spumosa
Paraspidodera uncinata
Strongyloides ratti
Strongyloides venezuelensis
Trichostrongylus retortaeformis
Trichostrongylus affinis
Graphidium strigosum

Nematodes (cont.)
Obeliscoides cuniculi
Nematodirus leporis
Nematodirus aspinosus
Spirura talpae
Protospiruria numidia
Protospiruria muricola
Mastophorus muris
Physaloptera clausa
Physaloptera erinacea

Nematodes (cont.)
Trichuris leporis
Trichuris sylvilagi
Capillaria gastrica
Capillaria bacielata
Capillaria annulosa
Capillaria intestinalis
Capillaria tavernae

Protozoa
Tritrichomonas muris
Tritrichomonas minuta
Tritrichomonas caviae
Trichomitus wenyoni
Tetratrichomonas microti
Monocercomonas caviae
Monocercomonas pistillum
Monocercomonas minuta
Monocercomonas cuniculi
Chilomitus caviae
Hexamastix caviae
Hexamastix robustus
Hexamastix muris
Enteromonas caviae
Retortamonas caviae
Retortamonas cuniculi
Chilomastix cuniculi
Chilomastix intestinalis
Chilomastix wenrichi
Chilomastix bettencourti
Giardia chinchillae
Giardia duodenalis
Giardia caviae
Giardia muris
Entamoeba histolytica
Entamoeba muris
Entamoeba caviae
Entamoeba cuniculi
Endolimax caviae
Endolimax ratti
Eimeria coecicola
Eimeria elongata
Eimeria exigua
Eimeria intestinalis
Eimeria irresidua
Eimeria magna
Eimeria matsubayashii
Eimeria media
Eimeria nagpurensis

Protozoa (cont.)
Isopora ratti
Eimeria caviae
Eimeria dolichotis
Eimeria nochti
Eimeria hasei
Eimeria ratti
Eimeria carinii
Eimeria falciformis
Eimeria ferrisi
Eimeria hindlei
Eimeria keilini
Eimeria krijsmanni
Eimeria musculi
Eimeria schueffneri
Eimeria neoleporis
Eimeria perforans
Eimeria piriformis
Eimeria stiedae
Eimeria miyairii
Eimeria nieschultzi
Eimeria separata
Cryptosporidium muris
Cryptosporidium parvum
Cryptosporidium wrairi

CIRCULATORY SYSTEM

Trematodes
Schistosomatium douthitti

Nematodes
Brugia tupaiae

Protozoa
Trypanosoma lewisi
Trypanosoma nabiasi
Leishmania donovani
Leishmania tropica
Leishmania aethiopica
Leishmania major
Leishmania mexicana mexicana
Leishmania mexicana amazonensis
Leishmania mexicana pifano
Leishmania braziliensis braziliensis
Leishmania braziliensis guyanensis
Leishmania braziliensis panamensis
Leishmania peruviana
Leishmania enriettii
Plasmodium berghei

Protozoa (cont.)
Plasmodium vinckei
Plasmodium chabaudi
Babesia microti
Babesia rodhaini
Eperythrozoon coccoides
Haemobartonella muris
Grahamella spp.

RESPIRATORY SYSTEM

Nematodes
Protostrongylus pulmonalis
Protostrongylus tauricus
Protostrongylus boughtoni
Protostrongylus sylvilagi
Protostrongylus oryctolagi
Angiostrongylus cantonensis
Angiostrongylus costaricensis
Pneumospirura rodentium
Crenosoma striatum

MUSCLES AND TENDONS

Nematodes
Trichinella spiralis
Sarcocystis cuniculi
Sarcocystis muris

URINARY SYSTEM

Nematodes
Capillaria papillosa
Capillaria prashadi
Trichosomoides crassicauda

Protozoa
Klossiella cobaye
Hammondia hammondi

LIVER

Protozoa
Hepatozoon muris
Hepatozoon musculi
Hepatozoon cuniculi
Hepatozoon griseisciuri

CENTRAL NERVOUS SYSTEM

Protozoa
Encephalitozoon cuniculi
Thelohania apodemi

SKIN AND SUBCUTANEOUS TISSUE

Arthropods
Gyropus ovalis
Gliricola porcelli
Trimenopon hispidum
Polyplax serrata
Polyplax spinulosa
Hoploplura acanthopus
Hoploplura captiosa
Hoploplura pacifica
Haemodipsus ventricosus
Spilopsyllus cuniculi
Ceratophyllus fasciatus
Xenopsylla cheopis
Oestromyia leporina (larva)
Cuterebra buccata (larva)
Cuterebra americana (larva)
Cuterebra lepivora (larva)
Cuterebra emasculator (larva)
Ornithonyssus bacoti
Allodermanyssus sanguineus
Echinolaelaps echidninus
Eulaelaps stabularis
Haemogamasus pontiger
Haemolaelaps casalis
Otobius lagophilus
Haemaphysalis leporispalustris
Haemaphysalis humerosa
Psorergates simplex
Psorergates oettlei
Cheyletiella parasitivorax
Cheyletiella farmani
Cheyletiella strandtmanni
Myobia musculi
Radfordia ensifera
Radfordia affinis
Trixacarus diversus
Trixacarus caviae
Notoedres muris
Notoedres budemansi
Notoedres douglasi
Psoroptes cuniculi

Protozoa
Besnoitia jellisoni
Besnoitia wallacei

MARINE MAMMALS

DIGESTIVE SYSTEM

Nematodes
Anisakis spp.
Contracaecum osculatum
Phocanema spp.
Uncinaria lucasi

CIRCULATORY SYSTEM

Nematodes
Dirofilaria immitis
Dipetalonema spirocauda

SKIN AND SUBCUTANEOUS TISSUE

Nematodes
Dipetalonema odendhali

RESPIRATORY SYSTEM

Arthropods
Halarachne spp.
Orthohalarachne spp.

Nematodes
Parafilaroides gymnurus
Parafilaroides decorus
Parafilaroides nanus
Parafilaroides prolificus

FOWL (INCLUDING WATER AND SEA BIRDS)

DIGESTIVE TRACT

Trematodes
Echinostoma revolutum
Echinoparyphium recurvatum
Hypoderaeum conoideum

Trematodes (cont.)
Notocotylus attenuatus
Catatropis verrucosa
Brachylaemus commutatus
Prosthogonimus pellucidus
Postharmostomum commutatus

Cestodes
Davainea proglottina
Raillietina tetragona
Raillietina echinobothrida
Raillietina cesticillus
Cotugnia digonopora

Cestodes (cont.)
Amoebotaenia sphenoides
Choanotaenia infundibulum
Metroliasthes lucida
Hymenolepis carioca
Fimbriaria fasciolaris

Nematodes
Heterakis gallinarum
Heterakis beramporia
Heterakis linganensis
Heterakis brevispiculum
Heterakis indica
Ascaridia galli
Subulura brumpti
Subulura differens
Subulura strongylina
Subulura suctoria
Subulura minetti
Strongyloides avium
Codiostomum struthionis
Capillaria caudinflata
Capillaria obsignata
Capillaria contorta
Capillaria anatis
Capillaria annulata
Trichostrongylus tenuis
Cyrnea piliata
Cyrnea colini
Hartertia gallinarum
Gongylonema ingluvicola
Gongylonema crami
Gongylonema sumani
Cheilospirura hamulosa
Dispharynx spiralis
Tetrameres americana
Tetrameres fissispina
Tetrameres confusa
Tetrameres mohtedai
Physaloptera gemina

Acanthocephala
Polymorphus boschadis

Protozoa
Histomonas meleagridis
Chilomastix gallinarum
Trichonomas gallinae
Tetratrichomonas gallinarum
Tritrichomonas eberthi
Eimeria acervulina

Protozoa (cont.)
Eimeria brunetti
Eimeria hagani
Eimeria maxima
Eimeria mivati
Eimeria mitis
Eimeria necatrix
Eimeria praecox
Eimeria tenella
Wenyonella gallinae
Cryptosporidium tyzzeri
Frenkelia clethrionomyobuteonis

LIVER

Protozoa
Histomonas meleagridis
Parahistomonas wenrichi

CIRCULATORY SYSTEM

Nematodes
Bhalfilaria ladamii

Protozoa
Trypanosoma avium
Trypanosoma gallinarum
Toxoplasma gondii
Leucocytozoon caulleryi
Leucocytozoon sakharoffi
Plasmodium cathemerium
Plasmodium relictum
Plasmodium circumflexum
Plasmodium durae
Plasmodium elongatum
Plasmodium fallax
Plasmodium gallincaeum
Plasmodium hexamerium
Plasmodium juxtanucleare
Plasmodium lophurae
Plasmodium polare
Plasmodium rouxi
Plasmodium vaughani
Haemoproteus columbae

Protozoa (cont.)
Haemoproteus danilewskii
Aegyptianella pullorum
Aegyptianella moshkovskii

UROGENITAL SYSTEM AND OVIDUCT

Trematodes
Prosthogonimus pellucidus
Prosthogonimus ovatus
Prosthogonimus macrorchis
Plagiorchis arcuatus

RESPIRATORY SYSTEM

Trematodes
Typhlocoelum cymbium

Nematodes
Syngamus trachea
Syngamus skrjabinomorpha

Arthropods
Cytodites nudus

SKIN AND SUBCUTANEOUS TISSUE

Trematodes
Collyriclum faba

Cestodes
Dithyridium variabile

Nematodes
Ornithofilaria fallisensis
Avioserpens taiwana

Arthropods
Callitroga hominivorax
Menopon gallinae
Menacanthus stramineus
Cuclotogaster heterographus
Lipeurus caponis
Goniocotes gallinae

Arthropods (cont.)
Goniodes gigas
Goniodes dissimilis
Ceratophyllus gallinae
Echidnophaga gallinacea
Dermanyssus gallinae
Ornithonyssus sylvarium
Ornithonyssus bursa
Argas persicus
Argas sanchezi
Argas radiatus
Argas miniatus
Argas reflexus
Ornithodorus savignyi
Amblyomma americanum
Haemaphysalis cinnabarina
Haemaphysalis leporispalustris
Haemaphysalis chordeilis
Amblyomma hebraeum
Cnemidocoptes gallinae
Cnemidocoptes mutans
Epidermoptes bilobatus
Epidermoptes bifurcata
Mégninia cubitalis
Pterolichus obtusus
Laminosioptes cysticola
Dermanyssus gallinae
Ornithonyssus bursa
Ornithonyssus bacoti
Ornithonyssus sylvarum
Syringophilus bipectinatus
Neoschongastia americana
Trombicula spp.

MUSCLES AND TENDONS

Protozoa
Sarcocystis rileyi

EYE

Nematodes
Oxyspirura mansoni
Oxyspirura parvovum
Oxyspirura petrowi

TURKEY, GUINEAFOWL AND PEAFOWL

DIGESTIVE TRACT

Trematodes
Brachylaemus commutatus
Plagiorchis megalorchis
Postharmostomum commutatus

Cestodes
Raillietina cesticillus
Raillietina georgiensis
Raillietina williamsi
Raillietina magninumida
Raillietina ransomi
Choanotaenia infundibulum
Metroliasthes lucida

Nematodes
Heterakis gallinarum
Heterakis meleagridis
Heterakis pavonis
Heterakis brevispiculum
Ascaridia galli
Ascaridia dissimilis

Nematodes (cont.)
Ascaridia numidae
Pseudoaspidodera pavonis
Pseudoaspidoderoides jnanendre
Subulura brumpti
Subulura differens
Strongyloides avium
Capillaria obsignata
Capillaria contorta
Trichostrongylus tenuis
Cheilospirura hamulosa
Dispharynx spiralis
Tetrameres americana
Tetrameres fissispina

Protozoa
Chilomastix gallinarum
Hexamita meleagridis
Histomonas meleagridis
Histomonas wenrichi
Trichomonas gallinae
Tetratrichomonas gallinarum

Protozoa (cont.)
Cryptosporidium meleagridis
Eimeria adenoeides
Eimeria dispersa
Eimeria gallopavonis
Eimeria innocua
Eimeria meleagridis
Eimeria meleagrimitis
Eimeria subrotunda
Eimeria gorakhpuri
Eimeria grenieri
Eimeria numida
Eimeria mandalis
Eimeria mayurae
Eimeria pavonina
Eimeria pavonis

LIVER

Protozoa
Histomonas meleagridis
Trichomonas gallinarum

CIRCULATORY SYSTEM

Protozoa
Haemoproteus meleagridis
Leucocytozoon smithi
Plasmodium griffithsi
Plasmodium durae
Aegyptianella pullorum
Aegyptianella moshkovskii

RESPIRATORY SYSTEM

Nematodes
Syngamus trachea

Arthropods
Cytodites nudus

SKIN AND SUBCUTANEOUS TISSUE

Trematodes
Collyriclum faba

Cestodes
Dithyridium variabile

Arthropods
Menacanthus stramineus

Arthropods (cont.)
Chelopistes meleagridis
Argas persicus

Arthropods (cont.)
Cnemidocoptes mutans
Freyana chanayi

Arthropods (cont.)
Dermanyssus gallinae
Ornithonyssus bursa

DUCK AND GOOSE

DIGESTIVE TRACT

Trematodes
Opisthorchis simulans
Echinostoma revolutum
Echinostoma paraulum
Echinoparyphium recurvatum
Hypoderaeum conoideum
Prosthogonimus pellucidus
Prosthogonimus macrorchis
Notocotylus attenuatus
Catatropis verrucosa
Typhlocoelum cymbium
Typhlocoelum obvale
Hyptiasmus tumidus
Apatemon gracilis
Parastrigea robusta
Cotylurus cornutus
Cotylurus platycephalus
Cotylurus flabelliformis
Cotylurus variegatus

Cestodes
Cotugnia fastigata
Hymenolepis lanceolata
Hymenolepis cantaniana
Diorchis nyrocae
Fimbriaria fasciolaris

Nematodes
Porrocaecum crassum
Contracaecum spiculigerum
Heterakis gallinarum
Heterakis dispar
Ascaridia galli
Capillaria contorta
Amidostomum anseris
Amidostomum skjabini

Nematodes (cont.)
Amidostomum cygni
Amidostomum acutum
Amidostomum simile
Epomidiostomum uncinatum
Epomidiostomum skrjabini
Epomidiostomum vogelsangi
Epomidiostomum orispinum
Trichostrongylus tenuis
Echinuria uncinata
Tetrameres fissipina
Tetrameres crami
Hystrichis tricolor
Eustrongyloides tubifex
Eustrongyloides papillosus

Acanthocephala
Polymorphus boschadis
Polymorphus botulus
Polymorphus magnus
Filicollis anatis

Protozoa
Trichomonas anatis
Trichomonas anseris
Cochlosoma anatis
Eimeria anatis
Eimeria anseris
Eimeria battakhi
Eimeria abramovi
Eimeria boschadis
Eimeria brantae
Eimeria bucephalae
Eimeria christianseni
Eimeria clarkei
Eimeria danailovi
Eimeria farri
Eimeria fulva
Eimeria hermani

Protozoa (cont.)
Eimeria kotlani
Eimeria koganae
Eimeria magnalabia
Eimeria nocens
Eimeria parvula
Eimeria saitamae
Eimeria schachdagica
Eimeria stigmosa
Eimeria somateriae
Eimeria striata
Tyzzeria anseris
Tyzzeria alleni
Tyzzeria pellerdyi
Tyzzeria perniciosa
Wenyonella anatis
Wenyonella gagari
Wenyonella philiplevinei

CIRCULATORY SYSTEM

Trematodes
Bilharziella polonica
Austrobilharzia variglandis
Trichobilharzia spp.
Dendritobilharzia spp.
Pseudobilharziella spp.

Protozoa
Plasmodium circumflexum
Haemoproteus nettionis
Leucocytozoon simondi
Aegyptianella pullorum
Trypanosoma calmetti

UROGENITAL SYSTEM

Trematodes
Prosthogonimus pellucidus
Prosthogonimus macrorchis
Prosthogonimus ovatus
Prosthogonimus antinus

Trematodes (cont.)
Prosthogonimus cuneatus
Prosthogonimus oviformis

Protozoa
Eimeria truncata
Eimeria boschadis

RESPIRATORY SYSTEM

Trematodes
Tracheophilus cymbius

Nematodes
Syngamus skrjabinomorpha
Cyathostoma bronchialis

Protozoa
Toxoplasma gondii

SKIN AND SUBCUTANEOUS TISSUE

Trematodes
Collyriclum faba

Nematodes
Avioserpens taiwana
Avioserpens mosgovoyi

Arthropods
Holomenopon leucoxanthum
Trinoton anserinum
Menopon gallinae
Mégninia velata
Anaticola crassicornis
Ceratophyllus garei
Argas persicus

MUSCULATURE

Protozoa
Sarcocystis rileyi

WATERBIRDS (SEE ALSO DUCKS AND GEESE)

DIGESTIVE TRACT

Trematodes
Clinostomum complanatum
Clinostomum marginatum
Diplostomum spathaceum
Prosthodiplostomum cuticula
Prosthodiplostomum minimum
Neodiplostomum perlatum
Neodiplostomum multicellulata
Cryptocotyle lingua
Cryptocotyle concava

Trematodes (cont.)
Cryptocotyle jejuna
Apophallus mühlingi

Cestodes
Ligula intestinalis
Schistocephalus solidus

Nematodes
Porrocaecum depressum
Porrocaecum angusticolle

Nematodes (cont.)
Porrocaecum aridae
Contracaecum microcephalum
Contracaecum osculatum

CIRCULATORY SYSTEM

Trematodes
Gigantobilharzia spp.
Trichobilharzia spp.
Dendritobilharzia spp.
Pseudobilharziella spp.

RESPIRATORY SYSTEM

Nematodes
Cyathostoma lari
Cyathostoma brantae
Cyathostoma variegatum

PIGEONS, DOVES, OSTRICH

DIGESTIVE TRACT

Cestodes
Cotugnia cuneata
Houttaynia struthionis

Nematodes
Ascaridia columbae
Ascaridia razia
Ornithostrongylus quadriradiatus

Nematodes (cont.)
Libyostrongylus douglassii
Physaloptera alata

Protozoa
Eimeria columbae
Eimeria columbarum
Eimeria labbeana
Eimeria tropicalis

SKIN AND SUBCUTANEOUS TISSUE

Arthropods
Columbicola columbae
Ceratophyllus columbae
Argas reflexus
Syringphilus columbae
Sarcopterinus nidulans
Mégninia columbae

CIRCULATORY SYSTEM

Protozoa
Plasmodium relictum
Haemoproteus columbae
Haemoproteus sacharovi
Leucocytozoon marchouxi

GAME BIRDS (PHEASANT, PARTRIDGE, GROUSE ETC.)

DIGESTIVE TRACT

Trematodes
Tetrameres pattersoni

Nematodes
Heterakis isolonche
Ascaridia compar
Trichostrongylus tenuis

Protozoa
Eimeria colchici
Eimeria dispersa

Protozoa (cont.)
Eimeria duodenalis
Eimeria langeroni
Eimeria megalostromata
Eimeria pacifica
Eimeria phasiani
Eimeria kofoidi
Eimeria lyruri
Eimeria procera
Eimeria augusta
Eimeria bonasae
Eimeria nadsoni
Eimeria tetricis

Protozoa (cont.)
Eimeria brinkmanni
Eimeria fanthami
Eimeria lagopodis
Eimeria coturnicus

RESPIRATORY SYSTEM

Nematodes
Syngamus trachea

Protozoa
Trichomonas phasioni

CIRCULATORY SYSTEM

Protozoa
Haemoproteus canachites
Haemoproteus lophortyx
Leucocytozoon bonasae
Leucocytozoon mansoni

SKIN AND SUBCUTANEOUS TISSUE

Arthropods
Dasypsyllus gallinulae
Mégninia phasiani

REPTILES AND AMPHIBIANS

DIGESTIVE TRACT

Nematodes
Tachygonetria spp.

CIRCULATORY SYSTEM

Protozoa
Leishmania adleri

RESPIRATORY SYSTEM

Arthropods
Entonyssus spp.
Entophionyssus spp.

Pentastomids
Porocephalus crotali
Armillifer armillatus

SKIN AND SUBCUTANEOUS TISSUE

Arthropods
Ophionyssus natricis
Aponomma spp.

Nematodes
Dracunculus globocephalus
Dracunculus ophidensis
Dracunculus alii

FISH

DIGESTIVE TRACT

Trematodes
Crepidostomum spp.
Stephanostomum baccatum

Cestodes
Proteocephalus ambloplitis
Triaenophorus spp.
Eubothrium spp.
Cyathocephalus truncatus
Caryophyllaeus fimbriceps
Caryophyllaeus laticeps
Khawia sinensis

Nematodes
Capillaria tomentosa
Capillaria tuberculata
Capillaria lewaschoffi
Capillaria cantenata
Capillaria eupomotis
Capillaria catostomi
Capillaria petruschewskii

Acanthocephala
Echinorhynchus salmonis
Pomphorhynchus laevis
Acanthocephalus spp.

Protozoa
Eimeria aurata
Eimeria carpelli
Eimeria ciprini
Eimeria subepitheliasis
Eimeria truttae
Ceratomyxa shasta

VASCULAR SYSTEM AND GILLS

Trematodes
Sanguinicola inermis
Sanguinicola klamathensis

Protozoa
Cryptobia borreli
Cryptobia brachialis
Cryptobia cyprini

Protozoa (cont.)
Nosema branchialis
Plistophora salmonae
Henneguya exilis
Myxosoma dujarine
Chilodonella cypruni
Chilodonella hexasticha
Trichodina spp.

LIVER

Protozoa
Plistophora cepedianae

MUSCLES, CONNECTIVE TISSUE AND CARTILAGE

Trematodes
Clinostomum spp. (metacercariae)

Protozoa
Plistophora hyphessobryconis
Plistophora macrozoarcidis

Protozoa (cont.)
Plistophora anguillarum
Glugea anomala
Glugea hertwigi
Glugea stephani
Ichthyosporidium giganteum
Myxosoma cerebralis
Myxosoma cartilaginis
Henneguya zchokkei
Thelohanellus piriformis

CENTRAL NERVOUS SYSTEM

Protozoa
Nosema lophi

REPRODUCTIVE SYSTEM

Protozoa
Plistophora ovariae
Thelohania baueri

SKIN AND SUBCUTANEOUS TISSUE

Trematodes
- *Gyrodoctylus elegans*
- *Dactylogyrus vastator*
- *Dactylogyrus extensus*
- *Benedenia* spp.
- *Discocotyle sagittata*
- *Diplozoon paradoxum*
- *Posthodiplostomum cuticula* (metacercariae)
- *Posthodiplostomum minimum* (metacercariae)
- *Neodiplostomum perlatum*
- *Neodiplostomum multicellulata*

Protozoa
- *Costia pyriformis*
- *Costia necatrix*
- *Ichthyophthirius multifiliis*
- *Cryptocaryon irritans*
- *Chilodonella cyprini*
- *Epistylis* spp.
- *Ambiphyra* spp.

Crustacea
- *Ergasilus* spp.
- *Salmincola* spp.
- *Achtheres* spp.
- *Lernaea* spp.
- *Argulus* spp.

Index